FLOOR PLAN MANUAL HOUSING

EDITED BY
OLIVER HECKMANN AND
FRIEDERIKE SCHNEIDER
WITH ERIC ZAPEL
FIFTH, REVISED AND
EXPANDED EDITION

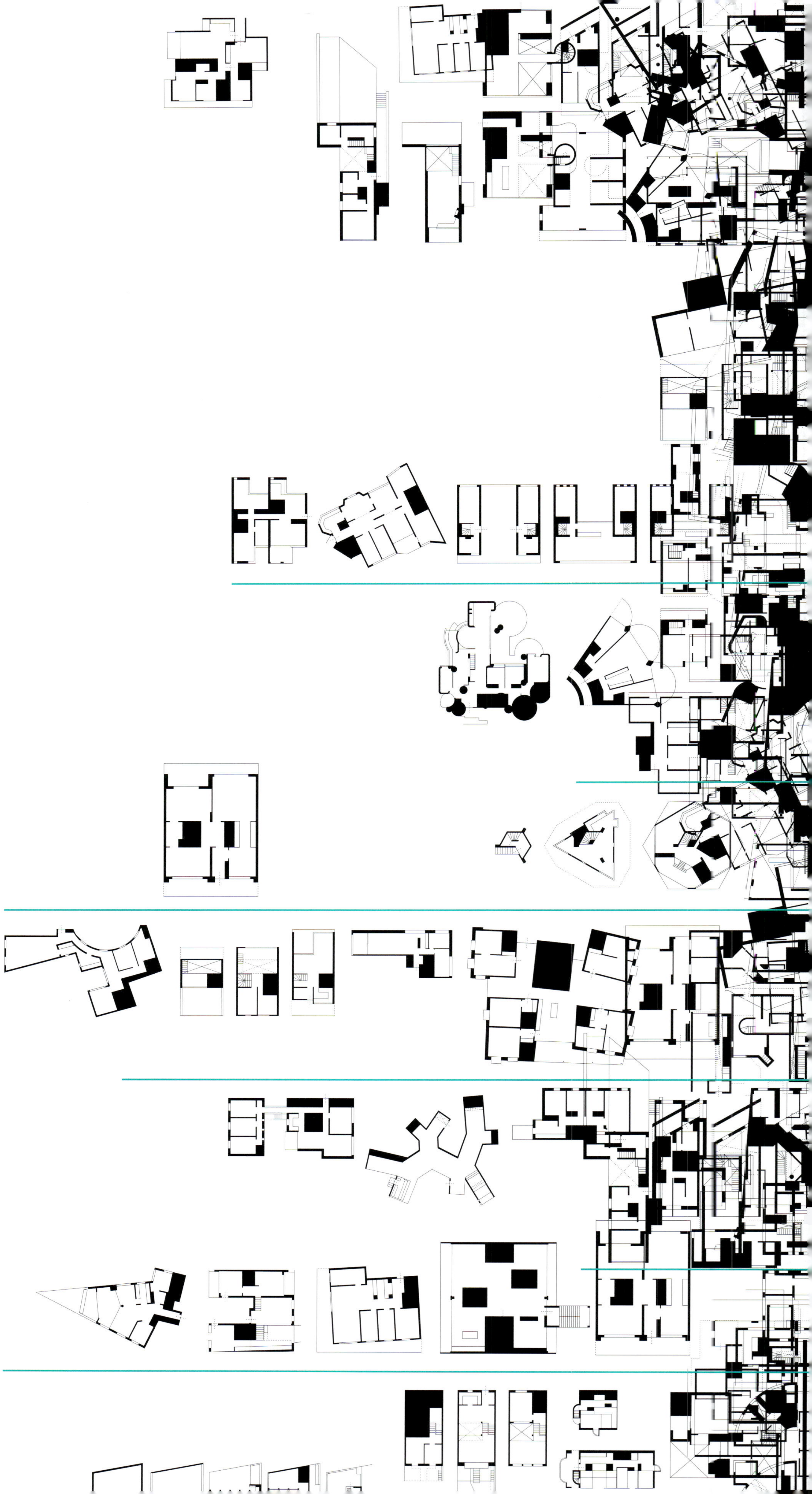

BIRKHÄUSER
BASEL

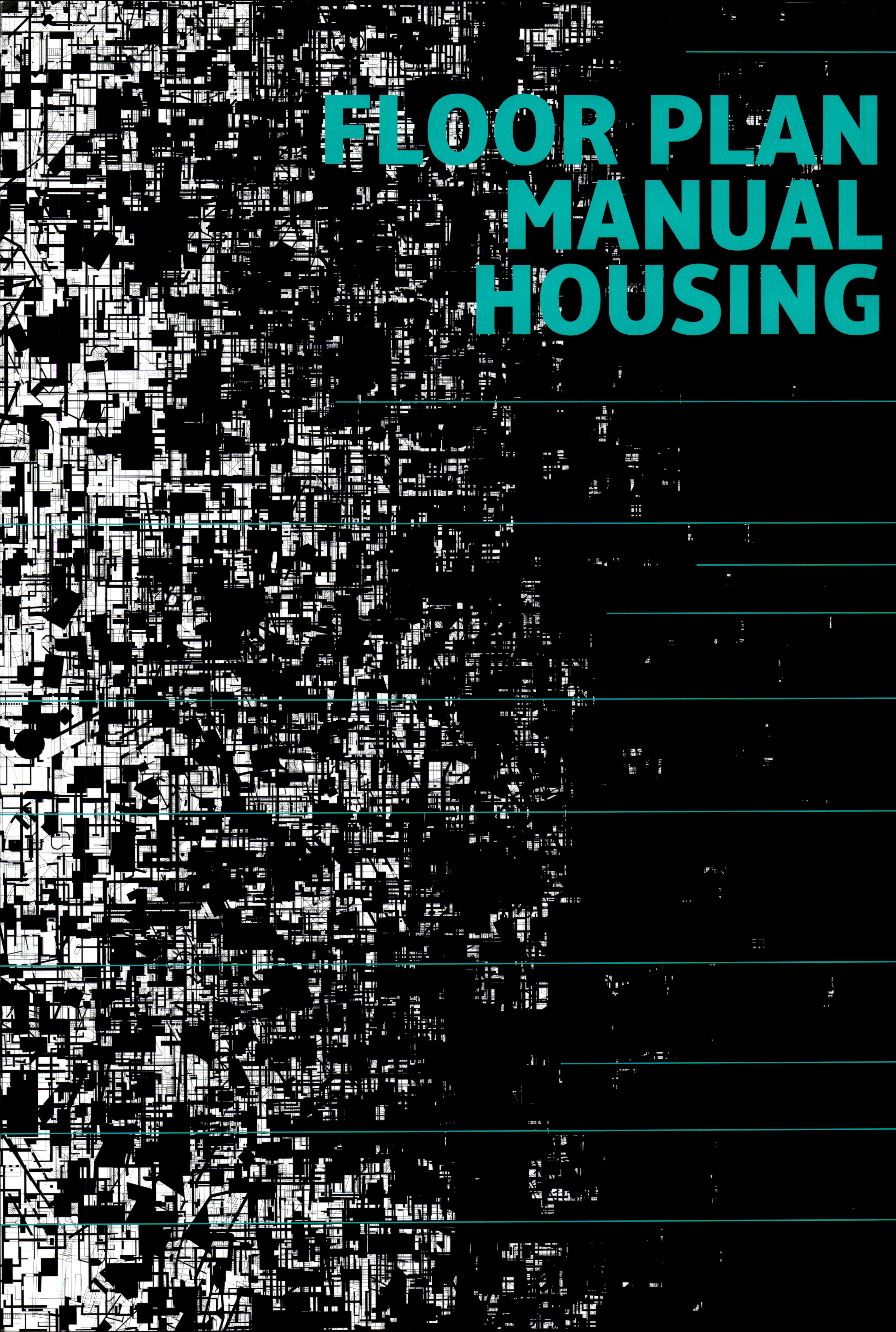

FLOOR PLAN MANUAL HOUSING

Content

Working with the Floor Plan Manual Housing

The Floor Plan Manual Housing is designed as a workbook. It documents and analyzes 161 international housing projects built since 1947. Its compact presentation of all projects, the range of the projects and the long time period covered by the selection of examples make the Floor Plan Manual Housing a tool for researching the latest developments in housing, contextualizing these developments through comparison with examples from the past 70 years. The systematic typological presentation of the projects allows readers to utilize the knowledge and ideas of others in a purposeful manner, thus finding inspiration for one's own work on floor plans.

A number of instruments in the book help provide an overview while offering focused approaches to reviewing the wealth of material. Thus, depending on one's mood or particular interest, different ways of using the book are made possible.

APPROACH A

The chart on the inside cover or insert

The navigation aid on the inside book cover (soft cover edition) or on the insert (hard cover edition) provides a systematic overview if you wish to search the collection of examples for specific floor plan ideas, access and circulation form, or for projects from a specific architectural practice. The chart categorizes the projects alphabetically from top to bottom by practice name and from left to right according to floor plan organization and means of access (cf. essays "The Floor Plan Idea", pages 48–53, and "The Path toward Access and Circulation", pages 42–47).

APPROACH B

Table of contents and introductions to typological groups
PAGES 4–5

The table of contents (like the project section) organizes the projects according to urban and morphological building types, each organized in its own chapter. Every chapter, in turn, is prefaced with a brief text describing the building type and the resultant demands for the floor plan. In this way, one can search and better compare the relevant projects for a given design task.

APPROACH C

The analytical essays
PAGES 8–57

The project section is preceded by six fundamental introductory essays, whose various focal points provide the reader and user with different possibilities to review and analyze floor plans. (see below: "The essays as working instruments")

APPROACH D

Floor plan diagram overview
PAGES 60–69

The overview of the floor plan diagrams of all examples can also be used as a visual table of contents: it allows the eye to travel and is intended to inspire a fresh look at the floor plans beyond the specific building tasks, names, or completion dates. One of the aims of this volume is to highlight the surprising and inspiring elements of floor plans, which at first glance often go unheeded as irrelevant or uninteresting.

The project selection

The focus of this documentation is on floor plans; therefore all key floor plans (or an exemplary plan for each project) are presented at the same scale of 1:200, thus enabling easy comparison. A north indicator at the bottom of the page identifies the orientation.

The projects are documented as a whole, by means of sections, site plans, small photographs as visual supplements, systematic additional information in the margins, and written descriptions. The descriptions fulfill a dual role: they are intended to guide the reader through the house or apartment; they also attempt to summarize the plan's underlying idea, thus freeing the reader from the concrete floor plan and inspiring him or her to form a personal interpretation of the concept.

The same is true for the diagrams, which in their abstraction – that is, the standardized presentation of an exemplary floor plan – allow the idea of the floor plan to become more evident and therefore comparable to others in the book. All diagrams are shown as figure-ground plans. The white area shows the open, undefined areas to better illustrate the spatial flow, while the black areas denote spatial boundaries and predetermined zones such as bathrooms and storage space. All diagrams are drawn at a scale of 1:500 and located at the top left corner of the first project page. Like a flipbook, they serve as a complementary search aid throughout the project section.

The publisher and the editors would like to thank all those architects who were kind enough to search for the plans – and the data – of buildings long completed. They deserve the merit for the accomplishment of an international "Floor Plan Manual Housing."

With regard to the project plans, we have retained the mode of representation chosen by each architect, for the design idea is always reflected in the visual language of the presentation.

Within each chapter, the projects are organized chronologically, in order to better trace the development of each building type.

The Floor Plan Manual features only built projects, although an argument could undoubtedly be made that unbuilt projects would be equally enriching for the design of floor plans. However, it was important that all projects "pass through the eye of the needle" called realization before a serious comparison can take place, because floor plan design is often subject to significant modification during the building process.

Emphasis was placed on a broad international scope of the selected projects in order to better reflect heterogeneous, worldwide developments in housing.

In addition to the objective selection criteria, a further stipulation was that each project selected is based on "exemplary floor plans." In our view, an exemplary floor plan is first and foremost distinguished by a good or excellent utilization of the given situation. Secondly, and equally important, is that the specific idea for a floor plan should be expressed with the greatest clarity possible, independent of whether the concept could be and indeed is applicable to all or only a small group of users. For the most part, we have selected universally applicable, easily transferable solutions, although it also seemed justified to include, here and there, several designs too unique to serve as references in other settings.

The essays as working instruments

The introductory texts provide the reader with various options for analysis. They demonstrate that one can only do justice to the complexity of what so prosaically is referred to as housing by layering different "ways of seeing."

The introduction "The Sweetness of Functioning is Architecture: On the Use of Floor Plans" focuses on the various ways of reading floor plans in general, whereby reading/use signifies the study and development of drafting the floor plan on the one hand and living in or making use of the built plan on the other. By shedding light on both of these "levels of reading," the text reveals the inherent sensuous nature of the seemingly abstract housing floor plan.

The "Chronicle of Developments in Housing" documents significant developments in housing from the industrialization in the 19th century to the year 2000. Its linear, chronological nature allows one to retrace the multifaceted parallel development of historic turning points, significant paradigm shifts, and newly developed buildings or building types created as a reaction thereto or creating new impulses.

The chronicle is complemented by a description of diverse "Challenges and Tendencies" in the global development of housing in recent years as we observe them today – without the benefit of hindsight.

The focus then shifts to the organization of the structure of a house, the ways in which the apartments are connected to one another and to their surroundings: "The Path toward Access and Circulation" describes the significance and potential of the access space and analyzes different access types.

The typology of the apartment, the search for the idea that underlies the specific arrangement of its individual rooms and their relationship to one another – be it linear, around a center, merging, or separate – is outlined in "The Floor Plan Idea."

While the "Floor Plan Idea" provides an analysis of built examples and suggests working with floor plan types as a design method, the essay "A Graphic Approach to Floor Plan Design" proposes methods of an abstract, graphic nature to first develop design intentions regarding spatial, functional, or social criteria without confining them to a concrete architectural form too early in the process.

The Floor Plan Manual Housing continues to be a work in progress. Thus we would like to once again issue an invitation to all readers and users of this manual who feel that a project they deem especially important – be it their own or a project created by someone else, old or new – is missing from this selection: please contact the publisher at www.birkhauser.com and share your knowledge for the next revised edition of this work.

"The Sweetness of Functioning is Architecture": On the Use of Floor Plans

By prefacing the term functioning with a set of sensations and employing the word "sweetness," Alison and Peter Smithson show the work of designing a floor plan in a different light. As architects who felt committed to modernism, they never questioned the necessity of functionality, which would after all be unthinkable in housing with its complex functional requirements. Yet they never relied solely on the functionality of the built structure; rather, they based their arguments on the pleasure of physically sensing that and how a house does function.

"... the light and the space and the air are one. Sniff the air, sense the space, know how to act." This expands the perspective: from working on a floor plan as an abstract and uninhabited concept to considering what the floor plan will become once it is in use. And conversely, how the work on a floor plan can be inspired if the person who designs it embraces all possible facets of use: the joy of exploring the potential of the floor plan, of resting in and moving through the space, and of engaging with the particular characteristics of a house – be it the sensory experience of its rooms or the manner in which they connect the inhabitant to his surroundings.

Ways of Reading

In his essay "Figures, Doors and Passages," the architectural theorist Robin Evans also explores the relationship between floor plans and their use by inhabitants. To this end, he compares the plans for two villas, one by John Webb, the other by Andrea Palladio. His description focuses less on the floor plans as such than on two different ways of reading them.

In John Webb's design for Amesbury House in Wiltshire (1661), the use to which the house is put seems secondary, for the architect has already anticipated and predetermined it in the plan. According to Evans, the process of industrialization and the separation of living and working led to the provision of privacy becoming the definitive topic in the debate on housing. The instrument of this intimacy is not the occupant of the house, but the house itself. Thus the plan for Amesbury House features a corridor and a two-tiered stairwell, which regulate all movement. In this case, the plan becomes a mechanism, capable of guiding its inhabitants to specific modes of behavior. One way of reading, therefore, is to test the spatial layout for its functionality and the systematic separation of private and public zones.

Such rules drove the design of floor plans in housing for a long time and, in addition to other factors, led to the extreme determination of floor plan structures. And this trend was further advanced during the modern era in particular, when the pursuit of economy and rationality led many architects

Ambient light, ambient air, no fuss about detail, awareness in a quiet way that the sweetness of functioning is architecture. ... In a real building, the light and the space and the air are one. Sniff the air, sense the space, know how to act. How to keep this sense of what is going on – where the light and air is coming from, how to get in and out .... that is the question...
A. + P. Smithson, in: Changing the Art of Inhabitation, London, Munich, 1994

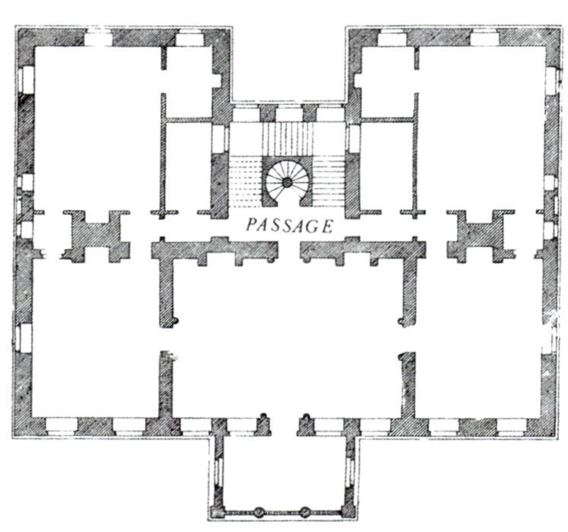

John Webb: Amesbury House, Wiltshire, 1661

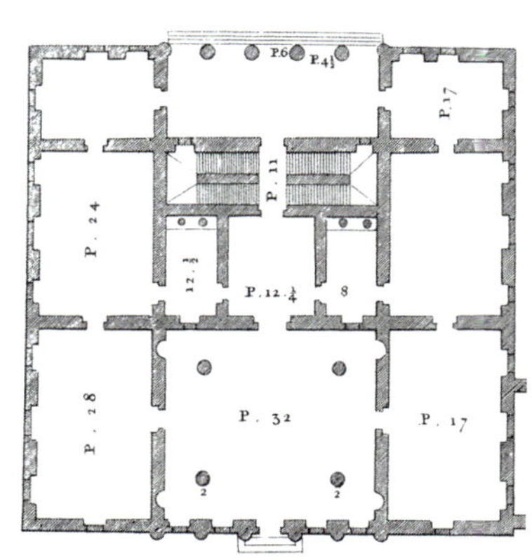

Andrea Palladio: Palazzo Antonini, Udine, 1556

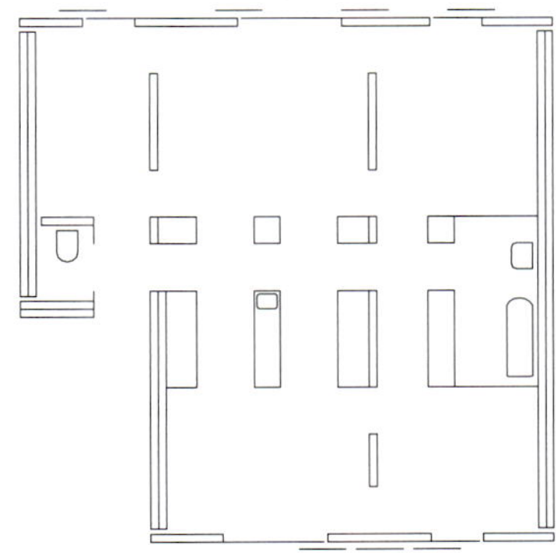

Riegler Riewe: Bahnhofstraße, Graz-Straßgang, 1994

to such functional and apparently logical solutions. The use of the house has only little relevance when verifying its functionality. The functions are predetermined to such a degree as to eliminate any surprises in the subsequent use.

In the second example cited by Evans, the Palazzo Antonini in Udine by Andrea Palladio (1556), the floor plan is only comprehensible through the history of how it was used. Evans uses a literary reference (a chambermaid was asked to guard one of the many rooms, thus temporarily separating it from the spatial continuum, in order to ensure privacy for a tête-à-tête) to explain that floor plans of this kind are greatly dependent on the ability of the occupants to control their boundaries. For all rooms in this plan are directly linked to one another as parts of an open sequence of spaces; they are equal in terms of structural prominence and are entirely indeterminate with regard to function or placement. Even the lavatories are connecting spaces.

Polyvalence

Thus the second way of reading operates on a completely different level. The reader speculates what may take place because the floor plan does not provide any clues as to what should take place. The indeterminate nature of the floor plan of the Palazzo Antonini turns out to be polyvalent: the potentialities of use are established, but only come to life through reading and using. In their residential building, Riegler Riewe have turned polyvalence into a design principle – not in the sense of Palladio, where it emerged more as a byproduct of his open, predominantly representative spatial order, but as a structured, very controlled and well-researched floor plan, whose numerous links between rooms of similar size allow for particularly diverse interpretations and functional allocations.

The activities of the residents were taken into account here. Instead of preemptively assigning each space a specific function such as living room, dining room, kitchen, bedroom and office, bathroom and WC, shared scenarios were devised in which the users relax, come together or seek out privacy, eat, cook, sleep, work, or wash and relieve themselves. The degree of interconnectedness between the rooms appears to be the most important aspect, as if the functions of the floor plan can only emerge once a spatial disposition of the volumes has been precisely arranged. Polyvalence, in working on a design, means creating spaces so rich and dense that they can be interpreted and occupied in a variety of ways.

Similarly, the floor plan for the Minus K House is composed of a homogenous matrix of equal-sized rooms, whose intended use is only hinted at through their linkages – either axially, on their sides, or diagonally with open corners to create larger areas.

[Architecture] during the last two centuries ... is employed more and more as a preventive measure; an agency for peace, security and segregation which, by its very nature, limits the horizon of experience – reducing noise-transmission, differentiating movement patterns, suppressing smells, ... veiling embarrassment, closeting indecency and abolishing the unnecessary; incidentally reducing daily life to a private shadow-play. But on the other side of this definition, there is surely another kind of architecture that would seek to give full play to the things that have been so carefully masked by its antitype; an architecture arising out of the deep fascination that draws people towards others; an architecture that recognizes passion, carnality and sociality.
"Figures, Doors and Passages," in: Robin Evans, Translations from Drawing to Building and Other Essays, London, 1997

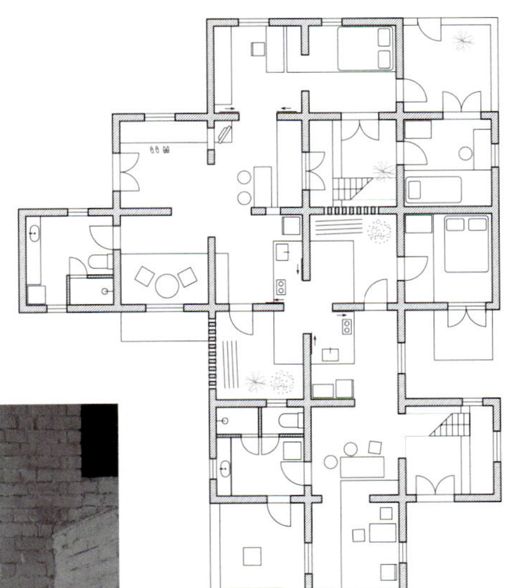

KUU: Minus K House, Nanhui, Shanghai, 2010

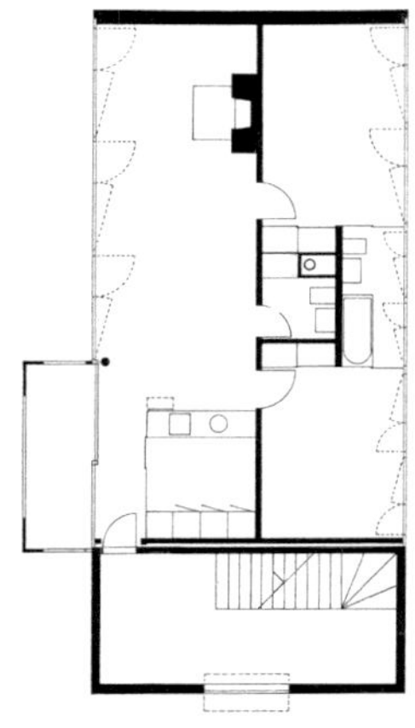

Peter Märkli with Gody Kühnis: Residential Building, Trübbach, 1989

Offer

In Peter Märkli's houses, functionality seems to play a subordinate role in the floor plan. He generates spatial tensions in his floor plan compositions and avoids anything that might be read as static and unambiguous; he allows the space to flow, creating the basis for the most varied forms of use. In his residential building in Trübbach, the apartments are organized by means of three volumes set off from one another and the middle wall. The kitchen is inserted and forms an entryway together with the loggia on the other side, while a fireplace at the far end of the room defines yet another locale by virtue of its corporeality and orientation. A sanitary core, around which the internal paths seem to circulate, is set back from the facade on the other side of the middle wall. There is no hallway or corridor. The subtle rhythmization of the areas – small-large-small on the entrance side, and large-small-large on the other side – evokes a flow of movement that blurs the boundaries and functions of the individual rooms, a quality that is further enhanced by the cohesively continuous facade. Although the floor plan establishes certain guidelines, it nevertheless remains polyvalent. The reader can promenade through such drawings as if on a virtual tour, measuring open spaces and defined areas of use, all the while speculating and deciding on how the house can be used.

Even more extreme are use-neutral and semi-finished buildings like the Kölner Brett by the architects b&k+ brandlhuber&kniess. The occupant is not only asked to establish a relationship with an existing spatial composition, he or she is even called upon to envision and then carry out its completion. This places a greater emphasis on use as an act of design. More than just reinterpreting the reading of a floor plan, here it remains to be completed.

To begin with, the building is use-neutral and the rooms are empty. The drawing of the spatial configuration is no more than an abstract system. Spatial manipulation, a linkage of flat, yet wide and tall, albeit narrow spaces, options for horizontal and vertical connections all contribute to form a spatial conglomerate, which offers sufficient density in any imaginable constellation.

Although the built units are empty and undefined, they are not devoid of atmosphere, for the concept, which is entirely diagrammatic to begin with, is defined by materiality and sensuality. Exposed concrete, massive oak planks, and tall wooden doors, which block the view behind the glazed

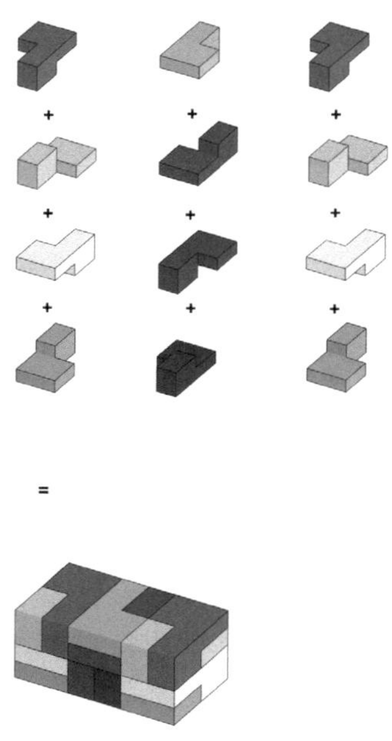

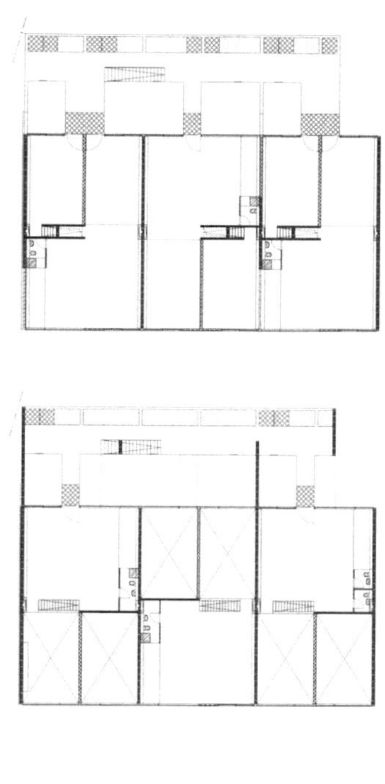

b&k+ brandlhuber & kniess GbR: Kölner Brett, 1999

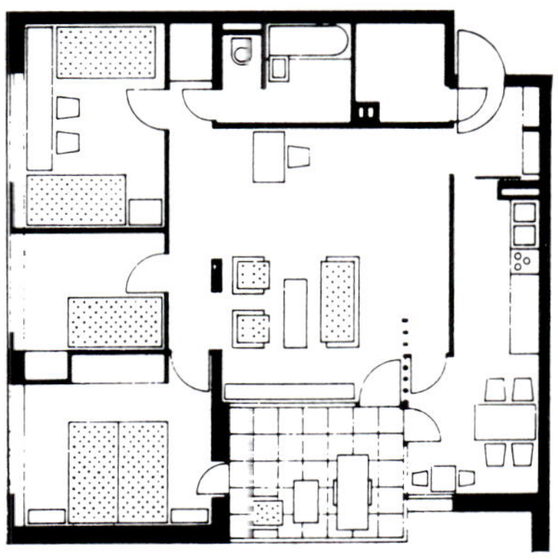

Alvar Aalto: Residential building in Hansaviertel, Berlin, 1957

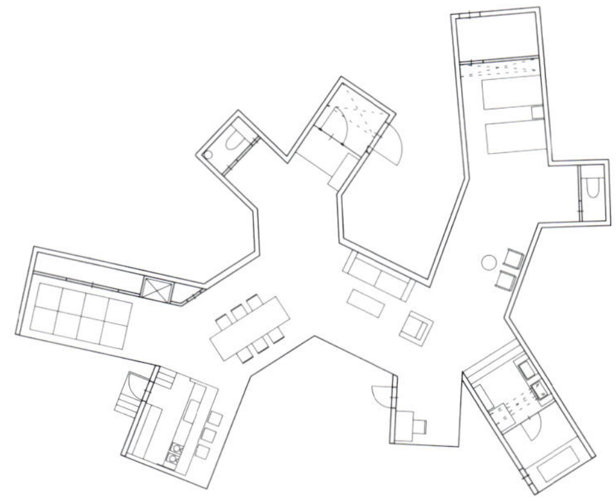

Sou Fujimoto Architects: House O, Chiba, 2007

facades at precisely those points where one walks through – all these elements define the character of the house. Given the emptiness and lack of definition, the occupant can seize upon these elements as a starting point of orientation.

Discretion

Evans' comparison between the floor plans of Webb and Palladio leads to another design question: was the guiding principle the separation or rather the connection of residents when the floor plan was being developed and its walls drawn? For instance, the categorical demand for the respect of privacy in the Amesbury House can be more subtly laid out in the floor plan design without such a strict spatial determination. An occupant who is able to connect with the space is also able to recognize openly formulated boundaries; the floor plan design may thus rely on the occupant's ability for discretion. Alvar Aalto's plan for a residential building in Berlin's Hansaviertel is particularly impressive to study, because such a balance is achieved: the focal point of the unit is clearly the idea of an open living room, – which he dubbed "the market square" – as the intersection of all movement. However, the rest of the apartment is so tactfully organized that no one ever feels too exposed. One of the doors to the private rooms is obscured by an equally wide support pillar, and to reach the others one first has to turn a corner; the dining area is part of the living room but formulated as a protected niche behind a wooden screen, thus slightly separated.

In House O, in which open spaces for living, eating, cooking, washing, and sleeping are all accommodated in a single large room, the branches in the floor plan point to the intimacy of individual areas – simply by virtue of preventing direct sightlines. The bed only comes into view after one has changed direction several times; the connection is delayed but not interrupted.

The residential units of SL Court are organized as split-levels and it is likewise not walls, but here the subtly offset levels that produce a feeling of safe distance. Only the bedrooms in the middle are given additional translucent sliding elements. One single space flows openly throughout the unit without completely dissolving the borders of the individual areas, enabling dense living in close proximity.

b&k+ brandlhuber & kniess GbR: Kölner Brett, 1999

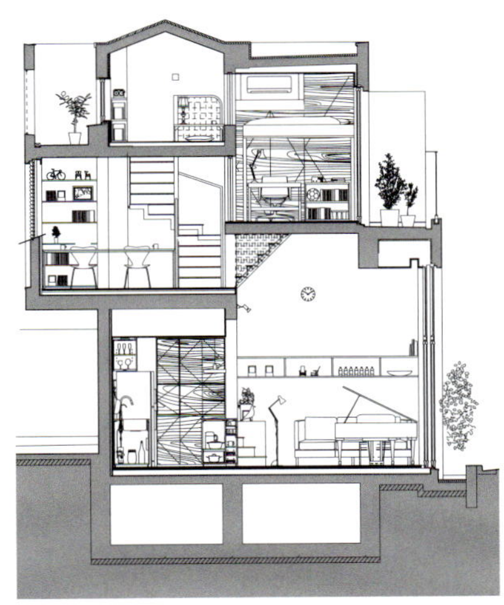

Field Design Architects: SL Court, Urawa, 2014

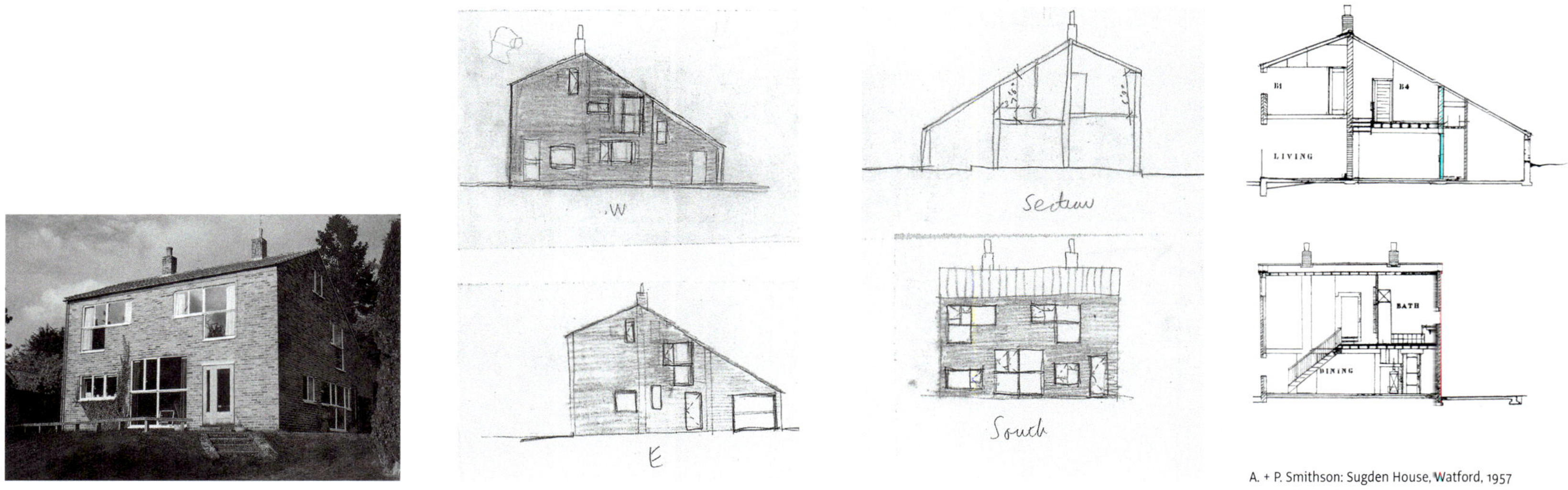

A. + P. Smithson: Sugden House, Watford, 1957

Pleasure

In the drawings "Small Pleasures of Life," Alison and Peter Smithson sketch an outline of how plain use can become a true pleasure. They cite the places where occupants can take possession of the house and explain how the house as a whole is enriched by the subtle definition of these places. Their observations seem to be rooted more in the experiences of an occupant who knows how to enjoy amenities like these than in the functional knowledge of an architect.

It must be noted that the highly specific window openings, in particular, are functionally predetermined in their sketches: they are arranged to direct the sightline and to clearly define the position of the inhabitant. There are spaces for sitting and looking outside, just as there are spaces for introspection. Rooms are not defined and delimited as dining room, bedroom, or kitchen; rather the architects extend an invitation to settle down in specific areas. And although one of the areas may be suitable for sitting on a sofa, that is by no means an indication that this space has been exclusively defined as a living room.

Alison and Peter Smithson's Sugden House is similarly distinguished by this kind of spatial definition of locales and the finely calibrated treatment of boundaries and transitions. This begins with the earth wall and the bench along its edge, which mark the ground level of the house. The house seems to have several entrances, giving equal value to all sides. The narrow metal sheets above the lintel seem to mark rather than shelter these entrance doors. The brickwork that surrounds the openings enhances the corporeal integrity of the house. In the interior, these openings inspire specific uses and activities; the inhabitant is connected to the outside world in different yet always unique ways.

In the house, the subtle treatment of the details characterizes the atmosphere of the whole. Take, for example, the additional step by which the bedrooms are raised: this gives them even greater privacy, while the living room below benefits from a higher ceiling and thus gains a greater spatial presence. The fireplace and the stairs define the large common area on the ground floor. At the same time, they also divide the large room and create a sense of shelter for the occupant rather than feeling "exposed" in a single large space. The bottom of the staircase is positioned in a manner that allows just enough space for lateral movement, while suggesting the path down the middle of the living room.

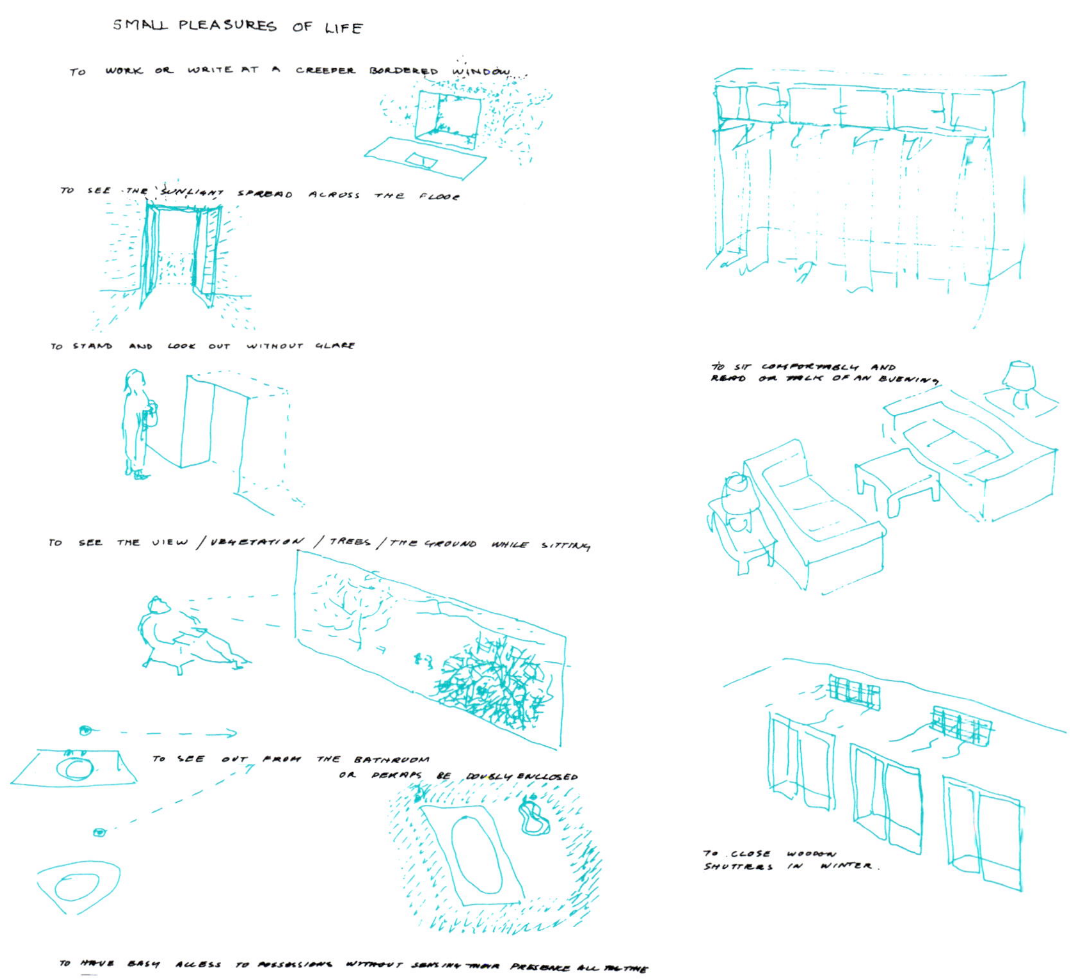

SMALL PLEASURES OF LIFE
to work or write at a creeper bordered window // to see the sunlight spread across the floor // to stand and look out without glare // to see the view / vegetation / trees / the ground while sitting // to see out from the bathroom / or perhaps be doubly enclosed // to have easy access to possessions without sensing their presence all the time // to sit comfortably and read or talk of an evening // to close wooden shutters in winter
A. + P. Smithson, in: Changing the Art of Inhabitation, London, Munich, 1994

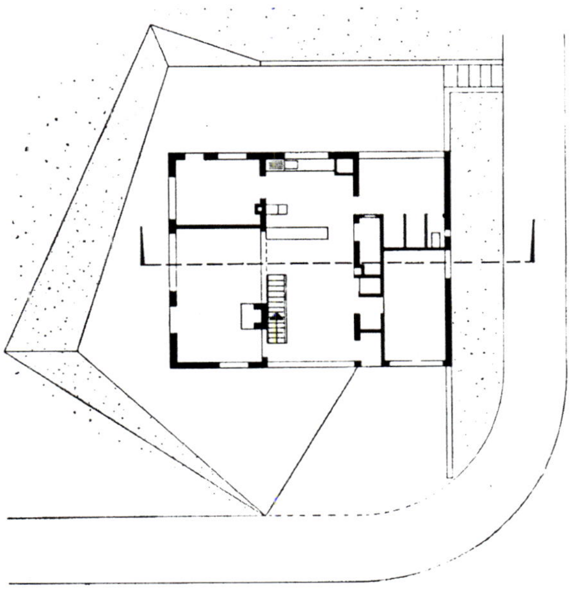

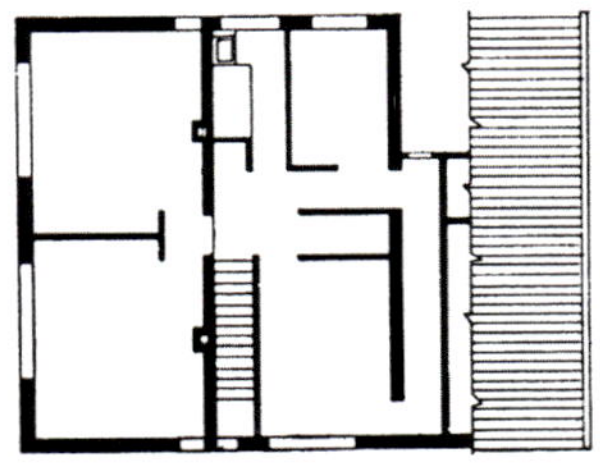

A. + P. Smithson: Sugden House, Watford, 1957

Although the floor plan is far from undetermined, it allows for great flexibility and the house thus achieves a balanced relationship between predetermined function and possible use. The design for the Sugden House "anticipates the matters of living," as Martin Steinmann* writes. The house seems to be imbued with a profound knowledge of what makes habitation enjoyable and how use makes the house into a home.

Similarly, Rudolf Olgiati's Witzig House could have served as a motif for Alison and Peter Smithson's sketches in that it is laid out with such a sense of comfort and scenic spaces, offering its inhabitants places for enjoying the surroundings. The funnel-shaped view from the dining area, the large window overlooking the valley, the sculptural cut-out window combined with a bench for taking in the mountain in one direction and facing the fireplace in the center of the room in the other direction: these spaces alternate between opening up expansive vistas and appearing enclosed and introverted. Occupants naturally use houses of this kind in a very intuitive manner.

Such floor plans require readers who take pleasure in deciphering their distinctive features. When encountering irritations or blank spaces, the question "how is this supposed to function" should be an expression of curiosity rather than a mere commentary.

This book addresses questions of this kind. Ultimately, the Floor Plan Manual Housing also provides an opportunity to read potential ways of living into the numerous examples it presents. The reader can peruse all the floor plans in this book, reading them simultaneously in a critical and curious, an analytical and imaginative manner. Even as bedtime reading.

* Literature
Martin Steinmann, "Das Haus als meine Welt – Zum architektonischen Denken von Michael Alder" (werk, bauen+wohnen 06, 2001)

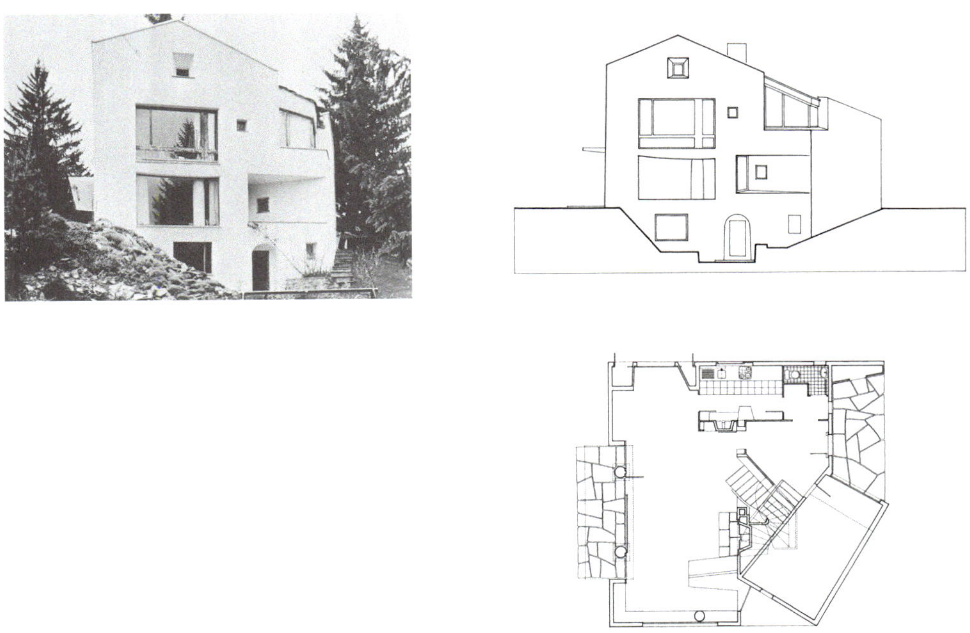

Rudolf Olgiati: Witzig House, Flims-Waldhaus, 1966

SMALL PLEASURES OF LIFE: to read in bed

A Chronicle of Developments in Housing

This timeline chronicles significant developments in housing from the industrialization of the 19th century to the year 2000. The passage of time appears to be the best nexus to visualize concurrent historical turning points and paradigm shifts, as well as buildings and newly developed building types emerging as a reaction thereto, or with the capability to set new standards around the world.

1850

Urban Housing Shortage

At the beginning of the 19th century, the industrialization led to an unprecedented migration of job seekers from rural regions to the factories in the rapidly growing cities in Europe and North America. Specific urban milieus arose, with the proletariat and the prosperous bourgeoisie in extremely different housing types and locations. In tenements developed on a massive scale, miserable living conditions were often the norm. Berlin's building code of 1853 had yet to define the apartment as an enclosed entity. Instead, rules determined the maximum allowable occupancy for individual rooms – single rooms with shared bathrooms and kitchens often accommodated entire families. For the most part, singles rented only a place to sleep.

Paternalist Workers Settlements/New Building Type: Row Houses

As an alternative to the generally miserable living conditions in the big cities – partly built for philanthropic reasons, and partly out of the fear of workers' riots – industrial workers settlements were realized in the immediate vicinity of production centers. The predominant housing types were row houses arranged in village-like layouts – often with garden plots to grow food to consume or sell as a source of supplementary income. The housing units in such settlements made a significant contribution to the development of the row house as a building type.

Workers Settlements/Row House
1850–1863
Saltaire, Bradford (UK) – Titus Salt (owner)
Settlement with row houses and communal facilities – built to the southwest of the factory to minimize exposure to harmful pollution.

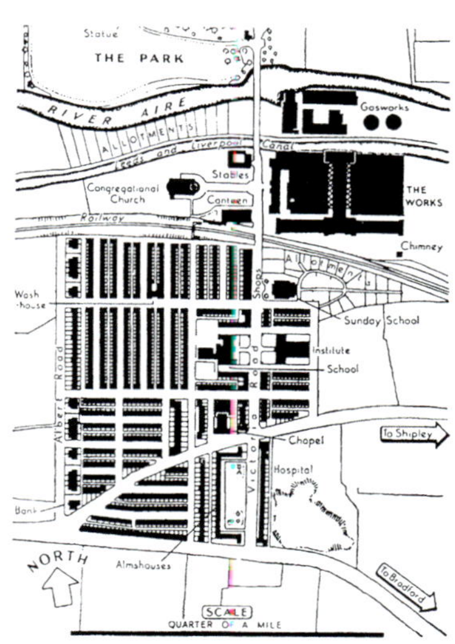

Workers Settlements
1859
Familistère, Guise (FR) – Jean-Baptiste André Godin
An ensemble of three buildings in an axial formation was designed and also cohabited by the owner of the nearby factory. Besides covered courtyards, which could be used for community events, other amenities included a kindergarten, school, bathhouse, and theater. Godin's ensemble followed the model of the Phalanstères as postulated by the utopian Charles Fourier, which had a deliberate resemblance to the palace at Versailles and proclaimed a cooperatively organized, collective vision of living and working.

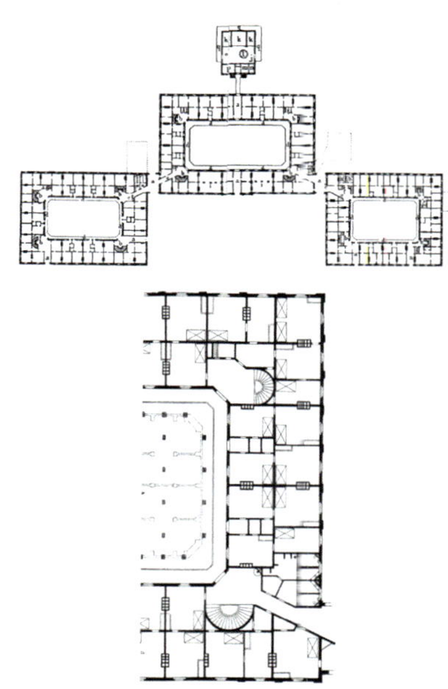

1889
Humane Housing Standards
Shortly before the advent of modern architecture, ideas on social reform led to improved living standards for the working class throughout Europe. While in England, private enterprises and philanthropic foundations operated freely in the market economy, in 1889, the Société Française à Bon Marché in France and the German Cooperative Law established the pillars for more humane living conditions.

Housing Types for the Urban Bourgeoisie
With the industrialization and the growth of cities, specific housing types arose for the urban bourgeoisie as well. Initially laid-out with palace-like suites of rooms and built for the upper class, more specific solutions in smaller units were developed for the growing middle class – with representative sequences of rooms along the street and, in many cases, separately organized entrances and areas for servants.

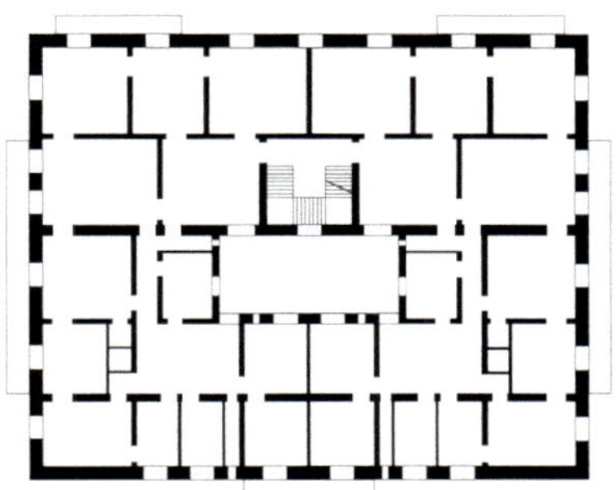

Housing Types for the Urban Bourgeoisie
Approx. 1870
Casa Blocco, Via Assarotti, Genoa (IT)
This building type emerged from the tradition of city palaces for wealthy families and was adapted to an urban, multi-story residential construction for the middle class. The two apartments per floor are laid out with similarly sized rooms partially offset to one another, revealing an almost use-neutral floor plan organization.

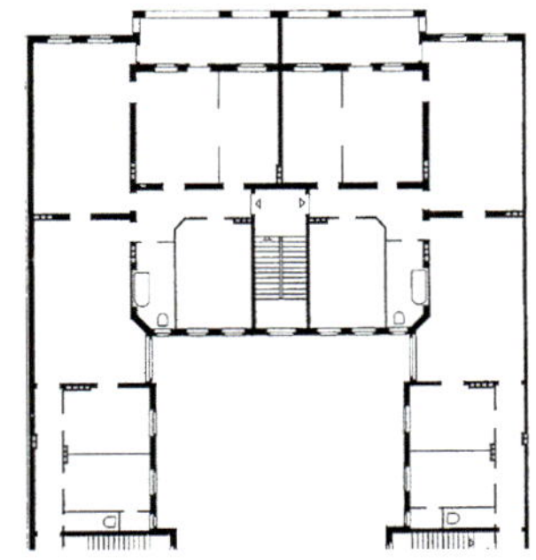

Housing Types for the Urban Bourgeoisie
1897
Brückenallee 4, Berlin (GER), Apartment Types for the Middle Class
In essence, most block edge residential buildings in Berlin are morphologically similar, with a load-bearing middle wall parallel to the street facade and a front and rear building with side wings. They often housed residents of diverse economic standings under the same roof. The wealthy looked out over the street, low-paid workers into generally dark courtyards. Apartments for the middle class and substantially larger ones for the wealthy were equipped with separate circulation routes for servants – with their own staircases, hallways, and wings; the living areas of the wealthy were linked in open enfilades.

1897

New Building Type: Apartment House (USA)
The row house dominated the streetscape of American cities until the early 19th century. A tenfold increase in the urban population between 1830 and 1870 led to rising land prices. Thus, despite a general distaste for multifamily dwellings so-called "French flats" were introduced and by 1875 regarded as an acceptable residential model for the middle class. Over time, these buildings expanded in size to cover entire blocks and were equipped with luxurious inner courtyards, penthouses and roof gardens in order to appeal to more wealthy segments of the population.

Apartment House/Housing Types for the Urban Bourgeoisie
1870
The Stuyvesant Flats, New York (USA) – Richard Morris Hunt
Manhattan's first multifamily dwelling had a two-part circulation core – with a main staircase for residents and a second for servants. While the representative rooms looked out over the street, a long corridor led to the rear of the building, where narrow courtyards provided the private chambers with some daylight.

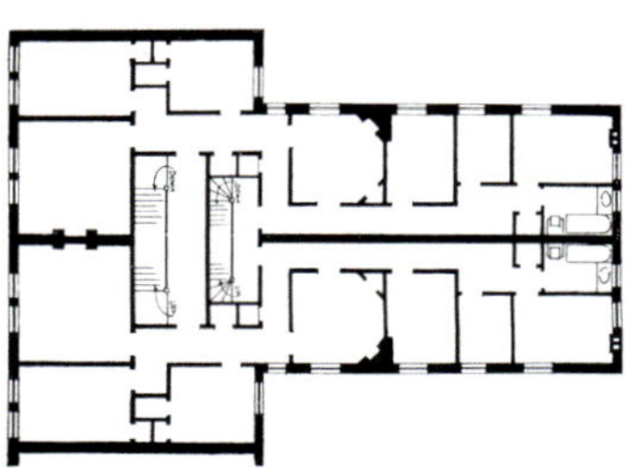

Urban Housing Shortage
1870
Perleberger Strasse 14, Berlin (GER)
Common building type for working families, with shared kitchens and WCs on the intermediate stair landings.

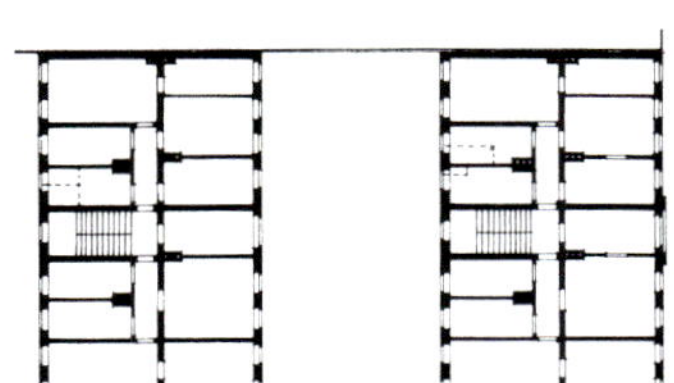

Workers Settlements
1879
Bournville, Birmingham (UK) – George Cadbury (Owner)
Spacious semi-detached house type in a workers village with recreational facilities, schools and its own train station.

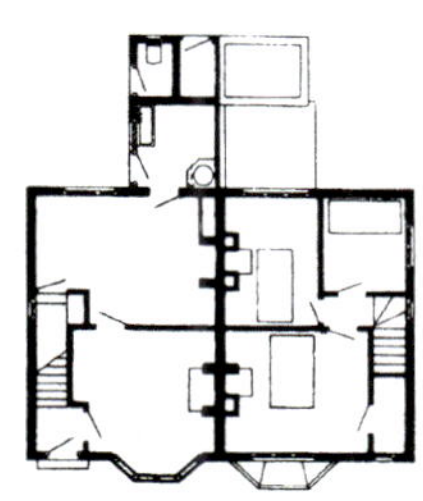

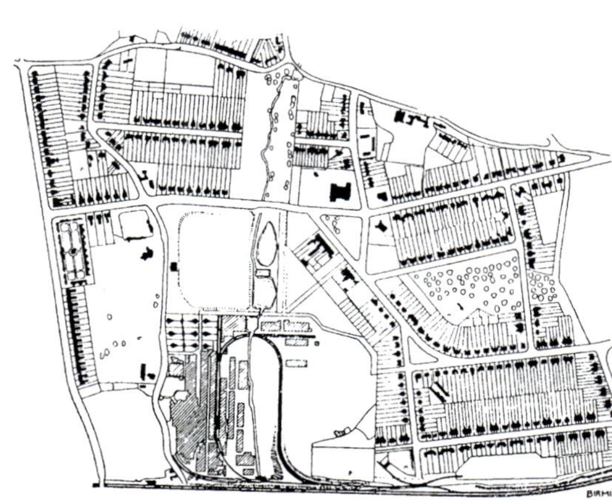

Apartment House/Housing Types for the Urban Bourgeoisie
1890–1930
Chicago courtyards (USA)
The organization of a U-shaped building around a courtyard open to the street became a typical convention for numerous, primarily three-story apartment buildings following the Great Chicago Fire in 1871. Double-loaded staircases are accessed directly from the landscaped courtyard, which serves as a semi-public transition from the street to the individual units.

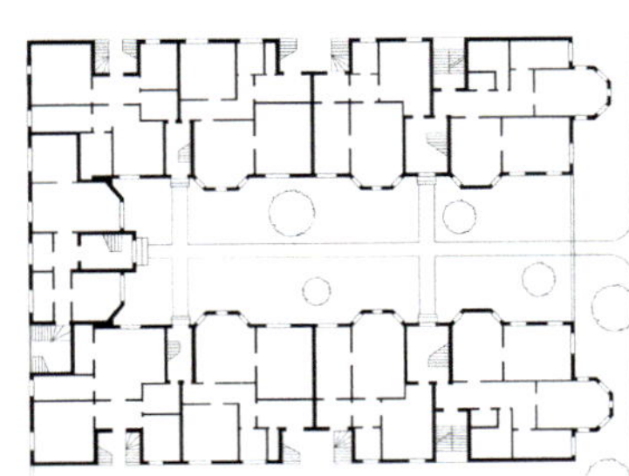

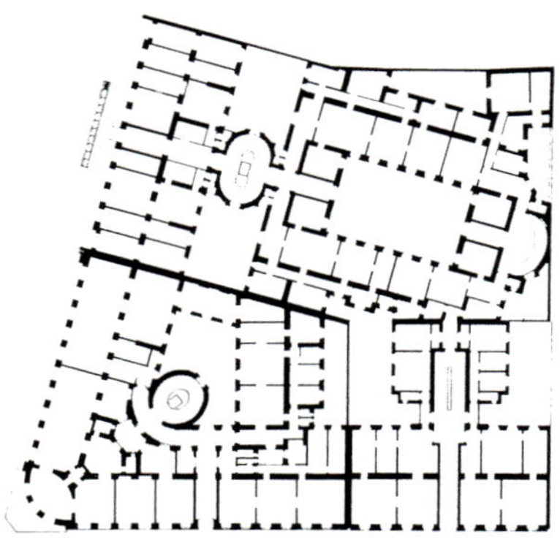

Housing Types for the Urban Bourgeoisie
1899
Linke Wienzeile, Vienna (AT) – Otto Wagner
Residential blocks for the upper middle class, with servant wings along very narrow courtyards.

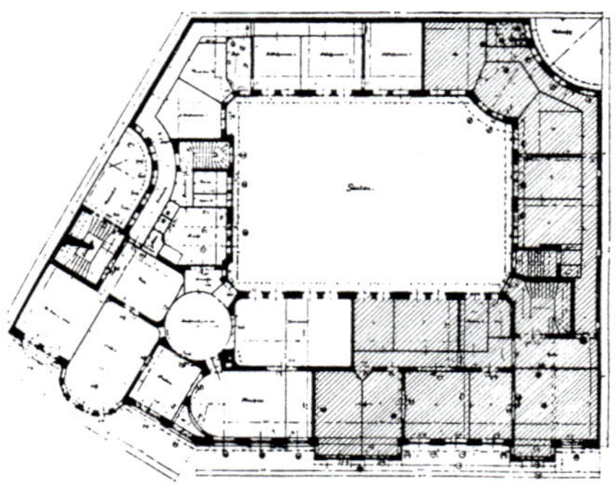

Housing Types for the Urban Bourgeoisie
Approx. 1900
Typical Block Edge Building Berlin (GER)
Apartment types for the upper class.

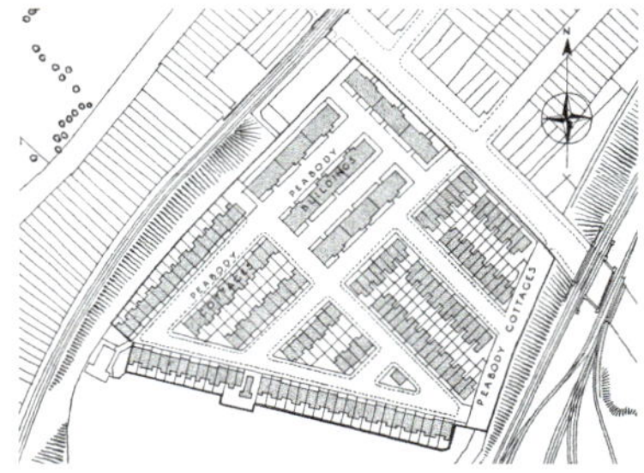

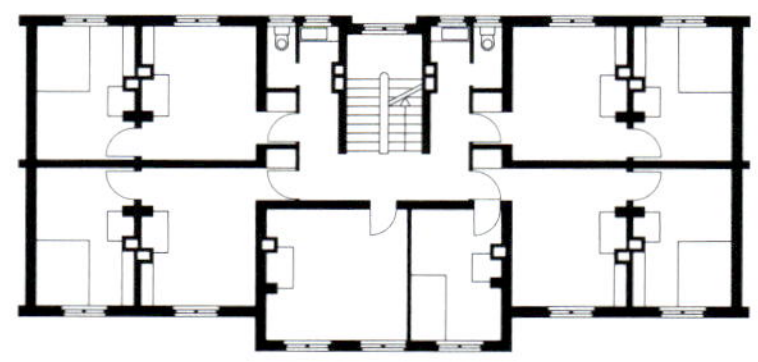

Humane Housing Standards
1902-08
Herne Hill Peabody Estate, London (UK) – Darbishire, Wallis, Wilkens
Developed as an inner-city housing type for working families, five 1–3-room apartments with shared bathrooms and washing rooms on each floor form a rational and cost-saving spatial principle, which was implemented in numerous residential buildings by the Peabody Trust over the following 40 years.

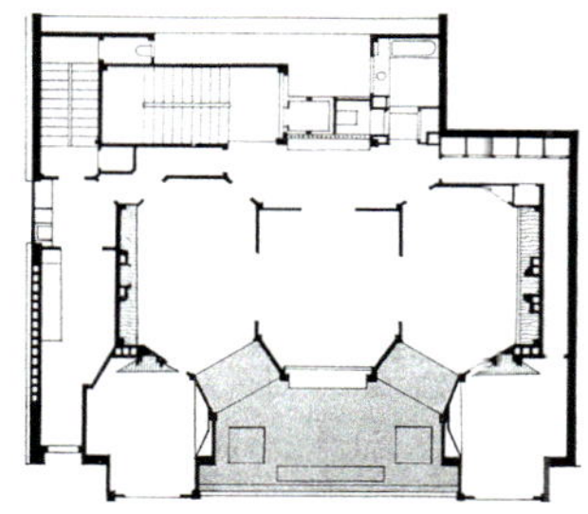

Housing Types for the Urban Bourgeoisie
1903
Rue Franklin, Paris (FR) – Auguste Perret
One of the first residential buildings erected in reinforced concrete. The living areas are generously linked to form an enfilade, the kitchen and servant staircases remain separate. Above the commercial ground floor, the building is U-shaped in order to bring in more natural light.

1899

Urban Housing Shortage
Approx. 1900
Bassena House Building Type, Vienna (AT)
In Vienna, the building code permitted the development of 85 % of a building site. The development of the Bassena House – named after a washbasin located in the hallway – at least provided each one-room family unit with its own kitchen in an anteroom.

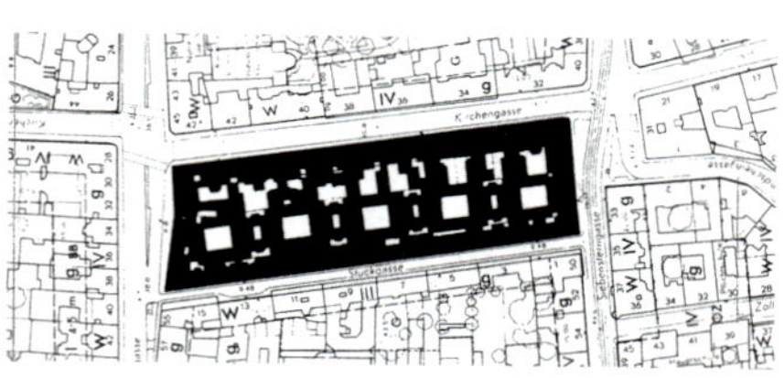

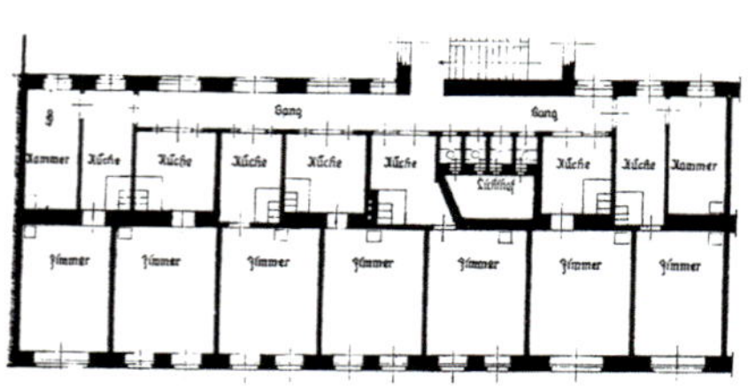

1902
Garden Cities of To-morrow, Ebenezer Howard

In light of miserable urban living conditions, a visionary idea designed to create a symbiosis between the advantages of rural and urban life. In a scheme of concentric rings, garden cities surround a central city beyond a green belt. The population is limited – if exceeded, further garden cities supplement the urban group. This influential model was implemented only in terms of typology (and drastically changed), while the economic and social strategies – simple building methods, collective forms of ownership, and the prohibition of land speculation – slipped into obscurity. The concept was influential above all in the later development of New Towns as well as suburban, single-family house settlements.

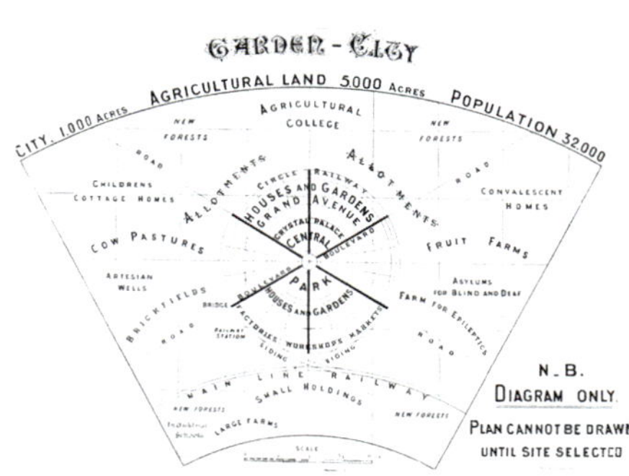

Garden Cities/Row House
1906
Hellerau Garden City (GER) – Riemerschmid, Muthesius, Tessenow
Muthesius' design for a 5.40 m-wide timber framed row house type with three rooms, an eat-in kitchen and annex with WC, stall, and laundry room.

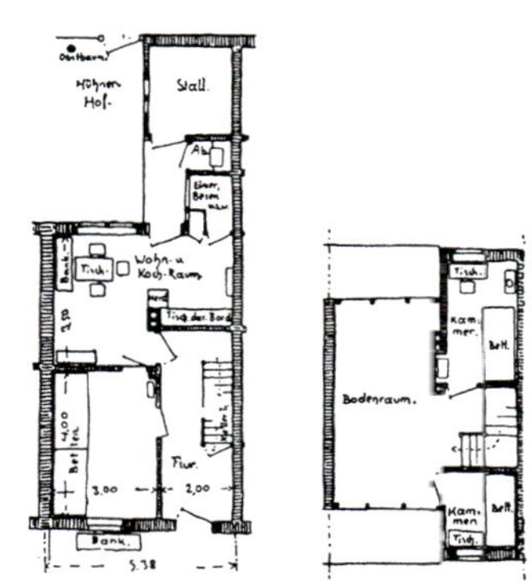

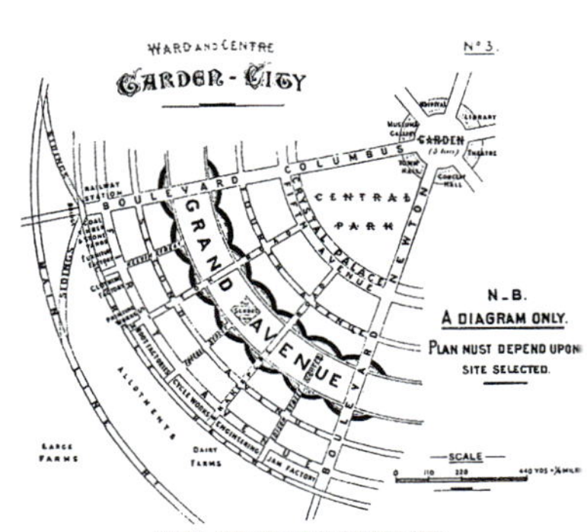

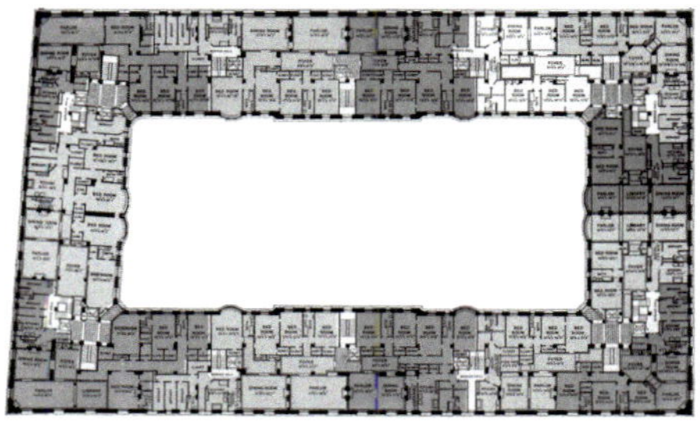

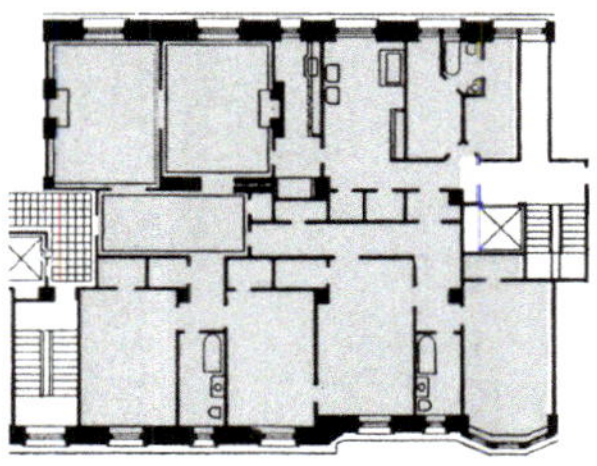

Apartment House/Housing Types for the Urban Bourgeoisie
1907
Belnord Apartments, New York (USA) – Hiss and Weekes
Occupying an entire city block, a massive cubic volume is organized around a spacious semi-public courtyard with a landscaped garden inspired by European models. The apartments are divided down the middle by a generous hallway space: representative living rooms (and servants' quarters) are oriented toward the loud street, the sleeping quarters toward the courtyard.

City Expansions

Large-scale city expansions helped alleviate the housing shortages. Influenced by contemporary social debates, considerable improvements in living standards were followed, such as the provision of public parks and community facilities as well as more modern apartments – spacious layouts for families and units specifically designed for single households – with much consideration given to the quality of light and air circulation.

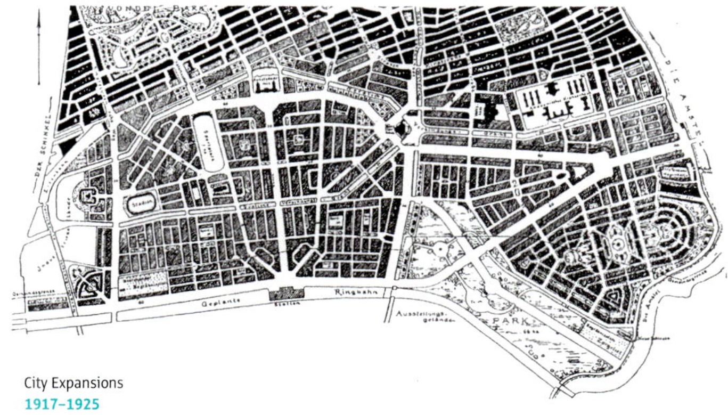

City Expansions
1917–1925
Amsterdam Zuid (NL) – H. P. Berlage
Berlage's second master plan for the expansion of Amsterdam placed important public spaces and buildings as accents at the junctures of the main axes flanked by long residential blocks. In order to achieve a mixed balance of residents, diverse apartment types were created – including minimal apartments for singles – in the slim, individually articulated buildings with large interior courtyards.

1918

Urban Housing Shortage
1906
National Hotel, San Francisco (USA)
"Rooming houses" were developed to fill a niche in the real-estate market: cell-like units are organized around light and air shafts and share a bathroom. In these buildings intended to house temporary renters, common shared spaces were reduced to an absolute minimum in order to maximize profits.

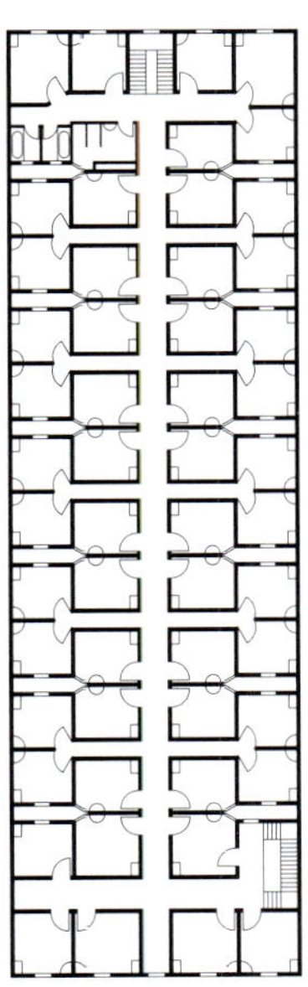

Garden Cities/Row House
1908
Letchworth Garden City (UK) – Unwin & Parker
A self-sufficient estate incorporating areas for farming, industry, a park and a residential quarter laid out in a star-shaped formation. The diverse row house types are remarkable for their shallow depth and dually oriented rooms.

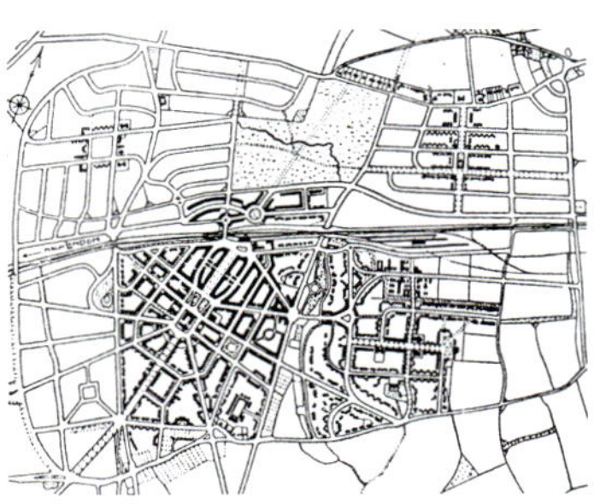

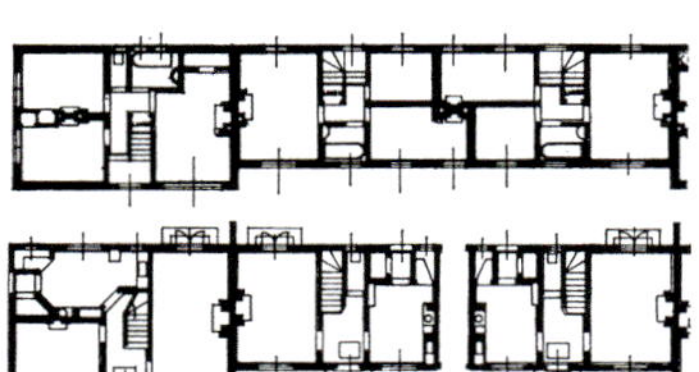

Affordable Housing

After World War I, economic crises and egregious housing shortages in Europe led to the vigorous pursuit of cost-effective mass housing solutions. The rationalization of production methods and a principally economic view of floor plan organization, building types, and their ensembles with the aim of standardization governed the debate on housing for many years to come.

Affordable Housing
1918
"On Economical Building: A Contribution on Housing Estates" – Peter Behrens, H. de Vries
This comprehensive study addresses, among other things, grouped buildings and back-to-back housing typologies in order to more effectively utilize building parcels. It presents strategies for typification and rationalization and encourages the use of affordable materials like concrete.

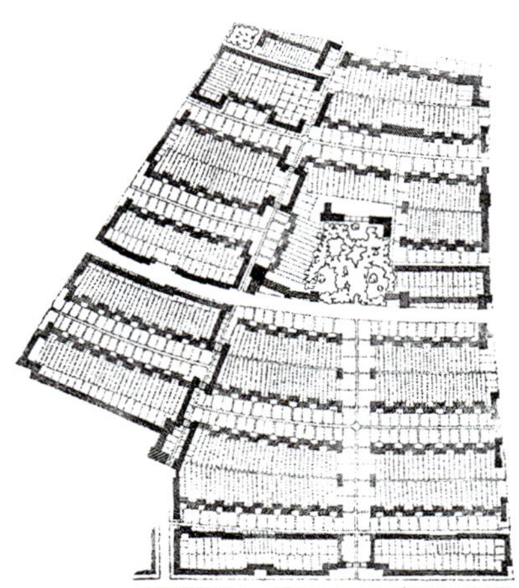

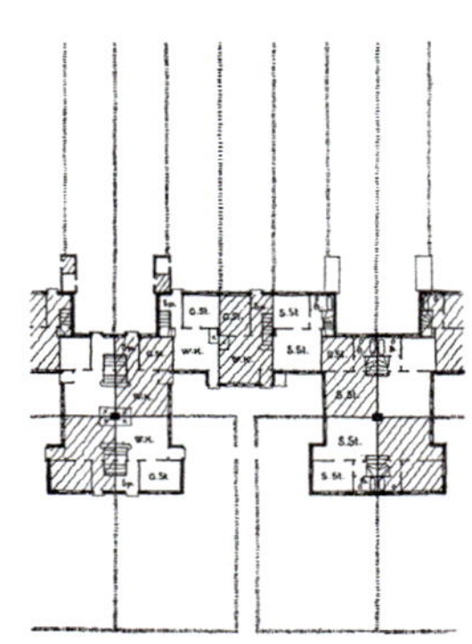

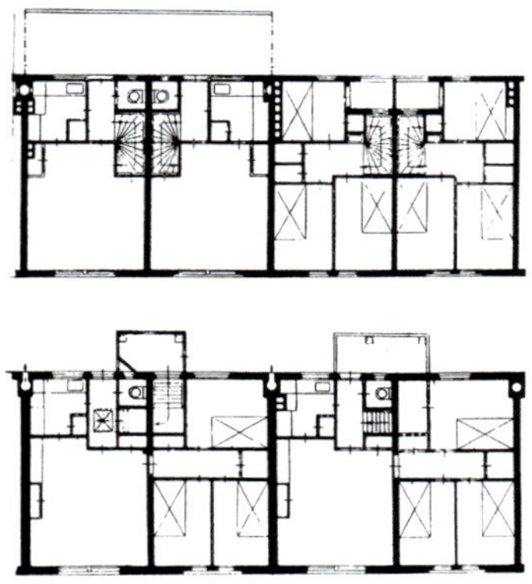

City Expansions
1919–22
Justus Van Effencomplex, Rotterdam-Spangen (NL) – Michiel Brinkman
As part of the development of Spangen Polder under the direction of city-architect J. J. P. Oud, where – inspired by Berlage – closed, mid-height block edge buildings were preferred over the garden city typology, here a symbiotic model was sought after. An urban block edge development encloses common courtyards with separate programs. The single-floor apartments on the ground and 2nd floors are accessed directly from the courtyard level, while the prominent, all-encompassing, and almost street-wide galleries on the third floor provide a second stage for neighborly interaction and access to the maisonette units.

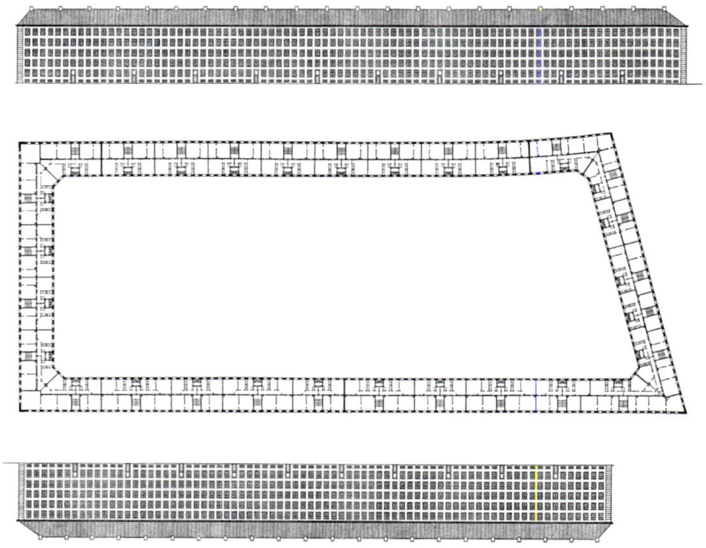

City Expansions
1923
Hornbækhus, Copenhagen (DK) – Kay Fisker
A "reform block" with roughly 300 apartments in an urban expansion zone surrounds a spacious, green courtyard. The homogeneity of its exterior facade lends it a monumental character. The living rooms face the street, while bedrooms and the kitchen – directly linked to the garden below by a second staircase – overlook the interior of the block.

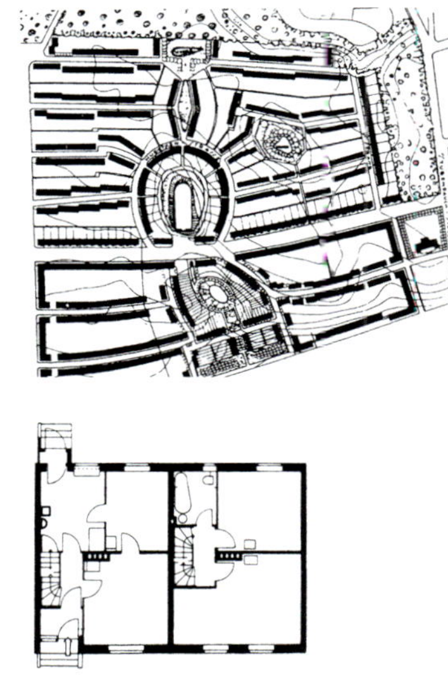

City Expansions
1925
Hufeisensiedlung, Berlin-Britz (GER) – Bruno Taut
One of 17 large housing developments built in Berlin between 1925 and 1933 with an intentional humanistic approach to modern architecture, this project saw the combination of single-story flats in the "horseshoe" with row houses. At the time, the flats were of ample size (living and dining room 24 m², kitchen 10 m², master bedroom 20 m²) and each with its own garden plot in the yard of the horseshoe.

1919

Affordable Housing/Row House
1920
"House with One Wall," Vienna (AT) – Adolf Loos
Loos patented his construction for this economic row house type: only the dividing walls between houses were bricked, the facades are made of wood.

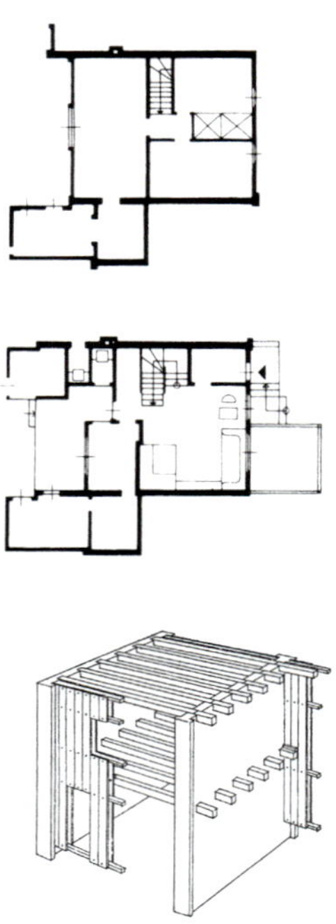

Affordable Housing/Garden City
1924–28
Sunnyside Gardens, New York (USA) – Clarence S. Stein, Henry Wright
Affordable housing development with simple flats and duplexes for families is a predecessor of garden cities in the USA. The layout of the estate adheres to the existing block structure, however, by not parceling out the individual lots, an open ensemble with offset buildings emerges where green spaces in front of and behind the houses connect to one another.

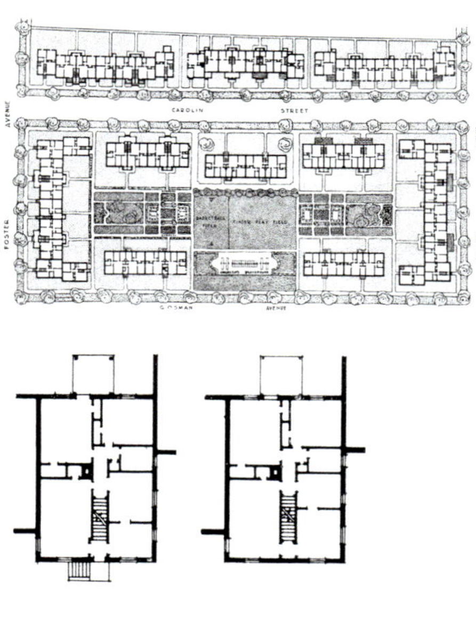

Row House
1925
Cité Moderne, Brussels (BE) – Victor Bourgeois
The design enriches the row house typology with a jagged row house development set at a 45° angle. Partially linked in their corners, rooms surround all sides of a central hall with staircase – day rooms as well as the kitchen and WC on the ground floor, bedrooms and bathrooms on the floor above.

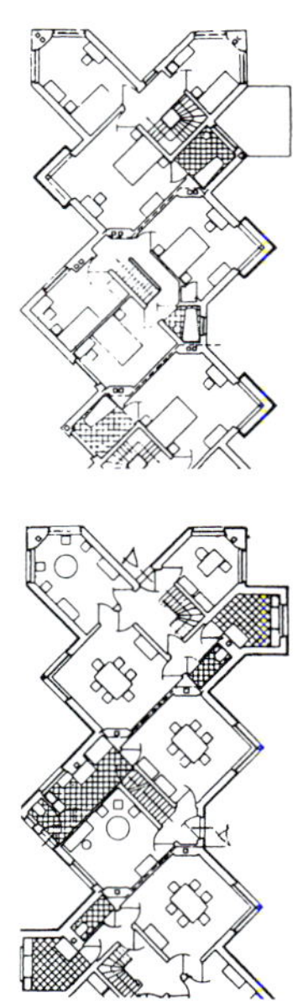

Affordable Housing
1925
Multifamily Dwelling – Ludwig Hilberseimer
Rational floor plan with combined kitchen and bath, living room as circulation space, and cabin-like rooms.

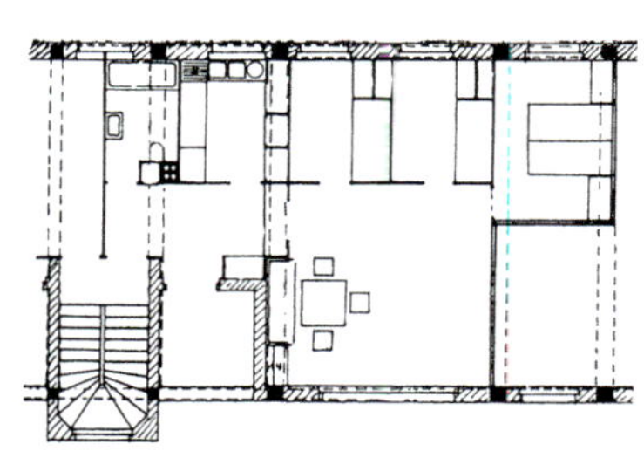

Row House
1925
Kiefhoek Row Houses, Rotterdam (NL) – J. J. Oud
Extremely compact, 4.20 m-wide house type for six people with minimal circulation space and no bathroom.

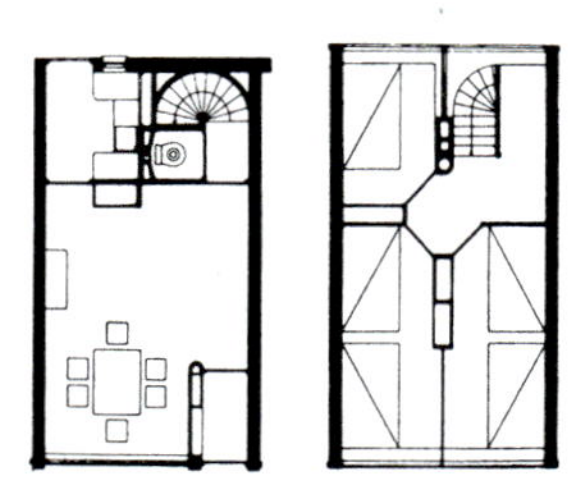

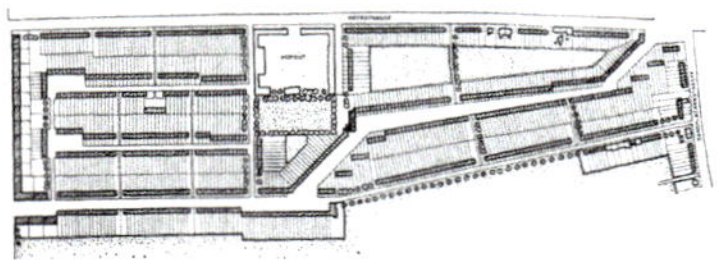

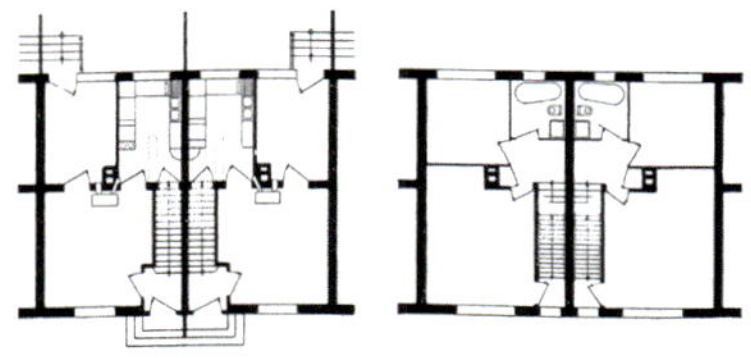

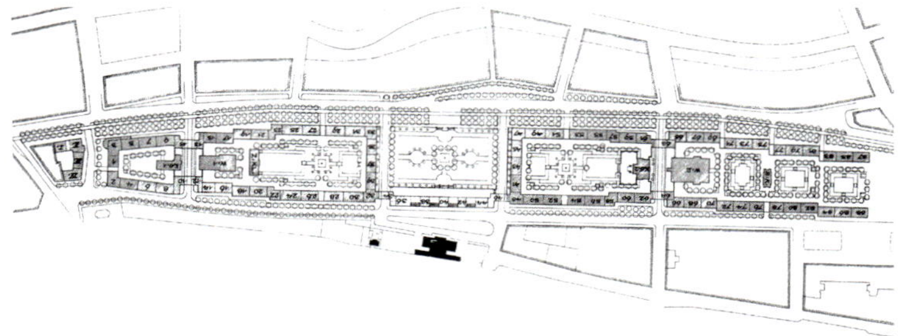

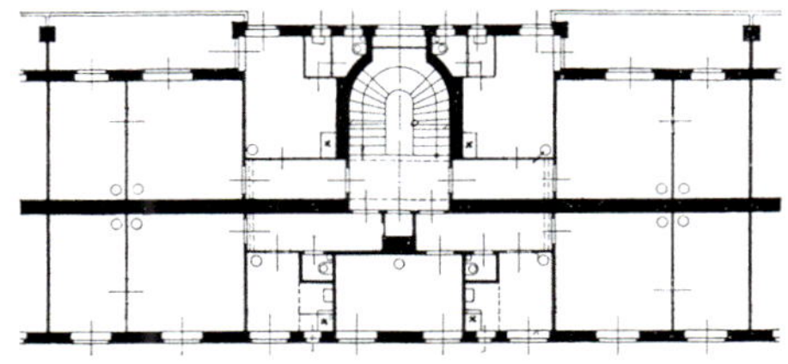

Building Systems

The development of innovative building elements, new concepts of prefabrication, and efficient, systematic building processes was influenced by many factors: massive housing shortages, a fascination with technical innovation such as Ford's new automobile production methods, new steel construction methods in American high-rises and industrial buildings, as well as the search for a rational aesthetic. According to Ernst May, the reduction in shape and size of building parts was intended to "have a beneficial impact on the simplicity and integrity of architectural design."

Building Systems/Affordable Housing
1926–30
Frankfurt Slab System, Floor Plans for the Praunheim Slab Type (GER) – Ernst May
Over 1,000 houses were built in Frankfurt using such industrialized assembly methods. The serial production of concrete building elements and the introduction of modern building methods were aimed at reducing building times and optimizing the construction of housing, so that even residents with modest incomes could manage to acquire affordable living space. The ground floor plan in Praunheim – part of the "first test group in Frankfurt" – followed a principle similar to all residential buildings in Frankfurt's housing program, in which a spacious living room is separated from a small, standardized kitchen outfitted with built-in furniture and equipment designed by Margarete Schütte-Lihotzky.

City Expansions
1927–30
Karl Marx Hof, Vienna (AT) – Karl Ehn
As an urban manifest of social housing, the blocks of this estate had large common courtyards with communal facilities such as laundry rooms, bathhouses, a library, shops, and restaurants. While the exterior is almost fortress-like in nature, the apartments are very compact and for the most part oriented in only one direction.

1927

Building Exhibitions and Model Estates

Starting in the 1920s and initiated by the "Deutscher Werkbund" (German Association of Craftsmen), among others, building exhibitions became testimonials to the "Neues Bauen" movement, which introduced Modernist architectural concepts – until then only discussed in international avant-garde circles – to the wider public. Together with model estates, they became an instrument of planning and building culture, which with often uniform concepts and projects created impulses for aesthetic, technical, and social changes.

Building Exhibitions and Model Estates
1927
Building Exhibition Weißenhofsiedlung Stuttgart (GER) – Ludwig Mies van der Rohe (master plan)
In contrast to more established conventions of the time, the master plan called for a layout of freestanding buildings with flat roofs, following the topography of the site.

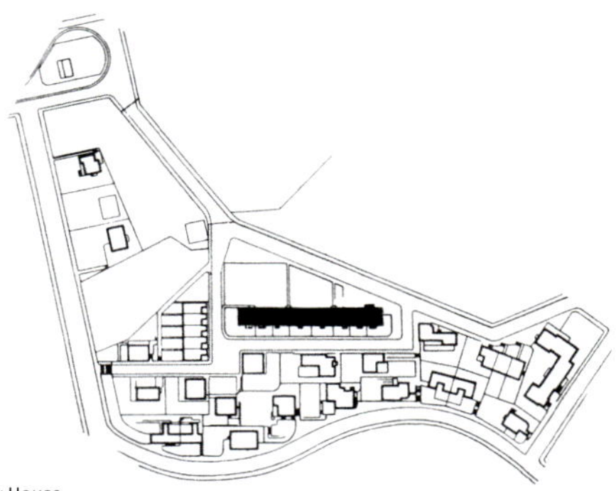

Building Exhibitions and Model Estates
1927
Weißenhofsiedlung Housing (GER) – Ludwig Mies van der Rohe
The steel-frame construction and the neutrality of the facades enable flexible floor plans arranged with movable walls located between the load-bearing columns to create space for four to six inhabitants. Possible configurations include cabin plans with the living room as circulation center.

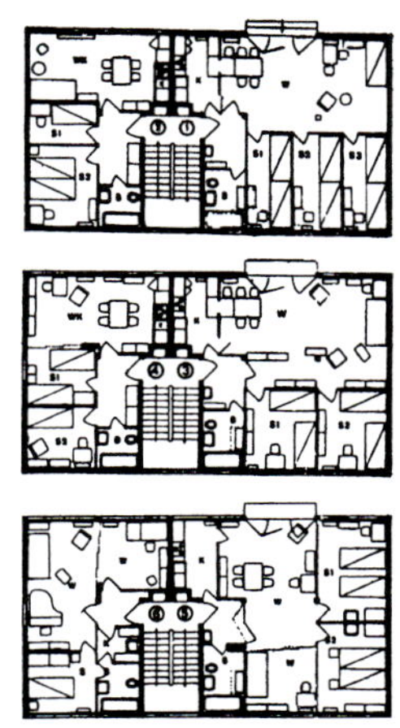

Building Exhibitions and Model Estates/Row House
1926
Bauhaus-Siedlung Dessau-Törten (GER) – Walter Gropius (master plan, row houses)
The row house types of this experimental estate project – on large parcels of land with enough room for keeping small animals or a vegetable garden – were built using serial construction processes and standardization methods and were intended as inexpensive housing for working families. An increase in building costs led to a further simplification of the later realized type "sietö IV": the split-level principle reduced the circulation space, baths were taken in the laundry room.

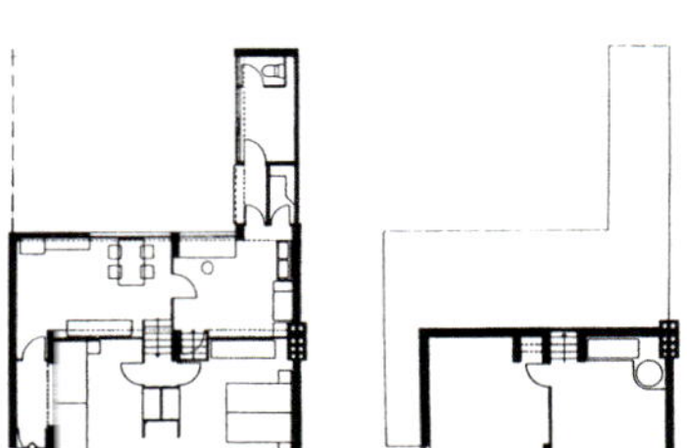

Building Exhibitions and Model Estates/Row houses
1927
Row houses, Weißenhofsiedlung Stuttgart (GER) – J. J. P. Oud
Housing type with low-lying annex for utility rooms on the stair landing, additional entrances from the rear yard, and a separate WC and bathroom (as connecting room).

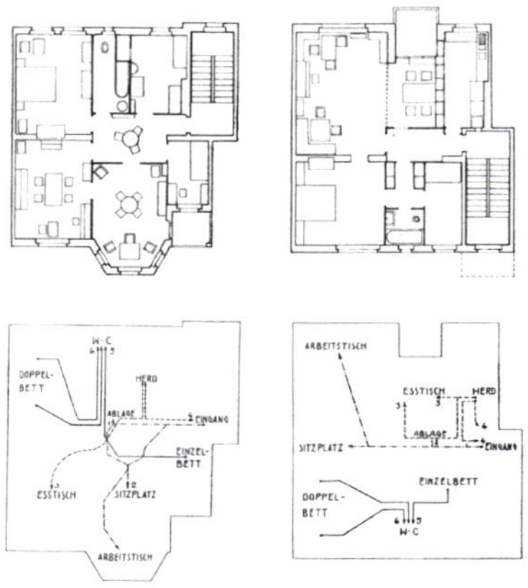

1928
Floor Plan Studies – Alexander Klein
Symptomatic of Modernist pursuits of systematically optimized housing types, such studies formed an important basis for their standardization. After extensive analyses of functional processes, movement flows, the placement of furniture, lighting conditions, and so on, Klein postulated a clear division of rooms into three groups – living, sleeping and utility – whose individual functions could in the best case be carried out with as little overlap as possible. The "Room Group" floor plan was presented in 1928 at the "Home and Technology" exhibition in Munich.

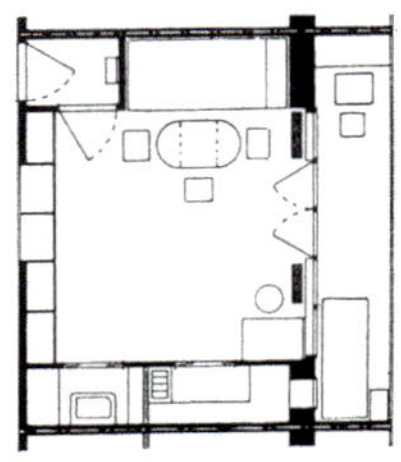

Floor Plan Studies
1928
Apartment for the Working Woman (GER), Margarete Schütte-Lihotzky
Apartment type presented at the "Home and Technology" exhibition in Munich.

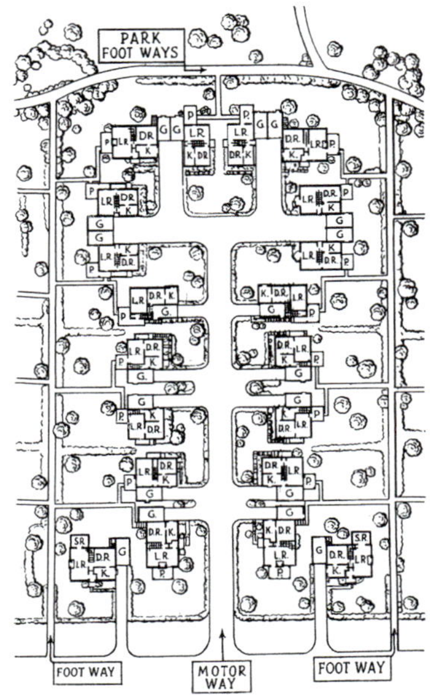

Garden Cities
1929
Radburn Estate (USA) – Clarence S. Stein, Henry Wright
Howard's Garden City concept served as a role model for the standardized, two-story detached homes – each with a garage, kitchen, living and dining room on the first floor and bedrooms on the second floor – arranged in groups around cul-de-sacs that flank a large public park. Semi-public paths connect the streets to this green space.

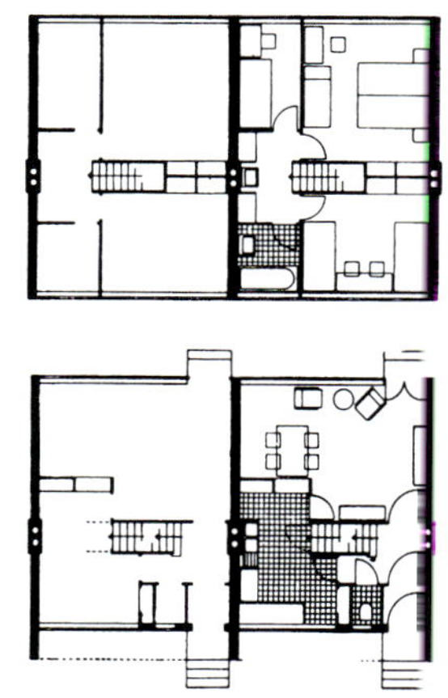

Building Exhibitions and Model Estates/Row House
1929
Dammerstock Estate Row Houses, Karlsruhe (GER) – Walter Gropius
Parallel slabs in a strict north-south orientation built according to "rational building methods." Half of the housing units are only 5.60 m-wide row houses, with transverse central staircases as organizing elements.

1928

New Building Type: Residential High-rises
The high-rise was developed in the USA in reaction to rising inner-city land prices following the Civil War. Made possible by innovations in steel-frame construction and the advent of new safety measures for elevators, they were initially conceived for commercial purposes. Around the turn of the century, residential high-rises emerged in a similar style. In part conceived as residential hotels to circumvent height restrictions for purely residential buildings, their shapes were for the most part derived from the lot structure and zoning laws. Starting in the late 1920s, varying models for different income groups led to the creation of both luxurious residences for the rich as well as inner-city housing for the middle class.

Residential High-rises
1928
Woodstock Tower, New York (USA) – H. Douglas Ives
Part of the world's first high-rise estate, Tudor City – a privately financed, speculative development aimed at offering office workers accommodation close to their place of employment – the building consists of over 400 hotel-like 1-room apartments ("Efficiencies") on its lower floors. More spacious apartments are located on the upper, terraced levels.

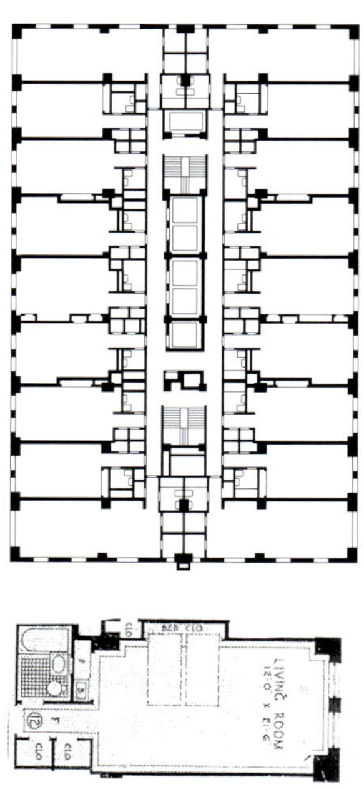

Residential High-rises
1929
San Remo Tower, New York (USA) – Emery Roth
The highest of Emery Roth's luxury high-rises on Central Park is comprised of two towers on a U-shaped, slightly terraced base covering the width of a city block. The maisonettes in the towers – with representative rooms below and sleeping quarters for residents and servants above – offer their inhabitants views in every direction and the feeling of living in a freestanding villa high above the city.

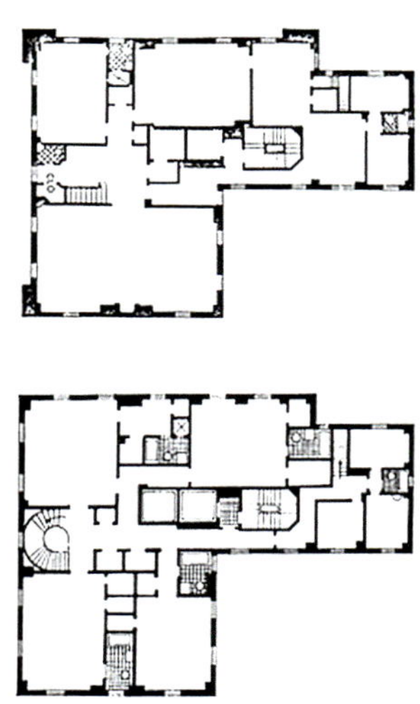

Residential High-rises
1929
1420 Lake Shore Drive, Chicago (USA) – Hooper & Janusch
9- and 11-room single-floor apartments and 16-room maisonettes are equipped with high-quality finishes, open fireplaces, and the latest innovations such as refrigerators. The floor plans exhibit a typical layout for the time period: large, representative rooms look out over Lake Michigan and are connected by a long corridor past the servants' quarters to the bedrooms at the rear.

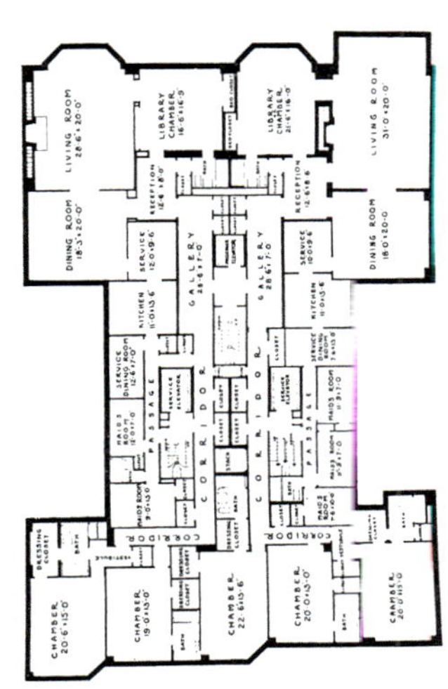

1929
Housing for the Minimum Subsistence Level

In 1929, an international group of architects under the direction of Ernst May posed the question of how to design an apartment for the minimum subsistence level at the second Congrès International d'Architecture Moderne (CIAM). 105 designs in a consistent graphic presentation with comparative values like apartment size, volume, and window surface area were collected to form a systematic floor plan typology for affordable housing.

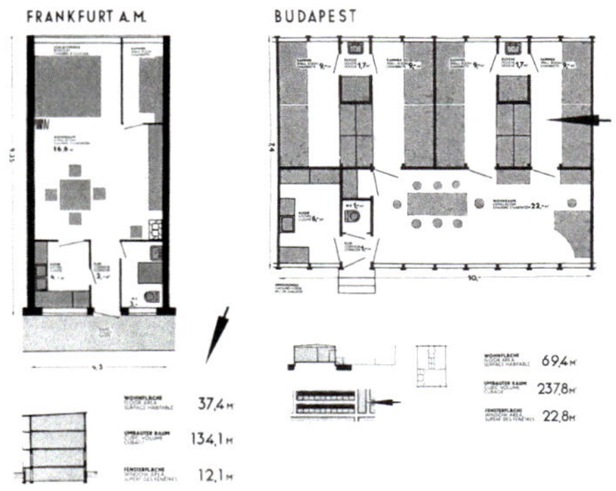

Housing for the Minimum Subsistence Level
1929
Apartment types Frankfurt a. M./Budapest

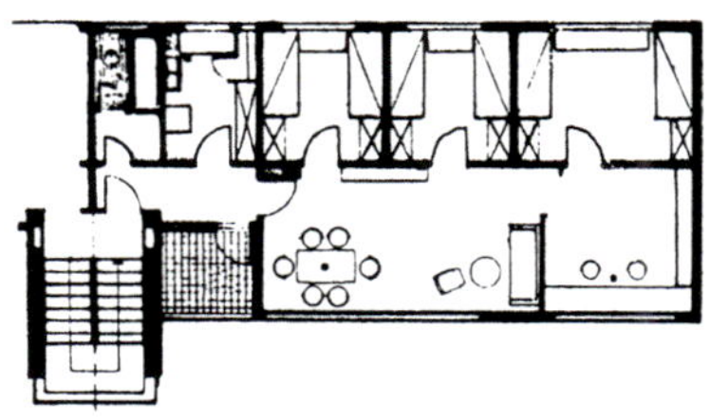

Housing for the Minimum Subsistence Level/Slab Housing
1930
Cabin plan, Kassel (GER) – Otto Haesler
Small individual rooms accessed through the living room.

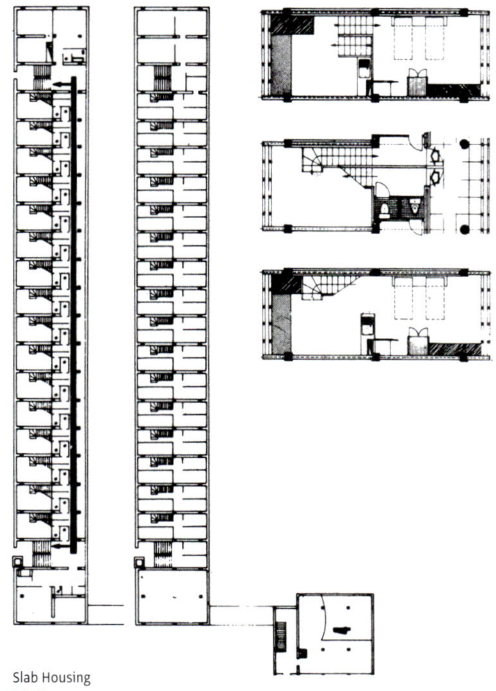

Slab Housing
1930
Narkomfin, Moscow (USSR) – Ginzburg & Milnus
A new model of society was proclaimed with the Russian Revolution, in which the traditional role of the family was questioned in favor of the collective. To facilitate the employment of women, kindergartens and canteens were incorporated into the building, while the private kitchens in the compact, split-level and duplex units of the residential slab were minimized. All collective spaces are accommodated in a separate building.

1930

Residential High-rises/City Expansions
1930–34
Les Gratte-Ciel de Villeurbane, Lyon (FR) – Môrice Leroux
Two long slabs, repeatedly interrupted by perpendicular, terraced residential towers, were created as part of a larger master plan for a new urban district. Various apartment sizes – but predominantly 3-room units – are efficiently designed to reduce the rent for the working-class residents. The high density of the configuration is moderated by setback roof terraces, public plazas, and a number of shops on the ground floor.

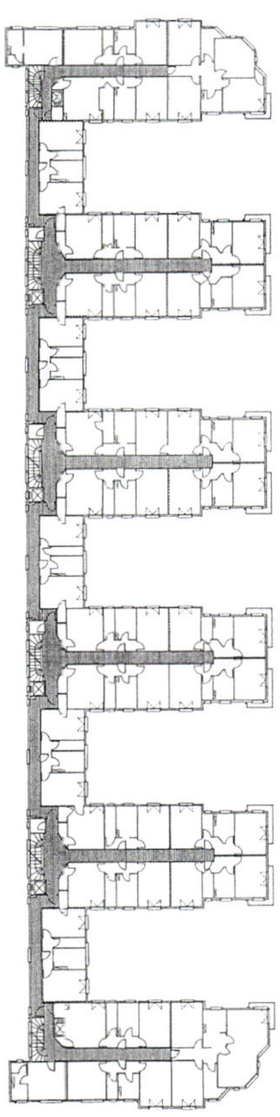

New Building Type: Slab Housing – Walter Gropius

Of the opinion that architecture should be practical and standardizable, Gropius established rules for a building type that was in part motivated by medical findings in the late 19th century concerning the effects of sunlight in preventing the spread of hazardous bacteria and the necessity of proper ventilation for healthy housing. The use of light studies resulted in precisely calculated distances between the parallel, open layout of the slab buildings oriented to the east and west. This configuration became an important instrument in modern urban planning. Although the scheme was implemented in row house developments as well, the dominant image of slab housing stems from multi-story, freestanding, linear buildings with shared access and circulation, be they vertical or horizontal.

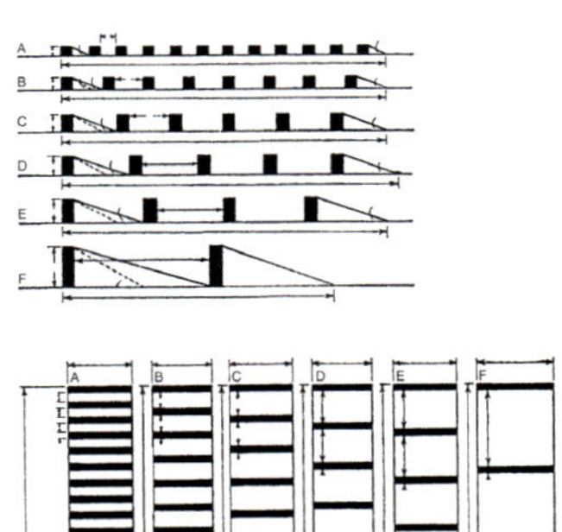

Slab housing/City Expansion
1930
Siemensstadt, Berlin-Jungfernheide (GER) – Hans Scharoun (master plan), Hans Scharoun, Walter Gropius, Hugo Häring (apartment types)
A strict east-west oriented layout of slab buildings for employees of the Siemens production facilities. Double-loaded staircases provide access to the apartments. In Scharoun's design (upper left), a living room spanning the width of the building – with a loggia, dining niche, and windows with flower boxes – serves as the central circulation space, from which a "slipper hall" leads to a group of bedrooms.

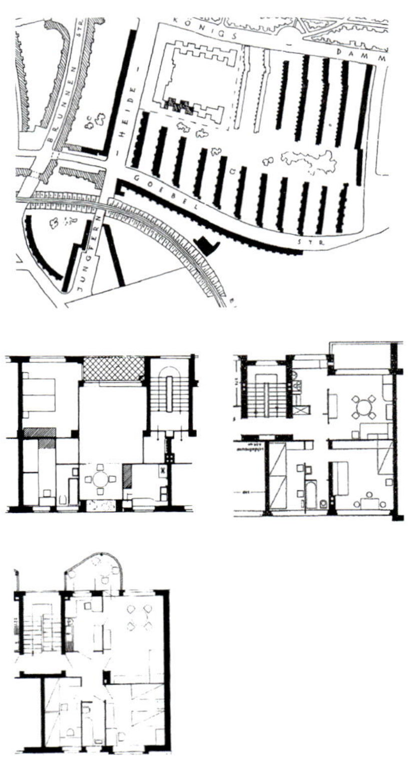

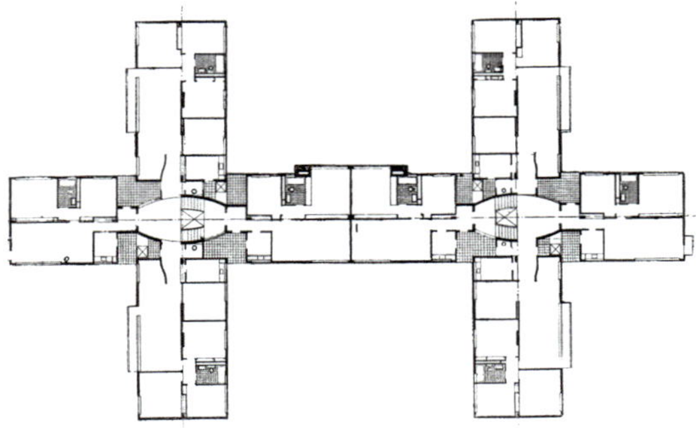

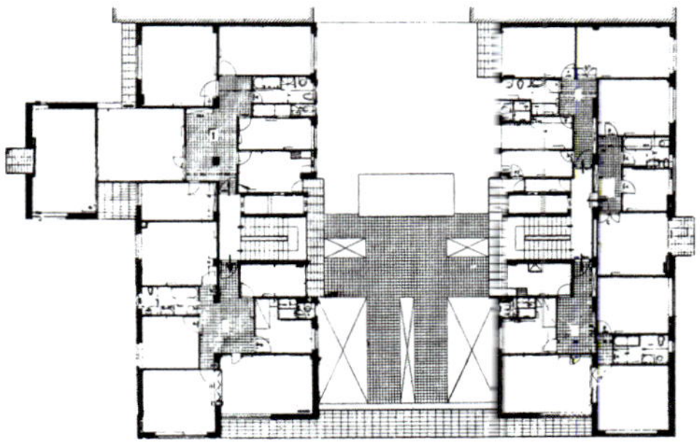

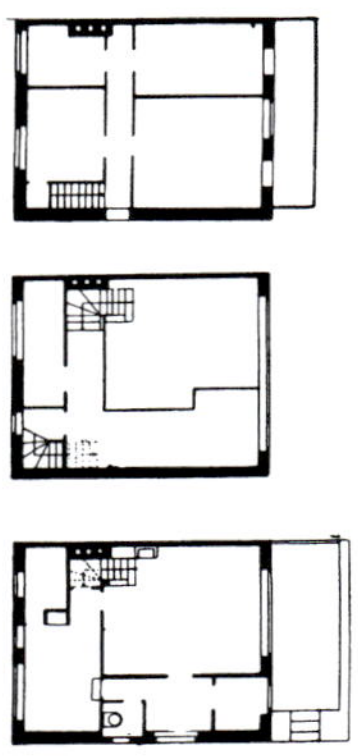

Building Exhibitions and Model Estates/Row House
1932
Werkbundsiedlung Vienna Row House (AT) – Adolf Loos
The "Raumplan" concept was adapted to this three-story row house: a double-height living room is bordered by a gallery with a built-in sofa.

1933–35
Highpoint Terrace, London (UK) – Lubetkin and Tecton
One the first Modernist residential buildings in England – where a ground floor raised on pilotis with a tea room and winter garden extends an invitation to linger – was commissioned by the industrialist Sigmund Gestetner for his office workers. The single-floor flats are accessed from a meandering circulation area. Due to the building's double-cross shape, most of the orthogonally organized units are exposed to natural light on three sides.

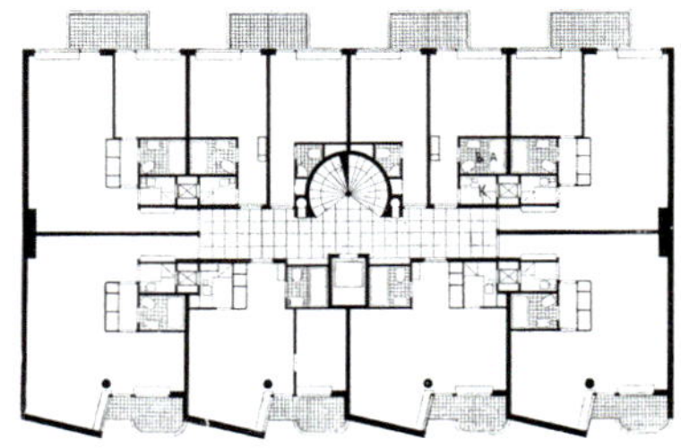

1935
Kollektivhus Stockholm (SE) – Sven Markelius
A residential building conceived to unburden working mothers. There is a day-care facility for children, meals can be ordered from the central kitchen and delivered to most apartments by dumbwaiter, a laundry chute leads straight to the serviced laundry room. Storage for sporting equipment, a restaurant, a reading room, and the roof terrace are all shared. Alternatively, small yet efficiently laid-out kitchens in each unit offer residents the possibility to cook for themselves. All ancillary rooms and bedrooms branch off from the open living room.

1935
Casa Rustici, Milan (IT) – G. Terragni, P. Lingeri
Modern architecture integrated into a historic block structure: two slab-like building segments, which – optimally cross-ventilated and oriented east-west – frame a courtyard that opens up to the city, are linked by a layer of balconies along the street to complete the block infill type. A spacious hall serves as the circulation center; almost all important living areas look out over the urban avenue.

1931

Building Systems
1931–34
Cité de la Muette, Drancy near Paris (FR) – Beaudouin, Lods, Mopin, Bodiansky
This project is regarded as the first systematically designed and erected large-scale housing estate, which – following the urban design principles of CIAM – merged industrially planned manufacturing cycles with customized construction methods. The distance between the steel-frame building structures was determined by the dimensions of the reinforced concrete facade elements produced on-site. Lightweight facade components were produced in the workshops of Jean Prouvé.

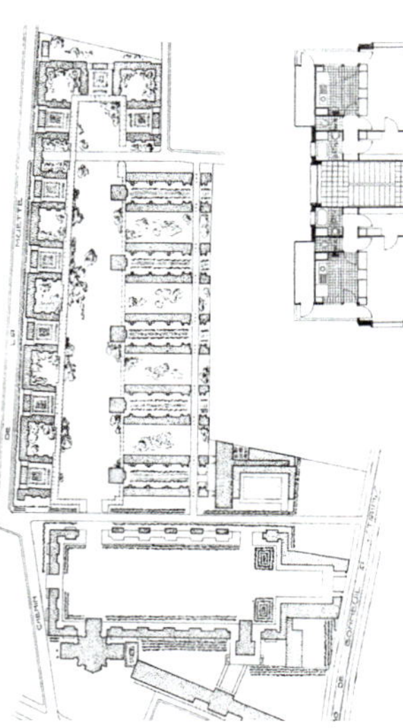

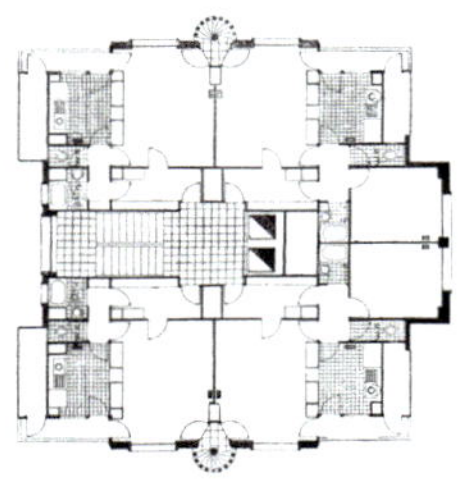

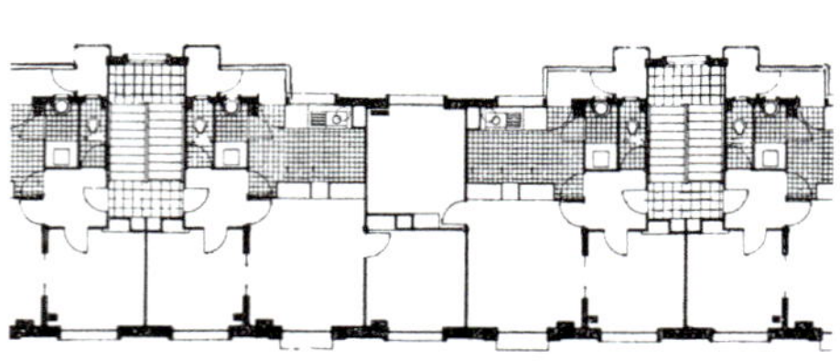

Building Systems/Slab Housing
1934
Bergpolder Residential High-rise, Rotterdam (NL) – Van Tijen, Brinkman & Van der Vlugt
An experimental residential high-rise realized as a steel-frame structure with wooden floors and a panel facade, oriented north-south to optimize the lighting situation of the living spaces. The efficient floor plan design aimed to minimize circulation areas; a sliding door enables flexible use of the rooms adjacent to the living room. The elevator stopped only at every second floor.

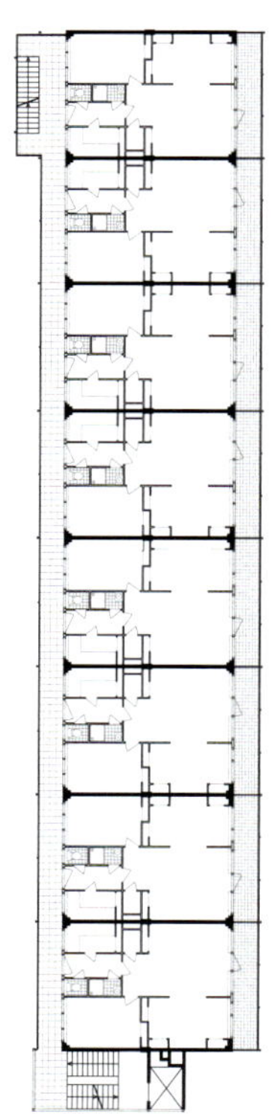

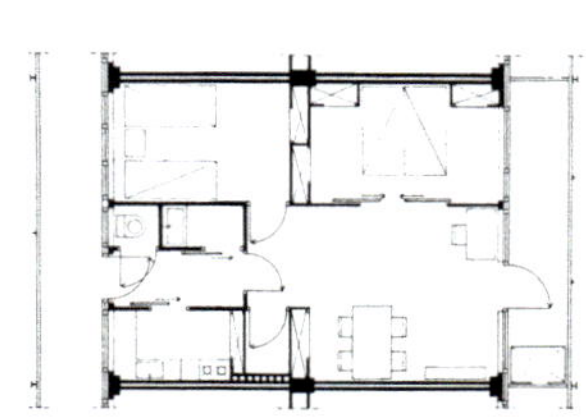

Building Systems
1935
Experimental House, Göteborg-Kallebeck (SE) – Erik Friberger
The only fixed elements are the reinforced concrete structure and the access and adjacent sanitary cores. The final layouts of the units were completed in their entirety by the residents.

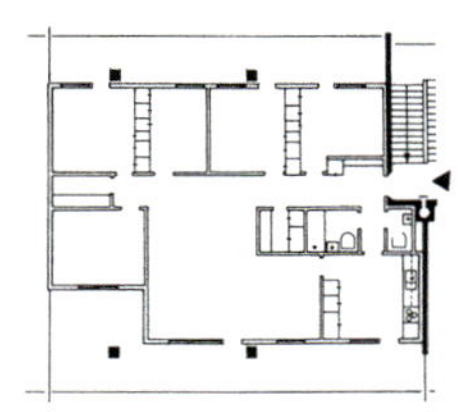

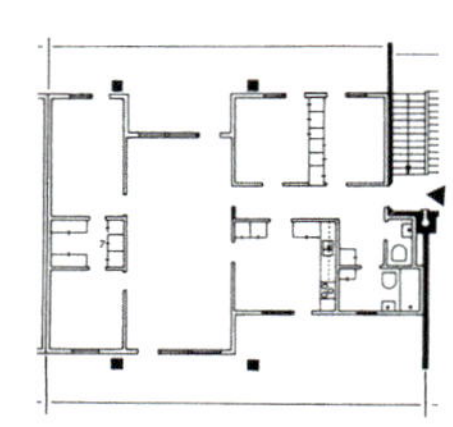

Housing in the Post-war Era

After the Second World War, many cities experienced an acute housing crisis. Housing developments therefore increasingly drew on the already established urban design and architectural repertoire of the Modernists – often at larger scales than before due to the housing shortage and at times built to inferior standards. Reconstruction efforts primarily followed the tenets of the landscaped automotive city, for which even completely intact urban quarters were razed. Traffic arteries linking growing suburban developments to the city exacerbated the situation.

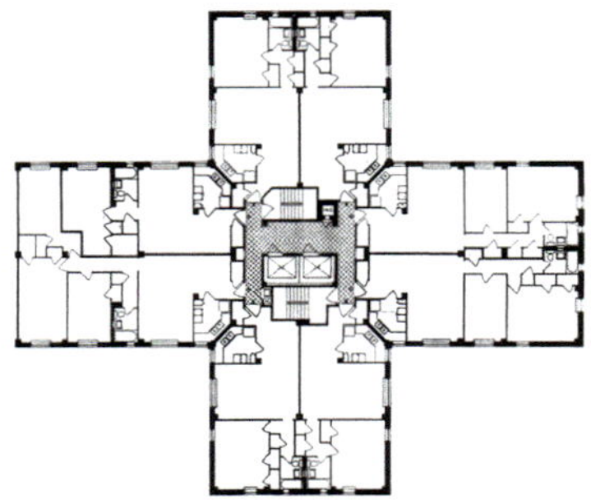

Housing in the Post-war Era
1943–49
Stuyvesant Town, New York (USA) – Irwin Clavan, Gilmore Clarke
Linked rows of austere, cross-shaped towers replaced 18 impoverished city blocks. Initially conceived as mass housing for war veterans, this privately financed large-scale estate provided its residents with affordable housing in park-like surroundings in the middle of the city.

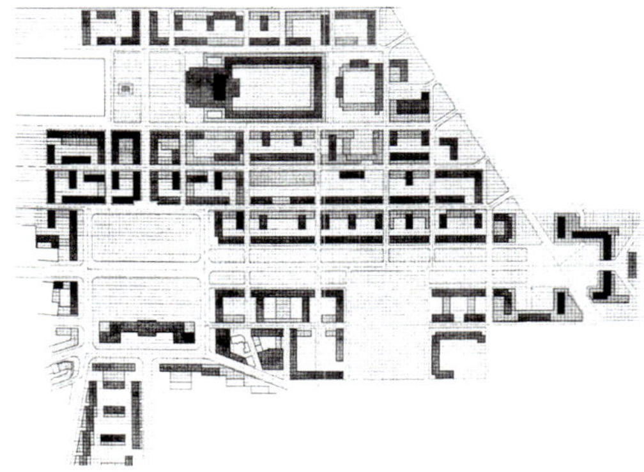

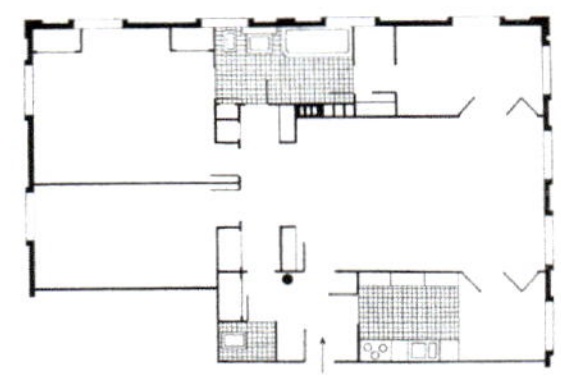

Housing in the Post-war Era
1945–1964
Reconstruction of Le Havre's City Center (FR) – Auguste Perret
The layout of the historic city center was readapted for the reconstruction and rebuilt with a uniform, representative building ensemble realized using methods taken from industrial mass housing. In the 4-room apartments, the living room forms the organizational and spatial focal point: work or dining areas can be flexibly connected along the street facade, bedrooms adjoin directly in the lateral axis. The – very innovative – mechanically ventilated bathrooms are accessed from three sides.

1945

Workers Settlements
1936–54
Sunila Workers Settlement, Kotka (FI) – Alvar Aalto
In this ensemble constructed over multiple decades, rows of three-story units are arranged along the southern slopes, gardens, and transport arteries of a valley. The building organization takes advantage of the topographical situation: 1-room apartments on the lower level are accessed from the south side, the 2-room apartments above either directly from the ground floor or by stairs from the north. The idea to provide each apartment with its own entrance instead of sharing a single staircase was based on sociological factors: privacy and identity of place despite high density was an important concern of the architect.

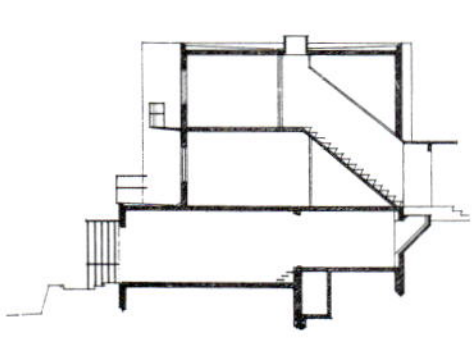

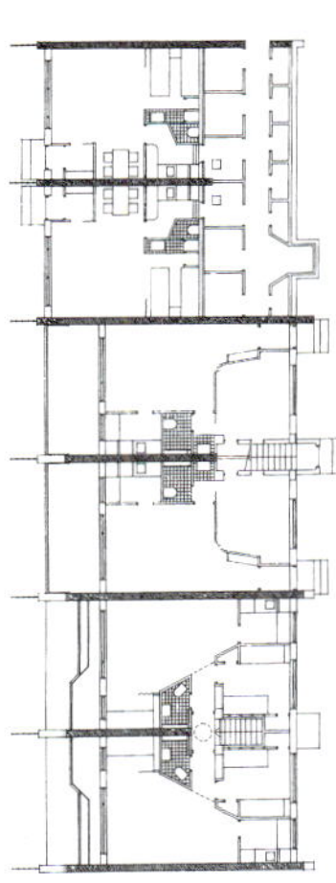

Urbanization Processes in Latin America, North Africa, and Asia

Concurrent with the developments in Europe and North America, many cities in North Africa, Latin America, and Asia were experiencing large population increases. This was partly due to the influx of political refugees, but above all else to large-scale economic migration to the urban centers. As the existing structures were not able to cope, large informal settlements arose, as did new housing policies and methods of dealing with the situation. Still under colonial rule, large-scale housing estates were initiated in North Africa as testing grounds for similarly large developments in France.

Urbanization Processes in Latin America
1945
El Silencio Residential Estate, Caracas (VE) – Carlos Raúl Villanueva
An urban renewal ensemble of housing blocks for residents with low incomes was erected to replace a rundown district. The apartments in Block 4 have three to four similarly sized linked rooms, which generate a nearly use-neutral floor plan open to interpretation.

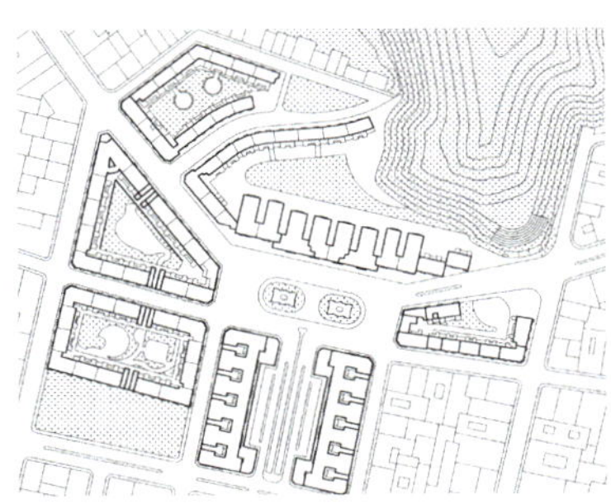

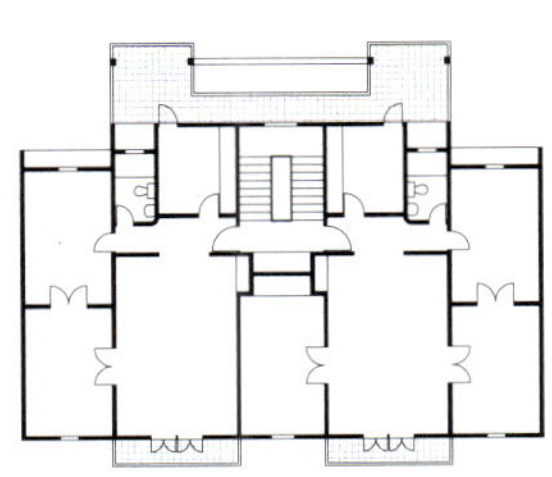

Urbanization Processes in North Africa
1945–48
New Gourna Village, Luxor (EG) – Hassan Fathy
Built to prevent expanding informal structures from encroaching on a historic burial site, the design for this settlement takes its cue from traditional social communities comprised of 10 to 20 families, each with its own designated area. The housing units are arranged around small courtyards. Utility rooms occupy the ground floor, bedrooms connect to a loggia on the upper floor that doubles as an outdoor living room.

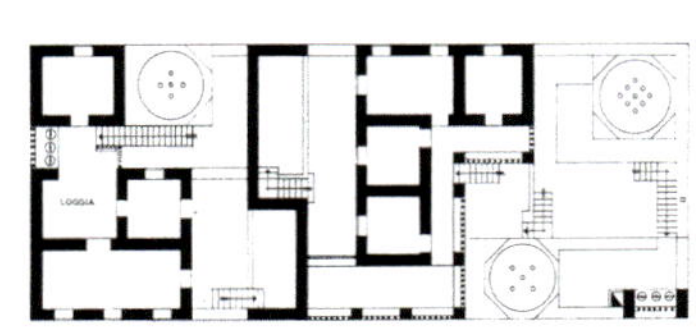

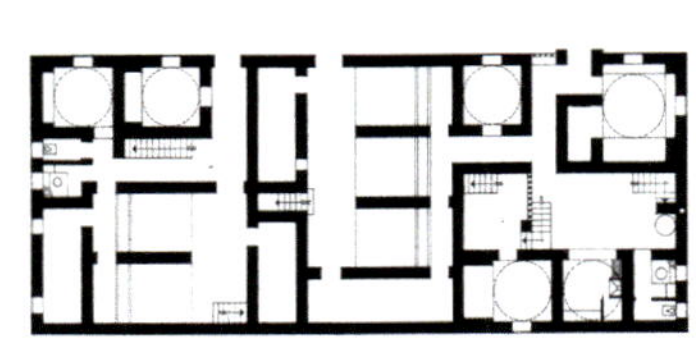

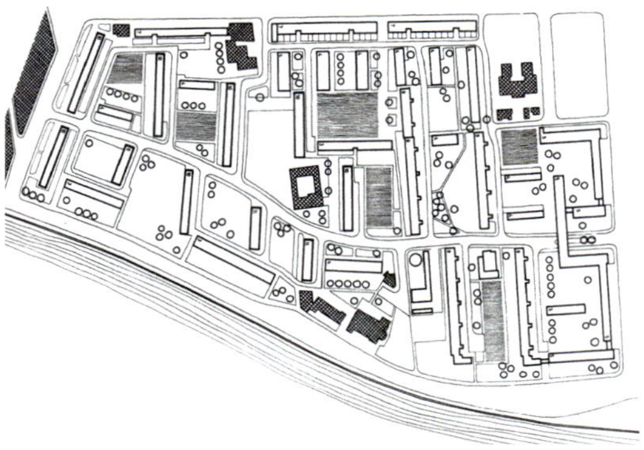

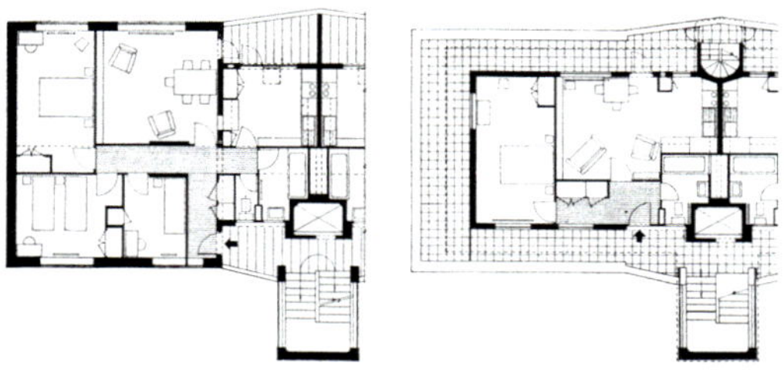

Housing in the Post-war Era
1946-62
Churchill Gardens, London (UK) – Powell and Moya
In this publicly funded residential estate, which replaced an urban neighborhood badly damaged during the war, a wide spectrum of housing types were on offer. In the tall slab buildings, external staircases lead to single-story flats with open kitchens and open galleries to stacked maisonettes.

Suburbia

Following the war, the erection of detached single-family home developments outside the city dramatically changed the face of domestic culture in the USA in only 25 years. Construction programs such as the Housing Act of 1949, the expansion of expressways, and new serial production methods coalesced in a building boom, which first and foremost made it possible for the middle class to move out of the increasingly unpopular, large cities. Despite contributing greatly to the overdevelopment of rural landscapes, the suburbs became a symbol of American lifestyle in the post-war era – and were copied around the world.

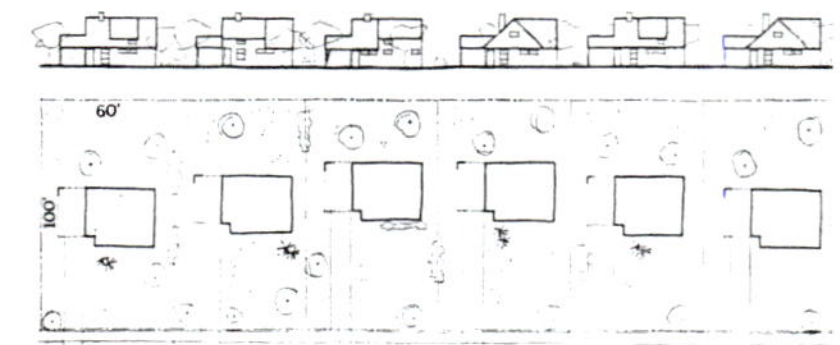

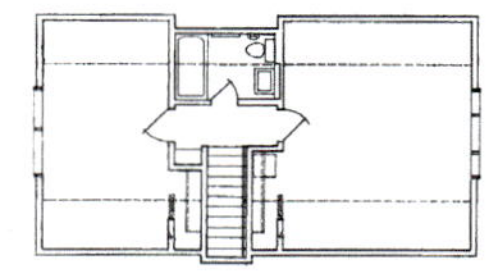

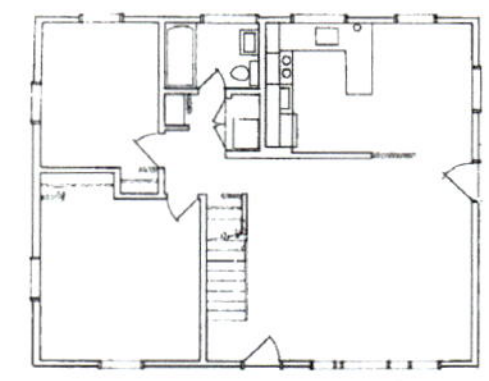

Suburbia
1947–51
Levittown, Long Island (USA) – Levitt & Sons
This is the first of more than 200 developments named after the proprietor and project developer. Designed and built by the same company, they optimized planning and construction processes ensuring short construction times. The residents could pick from a selection of catalogue models. The entrance of the housing types with unfinished but upgradable attics often led directly into an open living area and kitchen. In the one-story types with upgradable attics, the sleeping quarters were located to one side, in the two-story types on the floor above.

1946

New Towns Act (UK)

This program initiated the development in Great Britain of 27 new, autonomous cities over the coming 50 years. It was conceived in reaction to the acute housing shortage after the war and also to avoid further urban sprawl in the larger metropolitan regions – particularly London. New Towns were initially a modern interpretation of the garden city, but in the end were often realized as automotive cities. As a model, they were adopted worldwide.

New Towns
1947
Harlow (UK) – Frederick Gibberd
40 km outside London and built in a hilly, wooded, and sparsely populated landscape to alleviate the strain on industrial areas in the east of the metropolis, Harlow is divided into four districts, each conceived for 20,000 residents and 3 km in radius, which in turn are composed of multiple neighborhoods grouped around their individual central areas.

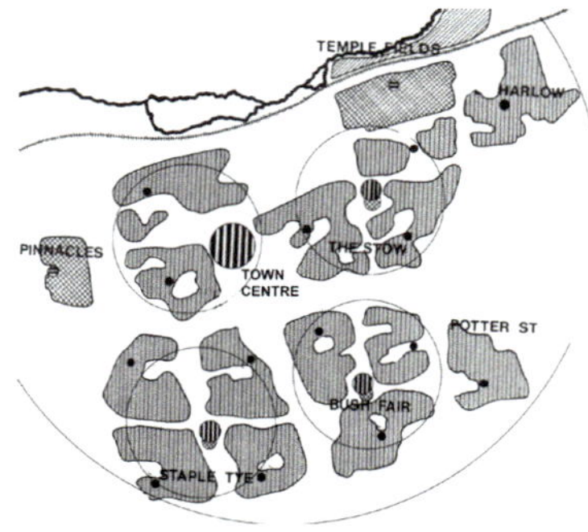

Housing in the Post-war Era/New Towns
1949–54
Tiburtino Residential Quarter, Rome (IT)
A modern urban expansion combined with the heterogeneous nature of village-like structures and vernacular architecture. Residential high-rises, terraced slabs, and row houses form a composition that follows the course of the streets. In the high-rises, the living room with loggia acts as a central circulation space, while in the lower slabs a corridor or hall divides the living and dining areas from the bedrooms. Offset and staggered building volumes, stepped access galleries, balconies, and rooftop gardens enliven the streetscape of this planned district.

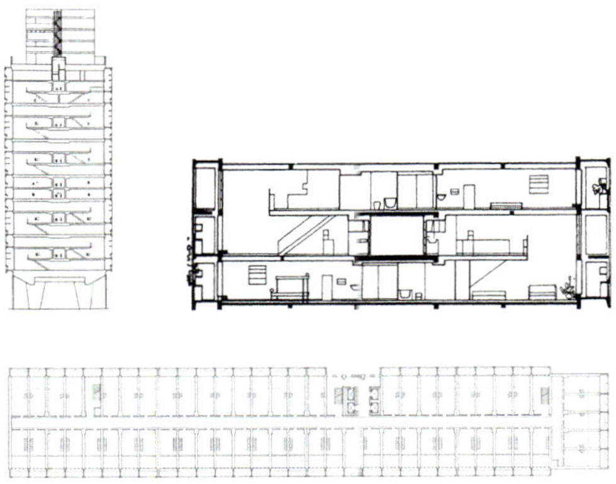

Housing in the Post-war Era
1952
Unité d'Habitation, Marseille (FR) – Le Corbusier
The Unité d'Habitation was conceptualized as an autonomous slice of city for 1,600 residents – the number Le Corbusier deemed necessary to generate a functioning urban community – with shopping streets, shared rooftop spaces as well as the "Rues Intérieures" on every third floor, from which different apartment types are accessed from both sides. The distinctive, very narrow maisonettes run the entire width of the building above and below the access floors. (see page 182)

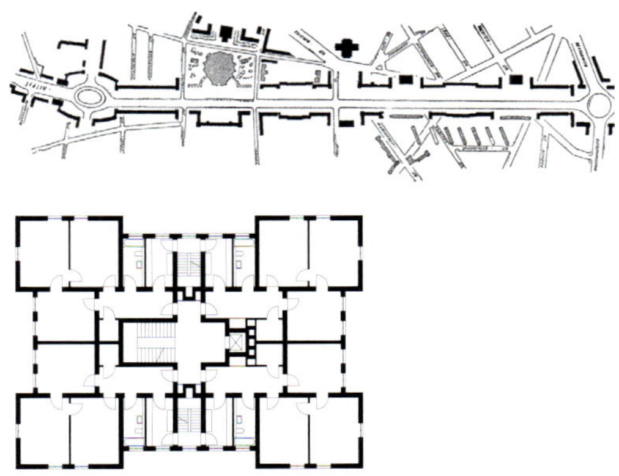

Housing in the Post-war Era
1952–60
Stalinallee, Berlin (GER) – Hermann Henselmann (master plan, high-rise)
As in other European capitals of the former Eastern Bloc, monumental, representative architecture as an expression of political power was adopted for public housing. "Residential palaces for workers" with roughly 3,000 apartments in blocks and towers offered many comforts to their inhabitants – such as high ceilings, spacious floor plans, and technological innovations including two-way intercom systems, telephone lines, and garbage chutes – which social housing in the West was not in a position to provide. (see page 229)

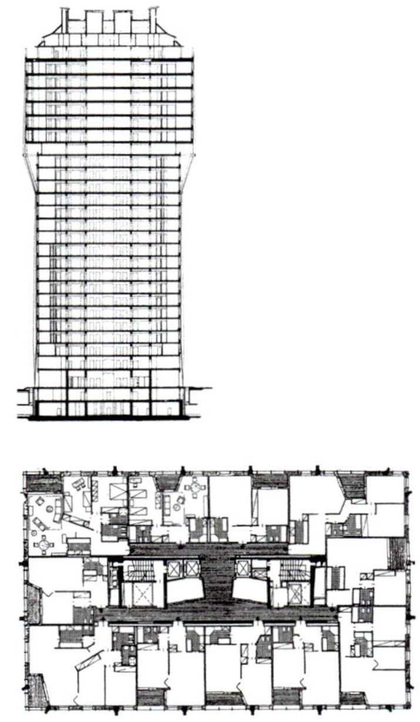

Housing in the Post-war Era
1954
Torre Velasca, Milan (IT) – BBPR
During the post-war reconstruction period, the multifunctional high-rise with commercial space on the ground floor, 18 office floors and eight residential levels at the very top strove to increase the density of the pre-existing city instead of replacing it. This position was pursuant to the contemporary debate in Italy, which endeavored to develop modern architecture in a contextual dialogue with the historic city instead of in opposition to it. A range of different apartment types was produced: studios, flats with spacious, corner living rooms, as well as maisonettes.

1954

Urbanization Processes in Latin America
1950–52
Pedregulho, Rio de Janeiro (BR) – Affonso Eduardo Reidy
This estate in an open landscaped park is regarded as an important example of Brazilian post-war Modernist social housing. The most prominent building is the curvilinear slab resting on pilotis that snakes its way through the topography of the site. Its mid-section remains open to the elements, is accessed from the upper slope and provides space for public programs such as a kindergarten or playground. The lower levels mostly contain very compact flats. Despite their modest size, the two levels of maisonettes above are visually linked by a connecting void between the floors.

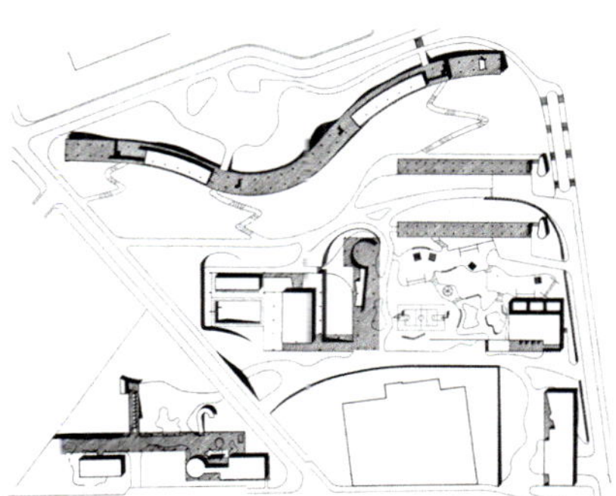

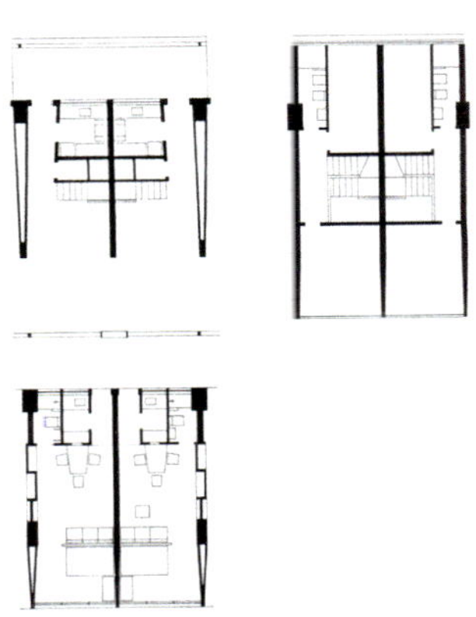

Slab housing
1952
Flats for Air France, Brazzaville (CG) – P. B. Berruet
Behind the adjustable brise-soleil, enclosed areas for the staircases, kitchens, and bathrooms alternate with living and sleeping areas open to both sides. By placing the circulation path along the facades, a buffer to the set-back living areas is formed, protecting them from direct sunlight and the tropical climate.

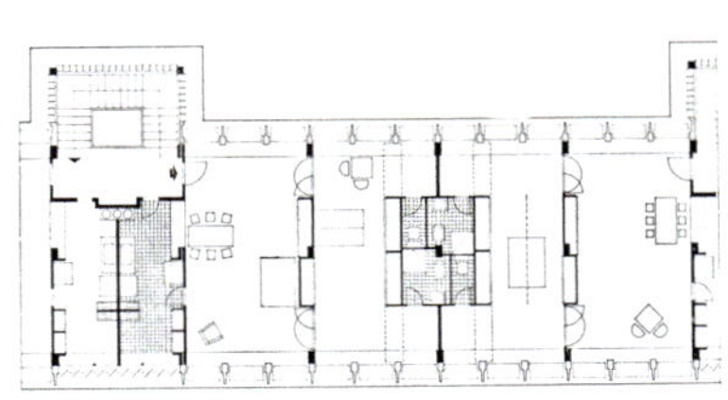

Urbanization Processes in North Africa
1954
ATBAT Collective Housing, Casablanca (MA) – Woods, Bodiansky, Candilis
As part of an estate for former slum inhabitants, three multi-story structures are based on the traditional courtyard houses of the region. The 2-room apartments in the slab building are arranged along patios that alternate every floor from one facade to the other. Completely walled-in but open to the sky, they form a climatic and social buffer by separating the interior rooms from the external access galleries and shielding them from direct sunlight. The apartments are cross-ventilated and the kitchen – a potential heat source – is located outside.

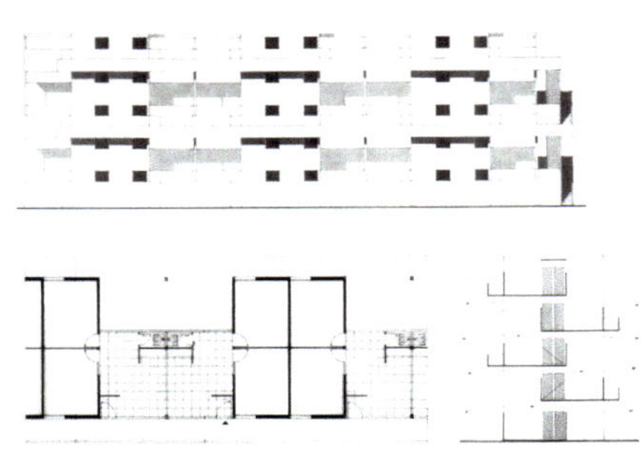

Urbanization Processes in North Africa
1954
Sidi Othman Apartments, Casablanca (MA) – André Studer
The slabs and high-rises are inspired by North African housing types. The compact units are organized around large living rooms and cantilevered patios. Stacked in an offset manner and each supported by a single column, these patios lend the facade its distinctive character.

New Building Type: Low Rise High Density

Criticism of urban sprawl due to the construction of detached single-family homes led to a search for alternative solutions. In contrast to other multifamily housing schemes, each unit was to be provided with private, outdoor spaces and a direct, preferably ground-level entrance from the public sphere, while interlocking or semi-detached constellations of similar, neighboring units simultaneously ensured a high ratio of land use. Besides more classical row house configurations, low-lying estates with cluster-like structures emerged that took advantage of back-to-back arrangements, patchwork-like patterns, or more complex, three-dimensional groupings. Lighting recesses in the building volume such as patios became distinctive design features.

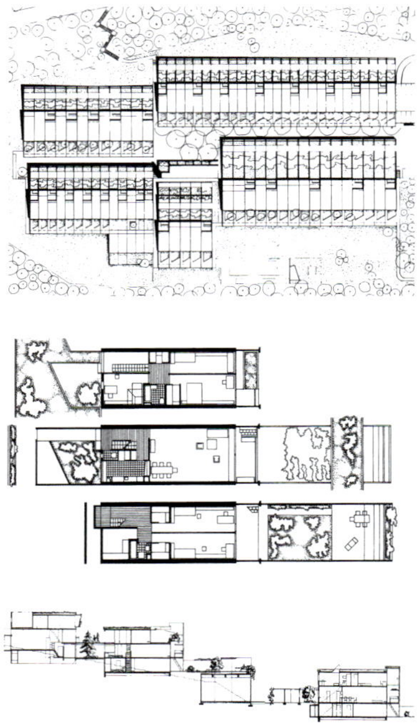

Low Rise High Density
1955
Halen Estate, Herrenschwanden (CH) – Atelier 5
In this dense ensemble on sloping terrain with very deep, primarily 4 m-wide, multi-storied housing units, the upper rows have gardens on the roofs of the unit below. Each house enjoys an unobstructed view. (see page 298)

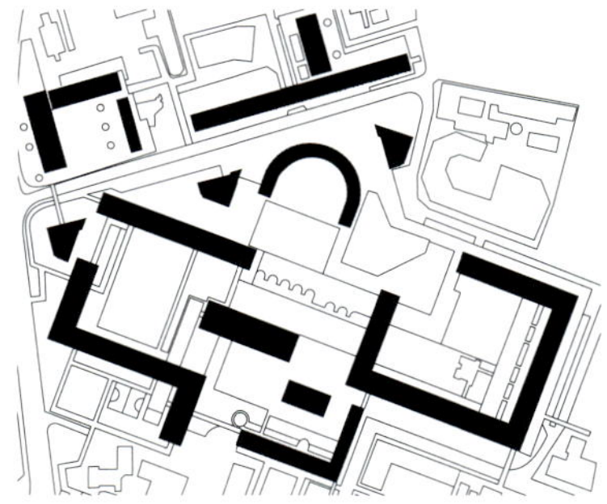

Housing in the Post-war Era
1955–83
Barbican Estate, London (UK) – Chamberlin, Powell, Bon, Arup
By rebuilding a site that had been vacant since the end of the war, an attempt was made to reinvigorate inner-city housing. The large-scale project offered the possibility of creating a unique ensemble by combining cultural and educational institutions as well as a variety of building types around generous outdoor green spaces. With over 100 different floor plans, a wide range of options were available, primarily geared toward upper middle-class residents. (see also 1974, Lauderdale tower)

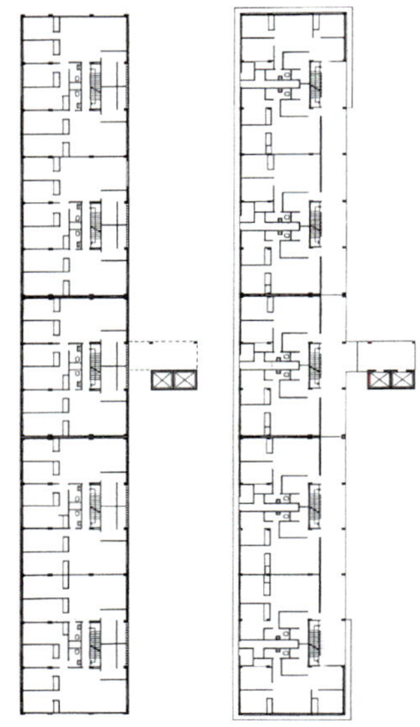

Urbanization Processes in Latin America
1955–57
23 de Enero housing development, Caracas (VE) – Villanueva
This residential district with a total of [illegible]08 apartments was regarded as an exemplary development scheme for the entire country. Elevators lead to external access galleries on every fourth floor, which connect to double-loaded staircases to the apartments on the other floors. At the front of the units, a living room and kitchen form day areas, while private rooms are arranged in a row at the rear.

1954

Mass Housing in Asia

In Asian cities like Singapore and Hong Kong, or in countries like South Korea, "emergency" housing types were initially implemented to deal with issues such as uncontrollable growth and the resulting informal settlements. From such immediate challenges, residential property developers arose who have since dominated large segments of housing construction. The supply of building land was partially monopolized and large-scale development given priority, as providing housing to the population became an important aspect of political and economic stabilization policies. Large housing estates were built not only in the form of New Towns, but also in part as replacements for older, more organically evolved, low-rise urban districts.

Mass Housing in Asia
1954
Shek Kip Mei Estate, Hong Kong (HK) – Resettlement Department
On Christmas day, 1953, a fire destroyed vast parts of an informal settlement in Shek Kip Mei inhabited by a large number of refugees from the Chinese Civil War. A relocation authority was initiated, marking the beginnings of a program of heavily subsidized housing construction under the later established Hong Kong Housing Authority. Initially, only emergency housing was built, but with improved political and economic stability, more permanent structures and increased standards for large portions of the urban population were introduced. The H-shaped block type realized in Shek Kip Mei in 1954 provided every family with only one room. Accessed by wraparound external galleries, they were arranged back-to-back, with the result that perforated dividing walls were necessary for cross-ventilation. Kitchens and bathrooms were shared.

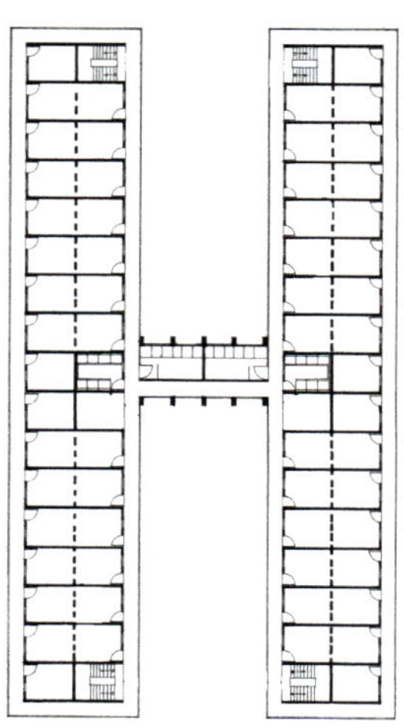

Housing in the Post-war Era
1955–85
Mikrorayon residential districts (USSR)
50 million apartments were built in the Soviet Union between 1955 and 1985. Using quantitative design charts, sun diagrams, and a multitude of standardized building elements, innumerable iterations of the so-called "Mikrorayons" were realized – prefabricated residential districts with clearly defined community facilities. The Mikrorayon system had an impact on almost all member states of the former Eastern Bloc and was later also influential in housing construction in the People's Republic of China.

Urbanization Processes in Latin America
1956–60
Brasília (BR) – Lucio Costa (master plan)
Brazil's new capital was built in only four years and adhered in every respect to the CIAM principles of a functionally segregated city. Each "Superquadra" contains roughly 600 apartments in nine to 11 slab buildings resting on pilotis above the open cityscape. Taken together, four such blocks and shared community facilities form a neighborhood zone. (see also 1959, SQS 308)

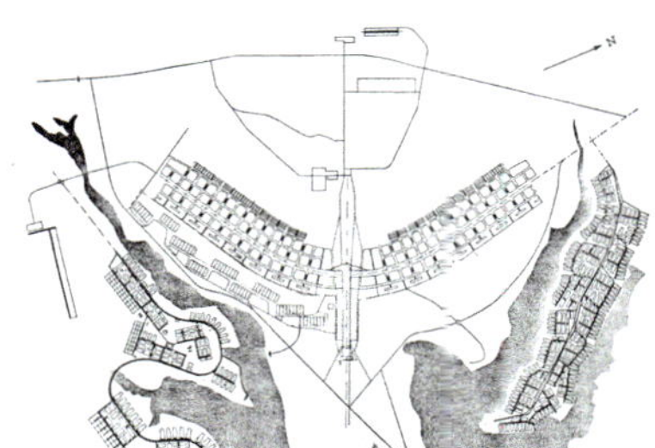

Housing in the Post-war Era

1957

Interbau, Hansaviertel, Berlin (GER) – Otto Bartning (overall design), Kreuer Jobst (master plan), Otto Senn (apartment building)

A model for the "city of tomorrow" – and a reaction in the West to the "Stalinallee" (see also 1952–60) – the ensemble of high-rises, slab and low-rise buildings was arranged in the middle of a landscaped urban park on a site where the urban block structure had been heavily damaged in the war. Renowned architects of the time like Le Corbusier, Aalto, van den Broek and Bakema, Gropius, Niemeyer or Senn designed and realized exemplary ideas on modern housing. In Senn's apartment building, three individual apartment types fan out around a central and natural lit circulation area. Each apartment has views in two directions.

New Building Type: Large-scale Estates with Space-enclosing Structures

Space-enclosing structures in taller and more meandering forms were implemented in the 1960s and 1970s to ensure the creation of protected inner spaces despite the high density of such projects on an urban scale – in contrast to the much-criticized open slab developments. Cityscapes accentuated by sporadic towers emerged, where spacious open areas free of traffic noise surrounded the buildings – large-scale sculptures for the society of the future. Many of these developments implemented cluster architecture consisting of similar floor plan segments arranged in rows, stacked, or turned to optimize natural lighting or views. As time passed, these large-scale developments were criticized as mono-functional and inhospitable bedroom communities, their buildings characterized as "residential silos" and "concrete fortresses."

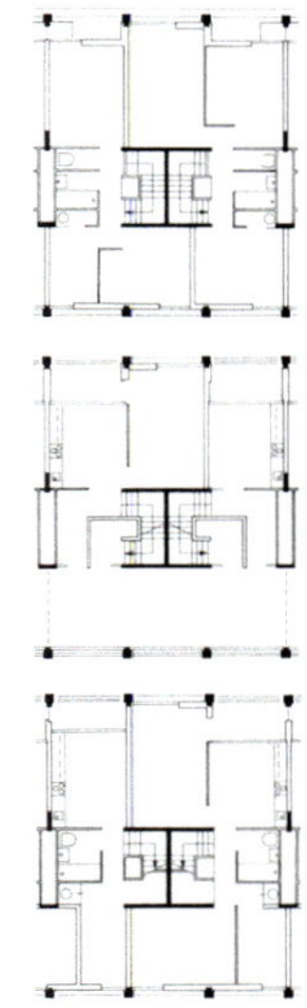

Large-scale Estates with Space-enclosing Structures

1957–61

Park Hill, Sheffield (UK) – J. L. Womersley

On a hill with a view over the city, the large-scale structure replaced an impoverished working-class district. The bent slabs follow the topography of the site. In the south, four-story building segments engage the surrounding lower buildings, while in the north, the building stands 13 stories tall. "Streets in the Air" on every third floor lead to apartment groups laid out across three bays and consisting of two studio apartments below and two maisonette units with open living areas above.

1958

Urbanization Processes in North Africa

1957

Climat de France, Algiers (DZ) – Fernand Pouillon

An elongated superblock with an inner courtyard surrounded by a high colonnade forms the center of the social housing district, composed of diverse building types derived partly from regional housing typologies. Above the abundant shops on the ground floor, 360 deep apartments stretch between the street facade perforated with small windows and the interior colonnade, which provides protection from direct sunlight.

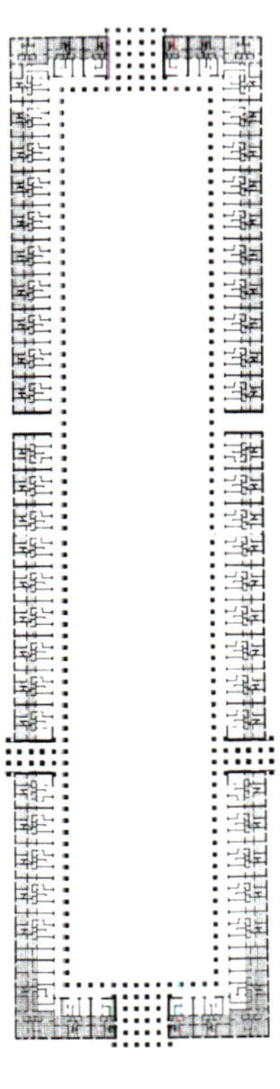

New Towns

1957

Gleadless Valley, Sheffield (UK) – J. L. Womersley

This development was built on three sides of a wooded valley outside the city at the same time as Park Hill. Varied building designs with approximately 4,500 apartments in total are divided into three "neighborhoods" connected by walkways, each with its own shops and schools. Row houses are in part accessed from a protruding terrace on their upper level, three-story units connected to form cross-shaped buildings.

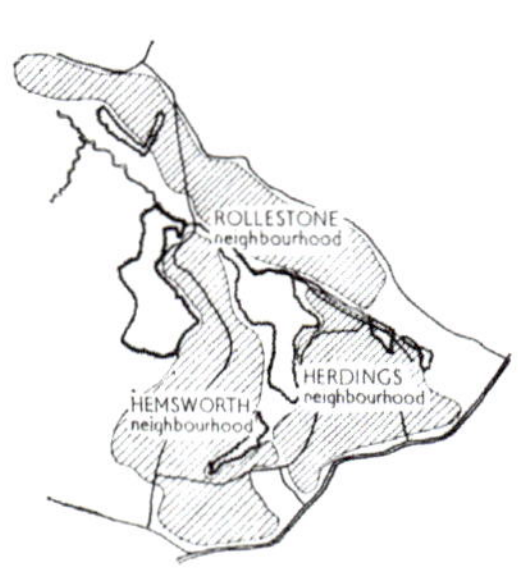

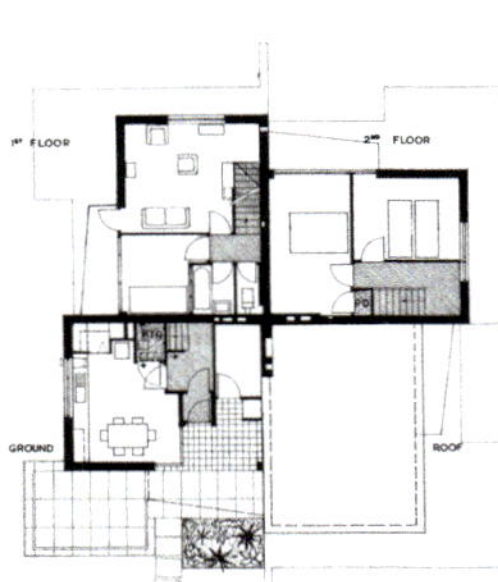

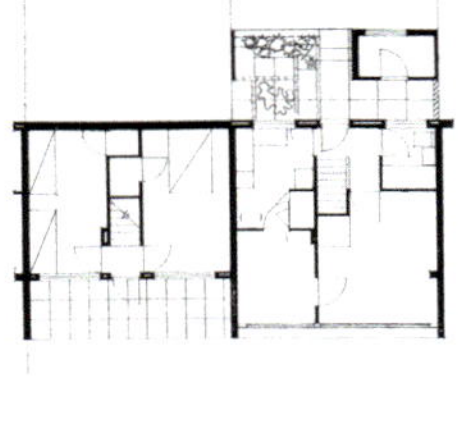

Urbanization Processes in Latin America/Residential High-rises

1957–66

Edificio Copan, São Paulo (BR) – Oscar Niemeyer

One of the tallest residential buildings of its time, this sinuous high-rise slab in the center of São Paulo was created with a diverse residential population in mind. The apartment types span from studio apartments occupying a single bay with sanitary rooms and kitchens in the darker, rear zone to 350 m² apartments with separate servants' quarters. The separate elevator core is connected with ramps and staircases to the main volume. Central corridors and access galleries behind the deep lamellae of the facade lead to the apartments entrances. In addition, a number of apartments are directly accessed from private elevators.

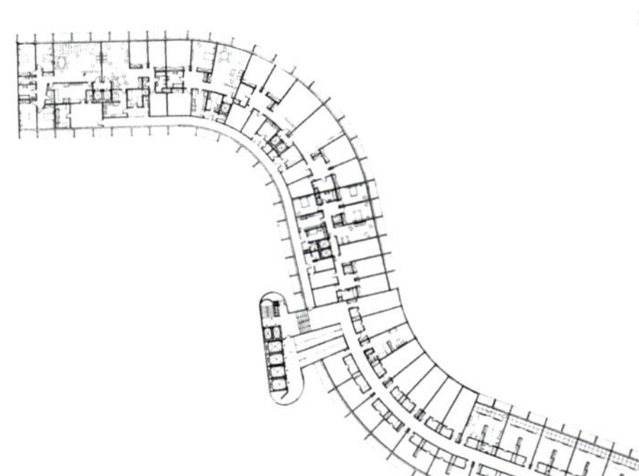

Mass Housing in Asia

1958

Harumi Apartments, Tokyo (JP) – Kunio Maekawa

Built by the Japanese Housing Association, this building remained exceptional in scale in Japan due to the traditionally small parcel structure. The high-rise slab combines a building structure inspired by Le Corbusier with traditional Japanese interior components, thus enabling a high degree of flexibility within the extremely compact apartments. (see page 187)

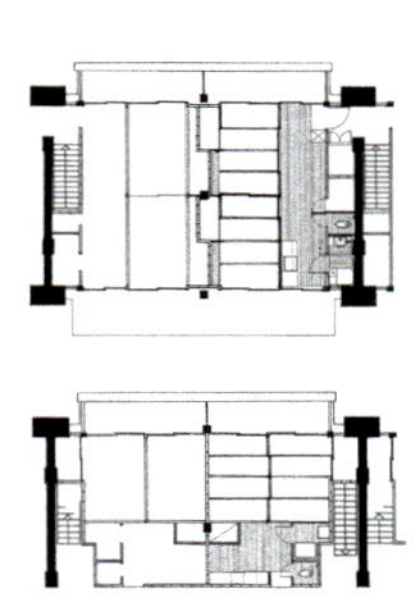

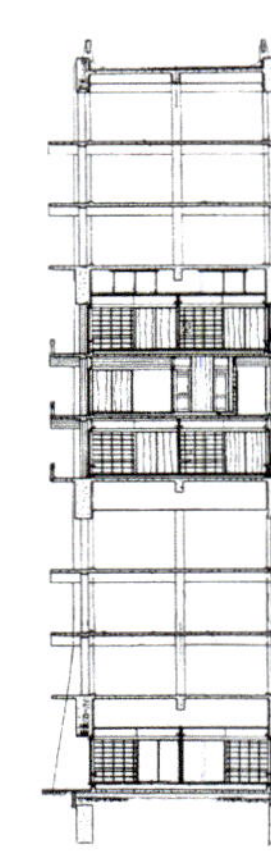

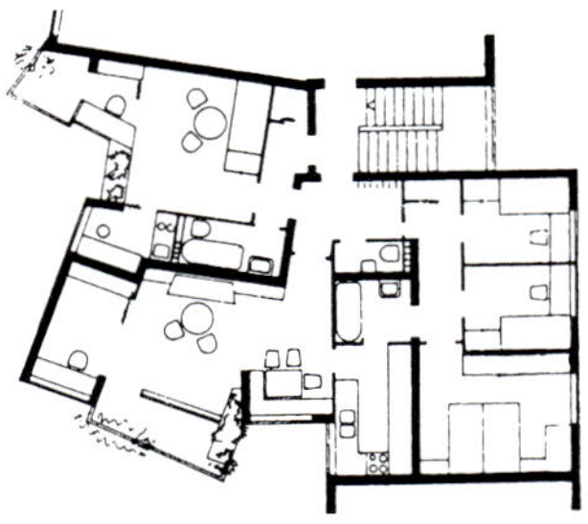

Housing in the Post-war Era
1958
Extension of the "Ring" Estate, Berlin-Siemensstadt (GER) – Hans Scharoun
So-called "residential homesteads" with community functions were formulated in conjunction with freely shaped slab buildings – as an extension and alternative to the more rigid structures from the 1930s. The "Symbiosis" apartment type combined a 4½ and a 1-room apartment with individual entrances, which could be combined or separated from one another. (see also 1930, Siemensstadt)

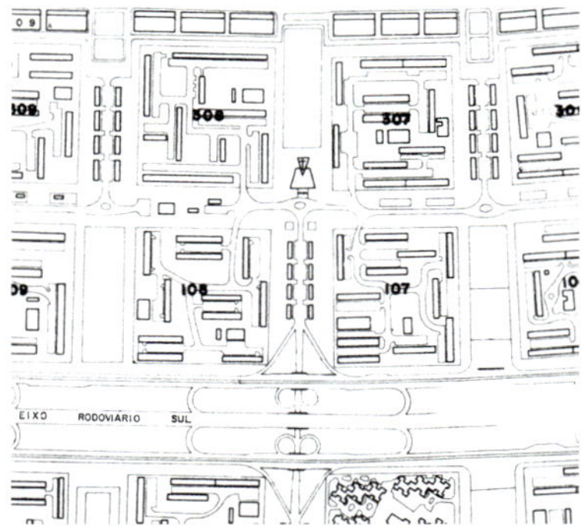

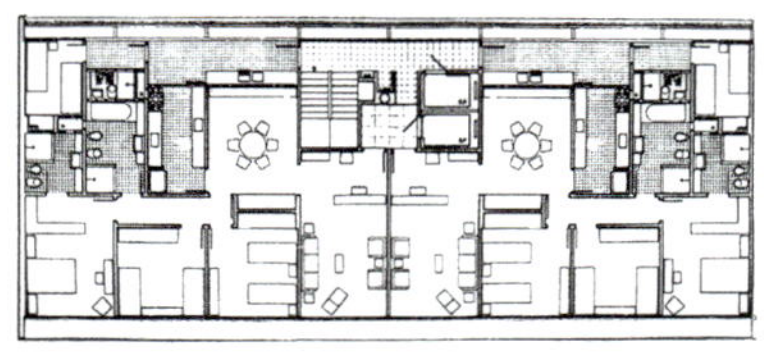

Slab Housing
1959
Slab type SQS 308, Brasilia (BR) – Marcelo Campello and Sérgio Rocha
Double-loaded staircases lead to apartments with separate service and servants' quarters along one of the long facades. The dining area adjacent to the kitchen is the center. The living room is offset to one side, and the bedroom wing with its own corridor on the other. Brise-soleil or loggias shield the units from the heat of the sun. (see also 1956, Brasilia)

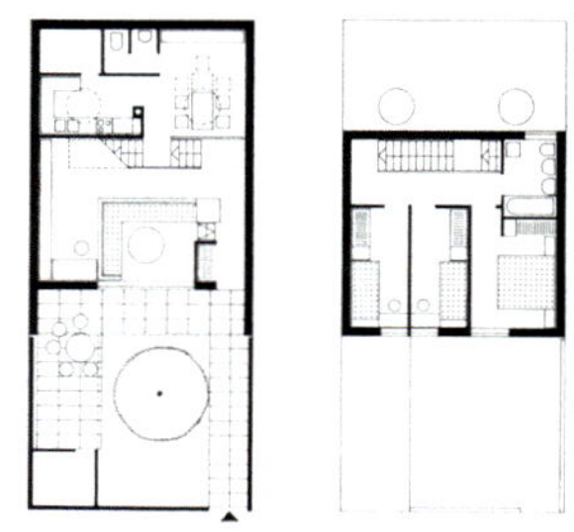

Low Rise High Density
1962–1995
Puchenau Estate, Linz (AT) – Roland Rainer
The result of over three decades of building and research into three topics: minimum parcel sizes, economic access systems, and passive solar thermal energy. From the four-story buildings with vertical access cores along the street to the one-story houses on the river, a formally and socially mixed estate design with a total of 984 housing units emerges that follows the cadence of the landscape.

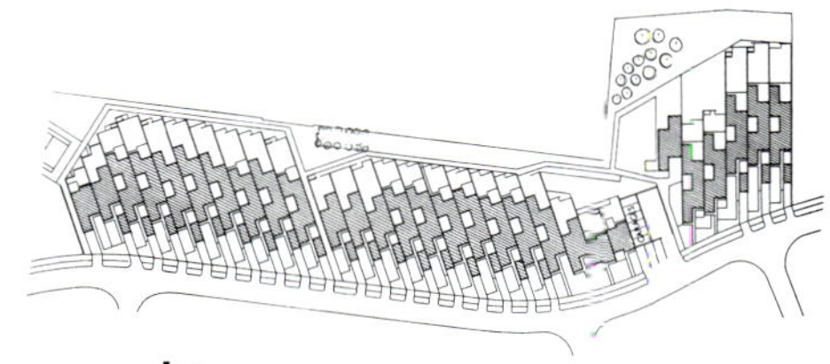

Low Rise High Density
1963
The Ryde, London (UK) – Phippen, Randall, Parkes
A diverse patio housing estate with high living quality despite minimal unit widths and a simple construction method. 28 staggered, one-story houses are structured by shear walls protruding into the exterior space, which divide the houses and their tightly interlocking outdoor spaces from one another. With the exception of the kitchen and bathroom, almost all interior rooms are enclosed with sliding walls and subsequently use-neutral. (see page 350)

1958

Mass Housing in Asia
1958
Hongshunli Commune, Tianjin (CN)
Due to an extreme housing shortage, individual living space in the young People's Republic of China was very limited, which – on top of dogma proclaiming the virtues of the commune over those of the individual – led to a search for compact housing arrangements. For this reason, all activities that could be organized collectively were practiced in the common areas. Residential units consequently consisted only of two bedrooms, the bathrooms were shared, communal canteens replaced individual kitchens, sewing rooms and nurseries were offered as additional amenities.

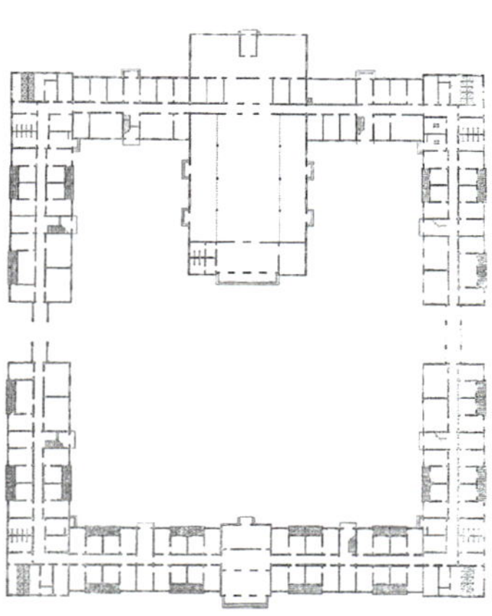

Mass Housing in Asia
Since 1960
Building and Apartment Types, Singapore (SG) – Housing Development Board
Between 1960 and 1965, the HDB built about 120,000 largely standardized apartment units in equally standardized housing blocks. Over the years, apartment standards, sizes, and variety increased and by 2000 more than 85 % of the population lived in HDB apartments. Subsidized home ownership and social mixing prevented the rise of many problems experienced in Western large-scale developments.

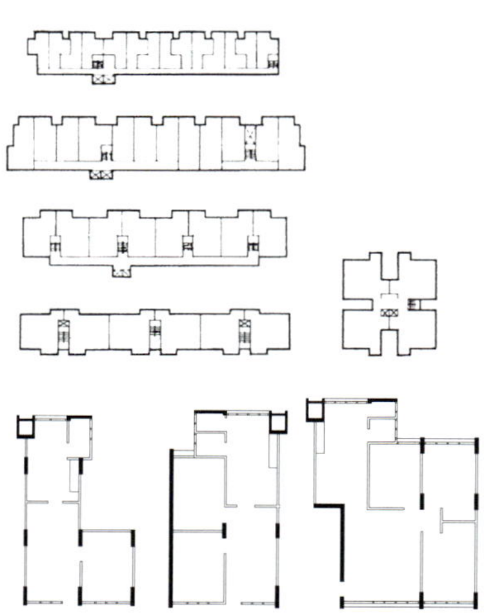

Alternative Concepts

During the 1960s, when the immediate housing crises of the post-war years had been alleviated and economies rebuilt, other aspects came to the foreground in Western societies. These were primarily incited by younger generations, who questioned the current practices and paradigms of society at large, but also within the architectural discourse of the time. The members of Team 10 countered the dogma of urban functional segregation with an alternative model that emerged from Candilis and Woods' well-known ethnographic studies of organically developed contexts in Casablanca, and which emphasized the relationship between housing unit, building, street, district, and city. Projects like Archigram's "Walking Cities," Constant Nieuwenhuys' "New Babylon" or Cedric Price's "Fun Palace" developed visionary concepts that anticipated social mobility and cultural change.

Alternative Concepts
1962
Open Building (NL) – N. John Habraken, SAR
Postulated as an "Alternative to Mass Housing," the concept proposed dividing the built environment into two distinct categories: a support system as the only part for which an architect would still be needed and an infill system placed in the hands of the residents so that housing could be adapted to individual needs. This approach was concerned with both the residential environment (SAR 73) as well as with the buildings themselves (SAR 65). A diagram from 1967 illustrates the principle of the twin structures: the support system is shown as parallel bands with predefined zones for the variable infill system comprised of common, service and circulation areas.

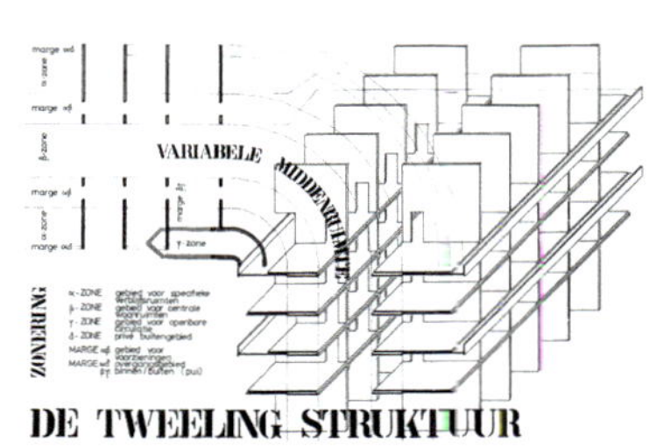

Large-scale Estates with Space-enclosing Structures
1963–1974
Märkisches Viertel, Berlin (GER) – W. Düttmann, H. C. Müller, G. Heinrichs
Space-enclosing, staggered, and linked buildings form large semi-courtyards; pre-existing detached single-family homes were left intact and complemented. This residential complex demonstrated a new trend toward increased vertical density in housing estates. (see also 1969, Wilhelmsruher Damm and page 272)

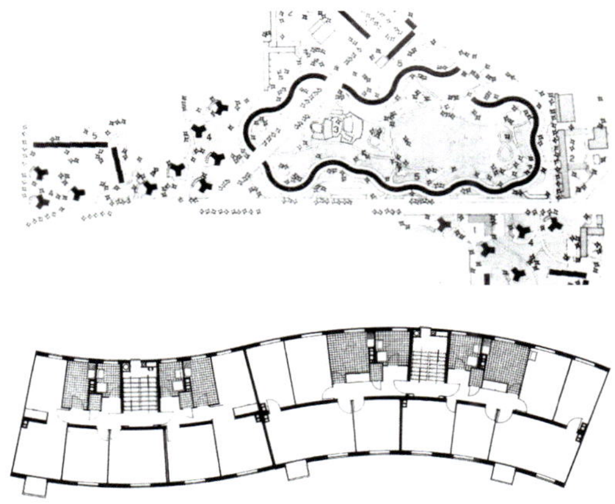

Large-scale Estates with Space-enclosing Structures
1966
Cité des Courtillières, Pantin (FR) – Émile Aillaud
As an alternative to the typical formal rigidity of large-scale complexes to date, a six-story ribbon-like slab building surrounded by star-shaped high-rises encloses an inner, landscaped park. The simple and economic floor plan with two groups of rooms meant it was possible to later combine them into larger apartments.

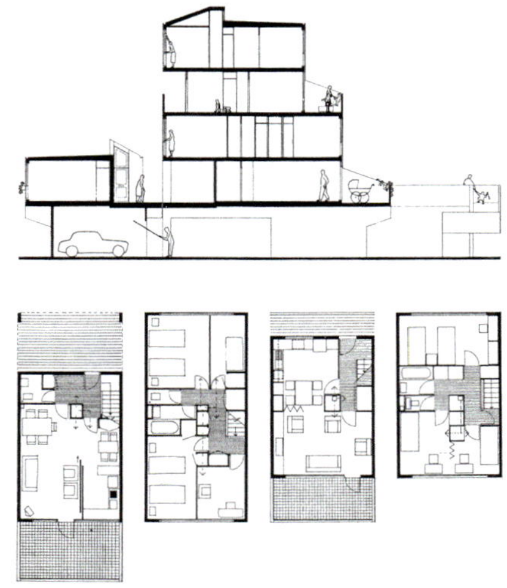

New Towns
1967–1970
Thamesmead New Town, London (UK) – Robert Rigg (GLC)
An ensemble of differentiated building types proposed as an alternative to the monotony of the early New Towns: residential high-rises increase the overall density, higher slabs along the river shelter lower-lying row houses with stacked maisonettes and community buildings. Most buildings are connected through a complex system of pathways raised one story above the ground – a typical model in which pedestrians are separated from automobile traffic, but also a protection measure against flooding. In the maisonettes, the kitchen, hall, and staircases form a buffer to the public space.

1967

New Building Type: Terraced Buildings

New ideas concerning terraced buildings as artificial landscapes emerged concurrent with the visions of Kenzo Tange, Walter Jonas, and Yona Friedman, who proposed city figures that hover above the ground and branch out horizontally and vertically. This new building type provided a chance to outfit every apartment with an open, outdoor space, while still generating a high level of built density. In cases where the terraced building did not follow an existing slope, the deep building interior on the lower floors was used for ancillary functions. Similar to other building types of the time, floor plan clusters were often implemented so that the overall expression was not defined by clear geometric volumes, but by a spatial structure that could seemingly continue on indefinitely.

Terraced Buildings
1963
Dubiner Apartments, Tel Aviv (IL) – Neumann, Hecker, Sharon
The intimacy of small-scale, vernacular forms of housing is combined with the efficiency of a terraced building on a sloping site. Stacked, hexagonal modules form the basic structure. A long corridor running along the terraces connects the bedrooms to the deep, polygonal living room and kitchen.

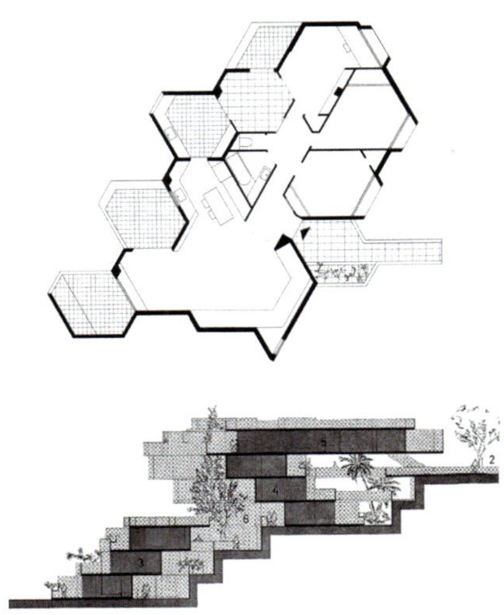

Terraced Buildings
1967
Habitat, Montreal (CA) – Moshe Safdie
The housing units of this experimental project created for the EXPO 67 consisted of 55 m² modules in a total of 16 different constellations. The basic module contains the entrance, living room, and kitchen; additional modules are attached vertically or horizontally. The interlocking configurations form a complex terrace-like structure with intriguing outdoor spaces and pathways. (see page 258)

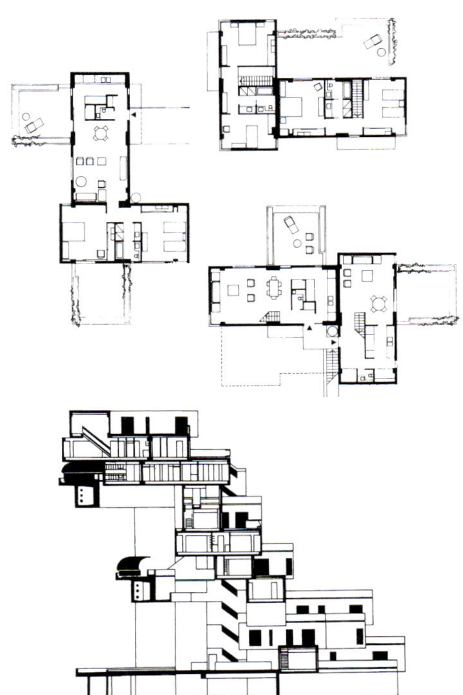

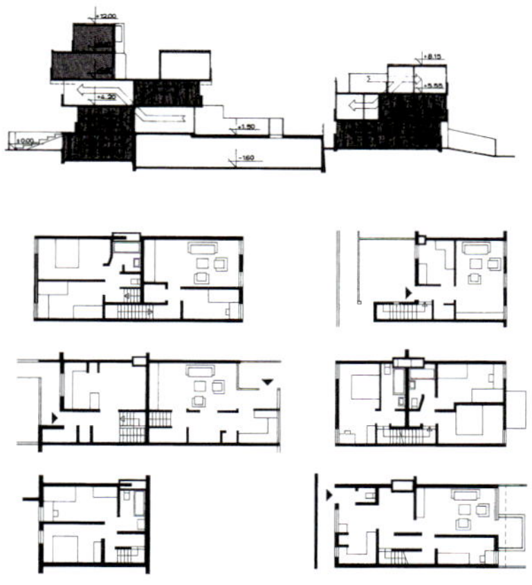

Alternative Concepts

1968–77

Marquess Road, London (UK) – Darbourne and Darke

As if along streets, apartments are strung along outdoor pathways on different levels; the rows are continually interrupted to create smaller groups of neighboring units. The apartments are primarily organized vertically: two- to three-story back-to-back maisonettes are accessed through private front yards. In section, their upper levels cross over to the other side to benefit from views in multiple directions. (see page 274)

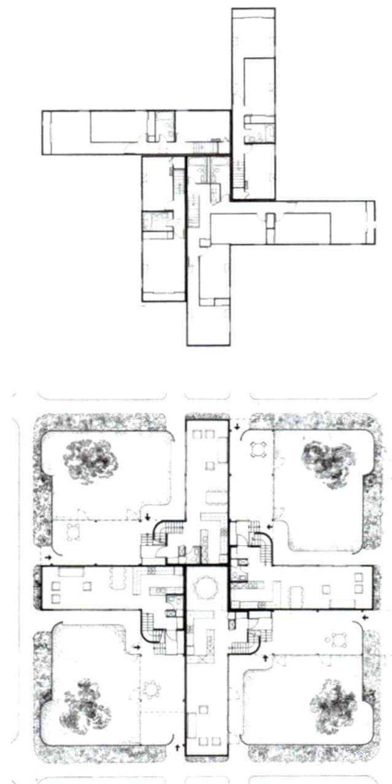

Low Rise High Density

1968

Oriental Masonic Gardens, New Haven (USA) – Paul Rudolph

An estate as a spatial system for low-income residents. Prefabricated container-like elements are arranged like a windmill and stacked to form two-story units with living areas below and bedrooms on the level above, which is rotated 90°, generating space for a garden in the resulting corners.

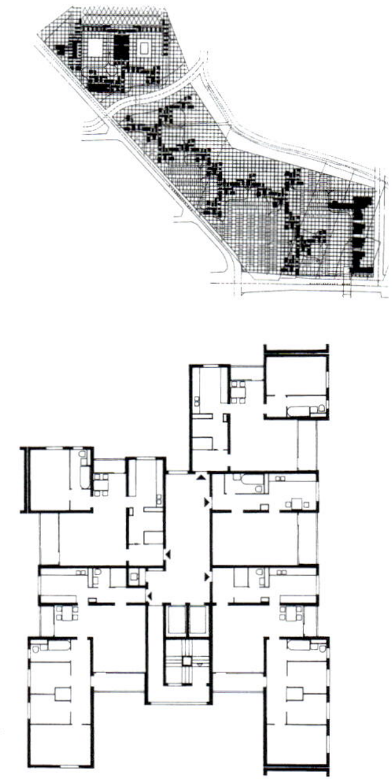

Residential High-rises

1969

Wilhelmsruher Damm 165–185, Märkisches Viertel, Berlin (GER) – O. M. Ungers

Service tracts (with entrance hall, kitchen, and WC) and bedroom tracts (with individual rooms, bathrooms, and a separate corridor) surround the predominantly L-shaped living rooms of the apartments. These floor plan clusters shape the external appearance of the buildings: connected groups of space-enclosing tower chains formed from stacked housing units. (see also 1963–74, Märkisches Viertel and page 273)

1968

Alternative Concepts/Low Rise High Density

1968–75

Proyecto Experimental de Vivienda (PREVI), Lima (PE) – Peter Land (master plan), Charles Correa/Candilis, Josic, Woods (housing type designs)

Architects from around the world conceived the layout and model houses for originally 1,500 low-rise units as an alternative to informal developments on the urban periphery. Their aim was the development of basic, prefabricated units for cost-effective "growing houses" – simple yet intelligent spatial structures that could be expanded by their inhabitants. In the design by Charles Correa, units with alternating narrow and wide modules interlock. The wide modules can accommodate larger rooms or be divided in two separate spaces, and the narrow modules ancillary functions like sanitary rooms. Staircases can be incorporated into the "side pockets." In the design by Candilis, Josic and Woods parallel rows of linked room sequences – with larger rooms or patios in the wide rows, smaller bedrooms or service spaces in the narrow rows – border each in such a way that different floor plan layouts are possible.

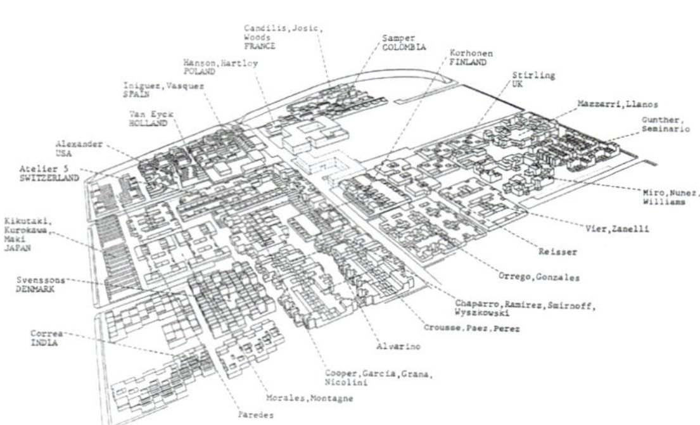

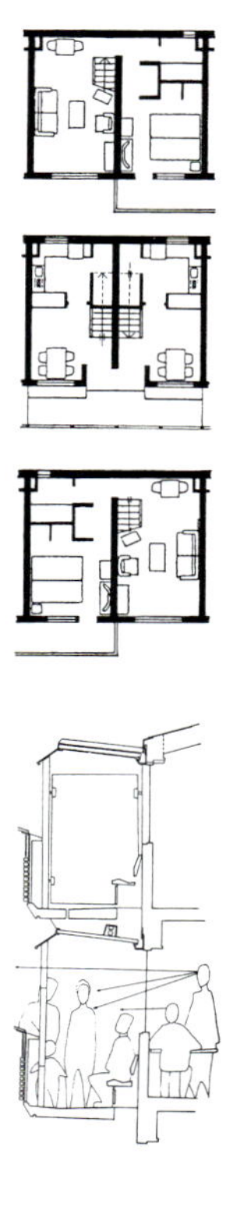

Alternative Concepts

1969–82

Byker Regeneration, Newcastle upon Tyne (UK) – Ralph Erskine

Instead of creating a master plan, a "plan of intent" formed the starting point for a mold-breaking participatory design process for both the public spaces and the apartment layouts. While 80 % of the housing units consist of low-rise row house types, the prominent Byker Wall wrapping around the northern edge of the estate was built to form a noise barrier to a planned thoroughfare. Very compact flats and maisonette apartments are accessed either from the ground floor or from external access galleries that double as social interfaces.

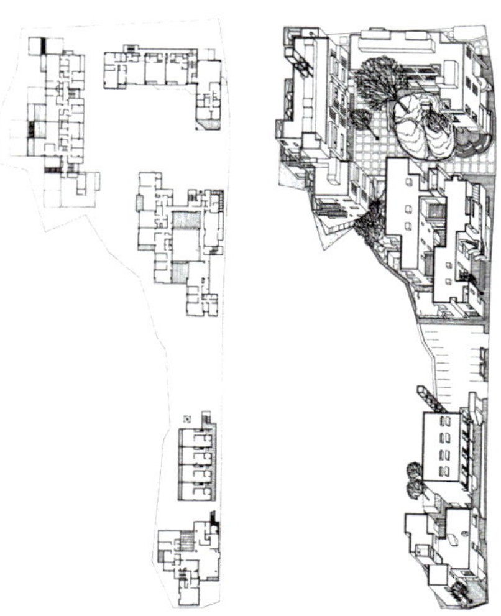

Alternative Concepts

1969–92

Daikanyama Hillside Terrace, Tokyo (JP) – Fumihiko Maki

This project was realized over multiple decades and makes use of Maki's concept of the "Group Form" as a strategy for urban change: a steadily, adaptively growing group of buildings along an eclectic series of public and semi-public spaces forge subtle transitions between the apartments and the city, between the private and the public. While the apartment floor plans of the first phase were simple and Modernist in nature, later types became more complex, adapting traditional Japanese principles for multilayered facades, making use of balconies and corridors as buffer zones, oriented toward calm courtyards.

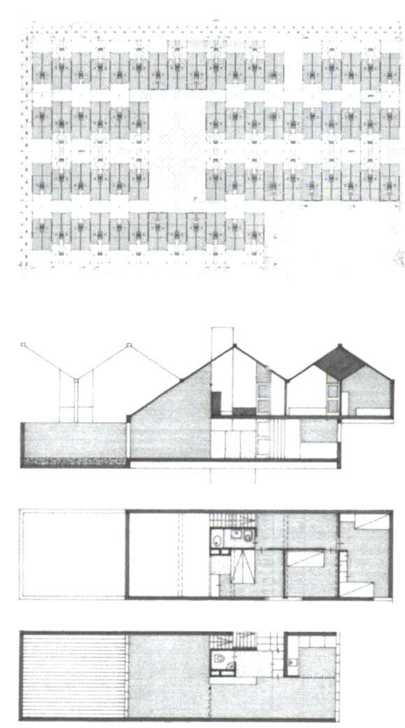

Alternative Concepts/Low Rise High Density

1971–73

Kasbah, Hengelo (NL) – Piet Blom

Under a raised patchwork-like cluster of houses, the concept called for the creation of an open urban space below, which residents could use for commercial and social initiatives, even at times adapting the space to their needs. Each apartment type consists of the same two-story basic unit to which additional modules can be connected: roof gardens, additional rooms above the paths and streets below or an entirely new unit. (see page 302)

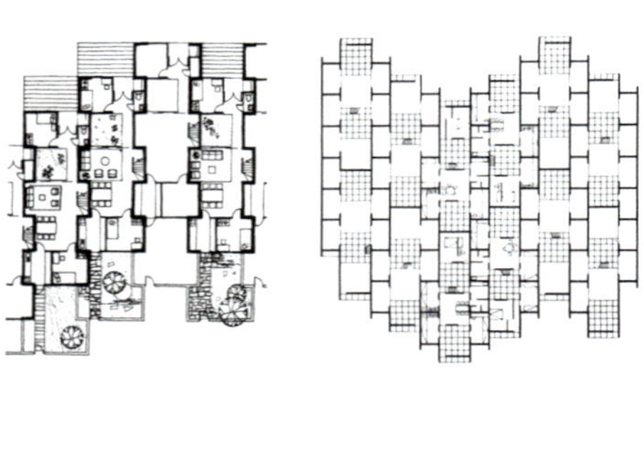

Asian New Towns

While fewer and fewer large-scale estates were realized in Western societies, mass housing in the growing industrial countries of Asia remains the dominant housing form to this day – often with high-rises as a high density building type. In contrast to the West, these configurations are not primarily intended for low-income renters, but instead were built for the growing middle class – predominantly with a model of home ownership in mind to help finance retirement. Cluster floor plans increasingly define the building shapes.

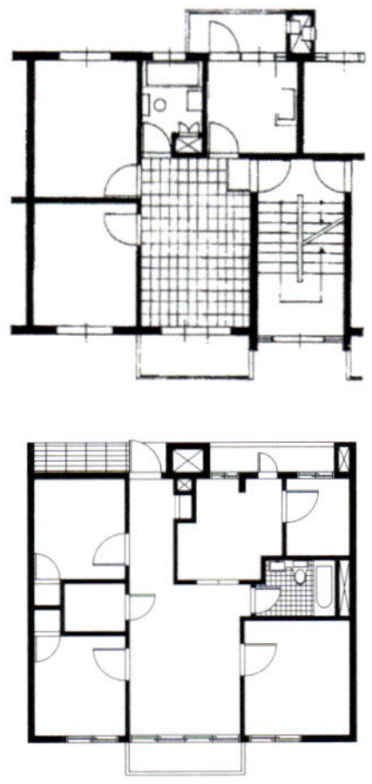

Asian New Towns/Residential High-rises
1970
Seoseoul Apartments, Ahyeon-Dong/Sibeom Apartments, Yeouido (SK)
Since the 1970s, residential high-rise estates have become the dominant form of housing in South Korea – realized by private developers instead of public entities and often replacing older, more organically evolved neighborhoods. By 2000, the number of residents in such developments rose to 50 % of the entire population. Thus, South Korea became the "Republic of Apartments." In most of the similarly laid-out apartments, the living room takes up the motive of the central courtyard in traditional Korean Hanok housing types.

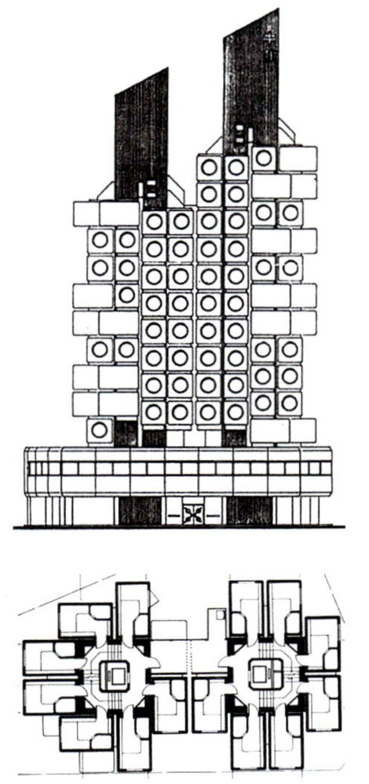

Residential High-rises
1972
Nagakin Capsule Tower, Tokyo (JP) – Kisho Kurokawa
Industrially prefabricated capsules were meant to be easily installable on the load-bearing cores, to be exchanged at any point or to be flexibly linked to one another. The compact units are only the size of four tatami mats, have integrated built-in furniture and bathing cells and were conceptualized as work-live cells.

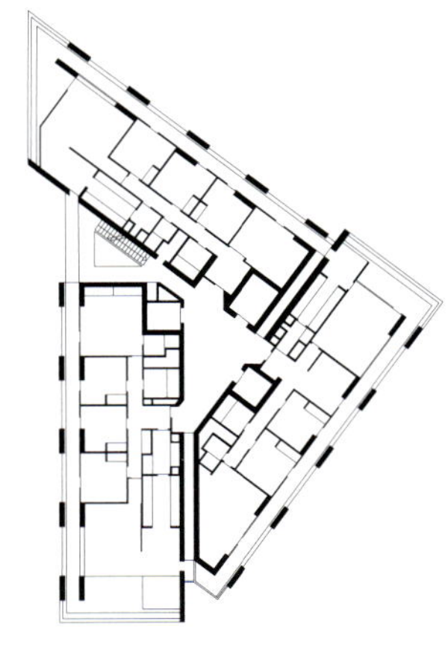

Residential High-rises
1974
Lauderdale Tower, Barbican Estate, London (UK) – Chamberlin, Powell, Bon
The three single-orientation, elongated 5-room apartments with corner living rooms opening up to loggias and escape balconies are functionally clearly structured and generate a cluster-like, triangular high-rise. (see also 1955–83, Barbican)

1974

Asian New Towns/Residential High-rises
1971–81
Sui Wo Court, Hong Kong (HK) – Palmer & Turner
Subsidized home ownership in an ensemble consisting of nine high-rises in Sha Tin, one of the New Towns inspired by British models and built in the New Territories. The cross-shaped cluster floor plan is laid out around stacked three-story halls accessed by elevator on their lower level, which connects every 36 apartments into a neighborhood. The units are for the most part accessed from secondary staircases and open into the living room, from which a separate corridor leads to the individual rooms.

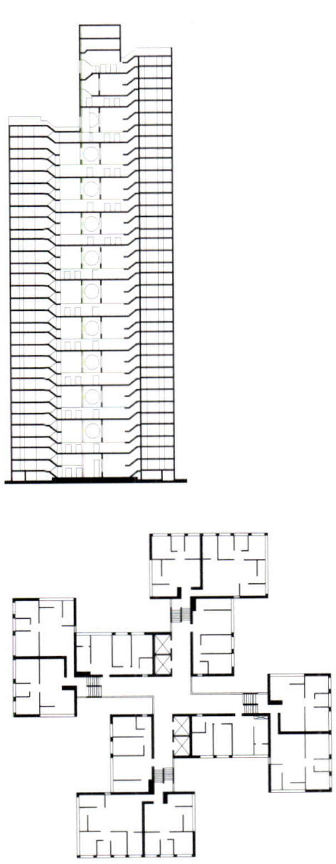

1972
Pruitt-Igoe, St. Louis (USA), beginning of demolition work

For the architectural theorist Charles Jencks, the demolition of the failed large-scale housing development Pruitt-Igoe (built 1951–56 – Minoru Yamasaki) signified the "death of modern architecture." But like many other such estates, its demise was not only the consequence of the anonymity of the architecture with its poor quality standards and lack of attention paid to communal issues. Originally designed for the middle class, its realization coincided with the exodus of better-off families to the newly built suburbs. These estates very quickly turned into ghettos, which in some cases eventually led to huge social problems accompanied by high vacancy rates.

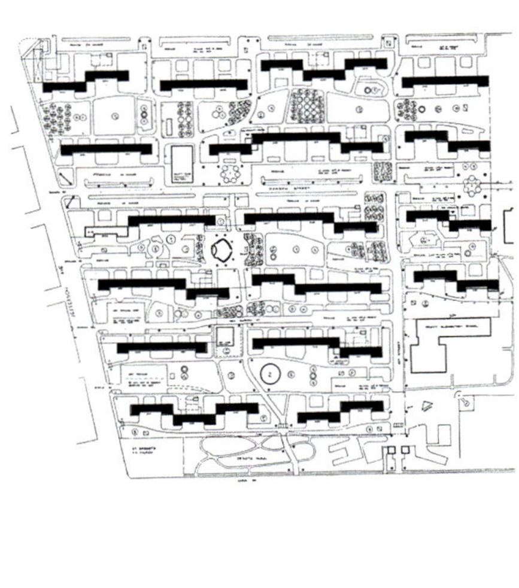

1973
Gallaratese II, Milan (IT) – Aldo Rossi
Up to this point, Rossi was only known as a theorist who spurred on important debates about the architectural relevance of the historic city. The elongated slab structures were part of a collaborative design for a new housing development. Clearly structured, two-tier apartment plans reveal a kitchen and bathroom placed along an external access gallery and a connecting hallway leading to all other spaces of the two-room apartments.

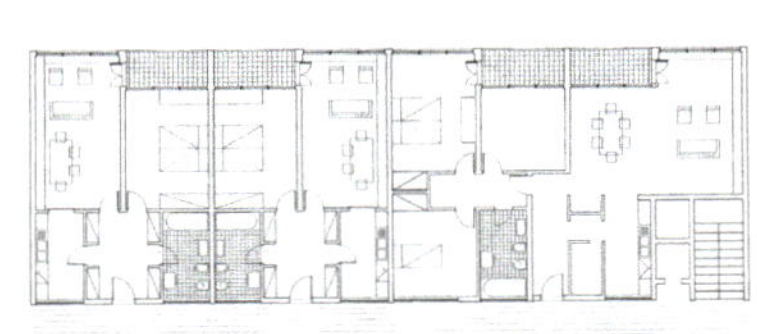

Residential High-rises
1974
Walden 7, Barcelona (ES) – Ricardo Bofill, Taller de Arquitectura
Tower segments laid out in a checkerboard grid and connected around courtyards to form a complex high-rise cluster. A network of circulation paths on every second floor unites the individual apartments composed of 30 m² modules into a community. Instead of walls and hallways, only built-in furniture and varying floor levels zone the flowing spaces of the units. (see page 240)

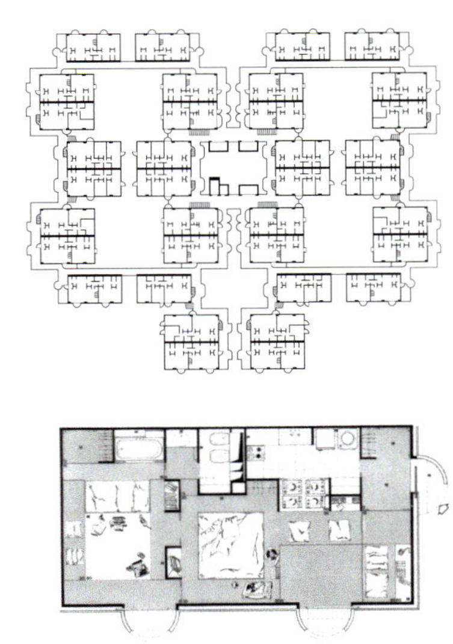

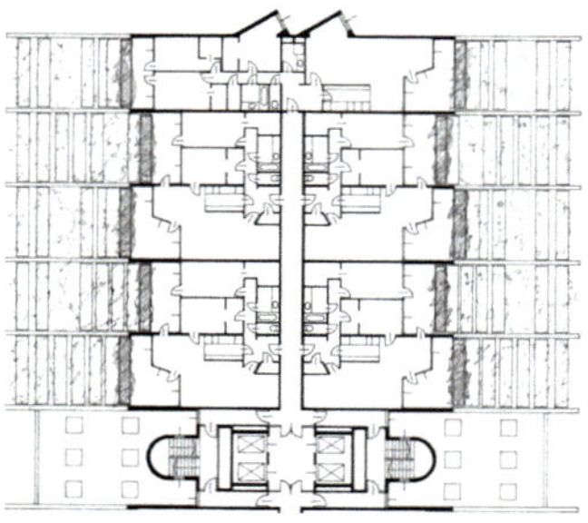

Residential High-rises/Terraced Buildings
1976–85
Alt-Erlaa Residential Park, Vienna (AT) – Harry Glück
Described as a "vertical garden city," this large-scale housing scheme with terraced high-rises, onto which the green surroundings seemingly continue to grow, are remarkably well-liked despite the interior corridors having no windows and the apartments being oriented to one side only. The numerous community spaces in the belly of the building nurture neighborly contact, while spacious terraces provide a replacement for one's own garden.

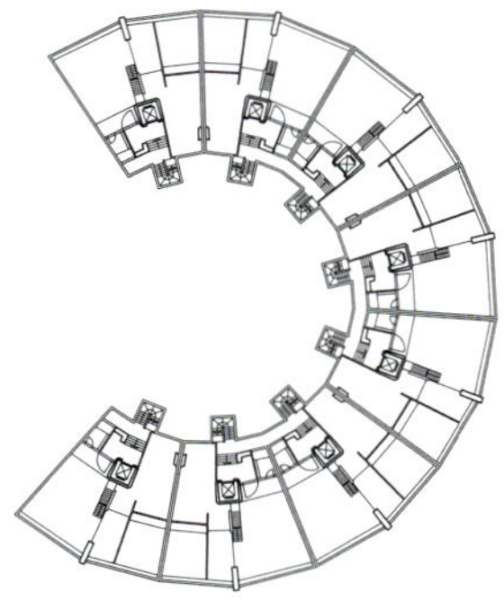

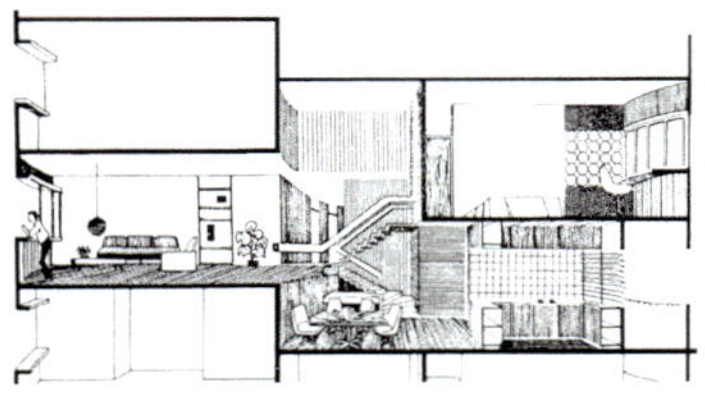

Residential High-rises
1976
Pearl Bank Apartments, Singapore (SG) – Tan Cheng Siong, Archurban
At 38 stories, the tallest residential high-rise in the city-state at the time and a response to a prolonged period of growth. The circulation system is located on the inner facade of the open cylinder and composed of external access galleries and branching secondary staircases. The varied apartment types are organized as split-levels, with open, dual-oriented living areas spanning between the inner circle and the exterior facade.

Postmodernism/Back to the City

With the increasing influence of Postmodernism and, particularly in Europe, the rediscovery of the city as an attractive place to live, small-scale and diverse solutions began to appear as did projects searching for urban and architectural historic continuities. Partially influenced by discussions around the European Architectural Heritage Year 1975 as well as growing criticism of post-war planning and the resulting deserted city centers, the long neglected historic districts were gaining more appreciation. Block infill projects and the modernization of existing buildings were becoming an option and apartment floor plans less schematic, as they had to take local circumstances into consideration.

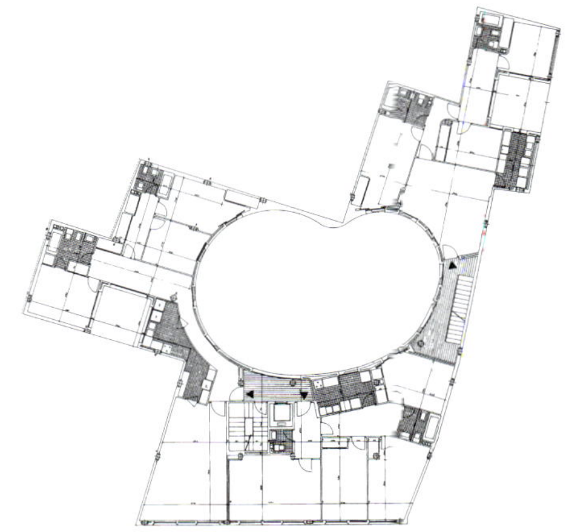

Back to the City
1976
Calle Doña Maria Coronel, Seville (ES) – Cruz, Ortiz
A kidney-shaped courtyard becomes a sculptural void on this irregular building site in the historic city center. Each apartment adapts very differently to the complex geometry, turning toward additional patios through crooked corridors, leading hallways around the courtyard or making use of the living room as an entrance and distribution center. The roof terraces and the courtyard are shared spaces. (see page 104)

1974

Low Rise High Density
1974
Galgebakken, Copenhagen (DK) – Storgård, Orum-Nielsen, Marcussen
A car-free estate providing a common site for diverse housing needs. Small groups of atrium houses, row houses and split-level housing types are bundled together with entrances, eat-in kitchens and at times play areas overlooking shared pathways or courtyards. For the most part, dually oriented living rooms open up to a garden at the rear. (see page 306)

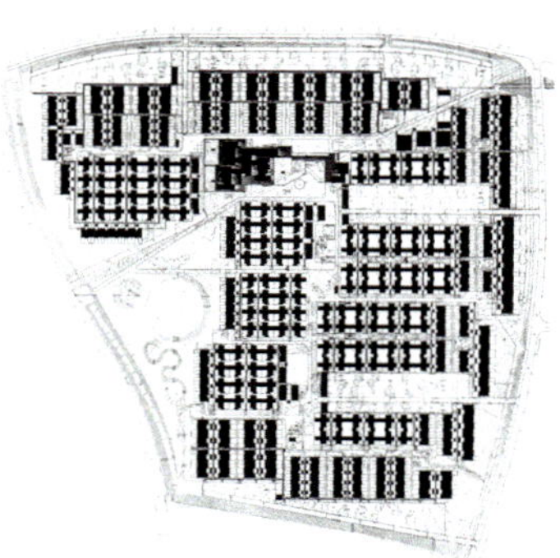

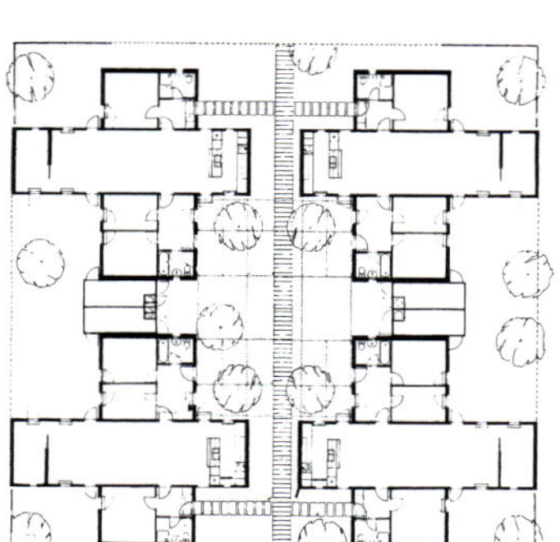

Second Demographic Shift: Diversification

Around the beginning of the 1970s, the so-called "second demographic shift," which covers both increasingly diverse familial relations – later and lower birth rates, higher divorce rates, and an increasing number of unmarried couples – as well as changing social values – like the increasing relevance of self-actualization and equal gender roles – also had an impact on housing concepts. Besides the classic nuclear family, diverse forms of cohabitation were developing, which were also often no longer conceived of as long-term solutions. Individual demands were formulated, which could no longer be satisfied by the dominant standard solutions of the time. Participation and self-building options changed approaches to design floor plan types were becoming more diverse and flexible.

Alternative Concepts/Diversification
1976
Diagoon Houses, Delft (NL) – Herman Hertzberger
The housing units were intentionally left incomplete: staircases, technical installations and the split-level-like diagonally offset floors were predetermined. The functional use and decision to enclose or leave the rooms open was left up to the discretion of the residents. (see page 352)

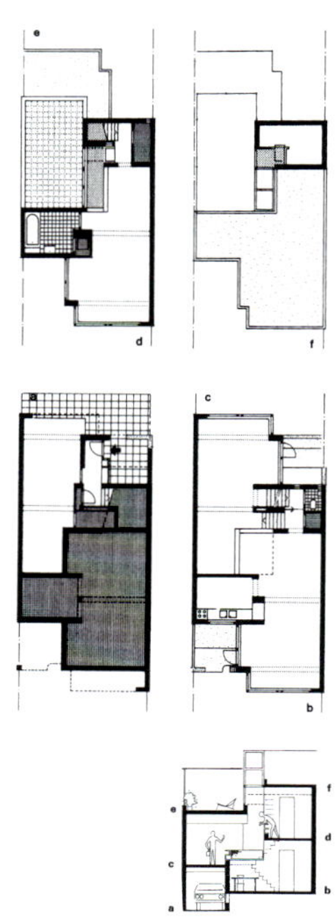

Low Rise High Density
1977
Quinta da Malagueira, Évora (PO) – Álvaro Siza
Affordable housing with a total of 1200 units arranged in a back-to-back configuration of L-shaped row houses with private patios. Residents are given a footprint of 8 x 12 m. The basic unit houses a kitchen, bathroom and living and bedrooms on the ground floor with the possibility to expand the surface area with a maximum of four additional bedrooms on the second floor.

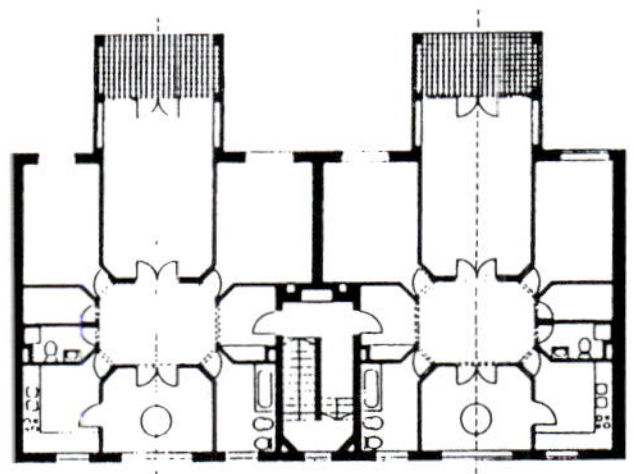

Back to the City
1979
Herter House, Munich (GER) – Hilmer & Sattler
A typical inner-city floor plan solution for the time, with a large hall as the spatial focal point.

Back to the City/ Building Exhibitions and Model Estates
1984–87
International Building Exhibition Berlin (GER)
Two conceptual topics stand for a new, exemplary approach to urban redevelopment. In contrast to the previous dominance of the automotive city and large-scale demolitions, which sacrificed many older, organically evolved urban structures, "critical reconstruction" sought to reconstruct and increase density in inner-city areas with the historic city layout as its basis. Cautious urban renewal – an approach which grew out of the protest milieu of the squatting movement – pursued the preservation and rehabilitation of older housing stock. It also intended to stabilize and further develop existing social and functional structures within the urban districts – using the instruments of participation, self-help and modernization measures carried out by the inhabitants themselves.

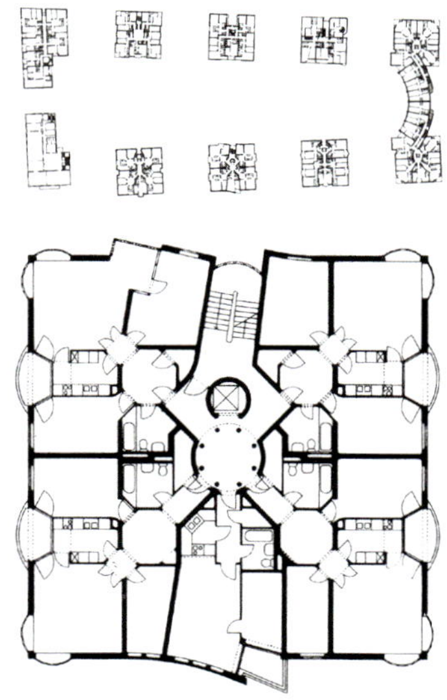

Back to the City/Building Exhibitions and Model Estates
1985
Rauchstraße, Berlin (GER) – Rob Krier (master plan), Hans Hollein (urban villa)
One of the most widely discussed urban compositions of the IBA was the solitary building ensemble on Rauchstraße, which transferred the suburban building type of the "Urban Villa" to the city center and adapted it for use as social housing. 10 solitary buildings with a footprint of 21 x 21 m each were designed by different architects, and together with the end building by Rob Krier, surround a large garden courtyard. In the diagonally laid- out apartments, an octagon-shaped hall, which connects to almost every other room, leads directly from the entrance into the corner living room.

1986

Alternative Concepts/Diversification
1977
Molenvliet, Papendrecht (NL) – Van der Werf, Werkgroep KOKON
This project tested the "Open Building" principle of support and infill systems as well as participatory design methods for both the residential environment and building structures. Grouped around four car-free courtyards – thus enabling the creation of small, intimate neighborhoods – the overall structure, on a 4.80 m grid which remained open in two directions, was first parceled according to individual space requirements and then built out in close consultation with the wishes of the residents. Its capability to be freely subdivided made it possible to create a variety of housing types from studio apartments to 6-room arrangements and the accommodation of both housing and commercial units. (see also 1962, Open Building)

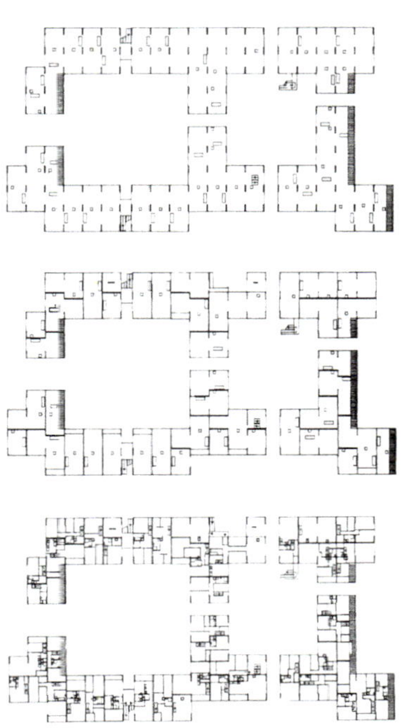

Alternative Concepts
1979
Odham's Walk, London (UK) – Donald Ball
Residents' protests in Covent Garden led to the design of a lively residential quarter instead of a planned wholesale redevelopment. While ateliers, workshops, and exhibition spaces on the ground floor anchor the complex to the surrounding streets, 2- to 6-room apartments above encircle multiple interlocking courtyards. The maisonette apartments at the top are accessed from an external gallery that winds its way around the courtyards.

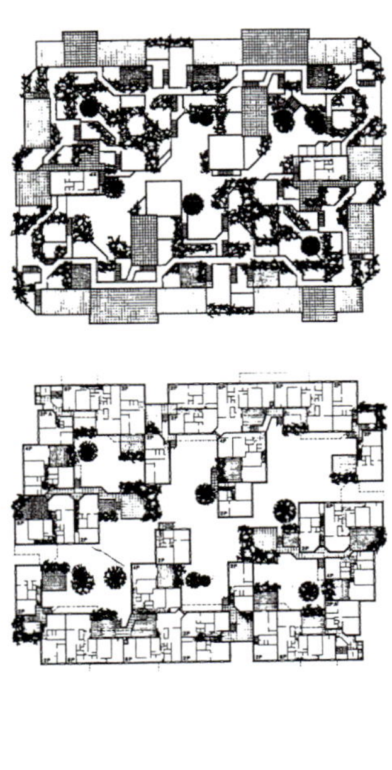

Residential High-rises
1983
Kanchanjunga Apartments, Mumbai (IN) – Charles Correa
A high-rise for wealthy residents is optimally laid out for the climate: both the building and the open, split-level floor plans are oriented to maximize cross ventilation from the dominant wind patterns. An outdoor layer of loggias – in the form of deeply drawn-in, double-height verandas at the corners of the building – help shield against direct sunlight and monsoon rains. (see page 242)

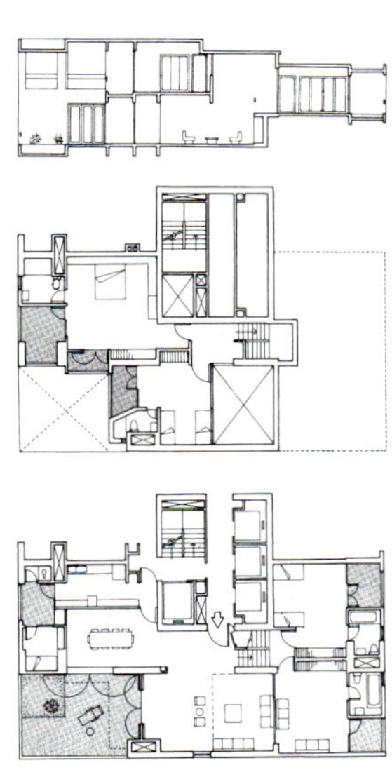

Back to the City
1986
Schrankenberggasse, Vienna (AT) – Rob Krier
This building with clear references to historic architectural elements combines a corner building with an infill typology. Symmetrically arranged living rooms in the axes of the entrances serve – particularly in the dually oriented corner apartments – as representative circulation centers. (see page 107)

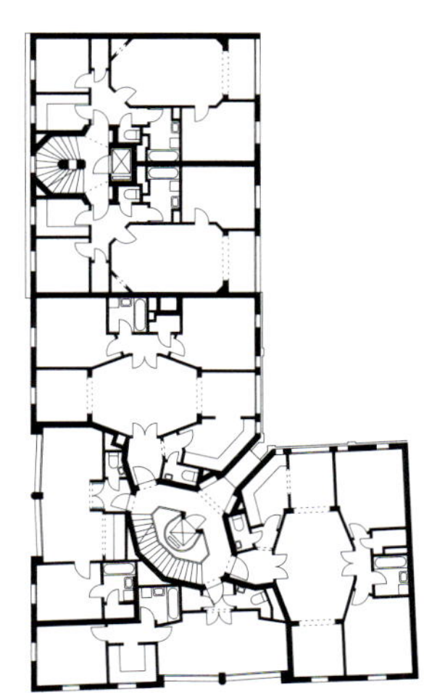

Urban Revitalization

Extensive, centrally located areas whose original purpose as industrial or transportation sites has become obsolete as a result of structural changes and technological innovation, were used for urban revitalization strategies. Projects such as IJplein, KNSM- & Java-eiland or Borneo Sporenburg in Amsterdam, HafenCity in Hamburg, Nyhavn in Copenhagen, Docklands in Melbourne or the port area of Toronto offered a unique opportunity to set a large-scale accent – not only for the reanimation of areas near the inner city, but also for new urban housing models.

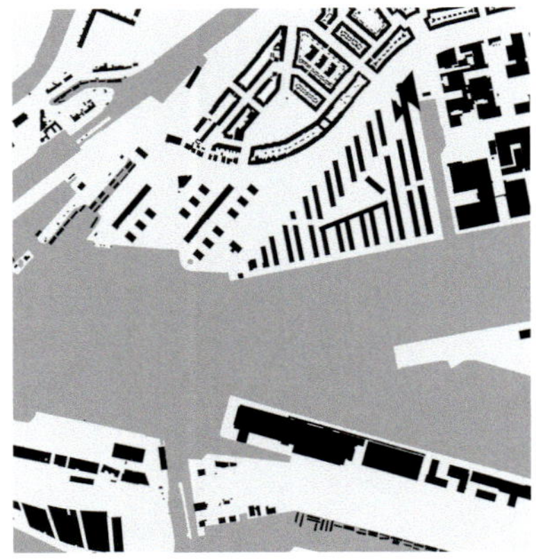

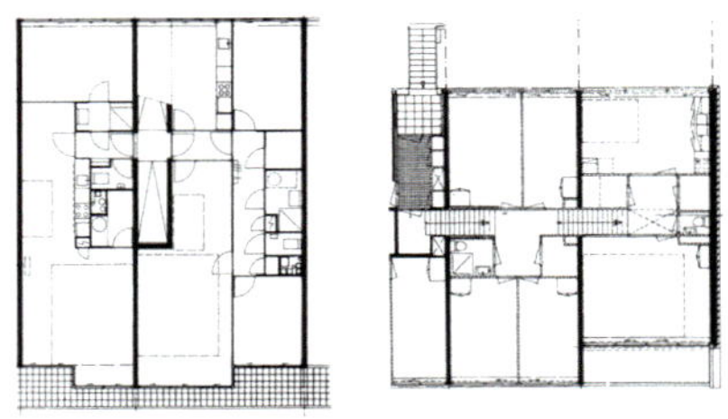

Urban Revitalization
1988
IJplein, Amsterdam (NL) – OMA (master plan, building design)
The original high-rise ensemble planned for a new urban district in a former shipyard directly across from the city center was changed to rows of parallel blocks due to protests. The floor plans are dominated by stairs – transversely connecting single-story flats at different junctures, and lengthwise within the maisonettes.

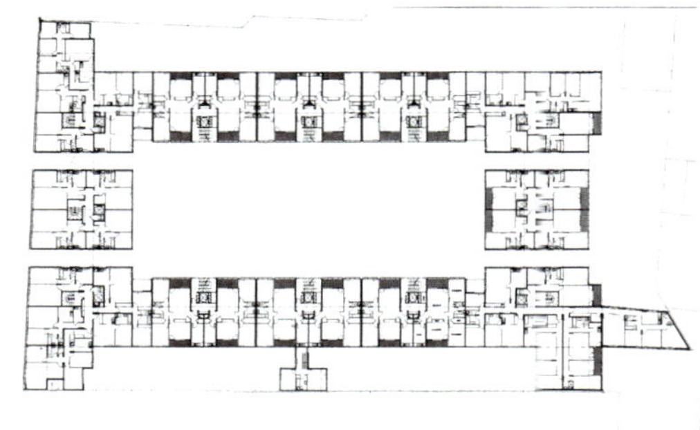

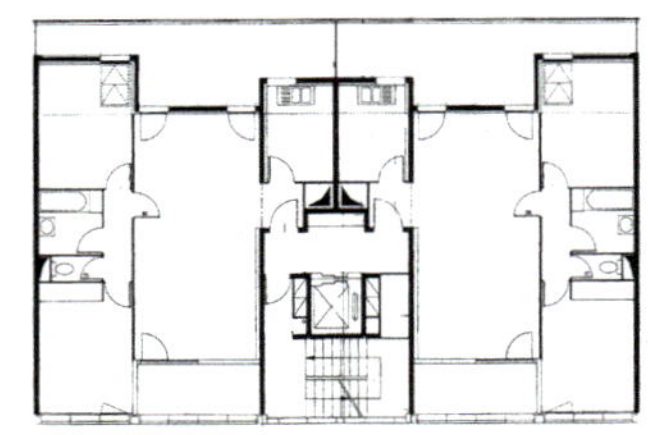

Back to the City
1991
Rue de Meaux, Paris (FR) – Renzo Piano
The complex extends across an entire block and is simultaneously influenced by both the surrounding block edge and courtyard building types to form a consistent urban ensemble. The floor plans are clearly structured: a dually oriented, divisible living room is located in the mid-section of the single-story flats, with the kitchen and access area on the one side, bedrooms with a secondary corridor and bathrooms on the other.

1987

1987
Avenue du Général Leclerc, Nîmes (FR) – Nouvel, Ibos
Wide wraparound galleries on every second floor provide access to the maisonettes and single-story apartments on the north side and function as balconies to the south. By setting the shear walls at 5 m intervals, the individual units can be reconfigured without any structural constraints, making it possible to create very diverse apartment types. In order to increase the amount of living space for each unit, standards were dramatically reduced, industrial building components used and walls left unfinished. (see page 194)

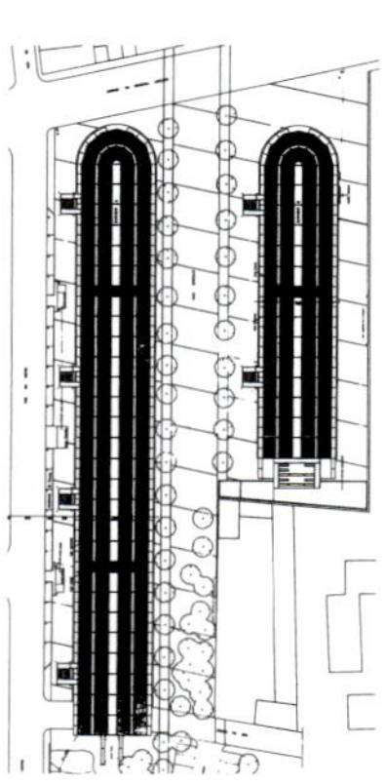

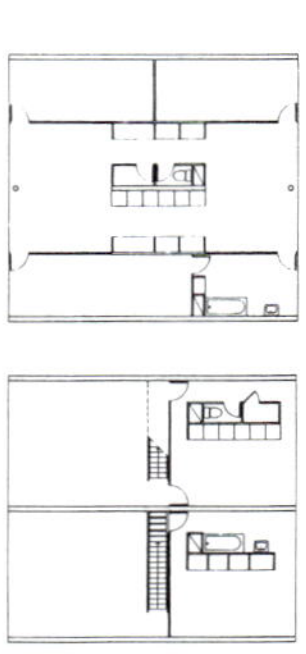

Low Rise High Density
1988–93
CIDCO Housing, Navi Mumbai (IN) – Raj Rewal
Part of a larger urban development scheme, the project is considered a model for affordable housing in a low-rise, pedestrian-friendly and dense building structure. A range of housing units brings together not only different sized families but also diverse income groups. Cubic volumes in a cluster-like composition form ensembles up to three stories in height around common courtyards with roof terraces at various levels.

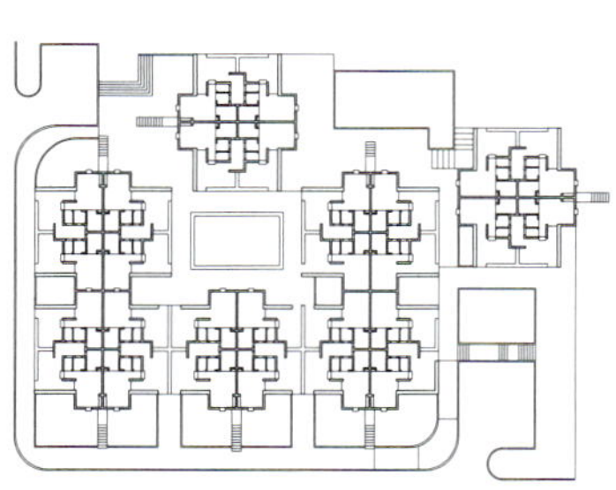

Housing at the End of the 20th Century

Housing at the end of the 20th century is exceptionally heterogeneous. Neoliberal patterns of thought influenced housing policies and in the 1990s led to the withdrawal of public funding from social housing in a number of Western societies. With this development, many restrictive measures were loosened and alternative approaches filled the resulting gaps. However, this also led to an increasing shortage of affordable housing. With the demise of Postmodernism – which was increasingly criticized as self-referential and purely formal – architects again increasingly turned towards more concrete problems and developments. They began to design more conceptually, and economic and social aspects were considered anew. Simultaneously, residential buildings outside of Europe and North America increasingly influenced the debate.

Building Exhibitions and Model Estates,
Low Rise High Density
1991
Nexus, Fukuoka (JP) – OMA
Three-story, compact low rise high density structure consisting of two blocks of L-shaped residential units with private patios accessed from a network of pathways on the ground floor. A roof terrace serves as a complementary outdoor space; wave-like shed roofs bring in additional natural lighting. (see page 308, see also 1962, Open Building)

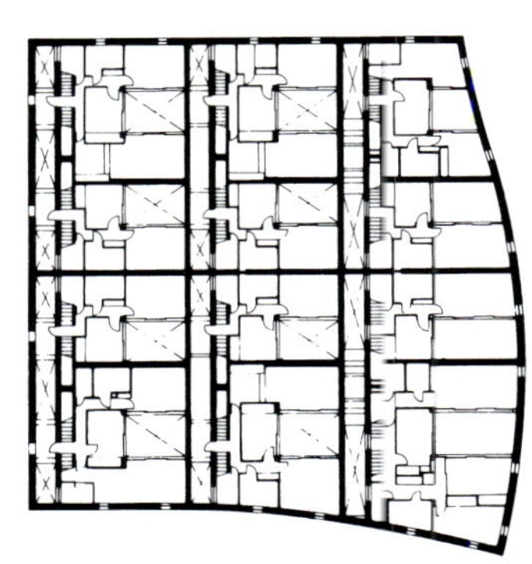

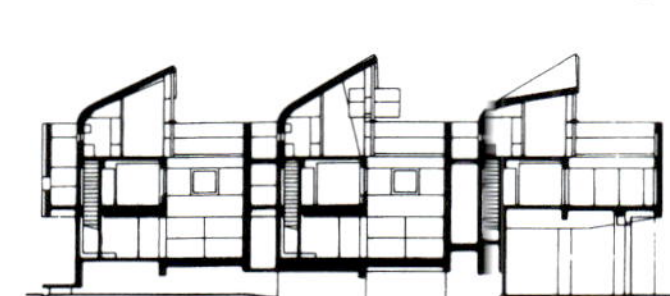

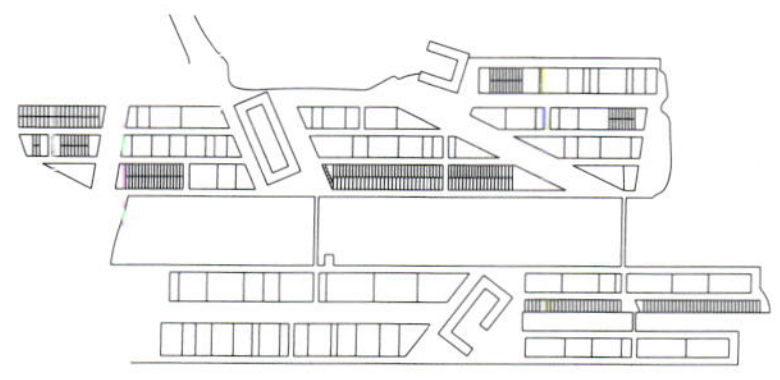

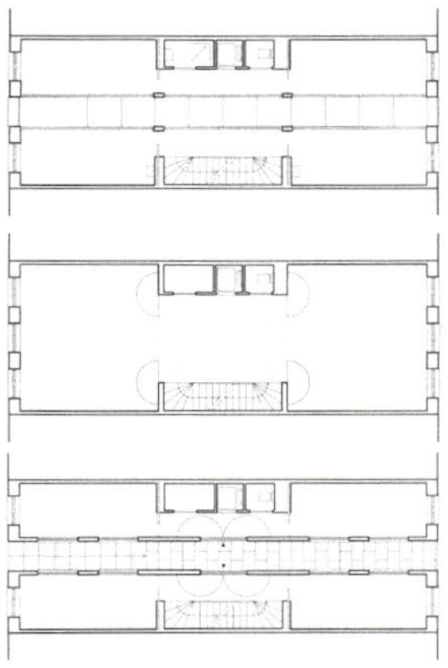

Urban Revitalization

1992

Borneo Sporenburg, Amsterdam (NL) – West 8 (master plan), Höhne & Rapp (row house)

The area was to be redeveloped to create an attractive, urban neighborhood for families near the city center. A high level of diversity was achieved through a mix of housing tenure models – 30 % public and 70 % privately financed residential units – and various building types – predominantly low rise high density structures combined with three larger multi-story apartment buildings. The owners of the 60 individual homes were allowed to choose their desired architect. (see page 121)

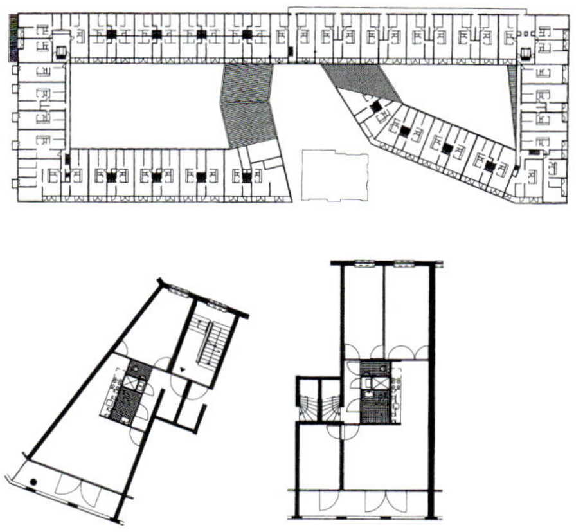

Urban Revitalization

1994

Piraeus, Amsterdam (NL) – Jo Coenen (master plan), Hans Kollhoff, with Christian Rapp (building)

Based on a design with conventional urban blocks, the building type was deformed to preserve an existing historic building on the site. The resulting complex figure brought about the development of a multifaceted access system and numerous apartment types. A consistent motive in the apartments is an inserted core with a kitchen and living room to one side, and a corridor to the other. (see page 86)

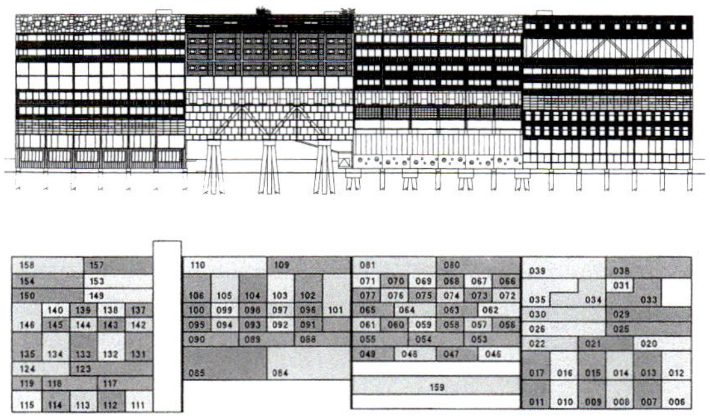

Urban Revitalization

1995–2003

Silodam, Amsterdam (NL) – MVRDV

This superblock is located in a harbor basin between the city center and Borneo Sporenburg and offers a multitude of apartment types, whose heterogeneity is reflected in the complex access system and diverse facade treatments.

2000

Alternative Concepts

1994

Next 21, Osaka (JP) – Utida et al.

A test model for sustainable housing designed according to the theories of John Habraken: an open frame structure similar to a shelving unit can be occupied with individualized and exchangeable housing units. (see also 1962, Open Building and page 156)

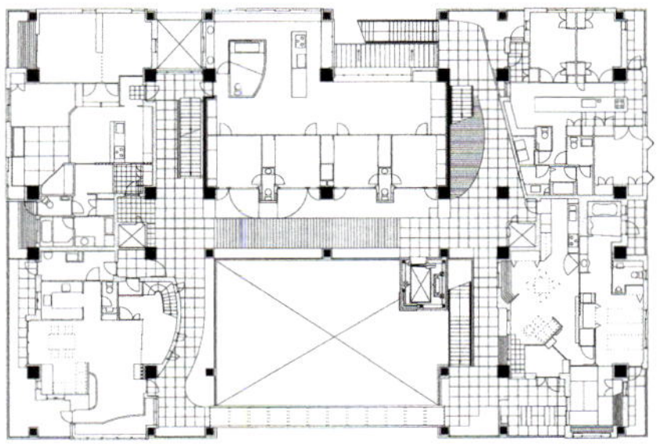

1994

Bahnhofstrasse, Graz (AT) – Riegler Riewe Architekten

An exemplary case of polyvalent housing, which can accommodate very diverse occupancies due to its versatile transverse and lengthwise circulation possibilities, the multi-part service zone in the middle and almost identically sized rooms. (see page 202)

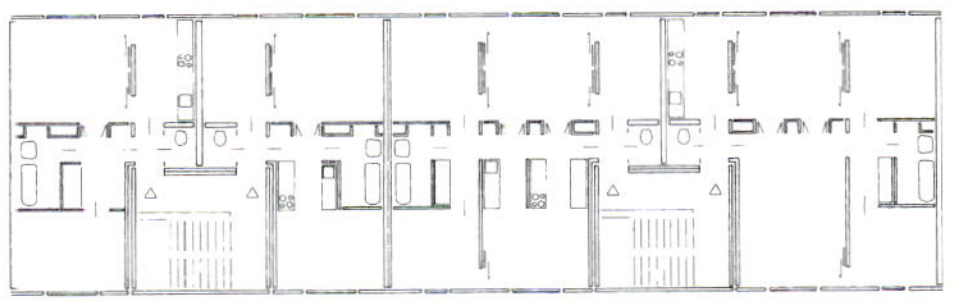

1998

Space Block Kamishinjo, Osaka (JP) – Kojima + Akamatsu

To improve the lighting conditions on the narrow infill site, the porosity of the building mass was modelled. The foundation of the design is a system of modular spatial blocks that are stacked, spring back and forth and create recesses to increase the surface area of the building's envelope, thus admitting more daylight. In this manner, spatially complex mini-apartments are generated in the interior. (see page 118)

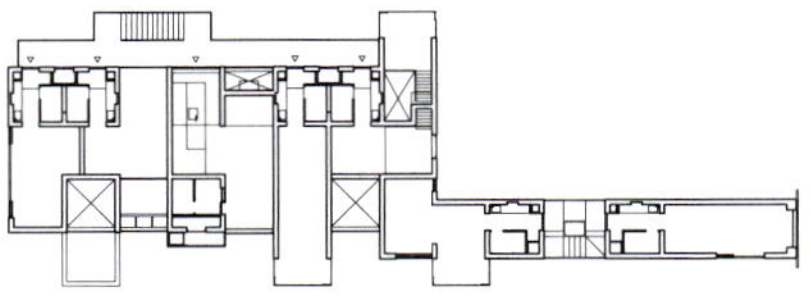

Alternative Concepts

1998–2000

Miss Sargfabrik, Vienna (AT) – BKK3

As a complement to the previously built project "Sargfabrik" – a converted factory site – a cooperative initiated, designed with a participatory approach and realized a building that sought to combine collective ways of living, communal facilities and work-life scenarios.

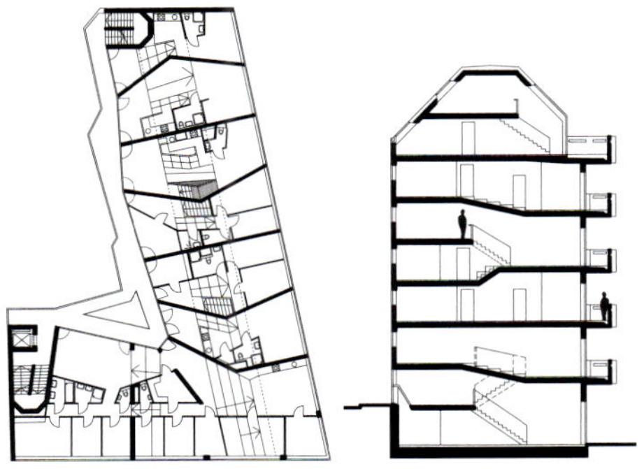

2000

Kitagata Residential Building (JP) – Kazuyo Sejima, Nishizawa

Rather than predefining housing units, various types of spaces (private room, tatami room, in part double-height kitchens and loggias oriented to two sides) can be combined and accessed lalong both facades from both a public and private walkway, making various cohabitation models possible. (see page 208)

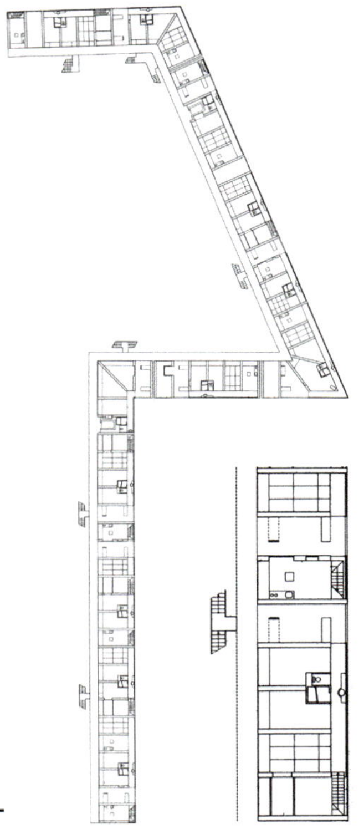

Challenges and Tendencies

In contrast to a review of the past – where the distance of time allows for a more accurate reflection on which historical, societal, and economic developments or design strategies challenged existing paradigms and contributed significantly to the development of housing – the present can only be described in tendencies. What is discussed here are some of the most important challenges facing urban housing today. Concrete examples are offered as illustrations of potential design solutions.

The Social Segregation of Cities

An increasing section of the population is threatened with exclusion from the real-estate market – whether as renters or buyers – because they simply cannot afford the rapidly accelerating cost of housing. This is not only due to the global financial crisis of 2007, but also to the increasing attractiveness of real estate as an investment, the often-cited gap in income development, and even factors like online vacation rental platforms that increasingly deprive cities attractive to tourists of sufficient residential space. The current scarcity of affordable housing is also further aggravated in Western industrial societies by the steady growth in the attractiveness of city life – with simultaneous high and still increasing numbers of empty properties in many rural regions. The situation is made more acute by processes of gentrification, where in well-located but often neglected neighborhoods, socially and ethnically diverse residents are pushed out by groups with higher incomes.

This development requires counter measures, that with more inclusive concepts attempt to stem the trend toward social segregation within cities – which on a societal level can also be viewed critically – by creating new models for affordable housing, whether in the form of new building types, new property development models, through the renewal and expansion of social housing programs, cautious forms of increasing inner-city density as well as renewed, contemporary research into "housing for the minimum subsistence level." The inhabitants to be considered are extremely heterogeneous and span from singles, couples, families, retirees, and immigrants to groups on the edge of society such as the homeless 1, 2, 3.

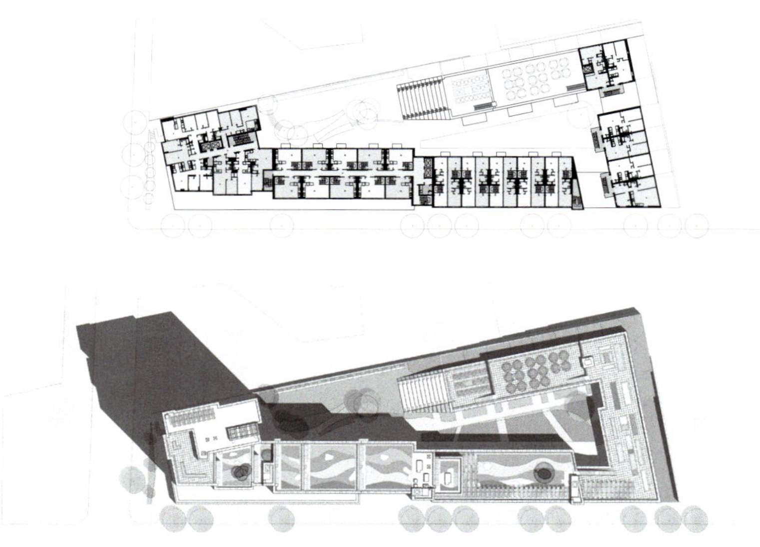

1 Dattner Architects, Grimshaw Architects: Bronx Social Housing, New York, 2012
The project, which was built as a model for "affordable, sustainable and reproducible" housing in a central location, combines three building types taken from the surroundings – a row of townhouses, a higher block with duplex units and a high-rise – to form a continuous building, the roof of which winds its way up from the courtyard to the uppermost roof terrace and can be used as a collective fruit and vegetable garden as well as for recreational activities. Small shops, a health center, work-life units and a fitness center enliven the juncture between the street and the courtyard.

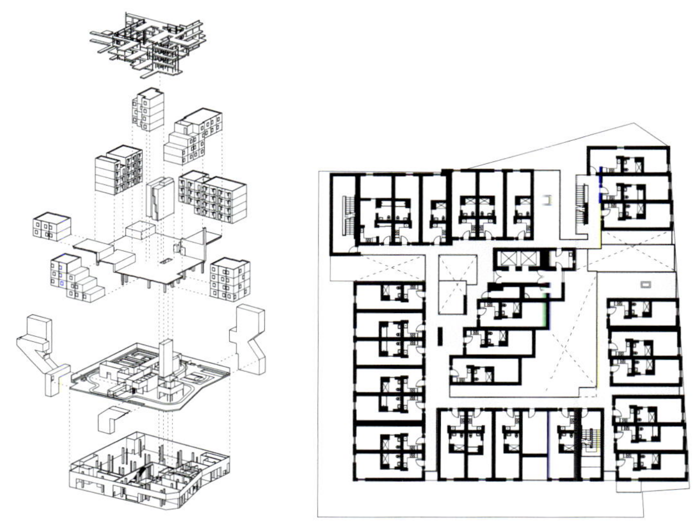

2 Michael Maltzan Architects: Star Apartments, Los Angeles, 2014
In this residential project for homeless people, the goal to foster a social interface with the immediate surroundings through a mix of uses was applied to the redevelopment of an existing parking structure. The ground floor is occupied by a health center for residents and neighbors, and the level above acts as a podium with communal gardens, jogging path and meeting places. The prefabricated residential units stacked above each have their own bathroom and kitchen and are connected in an open spatial fabric. Open voids offering views of both the city beyond and the podium below, and numerous staircases and external galleries are meant to foster contact.

4 HAHOH Haas Heckmann Architekten: Am Ottersgraben House, Zell am Harmersbach, 2007
A house with initially four compact apartments was to be built in such a way to keep open the option of completely different layouts. The scenarios included apartments for singles or couples, units for families as well as a multi-generational house comprising all floors. The two staircases on the ground floor serve as intersection for various connections: all adjacent areas – the units above and below, to the left and right as well as the stair landings – can be linked in a flexible manner.

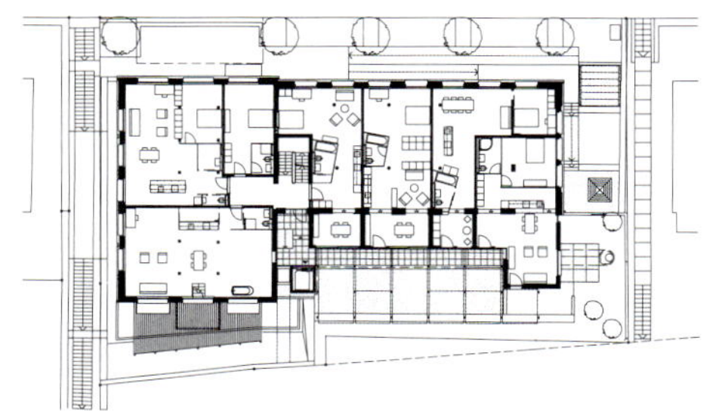

5 Archplan AG: Wohnfabrik Solinsieme, St. Gallen, 2002
Solinsieme – a combination of the words "solo" (alone) and "insieme" (together) – is a participatory project initiated by the residents, all in the second half of their lives, and combines individual housing with a more community-based lifestyle. An existing factory was converted and expanded with a new building, which acts as a collective intersection comprising access areas and sun decks. In total, the shared spaces make up almost 20 % of the building surface and also include studios, guest rooms, a roof terrace, a bike store and a community room with enough space for events.

DEMOGRAPHIC CHANGE: GENERATIONS

While the increasing quantitative imbalance between generations is above all a social and economic problem, specific alternative housing strategies are also required for the different generations – with foresight into the different residential forms and locations inhabitants prefer during each phase of life or the fact that future generations may simply be much smaller in number.

Families

In particular the family – with the nuclear family as the main object of housing standardization research for a long time – has become more diversified over the years. Besides the traditional model, there exist single-parent households, patchwork families, and multi-generational living situations, among many others. Familial relations and constellations are changing much quicker and more often than in previous times, thus increasing the demands on the flexibility of both the individual apartment floor plan, as well as the way in which adjacent units can be connected to or separated from one another 4.

Baby Boomers

As the baby boomer generation ages, the number of retirees is going to dramatically increase in the coming years. A growing elderly population in many societies will have to be countered with a considerable expansion of appropriate floor plan typologies and forms of living that are not only suitable for this demographic development, but also take into account that seniors today increasingly prefer to live in urban areas or have other, more diverse wishes and demands than earlier generations – for instance, a preference for collective or assisted living 5, 6.

Millennials

Among so-called millennials, who today range from 18 to 30 years old, a number of significant changes have also taken place: increasingly influential to where and how one lives are not only longer periods of education, a desire to first establish a career or the fact that marriage is more seldom, occurs later or not at all, while divorce rates increase, but also that in comparison to older generations they often earn considerably less and have fewer assets. While the more affluent segment often lives alone in 2- to-3-room apartments, reflective of the more general trend toward more living space per person, many millennials are increasingly excluded from the real-estate market. Breaking free of conventions and

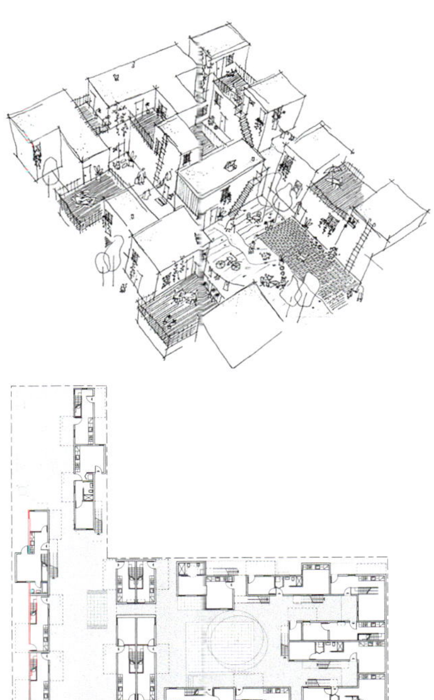

3 5468796 Architecture + Cohlmeyer Architects: Centre Village, Winnipeg, 2010
In this project realized for underprivileged families, multi-story residential units – each with its own entrance – surround a communal courtyard and space for everyday interaction. The units are comprised of very compact, narrow room modules connected to form vertical and horizontal spatial sequences, which generate apartment types for the various needs of the diverse inhabitants. The organization of both the building and the individual units aims to reduce inefficient and unattractive access spaces in order to reduce rental costs as much as possible.

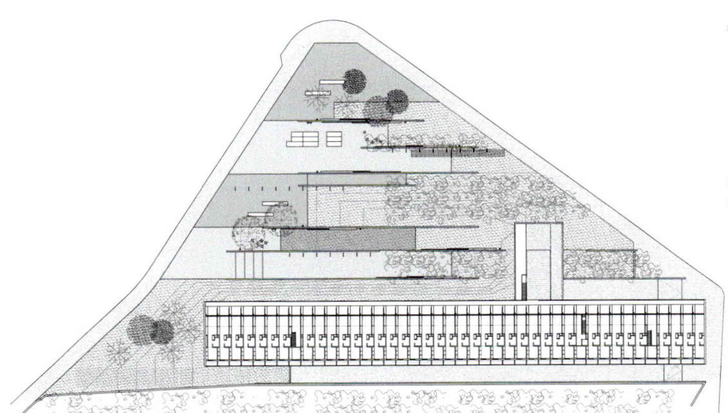

6 Javier García-Solera: Assisted living with day-care facilities, Santa Vicente del Raspeig, Alicante, 2005
Set in a new public parc, the project couples autonomous living for seniors with services and care above a day-care facility. The inserted bathrooms on both sides of the dividing walls cleverly zone the floor plans of the individual units: on one side they subtly narrow the open transition from the living room to the bedroom, and on the other they are inserted flush with the built-in furniture of the kitchen and storage spaces.

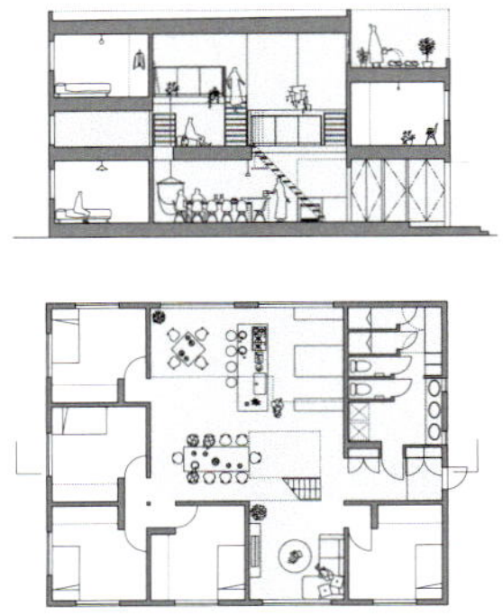

7 Naruse Inokuma Architects: LT Josai, Nagoya, 2013
LT Josai reacts to a growing social trend in Japan in which singles decide to share a living space temporarily or more permanently without knowing one another beforehand. Their private rooms surround an open, split-level common space spanning all floors. Even the bathrooms are shared.

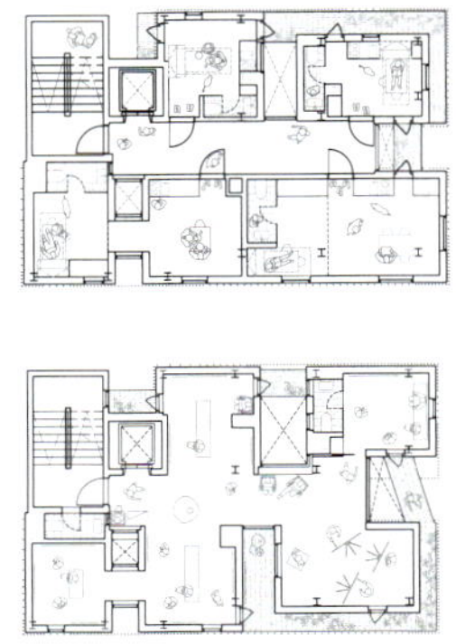

8 SsD Architecture: Songpa Micro Housing, Seoul, 2014
The spatial limitations of the micro apartments here are compensated for by semi-private open interstitial spaces that form shared thresholds through which multiple units can be combined. This matrix is complemented by "shared living rooms" that can be used by both residents and neighbors – a micro auditorium and a café in the lower level as well as a gallery and work spaces above the open ground floor.

9 narchitects: Carmel Place, New York, 2016
The project grew out of a publicly funded initiative to create affordable housing for the growing number of single-person households in New York. 22 of the 20–33.5 m^2 apartments are subsidized, 33 are offered at market price. The modular, prefabricated micro apartments are centrally located in Manhattan and complemented by service and other shared facilities.

increased flexibility has been helpful here: downsizing has become an acceptable option, housing has become more temporary or is fundamentally not thought about in permanent terms, establishing a household and a family is often postponed. The number of single-person residences is again on the rise and shared accommodation is – beyond one's time as a student – increasingly more attractive 7.

Micro Apartments

The development of micro apartments, based on the concept of serviced apartments from the hotel sector, is also an attempt to create living space in otherwise often unaffordable central urban areas. By implementing spatial scenarios that change throughout the day and built-in, multifunctional furniture like flexible tables, folding beds or compressible wardrobes – and at times a conscious removal of ancillary functions from the apartment or an increase in available resident services – they try to systematically minimize the required space for each individual. Their diverse manifestations are inhabited by singles, business travelers or commuters: common among all iterations, however, is that living here often remains impersonal and asocial, as residents can bring in little personal furniture and it is difficult to entertain guests 8, 9.

Outsourcing

While such developments are indicative of a new trend in many metropolitan areas, they are a long-established standard in the cities of Japan. The resulting building configurations are closely connected to the characteristics of their urban context, since original functions of living such as food preparation and eating have been in part "outsourced" to the public sphere. The forms of the urban and individual spheres complement and necessitate one another 10.

DEMOGRAPHIC CHANGE: DIVERSIFICATION

The increasingly heterogeneous nature of society has led to a broad multiplicity of ideas on possible housing forms, a multiplicity that still cannot be satisfied by the existing housing inventory 11, 12. But having adequate design ideas for these other modes of living is not only what matters, rather it must be accepted that the permanence of a chosen way of life – heretofore simply assumed as a constant – is hardly valid anymore.

The diversification of apartment size, layout and standards is an important criterion among all these trends when taking a variety of lifestyles and incomes into account. In or-

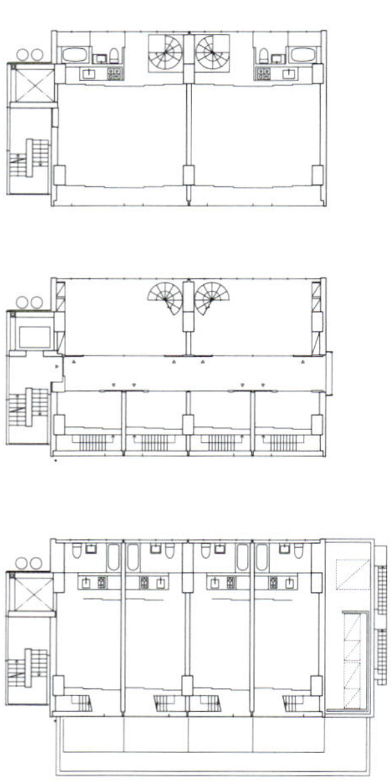

10 architecture WORKSHOP: Klarheit, Tokyo, 2008
On the entrance levels of the maisonette units, working spaces are lined up along the transparent central corridor. In the living spaces above and below, internal staircases, kitchenettes, bathrooms, and sliding doors serve as filters behind the floor-to-ceiling glazed facades.

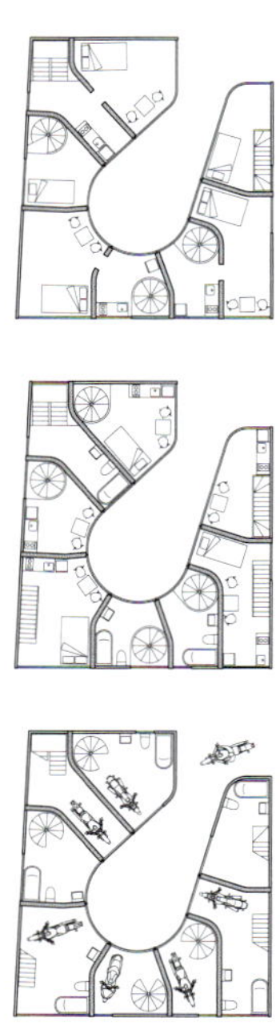

11 Nakae, Takagi, Ohno:
NE Apartment, Tokyo, 2007
House for motorcycle enthusiasts.

12 Riken Yamamoto & Field Shop: Local Community Area (theoretical proposal), 2012
The "Local Community Area" concept questions the sustainability of the nuclear family as the smallest social entity and instead proposes an urban community with roughly 500 residents. The relationships between the private and collective spheres are completely rethought: the family is no longer the only basis for shared economic activity or mutual and reciprocal care, and the exclusive rights to individual rooms are reduced to a minimum in favor of shared areas wherever possible. Two types of spaces are offered – for rent only, to ensure that constant adaptation to social changes remains a possibility. Open, transparent rooms can be used as commercial spaces, offices, studios, or as private verandas, forming an interface with the community. Enclosed units can be occupied as more conventional private households. Bathrooms, however, are shared, which is more efficient than providing each unit with its own.

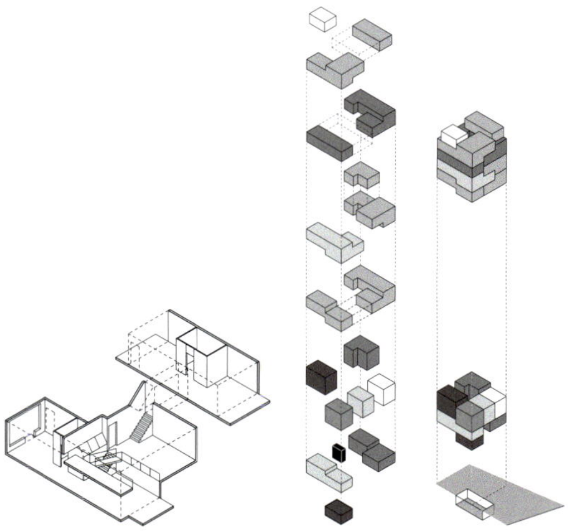

13 BARarchitekten: Oderberger Straße 56, Berlin, 2010
The units link with separate access points to the different stair landings enabling their division. The separable areas can be rented out, used for multi-generational living or as home offices.

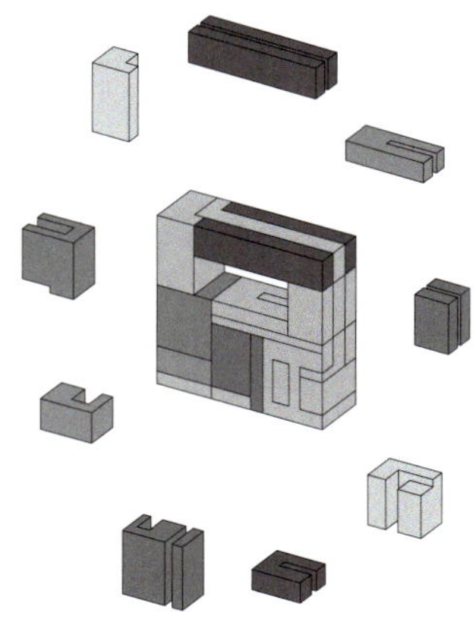

14 MVRDV & Blanca Lleó Associates: Mirador, Madrid-Sanchinarro, 2005
In Mirador, nine houses are stacked and compressed within each other– each with their own apartment types. Interwoven with a complex, vertical and horizontal access and circulation system, they form a "superblock" with an outdoor communal garden at a height of 40 m.

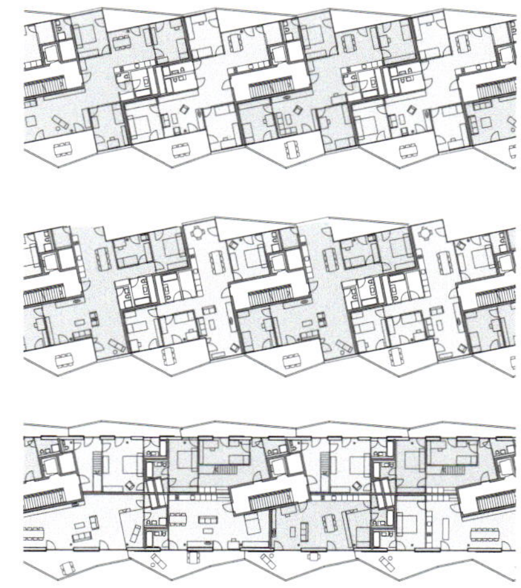

15 Matti Ragaz Hitz Architekten AG: Hardegg residential development, Bern, 2008
Shared flat typology (2nd floor)
Day/night area typology (3rd floor)
Open living room typology (4th floor)

der to potentially expand the choice, a broad range of different apartments – different typologies, sizes, floor plans, and standards – can be offered within one and the same project. Simultaneously, apartments with several access points allow for divisions or multiple uses 13, 14, 15.

The Dominance of the Generic

Many apartment floor plans still often functionally adhere to structures developed with the needs of the classic nuclear family in mind 16. In particular, large residential complexes realized by property developers show little interest in the creation of more contemporary floor plans – which may be due to potential buyers who purchase an apartment more as an investment preferring something non-specific, or that traditional patterns in mass housing are simply not questioned. Meanwhile, post-war housing estates are increasingly called into question due to the lack of flexibility in their floor plans, while the highly esteemed, spacious and less defined floor plans of mid- to late-19th century European residential buildings seem to offer more flexibility for changing living arrangements.

Deprogramming

Admittedly, too much diversification can also be viewed critically, at least when it results in extreme spatial and functional determination, which can only be adapted with considerable efforts – in contrast to office buildings, for example, where neutral floors are offered to accommodate the individual demands of each user. A small number of projects suggest a different path. Use-neutral buildings generate structures that incorporate the unpredictability of how the property will be utilized long-term, anticipating and allowing for uses besides housing, which are to a great degree vaguely defined and flexible and may therefore easily be reprogrammed 17.

Alternative models

It is often up to smaller projects to show new directions that not only follow different design strategies, but above all else endeavor to establish alternative models of cohabitation, connections between working and living, or even property development 18, 19.

Co-Housing

Co-housing or cooperative building ventures take advantage of niches resulting from diversification in the real-estate market, and react to the previously mentioned economic developments, which would otherwise preclude the par-

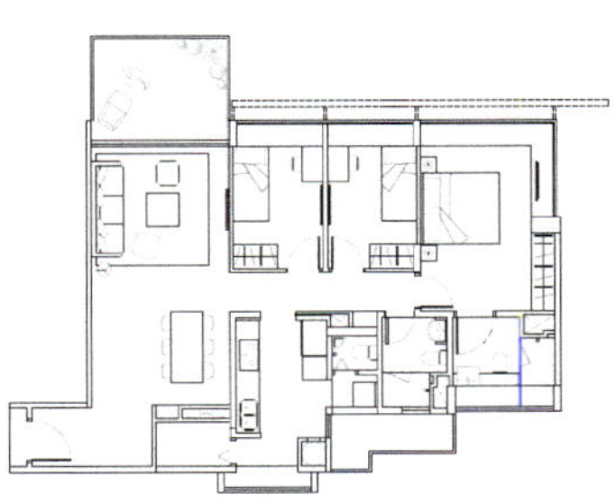

16 Generic floor plan type for the nuclear family
This inflexible floor plan is optimized for the nuclear family, and in similar fashion realized in many condominium projects in Asian cities. The unit is accessed through the living room, from which a hallway leads to extremely compact bedrooms on one side, a kitchen and bathroom on the other, and at its end a master bedroom with its own bathroom.

17 ANA: Multifunk, Amsterdam-Ijburg, 2007
The multifunctional building "Multifunk" is flexible both in its individual units and in its entirety – apartments can be transformed into offices and offices into apartments. All types of floor plans are possible: standard floor plans, duplexes, lofts, and commercial spaces. The only elements that are fixed are the load-bearing facades, the supply shafts, a series of void spaces with clerestory windows, and the access cores. Horizontal access, bathrooms and kitchens, loggias and additional stairs can be placed as needed. Natural ventilation, twin elevator shafts, and generous ceiling heights are further prerequisites for this flexibility.

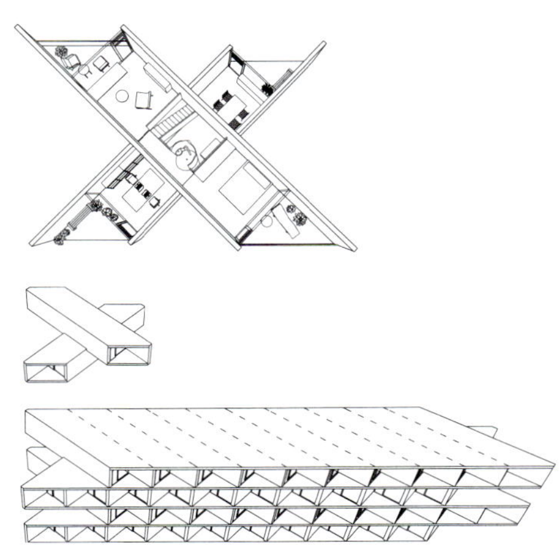

18 Brandlhuber + NiehüserS Architekten: 4-Direction Module, Hamburg 2013
Because the ideal orientation for housing in northern Europe is east-west, while a north-south orientation is preferred for work, this proposal calls for a spatial cross of two correspondingly oriented room modules.

19 Duplex Architekten: Hunziker Areal House A, Zurich, 2014
A model for collaborative living, in which individual and autonomous mini-apartments are inserted like a patchwork into much larger, shared flats.

20 Kaden Klingbeil Architekten: e3, Berlin, 2008
By setting the stairwell apart from the building and choosing a construction method in which two installation shafts are the only predetermined elements in the interior, a wide variety of floor plans could be realized in consultation with the residents.

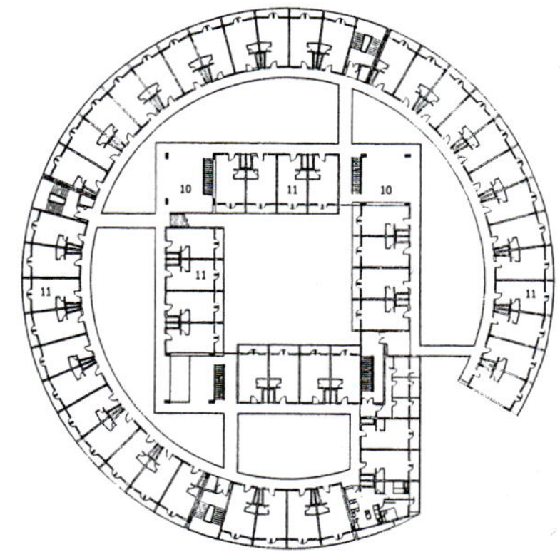

22 Urbanus: Tulou Residential Commune, Nanhai, Guandong, 2008
Typologically the project is premised on the traditional Tulou fortifications in southeast China; it was developed as a housing model for the numerous migrant laborers headed to China's growing megacities.

ticipants from finding affordable and adequate housing in their preferred locations. In co-housing design, users are no longer mere consumers of predefined forms of housing but become active participants in the creation of the floor plan – both in the act of inhabiting but also in the development of the buildings. Ideas on collective living are thus formulated. In these typically rather small collaborative housing projects, the inhabitants, while often assuming the economic risks for the project, have the opportunity to adapt the standard and size of their units to their income during the planning and the building stages. They are able to realize their individual concepts of living and to form a community, even prior to the completion of the building. Flexible floor plan structures with intelligently placed load-bearing, access and technical structures allow for individual floor plan solutions 20, 21.

FURTHER DEVELOPMENTS

Global Urbanization

According to a prognosis by the United Nations, in 2050, two-thirds of humanity will be living in cities worldwide. A significant portion of the global population growth will take place in urban areas, in particular in the so-called emerging economies of Africa, Asia and Latin America. Migration patterns from rural regions to urban areas are primarily motivated by economic factors and present challenges to the demographic, economic, and social structures of both cities and rural areas 22, 23.

Housing in growing and increasingly more permanent informal settlements and similar ways of living require strategies which – in lieu of demolition and forced resettlement of entire neighborhoods into places without any naturally evolved structures – build on legalization and participation, the stabilization of existing social networks, the creation of the necessary infrastructure to provide water supplies, sewers and drains, garbage removal, and more; strategies which employ simple means to design adequate and better quality housing forms with improved living conditions and economic possibilities 24.

Emergency Housing

The experience of recent years has shown how important it is to be prepared with adequate housing strategies for the enormous challenges posed by the immigration of refugees – whether their forced migration be due to natural catastrophes, violent conflict, economic reasons, or the consequences of climate change. Affordable, easily produced housing has to be created, but there is simultaneously a

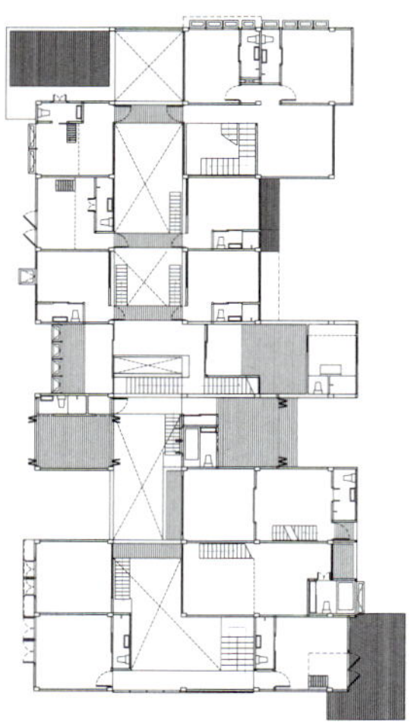

21 Case Architects: TEN Bangkok, Bangkok, 2009
Spurred on by a collectively organized loan, the building lot was divided into ten segments, which although individually designed, were developed in close collaboration with the neighbors. Inspired by a traditional Thai building type, the units enclose a common courtyard; kitchens remain open to the surroundings to provide better ventilation when cooking, facilities like the pool and laundry room are shared.

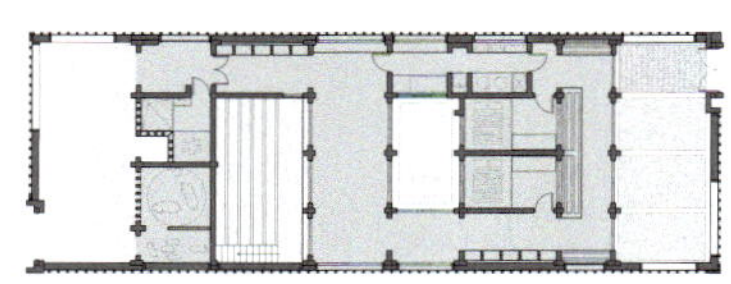

23 John Lin, Rural Urban Framework: Shijia House, Shijia, 2012
Against the backdrop of the massive population migrations to large Chinese cities, this house typology was developed as an alternative model to lend new vibrancy, sustainability and comfort to life in the countryside. By making use of natural resources, it sets an example for supply autonomy. The local artisanal production that takes place here contributes both to the village community and the local economy.

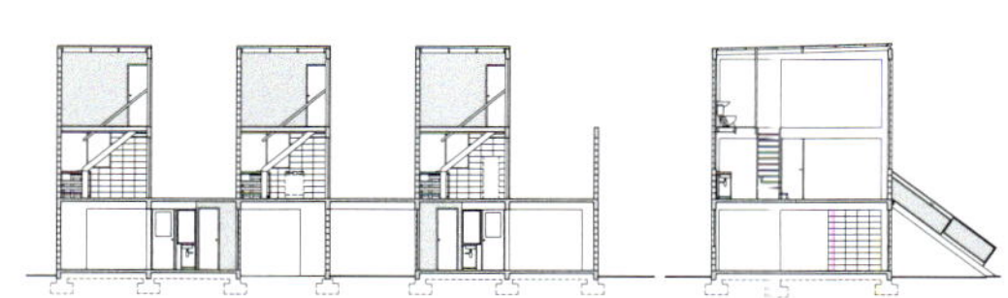

24 Elemental: Quinta Monroy, Iquique, Chile, 2004
This project for affordable housing was devised to preserve, legalize, and stabilize the vital economic and social networks of an informal settlement. The empty voids between the units not only accommodate self-built additions needed for changing demands, but also increase the value of the house in order to, for example, secure a loan to open a business or finance a child's education.

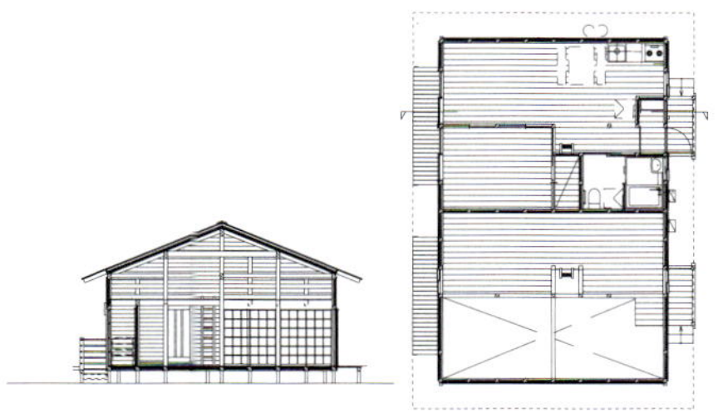

25 Kunihiro Ando + Satoyama Architecture Laboratory: Itakura Emergency Housing, Iwaki, 2011
In 2011, timber-framed houses were erected according to traditional building methods to house the victims of the Tsunami in Japan. The decision to use wood was above all made to bolster the local economy. After their completion, a local woodworker taught the residents how to build furniture from the leftover material, which was meant to both aid in the personalization of the individual houses as well as to strengthen bonds within the community. The housing units are organized on two levels under an exposed wooden roof construction. The upper level can be used for sleeping, living, or as storage.

26 Feldschnieders + Kister, Temporary Housing for Asylum Seekers, Bremen, 2015
12 units with a maximum of 16 people per floor in 2- to 4-person apartments, each with its own bathroom and kitchen, surround a shared courtyard, thus forming a carefully crafted transition from public to semi-public and private spheres. Four such courtyards enclose an open area with communal facilities.

need for more long-term solutions whose structures not only anticipate their disassembly or conversion, but also whose placement and layout avoid the formation of ghettos and encourage social integration 25, 26.

Asian Megacities

Given their enormous growth, Asian megacities can serve as a field for experimentation in path-breaking solutions on a large scale, to generate sustainable, alternative ways of building for the economic, ecological, and social challenges of increasing urban density. This applies both to historic quarters, which until now – despite being part of the local urban identity – were often demolished to make way for new buildings, as well as to the construction of New Towns, which have to establish new urban neighborhoods without benefitting from existing social networks 27, 28.

The complexity of all these trends cannot be tackled alone with static solutions. A design approach that aims to constantly identify questions, opportunities, and challenges in a spirit of curiosity is far more contemporary and sustainable. To this end, this book serves as a resource for reviewing floor plan solutions that have already – as an adequate answer to specific challenges – been invented, built, and inhabited.

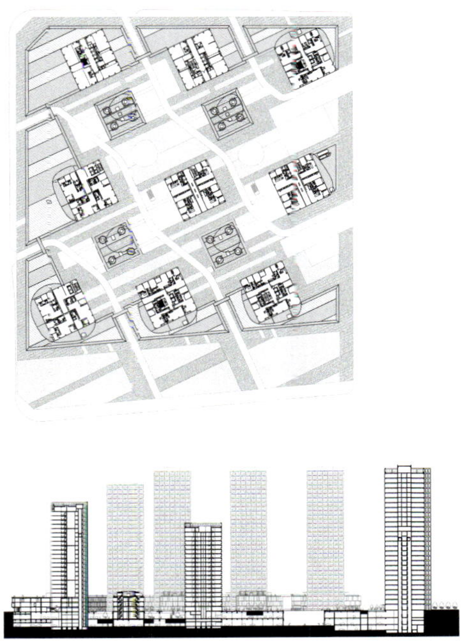

27 Riken Yamamoto & Field Shop, Jian Wai SOHO, Beijing, 2004
The multifunctional high-rise complex connects a field of towers, with home office units at their base, with an urban, multi-layered mesh of public passages incorporating shared and commercial functions – in stark contrast to the majority of Chinese residential estates, which often seem to strictly follow the Modernist model of the functionally segregated satellite city.

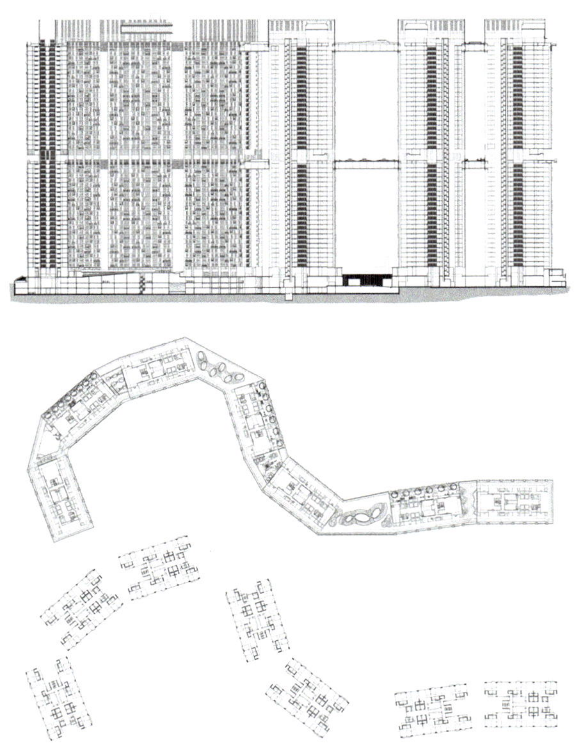

28 ARC Studio: Pinnacle@Duxton, Singapore, 2012
The project was developed as an innovative model for high-density public housing. The high-rise conglomeration is located in Singapore's Chinatown and is comprised of seven 50-story towers in a row, which are not only linked on the ground floor, but also by imposing sky bridges incorporating shared park-like arteries with playgrounds, a jogging path and other recreational facilities on the 26th and roof levels.

The Path toward Access and Circulation

Access describes the path from the public to the private sphere and the space it occupies, which begins with two thresholds: the first is the transition into the building; the second leads into one's own apartment. In between lies an entire sequence of spaces with horizontal and vertical connections, which can present a generous face to the public sphere, and either inviting or inhospitable in nature, may promote or discourage encounters among neighbors. In conjunction with open spaces such as balconies and terraces, the access also represents connection to the outside – it is, one might say, the prelude to a building. Similarly, it is a filter that controls and manages intimacy and allows for highly diverse lifestyles in close proximity to one another.

The word "access" is potentially ambivalent. To begin with, it simply describes an architectural space that allows residents to enter their private sphere; beyond architecture, however, it also describes a kind of taking stock, a step-by-step build-up of understanding and acquiring insight into complex contexts. According to the "promenade architecturale" postulated by Le Corbusier, which provides the basis for exemplary designs such as the Villa Savoye in Poissy (1929), the entire configuration of the design is derived from the path into and through the house. To the user, the "promenade architecturale" is revealed as a scenic sequence of spatial perceptions; however, it is also an element that establishes a hierarchy, a kind of unwritten user manual with which the structure of the house unfolds and becomes accessible to the user by virtue of his movement – step by step.

The Question of Function

But what is the function of this space if one thinks beyond its pure circulation role and wishes to explore its potential? The access space is a spatial and social buffer between a complex, anonymous public space and the intimate, individual environments of the residents. Sociologists have coined the term living environment for such spaces – an area that surrounds the residence, protects it while still being a part of it and at the same time providing an additional sphere of activity for the residents.

To regard the access space as no more than a thoroughfare is to misjudge its potential. After all, it offers ideal conditions for neighborly contact, for in contrast to the public space the number of local residents is manageable – an important prerequisite for people's willingness to socialize. Moreover, encounters take place out of necessity, as it were, and repeatedly; neighborly routines can therefore develop and become established. For children, especially, the access space also functions as an open space – located beyond the boundaries of the parental apartment but still sheltered. Thus it becomes part of their living environment in a most natural fashion – and it becomes a space in which they linger.

The question whether the access space is merely a passage space or whether one also enjoys spending time in it, is first and foremost in the hands of the residents. They might be able to live in the building without the access space as a social interface; however, when the space invites this type of use by virtue of its architecture and program, it is readily utilized as such. The invitation to linger is thus an offer with no strings attached. If it is accepted, the access space can become a place

... Inside: we enter, we walk around and the forms take on meaning, they expand, they combine with one another. Outside: we approach, we see, our interest is aroused, we stop, we appreciate, we turn around, we discover. We receive a series of sensory shocks, one after the other, varying in emotion. ... [And] the 'jeu' comes into play. We walk, we turn, we never stop moving or turning towards things ... Note the tools we use to perceive architecture: we have two eyes that can only look forward, we can move our heads from side to side and from top all the way to the bottom, we can turn our bodies or transport our bodies on our legs, all the while turning. The architectural sensation we experience stems from hundreds of different perceptions. It is the 'promenade', the movements we make that act as motor for architectural events.
Le Corbusier, cited in: Danièle Pauly: Le Corbusier: The Chapel at Ronchamp, 2008

In the film "Mon Oncle" by Jacques Tati (1958), the path is staged of a resident through an old Parisian apartment building leading up the stairs to his attic loft. In this case, the pathway space is no longer recognizable as a monofunctional access space with clear boundaries, but is entirely interwoven with the structure of the building, thus negating a clear hierarchy and separation between access space and accessed space. The viewer almost has the impression that the path to his apartment leads through all other apartments in the building; the access space thus gains an openness and boundlessness, as if all living areas were overlapping.

Guillermo Vázquez Consuegra: Calle Ramón y Cajal, Seville, 1987
Gallery with benches

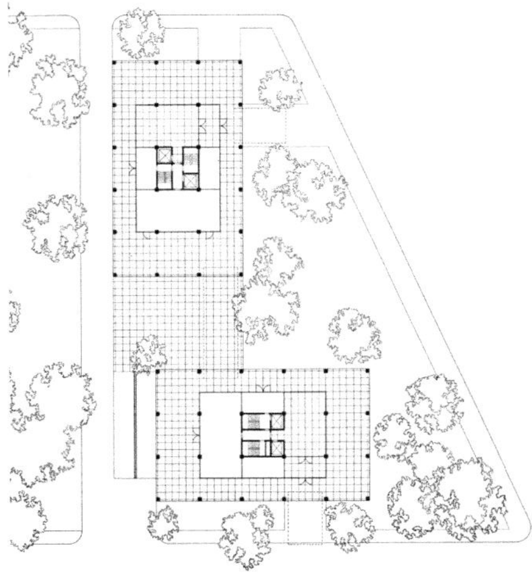

Ludwig Mies van der Rohe: Lake Shore Drive Apartments, Chicago, 1951
Ground floor with lobby

Heinrich Tessenow: Hellerau Garden City, 1911
Row house with wooden bench

for casual meeting, communal identification, and representation for all residents.

On the other hand, the access space is invariably in a tenuous relationship with the rest of the building, because efficiency dictates that it should occupy as little area as possible. This is an economic dilemma, especially evident in social housing developments where the dictum of economic efficiency has resulted in access spaces that are spatially, functionally, and qualitatively impoverished. Often, too many units are attached to a single access system. As a result, the access system becomes anonymous and inhospitable and is one of the reasons for the social problems arising in buildings of this kind.

Forms of Access and Circulation

Shaping the circulation in terms of space and program is one of the great challenges in housing. In this context, how to deal with the access space also depends on how habitation is understood and on the question of where habitation begins: in front of the building, behind the front entrance or only behind the apartment door?

Depending on the urban context and the building type, the access spaces vary in expanse and complexity: in the case of detached houses and larger housing estates, the houses are often surrounded by additional buildings, green spaces or common areas – and in some cases shared spaces within the buildings – which taken together can be regarded as part of this space and its potential. Apartment towers often feature lobbies in the entrance area, which deliberately delay the transition into the private sphere and invite residents to linger for a while. The access space tends to be more compact in urban infill developments, where details can play a key role: whether there is a canopy or not, whether the entrance door is recessed and thus sheltered from the elements, how the threshold is designed, how spacious, bright and inviting the space on the inside of the front door is, whether the landings are large enough to encourage social interaction, or even whether there are opportunities along the way to sit down for a brief chat.

Direct Access from Street or Courtyard

If apartments are directly connected with either the street or a courtyard, each resident has the privilege of having a personal entrance. Row houses, for instance, where multiple units have direct street access, provide each resident with an individual street address, often with a front garden or at the very least a front step. Tessenow articulates this area in a very subtle manner – running along multiple house fronts is a continuous stoop with room enough at the top for a bench. The space belongs to the individual unit and acts as a buffer despite being in a public setting – ideal circumstances for pleasant yet sheltered neighborly interaction.

In a courtyard, individual units are accessed from a semi-public open space either enclosed on all sides or part of an open building ensemble. This spatial arrangement forms a safe space for communal activities or events, which in turn may aid in community development.

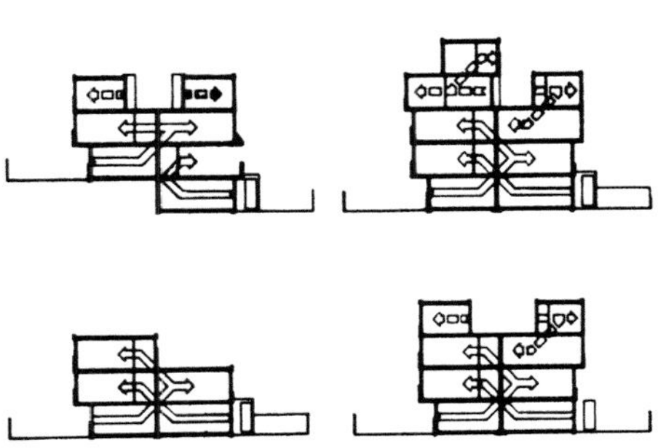

Darbourne & Darke: Marquess Road, London, 1976

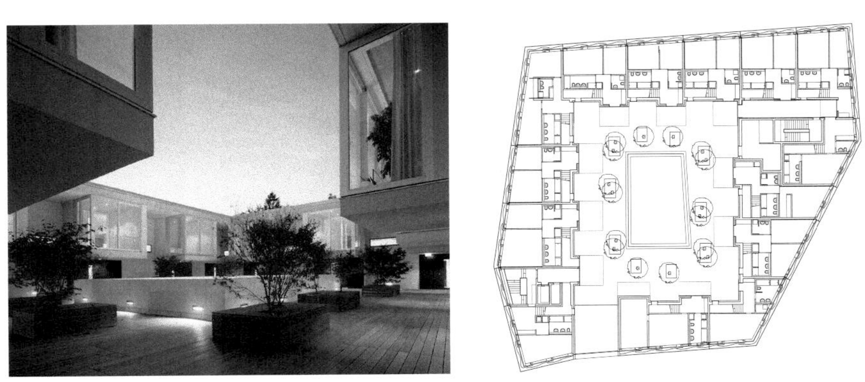

Burkhard Meyer: Falken residential and commercial building, Baden, 2006
In this project, residential units are stacked on top of commercial spaces. The entrances to the duplex units are lined up around an office atrium's skylight in the middle of the courtyard – a tranquil and quiet space for interaction.

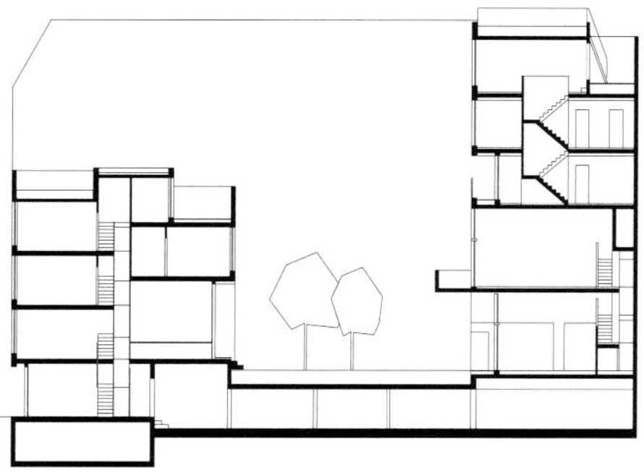

zanderroth architekten, BIGyard, Berlin, 2010, section
Townhouse (street access) & garden house (courtyard access)

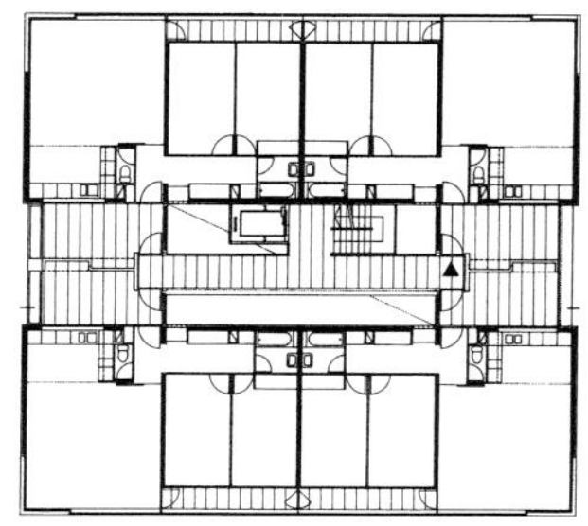

Henri Ciriani: Morgenstond, Dedemsvaartweg, The Hague, 1994
Four units per floor in open atrium, entries to apartments via bridge and private terraces

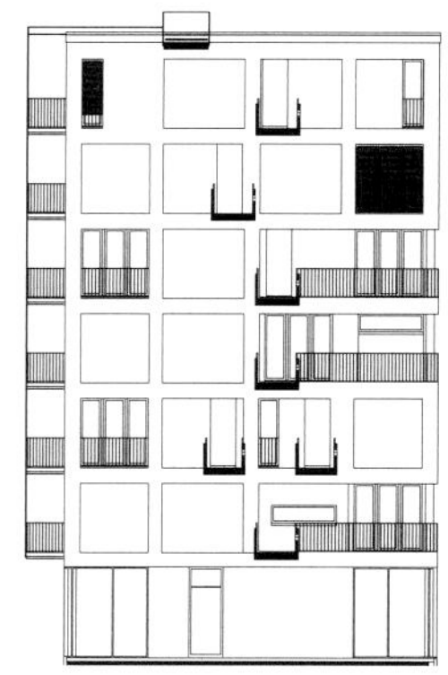

Kaden Klingbeil Architekten: e3, Berlin, 2008
Access to apartments via footbridges and recessed open spaces

In the project BIGyard, a townhouse typology forms part of a block edge development providing each unit with its own entrance and a room for potential commercial use despite its inner-city location. The garden houses run along the rear firewall and are accessed from the shared courtyard: eat-in kitchens form a buffer and threshold between the two.

Stairs

A set of stairs is probably the smallest common denominator among all residential building access systems. To begin with stairs are a mundane, functional building component; however, they also serve to stage the passage into the house. In baroque architecture, stairs evolved into complex, three-dimensional spatial configurations and became the stage for social encounters. The path towards the destination was artificially prolonged in order to endow the event of the encounter with time and space. In ordinary housing, stairwells can be designed in such a fashion as to consciously shape and publicly present the connection of residents with their environment.

In Álvaro Siza's housing development "Full Stop and Comma" in The Hague, the stairs define the character of the building to a large degree: access is via a portico – a wide set of stairs leads directly from the street to a loggia-like space, which is open towards the cityscape and from which all apartments are accessible through front doors exclusive to each unit.

Vertical Access Systems

The most commonly used stairwell in housing consists of several stacked flights of stairs that give access to a specific number of apartments at each landing: thus there are stairs that provide access to a single unit, two, three, and more units per floor. Horizontal access systems are preferred for buildings in cases where the number of units per level is too high. There are numerous subcategories to these models; there are buildings with internal or external stairs, which are combined with short horizontal corridors or housed in a separate building section. When they are constructed as a shaft, vertical access systems of this type can contribute to the structural integrity of a building and increase the design flexibility within the apartments.

The ten-story residential building by Henri Ciriani in The Hague demonstrates the possibilities of variation for this type. The apartments are organized into four units per floor; what is remarkable in this case is how the space around the sets of stairs expands into an open atrium reaching across all floors between the two building wings. On each floor, bridges connect to the landings and lead to the loggias projecting into the atrium, and it is these loggias that serve as entrance areas for the units. Beyond the loggias, the atrium opens onto the city, while the comings and goings of the residents become an event in its interior.

The same access type is interpreted in an entirely different manner in the project e3 by Kaden Klingbeil: for reasons of fire safety, the stairwell lies in a building component set apart from the apartment units. Because the building is comprised of a wooden structure, the escape route had to be constructed in reinforced concrete and set apart from the main building. The open gap gives each apartment a third facade and allows for more flexible interior spaces, thanks to the separation of

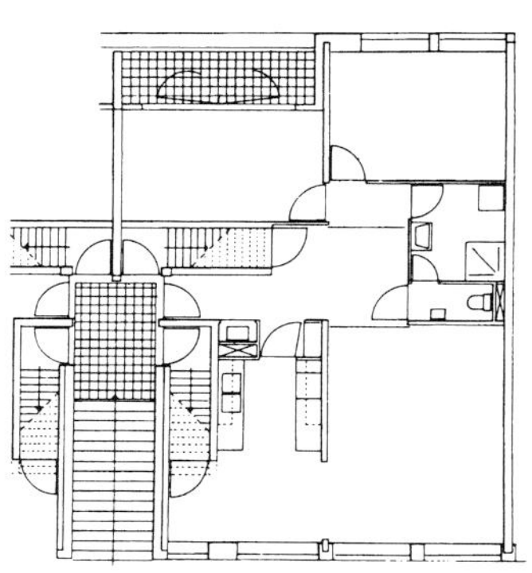

Álvaro Siza: "Full Stop and Comma," Schilderswijk, The Hague, 1988
Raised ground floor with six separate front doors

Rem Koolhaas refers to a "simple cartoon, ... which describes the ideal performance of a skyscraper," with the "skyscraper as a utopian instrument for the unlimited acquisition of new territory on a single urban site." In this cartoon, "a slender steel construction ... [supports] 84 horizontal levels, all of which are identical in size to the original lot. Each of these artificial levels is treated as a virgin piece of land – as if the other levels did not exist – in order to erect a self-contained empire with a single manor as focal point ..." (Rem Koolhaas, Delirious New York, 1978/1999).
For Koolhaas, this drawing is synonymous with a radical trend toward verticality in building, made necessary by the tremendous economic pressure on building lots in Manhattan, and made possible by innovative skeleton construction.

In 1853, Elisha Otis installed a spring-loaded brake, which would prevent the elevator cabin from falling and crashing in the event of a cable rupture. Due to the high crash risk, these vertical transport machines were at that time usually reserved for applications in mining and industry. The safety mechanism meant that access cores could reach previously unattainable building heights, reducing stairs to secondary escape routes and allowing for tremendous efficiency in stacking levels.
In 1854, Otis demonstrated the safety of his braking mechanism at the Crystal Palace World's Fair exposition in New York City.

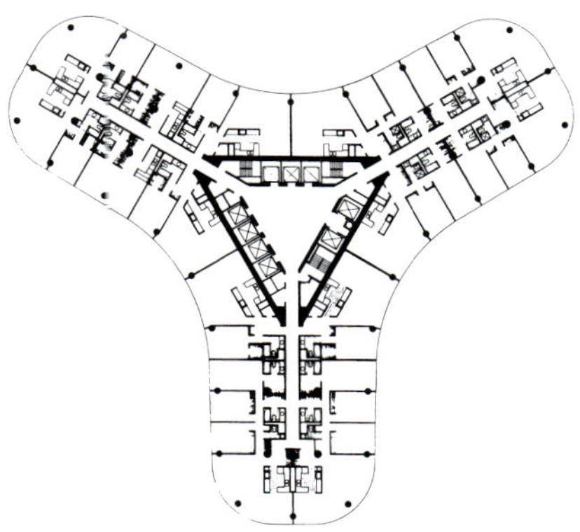

Schipporeit & Heinrich: Lake Point Tower, Chicago, 1965

A. + P. Smithson: Robin Hood Gardens, London, 1972 – "Streets in the Air"

the stairwell. Moreover, the access via a footbridge is pushed into the urban space and thus emphasized. On three floors, the apartments are accessed via generous, recessed open spaces, which enhances the sense of possessing one's own floor and home in the city.

Circulation as Separation –
The Invention of the Circulation Core

In the enclosed circulation cores of larger and higher buildings, vertical access systems are taken to the extreme: effective skeleton structures in steel or reinforced concrete and elevator facilities allow for extremely tall residential buildings with a steady repetition of the building's horizontal footprint. It is the use of elevators in access cores that radically reduces the amount of time that residents spend in this common interface. In this sense, the vertical circulation core is the symbol of an autonomous treatment of each floor and its inhabitants. The floors inhabited by "others" are merely perceived in a fleeting glimpse through the opening and closing elevator door; residents disembark only on their own floor.

There are, however, exceptions to the rule. At Citygate Tower in Vienna, a vertical incision in the building provides natural lighting to the central circulation core, opening it up to a view of the city. Additionally, shared and in part double-height spaces are inserted into this cut-out void every few floors and are used for instance as a media room, home theater, laundry room, play area or roof terrace.

Horizontal Connecting Spaces – "Rue Intérieure" and "Streets in the Air"

Central corridors such as the "Rue Intérieure" and external access areas such as galleries or "Streets in the Air" are horizontal access systems.

Corridor

A corridor is a horizontal passage located in a building's interior, which for the most part runs parallel with the building's facade and is lined on both sides by residential units. A groundbreaking example of this type of access system is the Unité d'Habitation, which emerged as a result of research on urban living. Le Corbusier established that it would require a minimum of 1,600 residents to create a functioning piece of urbanity and it was from this assumption that he derived the scale of the Unité – complete with retail strips, communal establishments and the so-called "Rues Intérieures." Located on every third floor, they provide access to the left and right to maisonette units, which occupy the entire width of the building on the floors above or below. In contrast to the galleries, these interior streets have virtually no relationship with the surroundings: running the entire length of the building, dark and with low ceilings, they failed to achieve the effect of a lively streetscape.

Van den Broek and Bakema's residential high-rise in Berlin also features an internal street. However, in this case it receives natural light from two sides and opens onto a two-story shared loggia on the south side with sun decks and playground terraces. The number of units along each corridor is manageable in this case. Spatially and programmatically, the access area is thus intentionally designed as a communal area.

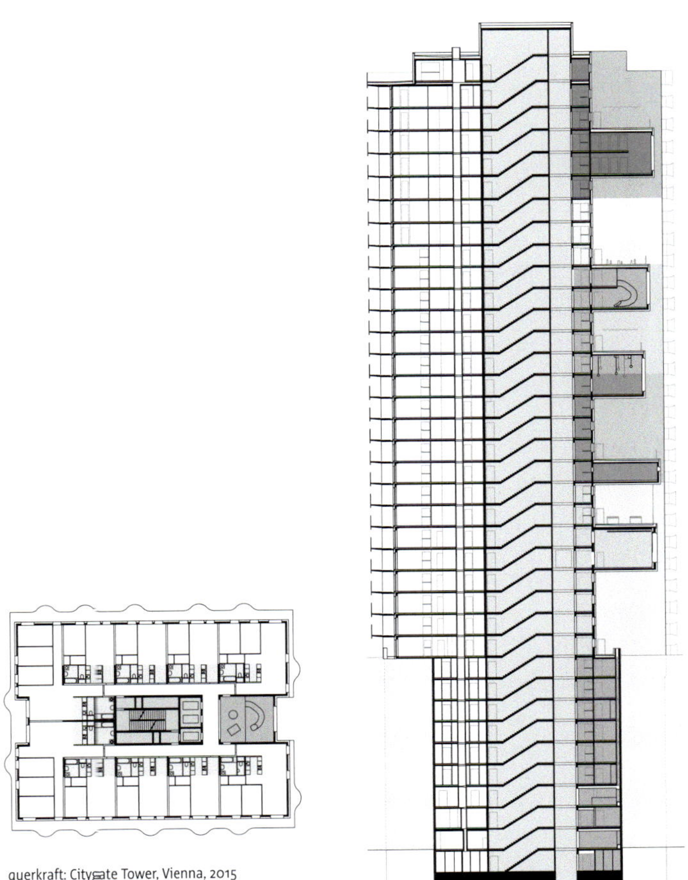

querkraft: Citygate Tower, Vienna, 2015
Section, typical floor plan

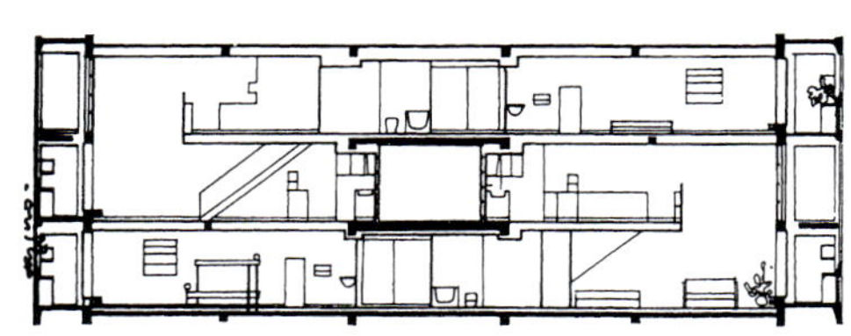

Le Corbusier: Unité d'Habitation, Marseille, 1947 – "Rue intérieure"

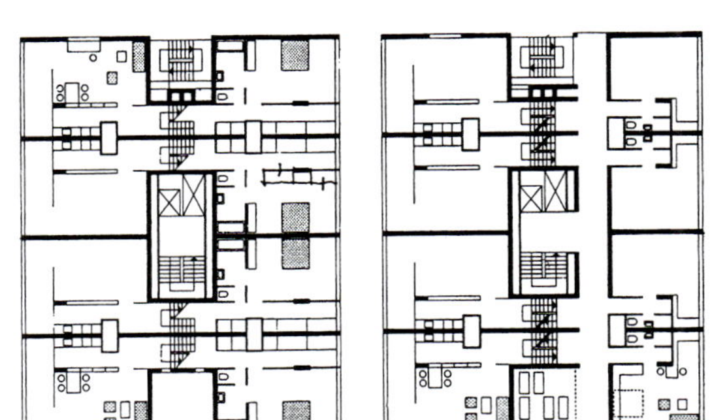

Van den Broek, Bakema: residential tower Hansaviertel, Berlin, 1958
Shared loggia at the southern end of internal streets

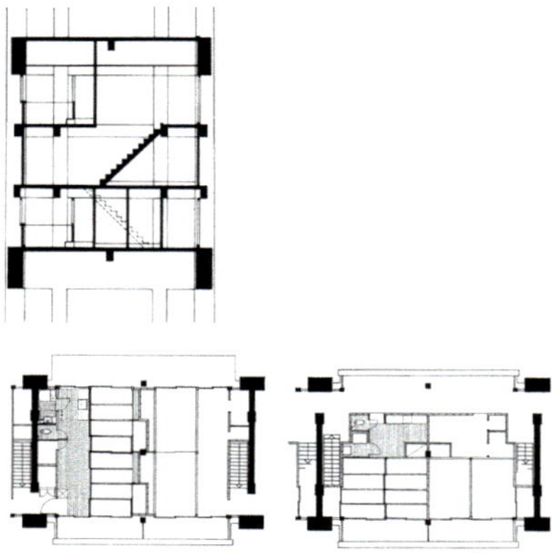

Kunio Maekawa: Harumi, Tokyo, 1958

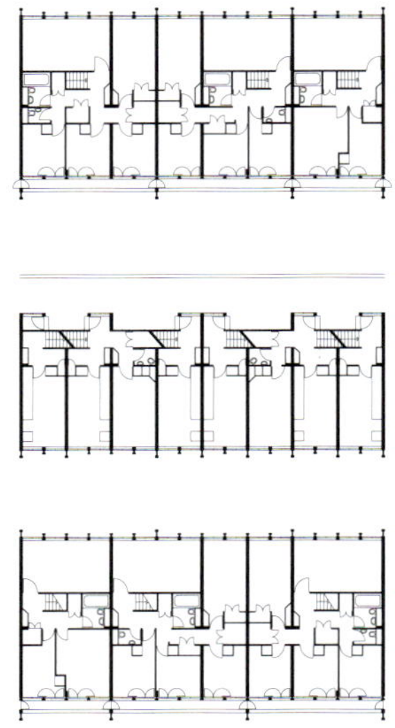

A. + P. Smithson: Robin Hood Gardens, London, 1972

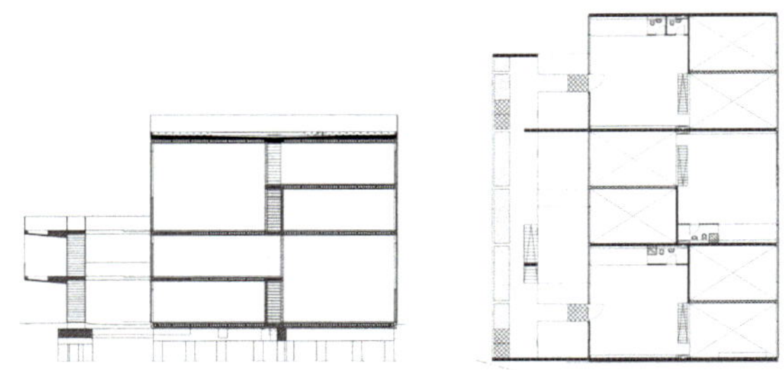

b&k+ brandlhuber&kniess GbR: Kölner Brett, Cologne, 1999

Access Gallery

An access gallery is generally speaking a path projecting from one side of a building, usually on the outside, and connected to one or several vertical circulation cores. Many of the built examples fall into a predictable pattern: the gallery is often located on the side disadvantageous for living spaces – depending on the geographic location, this may be the side with too little or too much sunlight. In heavily trafficked areas it may also serve as a noise barrier. It is sometimes set slightly lower than the floor level or else accompanied by horizontal windows set high into the wall to ensure privacy.

The kitchen, entrance hall, and sanitary block are often placed parallel to the gallery as an intermediary zone between it and the more private living areas, which run along the more attractive side of the building. This type of access is often employed in conjunction with maisonette units: within a spatial unit comprising several stories, galleries provide access to the entrance area on one level of the maisonette, which continues on the floor above or below. On this floor, front-to-back interiors are thus possible. On the lower floors, maisonette units can be arranged with ground floor entrances, thus allowing for access from private front gardens similar to a row house.

Alison and Peter Smithson adopted the term "Streets in the Air" – borrowed from the idea of the urban street based on traditional images – to describe their generous galleries.

They were proclaimed as a deliberate alternative to Le Corbusier's "Rue Intérieure" – the interior access corridors they criticized in the Unité d'Habitation. To the architects, the term "Streets in the Air" expresses their program: on the one hand, the building is interpreted as a small piece of city in its own right, and on the other hand, it illustrates the idea that access areas can serve as urban spaces for socializing with neighbors. Although their project Robin Hood Gardens in London ultimately revealed the limitations of access systems of this kind – the galleries proved to be inhospitable and anonymous due to the sheer number of units and the exposure to the noises of the highly trafficked surroundings – the spatial relationship between housing unit and gallery is nevertheless exemplary. Nearly 2 meters wide, the galleries expand into even wider bays in front of each entrance, creating a kind of forecourt. The apartment doors are set sideways into the resulting niche, further enhancing this entrance area. The apartment stairs lie parallel to the gallery and create a buffer between horizontal access and the housing unit. Internally, the stairs connect the common rooms, kitchen, and living rooms, which are deliberately located on different levels to avoid a strict separation of common and private spaces.

Taking the private sphere within an access space into consideration and providing each unit with its own anteroom can also become a unique quality of the common access space. In the Kölner Brett by b&k+ in Cologne, the access area is deliberately set apart as an independent sculptural element, with the result that access to the units is only possible via bridges. At the same time, the galleries are notable for their width and inviting spaces; they integrate balconies, patios, and even gardening areas.

Space-defining Access Systems

When horizontal access systems are not simply straight elements, but instead enclose spaces either singularly or with

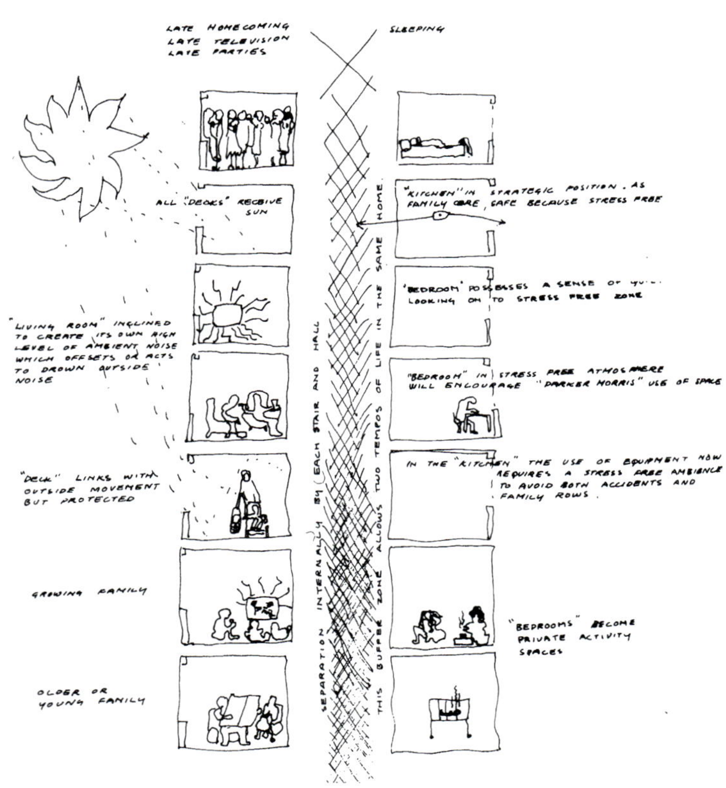

A. + P. Smithson: Robin Hood Gardens, London, 1972
Sectional sketch with reasoning behind disposition of access and apartments

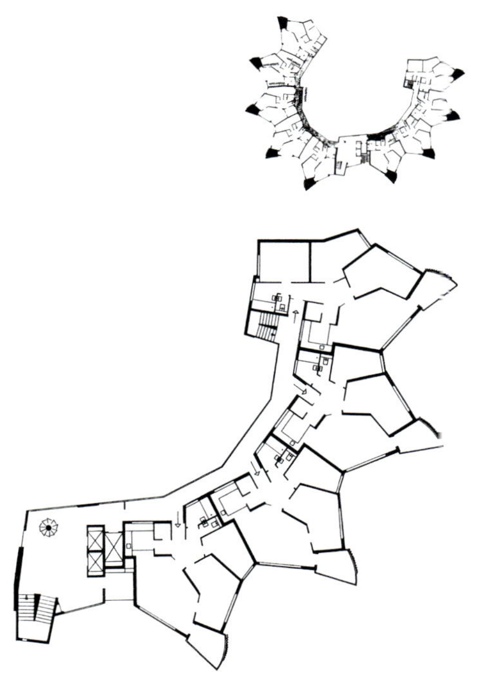

Scharoun: Romeo and Juliet, Stuttgart, 1959

The apartments are not merely accommodations strung along corridors, but lively habitats along "alleys" and open toward the external spaces in front of and behind the apartments. The "alleys" lead into the "street," the "street" unifies spaces, the main purpose of which is the development of community and the relationship with the outside.
Hans Scharoun, in: Jörg C. Kirschenmann, Eberhard Syring, Hans Scharoun, DVA, Stuttgart, 1993

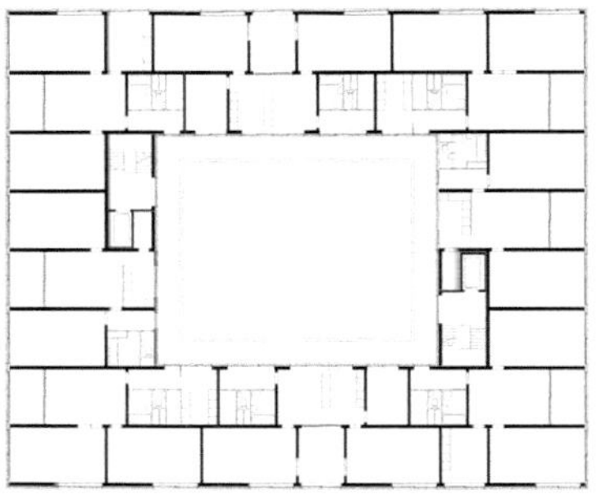

Diener & Diener Architekten: KNSM- & Java-eiland, Amsterdam, 2001

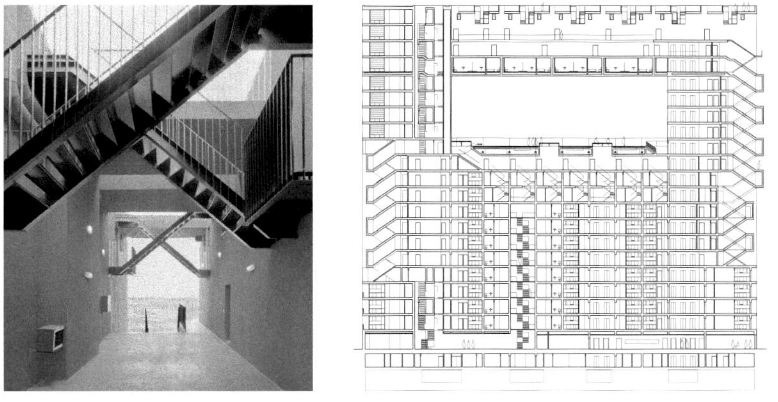

MVRDV & Blanca Lleó Associates: Mirador, Madrid-Sanchinarro, 2005
Internal street in open multi-story atrium, and longitudinal section with various access systems and 4-story open communal garden at a height of 40 m

an opposite, the impression of a shared communal space emerges that allows for visual and neighborly contact.

The galleries of "Juliet," part of Hans Scharoun's "Romeo and Juliet" twin housing complex in Stuttgart, can be overlooked and controlled from every vantage point thanks to their curved shape, which seems to almost form a circle and interior courtyard. By means of the intelligent shaping of the gallery balustrade and the exterior walls of the apartments, which usually run parallel to each other, each entrance lies in a triangular bay, enclosed on two sides and seemingly giving each unit an anteroom of its own. At the transition to the vertical access at the center, the gallery widens into a generous front-to-back space, with a large balcony facing towards the other side, inviting contact among neighbors.

Atrium

Similarly, an atrium uses access galleries on every floor that flank an open, naturally lit space and connect – whether covered or not – the horizontal and vertical access systems in a singular space, which also usually serves as a source of natural light. In the atrium-like access space of the residential building by Diener & Diener in Amsterdam, the vertical access elements are arranged at the peripheral gallery in such an intelligent way that one passes no more than one other unit on the way to one's own apartment. The placement of the floor-to-ceiling glazed eat-in kitchens at the transition between the public and private spheres lets life inside the apartment merge naturally with the communal life of the building as a whole.

Complex Access Systems

In some housing schemes, an intentional effort is made to go beyond a single access system; they are conceived as complex systems and as such permeate the spatial hierarchy of the entire building. One example for the combination of several access systems is the superblock Mirador; it integrates horizontal and vertical access routes, which – again and again – lead to communal spaces. Access cores project openly into the cityscape and lead to loggias cut into the corners. Internal streets provide access to maisonette units and, at the very top, become part of a multi-story open-air atrium. At the center of the high-rise, several access routes open simultaneously onto a dramatic five-story courtyard, which can be used by all residents as a common space offering an urban panorama.

In the high-rise cluster of Walden 7, a spatially complex network of pathways spans vertically and horizontally through a series of courtyards. Starting at the centrally placed elevators, numerous spatial situations and chances for neighborly encounters are experienced along the path to each individual apartment. The more generously dimensioned courtyards with entrances to the units broaden in section at their middle, thus allowing for a multitude of views.

In such cases the access system is simply a space so attractive that one happily accepts a delay en route to one's own apartment and is thereby drawn into the building's community – casually and only when desired.

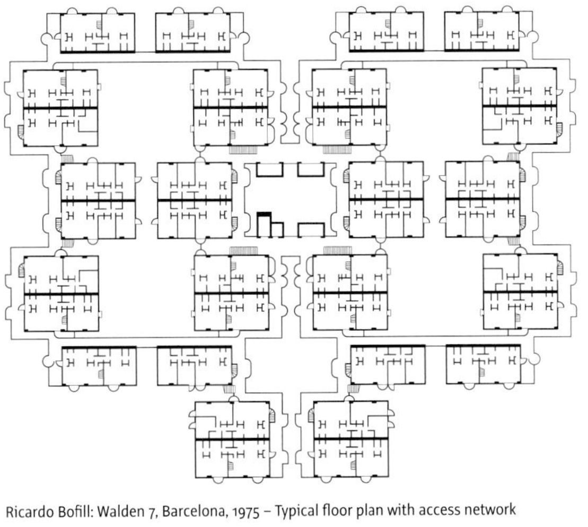

Ricardo Bofill: Walden 7, Barcelona, 1975 – Typical floor plan with access network

Pau Vidal: Torre Júlia, Barcelona, 2011
Typical floor plan with shared space, and section
Instead of banning the emergency staircases to a dark structural core, in Torre Júlia they are positioned in opposite corners of the building and intentionally left open – with generous staircases and balcony-like landings offering a view of the city. Roof terraces, the entrance lobby and well-lit, open corridors connect both staircases and the apartment entrances and in doing so create an inviting pathway that encourages residents to walk through the building and visit each other. Multiple floors are color-coded and share one of the three double-height community spaces. The top room is reserved for laundry while the others are left open to interpretation.

The Floor Plan Idea

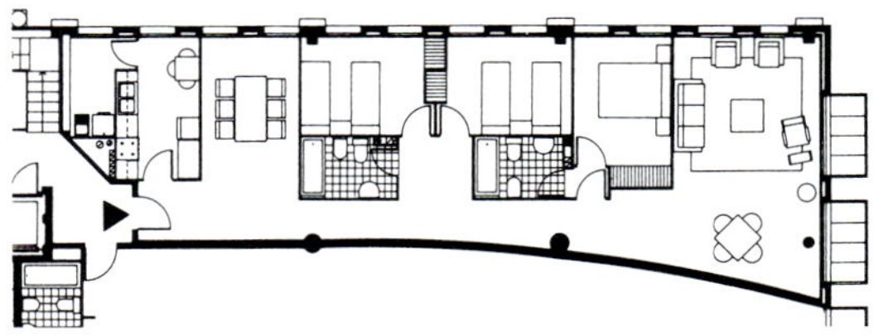

1

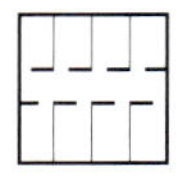

A floor plan idea reflects the interpretation of a certain notion of living. It is expressed in the internal organization, in the opening and closing of the rooms, in the connection and grouping of rooms, in the connection or isolation of functions and, last but not least, in paths and sightlines. Some floor plans subordinate everything to the spatial idea (cf. "Dividing Elements"), while others emphasize the link between the spaces, and optimize or celebrate the internal path (cf. "Floor Plan with Circular Path"). A floor plan may also articulate an apartment as a "space of social interactions": in this case, the spaces are assessed on a scale from highly communicative (social) to extremely private, and the floor plan is arranged accordingly. The result can be a largely balanced mix (cf. "Zoning") or a type that gives priority to privacy (cf. "Corridor"), or, conversely, a floor plan in which the communicative aspect of cohabitation dominates (cf. "The Living Room as Circulation Center"). These social interactions naturally influence how every floor plan is organized. However, since they cannot be assumed to be constant or universal in nature, new attempts are always being undertaken to render floor plans modifiable, open to multiple interpretations (cf. "Flexible Floor Plan", "Neutral Spaces") or even involve the future user in the design process from the beginning (cf. "User-defined Floor Plans").

The following classification is an instrument for reading and designing floor plans, whereby it must be noted that the pure type should not be mistaken for the best type: for truly exciting solutions most often lie at the intersections of these categories.

CORRIDOR/HALL

The apartment is organized along an axis, with the rooms arranged in sequence on one or both sides 1, 2. The advantage of this classic corridor type lies in the opportunity for simultaneous use of the rooms; completely enclosed and independently accessed, they offer all manner of freedom and flexibility for all kinds of users (families, co-op, etc.). The apartment can be accessed either in the axis or orthogonally 1, 7. In both cases, the layout of the apartment is instantly visible; a pleasant clarity is achieved. The end point of the axis is important, a common room in the best scenario, as is the case in example 1: the corridor widens in the direction of the living room. With the balcony in the sightline, spatial quality, visual experience and light intensity increase for the visitor, walking down the corridor becomes a pleasure. The width of the corridor determines to a great degree whether it is also suitable for other uses, for example, as a play area; natural light in a corridor, in particular, allows for a multitude of individual purposes 3. When corridors are no longer simply linear but widen into small bays or even entire rooms, for example into a wardrobe niche at the entrance door, an anteroom in front of bedrooms, or even a dining area 1, 4, the experience becomes enjoyable. The corridor is therefore wider and more generous in certain areas, but more importantly, it is structured and thus becomes interesting as a room in its own right. Given the linearity of the form, the corridor type quickly awakens a need for change in direction and widening of the space. Floor plans respond to this need in a variety of ways: with large, open living rooms oriented in all directions at the end of the corridor, or sightlines and room-to-room relationships set at right angles within the corridor 1, 3 and 6. In the classic type, encounters between inhabitants in the corridor are inevitable. A second corridor preceding the rooms and conceived as a loggia, for example, can alleviate this situation by affording a second access 5.

Similar to the corridor, a hall also provides access to all rooms simultaneously and the rooms are also individually usable. However, the hall is also distinguished by the fact that it welcomes the visitor in the form of a well-designed space that is often used to impress the visitor and allows for additional functions that the corridor cannot provide: one can set up a table, the space invites you to stay 7.

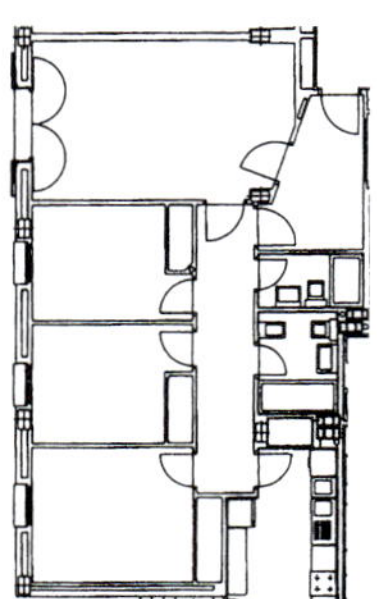

2

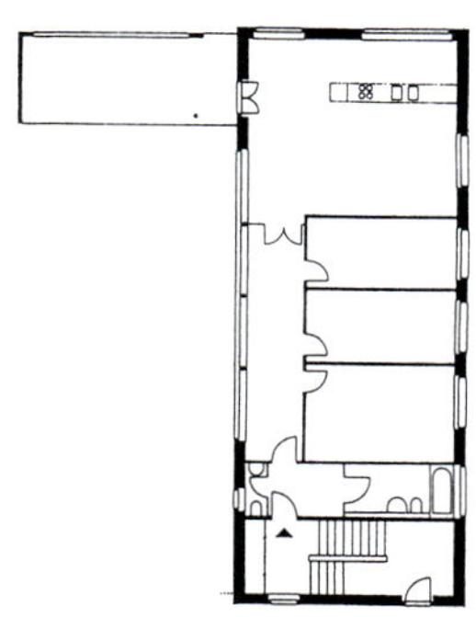

3

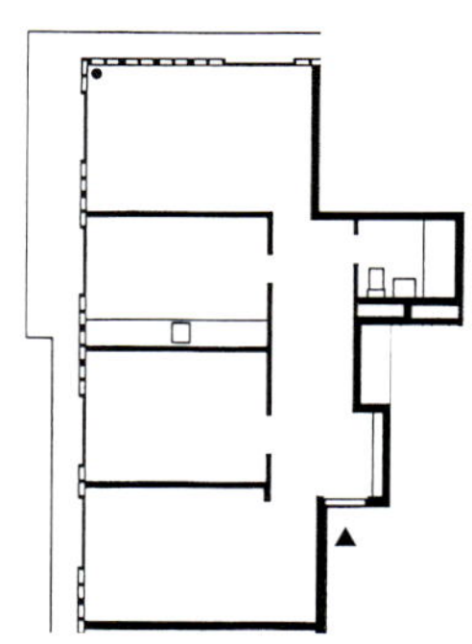

4

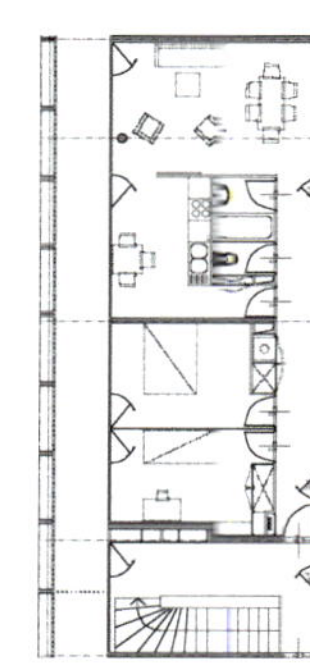

5

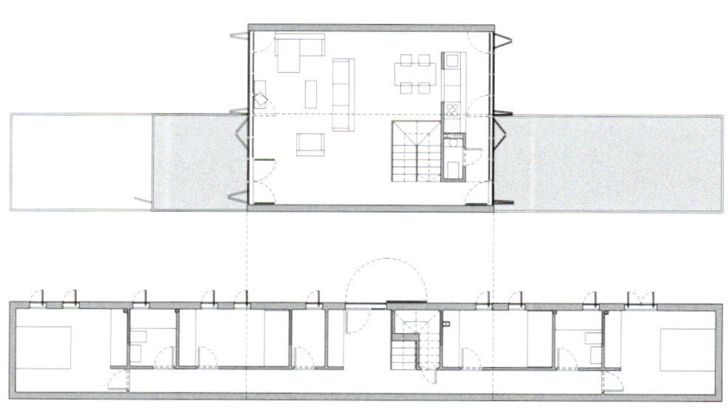

6

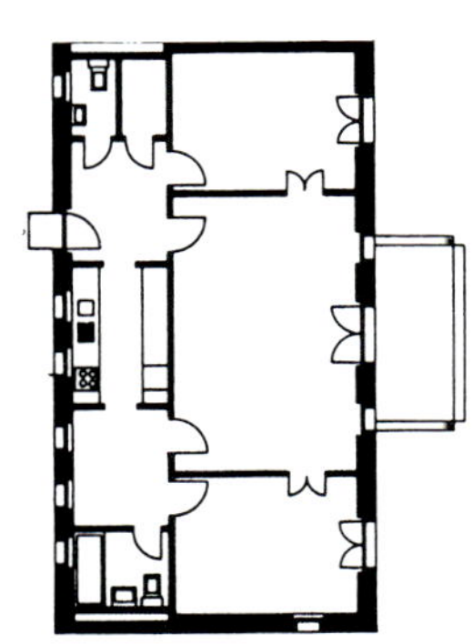

7

1 J. Martorell, O. Bohigas, D. Mackay: Villa Olimpica, Barcelona, 1991
2 Cruz, Ortiz: Carabanchel, Madrid, 1989
3 Michael Alder: Vogelbach, Riehen, Basel, 1992
4 Morger & Degelo: Müllheimerstrasse, Basel, 1993
5 Herzog & de Meuron: Rue des Suisses, Paris, 2000
6 Dosmasuno: Vellecas, Madrid, 2007
7 D. Schnebli, T. Ammann, W. Egli, H. Rohr: Baar, Zurich, 1985

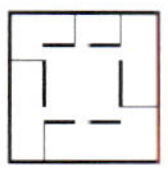

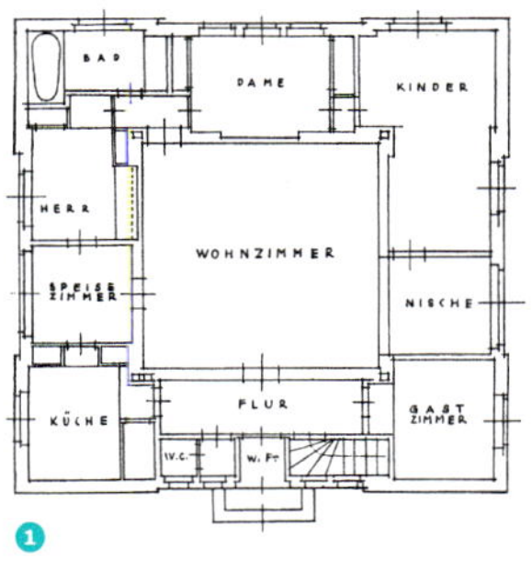

1

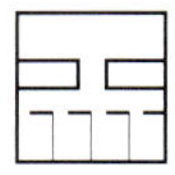

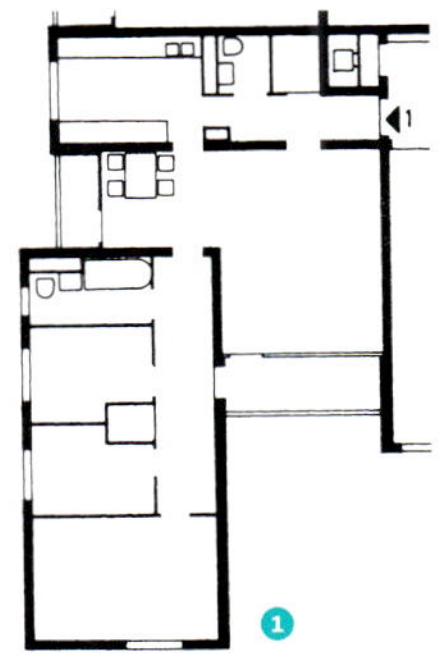

1

THE LIVING ROOM AS CIRCULATION CENTER/ FLOOR PLAN "WITHOUT CORRIDOR"

The floor plan fans out from and around the living room, which is both the center of the apartment and the distribution zone (circulation center). Nearly all paths lead through this room. The living room gains in floor area since less corridor space is required; sometimes, individual rooms are reduced in size in exchange for a larger living room. The concept is exceptionally conducive to communication, but it does restrict opportunities for privacy. Hilberseimer refers to this type of design as a "cabin system," Alvar Aalto called his all-purpose living room a "market square" emphasizing its lively nature. The apartment benefits from having a clearly defined center, while the remaining rooms are kept neutral with regard to use 1, 2, 3. The living room can also be defined as a large continuous hall determined by the spatial boundaries of the rooms 4. The paths through the living room must be designed in such a manner that the living quality is undiminished 5. Some cases feature a separate corridor with bathroom leading to the bedrooms. There is, of course, no need to place the living room in the geometrical center of the apartment 6.

ZONING

This type of floor plan clearly separates the different functional areas within an apartment: it differentiates between the common living area including living room, kitchen, and dining area, on the one hand, and the bedroom area with individual rooms and bathroom on the other 1. Studies or home office rooms can form a separate, third zone. The goal is to achieve an uninterrupted course of the individual functions, which can occur simultaneously and side-by-side; the individual member of the family or co-op is given as much freedom and privacy as possible. Each area has its own hallway, and the hallways are either gathered together at the entrance or arranged in sequence. The kitchen often separates the different zones; sometimes a bathroom core, which should, if possible, be accessible separately from both the common area and the bedroom area, fulfills this separating function 2, 3, and 4. However, the living area can also provide indirect access to the bedrooms by letting the living room hallway lead directly into the bedroom hallway 5. In order to achieve privacy even in the case of a direct link between living room and bedroom hallway, some plans offer a second path to the rooms, which may lead through the kitchen or even through the bathroom. The shape of the building can be chosen in a manner to ensure that a spatial separation of specific areas occurs naturally, for example, in L-shaped apartments, double rows 6, apartments arranged around an atrium 7, or, of course, maisonette apartments.

Examples from the 1990s often display a different type of zoning: the "service" spaces of an apartment (kitchen, bathroom, pantry, maisonette stairs) are bundled in a service core, with the living and bedrooms on the opposite side 8. This approach bundles the shafts and stacks the most noise-prone spaces one above the other. This zone is often employed as a buffer to the access area, especially when covered walkways or central corridors are used, and the rooms lie undisturbed. The latter – no longer squeezed between a variety of ancillary rooms – can therefore assume a clear, attractively designed form, frequently neutral (that is, flexible) in character.

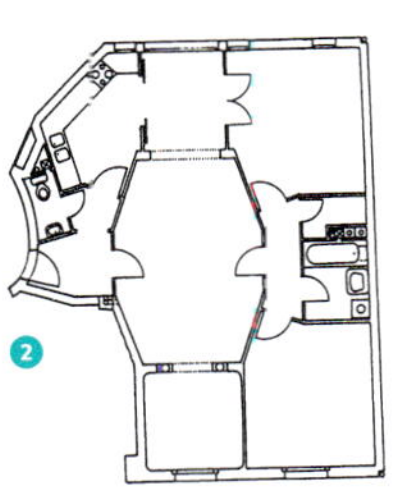

2

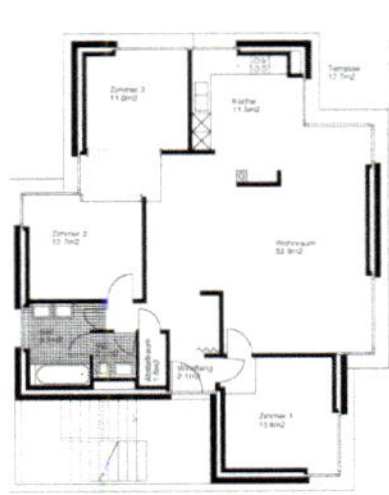

3

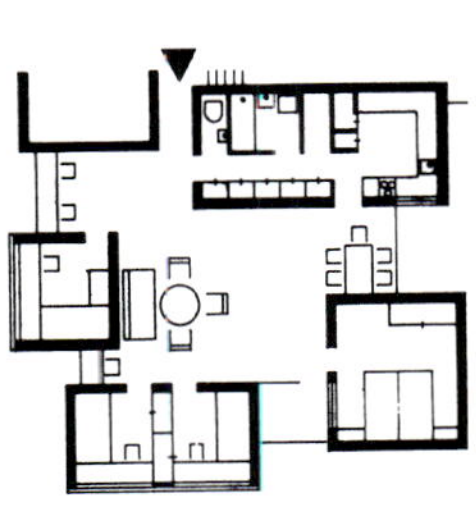

4

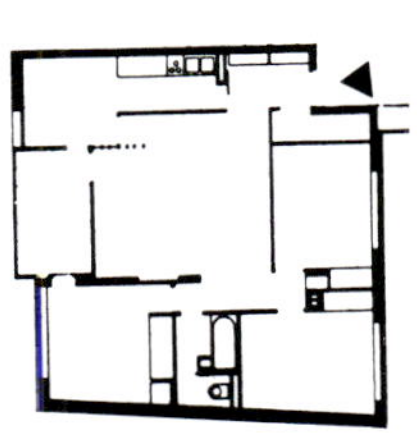

5

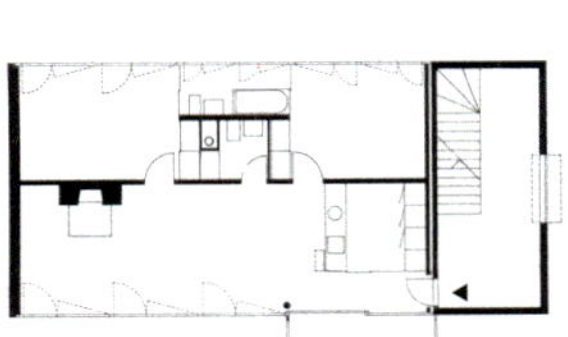

6

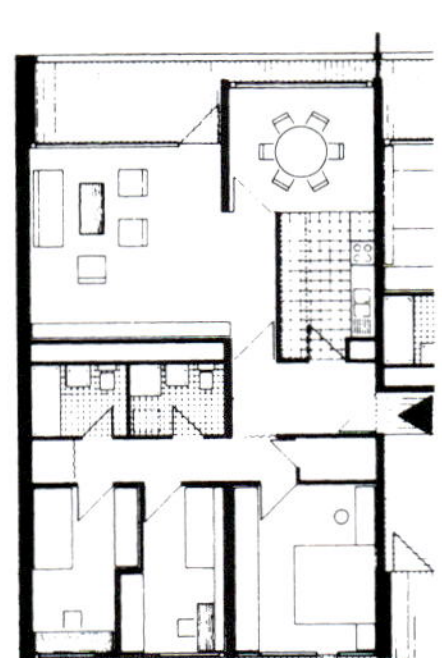

2

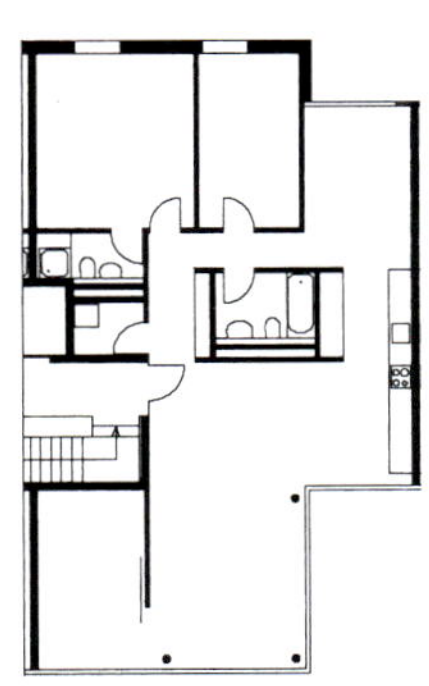

3

4

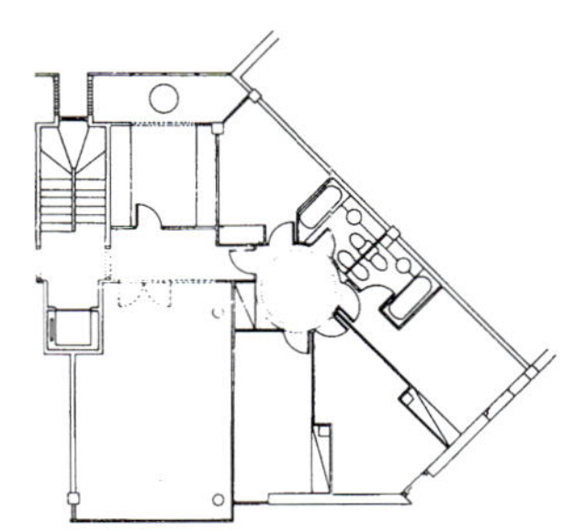

5

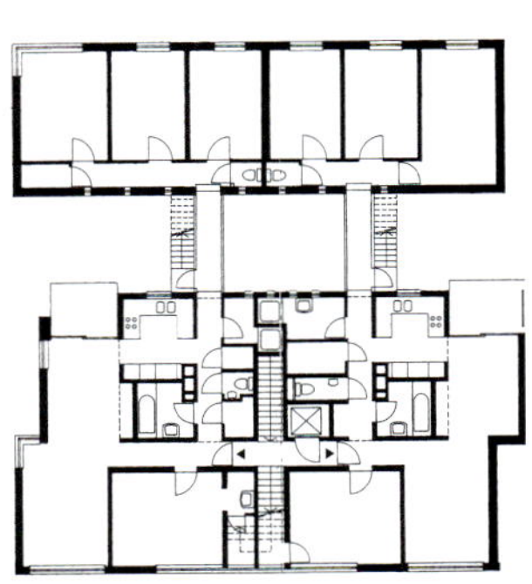

6

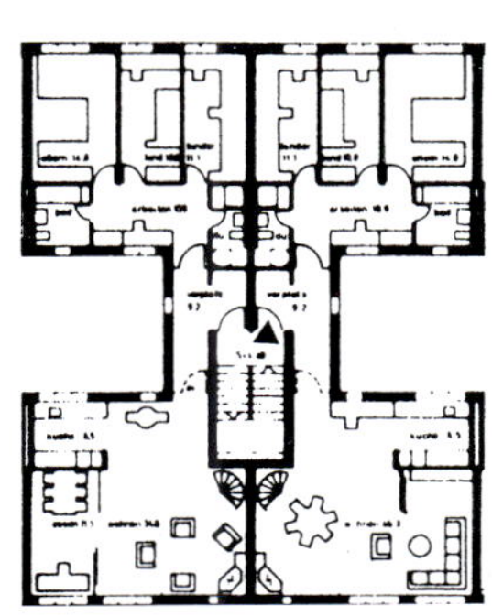

7

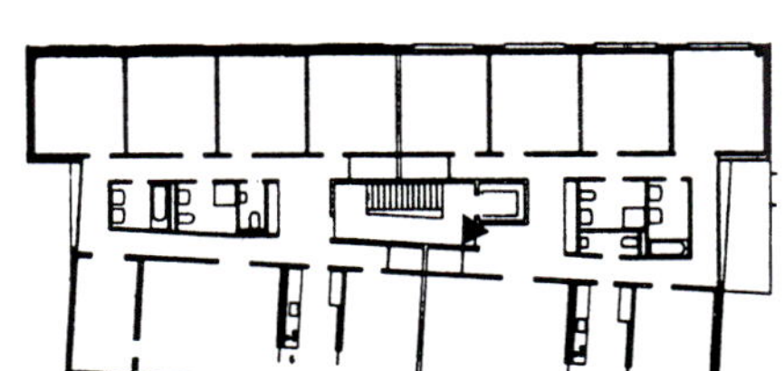

8

1 Georg Muche: House am Horn, Weimar, 1923
2 Rob Krier: Schrankenberggasse, Vienna, 1986
3 Peter Märkli: Mauerbach, House 1, Vienna
4 O. M. Ungers: Project "Cologne-Neue Stadt," 1962
5 Alvar Aalto, Paul Baumgarten: Klopstockstraße, Berlin, 1956–1957
6 Peter Märkli: Trübbach, Switzerland, 1989

1 O. M. Ungers: Wilhelmsruher Damm, Berlin, 1967–1969
2 Otto Jäger und Werner Müller: Hannibal, Stuttgart-Asemwald, 1969–1971
3 Kistler Vogt Architekten: Schüsspark Uno, Biel, 2004
4 Marc Langenegger: Bülachhof, Zurich, 2004

5 Carlos Ferrater: Villa Olímpica, Barcelona, 1991
6 Dercon, T'Jonck, van Broeck: Hogepontstraat, Gent-Scheldekaai, 1992–1996
7 Theo Hotz: Buchgrindel 2, Wetzikon, Zurich, 1979–1985
8 Diener & Diener: St. Alban-Tal, Basel, 1986

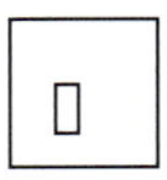

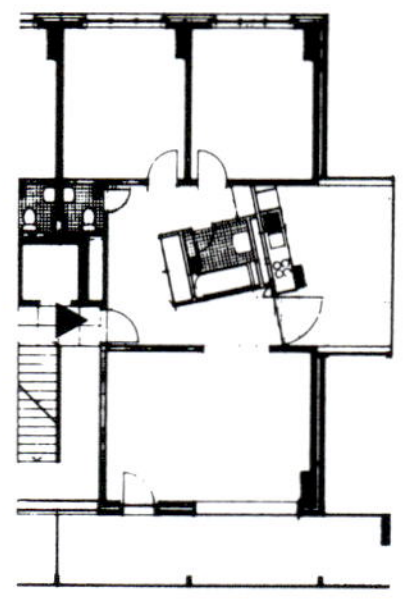

1

DIVIDING ELEMENTS

Visually, the apartment reads like a large, open space with prefabricated dividing elements (for example, installation core, maisonette stairs, wall pieces). Since the sense of space is derived from the large open space rather than the dividing elements, these apartments appear generous and open despite their real dimensions. To ensure that the core (with kitchen, bath, pantry) is perceived as a freestanding autonomous body, it is often rotated away from the axis 1. It guides the movements of the inhabitants, divides the paths or allows for a circular route. A pantry or storage room, a bathroom, or an installation core can also separate the corridors leading to living or bedrooms 1 or separate the living room from the bedroom wing 2, while a core with galley kitchen and bathroom can structure the link between living room and dining room 3. In example 4, a core with a galley kitchen, bathroom and integrated hallway structures the entire room: the only room partitions are sliding walls that can either enclose or open rooms, depending on the time of day or living configuration. In example 5, five large cores even accommodate small individual apartments (with bedrooms, kitchenettes and bathrooms). Through their juxtaposition to one another, spacious shared living areas are created with differently sized and shaped spaces suitable for various uses.

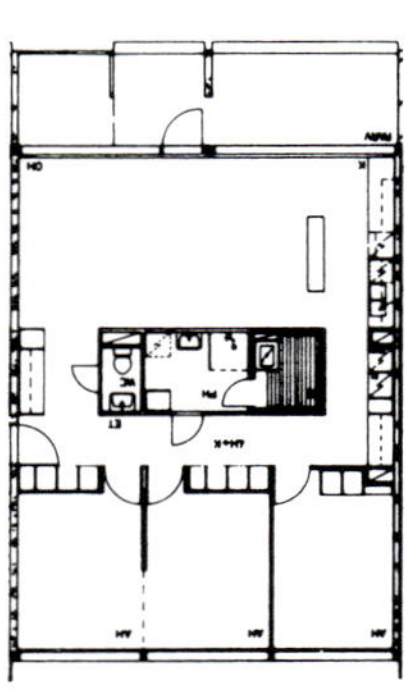

2

3

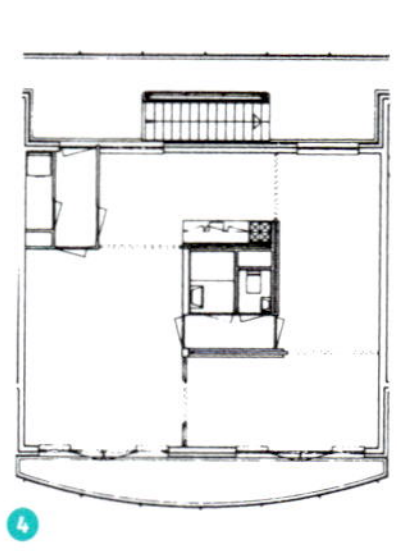

4

5

1 Diener & Diener: Riehenring, Basel, 1980–1985
2 Erkki Kairamo: Asunto-Oy Hiiralankaari, Espoo, Finland, 1982–1983
3 Guignard & Saner Architekten: Station Illnau, House A, 2010
4 M. Duinker, Machiel van der Torre: Dapperbuurt, Amsterdam, 1989
5 Duplex Architekten: Hunziker Areal, House A, Zurich, shared flat with five or six small, individual apartments, 2015

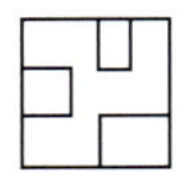

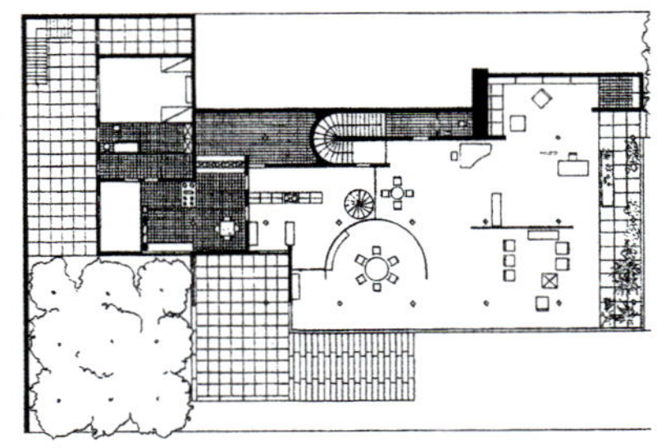

1

CONTINUOUS FLOOR PLAN

The continuous floor plan reduces room boundaries to a few walls, which are placed with great care in order to make the space dynamic and divide it into specific areas. The rooms are rarely separated from the traffic area and hardly at all from one another; they merge into each other, open up sightlines, extend an invitation to move forward 1. As a result, the apartments appear more generous and more open, the individual room always relates to the entire space. In examples with horizontal interconnectedness 2 and 3, living room, dining area, and kitchen blend with the corridors leading to the master and children's bedrooms; the rooms are set off from one another. In the case of vertical interconnection, this approach results in different room heights and room relationships, which increase the desire to move between spaces 4. The spatial flow is sometimes enhanced by the deliberate use of light, such as placing a source of light at the end of a path attracting the visitor to walk in that direction: in example 2, the source of light is provided in the form of the glass door leading to the garden; in example 4, it is the skylight, the entire house being arranged around a column of light. Example 5 is emblematic for the continuous floor plan. It is a promenade at its purest: a single, continuous space is structured through branches. Spatial relationships and sightlines are deliberately orchestrated through cleverly designed bends in the room branches; each branch ends in a quiet zone, even without a door as separation.

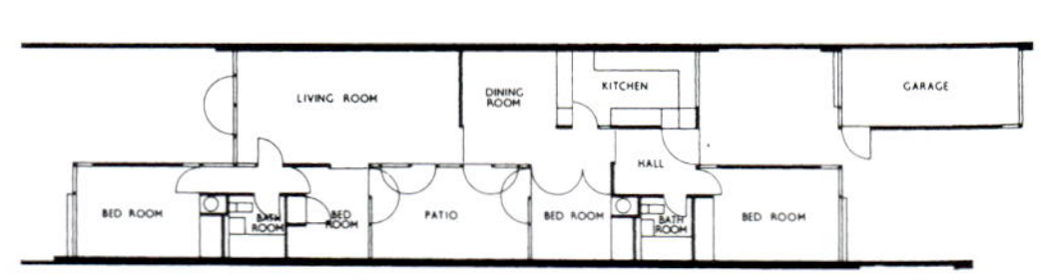

2

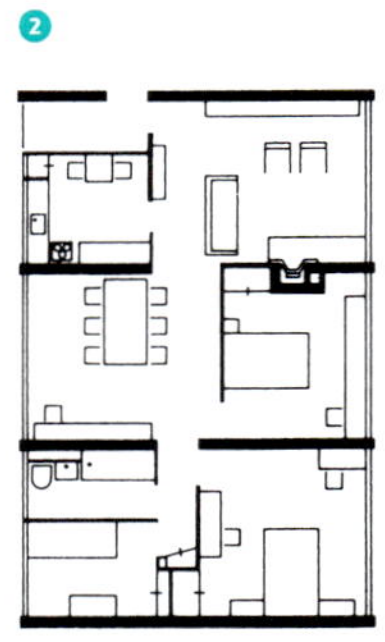

3

4

5

1 Ludwig Mies van der Rohe: Tugendhat House, Brünn, 1930
2 Peter Phippen: Hatfield, London, 1964
3 Eric Lyons: The Priory, Blackheath, London, 1957
4 Herman Hertzberger: Diagoon Houses, Delft, 1976
5 Sou Fujimoto: House O, Chiba, 2007

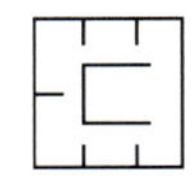

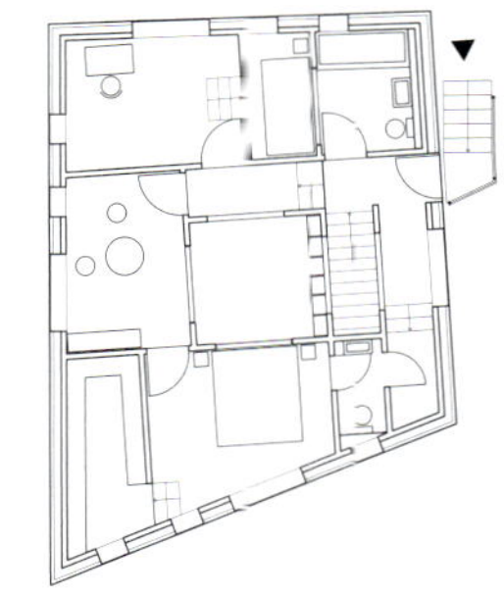

1

FLOOR PLAN WITH CIRCULAR PATH/ EMPHASIZING THE INTERNAL PATH

This kind of floor plan takes the path through the apartment as its theme. It creates the greatest number of functional and spatial relationships between different rooms, which can be experienced and actively utilized. Each space, each room can be reached via two or more paths in this type of floor plan. The multitude of possible paths allows the user to experience the apartment from ever-new perspectives, it appears to be more varied than it actually is. The path often circles around a center, for example, an atrium 1, 2. The visual contact with the other side ensures that the overall layout is clear at all times. The rooms can thus feel like "showcases" strung along the route. However, the circular path can also make the kitchen and sanitary zone into the center 3 or even the stairwell 4. The bedrooms can also lie at the center (cf. "Neutral or Ambiguous Spaces," bottom variant of 2), in which case they form the internal quiet zone, while at the same time functioning as a pass-through zone and filter. They can open toward the intimate side of the house or the common areas as needed. In examples 5 and 6, the individual rooms absorb the traffic area. The rooms are arranged in such a manner that the openings in the center of the wall surface allow for the "circular path." The downside of this arrangement is that it diminishes the floor area where furnishings can be placed and one can assume that these openings are left open or closed off by the users according to their individual needs.

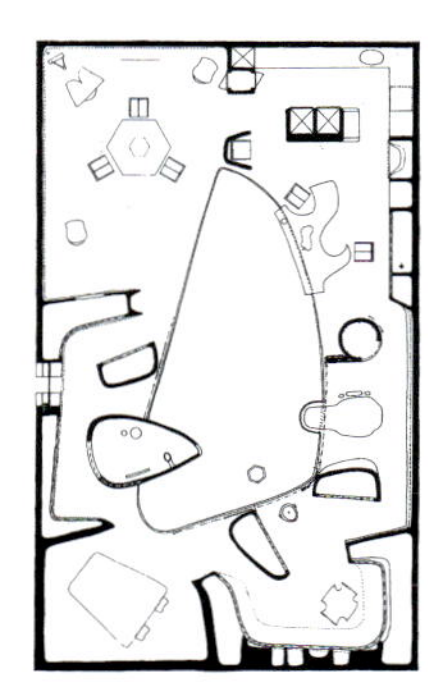

2

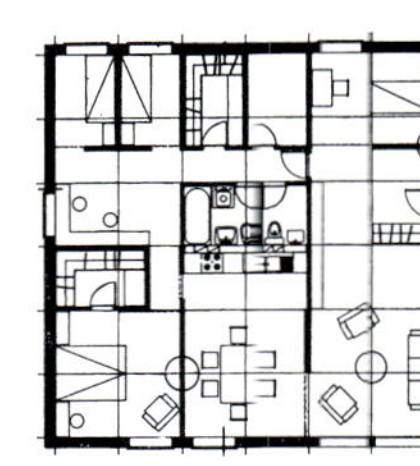

3

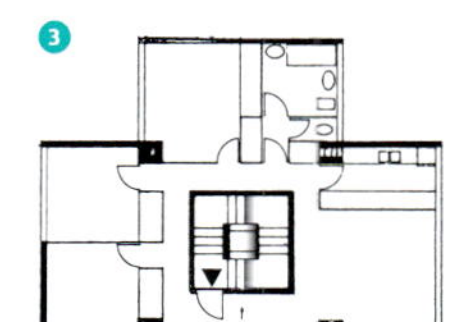

4

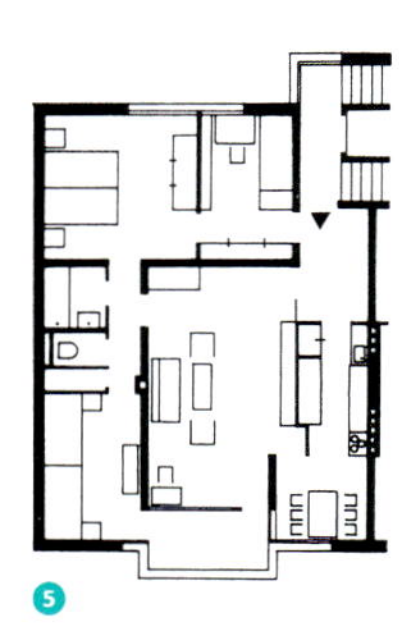

5

6

1 Kraus, Schönberg: House W, Hamburg, 2007
2 A. + P. Smithson: House of the Future, Ideal Homes Exhibition, London, 1956
3 Syndicat des Architectes de la Seine: System of flexible flat layout, Paris, 1960
4 Rem Suzuki: cruciformes, Paris, 1967
5 O. M. Ungers: Garthestraße 8, Riehl, Cologne, 1957
6 Martin Wagner: The Growing House, competition entry, Berlin, 1932

1

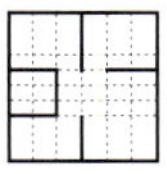

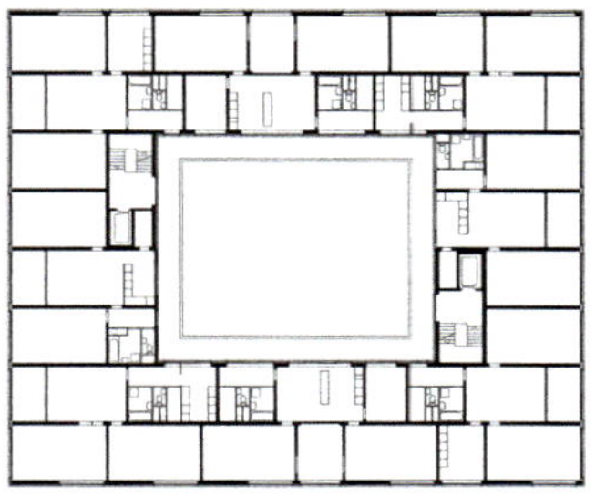

1

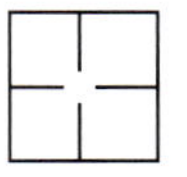

FLEXIBLE FLOOR PLAN

The constantly changing user requirements for an apartment (number of users, preference for close proximity/distance, live/work dynamic, financial situation) and the increasing diversity of forms of living in general have a tenuous relationship with the idea of real estate as an "immovable property," which – from an economic perspective – should be used by the same occupants for the longest possible time. In this case, the favored solution is conversion within the existing floor plan rather than a move to another apartment.

The approaches range from removable room dividers or additional rooms that can be connected to one apartment or another to changing the shape and size of the rooms by means of movable wall elements.

By means of special removable interior walls, the residents of example 2 are always able to fulfill their personal demands on the layout. The load-bearing exterior walls allow for a high degree of design flexibility (interior fittings play no structural role) and the shallow depth of the building guarantees adequate lighting for every alternative constellation.

"Connecting" rooms or groups of rooms require even less renovation effort: rooms with a neutral character (in apartment buildings, these are usually located across from the stairwell) can be added to individual apartments 3. Alternatively, the design is conceived to anticipate different room constellations for different living concepts within a house, so that several apartments can be easily connected or separated into individual units. Separate access to the different levels and room groupings is here the prerequisite, see example 3, where the position of the staircase at the entrance to the apartments allows four separate units to be combined to create a multi-generational house spanning across all floors.

While these concepts provide an answer to the needs of different life phases (and require building alterations), floor plans with movable wall elements can immediately react to the changing demands on an apartment over the course of a day. Such fixtures can take the form of sliding or folding doors that disappear into walls: in the daytime scenario, a large, multifunctional room is created by folding away wall elements, while at night they are used to close off individual rooms 1. The beds are stored during the day (here: under the raised hallway), similar to the Japanese living tradition. See example 4: the apartment size can be reduced to a minimum as there is no hallway, and no separate living or bedrooms otherwise occupied only part of the day; by day, the sliding doors of the bedrooms are opened to form a continuous living room. Personal items are stored in the built-in shelving units. A system of sliding and folding walls running both crosswise and lengthwise between short wall segments opens up the possibility to continually recombine the separate rooms in different ways, so that the apartment is open to reinterpretation, both over the course of a day and on the whole 5. Similarly versatile, yet without interior walls and consisting of only a central core and sliding walls is 6. The shape of a dividing element can also be more complex: rotating walls, closets, and room corners achieve surprising new spatial forms and connections 7.

NEUTRAL OR AMBIGUOUS SPACES

In this concept, the use of the space is not predefined by the size, form and arrangement of the rooms, which increases the opportunities for different uses to come into play for one and the same floor plan 1, 2, 3. There is no need to modify the apartment in order to meet new requirements (as is the case with the flexible floor plan); the occupant can and indeed must invent his or her own form of living. Personal requirements can be met by rearranging the interior. Occupants with different ideas on living can realize these ideas within the same floor plan model (greater mix of residents within one housing complex). To this end, the rooms need to be of a certain size and proportion and, ideally, with independent access. As a result, more floor area is required for corridors and rooms, and the best solutions for this type of floor plan are therefore found in more exclusive residential buildings. In most cases, the rooms are intentionally indistinct and able to accommodate a variety of uses. However, it would be interesting to explore yet another design approach to creating neutral, ambiguous spaces: instead of creating identical rooms with little to no character for the purpose of ensuring that they can accommodate the widest range of uses, the rooms could also be deliberately designed with distinctively different characteristics (room size and shape, light, etc.), thus inviting a very individual use – as is the case in many old buildings, which are once again popular 4.

2

3

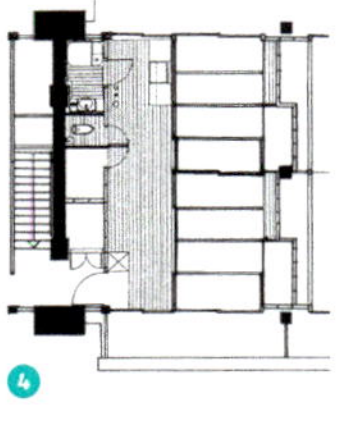

4

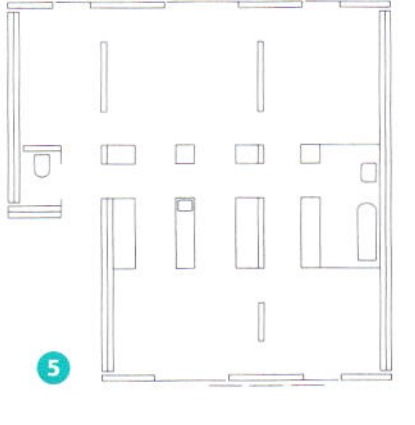

5

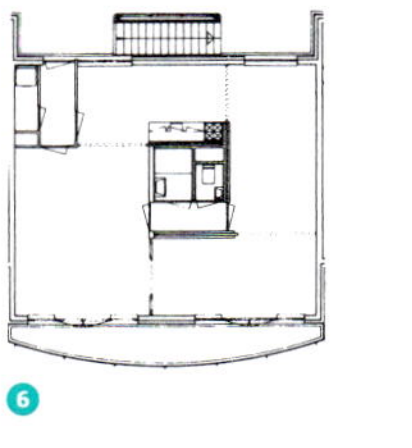

6

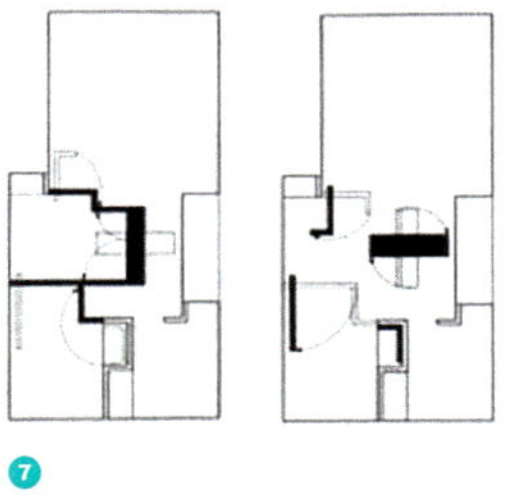

7

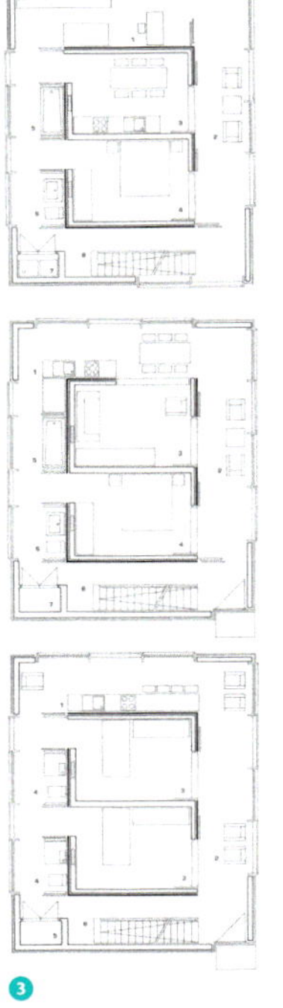

3

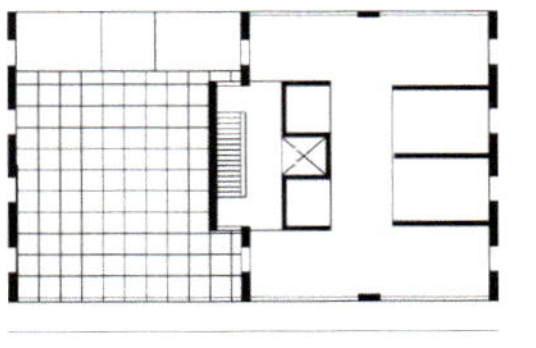

2

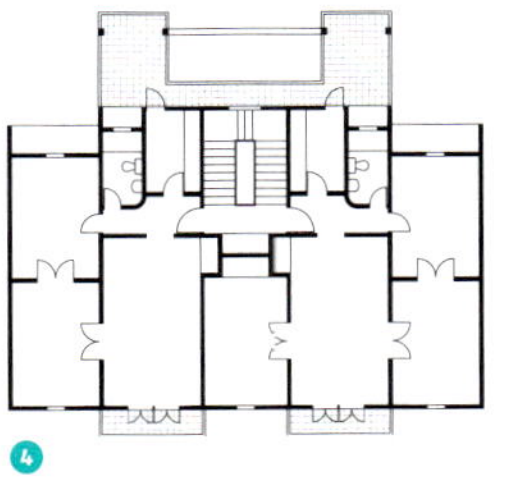

4

1 Aranguren Gallegos: Carabanchel, Madrid, 2003
2 Ludwig Mies van der Rohe: Weißenhof Estate, Stuttgart, 1927
3 HAHOH: House am Ottersgraben, Zell am Harmersbach, 2007
4 Kunio Maekawa: Harumi House, Tokyo, 1958
5 Riegler, Riewe: Bahnhofstraße, Graz-Straßgang, 1994
6 M. Duinker, Machiel van der Torre: Dapperbuurt, Amsterdam, 1989
7 Steven Holl: Nexus World, Fukuoka, Japan, 1986

1 Diener & Diener: KNSM- & Java-eiland, Amsterdam, 2001
2 Herczog Hubeli: Steinfelsareal, Zurich, 2002
3 heide von beckerath alberts: Wunschhaus #1, Hamburg-Sülldorf, 1999
4 Carlos Raúl Villanueva: El Silencio, Caracas, 1945

1

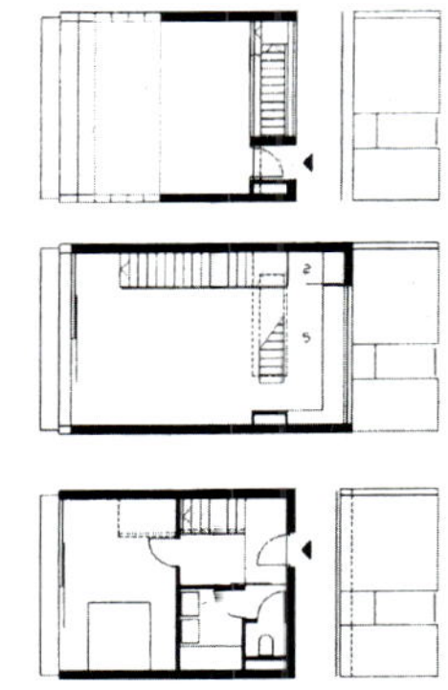

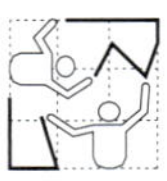

1

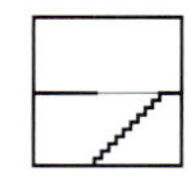

USER-DEFINED FLOOR PLAN

In the end, every resident is involved in determining his or her individual floor plan through use and furnishings, and in particular when it comes to flexible and neutral floor plans. However, in "user-defined floor plans," this potential to participate in the design is dramatically extended: for instance, in self-build scenarios, where the user "finishes" building the apartment, or in participatory floor plan design, where a floor plan is designed in cooperation with the architects. Both strategies provide the resident with a high level of personal identification with the apartment, since he or she helped design it, and the shape, size, furnishings, and price are suited to the user's current life situation. Because the structure of the house was devised to accommodate changes, the user is able to modify the floor plan again in the future – and in doing so stick around a bit longer.

In order to allow for such self-determination in a multi-unit residential building, architects generally design a basic building structure, which can facilitate numerous changes in the facade, interior walls and fittings due to the well-thought-out positioning of load-bearing elements, circulation areas, and building services. In smaller projects, architects can invent a personalized floor plan directly with input from the future users, for instance, if they form a Baugruppe – a joint building group – ahead of time and work together as one client. Beyond their individual apartments, members have a say in how the entire development and the shared spaces should be designed, and the opportunity to realize new living concepts. Example 2 illustrates the point: by inserting lightweight wooden walls on each level, individual floor plans can be created within the concrete skeletal structure. In order to minimize the coordination efforts with the future residents, architects can create a catalogue of options to choose from, which can be modified 3. Example 4 is an extreme case, in which a "residential shelving unit" is filled with individually designed apartments developed by the residents with the aid of multiple, independent architects. The result is a side-by-side juxtaposition of very different living concepts and styles (and social mix).

In self-build scenarios, the design is created so that expansion is not only possible, but encouraged: to some extent, the apartment/house is handed over in an "unfinished" state. Such a concept is particularly suited for low-budget situations. The architect specifies the structure and design in such a way that the user is free to finish the rest on his or her own, either in consultation with the architect, or with the aid of instruction manuals. In the case of the "growing house," only the circulation core (which is load-bearing and contains the building services) is handed over to the user. The house can then unfold step-by-step both vertically and horizontally 1. In example 5, two-story enclosures of space were handed over to the residents – kitchens, bathrooms, stairs, and gallery levels are added as desired; the accompanying handbook explains how the support system works and presents further examples of possible building elements. For example 6, a construction manual describes the building steps and techniques (for example, brick-laying), thus empowering the residents to complete the entire flat (including the exterior walls) in an affordable way.

MAISONETTE

The floor plan typology of a maisonette is characterized by the apartment stretching across several full floors. It is usually two stories high, rarely more. As a building type, the maisonette is above all dedicated to the idea of zoning: the two-story layout facilitates a spatial separation of specific functions. At the same time, two-story atria are often an impressive feature of maisonettes. The apartment with its connecting stairs, which can be designed with great flexibility, feels generous and spacious: one has a sense of being in a house of one's own. The stairs and the internal path can be staged in the manner seen in example 1, where the open stairs change direction on each floor and also serve as a room divider between living room and kitchen. Atria are also employed to merge the different levels: the clear height in the stairwell often widens so that the entrance area, kitchen, and dining area communicate with the bedroom wing, as seen in example 2. Conversely, this air well can also extend to the size of an entire room 3. But it can also be entirely separated from the stairs themselves, as in example 4, where a set-back dining gallery overlooks the living room below. A generous sense of space can also be achieved by not simply stacking the living levels between two partitions, but by arranging them in a staggered form. In this case, the apartment extends not only across several floors but also across several structural units 5. This multiplies the views and suggests a larger floor area than the apartment actually offers. In

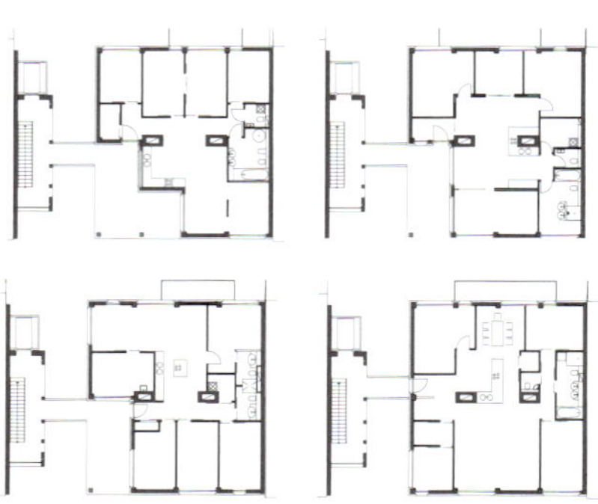

2

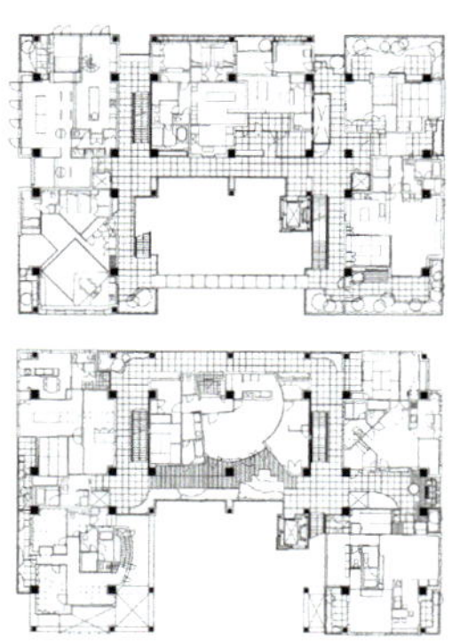

4

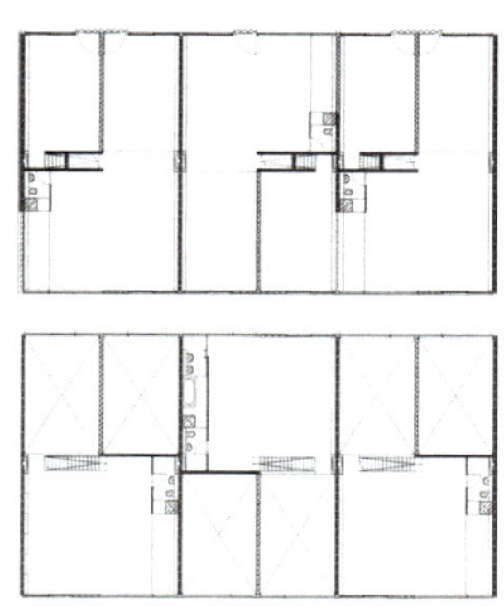

5

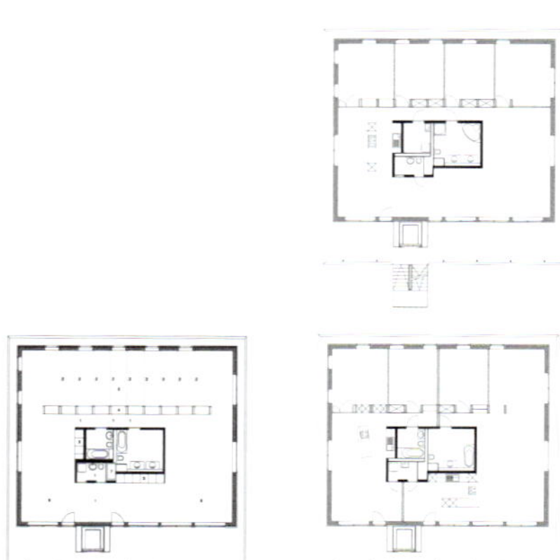

3

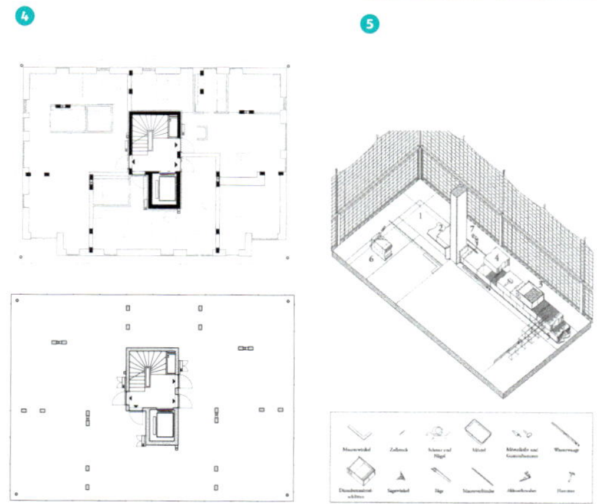

6

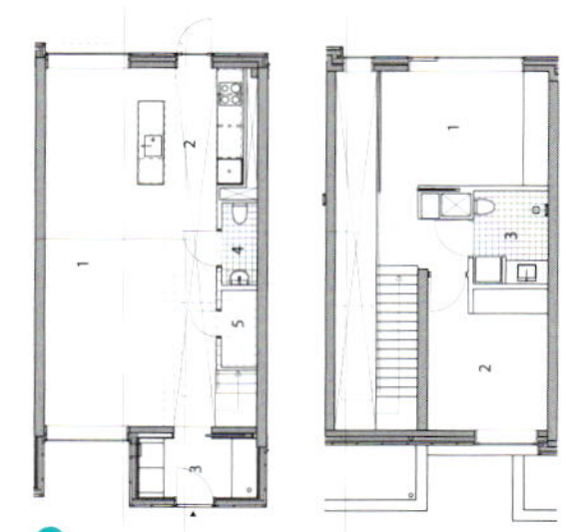

2

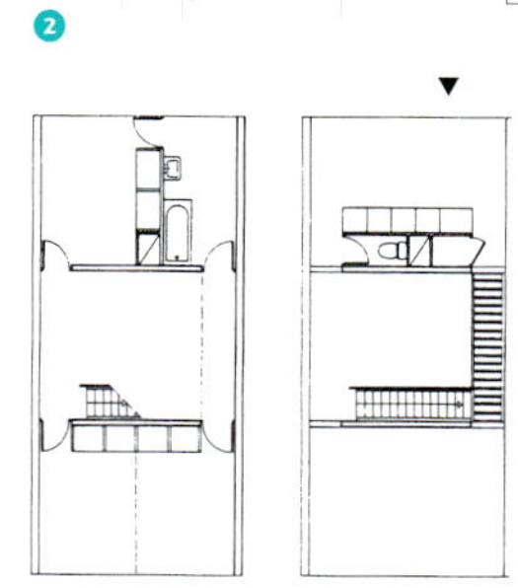

3

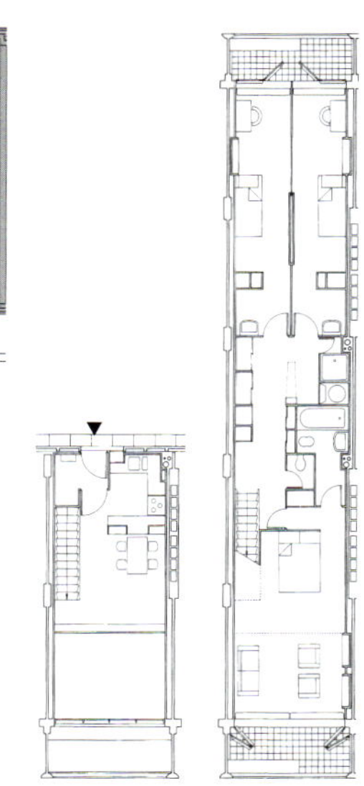

4

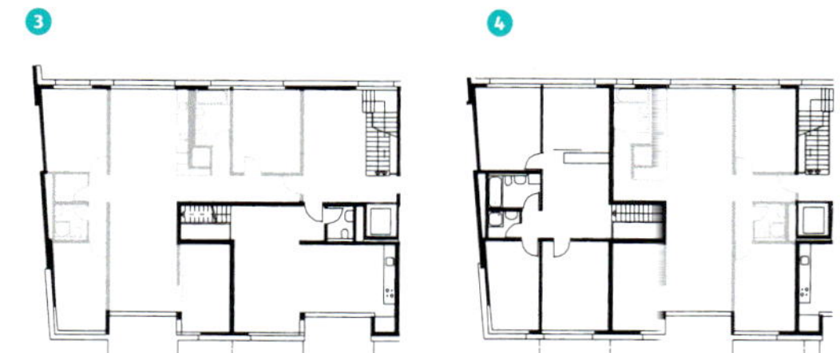

5

1 Elemental: Quinta Monroy, Iquique, Chile, 2004
2 Kaden Klingbeil: Esmarchstraße e3, Berlin, 2008
3 Haerle Hubacher Architekten: Balance Uster, Uster, 2001

4 Utida Shu-koh-sha Architectural and Urban Design Studio: Next 21, Osaka, 1993
5 b&k+ brandlhuber&kniess GbR: Kölner Brett, Cologne, 1999
6 BeL Sozietät für Architektur: "Grundbau und Siedler," Hamburg International Building Exhibition, 2013

1 Geiswinkler & Geiswinkler: Am Hofgartel, Vienna, 2004
2 Dorte Mandrup: LIVING 2006+, Ringsted, 2006
3 Nouvel, Ibos: Avenue du Général Leclerc, Nîmes, 1987
4 Le Corbusier: Unité d'Habitation, Marseille, 1947

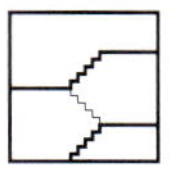

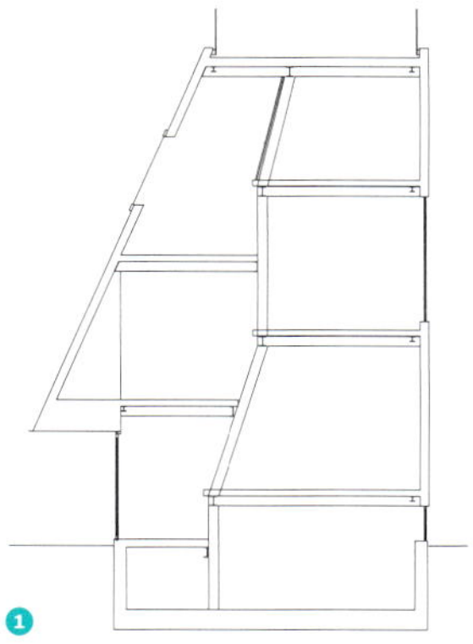

1

multi-story buildings, the maisonette type is often found in combination with a central corridor or covered walkway, because it allows for undisturbed front-to-back interiors despite the orientation to one side only on the entrance level. The entrance is usually located at the level where the common rooms are; kitchen, living, and dining room then function as a filter and protect the private sphere, even the stairs can serve as a buffer to the access path (see 6). However, the entrance can also be located at the bedroom level. This is often the case in center-corridor types, where only half of the building width is available at the entrance level. In this scenario, the stairs are designed and placed in a manner that draws the visitor directly into the living space 7.

The scissor-shaped floor plan, a variation of the "back-to-back, crossover"-type, is a unique form 8. In it the functions are bundled and stacked: a central corridor divides the house into two halves, for example, a living area that is more prone to noise and a quiet bedroom area. On one side, the corridor provides access to apartments with their living area at entrance level; on the other side it opens to apartments that have their sleeping area at this level; within each flat a set of internal stairs leads up to the next level and crosses the corridor to reach the other half of the house, where the second part of the apartment is located (with living room or bedroom, respectively). Each apartment is thus identical in structure and orientation, profiting equally from the advantages of the different sides of the building in terms of view or light.

SPLIT-LEVEL

Split-level is a floor plan type in which the apartment is usually distributed across at least three (often five to six) levels, staggered in half-level steps. The connecting stairs tend to be short, the levels small; what is at play here is a constant progression upward, the pleasure of how the spaces flow into one another as the path winds through the apartment – the qualities of the continuous floor plan. In extreme cases, each room, each function has its own level. Views into rooms and across spaces, sightlines of all kinds combined with a fine-tuned spatial separation characterize this type. Since one is constantly looking at the apartment from new positions, it can appear to be very diversified within a minimum of space. This type is often used when different functions must be accommodated in one apartment, for a room can be specifically designed in size and height for each of these functions and accordingly connected to a greater or lesser degree to the other activities in the home. Split-levels are most commonly used on narrow lots, for row housing, high-rises or in a building gap, and especially on hillsides where the topography can thus be utilized to best advantage. The internal stairs are centrally located, with landings of identical or different sizes for differing functions. The common rooms – following a sequence of entrance, kitchen, dining area, and living room – tend to transition seamlessly in a typical split-level plan. The spatial continuum is thus palpable and emphasized. At the same time, this generates a contradiction between deliberate openness and the need for withdrawing into a private sphere, to which each split-level floor plan must respond. One possibility is to place the stairs with oversized landings, which can serve for different work and living activities, apart from the main levels of the unit, separating them to a greater or lesser degree from these levels. The facing between the levels and the stairs becomes more and more closed with the increasing height of the building and the increasing need for intimacy called for by the uses (of the levels). Thus it is possible to optimize the degree of isolation provided 1. An atrium next to the stairs can further enhance the appeal of wandering through the apartment, as in example 2. Here, the atrium creates generous sightlines between the levels and the stairs all the way to the roof patio. The stairs, now merging with the main living areas, can also be closed off from these, creating a separate stairwell thanks to the careful placement of the stairs within the layout. Thus, when the requirements for use change, new individual rooms with separate access can be easily created (this is contrary to the prejudice that split-level types offer no flexibility). The split-level type can also be employed strategically, as in example 3. Here, several different split-level apartments unfold across three levels around the stairwell core of the multifamily home, whereby all three levels have a private entrance, some from the floor landing and some from the mezzanine landing. This independent access opens up new possibilities for subdividing the apartment and for different uses such as subletting, flats for adult children, in-law apartments, housing for a caregiver, or a home office space.

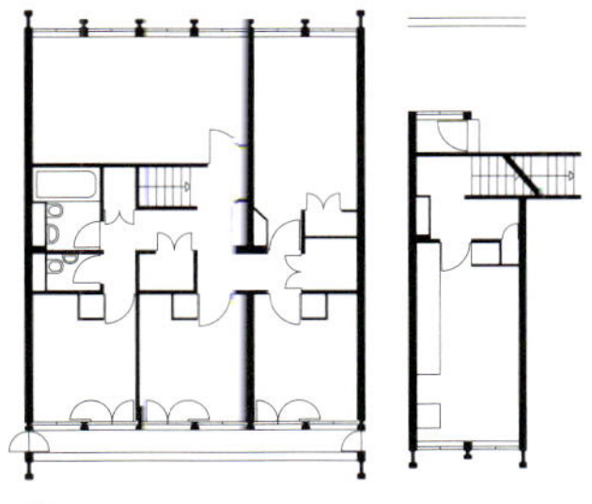

6

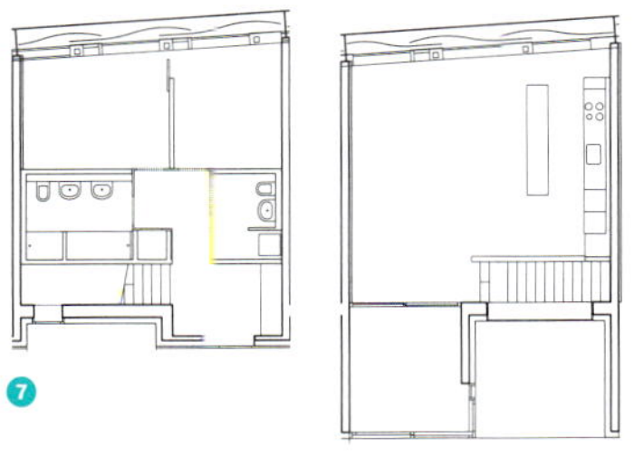

7

8

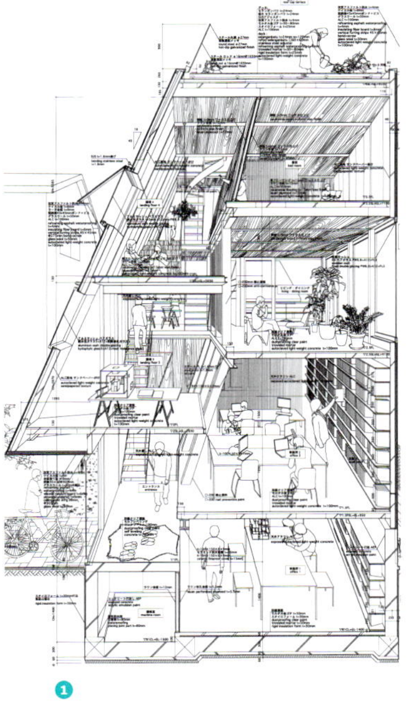

1

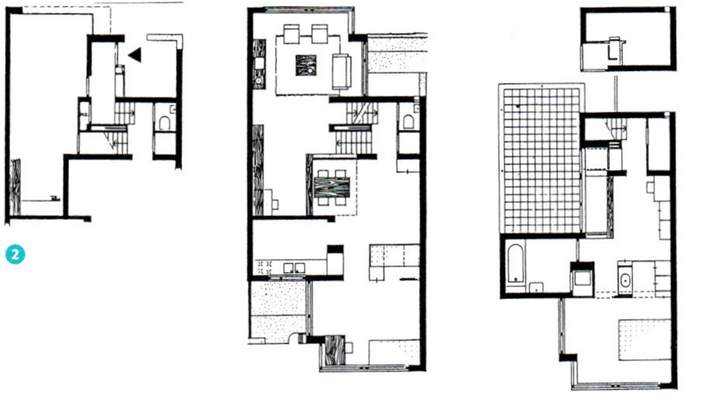

2

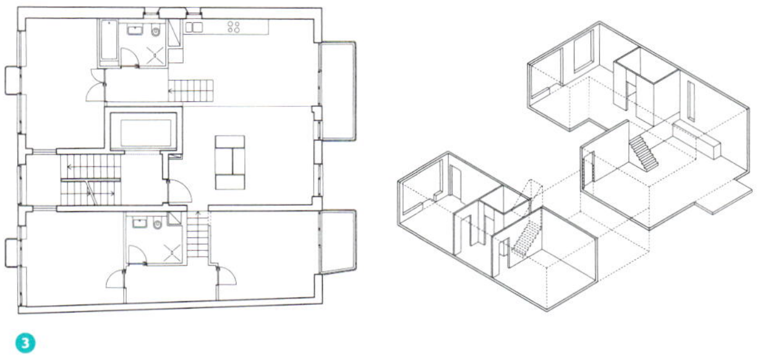

3

5 Pool architekten: Leimbachstrasse, Zurich, 2005
6 A. + P. Smithson: Robin Hood Gardens, London, 1972
7 Burkard Meyer: Falken House, Baden, 2006
8 CZWG: China Wharf, London, 1988

1 Atelier Bow-Wow: Bow-Wow House and Studiko, Shiujuku-Ku, Tokyo, 2005
2 Herman Hertzberger: Diagoon Houses, Delft, 1976
3 BARarchitekten: Oderberger Straße, Berlin, 2010

Oliver Heckmann

A Graphic Approach to Floor Plan Design

The Floor Plan Manual Housing builds to some extent on a kind of knowledge acquisition whose form is probably specific to designers: namely the ability to draw independent conclusions from the projects of others for use in one's own design work without simply copying them. Nigel Cross* argues that architectural knowledge is embedded in designs, and that architects are capable of deciphering projects and drawing conclusions from concrete examples about the underlying demands and design intentions. What can be deciphered here are essential aspects of the design, for instance, functional relationships, experiential qualities, or the relationship between an individual and the community – all of which are issues solved by the spatial configuration of a plan.

In this essay, an attempt is made to illustrate graphic methods of approaching some of these overarching spatial, functional, or social criteria, enabling their exploration at an early stage so they can be used to develop design intentions before they are confined to an architectural form. The following two themes, "Before Use" and "Before Form," describe two fields of application of this approach, the first morphological in nature, the latter more organizational. A definitive, all-encompassing strategy for designing a floor plan is not the goal here – and should continue to remain part of each individual design strategy.

BEFORE USE

The following studies investigate spatial aspects of floor plan designs that in principle can be defined independently of how these are later utilized. The example of the Palazzo Antonini by Palladio (see also page 8), but also mid- to late-19th-century European floor plans as well as emphatically use-neutral designs 1 show that to a certain degree functional specifications were simply not considered – or were intentionally avoided – and that floor plans have and can be developed on the basis of a spatial order alone.

Structuring the Depths of Buildings

Simple graphic studies on the spatial organization of a housing unit as it relates to the depth of the building 2 can help bring different aspects in relation to one another in such a way that an initial schema for the design of the floor plan can be formulated. The basis for such studies is formed by an initial assessment of the total living area within a grid as a placeholder for rooms or spatial units, which in a second step are laid out in spatial layers either parallel or perpen-

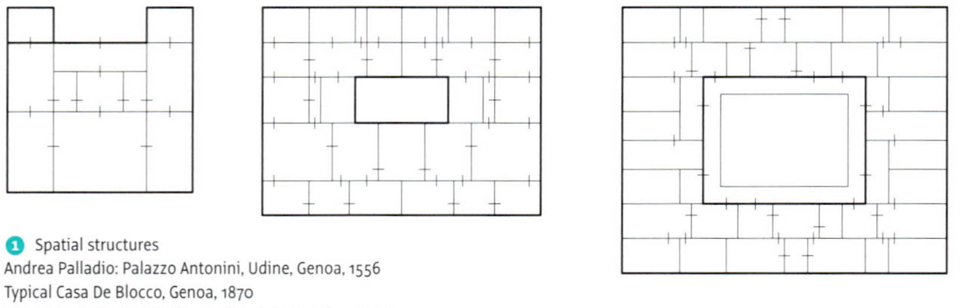

1 Spatial structures
Andrea Palladio: Palazzo Antonini, Udine, Genoa, 1556
Typical Casa De Blocco, Genoa, 1870
Diener & Diener: KNSM- & Java-eiland, Amsterdam, 2001

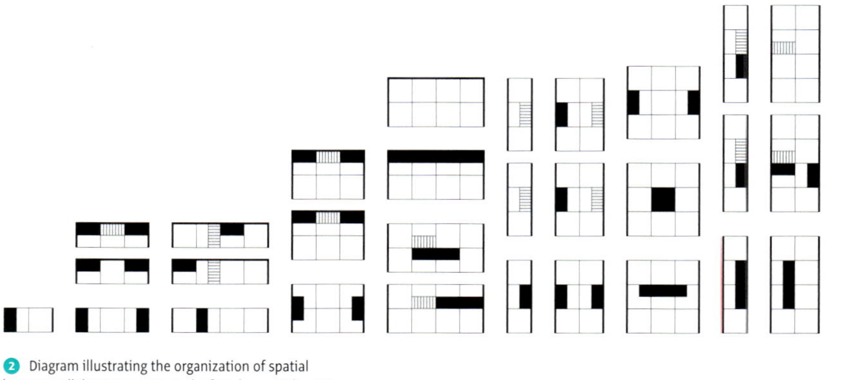

2 Diagram illustrating the organization of spatial layers parallel or transverse to the facade; spatial entities requiring no daylight are indicated

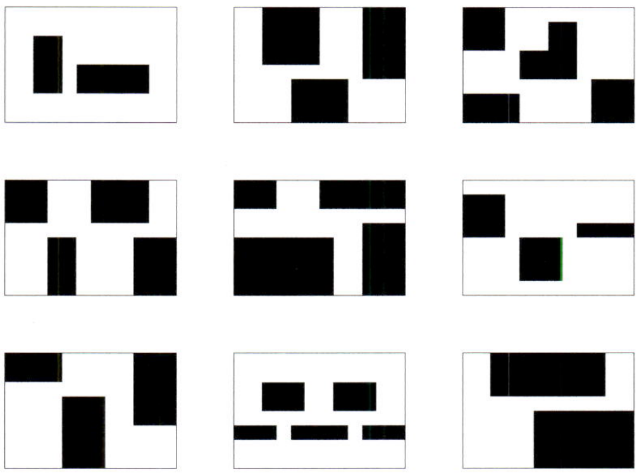

3 Figure-ground studies

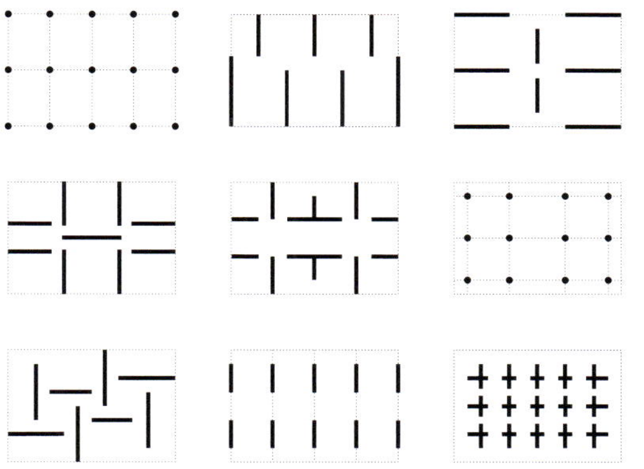

4 Studies on spatial structure

dicular to the facade, depending on the building depth. The following can then be illustrated in a basic schema:

— The depth of the units (as represented by the number of tiers, i.e., the spatial layers parallel to the facade), their width (as represented by the number of spatial layers transverse to the facade) as well as their height by the number of horizontal levels.

— The potential orientations of each unit: if, in the given context, the particular unit can be oriented on one, two, or even more sides to natural light and a view of the surroundings.

— The determination of spaces that need or can do without natural light – particularly relevant for units oriented to one side only or with an increased number of dark room tiers.

In the subsequent design steps, one can work out how the floor plan can potentially be organized and the functions allocated – whether, for instance, in a very deep floor plan the unlit groups can be used as hallways and enclosed bathrooms, how the sequence of rooms in an apartment with a single spatial layer can be arranged, or in which field the vertical access of a multi-story unit should be placed.

A Sequence of Rooms as a Graphic Composition

Diagrams in the style of figure-ground drawings can help develop ideas on composition, room sequences or experiences. They form spatial compositions of enclosed volumes and empty spaces with a specific rhythm and relationship to one another, as dualities of open and closed, light and dark 3.

Another aspect can be taken into account: black can be used for designating closed areas such as storage or sanitary rooms and white for common areas.

Even structural studies that make use of grids as well as linear or pointwise elements to develop structural strategies can simultaneously generate initial spatial hierarchies as basic, abstract systems 4.

Room Patterns

Schematic studies can develop patterns for spatial juxtapositions and the way they connect to one another. In a following step, each proximal relationship can then be formulated in more detail into direct connections – the more spaces adjoin a room (for instance, by offsetting them to one another) the greater the connective potential 5.

Such figure-ground and structural studies as well as patterns of spatial juxtaposition can generate schemes which, when subjected to a subsequent layering of functions and activities, can be post-rationalized, or more precisely formulated. Even at this stage, these schemes can work with approximate room sizes.

Spatial Experiences

Working with spatial sketches, which are often produced only at the end of the design, can help to conceive of floor plans in situational terms, for instance, to consider how visually permeable an apartment is, or how it can potentially forge a connection between its residents and the immediate surroundings. The sketches by Alison and Peter Smithson on page 12 show the value of such sketches 6.

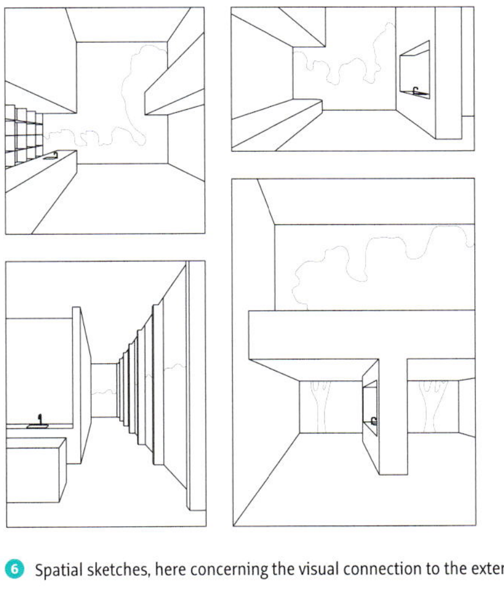

6 Spatial sketches, here concerning the visual connection to the exterior

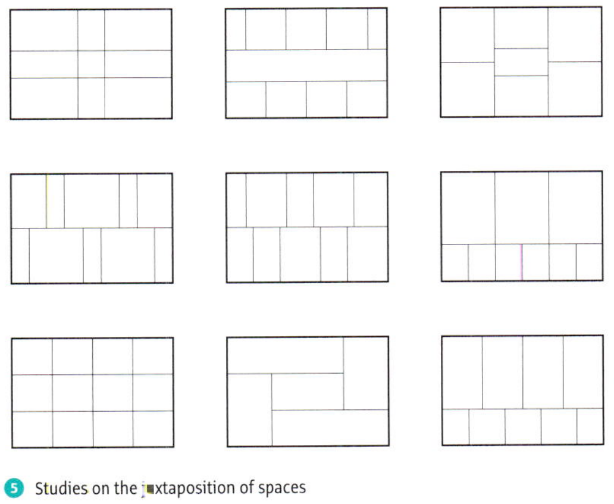

5 Studies on the juxtaposition of spaces

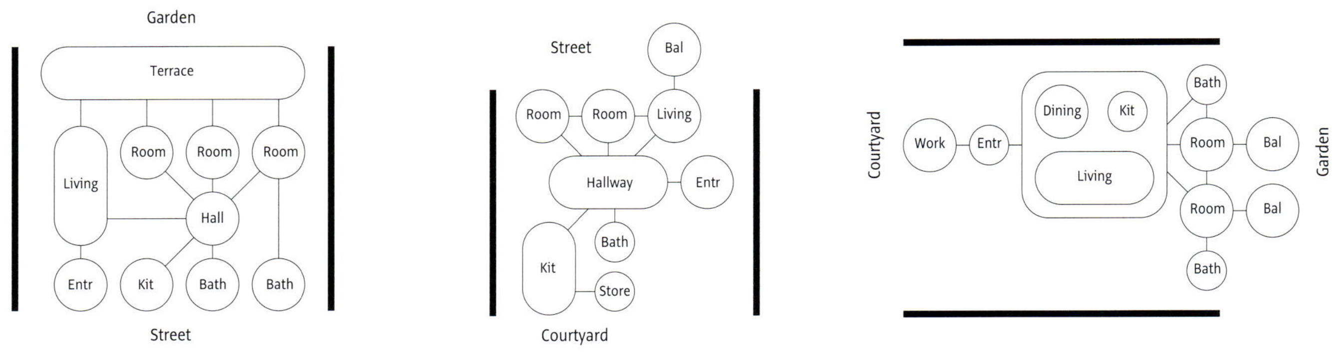

7 Examples of functional diagrams

BEFORE FORM: ACTIVITIES, FUNCTIONS, AND RELATIONSHIPS

Besides these rather spatial and morphological considerations, a floor plan can also be devised from purely abstract relational patterns with a view of, for instance, the organization of functions or activities, as well as the organization and prioritization of relationships between the individual and the collective.

Functional Diagrams

A convention of such preliminary analyses is the development of a functional diagram. This starts with taking inventory of the number and approximate dimensions of the rooms, their functions, and their resulting division into serving and served rooms, those requiring natural light and those that can do without. In a diagram, the relationships between the individual rooms to one another and to their surroundings are devised using connecting lines. The rooms are grouped and categorized hierarchically according to their functions, placed in either peripheral zones or in the center. Additionally, the interfaces with the surrounding context as well as the exterior orientation of the rooms to specific environmental conditions, such as direct sunlight, noise, or the public sphere, can be indicated. A spatial scheme begins to emerge, in which a preliminary organizational system is illustrated without too much formal or dimensional determination 7.

Activities instead of Functions

In a diagram by Kiyoyuki Nishihara*, a concept for room configurations is illustrated, which despite being based originally on traditional Japanese lifestyles, presents a transferrable alternative model to the functional diagram. The practice of flexible use in traditional Japanese housing – some of the rooms in his diagram are named only by their proximity to one another, not according to their functions – inspired him to take a closer look at the activities of the residents instead of focusing on the functions of the individual rooms. He defined six essential activities that cover the entire complexity of domestic life: gathering – eating – cooking – sleeping – working – washing and relieving oneself 8.

Localization of Activities

While Nishihara's graphic inquiries analyzed pre-existing situations, allocating activities can also form the basis for a floor plan design. Intentions can be formulated early on: rooms can be flexibly used for a number of activities; cooking, dining and gathering can be combined; all residents can work in one space, or in their own rooms. Simultaneously, these classifications allow conclusions to be drawn on how important certain rooms are for the overall structure, how open they should be and whether they are mono- or multifunctional 9.

The Integrity of Floor Plan Organization

The term 'integrity' is used by Julienne Hanson* to examine the floor plans of existing residential units as matrices of a spatially organized social framework. Emanating from the idea that the design of a spatial configuration always goes along with an attempt to formulate a specific balance between common and individual demands, the studies on integrity analyzed the level of connectedness between rooms within a unit and along its periphery, as it is precisely at these junctures that encounters are structured using spatial means.

Integrity as a Design Question

Such aspects – such as how the transition between public and private spheres is formulated, whether rooms are linked directly or through intermediate spaces such as corridors, whether they branch off in different directions or create cir-

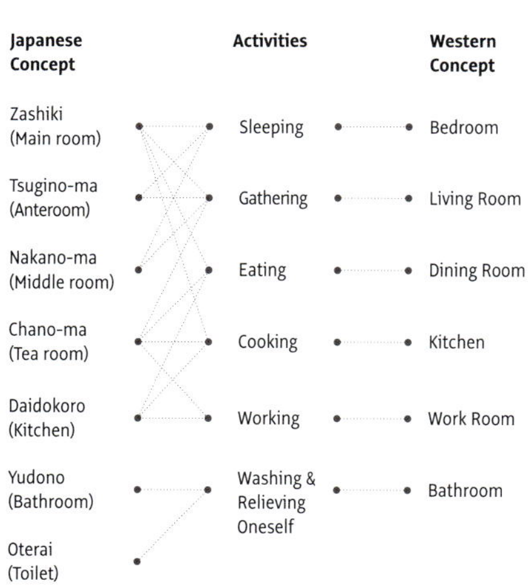

8 Nishihara's classification of six essential activities in comparison (redrawn from original)

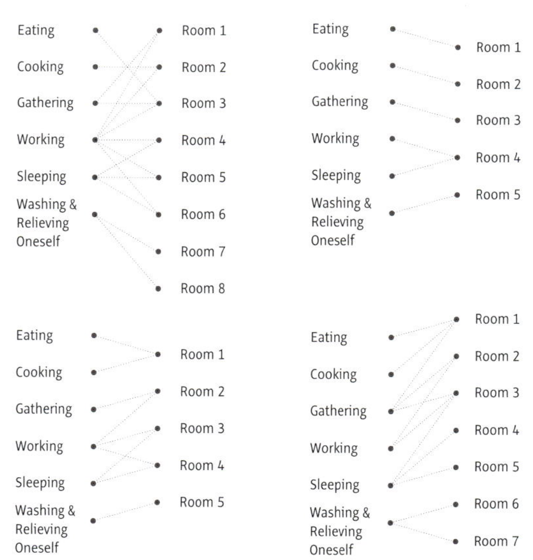

9 Examples of how the six essential activities are allocated among the rooms of different residential units

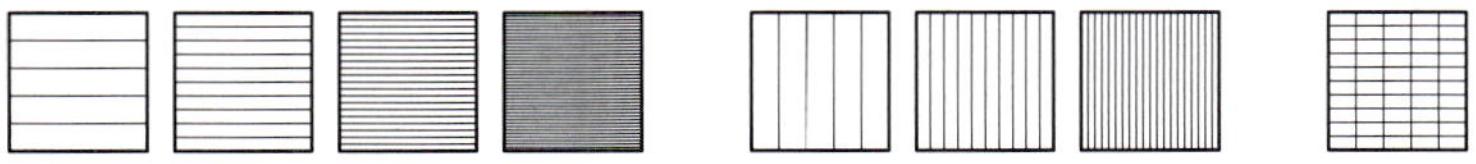

10 Public-private gradient, differentiated to indicate both indoor and outdoor spaces

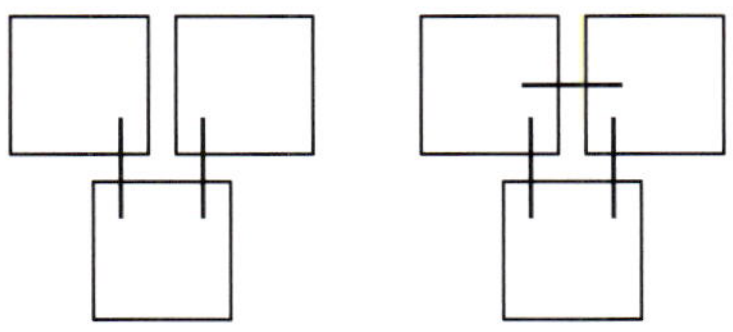

11 Branching or circular connections

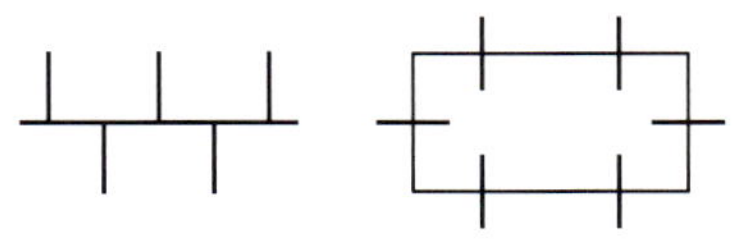

12 Circulation rooms: connections using living or intermediate rooms

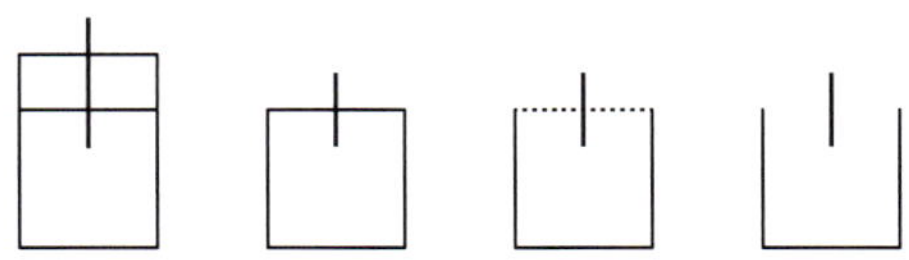

13 The permeability of thresholds

cular paths, or how thresholds are composed – can be taken into consideration before a concrete spatial design is drawn up in order to form clear intentions and ideas on cohabitation before a floor plan is devised. The aspects presented in the following examples and graphic codes, as well as the resulting diagrams on spatial integrity are for illustrative purposes only. Ultimately, this type of graphic instrument should be customized for each individual design strategy.

— Public-private Gradient
The gradual transition from more common, public areas to rather intimate private areas can be structured with a series of spatial situations. Differently spaced hatches can be used to indicate different levels of exposure (in this case: the more dense, the more private). When such transitions both within an apartment and in relation to its surroundings are indicated with lines in different directions, the design of such junctures can also be considered 10.

— Differentiation between Branching and Circular Spatial Connections
Connections that branch off – with only one link per room – threaten to turn rooms into isolated dead ends. The facilitation of circular paths with multiple connections per room increases both the social potential and the functional flexibility of an apartment, but can possibly come into conflict with the demands of individual intimacy 11.

— Circulation Spaces
A supplementary differentiation can help to decide whether the spatial distribution should be organized using actual rooms (the living room, for example) or intermediate spaces (such as corridors) 12.

— The Permeability of Thresholds
Depending on the relationship between the shared and the private sphere, transitions can be formulated using buffer zones, closed walls, variable or somewhat hidden connections, or even open ones (like a niche adjacent to another room) 13.

Diagram for the Assessment of Spatial Integrity
Combining such aspects and their schematic organization into a single diagram can help to visualize a matrix of cohabitation, which in a further step can be transformed into a floor plan design 14.

VERIFICATION

Working with diagrams can also take place while more concrete design work on a floor plan has already begun – as a verification measure. In this example, one aspect is considered: the examination of possible alternative occupancies within a unit for different resident constellations 15. The type and number of connections, assessment of privacy, and the degree to which rooms are predetermined can be adjusted depending on the results of such verification measures.

It is always helpful to transfer projects – even ones at a much more advanced stage in the design process – back into such abstract graphic illustrations in order to reassess important qualities such the organization of the community or the structure of spatial connections.

* Literature
Nigel Cross, "Designerly ways of knowing", Design Studies, vol. 3, no. 4, October 1982, pp. 221–227.
Kiyoyuki Nishihara, Japanese Houses: Patterns for Living, Tokyo 1967.
Julienne Hanson, Decoding Homes and Houses, Cambridge 1998.

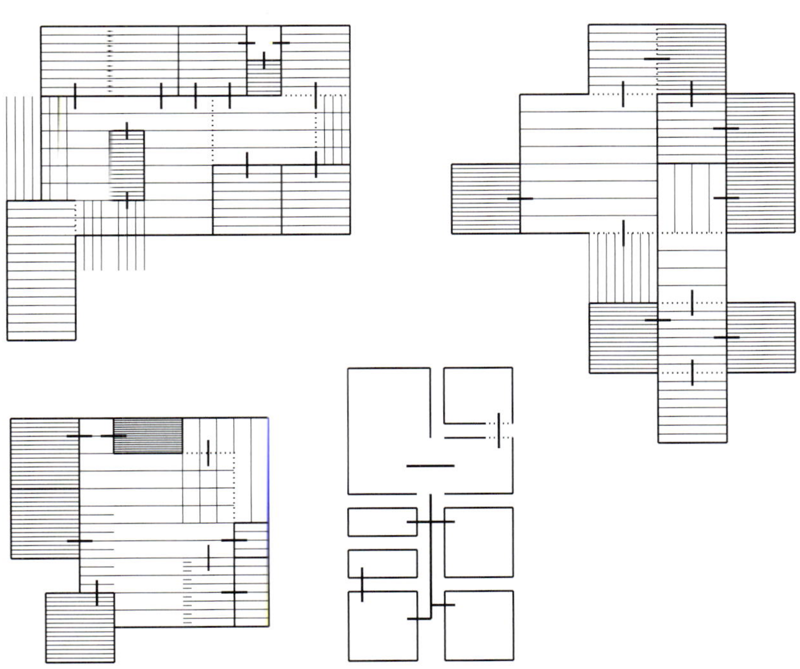

14 Sample diagrams for the assessment of spatial integrity

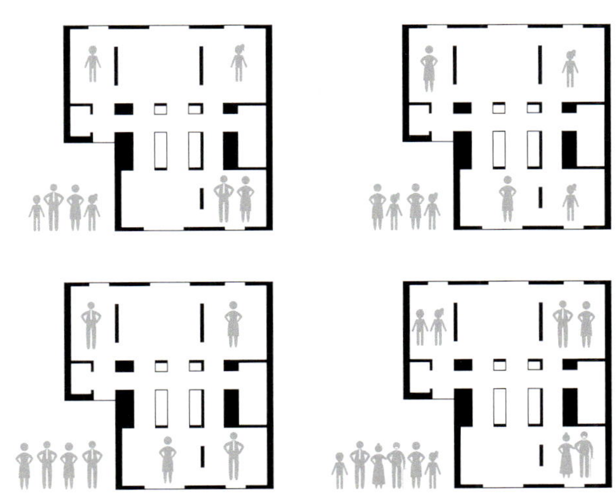

15 Potential occupancy configurations for a residential unit

PROJECTS

Floor Plan Diagrams of all Projects

Scale 1:500, with page references of the respective projects

186

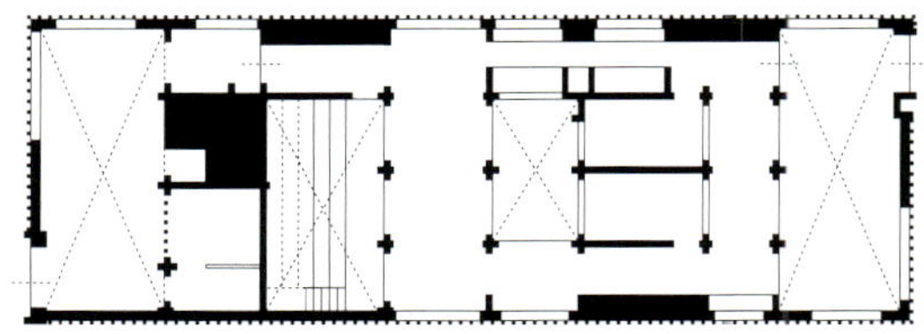

337

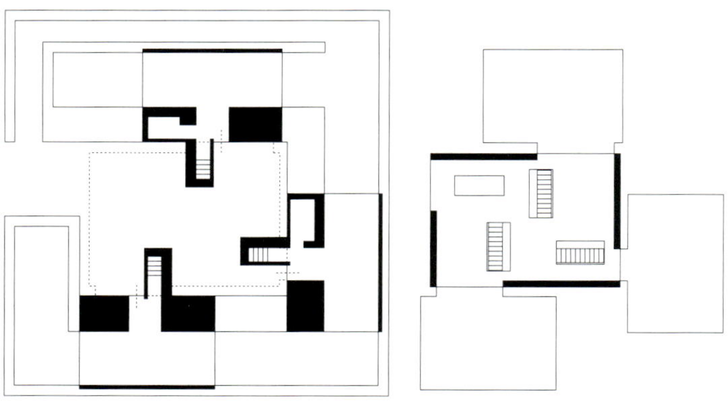

330

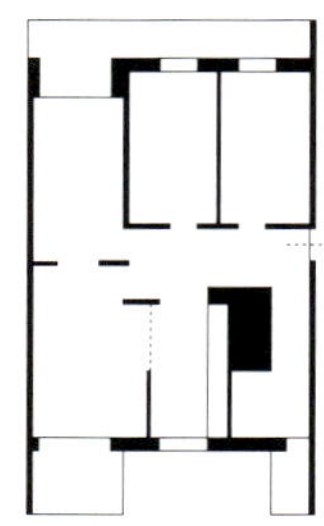

82

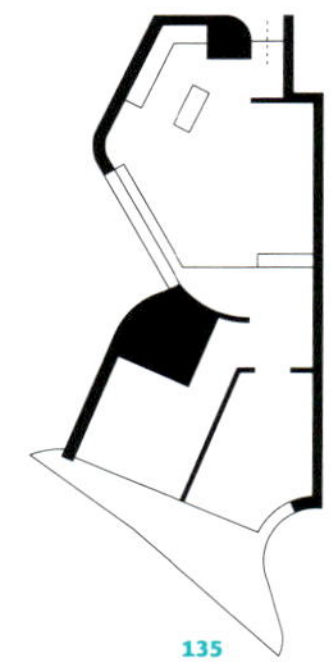

135

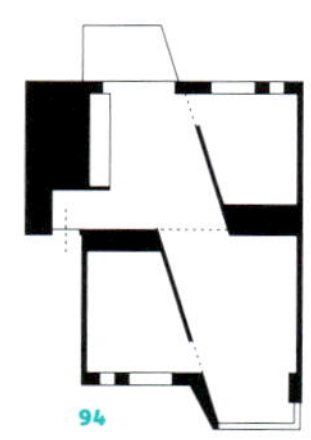

94

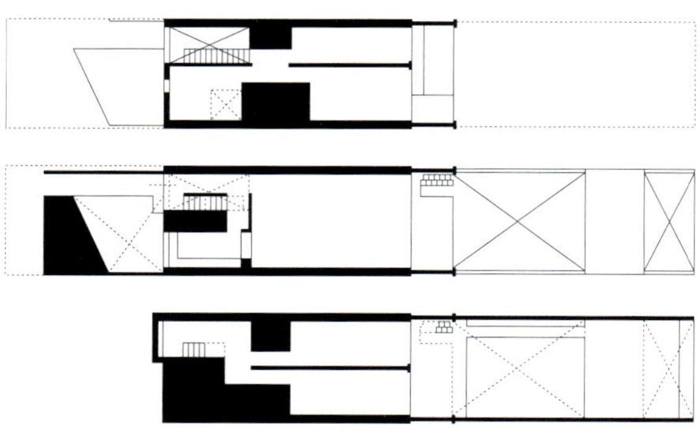

298

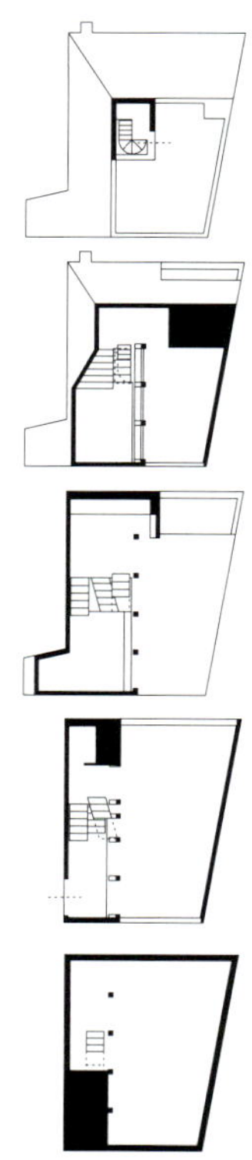

122

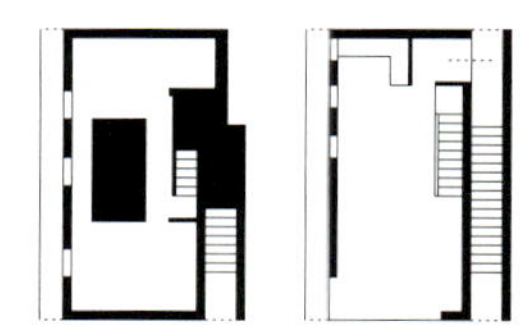

266

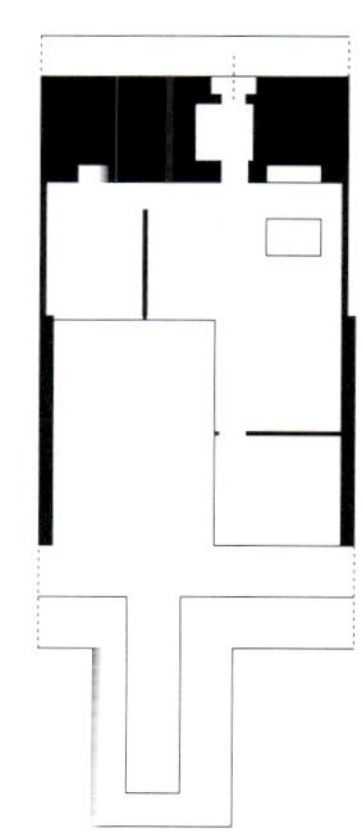

268

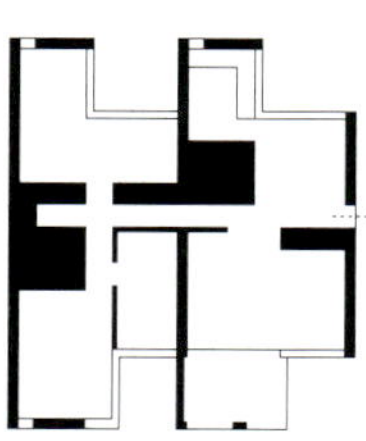

278

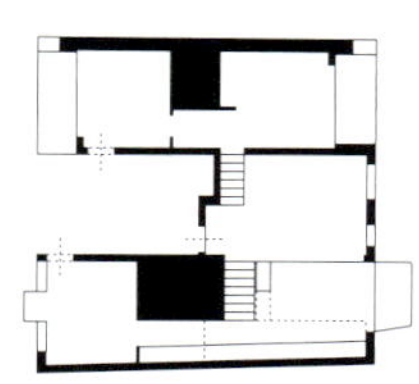

126

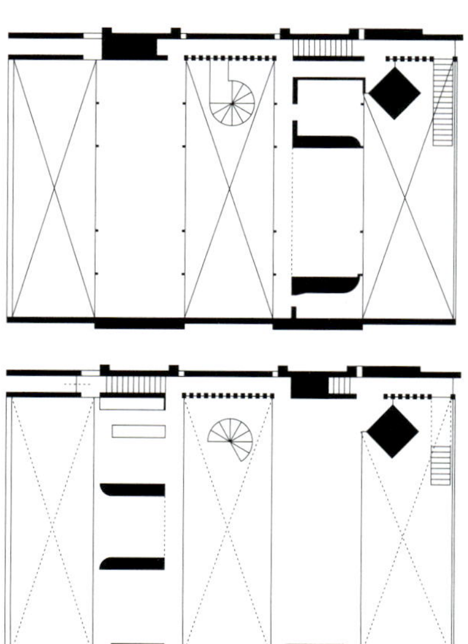

324

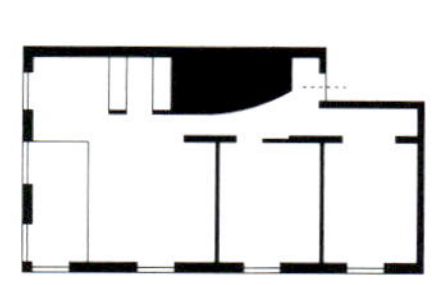

196

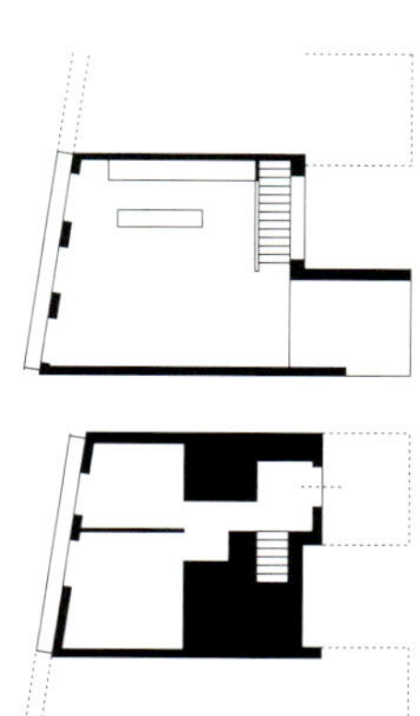

186

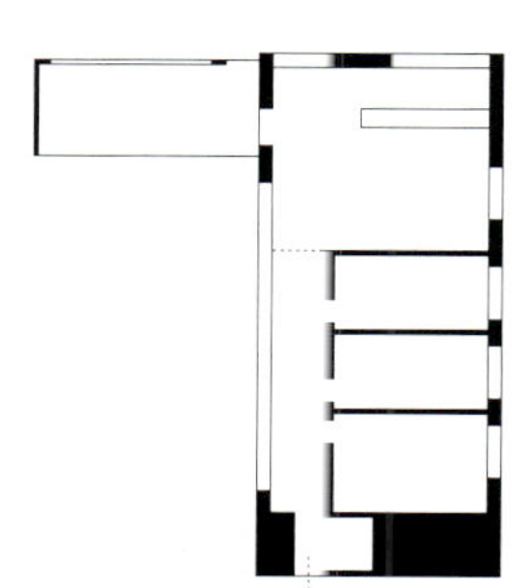

280

Floor Plan Diagrams of all Projects

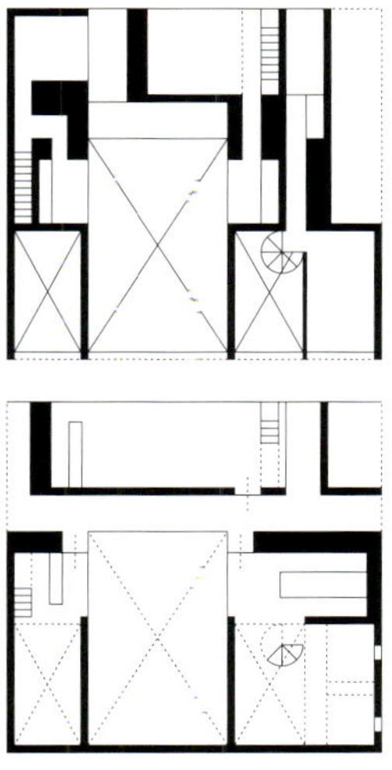

164

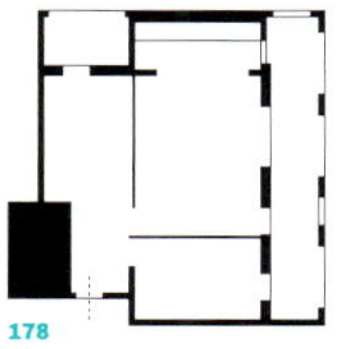

178

294

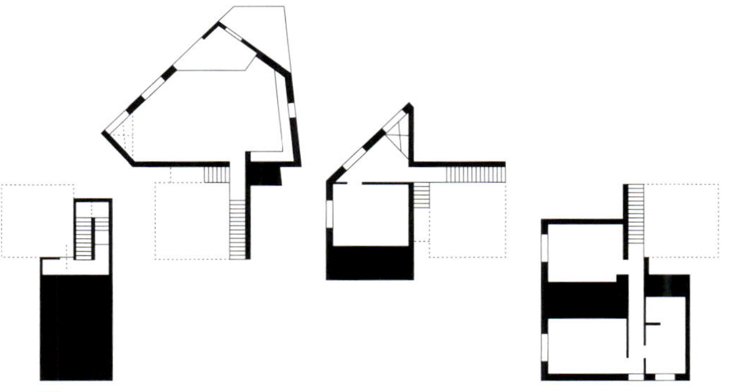

174

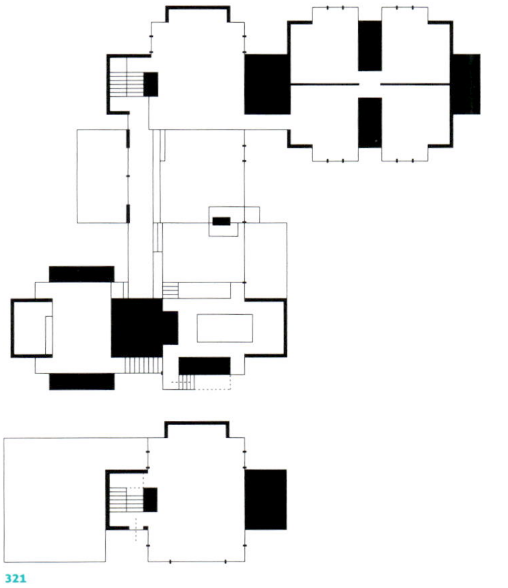

321

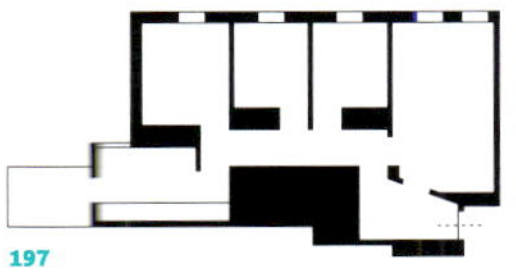

197

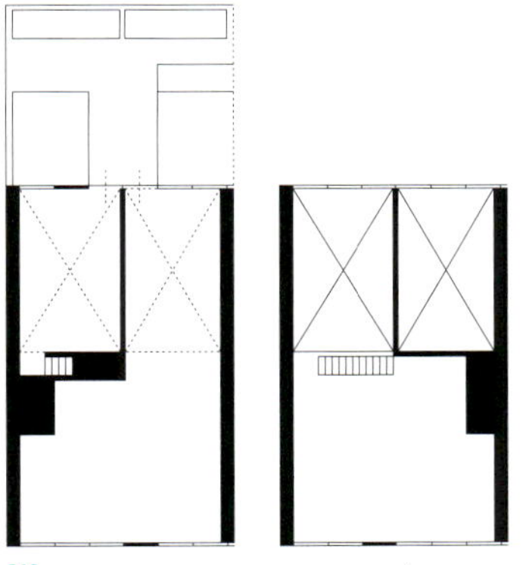

210

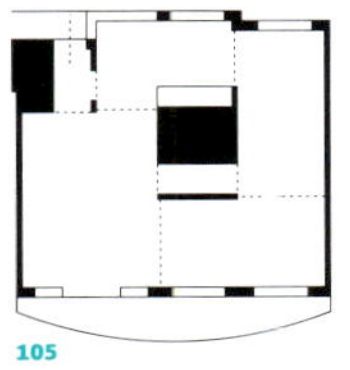

105

104

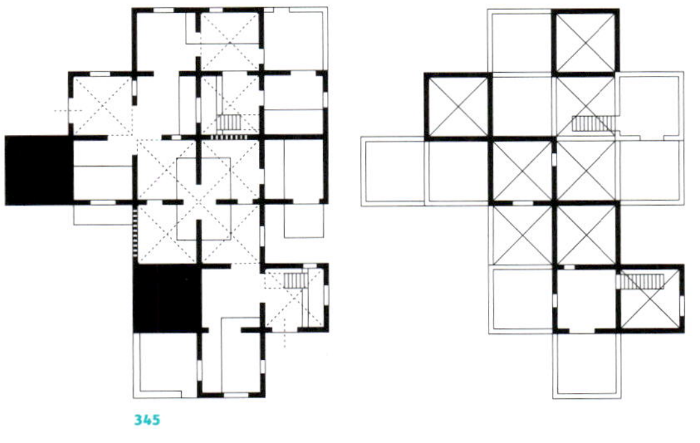

345

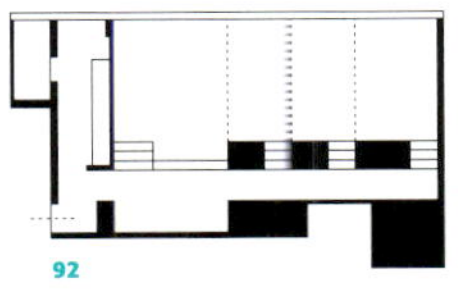

92

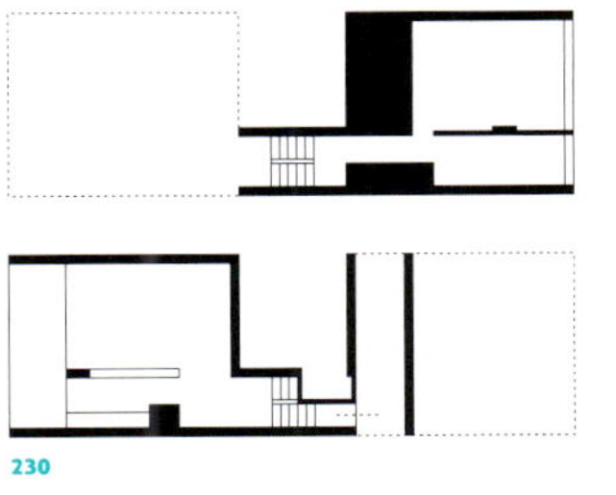

230

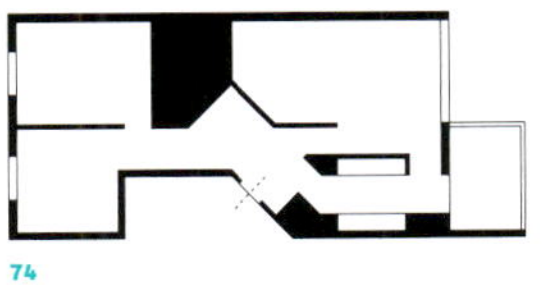

74

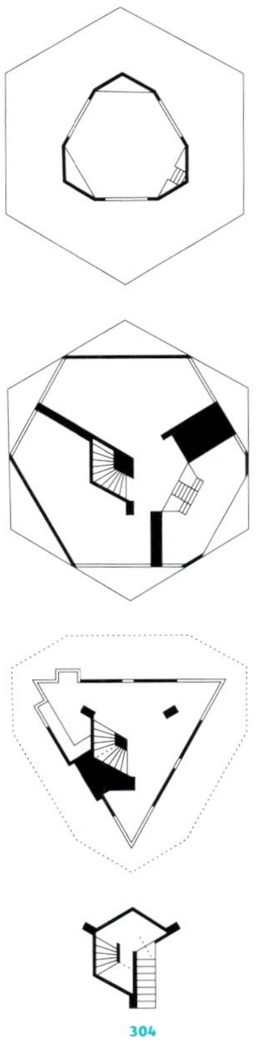

304

134

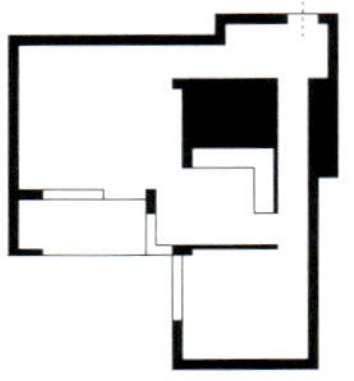
272

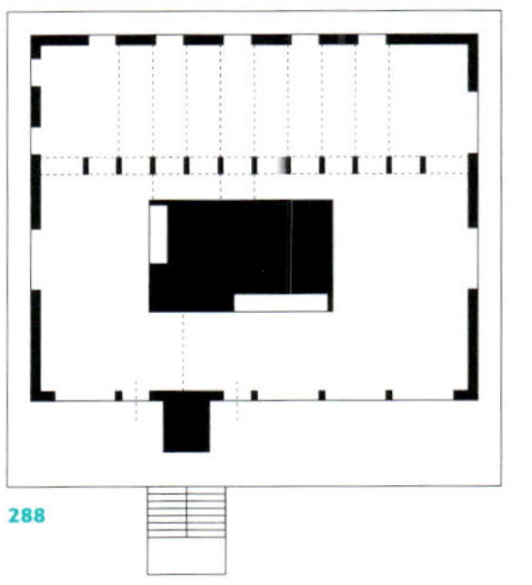
288

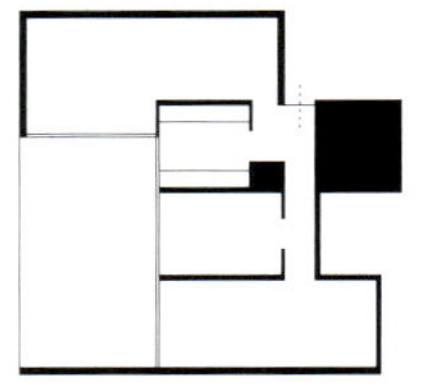
260

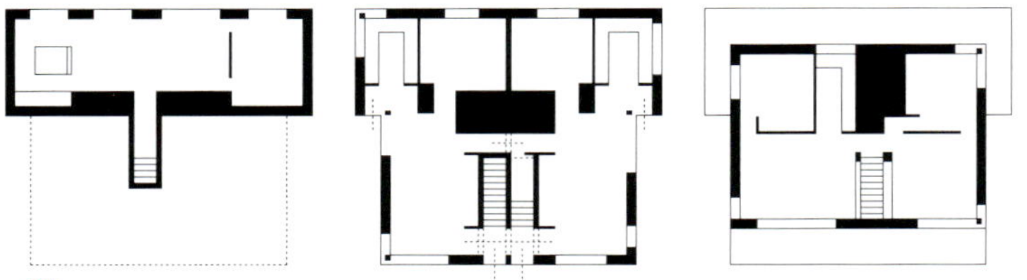
334

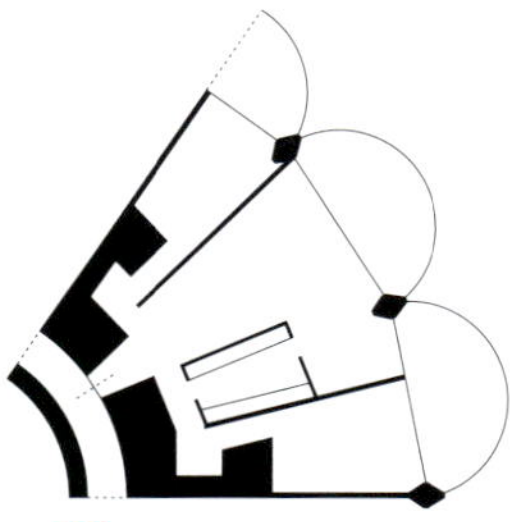
233

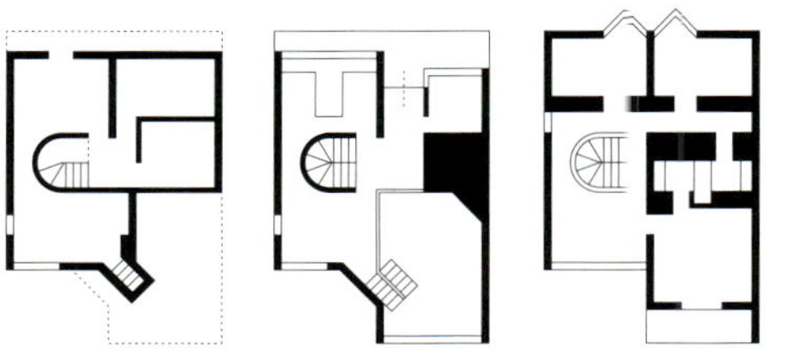
354

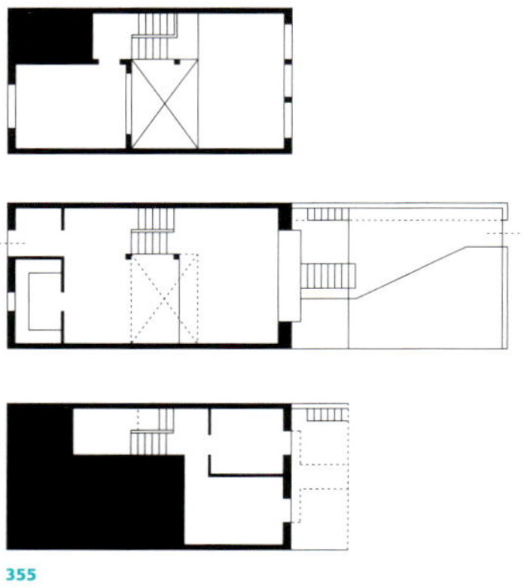
355

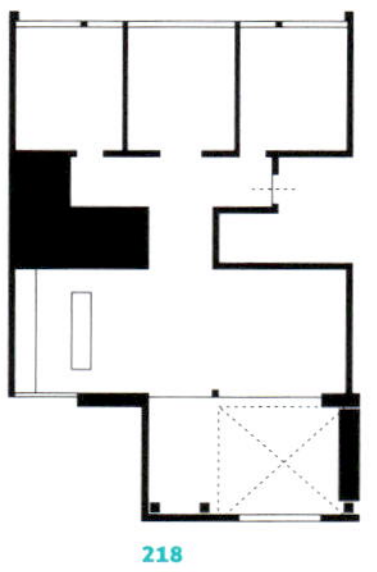
218

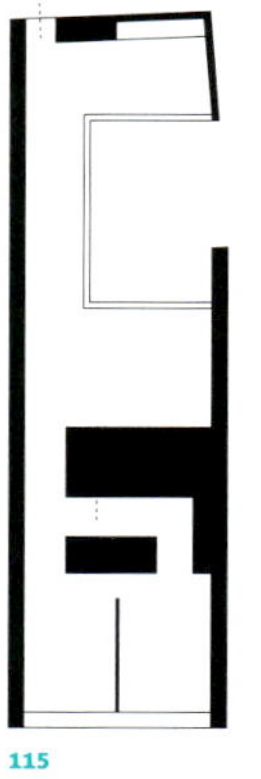
115

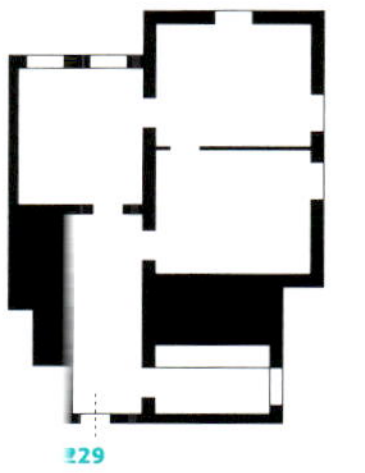
229

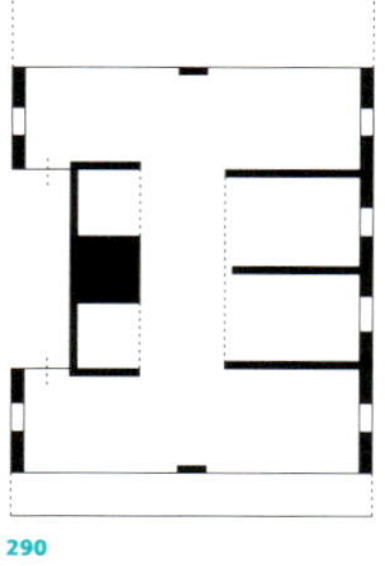
290

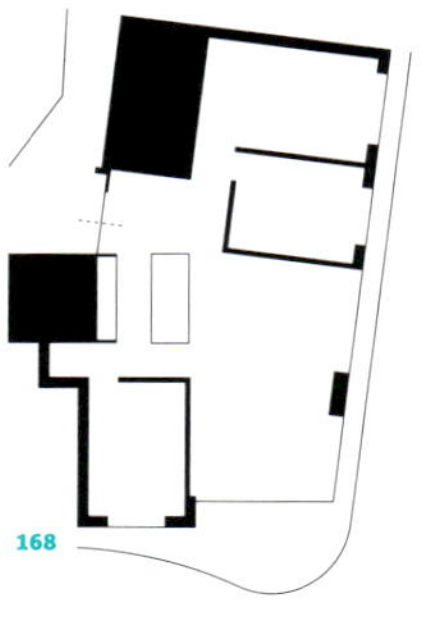
168

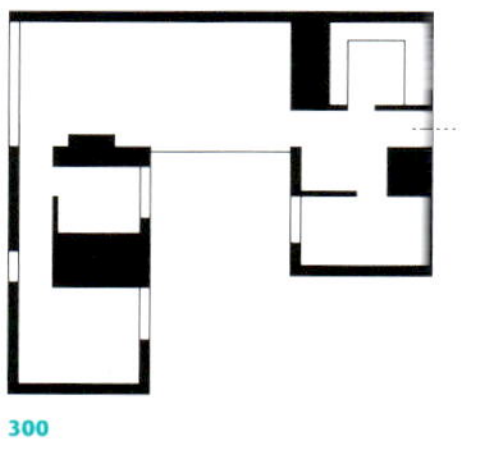
300

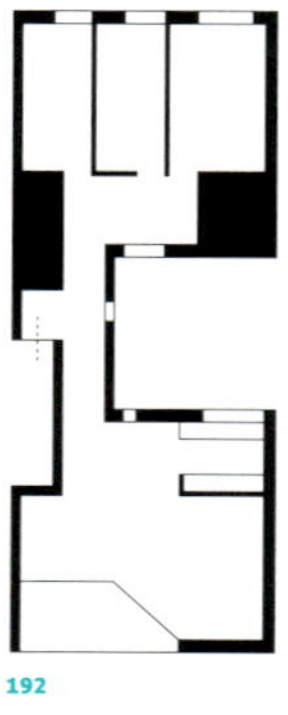
192

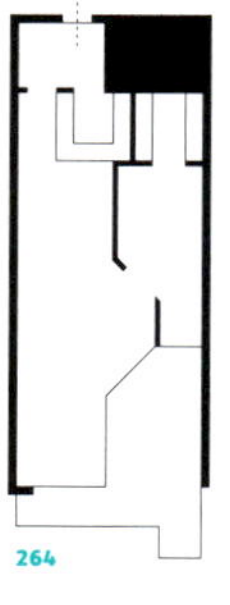
264

340

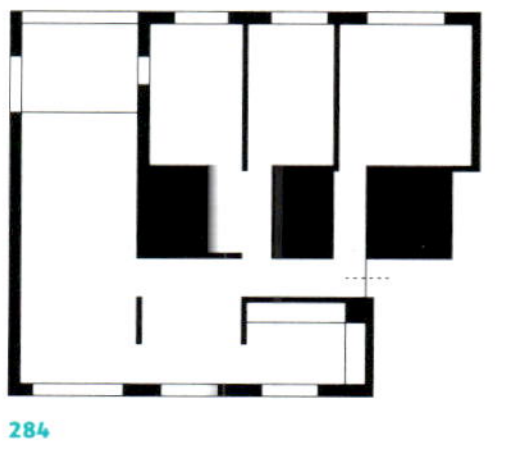
284

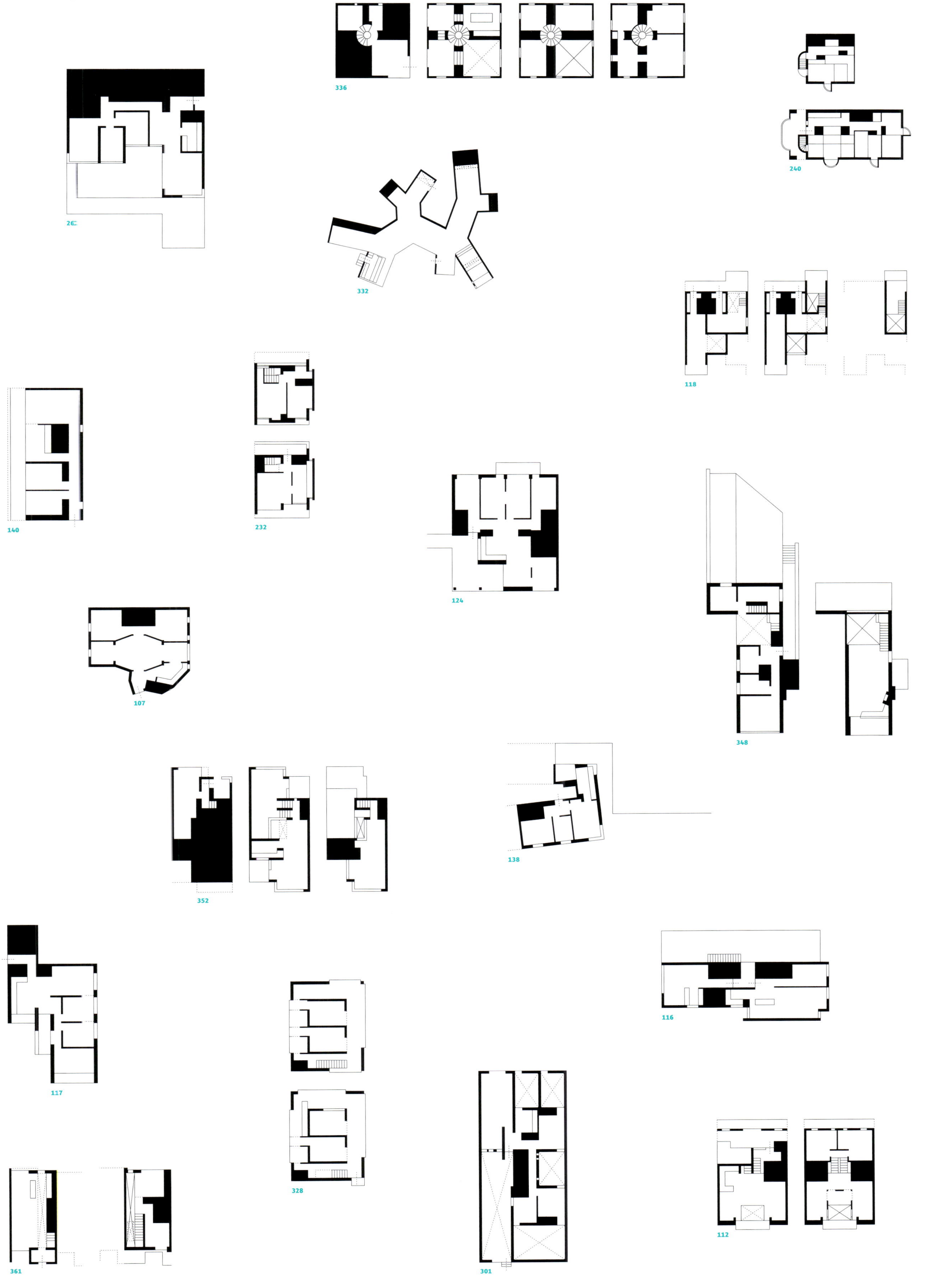
336
26
240
332
118
140
232
124
348
107
352
138
116
117
328
112
361
301

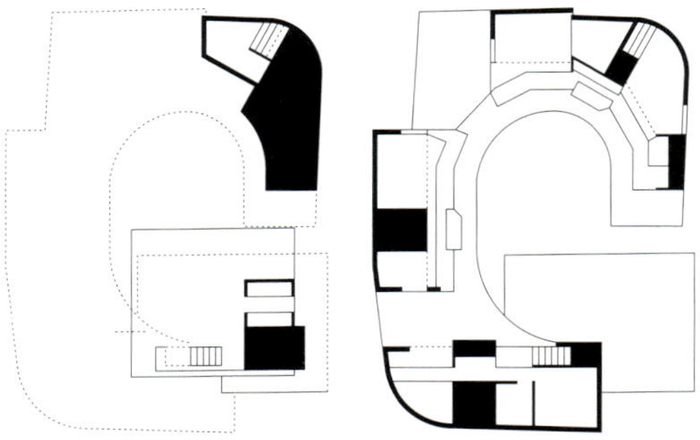
325

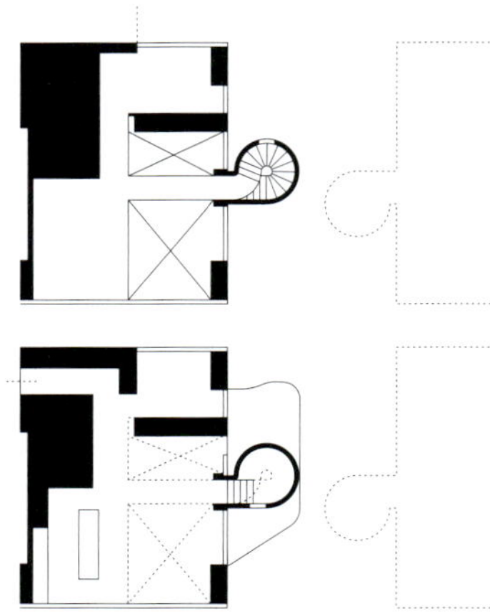
250

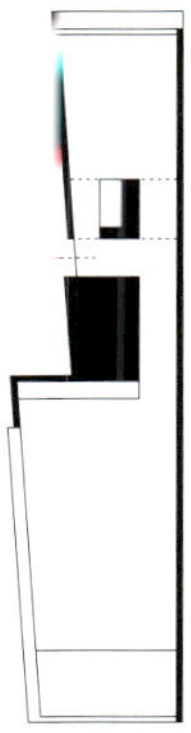
120

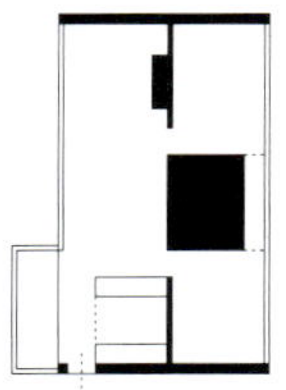
323

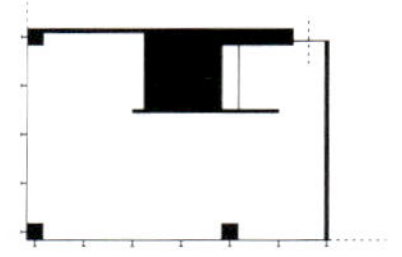
228

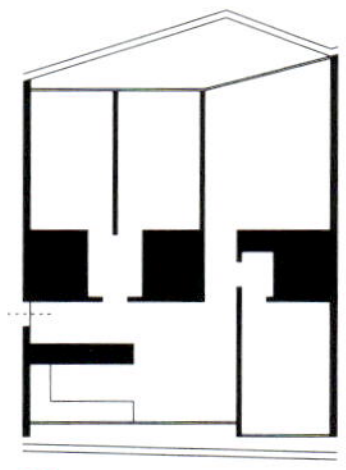
214

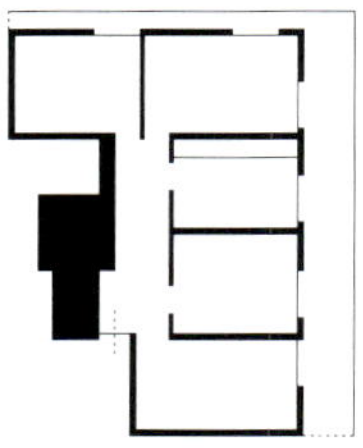
84

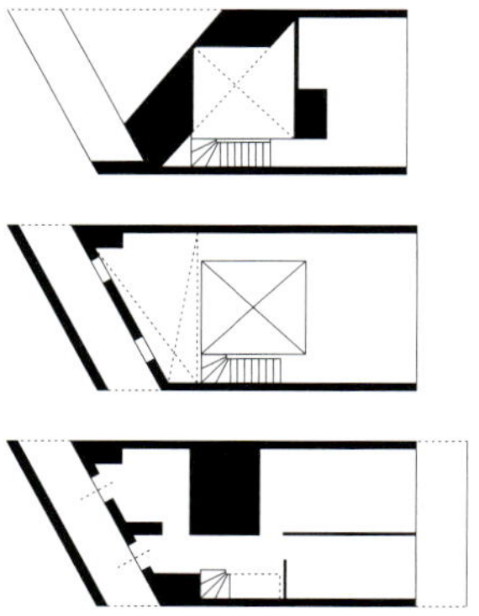
172

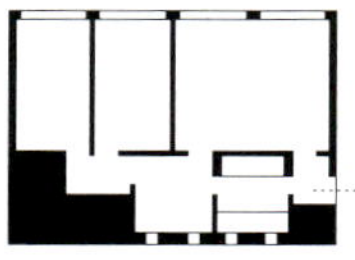
212

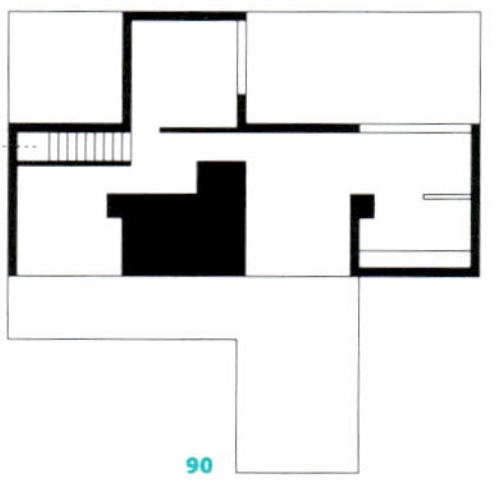
90

108

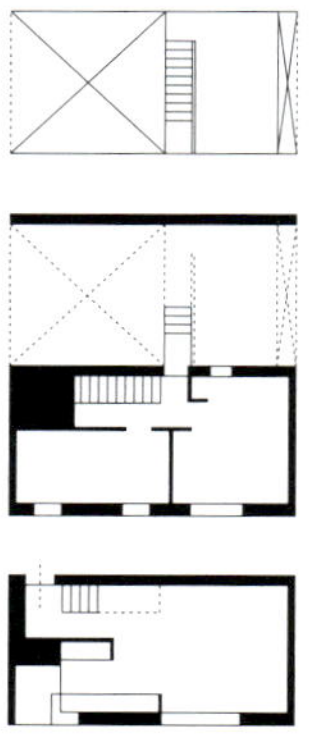
248

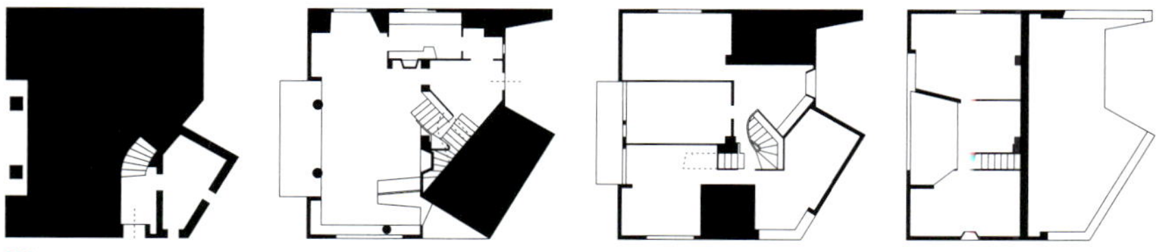
320

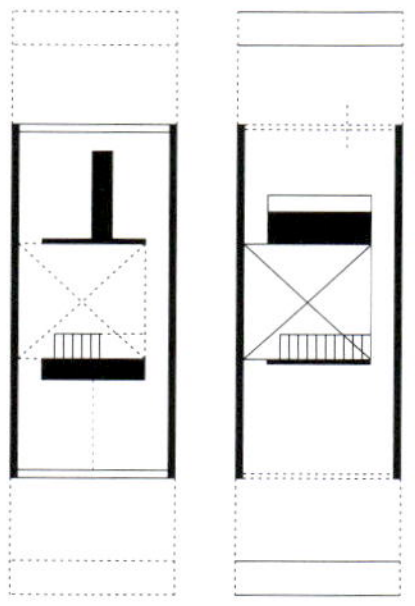
194

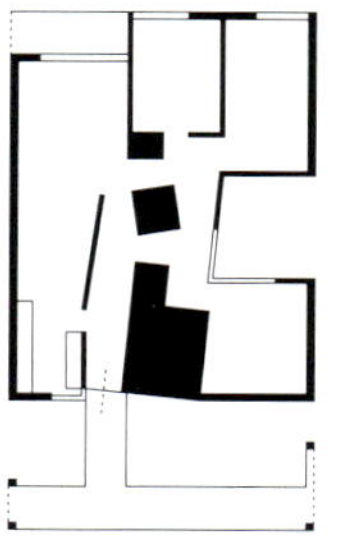
80

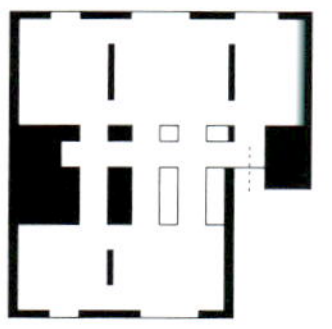
202

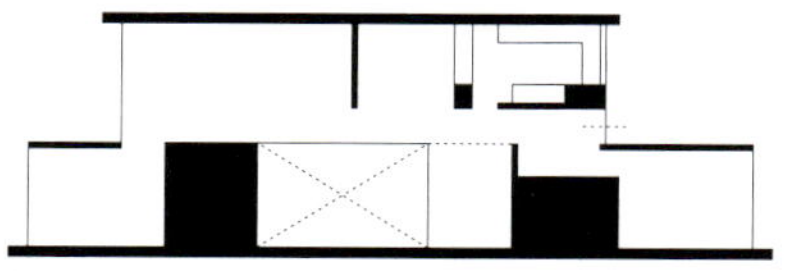
350

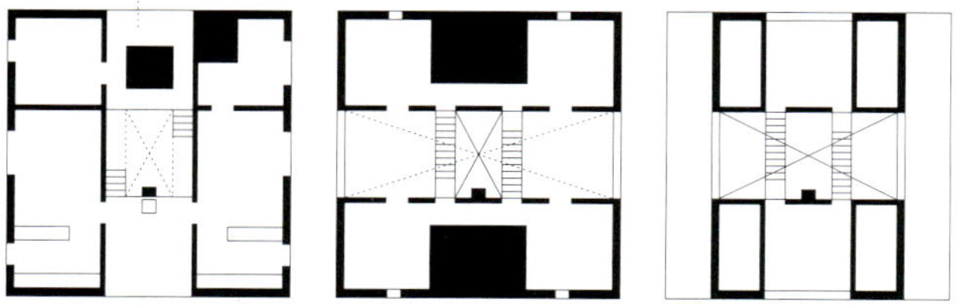
344

Floor Plan Diagrams of all Projects

238

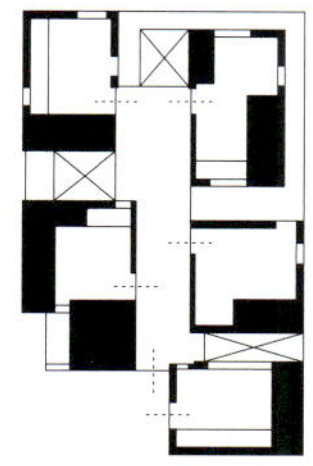
129

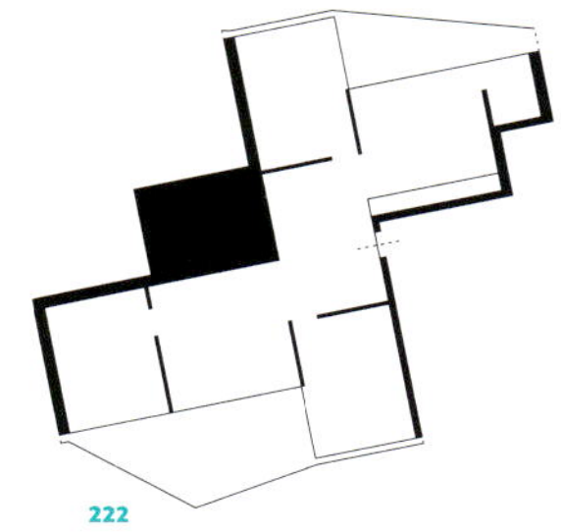
222

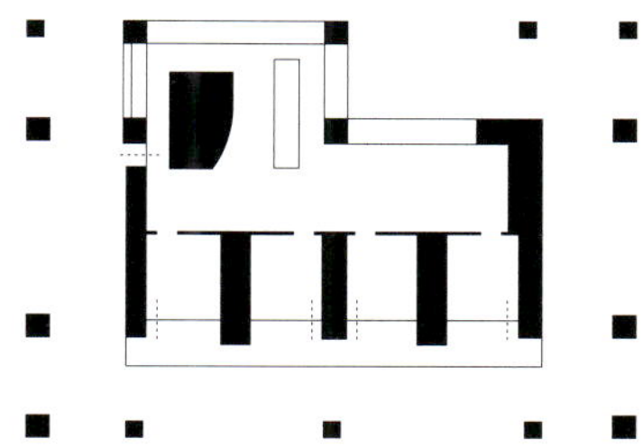
156

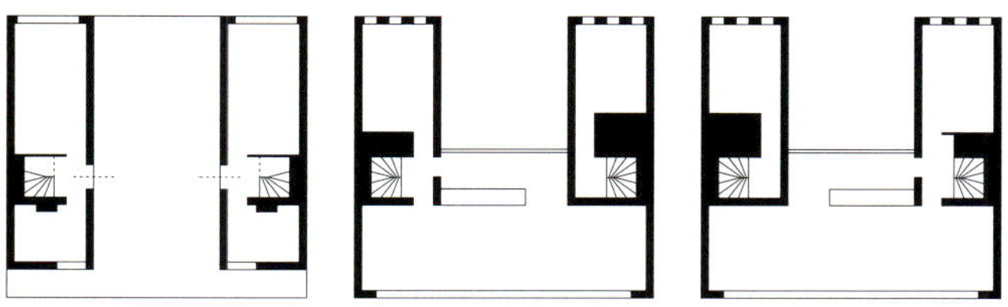
356

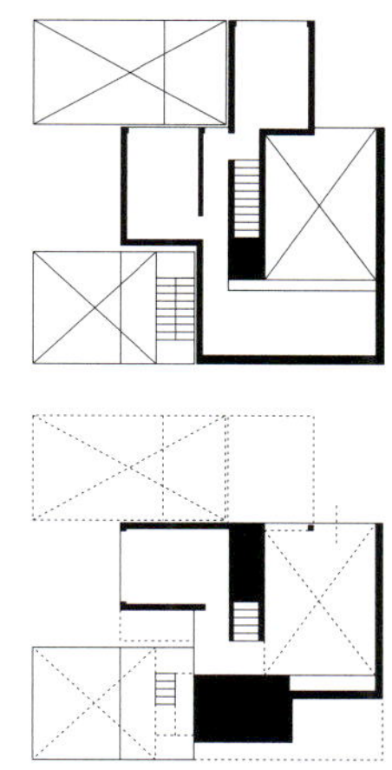
310

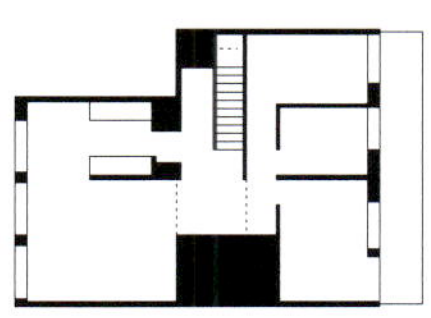
78

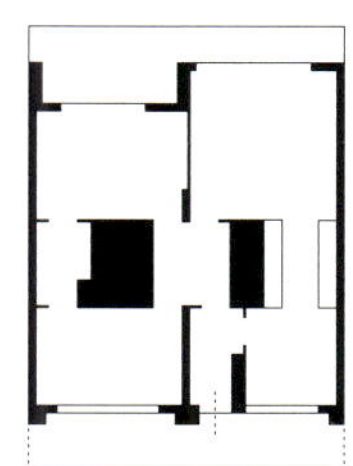
96

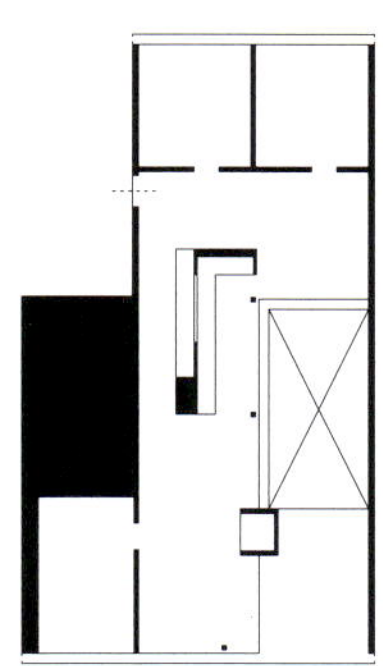
220

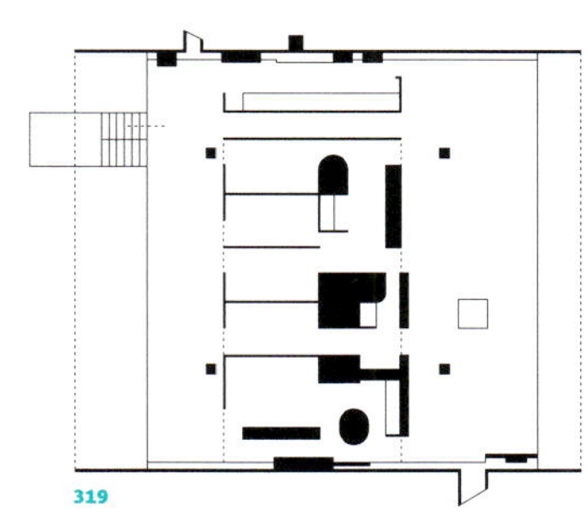
319

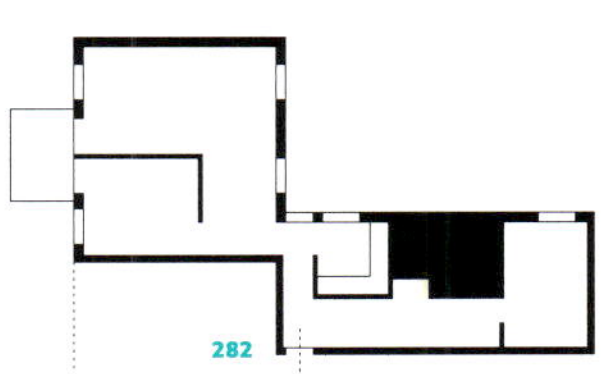
282

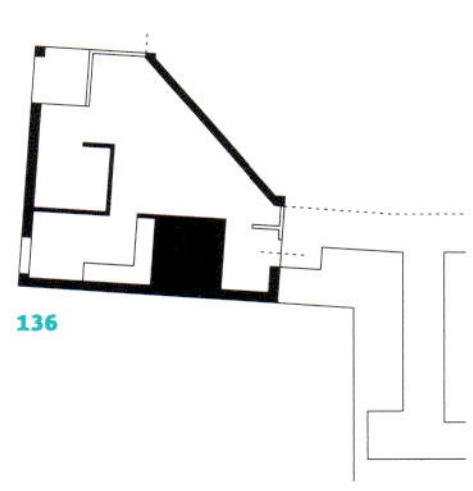
136

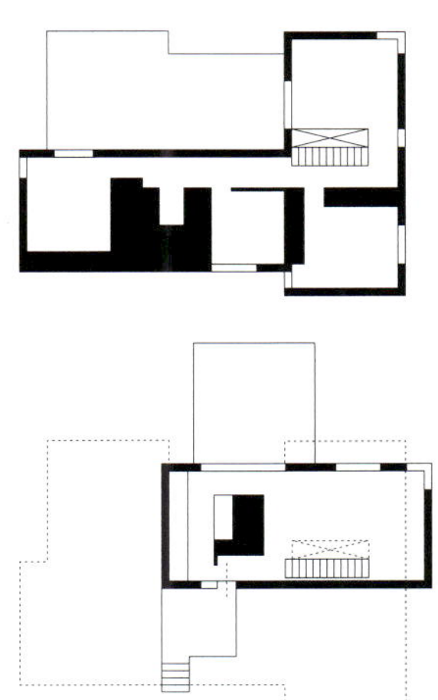
258

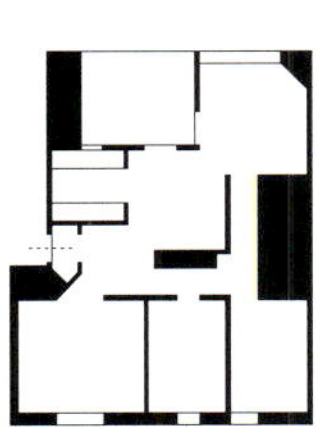
196

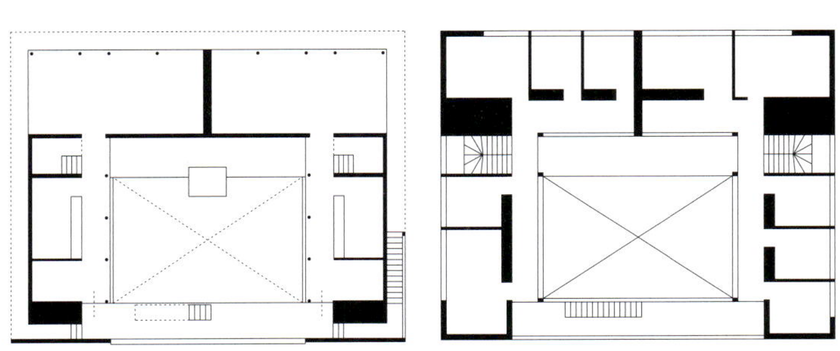
341

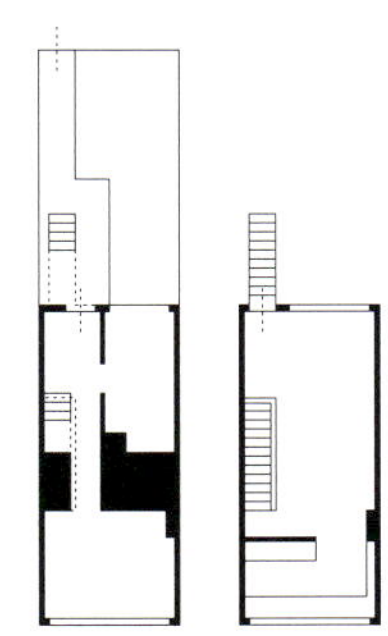
110

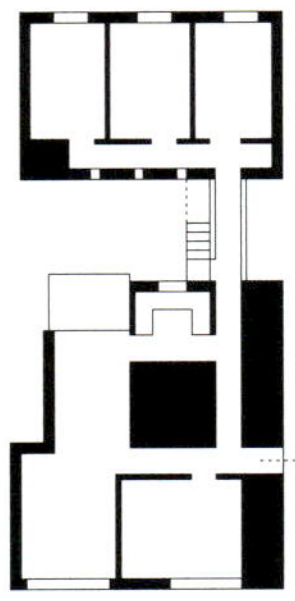
206

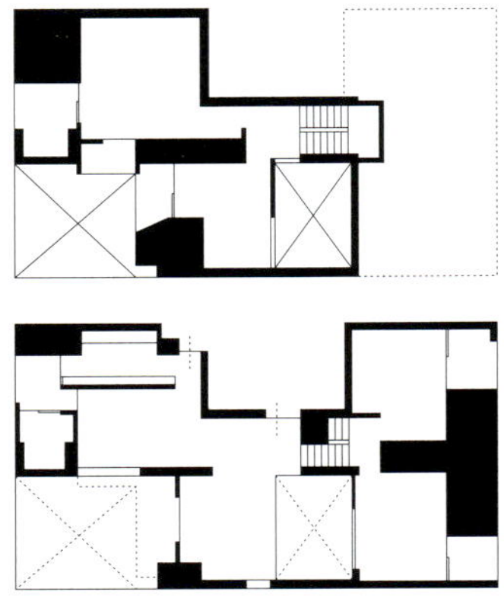
242

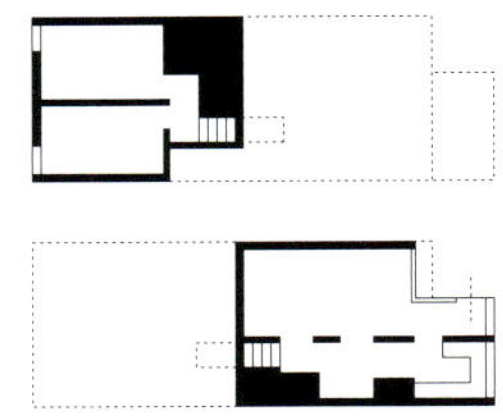
274

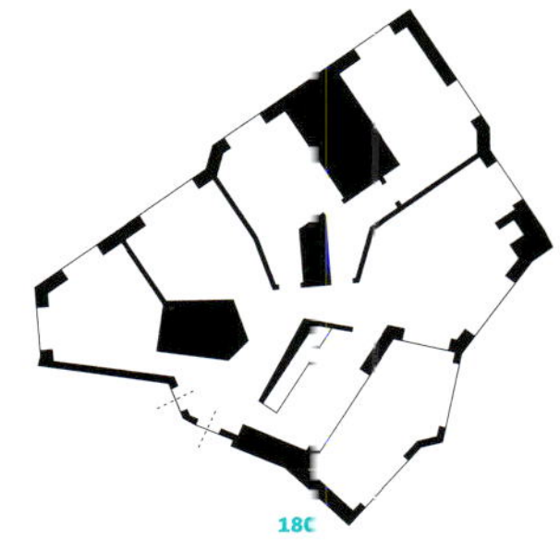
180

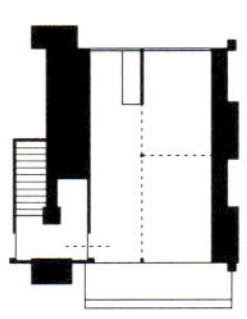
187

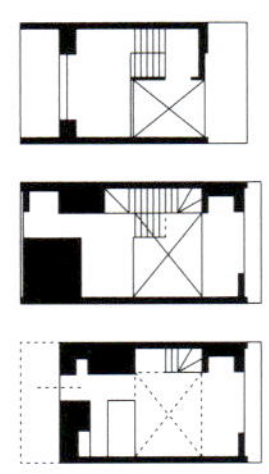
130

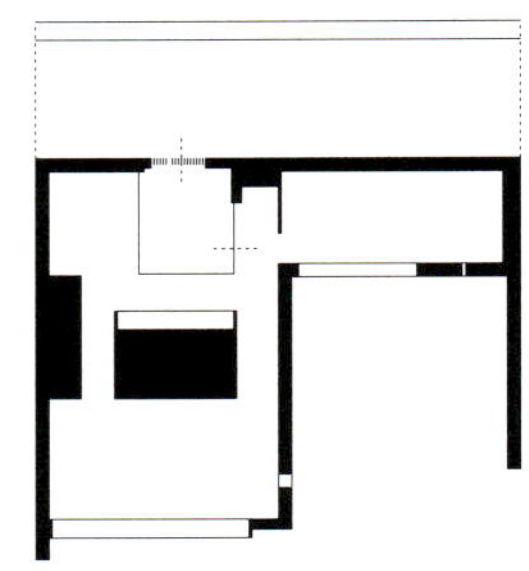
142

198

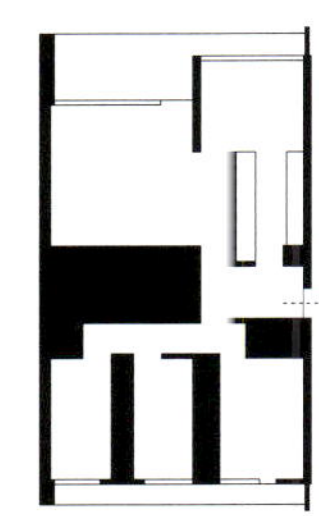
198

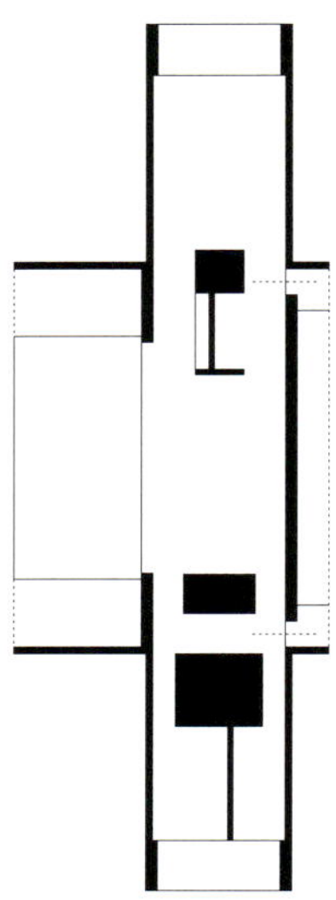
162

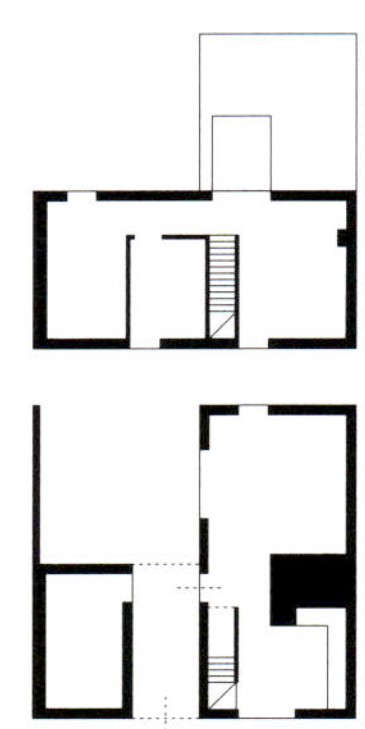
314

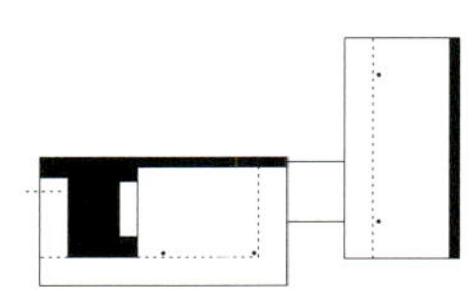
292

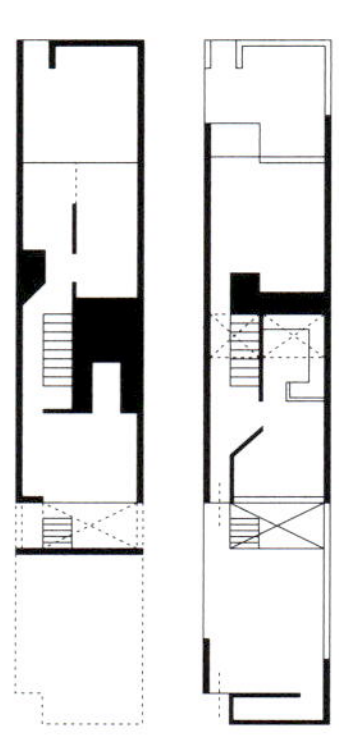
276

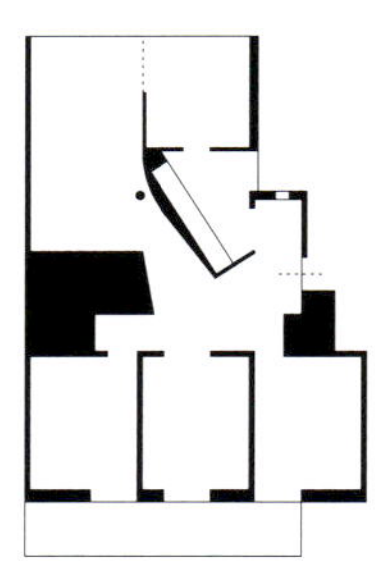
76

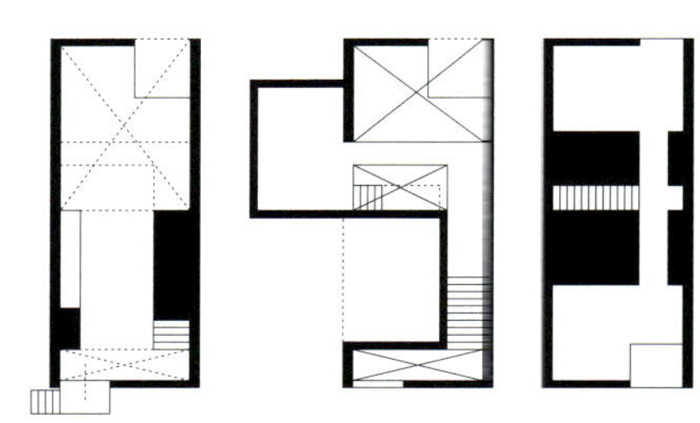
170

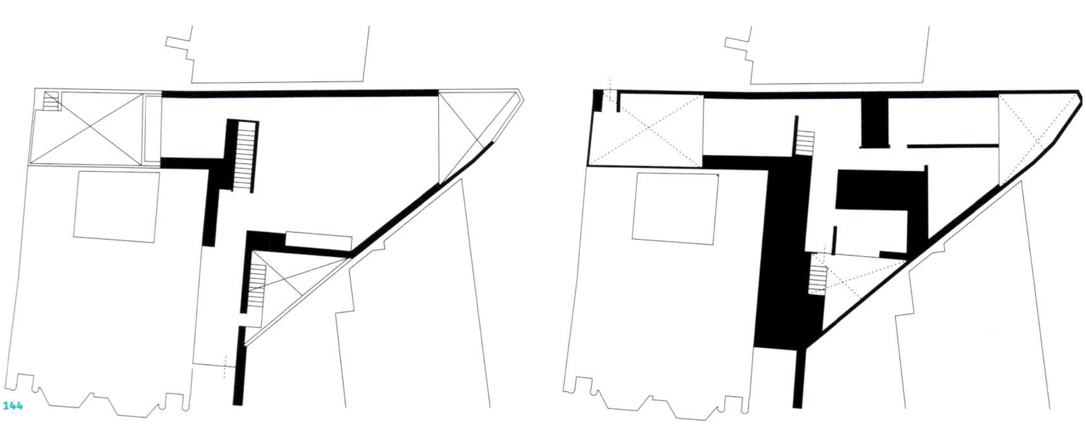
144

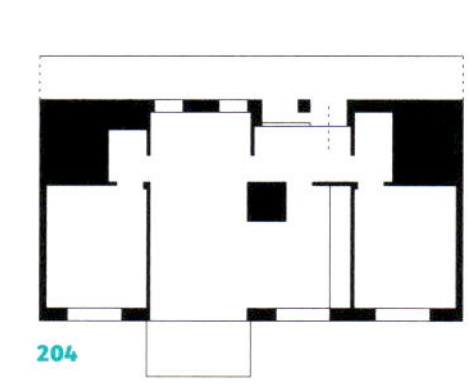
204

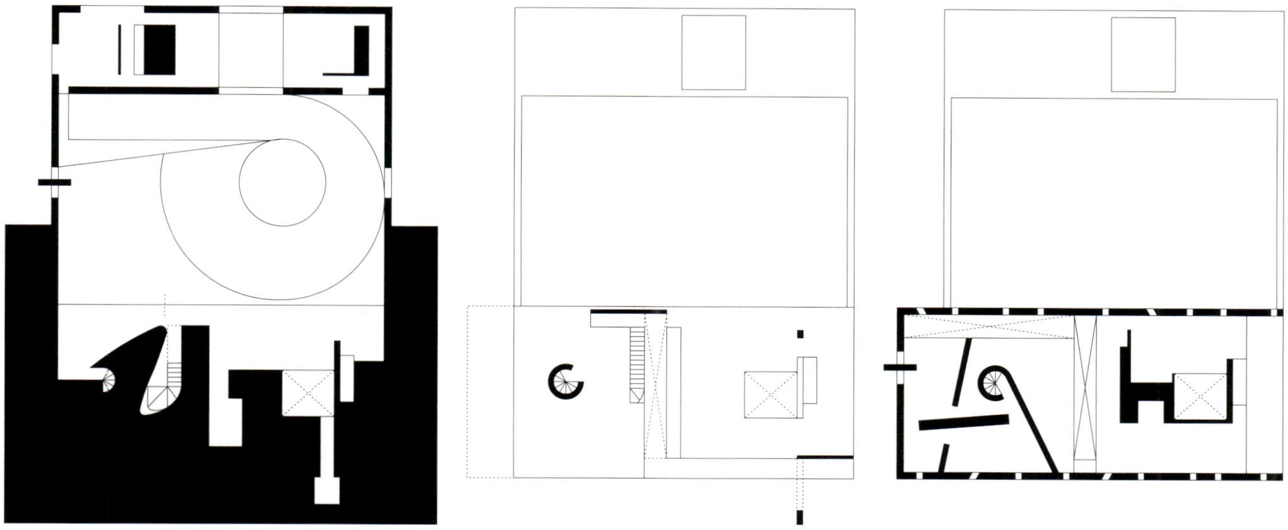

326

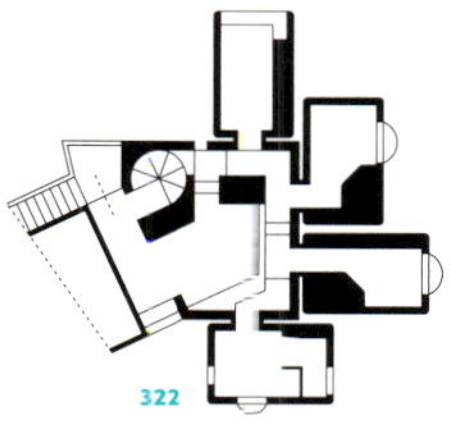

322

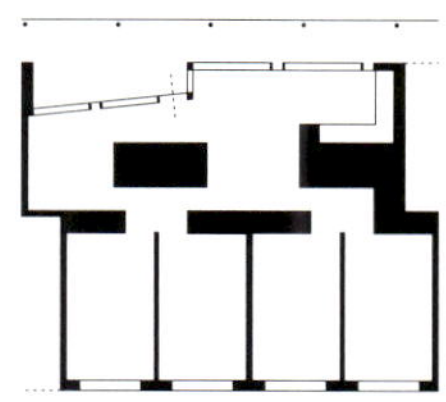

216

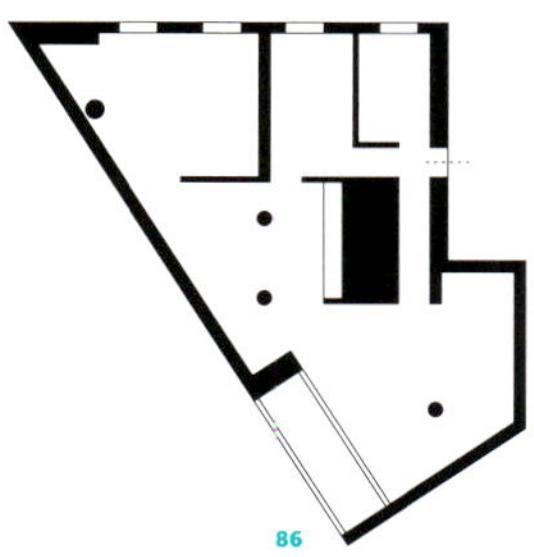

86

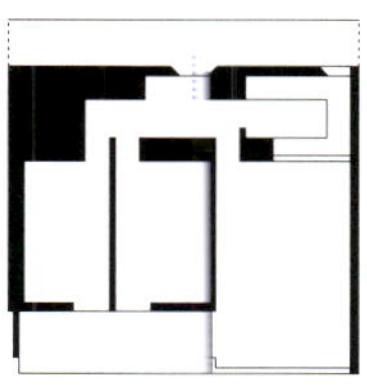

261

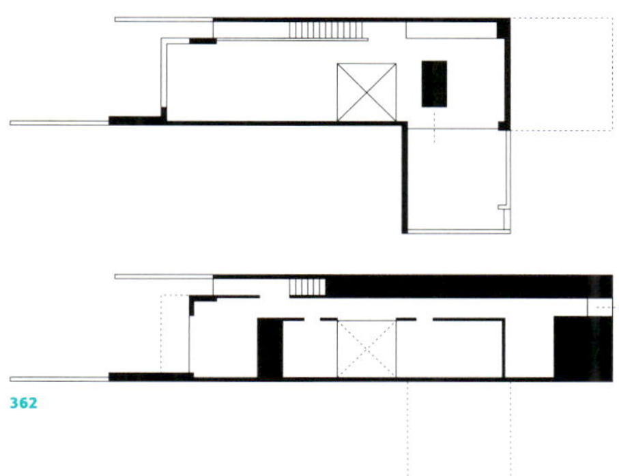

362

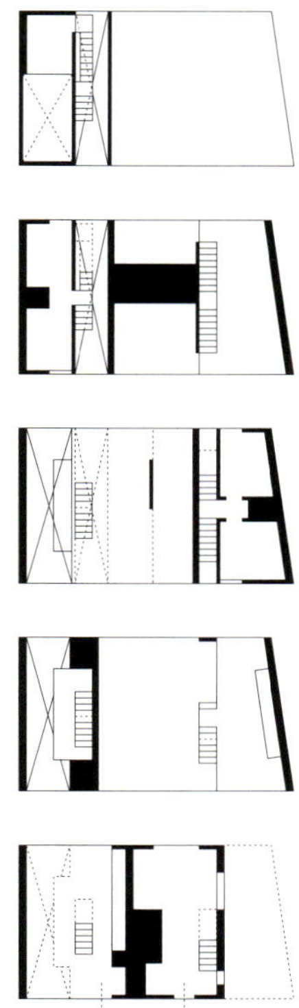

342

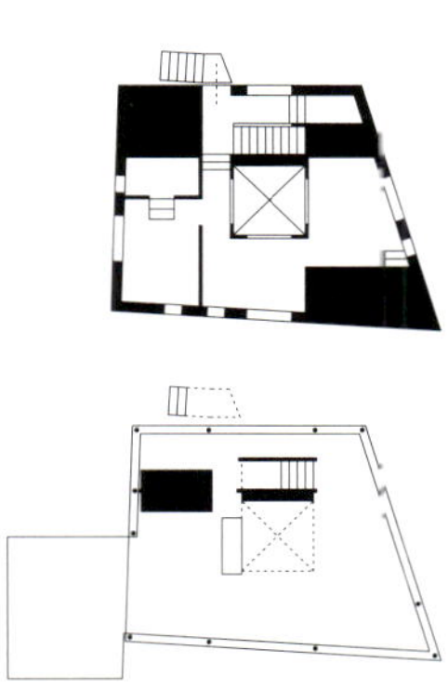

333

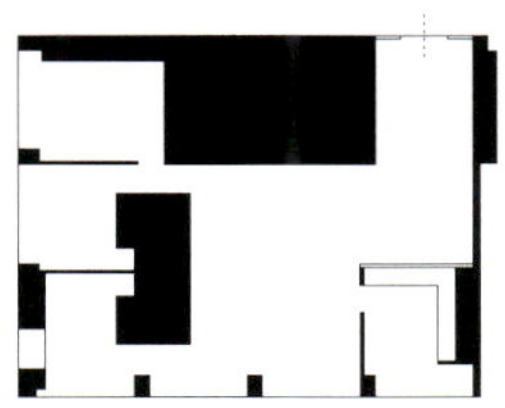

246

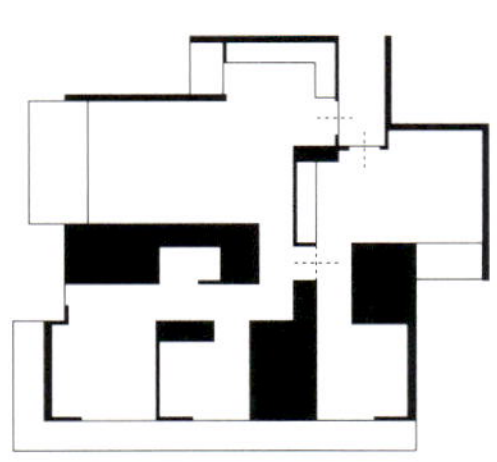

254

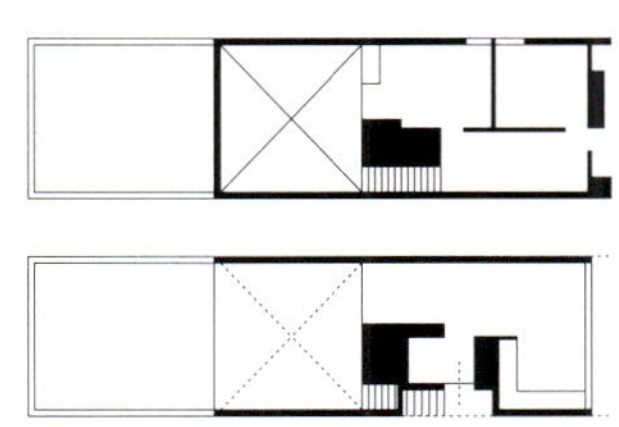

302

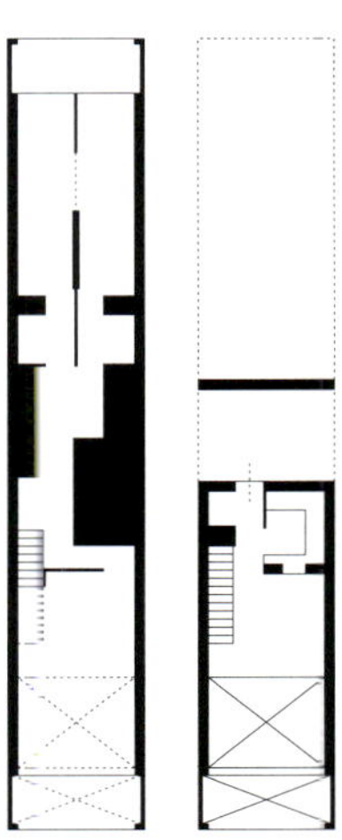

184

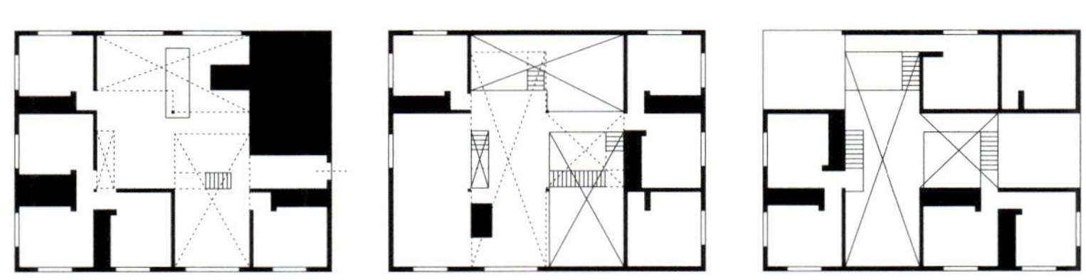

128

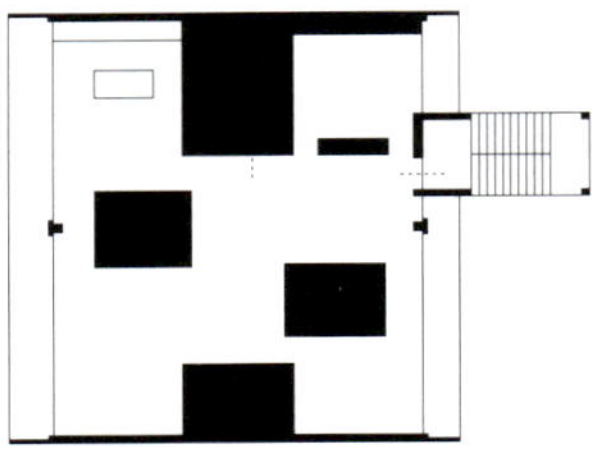
100

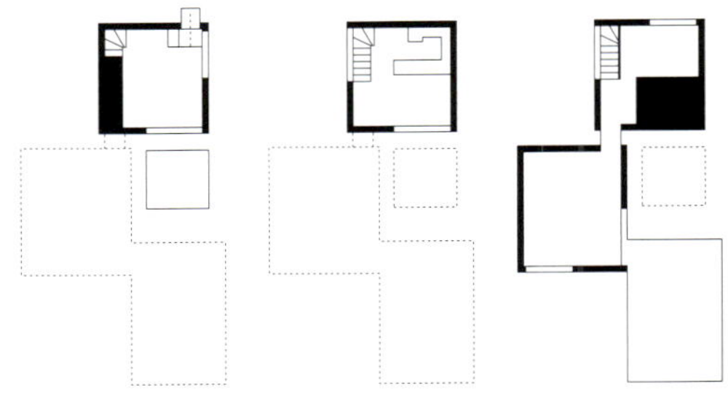
312

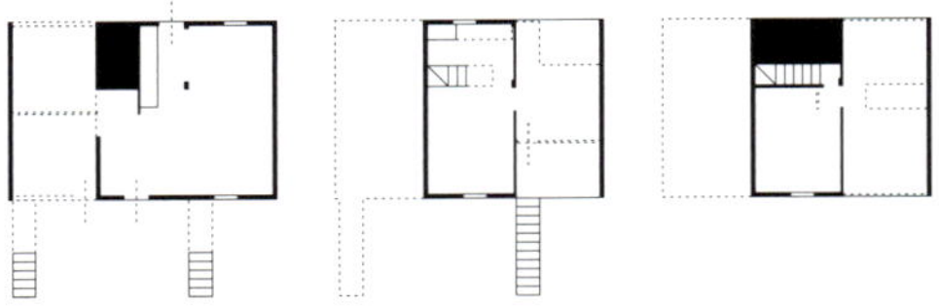
360

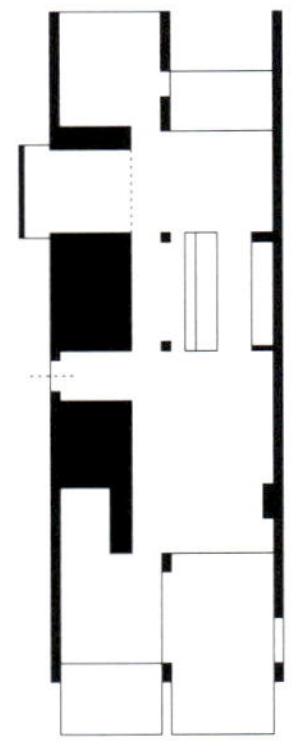
286

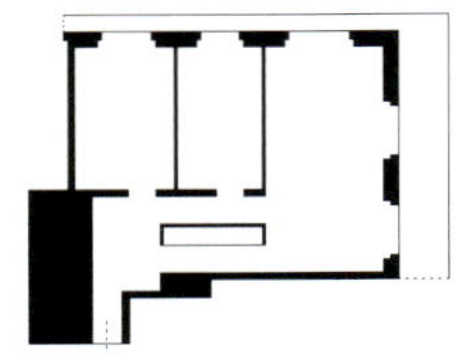
159

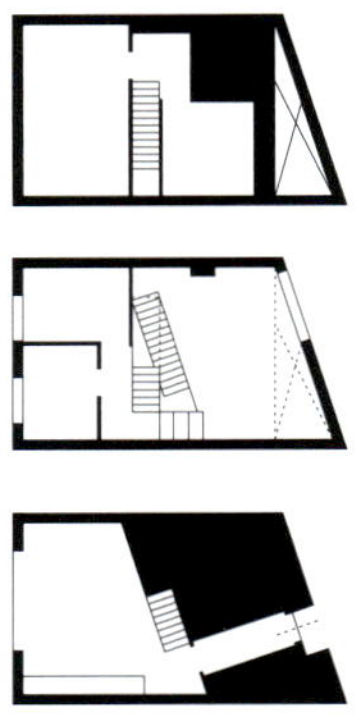
364

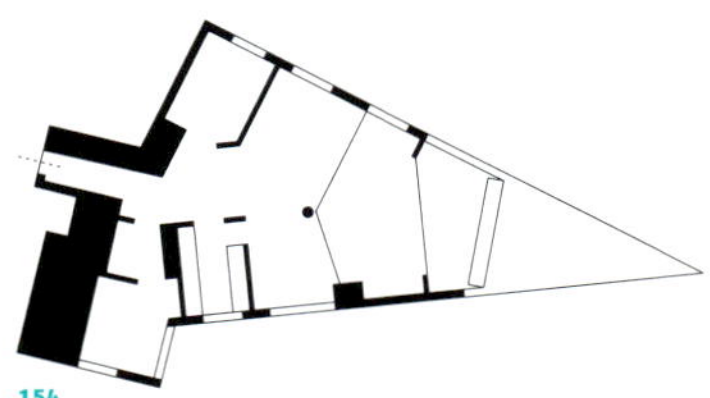
154

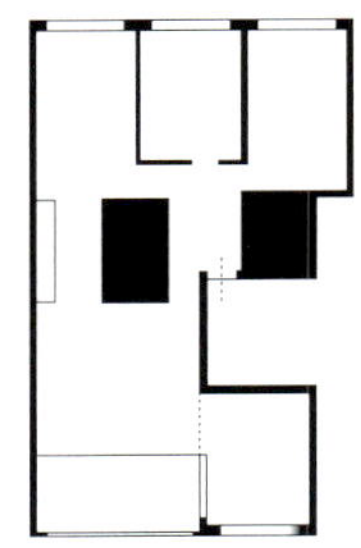
224

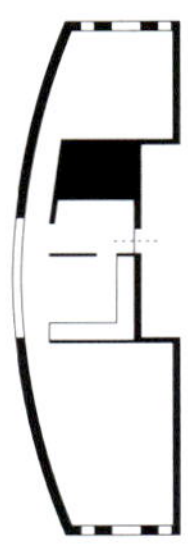
114

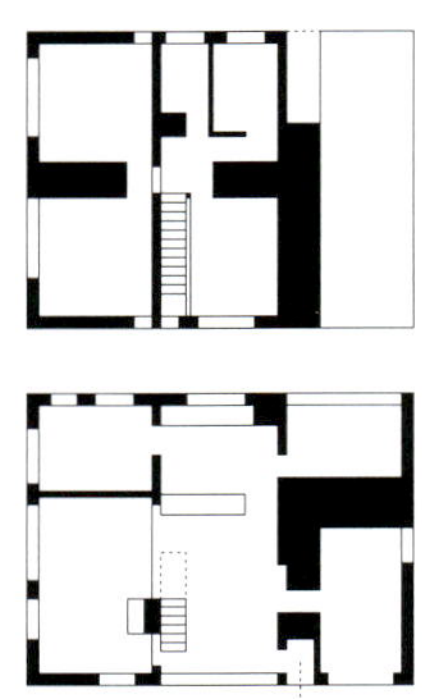
318

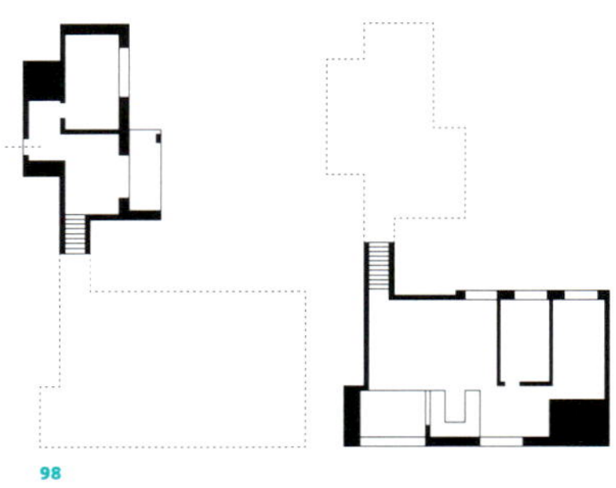
98

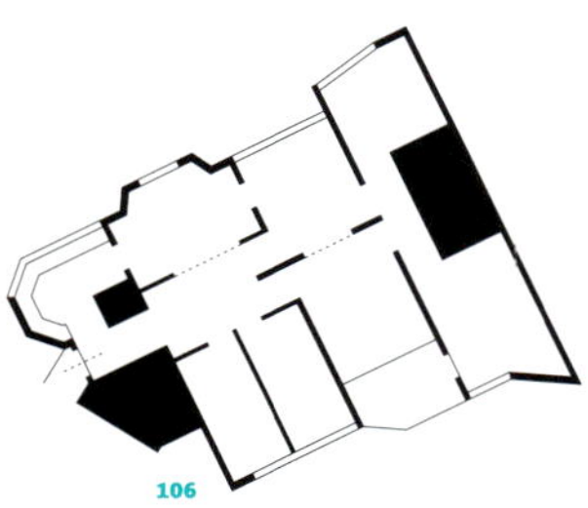
106

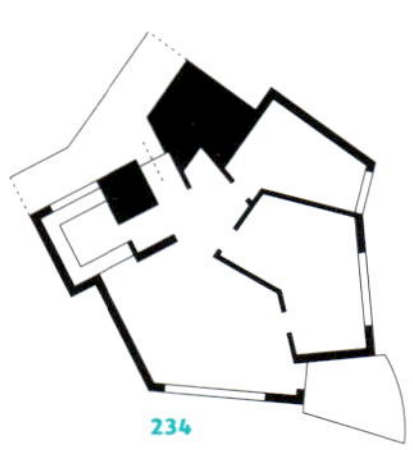
234

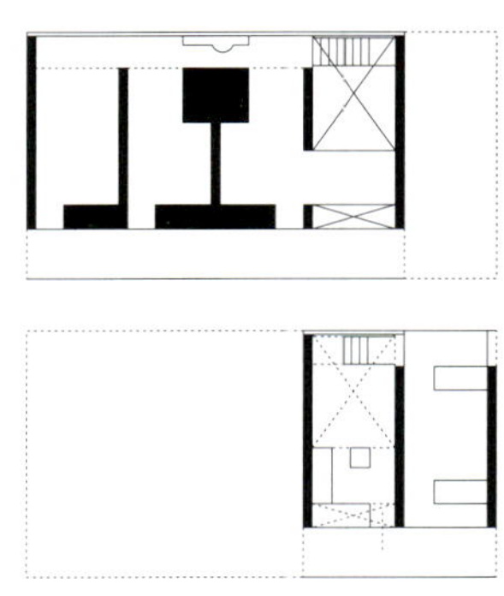
208

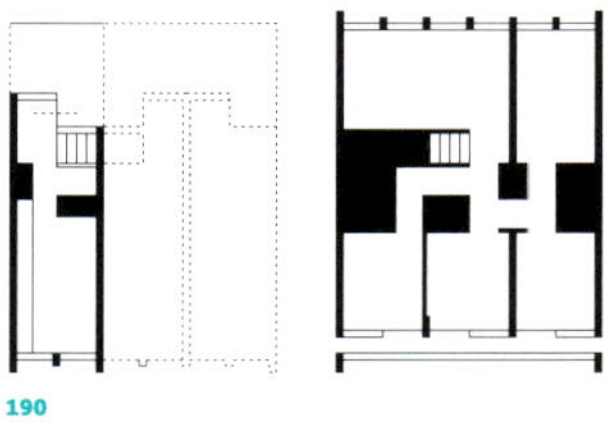
190

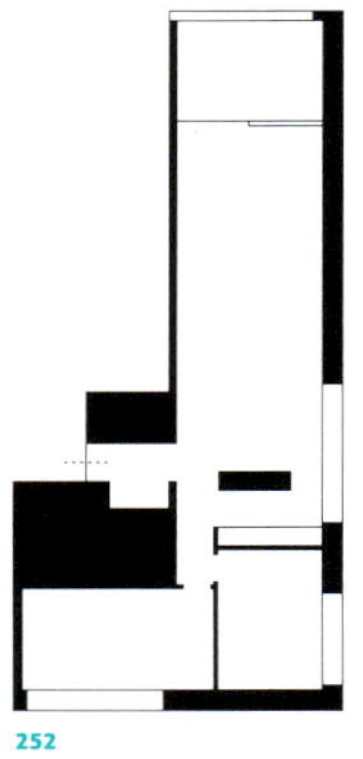
252

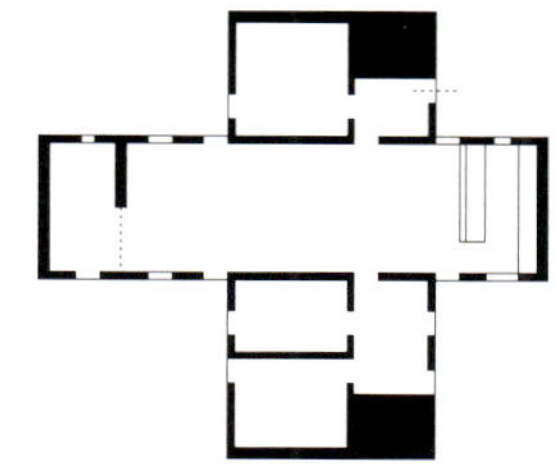

306

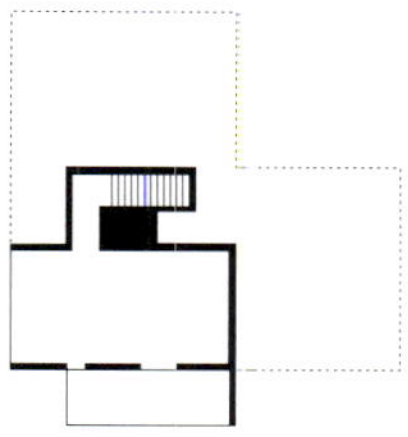

176

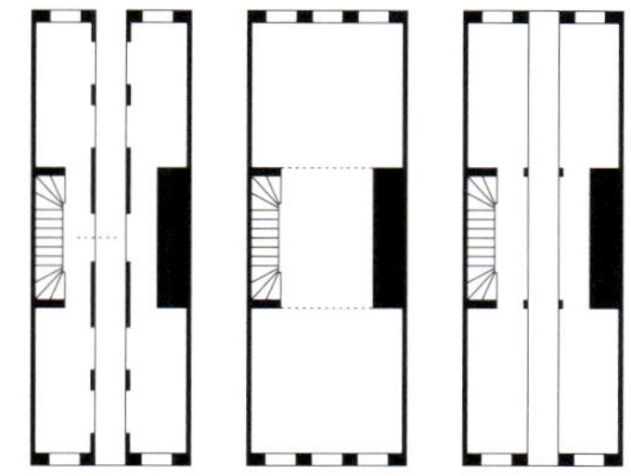

121

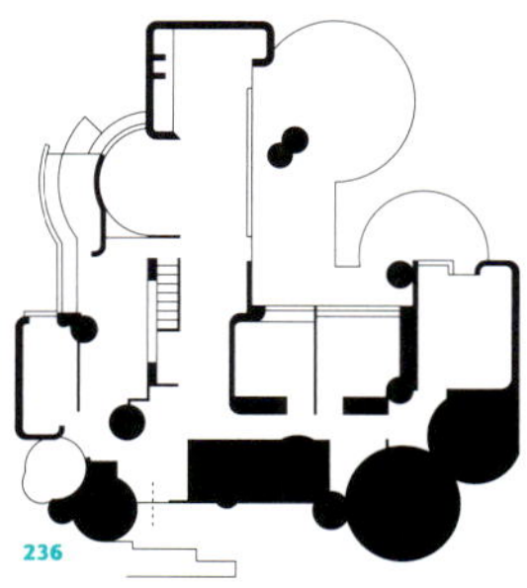

236

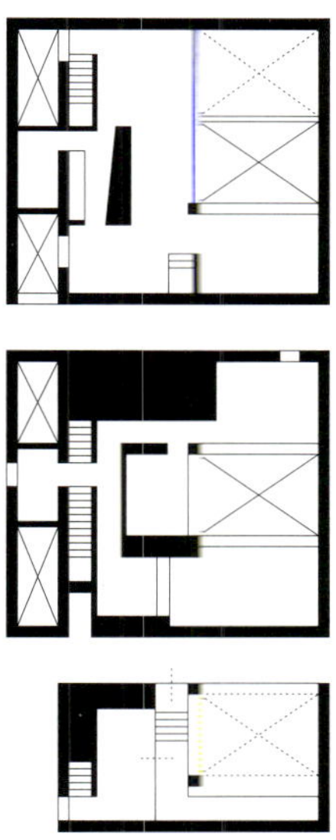

308

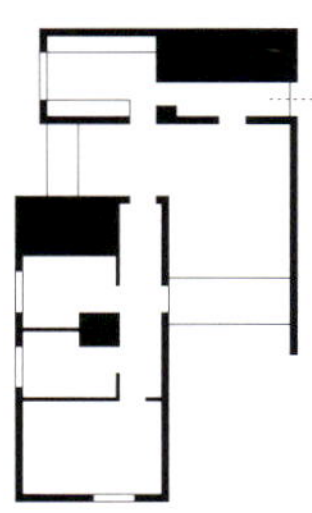

273

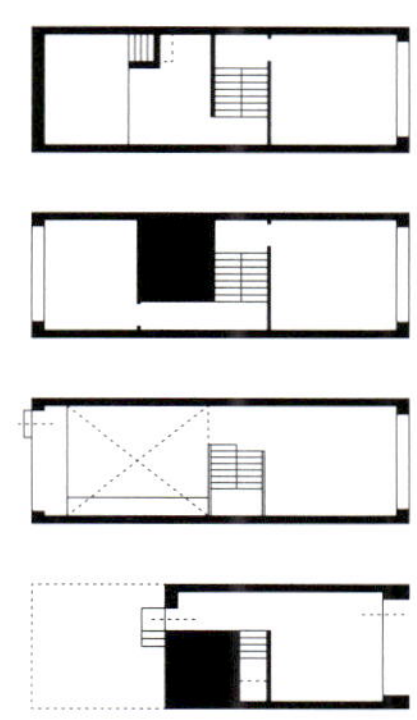

146

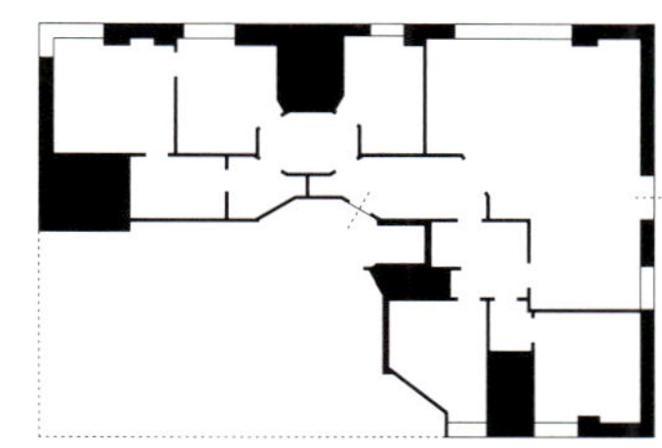

152

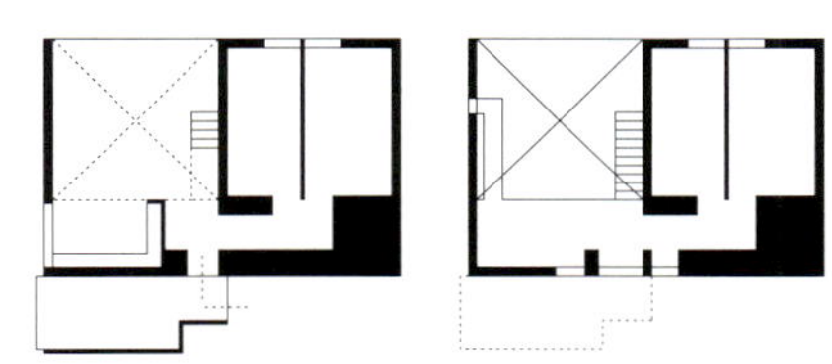

244

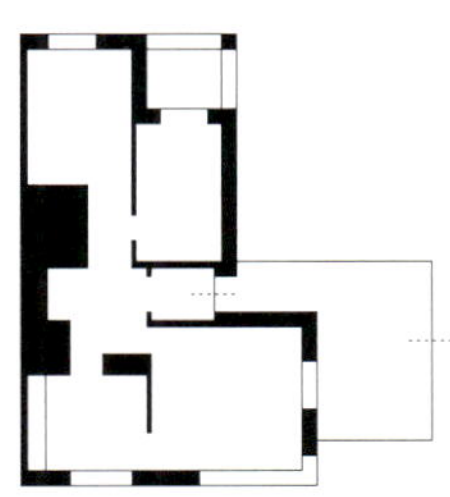

88

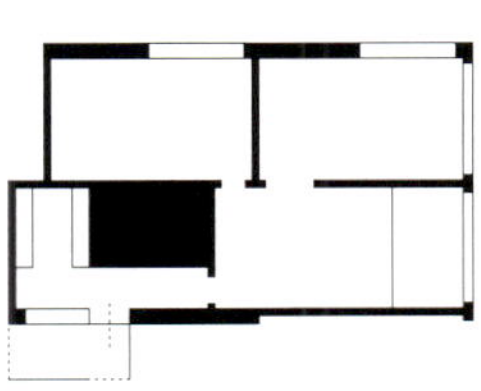

160

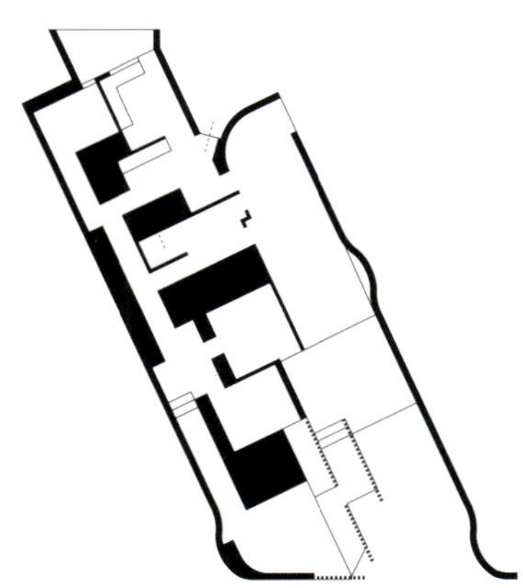

72

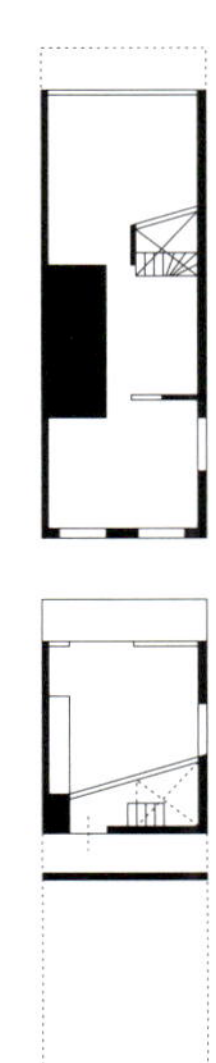

200

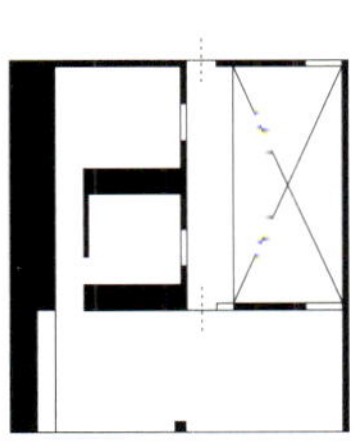

148

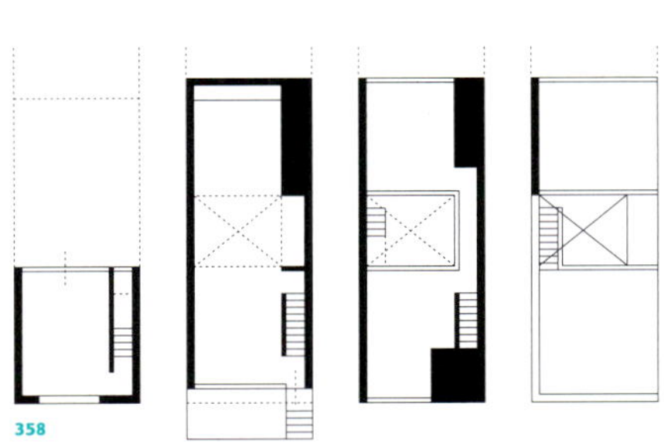

358

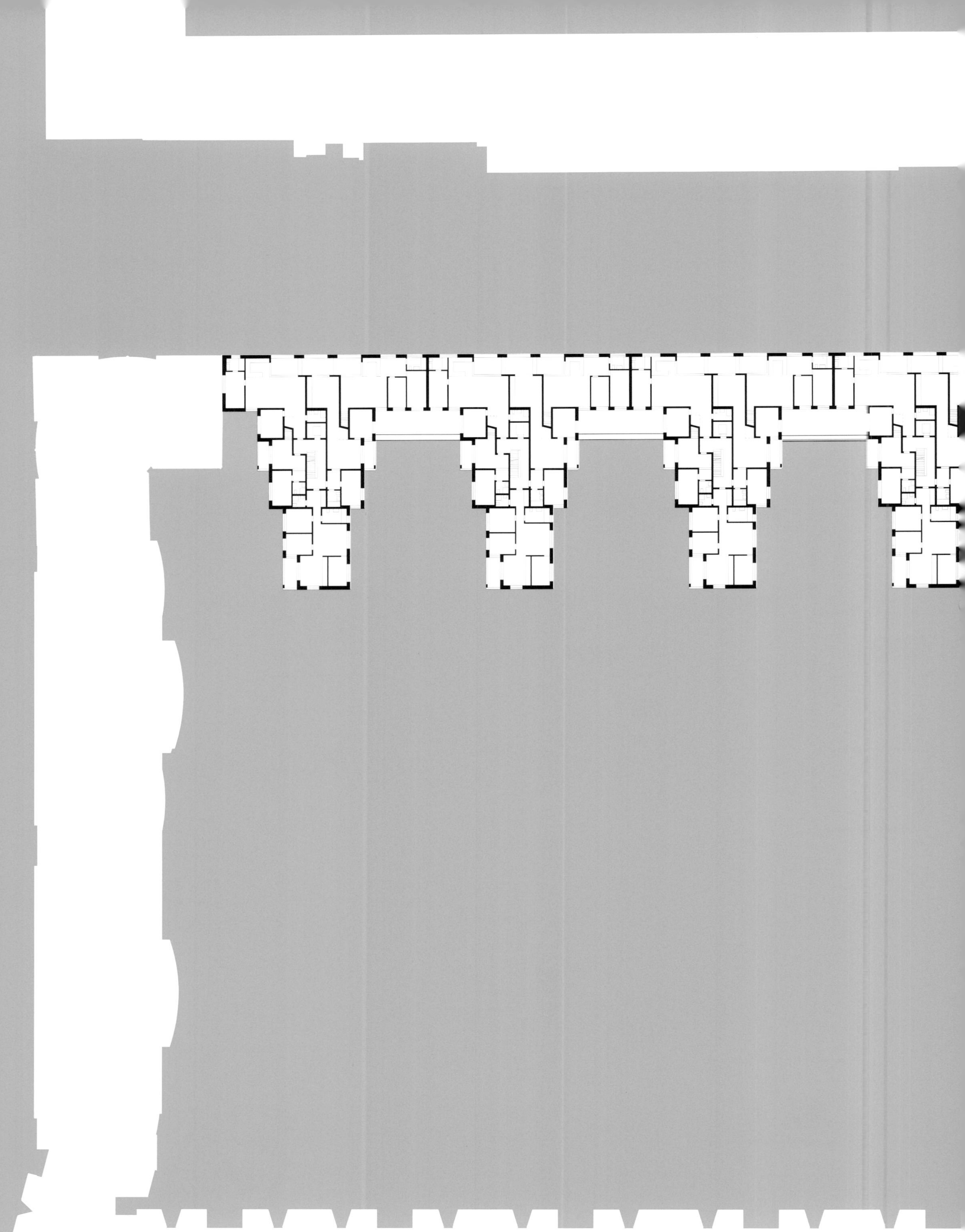

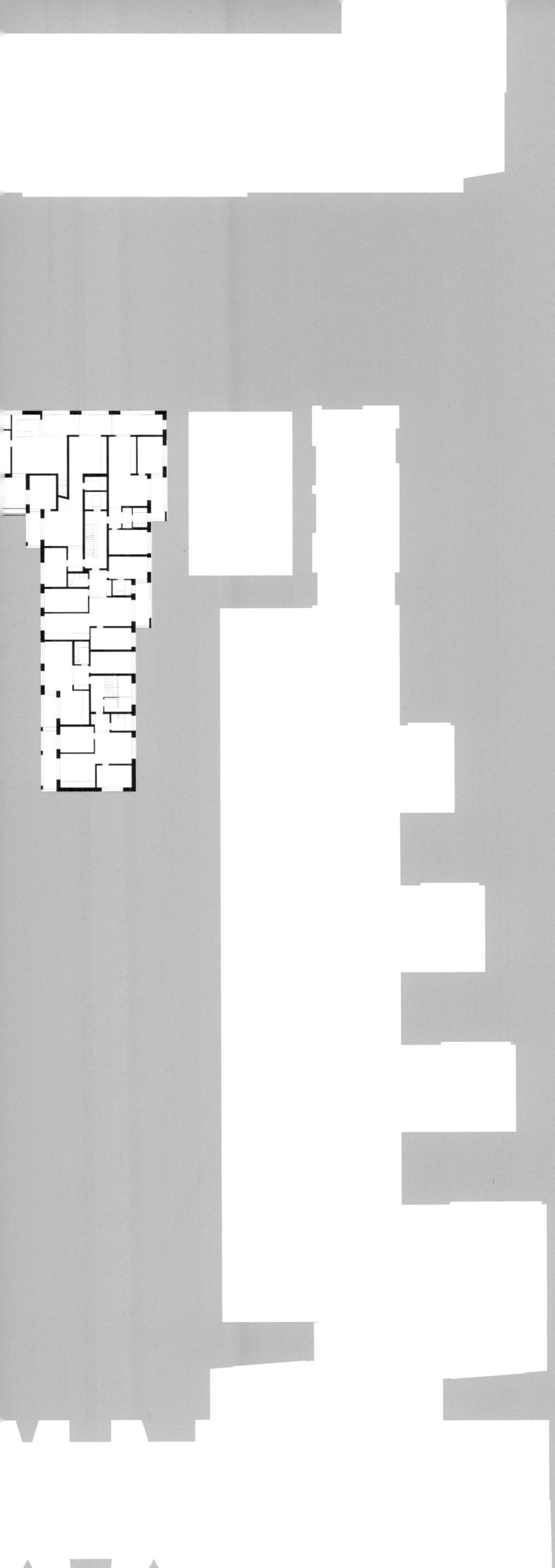

1.1
BLOCK EDGE

The urban block forms the common context for the following chapters on block edge, urban infill, and firewall buildings. While in European cities, the closed block edge remains the dominant form, urban blocks can also be occupied with buildings offset from one another along the street or even in larger ensembles. For the most part, city blocks and their buildings are not developed simultaneously, and often the building parcels in a block belong to multiple owners. The buildings within a block are therefore often very different from one another, even if they have similar dimensions.

Block-edge developments close one or more sides of an existing urban block. The building design is thus free and "unfree" at the same time: "unfree" with regard to the specified block, its street layout, orientation, height, building depth, and the size of the courtyard, yet free with regard to the decision whether the line of the street front should be preserved or structured with a setback, broken up by staggered volumes or whether the block interior should be opened through a gap.

A building in an urban block has essentially two orientations: to the street and to the courtyard, whereby factors like orientation, view, and exposure to noise are often at odds with one another. Due to the lack of light and insufficient separation from the street, the ground floor is reserved for commercial uses whenever possible; alternatively it is laid out as a maisonette and combined with a private green space.

In several situations, block-edge developments deviate from existing urban dimensions, for example, when they occupy entire streets or even blocks within a context of otherwise individually developed parcels. In this way, they offer the opportunity to experiment with contemporary housing solutions within the existing urban layout. Since they are greater in scale than infill developments, it is possible for each project to develop patterns of its own and to work with more complex access systems. This building type allows for all types of floor plans and makes it possible to play with a broad range of floor plan types in one project.

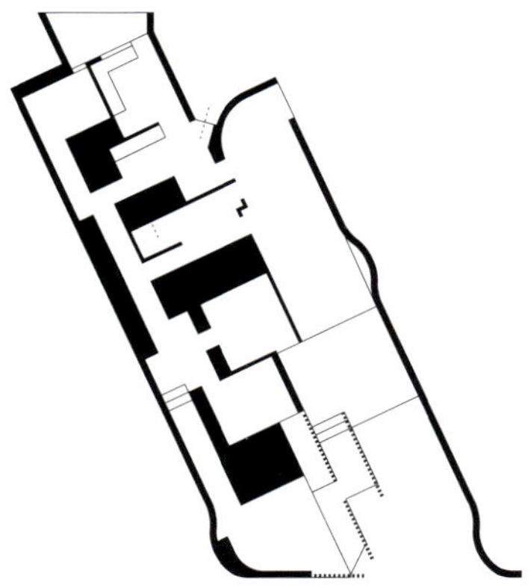

Girasol | Coderch / Valls | 1966

Instead of creating a large courtyard, the street facade was dissolved into a rhythm of rounded, solid projections and open perforations, which allow daylight and the life of the city to penetrate deep into the apartments like a filter. Thanks to the rotation, the building appears to be composed of individual stacked volumes, thus setting it apart from the constraints of the urban block. The retail spaces on the ground floor are followed by a mezzanine level with vegetation: it serves as a horizontal circulation level to the seven (or eight) apartment levels and also acts as a calming filter. The floor plan is divided into three sections: the service area with separate access lies to the rear, followed by the elongated living area – for which the central elevator offers a formal prelude – and parallel to the this space a sequence of individual rooms with bathrooms. The front-to-back living area reaches from the patio on the courtyard side to the street facade, where it is extended by a room-size exterior space. Depending on weather conditions, this space can be opened toward the outside or toward the interior, entirely at the decision of the users. Raised by a few steps and at right angles to this space are the loggias in front of the individual rooms. They are stepped back to ensure that all interior spaces have a connection with the outside. The dividing walls between the flats are the characteristic feature. They are found throughout the apartments and continue on the outside, materializing in the form of a ceramic skin, stretching into a protective curve, followed by a fluid transition into a homogeneous layer of light, wooden venetian blinds. The result: from the street perspective, there is very little evidence of the residential activities, which remain discrete, almost hidden behind an envelope of material and form.

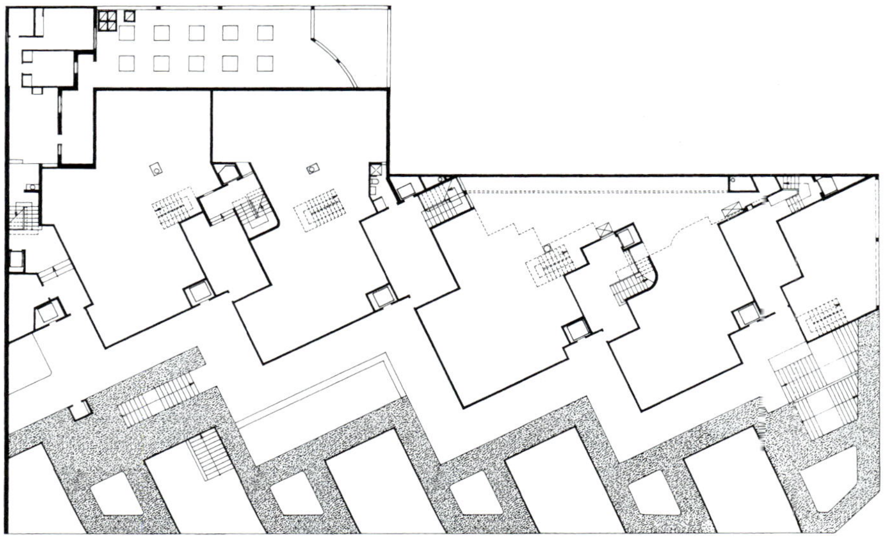

1

3

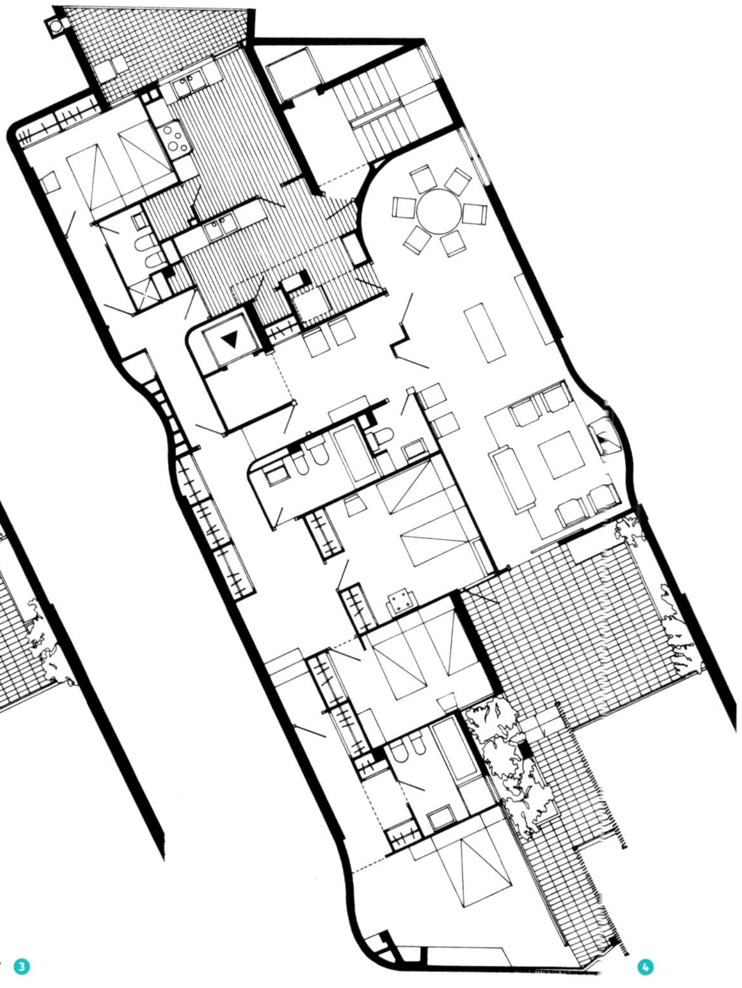

4

1 Open, planted access level on 2nd floor 1:500
2 Standard floor 1:500
3 5-room apt. and staff apartment 1:200
4 4-room apt. and 1 staff room 1:200
5 3-room apt. and 1 staff room 1:200
6 West elevation 1:500

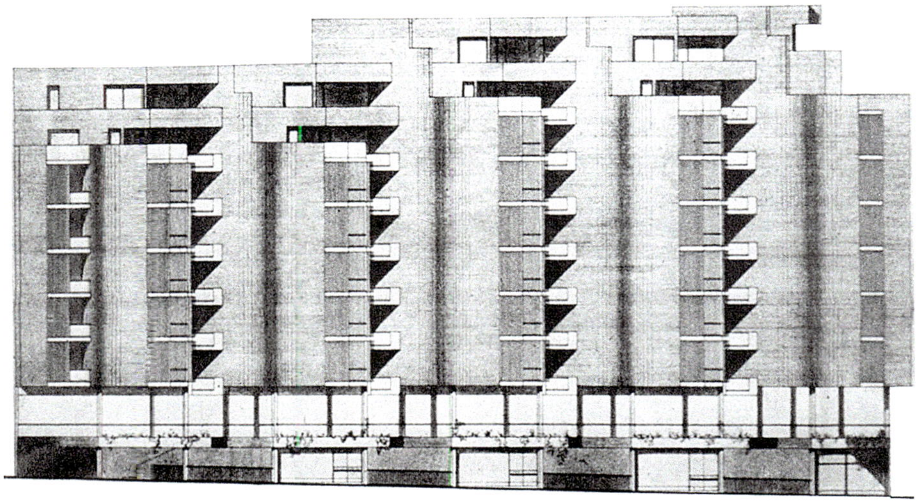

6

1.1
BLOCK EDGE

2

5

Building type
9/10 stories
perimeter block
with cut in atria
facing NE/SW

Date of construction
1964–1966

Number of units
38

Site of units
5 (+2)-room apts., approx. 190 m²
4 (+1)-room apts., approx. 148 m²
4 (+1)-room apts., approx. 170 m²
3 (+1)-room apts., approx. 147.5 m²
6 (+1)-room apts., approx. 166.5 m²
(number of staff rooms
indicated in parentheses)

Area per user
24–37 m²

Building depth
approx. 16–28 m

Access
2nd floor as open
horizontal circulation level,
single-loaded stairs to
service area of apartments,
elevators as main direct access

Open spaces
room-size loggia
in front of living room,
loggias in front of bedrooms

Parking
underground garage

Architect
J.A. Coderch / M. Valls
Barcelona

Location
Madrid
Spain

Referring to
floor plan 1:200

Bläsiring | Diener & Diener | 1981

The three-winged block-edge development does not connect to the adjacent buildings, thereby creating a passage for an avenue of trees, which precedes the block to the north (see page 76). The short end sides are occupied by corner maisonettes and are thus articulated by openings that overlook the passage on one side and the city on the other, entirely different than would be the case with a standard compartment wall. In the north-south facing standard floor plan **A**, the hallway is characterized by the intersection of two directions, one orthogonal and the other diagonal. The hallways, thus connects entrance and living room, living room and sleeping area and also links bathroom, WC, and kitchen to the center area. In the corner type with central access, the diagonal divides four units, but it also characterizes the larger apartments in a subtle way **B**. By turning into the diagonal, the dining area expands the living room, and in the other direction it links up with the kitchen and the loggia. The hallway is thus the focal point of a circular internal path. One of the small one-room apartments **C** overlooking the street can be integrated to create a larger unit. The flats in the east-west wings **D** are laid out on a cruciform plan with a central hall. In the longitudinal direction, this hall is defined by two room enclosures in front of the sanitary- and the stairwell core, respectively, and parallel hallways on both sides along which the rooms are lined up. Crosswise, two identical verandahs link the spacious hall to the street and the courtyard. The floor plan is symmetrical in structure, but allocated to different functions.

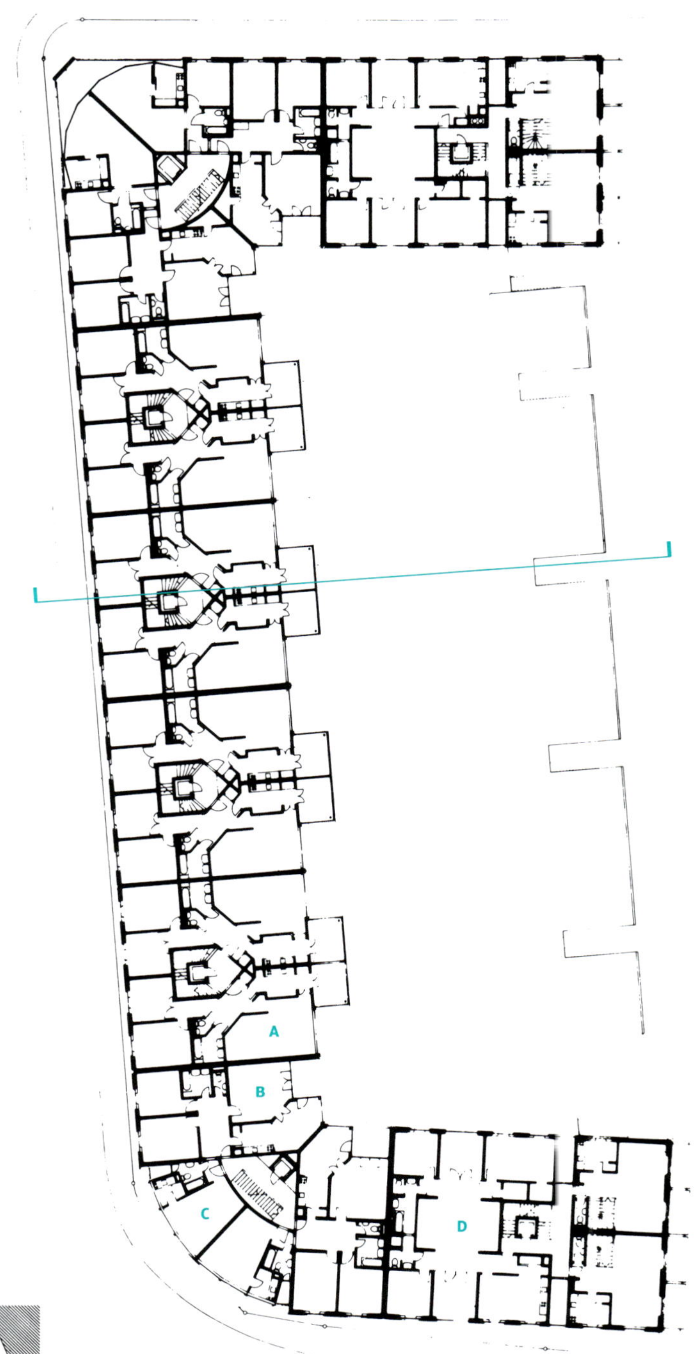

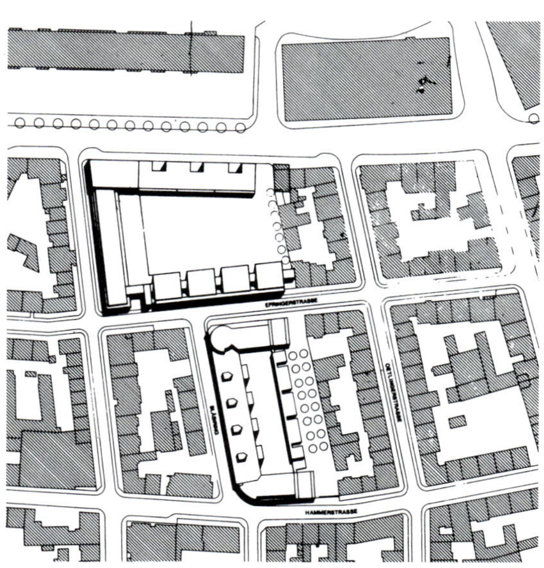

1 3-room apartments type **A** 1:200
2 1- and 3-room apartments type **B**, **C** 1:200
3 6-room apartment type **D** 1:200

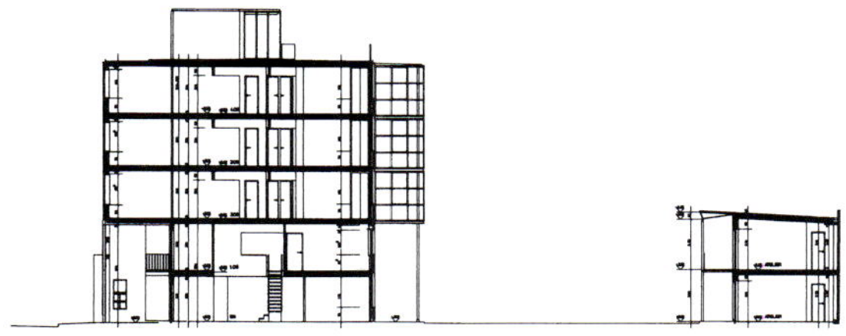

1.1
BLOCK EDGE

Building type
U-shaped development
5 stories
facing N/S and E/W

Date of construction
1978–1981

Number of units
88

Size of units
1-room apts., 35–46 m²
2-room apts., 52 m²
3-room apts., 85–98 m²
5-room apts., 130 m²
5-room mais., 137 m²
6-room apts., 147 m²
studios, 29 m²

Area per user
24.5–41.5 m²

Building depth
14.7 m

Access
multiple-loaded interior stairwells,
maisonettes direct from forecourt,
studios through gallery

Open spaces
shared rooftop terrace,
garden in inner court

Parking
91 parking spaces in underground garage beneath inner court

Architect
Diener & Diener

Location
Hammerstrasse/Bläsiring
Basel
Switzerland

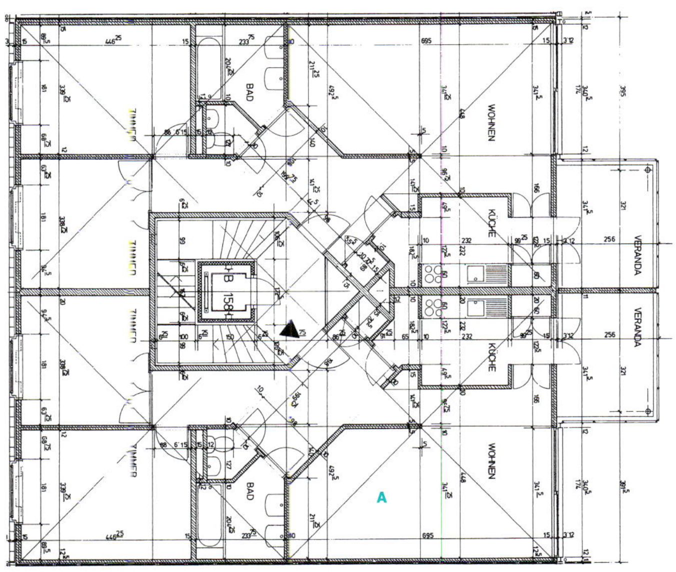

1

2

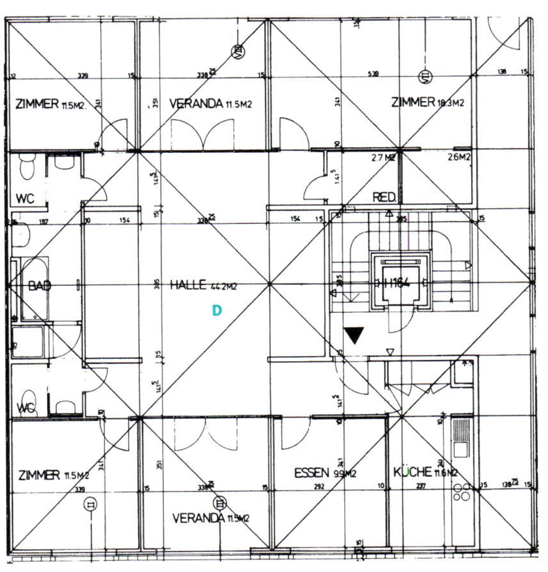

3

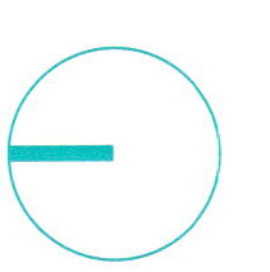

Referring to floor plan 1:200

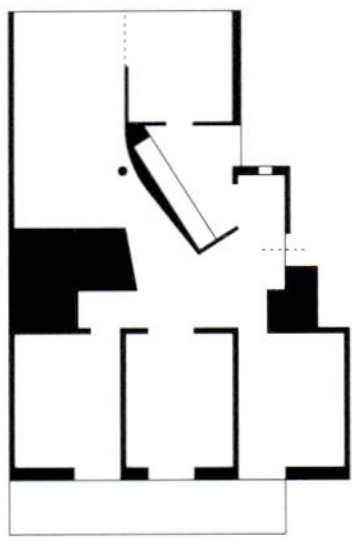

Riehenring | Diener & Diener | 1985

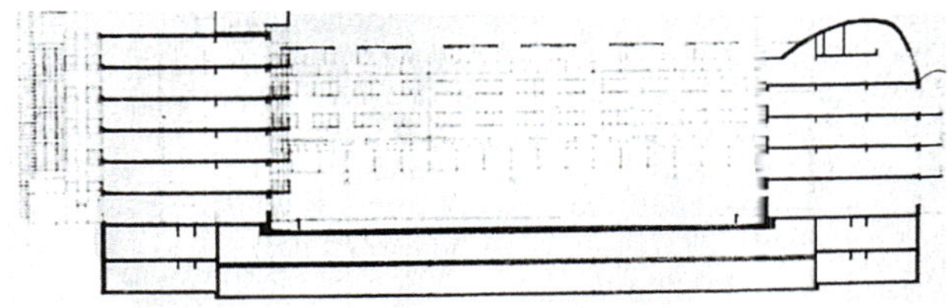

The block-edge completion is adapted to the texture of the city and makes reference to the existing parcel structure in the structure of its facade and plan. A footpath between the new and the old development connects to the arcade of the block on the Bläsiring (see page 74). Office areas at the corners and retail spaces on the ground floor complete the scheme. Each of the three block edges is characterized by a unique floor plan type. The north-south facing maisonettes **A** are arranged around an open atrium. On the lower level, the rooms located on the courtyard and street side form a representative enfilade in combination with the atrium. The sanitary zones and the individual rooms on the upper level are clearly separated from this spatial sequence. In the east-west facing five-room apartments **B**, the diagonally placed kitchen and the sanitary zone divide a continuous common room into entrance hall, a hallway facing towards the individual rooms overlooking the street, and living room and dining area. On the courtyard elevation, the building volume is characterized by recesses toward which the stairwell, hallways, and kitchens are oriented. Within the three-room apartments **C** in the block edge on the opposite side a slightly rotated core with bathroom, storage space, and kitchen row divides the living and sleeping area hallway. The building volume is recessed on two sides. The kitchen is thus lit from two sides and is transformed into a unique space, although it is located at the far end of the unit. Kitchen and living room are positioned in a manner to create a continuous spatial sequence from one end of the building to the other.

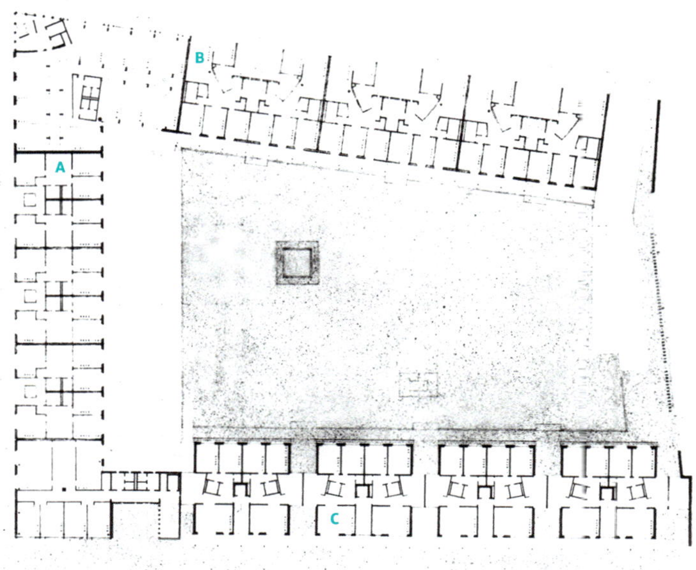

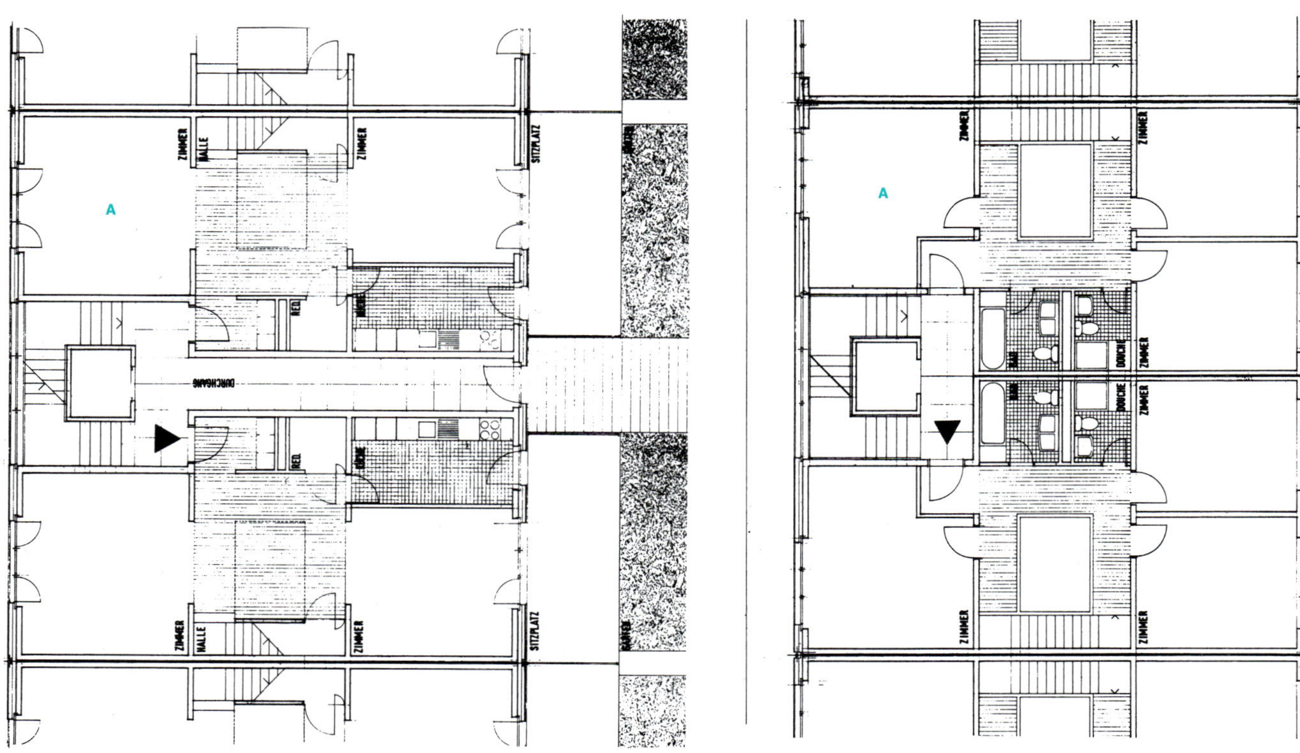

1

1 N/S wing: 5-room maisonette type **A** 1:200
2 E/W wing: 5-room apartments type **B** 1:200
3 E/W wing: 3-room apartments type **C** 1:200

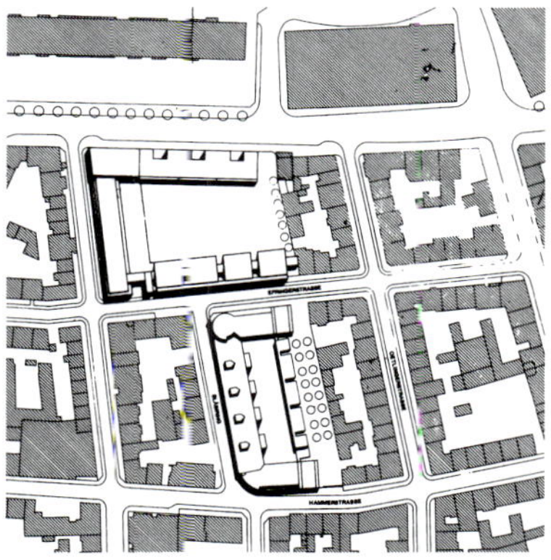

1.1
BLOCK EDGE

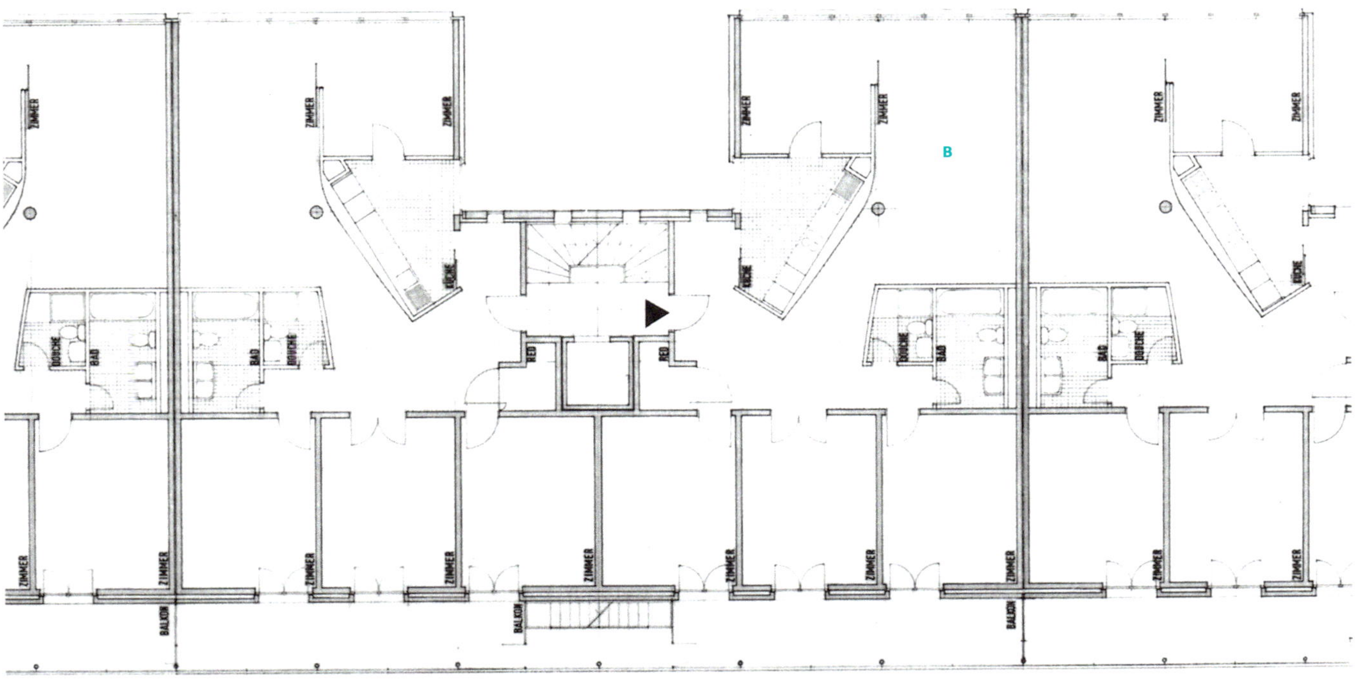

2

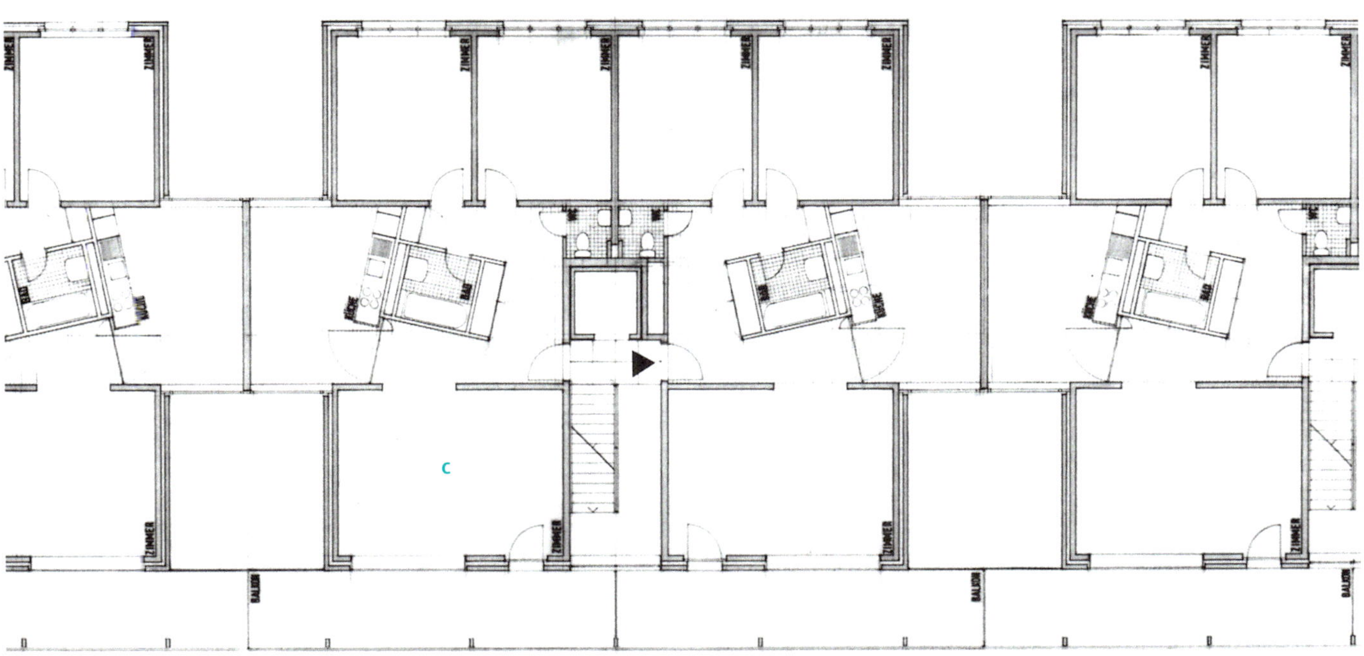

3

Building type
three-wing block
5/6 stories
facing N/S and E/W
with offices and stores

Date of construction
1982–1985

Number of units
74

Size of units
3-room apts., 81–100 m²
5-room apts., 127 m²
5-room mais., 139–146 m²

Area per user
25.5–30 m²

Building depth
15.4 m on Riehenring side,
otherwise 14.2 m

Access
double-loaded
stairwells,
maisonettes,
offices with
separate stairwells

Open spaces
rooftop terrace
(on Riehenring side),
trees in inner court

Parking
307 parking spaces in
underground garage
beneath inner court

Architect
Diener & Diener
Basel

Location
Riehenring/
Amerbachstrasse/
Efringerstrasse
Basel
Switzerland

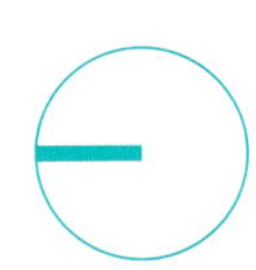

Referring to
floor plan 1:200

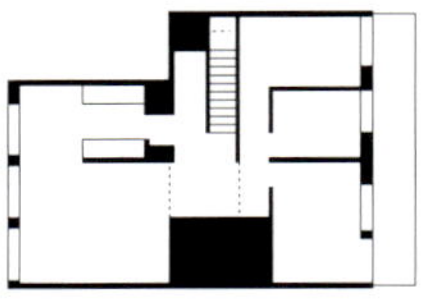

Full Stop and Comma | Siza | 1988

The hope here was to breathe new life into a decrepit working class neighbourhood by introducing new and international architecture. Siza and Castanheira studied the structure of the quarter and reinterpreted it in their design. They borrowed building heights, materials (various colors of brick) and access type from the surroundings. The closed building form is broken at the corners to permit inner courtyard entry and space for low corner shops. Typical access is here the "portico"; steps lead directly from the street to a landing from which all but the ground floor apartments are reached by individual interior stairways. This means that each apartment door can be seen from the street, an important safety feature for the residents. Apartment doors open into a corridor which leads to kitchen, living room, bathroom, and a further bedroom hallway. This hallway gives access to either two or three bedrooms and a bath. The floor plans were tailored to suit the customs of practicing Muslims, as they represent the majority of the project's residents. It was therefore necessary to provide for corridors through which women could move freely between bath and bedroom unseen by men in the living room. Hence the strict separation of living and sleeping areas and the interior door between bath and toilet. By opening sliding walls, other residents can expand the living room into the entrance hall.

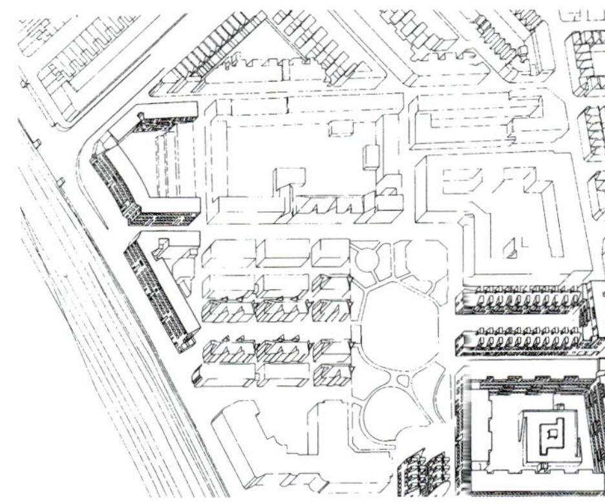

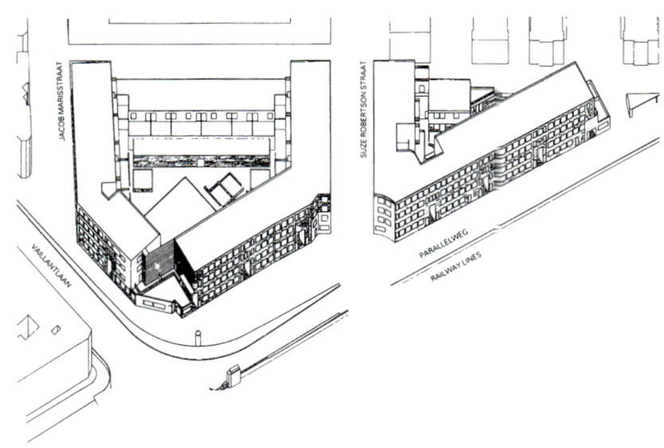

1

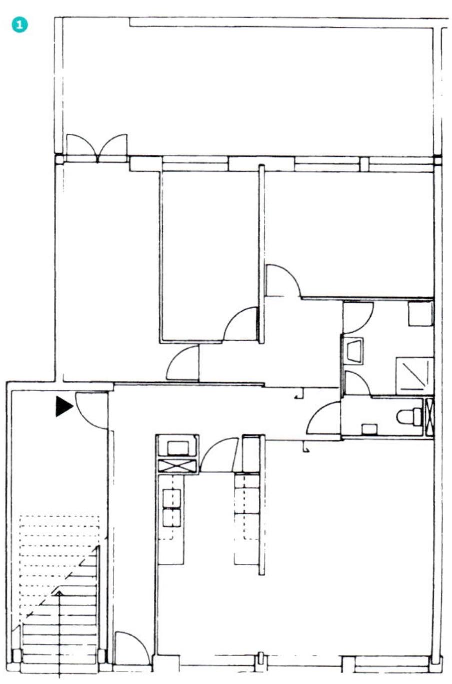

2

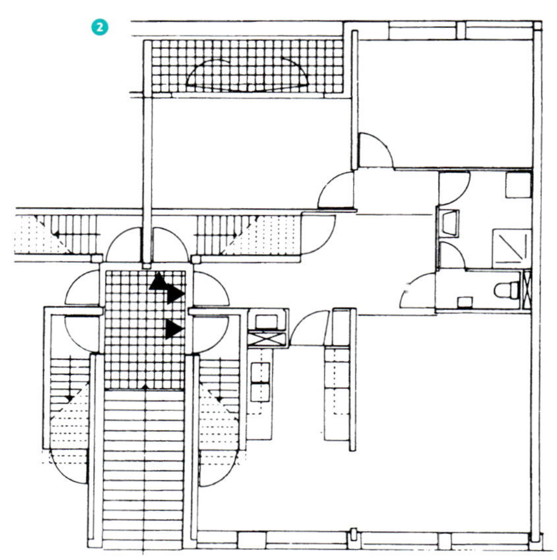

1 Ground floor: 4-room apartment with garden
2 2nd floor: 3-room apartment with loggia
3 3rd floor: 4-room apartment with balcony
4 4th floor: 4-room apartment with balcony
5 General layout ground floor 1:500

1.1
BLOCK EDGE

5

Building type
block housing
2–4 stories
street shops

Date of construction
1983–1988

Number of units
106

Size of units
2-room apts., 51/56/61 m² (11 units)
3-room apts., 61/75 m² (18 units)
4-room apts., 85/90 m² (59 units)
5-room apts., 110 m² (4 units)
5-room mais., 97 m² (4 units)
6-room apts., 130 m² (4 units)

Area per user
19.5–30.5 m²

Building depth
12 m

Access
"portico" accesses
6 apartments directly,
ground floor apartments
have street entry

Open spaces
private gardens,
loggias, balconies,
common courtyard

Parking
parking spaces on street,
in courtyard garages

Architect
Álvaro Siza with
Carlos Castanheira
Porto
in collaboration with
van den Broek and Bakema
The Hague

Location
Jakob Marisstraat / Vaillantlaan
Parallelweg / Suze Robertson Straat
The Hague-Schilderswijk
The Netherlands

3

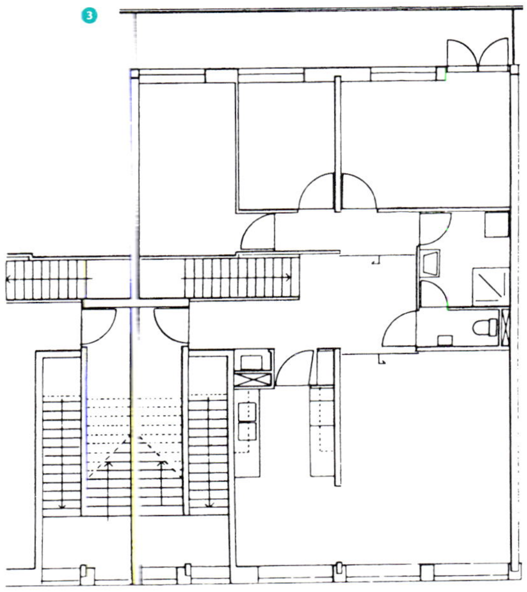

4

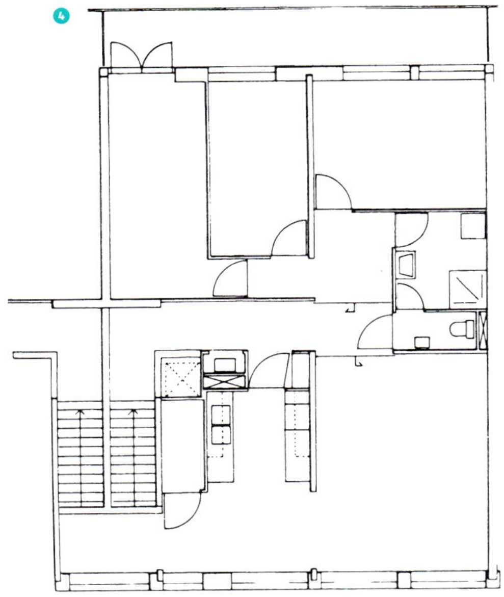

Referring to
floor plan 1:200

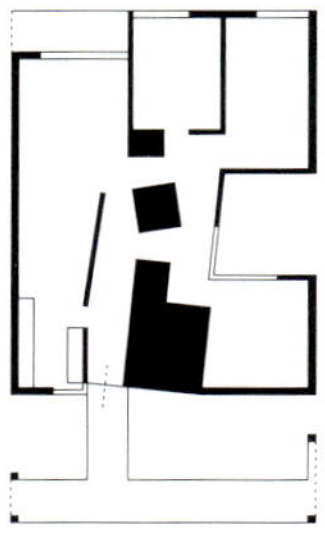

Brunnerstraße | Richter | 1990

Building row with front gallery behind noise protection wall of large glass panels. The gallery also provides a link to the housing units at the end of the row. Straight-run stairs, parallel to the facade, provide access to the floors; the apartments can be reached by walkways and are thus separated from events on the gallery. The entrance doors, turned slightly inward, allow for visual contact into the access area through the kitchen windows situated at the corner. The organization of the apartment is essentially the same despite varying apartment sizes: located along a slanted hallway are, first, the kitchen, then the wedge-shaped living area. In the hallway, a diagonally inserted yellow cube (storage room) separates the living and sleeping areas and directs circulation toward the living room. A second door behind the cube provides a direct short connection between living and sleeping areas (second path). Upon entering the apartment, a visitor can see through it all the way to the loggia; floor-to-ceiling sliding windows admit light and open the room up to the outside. On the second floor the apartments are organized around inserted interior courtyards. These interior courtyards and the rooftop terraces on the third floor are marked on the courtyard side as recesses in the built volume. The crosswall/lightweight structure enables changes after the fact: only one column, one wall, and the installation shaft are fixed inside an apartment.

1

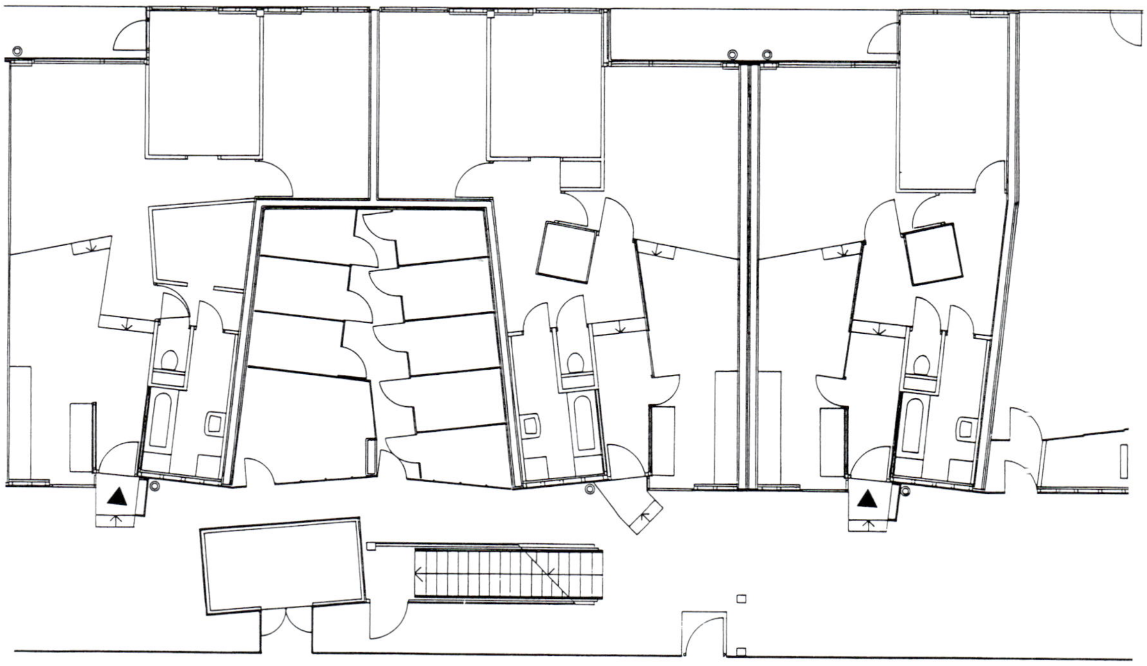

2

1 Layout of ground floor and second floor
2 Ground floor: section of floor plan with 2- i.e. 3-room apartments 1:200
3 2nd floor: section of floor plan with 4-room apartments 1:200

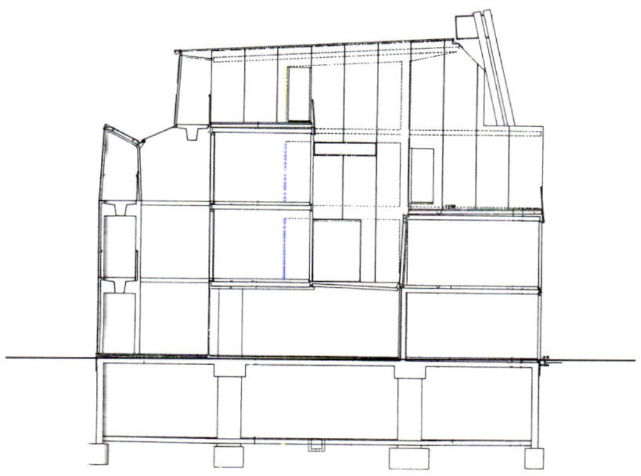

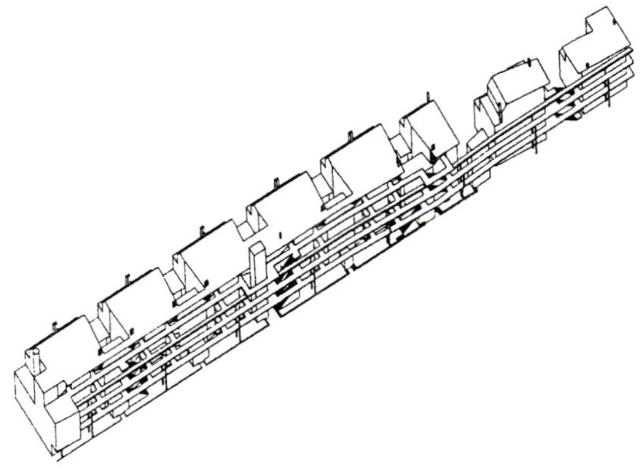

1.1
BLOCK EDGE

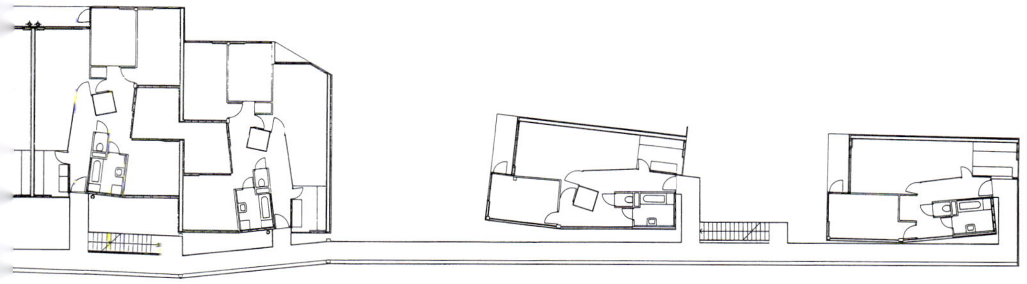

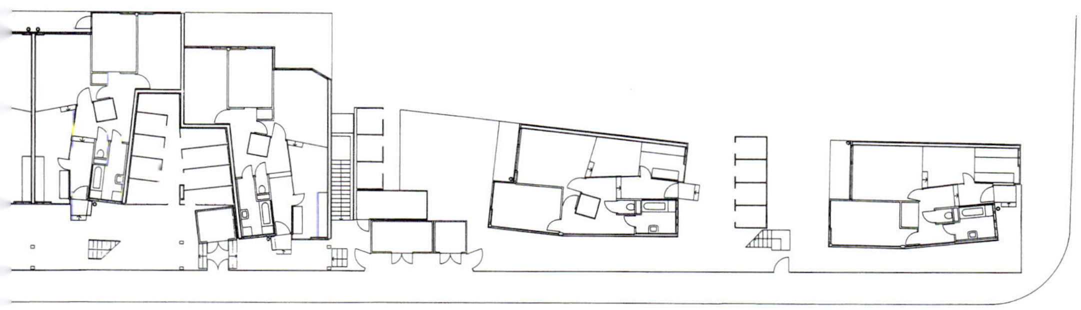

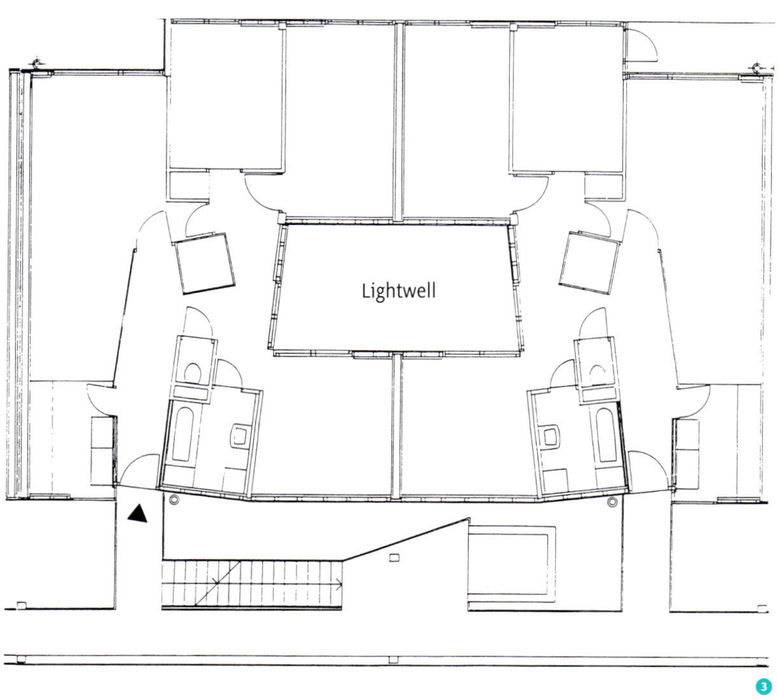

Building type
building row with
2 separate housing units
4 stories
E/W

Date of construction
1990

Number of units
62

Size of units
2-room apts., 61 m²
3-room apts., 74–88 m²
4-room apts., 85–106 m²
(4 apartment types,
one type per floor)

Area per user
24–30.5 m²

Building depth
17 m

Access
glass-paneled gallery
with catwalks

Open spaces
gardens on the rear side
of the buildings,
inserted inside courtyards
rooftop terraces

Parking
subterranean,
1 place per unit

Architect
Helmut Richter
Vienna
with
Anne Hengst
Gogo Kempinger
Bert Dorfner

Location
Brunnerstraße
Vienna
Austria

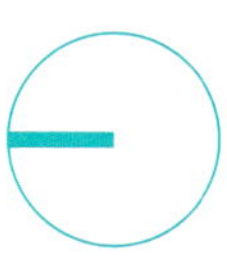

Referring to
floor plan 1:200

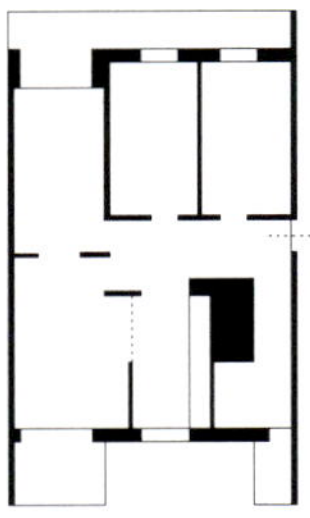

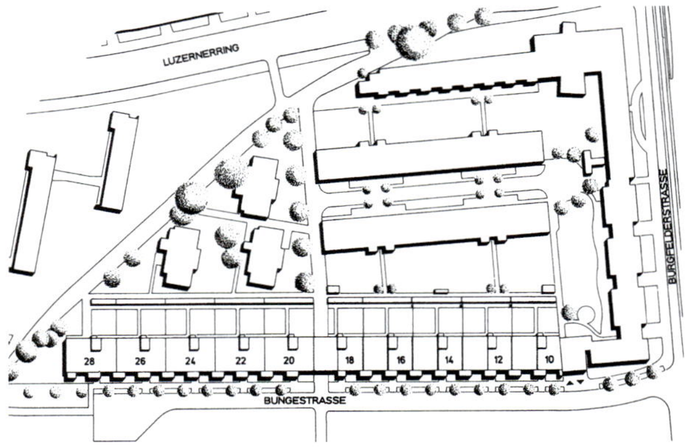

Bungestrasse | Alder | 1993

The five-story apartment building along the road has a supporting structure composed of crosswalls and columns, which provides a division for the dwellings. According to their size they take up one, two, or three sections, whereby one section corresponds to the width of a room. Studios and lofts are available as variations on the apartments. Each stairwell has a lift and a generously proportioned entrance area (with a communal room!), and leads to two dwelling units per story. As a rule, a 3½- and 4½-room flat are located next to each other, each taking up 2 or 3 sections respectively. The former makes use of the width of the stairwell as an "additional" room. The design also allows freedom of choice within each flat. Walls can be placed in any of a number of predetermined locations: at the expense of the living room, for example, the kitchen can be extended to become a dwelling kitchen, and the walls separating the bedrooms can be placed in a space-defining manner. But the uniform kitchen/bathroom unit next to the entrance complete with bath, WC, and réduit remains fixed. All rooms can be reached directly from the large corridor. The bedrooms face the courtyard, the main rooms the road. Apart from the kitchen, all rooms have access to balconies – even the bathroom. This makes the flat seem very open. On the ground floor commercial use units alternate with 3-room apartments with private front gardens facing the street. A roofed area for locking up bicycles shields the back of the communal garden from the paved inner yard. The communal roof terrace affords a magnificent view of nearby Alsace.

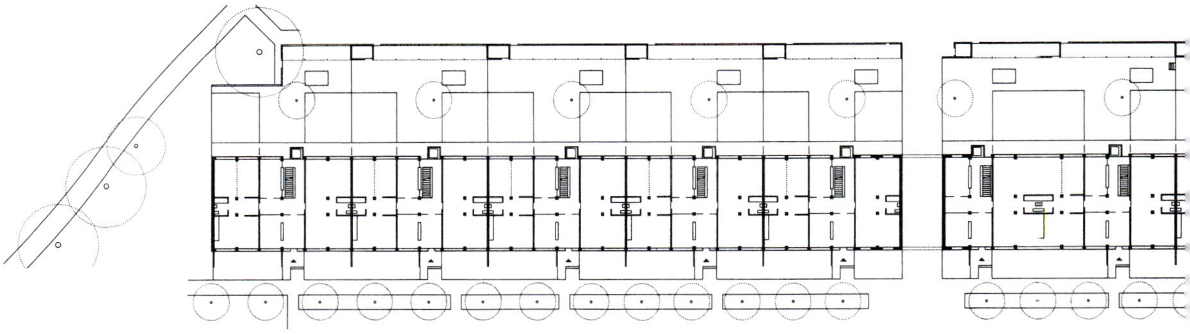

1

3

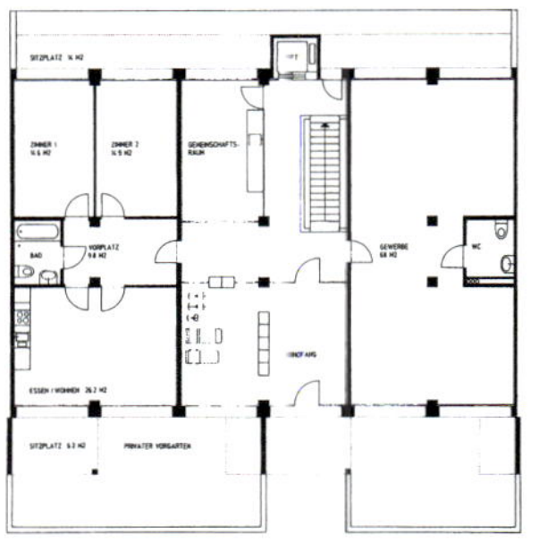

2

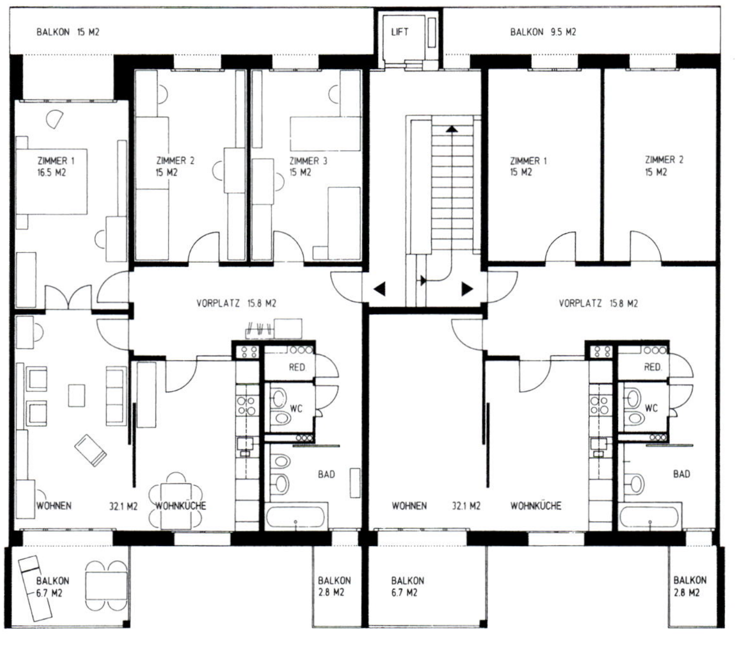

1 Typical floor plan with different options for the wall positions
2 2nd to 5th floor with 3½- and 4½-room apartments with large kitchen 1:200
3 Ground floor with commercial use and 3-room apartment
4 2nd floor with studio and 3-room apartment
5 3rd to 5th floor with loft and 3-room apartment
6 2nd to 5th floor with 3- and 4-room apartments with enlarged living room 1:200

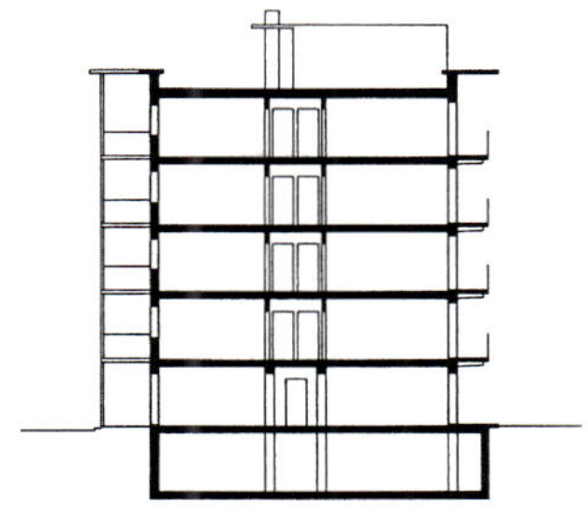

1.1 BLOCK EDGE

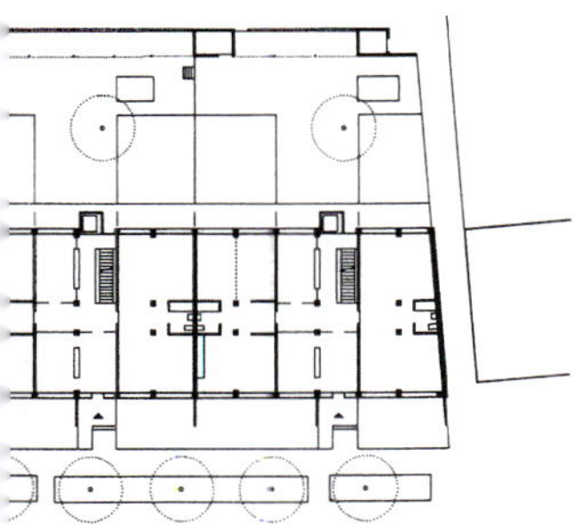

Building type
row in block perimeter
5 stories
facing NW/SE
with commerical use
on ground floor

Date of construction
1991–1993

Number of units
99 (including studio)

Size of units
3-room apts., 70 m² (19 units)
3½-room apts., 87 m² (36 units)
4½-room apts., 103 m² (36 units)
lofts, 87/105 m² (7 units)
studio, 40 m² (1 unit)

Area per user
23.5–48 m²

Building depth
13 m

Access
double-loaded stairwells

Open spaces
garden or three balconies
on two sides
(19/24.5 m² in total)

Parking
290 underground spaces
under communal court yard

Architect
Michael Alder
Hanspeter Müller
and Andy Hindemann
Basel

Location
Bungestrasse 10–28
Basel
Switzerland

4

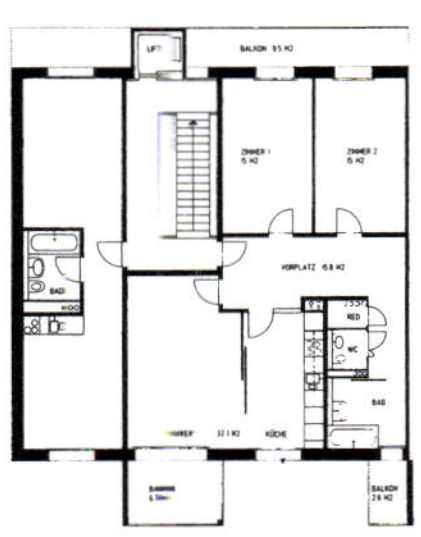

5

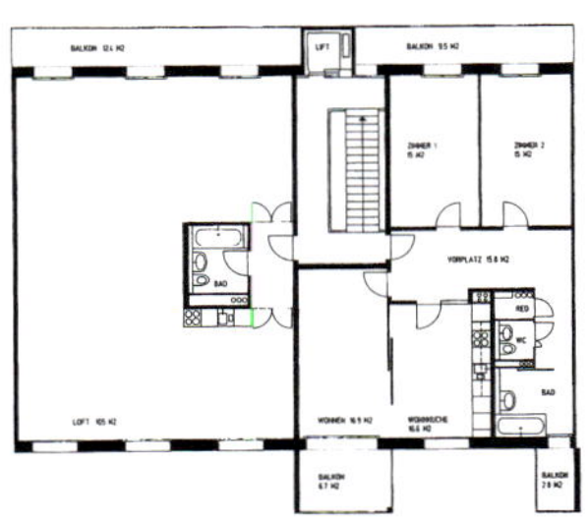

6

Referring to floor plan 1:200

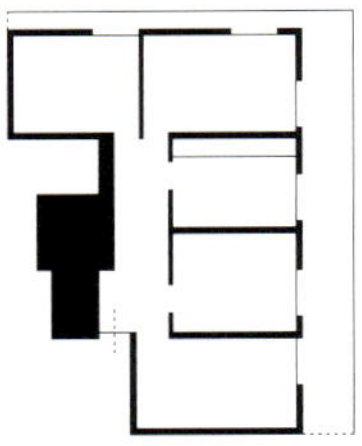

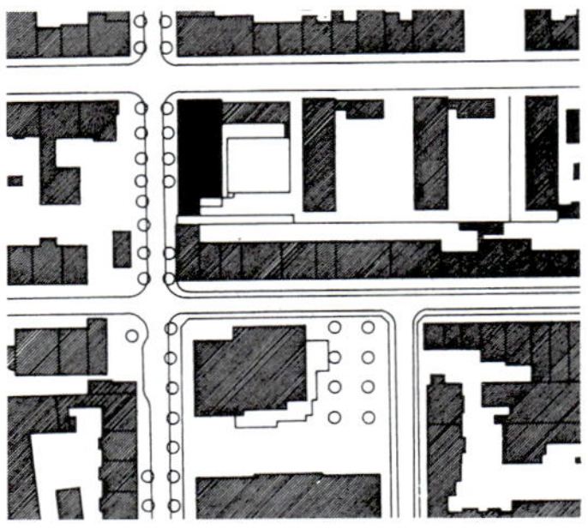

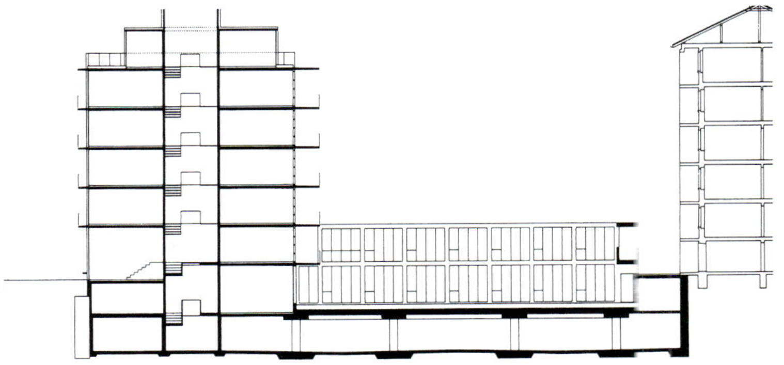

Müllheimerstrasse | Morger & Degelo | 1993

The plan closes the corner lot with a row, and the circumferential balconies mark the end of the street. The building is connected to the shallower neighbouring house in the street by a narrow structure with terraces (running all the way from the street through to the courtyard). The apartment floor plans are wrapped around two stairwells so that all apartments can claim two orientations or views (street/courtyard or street/corner or courtyard/corner). The stairwells are coiled around a lightwell that runs from top to bottom through the building; they begin in the raised ground floor with a large entry hall, almost one and a half stories high, and open out onto long landings on the floors. Almost every apartment has a special feature: corner living rooms with windows on two sides (at the end of the row); clear separation of living and sleeping according to two points of the compass (central apartments); terraces extending from front to back of building (apartments connected to the neighbouring house); circumferential rooftop terraces (apartments on sixth floor); etc. All windows are from floor to ceiling; the circumferential balconies are 60 cm wide on the street side (exception: in front of corner apartment) and 1.75 m wide on the courtyard side.

1

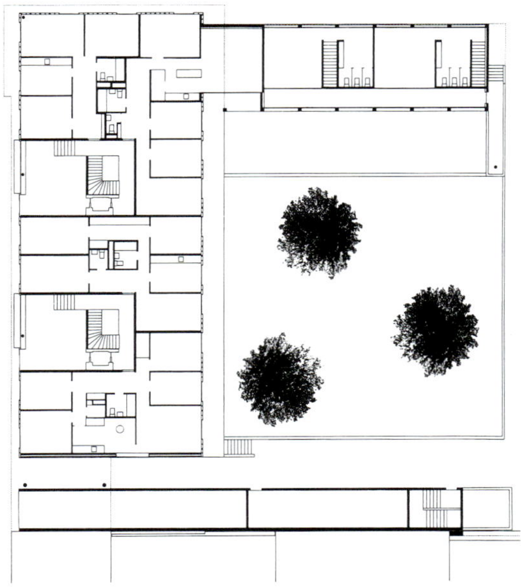

2

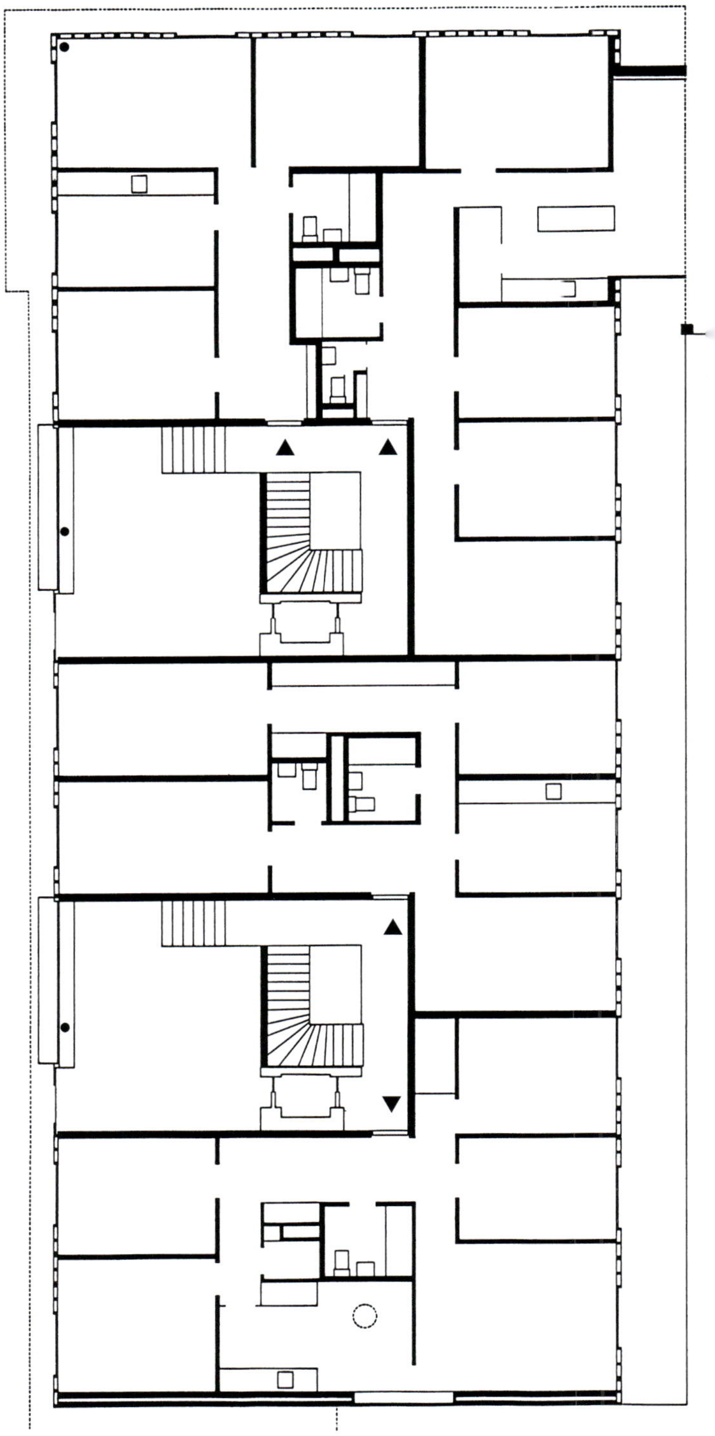

1 Ground floor layout: apartment building and kindergarten
2 Ground floor plan 1:200
3 Typical floor plan, 2nd to 5th floors 1:200
4 Roof floor layout

1.1
BLOCK EDGE

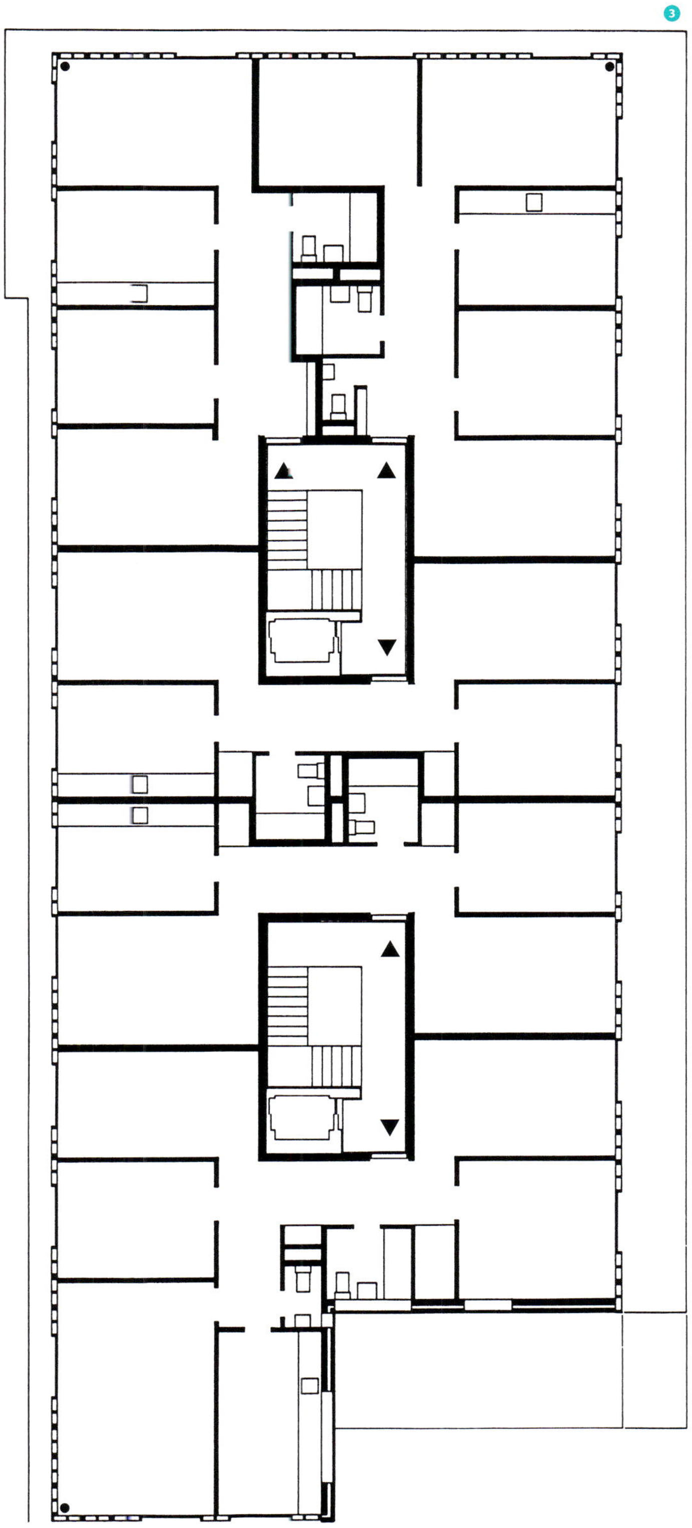

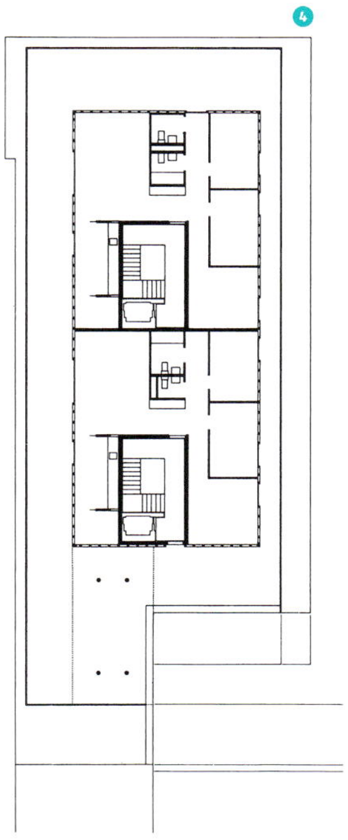

Building type
corner building as end of row
6 stories
facing E/W and N
(block completed by kindergarten)

Date of construction
1993
(winning competition entry: 1989)

Number of units
26

Size of units
3-room apts., 74.6 m² (12 units)
4-room apts., 99.5 m² (7 units)
5-room apts., 107/117 m² (5 units)
4-room barrier-free apts. (2 units)

Area per user
22.5–25 m²

Building depth
15 m

Access
double- and triple-loaded stairwells

Open spaces
courtyard with trees and fountain
(sunken area for kindergarten)
balconies
rooftop terraces

Parking
in basement story
(50 spaces)

Architect
Morger & Degelo
Basel

Location
Müllheimerstrasse
Basel
Switzerland

Referring to floor plan 1:200

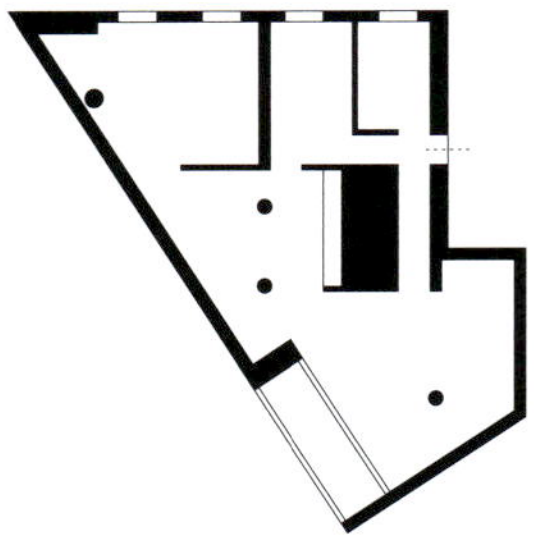

Piraeus | Kollhoff | 1994

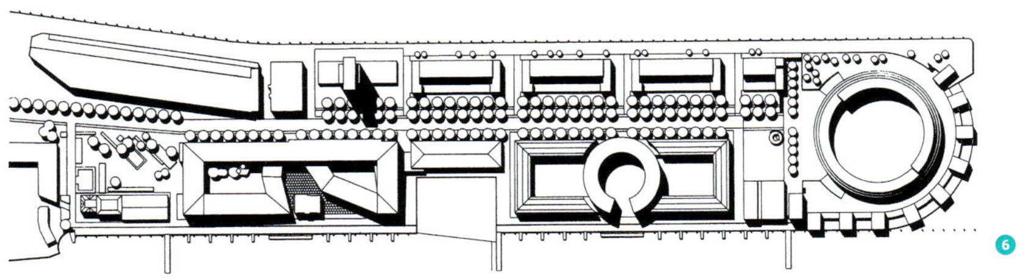

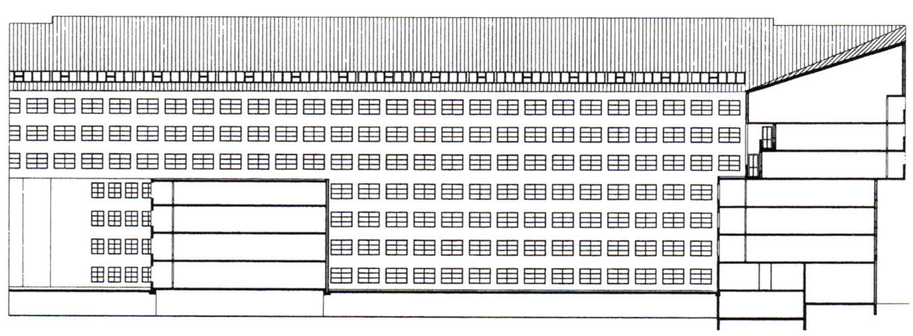

This sculptural superblock was based on an urban plan by Jo Coenen, whose master plan had envisioned a more conventional block-edge development at this site. However, the existing building on the water's edge, which was to be preserved, acted as a catalyst for the metamorphosis of this fundamental urban type. The block edge skirts around the old building, it is angled in plan and section, thus establishing a distinctive urban volume. The building opens and turns toward the surroundings in a wide variety of ways, resulting in brightly lit rooms, improved views of the city and the water, and rooms that are up to 8 m high in the attic story. From the ground to the 4th floor the internal organization of the building is almost exclusively realized with double-loaded stairwells. Covered walkways, serviced by only two lifts, complement the provisions for access; they alternate on the various levels between courtyard elevation and street/waterside elevation and also articulate the image of the facades. There are 143 floor-plan options for the 304 apartments, in part due to the complex shape of the plan. However, despite the multitude of plans, the apartments are invariably divided by the same core consisting of hallway, WC, bathroom, storage room, shaft, and linear kitchen. This core always results in a circular internal path with the bedrooms facing in one direction and the living area with loggia or balcony facing in the other.

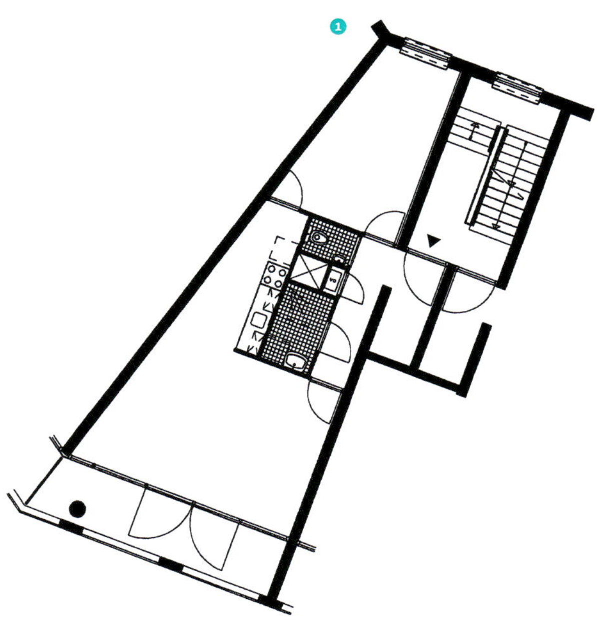

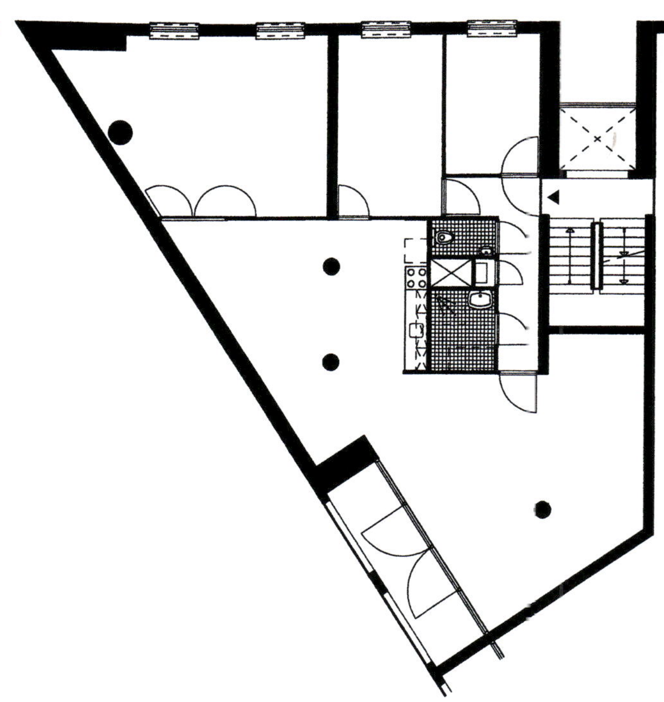

1 3rd floor: 2-room apartment 1:200
2 3rd floor: 4-room apartment 1:200
3 5th floor: 4-room apartment 1:200
4 5th floor: 4-room apartment 1:200
5 3rd floor: 2-room apartment 1:200
6 Master plan by Jo Coenen 1:7500
7 Layout of 3rd, 5th, and 7th floor 1:1500

1.1
BLOCK EDGE

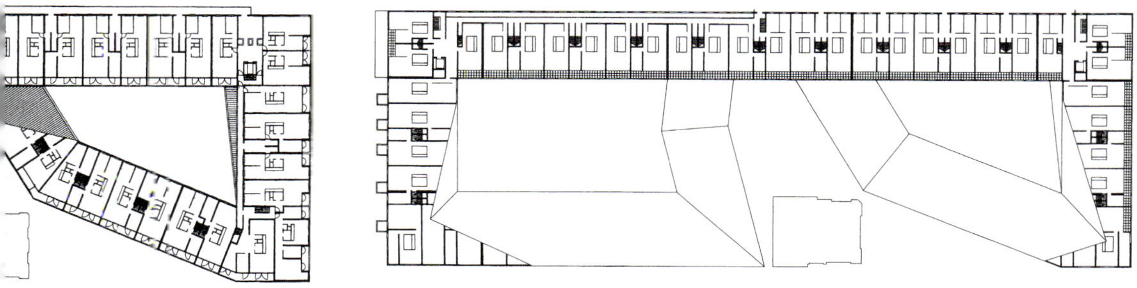

7

Building type
superblock
4–8 stories

Date of construction
1989–1994

Number of units
304

Size of units
2-room apts., 52–87 m² (64 units)
3-room apts., 71–106.5 m² (143 units)
4-room apts., 85–130 m² (75 units)
5-room apts., 110–153 m² (18 units)
5-room apt., (with atelier), 288 m² (1 unit)
6-room apts., 187 m² (3 units)

Area per user
21–57.5 m²

Building depth
13.5/14.5 m

Access
various:
Ground floor–4th floor
double-loaded stairwells,
5th–9th floor glazed
or roofed, open galleries

Open spaces
glazed loggias, balconies

Parking
underground garage

Architect
Hans Kollhoff
with Christian Rapp
Berlin

Location
Levantkade 8
KNSM-Eiland
Amsterdam
The Netherlands

4

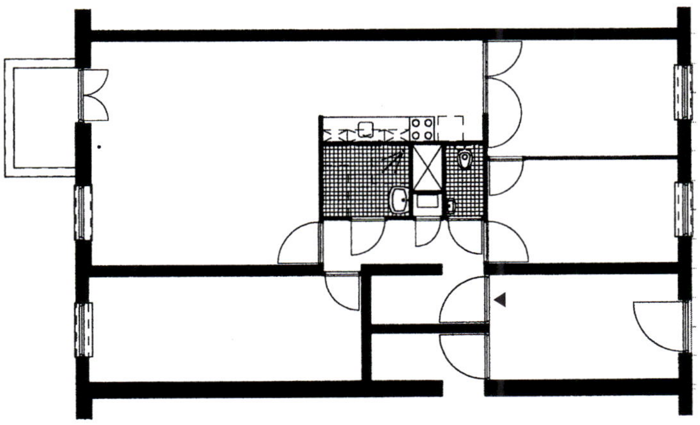

3

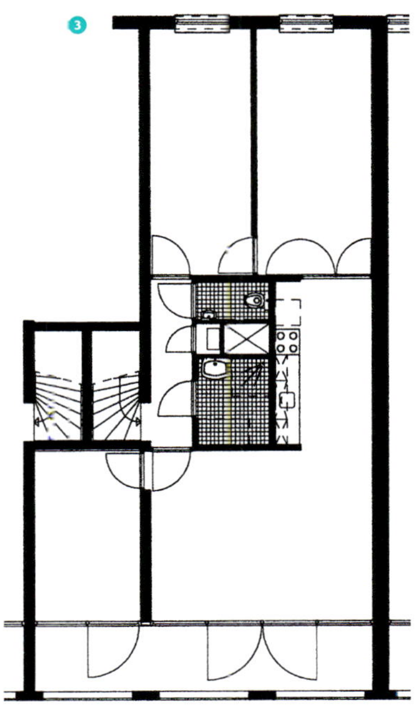

5

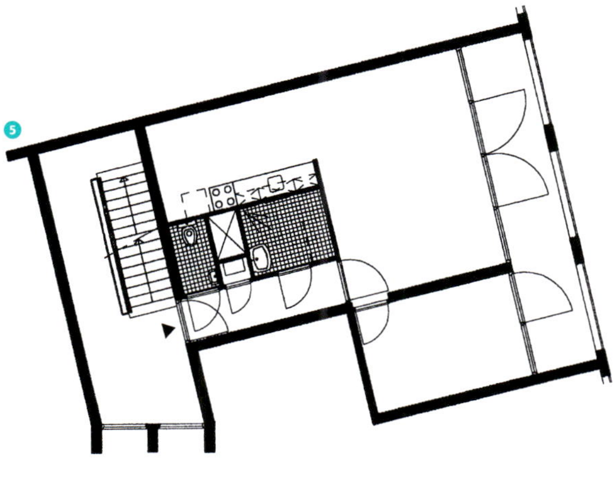

Referring to
floor plan 1:200

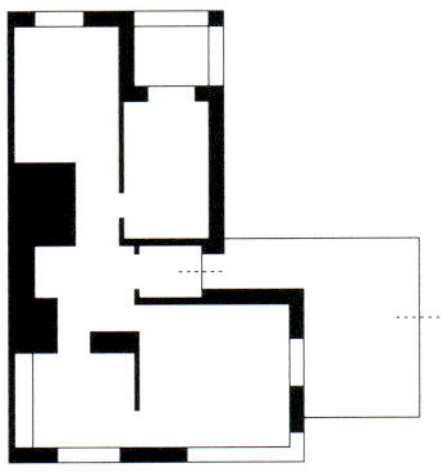

Sihlhölzlistrasse | Spühler | 1995

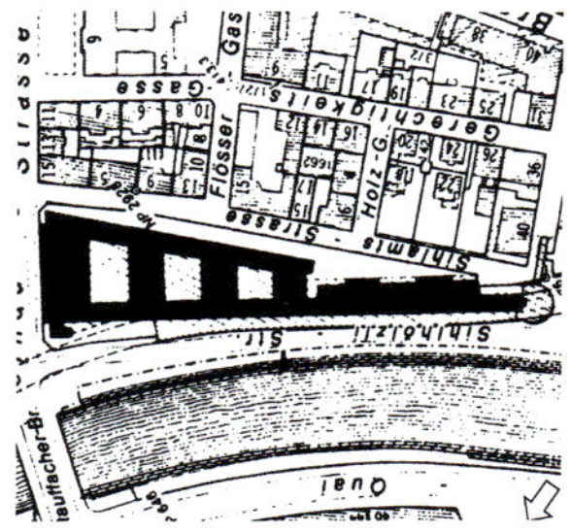

Construction on a long triangular plot at the waterside. Two narrow linear blocks run lengthways. Lateral buildings between these blocks create interior courtyards, which grow shorter toward the end of the complex and are open to the south. Towards the river the facades are smooth and 5 stories high; towards the residential road they are 3 and 4 stories high. There is an administration building at the base of the triangle. The ground floor contains shops, offices, and community services (kindergarten, crib and crèche). Each of the stairwells in the long rows (with individual elevator) provides access to just two apartments per landing. On the standard floor the apartment door leads, surprisingly, first to the large loggia, which is simultaneously an entrance and an open space, and serves as a buffer zone. The living space is adjacent. The apartments bend sharply around the interior courtyard, with their main rooms (living room, kitchen, dining area) facing the road, and the bedrooms facing the yard. The bend is the central point between the entrance, living room, and private sleeping corridor. The auxiliary rooms lie along the seam where the apartments connect, acting as a service spine serving both sides. The shape of the plot has endowed the lateral buildings with a variety of length, and thus also the apartments. The longer the building, the more rooms there are along the corridor. In part the apartments interlock at their (wide) ends. As a result there is a whole range of apartment sizes. The longest side building contains four large apartments serving as communal flats.

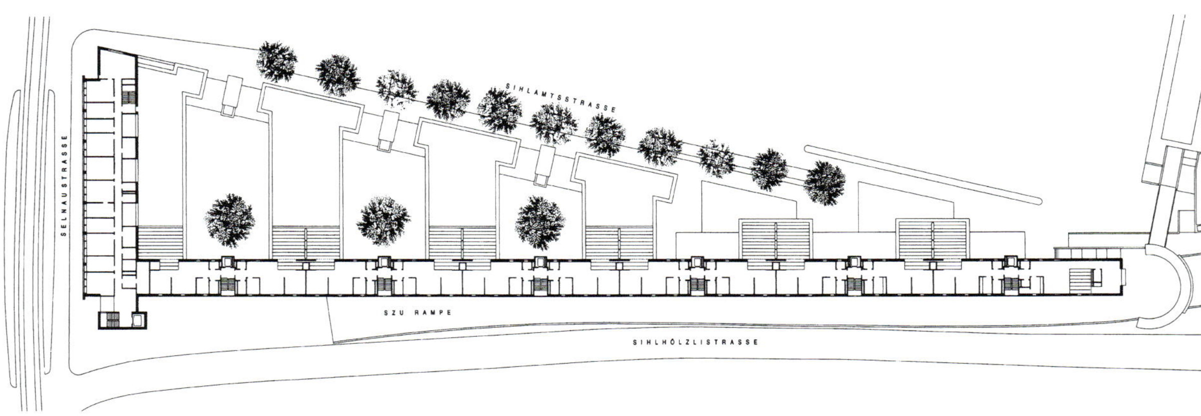

2

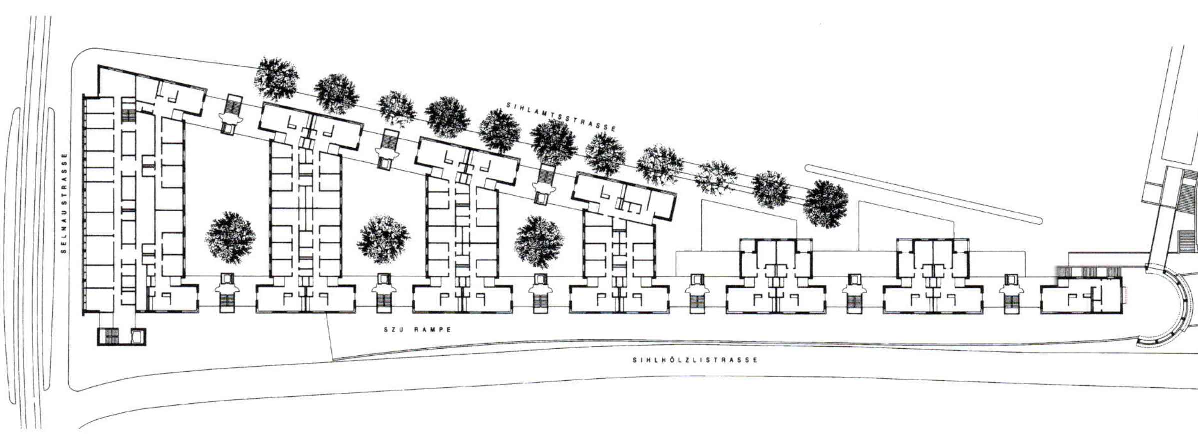

1

1 3rd floor of overall complex
2 4th floor of overall complex
3 3rd floor with 4½- and 5½-room apartments 1:500
4 3rd floor with 3-room apartments 1:200
5 5th floor with large 2-room apartments 1:500

1.1
BLOCK EDGE

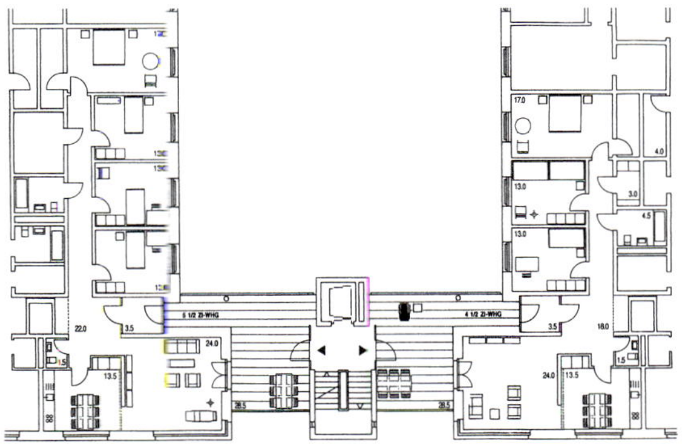

3

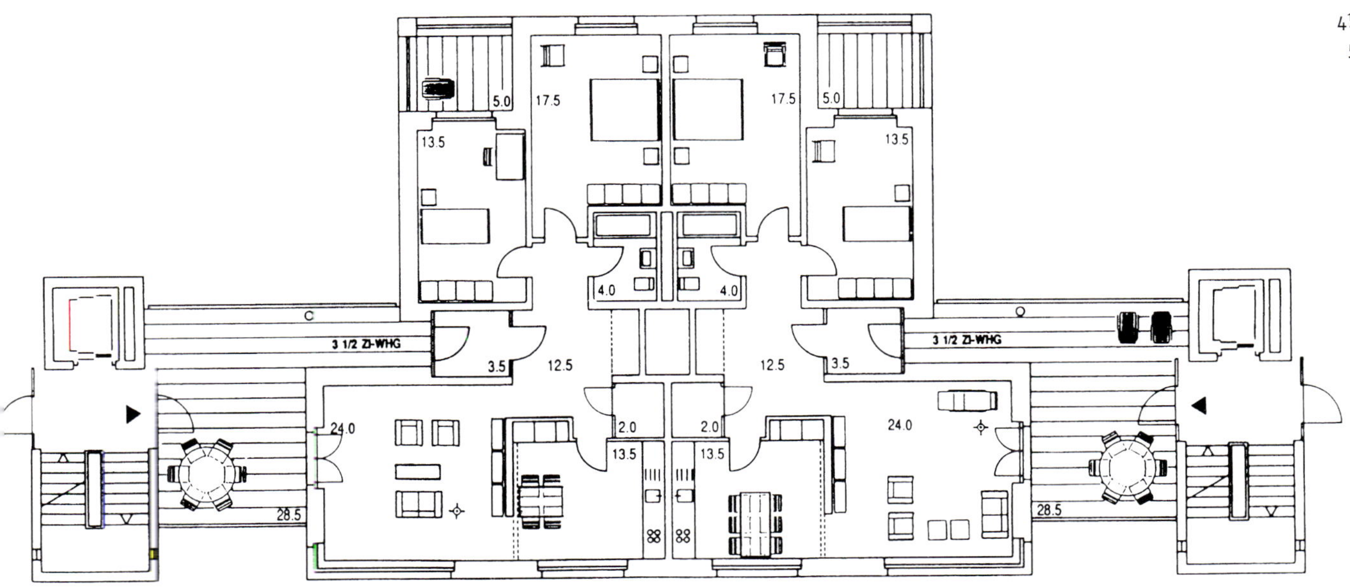

4

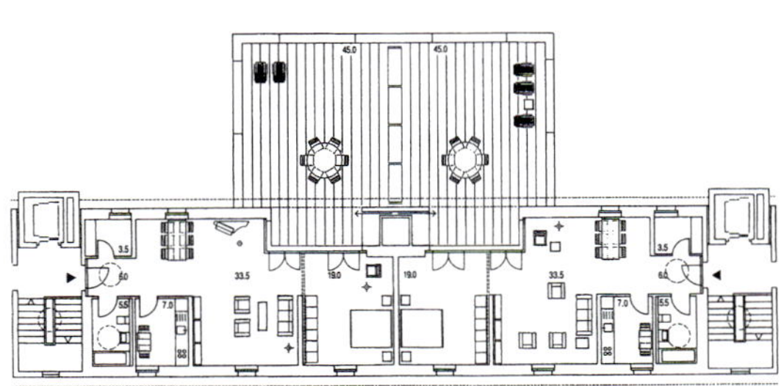

5

Building type
2 rows along road with lateral buildings creating courtyards, with offices, shops, administration
4–5 stories
facing NW/SE

Date of construction
1992–1995
(competition 1985, 1986, 1987)

Number of units
64
(incl. 4 large apartments)

Size of units
1½-room apt., 53 m² (1 unit)
2½-room apts., 74–100 m² (15 units)
3½-room apts., 90/94 m² (14 units)
4½-room apts., 112–145 m² (23 units)
5½-room apts., 127/136 m² (9 units)
6½-room apt., 177.5 m² (1 unit)

Area per user
26.5–53 m²

Building depth
7.8/14.8 m

Access
double-loaded stairwell with lift

Open spaces
spacious loggias and roof terraces provide access to the apartments

Parking
64 parking spaces in two-storied underground car park

Architect
Martin Spühler
with David Munz
and Bruno Senn
Zurich

Loction
Selnau-, Sihlamt-,
Sihlhölzlistrasse
Zurich
Switzerland

Referring to floor plan 1:200

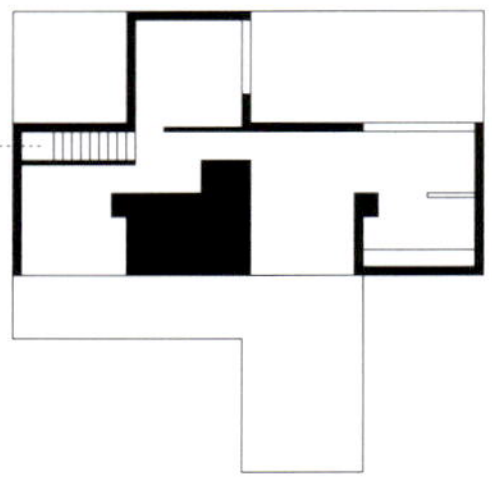

Hollainhof
Neutelings Riedijk | 1999

The "Hollainhof" subsidized housing project is located in the center of Ghent between the street and the riverside promenade. An interior courtyard has been created by concentrating the development along the edges of the lot. This open space provides tranquility and seclusion for the complex. The verdant courtyard offers access to most apartments, with the exception of a few ground-floor units facing the street. The rows consist of 5 and 9 individual cubes, respectively, which are linked via access elements or auxiliary rooms. Each cube is based on a square grid of 4 × 4 units. Each unit, in turn, can be occupied by a room or a terrace area. Living rooms generally extend across two side-by-side or sequential units. The various combinations of these building blocks result in a wide variety of apartment types, which are suitable for differing tenant groups. By joining different apartment types, each house is given a unique shape and the entire row presents an animated silhouette. The street elevation, alone, presents a cohesive image provided by a long wall panel with vertical slits. It acts as a visual and acoustic shield between the street and the covered walkways. These decks and their small front gardens create a buffer zone for the flats and maisonettes on the second and third floors. On the riverside elevation, the cubes are composed of row houses with private gardens and the complex configuration of flats and maisonettes above, which are accessed via exterior stairs.

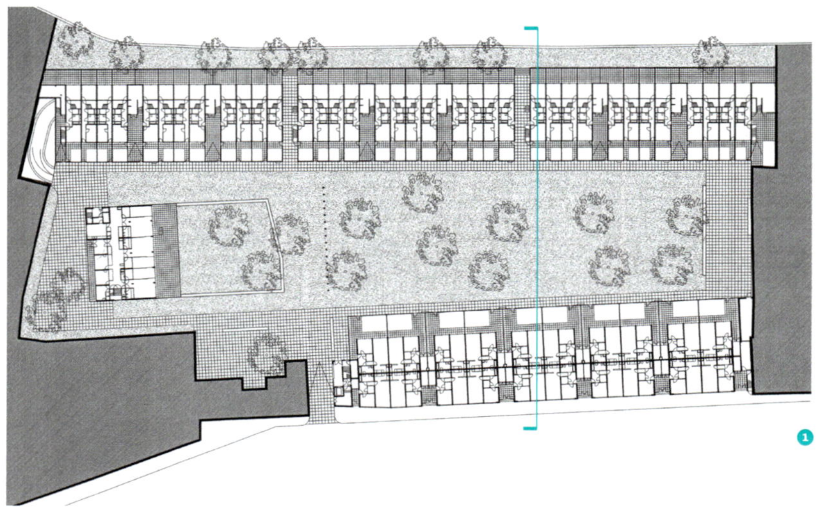

1

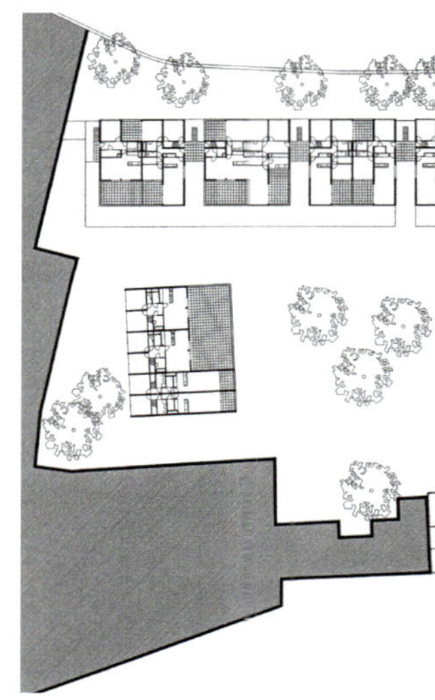

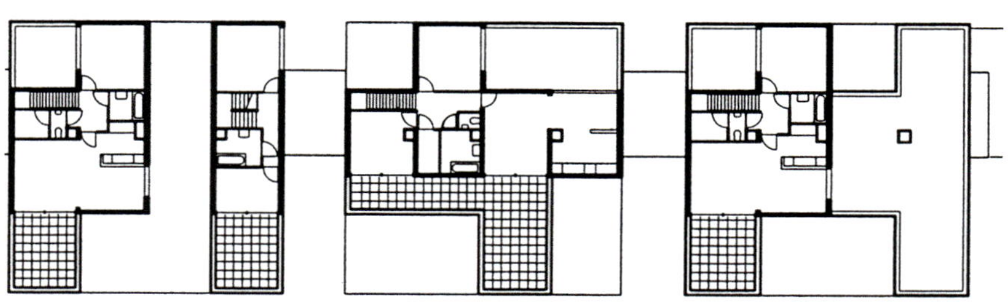

5

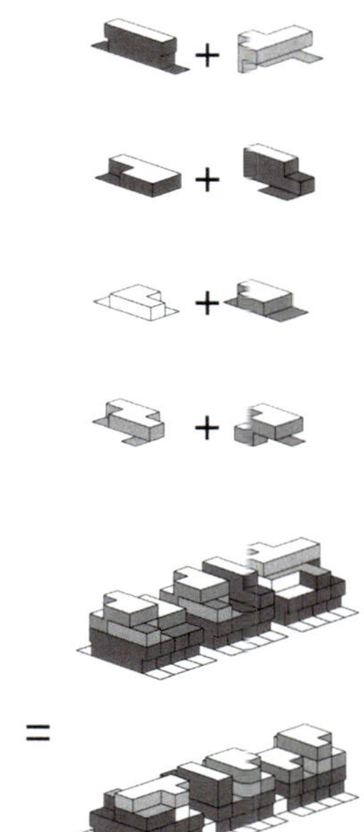

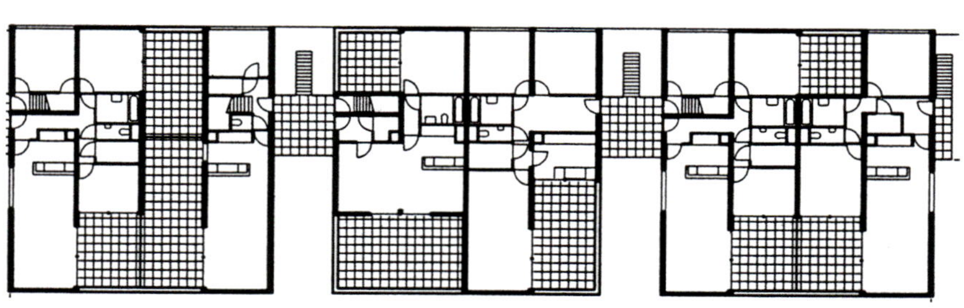

4

1 Ground floor of overall complex
2 3rd floor of overall complex
3 4th floor of overall complex
4 Floor plan detail, riverside row 3rd floor 1:500
5 Floor plan detail, riverside row 4th floor 1:500
6 Lower and upper level of one of many building cubes variants 1:200

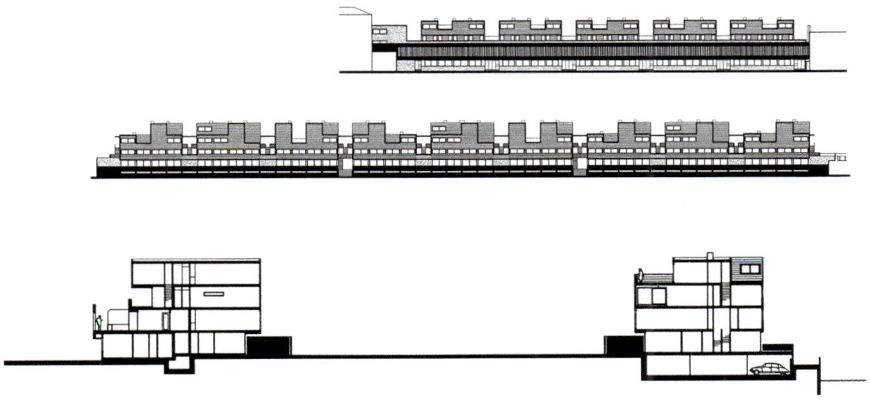

1.1
BLOCK EDGE

2

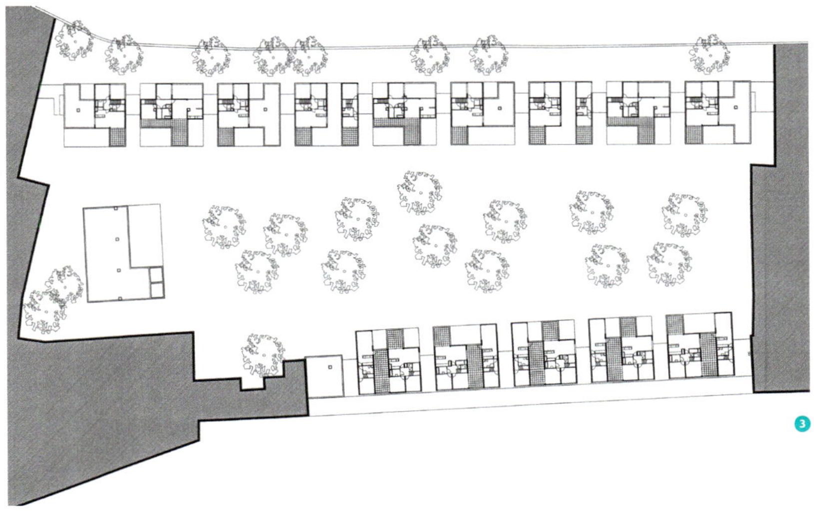

3

Building type
2 parallel rows of differing building cubes along street and river, resp.
4 stories
facing NE/SW
daycare in courtyard building

Date of construction
1993–1999

Number of units
123

Size of units
14 unit types
2-room mais., approx. 60 m² (5 units)
3-room mais., approx. 73 m² (14 units)
4-room mais., approx. 92 m² (38 units)
2-room apt., approx. 54 m² (55 units)
3-room apt., approx. 68 m² (11 units)

Area per user
22.5–30 m²

Building depth
15–17 m/14–15.5 m

Access
direct access from courtyard or street,
exterior stairs
covered walkways

Open spaces
private gardens
generous roof terraces
interior courtyard with greenery
public promenade along the Schelde river

Parking
underground garage with 90 parking spaces,
entrance ramp behind daycare

Architect
Neutelings Riedijk Architecten
Rotterdam

Location
Ghent
Belgium

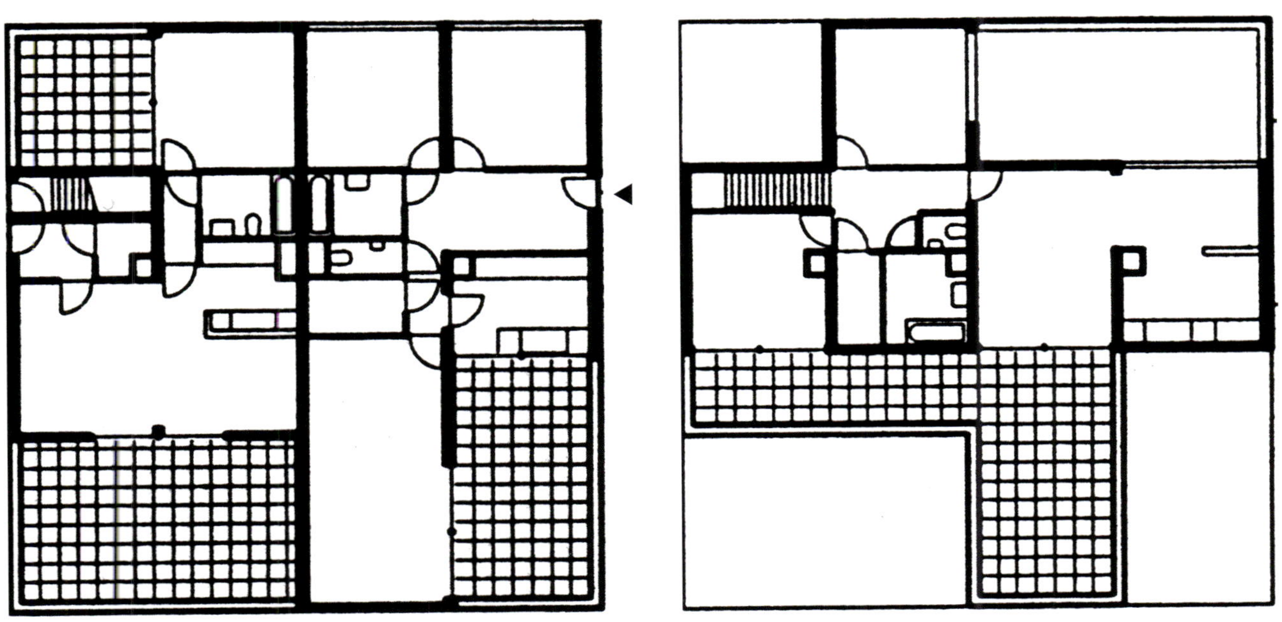

6

Referring to floor plan 1:200

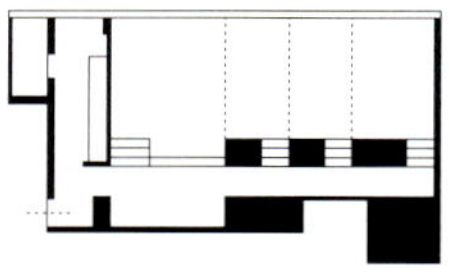

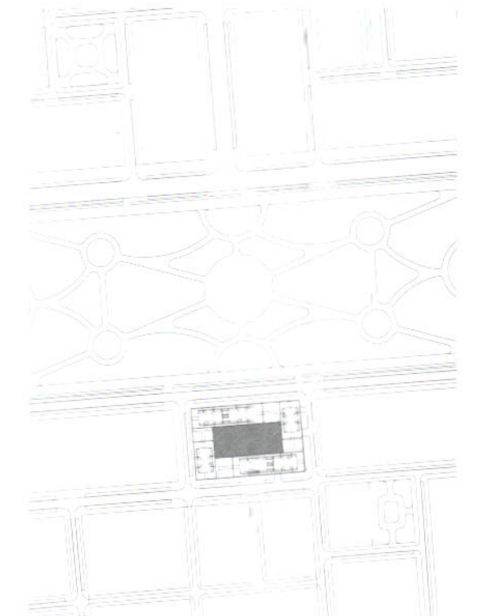

Carabanchel | Aranguren & Gallegos | 2003

Two requirements define the social housing project in Carabanchel: rationalization of the elements with a view to keeping construction costs low, and the need for a high degree of flexibility. The development consists of blocks that are lined up in sequence. Each block arrangement occupies an entire city block and creates a large, private inner courtyard. The standardization of the building elements and the rational layout of the floor plans are applied to both the exterior and the interior.

All residential units are organized in the same manner: kitchens and sanitary blocks, located at the center partition wall between units, create a fixed, raised core, while the remaining space of the apartment is arranged along the facade and interpreted as a flexible space. Various elements – sliding walls, built-in cabinets, and niches for bed storage recessed into the floor in the corridor – make it possible to transform this open space which is bathed in natural light. The raised dining area expands this field, adds space to the living room and serves as an interface with the other areas. A subtle transition of four steps between corridor, kitchen, baths, and rooms separates the fixed from the flexible spaces. It also supports the idea of habitation as an activity on the part of the residents: that is, room configurations that differ depending on use and time of day.

2

1

1. 2nd and 3rd floor of the whole complex 1:500
2. Cross- and longitudinal section 1:500
3. Different apartment types on ground floor and 2nd i.e. 3rd floor
4. Apartment type **A**, floor plan, day and night situation 1:200
5. Apartment type **B**, floor plan, day and night situation 1:200
6. Apartment type **B**, longitudinal section 1:200
7. Apartment type **B**, cross sections 1:200

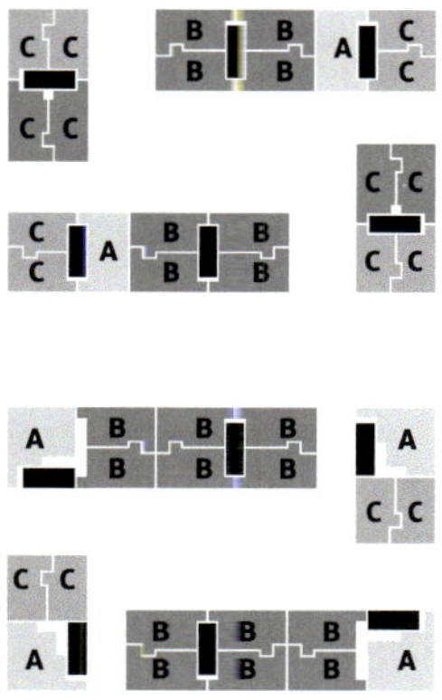

3

1.1
BLOCK EDGE

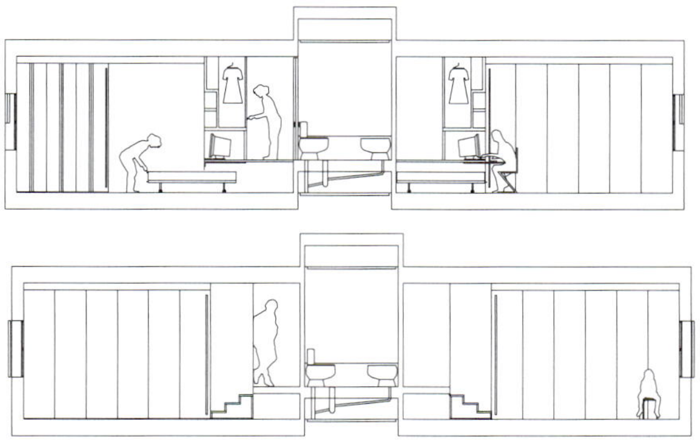

7

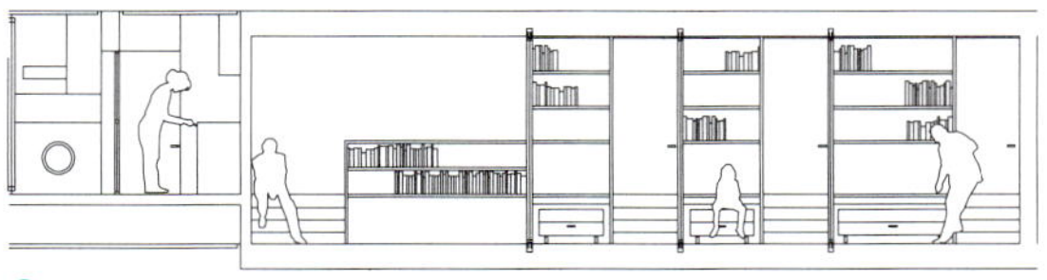

6

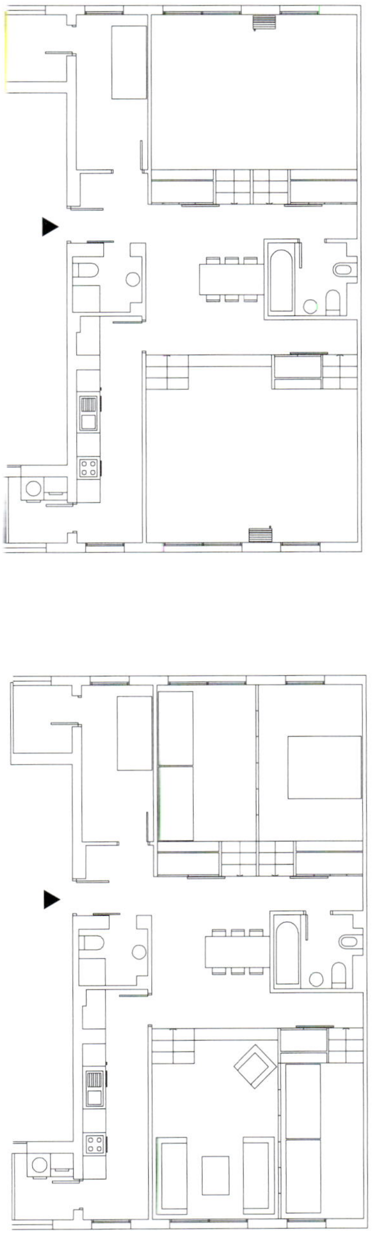

4

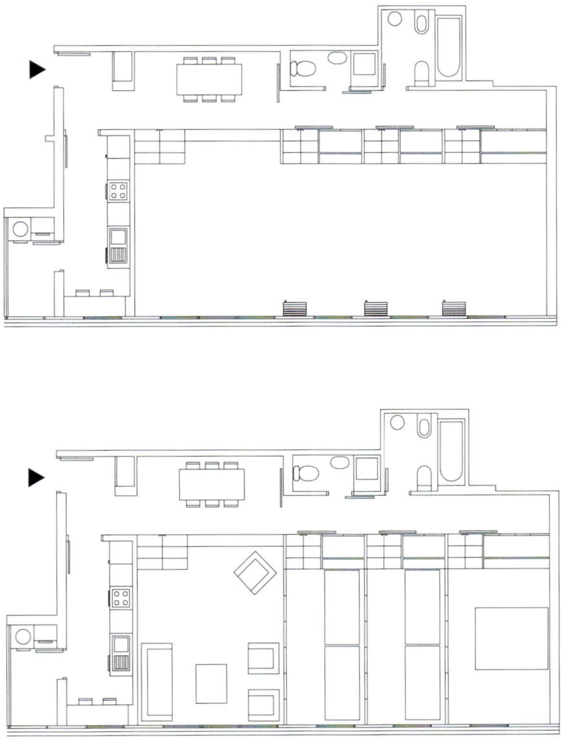

5

Building type
open block
4 stories
facing W/E

Date of construction
2001–2003

Number of units
64

Size of units
5-room apts., type **A**: 106 m²
4-room apts., type **B**: 90 m²
3-room apts., type **C**: 73 m²

Area per user
18.5–22.5 m²

Building depth
13.5 m

Layout
four slabs form an open block, triple- and quadruple-loaded stairwell

Open spaces
rock garden in interior courtyard, kitchen balconies, roof terraces

Parking
no parking on lot

Architect
Aranguren & Gallegos Arquitectos SL
Madrid

Location
Ensanche 6
Madrid-Carabanchel
Spain

Referring to floor plan 1:200

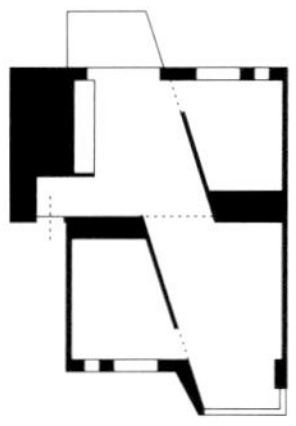

Østerbrogade | C. F. Møller | 2006

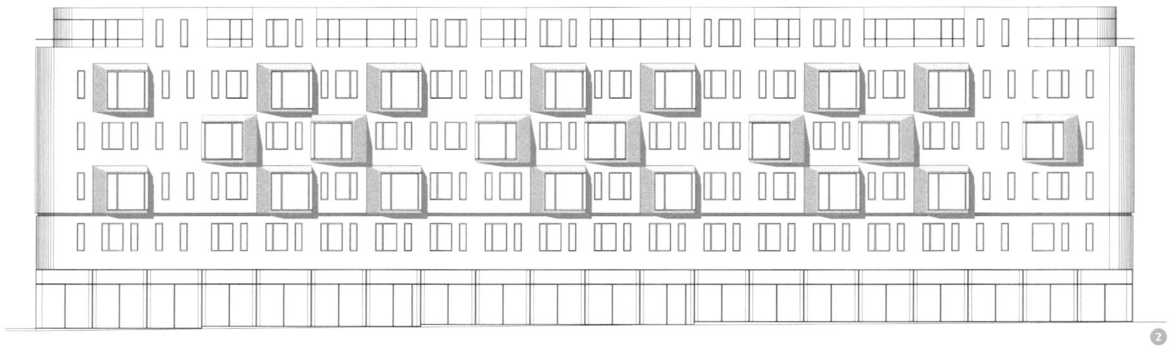

The building is sited along the full length of a triangular, inner city block, defining the corners of the block with its distinctive curves. The standard floor plan of the apartments is defined by an open-plan area that runs on a diagonal from front to back: from the south-facing balcony on the courtyard side to the bays on the street side. Although the living area is only 2 m wide, it opens diagonally into the kitchen and the living room, respectively. On the other sides, the two individual rooms are linked along the facades with sliding doors, making it possible to experience the full width of the apartment when these are opened. The floor-plan pattern that emerges – also by means of the seamless integration of built-in cabinets and kitchen counters, the placement of the bathroom behind the access core, and the option of partitioning the large living area – creates an ambiguous, open space with natural zones for access, eating, cooking, and living. Due to the alternating crosswise arrangement of the floor plans from one level to the next, the bays also appear in an alternating fashion on the exterior, creating a dynamic external expression that reflects the diagonals in the internal layout, thus informing the sculptural character of the entire building.

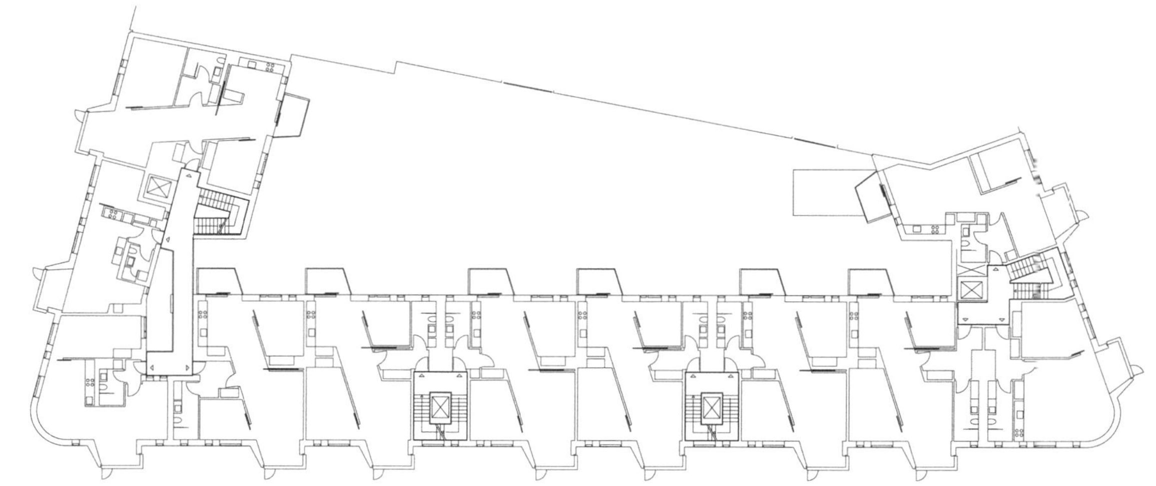

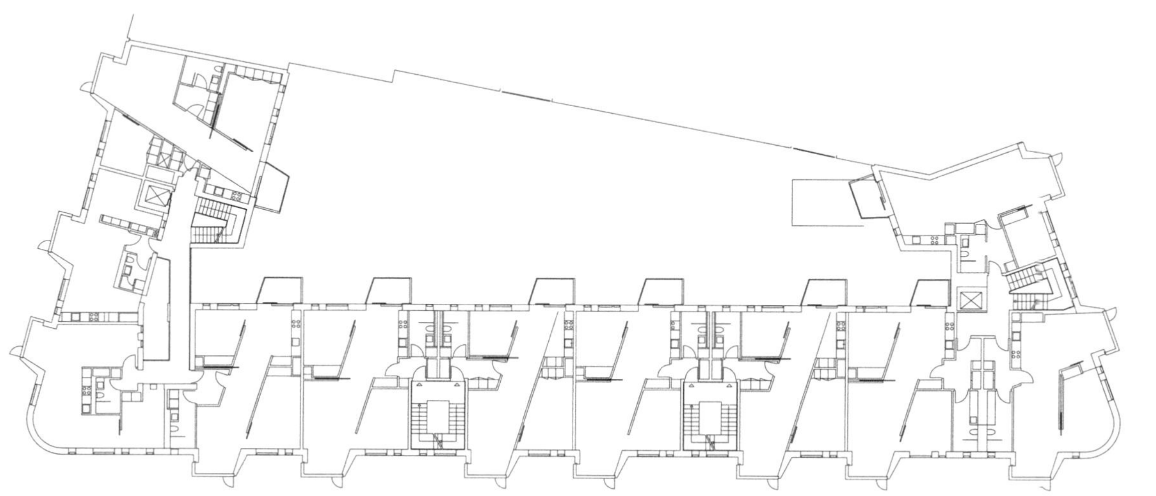

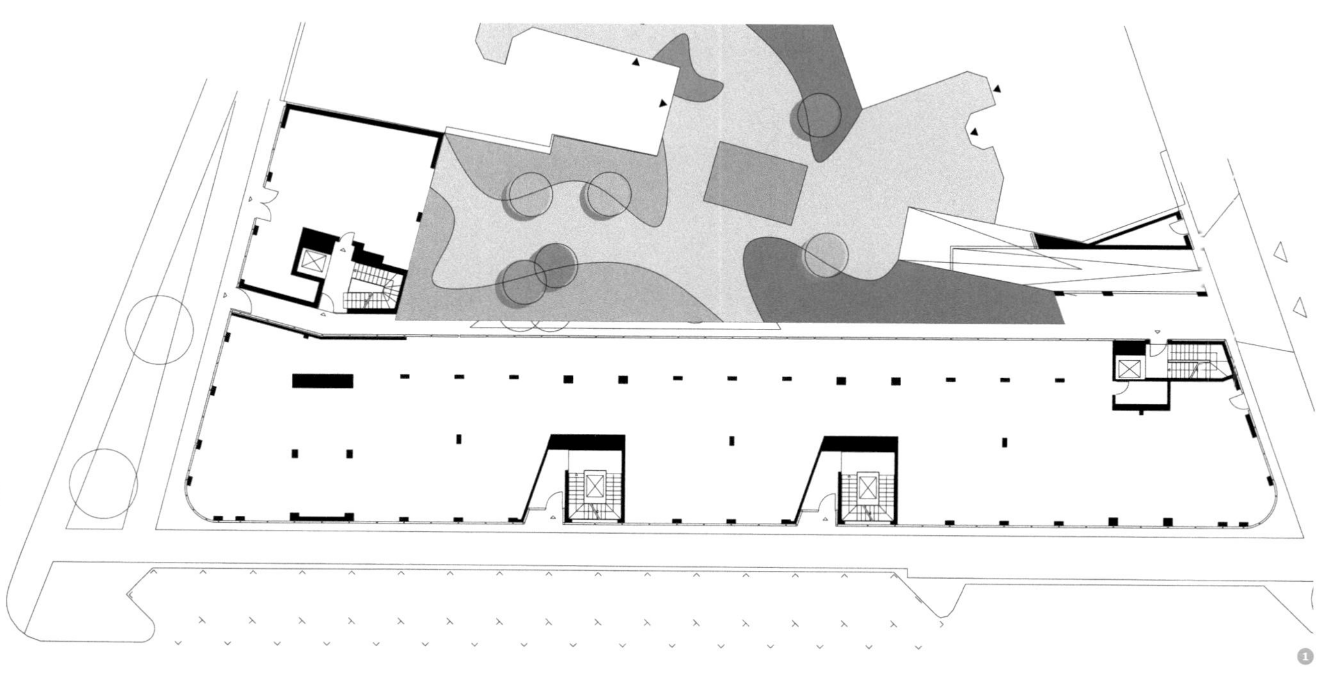

1. Total floor plan, ground floor, 2nd floor, 3rd floor 1:500
2. North elevation 1:500
3. Floor plan section of typical floor plan with 3-room apts. 1:200
4. Furnishing suggestion
5. Cross section 1:500

1.1
BLOCK EDGE

Building type
C-shaped development
6 stories with retail on ground floor
facing NW/E, E/W, WS/EN

Date of construction
2004–2006

Number of units
52

Size of units
3-room apts., 95 m²

Area per user
31.5 m²

Building depth
12.5 m

Access
double-loaded stairwell

Open spaces
balconies

Parking
underground garage with
29 parking spaces

Architect
Arkitektfirmaet C. F. Møller
Århus

Location
Østerbrogade 105
Copenhagen-Østerbro
Denmark

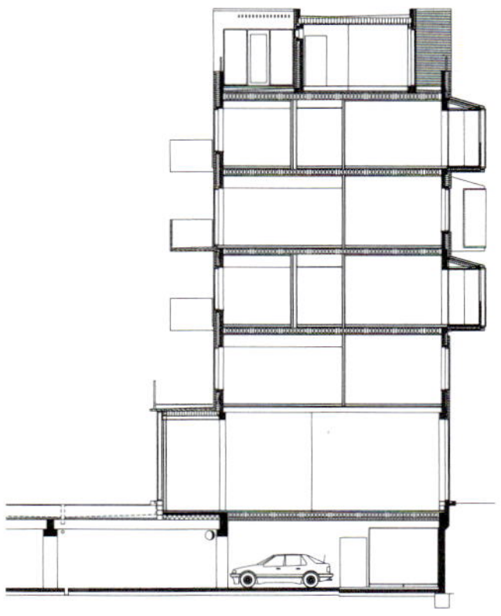

5

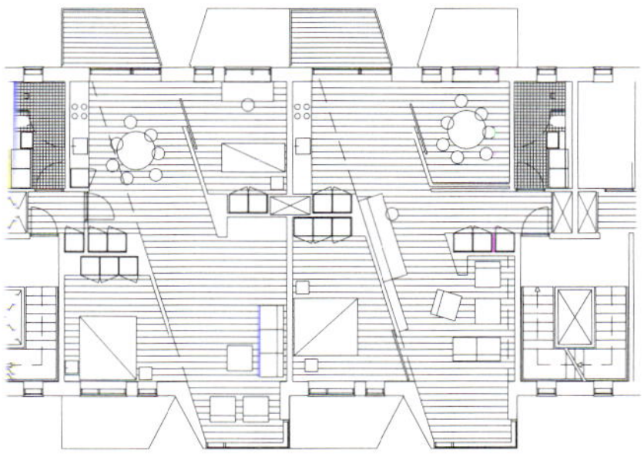

4

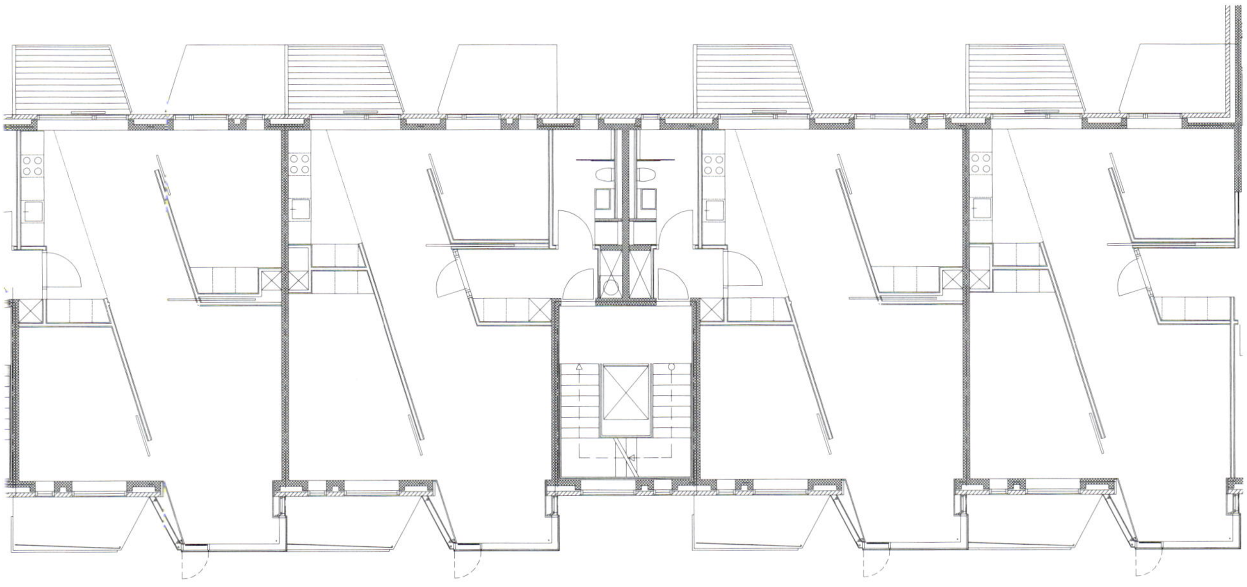

3

Referring to
floor plan 1:200

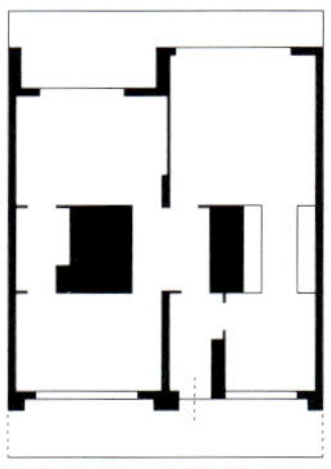

Staalmanplein | Office Winhov | 2010

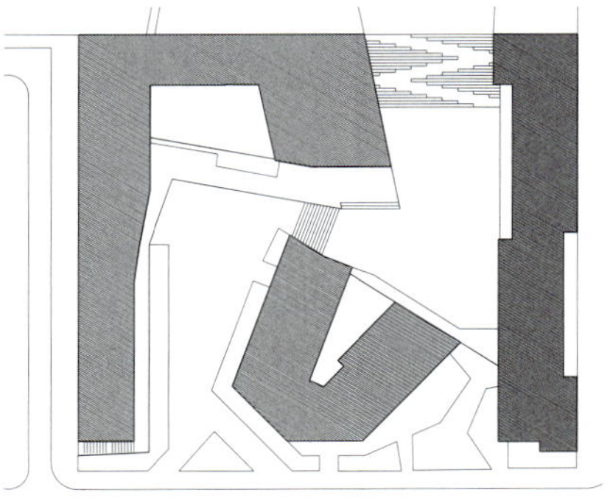

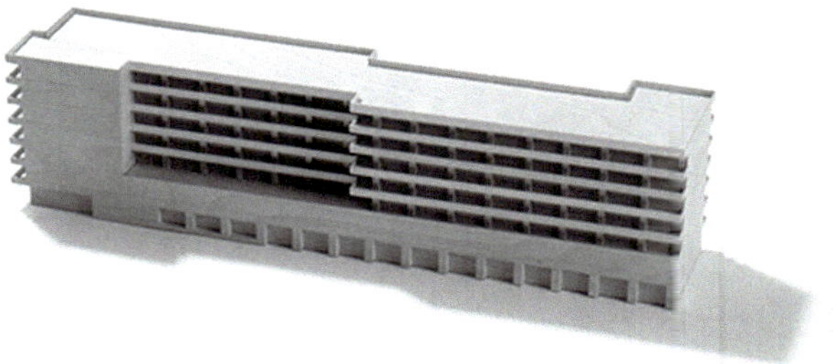

In a post-war development on the outskirts of Amsterdam, two linear slab blocks form the east and west edges of an urban block ensemble framing a raised plaza above a parking garage in between, on which a mosque, a solitary residential building, and a school are located. This plaza opens to the north towards a public park and the sporting facility of a school via a generous stairway. The distinctively articulated plinth of the eastern block edge building mediates between the different levels of the street and the raised courtyard. It houses spacious entrance halls open to both sides, which lead to the half-sunken gymnasium and the apartments above. At the building ends, individual entrances to the duplex units (Types **A** and **C**) lead directly to the street, their transparent facades also enabling their usage as commercial spaces.

The floor plans of some of the flats (Type **D**) follow a simple but effective principle of division into three sections. The mid-section – a row of storage spaces, built-in furniture, inserted cores with bathrooms and WCs, and circulation areas – is offset to the sequences of rooms along both facades in such a way as to allow numerous crosslinks: from the living room to the dining space through the kitchen, from the entrance hall to the three rooms through a corridor or – in some apartments – from room to room through a shared closet. A certain degree of flexibility is maintained. For instance, the rooms can be combined by removing the non-structural dividing wall between the living room and bedroom facing the courtyard. And as the dining area has its own access point, it can be separated from the kitchen and used for other purposes. The principle of multiple links on both sides of an inserted core is also implemented in the 2-room apartment (Type **B**).

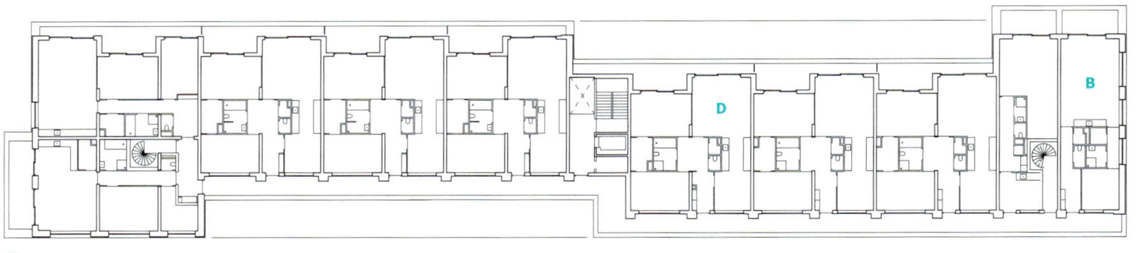

3

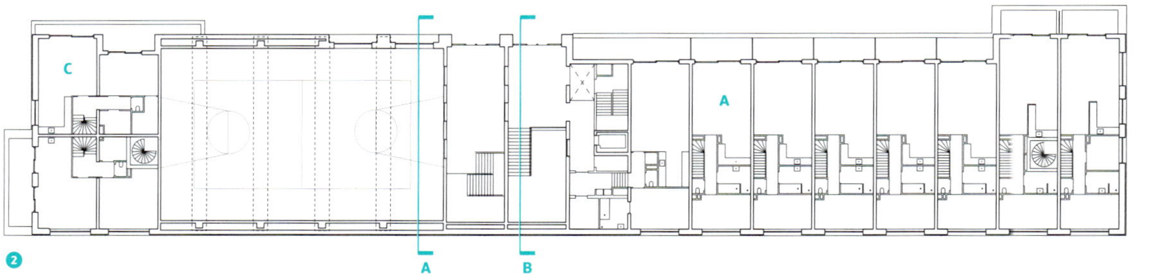

2

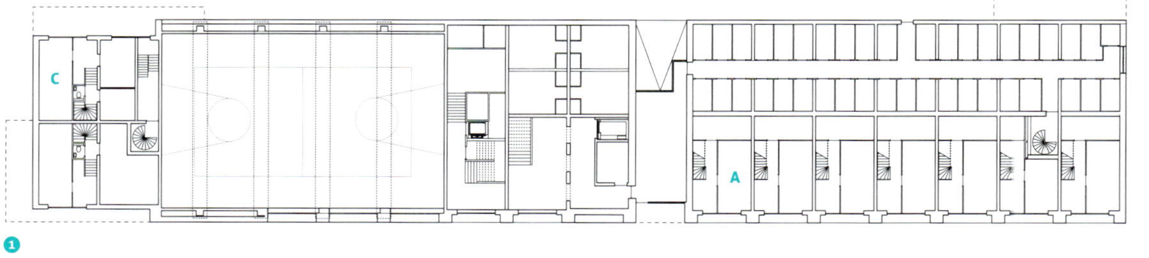

1

1 Street level floor plan
2 Courtyard level floor plan
3 Typical floor plan
4 Cross section with gymnasium 1:500
5 Cross section with stairs 1:500
6 3-room mais. Type **A** 1:200
7 2-room apt. Type **B** 1:200
8 4-room mais. Type **C** 1:200
9 3- to 4-room apt. Type **D** 1:200

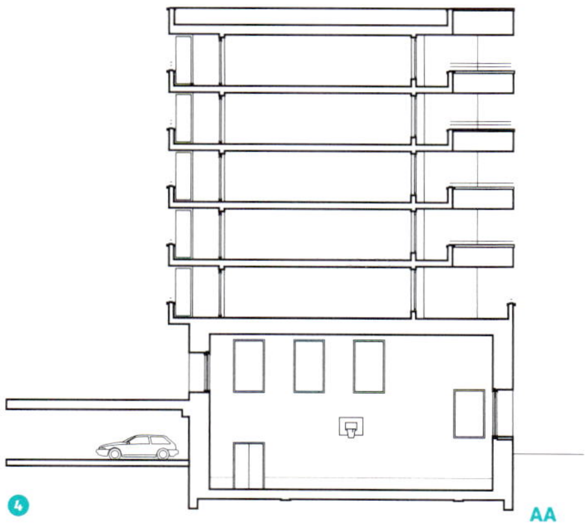

4

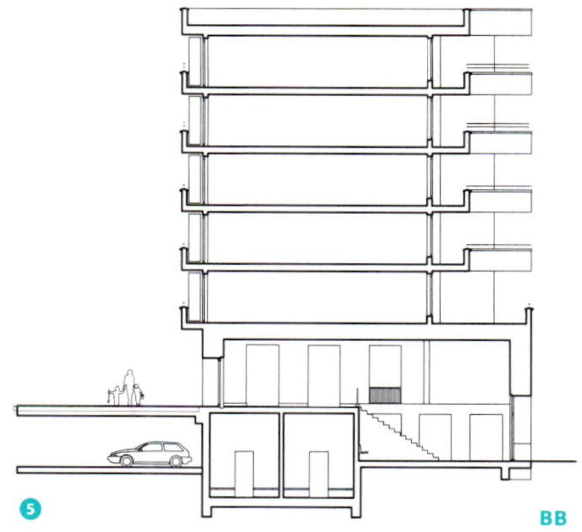

5

1.1
BLOCK EDGE

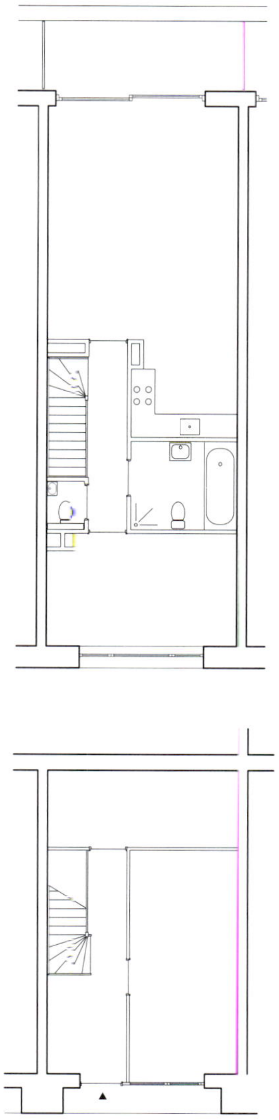

6

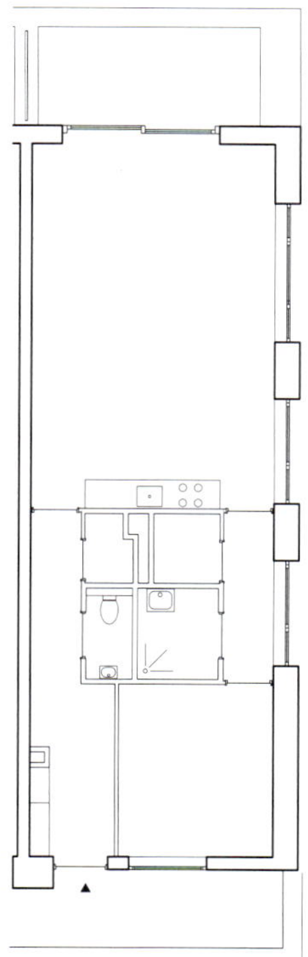

7

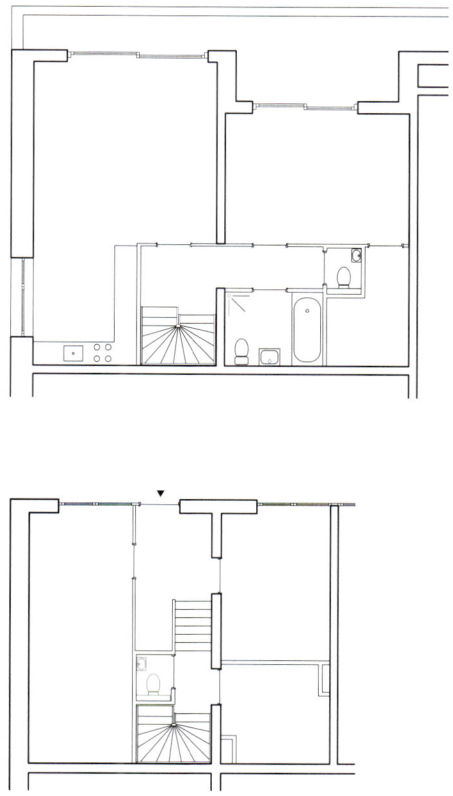

8

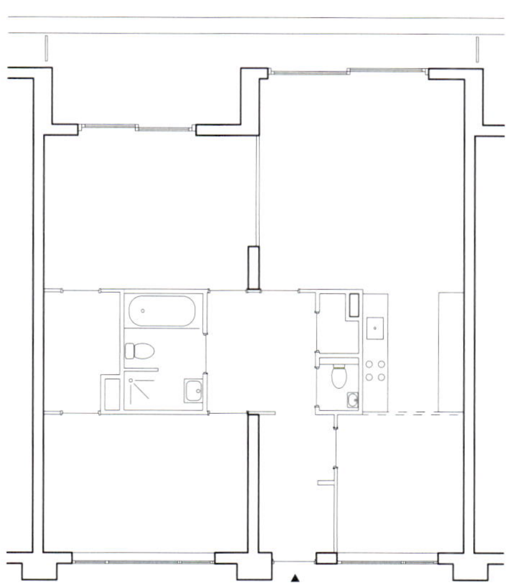

9

Building type
block edge slab
with gymnasium at street level
and public courtyard
one floor above
6/7 stories
facing E/W

Date of construction
2009–2010

Number of units
60

Size of units
studios, 57 m² (5 units)
2-room apts., 69–88 m² (6 units)
3-room apts., 79–97 m² (10 units)
3-room mais., 99–121 m² (8 units)
3–4-room apts., 102 m² (30 units)
4-room mais., 125 m² (1 unit)

Area per user
25.5–57 m²

Building depth
12–15 m

Access
external access gallery, duplex units
with street access

Open spaces
loggias, balconies

Parking
garage below public courtyard

Architect
Office Winhov
Amsterdam

Location
Ottho Heldringstraat
Staalmanplein
Amsterdam Slotervaart
The Netherlands

Referring to
floor plan 1:200

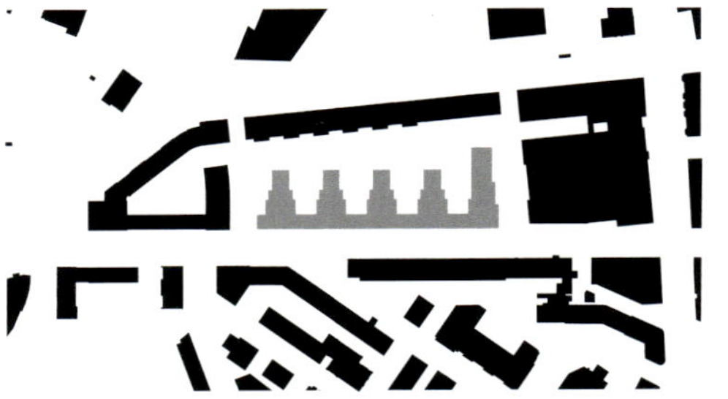

1
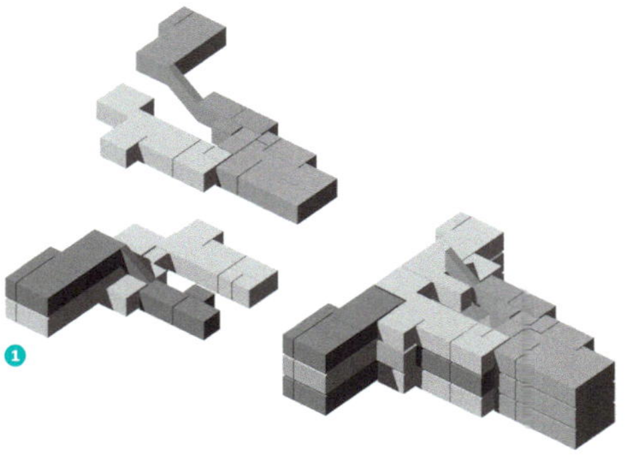

Brunnmatt-Ost | Esch Sintzel | 2013

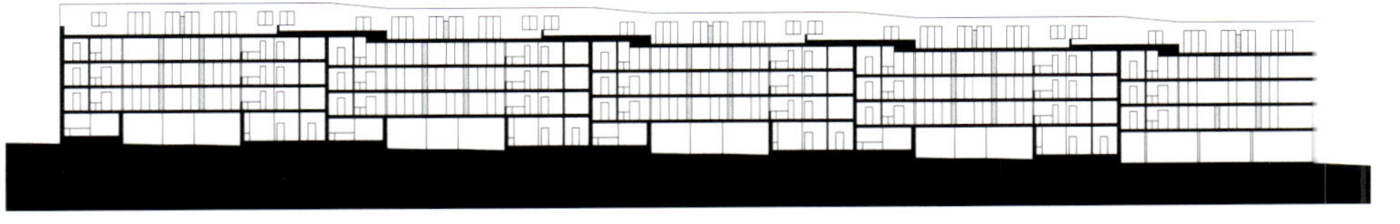

6

Intent on orienting as many apartments as possible in both directions – both toward the bright and lively but also loud street as well as the quiet, green courtyard – a block edge building is connected to five side wings. In the T-shaped segments, where the front building transitions into the tapered, dually oriented side wings, different apartment types emerge from layered sequences of rooms. Strings of connected and side-by-side spaces run from the front to the back of the apartment. They are at times passages from one room to another, at other times enclosed spaces and open up toward either the street or the courtyard.

In the single-story units (Type A and B) a room-sized loggia, eat-in kitchen, and bathroom are lined up along the street facade. Following in the layer behind on the courtyard side is a large living room and one or two bedrooms. At the juncture of the front building and the side wing, a second bedroom oriented toward the side courtyard, a bathroom, and an open hallway are lined up with the apartment entrance.

This spatial system is similarly applied in the front section of the duplex units (Types C and D). Instead of a hallway, however, a staircase descends or ascends into the side wing, where a further living room leads to an entrance area, from which an additional room and bathroom branch off.

This spatial arrangement in the back area is again repeated in the smaller single-story units (Type E). In the front building, a narrow connecting hallway, analogous to the duplex staircase, leads into an open eat-in kitchen followed by a loggia.

Complementing these similarly laid-out and cleverly interlaced units oriented in all directions are the apartments (Type F) at the rear, which look out on three sides toward the courtyard. Here the living room is a meandering circulation space at times looking out onto one courtyard or together with the kitchen toward another.

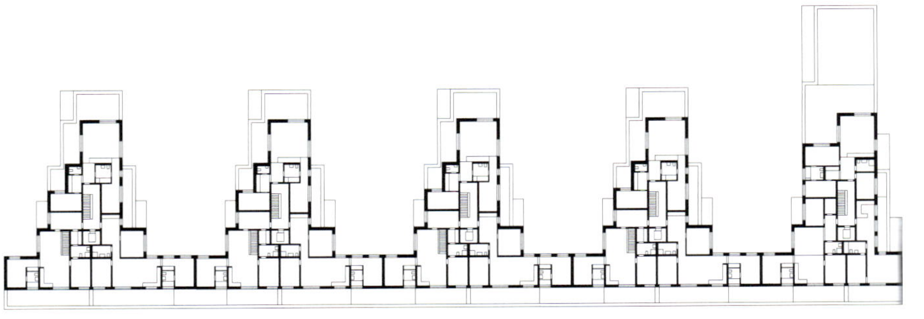

5

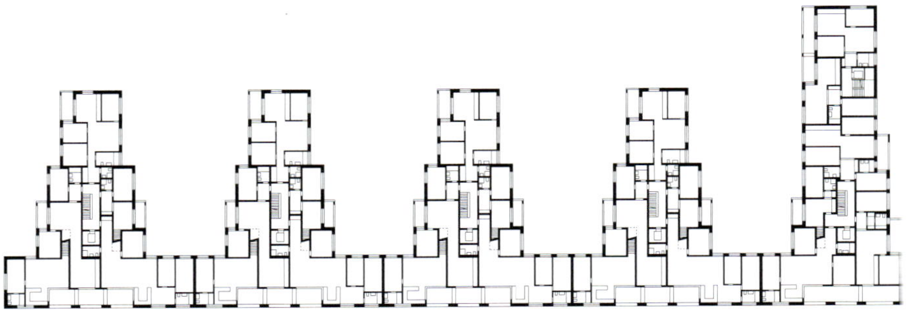

4

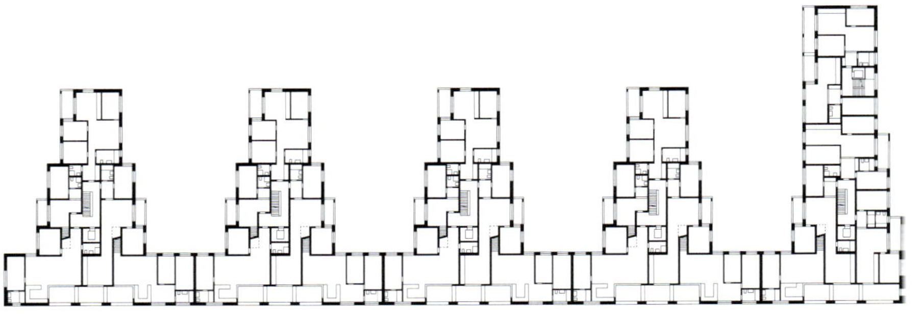

3

2

1 Combination principle of residential units
2 Ground floor 1:1250
3 2nd and 4th floor 1:1250
4 3rd floor 1:1250
5 Top floor 1:1250
6 Longitudinal section 1:1250
7 Cross sections 1:1000
8 2nd floor 1:200
9 3rd floor 1:400

1.1
BLOCK EDGE

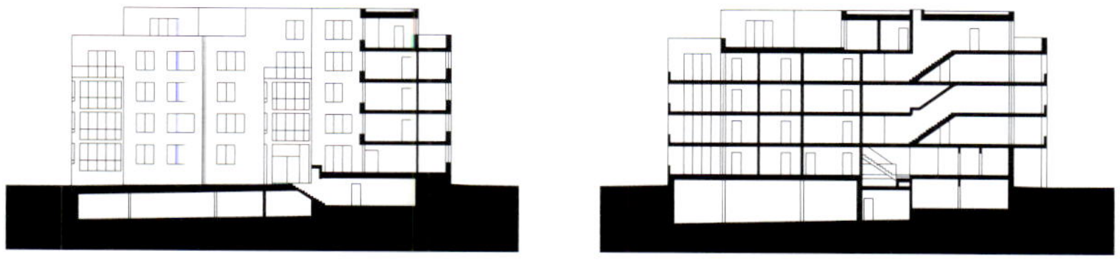

7 AA BB

Building type
comb-like block edge development
with setback levels
5 stories
facing in all directions

Date of construction
2009–2013

Number of units
95

Size of units
2½-room apts., 51–77 m² (5 units)
3½-room apts., 80–117 m² (37 units)
3½-room mais., 109 m² (5 units)
4½-room apts., 104–124 m² (31 units)
4½-room mais., 130 m² (4 units)
5½-room apts., 124 m² (4 units)
5½-room mais., 140–154 m² (9 units)

Area per user
25–39 m²

Building depth
8.5–18 m

Access
quadruple-loaded
staircases in the middle
axes of the side wings

Open spaces
terraces, loggias, balconies,
roof terraces

Parking
limited parking on side street

Architect
Esch Sintzel Architekten
Zurich

Location
Schwarztorstrasse 102–110
Bern
Switzerland

9

8

Referring to
floor plan 1:200

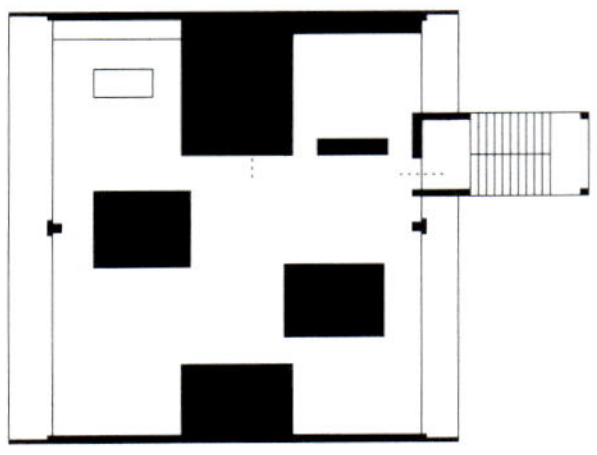

Am Lokdepot | ROBERTNEUN™ | 2013

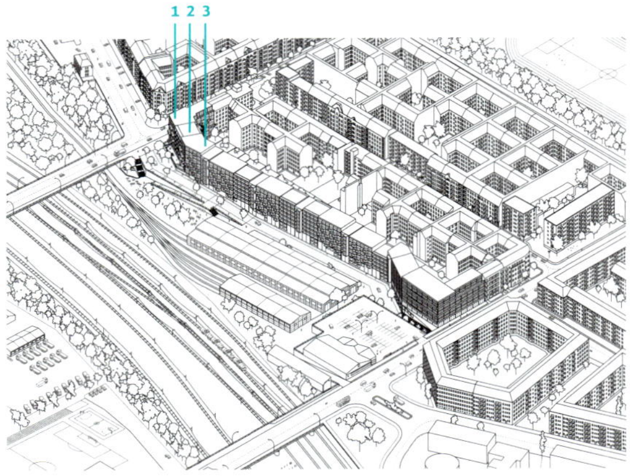

The buildings are part of a larger conversion scheme for a former rail yard on the edge of a new park, which is intended to contribute to urban densification. The conceptual and aesthetic inspiration for the residential development was industrial buildings, whose structural and robust nature offers spatial possibilities for both living and work configurations that can be interpreted individually and rearranged by the user.

In building **3**, the units are arranged around a room-sized central hallway in such a way that almost every other field within the spatial grid of the apartment is directly accessible. However, the flexibility in use for each room stems not only from this spatial configuration: by implementing a concrete skeletal structure, almost all the vertical planes can be filled in or changed at will – either as dividing walls or built-in furniture. The level of material transparency between adjacent rooms is similarly open to interpretation. In addition, the 1½-height spaces along the facade form particular focal points for the apartment.

The loft-like apartments in building **2** are each laid out over an entire floor and are directly accessed by elevator or external staircase. The placement of two interior mechanical cores divides the apartments into three parallel zones, providing a sense of how the floor plan could be organized without specifically defining the use of each space. For example, the hallway could be placed in the middle with enclosed functional zones at either end in order to leave the rooms along the facade as open as possible. Or more cores could be inserted on the apartment perimeter to create differently sized zones, while still ensuring free circulation within the apartment.

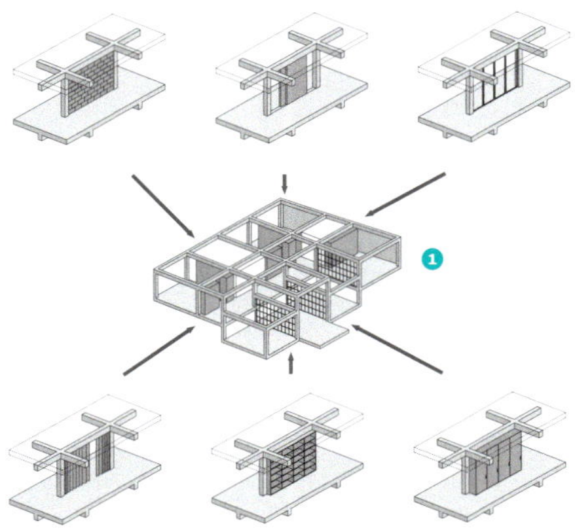
1

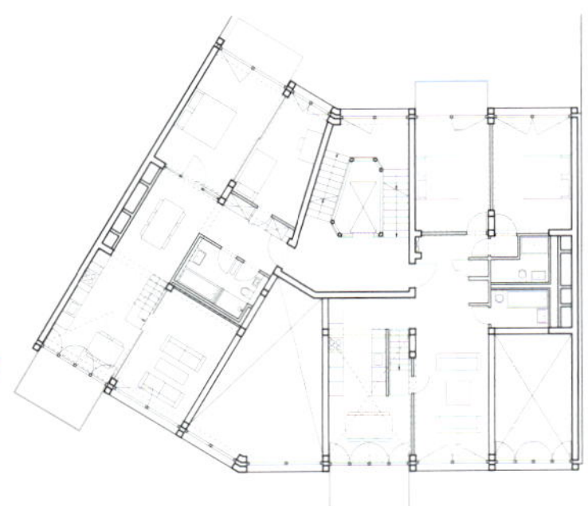
4

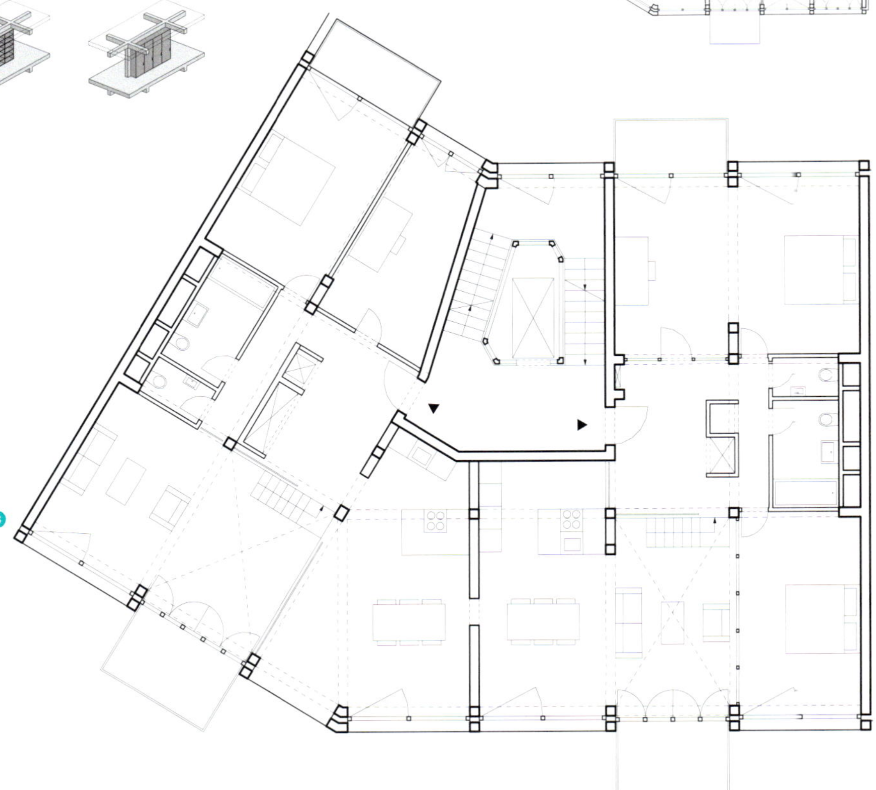
3

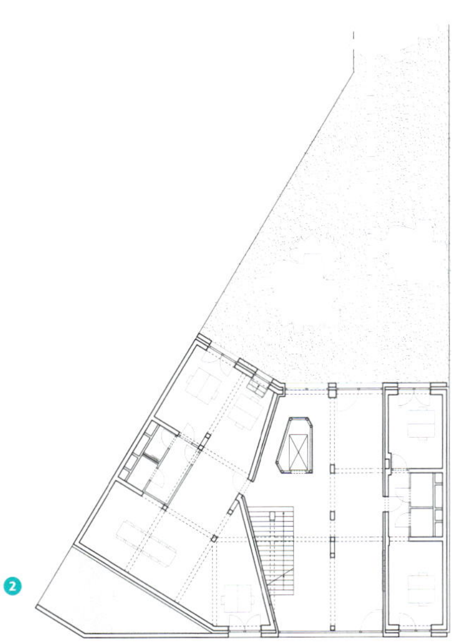
2

1 Structural shell/infill principle in buildings **1** and **3**
2 Building **3**: ground floor 1:500
3 Building **3**: 2nd floor 1:200
4 Building **3**: 3rd floor 1:500
5 Building **2**: ground floor 1:500
6 Building **2**: floor plan variation 1:200
7 Building **2**: floor plan variations 1:500
8 Cross section building **2**
9 Building **1**: typical floor 1:500
10 Building **1**: overview of apartment volumes

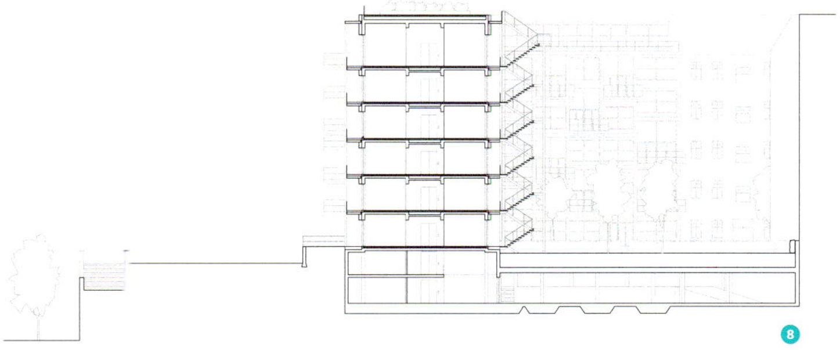

8

1.1 BLOCK EDGE

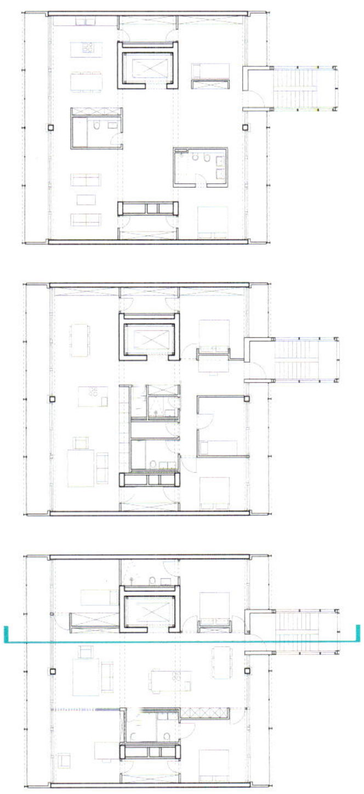

7

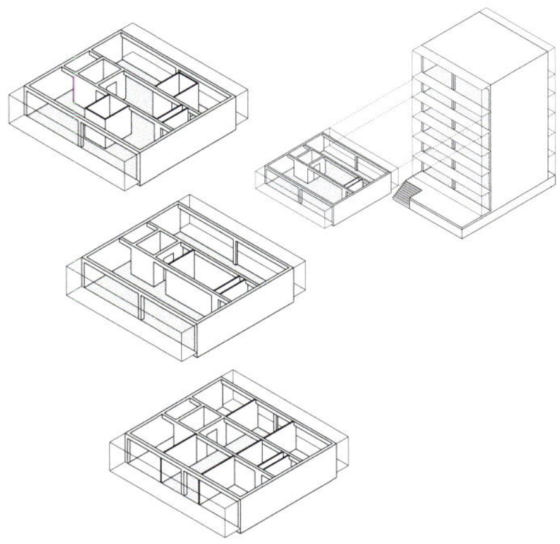

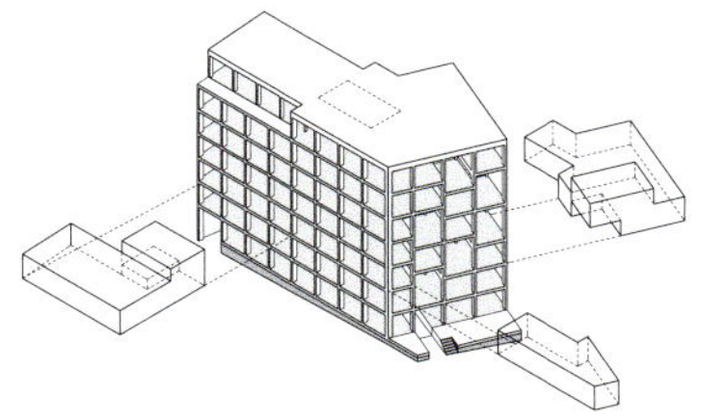

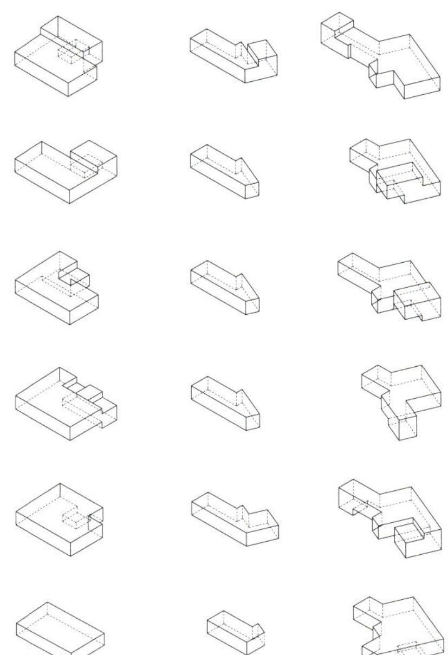

10

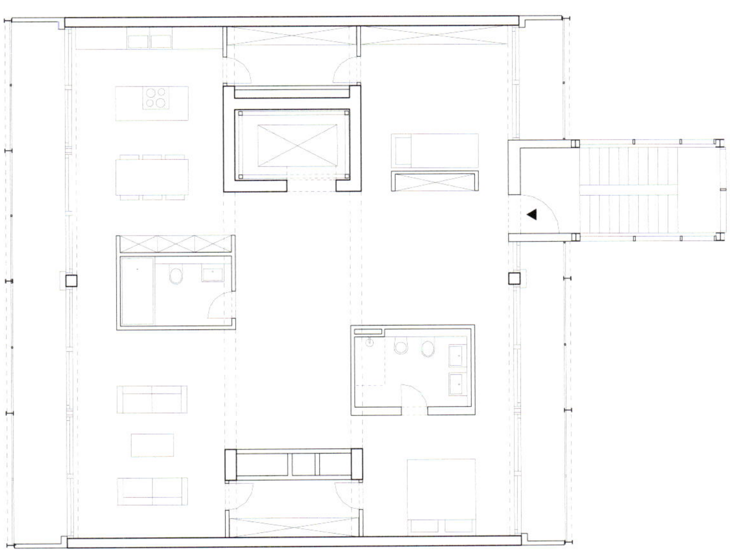

6

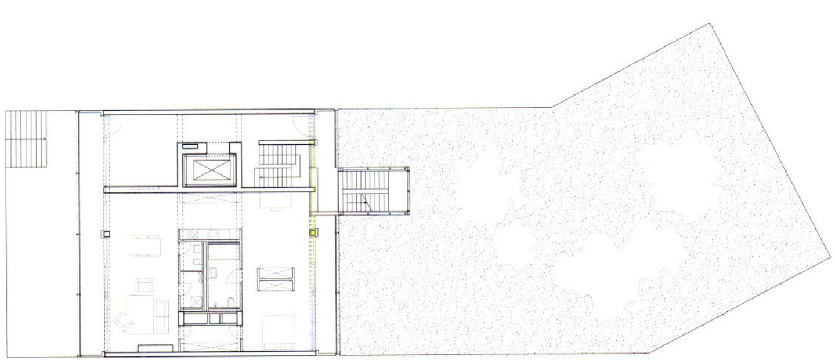

5

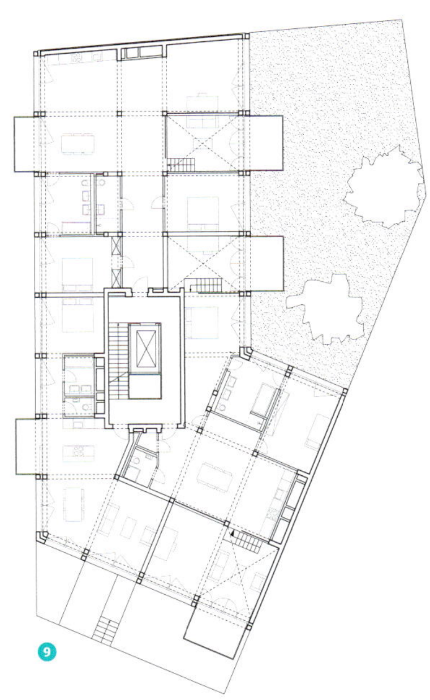

9

Building type
block edge development
ensemble
6–7 stories
primarily facing E/W

Date of construction
2007–2013

Number of units
39

Size of units
building 1: 53–147 m² (19 units)
building 2: 165 m² (6 units)
building 3: 78–135 m² (14 units)

Area per user
27–165 m²

Building depth
15 m

Access
single-, double- and
triple-loaded staircases

Open spaces
balconies, loggias and
roof terraces

Parking
underground garage

Architect
ROBERTNEUN™
Architekten GmbH
Nils Buschmann, Tom Friedrich
Berlin

Location
Am Lokdepot 1–14
Berlin
Germany

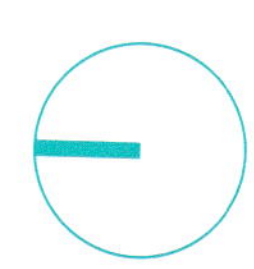

Referring to floor plan building **3**
1:200

Referring to floor plan building **2**
1:200

1.2
URBAN INFILL

The building scale and complexity of floor plans of these developments can vary greatly. Urban infill is a building task with numerous stipulations and requires an intense examination into the built environment and its characteristics. Yet precisely these narrow specifications sometimes inspire experimental floor plan solutions.

While in Europe, the context for this building type is dominated by the closed urban block, so that urban infill projects are for the most part bordered by two firewalls, they can also be set apart from one another in more open blocks, either along the street or in the block's interior.

Approaches to the building task can vary greatly and are often radical in nature: examples include complete coverage of the infill lot combined with an atrium; offsetting the infill volume on one side and thus creating an opening to the courtyard; complete isolation of the volume from all surrounding firewalls; or dissolving the new development into many small volumes, which seamlessly adapt to the existing configuration and every possible gap within it. Corner buildings, on the other hand – having the advantage of two street-facing facades and an increased urban presence – assume a more prominent place in the urban plan. The disadvantage of a poorly lit inner corner, however, demands specific solutions – such as the placement of the access core at this point.

Given their manageable scale, urban infill projects create opportunities for innovation not only in floor plan design but also in construction, energy concepts, or financing methods. Thus, many urban dwellers discover the idea of a joint building venture: they assume the role of property developer with the advantage of being able to participate in the design of the building and their own apartment from the ground up as future tenants.

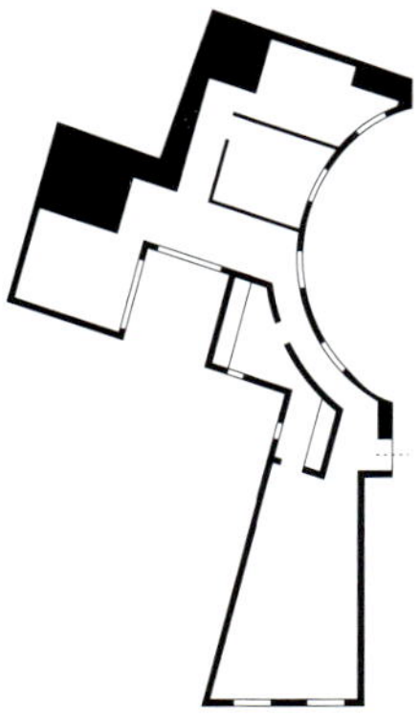

Calle Doña Maria Coronel | Cruz, Ortiz | 1976

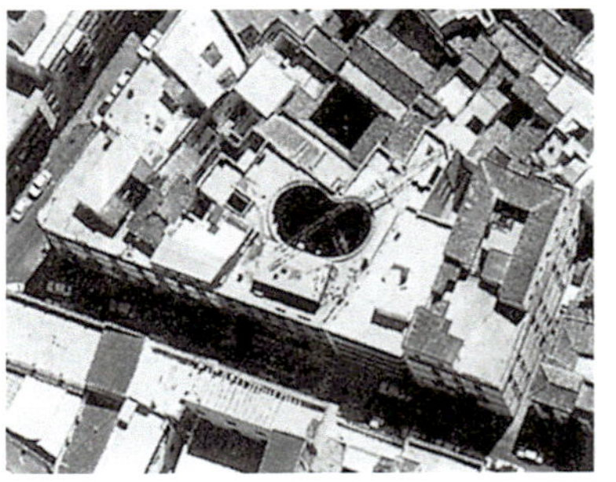

Maximum-utilization urban infill in the historical old town of Seville. Not only was the block perimeter closed here, but the 500 m², irregular lot was completely built over. Only one patio, whose basic area had to amount to 25 percent of the property area, was left free. The built volume, which branches out on three sides with the kidney-shaped patio in the center, made it impossible to give the three apartments equal value: they had to be adapted. One of the four-room apartments has a living room that serves as hallway and dining area; the bedrooms are located along twisted corridors. Two of the bedrooms and the kitchen are opened up to the neighbouring courtyard through a lightwell. One apartment is organized along a long hallway around the patio. Accessible to all residents: the roof terrace. The patio can be covered with a cotton tarpaulin.

Building type
block infill with patio
4 stories

Date of construction
1974–1976

Number of units
11

Size of units
4-room apts., 100–110 m²

Area per user
26.5 m²

Building depth
irregular

Access
single- and double-loaded stairwells

Open spaces
roof terrace

Parking
garage in basement

Architect
Antonio Cruz
Antonio Ortiz
Seville

Location
Calle Doña Maria Coronel
Seville
Spain

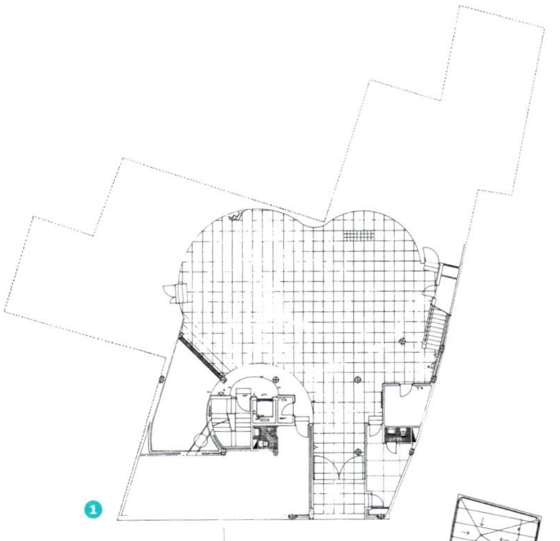

1

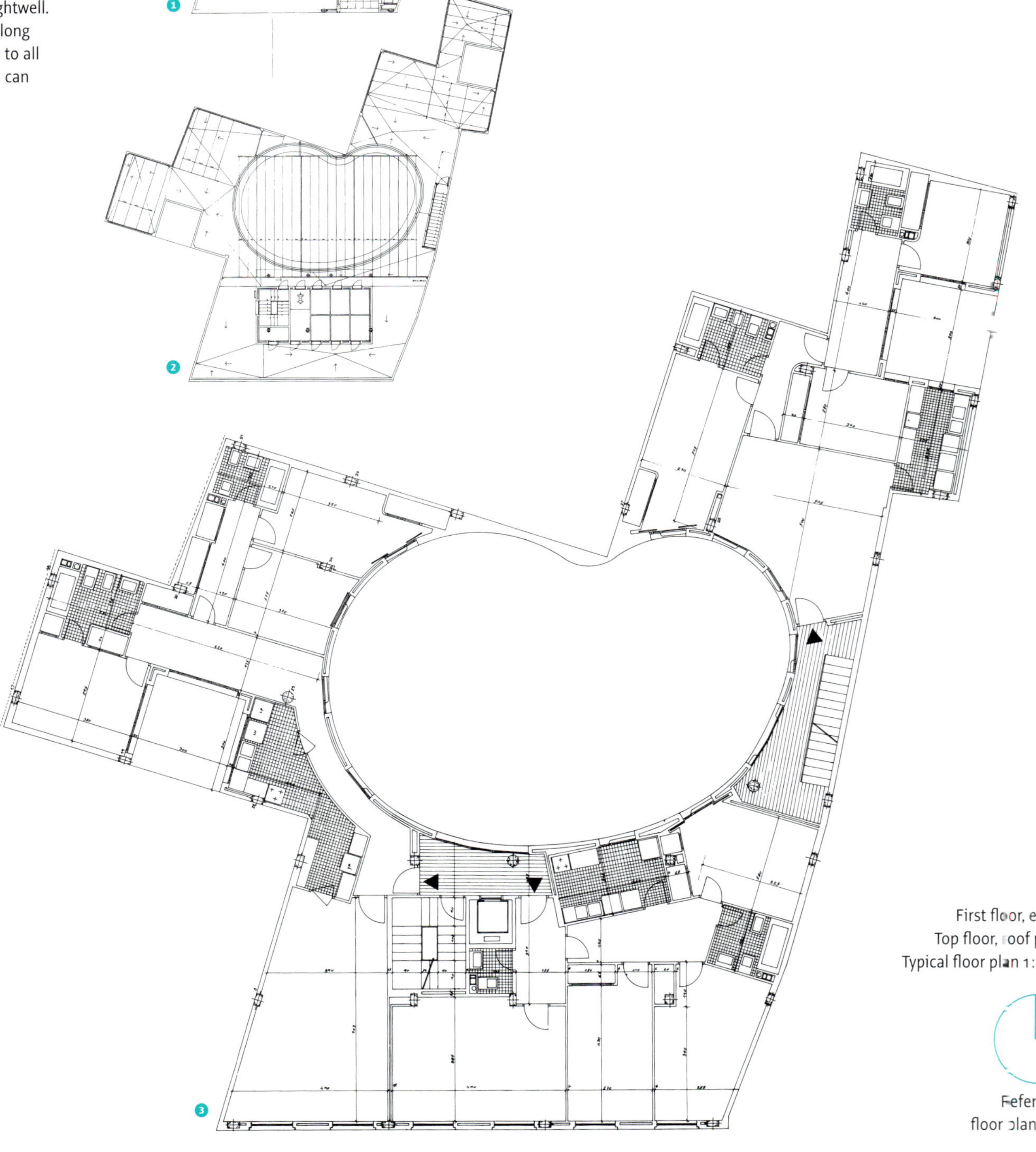

2

3

First floor, entry 1
Top floor, roof plan 2
Typical floor plan 1:200 3

Referring to floor plan 1:200

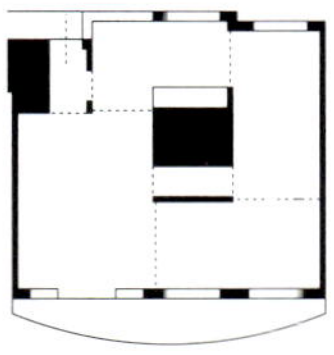

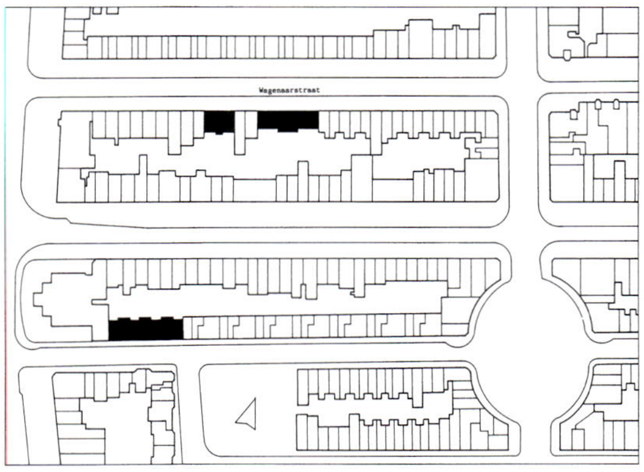

Wagenaarstraat | Duinker, van der Torre | 1989

1.2 URBAN INFILL

Open-plan approach to urban infill. The shallowness of the building and the generous use of glass (glass facade to courtyard, corner windows, glass stairwell to street) are designed to channel as much light as possible into and through the building while illuminating the street and block interior at the same time. The glazed stairwell, in which the staircase lies parallel to the facade, dramatizes the coming and going of residents. The roughly square floor plans are arranged around a compact core – with bath, WC, forecourt, kitchen row; they can be divided by sliding walls or used as freely flowing space. The sliding walls disappear entirely into their wall pockets.

Building type
residential building
5 stories
facing NNW/SSE

Date of construction
1989

Number of units
total of 49
(here 15)

Size of units
1- or 2-room apts. (9 units)
3- and 4-room apts.
(depending on configuration)
approx. 85 m² (40 units)

Area per user
21.5–85 m²

Building depth
10/12 m

Access
triple-loaded stairwells

Open spaces
apartment-wide balconies

Parking
on street

Architect
Margret Duinker
Machiel van der Torre
Amsterdam

Location
Wagenaarstraat/
Van Swindenstraat
Amsterdam-Dapperbuurt
The Netherlands

1

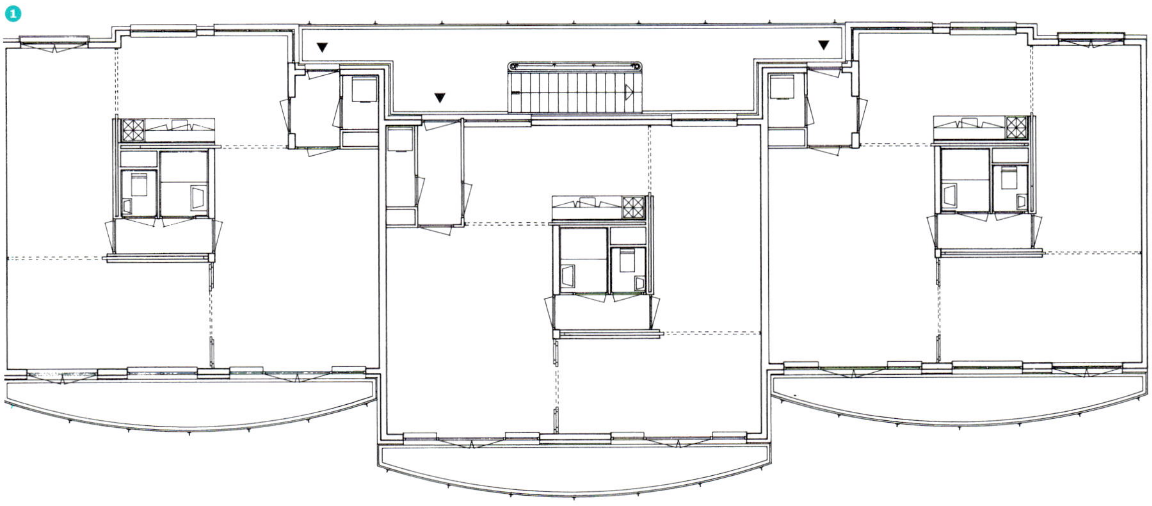

2

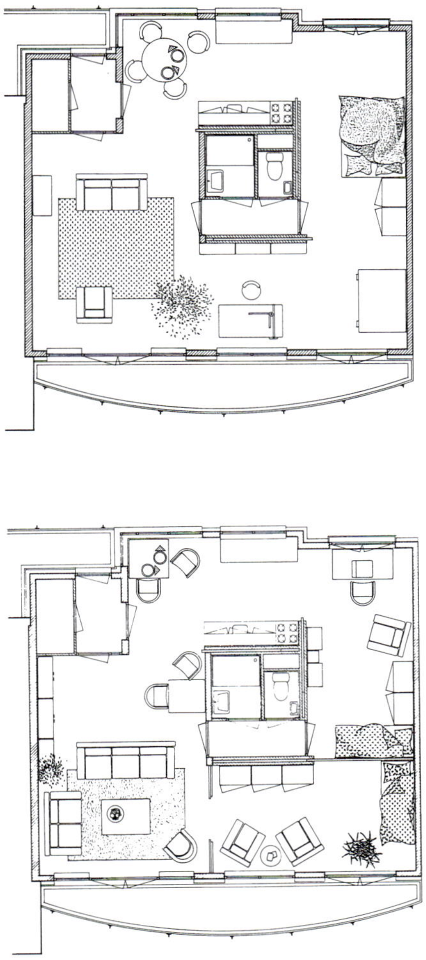

Typical floor plan 2nd–5th floor 1:200 1
2 furnishing proposals 1:200 2

Referring to
floor plan 1:200

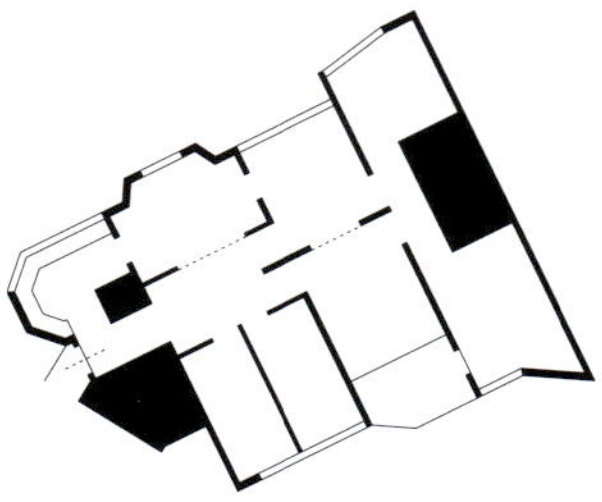

Elberfelder Straße | Uhl | 1981

A corner building with wings of different lengths, which makes the corner visible by means of a wide through passage to the block interior (via steps that first lead up and then down into the courtyard). The corner symmetry is emphasized by the U-shaped heads of the wings and through the connected oriels. The kitchens feature a band of windows around the corner, providing a view into both streets. Symmetry is key – in terms of access and internal layout: the living room extends from the street to the courtyard (room depth roughly 8 m plus balcony) and can be divided by sliding wall panels or double doors. The bedrooms are reached along a separate hallway from the living room.

Building type
corner building with through passage to courtyard
6 stories
facing NW/SE and W/E

Date of construction
1977–1981

Number of units
11

Size of units
3-room apts., approx. 85 m²
5-room apts., approx. 135 m²

Area per user
27–28.5 m²

Building depth
11/12.5 m

Access
double-loaded stairwell

Open spaces
green courtyard

Parking
underground parking

Architect
Johannes Uhl
Berlin
with Joachim Nowak,
Wolfgang Müller, Andreas Essig

Location
Elberfelder Straße
Alt-Moabit
Berlin-Tiergarten
Germany

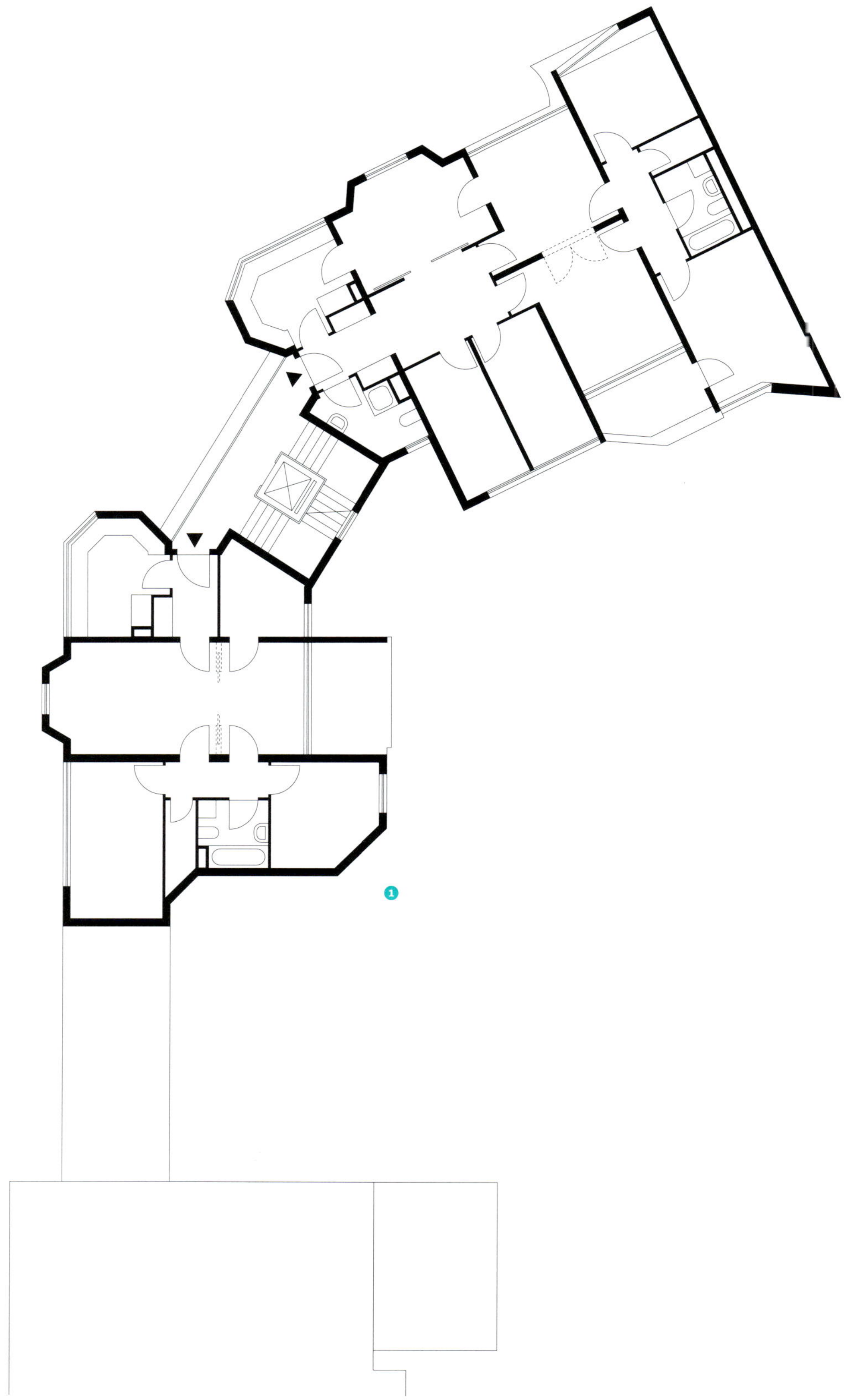

Typical floor plan 1:200 1

Referring to floor plan 1:200

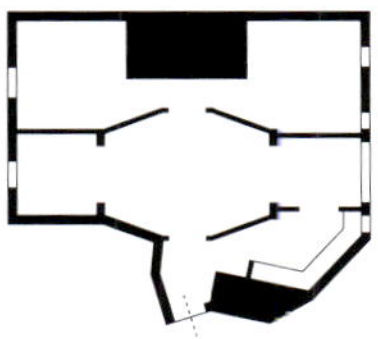

Schrankenberggasse | Krier | 1986

Two individual buildings can be distinguished. The floor plans of the buildings are based on a symmetrical scheme. The center of the corner building contains a formally defined stairwell placed on the bisecting line; the two- and three-room apartments are grouped around the stairwell, again symmetrically. The tiny entryway leads directly into the large living room, which extends from the street to the courtyard in the large apartments; in the case of the smaller units, the living room stretches parallel to the street. Both living rooms are developed symmetrically to the entrance; they function as center and distributor for the apartment. In the adjacent house, a corridor continuing around the corners connects nearly all rooms. The living area, which is once again octagonal, leads to the winter garden and bedroom.

Building type
urban residence
8 stories
facing E/W and N/S

Date of construction
1983–1986

Number of units
43

Size of units
2-room apts. (15 units)
3-room apts. (15 units)
4-room apts. (6 units)
5-room apts. (7 units)

Area per user
23.5 m²

Building depth
12 m

Access
quadruple- and double-loaded stairwells

Open spaces
courtyard

Parking
on street

Architect
Rob Krier
Vienna/Berlin
with Markus Geiswinkler

Location
Schrankenberggasse / Puchsbaumgasse
10th District
Vienna
Austria

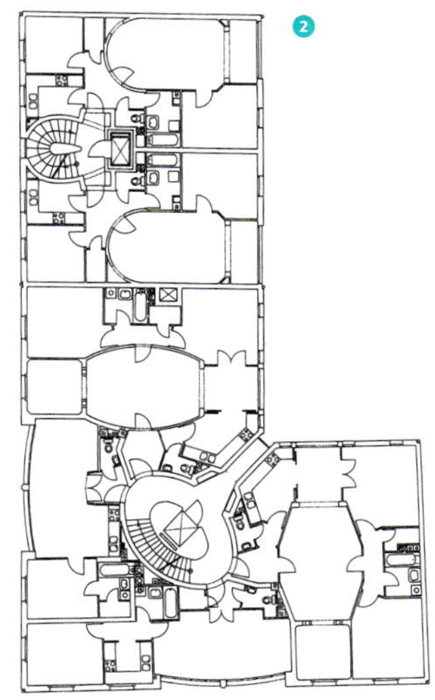
2

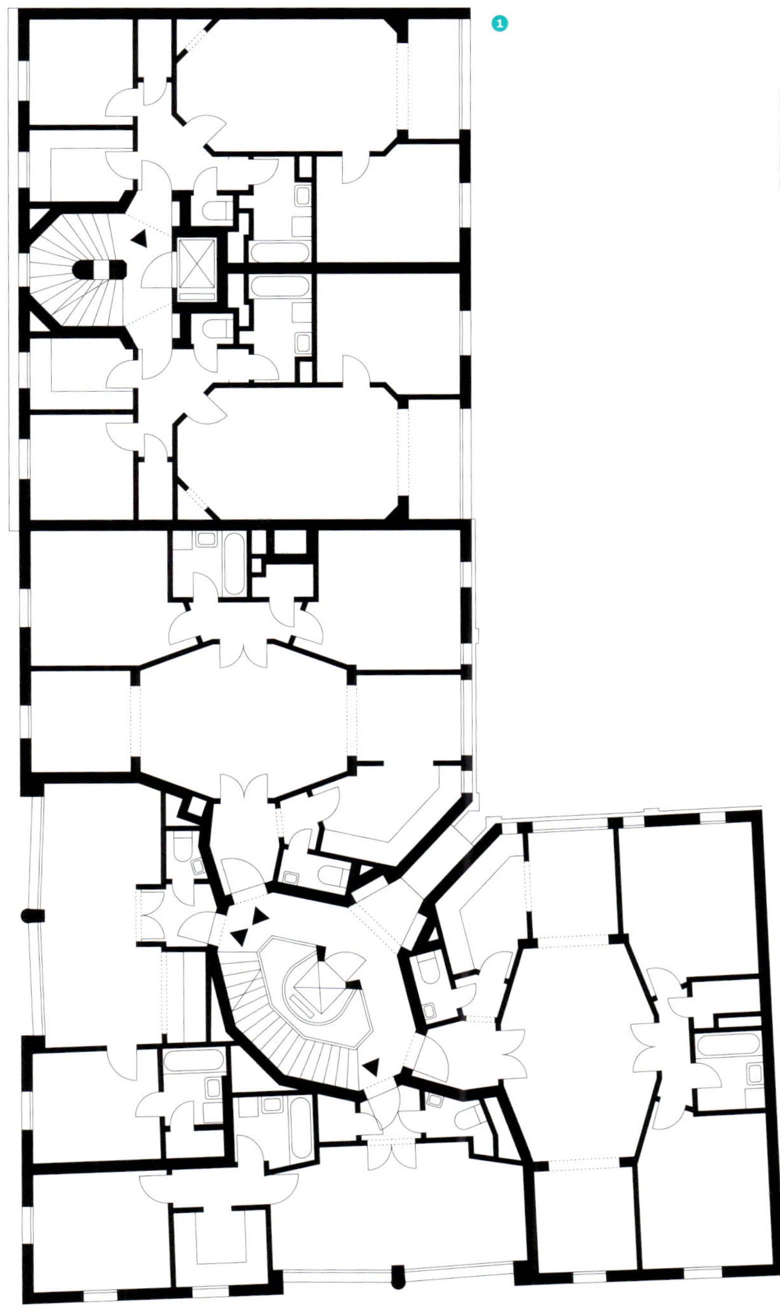
1

Typical floor plan 1:200 1
(for comparison the ideal 2
version in reduced scale)

Referring to
floor plan 1:200

Admiralstraße | Nylund, Puttfarken, Stürzebecher | 1986

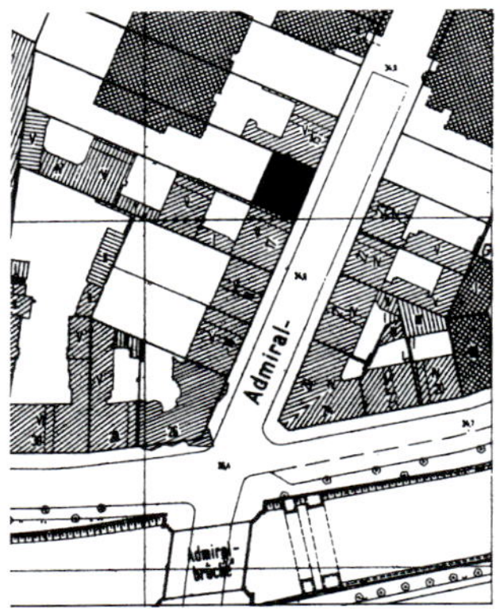

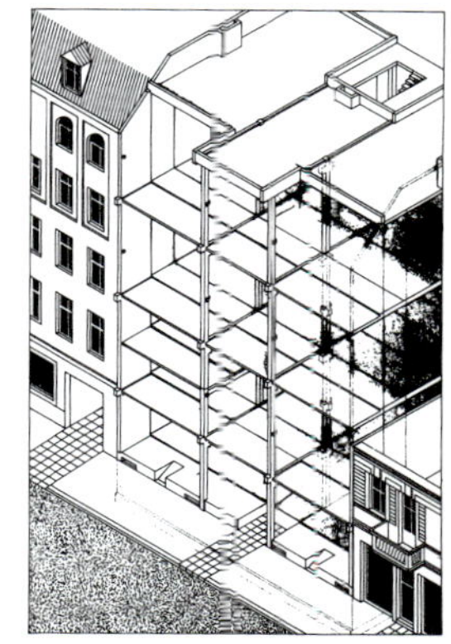

One goal of the 7-story "housing shelf," which was realized as part of an international building expo in the 1980s, was to stem the exodus toward home ownership on the periphery and the resulting urban sprawl, and to encourage residents to remain in their urban quarter. Collaborative planning, building, and living were part of the program for the cooperative. The primary structure of the house consists of a framework of prefabricated reinforced concrete components: when the shell was completed, it had the appearance of a shelf with (open) compartments – hence the nickname. The future residents then "inserted" the 2-story apartments into this framework. Realized in wooden skeleton construction and built in part by the residents themselves, the idea originated in the concept of creating stacked rowhouses right in the city. The individual floor plans were determined by the users themselves in consultation with the architect. While the room divisions are entirely different in the twelve units, the floor plan concepts remain conventional: either with a front-to-back open living room or with a kitchen separated from the living room and bedrooms on the upper level. In some units, the bathtub is accessible from two sides. All units have the benefit of generous winter gardens or loggias and small balconies as well. The planted roof patio is for shared use by all residents. Any decisions with regard to modifications to the building are made jointly by the cooperative.

1

1 Isometric views of different floor plan layouts
2 3 Layouts of maisonettes as designed by the inhabitants, 4th/5th floor 1:200
4 5 Layouts of maisonettes, 6th/7th floor 1:200

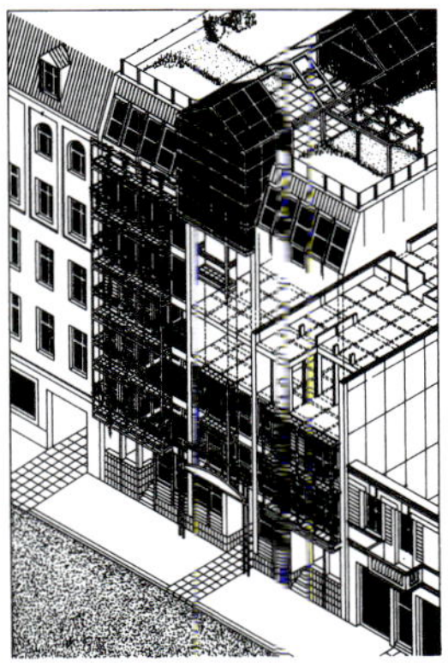

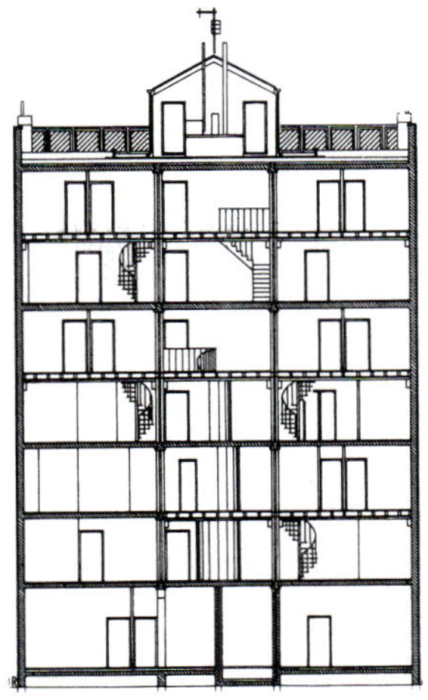

1.2
URBAN INFILL

Building type
"shelf house"
7 stories
facing NNW/SSE

Date of construction
1986

Number of units
12

Size of units
1- and 2-story units
designed by residents

Area per user
nondeterminable

Building depth
13 m

Access
triple-loaded stairwell

Open spaces
roof gardens
balconies
winter gardens
greenhouse

Parking
parking spaces in courtyard

Architect
Kjell Nylund
Christof Puttfarken
Peter Stürzebecher
Berlin

Location
Admiralstraße 16
Berlin-Kreuzberg
Germany

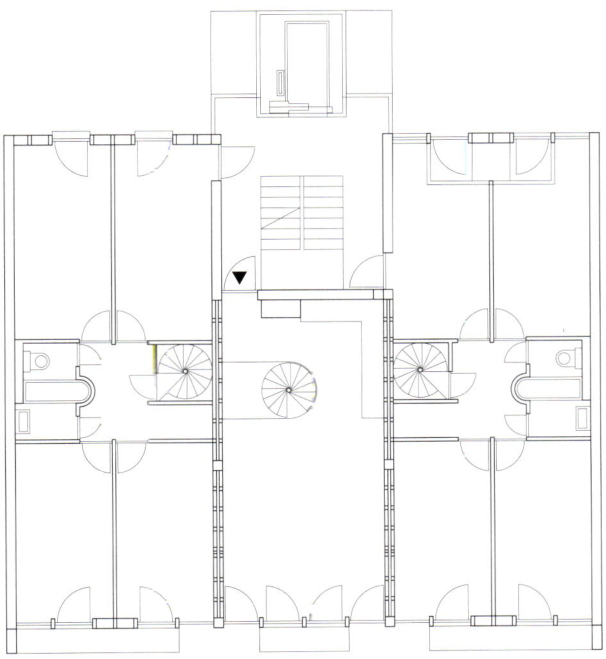

3

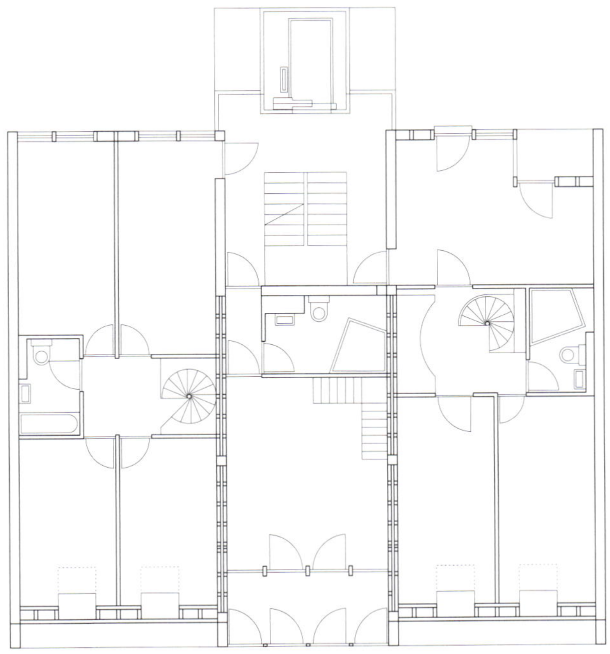

5

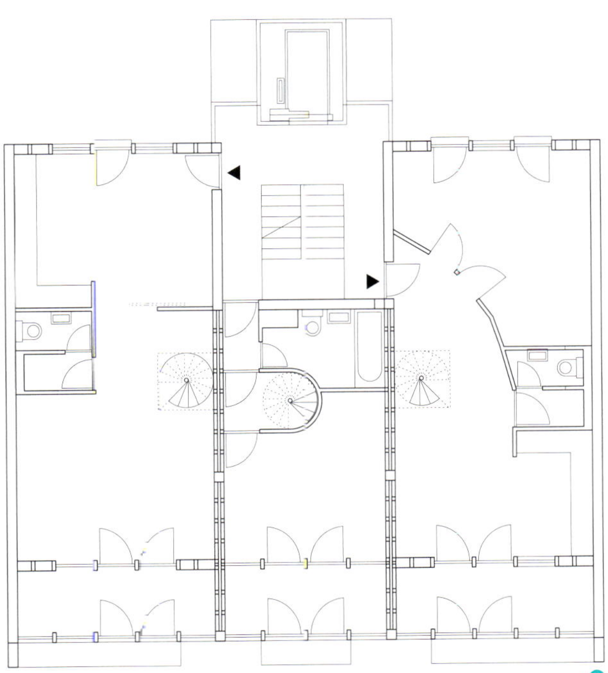

2

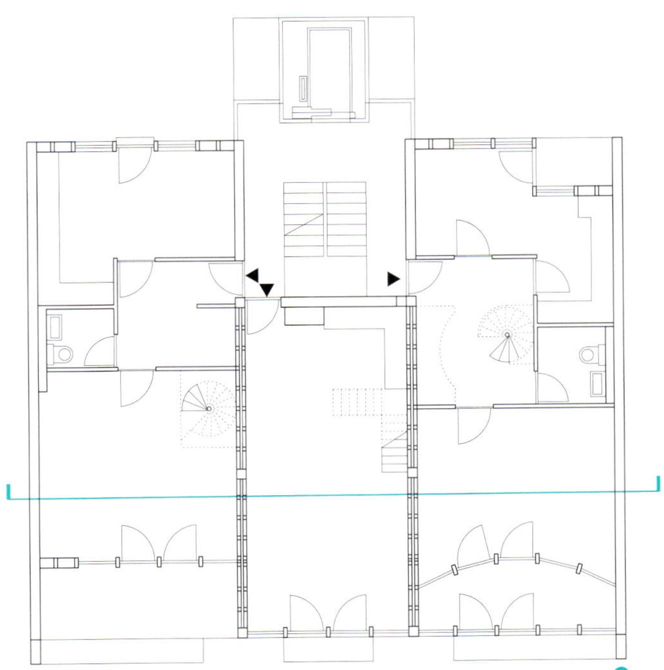

4

Referring to
floor plan 1:200

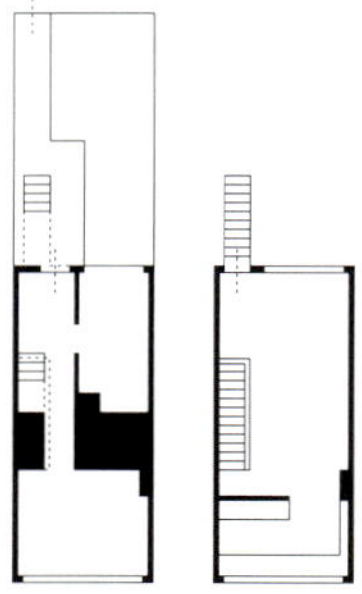

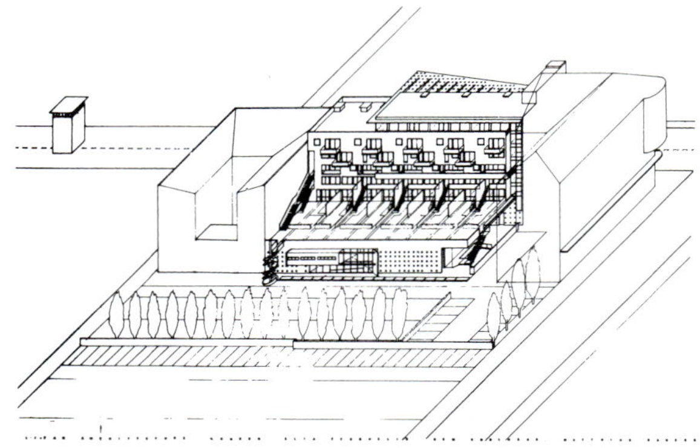

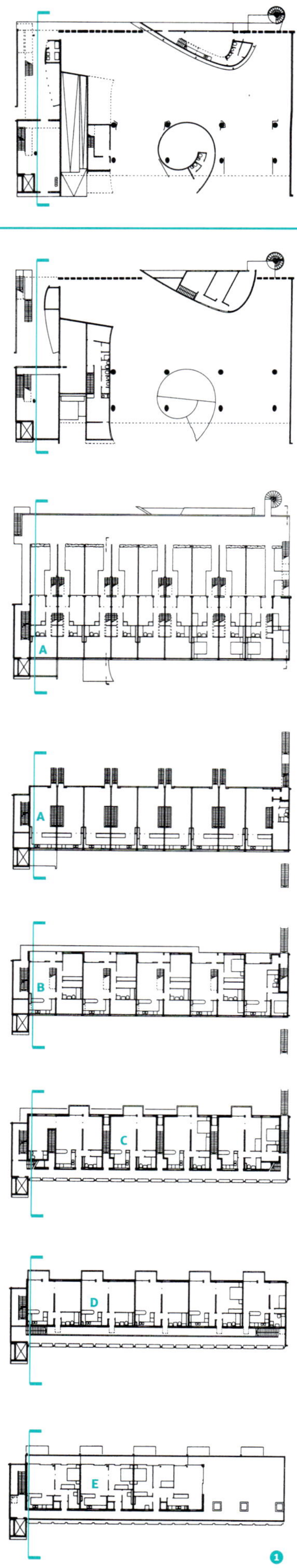

Friedrichstraße | OMA | 1989

A display of different kinds of circulation: The lower building section – originally a garage with a turning loop for the buses of the Allied Forces at the former Checkpoint Charlie – occupies the entire lot and forms a plateau on which the six-story apartment building is erected. Levels 1 and 2 are occupied by maisonettes with the luxury of private front gardens that are directly accessible from the plateau. Both interior and exterior stairs lead to the upper level. The four upper apartment floors are accessible via galleries, two of which as part of a two-story gallery clad in glass louvers, and another gallery that is open beneath a cantilevered roof projecting far into the urban space. The simple apartments along the galleries utilize and emphasize the east-west orientation: front-to-back east-west living rooms with galley kitchens or kitchens facing east, living rooms with a loggia or balcony facing west. The lower gallery also provides access to the apartments on level 3 via stairs inside the apartments. They are comparable to the other apartments; however, a wall parallel to the stairs separates a corridor, which connects the bedrooms, bathroom, and a storage room and loops around to connect with the living room. The penthouse apartments feature one extra study or bedroom overlooking the short gallery and have the added benefit of a roof terrace.

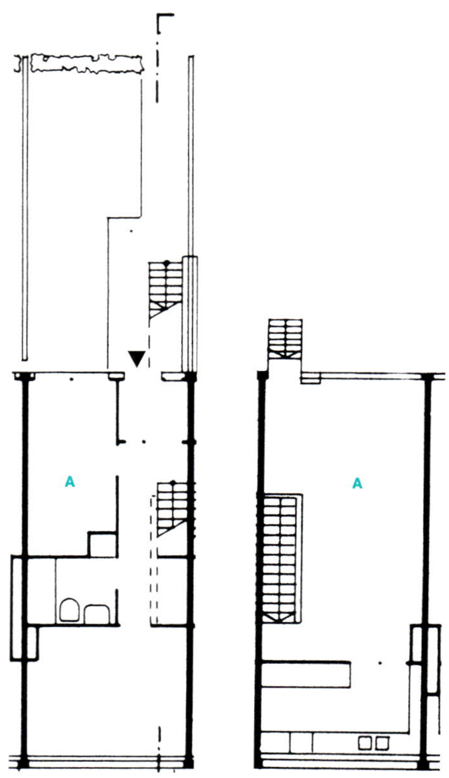

2

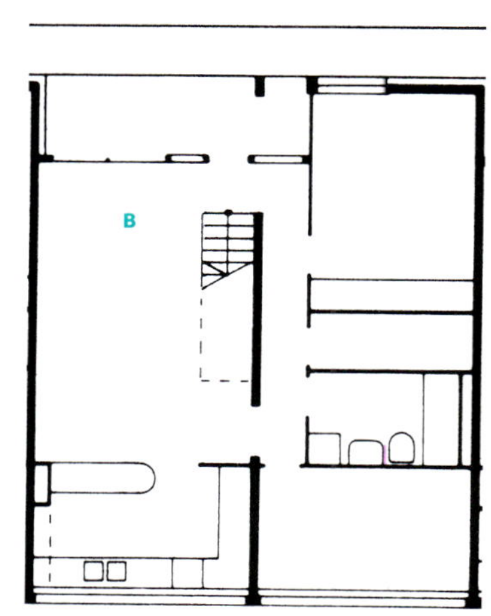

3

1

1 General layout: Ground floor–6th floor, from top to bottom
2 3-room maisonette, 2nd and 3rd floor 1:200
3 3-room apartment (entrance on level above), 4th floor 1:200
4 2-room apartment, 5th floor 1:200
5 2-room apartment, 6th floor 1:200
6 2-room apartment, 7th floor 1:200

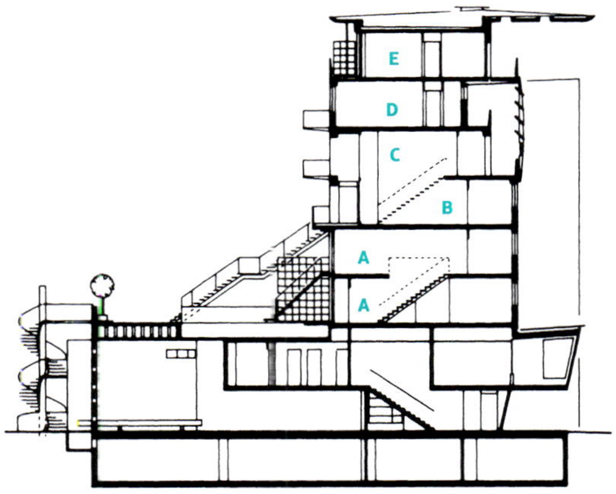

1.2
URBAN INFILL

Building type
8-story bar
facing E/W

Date of construction
1989

Number of units
26

Size of units
3-room mais., 73 m² (8 units)
6-room mais., 120 m² (1 unit)
3-room apts., 73–83 m² (7 units)
2-room apts., 48–59 m² (10 units)

Area per user
20–27.5 m²

Building depth
11 m

Access
maisonettes on 2nd
and 3rd floor from garden,
other floors from
gallery on street side

Open spaces
maisonettes have gardens
above the ground floor,
roof terraces

Parking
underground garage
and open-air parking

Architect
Office for
Metropolitan Architecture – OMA
Matthias Sauerbruch
Elia Zenghelis
Berlin
with
Dirk Alten
Barbara Burren
Eleni Gigantes
Reni Keller
Axel Wall

Location
Friedrichstraße 207/208
Berlin-Kreuzberg
Germany

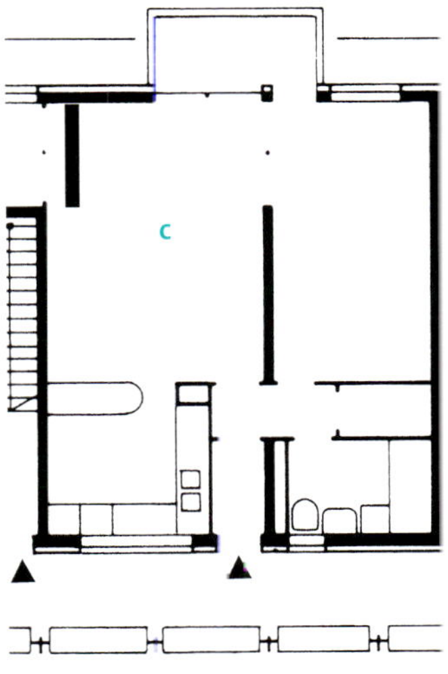

4

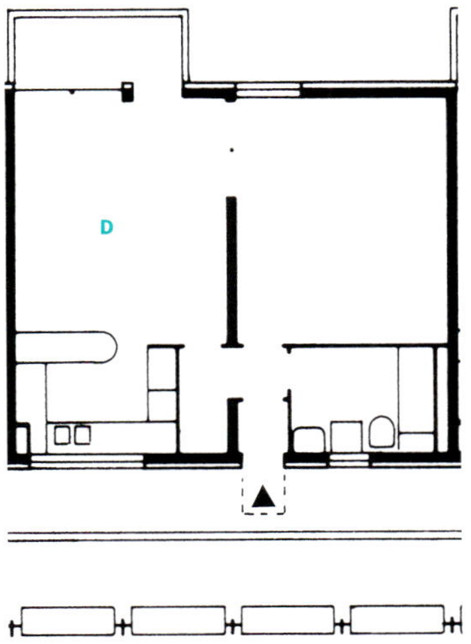

5

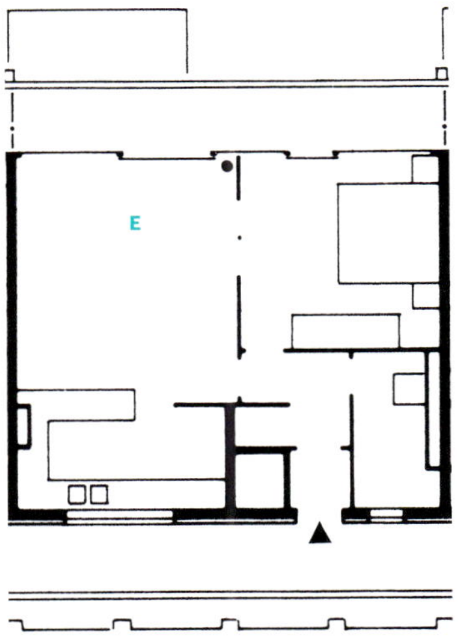

6

Referring to
floor plan 1:200

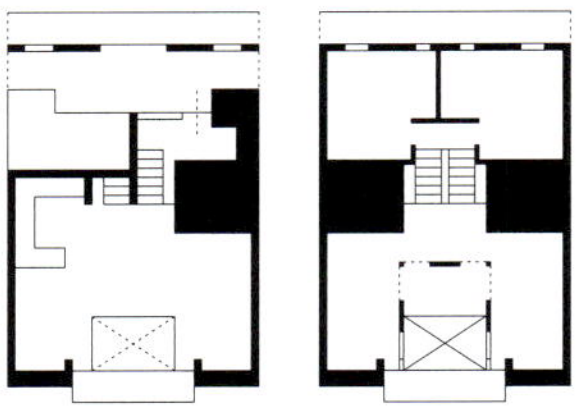

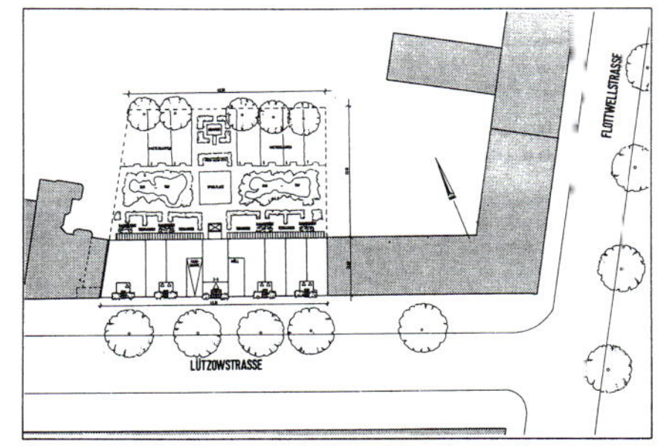

Lützowstraße | IBUS | 1989

Floor plans of the energy-efficient building establish protective zones: access from the north as climatic buffer, entrance, sleeping, and auxiliary rooms, living rooms and kitchen, winter garden from north to south. On ground floor/first level: two-story maisonette with separate entrance from street, treated as row house, ground floor on street side raised by a half-story. Continuous from south to north: kitchen, dining area, two-story living room. Spatial sequence lit from two sides. Terrace on the north side, screened off by trellis from semi-public open spaces. Sleeping area toward southwest. Inner bath as warm core. The galleries on level 2½ and 5½ provide access upward and downward to split level maisonettes. The following levels are combined: level 1½/2/3½/4; level 4½/5/5½ and level 5½/6/6½ with roof terrace on seventh floor. Storage space and closet next to covered walkway. On the street side (south-west), staggered arrangement of bathroom, kitchen, and living room with connected dining area and winter garden. The bedrooms are located on the north-east side, overlooking the courtyard. They are lit and ventilated from the glazed escape balcony; solar collectors integrated into the facade preheat air, which is blown into the cavity ceilings. The ceilings emit heat into the room as needed in a delayed fashion.

2

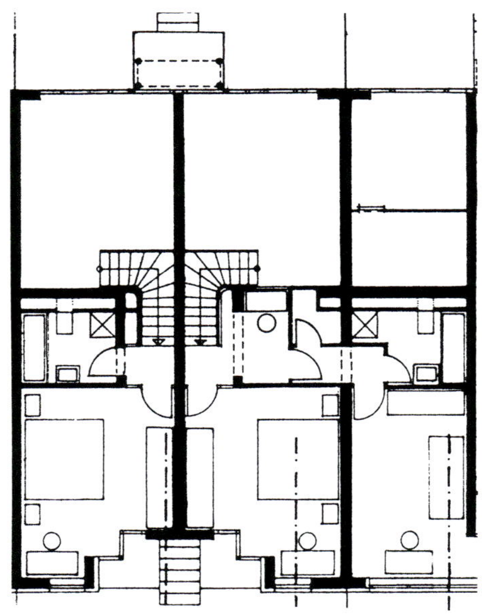

4

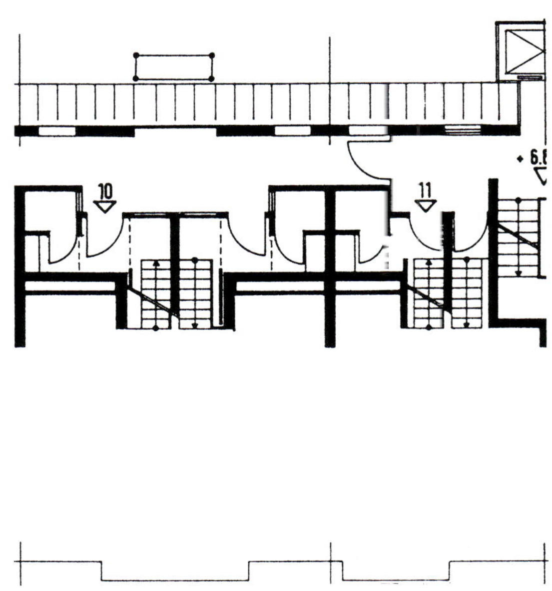

1

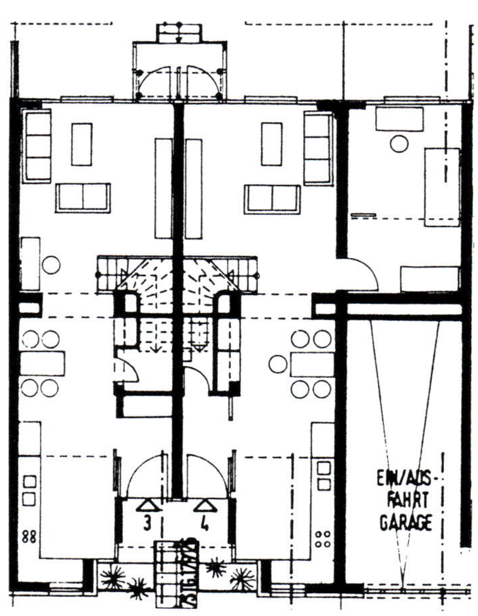

3

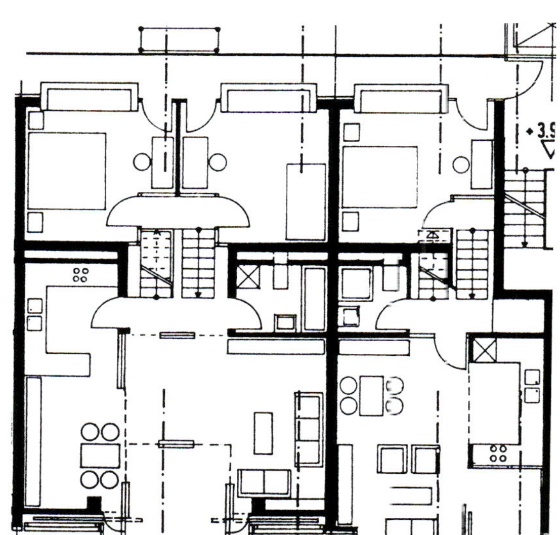

1 Maisonettes with entry from garden, level 0, 1:200
2 Maisonettes level 1, 1:200
3 Split-level apartment, level 1½ and 2, 1:200
4 Lower gallery with entry for split-level apartment, developing downwards, level 2½, 1:200

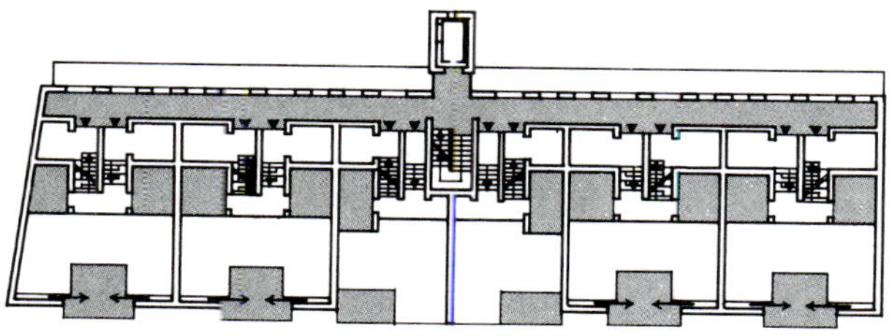

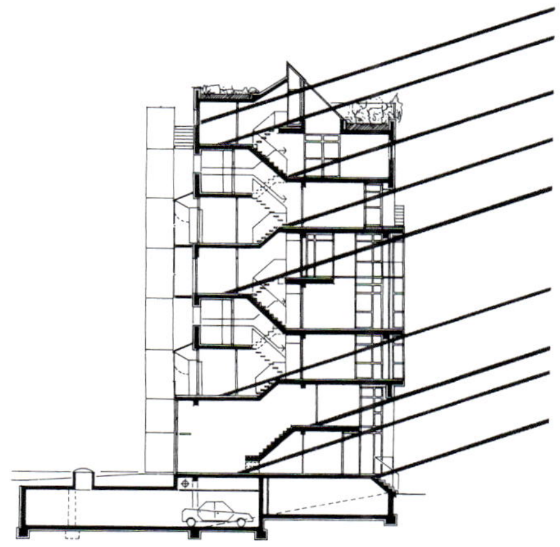

1.2
URBAN INFILL

Building type
row development
in block perimeter
7 stories
facing NE/SW

Date of construction
1989

Number of units
31

Size of units
1½-room apts., 48/52 m²
2-room apts., 58/62 m²
3-room apts., 86 m²
4-room apts., 96 m²
5-room apts., 116 m²

Area per user
23–50 m²

Building depth
11.5 m

Access
ground floor maisonettes
from street,
split-level apartments from
gallery on the north side

Open spaces
tenant gardens
terrace
rooftops
green block interiors

Parking
underground garage (16 spaces)

Architect
IBUS (Institute for Architectural,
Environmental, and Solar Research)
Hasso Schreck
Gustav Hillmann
Joachim Nagel
Berlin
with Michael Güldenberg
Peter Kempchen

Location
Lützowstraße 5–7
Berlin-Tiergarten
Germany

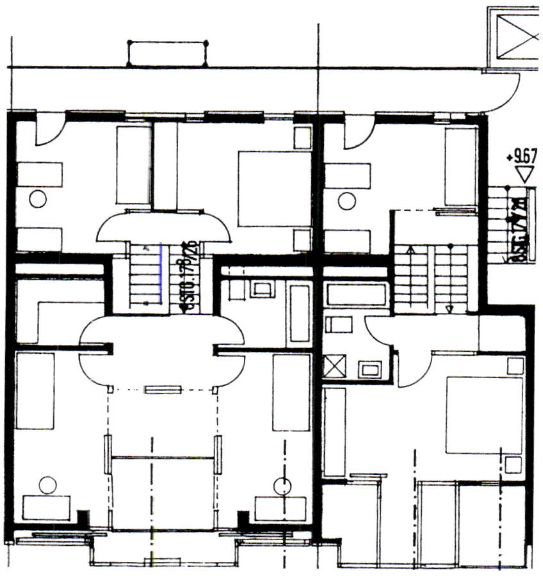
6

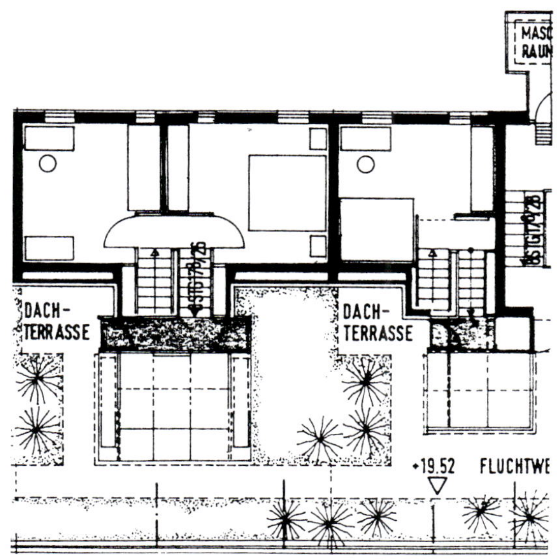

8

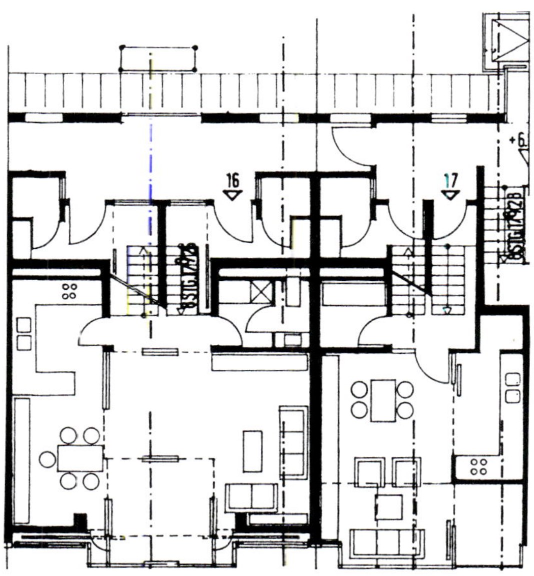
5

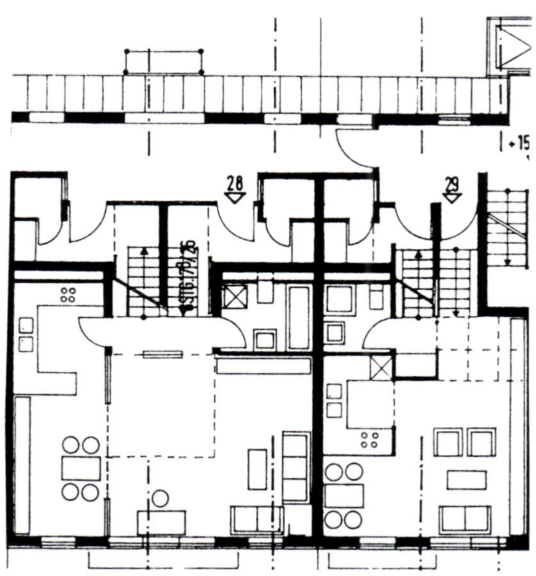
7

5 Lower gallery with entry for split-level apartment, developing upwards, level 2½ and 3, 1:200
6 Level 3½ and 4, 1:200
7 Upper gallery with entry for split-level apartment with roof garden, level 5½ and 6, 1:200
8 Level 6½ and 7, 1:200

Referring to
floor plan 1:200

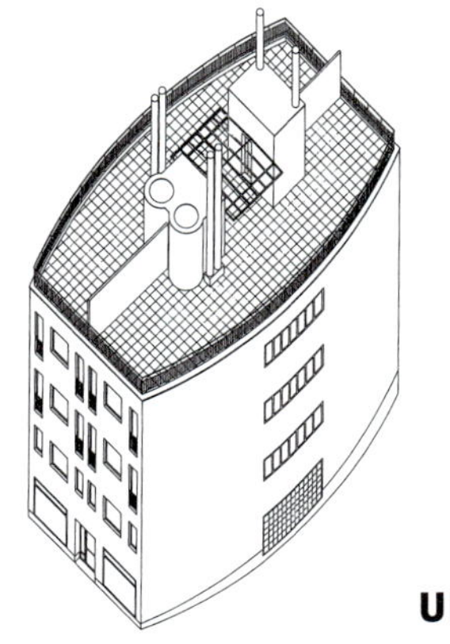

Alte Zürcherstrasse | Schnebli / Ammann | 1993

The building site is narrowed by the two adjacent buildings that reach deep into the courtyard. The new building is set independently into the courtyard; its sides are concave to make the gap seem larger. This unusual structure was made possible by an exception in the building code (for architectural ensembles). The apartments on the third and fourth stories, above store and offices, each occupy one half of the longitudinally divided building. The living rooms open up to the east, to the street; bedrooms face the courtyard to the west. The open kitchens in the center of the apartments are lit through glass-brick walls. The rooms are connected by a minimal hallway along the curved outer wall. Each of the fourth-floor apartments has a rooftop terrace.

Building type
freestanding
4 stories
facing E/W

Date of construction
1987–1990

Number of units
4

Size of units
2-room apts., 60 m² (4 units)

Area per user
40 m²

Building depth
17 m

Access
double-loaded stairwell

Open spaces
courtyard
rooftop terraces

Parking
on street

Architect
Dolf Schnebli /
Tobias Ammann Partners
Zurich
with Isidor Ryser, M. Meili

Location
Alte Zürcherstrasse 13
Baden
Switzerland

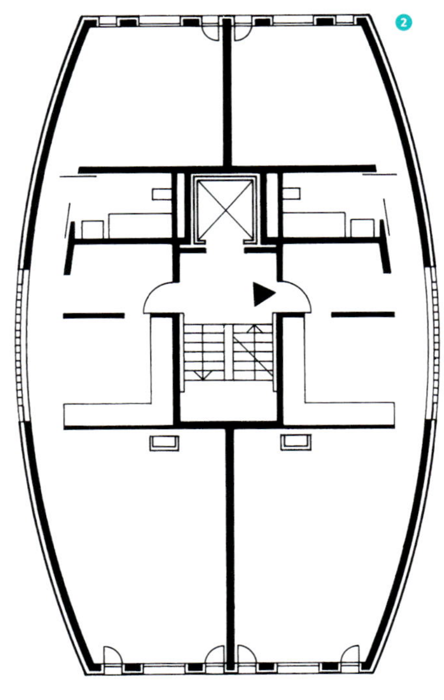

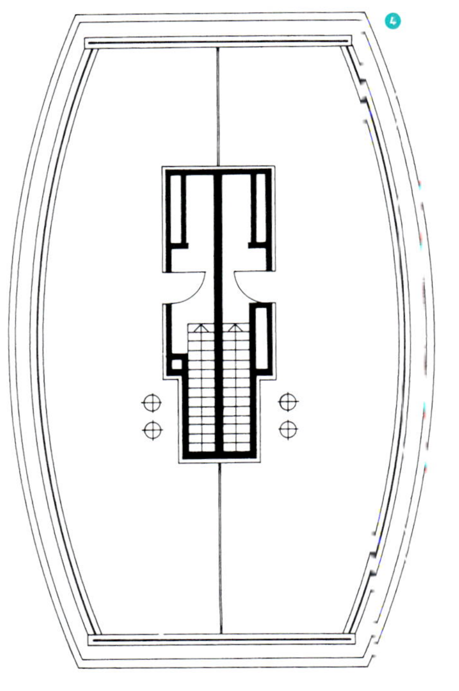

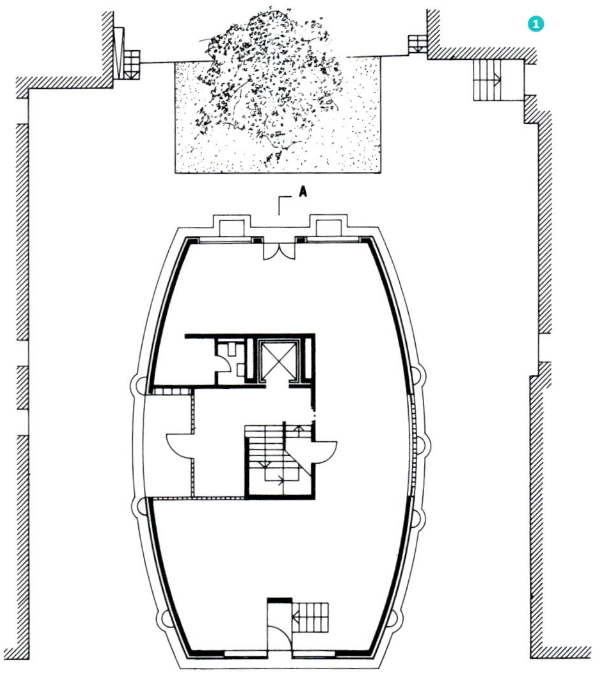

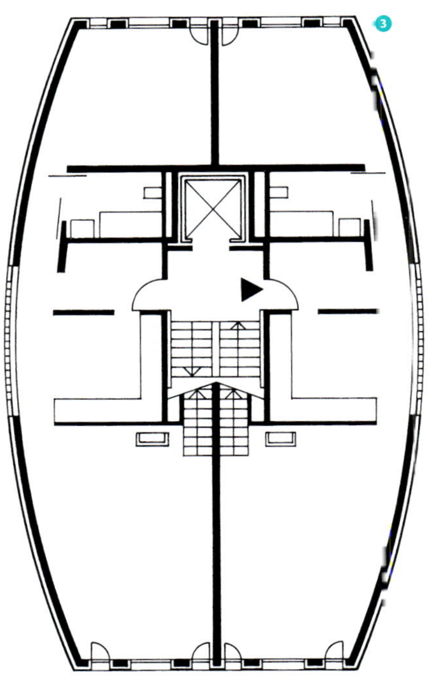

Ground floor with store (1)
3rd floor: 2-room apartments 1:200 (2)
4th floor: 2-room apartments 1:200 (3)
Rooftop terraces 1:200 (4)

Referring to floor plan 1:200

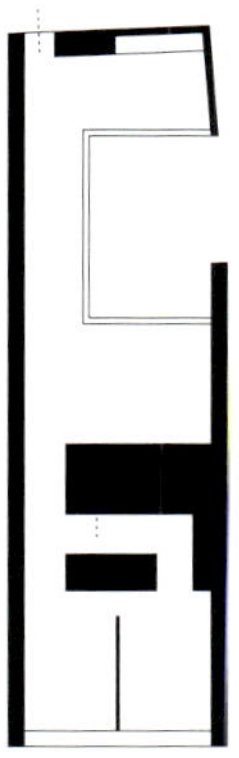
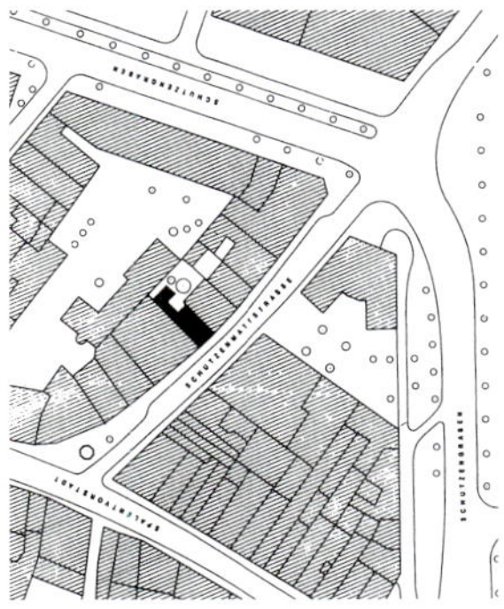

Schützenmattstrasse | Herzog & de Meuron | 1993

1.2 URBAN INFILL

The deep and narrow building form is a result of the medieval lot. An elevator provides direct access to the apartments, while an open staircase at the end of the lot also serves as a loggia on each level. The apartments are divided by the elevator shaft and the sanitary block. On the street side, the bedrooms lie behind a fully glazed facade covered by a moveable curtain of perforated latticework. The living room and kitchen to the rear of the building are grouped around a central light well that opens onto the adjacent southeastern lot and is horizontally stepped back on each floor. The 6th and 7th floors accommodate a maisonette: living area below (kitchen integrated into core, flexible use of all rooms), bedrooms above (large connecting ensuite bathroom). Narrow patios overlooking the street and the courtyard.

Building type
mixed-use residential and commercial building, 7 stories facing SW/NE

Date of construction
1991–1993

Number of units
4

Size of units
3-room apts.: 100 m² (3 units)
4-room maisonette: 170 m² (1 unit)

Area per user
33.5–43.5 m²

Building depth
26 m

Access
single-loaded staircase
direct elevator access

Open spaces
loggias, patios

Parking
street

Architect
Jacques Herzog, Pierre de Meuron
Basel
with
Dieter Jüngling and Andreas Stöcklin

Location
Schützenmattstrasse 11
Basel
Switzerland

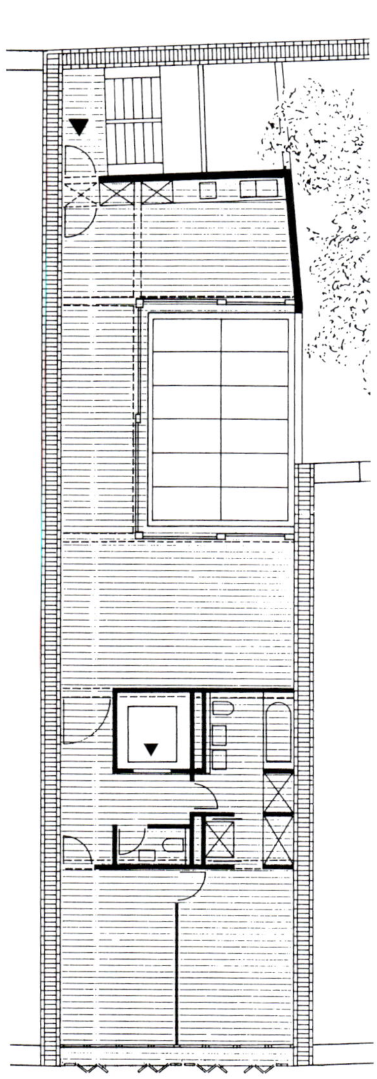
1

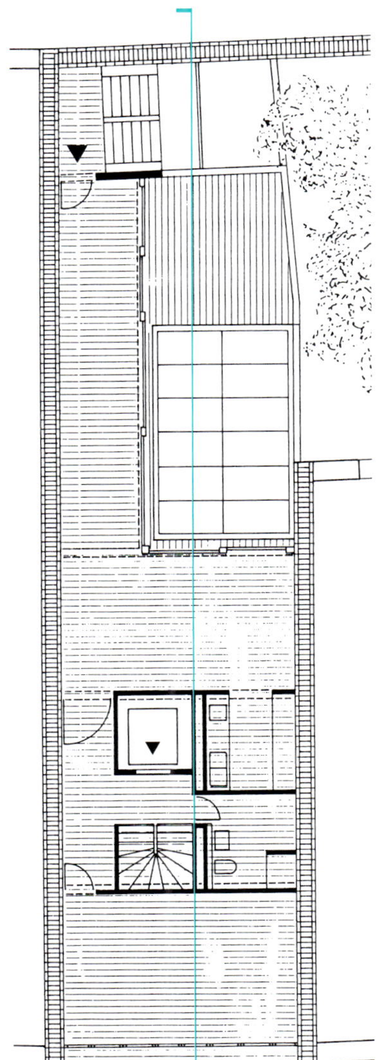
2

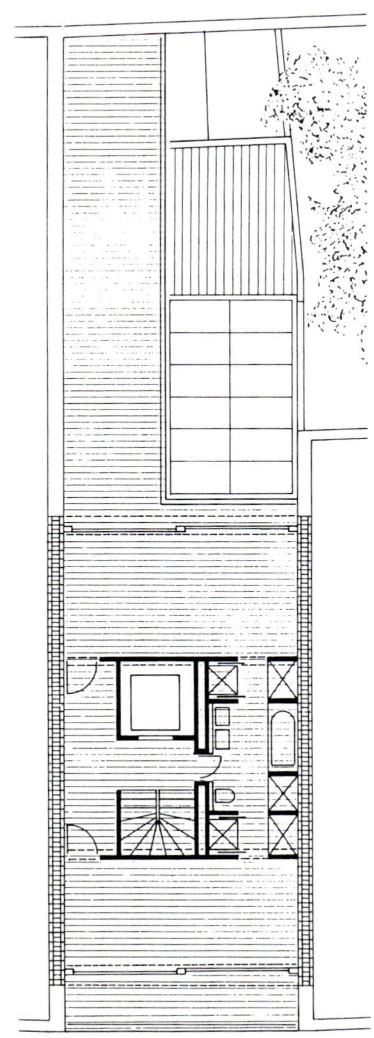

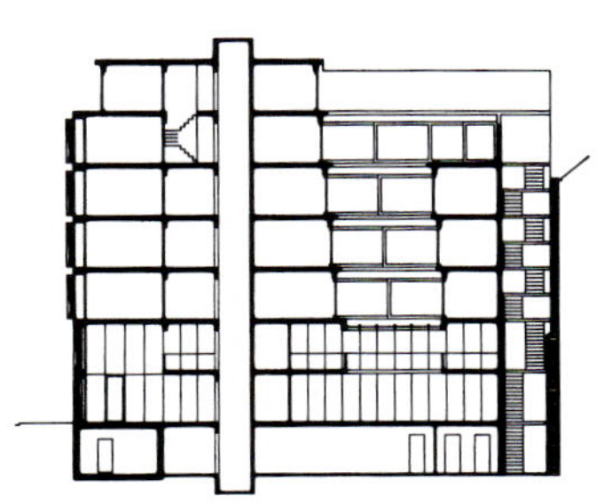

3rd, 4th and 5th floor: 1
3-room apartment 1:200
6th/7th floor: 2
4-room maisonette 1:200

Referring to floor plan 1:200

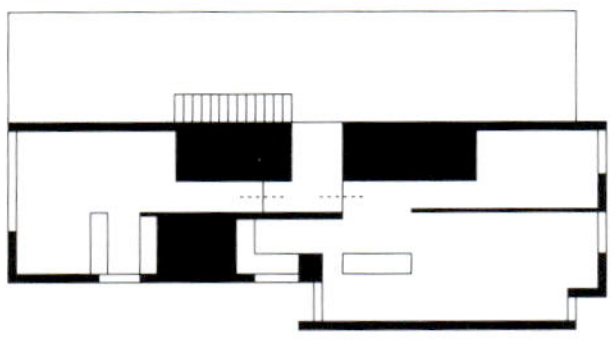

Rue de l'Ourcq | Gazeau | 1993

The design of this infill construction – which opens up more than it closes – was inspired by its dense, heterogeneous surroundings. A slit divides the block into two volumes (3.5 m/7.5 m) both of which stand tall toward the street and short in the back. In this way sunlight and the life of the street reach the block's interior; apartments can orient themselves to more than two sides. A generous stairwell spans the gap between the two building halves, connecting them together. The open stairs have giant wooden platforms that reach into the block's interior, and are at once terraces and meeting places for the residents. The first two floors to the rear hold larger apartments distributed over several levels. Along the street front, modest studios and 2-room units are stacked up to the eighth floor on top of the ground floor shops. The apartments generally open onto a corridor that leads to a bedroom and bath on one side, and living room, kitchen, and toilet on the other.

Building type
infill construction with gap over entire plot's length
7 stories
facing NE/SW

Date of construction
1993

Number of units
26

Size of units
studios, 35 m² (13 units)
2-room apts., 55/58/62 m² (11 units)
3-room apt., 60 m² (1 unit)
4-room apt., 77 m² (1 unit)

Area per user
19–35 m²

Building depth
3.5/7.5 m

Access
double- and triple-loaded stairwell, generous open access in building slit, elevator

Open spaces
large terraces often part of access
communal garden

Parking
31 underground parking spaces

Architect
Philippe Gazeau
Paris

Location
46, rue de l'Ourcq
Paris
France

1
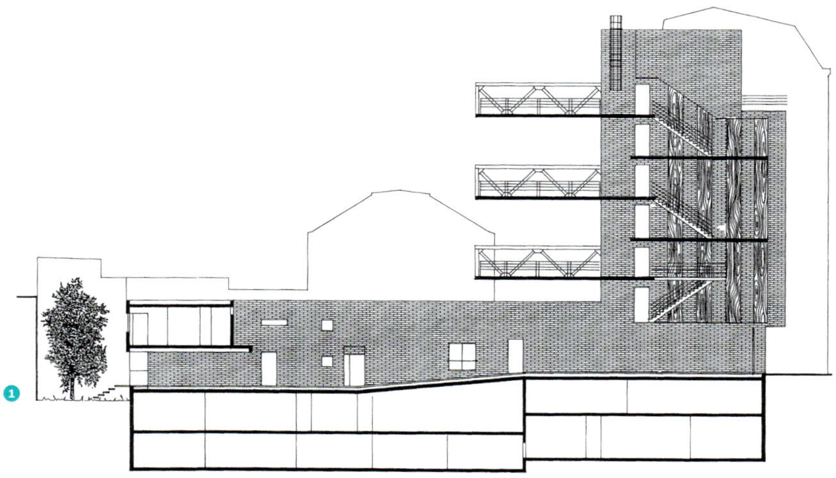

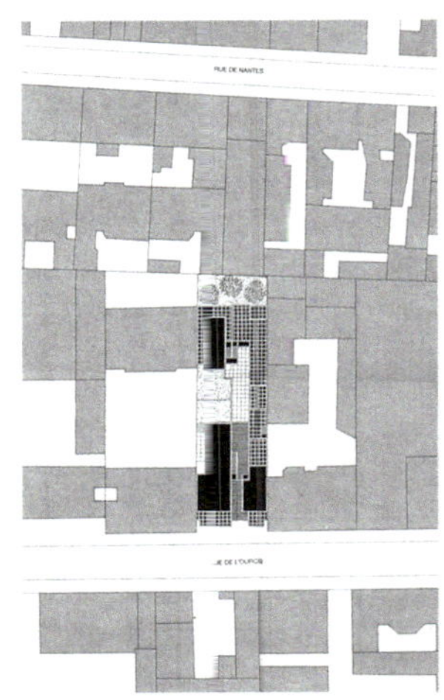

2
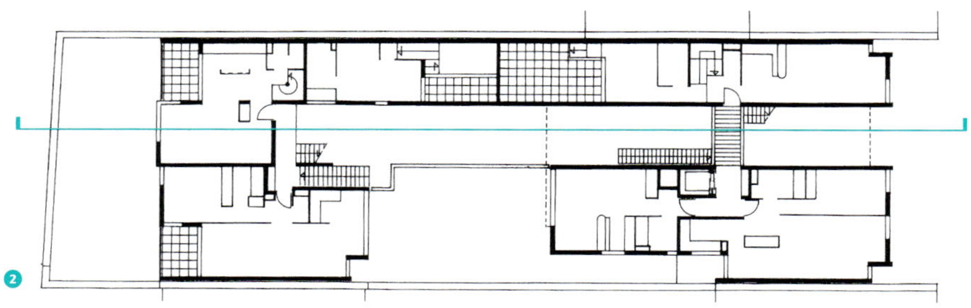

3
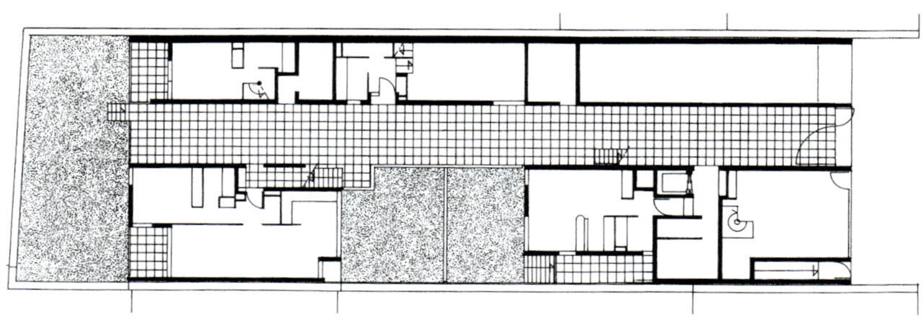

4
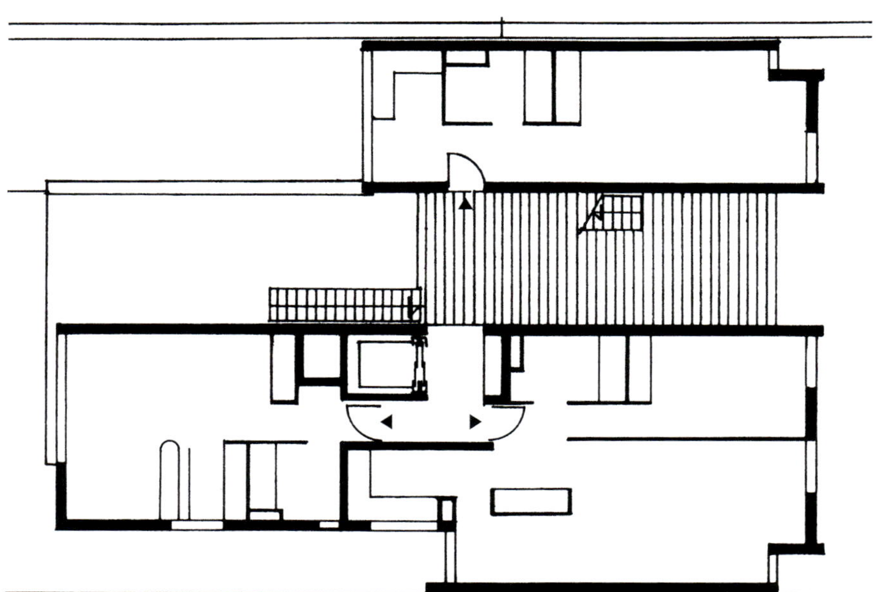

Section cutting through the generous stairwell 1:500 1
Ground floor 1:500 2
2nd floor 1:500 3
4th floor 1:200 4

Referring to floor plan 1:200

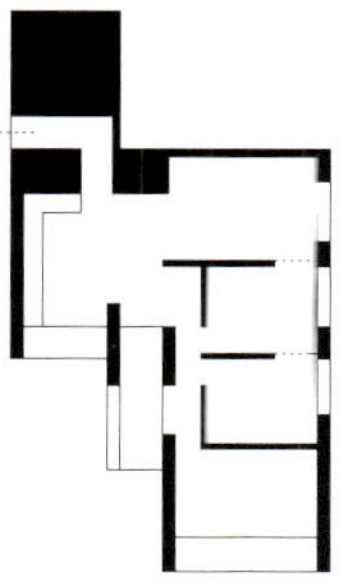

Schlesische Straße | Léon, Wohlhage | 1993

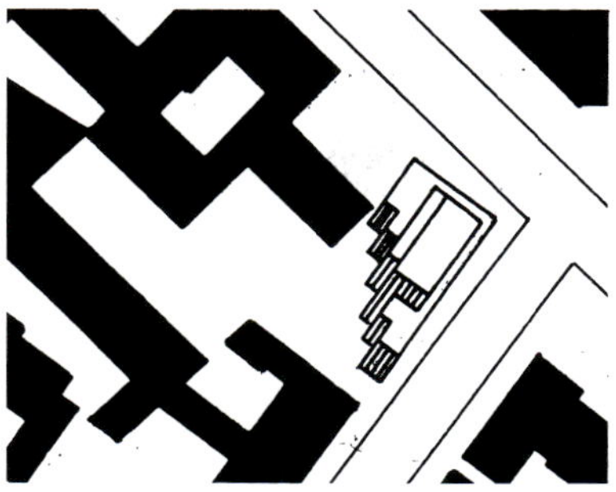

The structure with slanted glazing on the courtyard side and a punctuated masonry wall on the street elevation is conceived as a freestanding corner building. For a social housing development, the standards are of extraordinarily high quality, made possible by a compensatory model: in return for increased utilization of the lot, the client financed conservatories, ecologically sustainable exterior walls, and interior sliding walls. Parallel to the longitudinal side, the plan is divided into static zones: living areas, sanitary areas, and central corridor. The zones are stepped back. Two apartments per floor are furnished with two conservatories each. The other three apartments are oriented toward the street and open onto a shared balcony. In these units the rooms are kept functionally neutral and are arranged side-by-side; sliding walls make it possible to combine two rooms into one. The two-bedroom apartments with kitchen block can be opened up into one single open-plan space.

Building type
freestanding corner building, 7 stories facing NE/SE/SW, ground floor and part of second floor dedicated to commercial use

Date of construction
1993

Number of units
28

Size of units
2-room apts., 65 m² (11 units)
3-room apts., 73/82/93/97 m² (9 units) (including barrier-free units for the disabled)
4-room apts., 90/95/98 m² (4 units)
4-room mais., 92/98/103 m² (4 units)

Area per user
21.5–32.5 m²

Building depth
18 m (staggered length: 38.5–18 m)

Access
central corridor with one straight-run stair per floor, access to five apartments per floor

Open spaces
green space in courtyard with fountain

Parking
underground garage

Architect
Hilde Léon, Konrad Wohlhage, Berlin with Ebba Zernack, Nina Lambea, and Klaus Wiechers

Location
Schlesische Straße/Taborstraße
Berlin-Kreuzberg
Germany

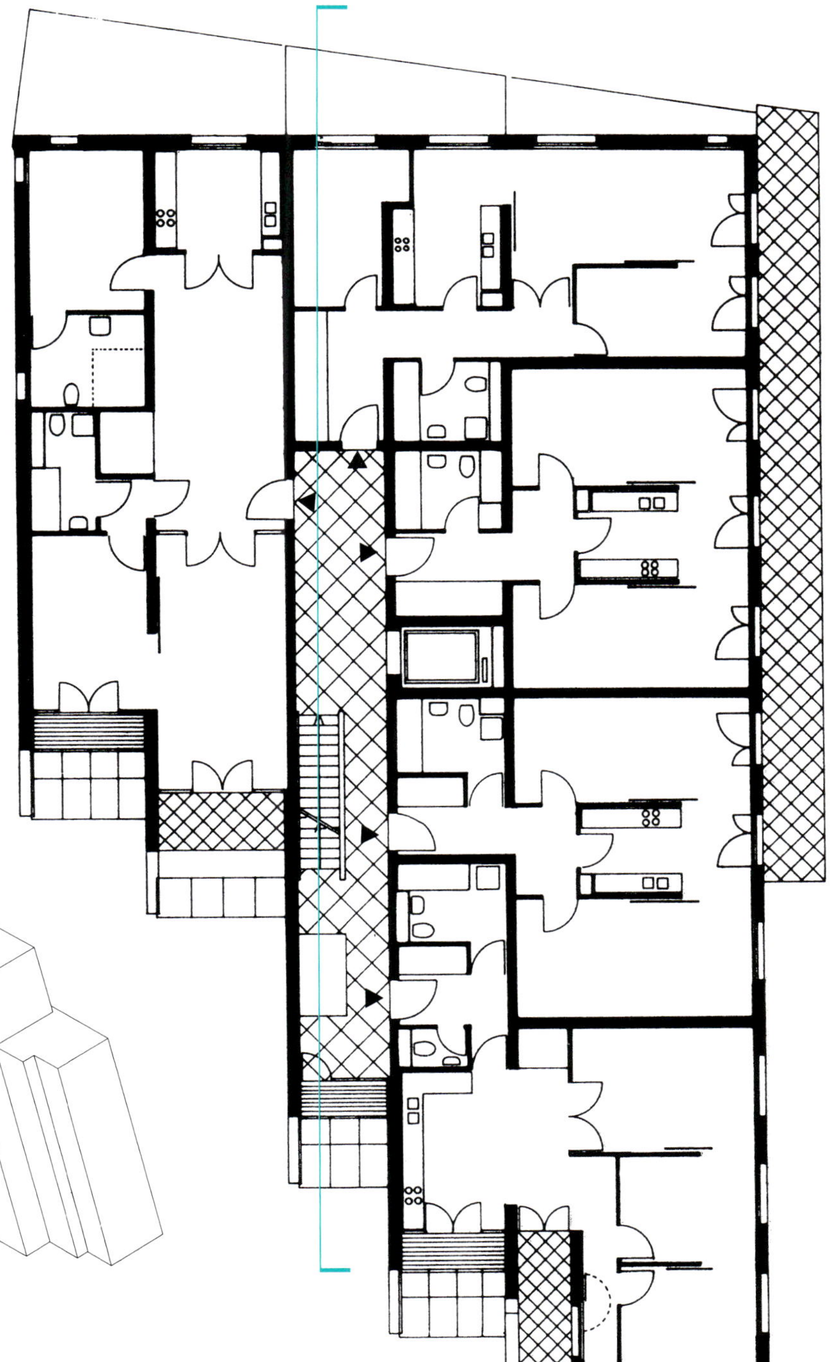

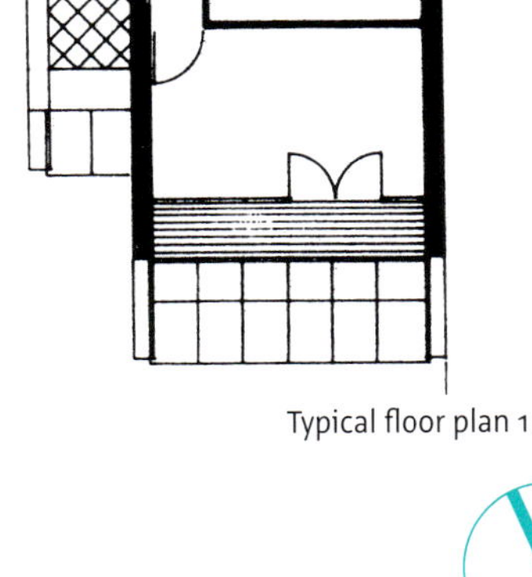

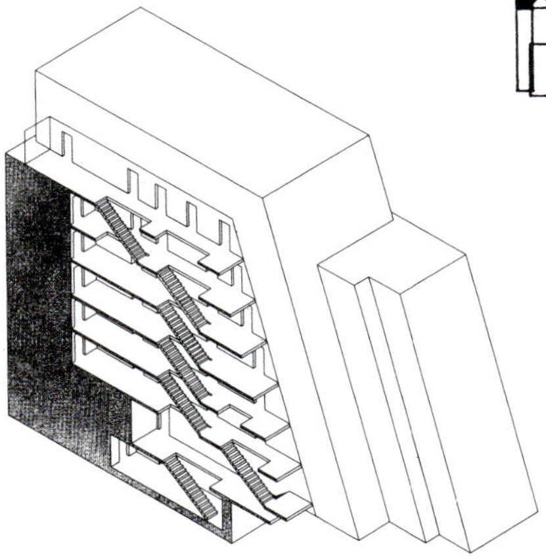

Longitudial zoning: functional

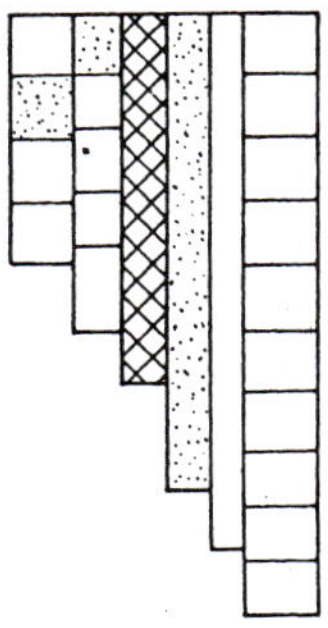

Longitudial zoning: structural

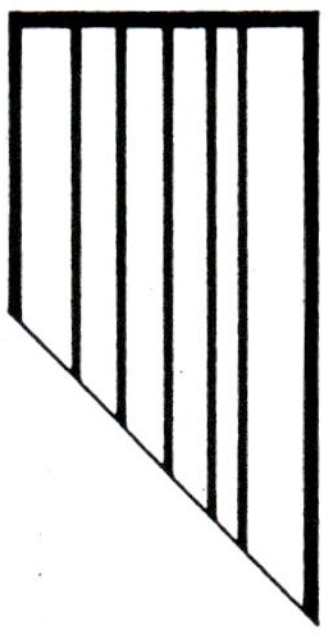

Functional flexibility through nonbearing cross walls

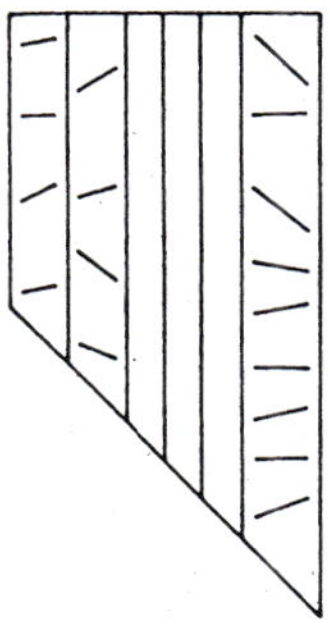

Typical floor plan 1:200 1

Referring to floor plan 1:200

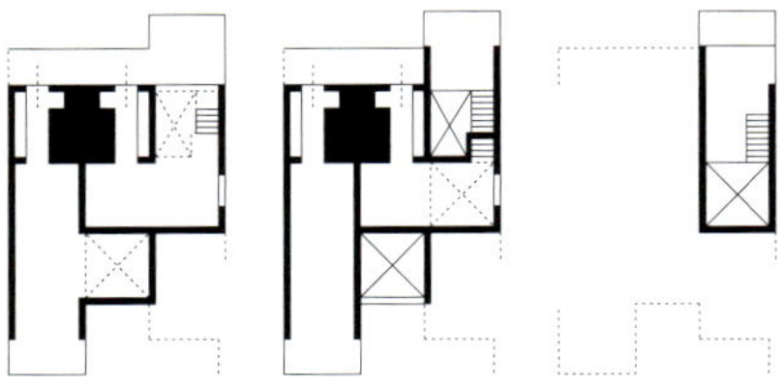

Space Block Kamishinjo | Kojima + Akamatsu | 1998

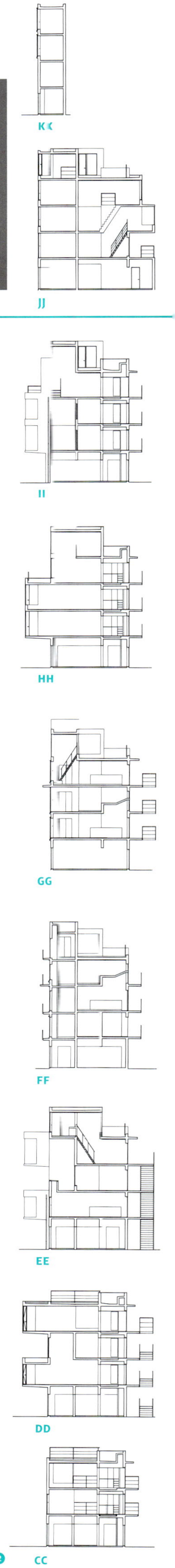

The building closes a small building gap that runs crosswise through an urban block. This situation permits neither front nor rear buildings, and no courtyard space for natural light; instead, a dense, homogeneous spatial structure composed of "space blocks" runs from one end through to the far end. Basic Space Blocks is a system of modular space blocks with an edge length of 2.4 m, which are placed front-to-back, side-by-side or on top of one another, thereby creating complex, three-dimensional, stacked housing configurations. Natural light and views are provided for the apartment interiors by manipulating this spatial construct. This results in miniapartments, which offer surprising spatial qualities in a continuum of three-dimensional situations despite their limited floor area: some are single-level but with two-story atria in one segment, which allows light to fall into the apartment from the other side. At other times, stairs set into similar atria provide access to generous roof patios, or else a unit is oriented in the opposite direction on the second level and connected on that side to another space module. Some modules are cantilevered, simply to open the view into the street perspective. The building was the first in a series of similar projects in urban contexts with extreme density. The "porosity factor" – which measures the permeability of the space blocks to light and air – emerged as the conceptual leitmotif for the design method.

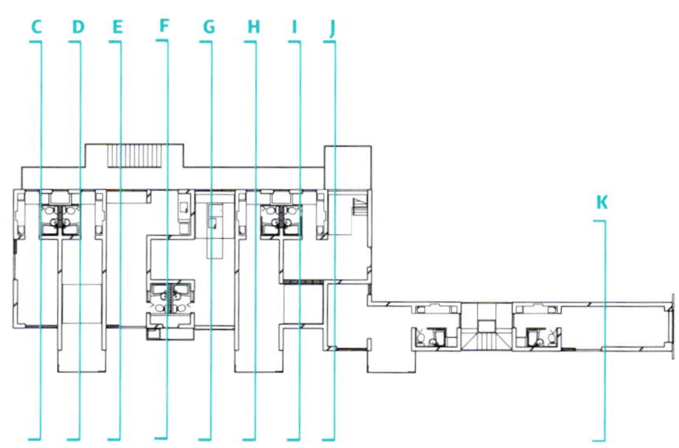

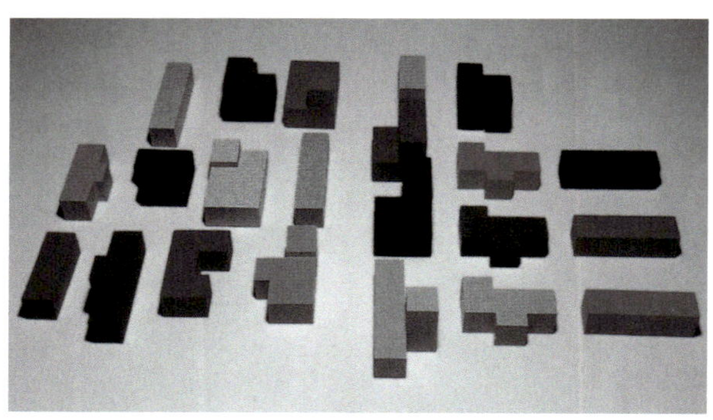

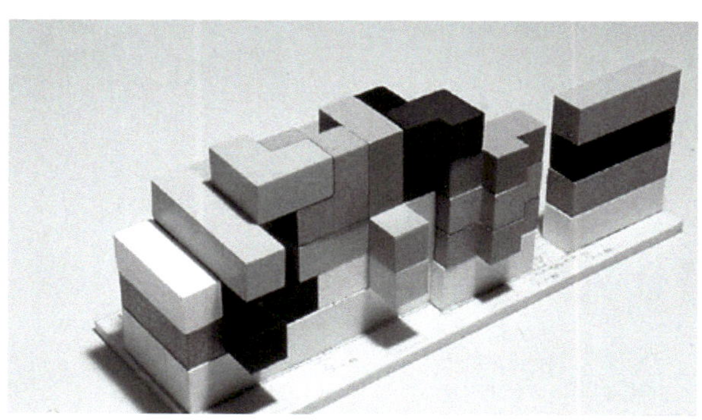

❶ Concept model: the units composed of "space blocks" and their combinations
❷ Sequence of cross sections 1:500
❸ 2nd floor 1:200
❹ 3rd floor 1:200
❺ 4th floor 1:200
❻ 5th floor 1:200

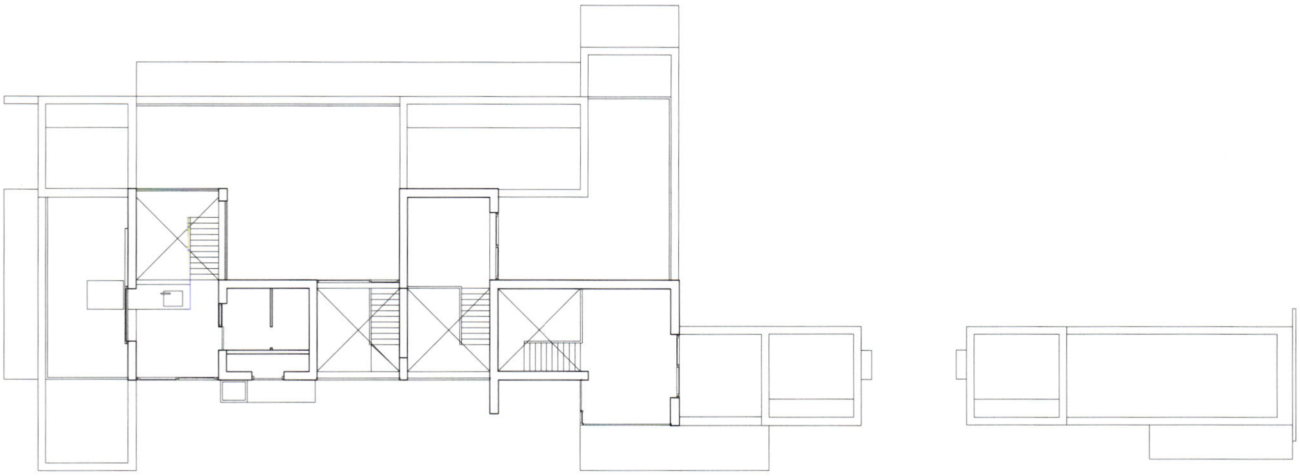

6

1.2
URBAN INFILL

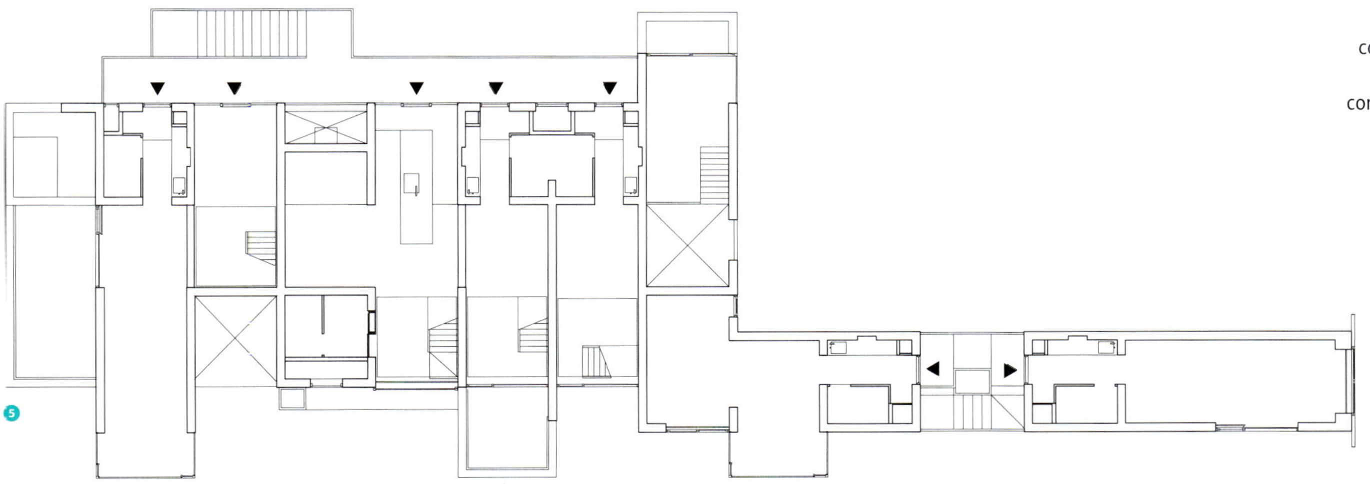

5

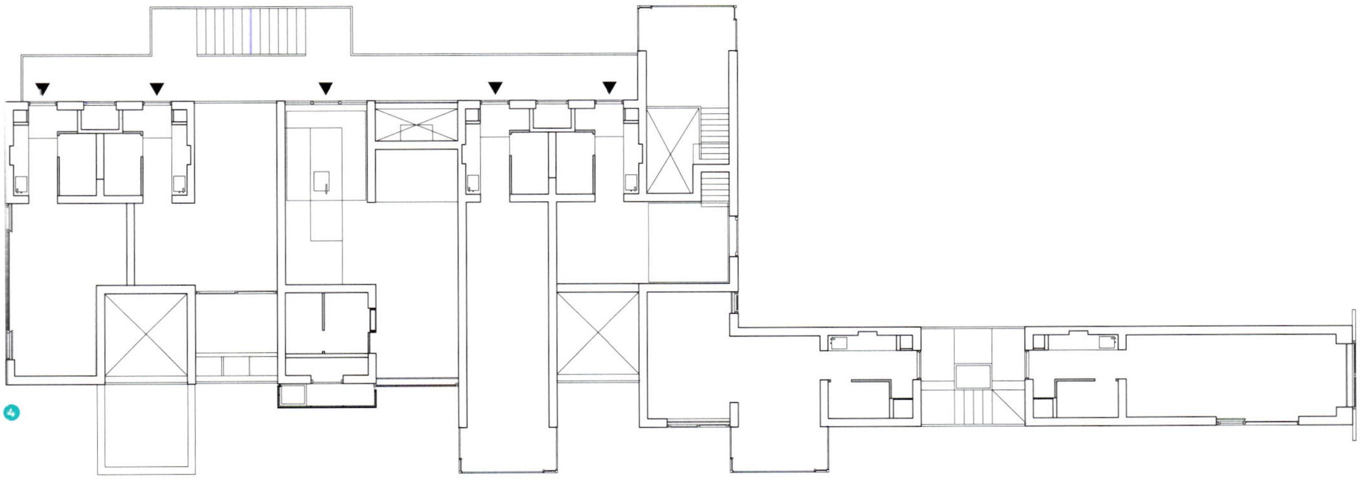

4

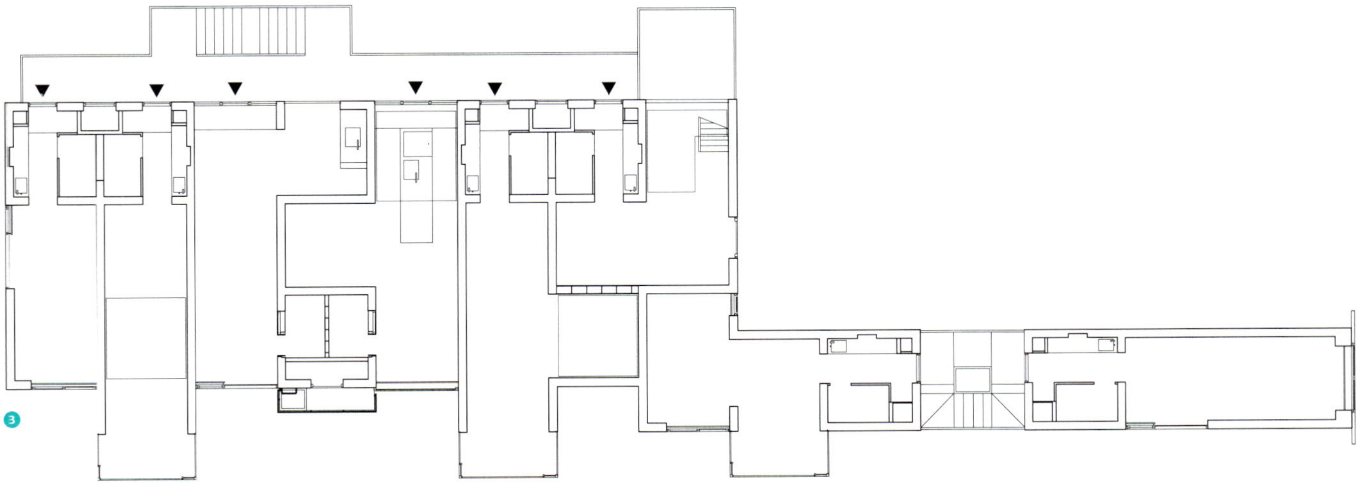

3

Building type
conglomerate of spatial blocks
with miniapartments
commercial use on ground floor
5 stories
facing SW

Date of construction
1998

Number of units
22

Size of units
miniapartments, 17–35 m^2

Area per user
17–35 m^2

Building depth
2.6–10 m

Access
covered walkway

Open spaces
roof patios

Parking
no parking on lot

Architect
Kazuhiro Kojima +
Kazuko Akamatsu
Tokyo

Location
Higashi-Yodogawa
Osaka
Japan

Referring to
floor plan 1:200

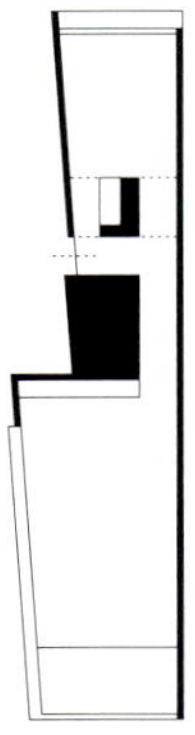

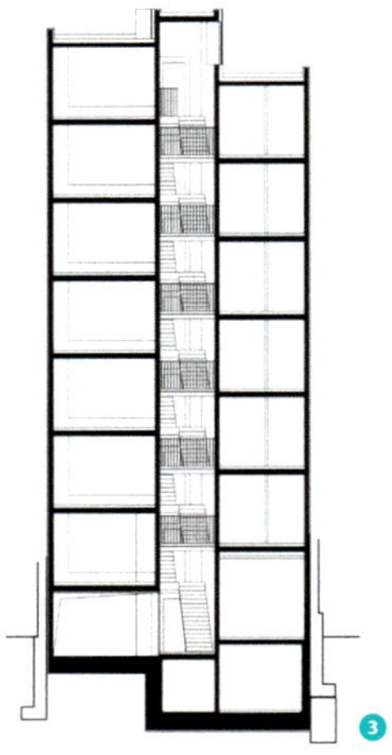

3

Lychener Straße |
Nägeli, Zander | 2000

1.2
URBAN INFILL

A small, very narrow building gap fronted by a square and with a garden to the rear. In response to the site, the building is divided lengthwise; the two identical halves are set off from one another in plan and section to leave the block edge open and to allow light into the units. Interior and exterior are intertwined. The space between the party walls remains perceptible. The apartments are oriented both toward the square and to the street. Floor-to-ceiling glazing in the facades extends part of the living space in a visual fashion across the entire width of the gap, the other section – raised by 40 cm – mediates between interior and exterior and forms a "panoramic window." The private area of the apartment is divided into hallway, bathroom, and bedroom by a built-in unit with sliding doors.

Building type
mixed-use building
7 stories
facing SE/NW
bar on ground floor, office on first floor

Number of units
12

Size of units
2-room units, 110 m² (10 units)
2-room units, 100 m² (2 units)

Area per user
52.5 m²

Building depth
36 m (apartment 24.7 m)

Access
single-loaded stairwell
split-level access

Open spaces
communal garden,
balconies, terraces

Parking
no parking on lot

Architect
zanderroth architekten gmbh
Christian Roth, Sascha Zander,
Kirka Fietzek, Diana Gunkel,
Guido Neubeck, Konrad Scholz,
Lutz Tinius

Location
Lychener Straße
Berlin-Prenzlauer Berg
Germany

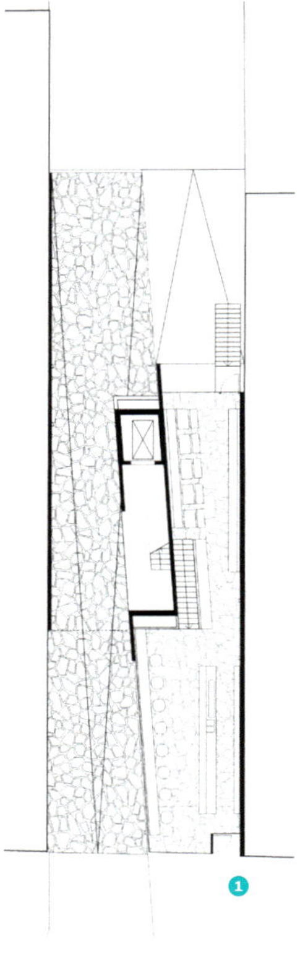

1

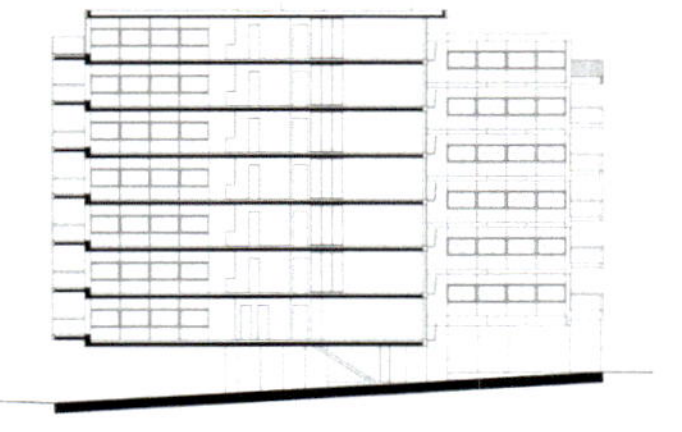

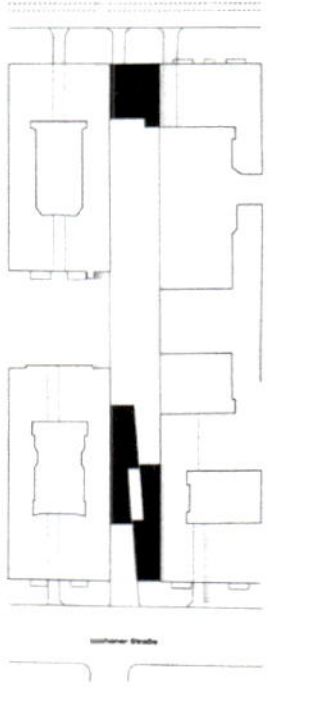

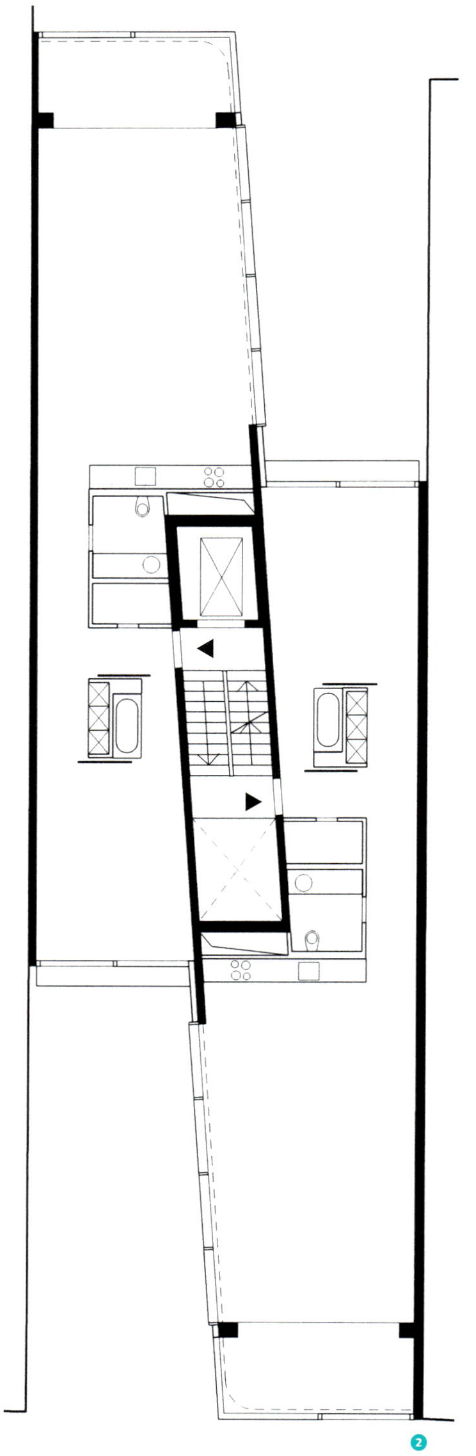

2

Ground floor 1:500 1
Standard floor 1:200 2
Cross-section 1:500 3

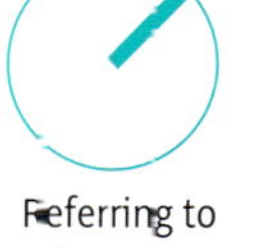

Referring to
floor plan 1:200

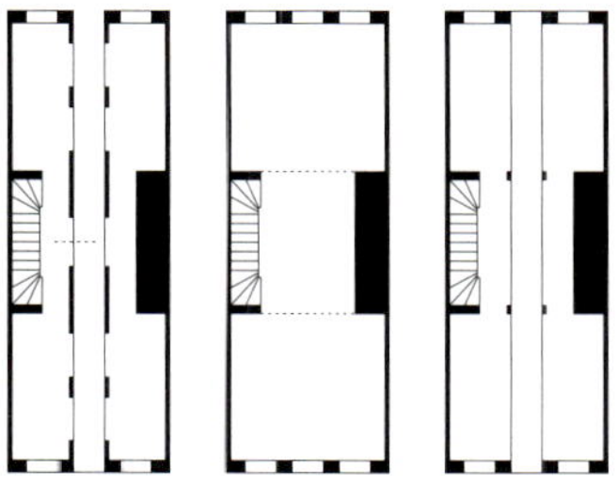

Santen House | Höhne & Rapp | 2000

The house is located in Borneo-Eiland, a new urban development area in Amsterdam, whose parcels were sold to private individuals and their architects with the proviso of adhering to the row house typology. This home divides a "quasi public" passage way, which links the road and the waterway. The entrance is located in this passage way, with a work area occupying the other side. Both halves of the building are united on the second floor (open living zone). On the third floor, however, an elongated terrace – which echoes the passage below – serves as a dividing element between bedrooms (master bedroom, children's bedroom[s]). The middle of the house is occupied on one side by the service rooms, to the left by a spiraling staircase, and to the right by the sanitary rooms. Qualitatively, this design is remarkable for the idea of drawing the exterior space crosswise through the house, with all the advantages (e.g., lighting) and disadvantages (e.g., narrow rooms) this entails.

Building type
single-family house
3/4 stories
facing N/S

Number of units
1

Size of units
291 m² (including basement area)

Area per user
42 m²

Building depth
16 m

Access
internal stairwell

Open spaces
terrace on 3rd floor
internal passage way
leads to waterfront

Parking
no parking on lot

Architect
Höhne & Rapp
Amsterdam

Location
Borneo
Amsterdam
The Netherlands

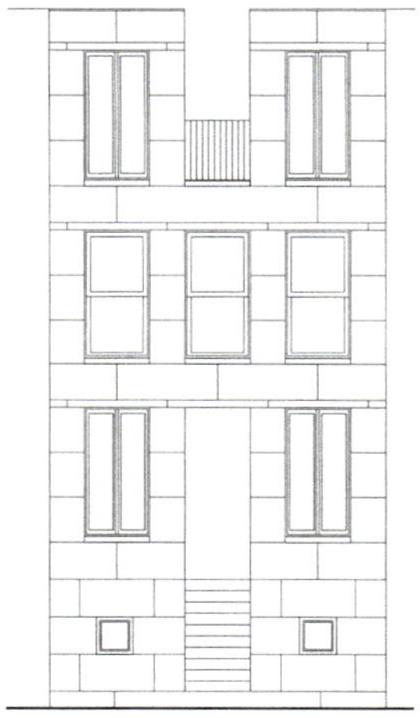

5

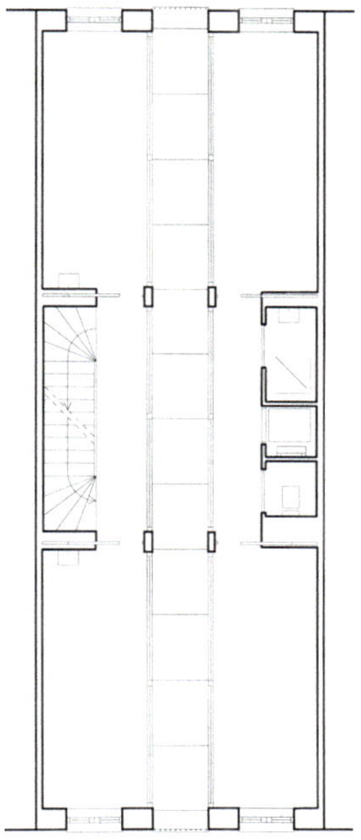

4

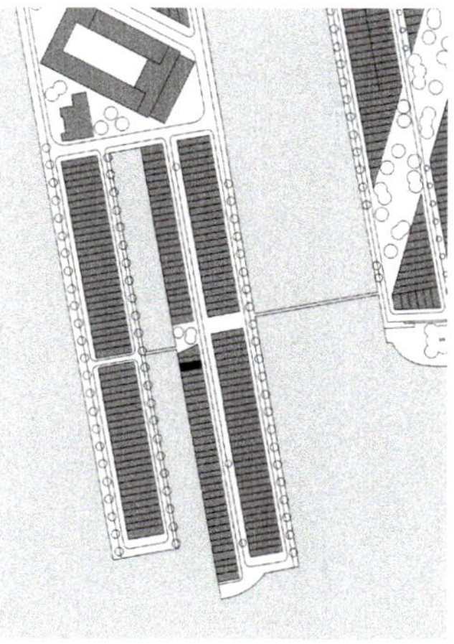

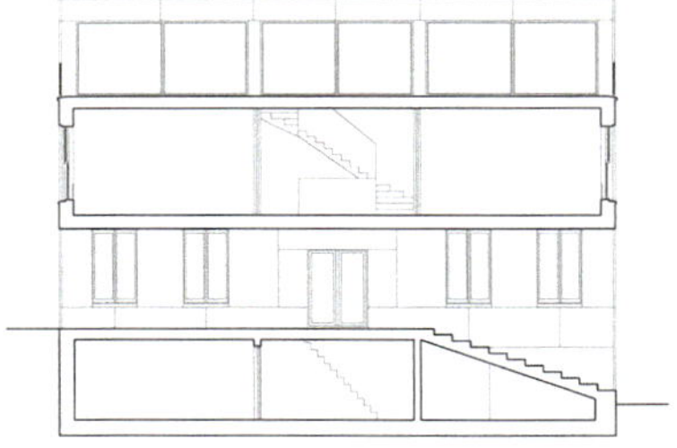

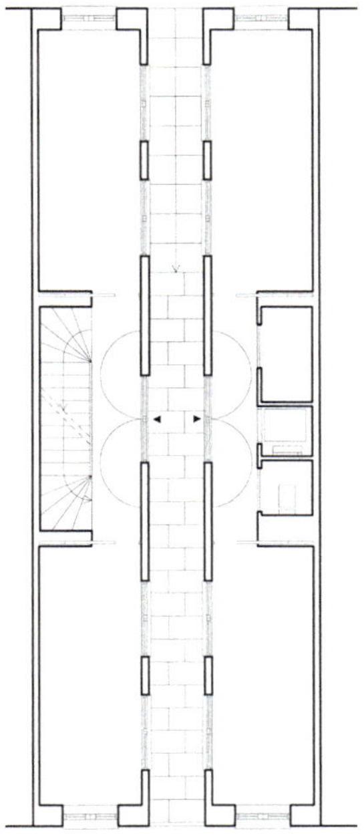

2

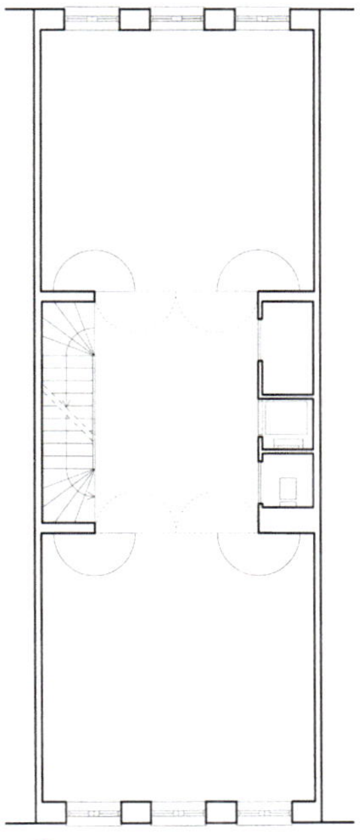

3

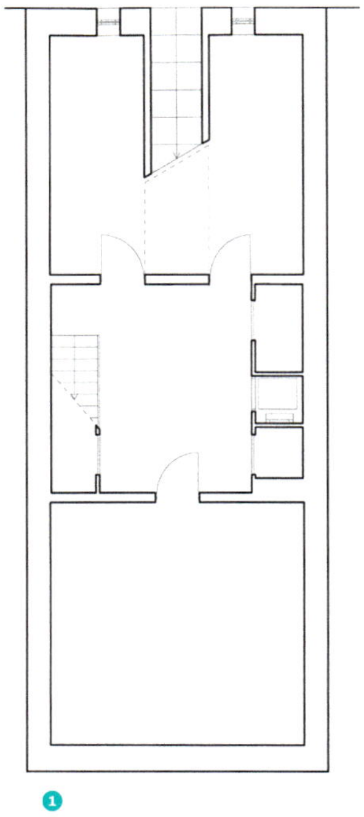

1

Basement 1:200 1
Ground floor 1:200 2
2nd floor 1:200 3
3rd floor 1:200 4
Elevation on water 5

Referring to
floor plan 1:200

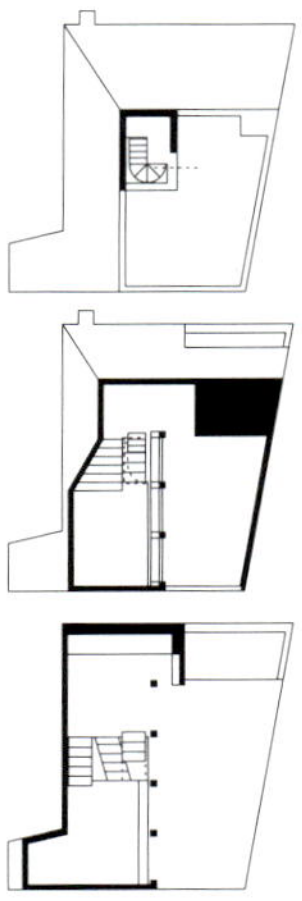

House & Atelier Bow-Wow | Atelier Bow-Wow | 2005

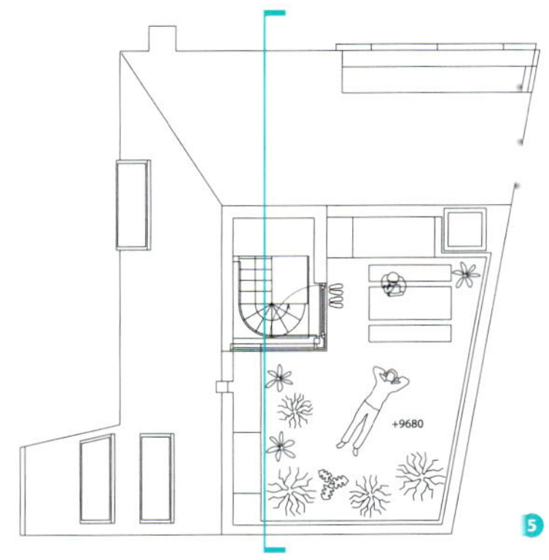

5

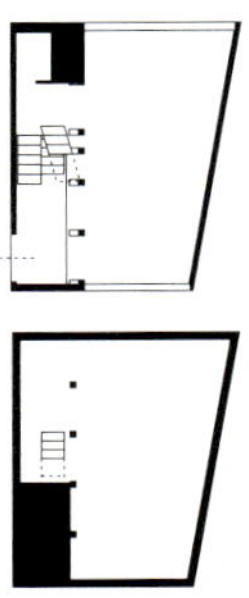

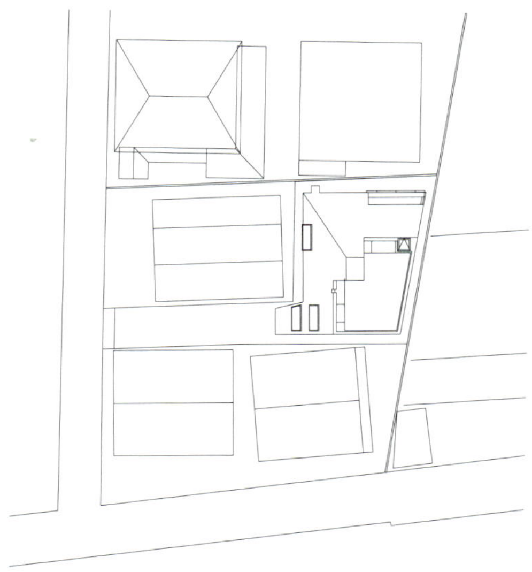

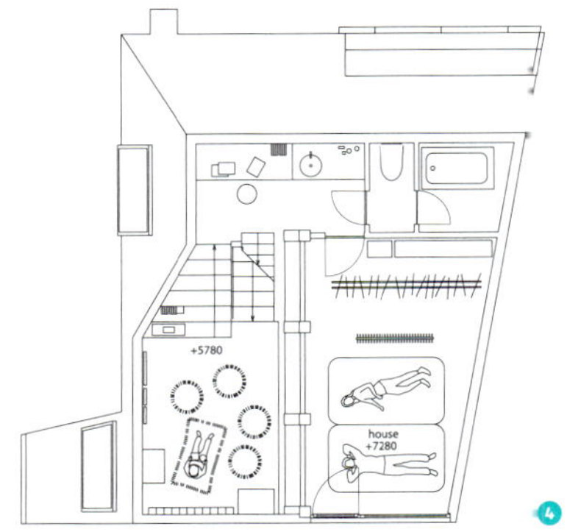

4

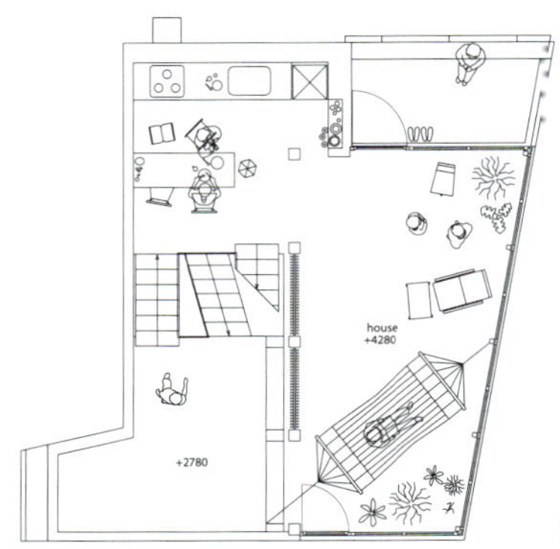

3

The sculptural shape of the minihouse evolved out of the interplay of several parameters, such as the maximum allowable building volume, rules regarding distance, fire, and earthquake safety regulations, the mandate that only a part of the sky may be obstructed from the street perspective, as well as the hybrid program for the interior. The work/live building consists of open levels, progressing from floor to floor in the manner of a split-level and terminating in the roof patio. The result is a spatial continuum, where working (basement and mezzanine) and living (upper floors) transition into one another without clear spatial separation; with the exception of the bathrooms and WCs, there are no room enclosures. Intimacy is indicated by subtle means: for example, the heating element which reaches through the full height of the building, permits or blocks visual sightlines, the functional definition of the mezzanine hints at the use assigned to the following floor, the translucent wall elements slow down sight and movement alike. To one side of the entrance, the upper floor of the living area is clad in the same wood elements as the suspended ceilings. The small windows in the internal facade seem like an ironic nod to conventional living. Despite the immediate proximity, subtly formulated boundaries and the staggered levels promote a sense of spatial separation between the different areas. This corresponds with the users' wish for a house in which the transition from public to private space is fluid and not abrupt in nature.

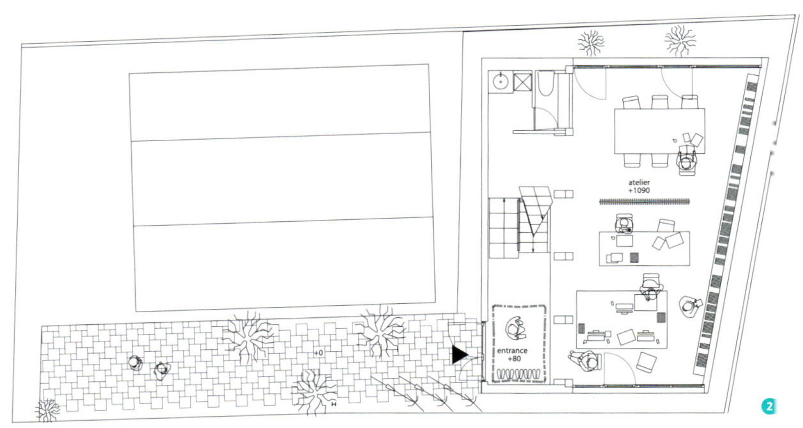

2

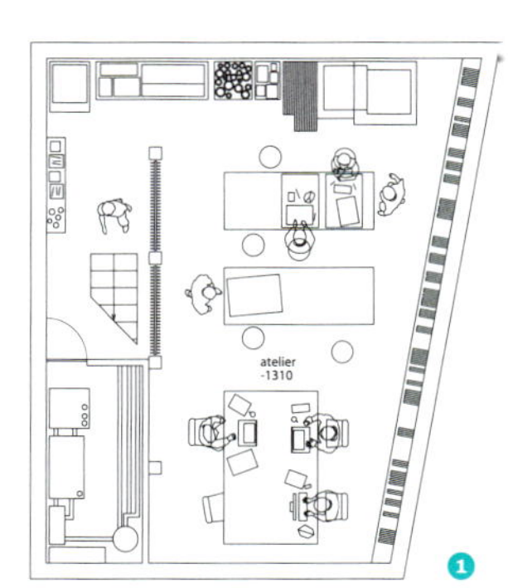

1

1 Lowered ground level with studio 1:200
2 Raised ground floor with studio, entrance on intermediary level 1:200
3 2nd floor with kitchen and living area, studio (model archive) on intermediary level 1:200
4 3rd floor with sleeping area and bathroom, living area on intermediary level 1:200
5 Roof terrace 1:200

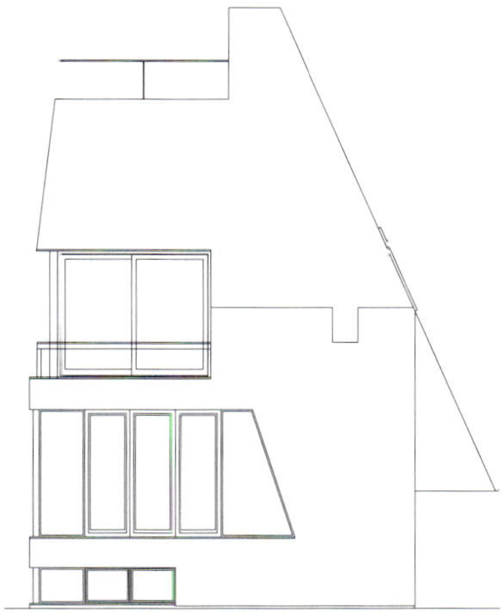

6

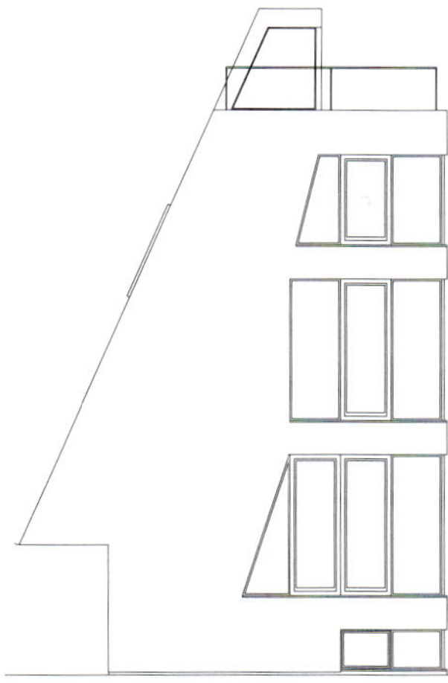

7

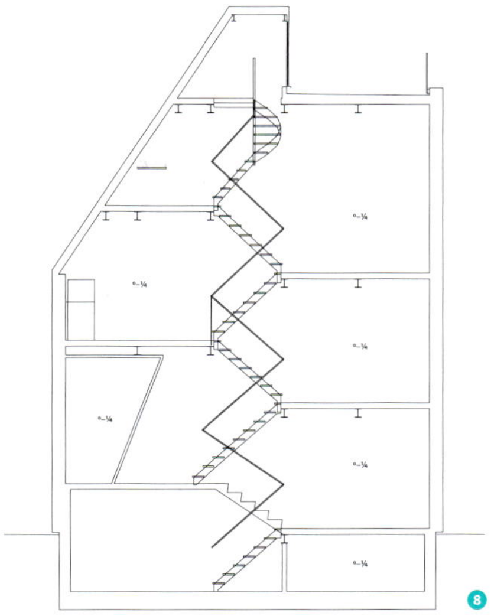

8

1.2
URBAN INFILL

Building type
detached, second-tier infill
4 stories with basement
facing NW/NE/SE/SW

Date of construction
2005

Size of units
219 m²

Area per user
109.5 m²
(incl. office areas)

Building depth
9 m

Access
split-level
with generous landings

Parking
no parking on lot

Architect
Atelier Bow-Wow
Tokyo
Yoshiharu Tsukamoto
Momoyo Kaijima
Shun Takagi

Location
Shinjuku-ku
Tokyo
Japan

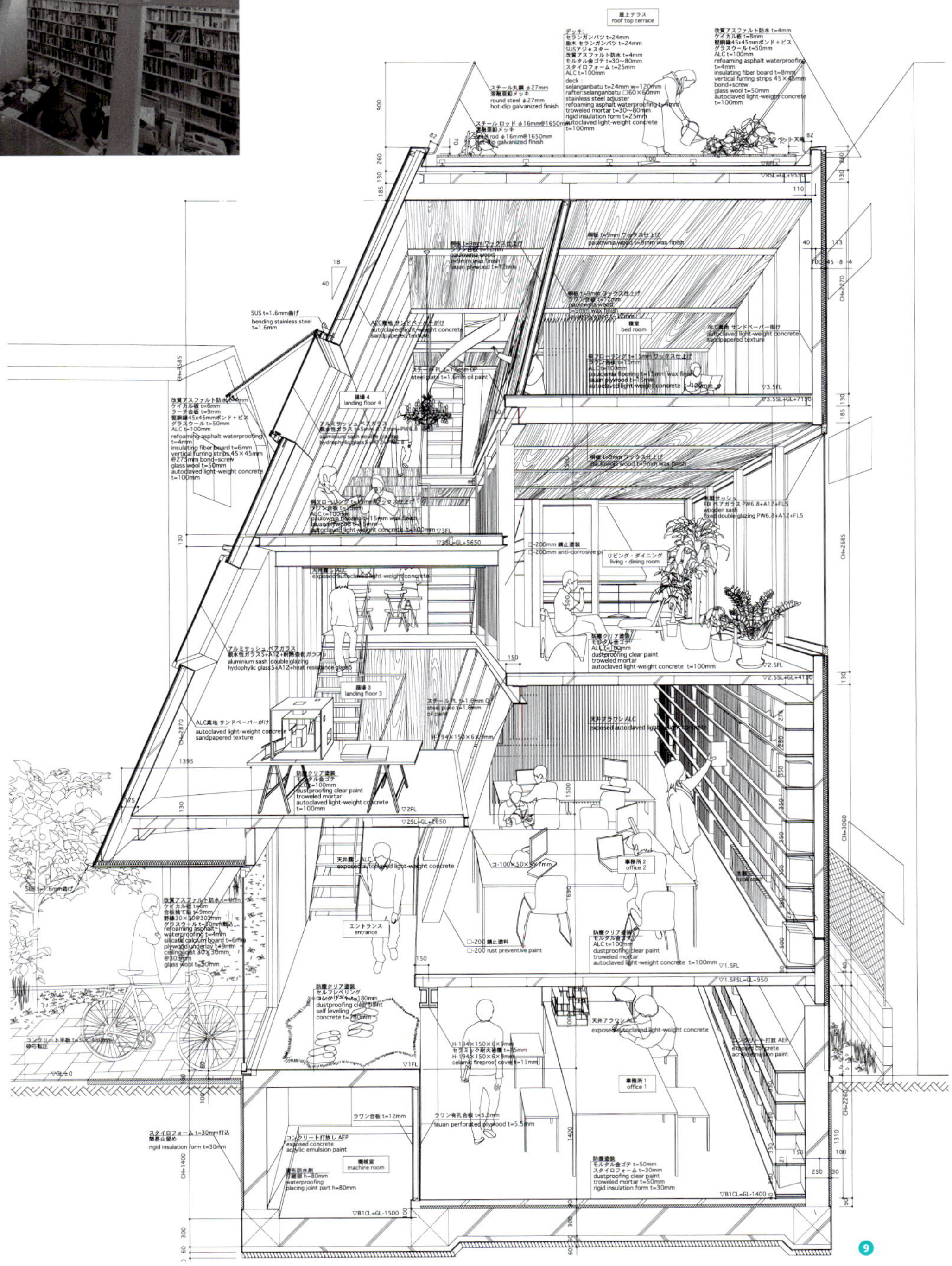

9

6 North elevation 1:200
7 South elevation 1:200
8 Longitudinal section 1:200
9 Sectional drawing

Referring to
floor plan 1:200

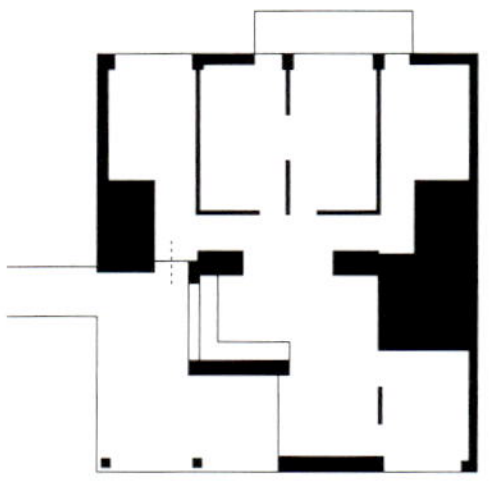

e3 | Kaden Klingbeil Architekten | 2008

The project was realized by a housing co-operative with the future residents assuming the role of client. This allowed them to influence the building costs, implement individual floor plan preferences, and form a cooperative community of residents even prior to completion. The group chose a construction method that is amenable to flexible floor plan design while also being ecologically and economically sustainable: a seven-story post-and-beam wood construction with reinforcing solid wood walls and composite wood-concrete ceilings that freely span the space with a central girder that is flush with the ceiling; the only elements in the interior space of the room are two installation shafts. Completely different floor plans with open living areas were developed in consultation with the users. While some of the living areas are fully open from front to rear and others are oriented towards the courtyard or the access, all are wrapped around the individual rooms located at the corners or the courtyard i.e. street elevation.

The access core is separate for reasons of fire protection: since the apartment building is a wood construction, the escape route had to be built in reinforced concrete and set apart from the building. The resulting gap creates a "third facade". Transferring the stairwell to the outside also helps to facilitate flexibility in the interior and shifts access to each apartment, via 3-m-long footbridges, into the urban space. On three floors, the access route leads across a generously proportioned incised open space, further enhancing the feeling of owning an entire level and hence a home of one's own in the city.

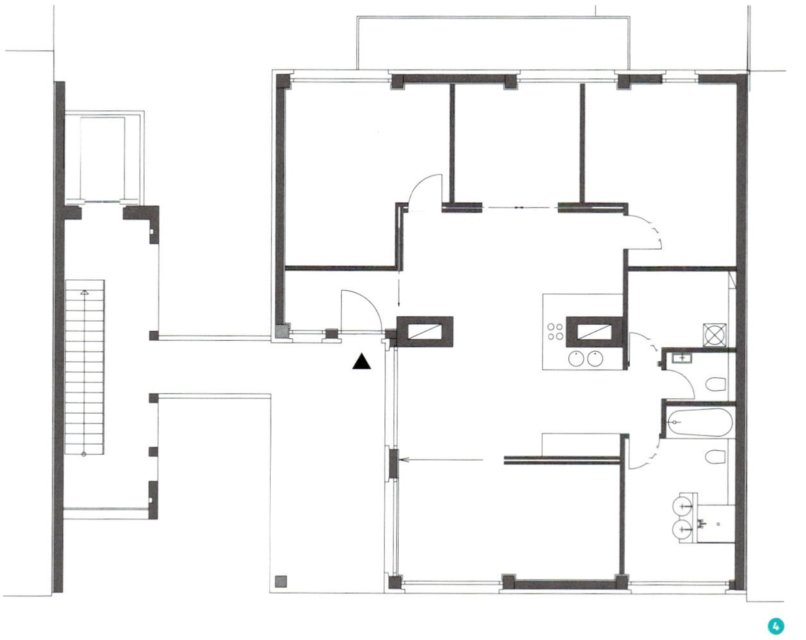

4

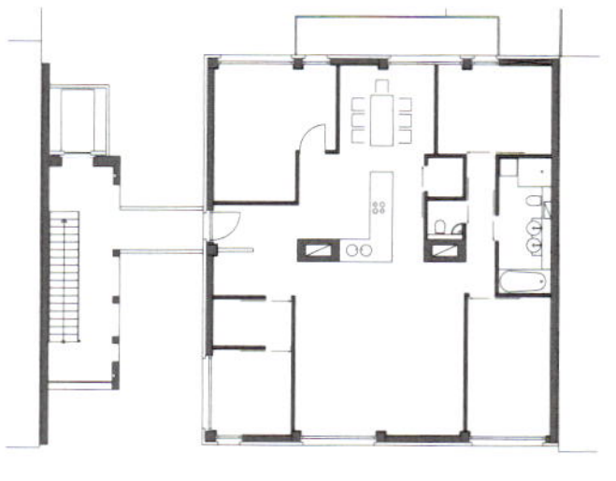

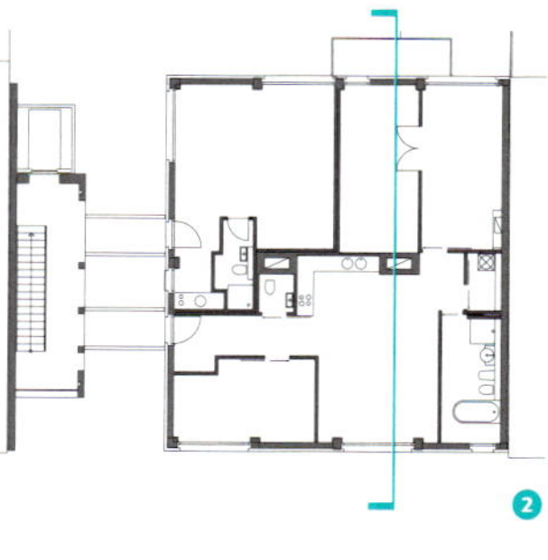

2

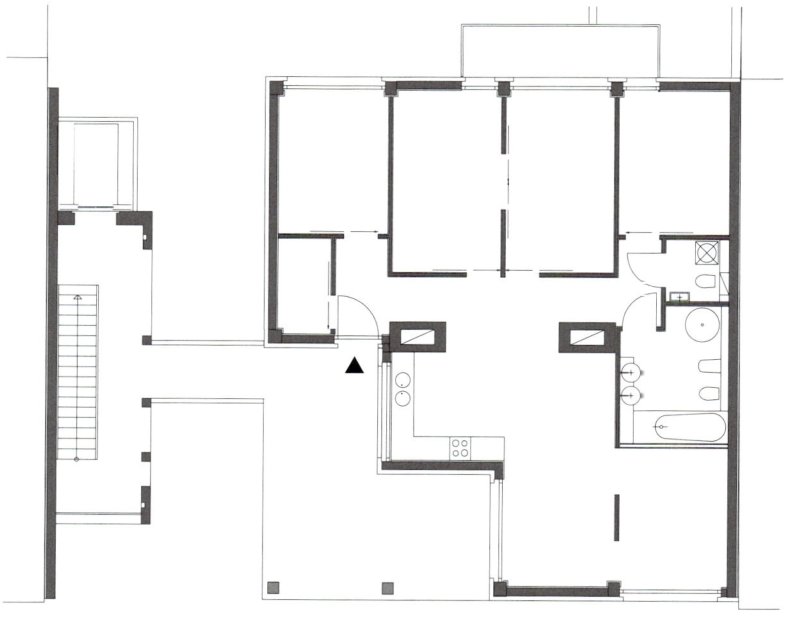

3

1

1. Floor plans developed in consultation with residents, ground floor 1:500
2. 3rd floor, 6th floor 1:500
3. 2nd floor: 6-room apt. with living area with three partitions and corner terrace 1:200
4. 4th floor: 5-room apt. with central, expandable living area and corner terrace 1:200
5. 5th floor: 6-room apt. with living room and working space, overlooking garden, corner terrace 1:200
6. Attic story: 6-room apt. with front-to-back living room and balcony overlooking garden 1:200
7. Garden elevation 1:500
8. Cross section 1:500
9. Side view, section through connecting bridges 1:500

7

8

9

1.2
URBAN INFILL

Building type
low-energy apartment building
wood construction
7 stories
facing NW/SE

Date of construction
2007–2008

Number of units
7

Size of units
2nd floor: 6-room apt.: 118.5 m²
3rd floor: 4-room apt.: 107.5 m²
1-room apt.: 40.5 m²
4th floor: 5-room apt.: 130 m²
5th floor: 6-room apt.: 131.5 m²
6th floor: 5-room apt.: 149.5 m²
attic story: 5-room apt.: 145.5 m²

Area per user
23.5–40.5 m²

Building depth
13.8 m

Access
single-loaded staircase
(except for 3rd floor),
separate open stairwell

Open spaces
loggias
balconies
bridges

Parking
no parking on lot

Architect
Kaden Klingbeil Architekten
Berlin

Location
Esmarchstraße
Berlin-Prenzlauer Berg
Germany

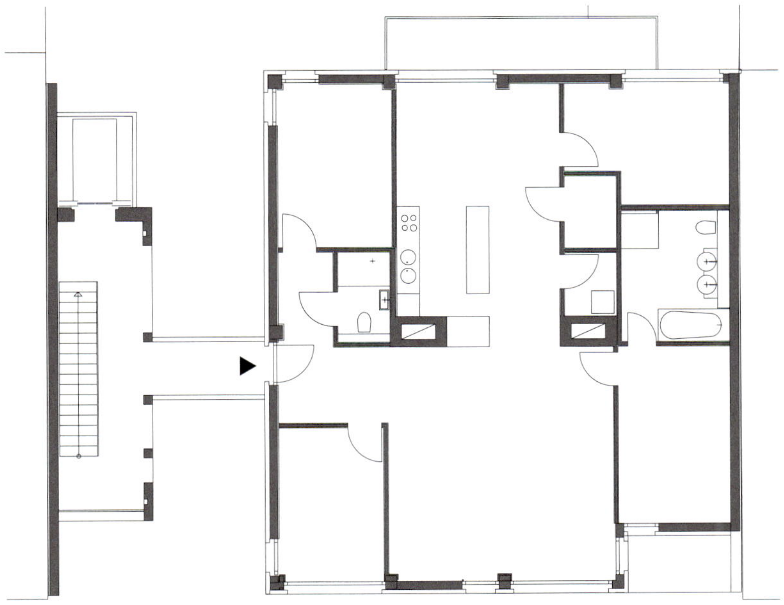

6

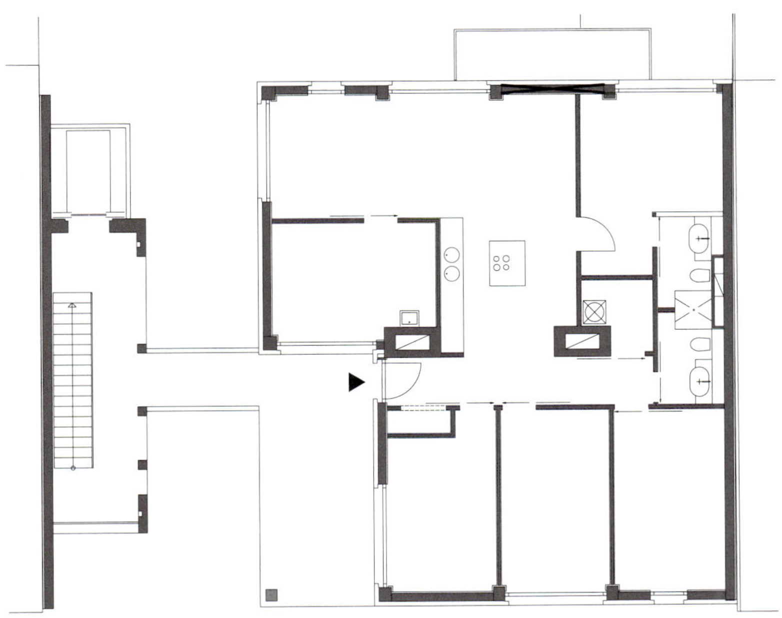

5

Referring to
floor plan 1:200

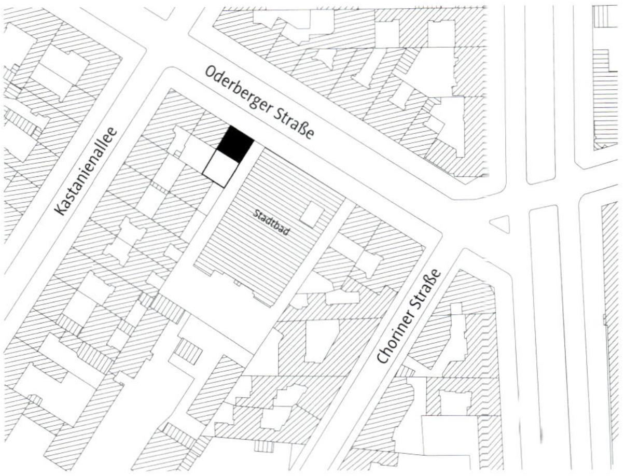

Oderberger Straße | BARarchitekten | 2010

The hybrid building is conceived as an interface for a variety of different uses. Ground floor and basement with an inserted light well accommodate retail and event areas. The 2nd and 3rd floors above contain two-story work/live studios. The mixed-use housing in this "urban plinth" is aimed at promoting the formation of local structures, social integration and cooperation within and outside of the building.

From the 4th floor onward, the building is distinguished by a highly differentiated spatial structure, creating flexible units with the character of a very unique "house within a house."

The varying floor heights – on both the courtyard and the street elevation – as well as the cascading offset levels parallel to the facades, create complex unit types with up to four different levels.

The internal room sequences border onto the stairwell – either on two or on three sides – and simultaneously connect with different floor and mezzanine landings, making it possible to subdivide the units. The separable units can be rented out, utilized for multigenerational living or as home offices, and also serve as small flats for caregivers or in-house help.

On the narrow, south-facing courtyard side, ceilings soar up to 4.3 m: above the kitchens, the space on this side of the building includes a gallery for use as a library or an additional open space. A guest apartment is available in the attic story, while the roof patio serves as a common outdoor space.

C

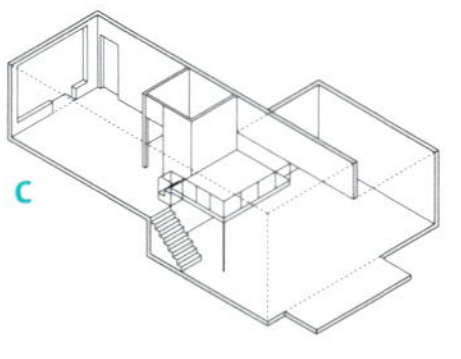

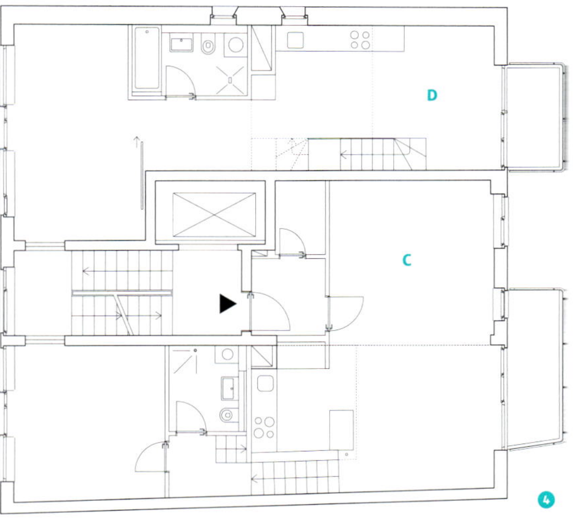

4

B

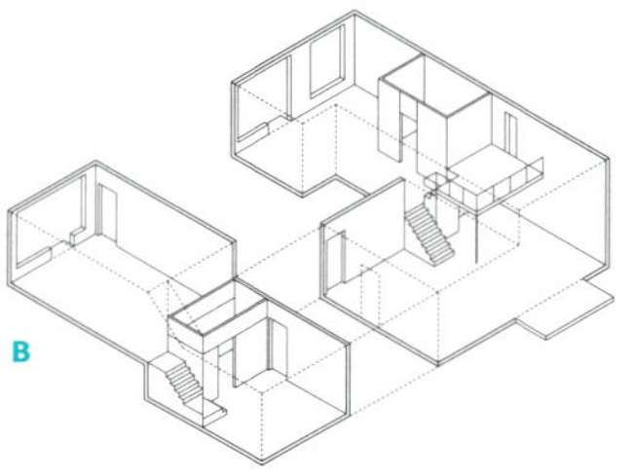

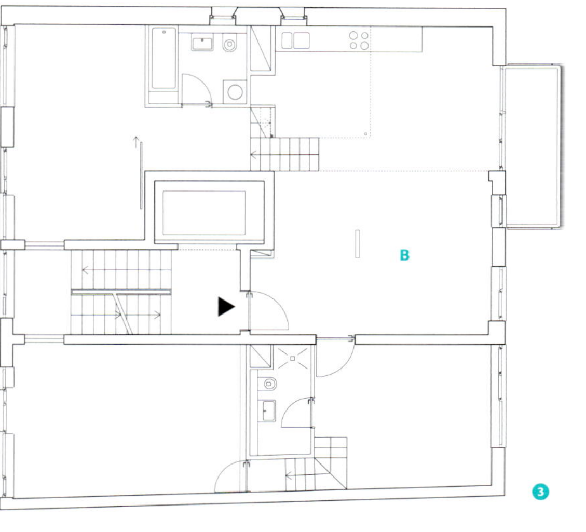

3

A

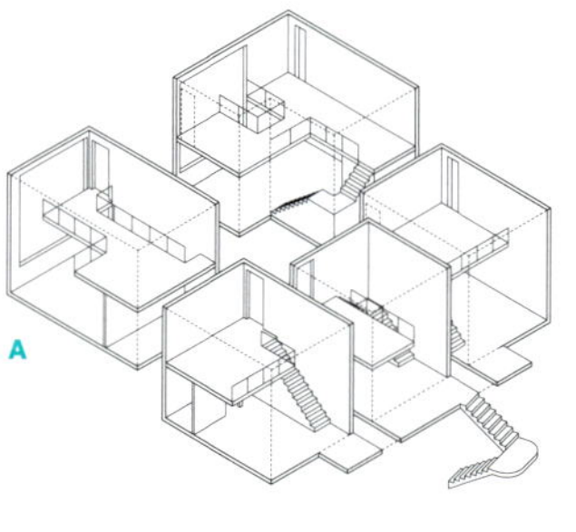

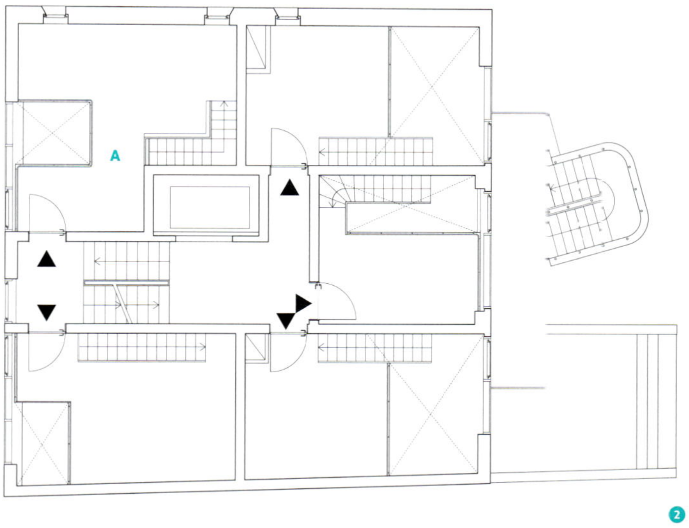

2

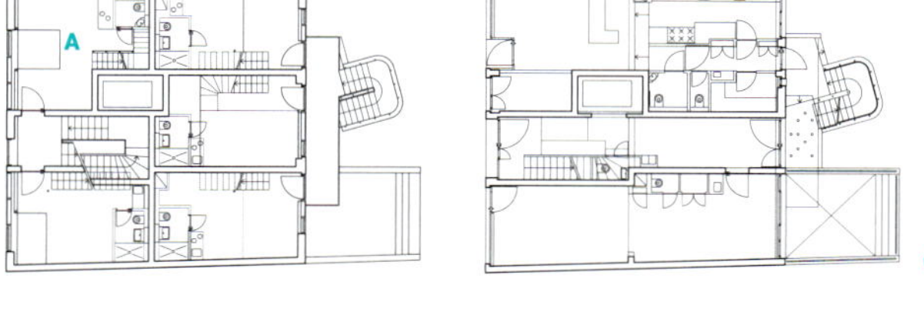

1

1 Floor plans of ground floor, 2nd floor 1:500
2 3rd floor: work/live studios **A** 1:200
3 4th floor: apartment **B** 1:200
4 5th floor: apartments **C** and **D** 1:200
5 6th floor: apartments **D** and **E** 1:200
6 7th floor: apartment **E** 1:200
7 8th floor: apartment **F** 1:200
8 Isometric views of various apartment volumes and their composition
9 Cross- and longitudinal section 1:500

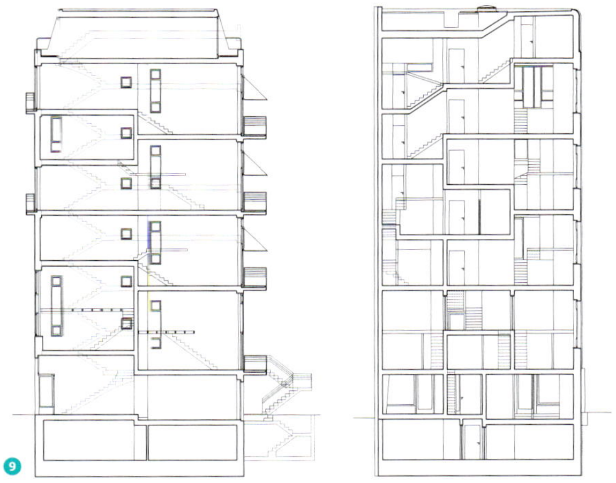

9

1.2
URBAN INFILL

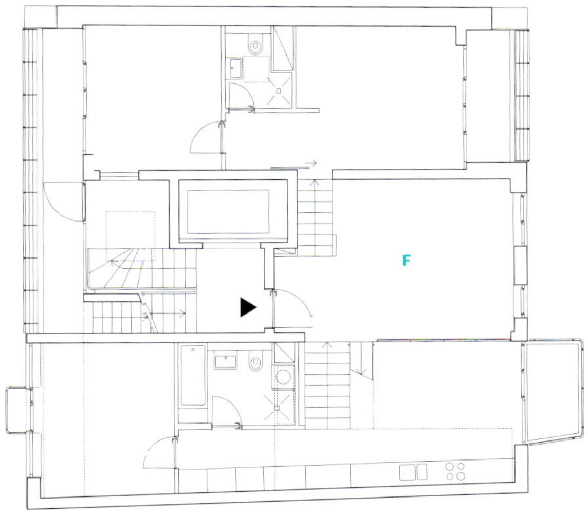

7

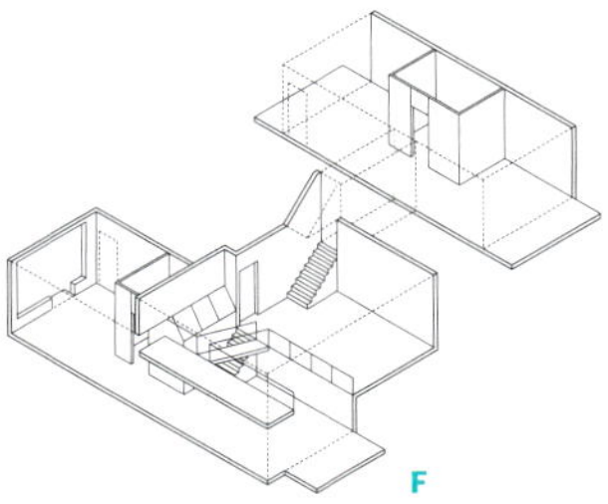

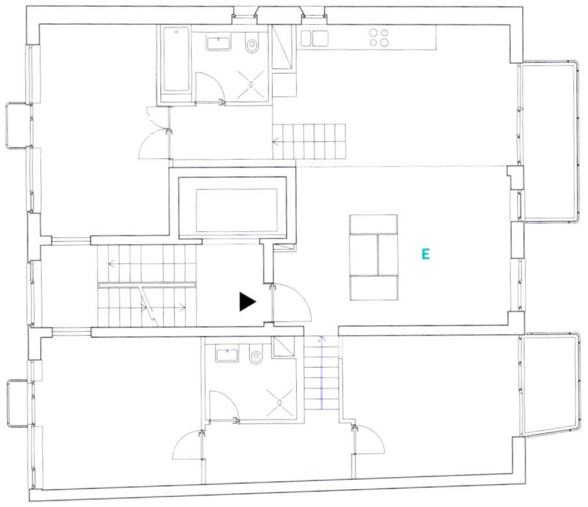

6

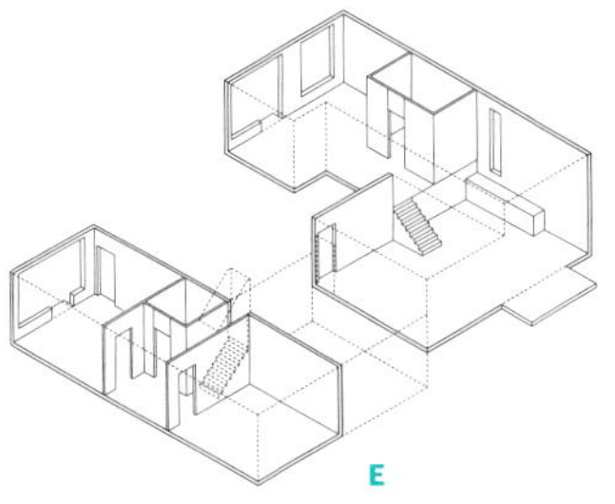

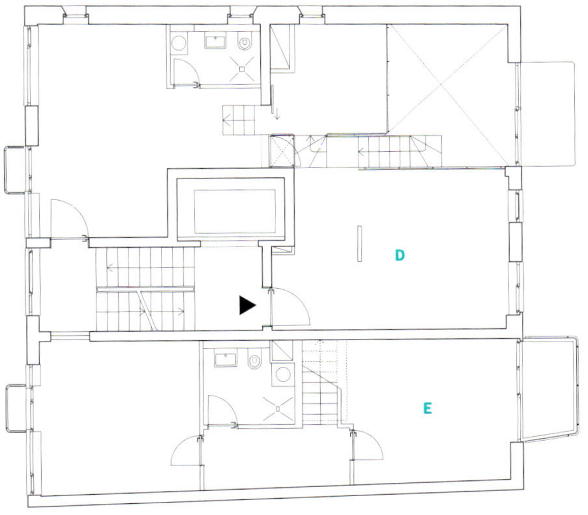

5

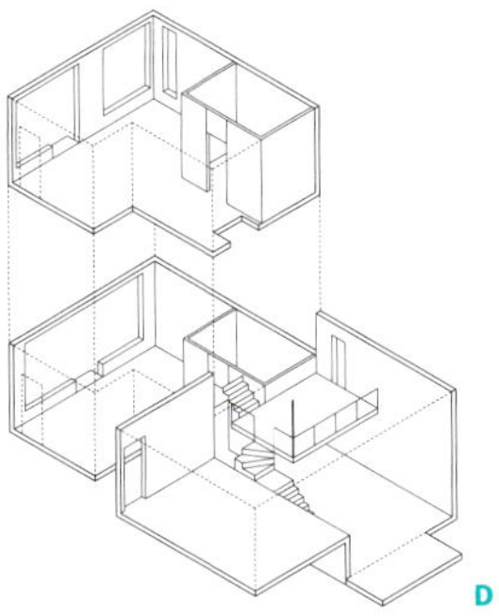

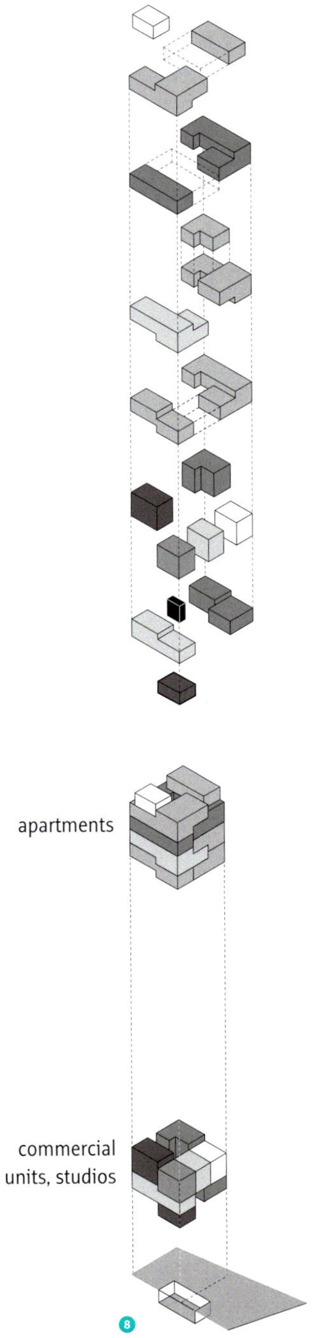

8

Building type
mixed-use residential building with commercial units, minigallery and music room on ground floor and basement

Date of construction
2008–2010

Number of units
10 to 14 residential units (depending on subdivision of the units)

Size of units
2nd–3rd floor:
studios (work/live studios), 35–45 m² (5 units)
4th–8th floor:
2–3-room apartments, 75–85 m² (5 units)
additionally 4 flexible apartment units with 1–2 rooms, 35–45 m²
guest room (18 m²)

Area per user
35–45 m²

Access
single- and double-loaded stairwell, with additional access via mezzanine landings

Open spaces
balconies, courtyard and roof patio for common use

Parking
no parking on lot

Architect
BARarchitekten
Berlin
Antje Buchholz
Jack Burnett-Stuart
Michael v. Matuschka
Jürgen Patzak-Poor

Location
Oderberger Straße 56
Berlin–Prenzlauer Berg
Germany

Referring to floor plan 1:200

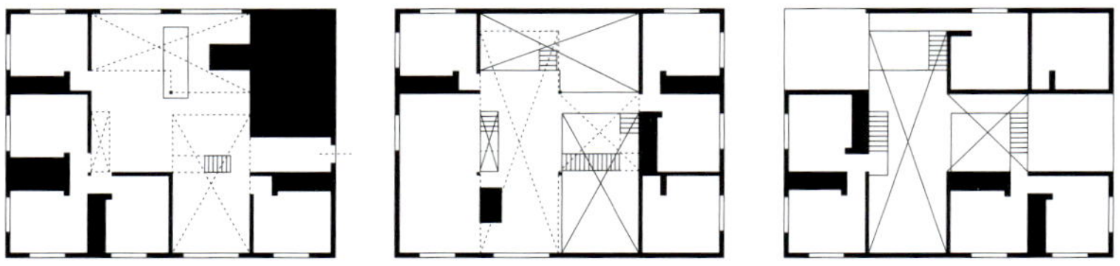

LT Josai | Naruse Inokuma | 2013

1.2 URBAN INFILL

The project reacts to a growing trend in Japan in which singles decide to live together without knowing one another beforehand. The building is structured in a manner to help the residents overcome initial shyness, fostering both interaction and exchange in variable intensities, while also providing each resident with a more private spatial sphere in which to withdraw. The large dining table, for instance, encourages contact between many residents. Various niches are more suitable for spending time alone or in small groups. Private and shared spaces interlock to form a three-dimensional system with offset levels, subtly situated staircases, and numerous visual connections flanked by private rooms. Wall segments in the private rooms, placed perpendicularly behind the door, form an intermediate space and visual buffer to ensure that the interior of these rooms cannot be directly seen from the shared areas. All residents share a bathroom centrally located on the ground floor.

Building type
freestanding cubic building volume
2 stories
facing in all directions

Date of construction
2013

Number of units
1

Size of units
310 m²

Area per user
24 m²

Building depth
10.9 m

Access
open staircases in multi-level collective living room

Open spaces
garden, roof terrace

Parking
parking spaces along street

Architect
Naruse Inokuma Architects
Tokyo

Location
3 Chome Josai
Nagoya
Japan

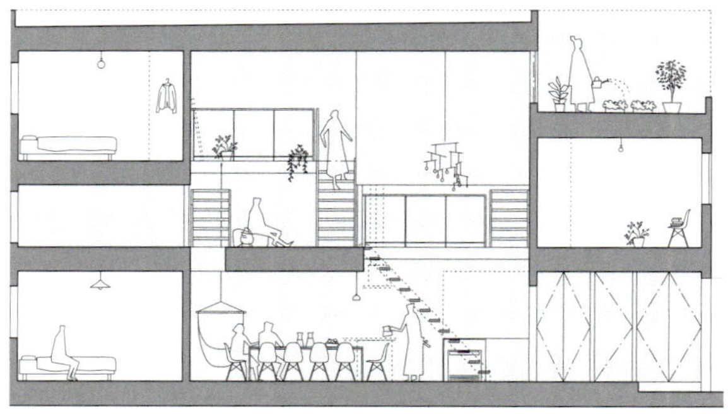

4

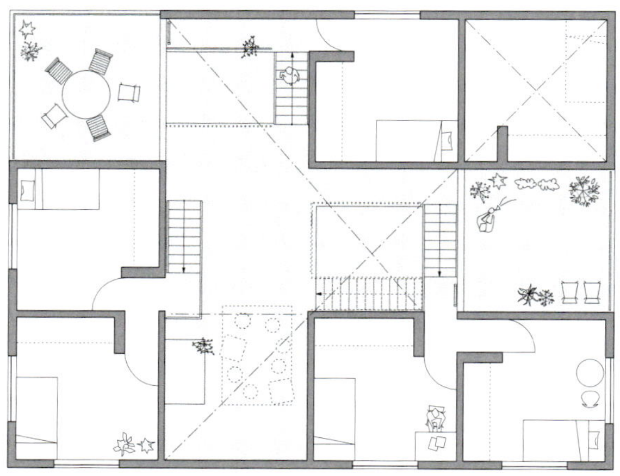

3

5

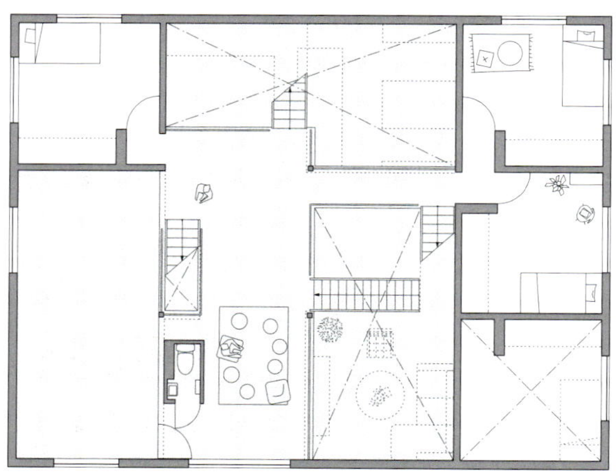

2

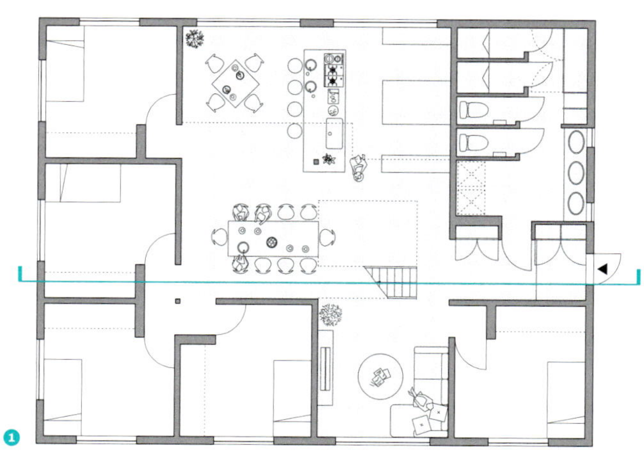

1

Ground floor with eat-in kitchen 1:200 1
2nd floor with "rug space" 1:200 2
Upper level with private rooms and roof terraces 1:200 3
Longitudinal section 1:200 4
Interwoven spatial volumes of private and common areas 5

Referring to floor plan 1:200

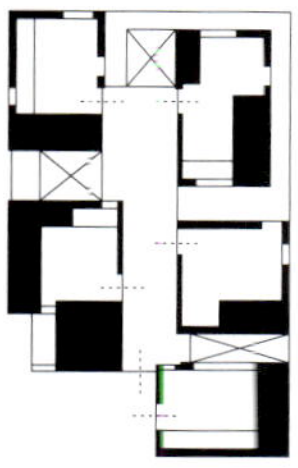

Songpa Micro-Housing | SsD, Park, Hong | 2014

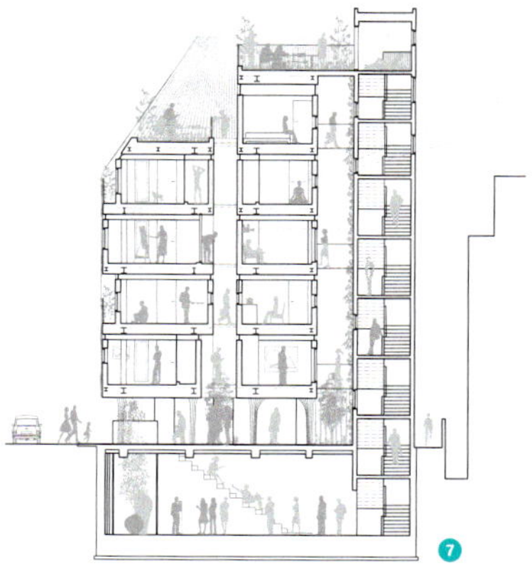

7

This project proposes a new building type in recognition of the criticism that micro-apartments are often only temporary in nature and lacking in social potential. The spatial finiteness of the units – each outfitted with foldable furniture and built-in storage spaces – is compensated by open access spaces running between them. To a certain degree, this softens the divisions between private and public, transforming them into a social fabric stretched between the residents. Simultaneously, the spatial system allows the combination of individual units across the interstitial spaces. This flexibility is intended to enable the residents to live much longer in the building regardless of changing personal conditions. The lower levels form an interface: the open space at street level can also be used by visitors. A micro-auditorium extends this space downward and leads to a café and gallery. The units on the 2nd floor can also be easily converted into work or gallery spaces.

Building type
freestanding open building with inserted micro-apartments above an open ground floor and public spaces below grade
3–7 stories
facing NW/SE/NE/SW

Date of construction
2014

Number of units
14

Size of units
micro-apts., approx. 10–22 m²

Area per user
14 m²

Building depth
15 m

Access
naturally lit, flexibly divisible corridor

Open spaces
open interstices between apartments and the external facade lamellae
shared roof terrace, open ground floor spaces

Parking
open parking spaces on ground floor

Architect
Jinhee Park, John Hong, SsD, New York/Seoul

Location
Songpa-Gu, Seoul, South Korea

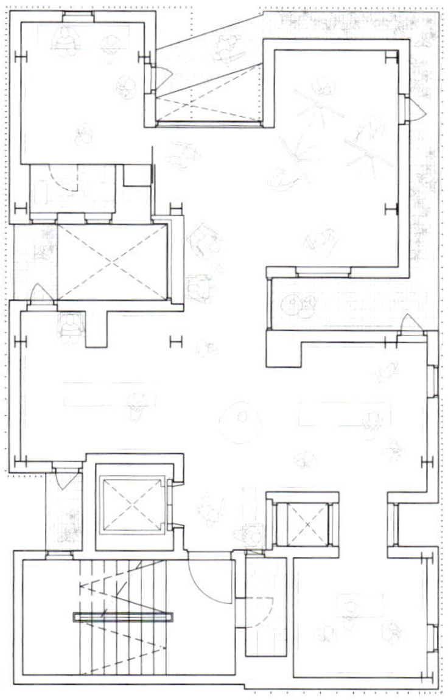

3

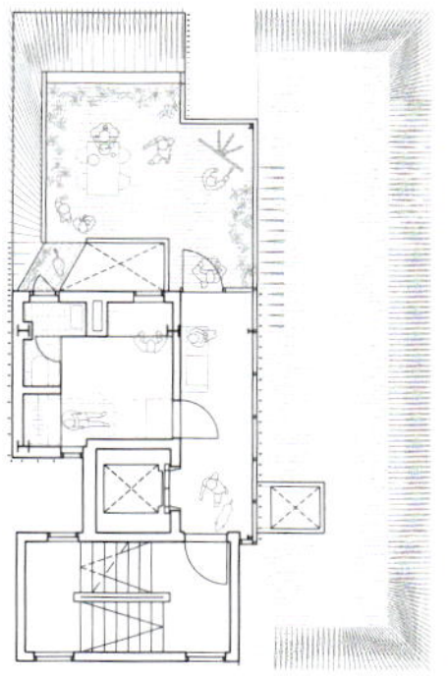

6

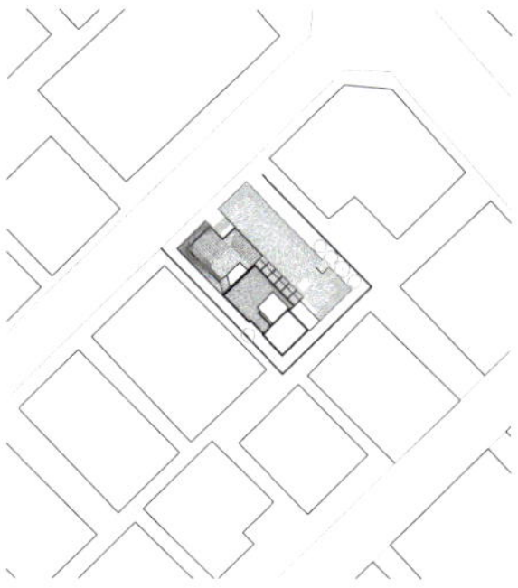

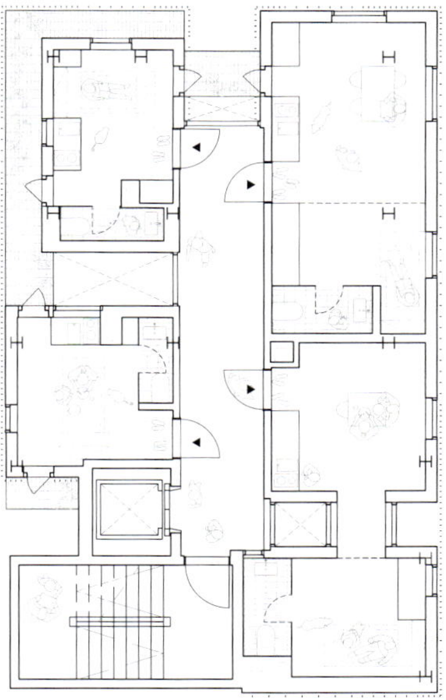

5

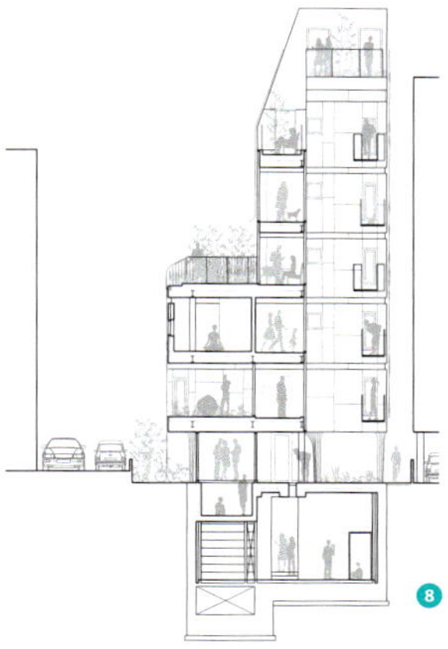

8

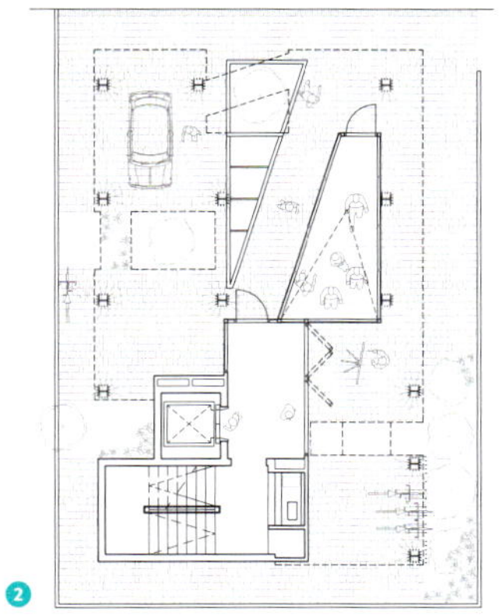

2

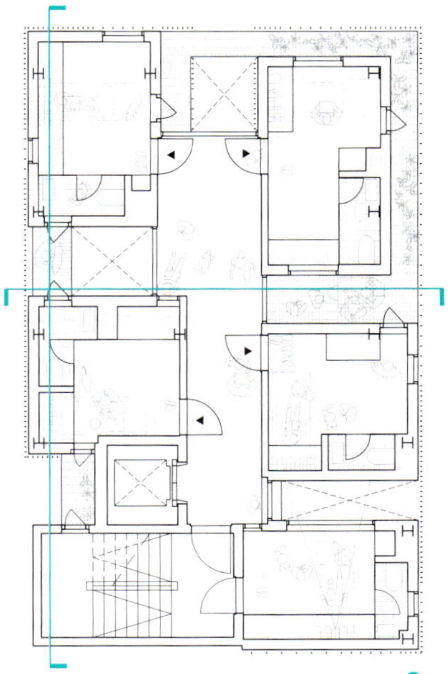

4

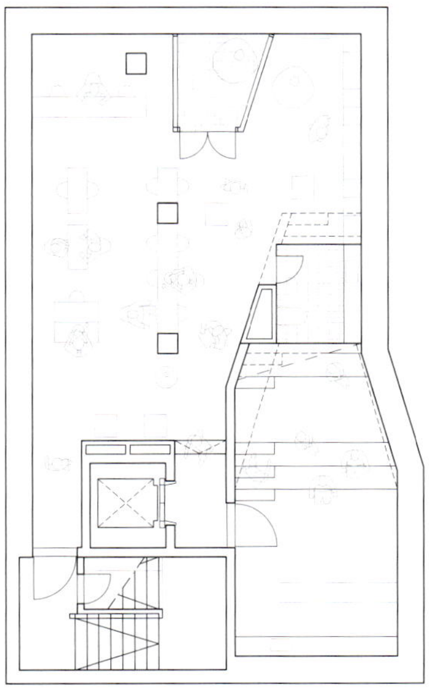

1

Basement level with café and micro-auditorium 1:200 1
Open ground floor with upper access to micro-auditorium 1:333 2
2nd floor, gallery and working space option 1:200 3
2nd floor, micro-apartments option 1:200 4
3rd floor 1:200 5
5th floor 1:333 6
Longitudinal section 1:500 7
Cross section 1:500 8

Referring to floor plan 1:200

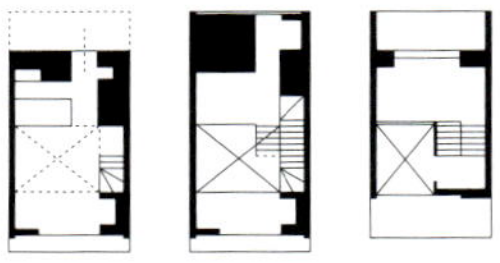

SL Court | Field Design | 2014

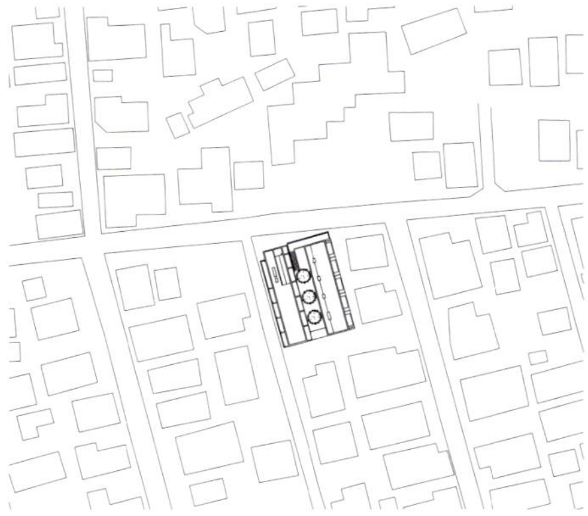

Occupying the corner of an open block edge development, the building consists of two parallel rows framing a shared courtyard. In the street side row, residential units for families extend through the entire height of the building, while two much smaller apartments for singles or couples are stacked on top of one another on the other side. The building's section reveals its unique spatial organization: both rows are divided along their entire lengths and organized as split-levels. Within the same height as a conventional two-story building, up to seven different intermediate levels are arranged – each with very different ceiling heights from 1 to 4½ m and with floor areas between 4½ and 25 m², cut to size for various activities and uses. The implementation of offset levels instead of dividing walls provides a feeling of protective distance. Only the bedrooms in the middle are given additional translucent sliding elements. One single space flows openly throughout the units without completely dissolving the borders of the individual areas, enabling dense living in close proximity.

Type **C** is laid out over four levels for a family with one child. It begins with an extremely compactly arranged living area, whose different ceiling heights subtly zone the space. Located on the level above is a small bridge-like workspace with view below into the living room as well as to the child's room a half-story above. At the top, another level provides space for the parent's bed, which lies at almost the exact height of the child's bunk. The ceiling here is so low that the small roof patio/loggia can only be reached by crawling out over the bed.

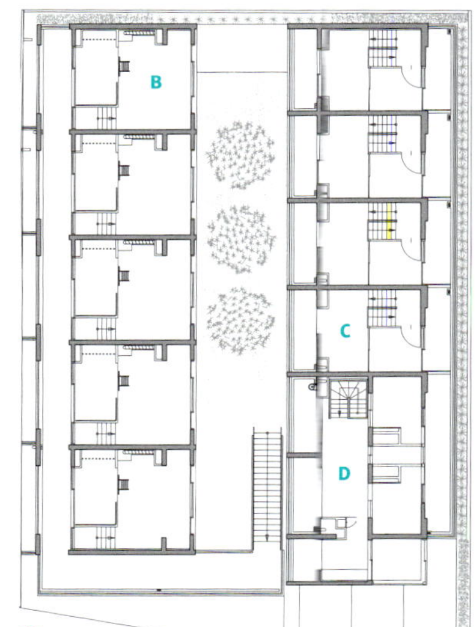

4

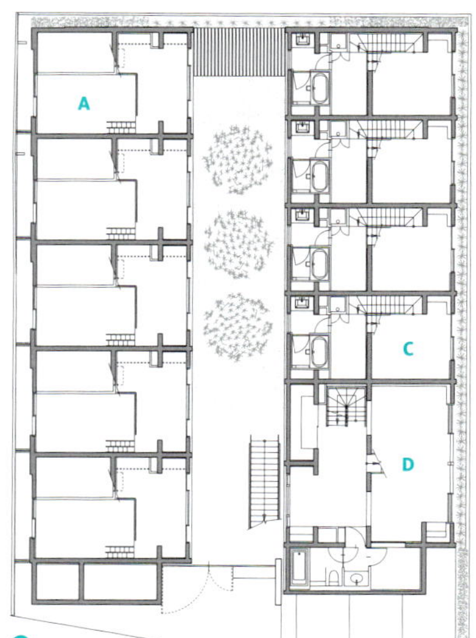

2

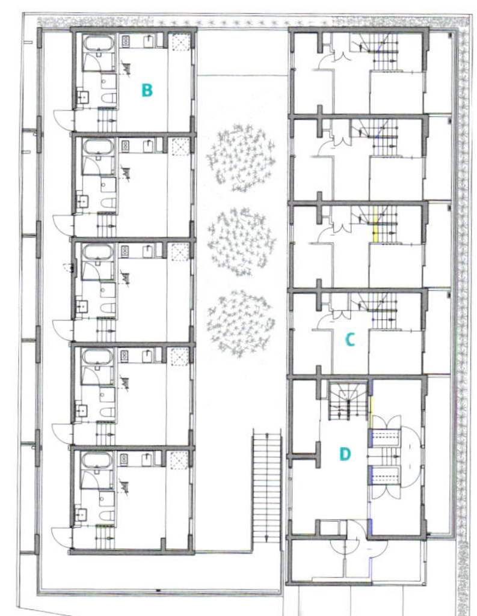

3

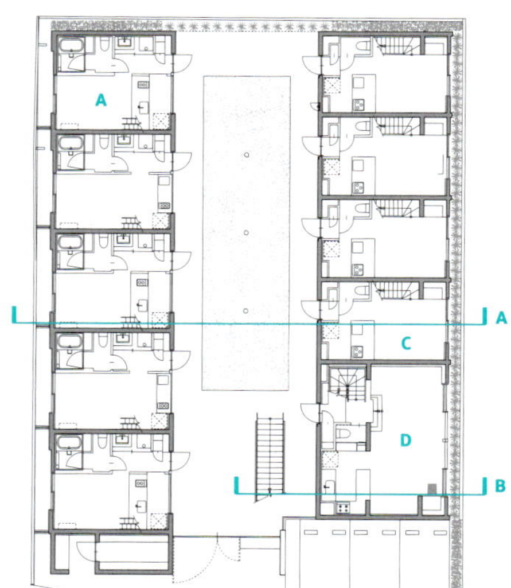

1

1 Ground floor 1:400
2 Intermediate levels above ground floor 1:400
3 2nd floor 1:400
4 Intermediate levels above 2nd floor 1:400
5 Floor plans type **A** 1:200
6 Floor plans type **B** 1:200
7 Floor plans type **C** 1:200
8 Sections types **A**, **B**, **C** 1:200
9 Floor plans type **D** 1:200
10 Section type **D** 1:200

1.2
URBAN INFILL

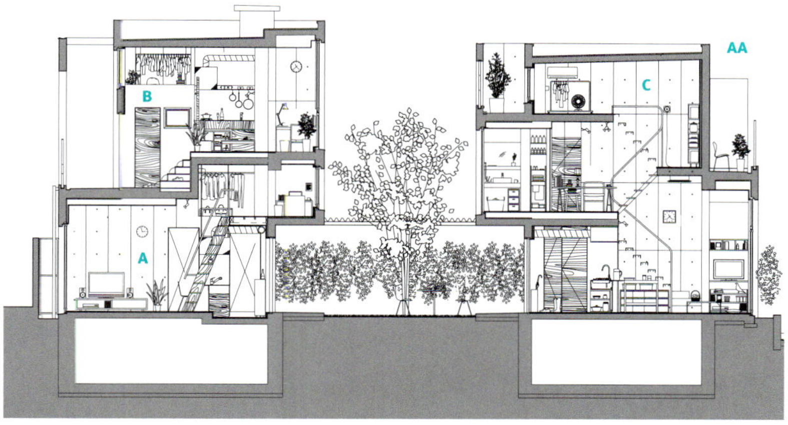

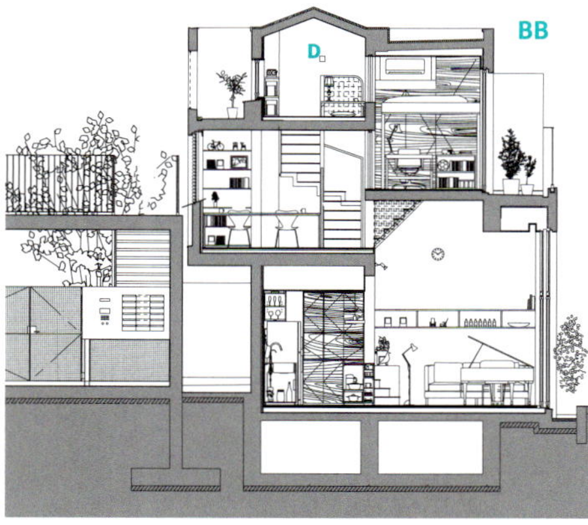

10

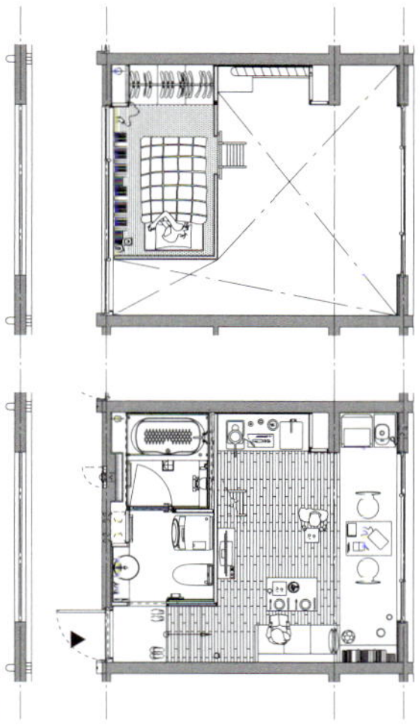

6

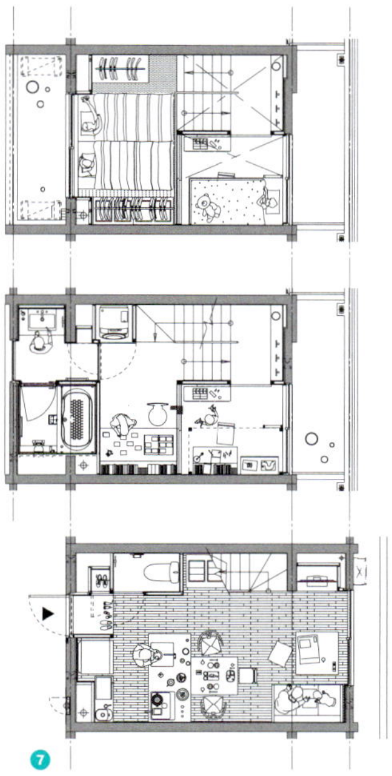

7

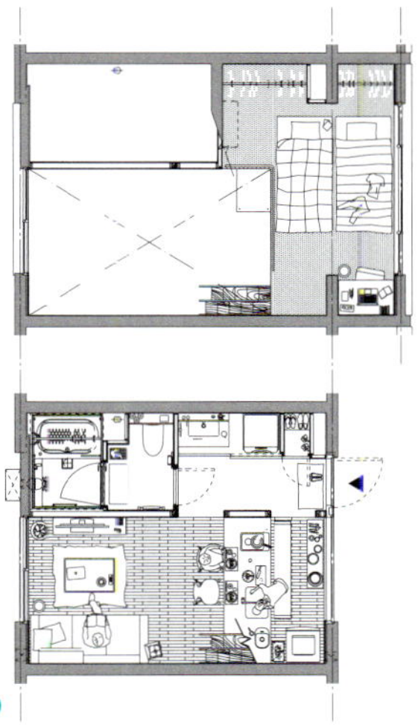

5

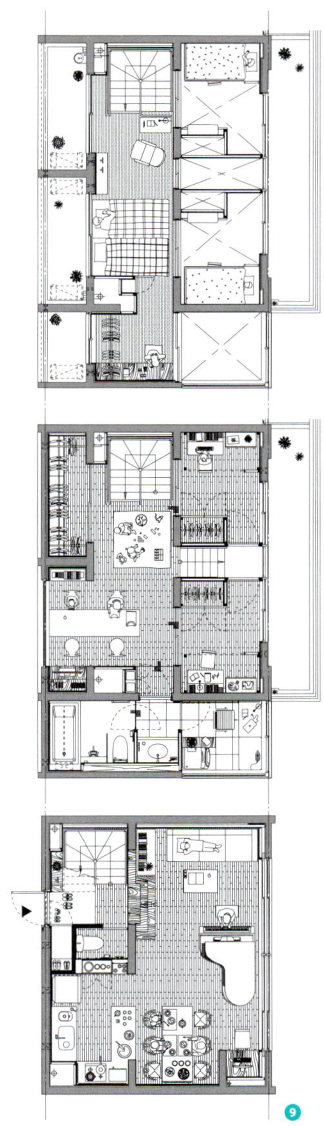

9

Building type
corner building on an open urban block
2 parallel bars flanking a courtyard garden
height equivalent to 4 full stories
facing E/W

Date of construction
2014

Number of units
15

Size of units
Type **A**: 1-room apts. with raised level, 26 m² (5 units)
Type **B**: 1-room apts. with raised level, 24 m² (5 units)
Type **C**: 3-room apts. with raised level, 45 m² (4 units)
Type **D**: 4-room apt. with raised level, 95 m² (1 unit)

Area per user
22–26 m²

Building depth
7.5 m

Access
direct access from courtyard garden
external access gallery for micro-apartments on upper levels

Open spaces
roof terraces and minimal rear gardens
common courtyard garden

Parking
limited parking near courtyard entrance

Architect
Field Design Architects
Tokyo

Location
Urawa-ku
Saitama
Japan

Referring to floor plan 1:200

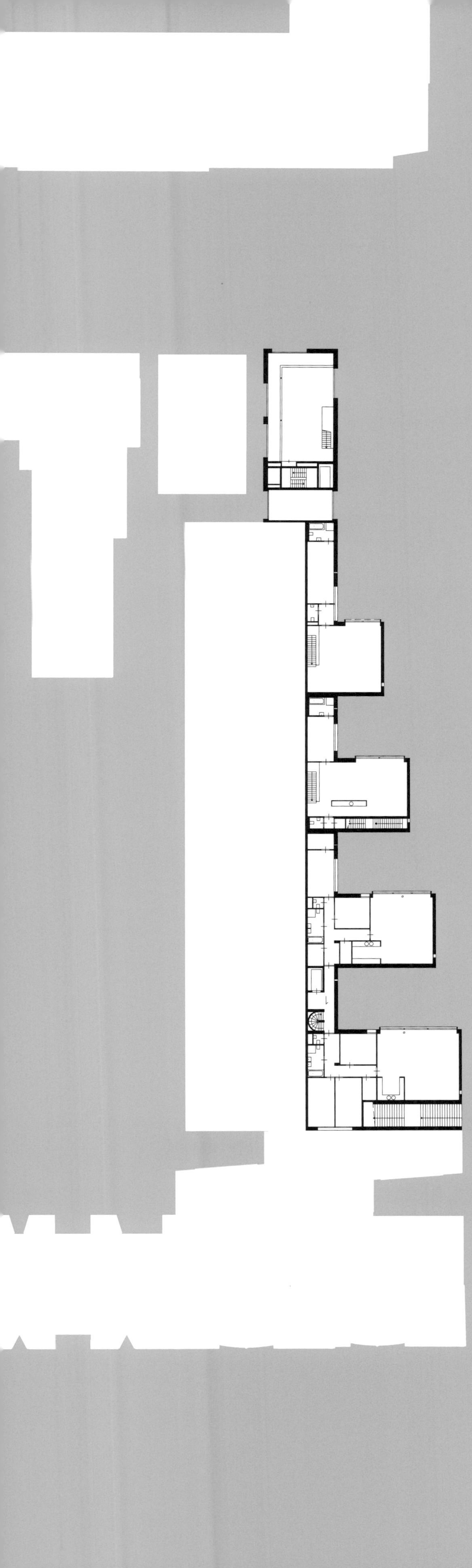

1.3 FIREWALL BUILDING

Building lots between firewalls are often the result of wounds and breaks within the historic pattern of a city; they are not intended as part of an ideal urban plan. Firewalls are created, for example, when new traffic arteries create residual plots. In this case, the expectation of their development is to mediate aesthetically between enclosed city blocks and the open modern city. Firewall developments always have the opportunity to make a virtue out of necessity and to turn the disadvantage of the necessarily one-sided orientation into developing a unique housing type: the structure can take on an additional dimension through folds and undulations in the facade or by dissolving the whole into individual sections, offering variety in views and lighting possibilities. Special attention in the design can be given to the interface between firewall and new development: thus the new building can be set apart providing the building sections adjacent to the firewall with more air and light.

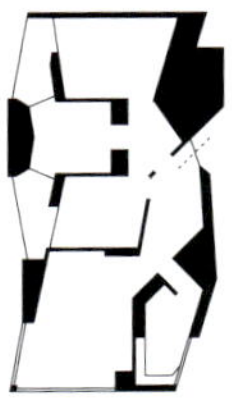

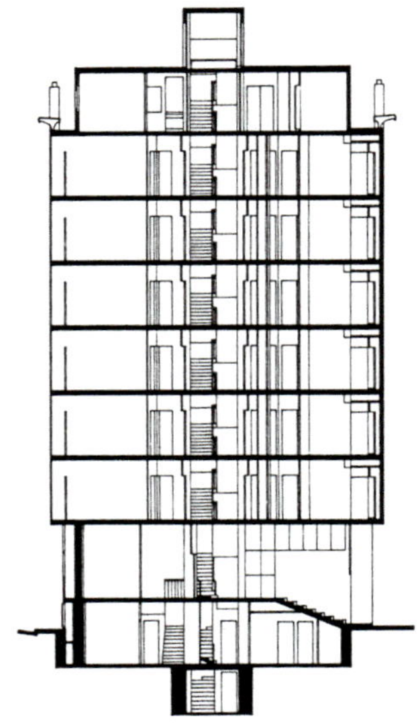

I.S.M. House | Coderch | 1951

1.3
FIREWALL BUILDING

The residential building for the families of fishermen forms the head of a row in the harbor district. The geometry of the ground plan with its two mirror-image apartments is as animated as the undulating facade. The corridor leads in a sweeping diagonal across the apartment and straight ahead into the living room to a corner which is dissolved into window elements allowing the space to be flooded in light and offering a view onto the street space. A passage at a right angle in front of the living room leads to the kitchen with terrace. The bedroom wing with ensuite bath is located on the other side of the main corridor; the bedrooms are reached via a separate corridor and are connected on the exterior by loggias, which are protected from the elements with louvered shutters. The movable louvers, which allow the occupants to regulate the ingress of light, further animate the facade. The modest apartment size is amply compensated by the optimized use of all rooms, for which the furnishings were custom designed with great attention to detail.

Building type
head building at compartment wall
8 stories with retail stores on ground floor
facing E/W/N

Date of construction
1951

Number of units
13

Size of units
3-room apts., approx. 74 m² (12 units)
3-room penthouse apt., approx. 51 m² (1 unit)
2-room caretaker apt., approx. 48 m² (1 unit)

Area per user
17–24.5 m²

Building depth
12.2 m/15.1 m

Access
double-loaded stairwell,
internal stairwell lit from private loggias

Open spaces
loggias, roof terrace

Parking
no parking spaces

Architect
J. A. Coderch with Manuell Valls, Barcelona

Location
I.S.M. House at the Paseo Nacional, Barceloneta
Barcelona, Spain

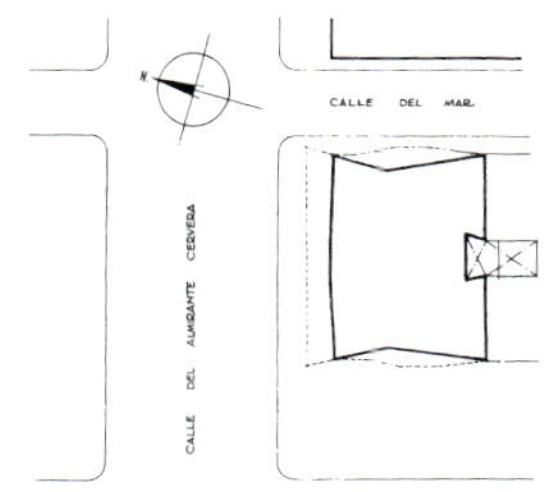

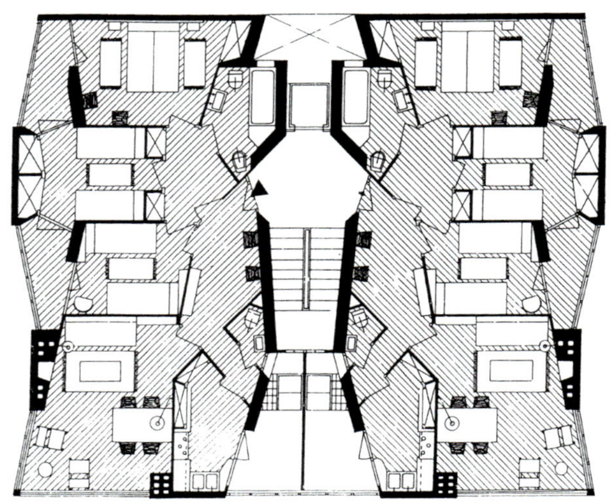

3

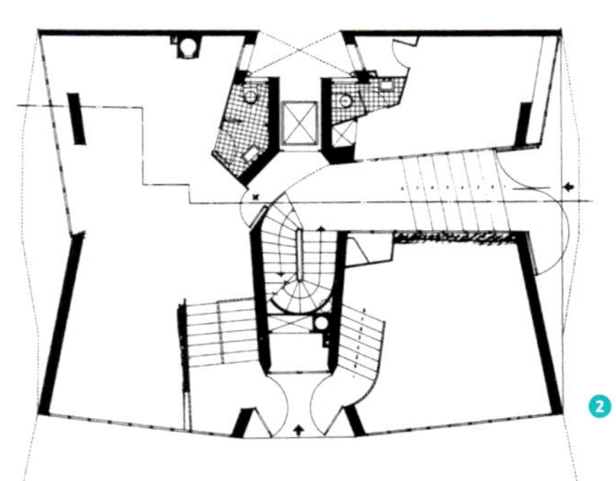

2

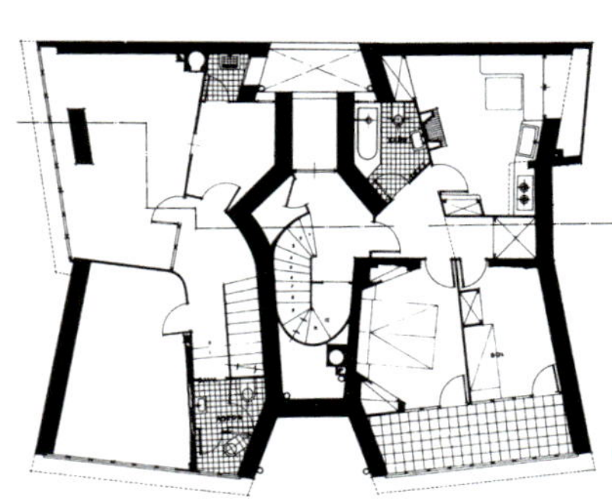

1

Basement with stores and storage spaces 1
Raised ground floor with caretaker apartment and commercial spaces 2
Standard floor with two 3-room apartments 1:200 3

Referring to floor plan 1:200

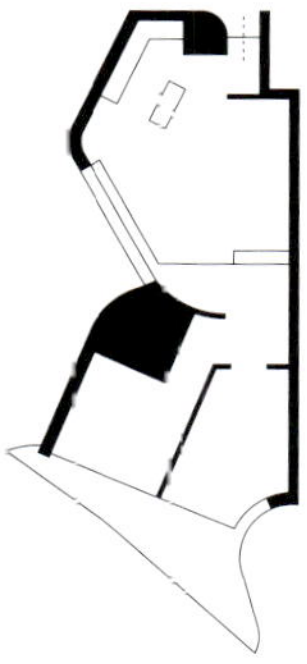

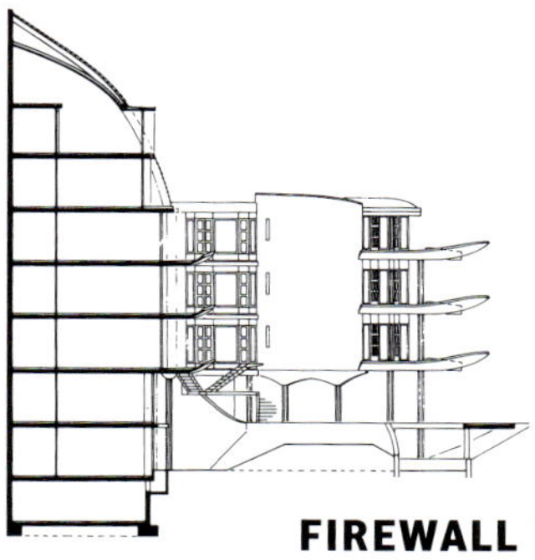

Fraenkelufer | Baller | 1984

1.3 FIREWALL BUILDING

This building along a firewall makes use of organic forms. All apartments open up toward the garden courtyard; the entrance zones with elevators and stairwells are placed against the end wall. The division of the apartments does away with hallways and lobbies, the soft outlines and height differences of the floor plans create zones that can and must be interpreted by the resident. 2-room apartments: the kitchen (as counter/bar) is lit and ventilated indirectly via the living room. The rounded area adjacent to the kitchen can be used as a cloakroom. The separation or connection of sleeping and living areas is possible. 4-room apartments: the kitchen is part of the large living room/throughway (the traditional "Berlin room"). From there, on the second and third floors direct access to the garden is provided via spiral staircases. 6-room maisonettes of exceptional size on the basement story/ground floor and on the 5th and 6th floors: A spiral staircase connects the two levels, steps and galleries structure the space. They have direct access to the garden or the roof terraces.

Building type
courtyard building along firewall
6 stories
facing S

Date of construction
1982–1984

Number of units
48 units in courtyard buildings
40 units in the gate buildings

Size of units
2-room apts., approx. 64 m²
3-room apts., approx. 93 m²
6-room mais., approx. 167 m²

Area per user
27.5–32 m²

Building depth
approx. 8/18 m

Access
quadruple- and
double-loaded stairwells

Open spaces
balconies, garden in inner court

Parking
in basement story with garden above

Architect
Inken und Hinrich Baller, Berlin

Location
Fraenkelufer
Berlin-Kreuzberg, Germany

3
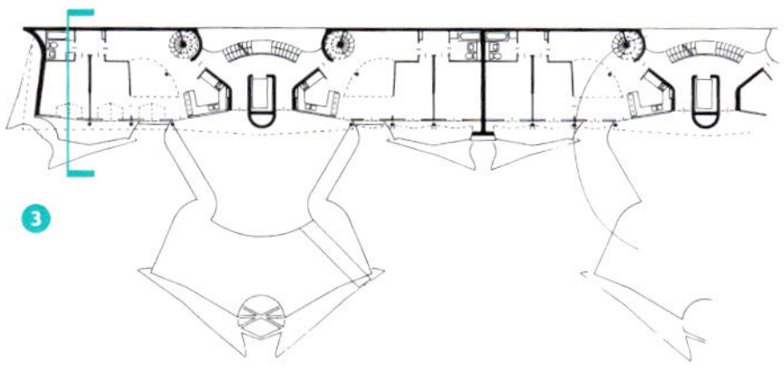

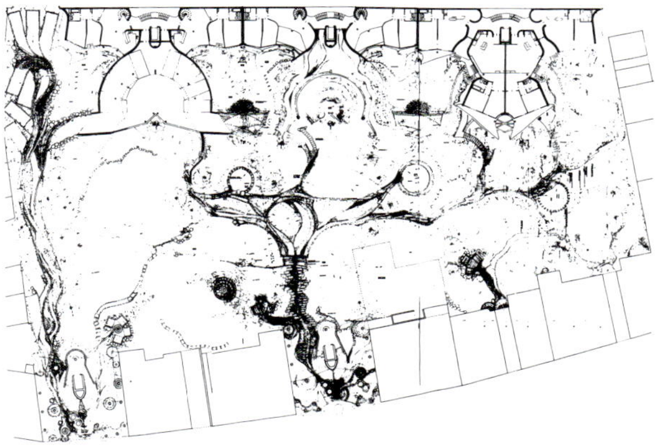

2
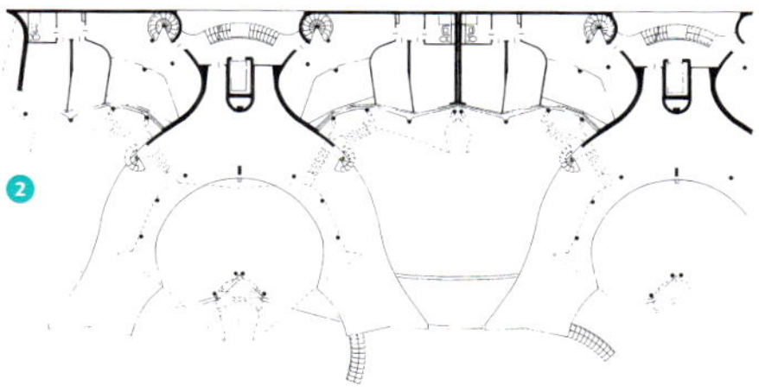

1
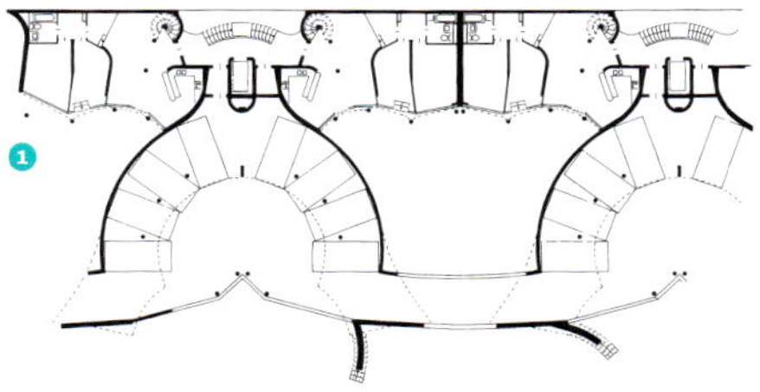

4
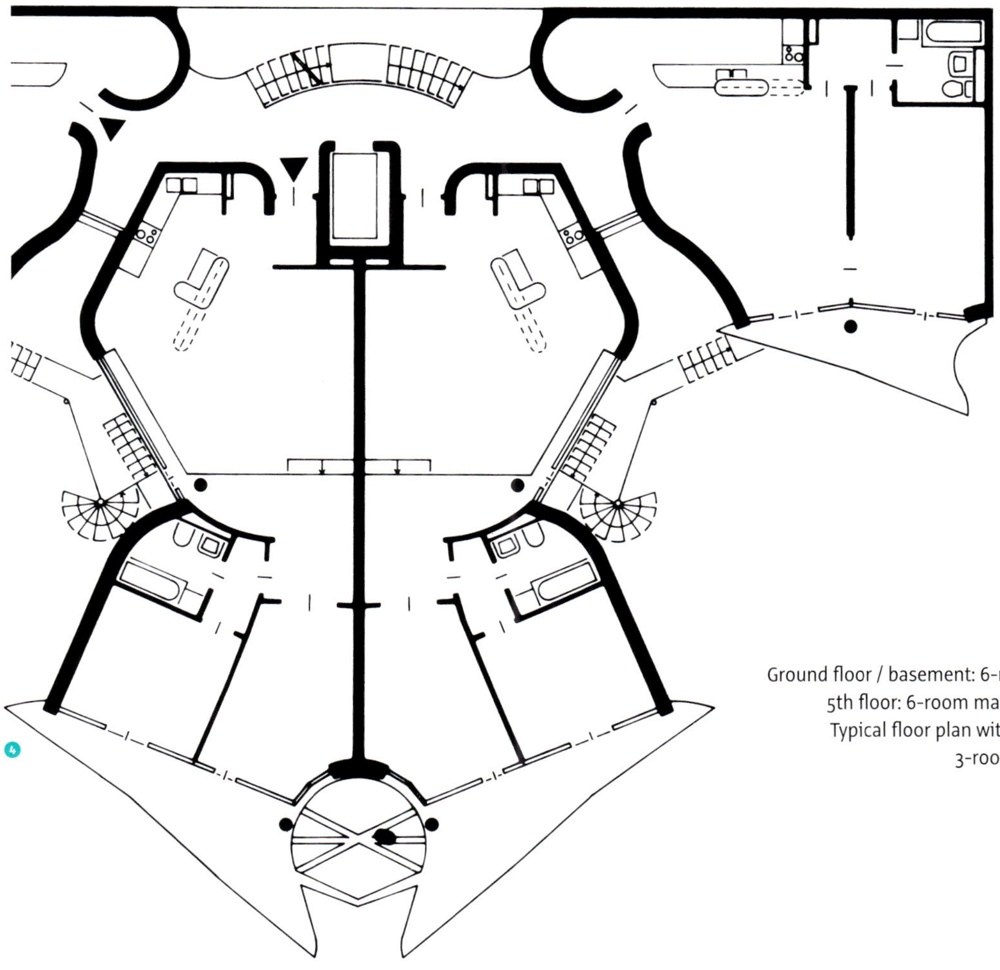

Ground floor / basement: 6-room mais. 1 2
5th floor: 6-room mais., lower level 3
Typical floor plan with 2-room and 4
3-room apts. 1:200

Referring to
floor plan 1:200

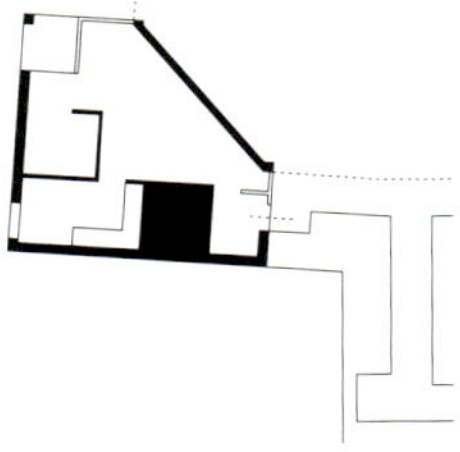

Köpenicker Straße | Steidle | 1985

The project consists of three different structures: the conversion of the existing transverse wing, the four-story row along the firewall (documented here), and a building on Köpenicker Straße. The new row is set off from the firewall roughly by the depth of the building; the intermediate space was transformed into a glassed-in access hall with ramps in the length of the building. The plan of the building, legible as a meandering pattern, is comprised of two-apartment modules (apartments are divided by diagonal walls); the glass surfaces in front of the sleeping and living areas form another meandering pattern in plan, developed around the facade recesses. The apartments are accessed via a short entry hallway; kitchen and small bedroom (with bed niche) can be entered from the living room. In the apartments along the access hall, the kitchens and baths are lit and ventilated via the hall; the outside-oriented apartments have baths located at the core, but kitchens with large windows to the park.

2

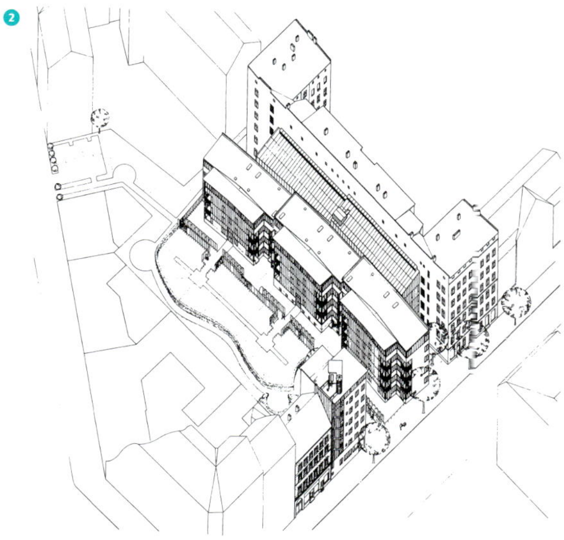

1

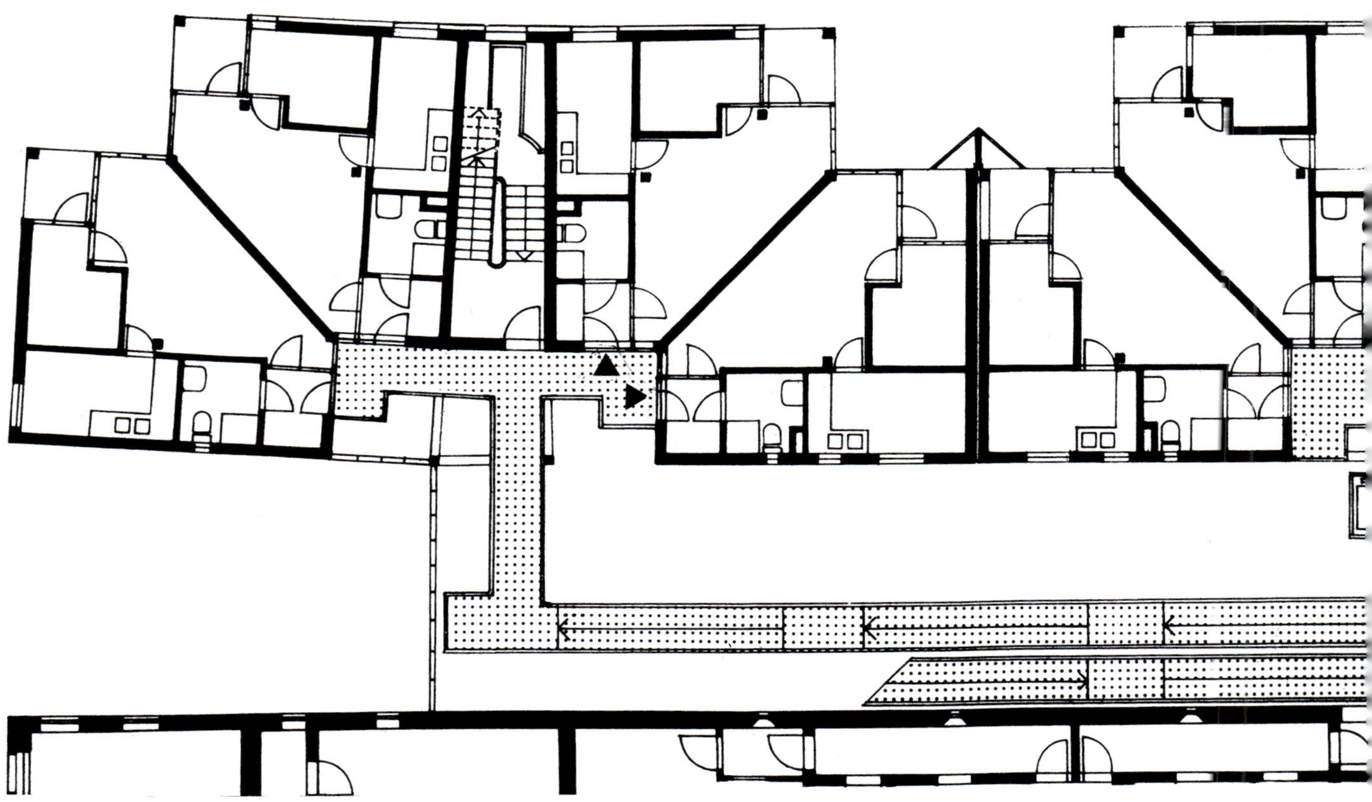

1 Typical floor plan 1:200
2 Isometric view of overall project
3 General layout of overall project

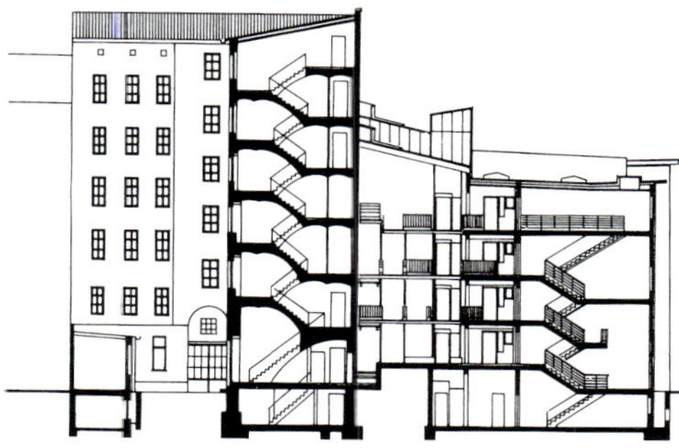

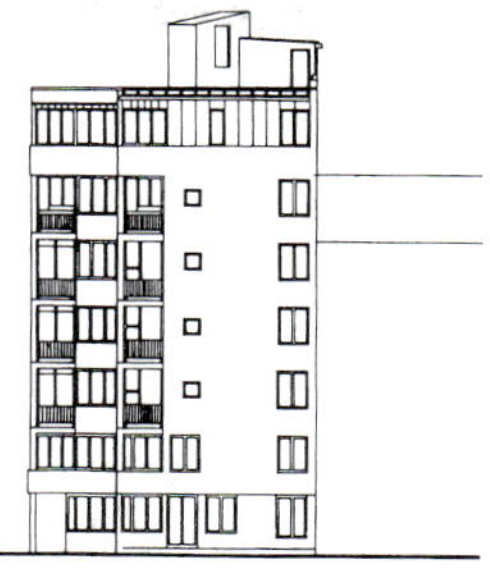

1.3
FIREWALL BUILDING

3

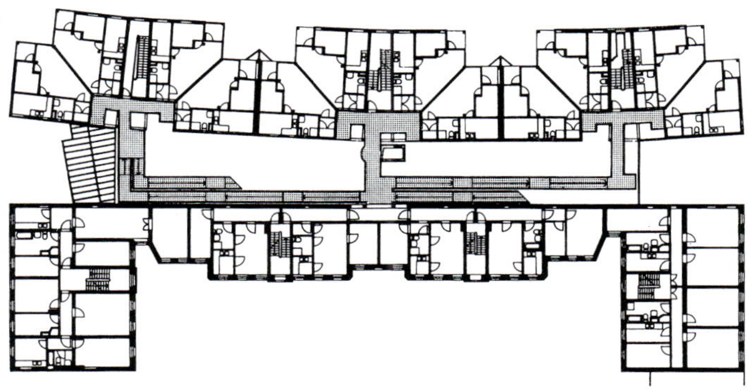

Building type
home for the elderly
along access hall
4 stories
facing SE/SW

Date of construction
1984–1985

Number of units
48
(only firewall building)

Size of units
1½-room apts., approx. 59 m²

Area per user
29.5–59 m²

Building depth
8/11.7 m

Access
quadruple-loaded stairwells,
access hall
with ramps

Open spaces
balconies, geometrical
garden in courtyard

Parking
open-air parking

Architect
Otto Steidle
Munich

Location
Köpenicker Straße
Berlin-Kreuzberg
Germany

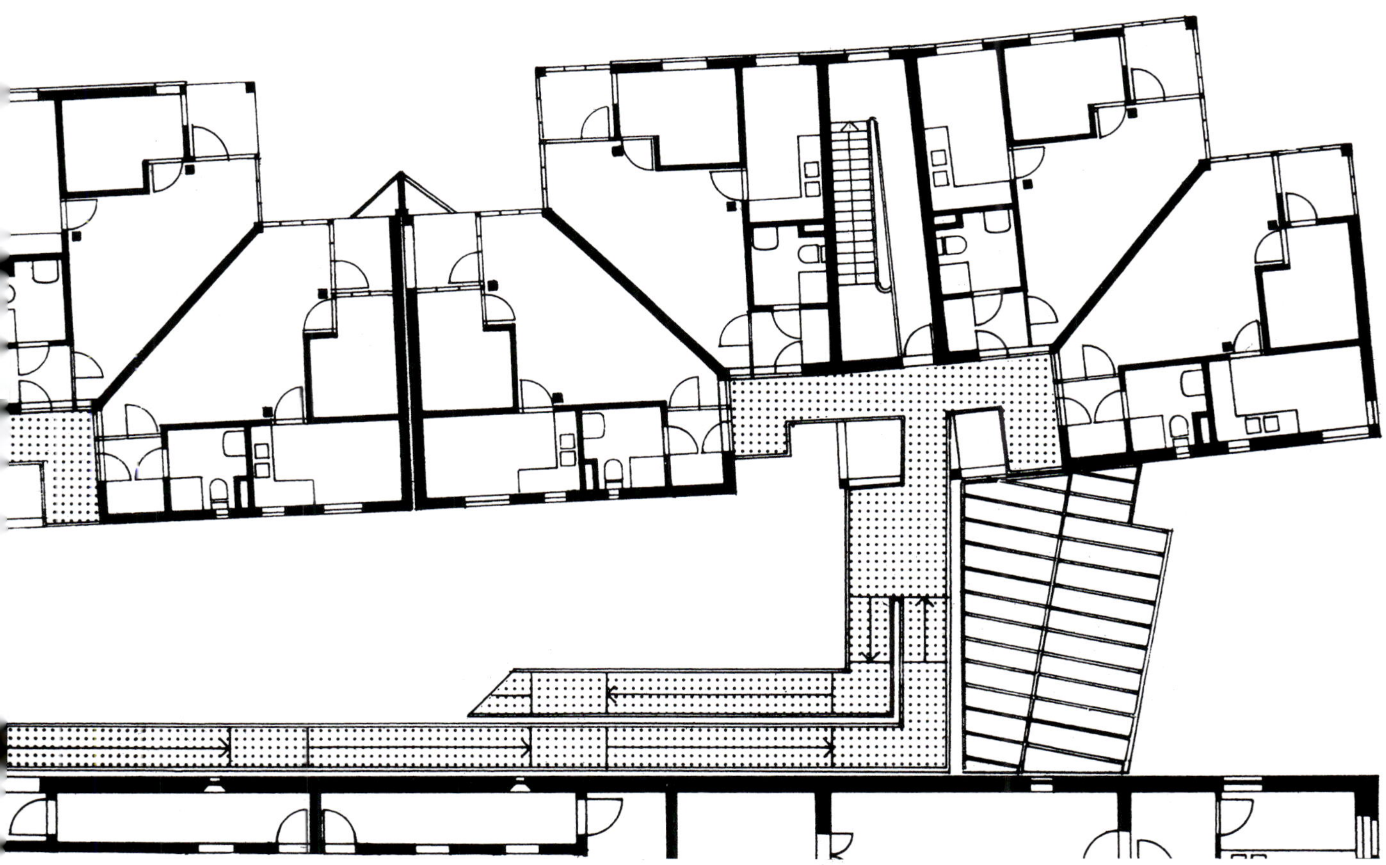

Referring to
floor plan 1:200

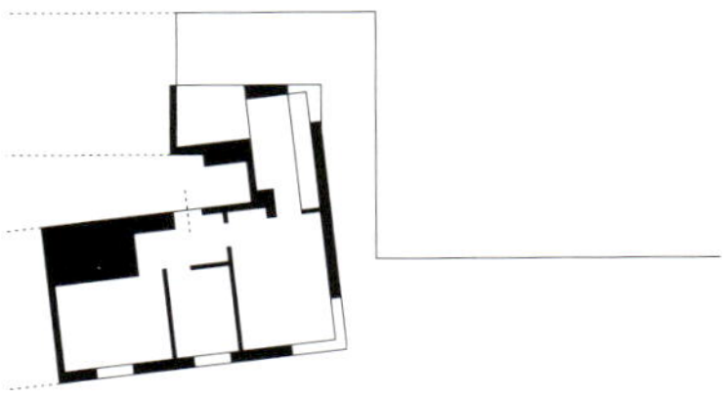

Carrer Carme / Carrer Roig | Llinàs | 1994

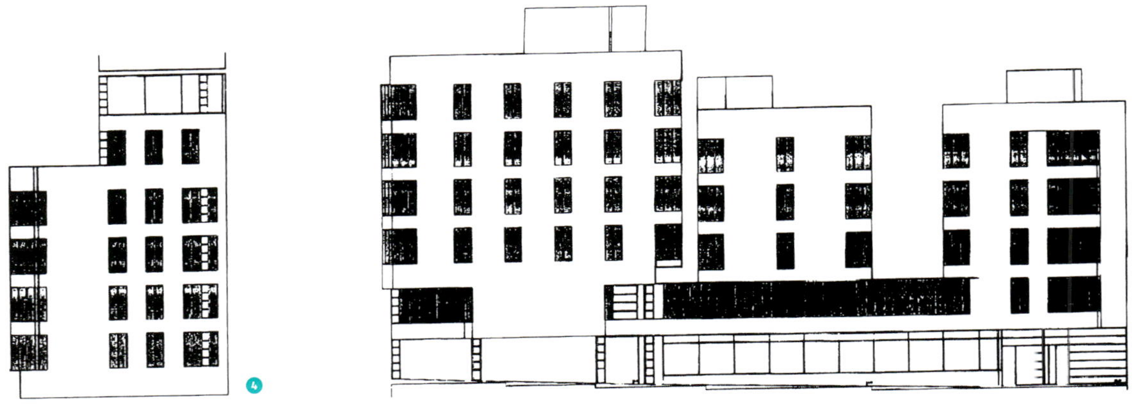

A firewall construction on a narrow (4-m-wide) street. The building's mass was broken down into three individual blocks. This was done not so much out of the architect's concern for the apartments' qualities, but rather to seize the chance to bring light into the shaded Ciutat Vella quarter. The boxes of varying heights sit atop a shared base which tapers such that it pours the life of the main street into the side street like a funnel. Shops (ground floor), entry to the underground garage, an office (second floor), and two entries to the upper stories occupy this shared base. From the entry on the ground floor, one has two stairwells, each with elevator. These split into three on the second floor. The apartments are reached by these double- and triple-loaded stairwells. The units are all compact 3-room apartments of similar composition. The living rooms are set in the corner and open onto the street, underscoring the new – mainly visual – relationship established by the base beneath. The living rooms give access to the narrow kitchens, while all other rooms are reached by a short hall. Each of the three blocks is turned and situated such that slits of light are created between it and the firewall. Some spaces are lit only from small loggias. In most cases these loggias are attached to the bedroom; while others adjoin the kitchen.

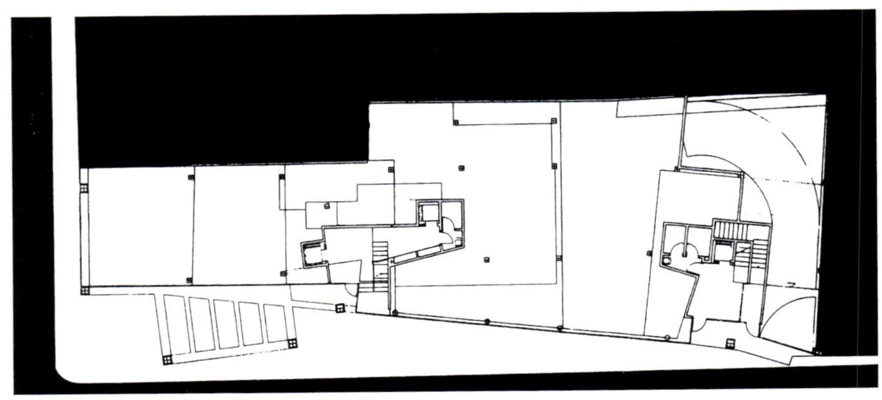

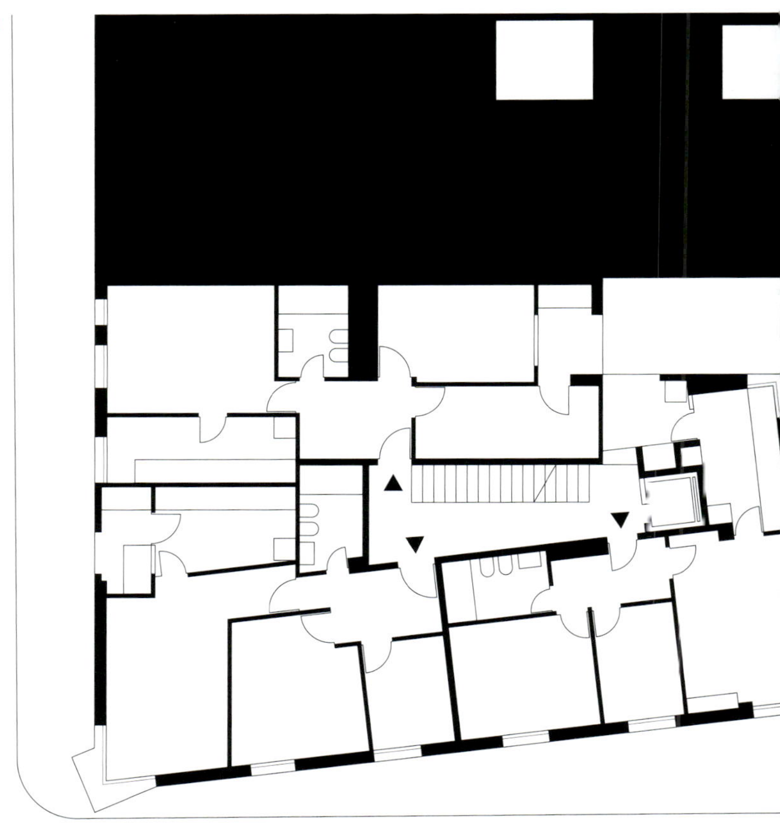

1. Ground floor 1:500
2. 2nd floor 1:500
3. Typical floor plan 1:200
4. Elevations from main street and from side street 1:200

1.3
FIREWALL BUILDING

2

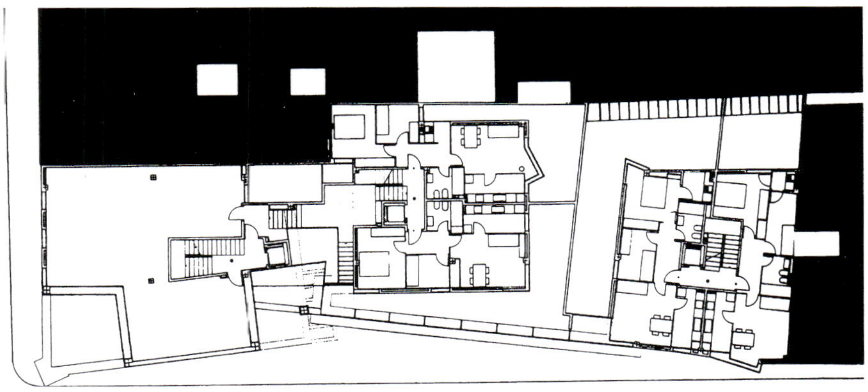

Building type
firewall construction
in three individual blocks
5–6 stories
facing NW/NE

Date of construction
1989–1994

Number of units
28

Size of units
3-room apts., 52–62 m²

Area per user
17.5–20.5 m²

Building depth
9–13.5 m

Access
double- and triple-loaded
stairwells, 2 elevators

Open spaces
small loggias

Parking
underground garage

Architect
Josep Llinàs
Barcelona
with
E. Monte, J. Vera

Location
Ciutat Vella
Carrer Carme/Carrer Roig
Barcelona
Spain

Referring to
floor plan 1:200

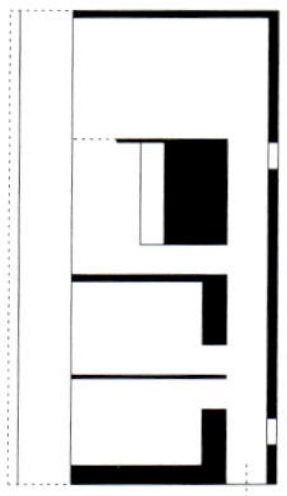

Rue des Suisses | Herzog & de Meuron | 2000

Densification scheme in a Parisian urban block: a measure that included closing two building gaps and several new constructions at the center of the block. The architects reprise several building typologies already existing in this urban quarter and the charming mix they establish, formulating them in a more rigorous vocabulary that is entirely their own. On two sides of the block, two seven-story buildings fill the gaps. Passing beneath these buildings, one enters an inner courtyard, which is dominated by a three-story elongated building that is linked to the adjacent buildings via low, house-like extensions. Together with the elongated slab, the extensions surround picturesque gardens to the rear for the ground-level apartments. Small, two-story houses are nestled against the boundary wall on the opposite side, dividing the elongated courtyard into intimate spaces. The most interesting floor plan is found in the long building: all rooms are strung in a row on the courtyard side, linked by a corridor, which transitions into the living room. The same rooms are also linked on the exterior by a verandah. Thus, each room has two direct links to the outside. These loggias on the courtyard side are covered by wooden roller blinds (reminiscent of a bureau), which the occupants can open and close as needed as protection against incident sun or for privacy. The small projecting element functions as a loggia; on the third floor where large roof gardens stretch between the apartments, it is reduced to a small balcony. Like the long building, the facades of new structures closing the gaps are also clad in an external facade layer. In this case, the layer consists of perforated, individually foldable and sliding aluminum sheeting, which is also an understated reference to the narrow courtyard entrances.

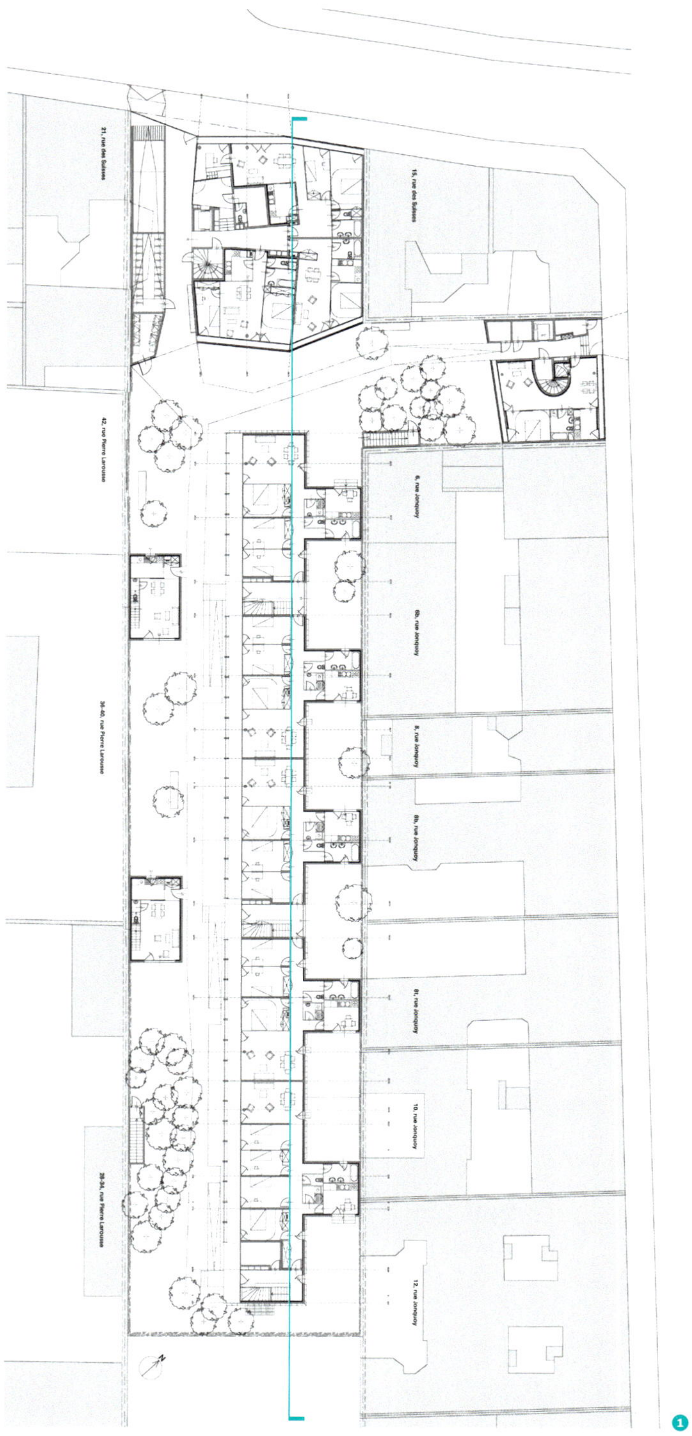

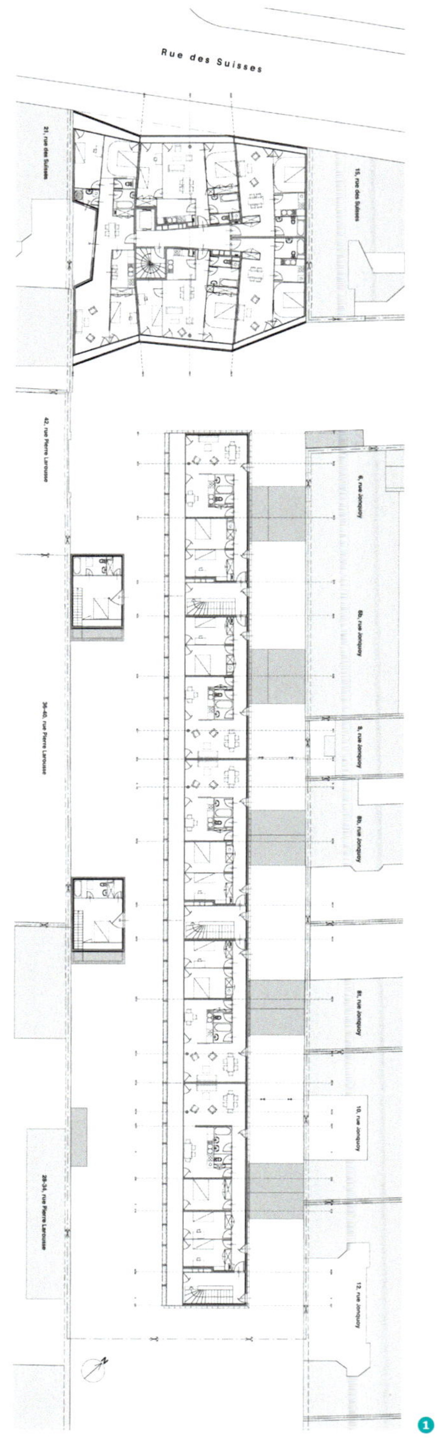

1 Overall plan of complex: ground floor, 2nd floor, 3rd floor (from left to right)
2 Floor plan of 3-room apartment, long apartment house 1:200
3 Section of long building with loggias behind curved louvered profiles
4 Section through building gap closure and house-like extension of block development

1.3
FIREWALL BUILDING

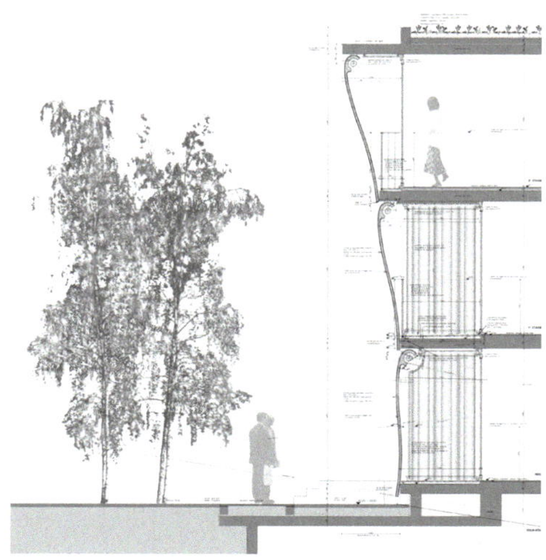

❸

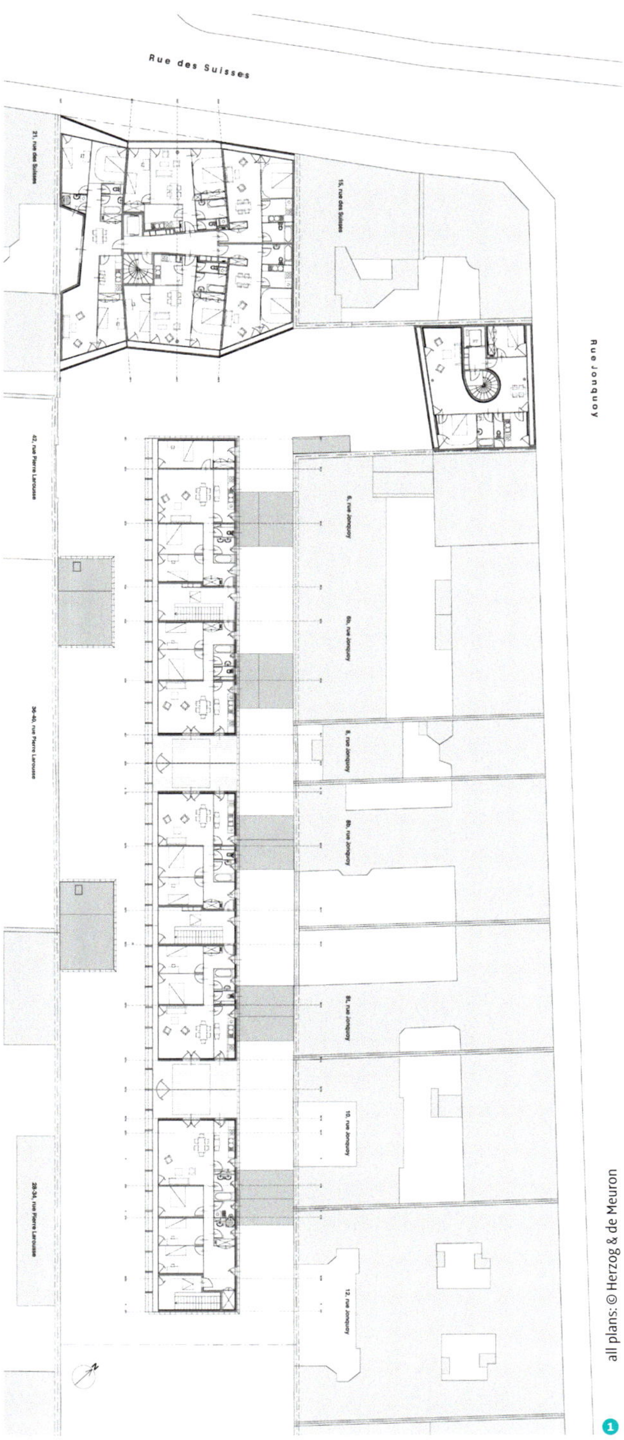

all plans: © Herzog & de Meuron

❶

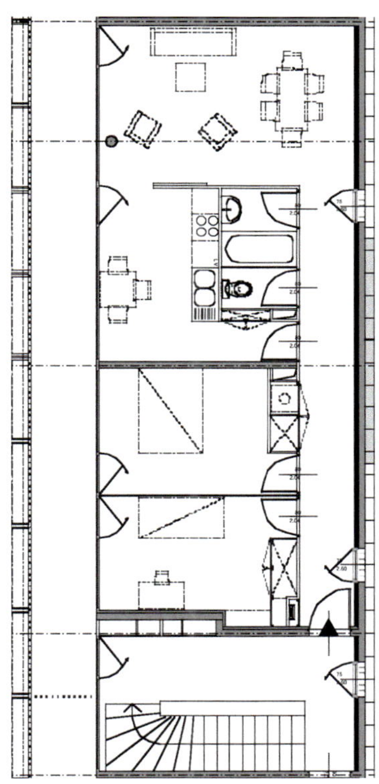

❷

Building type
urban densification project
two urban infill buildings and three developments along fire walls
in courtyard: 2/3/7 stories
facing NW/SE and NE/SW

Date of construction
2000

Number of units
57, plus 2 single-family homes

Size of units
total living area long building: 2,055 m²
house Suisses: 2,800 m²
house Jonquoy: 561 m²
different apartment types and sizes:
2-room apts. (14 units)
3-room apts. (28 units)
4-room apts. (14 units)
5-room apt. (1 unit)
2-bedroom houses (2 units)

Area per user
30.5 m²

Building depth
12–18 m (Suisses)
7.2–8.5 m (Jonquoy)
6.5 m (long building)
4 m (small houses)

Access
1-, 2- and 5-strung, respectively, direct access for single-family homes

Open spaces
urban infill:
very narrow balconies
long apartment house:
small gardens and loggias to the rear on ground floor, loggias on 2nd floor, terraces on 3rd floor,
semipublic, planted courtyard

Parking
underground garage
beneath entire courtyard area

Architect
Herzog & de Meuron
Basel

Location
Rue des Suisses
14ème Arrondissement
Paris, France

Referring to floor plan 1:200

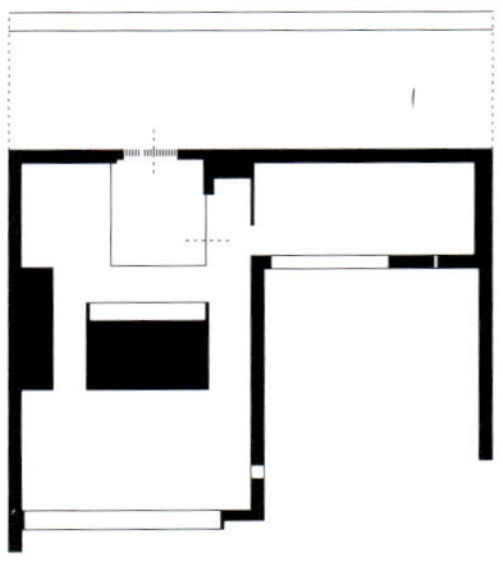

Pieter Vreedeplein | Bedaux de Brouwer | 2007

The development on a triangular lot is an addition to a new shopping mall, mediating between it and the historic parcel development of the old town. The structure is based on the grid of the underground garage beneath it. Arranged in stacked segments and surrounding four patio courtyards in the manner of a comb, the structure of the development creates a multitude of different apartment types and sizes. Despite the increasing building depth, the inserted patios ensure that natural light flows into all interiors. Each patio offers a private interstitial space between the building and the city, shared among three units. Thanks to the staggered layout, the patios also open up a vista in the direction of the square to the south. The L-shaped apartments are laid out in different plans: as maisonettes around a private courtyard; as single-level apartments with individual access stairs from the street; or as a row of penthouse bungalows reached via a footbridge from the shopping mall and inserted roof terraces. The floor plans reveal that all access areas, kitchens, and sanitary blocks are situated in the corner without natural light, while the living rooms and individual rooms all face towards the patio and the street across the full width of the space. An apartment tower forms the southern completion of the complex, creating a new visual accent for the urban square. In the tower, four maisonette units with lightwells along the facade and cantilevered two-story loggias on alternating sides are stacked one above the other, facing toward the cityscape.

1

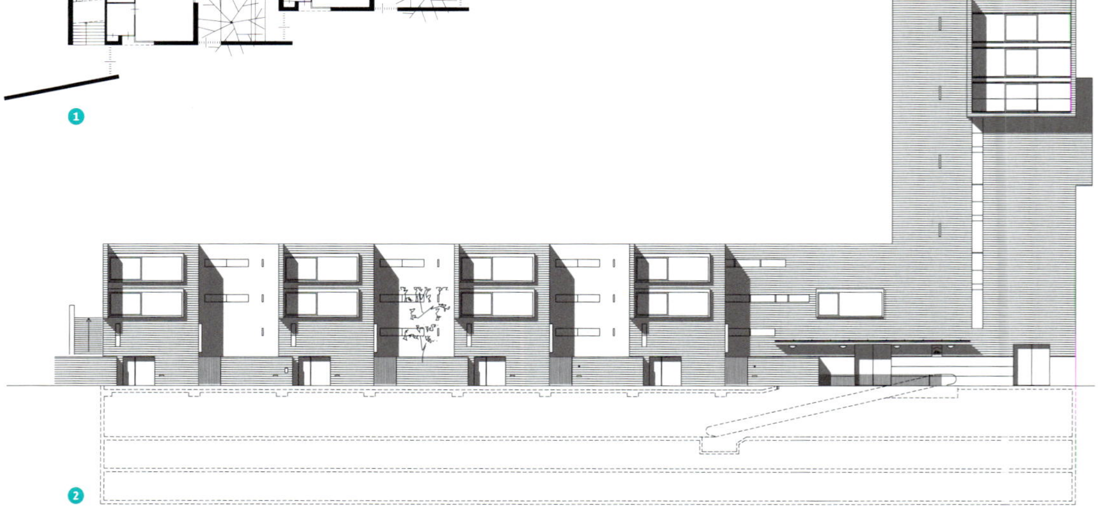

2

1 Ground floor – 4th floor 1:600
2 West elevation 1:600
3 Ground floor: 4-room maisonette, lower level, access to unit on 3rd floor 1:200
4 2nd floor: 4-room maisonette, upper level 1:200
5 3rd floor: 2-room apartment, access on ground level 1:200
6 4th floor: 2-room apartment with access via patio and roof passage on adjacent shopping mall 1:200
7 Section of apartment tower 1:1000

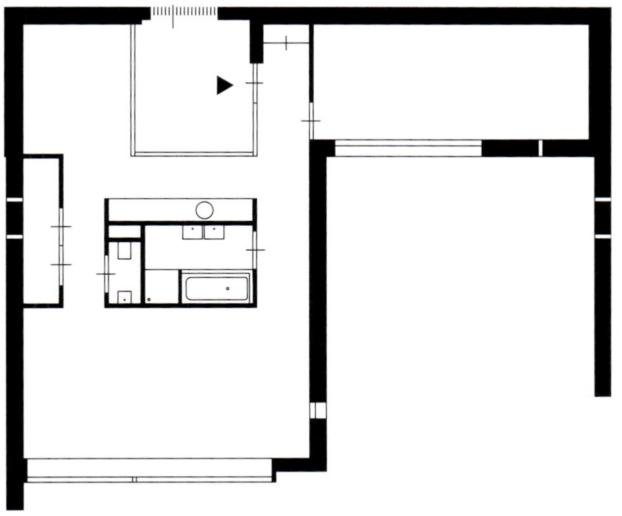

6

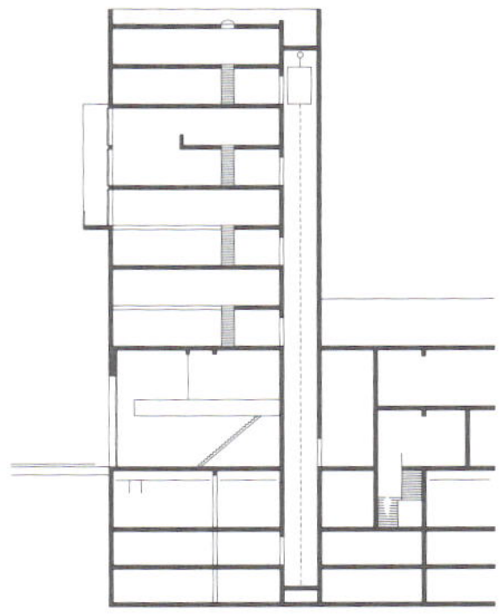

7

1.3
FIREWALL BUILDING

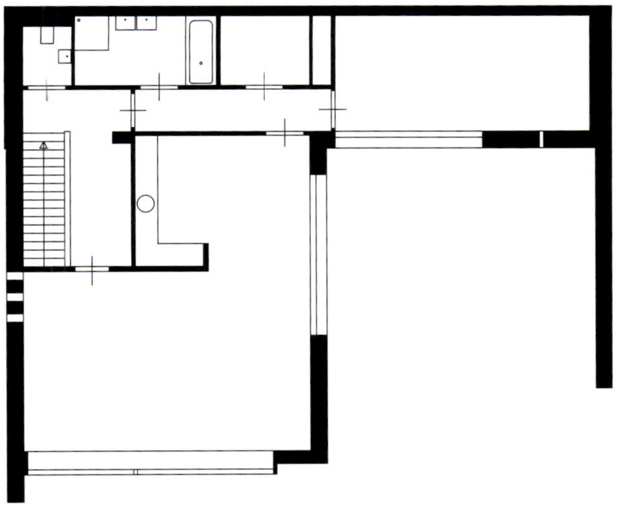

5

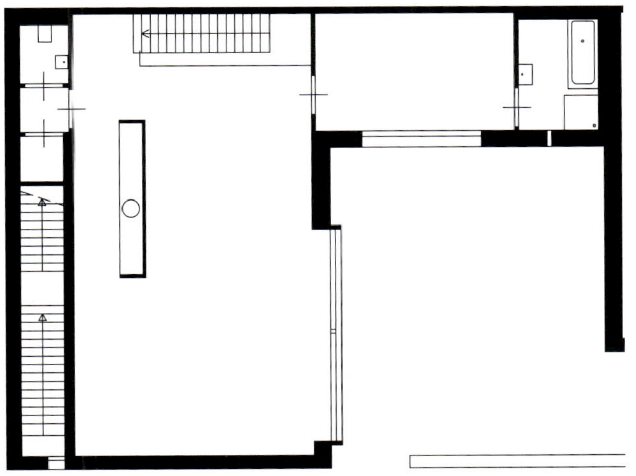

4

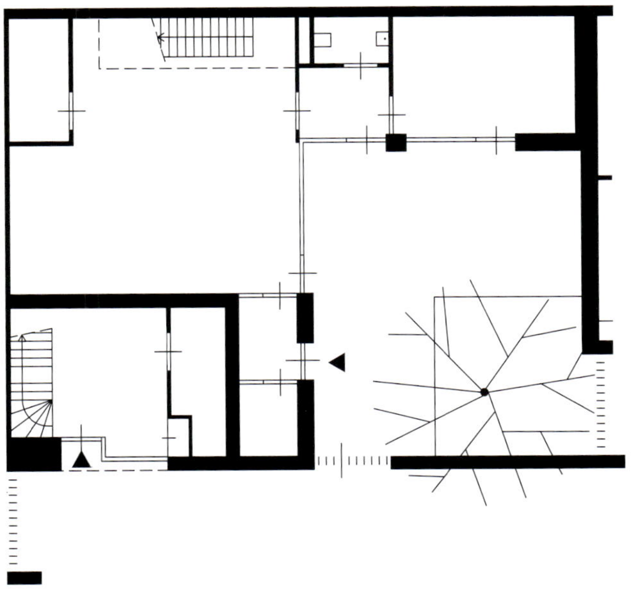

3

Building type
firewall development with
comb-shaped building layout
with patio courtyards and
apartment tower
4 stories
tower 10 stories
facing SSW/WNW
tower also facing ESE

Date of construction
2007

Number of units
18

Size of units
2-room apts., 73–116 m² (4 units)
3-room apts., 94–109 m² (4 units)
4-room apts., 92–94 m² (2 units)
4-room mais., 81–158 m² (4 units)
3-room mais., 137 m² each tower
(4 units)

Area per user
20–58 m²

Building depth
8.5–18 m

Access
direct access via patios,
covered walkways on 5th floor,
single-loaded staircase
in apartment tower

Open spaces
garden in patio courtyard
roof patios
balconies

Parking
underground garage

Architect
Bedaux de Brouwer Architecten BV BNA
Jacq. de Brouwer (project architect)
Peter Keijsers (coarchitect)
Koen de Witte

Location
Ijzerstraat/Pieter Vreedeplein
Tilburg
The Netherlands

Referring to
floor plan 1:200

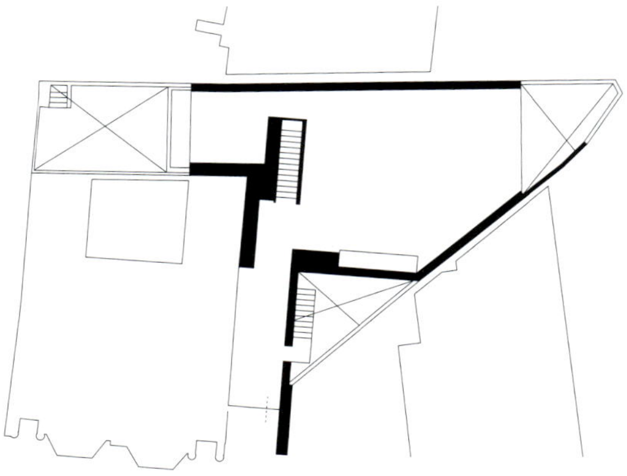

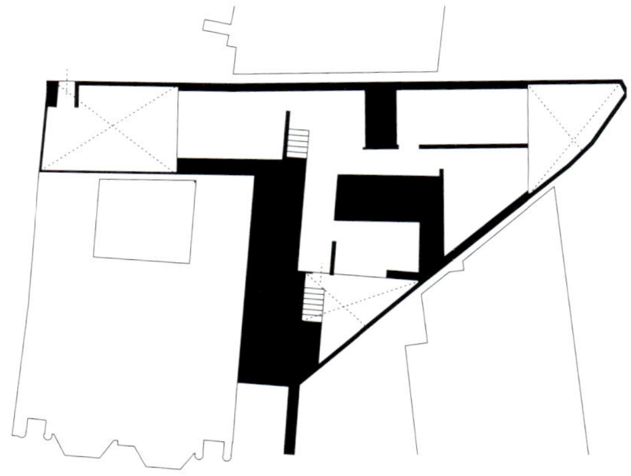

Brick House | Caruso St John | 2005

The shape of the house evolved entirely in response to the unique characteristics of the site. Hemmed in by a homogeneous Victorian row and a heterogeneous rear building, the lot is a residual building gap in the shape of a triangle. Devoid of conventional facades and windows, it lies between three fire walls and is entirely inward in orientation. With its enclosing brick walls, three patios, and sculptural roof, the house is only ever visible as a fragment from the outside, without a defined front or rear. Access is through the carriageway of the adjacent Victorian terrace and via a ramp that leads past a patio with exterior stairs and directly into the living area, which occupies nearly the entire upper level. This space is defined entirely by form and material; it is bounded by stairs leading to the private rooms downstairs, the galley kitchen, the large patio at the far end of the room, and the sculptural, reinforced concrete roof, whose pronounced three-dimensional differences in height and skylights create subtle accents in the individual areas. On the lower level, the cruciform hall and block-like bathrooms and storage rooms create the voids for the private rooms, whose walls transition almost seamlessly onto the patios thereby extending the interior to the exterior. Here, too, the consistency in material and the dark built-in furniture once again serve to emphasize the fluid sculptural character of the space. The same plasticity is found on the three patios, which bathe the rooms in a soft, muted light.

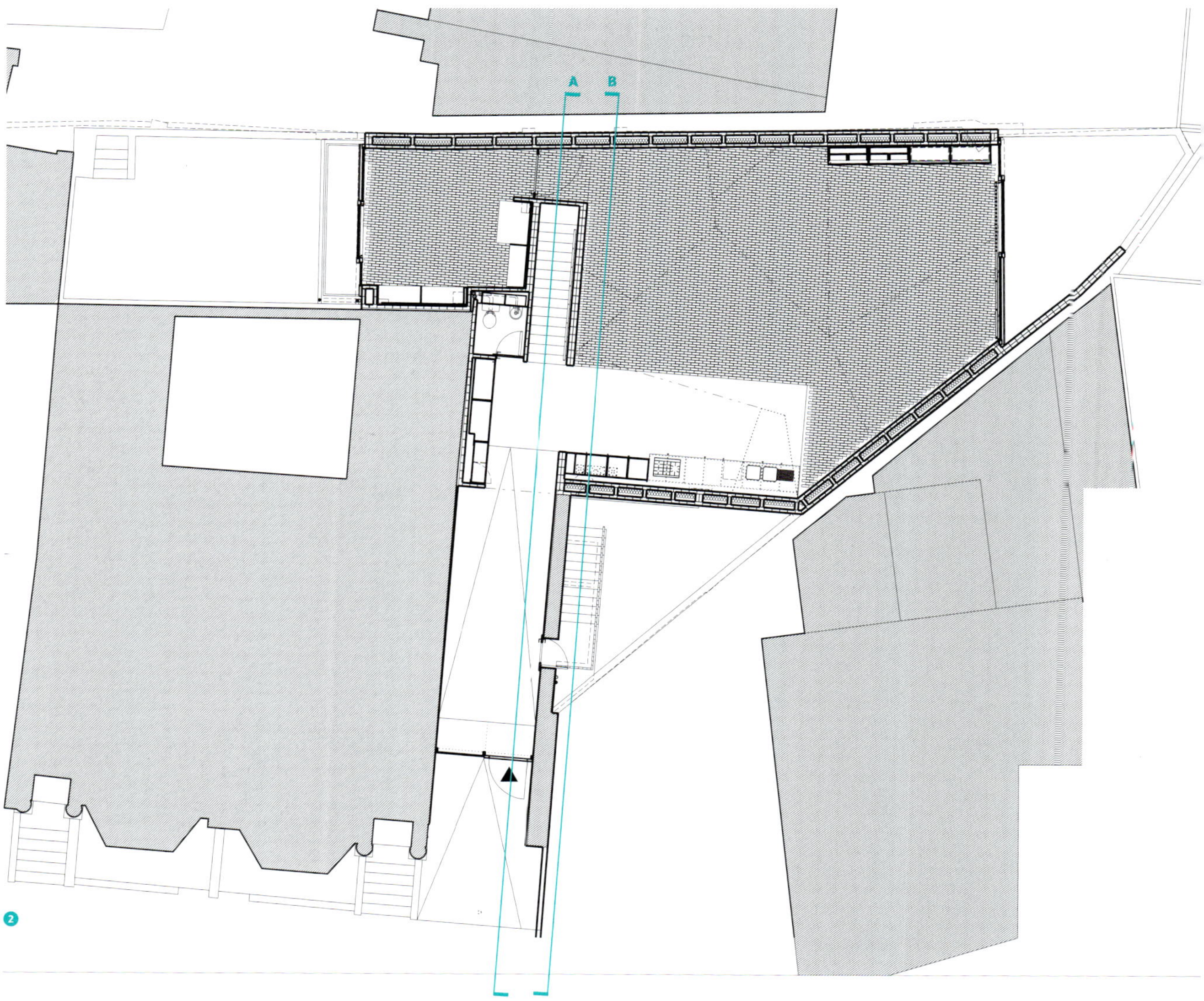

2

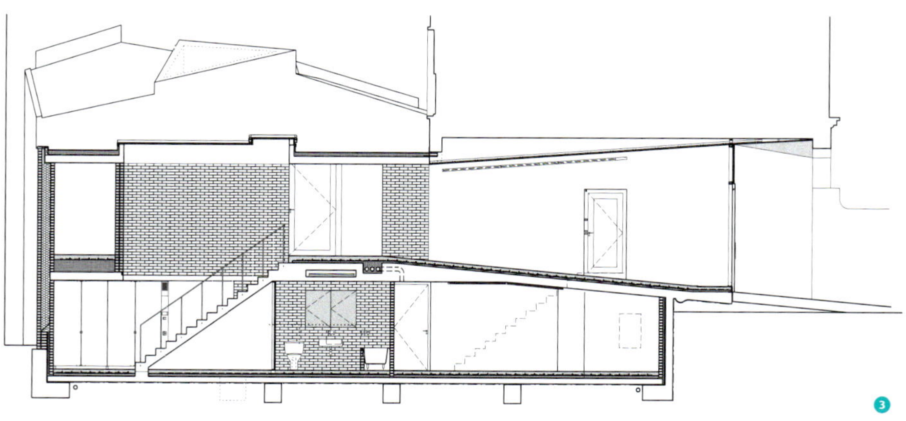

3

1 Level with entrance and living area (raised ground floor) 1:200
2 Level with private rooms (lowered ground level) 1:200
3 Cross section with entrance ramp and stairs to private rooms 1:200
4 Cross section of patio and living area 1:200

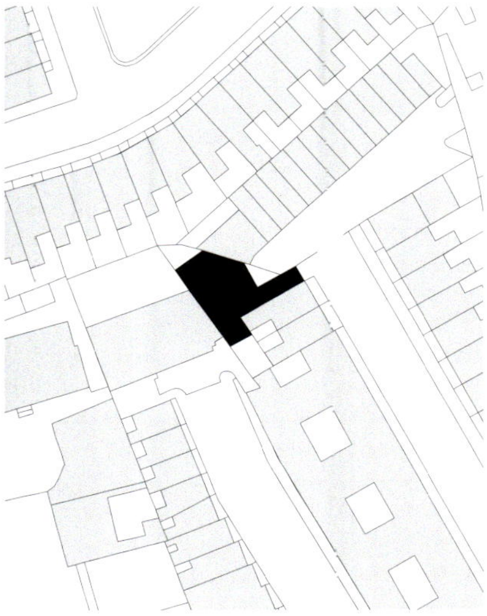

1.3
FIREWALL BUILDING

Building type
patio house on residual lot
between three fire walls
2 stories (raised ground floor
and lowered ground level)
facing NW/SE/SW

Date of construction
2005

Number of units
1

Size of units
approx. 330 m²

Area per user
66 m²

Building depth
10–20 m

Access
through carriageway in
adjacent building and via ramp

Open spaces
patios
balcony

Parking
no parking on lot

Architect
Caruso St John Architects LLP
project architect:
Rod Heyes

Location
London
United Kingdom

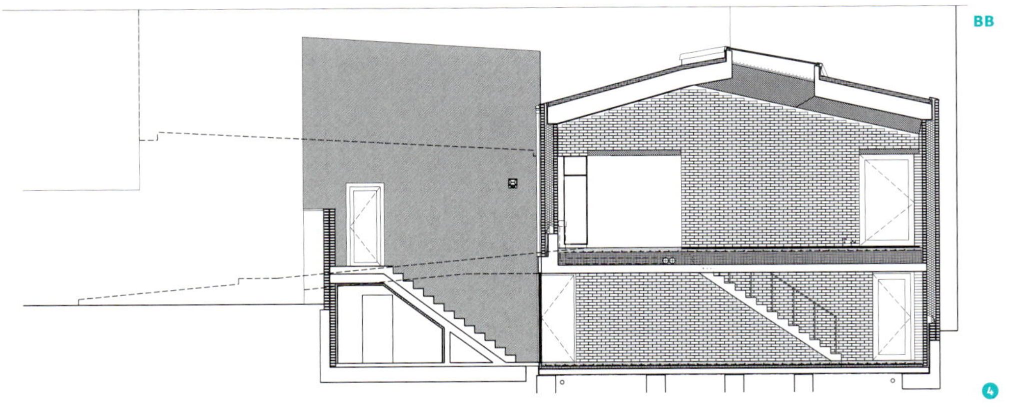

Referring to
floor plan 1:200

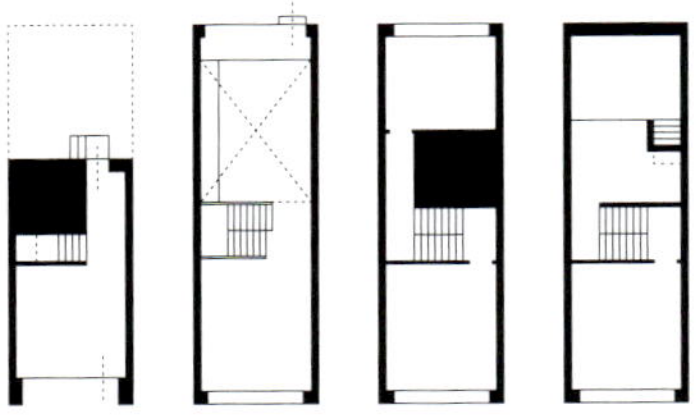

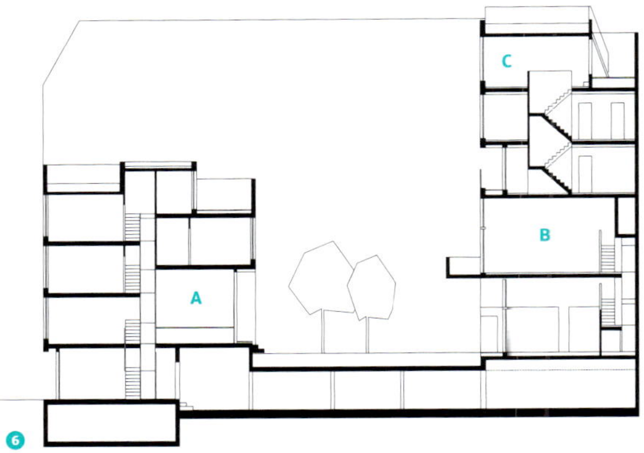

6

BIGyard | zanderroth | 2010

The inspiration for this cooperative building venture was the idea to connect the ideals of a single-family home with the advantages of living in a dense, urban context. Thus, the spatial organization of the individual building types and the complex as a whole provides numerous neighborly junctures, while also ensuring the ability to retreat into more private areas. The "townhouses" facing the street (**A**) are organized as split-level units and open up to their surroundings at multiple points: the entrance and optional commercial spaces are connected to the street, the raised eat-in kitchen is directly linked to the courtyard, while the dual-level roof terraces provide views into the surrounding neighborhood. The firewall building consists of a row of residential units stacked on top of each other. These so-called garden houses (**B**) are accessed directly from the courtyard, where their eat-in kitchen is also located. Situated on the more secluded floor above is the living room. The penthouses (**C**) above these units are accessed mid-level through a corridor to the rear. The living room is located on the top floor and has access to a private roof terrace above the fray.

Communal activities foster the feeling of a neighborhood: the garden has not been parceled out, thus offering spaces to meet and play, and a shared roof terrace with an outdoor kitchen can be used by residents in the warmer months. There are also a sauna and guest apartments.

The configurations of the block edge and the firewall building can be read as a reaction to the disadvantages of the lot, which faces north and is surrounded by the high walls of the neighboring properties. For this reason, the townhouses on the street are only 4 stories high to allow enough light to reach the garden houses. The 3-story garden houses along the firewall are built on top of the underground garage and therefore raised an entire level. Stacked above, the penthouses are exposed to natural light from the courtyard as well through a patio to the rear.

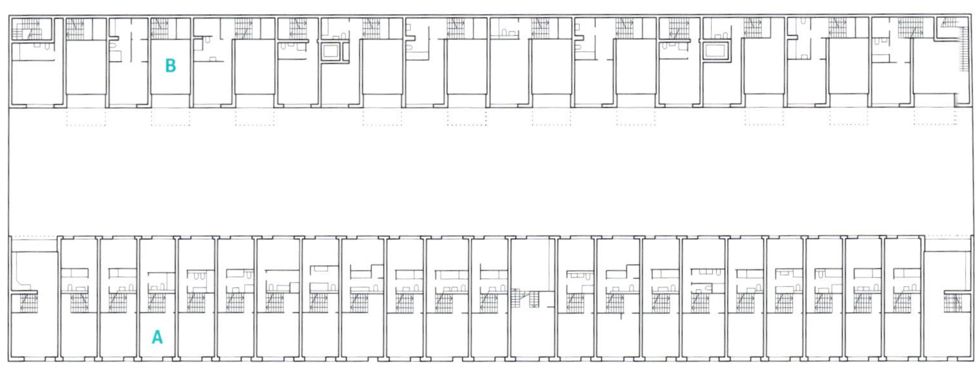

2

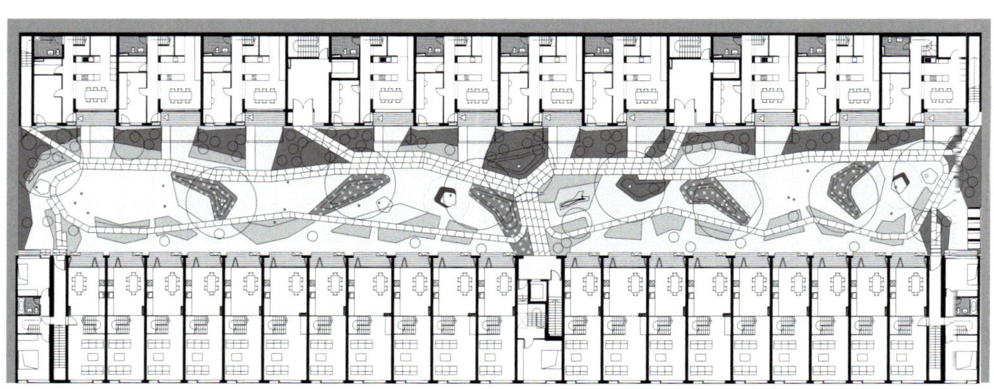

1

1 Garden level, entire ensemble (2nd floor) 1:750
2 3rd floor, entire ensemble 1:750
3 Townhouse floor plans **A**: Ground floor with street access and parking garage to the rear, upper split levels offset between street and courtyard facades 1:200
4 Garden house floor plans **B**: 2nd floor with courtyard access, upper split-levels offset to the left and right 1:200
5 Penthouse floor plans **C**: 5th floor, 6th floor with access corridor, 7th floor with stairs to upper roof terrace 1:200
6 Cross section through the split-level floors of the garden house 1:500
7 Cross section through the raised living room of the garden house 1:500

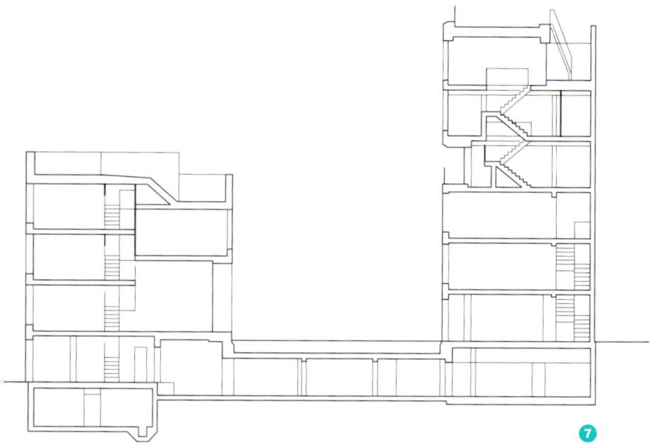

7

1.3 FIREWALL BUILDING

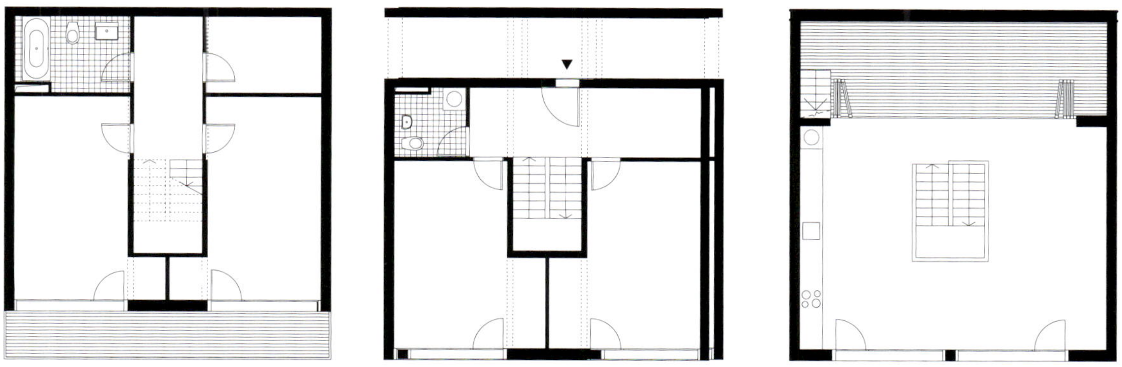

5

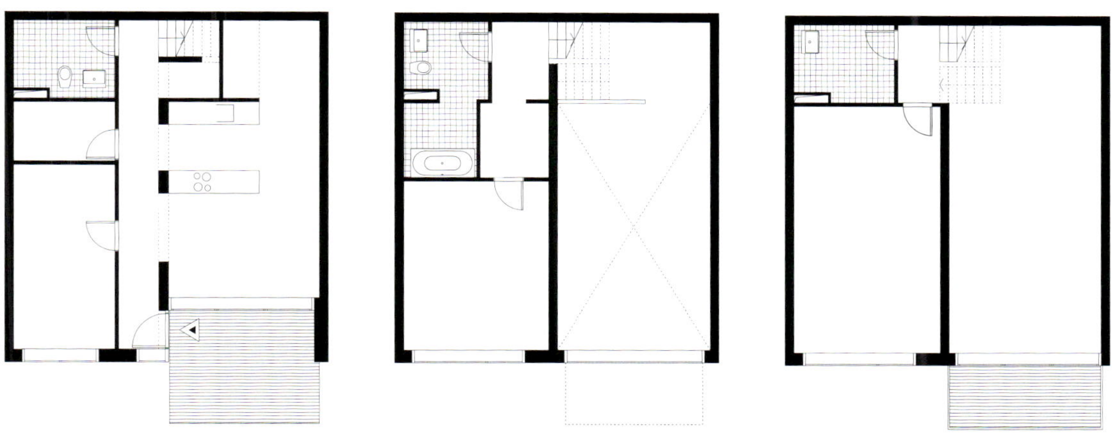

4

3

Building type
row house types
in block edge and
firewall configuration
4 and 6 stories
facing N/S

Date of construction
2007–2010

Number of units
45

Size of units
townhouses, 130–155 m²
(23 units)
garden houses, 139–162 m²
(10 units)
penthouses, 65–156 m²
(12 units)

Area per user
39 m² (entire development)

Building depth
9 m (firewall) /
12 m (block edge)

Access
direct access from street
or courtyard,
corridor along firewall for the
upper floor penthouses

Open spaces
terraces, balconies, private
and shared roof terraces,
garden courtyard above
underground garage

Parking
underground garage at street
level below garden courtyard

Architect
zanderroth architekten gmbh
with Kirka Fietzek,
Diana Gunkel, Guido Neubeck,
Christian Roth, Konrad Scholz,
Lutz Tinius, Sascha Zander
Berlin

Location
Zelterstraße 5–11
Berlin
Germany

Referring to
floor plan 1:200

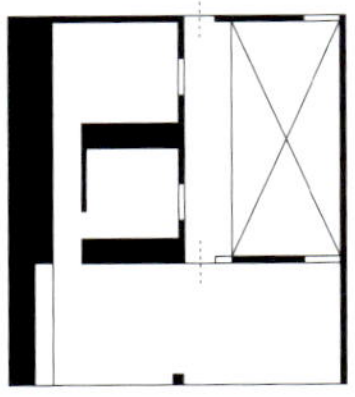

8 Octavia | Stanley Saitowitz | Natoma Architects | 2014

Located on a steeply inclined road, the building occupies one complete side of a city block along a firewall and marks the transition from a freeway to the urban block structure of San Francisco. The building code requirement of a rear courtyard was translated into the final meandering figure with its four vertical courtyard voids. In section, these spaces are alternatingly oriented outward or inward, at once organizing the building volume and increasing the surface area of the facade, and thereby the amount of natural light to which each apartment is exposed.

Simultaneously, these well-lit courtyards are intended as meeting places for residents after walking through the long access corridor along the firewall. Most apartments are oriented in some way toward the courtyard. Some units are accessed across bridges in the void spaces, while spiral staircases here connect the penthouses to private roof terraces.

The residential units are organized around the courtyard on one (Type **I**), two (Type **L**), or three sides (Type **U**). In the smaller Type **I** apartments, closet space, bathrooms, and kitchen units are organized within a "pod," which divides the studio apartment into dining, living, and sleeping areas with the aim of creating loft-like open floor plans. In the larger Type **L** units, the secondary spaces are bundled into a thickened wall running along the apartment perimeter and through all living spaces, thus complementing each space with the necessary service elements, such as a kitchen unit or bathroom, and leaving the rest of the apartment as unencumbered as possible.

The individual facade panels can be adjusted manually to control the amount of sunlight, and the view into and out of the apartment as well as reduce traffic noise.

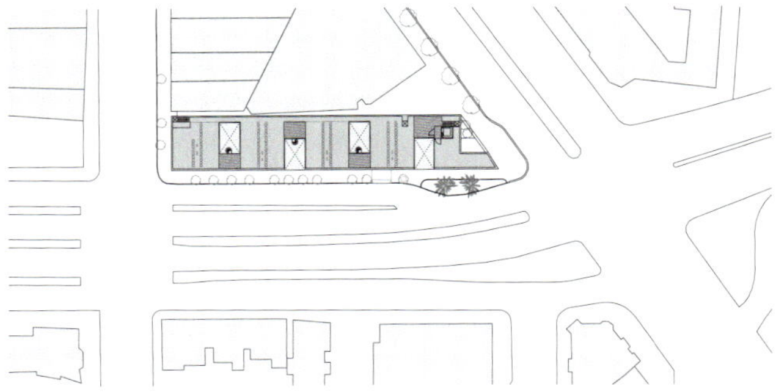

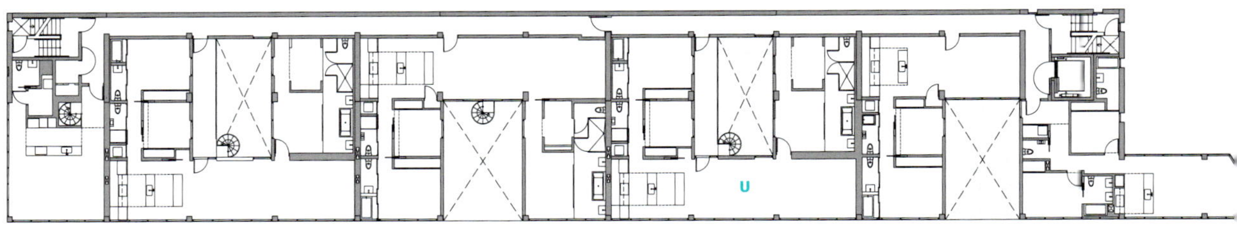

4

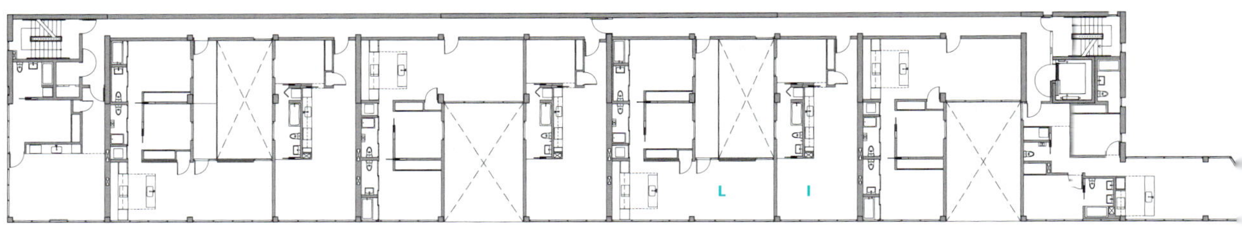

3

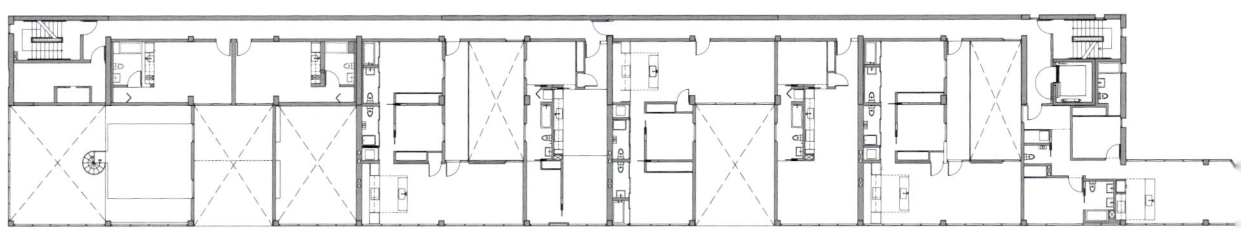

2

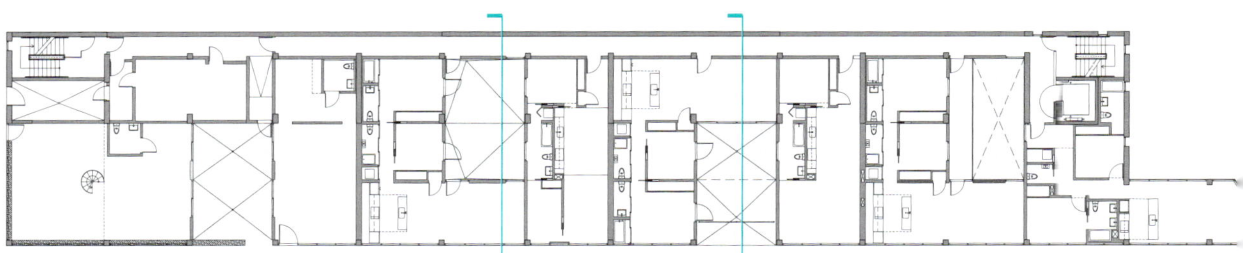

1

1 Total floor plan, 3rd floor 1:500
2 Total floor plan, 4th floor 1:500
3 Total floor plan, 6th floor 1:500
4 Total floor plan, 9th floor 1:500
5 West elevation 1:800
6 Cross sections 1:800
7 Longitudinal section 1:800
8 3-room apt. type **L** 1:200
9 2-room apt. type **I** 1:200
10 4-room apt. type **U** 1:200

5

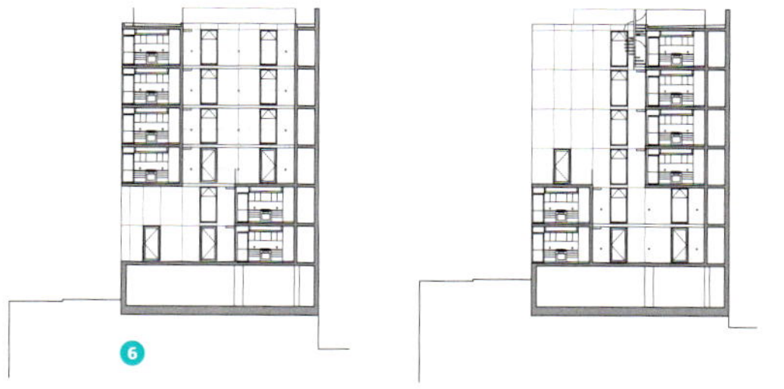

6

7

1.3
FIREWALL BUILDING

Building type
firewall building completing a block edge on an inclined road, meandering form around alternating courtyard voids
5–8 stories
facing W

Date of construction
2014

Number of units
47

Size of units
2-room apts., 67 m² (9 units)
3-room apts., 93 m² (35 units)
4-room apts., 164 m² (3 units)

Area per user
31–41 m²

Building depth
13.8 m

Access
rear corridor along firewall, connecting bridges through courtyard voids

Open spaces
shared and private roof terraces, terraces in courtyards

Parking
underground garage

Architect
Stanley Saitowitz | Natoma Architects
San Francisco

Location
8 Octavia Street
San Francisco
USA

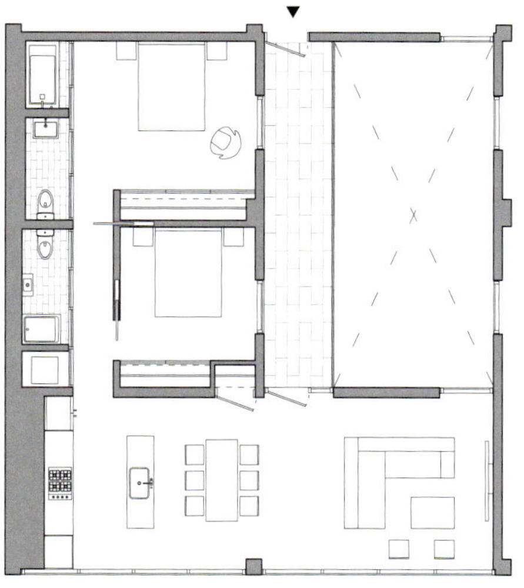

8

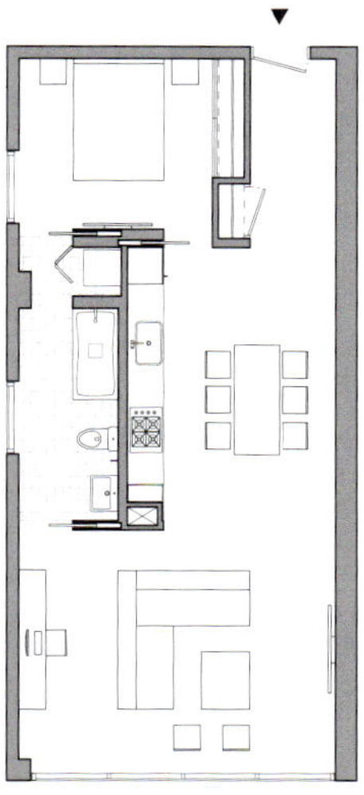

9

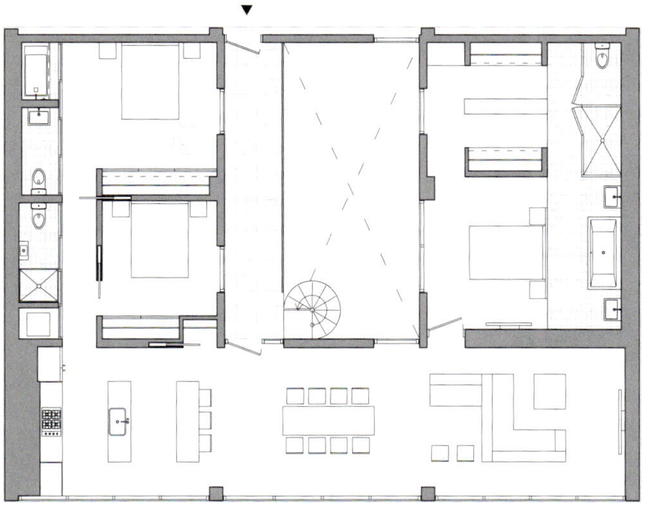

10

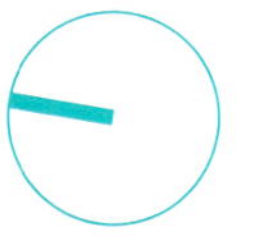

Referring to floor plan 1:200

2.1
SOLITAIRE

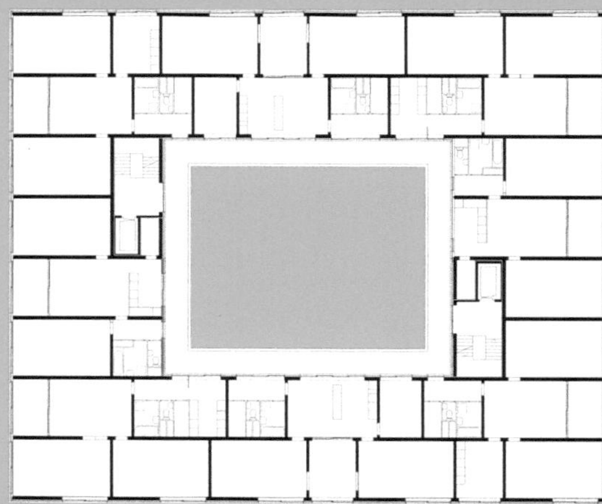

Solitaires indicate something singular, sometimes unique. They are fully detached and receive natural light from all sides; exterior form, orientation, and height can be determined with great freedom. All these factors offer ideal conditions for creating and exploring exemplary housing models. Solitaires differ primarily from one another with regard to scale: the small, suburban multifamily house or the urban villa in an open city block are distinguished for their clarity, their relationship to public and private green, the occupants' identification with their home and with their neighbours. The advantages of the single-family house can be realized here within the more economical multifamily house. Large urban solitaires – the courtyard house, the atrium house, or the small urban block – are configurations with more complex access systems and internal hierarchies, shared internal and external spaces, and a great variety of apartment types. A decisive factor for the interpretation as a solitaire is its urban presence: in other words, whether the building appears as an autonomous object. Naturally, in doing so it can be integrated into an urban planning concept that allocates an entire urban block with a more or less stringent canon of regulations to a series of solitaires.

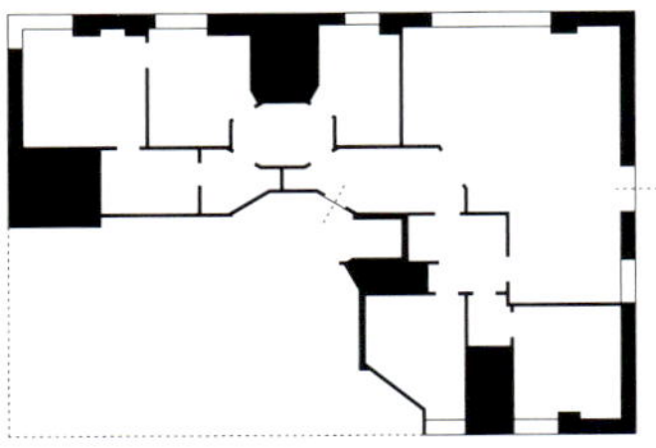

Piazza Carbonari | Caccia Dominioni | 1961

The 7-story apartment building is a free-standing urban structure in the vicinity of a large roundabout. The sculptural appearance of the building is not only an end in itself: the cubature and the gabled roof profile emerged as a compromise in response to the requirements set by the development guidelines and a desire for maximum utilization and height. The windows, which appear to be randomly placed, flush with the homogeneous ceramic skin and wrapped around all four facades like a pictorial composition that seems to ignore the structure of the floors, are in fact essentially a reflection of the internal organization of the building. The leitmotif for this project are stacked villas, each of which occupy an entire floor. Each villa has a unique floor plan: with openings that serve rooms with different dimensions and uses. The skeleton construction allows for a very free floor plan design, which connects open spatial sequences with a series of smaller and more private spaces – accessed through an array of various corridors and hallways, around which other rooms are bundled in turn.

These conjunctions of spaces form a floor plan, in which not everything is immediately revealed. The spatial sequences are only deciphered through the movements of the users, successively and in fluid changes of direction.

1

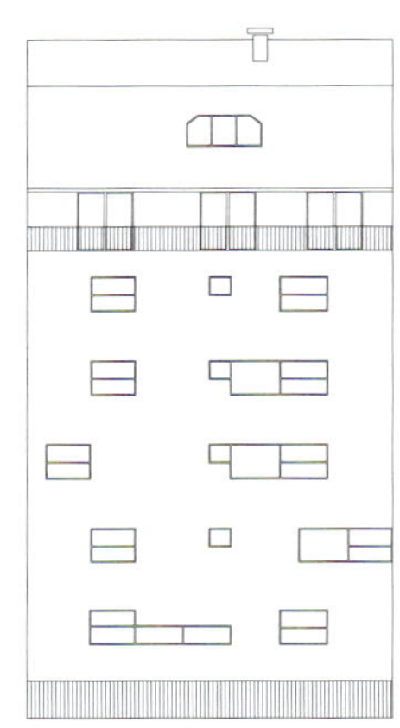

2

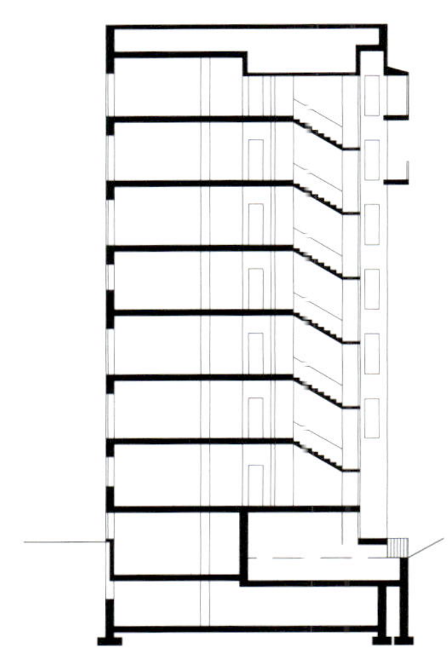

3

4

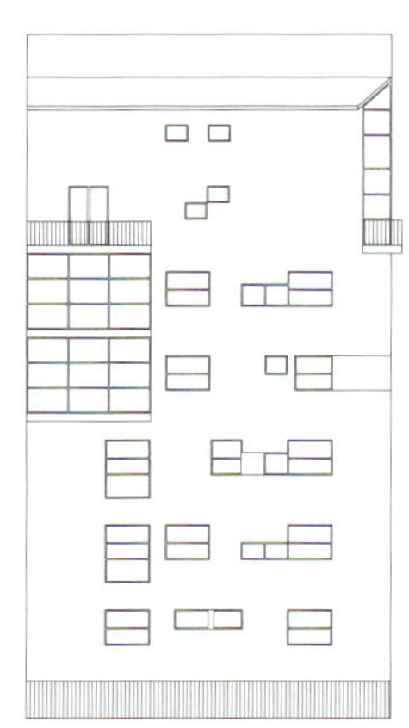

5

1 Northeast elevation
2 Southeast elevation
3 Cross section 1:500
4 Southwest elevation
5 Northwest elevation
6 Raised ground floor 1:200
7 Basic floor plan for standard floor 1:200
8 Attic story 1:200
(all drawings from the design stage)
9 Lot with grounds

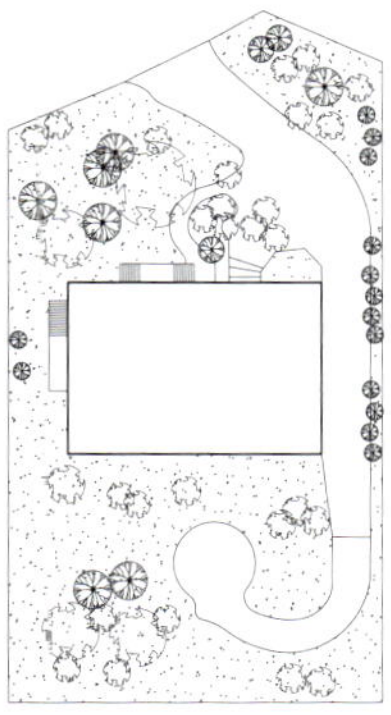

9

2.1
SOLITAIRE

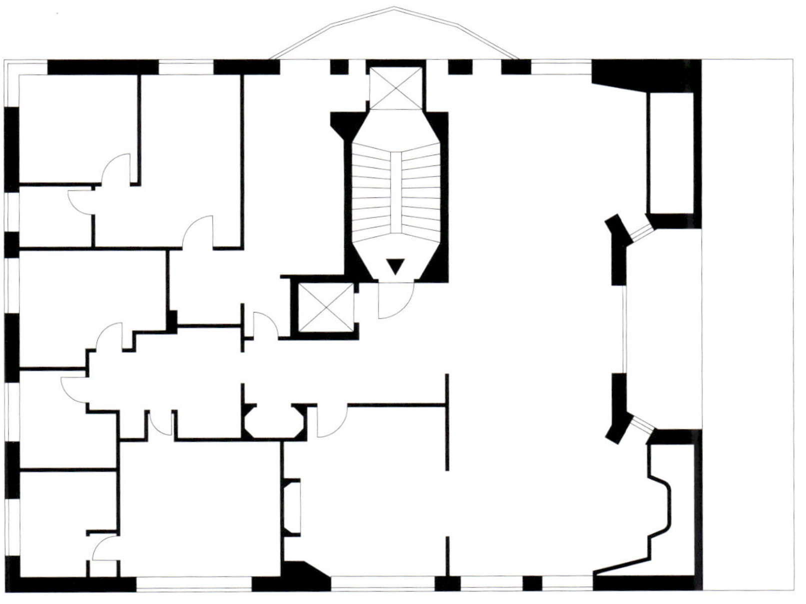

8

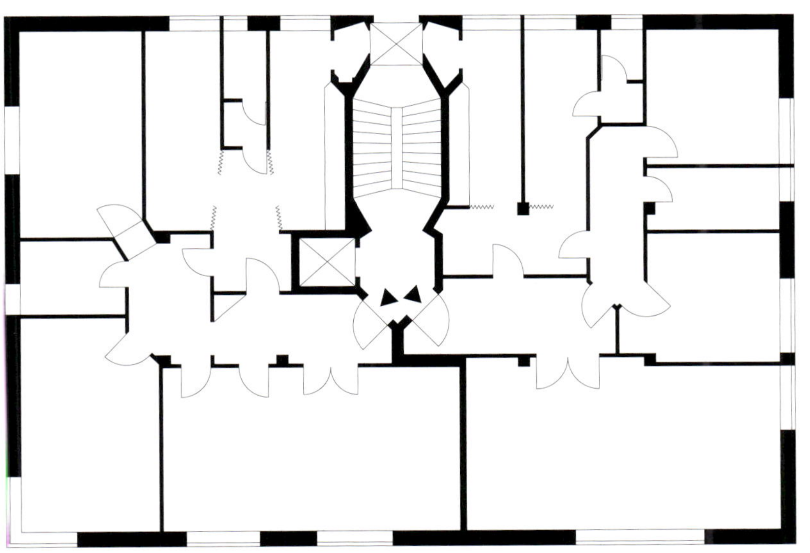

7

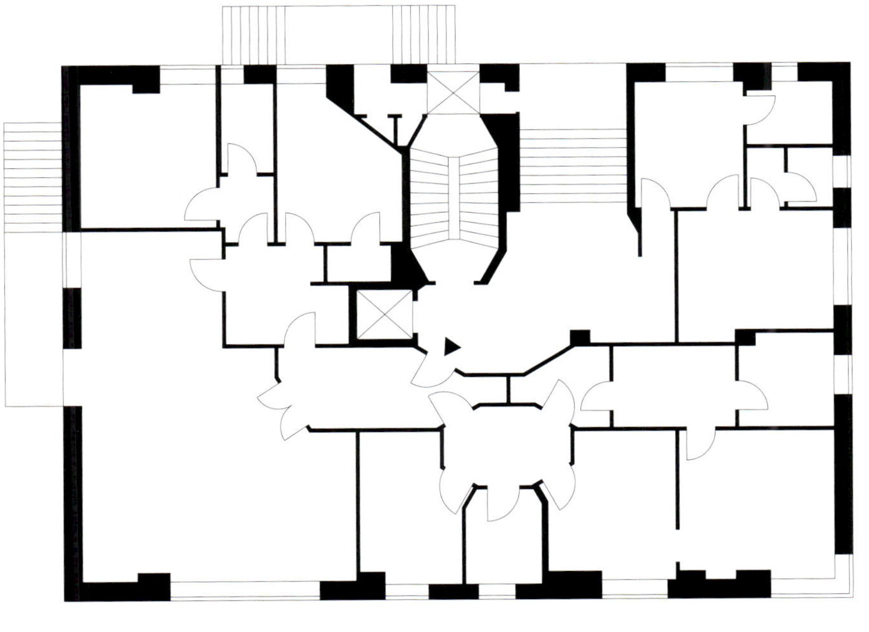

6

Building type
sculptural volume
with vividly composed,
ceramic-clad perforated facade,
skeleton construction
with individual floor plans
facing in all directions

Date of construction
1960–1961

Number of units
7

Size of units
raised ground floor, approx. 170 m²
standard floor, approx. 238 m²
attic floor, approx. 193 m²

Area per user
nondeterminable

Building depth
approx. 14 m

Access
single unit per floor with
one internal elevator,
in part opening directly
into the apartments and
one external elevator,
flush with the facade and
leading directly to the kitchens

Open spaces
large, shared garden
some units with roof terraces
and balconies

Parking
underground garage

Architect
Luigi Caccia Dominioni

Location
Milan
Italy

Referring to
floor plan 1:200

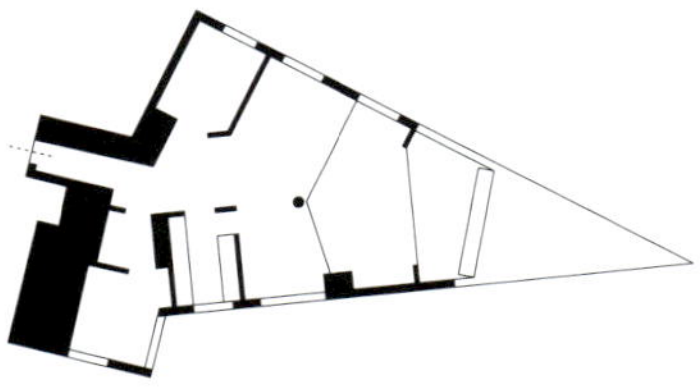

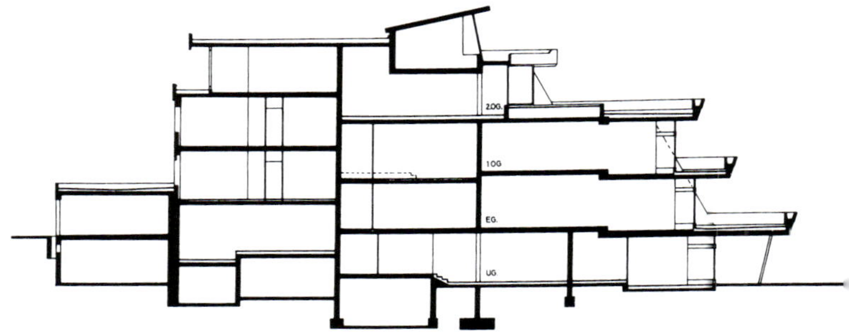

Wallotstraße | Schudnagis | 1972

The complex floor plan design evolved by rotating all apartments toward the shoreline and the landscape. Despite angled walls, a winding building fabric, and the stepped, pointed terraces, the layout of the floor plans is essentially economical and functional: the rooms fan out from a corridor in all directions, thus facing toward the various views. The most diverse apartment sizes and layouts are unified under one roof: from the 25 m² 1-room apartment with minimal kitchen nook to the 156 m² apartment, which echoes the landscape outside in its sequence of terrace, patio, flower boxes, and seamless transitions between living spaces in the interior. The living rooms are extended upward and create space for a rooftop studio. The slope site is not only reflected in the apartments at the rear elevation, stepped back by one half story, it is also thematicized in that the living room that overlooks the water (with the pointed terrace) is always slightly lower than the rest of the rooms or by virtue of other changes in floor level, which add interest to the room layout. The design provides many built-in pieces of furniture and an open fireplace in the living room of the larger units. The garden floor accommodates garages, an indoor pool, and a sauna.

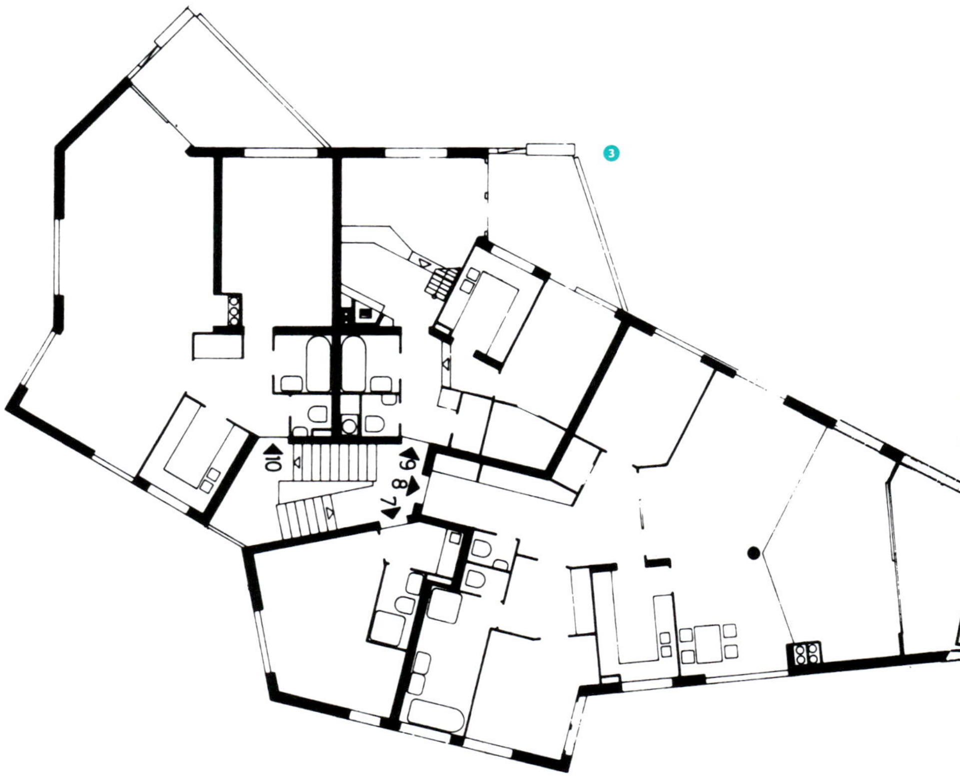

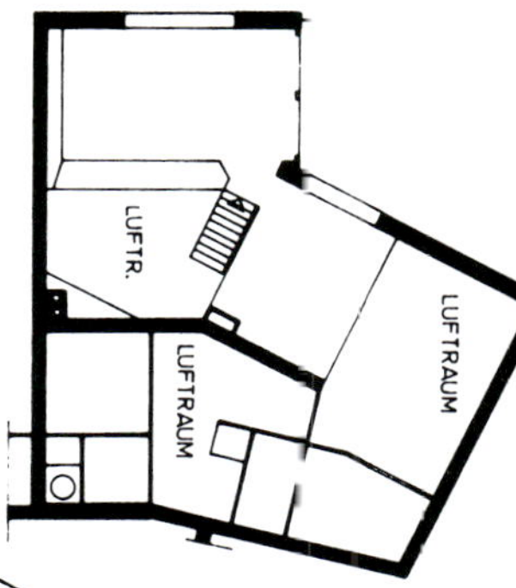

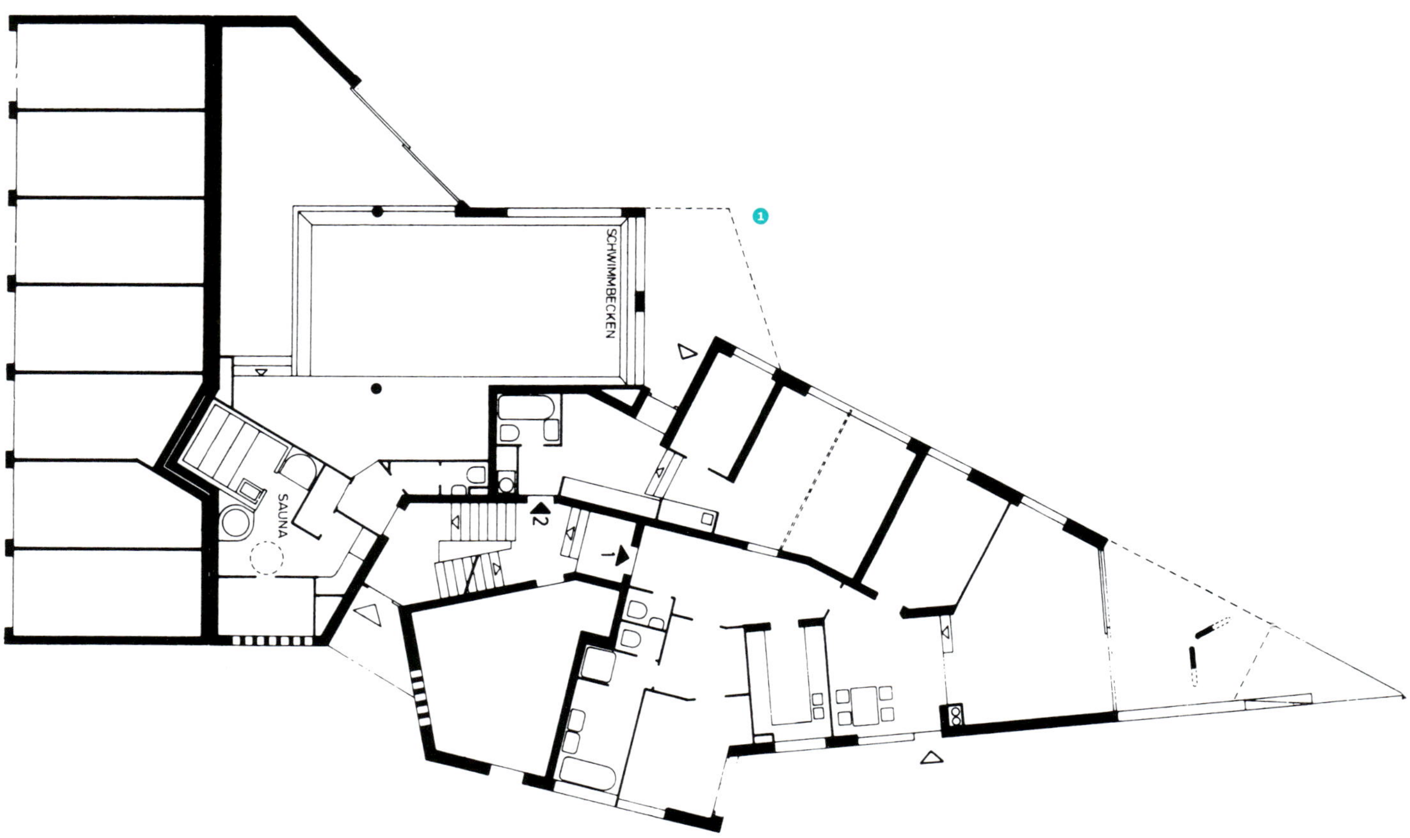

1 Ground floor 1:200
2 Second floor 1:200
3 3rd floor 1:200
4 4th floor 1:200

2.1
SOLITAIRE

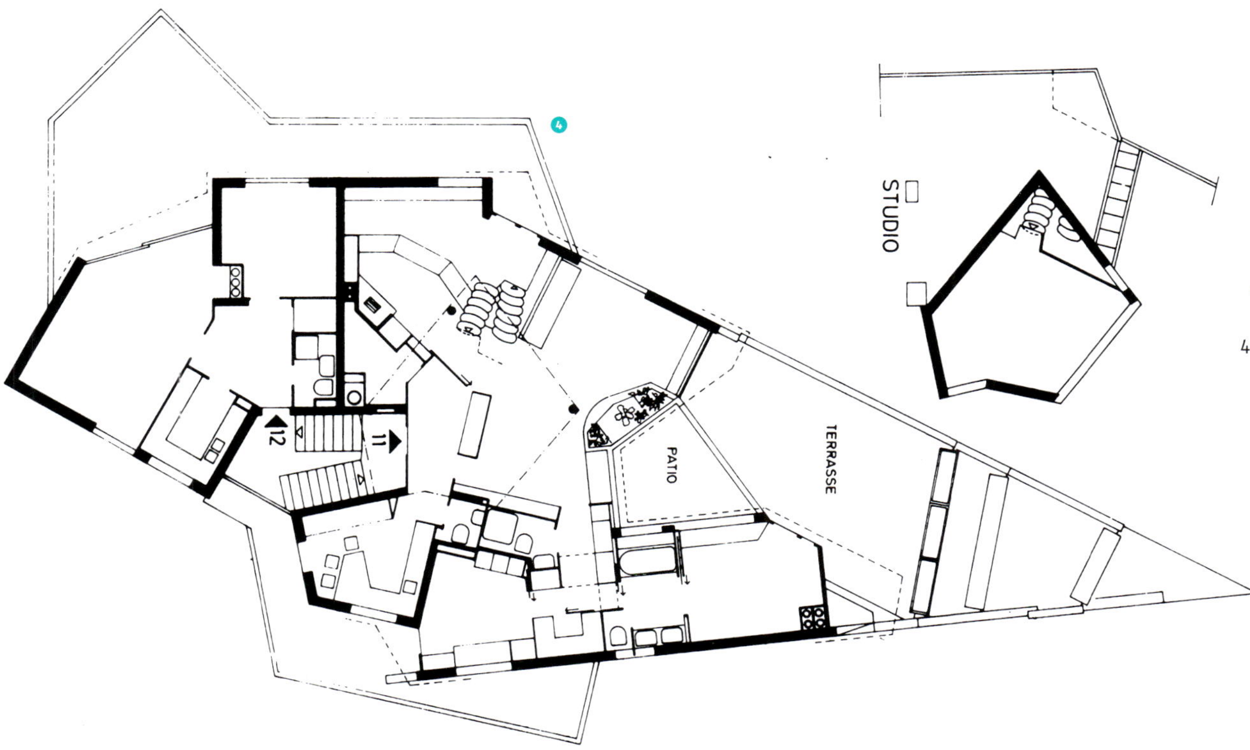

Building type
multiple-family house at a lake
4 stories
facing E/W

Date of construction
1972

Number of units
12

Size of units
1-room apts., 25 m² (2 units)
2-room apts., 56–60 m² (3 units)
3-room apts., 87 m² (2 units)
4-room apts., 126–135 m² (3 units)
4½-room apt., 156 m² (1 unit)
1½-room apt. can be separated

Area per user
25–39 m²

Building depth
varied

Access
double- and quadruple-loaded stairwell

Open spaces
garden
rooftop terraces
pool

Parking
7 basement garages

Architect
Heinz Schudnagis
Berlin

Location
Wallotstraße 9
Berlin-Wilmersdorf
Germany

Referring to
floor plan 1:200

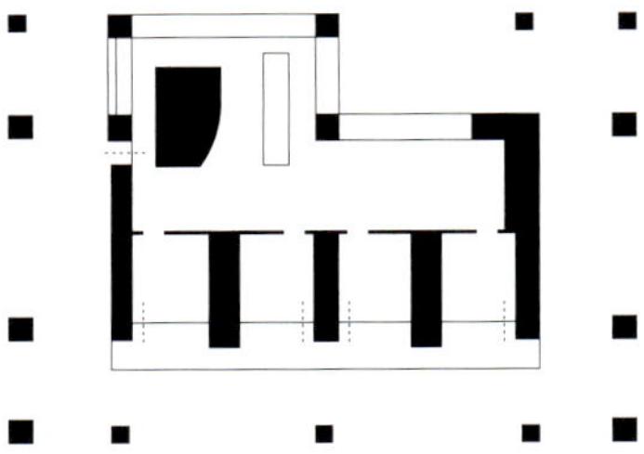

Next 21 |
Utida et al. | 1993

This building was intended as an experiment in spatial, technical, and conceptual strategies to explore the development of flexible apartment buildings able to react long-term to the changing preferences of residents. The basic principle was formed by the "Open Building" concept of the Dutch architect John Habraken in which a building is understood as a composite of two systems: the support system forming a stable and shared property and an infill system, which is individually owned and interchangeable. Here the concept was realized with a series of constituent subsystems: a skeletal structure based on a multi-layered grid, a modular facade system and technical infrastructure that can be easily renewed at any time.

U-shaped levels are laid out to enable the creation of completely different apartment types and sizes, and remain open to the neighboring buildings, thus providing room for green spaces. In contrast to the layouts of conventional apartment buildings, which need to be standardized in order to anticipate the potential demands of future users, the conditions here proved congenial to having 18 residential units designed by 13 different architects in direct contact with the future residents of each.

The individual designs are fundamentally different from one another. Some (apartment **A**, for example) are reminiscent of the spatial character of a single-family home: they are individually designed and comprised of two stories with wraparound facades. Others (**B**, for example) connect open living spaces with individual cabinet-like private rooms. And still others (**C**, for example) incorporate traditional Japanese elements such as sliding walls to create flexibility of use. The first unit conversions have successfully illustrated the ability of this open framework to incorporate changes.

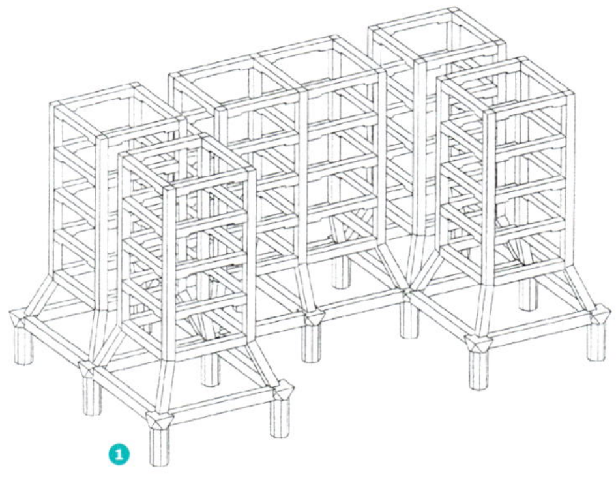

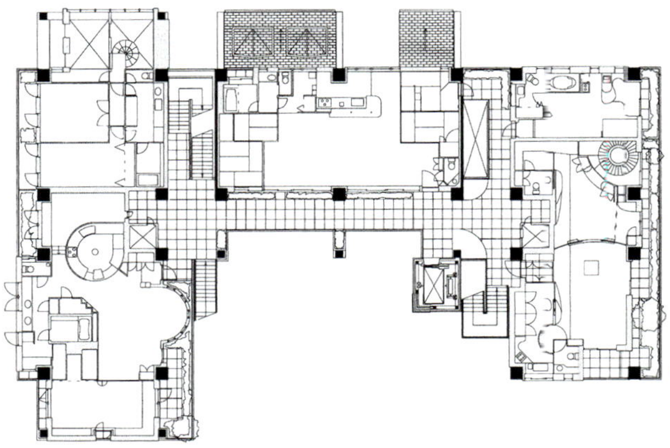

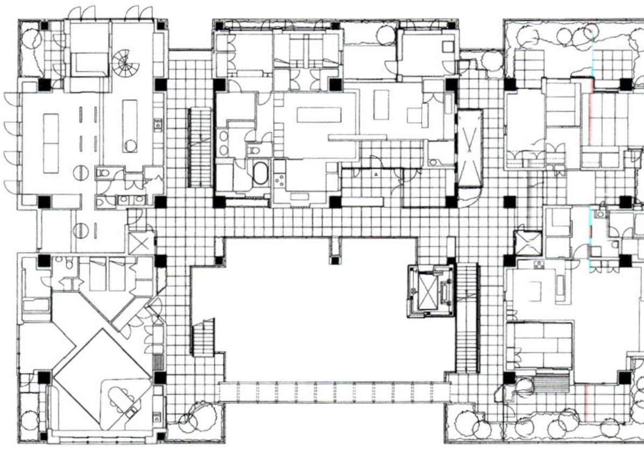

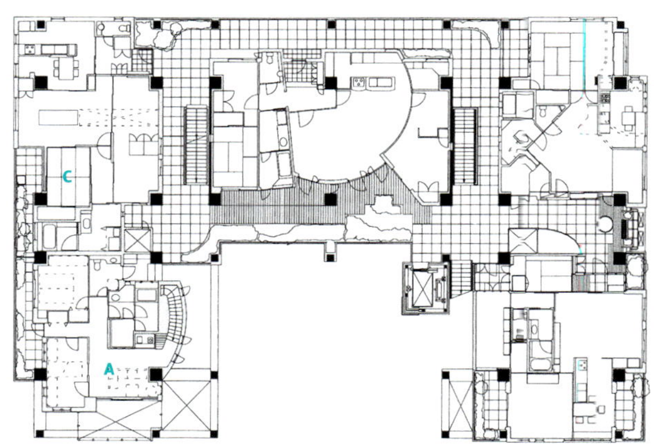

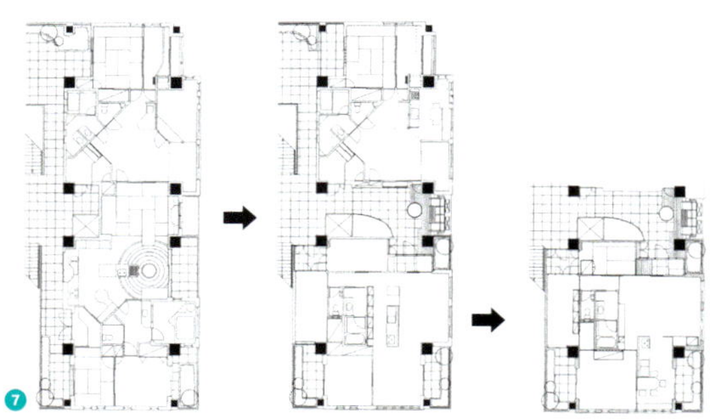

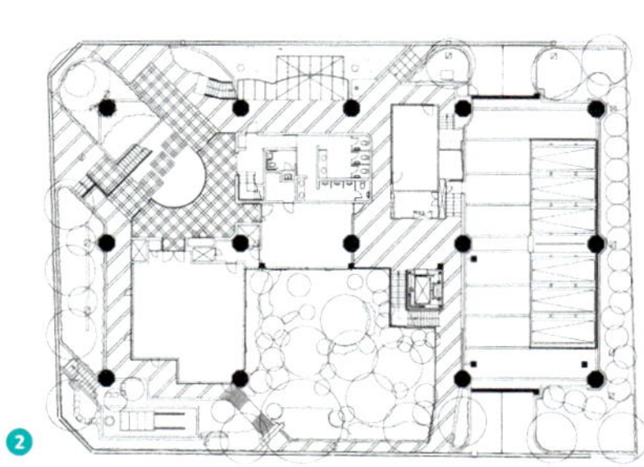

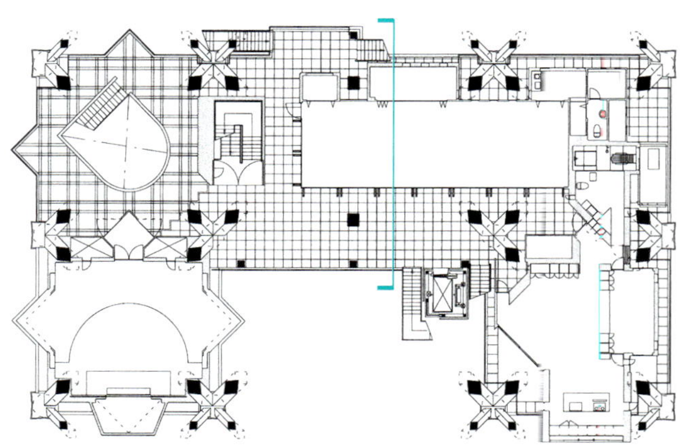

1 Load-bearing structure as part of the support system
2 Ground floor 1:1000
3 2nd floor 1:500
4 4th floor 1:500
5 5th floor 1:500
6 6th floor 1:500
7 Infill system: renovation phases of a unit on the 4th floor
8 South elevation 1:500
9 West elevation 1:500
10 Cross section 1:500
11 Raised floors for technical installations, table-like structure within residential unit for mounting air-conditioning unit on suspended ceiling
12 3rd floor 1:200

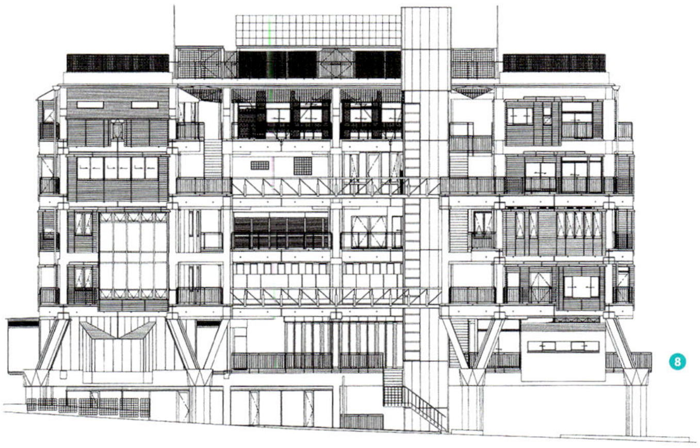
8

9
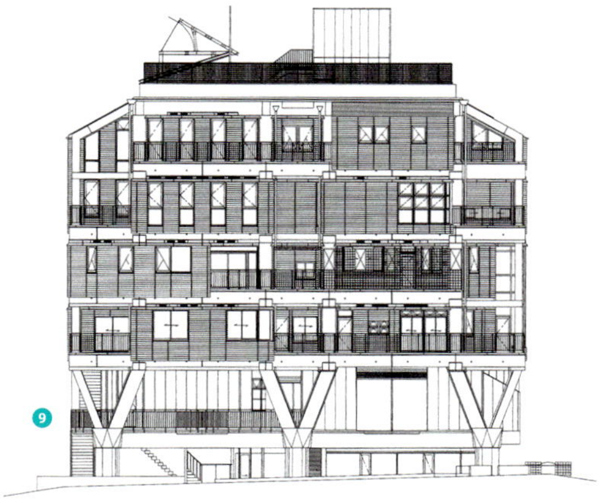

2.1
SOLITAIRE

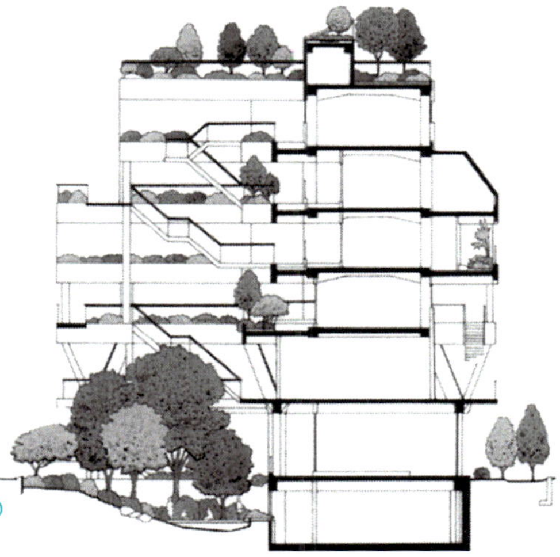
10

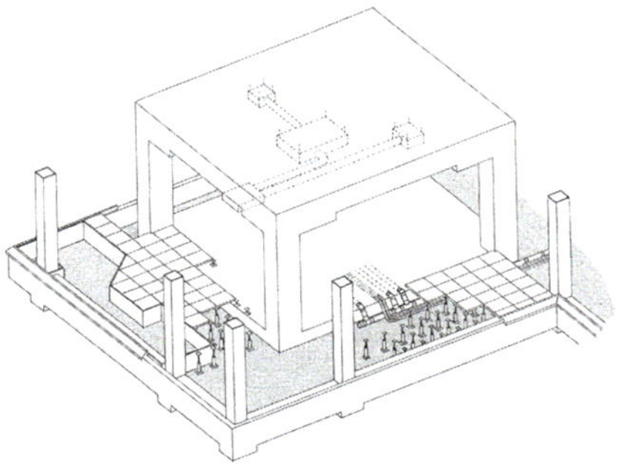
11

Building type
Open Building: U-shaped, open skeletal frame construction ("support system"), individually designed, modifiable or replaceable residential units ("infill system")
6 stories
facing in all directions

Date of construction
1993

Number of units
18

Size of units
1- to 5-room apts., incl. duplex units, approx. 58–187 m²

Area per user
20–58 m²

Building depth
11.8/13.9 m

Access
open levels between residential units

Open spaces
loggias, balconies, winter gardens, common spaces on open ground and second floors, as well as the roof terraces and access spaces

Parking
parking system on ground floor and below grade

Architect
building design, supervision:
Yositika Utida
Shu-koh-sha Architectural and Urban Design Studio
Tokyo

Location
Shimizudanicho, Osaka
Japan

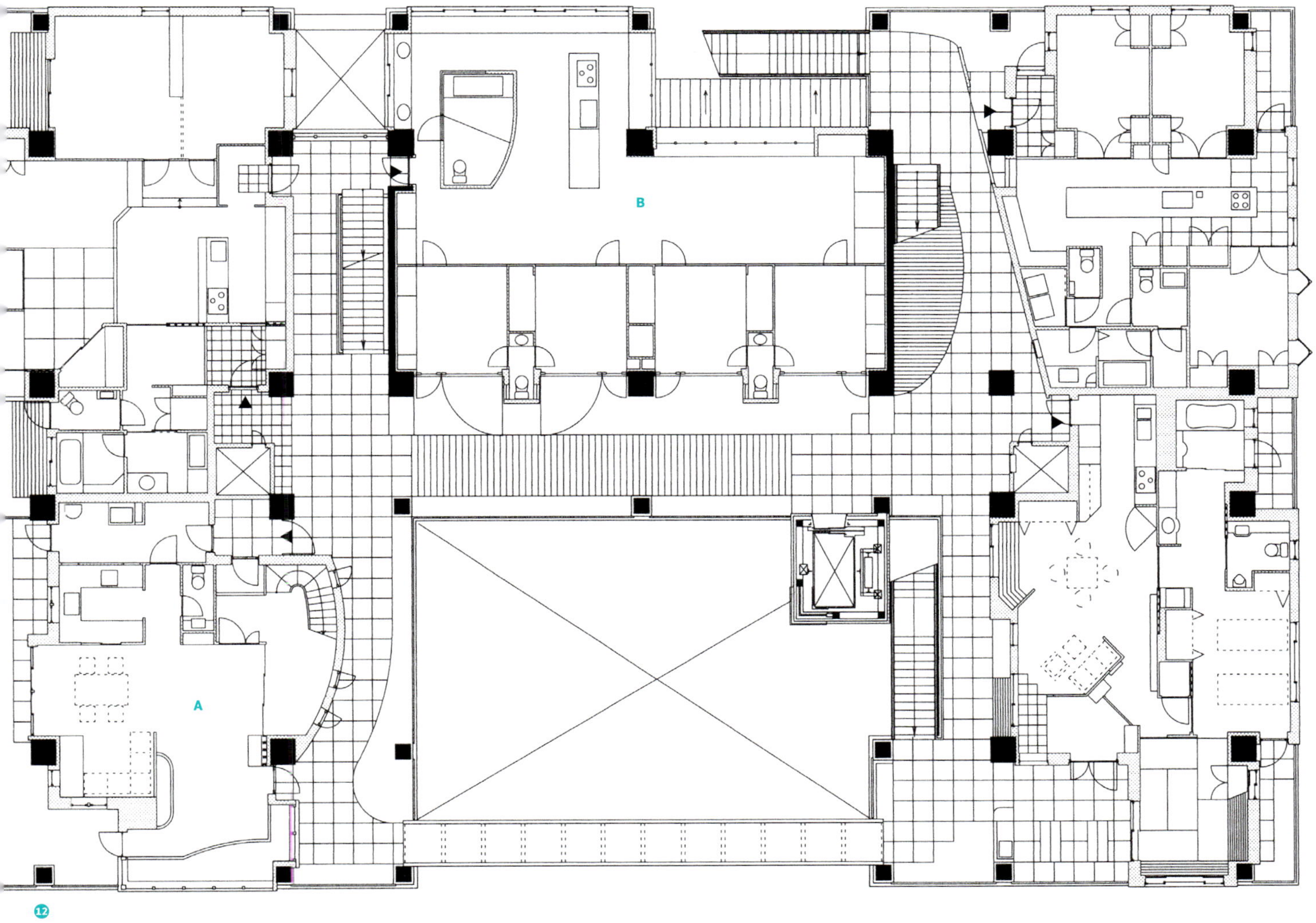

12

Referring to floor plan 1:200

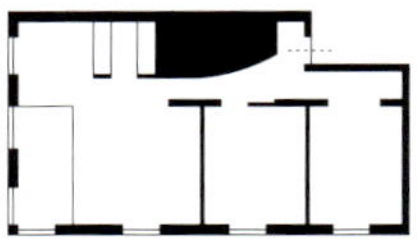

Kapellenweg | Baumschlager Eberle | 1996

2.1
SOLITAIRE

Three houses situated relatively close to each other. The facade, evenly broken down by floor-high windows, makes it impossible to deduce the true inner arrangement of the house. A central stairwell with skylights allows access to four apartments on every landing. The symmetrically placed doors conceal, surprisingly, apartments of different sizes: made possible by shifting a separating wall within the pattern established by the windows – one room per window – by one room module. Thus, two 3-room apartments become a 2- and a 4-room apartment. The bedrooms are aligned along a corridor remaining private, whereby the round bathroom guides the visitor straight into the living space. The living rooms with kitchens and loggias are situated in the four corners of the building. On the ground floor are three 2-room apartments and various kinds of ancillary rooms.

Building type
3 houses
3 stories
orientation to all sides

Date of construction
1994–1996

Number of units
11 in every house

Size of units
in every house:
2-room apts., 65 m² (5 units)
3-room apts., 73 m² (4 units)
4-room apts., 90 m² (2 units)

Area per user
22.5–32.5 m²

Building depth
14.7/26.7 m

Access
central stairwell

Open spaces
loggias

Parking
33 underground parking places
14 open places

Architect
Baumschlager Eberle
Lustenau

Location
Kapellenweg
Feldkirch-Tosters
Austria

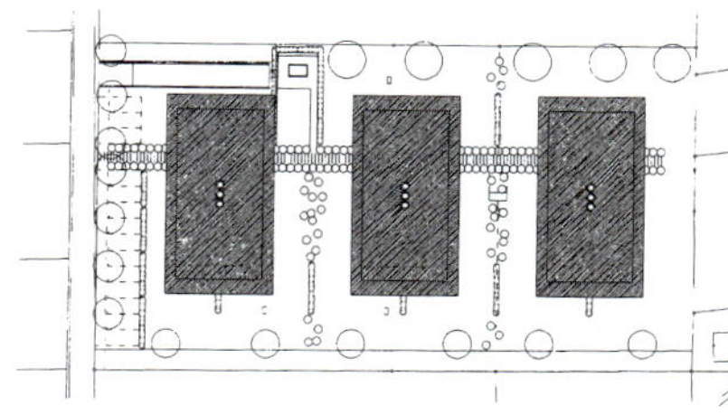

1
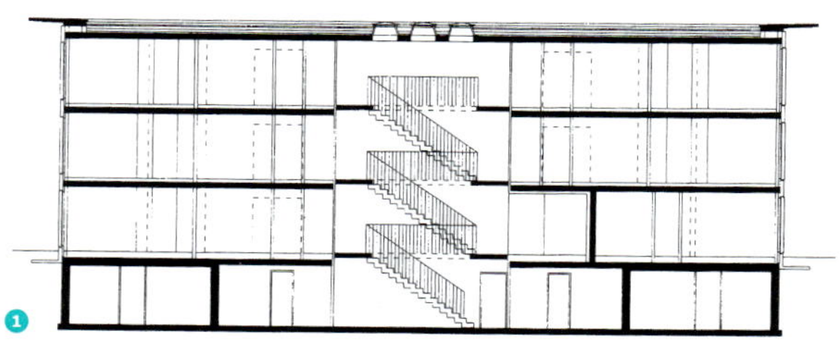

2
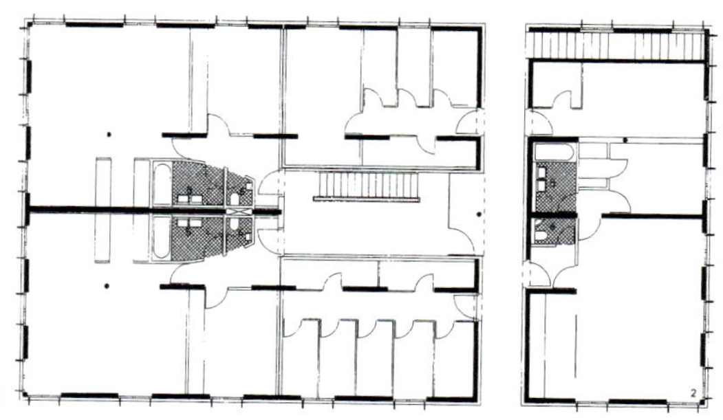

3
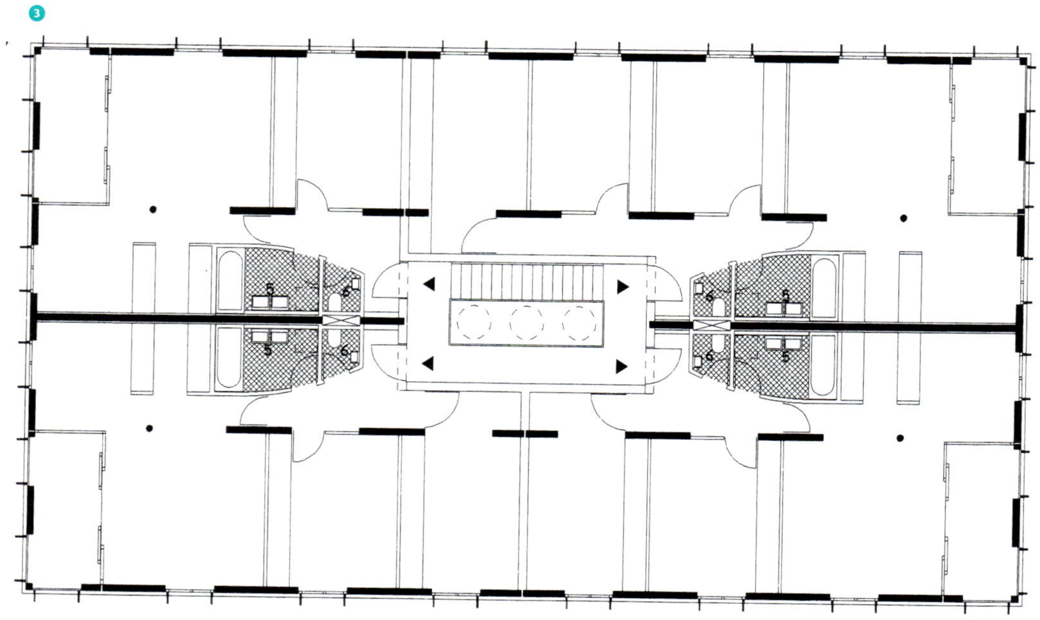

Site plan 1:2000
Section 1:400 1
Ground floor with ancillary rooms 1:400 2
Typical floor plan 1:200 3

Referring to floor plan 1:200

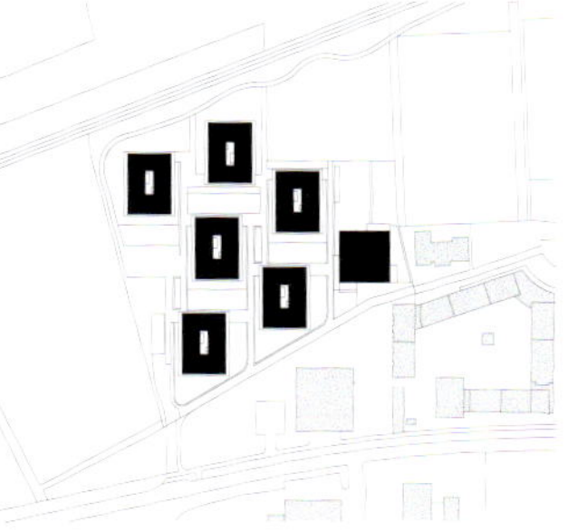

Lohbach | Baumschlager Eberle | 2000

The dense residential complex consists of six compact, cube-shaped residential buildings in an attractive green space, each with an internal circulation and a prominent wraparound balcony layer, which can be completely closed off to its surroundings with copper screen elements. The buildings are offset to one another in a checkerboard pattern so as not to hinder outside views. A network of walkways, small courtyard-like spaces, and gardens on the ground floor form a communal area in the interstitial spaces. The apartments are accessed through a glazed atrium in which the elevator and staircase cores are situated. The sculptural character of this space stands in stark contrast to the rather modular outer appearance of the buildings. Within the apartments, a layer of entrance halls, sanitary, and storage rooms border the atrium, forming a transitional zone that serves both as a spatial buffer and a constructive element. Because the load-bearing walls run parallel and not perpendicular to the facades, the second layer of bedrooms and living areas can be subdivided as desired.

Building type
ensemble of 6 identical cubic buildings
7 stories
facing in all directions

Date of construction
1998–2000

Number of units
297

Size of units
1-room apts., 43 m², balcony 8 m² (34 units)
2-room apts., 54 m², balcony 12 m² (82 units)
3-room apts., 85 m², balcony 19 m² (131 units)
4-room apts., 98 m², balcony 20 m² (50 units)

Area per user
20–24 m²

Building depth
27.9 m

Access
central atrium

Open spaces
wraparound balcony zone with closable folding shutters

Architect
Baumschlager Eberle
Lustenau

Location
Technikerstrasse/
Franz-Baumann-Weg
Innsbruck
Austria

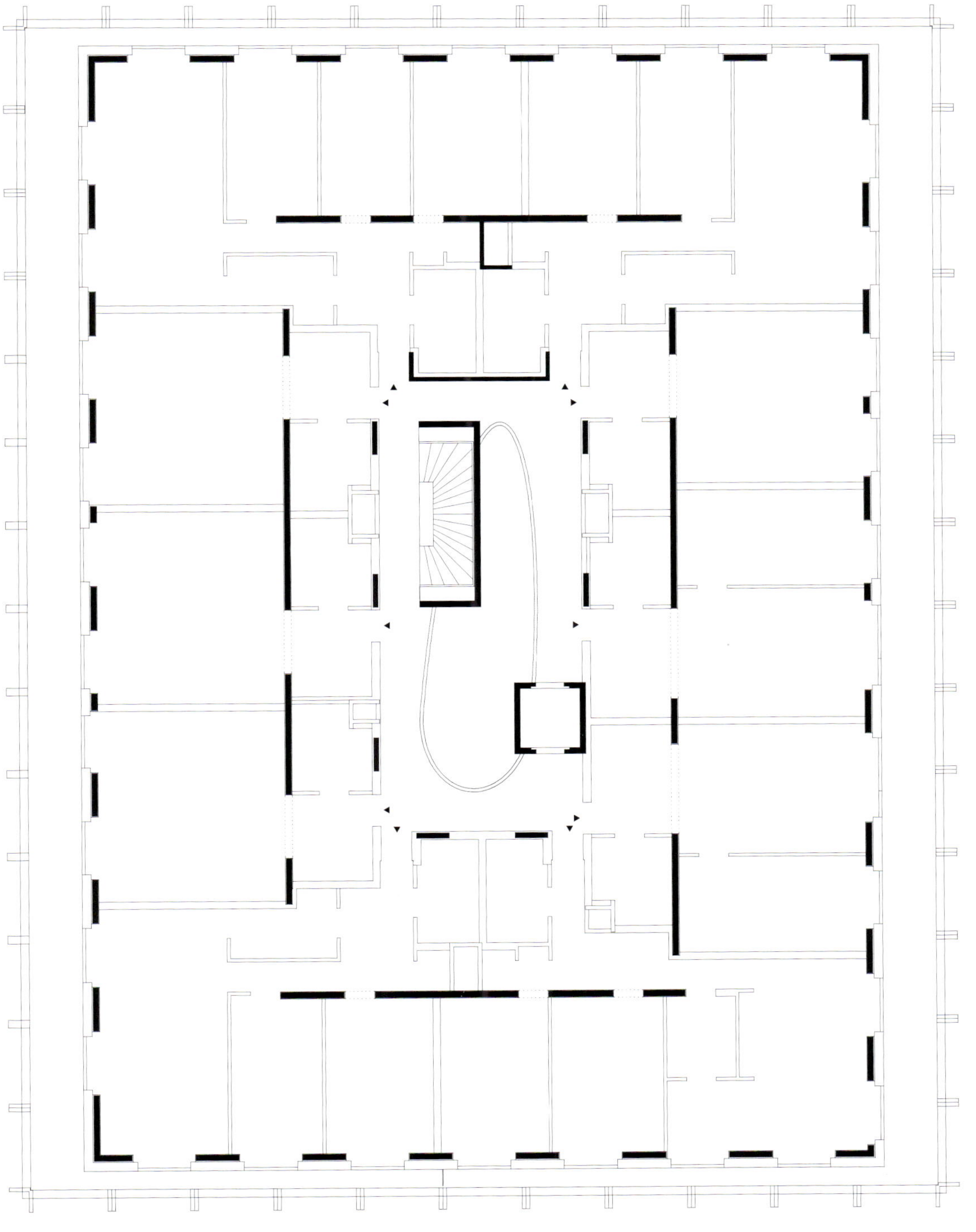

2

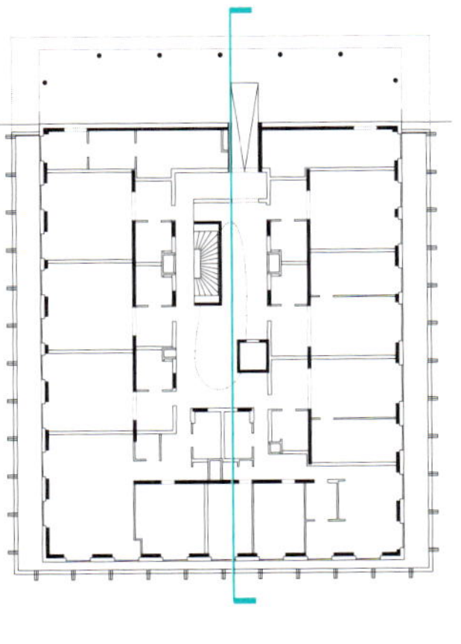

1

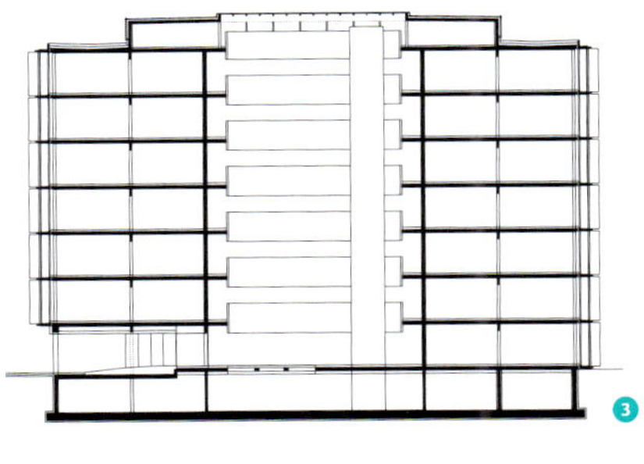

3

Ground floor 1:750 1
Typical floor 1:200 2
Longitudinal section 1:750 3

Referring to
floor plan 1:200

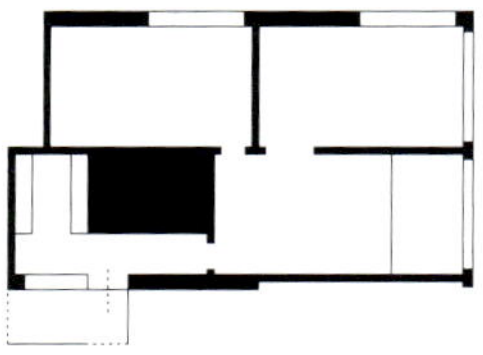

KNSM- and Java-Eiland | Diener & Diener | 2001

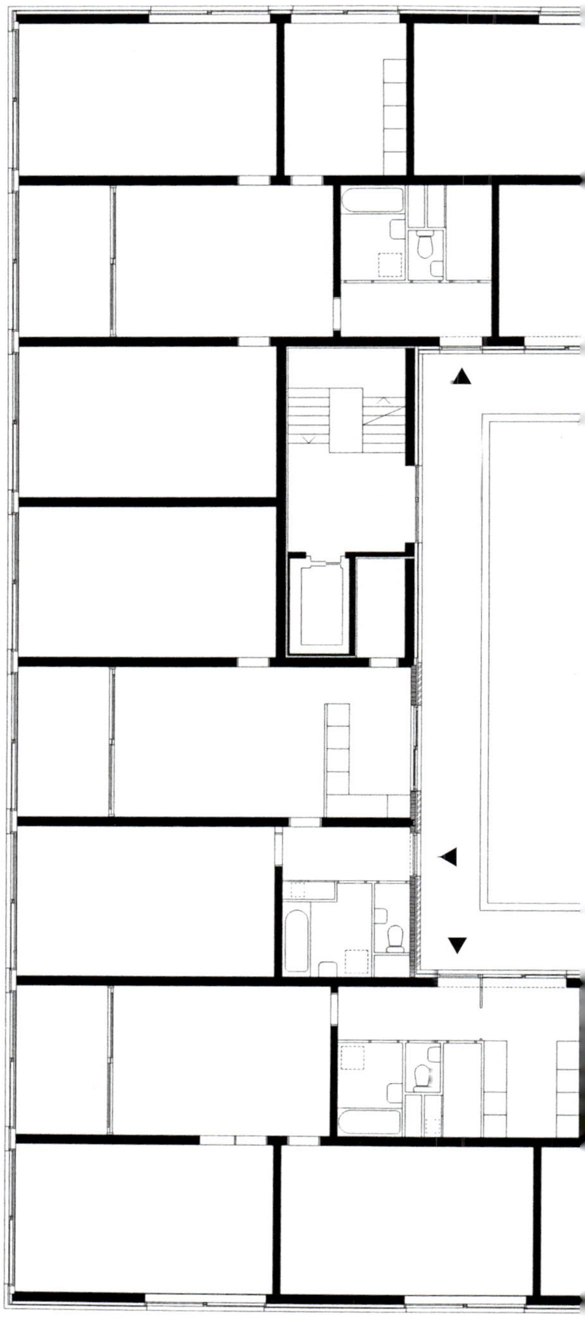

Two buildings, the long and the courtyard building, by Diener & Diener Architects bring structure into the transitional area of the new housing district in Amsterdam on the two peninsulas KNSM and Java, whose heterogeneous urban concepts converge at precisely this location (see also Kollhoff, page 70). The courtyard house with its unique plan is a cubic volume, whose eastern elevation is cantilevered across the access road. The concentric organization of the courtyard type and the orientation of the building fabrics in the axis of the peninsula have informed the building. The volume seems to be simultaneously in the grip of static and dynamic forces. This delicate balance is continued in the apartments, which are not hierarchically ordered: with a long rectangular plan, with front or lateral windows and doors depending on location, the individual rooms convey a sense of tranquility and animation at the same time – much like the entire complex. This is a residential dwelling with a distinctly public character, which is conceived to also embrace ways of living outside of the traditional family model. The apartments, usually 8 per story, are generally composed of three large rooms. There are four types, two for corner units and two for units located at the core (longitudinal/crosswise). The latter feature a central live-in kitchen, which is equipped with floor-to-ceiling sliding elements for opening onto the glazed loggia, thereby linking the living within the unit with the communal space of the entire building, the courtyard.

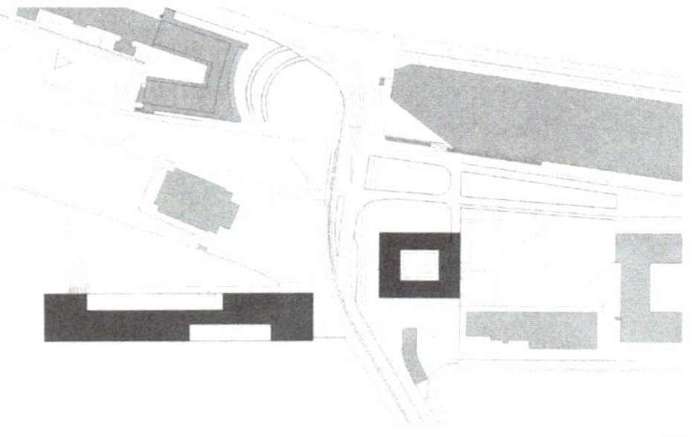

1

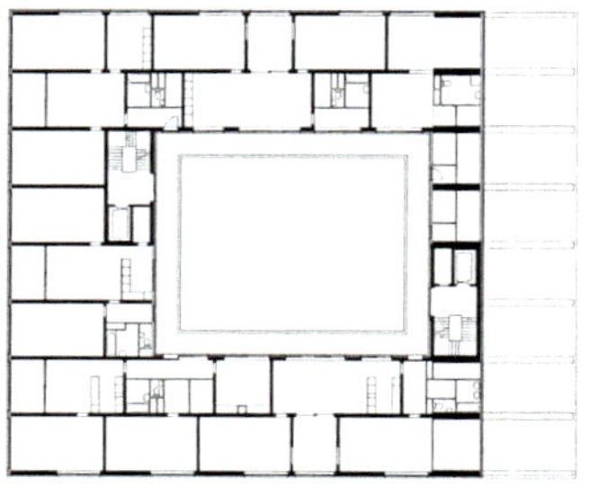

3

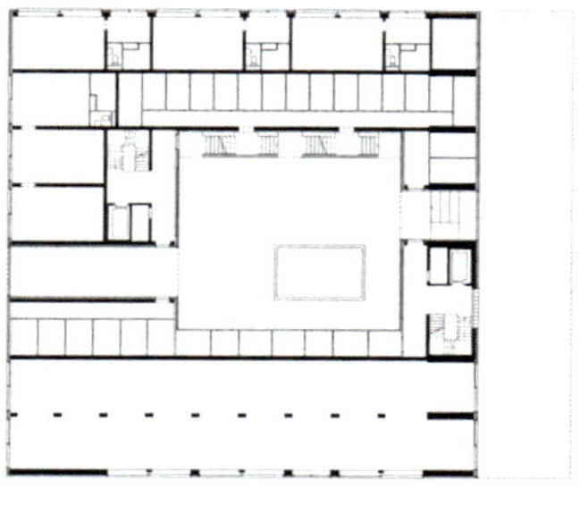

2

1 Site plan courtyard house, long building on peninsula KNSM/Java with connecting pier to Sporenburg
2 Ground floor with shops and commercial spaces 1:1000
3 2nd floor, mixed use (residential/commercial) 1:1000
4 Typical floor plan 1:200
5 Section 1:1000

2.1
SOLITAIRE

Building type
courtyard building with retail and commercial use on ground level
7 stories
facing in all directions

Date of construction
1999–2001

Number of units
45

Size of units
different 3-room and 4-room apartments, 103 m^2–125 m^2

Area per user
26–41.5 m^2

Building depth
34.2 m

Access
covered walkway in interior courtyard,
2 stairwells with lift

Open spaces
glazed loggias
interior courtyard

Parking
180 parking places in underground garage beneath long building

Architect
Diener & Diener Architects
Basel

Location
Amsterdam
The Netherlands

4

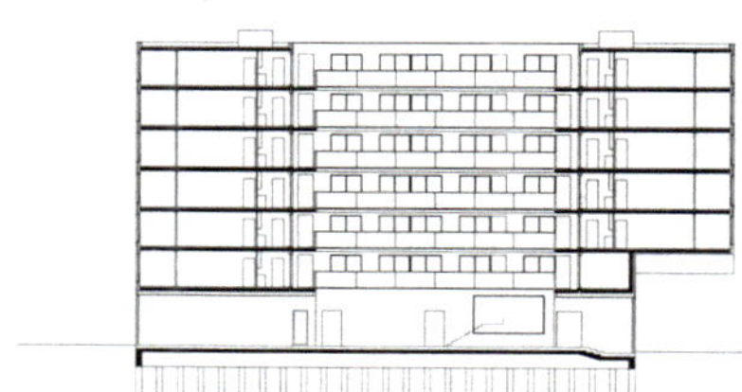

5

Referring to floor plan 1:200

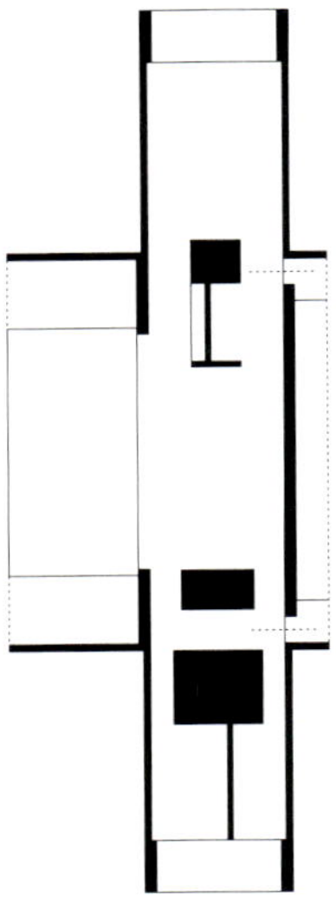

Botania | De Architekten Cie./ van Dongen | 2002

From the outside, the residential block on the water's edge appears uniform and monolithic. The interior courtyard, however, opens the view onto a complex and typologically interesting structure. A series of 33-m-deep "bridge" apartments, which are strung in a row from one exterior facade to the other, are staggered each by one crosswall depth to the rear and by one story to the top, the roof of one unit thus constituting the terrace of the unit above. This cascade of housing units divides the block interior into two distinct sections: as the area of each stepped terrace widens from floor to floor, the width of each atrium-like interior beneath the "bridge" apartments diminishes in equal measure. This sectional configuration is what makes this project so attractive. Parallel to these tiered levels, cascading stairs lead to the wraparound galleries that link all units. The floor plan arrangement of the standard apartments oriented to one side is rather more conventional: the service zone lies parallel to the corridor, next to it is a hallway, which provides access first into the common area and then to the private spaces. The upper units offer roof terraces accessed via the loggias. The 100-m²-large living area of the bridge apartment, on the other hand, is only defined in very broad strokes: a kitchen/WC core divides it into a zone oriented toward the exterior facade and another facing the generous terrace. A glazed loggia completes the interior at the far end and, on the opposite, lie the bedrooms with their own corridor and sanitary zone.

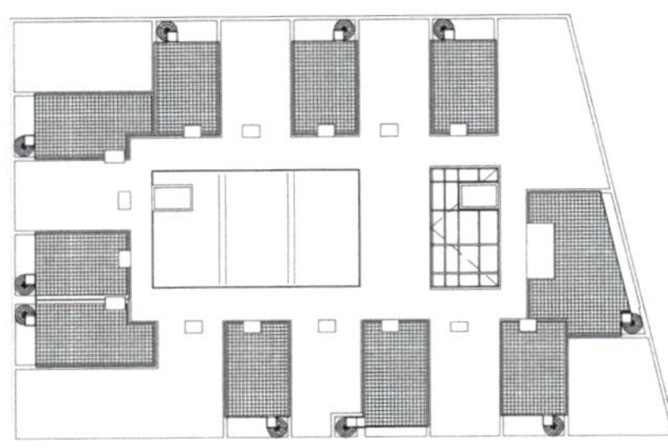

❷

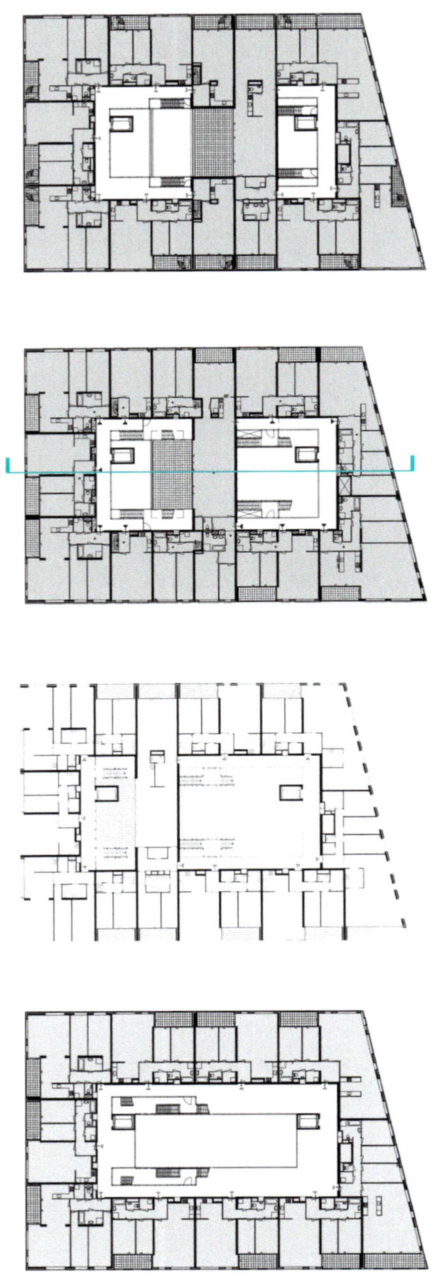

❶

❶ From bottom to top:
overall plan 2nd–5th floor 1:1000
❷ Roof with terrace 1:1000
❸ Longitudinal section 1:500
❹ Floor plan detail 1:200

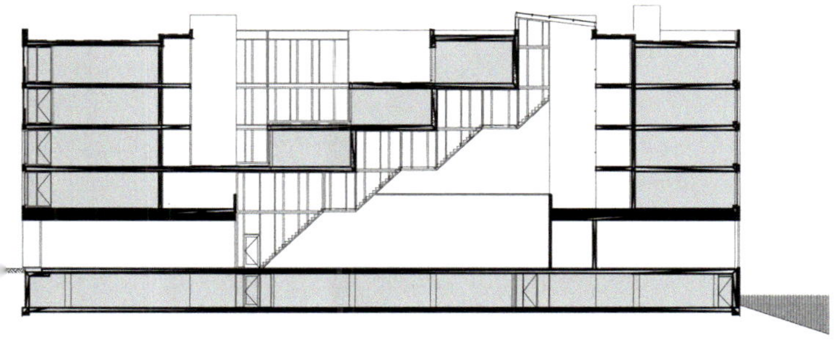

3

2.1 SOLITAIRE

4

Building type
mixed-use apartment building
5 stories
facing in all directions

Date of construction
2002

Number of units
40

Size of units
3-room apts. (9 types)
84/98/100/102/137/155/158 m²
(33 units, of which 4 maisonettes)
4-room apts. (5 types)
108/111/118/120/123 m² (7 units)
in addition: 1000 m² retail area
on ground floor

Area per user
28–52.5 m²

Building depth
33.6 m (crosswise)

Access
2 cascading stairs access inner
covered walkway in the
central circulation lobby

Open spaces
glazed loggias, roof terraces
(roof penthouse apartments)

Parking
underground garage with
24 parking places

Architect
De Architekten Cie./
Frits van Dongen
Amsterdam

Location
Amsterdam
The Netherlands

Referring to
floor plan 1:200

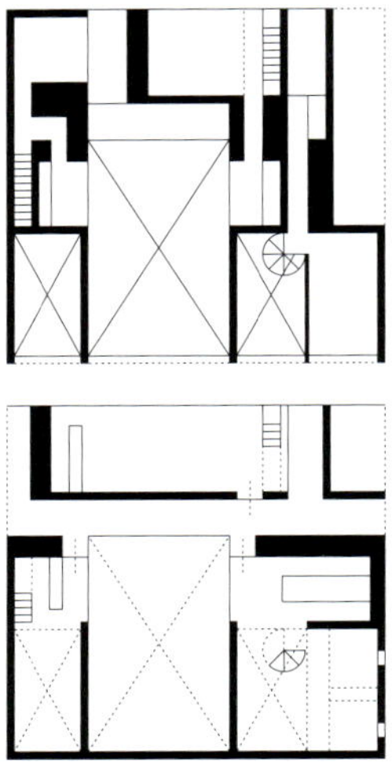

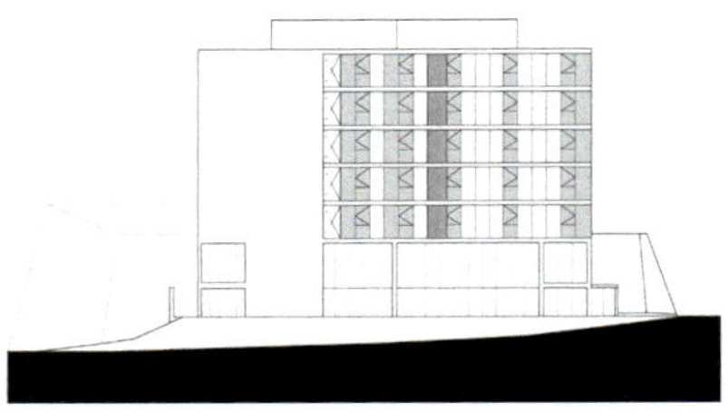

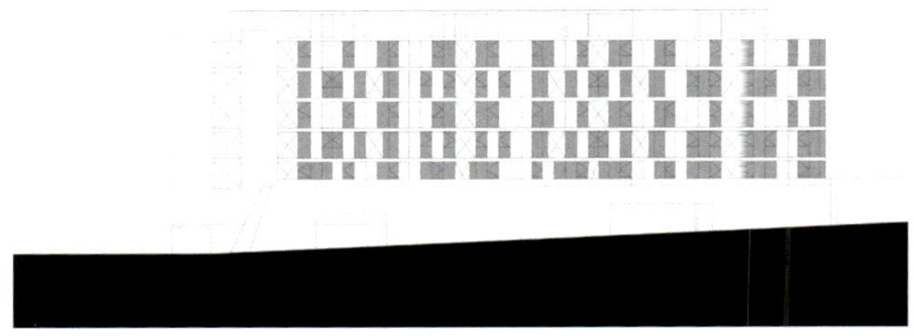

Eda Housing | Chiba | 2005

The dense, typologically interesting complex of buildings in a suburb of Tokyo is located on an open site near the train station. Due to the location, the brief was to seek a structure that would be suitable both for single commuters and for families. The building consists of an envelope and a core, or rather many cores, which are arranged in a chessboard-like pattern in the interior. The envelope and core are linked by an interior corridor, which wraps around the interior in the form of a ring on each second level, thereby providing access to all units. The corridor is lit from both sides, from the outside via the series of loggias, and from the inside through the rhythmic disposition of the cores. The envelope consists of elongated, narrow, bright living areas with attached loggias oriented toward the urban context, while the core zone contains two-story living rooms overlooking the tranquil interior courtyards. The view into the garden from the kitchen – located behind the corridor – is sideways, while windows set high into the wall light the rather more introverted living space from the front. The service zones are housed between envelope and core: the access corridor, or – in those positions where the maisonettes extend across the corridor – the installation cores. The apartments are composed of the different sections of the horizontal space (the envelope) and the vertical space (the cores) into a multiplicity of forms. This fabric of different sections and types is intended to give the occupants a sense of individuality and distance from the next-door neighbour despite the density of the complex.

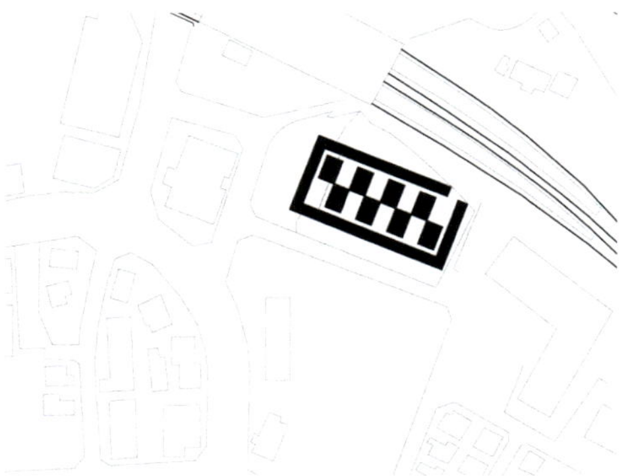

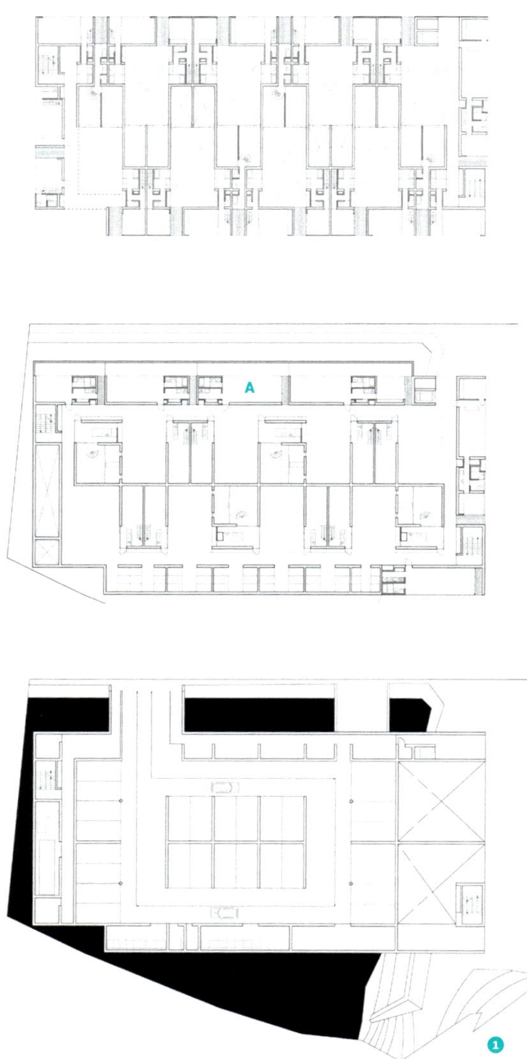

1

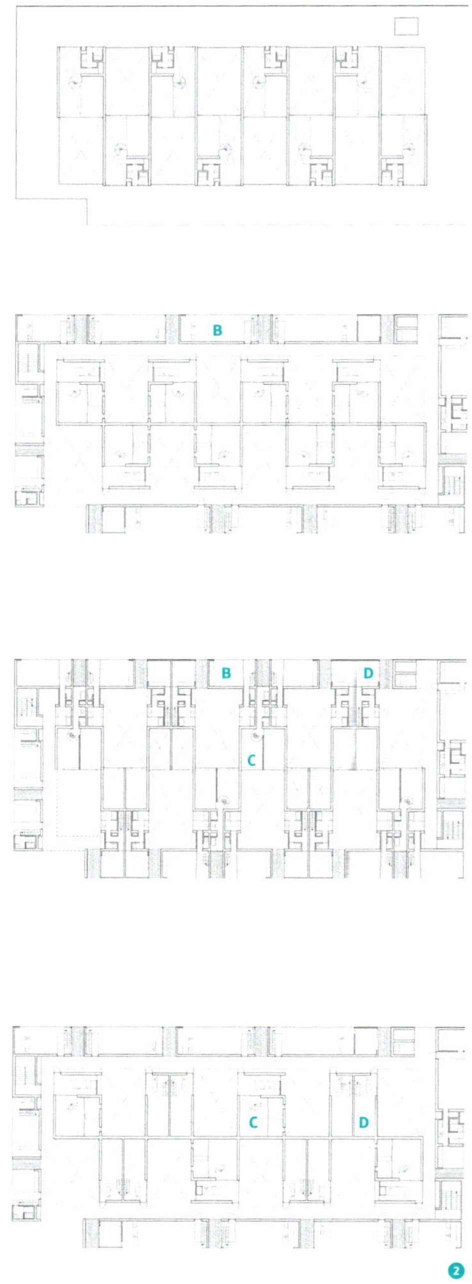

2

1 From bottom to top: 2nd below-ground level with garage, entrance level, 2nd floor 1:1000
2 From bottom to top: 3rd floor–6th floor 1:1000

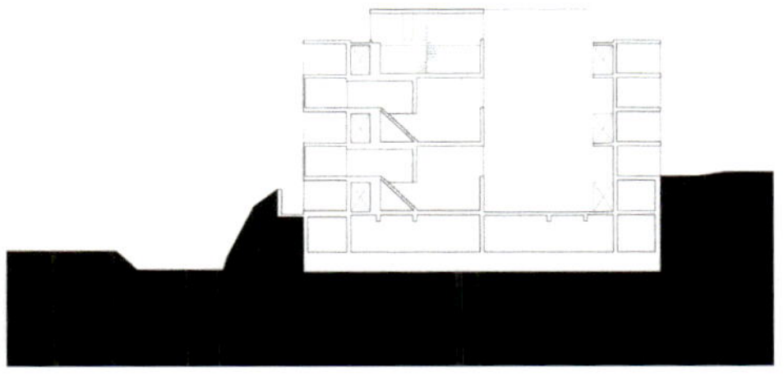

2.1
SOLITAIRE

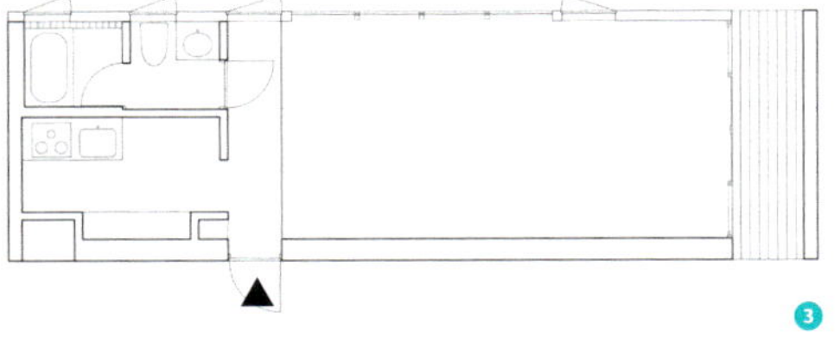

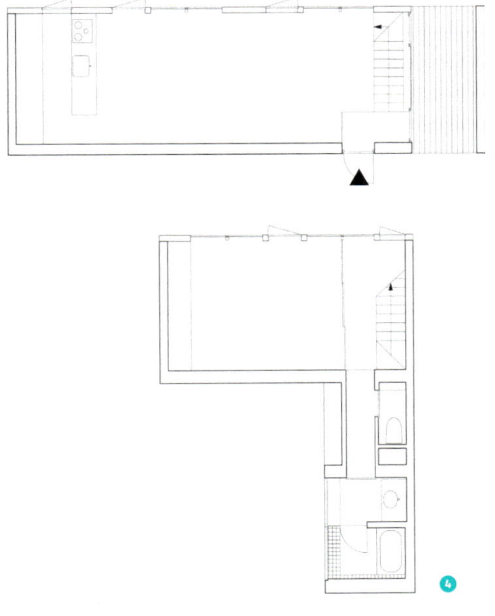

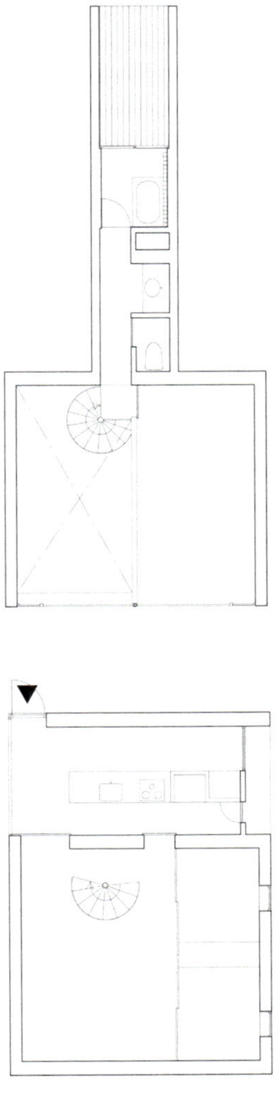

Building type
dense housing and commercial complex
8 stories
facing in all directions

Date of construction
2005

Number of units
62

Size of units
8 different types of flats and maisonettes:
1-room apt. **A**, 29.5/39/44 m^2 (5 units)
2-room mais. **B**, **C**, **F**, **G**,
49.5–54/65/70.5 m^2 (36 units)
2-room apt. **H**, 63 m^2 (5 units)
4-room mais. **D**, **E**, 78.5–80 m^2 (16 units)

Area per user
20–44 m^2

Building depth
29 m

Layout
courtyard building with
chessboard-like core,
vertical division in underground garage,
commercial and residential use,
ring-shaped interior corridor access

Open spaces
private loggias
gardens in interior
courtyard

Parking
28 parking spaces in
underground garage,
plus additional parking spaces
on north side

Architect
Chiba Manabu Architects
Tokyo

Location
Yokohama
Kanagawa, Japan

Referring to
floor plan 1:200

3 1-room flat, type **A** 1:200
4 2-room maisonette, type **B** 1:200
5 3-room maisonette, type **D** 1:200
6 2-room maisonette, type **C** 1:200

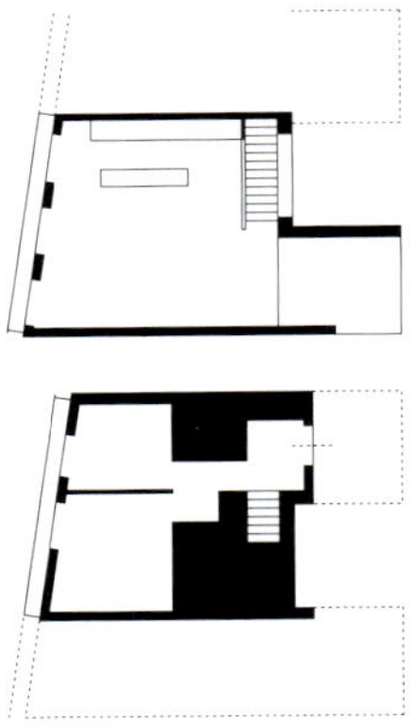

Falken |
Burkard Meyer | 2006

This structure is primarily used as an administration building. Behind the double skin glass facade fitted with uniform drapes, the residential maisonettes located on the two top floors, are barely decipherable as such from the outside. On the elevation overlooking the city, the volume has the appearance of a crystalline and homogeneous sculpture. However, a ring-shaped completion of fourteen maisonettes arranged around a shared courtyard is situated above the office atrium and beneath a cantilevered roof. On one facade, the continuous wooden deck opens up with a two-story loggia to the city; otherwise, this space exudes a tranquility that is almost reminiscent of a cloister. At the entrance to the apartments, two slightly offset lobbies – with the bottom landing of the stairs and the bathrooms to one side and the rooms i.e. apartment doors to the other side – separate the public and private areas of each unit. The open-plan living area is located on the upper level and reached via a set of single-flight stairs on the courtyard side, which defines the kitchen and dining areas. The conservatories, which are almost room size, extend the open-plan living area and are cantilevered over the courtyard. They also shelter the entrances below and give each unit a unique "face". All rooms set along the outer facade benefit from a panoramic view of the city through floor-to-ceiling windows.

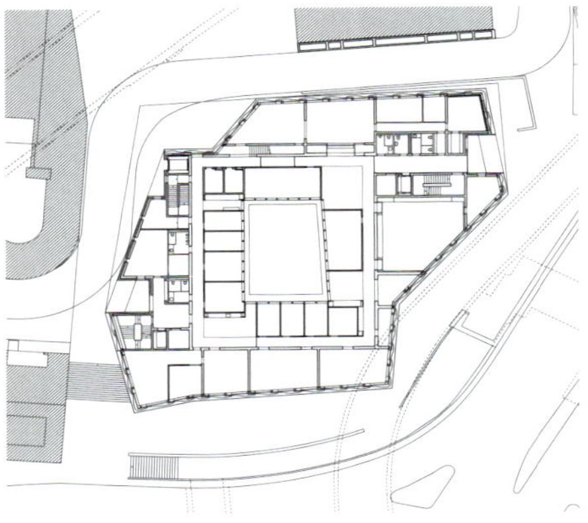

❶

❸

❷

❶ Site plan, 2nd floor with offices
❷ 5th floor: courtyard with entrance level to maisonettes 1:500
❸ 6th floor: living area of maisonettes 1:500
❹ 3-room maisonette apartments, entrance and private rooms 1:200
❺ 3-room maisonette apartments, living area 1:200
❻ Section of atrium and courtyard 1:750
❼ Southeast elevation 1:750

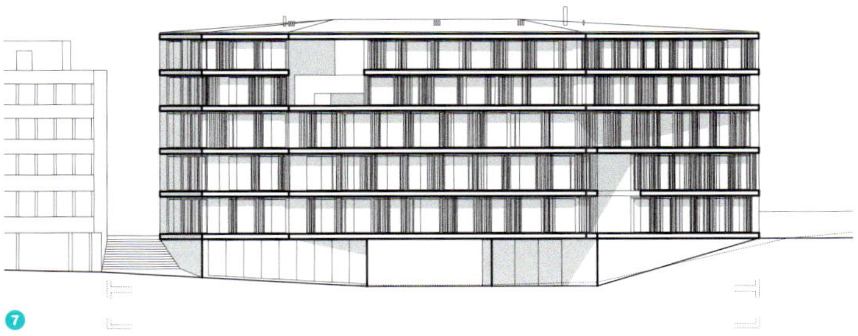

7

2.1 SOLITAIRE

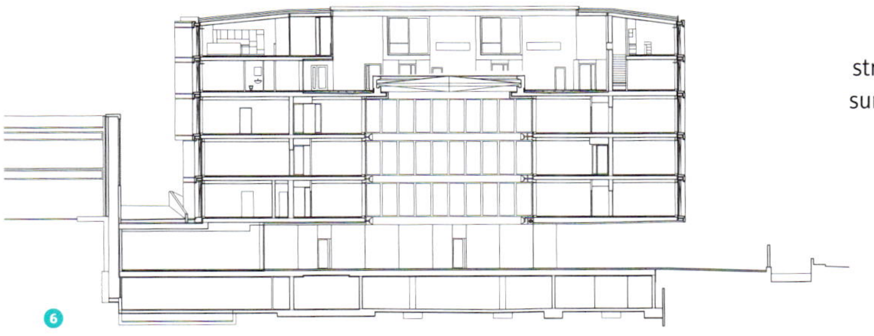

6

Building type
freestanding sculptural urban structure with 2-story maisonettes surrounding a courtyard set on top of a 4-story office wing
6 stories
facing NW/NE/SE/SW

Date of construction
1999–2006

Number of units
14

Size of units
2- and 3-room maisonettes, 90–140 m²

Area per user
30–46.5 m²

Building depth
7–16 m
(residential level)

Access
access core and gallery in courtyard

Open spaces
conservatories, shared courtyard with loggia overlooking the city

Parking
underground garage

Architect
Burkard Meyer Architekten BSA
Baden

Location
Baden
Switzerland

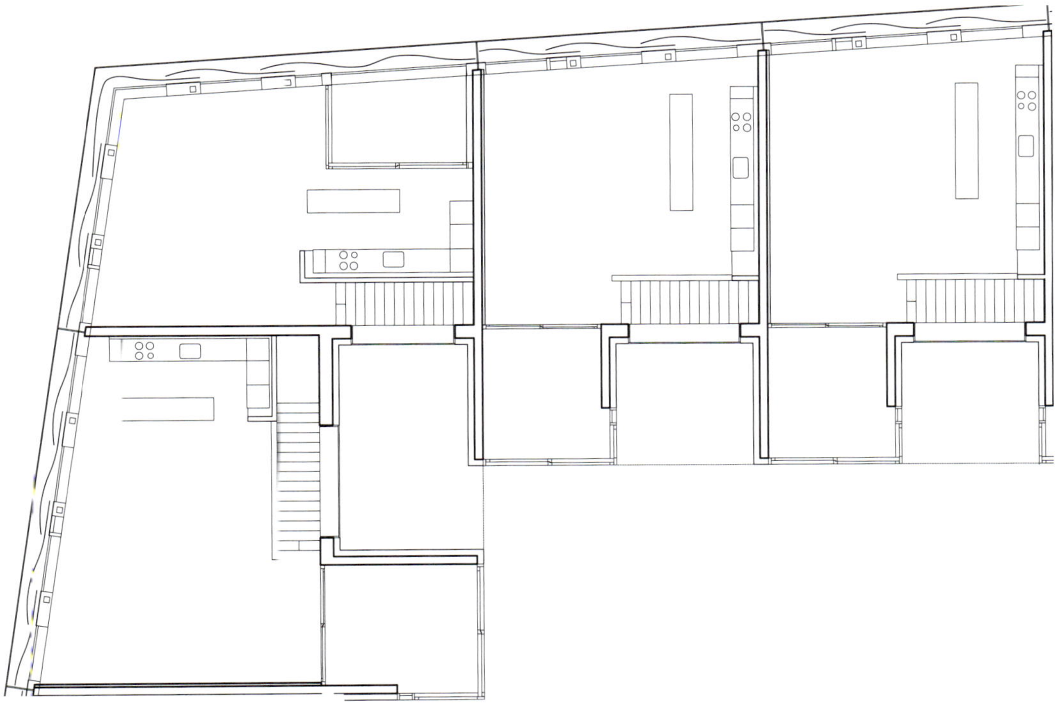

5

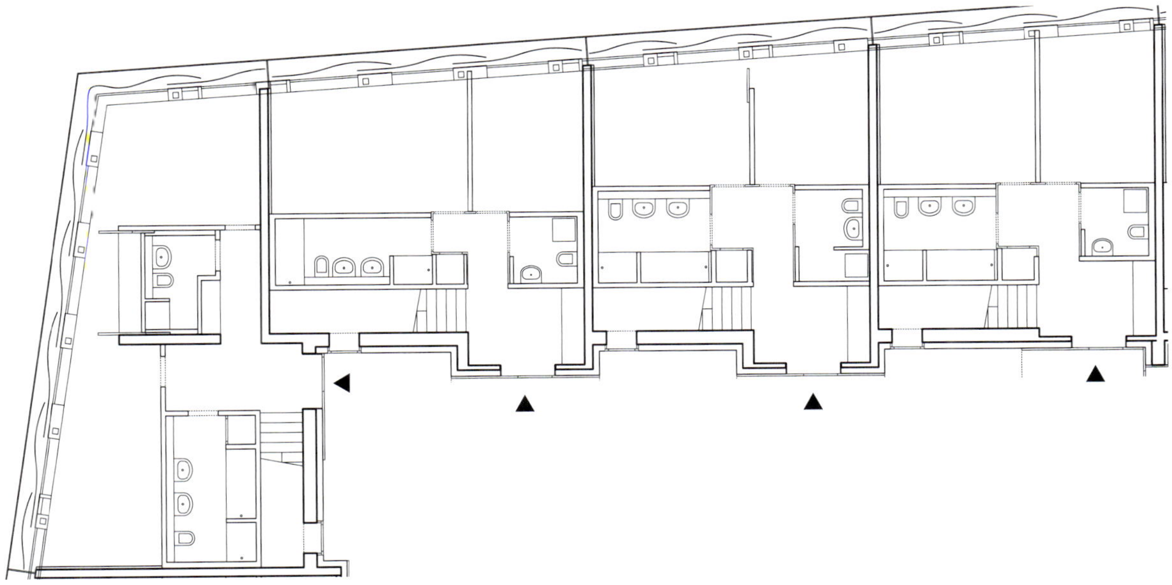

4

Referring to floor plan 1:200

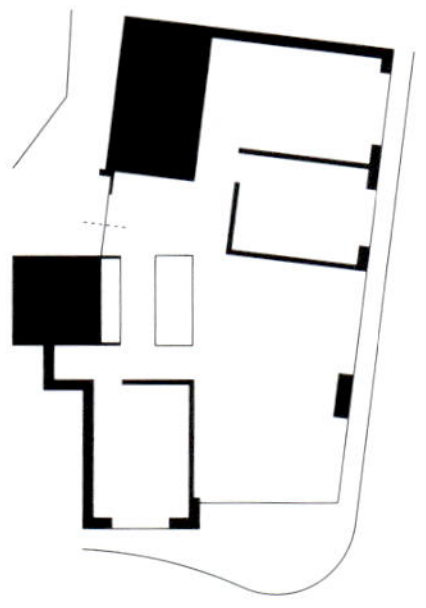

Rondo | Graber Pulver | 2007

The irregular pentagonal form of the building was determined by the shape of the lot. In plan, the building layout is organized into ring-like layers with a generous, top-lit access and circulation atrium at the core. The space-defining cascading stairs within this atrium provide access to up to five units per floor. This strikingly sculptural space functions as a neighbourly intersection and also gives each apartment a unique "address" thanks to the staggered stairs and alternating mezzanine sections. All unit entrances are located in the corners of the atrium; in the apartments' interior, the space opens up, widening like a fan toward the outer corners and extending out onto the rounded balcony bays. The kitchen units are placed near the entrance – the first element in a series of ancillary rooms with bathrooms and storage rooms – offering anyone who enters an immediate focal point much like a counter. Bedrooms follow to the left and right, in part as a kind of private block inserted into the communal space, in part strung along small corridors on both sides. The building is clad in a layer of undulating, flexible metal mesh, which is permeable to light while providing sun protection and a visual screen for privacy.

1

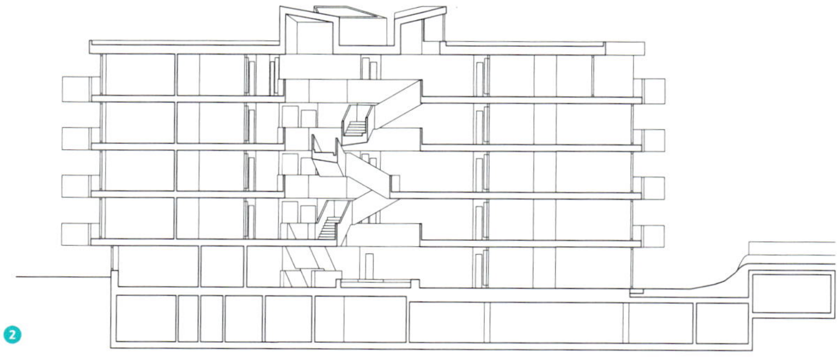

2

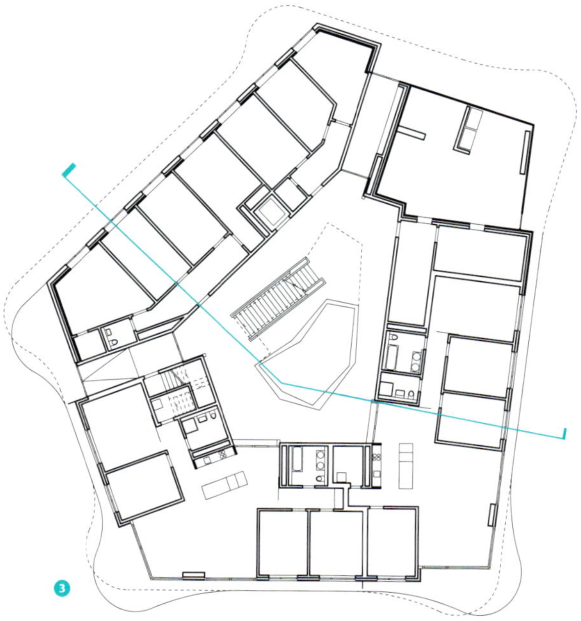

3

4

1 Southwest elevation 1:500
2 Section 1:500
3 Ground floor 1:500
4 Attic story 1:500
5 3rd floor: five corner units 1:200

2.1 SOLITAIRE

Building type
residential building with central atrium
5 stories
facing in all directions

Date of construction
2005–2007

Number of units
22 condominium apartments

Size of units
2½-room apt., 61 m^2 (1 unit)
3½-room apts., 101–109 m^2 (5 units)
4½-room apts., 127–138 m^2 (8 units)
5½-room apts., 142–157 m^2 (8 units)
3 i.e. 4 studios on ground floor

Area per user
30.5–39.5 m^2

Building depth
approx. 22–34 m

Access
central atrium

Open spaces
wraparound balconies, widening into rounded bays at the corners

Parking
underground garage

Architect
Graber Pulver Architekten AG
Zurich/Bern
project manager:
Alexander Huhle
team:
Susana Elias Robles
Yvonne Urscheler Lofteröd
Marcel Weiler

Location
Zurich
Switzerland

5

Referring to floor plan 1:200

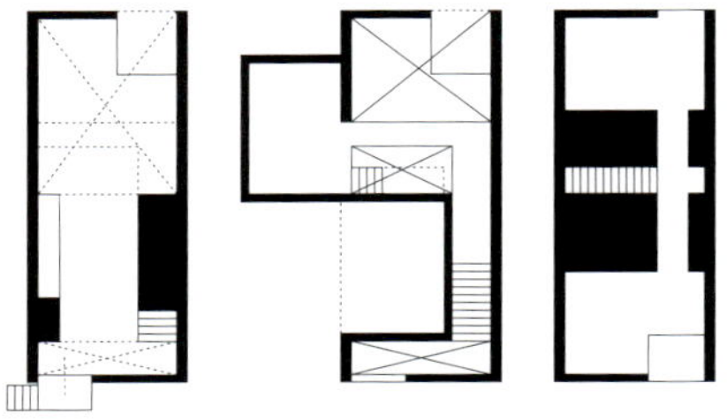

Willoughby 7917 | LOHA | 2008

Six maisonette apartments, front-to-back across all three levels, feature two-story open spaces on both sides and terminate in a floor entirely reserved for individual rooms on the third level. However, what sets this building apart is the transformation of the standard layout of comparable maisonettes: on the mezzanine level, the interior is liberated from the corset of the two dividing walls and is pushed into the adjacent apartment. On the one hand, this has the effect that the open living space is organized into three areas – soaring across two stories height at the two facades and low in the middle, where kitchen and WC are located. On the other hand, this solution creates a unique space on the mezzanine between the open living area below and the closed, predetermined individual rooms above, with sightlines in all directions and suitable for a variety of flexible uses. The cave-like penetration into the adjacent unit creates a complex floor plan that defines the entire volume. Two apartments are combined into a unit, in which the inserted rooms are oriented in opposite directions on one or the other side of the building. In this manner, the living areas too change in orientation with the inserted open patio on one side. The two penthouse apartments occupy the entire length of the building and are entwined in a meandering manner. Each unit is oriented toward two sides and surrounds one, i.e. two patios. The penthouse units are accessed from a shared roof patio, while a second entrance from the fire escape allows for divisions across the middle of the plan.

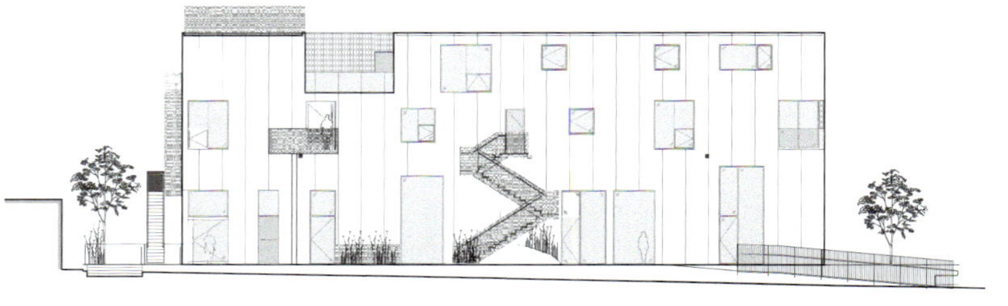

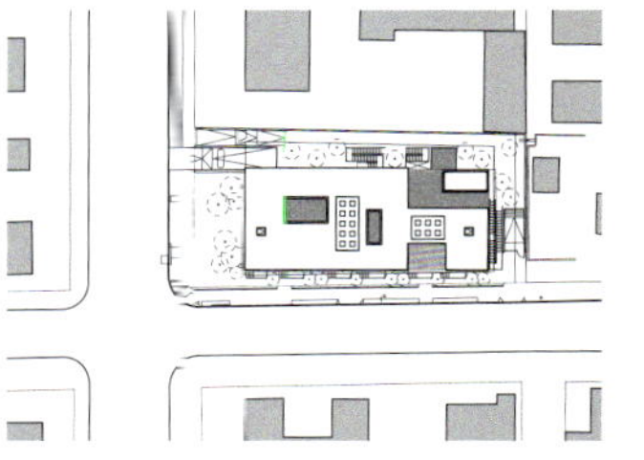

2

1

1 Design variant (not realized): floor plan layout for ground floor, 2nd floor, 3rd floor, attic story 1:500
2 South elevation (realized) 1:500
3 Realized total floor plans: ground floor, 2nd floor, 3rd floor, attic story, roof aspect 1:500
4 Two intertwined triplex units on mezzanine level: ground floor, 2nd floor, 3rd floor 1:200
5 Longitudinal sections of apartment stairs and inserted rooms 1:500
6 Cross section 1:500

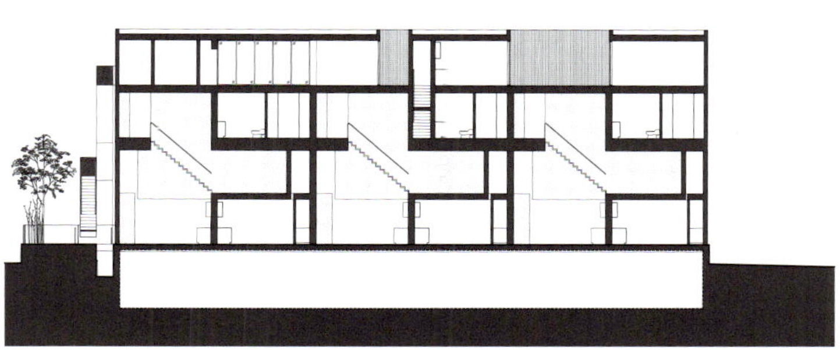

5

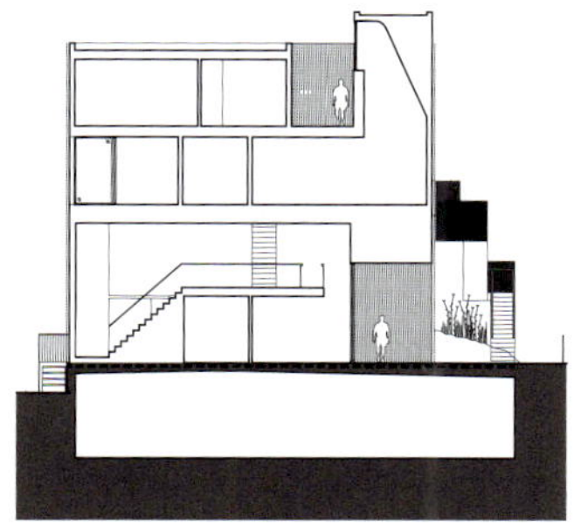

6

2.1
SOLITAIRE

Building type
freestanding slab
4 stories
facing N/S

Date of construction
2008

Number of units
8

Size of units
4-room maisonettes on 3 levels, approx. 156 m² (6 units)
penthouses, approx. 146/148 m² (2 units)

Area per user
39–49.5 m²

Building depth
13.8 m

Access
maisonettes with direct ground-floor access, penthouse access via external stairs and elevator

Open spaces
2-story loggias
roof patios

Parking
no parking on lot

Architect
LOHA
(Lorcan O'Herlihy Architects)
project architect:
Pierre De Angelis (PD)
Sabrina Schmidt-Wetekam

Location
7917 Willoughby Avenue
West Hollywood
California
USA

3

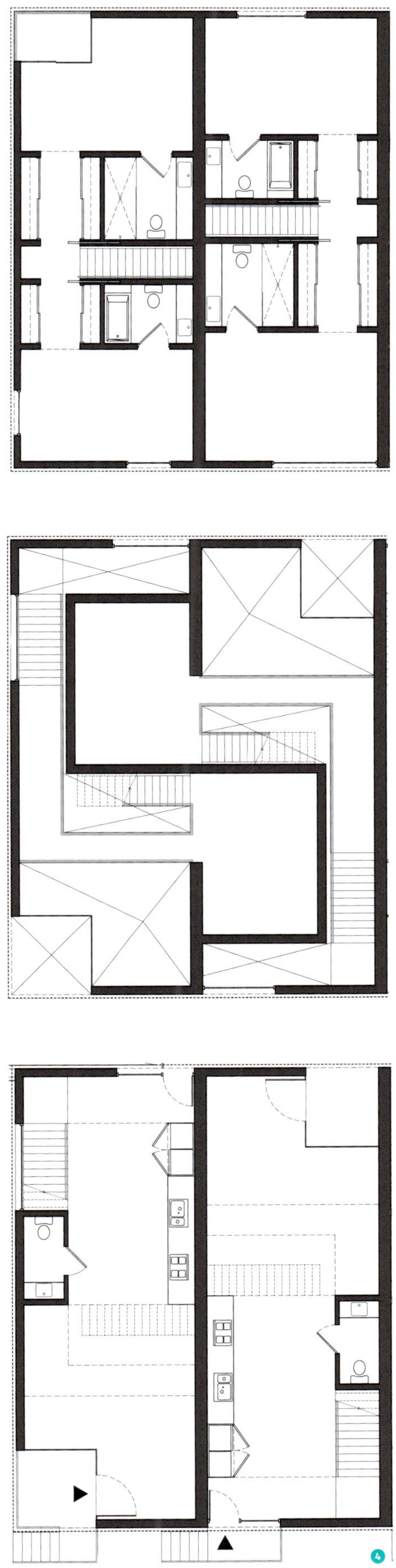

4

Referring to floor plan 1:200

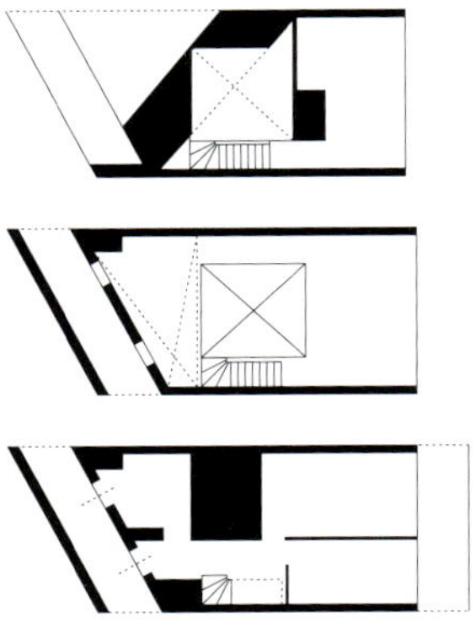

Funen Blok K | NL Architects | 2009

The complex is one of sixteen blocks, all of which were to adhere to a common rule: to rise to two-and-a-half stories on a nearly square plan; the top level was to be half living area and half roof terrace and garden. The inventive interpretation of these stipulations forms the parameters for this design.

The volume is divided into ten units of 633 m³ each, accessed via a diagonally inserted "minicanyon." The units are turned inside out; in this manner, corridors, stairs and storage rooms do not take up valuable facade areas, instead the large rooms benefit from all available natural light. The undulating roof, with its extreme height fluctuations between 5 and 15 m, nevertheless achieves the stipulated two-and-a-half stories on average. As a result of the diagonal section, the different widths, and the roof profile, the units have very different cubatures. Although all have the same spatial volume – units with shallow depth have high rooflines and vice versa –, apartment size and number of stories do vary. What is unique, however, is that the interplay of these parameters creates very different room layouts, each of which has special characteristics: tall and narrow i.e. low and deep rooms, roof terraces accessed from a patio on a lower level, there are living rooms that soar towards the light, with galleries along the facade, or patios that are pushed into the double-height space. The roof terraces perforating the undulating green carpet of the roof surface are the defining element – as a fifth elevation visible from all the directions, from the apartments as well as from the city.

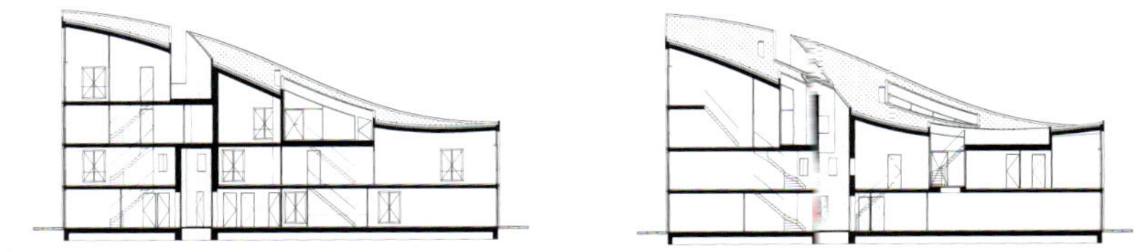

1

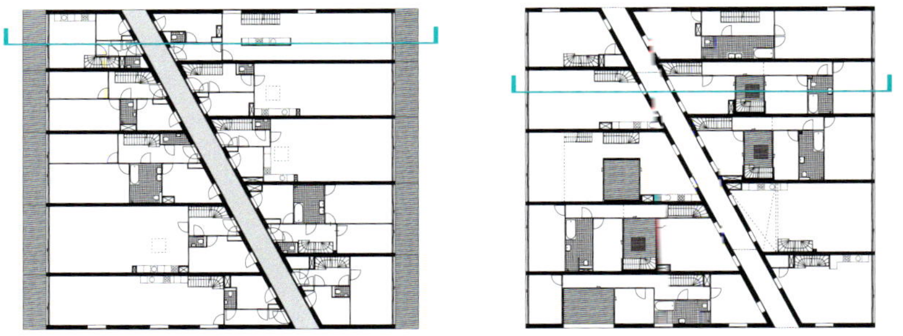

2

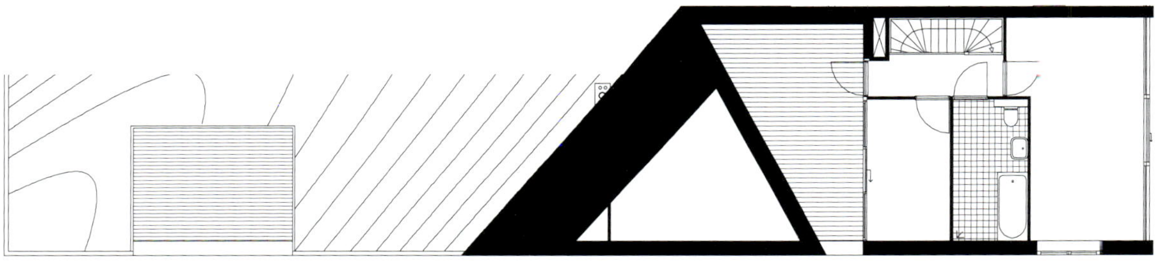

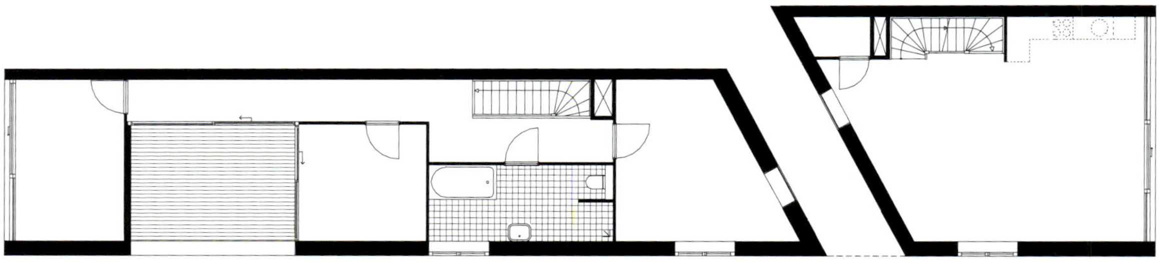

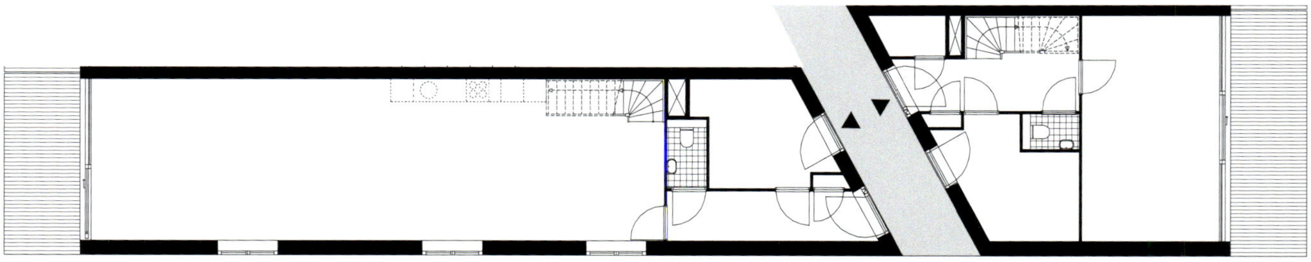

3

1 Sequence of long. sections with "minicanyon" 1:750
2 Total floor plans: ground floor, 2nd floor, 3rd floor (on two sectional levels), roof 1:750
3 Different units with 2- i.e. 3-story, 4-room apartments 1:200
4 West and south elevation 1:750

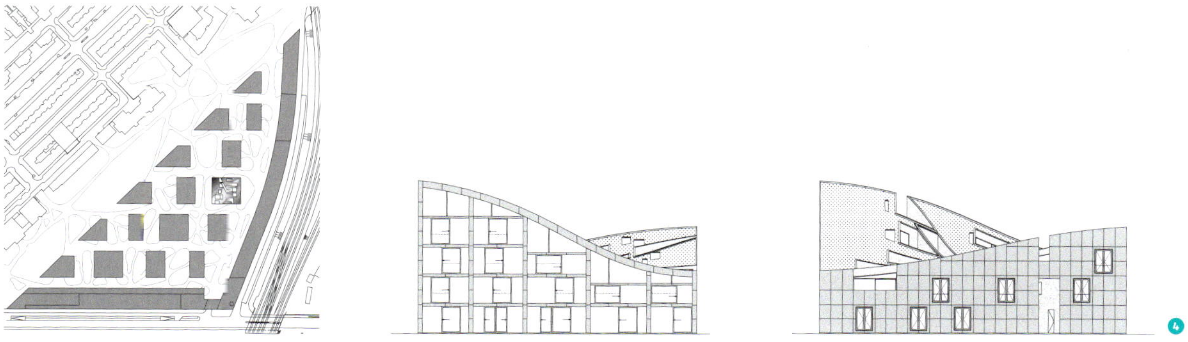

2.1 SOLITAIRE

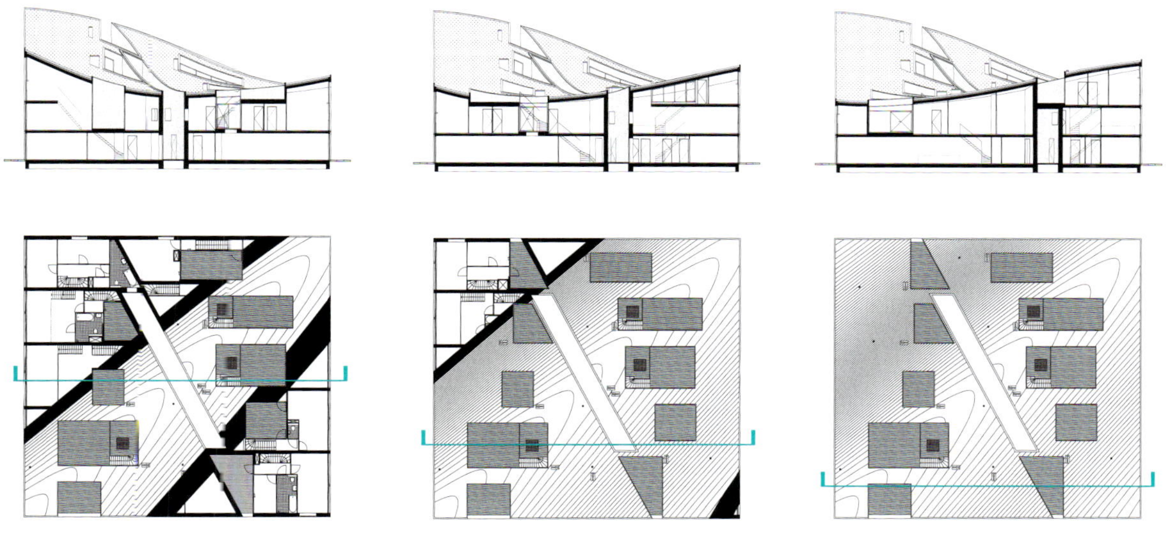

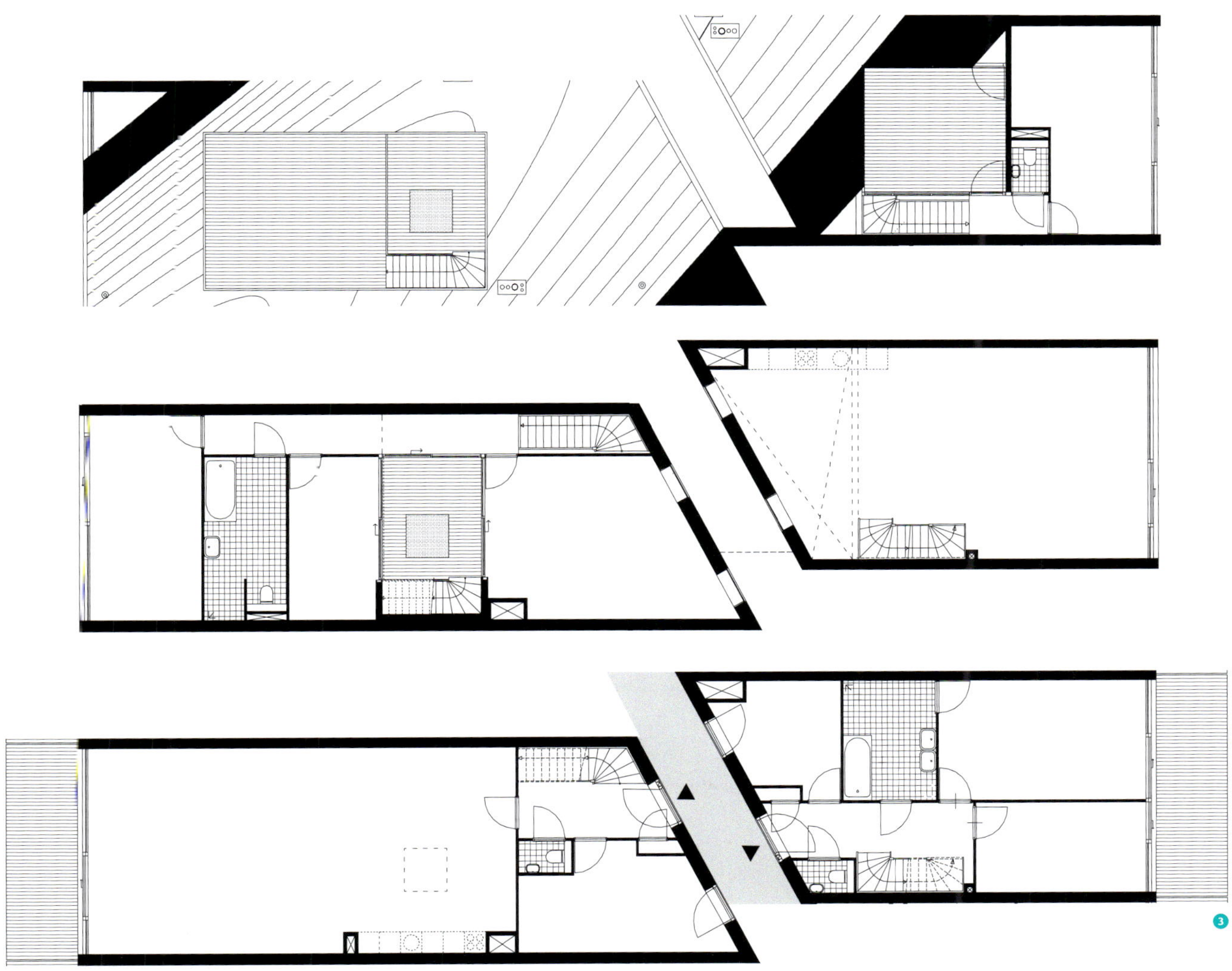

Building type
apartment block divided into 10 units with different unit depths and number of stories, undulating roof landscape, 2.5 stories on average, facing N/E/W/S

Date of construction
2006–2009

Number of units
10

Size of unit
3-room apts., 147/160 m² (2 units)
with patios 158/170.5 m²
4-room apts., 127–180.5 m² (7 units)
with patios 142.5–204 m²
5-room apt., 142 m² (1 unit)
with patio 153 m²

Area per user
28.5–53.5 m²
(with patios 30.5–57 m²)

Building depth
27.7–30.5 m

Access
diagonally inserted aisle ("minicanyon")

Open spaces
roof terraces

Parking
no parking on lot

Architect
NL Architects
Amsterdam
Pieter Bannenberg, Walter van Dijk
Kamiel Klaasse, Mark Linnemann
Associates:
Caro Baumann, Jennifer Petersen
Niels Petersen, Holger Schurk
Misa Shibukawa,
Rolf Touzimsky

Location
Amsterdam
The Netherlands

Referring to
floor plan 1:200

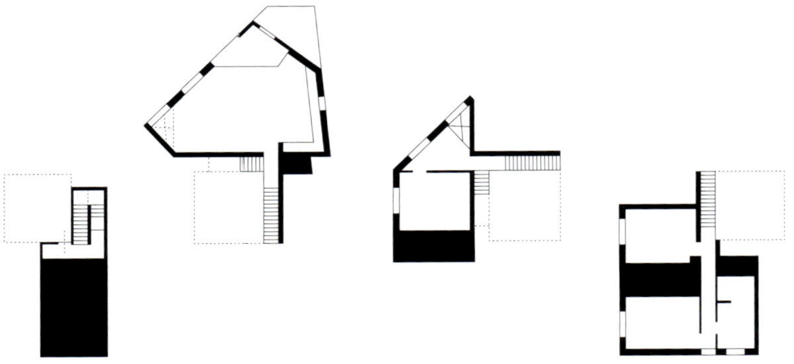

Urban Villa 4 in 1 | 2b architectes | 2011

The crystalline shape of this urban villa follows the geometry of the plot and is embedded in the terrain, surrounded on three sides by a retaining wall. On the fourth side, a sunken level leads to a four-story glazed atrium belying the unique structure of this building and forming a common point of reference for all residents through which each unit can be accessed, one per side. The apartments are laid out over multiple levels connected by an internal staircase running from one corner to the next, wrapping around the central atrium like superimposed spirals. This interlocking relationship combines the advantages of shared living within a dense urban form with the conceptual characteristics of a freestanding home. The rooms of each unit are distributed over all levels throughout the building, providing views in all directions. Each unit has its own garage next to the entrance and a private garden plot.

The organizational uniformity in combination with the complex geometric shape facilitates a wide range of sizes, functions, and spatial situations. This results in double-height rooms, voids in the building corners where rooms overlap, as well as apartments that distribute eat-in kitchens and living spaces over two levels or connect them within a single room. One of the units is used as an office.

On the facade, the inner layers of the building are expressed by the varied finishes of the concrete surfaces accentuating the outline of each apartment, whereas the sculptural character of the entire building is emphasized by a free composition of openings.

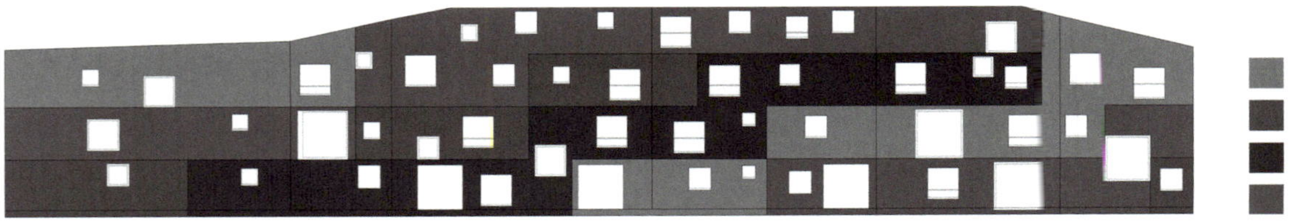

1

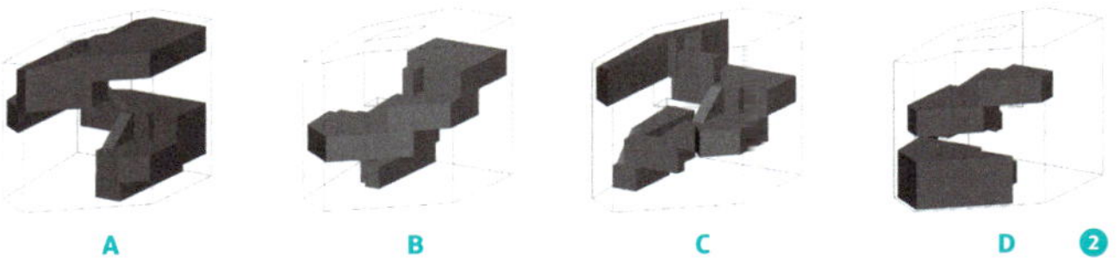

2

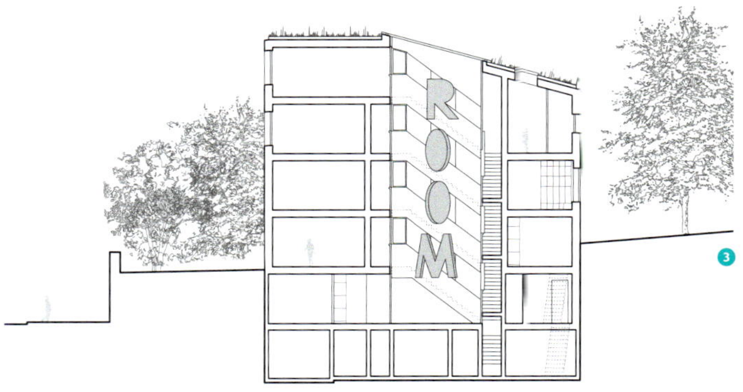

3

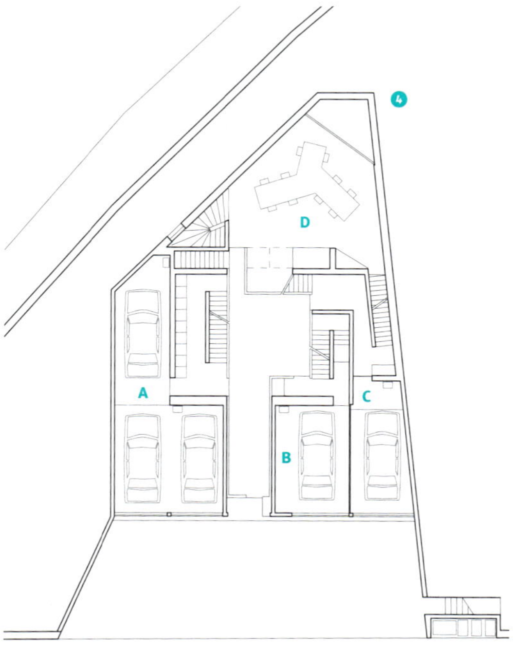

4

5

1 Facades with accentuation of the four residential units
2 Apartment volumes
A: 7½-room apt.
B: 5½-room apt.
C: 5½-room apt.
D: 4½-room apt.
3 Section 1:400
4–8 Ground floor, 2nd – 5th floor 1:400
9 **B1** – **B4** correspond to ground floor, 2nd-4th floor 1:200

2.1
SOLITAIRE

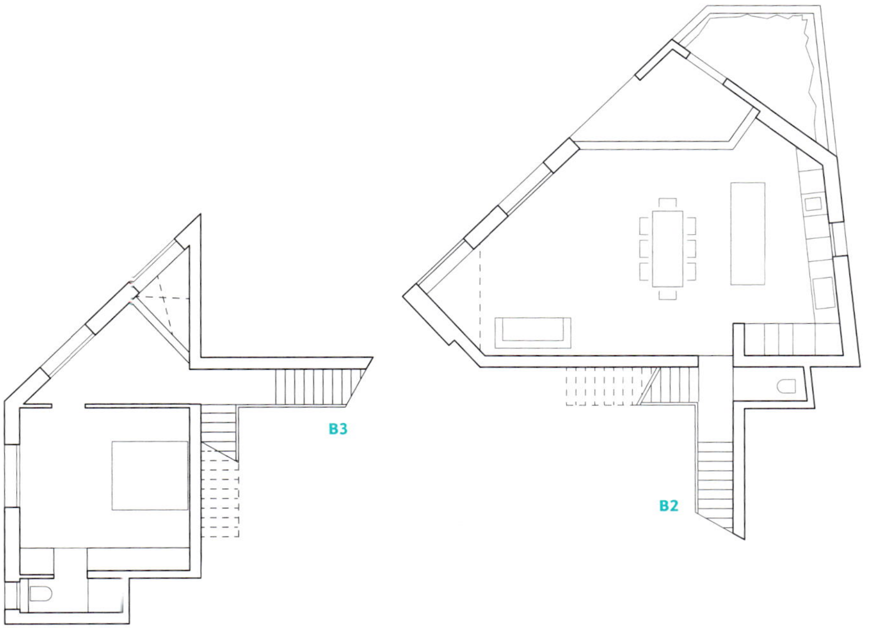

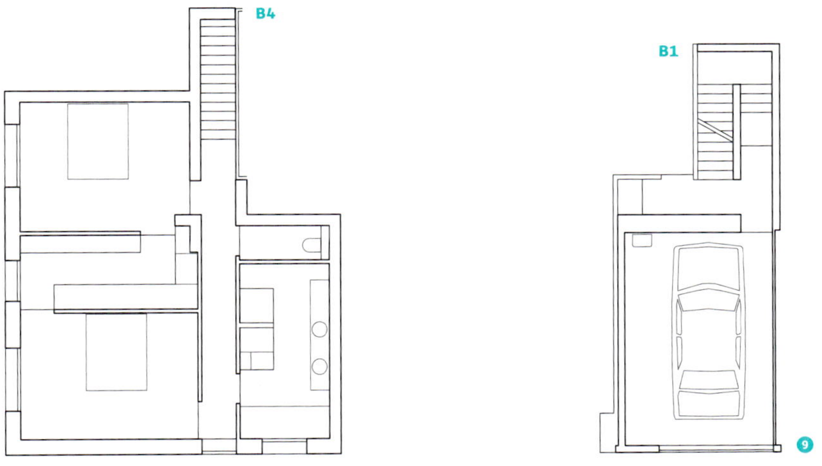

Building type
sculptural multifamily building
on a slope in villa district
3–5 stories
facing in all directions

Date of construction
2009–2011

Number of units
4

Size of units
A: 7 ½-room apt., 288 m²
B: 5 ½-room apt., 182 m²
C: 5 ½-room apt., 168 m²
D: 4 ½-room apt., 151 m²

Area per user
34–41 m²

Building depth
12–21.4 m

Access
apartment entrances and staircases on
each side of a shared atrium

Open spaces
loggias, garden plots

Parking
individual garages

Architect
2b architectes
project team: Ph. Béboux, S. Bender,
T. Borges, G. Warnking, C. Jung,
G. A. Lili, M. Baumgartner, F. Ben-Amor,
J. Blanc, A. Clement, P. Gabrielli,
S. Grimm, E. Quaresina
Lausanne

Location
Avenue de Beaumont 22a
Lausanne
Switzerland

Referring to
floor plan 1:200

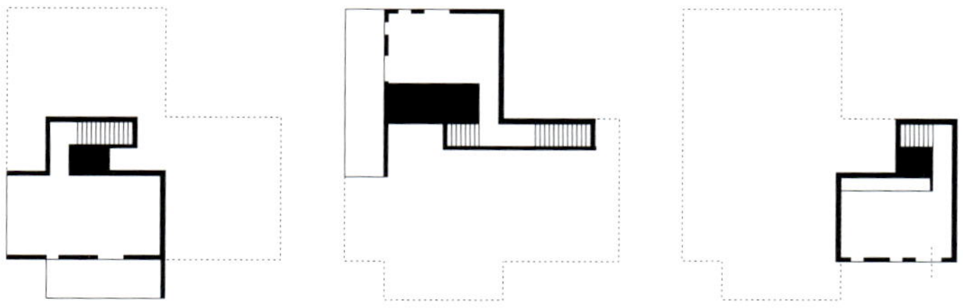

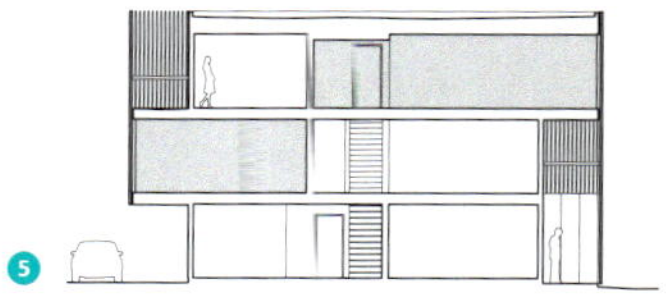

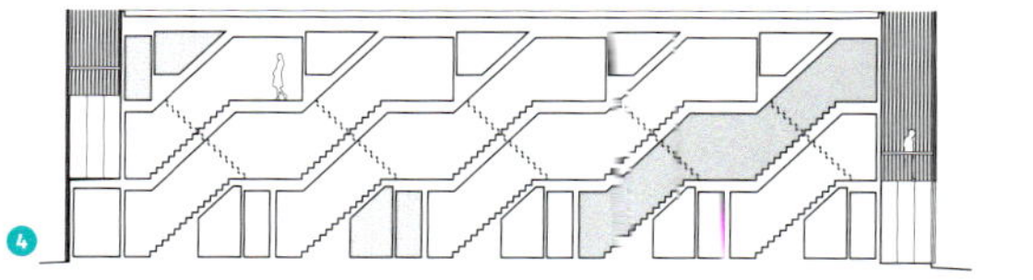

Bloc_10 | 5468796 Architecture | 2011

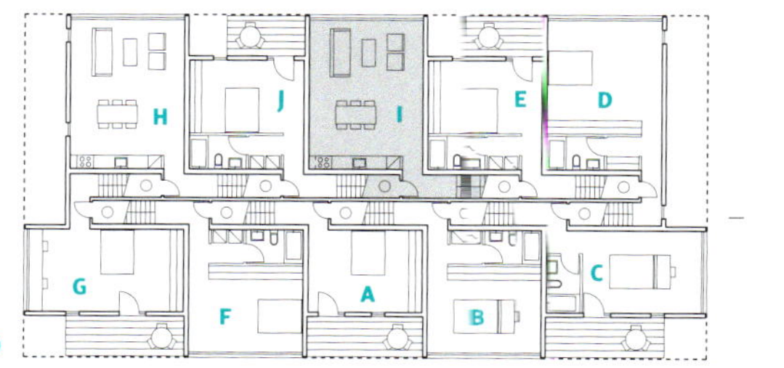

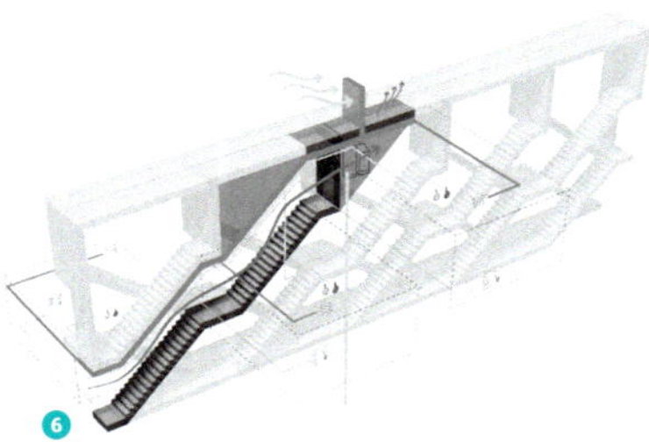

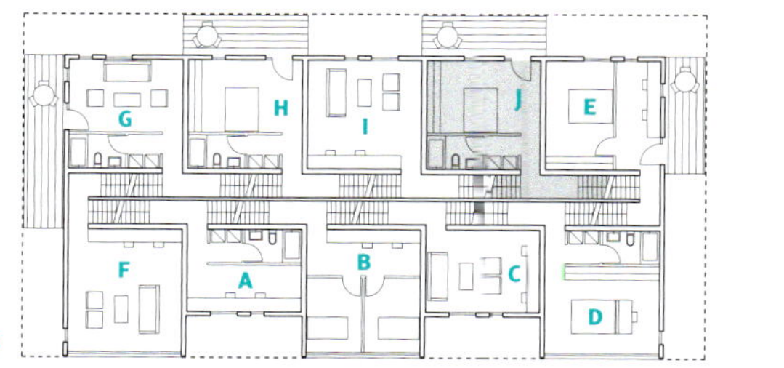

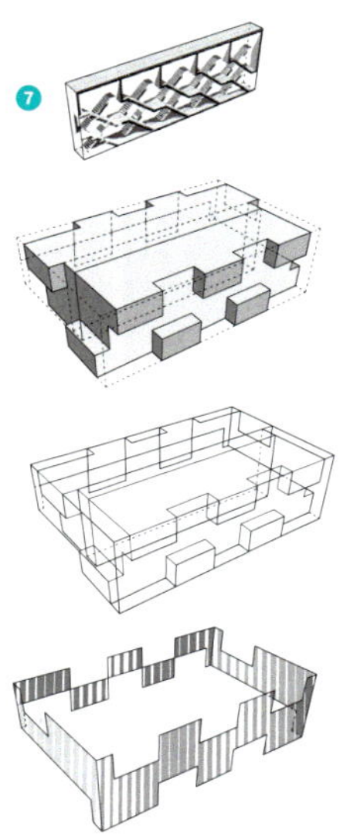

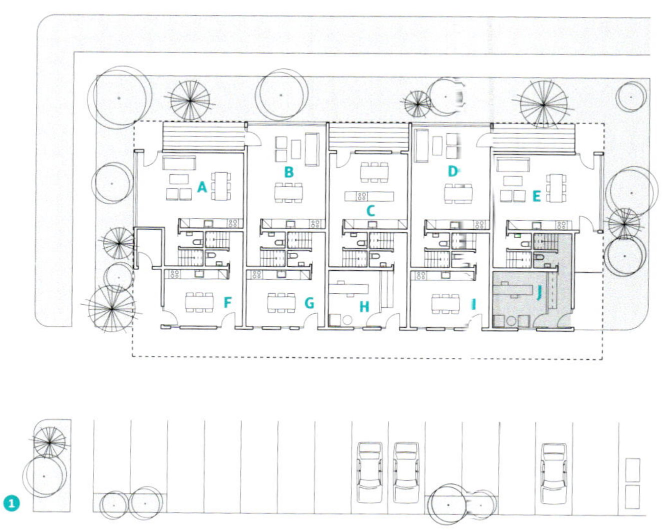

The apartments in this project were designed according to the "white box" principle: an unfinished unit with the essential sanitary, heating, and electric installations is purchased, after which each buyer can then determine the surface materials based on personal taste, the available budget, and how the spaces on the three levels will be used. Office spaces can for instance be built out on the ground floor, while the living and dining rooms are spread over two floors or combined. The top floor can be used as either a living room or bedroom.

In contrast to the standard convention where each level of a residential unit is directly above the other, or where all rooms are spread out on a single floor, the three levels of each apartment spring from one side of the building to the other, forming a linear cascade of spaces from one building segment to the next. They are connected by a series of interior staircases above and behind one another, which form a sort of backbone down the midsection of the entire building. The result is very different unit types with surprising spatial organization and exterior views.

Complementing the inner access core – integrating the technical infrastructure that enables the flexible placement of sanitary rooms within each apartment segment and on all levels – and the 3-story units, which wind their way around one another, is another spatial layer. The difference between the permitted building volume and the interior volume is used to create a series of cut-out open spaces clad with wooden lamellae across most of the facade surface. Depending on the point of view, the building looks either like a solid sculptural object or an open structure.

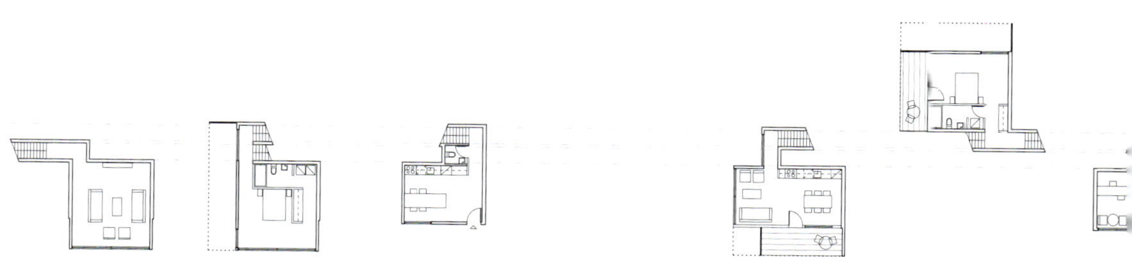

1 Ground floor 1:500
2 2nd floor 1:500
3 3rd floor 1:500
4 Longitudinal section with interior staircases above and behind one another 1:500
5 Cross section 1:500
6 Access section with integrated building services
7 Building structure
8 Floor plan options 1:500
9 Apartment J 1:200
10 Apartment G 1:200

2.1
SOLITAIRE

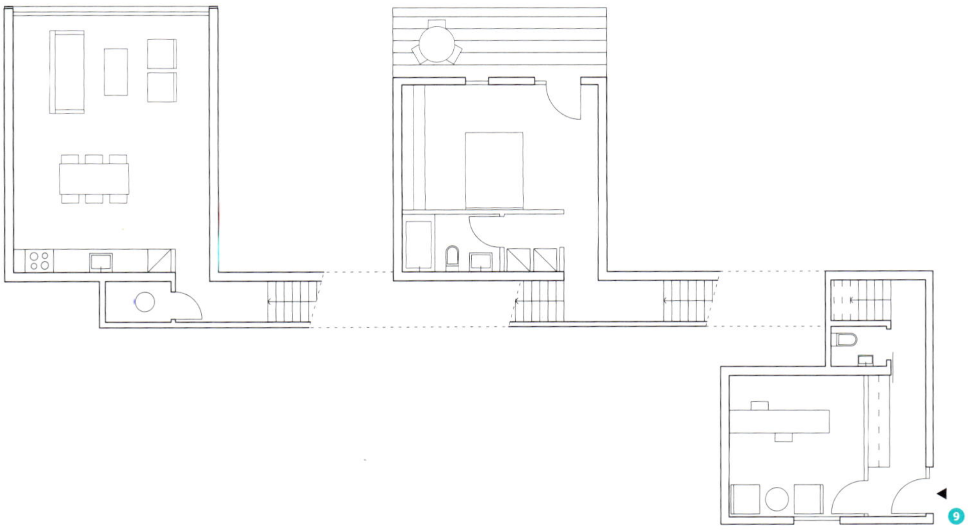

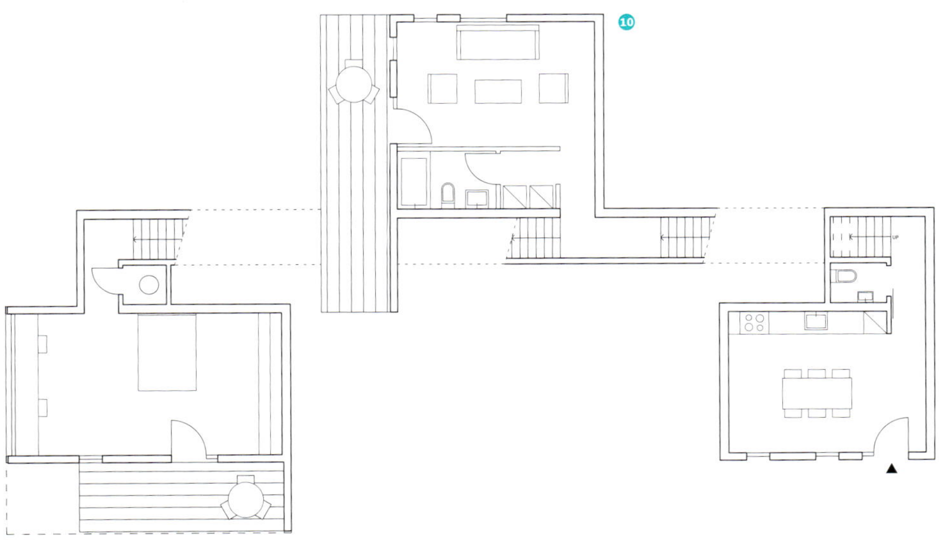

Building type
free-standing slab on corner lot of an open block with detached houses
cut-out open spaces behind facade lamellae
3 stories
facing in all directions

Date of construction
2011

Number of units
10

Size of units
3- to 4-room apts., 88–118 m²

Area per user
36–59 m²

Building depth
13.6 m

Access
direct street and courtyard access to each unit, middle section, staircases inside apartments arranged on top of and behind one another

Open spaces
terraces, loggias, balconies

Parking
open parking spaces behind building

Architect
5468796 Architecture
Winnipeg

Location
1530 Grant Ave.
Winnipeg
Canada

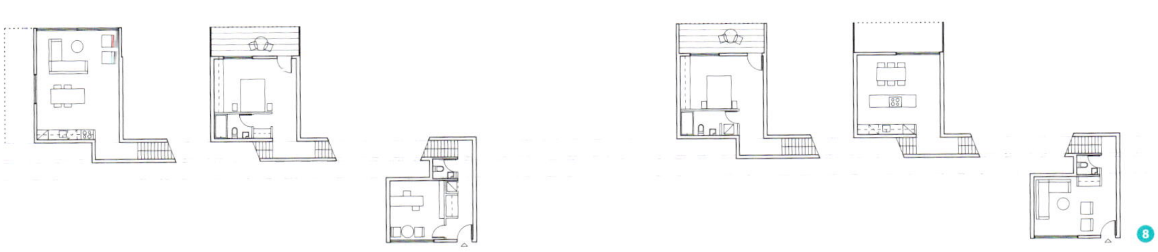

Referring to floor plan 1:200

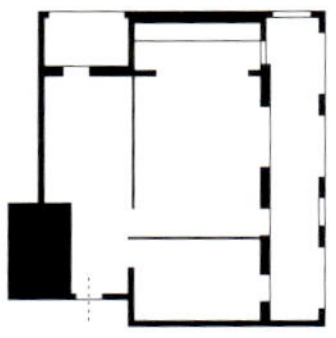

86 Dwellings | Eric Lapierre Architecture | 2013

Located on a corner lot of an important east-west thoroughfare in Lyon, two solitary volumes are offset from one another to form two small squares – one towards the heavily trafficked intersection, the other towards the quiet block interior. The floor plan of the 20 m deep volumes is characterized by several concentric spatial layers wrapped around one another: a narrow access corridor is located in the middle axis followed on both sides by a layer containing the stairwells, elevators and entrance halls, and bathrooms of the individual apartments. Following this is a ring of connected living spaces with predominantly two access points each, opening up their potential use to interpretation. Surrounding these interior layers is a double-skinned facade, which while providing a traffic noise buffer also imparts the buildings with their monolithic character. Loggias and the kitchens, which remain open to the living space, occupy the space within the double facade.

Within this strict organizational system, various divisions and connections, set either parallel or perpendicular to the structural layers, allow the creation of differently sized apartments and floor plans: small and large apartments, duplex units, some of which have double-height living rooms and loggias, and on the top floor, apartments with patios located on the buildings' perimeter. On the sides of the patios, small spaces are created between the double facades, with views in almost every direction.

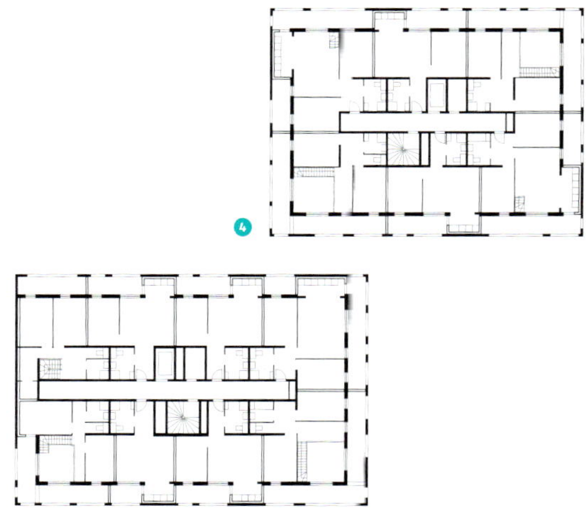

4

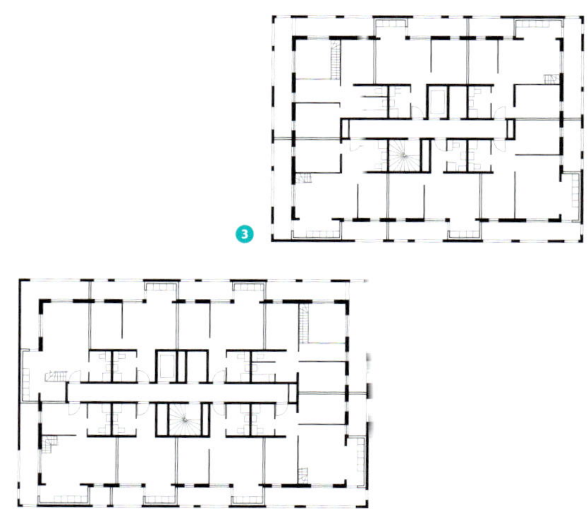

3

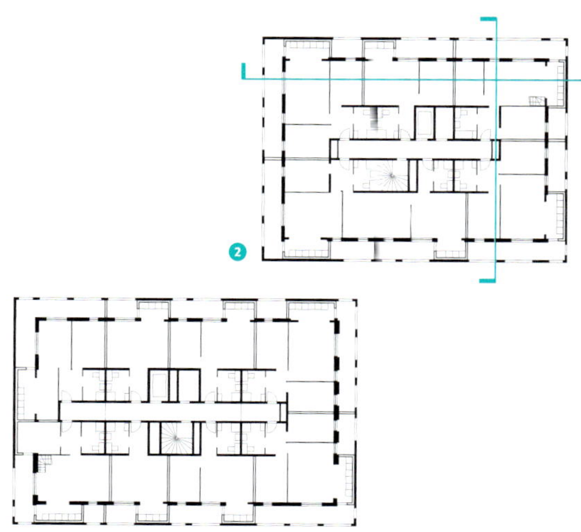

2

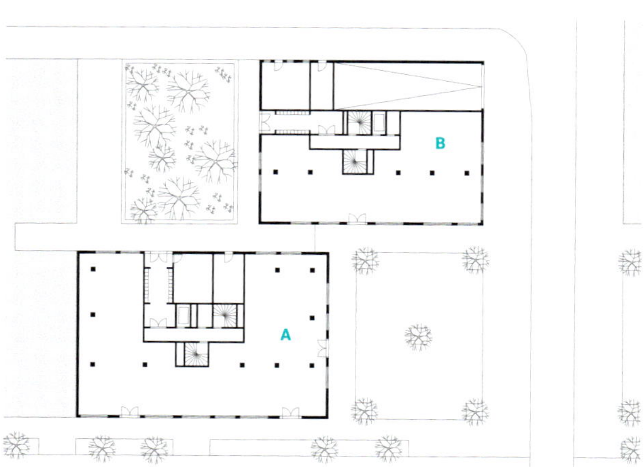

1

1 Total floor plan, ground floor 1:1000
2 Total floor plan, 4th floor 1:750
3 Total floor plan, 6th floor 1:750
4 Total floor plan, 7th floor 1:750
5 Floor plan House B, 7th floor 1:200
6 Total floor plan, 8th floor 1:750
7 Cross section of the facade 1:300
8 Cross and longitudinal sections 1:750

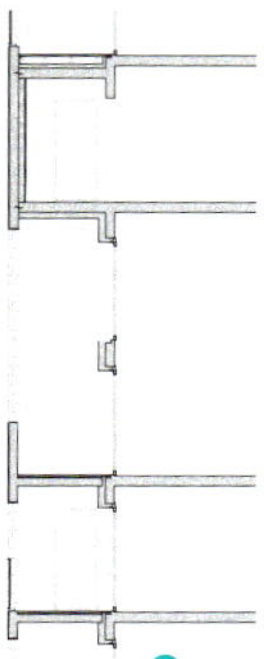

7

8

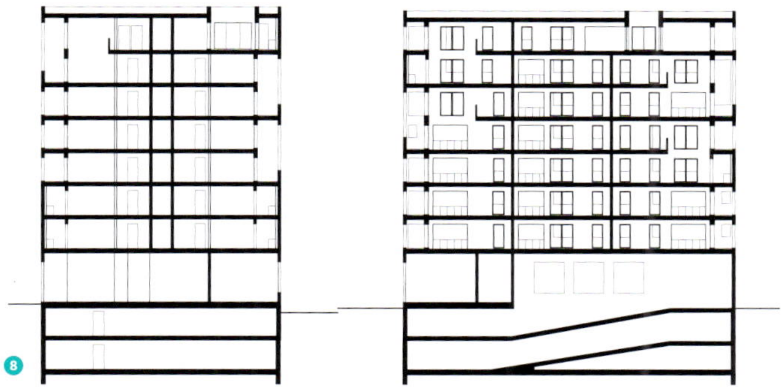

2.1
SOLITAIRE

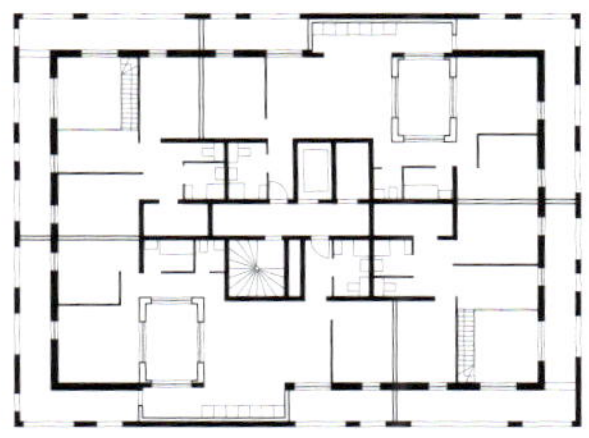

6

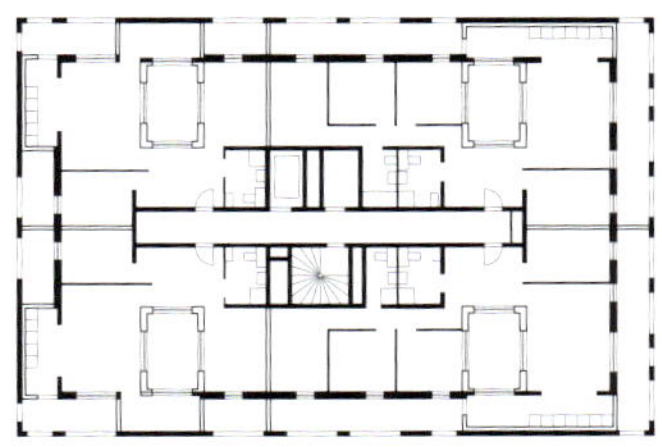

Building type
two solitary buildings offset to one another in an urban block
8 stories
facing NW/SE/NE/SW

Date of construction
2010–2013

Number of units
86

Size of units
1-room apts., 36 m²
2-room apts., 45 m²
3-room apts., 65 m²
4-room apts., 93 m²
5-room apts., 104 m²

Area per user
21–36 m²

Building depth
20.6 m

Access
double-loaded corridor with 4 to 7 apartments

Open spaces
single- and double-height loggias, roof patios

Parking
underground garage

Architect
Eric Lapierre Architecture
Paris

Location
4–6 rue de l'Égalité
Lyon
France

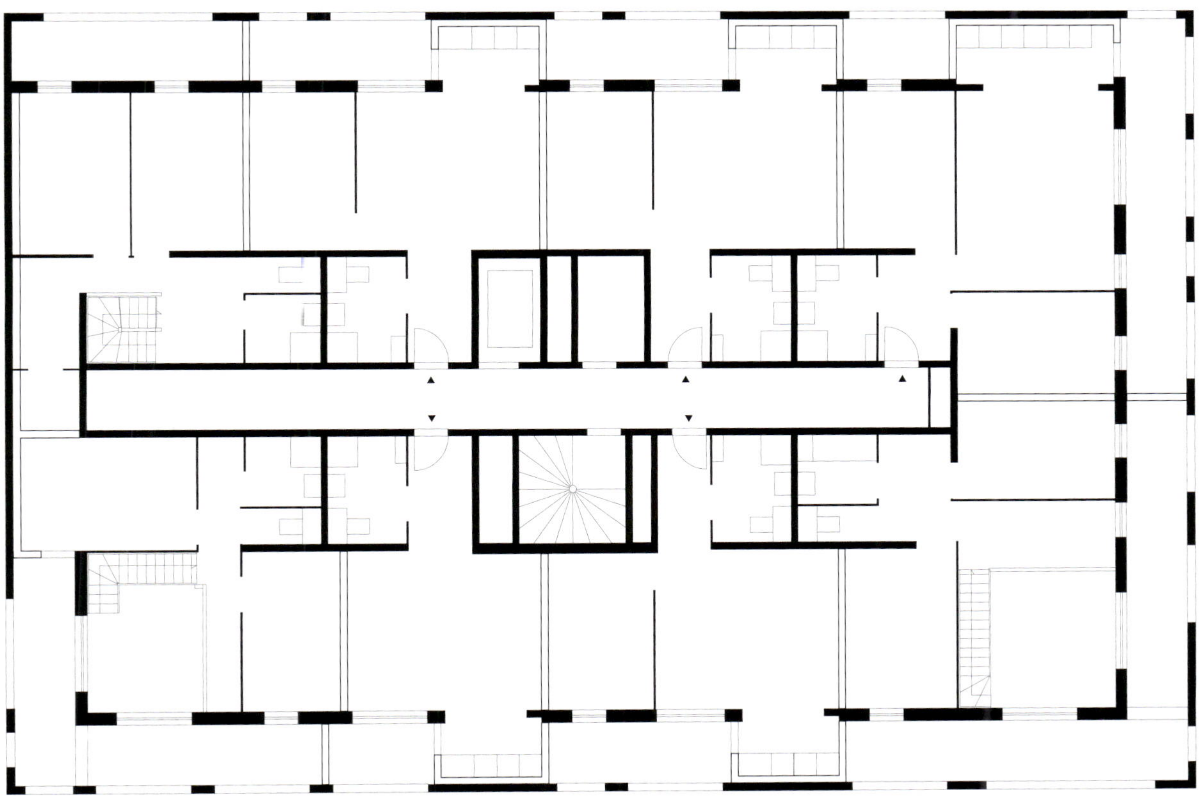

5

Referring to floor plan 1:200

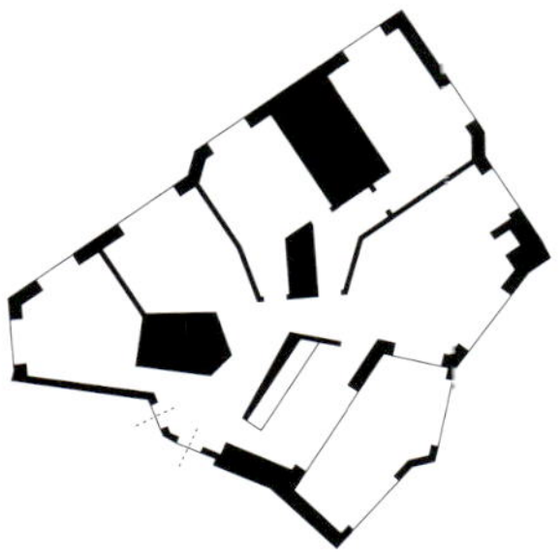

Steinwiesstrasse/Irisstrasse | Edelaar Mosayebi Inderbitzin | 2014

Located in a wooded urban district with open blocks, this ensemble consists of three very different structures. While the renovated villa and the atelier house are reflective of the surrounding heterogeneous building types, the residential building in the center adheres to its own design language. Formally inspired by the protected tree stock of the vicinity as well as the structure of rock formations, the built figure is embossed with vertically alternating surfaces and shaped to leave enough room for the surrounding trees. The result is a floor plan with four multifaceted wings, which appear to grow into the clearings.

The resulting amorphous shape continues on the inside and is translated in the building's footprint – with one apartment in each wing – into a cluster of adjoined, skewed polygons. These combine to create a succession of open spatial linkages which widen and narrow, allow diverse views into the surroundings or connect to neighboring rooms. A spatial path thus emerges in each of the four, slightly vertically offset apartments. The path begins at an entrance hall directly accessed by elevator; it is at times angled toward one or the other view of the garden while leading past an initial living space, then narrows near the kitchen and opens at the end into a living room lit naturally from three sides – one of which contains a loggia. Enclosed "chambers" with intentionally smaller doors lie along this path and join to form separate groups of rooms – for instance with their own corridor, bathroom, dressing room, or as connected bedrooms.

1

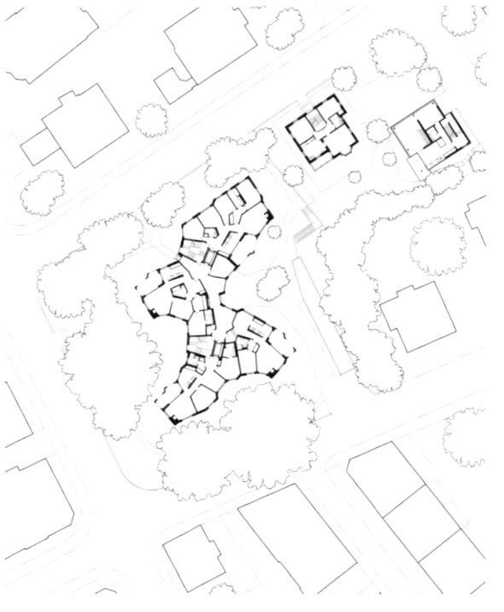

2

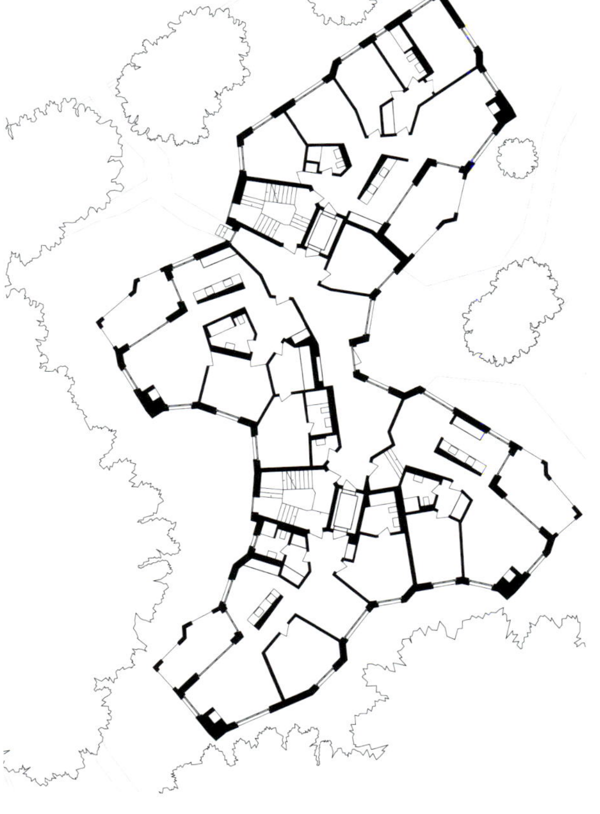

3

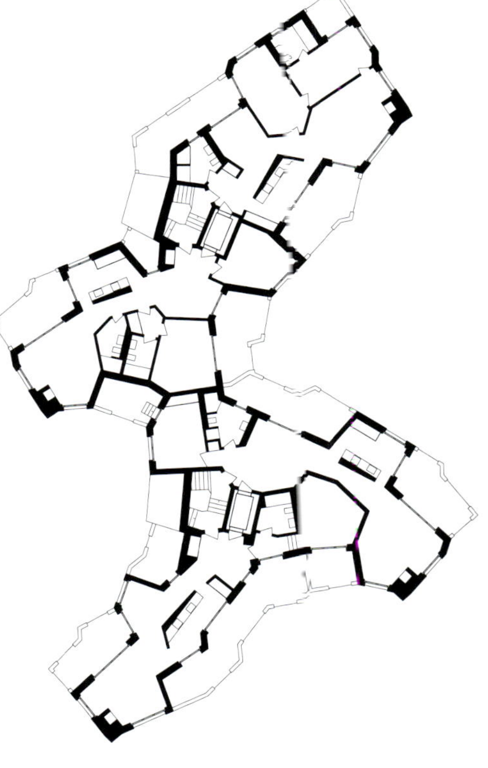

1 Southeast elevation 1:500
2 Ground floor 1:500
3 4th floor 1:500
4 2nd floor 1:200

2.1
SOLITAIRE

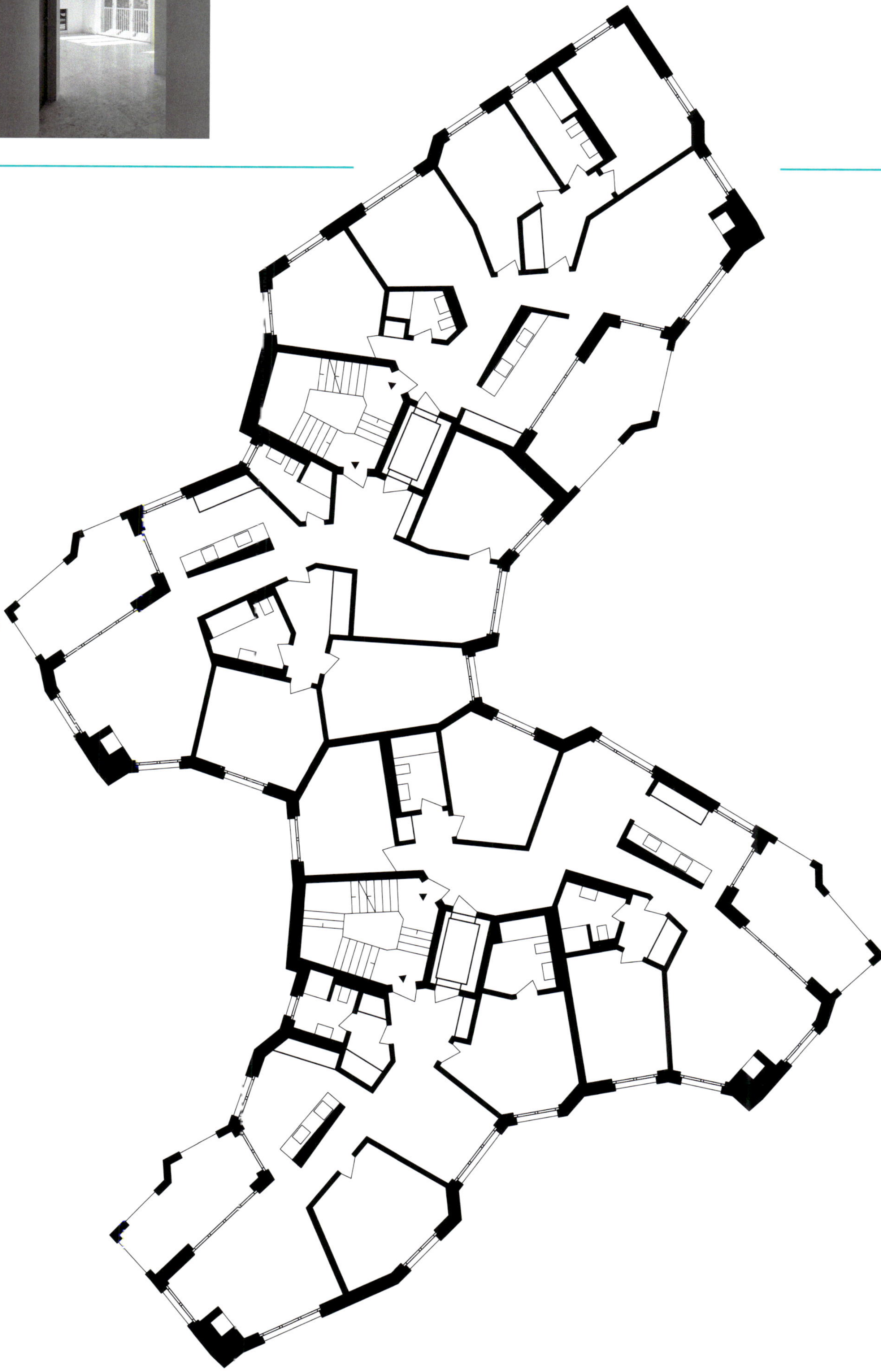

Building type
sculptural building volume branching out into three clearings as part of an ensemble in a villa district
4 stories
facing in all directions

Date of construction
2014

Number of units
16

Size of units
2 ½-room apts., approx. 83 m² (3 units)
3 ½-room apts., approx. 104 m² (6 units)
4 ½-room apts., approx. 129 m² (7 units)

Area per user
34 m² (incl. hobby room)

Building depth
8.4–15.6 m

Access
double-loaded staircase with entrances on offset landings, direct elevator access to each unit

Open spaces
loggias, roof terraces

Parking
underground garage

Architect
Edelaar Mosayebi Inderbitzin Architekten AG ETH SIA BSA
Ron Edelaar, Elli Mosayebi, Christian Inderbitzin
project team: Theres Hollenstein (project manager), Philipp Schneider (construction manager), Michael Reiterer, Nina Villiger
competition team: Jonathan Roider, Michael Reiterer, Lukas Prestele
Zurich

Location
Steinwiesstrasse/Irisstrasse
Hottingen, Zurich
Switzerland

Referring to floor plan 1:200

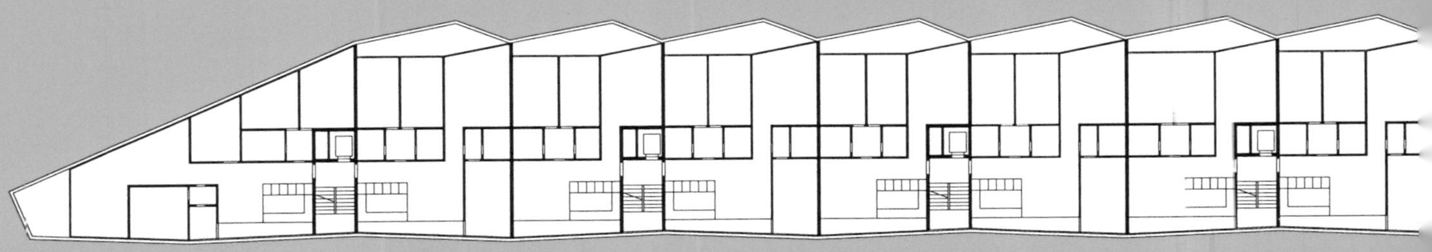

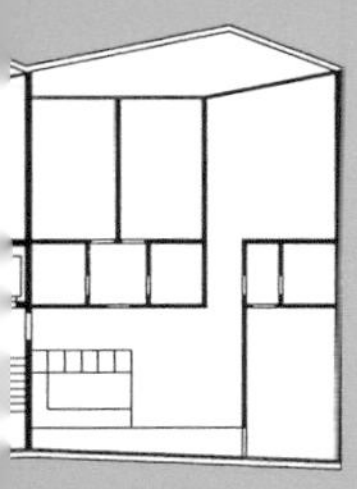

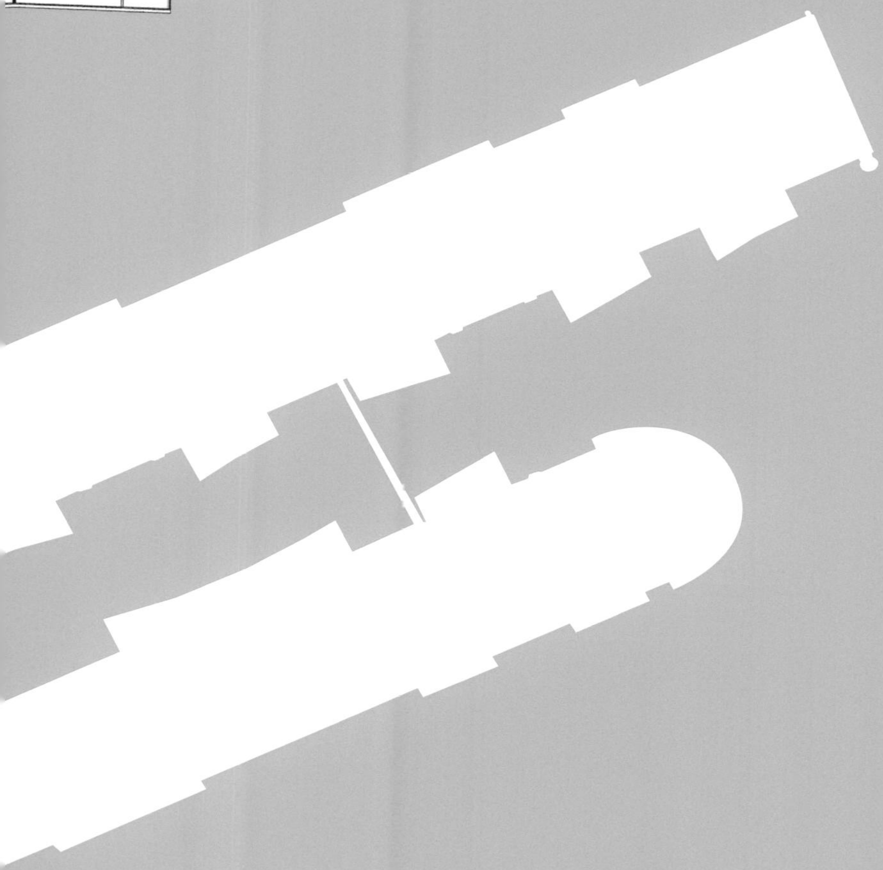

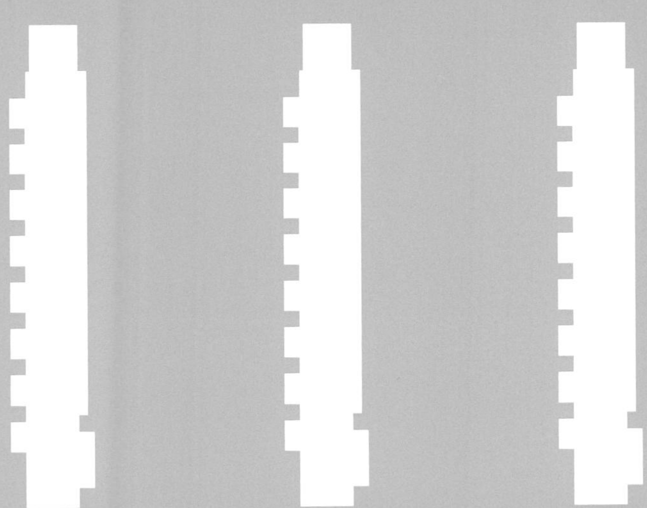

2.2
LINEAR BLOCK / SUPERBLOCK

This chapter combines the typologies linear block and superblock, which are similar in appearance. As an urban planning leitmotif of modernism, which promoted less dense developments with light, air, and sun, linear blocks were the antithesis to the tenement city and their enclosed block-edge developments in the late 19th century. For a long time, linear block developments were the ideal solution for societies that had a pressing need for inexpensive housing on a large scale. Owing to their serial construction method, linear blocks are characterized by systematic floor plans, which nevertheless allow for some variation within the rhythm of the dividing walls and the changes from floor to floor (one-level apartment, maisonette, triplex or split-level type). A variety of access systems, such as landings, galleries, or internal streets are possible. Ideally with an east-west orientation, they occupy tailor-made urban spaces with dedicated traffic areas and green spaces and are often accompanied by functional segregation: the occupant works, shops, and spends his or her leisure time in different locations; the estate becomes effectively a "dormitory town." Linear blocks were the dominant building type in large estates for a long time, arranged in serial rows facing one another and forming open, often monotonous urban spaces.

The superblock is larger in scale, more complex than the linear block and nearly always a solitary structure; it is a massive, large volume whose dominant, often sculptural presence can structure an entire urban situation. Multilayered variations characterize the interior of the complex fabric: different apartment sizes and types for different housing requirements complement one another, accompanied by interlinked vertical and horizontal access spaces. Common areas and services not only structurally shape the residential community inside the building, but also the exterior of the superblock. This is a case of unity in diversity: the superblock is thus a quintessentially urban building form. Introduced in the 1950s – at the time usually as a series of always identical apartments arranged along a corridor or internal access street – it has continually evolved up to the present time, and has now been rediscovered as an urban building block appreciated for its unlimited variety. In the meantime, linear block and superblock are no longer employed solely for municipality-owned social housing; they are now being constructed – sometimes in single and shorter form – for all social strata and utilized for urban densification and agglomerations.

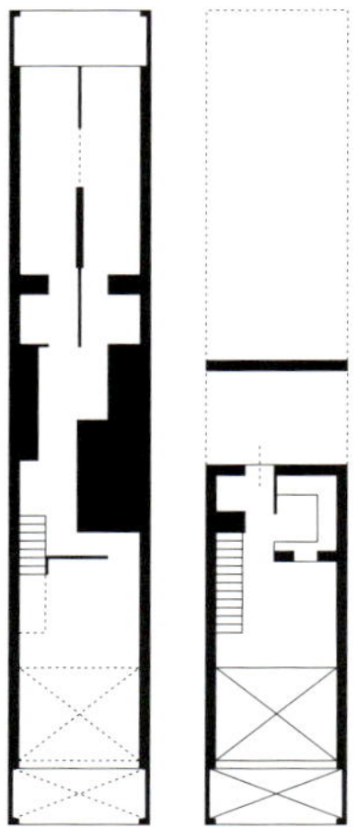

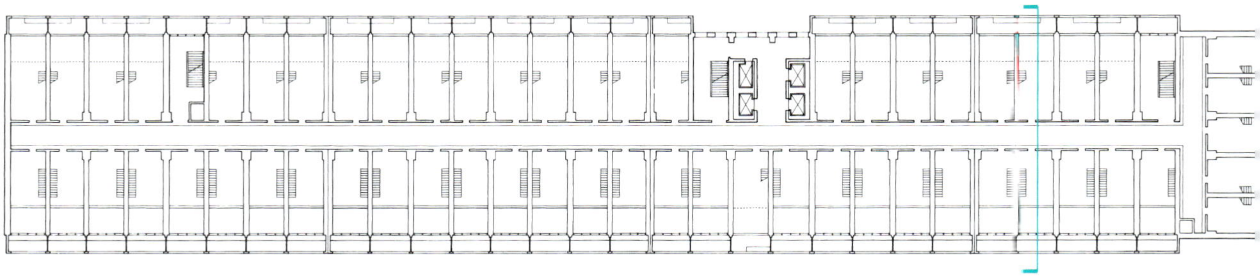

Unité d'Habitation | Le Corbusier | 1947

The "machine for living" model: prototype of an autarchic housing unit as module for a new town-planning scheme. The building is set freely into the landscape and oriented entirely according to the course of the sun: clear east-west orientation. Optimization of floor plans according to Corbusier's "Modulor" measurements and laws of insolation. A central access corridor (every two or three stories, without natural lighting) is surrounded by four basic spatial modules (criterion of industrial construction) that are varied in up to three levels and three parallel units. This allows a variety of apartment sizes and layouts, a total of 23 floor plan types were developed for this project. All apartments (except for the 1-room apts.) are maisonettes. The great room depth between the two only 3.66-m-wide crosswalls is utilized to maximum advantage; the apartments follow consistent zoning (relationship between incidence of light and depth of room). Living rooms or joint play areas open up toward outside (living rooms are always two stories, opening out to continuous terraces); bedrooms (minimized) on upper floor (galleries) or on setback story below; sanitary rooms and kitchens toward central corridor. Through-living spaces, both horizontal and vertical. Large common terrace on roof (recreation facilities). Comprehensive service facilities (hotel, businesses, offices, and restaurants) on two full stories (level 7/8), like gym, pool; also crèche and kindergarten.

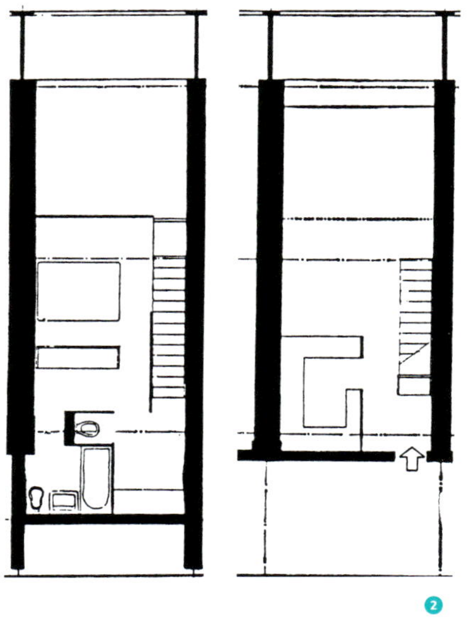

2

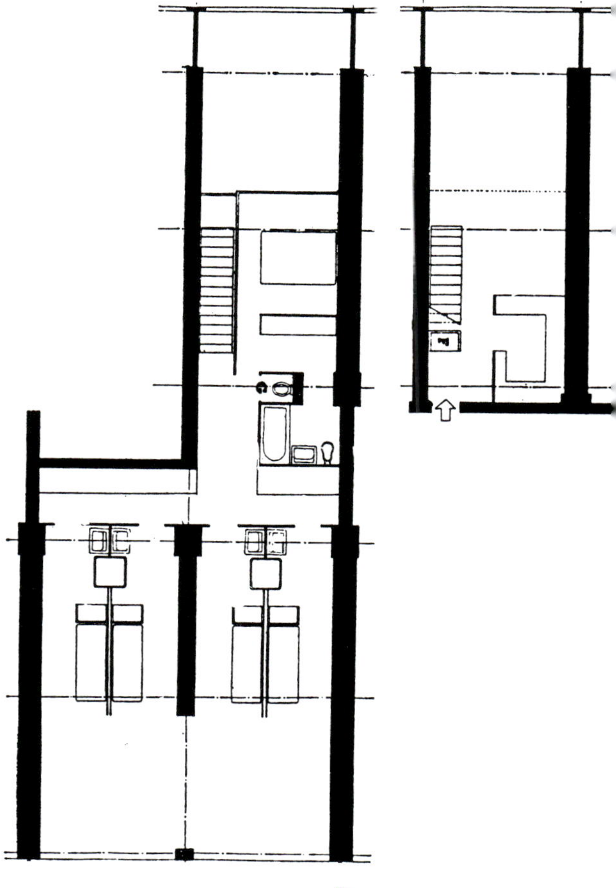

3

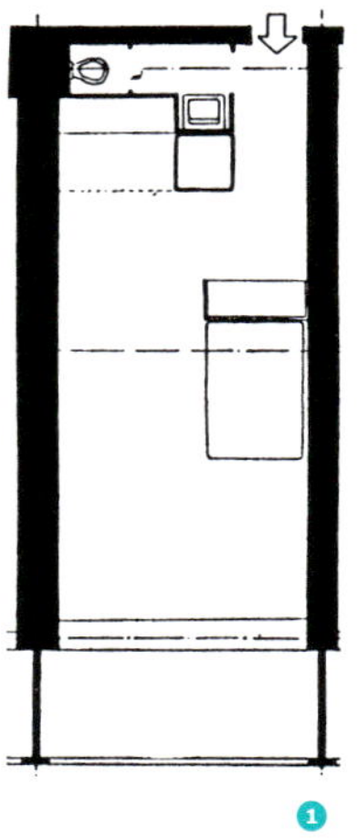

1

1 Type B: 32.5 m²
Single-person flat 1:200

2 Type C: 59 m²
Maisonette for 2 persons 1:200

3 Type G: 137 m²
Maisonette for family with 4–8 children 1:200

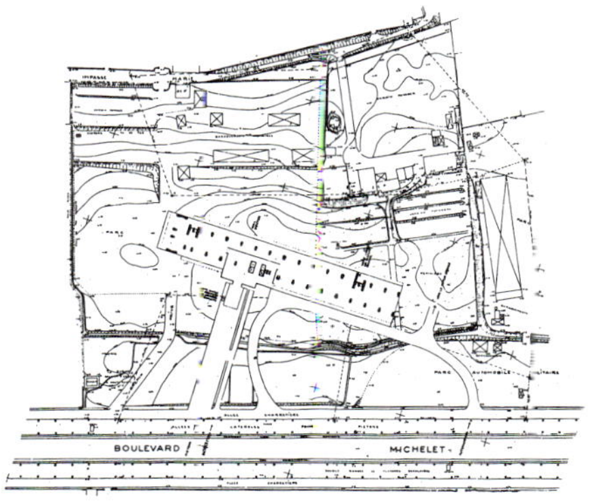

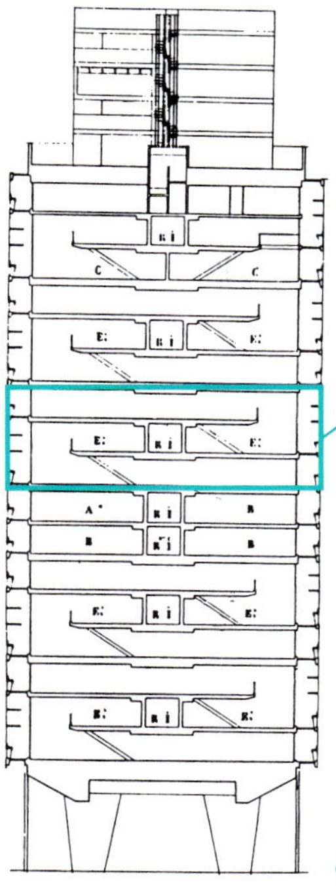

7

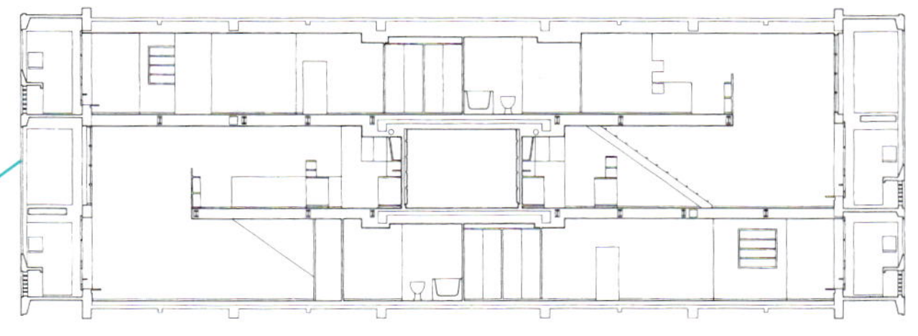

8

2.2
LINEAR BLOCK / SUPERBLOCK

Building type
Unité d'Habitation
18 stories
facing E/W

Date of construction
1946–1947

Number of units
337

Size of units
1-room apts., 32.5 m²
2- to 5-room mais.,
59/98/137 m²
(23 types of floor plans)

Area per user
23–32.5 m²

Building depth
24.4 m

Access
interior access roads on levels 2/5/7/8/10/13/16

Open spaces
loggias
rooftop terrace with swimming pool
park

Parking
on street level under the building

Architect
Le Corbusier
Paris

Location
Boulevard Michelet
Marseille
France

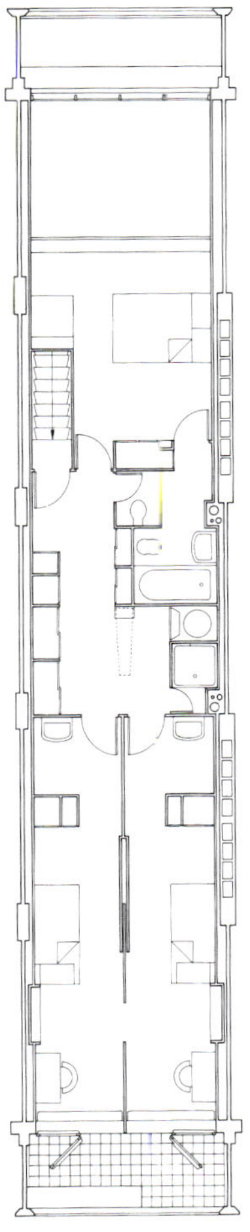

4

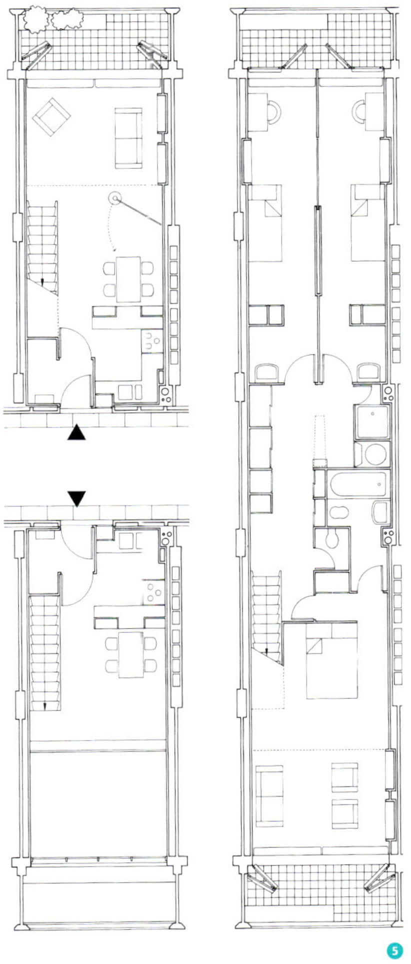

5

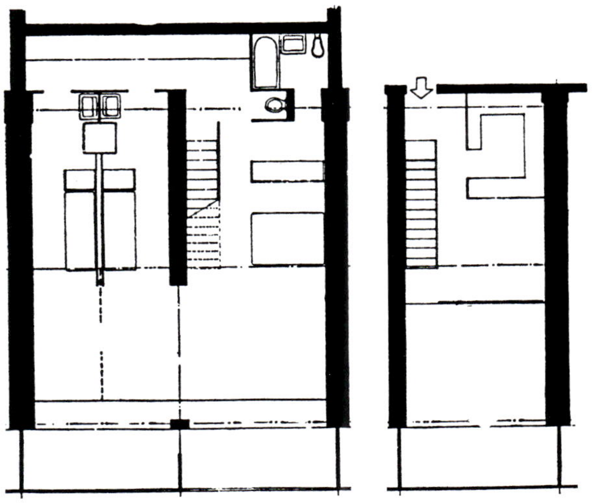

6

4 5 Type **E2**: 98 m²
Typical maisonette for family with 2–4 children 1:200
6 Type **E1**: 98 m²
Maisonette for family with 2–4 children 1:200
7 Section showing central corridors
8 Section through typical maisonette

Referring to floor plan 1:200

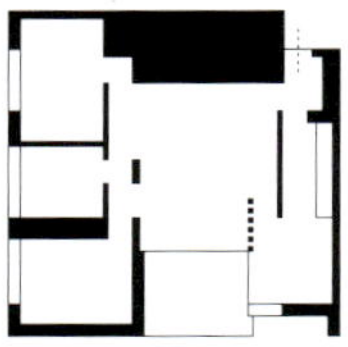

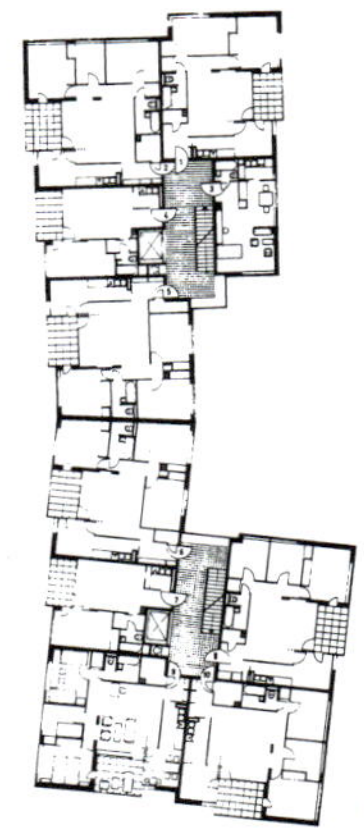

Klopstockstraße |
Aalto, Baumgarten | 1957

What makes these floor plans unique is their living space, which forms the center of the apartments (the "marketplace of the family," according to the architect). From there access to the sleeping-room hallway, which can be optionally incorporated into or separated from the living space, by means of folding doors. Deep loggias in front of living spaces, also accessible from sleeping and eating areas. Small private rooms at the cost of the large living space. Economical quintuple-loaded stairwells with large landings, elevator, straight-run stairs. 10 apartments per story, 8 apartments open to two points of the compass. Sauna, sundeck on roof; laundry, hobby room in basement.

Building type
residential slab
8 stories
facing E/W

Date of construction
1956–1957

Number of units
84

Size of units
1- and 2-room apts., 38–45 m^2
$3^{1}/_{2}$-room apts., 77 m^2
$2^{3}/_{2}$-, $3^{2}/_{2}$-, $4^{1}/_{2}$-room apts., 83–90 m^2

Area per user
22.5–38 m^2

Building depth
approx. 22/10.5 m

Access
quintuple-loaded stairwells

Open spaces
rooftop terrace with sauna
park-like surroundings

Parking
open-air parking

Architect
Alvar Aalto
Munkiniemi
Paul Baumgarten
Berlin

Location
Klopstockstraße 30, 32
Hansaviertel
Berlin-Tiergarten
Germany

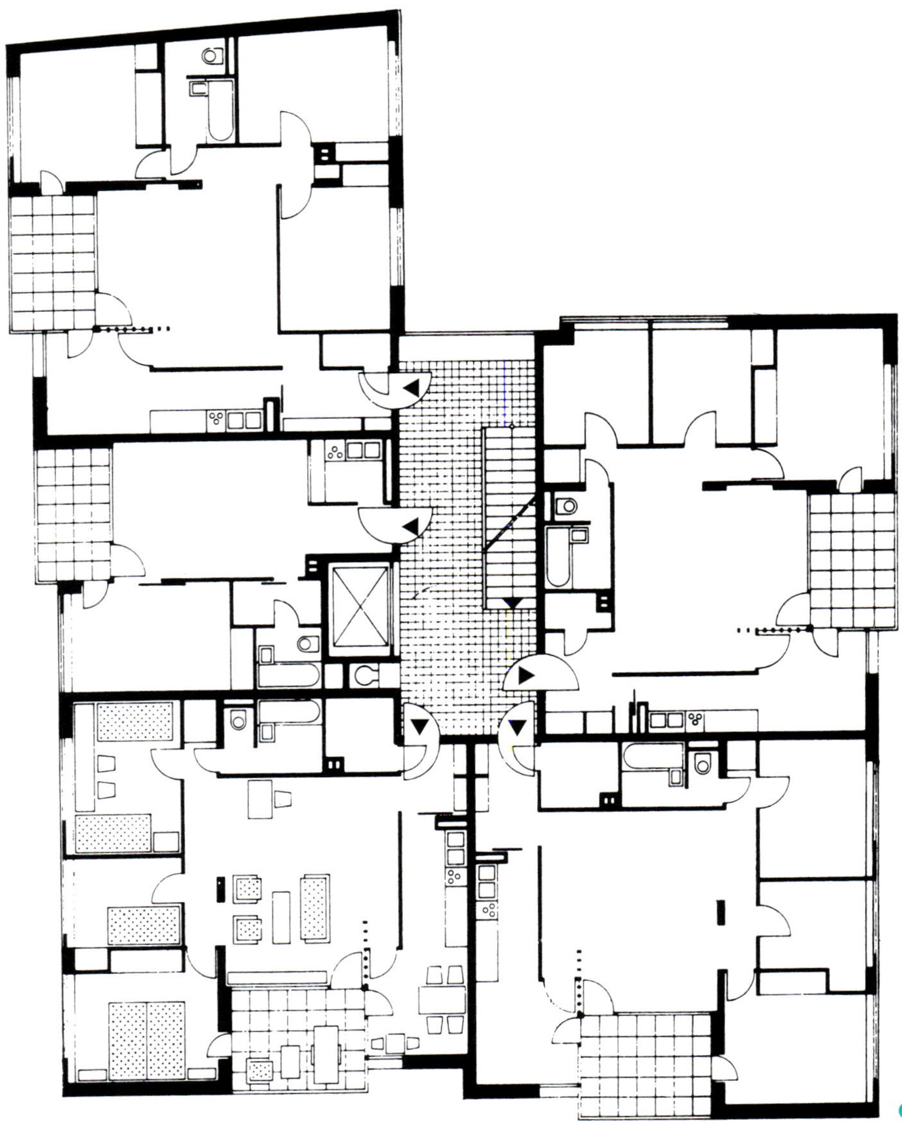

Typical floor plan 1
Section of 2
typical floor plan 1:200

Referring to
floor plan 1:200

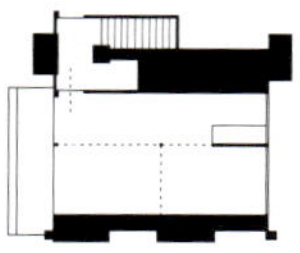

Harumi Apartments | Maekawa | 1958

This slab unites a building type inspired by Le Corbusier with Japanese domestic culture. On every third level of the exposed concrete structure there is a spacious outdoor access gallery from which perpendicular staircases lead to the floors above and below. The use of built-in wooden carpentry and the clear division of the floor plan are inspired by traditional Japanese building types, in which a kitchen and dining area have wooden floors, while tatami mats line the floor of the sleeping and living areas. Sliding walls and storage nooks for futons allow different usage scenarios. On the access levels, the units are organized parallel to the volume and entered through the kitchen, from which the WC, bathroom area, and two rooms are accessed. On the floors above and below, they are organized perpendicular to the volume. Here, the unit is divided into four parallel zones: one strip is occupied by storage spaces, bathroom, WC, and the shared staircase, followed by a hallway-like space with a mini-kitchen and wide enough to double as a living room. In the adjacent zone, two rooms can be connected to one another or the hallway by sliding walls. The meandering dividing wall creates niches for furniture or storage.

Building type
residential slab
10 stories
facing NW/SE

Date of construction
1958

Number of units
170

Size of units
3-room apts., 30–43 m²

Area per user
10–14.5 m²

Building depth
10.5 m

Access
external access gallery every 3rd floor

Open spaces
balconies

Parking
on forecourt

Architect
Kunio Maekawa
Tokyo

Location
Harumi, Tokyo, Japan

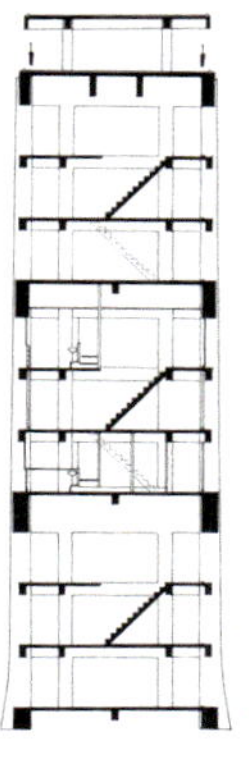
3

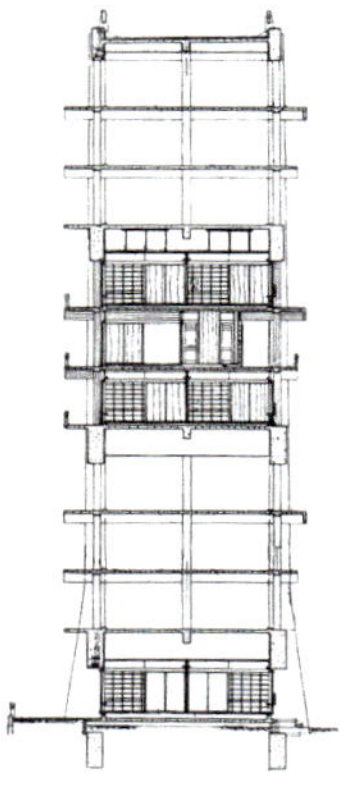
4

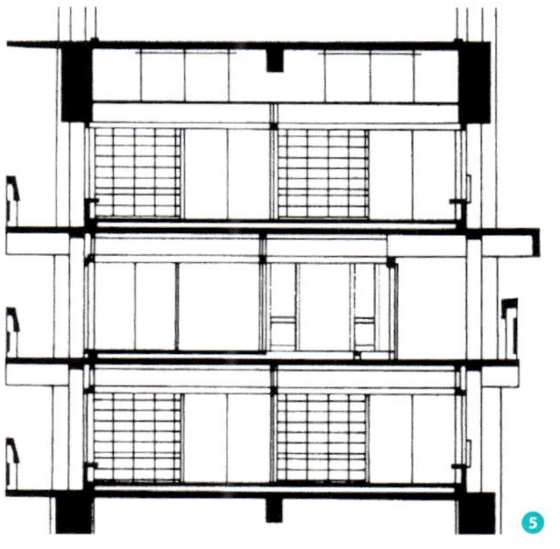
5

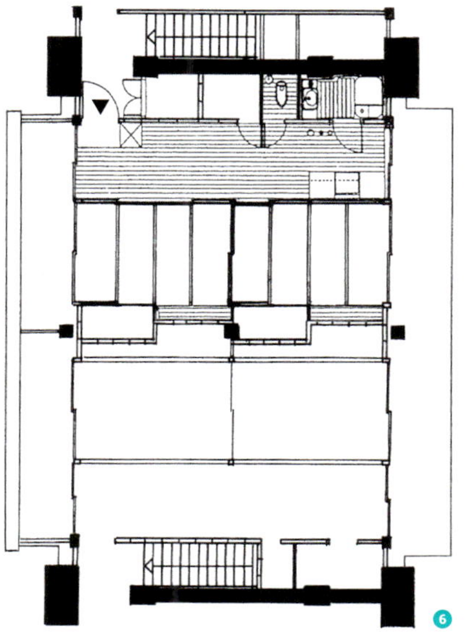
6

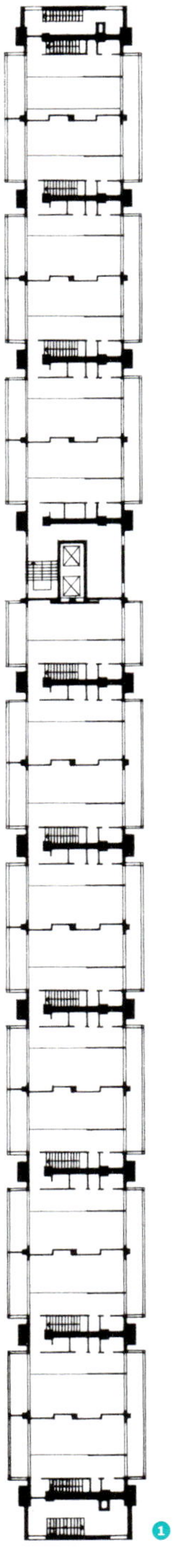
1

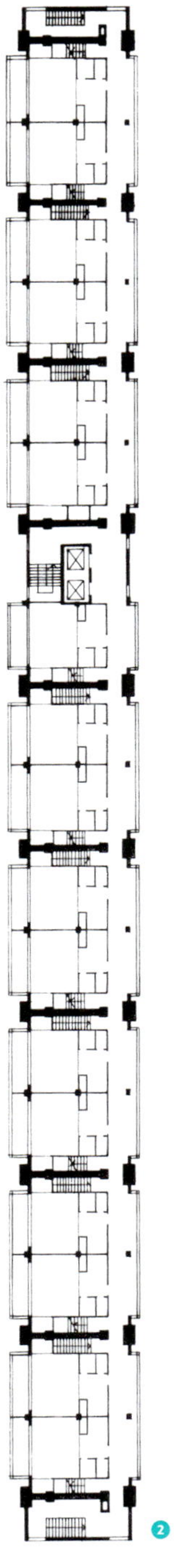
2

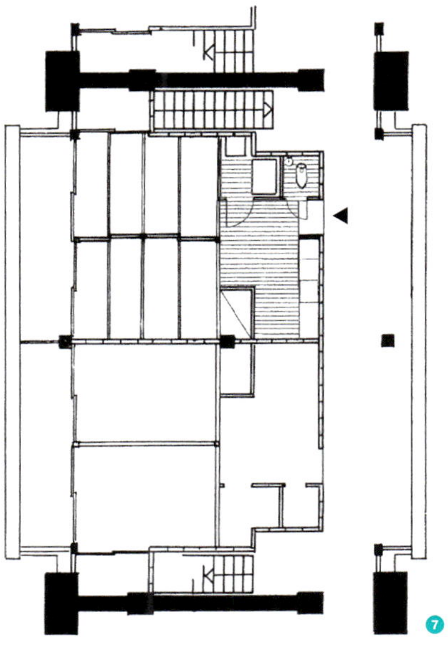
7

Typical floor 1
above and below
access level
Typical floor on level 2
with with access gallery
Cross section with 3
stairs branching off
open access gallery
Cross section residential units 4
Cross section of residential 5
floors with access gallery in
the middle 1:200
Two standard units 6
above and below
access levels 1:200
Two standard units on level 7
with access gallery 1:200

Referring to
floor plan 1:200

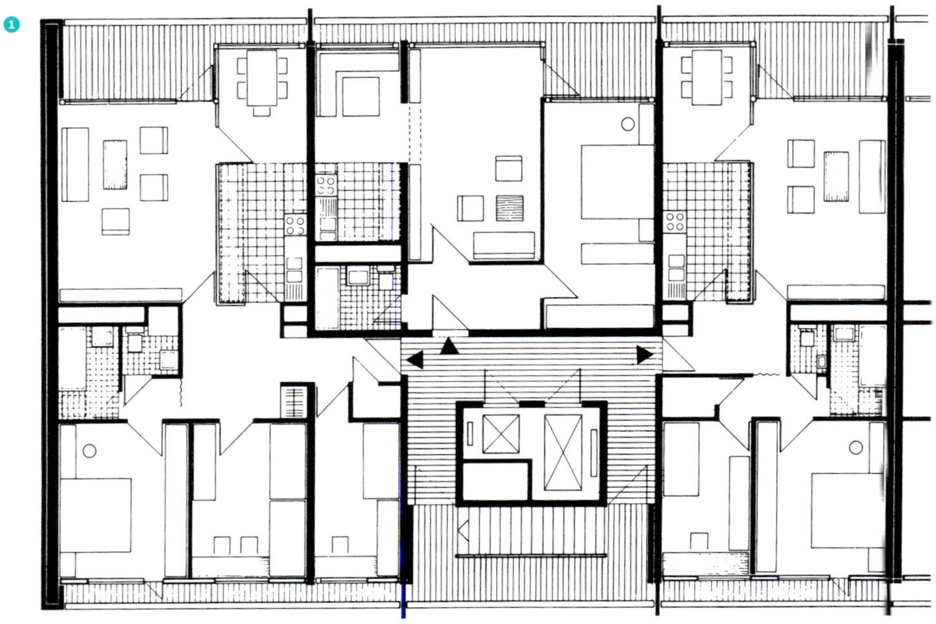

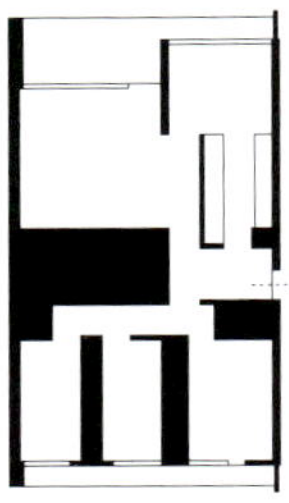

Hannibal | Jäger, Müller, Wirth | 1971

Access to buildings by means of double-height entry vestibules ("hotel atmosphere"). In the suspended mezzanine story, small apartments for the elderly with hallway entries. Elsewhere, double- and triple-loaded stairwells around open elevator and stair cores. Free plan layouts are possible since only the outer walls and shafts are load-bearing (21 different types of floor plans, additional changes by tenants also possible). Generally: clear separation of living and sleeping zones, both always directly accessible from entrance hall. Organization of living room, kitchen, dining area always handled differently: kitchen between living and dining areas; separate kitchen with dining area in living room; linear kitchen with dining area at the end, parallel to living room; or central kitchen as focal point of plan, with access to balcony. The bedrooms are always lined up along a separate corridor, bath and WC are invariably separated. Each apartment has one or two, often large, balconies. Maisonettes, terrace apartments, and studios on roof level.

Lower levels of maisonettes have bedrooms, baths, kitchens, and an expanded double-height dining area (small living room). The stairs between bath and dining area open above into a long gallery leading to a large living room (or living/bedroom).

A number of sports and leisure facilities: on ground floor level laundry rooms, common rooms, and daycare center. Semi-public sun terraces, indoor pool, sauna and restaurant on roof level.

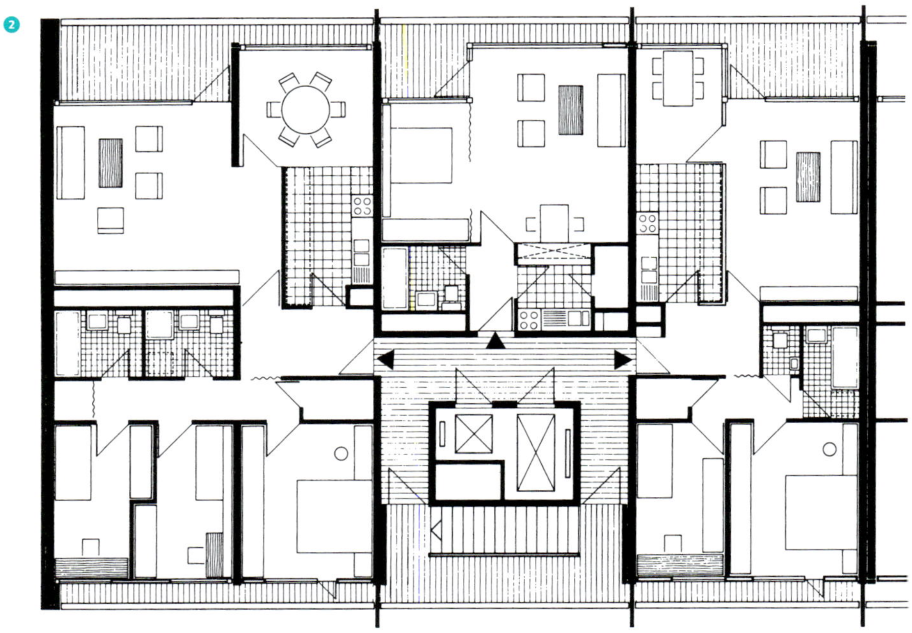

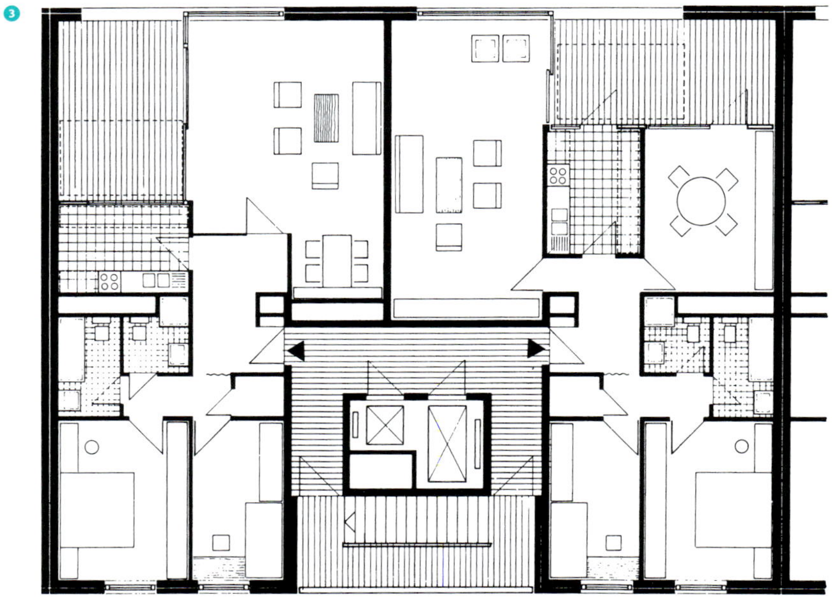

1 Bldg. A, No 5, floors 2–21: 4-, 1½- and 3-room apts. 1:200
2 Bldg. A, No 1, floors 2–21: 4-, 2, and 3-room apts. 1:200
3 Bldg. A, No 2, roof story: 3-room apts. 1:200

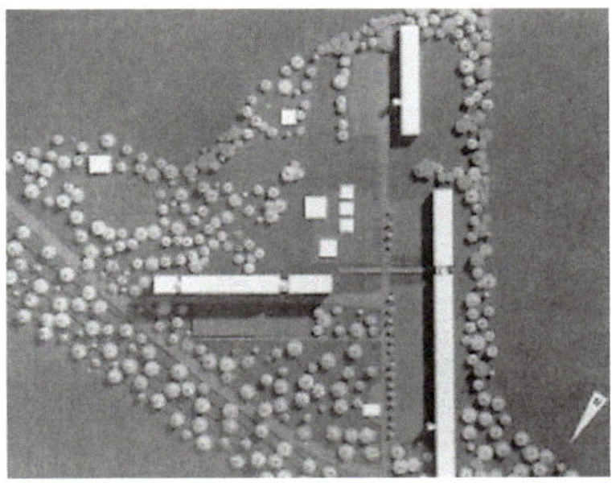

❹

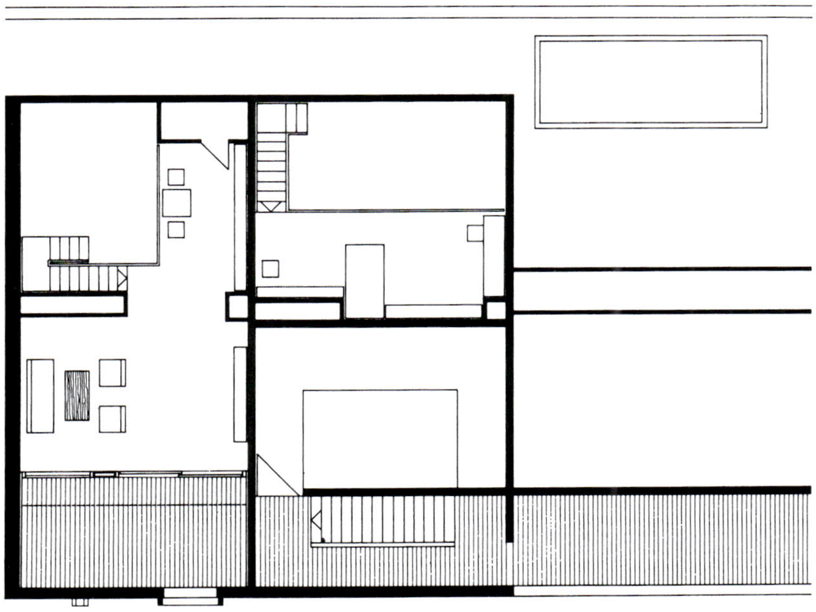

2.2 LINEAR BLOCK / SUPERBLOCK

❺

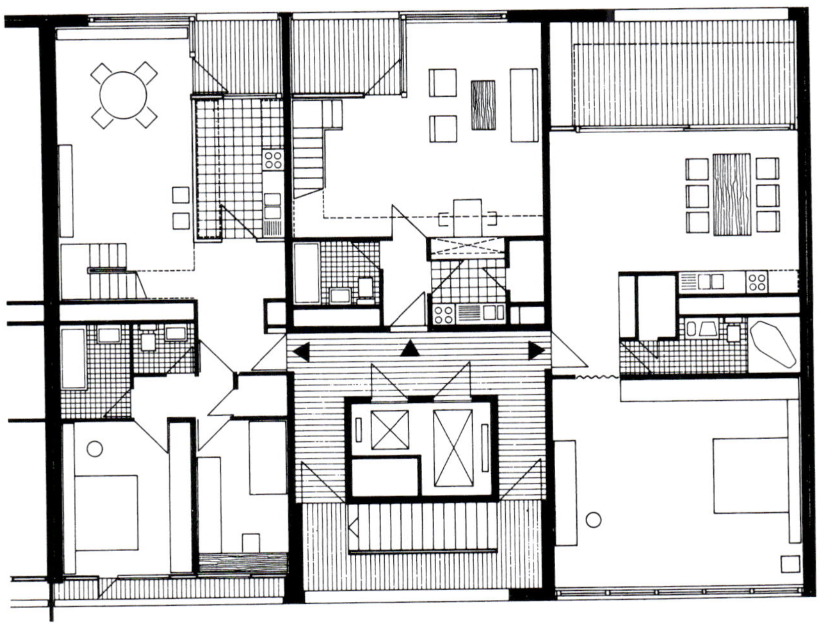

❻

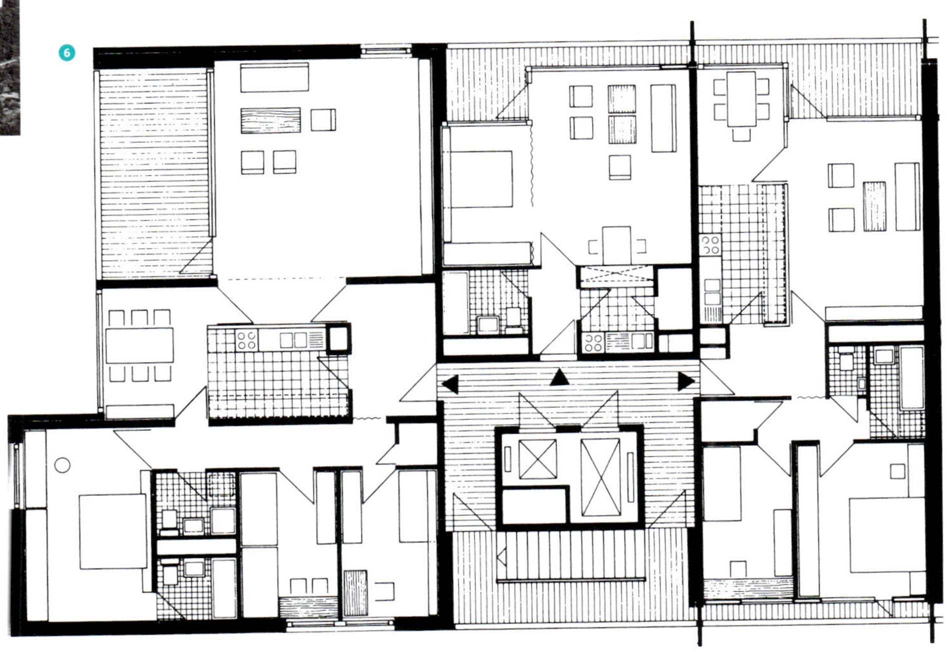

Building type
three 22/24-storied apartments slabs
Bldg. A facing NW/SE
B and C facing NE/SW

Date of construction
1969–1971

Number of units
1143

Size of units
1-room apts., 41–45 m²
1½-room apts., 45–47 m²
2-room apts., 66 m²
3-room apts., 80–89 m²
4-room apts., 104 m²
5-room apts., 130–139 m²
6-room apts., 155 m²

Area per user
26–44.5 m²

Building depth
14/15 m

Access
double- and triple-loaded stairwells

Open spaces
park and water areas, balconies

Parking
mostly with underground parking, some open-air parking spaces

Architect
Otto Jäger
Werner Müller
H. P. Wirth
Stuttgart

Location
Stuttgart-Asemwald
Germany

❹ Bldg. **c**, No 4, 24th floor: upper level maisonette, upper level studio 1:200
❺ Bldg. **c**, No 4, 23th floor: 4-room maisonette, 2-room studio, 2-room apt. 1:200
❻ Bldg. **c**, No 1, floors 2–22: 4-, 1½-, 3-room apt. 1:200

Referring to floor plan 1:200

2.2

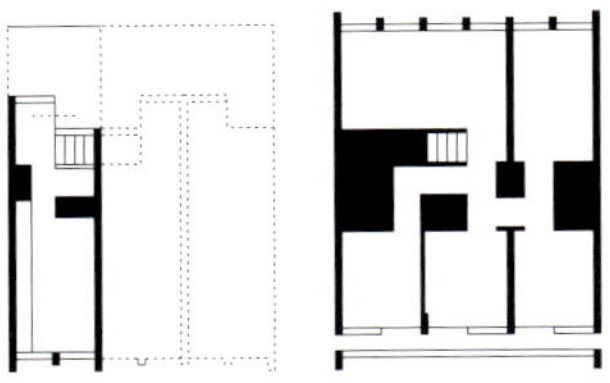

Robin Hood Gardens | Smithson | 1972

The leitmotif of the complex are the so-called "streets in the sky," broad covered walkways designed to create a neighbourly urban space for socializing. Located between two major thoroughfares, Robin Hood Gardens is an island in a sea of traffic. The complex consists of two long, sculptural blocks – one of seven stories, the other of ten –, forming a protective enclosure around a landscaped green area with a small hill made from construction rubble and sheltered depressions as children's play areas. The bends in the blocks are utilized for access and in part as storage areas. The walkways provide access to apartments in units comprised of three levels, in which various types of maisonettes alternate with the apartments and are continued on the floor above or below. Some two meters in width, the walkways are even wider in front of the entrances, which are set off to the side, creating niche-like anterooms. In the interior, the maisonette stairs run parallel to this entrance area, thereby creating a buffer zone for the adjacent combined kitchen and dining area. On the upper level, the stairs lead through the corner of the living room into a corridor, the meandering layout of which subtly separates public from private spaces. On this level, the bedrooms are largely oriented toward the quieter courtyard and linked via a narrow balcony. The garden or ground floor contains apartments for seniors and families with direct access from the outside.

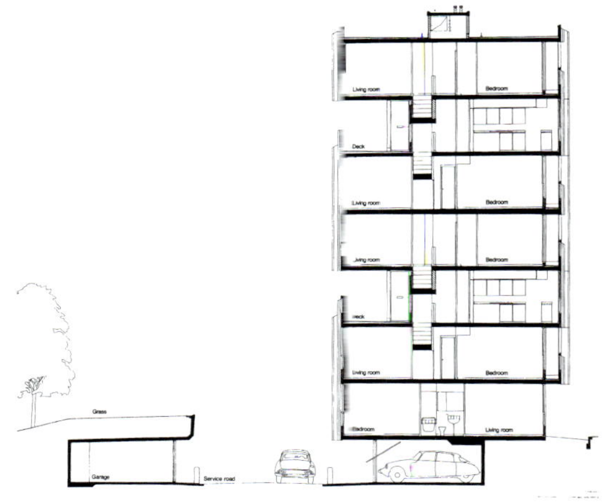

3

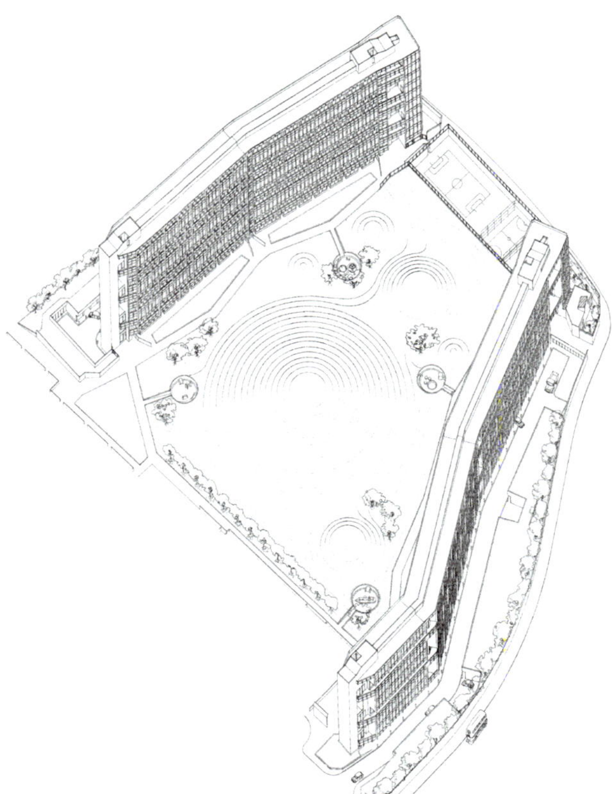

1

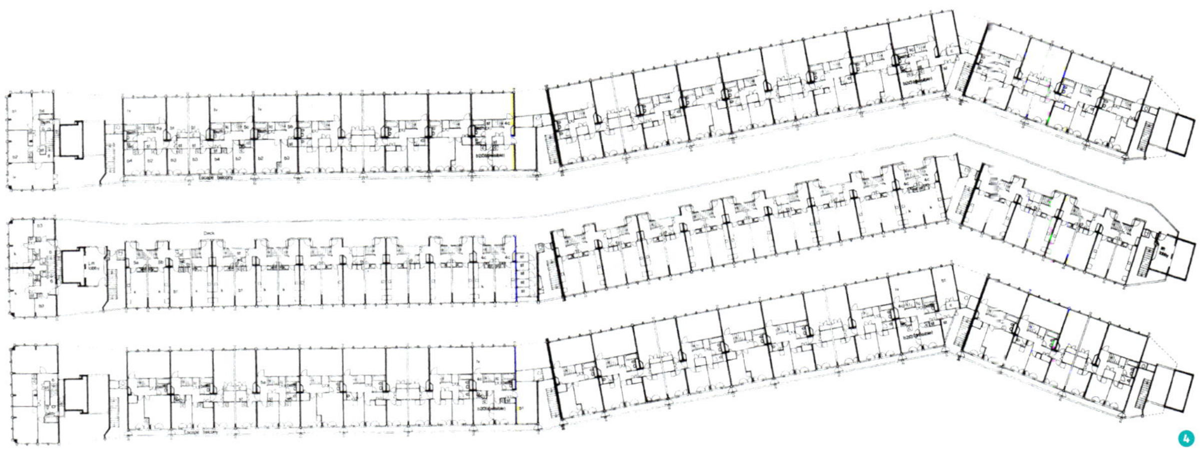

4

1. Axonometric drawing with landscaped park, to the right: Blackwall Tunnel South Block (BTS Block), to the left: Cotton Street Block (CS Block)
2. Section through CS Block maisonette, 6th, 7th floor
3. Section through CS Block with garages
4. Floor plan CS-Block unit on 3 levels with "street in the air" in the middle
5. Floor plan sections with 4- and 5-room maisonettes **A–F** on 3 levels 1:200
6. Site plan of the six originally planned slabs

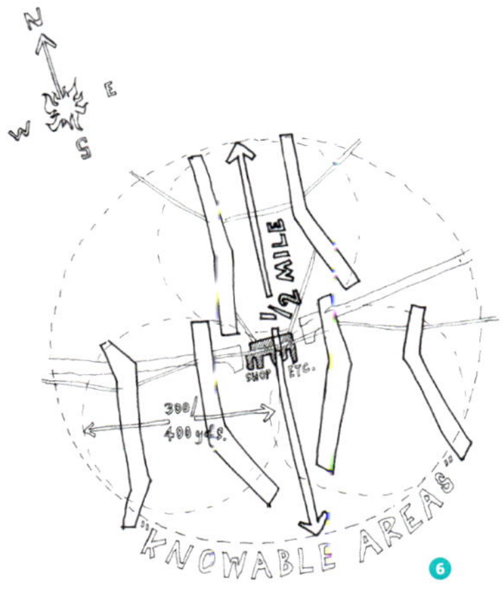

6

2.2 LINEAR BLOCK / SUPERBLOCK

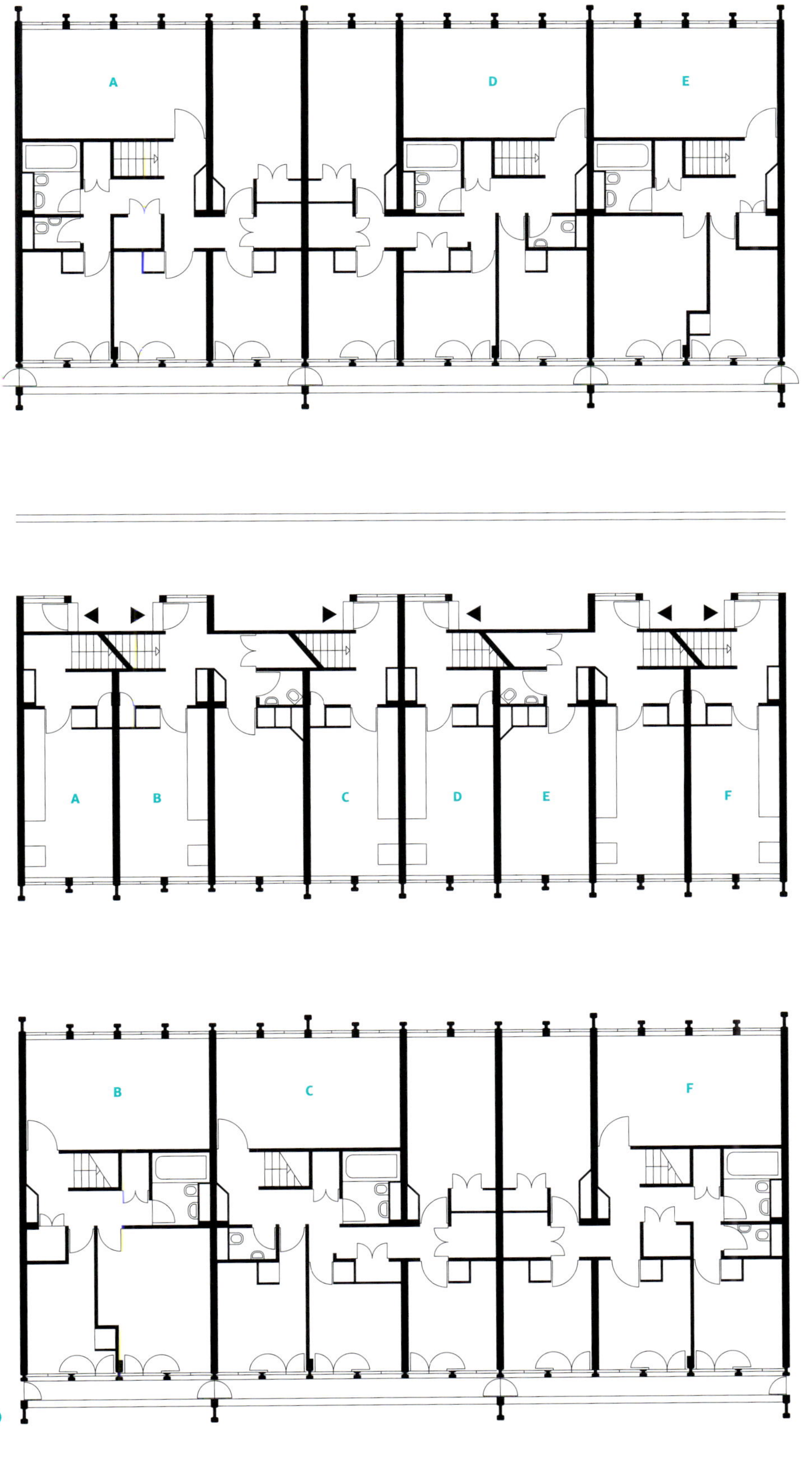

5

Building type
2 housing slabs framing an open space, businesses integrated into the buildings
7 and 10 stories, respectively facing E/W

Date of construction
1966–1972

Number of units
213 (total of 15,670 m²)

Size of units
6 different types
e.g.: garden flats for families or seniors,
4-room mais., approx. 100 m²
5-room mais., approx. 112 m²

Area per user
22.5–25 m²

Building depth
approx. 11.2 m

Access
direct access on ground floor, then covered walkways on every third floor for maisonettes above and below

Open spaces
generous covered walkways
hilly landscape for shared use created with construction waste between the buildings with playgrounds in the depressions

Parking
143 parking spaces underneath the building, guest parking

Architect
Alison + Peter Smithson
London

Location
London
United Kingdom

Referring to floor plan 1:200

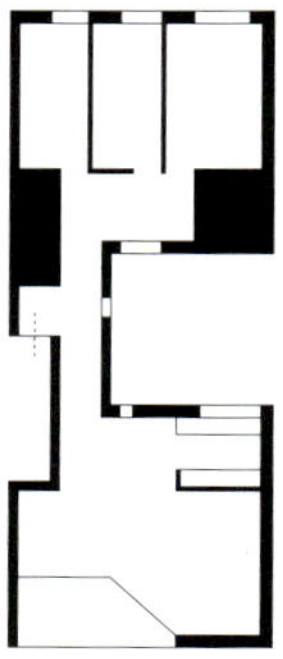

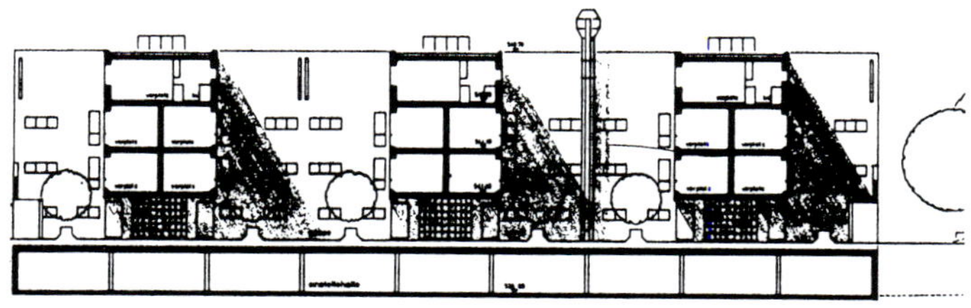

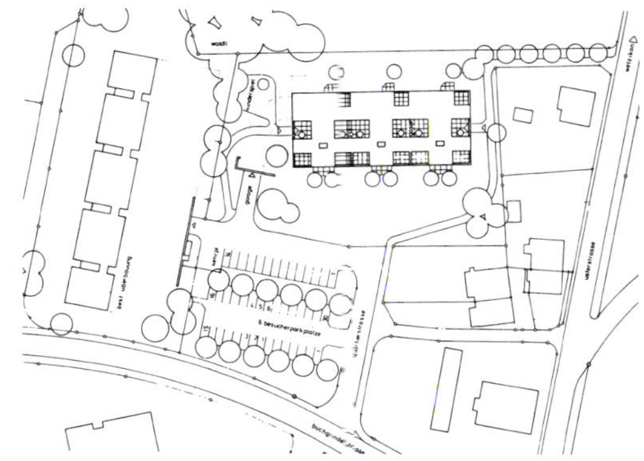

Buchgrindel II | Hotz | 1985

Compact double row as an autonomous urban module: open spaces and rooftop terraces are cut into the built volume from outside. Inside is a residential street that runs along the longitudinal axis of the building, closed to the outside by gates and offering a starting point for the stairwells. On the upper stories, bridges link the two blocks, thus creating lightwells. A mixture of apartments: the small apartments with open spaces cut into the building volume are located on the ground floor. The upper floors contain larger apartments, which extend from one side to the other. Work area, dining area, kitchen, and bath are located on the lightwells, with living and sleeping to the outside. The apartments on the third floor have two-story high living rooms with access via spiral staircases to large rooftop terraces. On the fourth floor, the living room stretches from the stairwell core across the bridge all the way to the exterior wall. It integrates the entrance area, kitchen, and dining area and opens toward the sleeping zone on the one side, and toward the building corner with recessed terraces on the other. Communal secondary and storage rooms between the apartments are located on the ground floor.

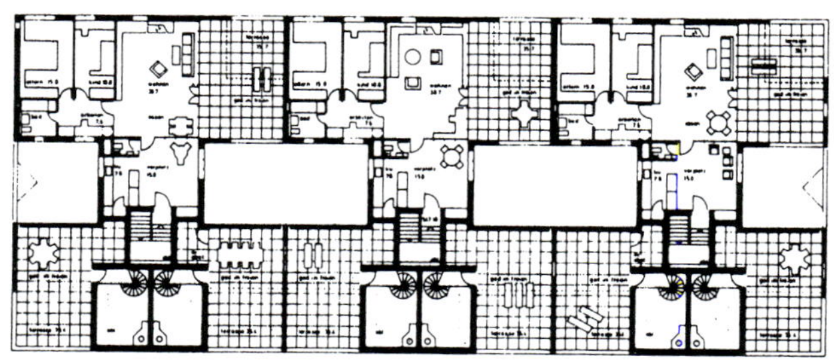

4

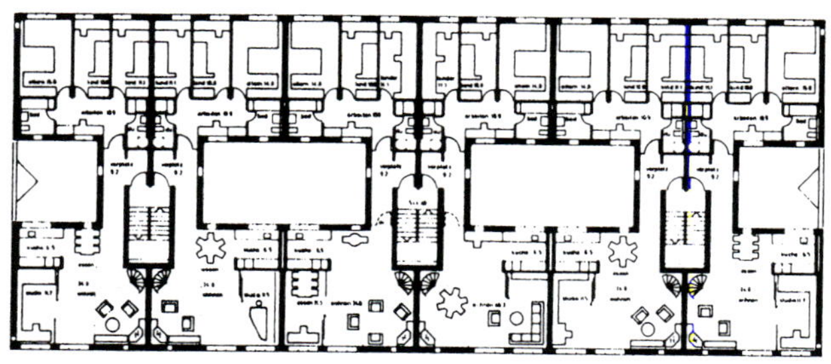

3

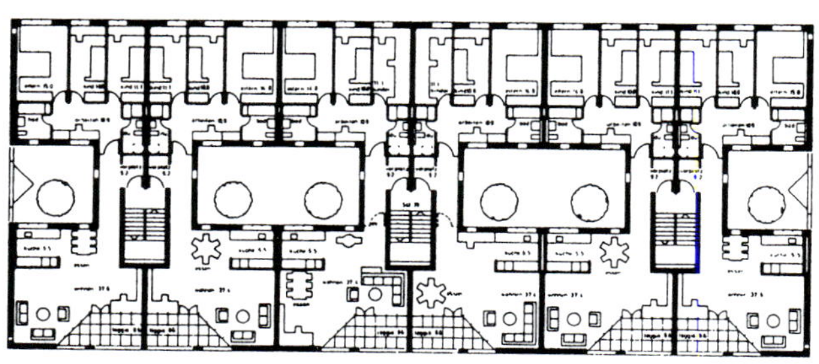

2

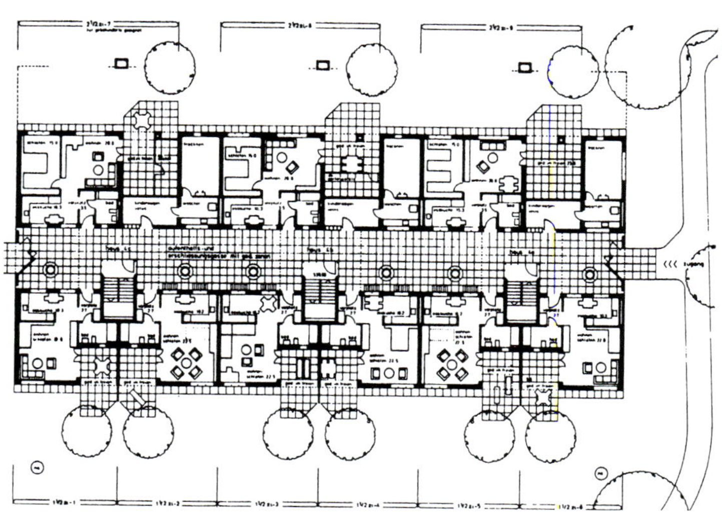

1

5

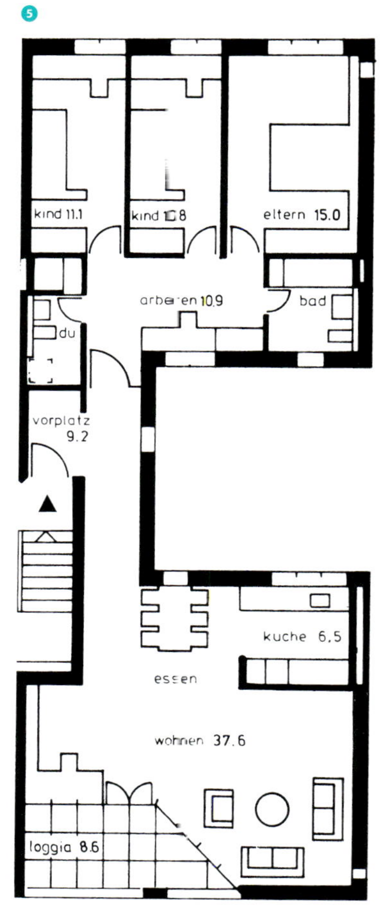

1–4 General layout ground floor–4th floor
5 4½-room apt., 2nd floor 1:200
6 5½-room apt., 3rd floor with roof terrace on 4th floor 1:200
7 3½-room apt., 4th floor with roof terrace 1:200

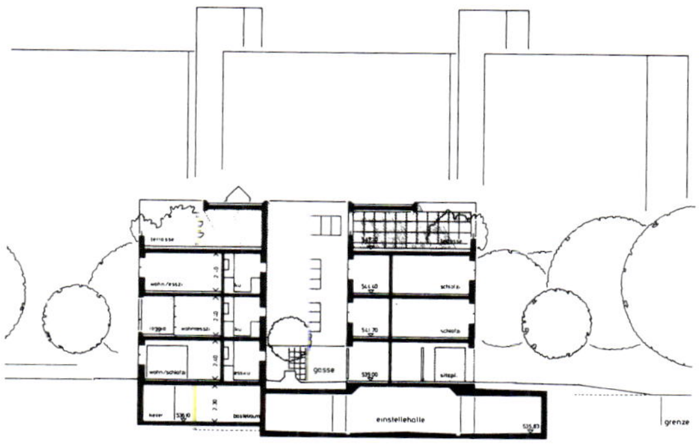

2.2
LINEAR BLOCK / SUPERBLOCK

Building type
connected blocks
with inside courts
4 stories
facing E/W

Date of construction
1985

Number of units
24

Size of units
1½-room apts., 45 m² (6 units)
2½-room apts., 57 m² (3 units)
3½-room apts., 99 m² (3 units)
4½-room apts., 116 m² (6 units)
5½-room apts., 126 m² (6 units)

Area per user
25–45 m²

Building depth
21 m

Access
internal pedestrian zone
with 3 stairwells,
double-loaded stairwells

Open spaces
access lane
play zone
ground-floor garden seating area
loggias
rooftop terraces

Parking
garage on basement story

Architect
Theo Hotz AG
architects and planners
Zurich

Location
Buchgrindelstrasse 4
Zurich-Wetzikon
Switzerland

6

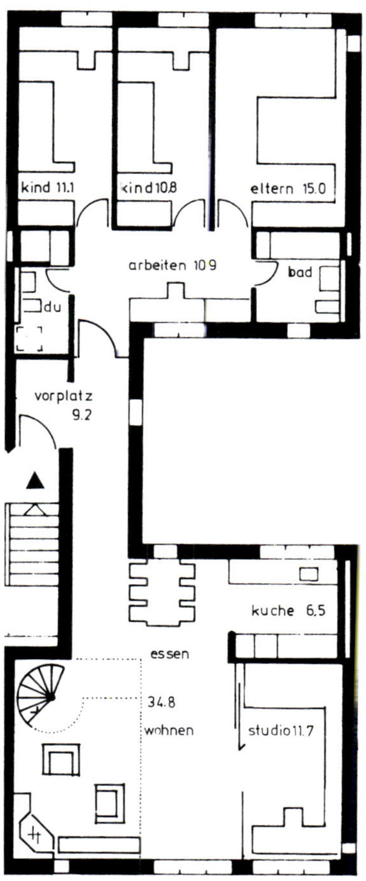

7

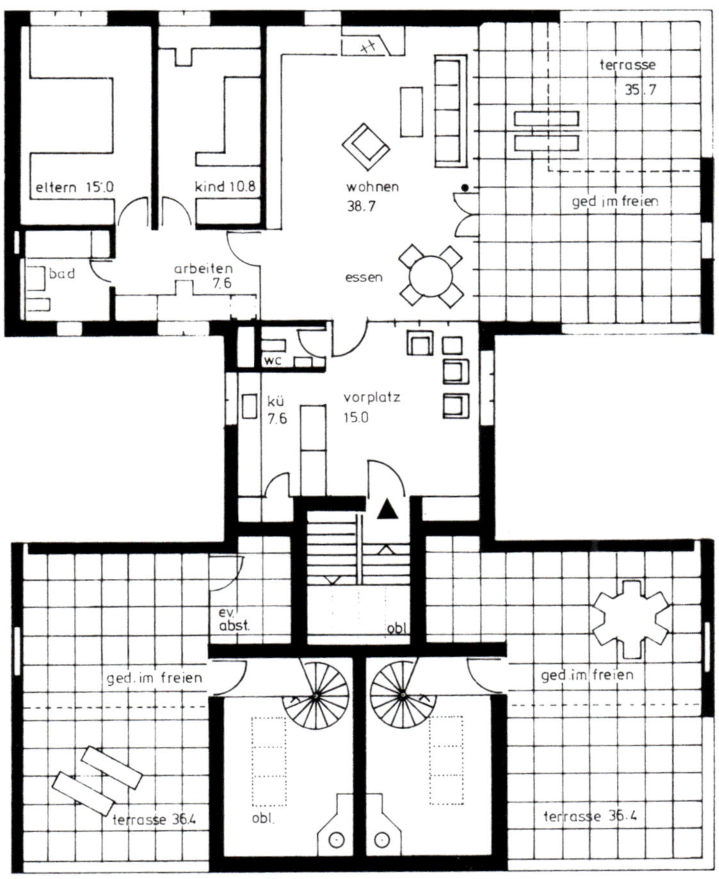

Referring to
floor plan 1:200

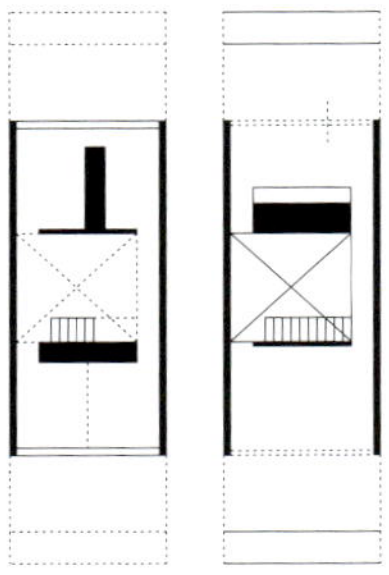

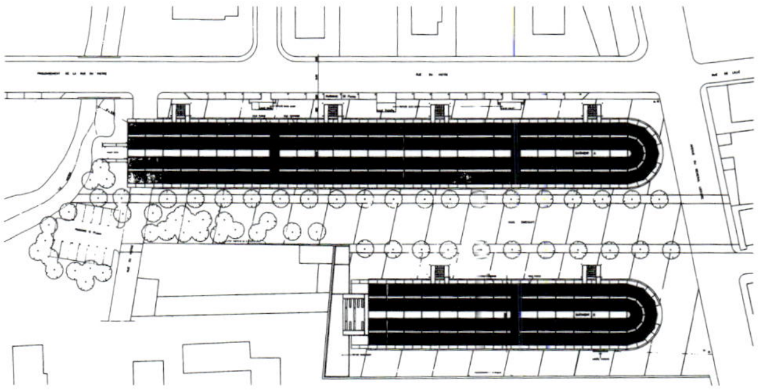

Avenue du Général Leclerc | Nouvel, Ibos | 1987

Two long, ship-like buildings on a mediterranean square with large plane trees. The open basement serves as a covered parking lot, the cars, hidden on the level lowered by a half story, are no longer visible from the grove of plane trees. The building is composed of two maisonettes, i.e. double floors, stacked one above the other, and an attic story. Every second story has 3-m-wide, perimeter galleries, which serve as covered walkways on the north side and as balconies on the south side. The fabrics are divided into 5-m-wide crosswall sections. The maisonettes can be completed without further structural constraints, the only fixed component being the location of the installation shaft. Aside from a few interior walls, the otherwise undivided space is defined by the bathroom and kitchen core, often set into the space, and the maisonette stairwell. Most of the apartments occupy one crosswall width and are developed vertically as two- or three-story maisonettes. Bathrooms and kitchens lie mostly on the north side, the living areas, many two-stories high, open onto the balcony. The building is executed as a rough concrete structure, made inhabitable with simple, prefabricated elements adopted from 1980s industrial architecture: galvanized folding doors open the living space onto the terrace, the entrance doors are made of steel sheet, a red alarm button serves as a door bell.

1

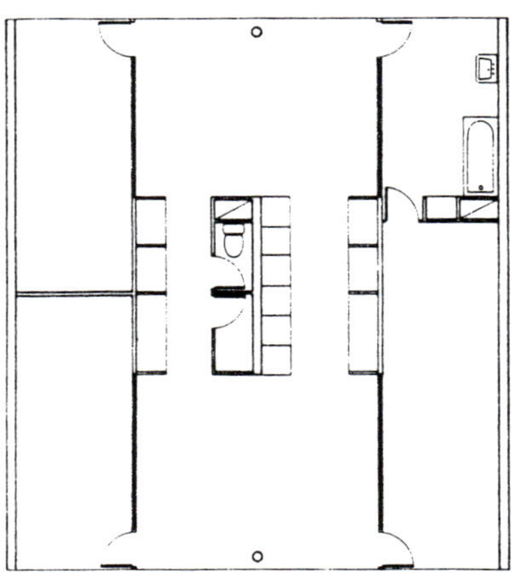

2

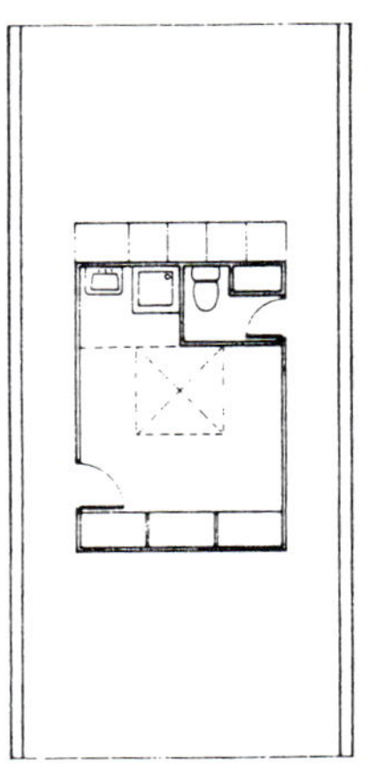

3

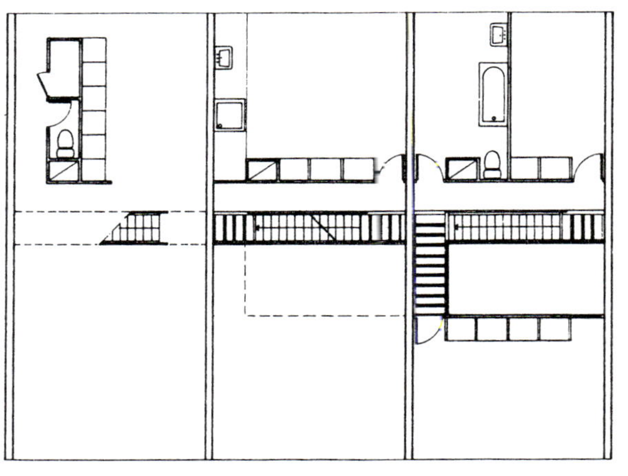

5

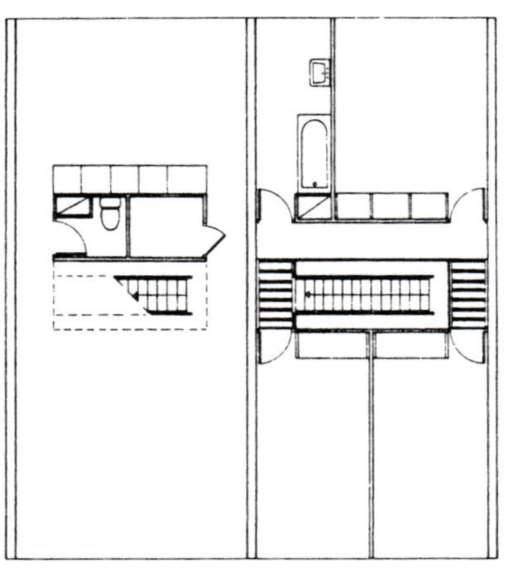

5

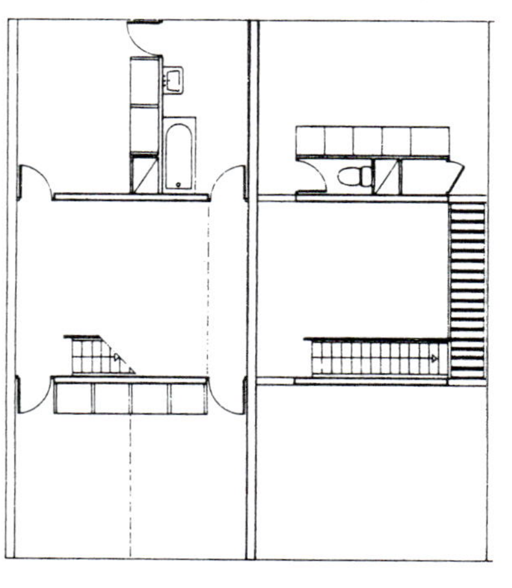

5

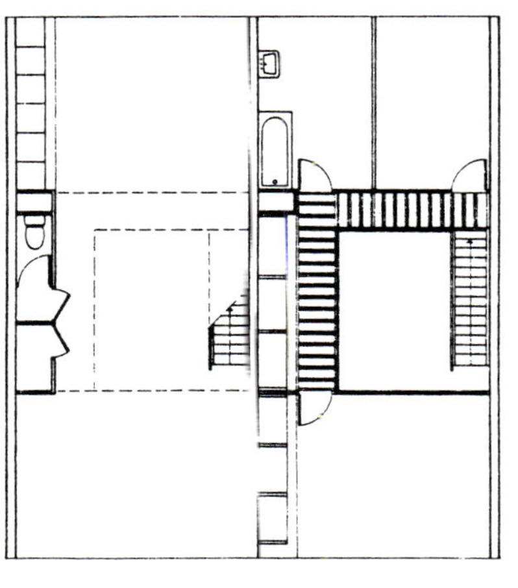

1 4-room apartment, one level 1:200
2 2-room apartment, one level 1:200
3 6-room maisonette, 3 levels 1:200
4 5-room maisonette, 2 levels 1:200
5 4-room maisonettes, 2 levels 1:200
6 3-room maisonette, 2 levels 1:200
7 2-room maisonette, 2 levels 1:200
8 2-room apartments, one level 1:200

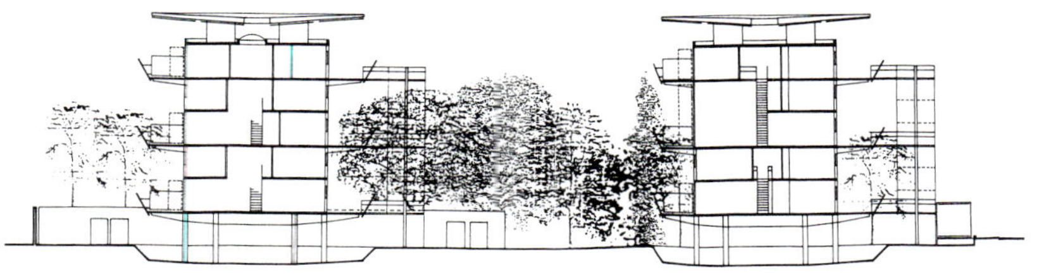

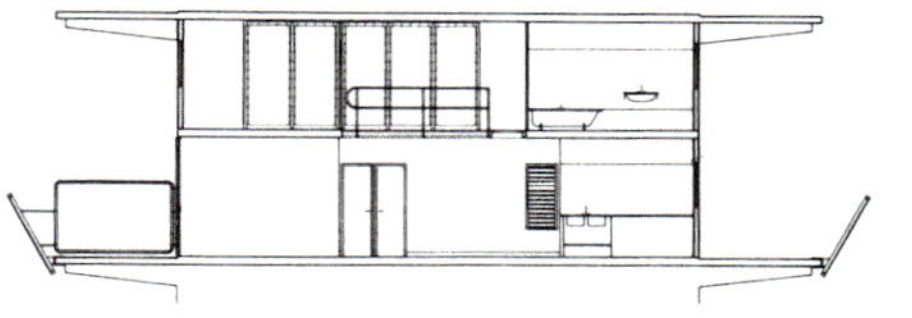

2.2
LINEAR BLOCK / SUPERBLOCK

4

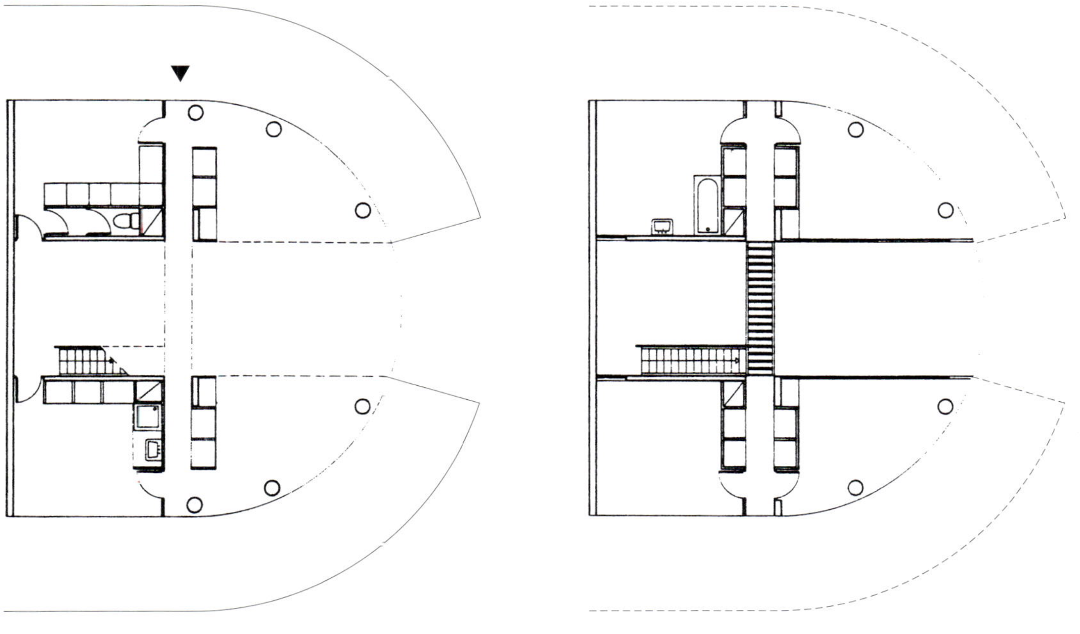

6

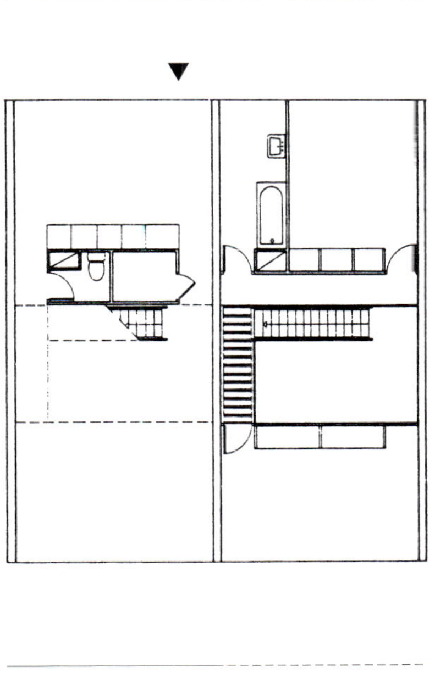

7

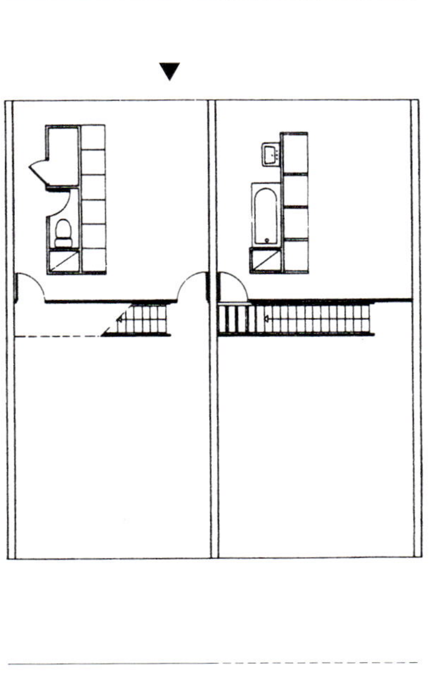

8

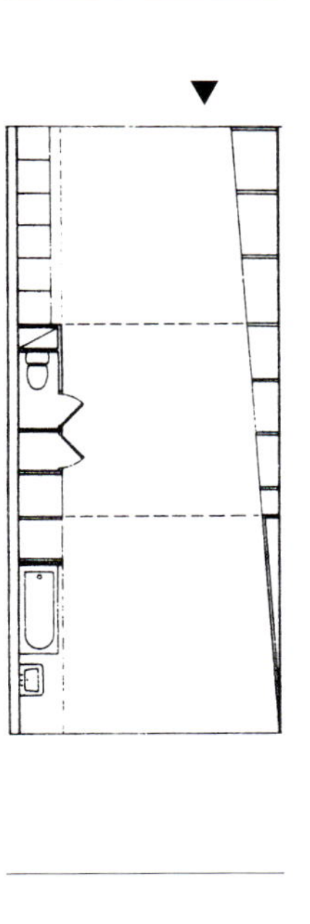

8

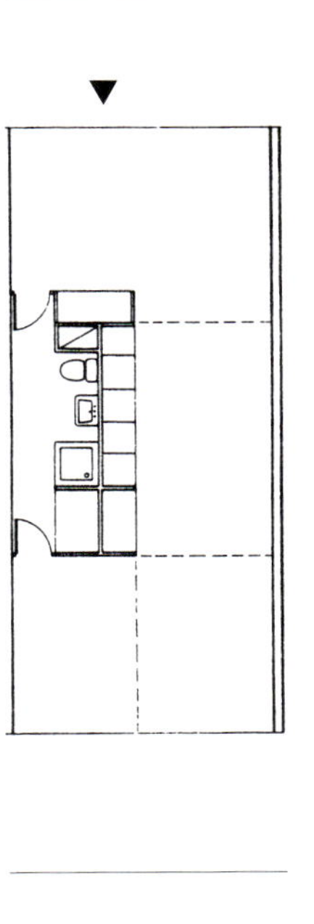

Building type
2 5-story bars
facing N/S

Date of construction
1987

Number of units
114

Size of units
2-room apts., 52/73 m²
3-room apts., 89/97 m²
2-/3-/4-/5-/6-room
maisonettes 100/116/170 m²
(17 different types)

Area per user
20–42.5 m²

Building depth
18 m

Access
gallery
steel stairs
on north side

Open spaces
rows of plane trees
between buildings,
balconies

Parking
parking spaces in sunken
ground floor

Architect
Jean Nouvel, Jean-Marc Ibos
Paris
with
Jean-Rémy Nègre
Frédéric Chambon

Location
Avenue du Général Leclerc
Nîmes
France

Referring to
floor plan 1:200

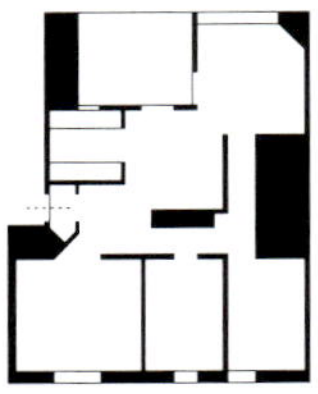

Mas Abelló Reus | Tusquets Blanca | 1988

Each of the four buildings is designed as a symmetrical urban villa, with four completely identical apartments per story. From the street side the buildings appear closed, everything that leaves an imprint of its particular function on the facade (kitchens, stairwells, balconies, etc.) has been placed on the garden side. The floor plans are developed in two directions: one parallel to the street with an optional living room/reception area and two additional bedrooms; the other, more dominant plan, in the diagonal line leads out to increasing brightness – with lobby, dining area, loggia, and a room overlooking the garden. Corridor segments around the dining area connect all rooms. The functions of the spatial sequences can be freely determined.

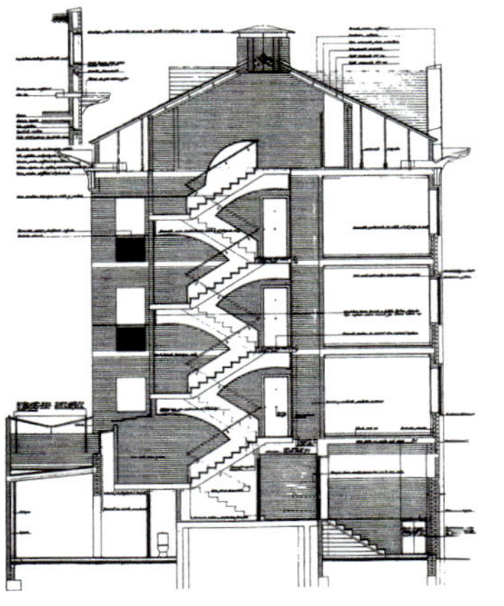

Building type
4 urban villas on block perimeter
4 stories
facing E/W

Date of construction
1986–1988

Number of units
48

Size of units
5-room apts., approx. 100 m²

Area per user
25 m²

Building depth
11 m

Access
double-loaded stairwells

Open spaces
terraces, gardens

Parking
basement

Architect
Oscar Tusquets Blanca
Barcelona
with
Carlos M. Díaz
Charles Bassó

Location
Tarragona
Spain

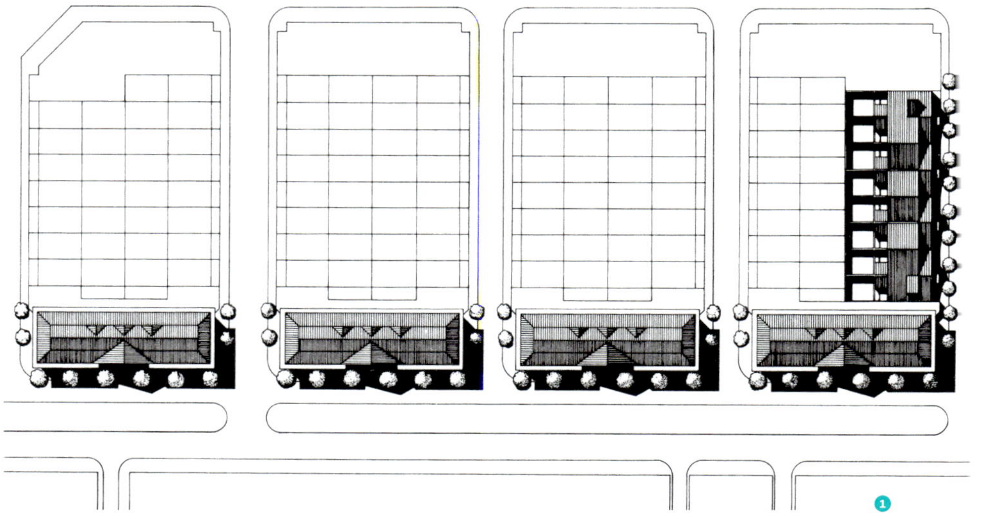

1

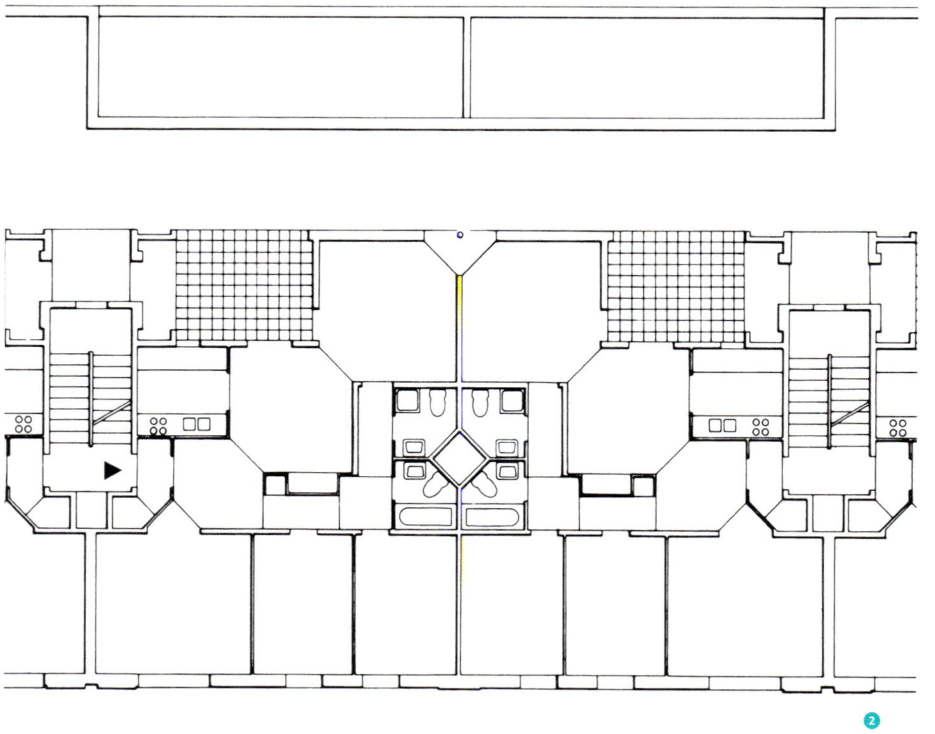

2

Site plan with 1
4 houses
Section of 2
typical floor plan
with identical
5-room apts.
1:200

Referring to
floor plan 1:200

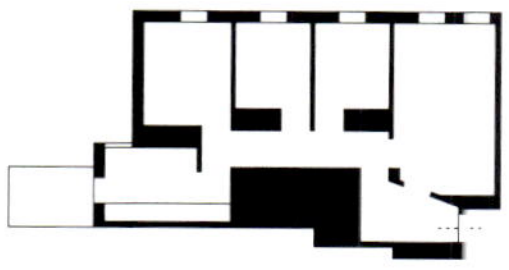

Carabanchel |
Cruz, Ortiz | 1989

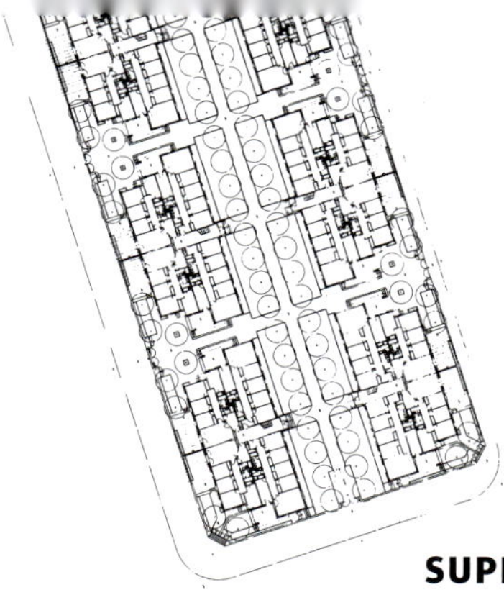

Ten identical multifamily buildings along a private path: despite the repetition of the symmetrical housing type, the complex still appears varied and open. Thanks to the staggered building ends, no two floors are alike. One room of the preceding apartment serves as a – partially covered – terrace for the unit above. An apartment is situated in each building corner, with rooms arranged along a corridor. An entrance foyer leads to the corridor; the bathrooms and kitchen are lined up along the interior, the bedrooms on the exterior. The living room is straight ahead. Both kitchen and living room open onto the terrace. In the 4-room apartments on the back side, the living room is located directly off the foyer space. The corridor is transformed into a private space. A tree-lined boulevard between the rows serves as a communal green space, and the roof terrace is for shared use.

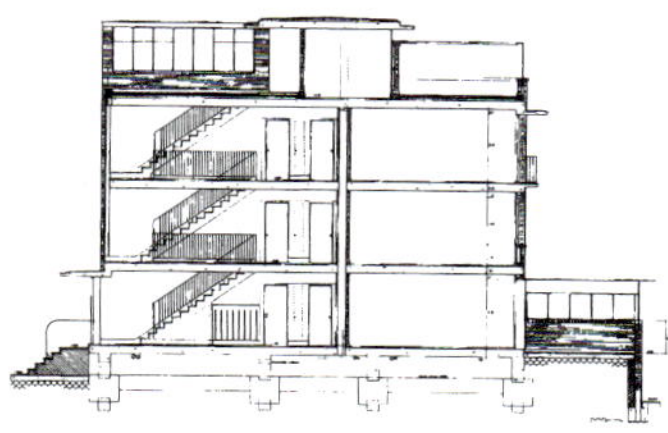

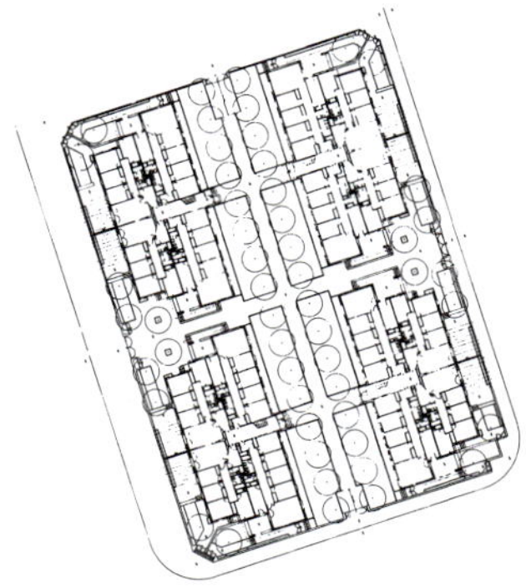

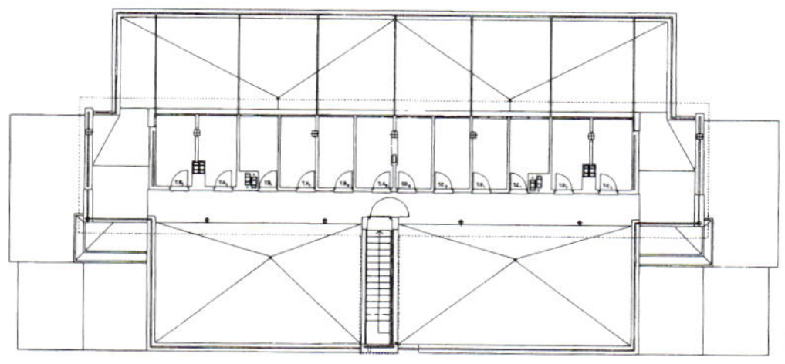

4

Building type
10 multifamily houses
3 stories
facing NW/SE

Date of construction
1986–1989

Number of units
120

Size of units
each house:
5-room apts., 90 m² (2 units)
4-room apts., 75/77.5 m² (3 units)
3-room apts., 59.5 m² (2 units)

Area per user
18–20 m²

Building depth
14.6 m

Access
quadruple-loaded stairwell

Open spaces
shared terraces, gardens, shared green between the houses

Parking
external parking

Architect
Antonio Cruz
Antonio Ortiz
Sevilla

Location
General Romero Basart
Madrid
Spain

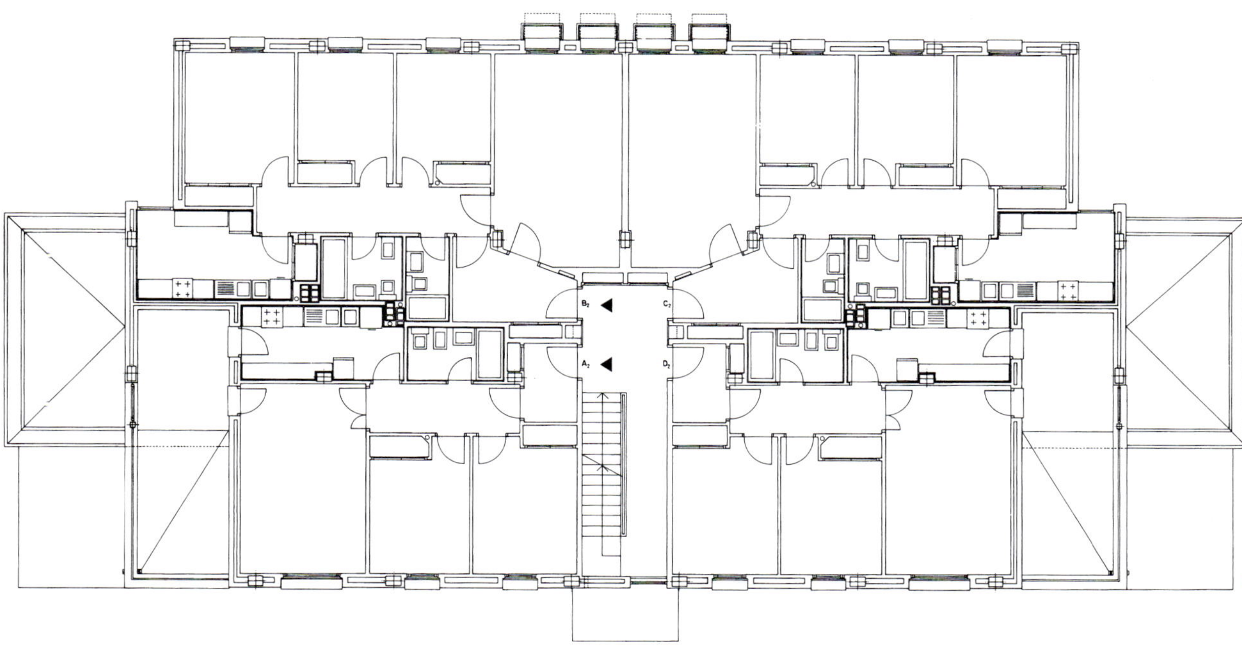

3

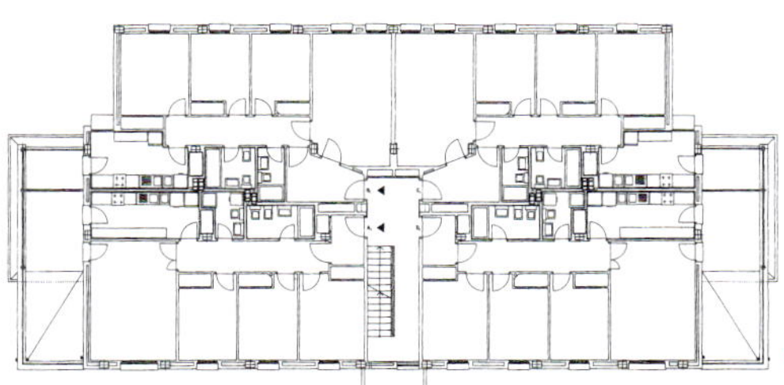

2

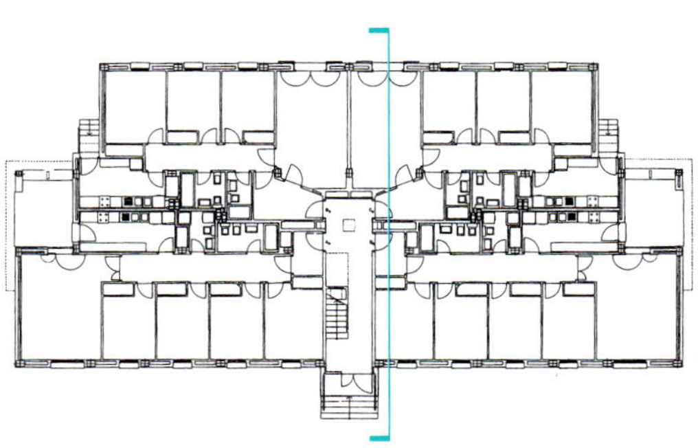

1

Ground floor with 4- and 5-room- 1
apartments 1:500
2nd floor with four 4-room- 2
apartments 1:500
3rd floor with 3- and 4-room- 3
apartments 1:200
Roof with storage spaces and communal 4
roof terraces 1:500

Referring to
floor plan 1:200

5

Nexus World | Holl | 1991

The aim of the housing strategy: variety of apartment types and, in each individual apartment, great flexibility of spatial divisions by means of rotating wall sections and shelving units. (Spatial division according to times of day: bedrooms become part of the living room/spatial division according to periods of life: parts of the apartment can be separated off.) The problem of the site: the street is in the south. Hence the comb-shaped plan, with inner courtyards on the south side (the courtyards are filled with water basins that reflect the light and are intended to symbolize peace and tranquility). Covered walkways with differing orientation, south or north, in part enclosed and in part open; stairwells located at both ends of the building. Basically there are two arrangements for the maisonettes: 1. The maisonettes have their living areas in the individual wings; the sleeping area, set off by one floor and pushed toward the back, lies in the continuous bar (light from east, south, west, north). 2. The maisonettes are stacked on top of each other in the continuous bar (light from south and north). Meandering paths with intermediate landings i.e. stairs lead from the entrances past one or two bedrooms to the central living area, which frequently provides direct access to two additional bedrooms.

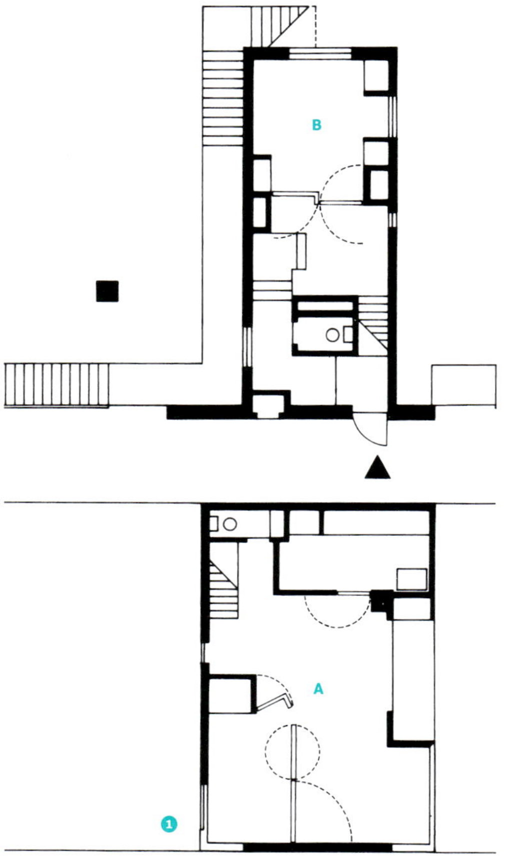

1

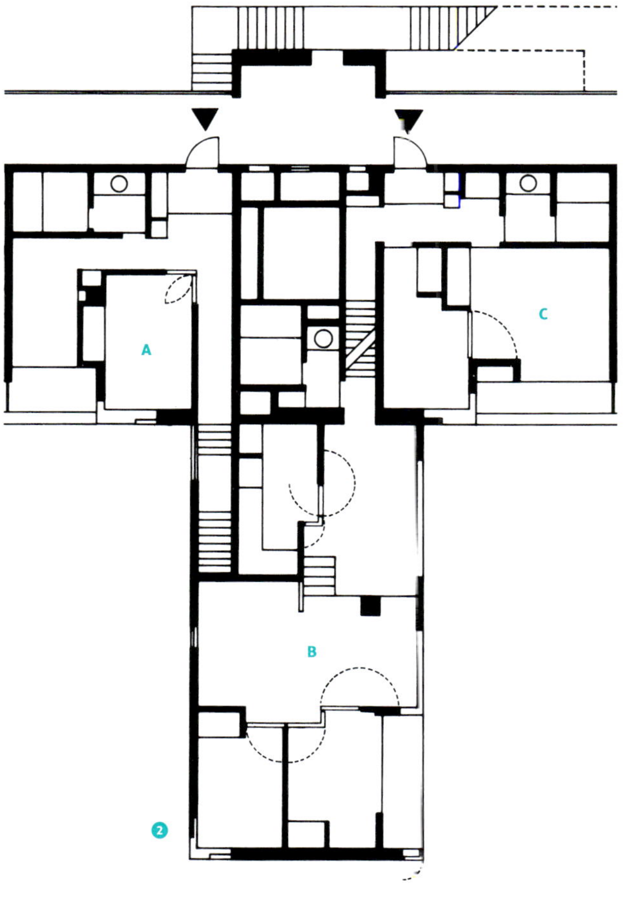

2

1 2 Floor plan sections 2nd and 3rd floor: 3 maisonettes **A**, **B**, **C** (only lower level) 1:200
3 4 Floor plan sections 4th and 5th floor: 4 maisonettes **C** (upper level), **D**, **E**, **F** 1:200
5 Schematic plans of 2nd, 3rd, 4th, 5th floor from bottom to top
6 Spatial division transformed in three steps

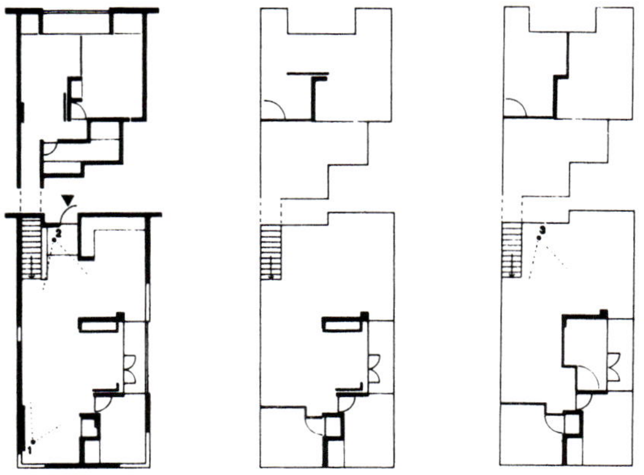

6

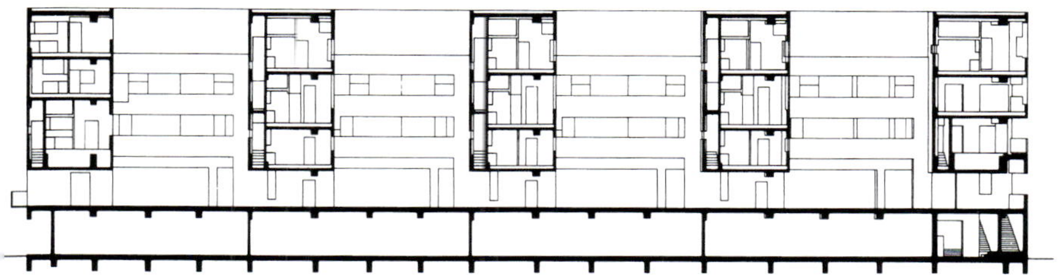

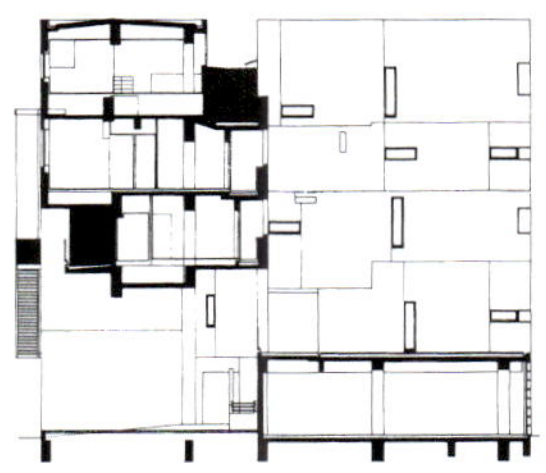

2.2
LINEAR BLOCK / SUPERBLOCK

Building type
comb-shaped plan
4 residential floors over stores
facing NW/SE, NE/SW

Date of construction
1991

Number of units
28

Size of units
2-/3-room apts.
73.5–110 m² (25 units)

Area per user
36.5 m²

Building depth
10/21 m

Access
galleries
(outside/central)

Open spaces
rooftop gardens

Parking
street space

Architect
Steven Holl Architects
New York

Location
Fukuoka
Japan

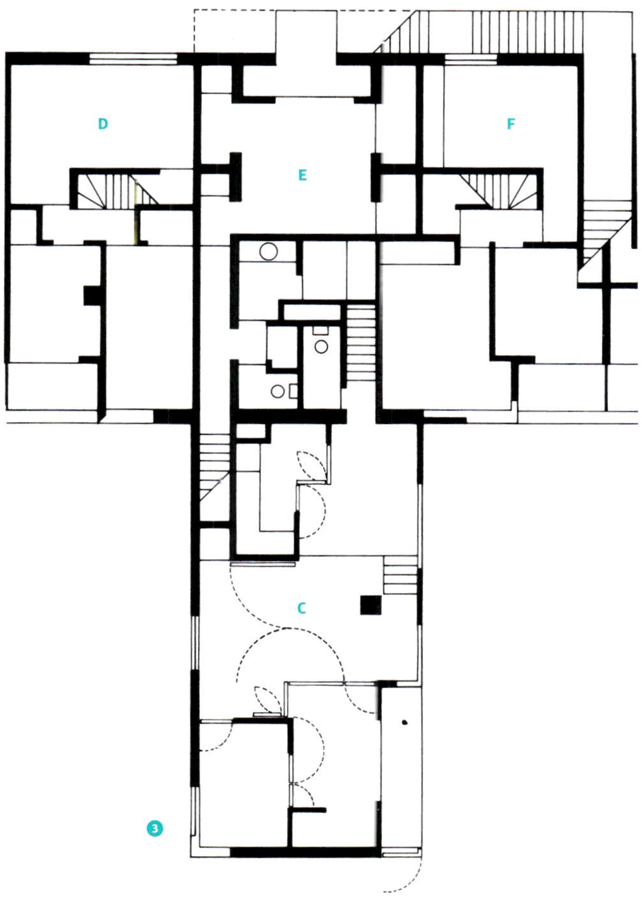

3

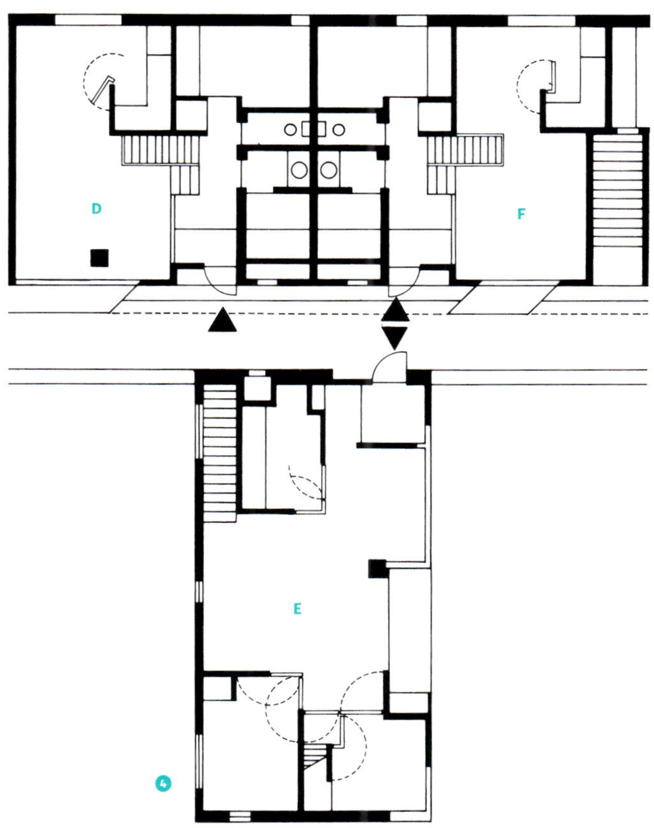

4

Referring to
floor plan 1:200

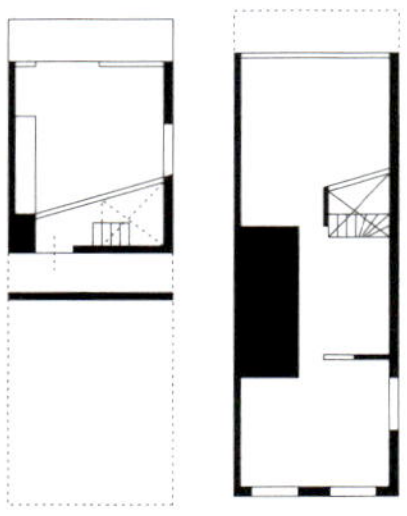

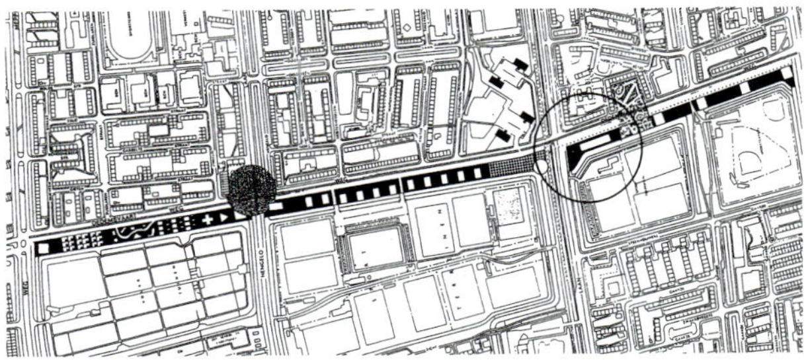

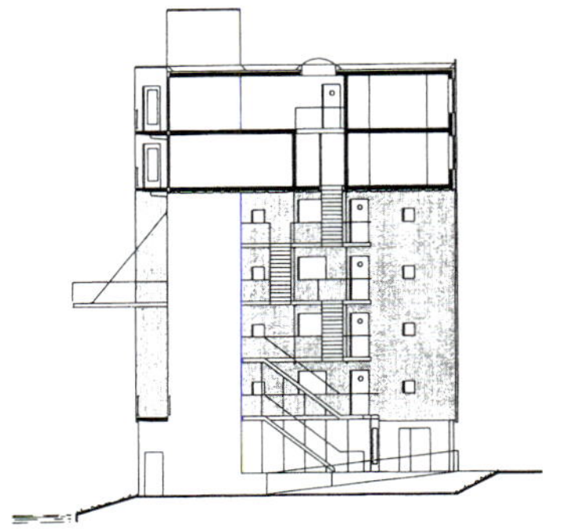

K25 | Zaaijer, Christiaanse | 1992

The project is located on a strip of the city on which The Hague's social housing festival was held. The festival's urban plan asked for a maximum of transparency – paradoxically – from the superblock, as this specific site marks the border between a residential and a green zone. Zaaijer and Christiaanse, who also conceived the master plan, recognized the challenge and put the bar on pilotis (with parking below) and gave their building two distinct sides: one, of stone, to the street; the second, dematerializing in glass and balconies facing the canal. Moreover, they perforated the bar with 3 openings in which the circulation and communal terraces are visible from the street. These perforations divide the very deep building into three sections, each with its own access. The entire building has an abundance of offerings: various apartment types for different resident groups; various forms of access (central corridor on the 4th and 6th floor, gallery on the 2nd floor) as well as gigantic communal terraces and indoor spaces for flexible use according to the desires of the residents. These can also be joined with the apartments to permit different living arrangements. Like the building as a whole, the floor plans of the individual units offer tremendous flexibility. The bath and kitchen complex is as compact as possible, there are as few walls as possible. The resident is called upon to design his/her own living environment. The apartments on the second floor are either continuous (1½ crosswall widths) or developed around the corner at the head of the block. The maisonettes adjoin the central corridor (1 crosswall width). At access level, they feature one room (kitchen with dining area, or bedroom with bathroom); on the level above or below, respectively, they offer a continuous living space (with kitchen or a generous sleeping area).

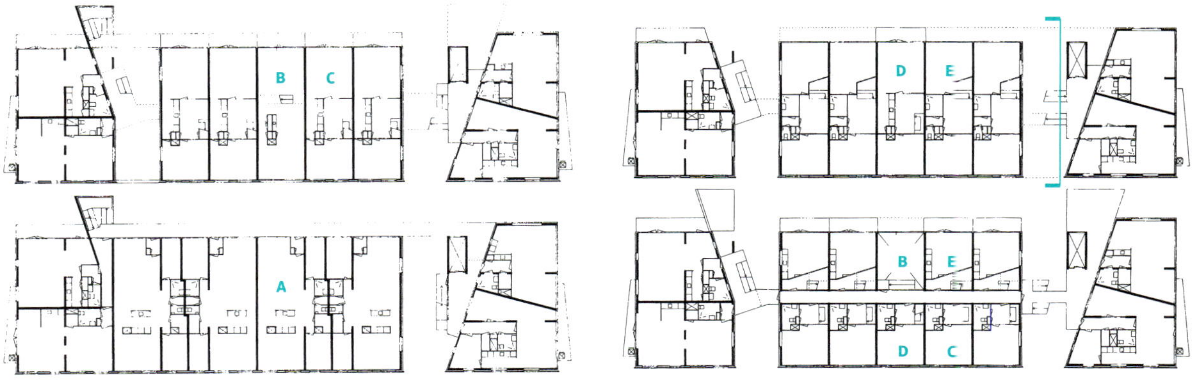

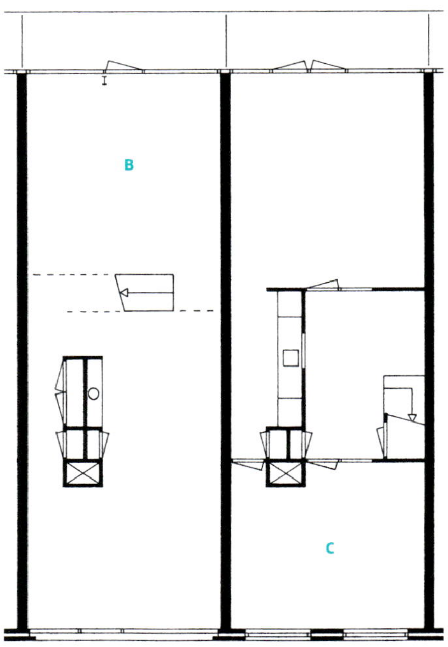

2

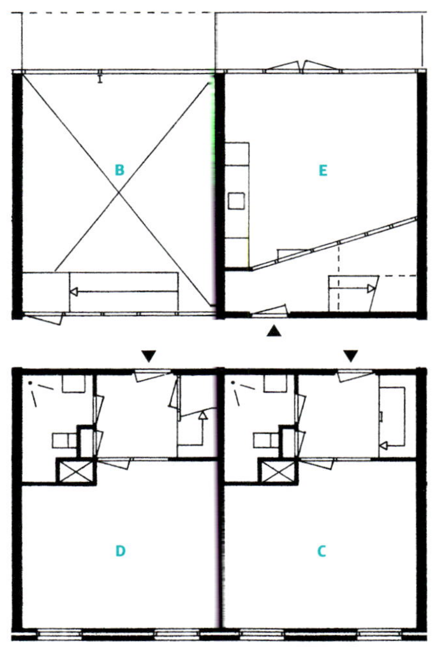

3

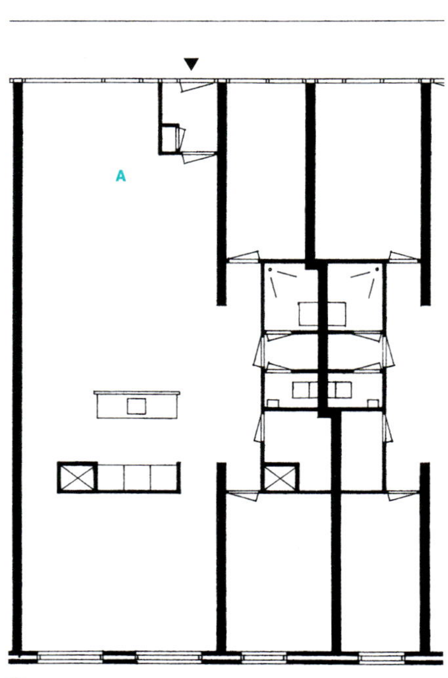

1

1 2nd floor: 4-room apt. **A** 1:200
2 3rd floor: communal space **B**, 3-room mais. **C** 1:200
3 4th floor: three 3-room mais. **C**, **D**, **E** 1:200
4 5th floor: two 3-room mais. **D**, **E** 1:200
5 6th floor: 2-room apt. **F**, 3-room mais. **D**, **E** communal terrace **G** 1:200
6 7th floor: communal terrace **G**, 3-room mais. **D** 1:200
7 General layout 2nd/3rd, 4th/5th, 6th/7th floor (starting at the bottom)

2.2
LINEAR BLOCK / SUPERBLOCK

Building type
superblock
6 stories
facing NE/SW

Date of construction
1988–1992

Number of units
44

Size of units
11 unit types
2-room apts., 56–65 m² (12 units)
3-room apts., 69–85 m² (15 units)
3-room mais., 97–99 m² (13 units)
4-room apts., 103 m² (4 units)

Area per user
23–32.5 m²

Building depth
14.6 m

Access
gallery on second floor, short center corridor on 4th and 6th floors

Open spaces
balconies, additional communal spaces and terraces

Parking
on street level under the building

Architect
Art Zaaijer
Kees Christiaanse
with Han van De Born
Rotterdam

Location
Dedemsvaartweg 893–981
The Hague
The Netherland

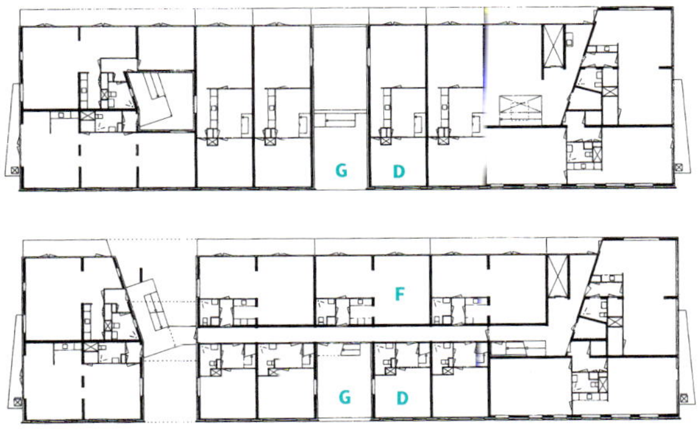

7

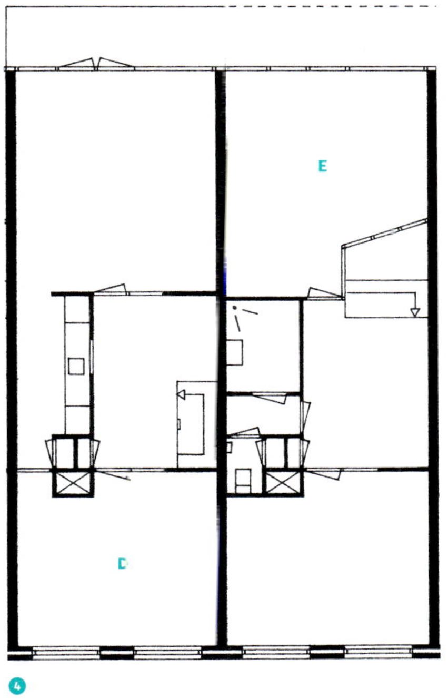

4

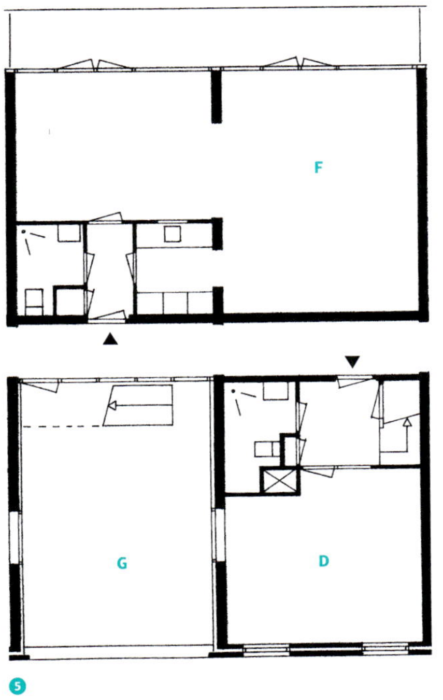

5

6

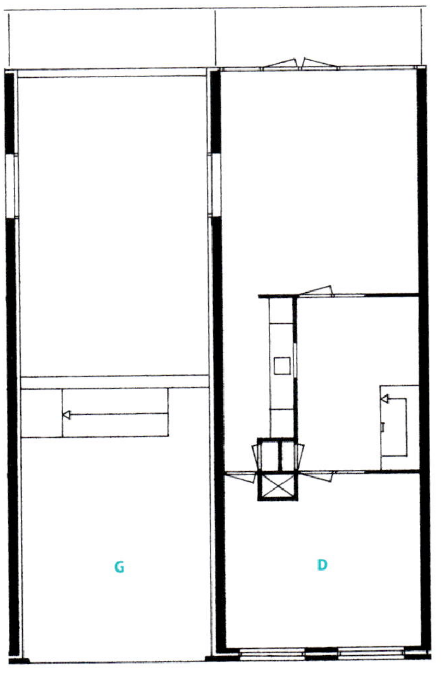

Referring to floor plan 1:200

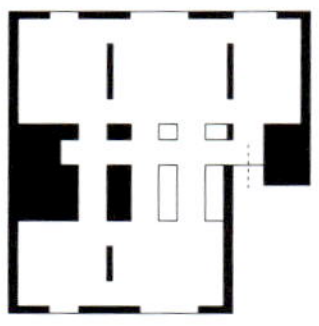

Bahnhofstrasse |
Riegler, Riewe | 1994

A concrete bar of precise and uncompromising structure which offers great flexibility within its rules. The extruded-looking building is not subordinate to its haphazard surroundings. Rather, it sits autonomously in the landscape. The smooth facade is unfettered by loggias or balconies. It comes alive instead through the animated image of floor-high sliding shutters. On the east elevation, these shutters are made of expanded metal and provide visual protection; on the west side, they are made of metal frames with nylon fabric and filter sunlight. They conceal French windows. Opening to 180°, these windows transform the living room into a loggia. Basically, there exists only one, albeit extremely flexible, apartment type in the building – in smaller and larger versions: the rooms offer a functional openness thanks to their size, shape, and individual entrance area. The plan can be read in two ways with the spaces ordered lengthwise or transversely into three zones (4½-room apt.). Lengthwise: living rooms and bedrooms along the two outer sides, the service core (hall, kitchen, bathroom) in the middle. Transverse: first axis, a half-bedroom (7.5 m^2) and guest WC; second axis, living room and kitchen; third axis, two bedrooms and bathroom. The spaces are connected lengthwise by wide sliding doors and transversely by folding doors. The spaces can thus be linked with complete flexibility. There is one additional variation: occupants can request that the kitchen be located in the small room on the west side.

2

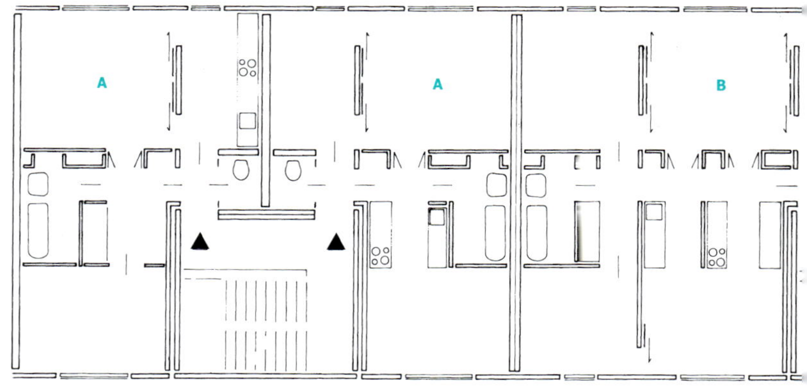

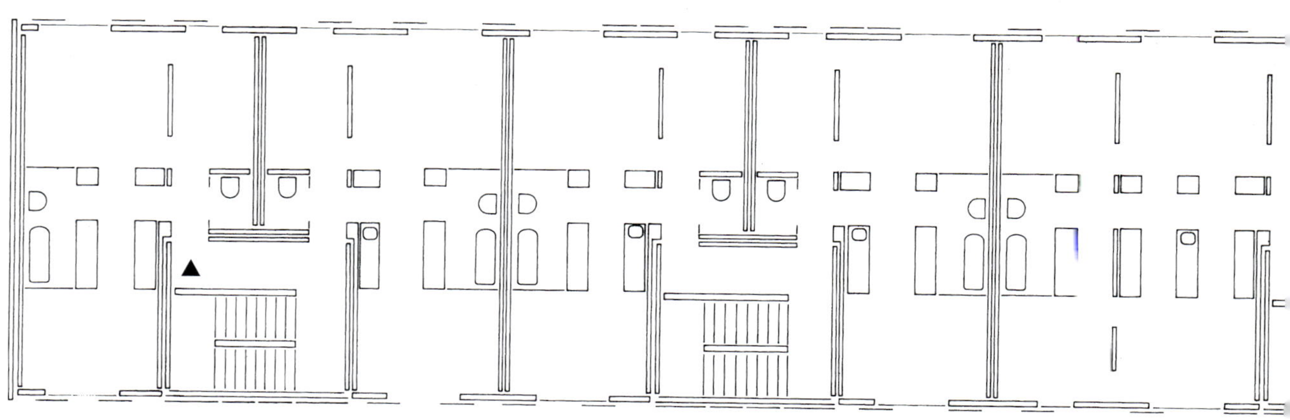

1

1 Typical floor plan 1:200

2 Type **A**: 2½-room apartment, with kitchen in the small room or in the center 1:200
Type **B**: 4½-room apartment, with kitchen in the center or in the small room 1:200

2.2
LINEAR BLOCK / SUPERBLOCK

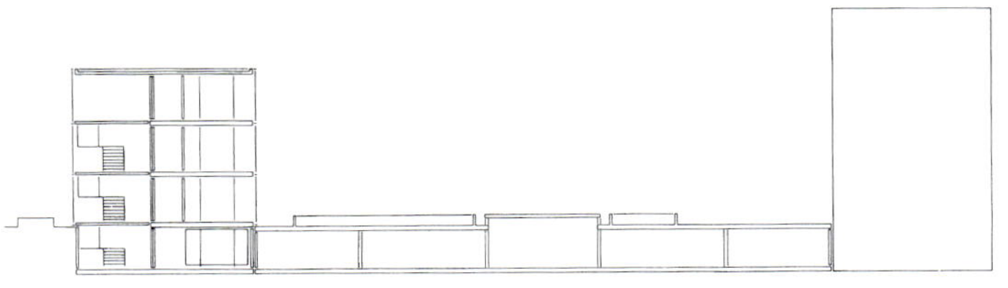

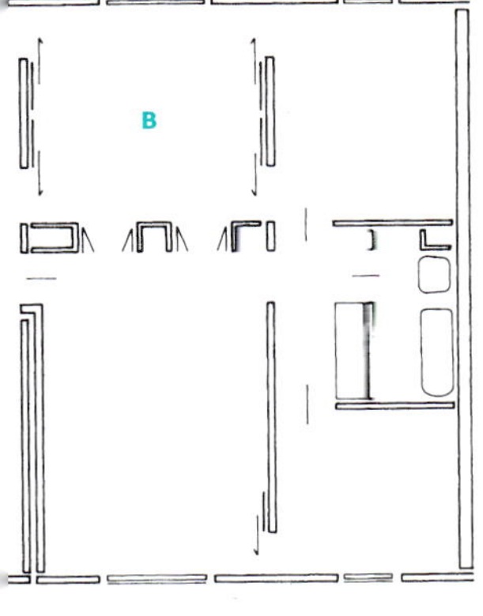

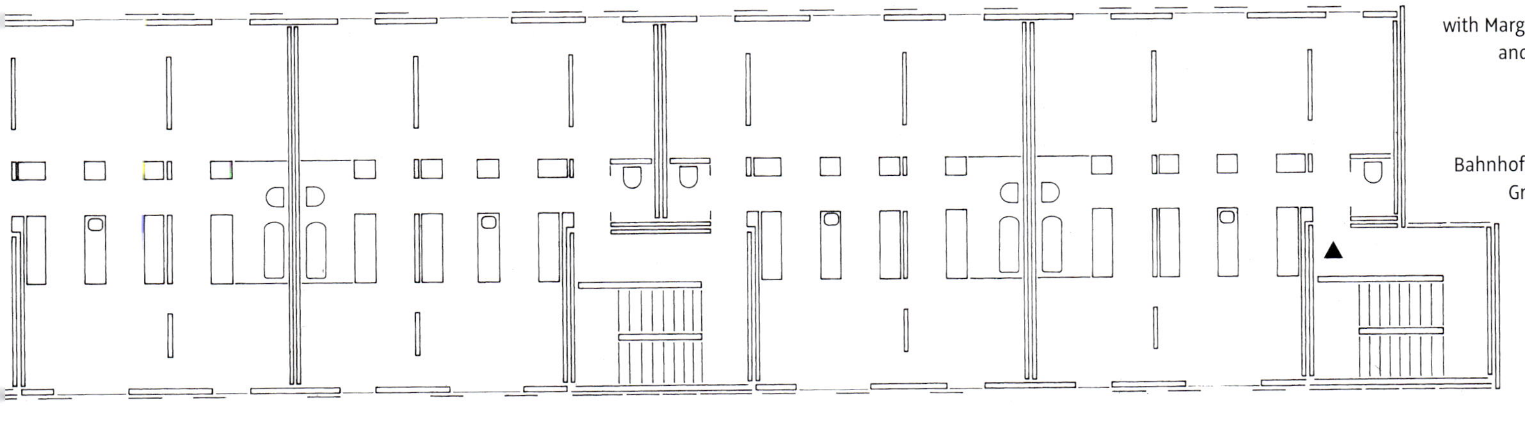

Building type
bar
3 stories
facing E/W

Date of construction
1992–1994

Number of units
24

Size of units
2½-room apts., 46 m² (6 units)
4½-room apts., 76 m² (15 units)
5½-room apts., 96 m² (3 units)

Area per user
19–23 m²

Building depth
10 m

Access
double-loaded
stairwell

Open spaces
living spaces with floor-high windows become loggias, public open spaces

Parking
subterranean garage

Architect
Florian Riegler
Roger Riewe
with Margarethe Müller
and Birgit Theißl
Graz

Location
Bahnhofstraße 10 a–e
Graz-Straßgang
Austria

Referring to floor plan 1:200

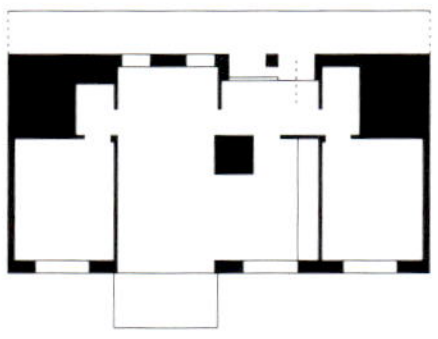

Tyroltgasse | Kovatsch | 1994

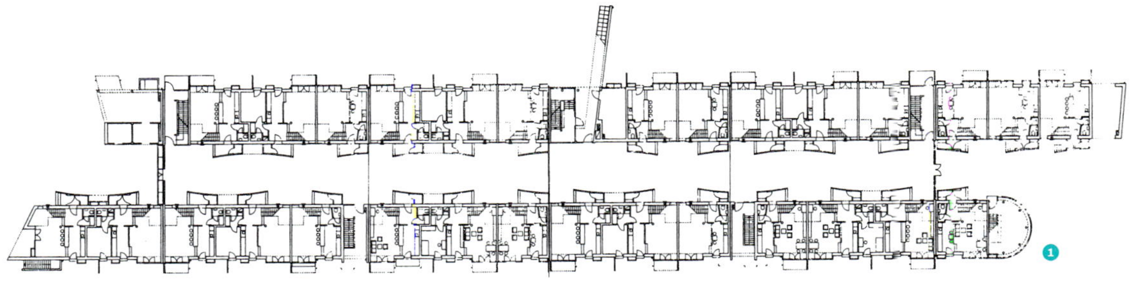

1

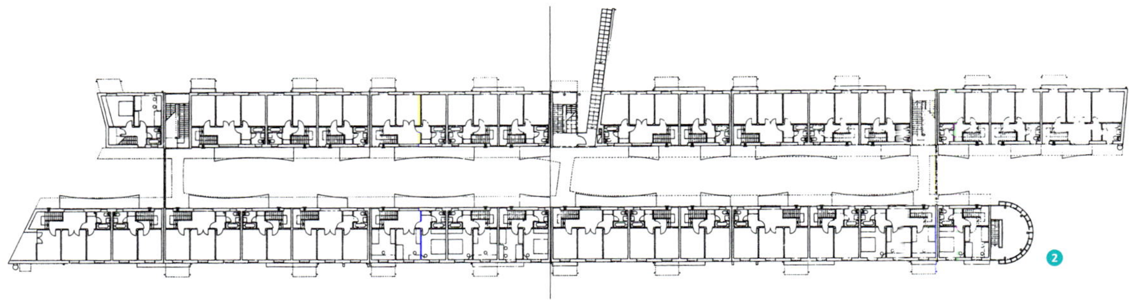

2

The clear, structure-giving form of a 110-meter-long double row is placed within a haphazardly developed residential area. A glass-roofed hall joins the two parallel north-south rows. It serves as a means of access and communal room. The hall's strong linearity is softened by the maisonettes' entrance porches (first and second floor), their colored boxes pushing forwards at an angle into the corridor. The maisonettes' entrances and bathrooms lie towards the passageway with all other rooms facing their own individual gardens. Two steps elevate and separate the living room from the kitchen (and, depending on the plan, an additional bedroom). The stairs have a high window towards the hall and lead up to further bedrooms and the bathroom. The third-floor apartments are reached via slightly curved access galleries which cantilever into the hall and are linked by bridges at three points. These single-story apartments consist of a string of identical rooms, kitchen, living room or bedroom, and a zone of unspecified use in front facing the hall. This area is used as access to the rooms, but it is also broad enough for working or playing. The living room – not separated from the kitchen – can carry over into this zone to the corridor, and across to the garden via the balcony in the opposite direction. Also at the disposal of the residents, who have deliberately chosen a form of communal living, is a range of other rooms and workshops at each end of the row.

3

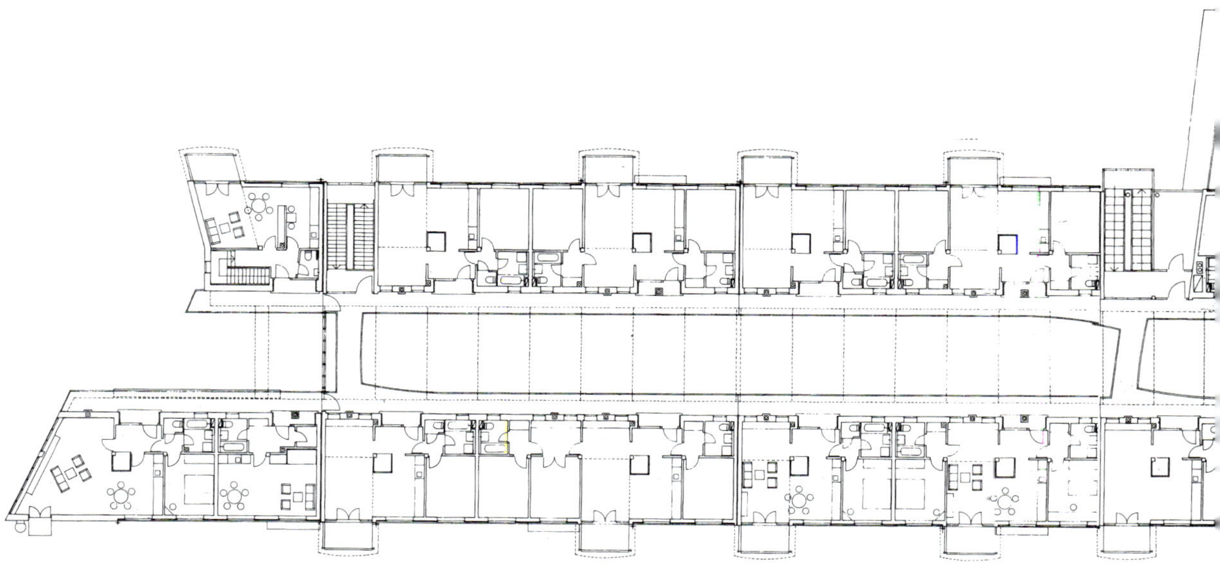

4

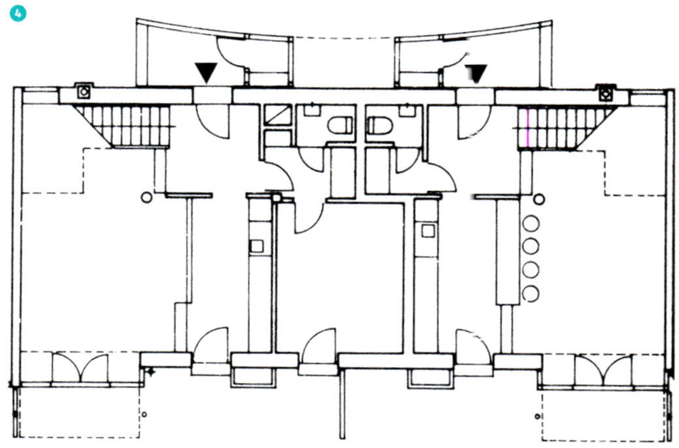

1 Ground floor 1:1000
2 2nd floor 1:1000
3 3rd floor 1:500
4 Ground floor and 2nd floor with 3- and 5-room maisonette 1:200
5 3rd floor with 3-room apartment 1:200

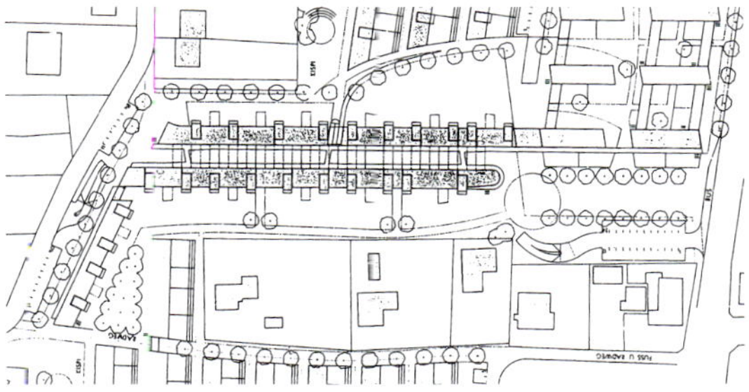

2.2 LINEAR BLOCK / SUPERBLOCK

Building type
double row with central hall
(access/communal functions)
3 stories
facing E/W

Date of construction
1994

Number of units
50

Size of units
1½-room apt., 38 m² (1 unit)
2-room apts., 61–65 m² (10 units)
3-room apts., 76–81 m² (6 units)
3-room mais., 76–81 m² (11 units)
4-room apts., 93–103 m² (3 units)
4-room mais., 93–103 m² (19 units)

Area per user
23–38 m²

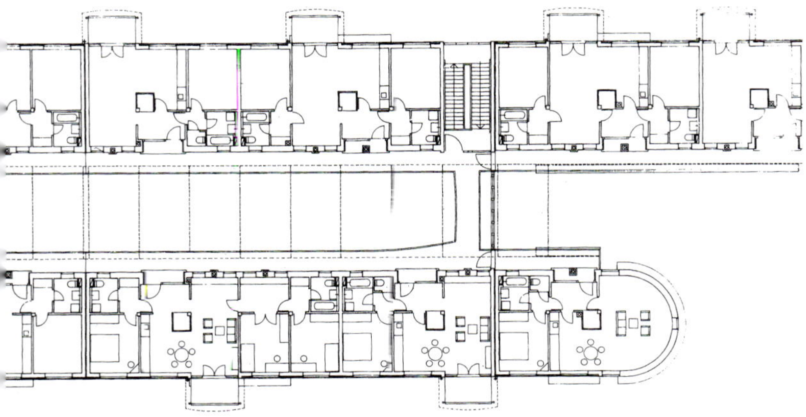

Building depth
7.25 m (individual row)
8 m (hall)

Access
maisonettes:
ground floor
from roofed hall
one-story apartments:
access gallery

Open spaces
gardens
balconies

Parking
underground garage

4

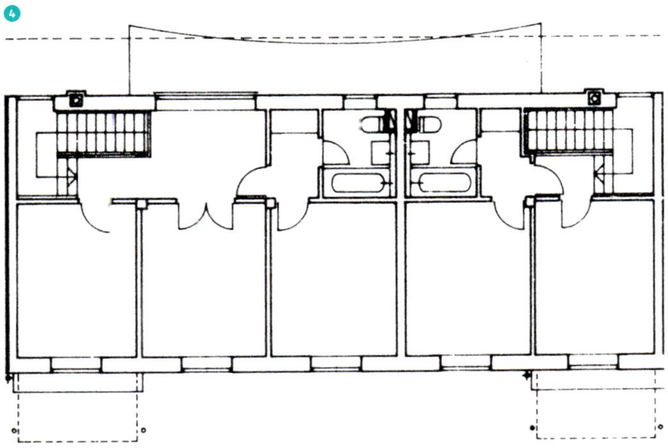

5

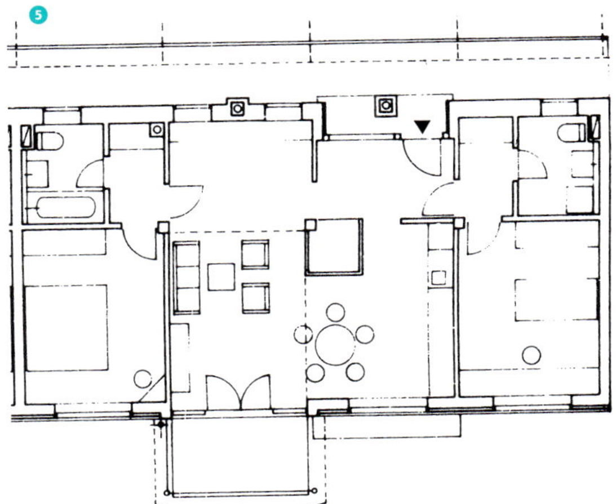

Architect
Manfred Kovatsch
with Helmut Bielenski
Gerhard Breu
Munich

Location
Tyroltgasse
Graz
Austria

Referring to
floor plan 1:200

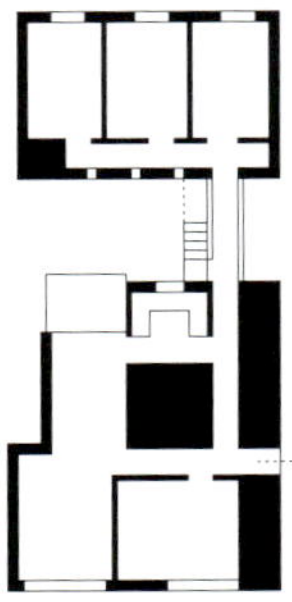

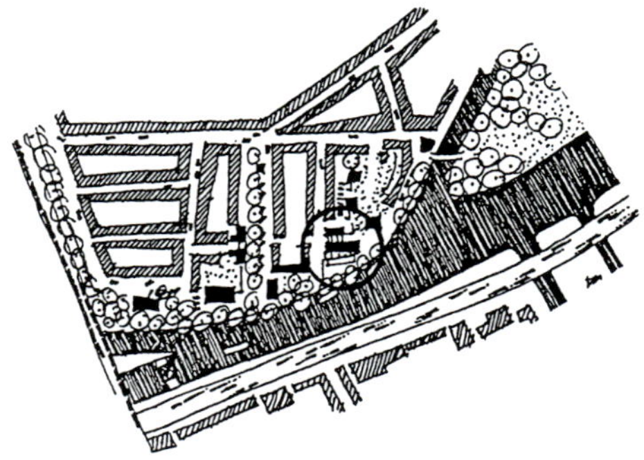

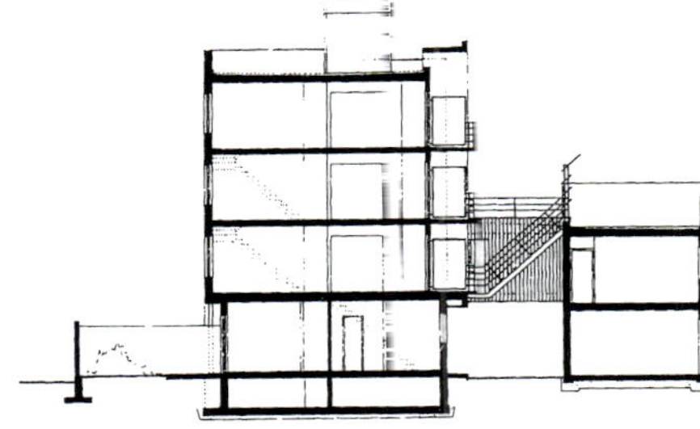

Hoge Pontstraat | Dercon, T'Jonck, Van Broeck | 1996

Double row. This is part of an urban plan intended to refurbish the impoverished Scheldekaai industrial area along the banks of the Nederschelde. Although the ambitious urban planning scheme won the architects the competition, it has been realized only partially and thus the building stands alone at present. It is divided into a principal row and a secondary row. The secondary row stacks garage and storage spaces at the bottom, children's rooms above, and is topped by spacious roof terraces. The colorful principal row contains all remaining functions, the circulation area, and other apartments. The generous access (three stairwells and two lifts) is oriented towards the semipublic alley between the rows. Small apartments with gardens, ideally suited for the elderly, are arranged on the ground floor. Large five-room apartments for families with children are on the second floor. A master bedroom, living room with balcony, kitchen, storage, and guest WC are grouped around the central bathroom. The three separate children's bedrooms (with WC), all lined up side by side above the garages, are reached via a wood-clad bridge. The roof terrace is, in turn, reached by a stair parallel to this bridge. On the third floor, one finds 2- and 3-room apartments with identical layouts. Their bridges open onto large roof terraces. Above are similar 3-room apartments with balconies. The western row, without lift, accommodates 2-room apartments (for singles) with large kitchens and two smaller rooms.

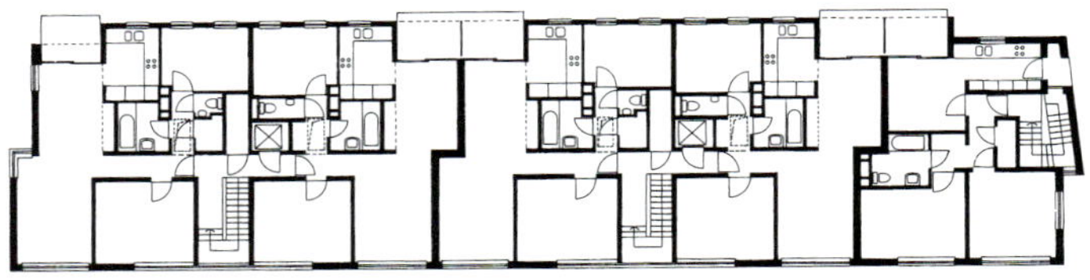

4

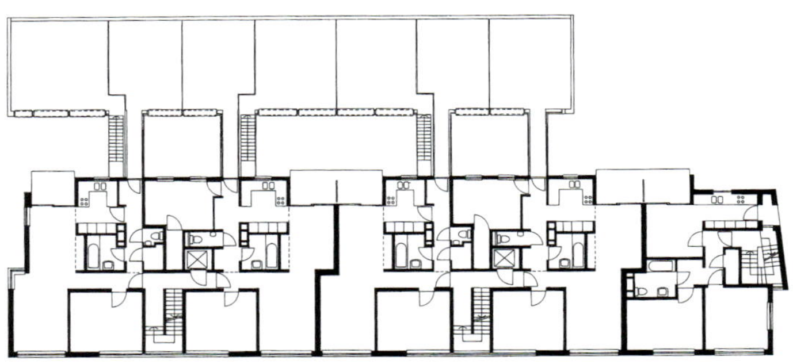

3

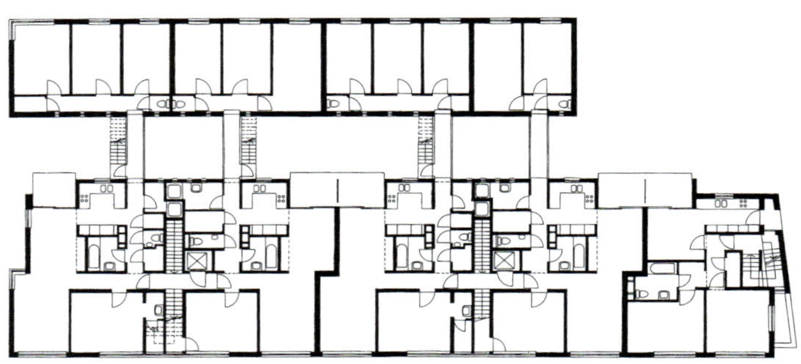

2

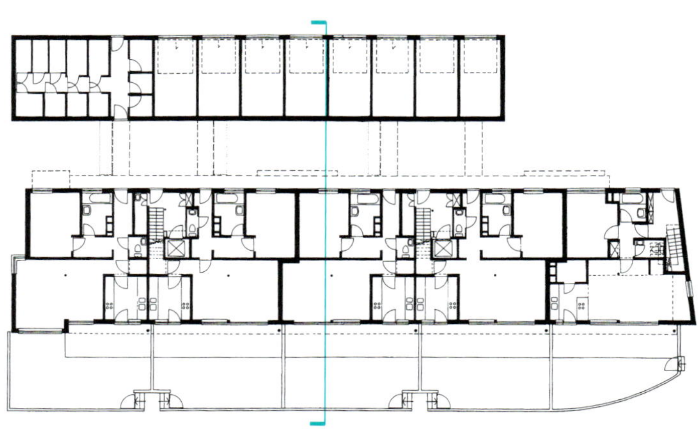

1

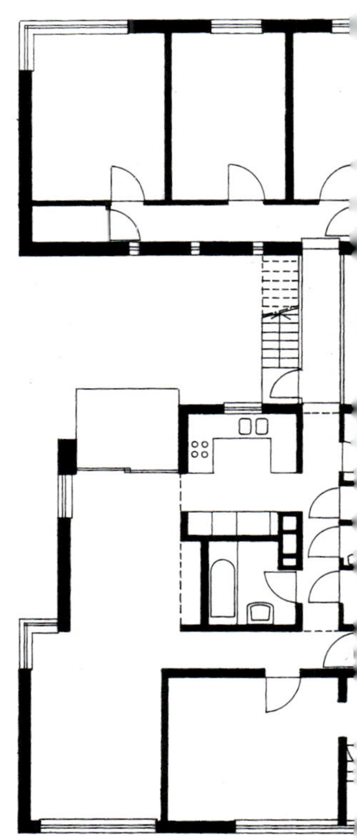

1 Ground floor: 2-room apts. with garden 1:500
2 2nd floor: 2-, 4- and 5-room apts. with roof terraces on level above 1:500
3 3rd floor: 2- and 3-room apts. with roof terraces 1:500
4 4th floor: 2- and 3-room apts. 1:500

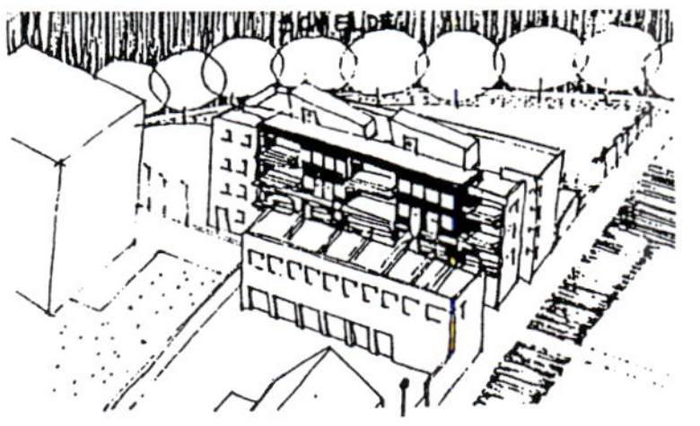

6

2.2 LINEAR BLOCK / SUPERBLOCK

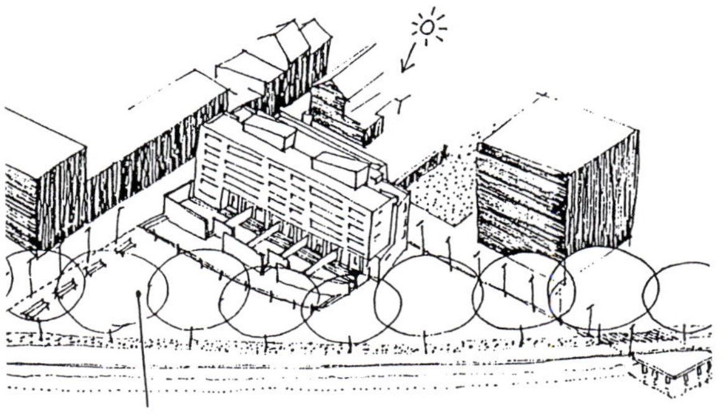

6

Building type
double row, 2 and 4 stories
joined at second level
facing N/S

Date of construction
1992–1996

Number of units
20

Size of units
2-room apts., 54.5/60.5/62.5 m² (10 units)
3-room apts., 71–75.5 m² (6 units)
4-room apt., 97 m² (1 unit)
5-room apts., 110.5/113 m² (3 units)

Area per user
22.1–31.5 m²

Building depth
10.3 m (main row),
5.6 m (secondary row)

Access
3 single- and double-loaded stairwells,
2 elevators

Open spaces
garden, rooftop garden, balconies, communal garden

Parking
garage (8 spaces)
open parking spaces

Architect
Bruno Dercon
Pieter T'Jonck
Leo Van Broeck
samenwerking
ir.architecten
Leuven

Location
Hoge Pontstraat
Ghent-Scheldekaai
Belgium

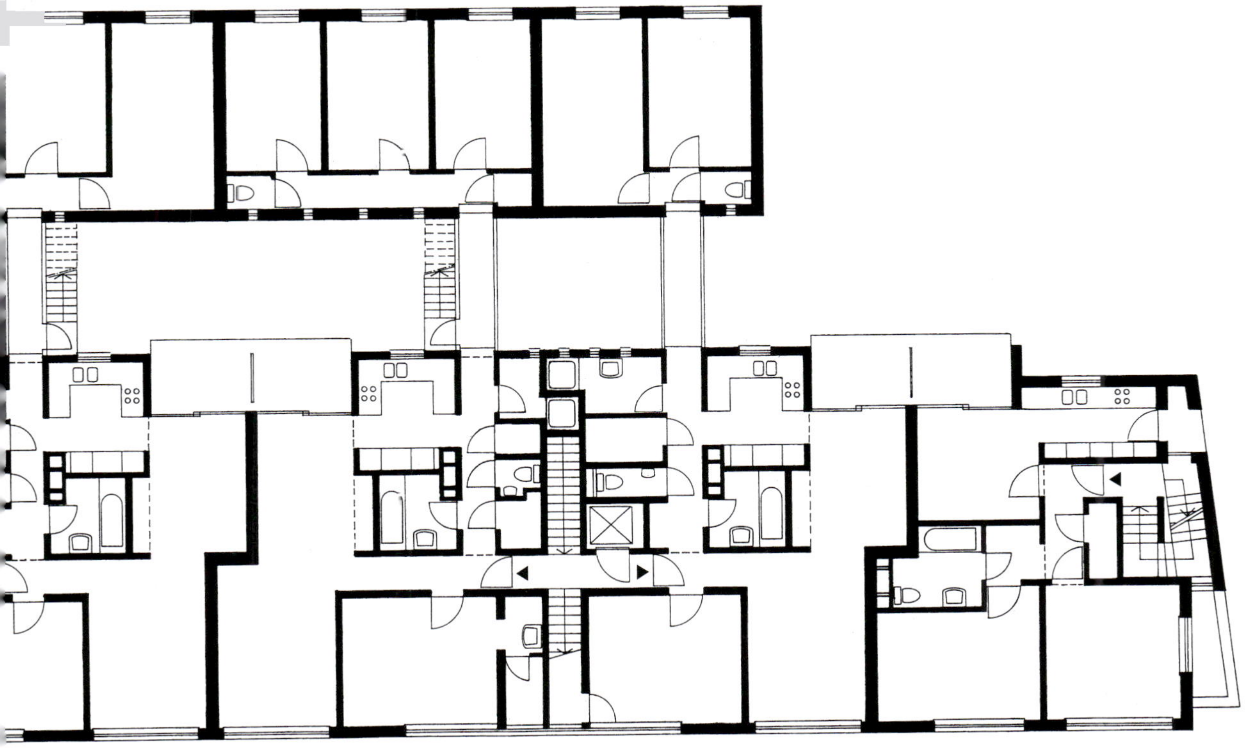

5

5 2nd floor: 2-, 4- and 5-room apts. with roof terraces on level above 1:200
6 The building in the projected urban design, which formed the basis for the design

Referring to floor plan 1:200

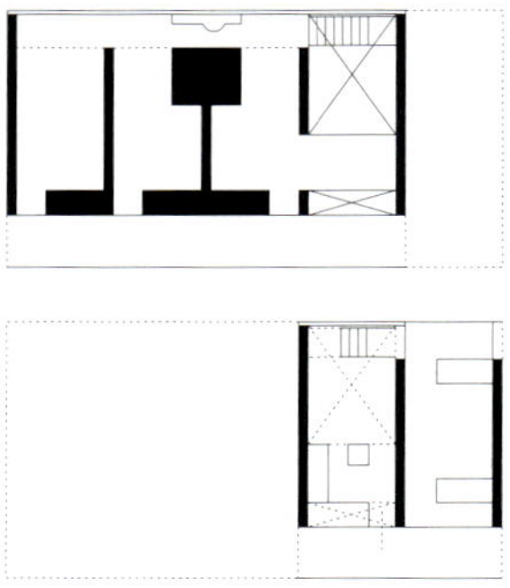

Kitagata | Sejima, Nishizawa | 1998

This housing development bent along the contour of the lot is part of a large housing scheme built by four women architects in Gifu. The structure is raised on pilotis, providing space at ground level for parking and allowing access to the complex from all sides. The crosswall type of construction determines the structure of the building in its entirety – with each crosswall unit corresponding to one room regardless of the allocated use. Each apartment consists of a combination of eat-in kitchen and living room, loggia, and a varying number of rooms on one or two levels. Half of the units even feature a two-story space. All rooms in each apartment are linked via a "sunlit corridor" (with washstands), whose floor-to-ceiling windows look out onto the landscape. The apartments, and all private rooms individually, are provided with access from the continuous covered walkway on the side of the building that is exposed to the elements, in other words, each of the private (or bed-)rooms is accessible from two sides (suitable for renting). The loggias – true "open spaces" – perforate the facade and, in combination with the extremely shallow building depth (lighting, ventilation), contribute to the pronounced light character of the building. The variations in plan and elevation create a stimulating facade image on both sides. The access side with its diagonal strings of fire-escape stairs and the mesh wire skin, as well as the glazed side facing the sun, also make the movements of the inhabitants perceptible from the outside, thus bringing more movement into the facade.

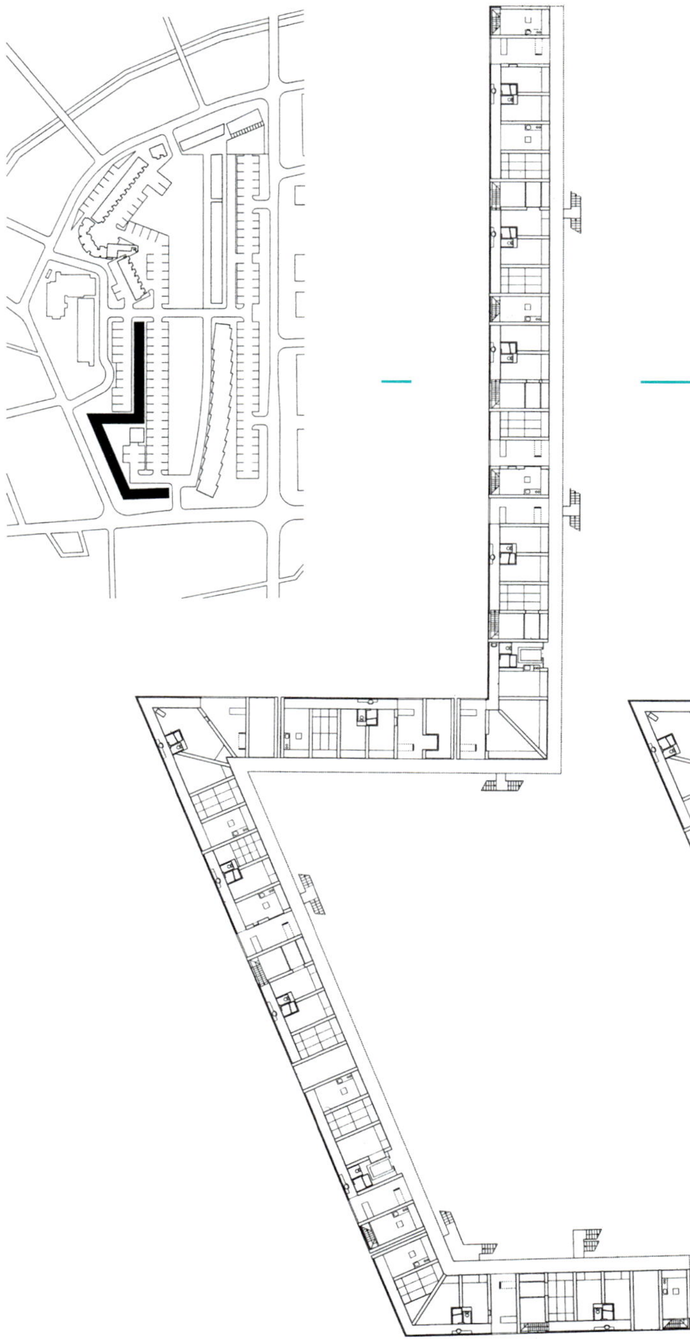

1

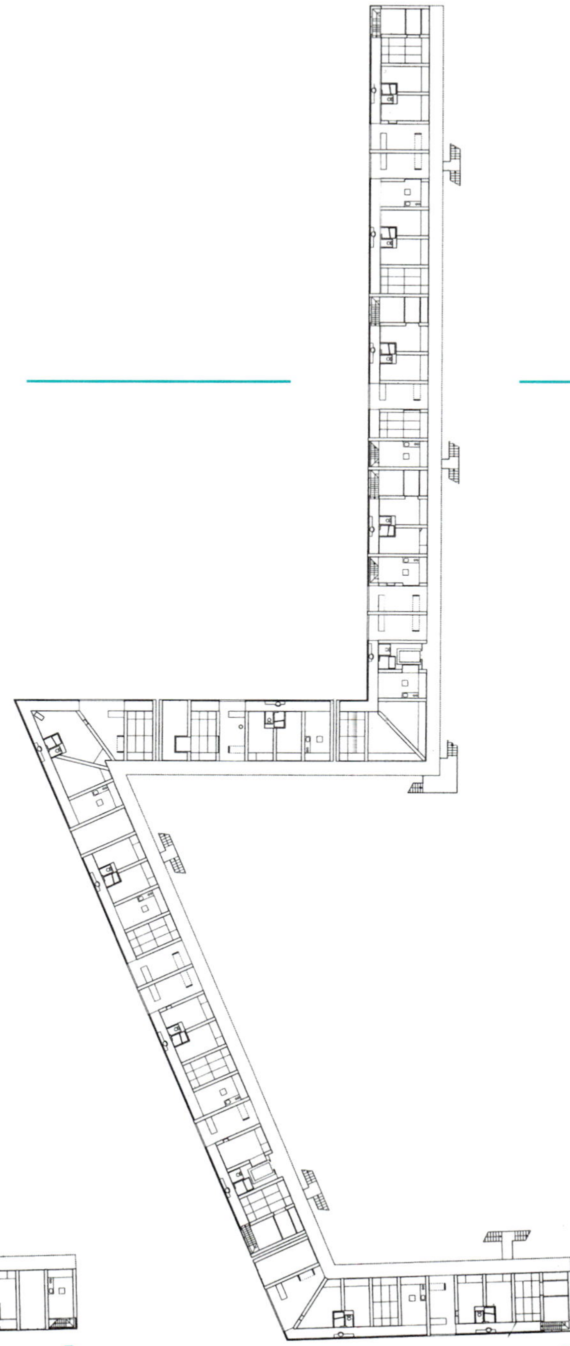

2

3

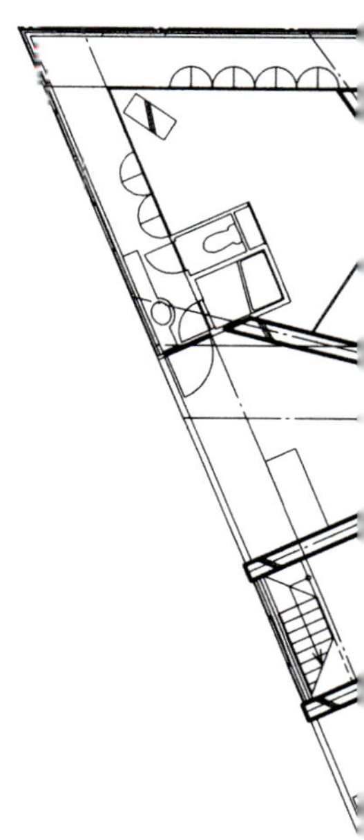

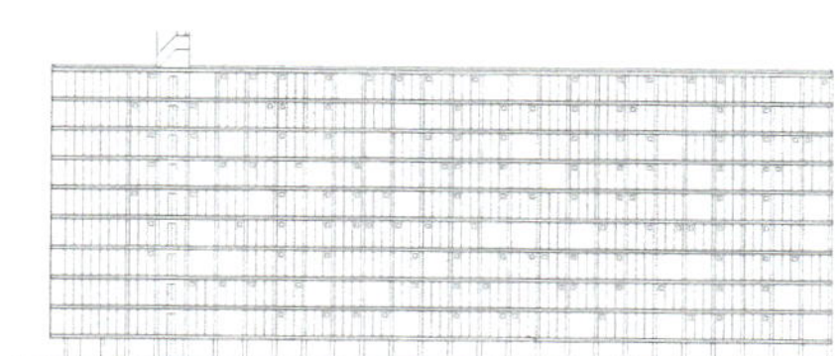

4

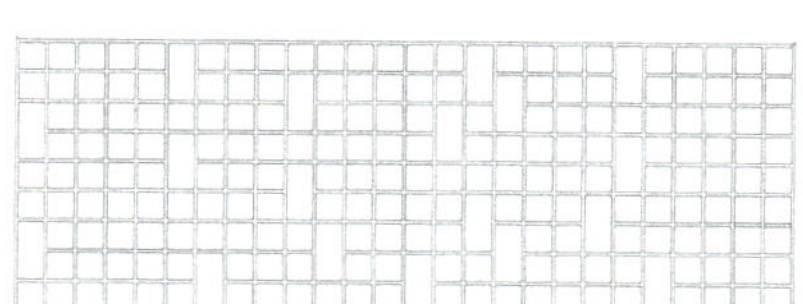

5

6

1. 4th floor of entire building
2. 5th floor of entire building
3. Detail section of standard floor 1:200
4. Detail: elevation of north facade with distribution of open loggias within the enclosed facade
5. Detail: longitudinal section with distribution of apartments on one i.e. two levels
6. Detail: longitudinal section with distribution of one i.e. two story
7. South facade (detail)
8. North facade (detail)

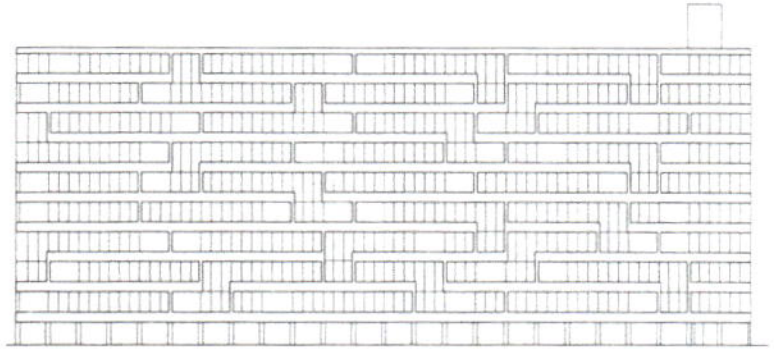

7

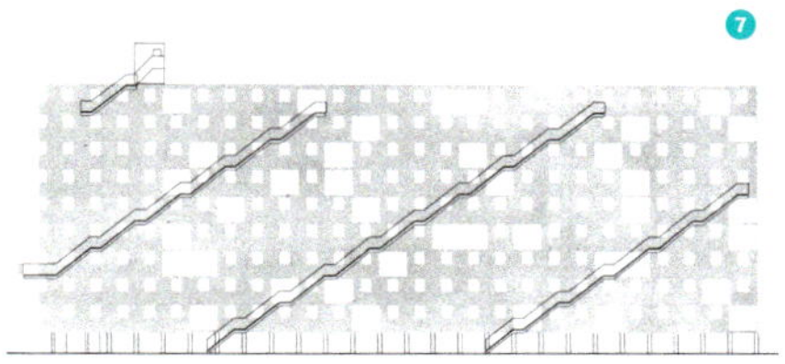

8

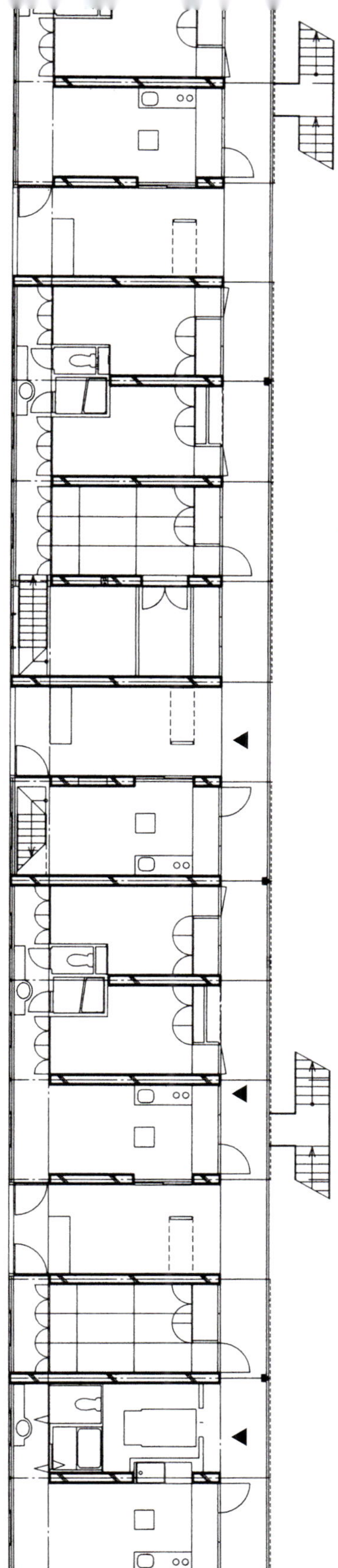

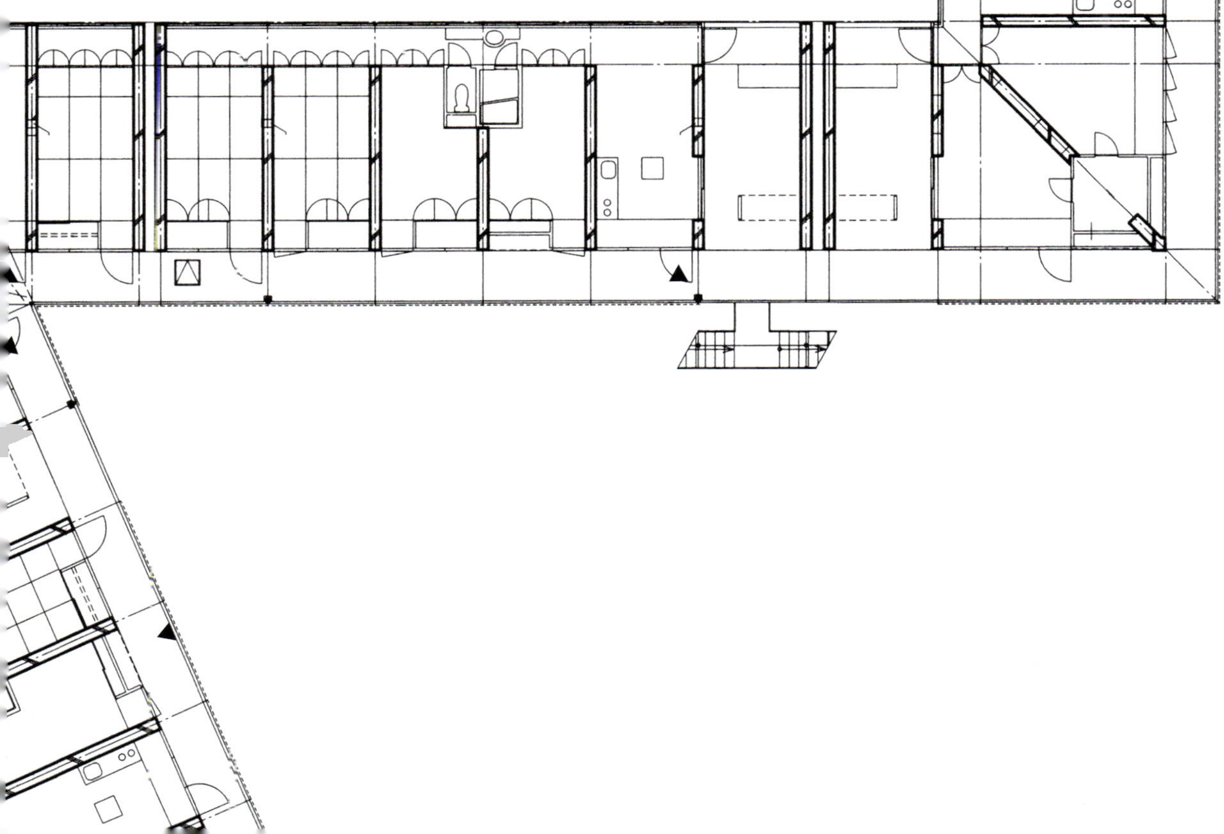

2.2 LINEAR BLOCK / SUPERBLOCK

Building type
10-story housing slab arranged along the contour of the lot boundary facing N/S and E/W

Date of construction
1994/1998 (1st/2nd construction phase)

Number of units
107

Size of units
2/3 flats, 1/3 maisonette units of varying sizes

Area per user
nondeterminable

Building depth
approx. 7.2 m

Access
covered walkways on each level running the full length of the building

Open spaces
generous private loggias

Parking
open parking spaces at ground level below raised slab on pilotis

Architect
Kazuyo Sejima
Ryue Nishizawa
Tokyo

Location
Gifu
Japan

Referring to floor plan 1:200

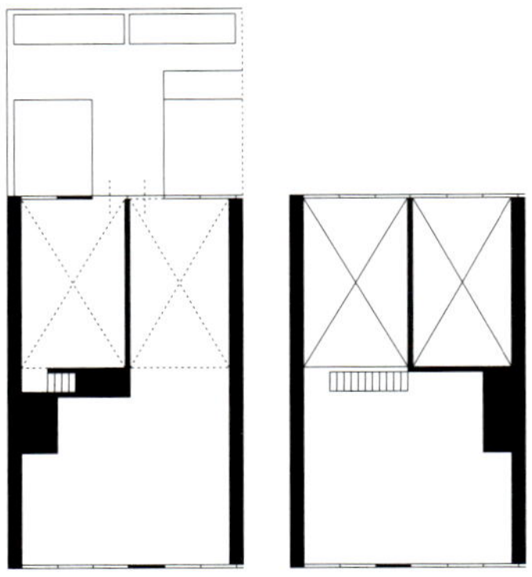

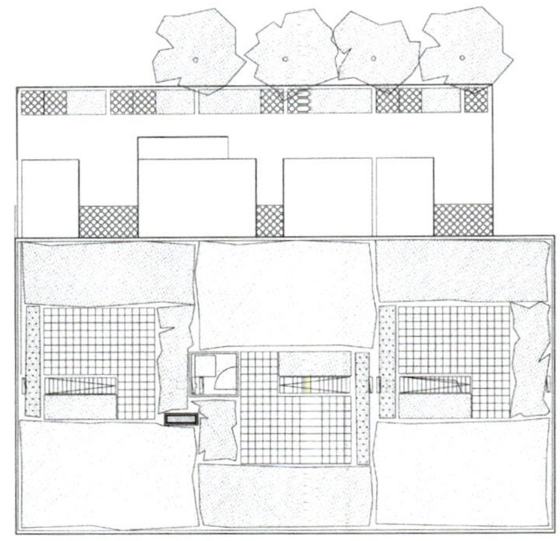

4

Kölner Brett | b&k+ brandlhuber&kniess GbR | 1999

In a quarter characterized by converted factories, a new building has been created, which offers a deliberate interpretation of the appeal and advantages of such converted structures and utilizes them for its own purposes. The architects have formulated a conscious response to the changing lifestyles and the growing integration of living and working. The building is composed of 12 identical modules, consisting of a 1-story section and a 2-story section (optional 2nd level). By rotating and creating a mirror image of this basic unit, they have achieved 8 different modules of tremendous spatial variety, which can be further increased by connecting individual modules. The individual volumes are completely usage-neutral – they are handed over without kitchen and bathroom – and are so inspiring in their spatial arrangement that they invite immediate occupation. Buyers are offered a variety of options. Thus, there were three different sanitary units for installation in the 1-story section; a detailed handbook explains where the attachment points are located, from which galleries can be suspended as desired, among other things; steel stairs are available upon request. However, the buyers were also free to complete the interiors entirely on their own. Due to soil contamination, the site had to be fully sealed and open spaces were therefore integrated on the roof or on top of the access structure to the rear of the building. The wide access bridges of this attached structure feature integrated planted borders and serve as additional semipublic balconies or terraces.

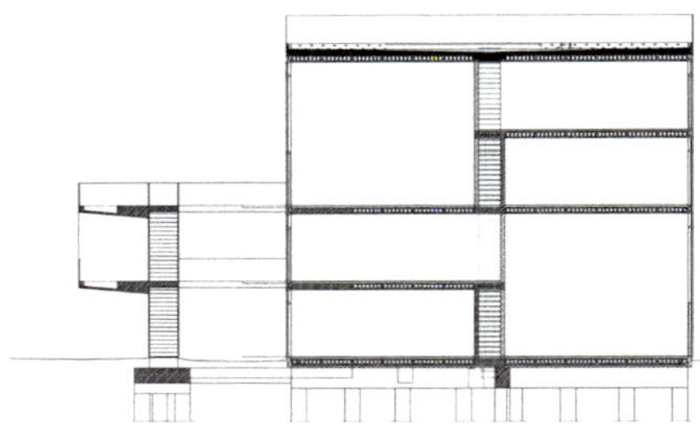

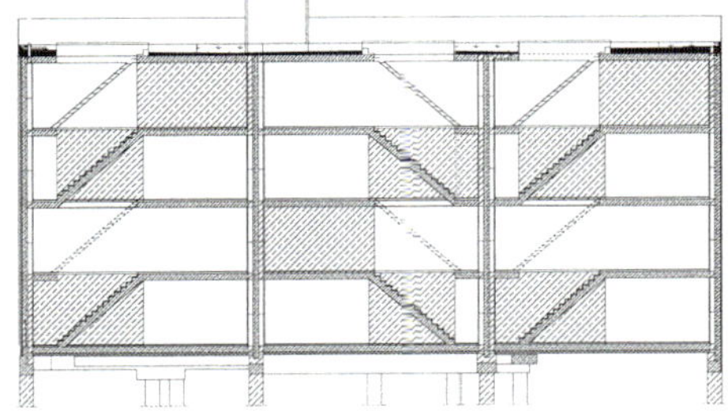

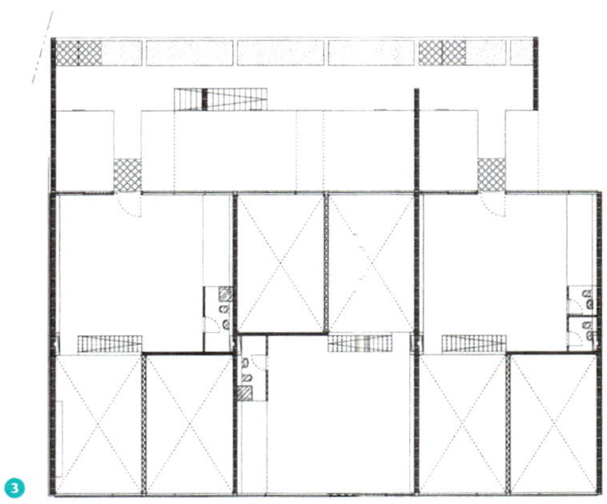

3

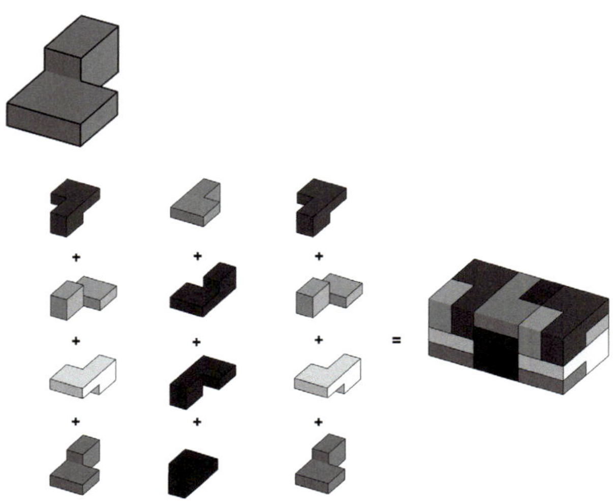

1

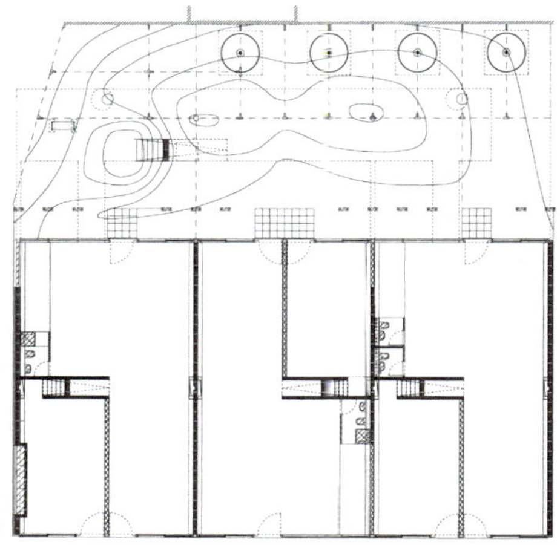

2

1 Basic spatial module and options
2 Ground floor with 6 units 1:500
3 2nd floor with upper part of 6 units 1:500
4 Roof level with roof gardens 1:500

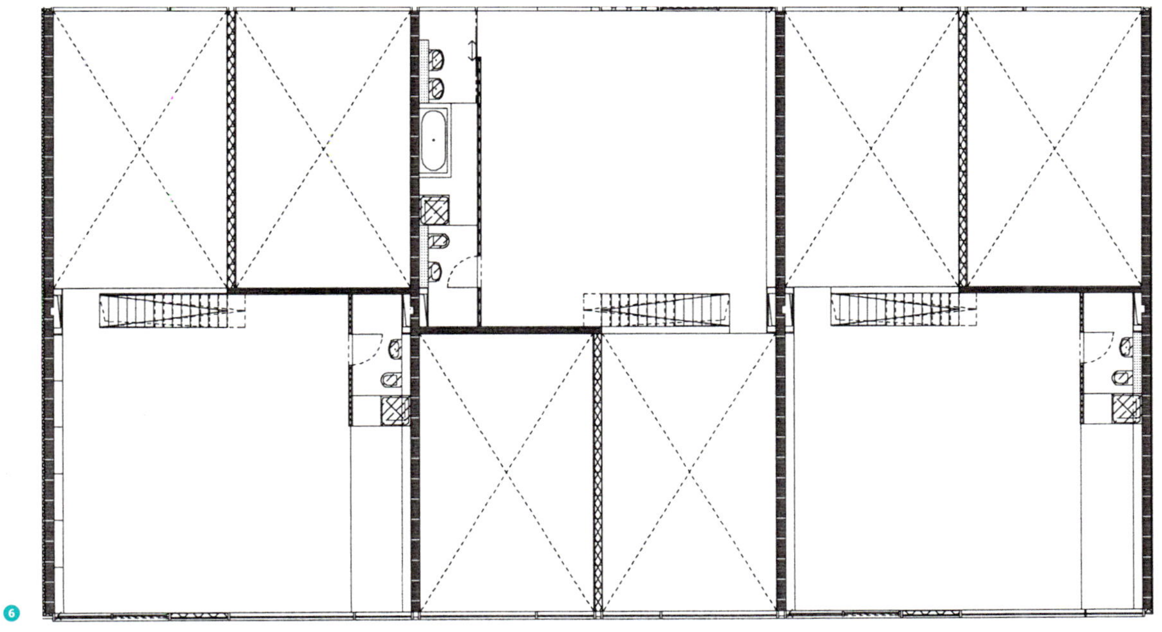

2.2
LINEAR BLOCK / SUPERBLOCK

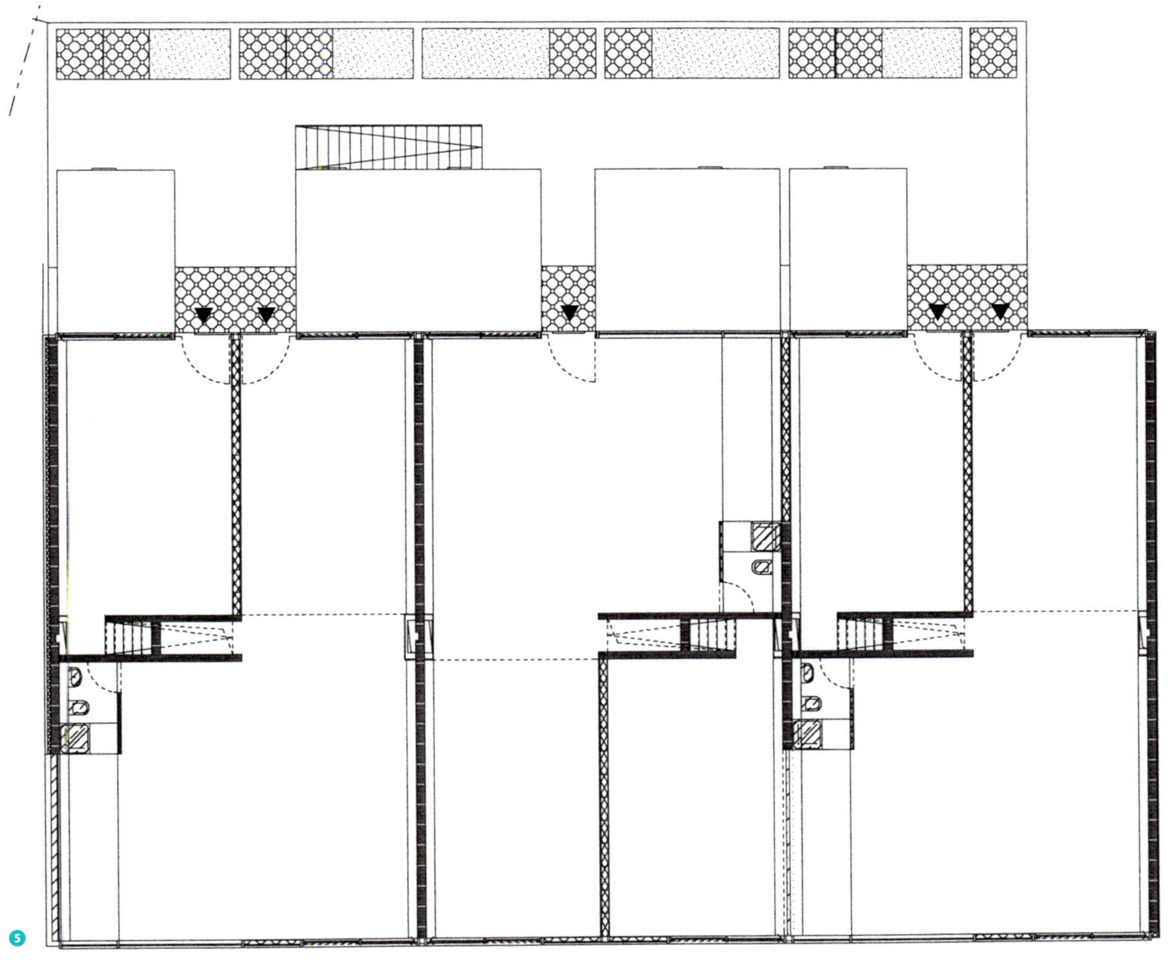

Building type
apartment and studio building
4 stories
facing NE/SW

Date of construction
1997–1999

Number of units
12

Size of units
8 different spatial modules of identical size: min. 110 m²
approx. 30 m² open space
(two-story, extendable as needed)

Area per user
nondeterminable

Building depth
15.75 m

Access
ground floor: direct access
on 2nd and 3rd floor:
fronted by access structure
with partially covered walkways

Open spaces
access structure with
terraces and balconies,
roof gardens

Parking
10 parking spots beneath
access structure

Architect
b&k+ brandlhuber&kniess GbR
Cologne

Location
Cologne
Germany

Referring to
floor plan 1:200

5 3rd floor with lower part of 5 units 1:200
6 4th floor with upper part of 5 units 1:200
(middle 2 units are connected)

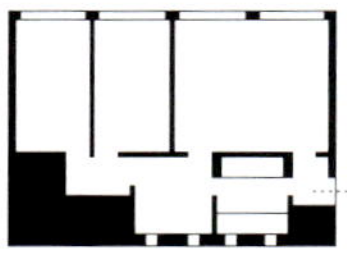

Maia I | Rocha | 1999

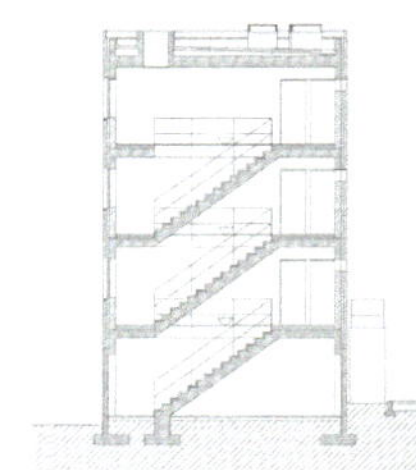

The subsidized housing program called upon the architects to manage with a significant reduction in funds – a topic that was treated as the defining design theme: limitation to two room sizes (single/double), two apartment sizes, two window sizes, etc. The rigidity of this repetition invests the slab with a presence on the outside, which is intended to promote an urban restructuring of the sprawling suburb. The floor plans are characterized by a pronounced simplicity. They are divided lengthwise in two: one row of living and bedrooms runs parallel to a sequence of service rooms. A small hallway provides access to the living room, and also, in a straight line to the galley kitchen, which leads to another hallway to bedroom, bathroom, and laundry room. The second hallway is large enough to accommodate a dining area or to be used for other purposes. The division of the floor plan in two is also visible on the facade. On the street elevation, the latter is perforated with small, square windows, which light the service zone, while the brick facade overlooking the garden is structured with window ribbons running the length of the building. These ribbons consist of an equal number of opaque and transparent areas; it is interesting to note that in this case it is the timber elements that are the movable components. The basement houses storage areas, as well as a common room in the above-ground area of the concrete plinth fronted by a terrace. This floor plan model was applied to a variety of building projects in the community, with strong variations in the length and height of the slab.

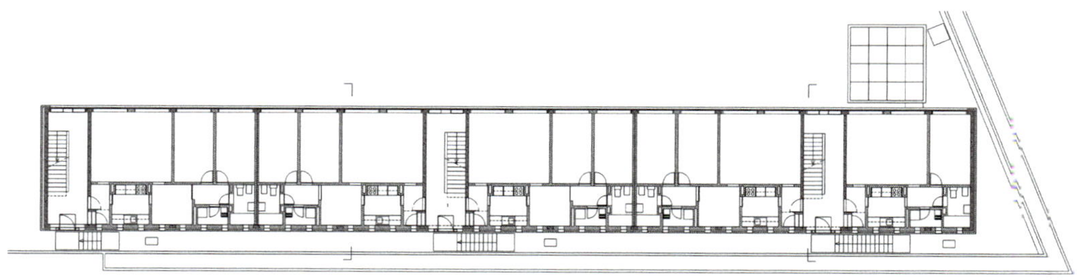

2

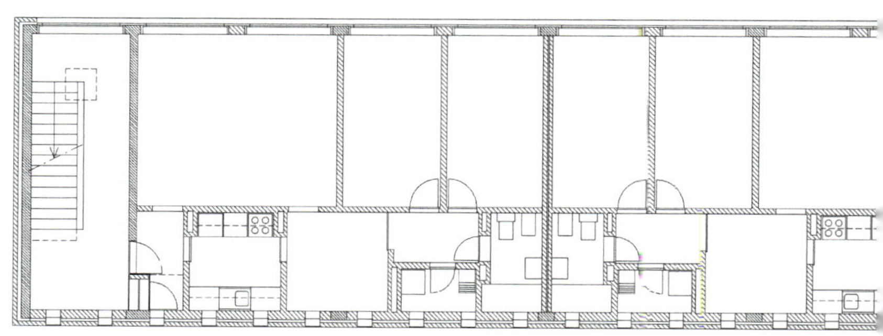

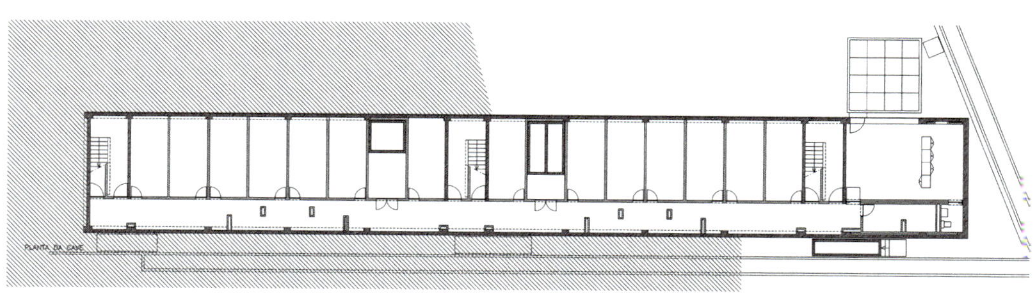

1

1 Basement 1:500
2 Ground floor 1:500
3 Standard floor 1:200

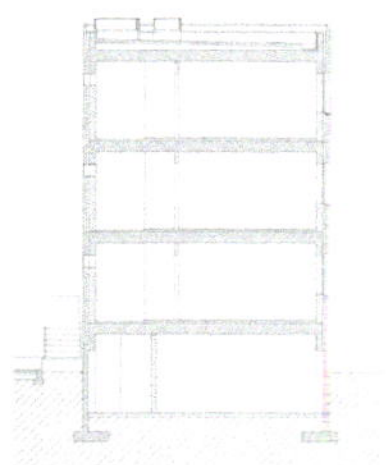

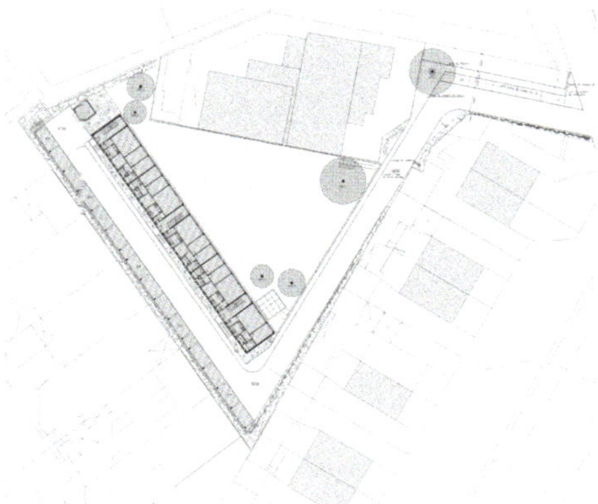

2.2
LINEAR BLOCK / SUPERBLOCK

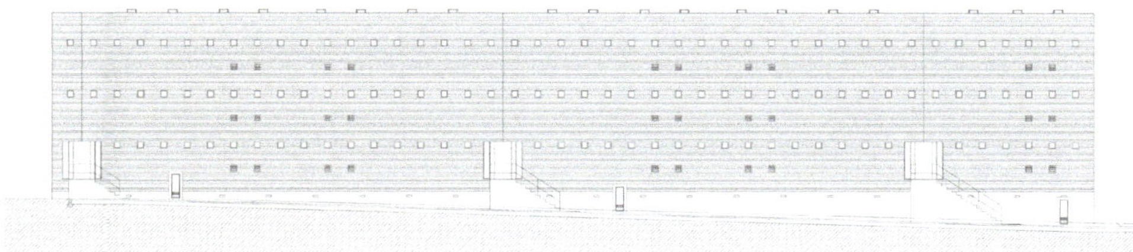

Building type
housing slab
3–4 stories
facing NW/SE

Date of construction
1999

Number of units
15

Size of units
2-room apts., 62.5 m² (3 units)
3-room apts., 81 m² (12 units)

Area per user
27–31.5 m²

Building depth
7.75 m

Access
single- and double-loaded stairwells

Open spaces
large, shared garden area for common use

Parking
15 garages on opposite side of road

Architect
João Álvaro Rocha
Porto

Location
Maia
Portugal

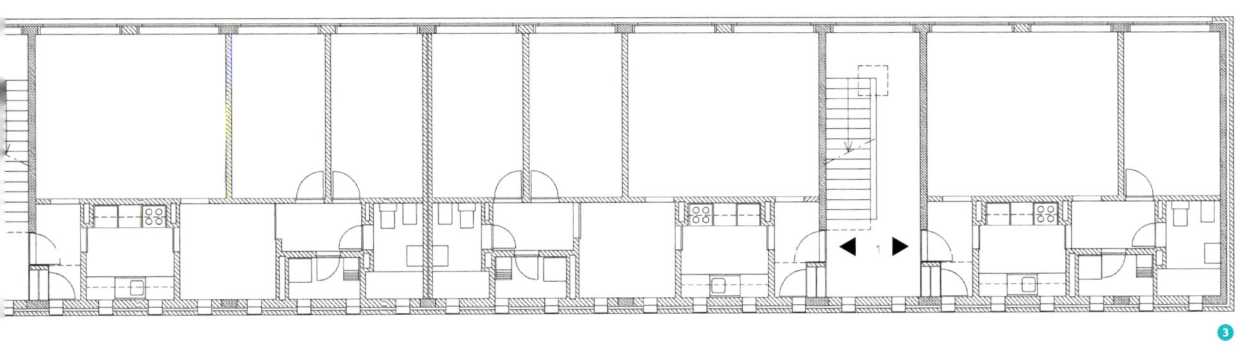

Referring to floor plan 1:200

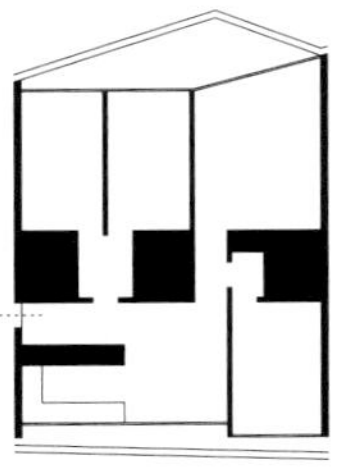

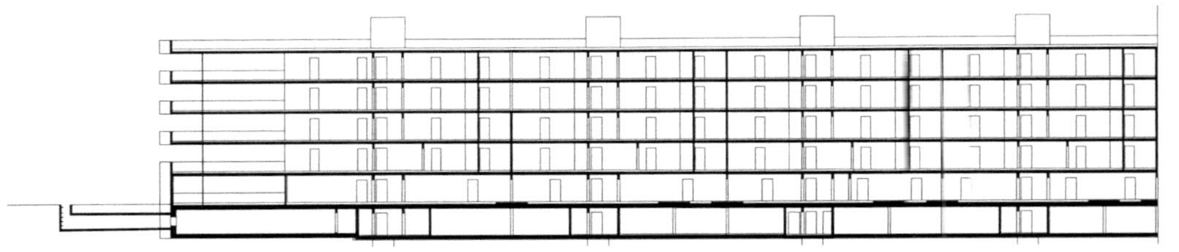

St. Alban-Ring | Morger & Degelo | 2002

A building slab located on the boundary between two very different urban areas: on one side, a street followed by an elevated highway and interregional railway line; on the other side, a townhouse district with an adjacent public park. The building has been pushed right up to the edge of the street to leave as much space as possible on the more attractive side overlooking the park. Each standard floor contains alternating 3- and 4-room apartments, with the stairwell occupying the space of what would be the fourth room in the smaller units. The interiors are characterized by a pronounced shift in direction: one enters into an elongated space with kitchen and dining area parallel to the street. At a right angle, a narrow passage leads from this space to the generously proportioned living room overlooking the park on the other side. This is where the large balcony is located, onto which the bedrooms also open. The result is a surprising diagonal relationship between the two shared living areas, connected via the small passage. The private spaces, which surround and shape this shared open area, are preceded by a separate hallway with bathroom. The building was partially lowered on the park side to accommodate apartments that are oriented toward one side only and hobby/recreation rooms in the basement. The polygonal shape of the balconies invests the garden facade with an undulating movement, which is continued – in a more restrained fashion – on the street side.

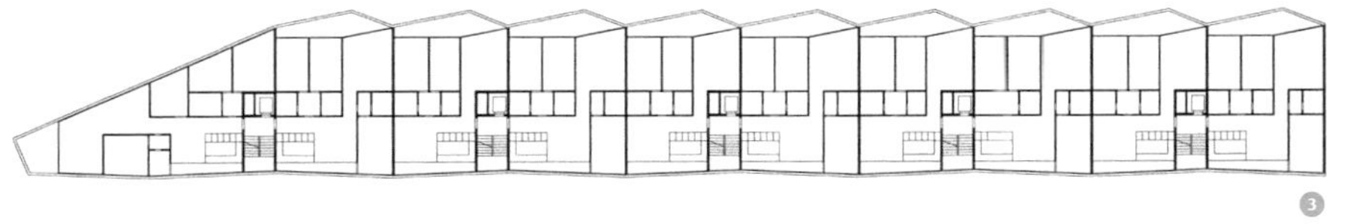

3

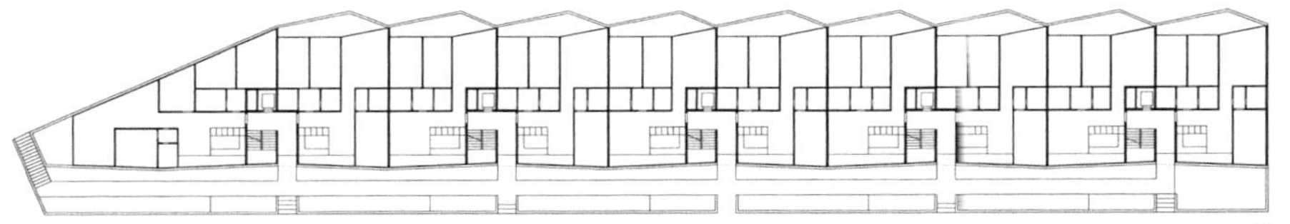

2

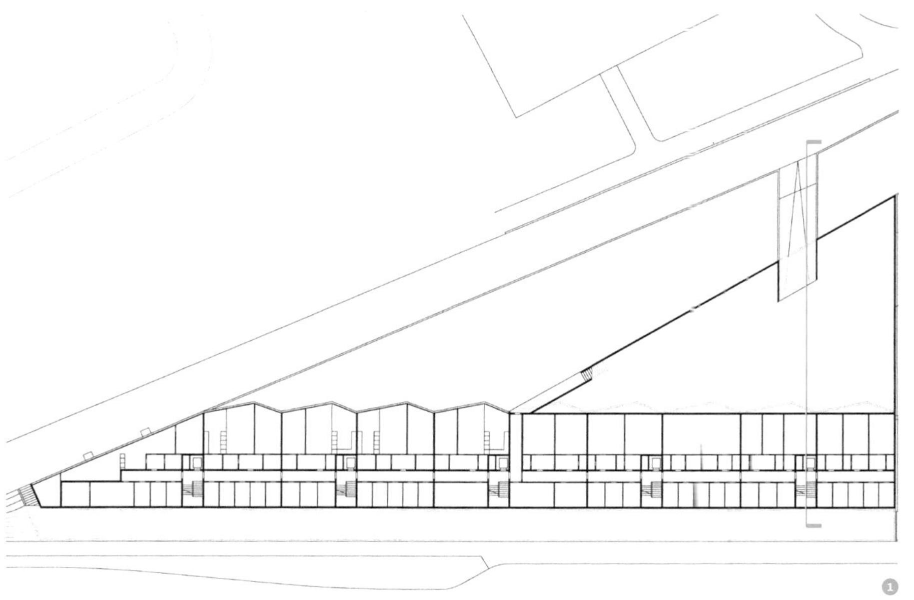

1

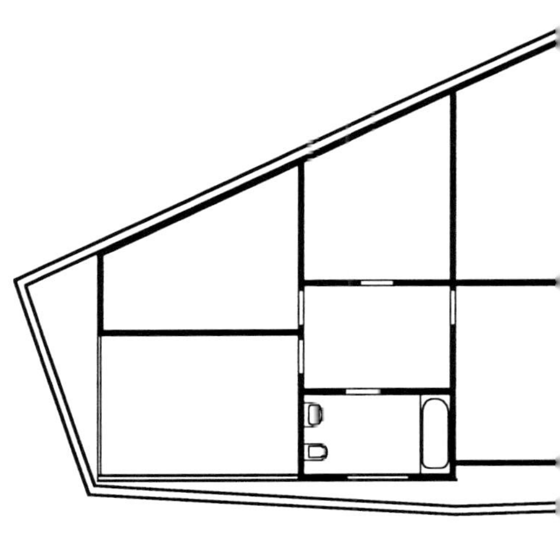

1 Basement with apartments and recreation rooms 1:1000
2 Entrance level 1:1000
3 Standard floor 1:1000
4 Detail standard floor 1:200

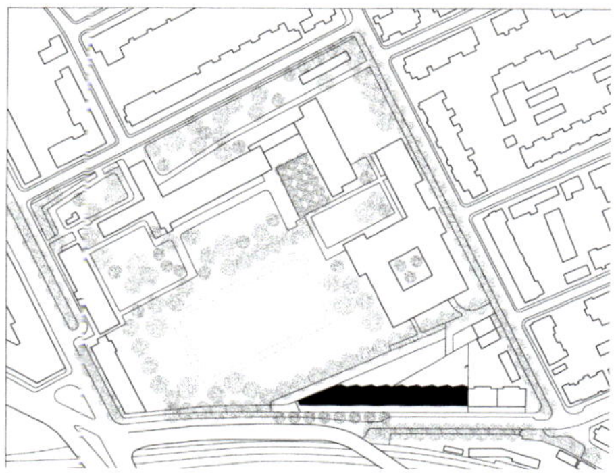

2.2
LINEAR BLOCK / SUPERBLOCK

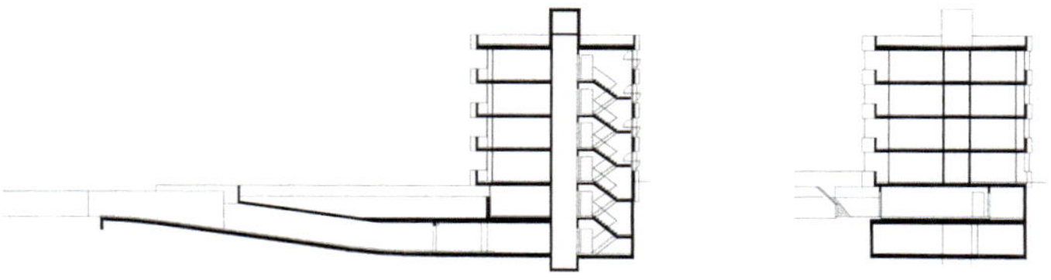

Building type
5-story housing slab
facing NW/SE

Date of construction
2002

Number of units
45

Size of units
2½-room apts., 70–80 m² (5 units)
3-room apts., 97 m² (16 units)
4-room apts., 111 m² (5 units)
4-room apts., 119 m² (15 units)
5-room apts., 135 m² (4 units)

Area per user
27–40 m²

Building depth
11.9 m

Access
Double-loaded stairwell

Open spaces
balconies, communal garden
with playground,
11 hobby/recreation rooms

Parking
underground garage with
43 parking spaces

Architect
Morger & Degelo Architekten
Basel

Location
St. Alban-Ring
Basel
Switzerland

Referring to
floor plan 1:200

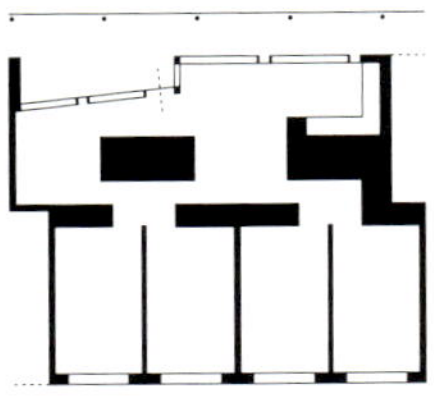

Bülachhof | Langenegger | 2004

Although this residential development is conceived as an expansion to an existing student housing complex, its plan can easily be applied to other residential developments in different contexts. There is a clear separation – both in the internal organization and in the external appearance – between a closed sequence of individual rooms and an open progression of living spaces along the light covered walkway. The spatial composition of the living room is exemplary: the contours of the walkway, the inserted prefabricated sanitary cores, and the built-in closets in the individual rooms projecting into the living room, create designated zones that are differentiated in a completely unforced manner. The bays of the walkway become an exterior space for socializing; in the interior, the kitchen plan resembles a niche; the large dining area, crosswise to the kitchen, functions as the center; an open vestibule, the sitting area and anterooms that expand into two rooms each, are screened off by the sanitary core. The living area promotes visual contact to and from the outside, allows for ease of circulation and simultaneously inspires inhabitants to put the various zones to the uses for which they were intended. Access via the walkways is designed to promote communication among residents. The stairwells at the far ends of the buildings feature open entrance zones, and the basement level of each building accommodates a party room and other common areas. The lower-lying roof areas serve as large patios for all residents. Lilac bushes between the buildings create a filter between the individual spaces, each row with its own hue and perfume.

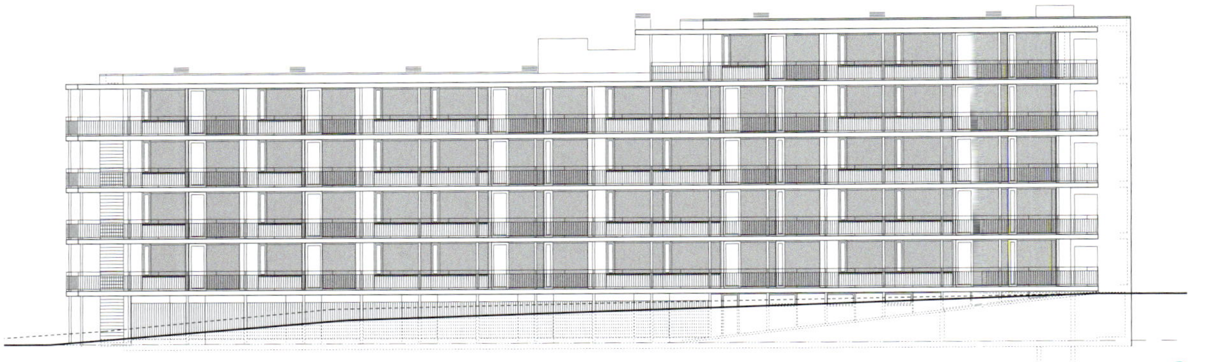

1

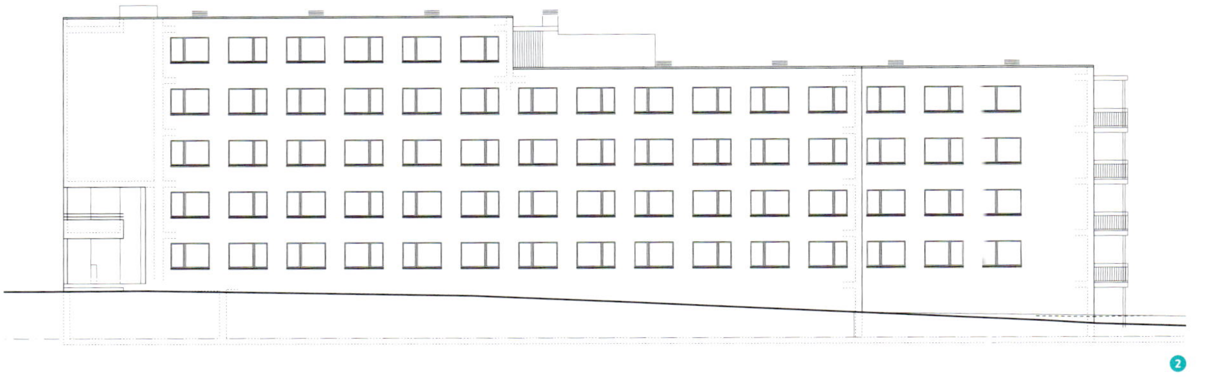

2

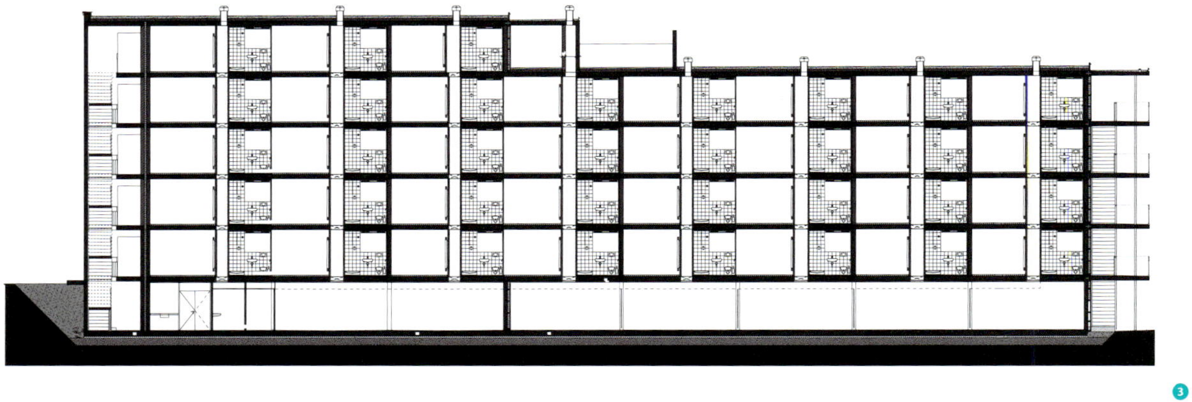

3

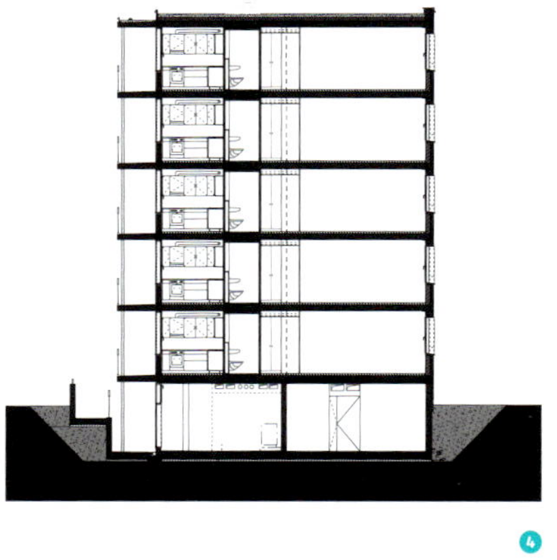

4

1. West elevation with walkways and living areas 1:400
2. East elevation with bedrooms 1:400
3. Longitudinal section with prefabricated sanitary blocks 1:400
4. Cross section 1:400
5. Total floor plan of ground floor with 3- and 5-room apartments 1:400
6. Typical floor plan detail: 5-room apartment and 3-room apartment with gallery, communal area with inserted sanitary block, and sequence of bedrooms 1:200

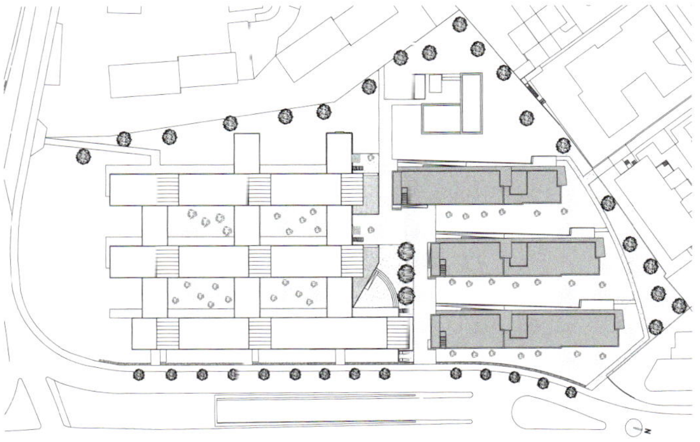

2.2
LINEAR BLOCK / SUPERBLOCK

Building type
5- to 6-story residential building facing W/E

Date of construction
2002–2004

Number of units
71

Size of units
3-room apts., approx. 54 m² (28 units)
5-room apts., approx. 110 m² (43 units)

Area per user
27–27.5 m²

Building depth
12 m

Access
covered walkway

Open spaces
roof patios for common use, walkway widening into bays that serve as balconies

Parking
no parking on lot

Architect
Marc Langenegger
Architekt EPFL SIA
Bern

Location
Bülachhof
Zurich
Switzerland

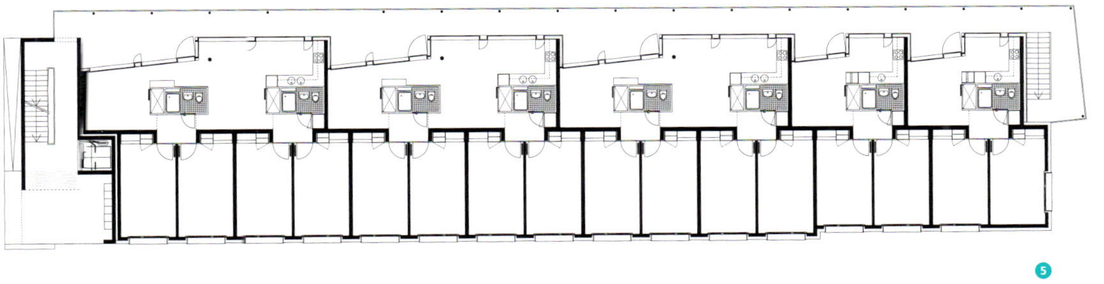

5

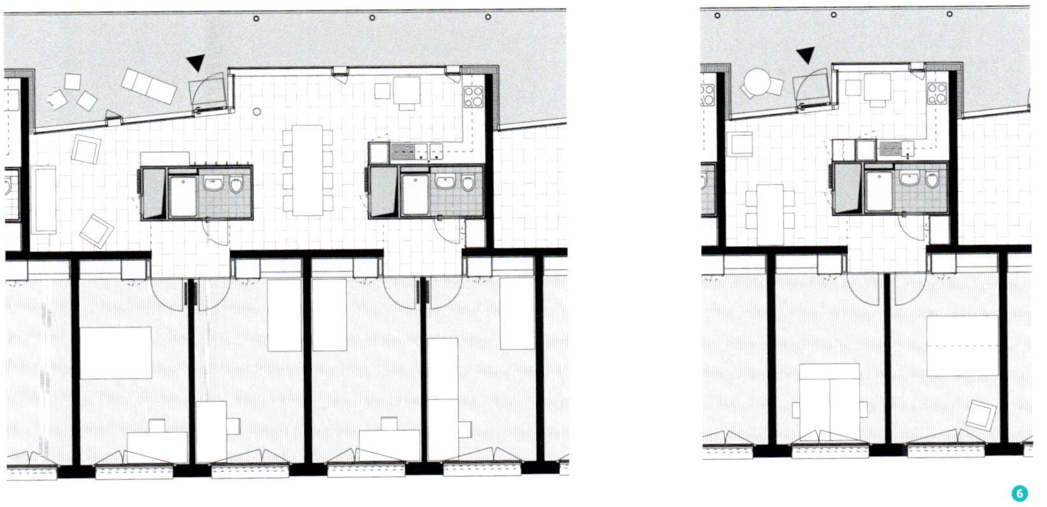

6

Referring to floor plan 1:200

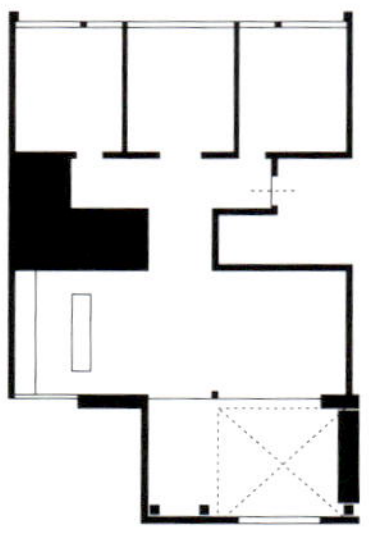

Paul-Clairmont-Strasse | Gmür & Steib Architekten AG | 2006

The two-story balcony patios staggered on the southwest facade are the most distinctive feature of these apartments. In combination with the eat-in kitchen and the living rooms, these large, sheltered spaces provide a bright and spacious living area. Two-meter-high balustrades transform them into intimate patios by providing shelter from the elements and a visual screen toward surrounding high-rises. A cutout in the man-high balustrade frames a picture-book vista onto the landscape near the Uetliberg mountain. The floor plans are divided into four parallel layers. On the one side lies a continuous sequence of individual rooms, all identical in size, with every second apartment on the ground floor featuring one additional room. Stairwells and bathrooms are contained in a strip located at the core, while living areas and the balcony patios are situated along the front of the building. The kitchens are recessed as if in a niche, providing visual contact with the outside with a strip window above the patio of the apartment below. The four spatial layers are linked with each other via a nearly symmetrically arranged central axis. On the ground floor, the five stairwells are connected by a public walkway with two access points; this floor also accommodates meeting areas of the housing complex: additional studio spaces for rent, live-in studios, a common room, and storage rooms for strollers and bicycles.

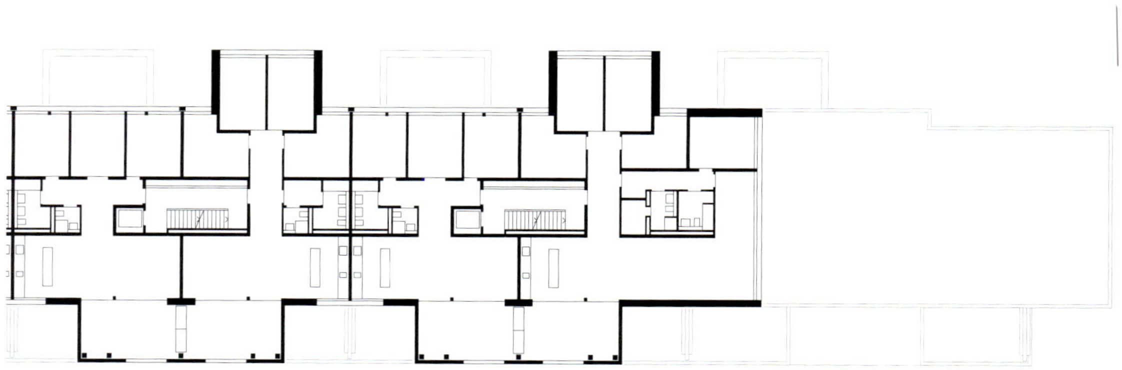

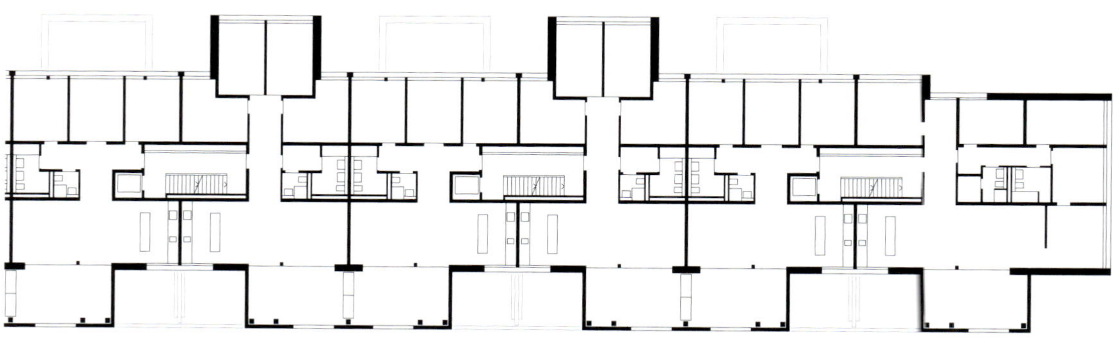

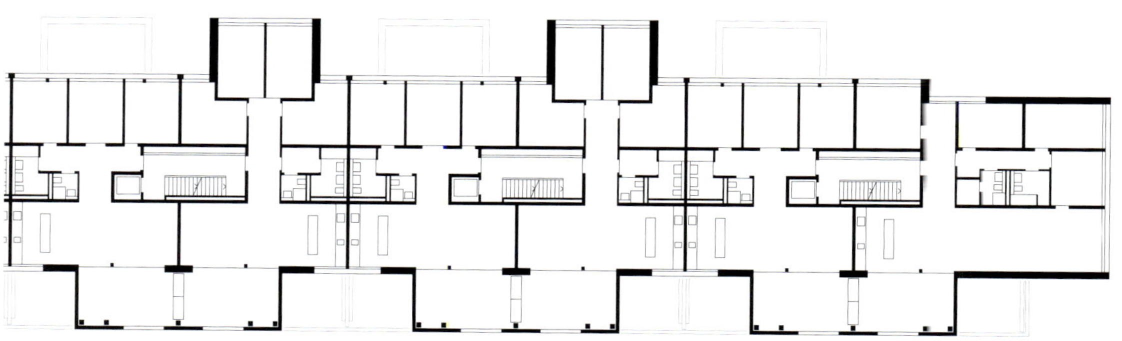

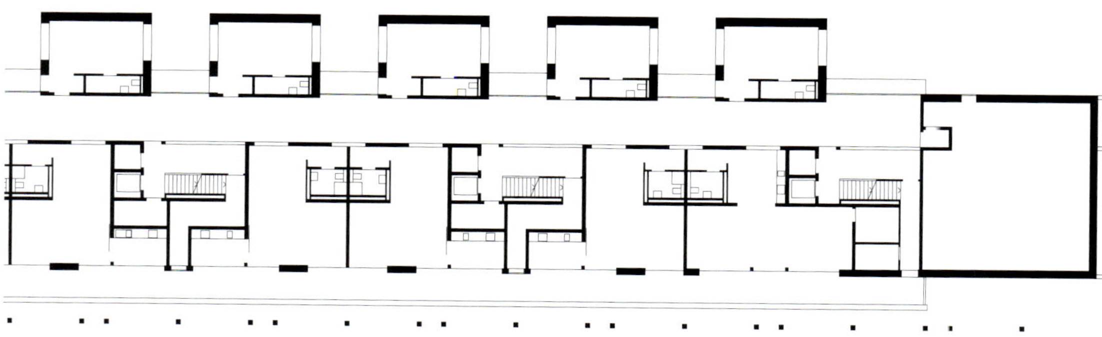

1

1 Floor plan detail ground floor: 3rd to 5th floor 1:500
2 3rd floor: 4- and 5-room apartments 1:200
3 Furnishing suggestion
4 Cross section 1:500

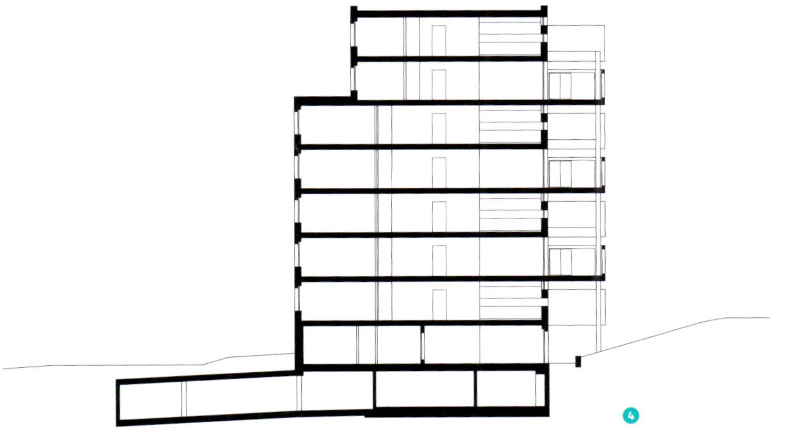

4

2.2 LINEAR BLOCK / SUPERBLOCK

Building type
residential slab with medical practice, kindergarten, studio, hobby and community rooms on ground floor
4 to 8 stories
facing NE/SW

Date of construction
2000–2006

Number of units
49

Size of units
4½-room apartments, 113–116 m^2 (23 units)
5½-room apartments, 135–138 m^2 (20 units)
6½-room apartments, 161–186 m^2 (6 units)

Area per user
27–31 m^2

Building depth
17–21 m

Access
5 double-loaded staircases connected by public walkway on ground floor (rue intérieur)

Open spaces
2-story balcony patios

Parking
underground garage

Architect
Gmür & Steib Architekten AG
Zurich
Patrick Gmür, Jakob Steib
Nicole Deiss, Barbara Ruppeiner
Michael Geschwentner

Location
Paul-Clairmont-Strasse
Zurich-Wiedikon
Switzerland

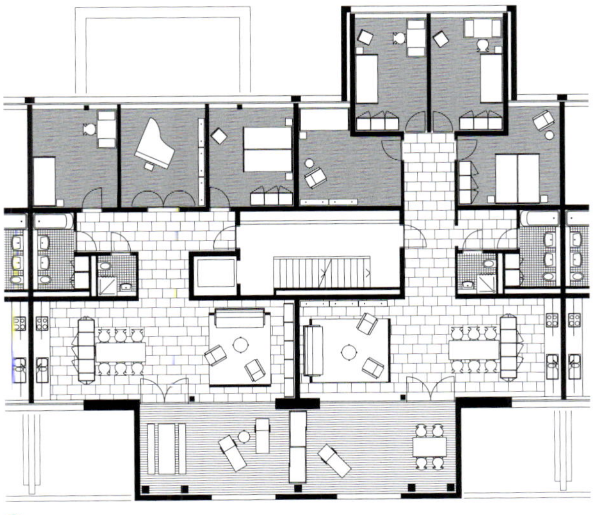

3

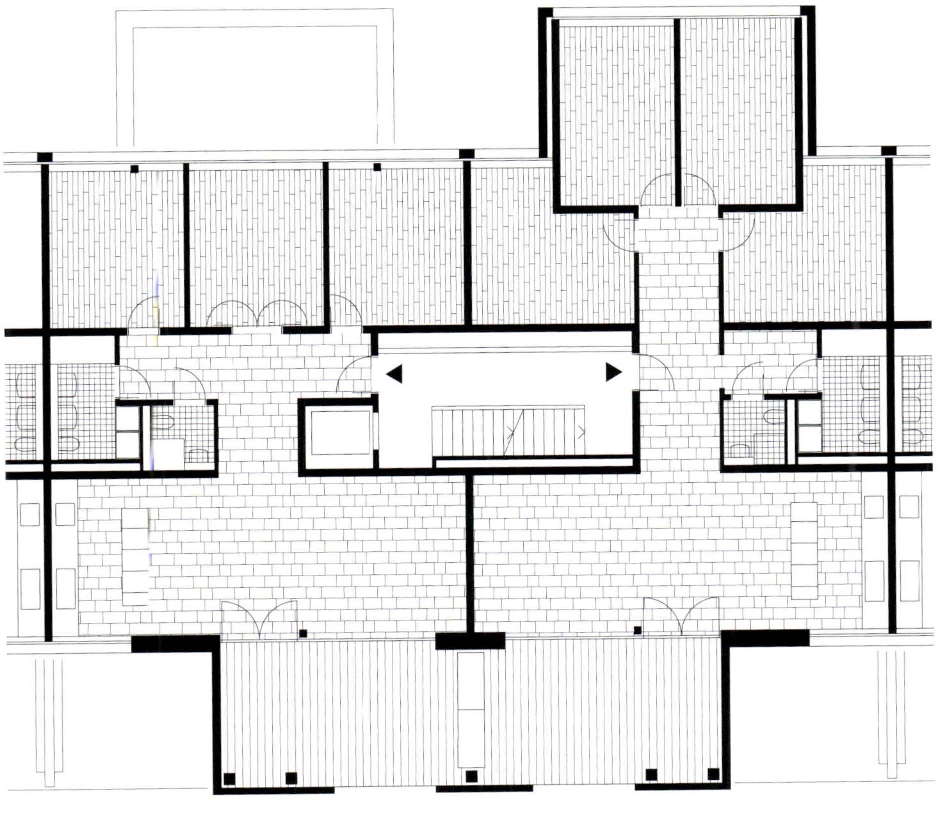

2

Referring to floor plan 1:200

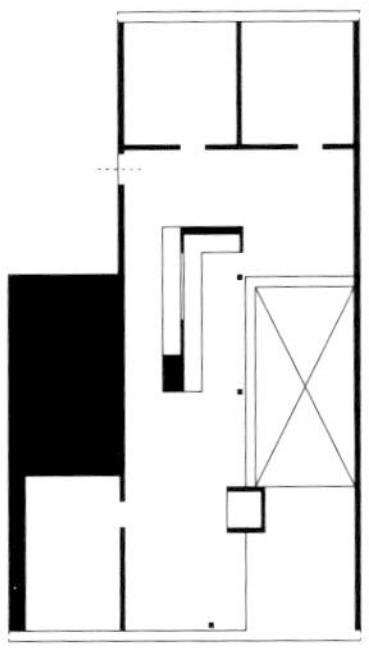

Rheinresidenz | Neff Neumann | 2006

The goal of this urban planning concept – leaving sufficient space on one side of a historic city wall and formulating a spatial border to a 1950s development on the other side – yielded a simple, albeit very deep volume (21 m building depth). The floor plan layout responds to this building depth by featuring narrow enclosed and open atria inserted in the volume, and loggias along the west facade, which are open on two sides.

The open living area forms an L-shape around these open spaces: the inserted block with kitchen counters and built-in closet reprises the shape and defines areas for access, cooking, dining, and living. The bedrooms and bathrooms – again in an L-plan – adjoin this area. The arrangement of living area, atrium, and loggia allows for complex sightlines through different spatial layers, both horizontal and vertical, resulting in a stimulating ambiguity in the perception of the spaces, be they interior or exterior, accessible or nonaccessible. In each second segment, the access cores occupy the space of a bedroom and bathroom: thus, although the apartments appear identical from the outside, they are of different sizes on the inside and are accessed from different sides.

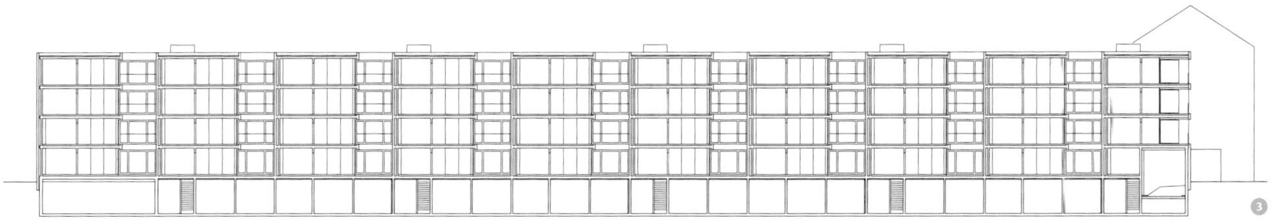

3

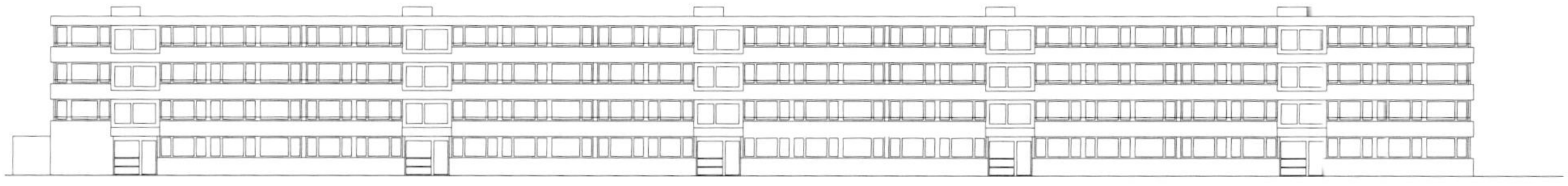

2

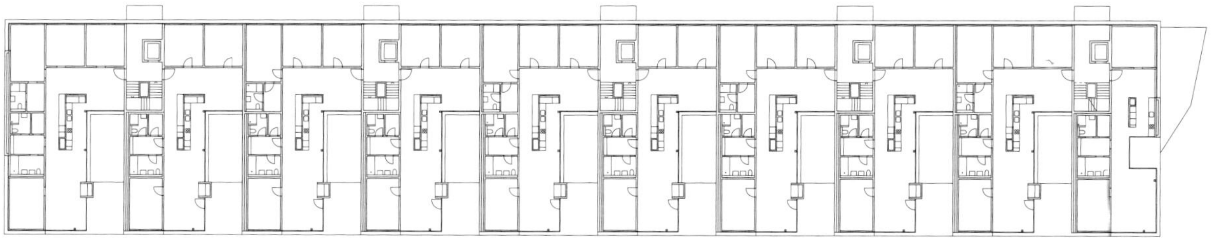

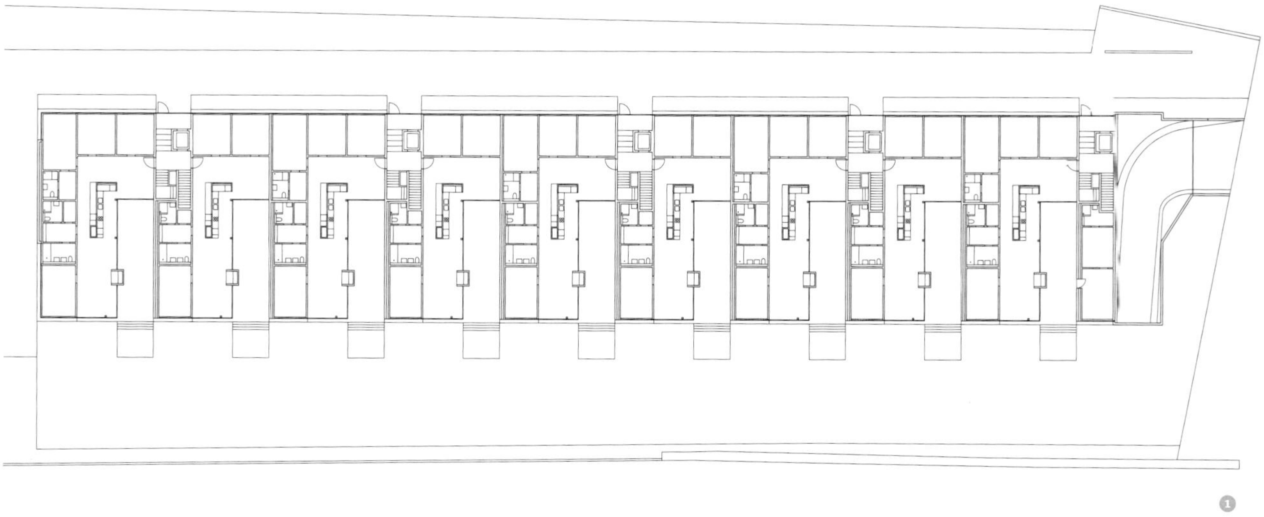

1

1 Total floor plan for ground floor and standard floor 1:750
2 East elevation 1:750
3 Longitudinal section/west elevation 1:750
4 Cross section/south elevation 1:750
5 Standard floor plans for 5½- and 4½-room apartment 1:200

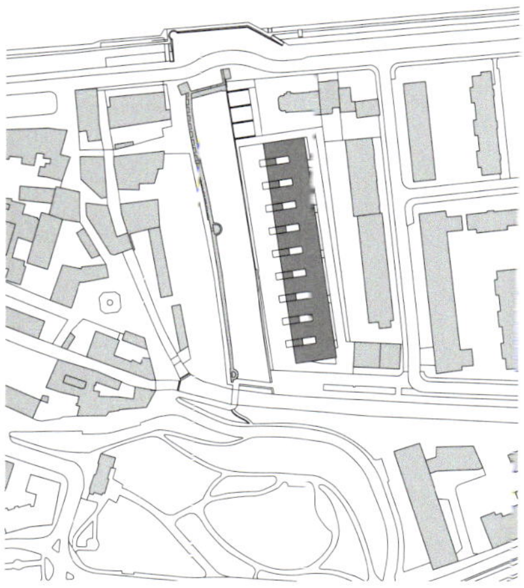

2.2
LINEAR BLOCK / SUPERBLOCK

Building type
deep row-house development with inserted atria and loggias
4 stories
facing WSW/ENE

Date of construction
2002–2006

Number of units
39

Size of units
5½-room apts., 166 m² (20 units)
4½-room apts., 139 m² (16 units)
3½-room apts., 105 m² (3 units)
studio (supplementary to 5-room apartment), 32 m² (1 unit)

Area per user
33–35 m²

Building depth
21 m

Access
double-loaded stairwells

Open spaces
Loggias

Parking
underground garage

Architect
Neff Neumann Architekten AG
Zurich
partners-in-charge:
Barbara Neff
Bettina Neumann
project management:
Philippe Vaucher
Florian Kirfel

Location
Basel
Switzerland

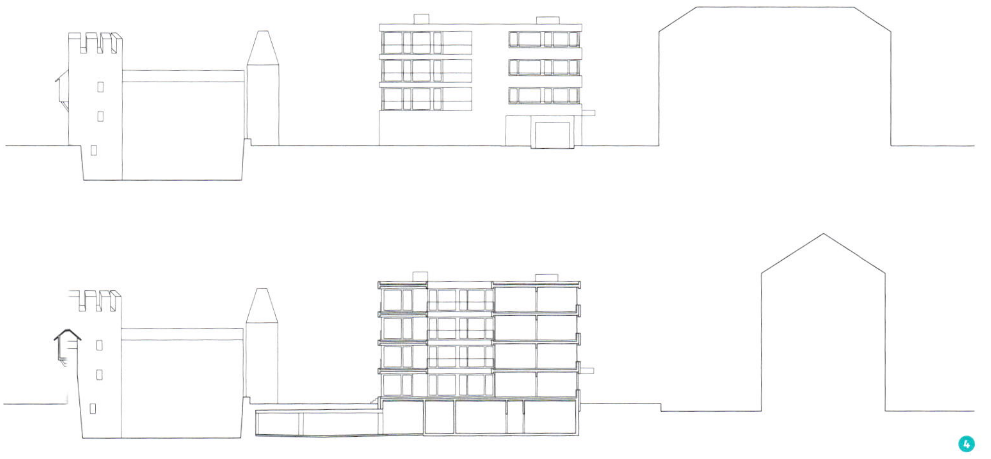

4

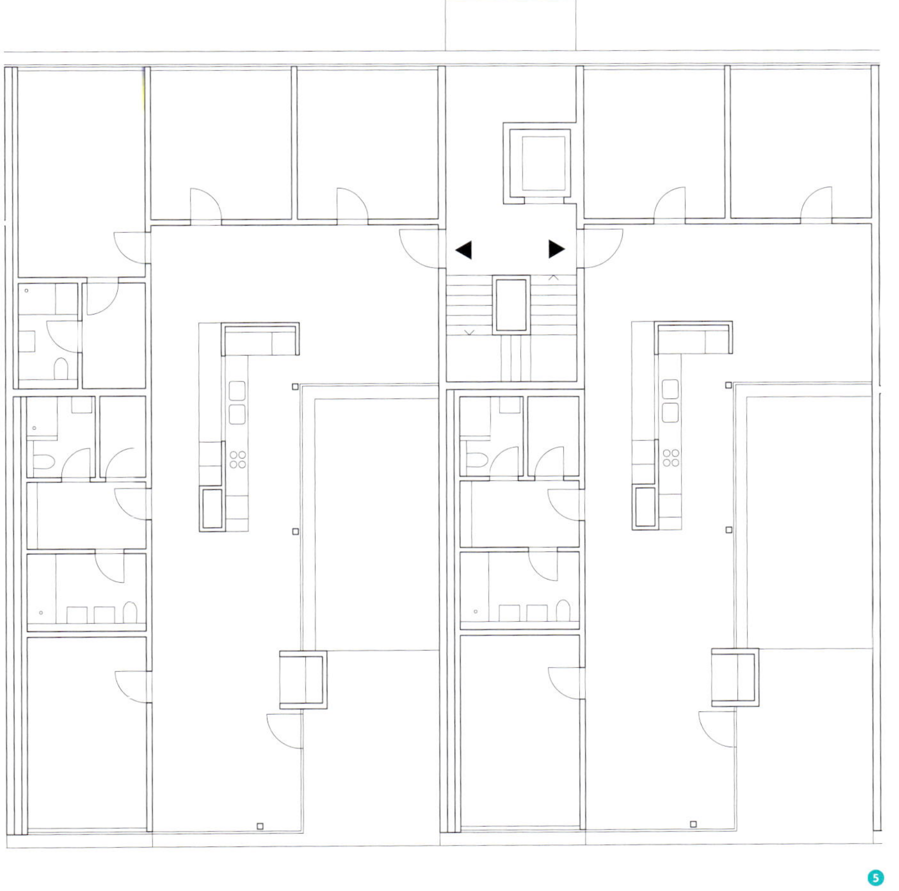

5

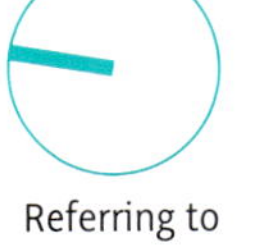

Referring to floor plan 1:200

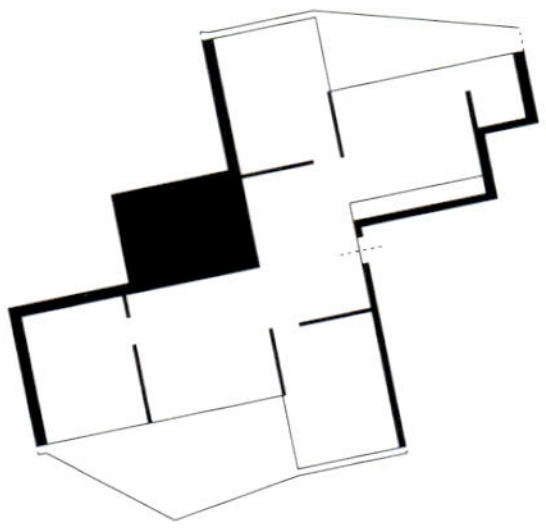

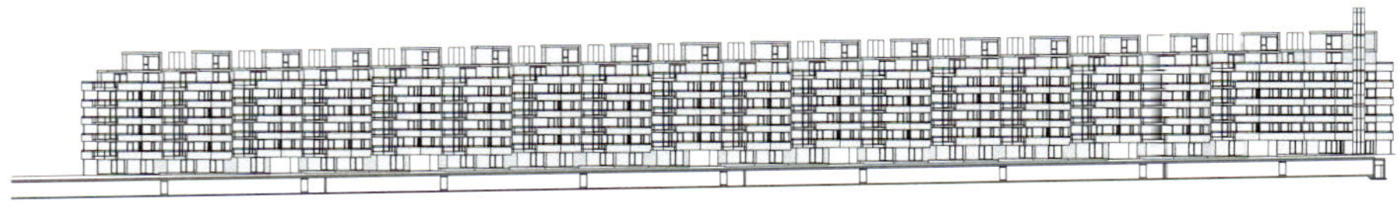

Hardegg | Matti Ragaz Hitz | 2008

The long building (225 m long), six freestanding buildings and a building with kindergarten and daycare facilities are conceived to give the former industrial site a new urban identity. The great variety of apartment types from studios on the ground floor to loft apartments and various duplex types is geared toward appealing to a wide range of potential residents, creating a sense of urban community. On the standard floors of the long building, the apartments are rotated away from the axis, dynamically oriented toward the exterior with balconies on both sides. This rotation results in a cruciform layout, which lends itself to a variety of floor plans. In one variation, three equally sized rooms form a sequence across the depth of the units: the bedrooms on the facade side open onto the balcony at right angles, while the open room at the center functions as interface and intersection. The longitudinal sides of this space adjoin the bathrooms and stairwell and, with a set back, lead into two additional rooms, which open onto the balconies on the far side. This results in a diagonal layout to the open, loft-like living room with the bedrooms at the end. In another variation, the rooms create a front-to-back continuous space followed by a series of bedrooms along side corridors with a similar set back and cruciform room disposition as above. All living areas are open from balcony to balcony, generously oriented toward the environment. The two crested attic stories accommodate apartments and large maisonettes with front-to-back roof patios.

The single-floor and duplex units in the freestanding buildings are open on three sides with loggias at the corners.

5

4

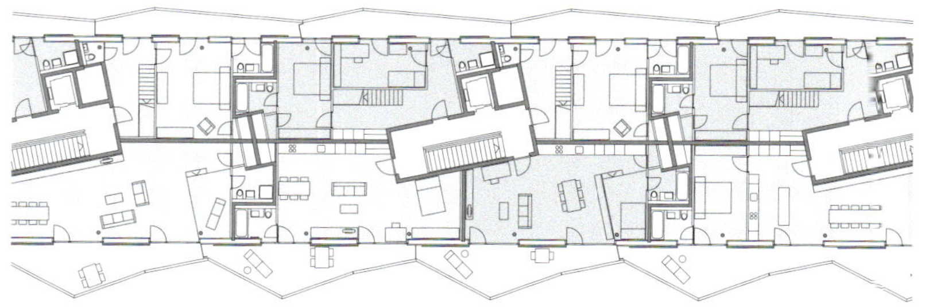

3

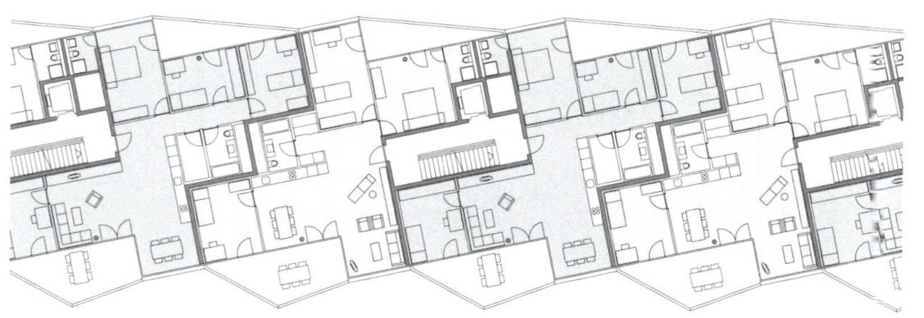

2

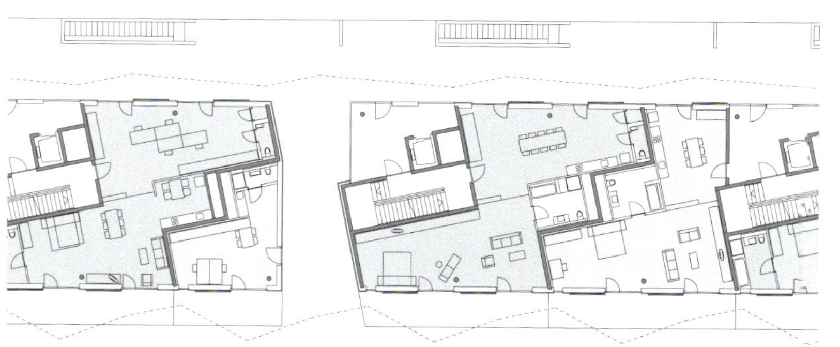

1

1 Floor plan section long building, ground floor: typology "loft" 1:500
2 Floor plan section long building, 3rd floor: typology "day/night area" 1:500
3 Floor plan section long building, 6th floor: typology "loft and maisonette" 1:500
4 Floor plan section long building, attic story: typology "loft and maisonette" 1:500
5 Site plan with long building, individual buildings, and kindergarten and daycare building 1:2000
6 Floor plan sections of long building, 4th floor, typology "open living" area 1:200
7 Floor plan sections of long building, 2nd floor, typology "communal living" 1:200
8 Freestanding buildings 4th floor, 5th floor/3rd floor, 6th floor 1:500
9 Section of freestanding building and long building 1:1000

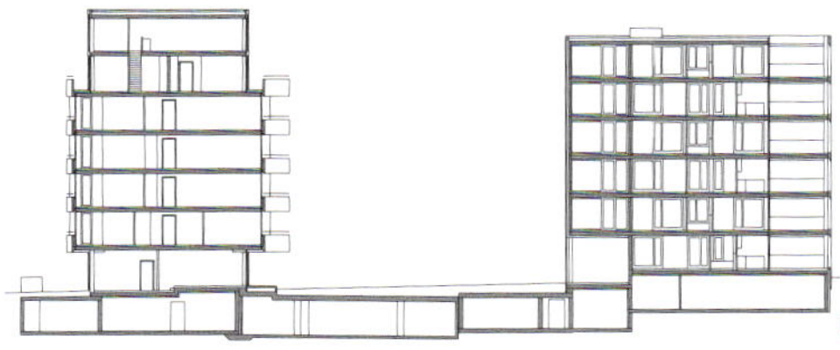

9

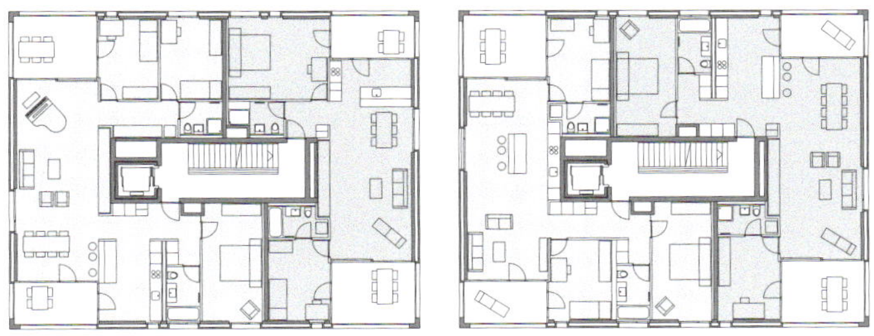

8

Building type
7-story long building (225 m long) with rotation, facing N/S, 6 freestanding buildings, 6 stories, facing in all directions freestandig buildings with kindergartens

Date of construction
2004–2008

Number of units
long building: 120 apts., freestanding buildings: 66 apts.

Size of units
long building:
type "loft":
studio, 36 m²
apt.-studio, 91/117 m²
type "communal living":
4½-room apts., 116 m²
5½-room apts., 138 m²
type "day/night area":
4½-room apts., 116 m²
5½-room apts., 138 m²
type "open living area":
4½-room apts., 116 m²
4½-room apts., 138 m²
type "loft + maisonette":
4½-room maisonettes, 129 m²
3½-room maisonettes, 112 m²
lofts, 56/60 m²

Area per user
27.5–60 m²

7

Building depth
long building: 15 m
freestanding buildings: 14–16 m

Access
double-loaded stairwells

Open spaces
balconies, roof patios

Parking
garage

Architect
Matti Ragaz Hitz Architekten AG
Liebefeld-Bern

Location
Hardegg
Bern
Switzerland

6

Referring to floor plan 1:200

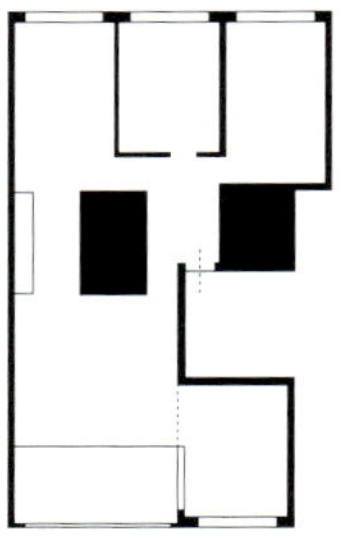

Station Illnau | Guignard & Saner | 2010

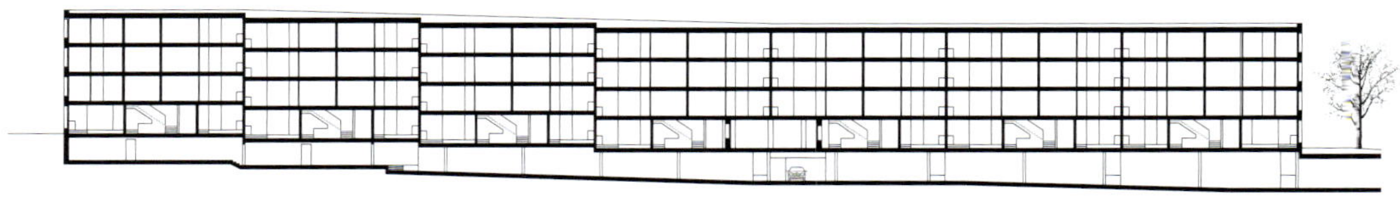

❹

Despite its location between a train station and village center, this mixed residential and commercial ensemble is molded by the landscape. It consists of two linear slab blocks and a solitary building nestled into the sloped, meadow-like topography. A staircase descends from the train station to pathways leading to the building entrances. One of the paths pierces the long slab, continuing on eastward towards the village.

In the flats in the long slab (Building **A**), the enclosed rooms and an inserted core define the boundaries for an open but subtly zoned living area. The lengthwise orientation of the staircase, the position of the bathroom core, and the arrangement of the individual rooms and loggia create an open sequence of rooms offset to one another – with a large hallway, living and dining areas connected by the kitchen, and an additional room which can be opened by a sliding door.

On the top floors of the solitary building (Building **C**) recessed into the slope, the apartments are arranged around one of the building's corners or open to three sides and laid out in a similar fashion. Between the inserted core and the encompassing walls, a corridor, foyer, kitchen, and two living spaces are formed. The open corners of the core, created by bundling two bathrooms and a storage closet, help to mark the boundaries of the different areas without inhibiting the open and flowing space.

Within the shorter slab (Building **B**), a central corridor leads to studios on one side, duplex apartments on the other, and once more flats with inserted cores at either end.

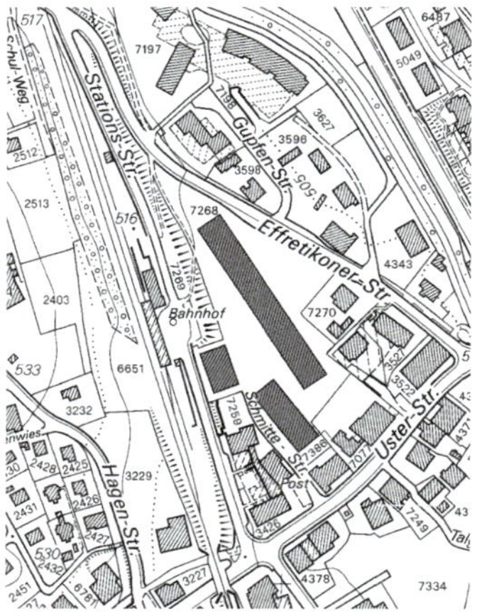

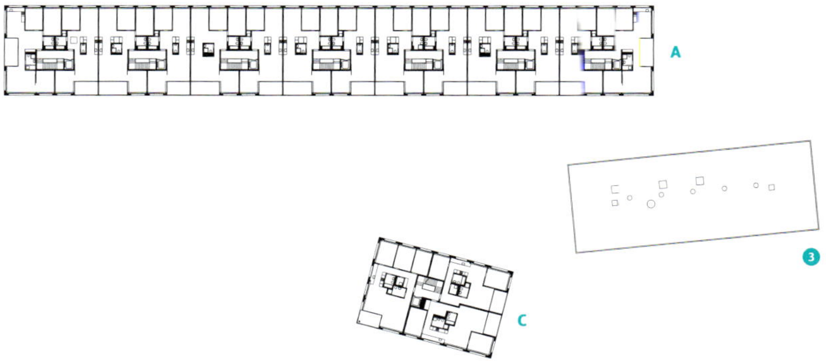

❸

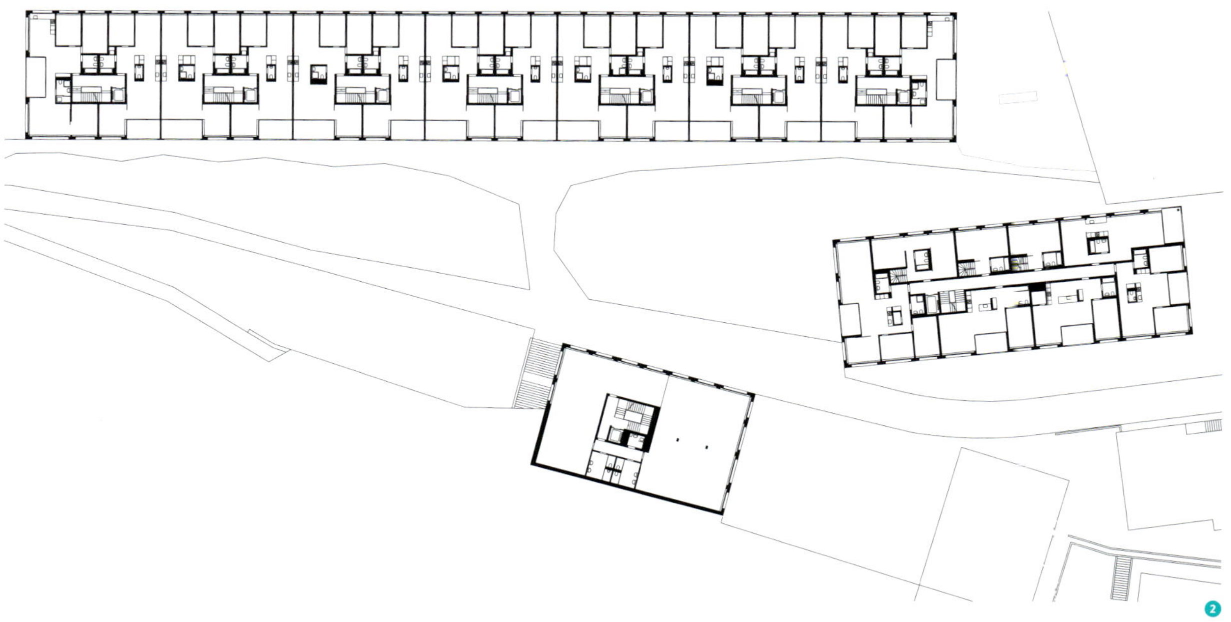

❷

❶

❶ Ground floor, entire ensemble 1:1500
❷ 2nd floor, entire ensemble 1:1000
❸ 4th floor, entire ensemble 1:1500
❹ Longitudinal section building A 1:1000
❺ Cross section building A, elevation building C 1:1000
❻ Building **C**: 4 ½-room apt. 1:400
❼ Building **C**: 3 ½-room apt. 1:400
❽ Building **A**: 4 ½-room work-live or commercial apt. (ground floor) 1:400
❾ Building **A**: 4 ½-room apt. and 3 ½-room apt. (typical floor) 1:200
❿ Building **A**: 3 ½-room work-live or commercial apt. (ground floor) 1:400
⓫ Building **B**: 3 ½-room maisonette 1:400
⓬ Building **B**: 2 ½-room apt. 1:400
⓭ Building **B**: 3 ½-room apt. 1:400

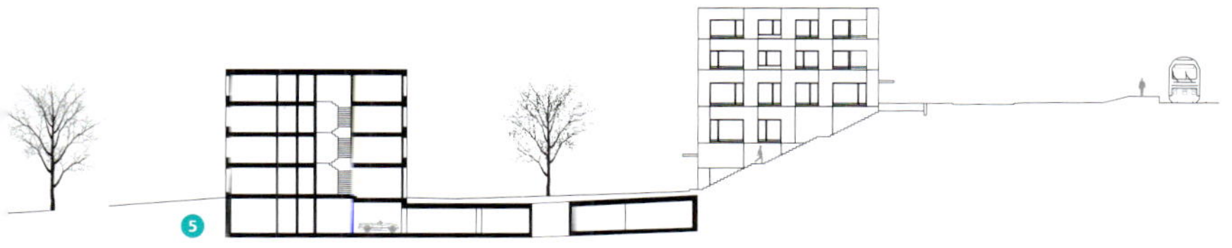

5

2.2 LINEAR BLOCK / SUPERBLOCK

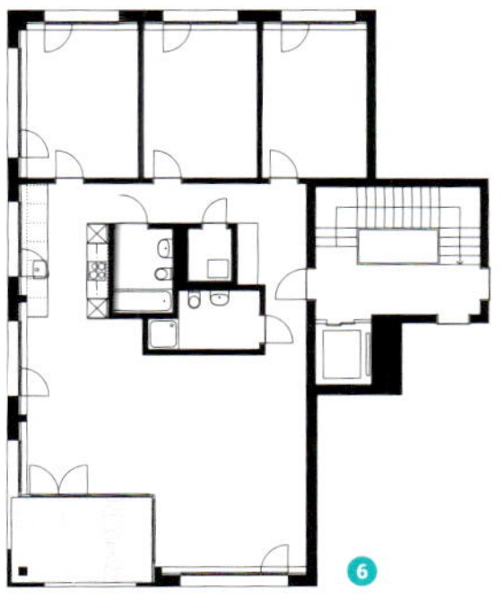

6

7

8

9

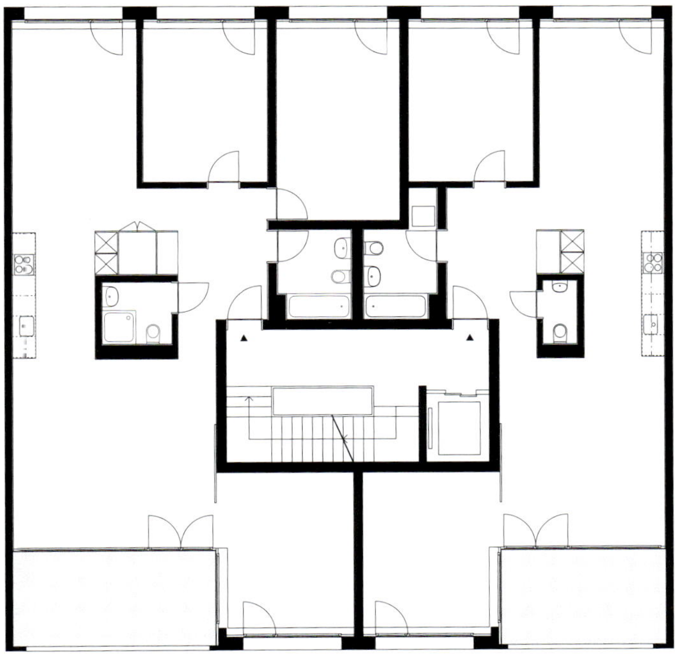

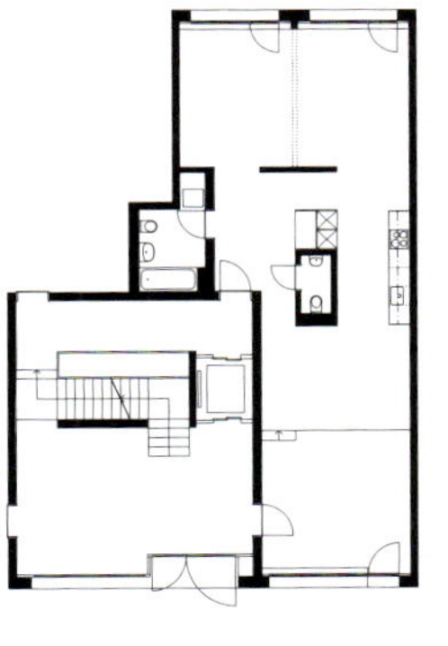

10

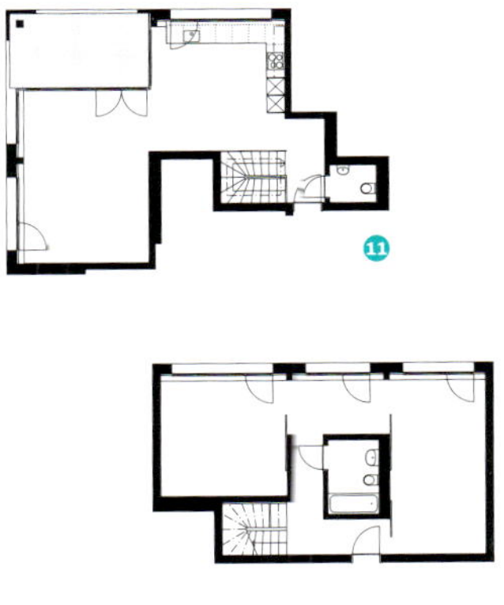

11

12

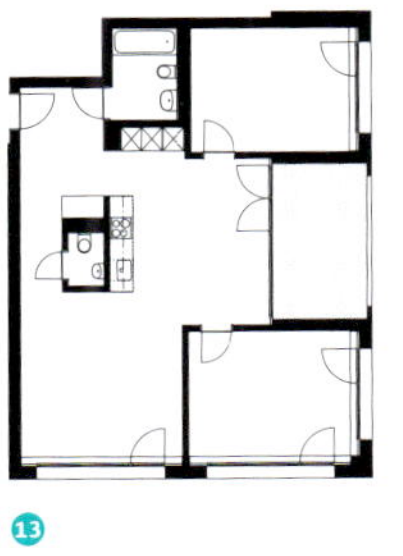

13

Building type
picturesque building ensemble comprised of two slabs and a solitary structure
slabs 3–4 stories
facing NE/SW
solitary structure on a slope, 3 or 5 stories
facing in all directions

Date of construction
2007–2010

Number of units
74

Size of units
building **A**
ground floor work-live apts./commerical units
1 ½-room apt., 48 m² (1 unit)
2 ½- / 3 ½-room apts., 89.5 m² (6 units)
3 ½- / 4 ½-room apts., 117 m² (6 units)
upper floors
3 ½-room apts., 99 m² (21 units)
4 ½-room apts., 126 m² (21 units)
building **B**
2 ½-room apts., 75.5–80 m² (6 units)
2 ½-room mais., 82–91 m² (2 units)
3 ½-room apts., 86–93 m² (3 units)
3 ½-room mais., 113 m² (1 unit)
4 ½-room apts., 129 m² (1 unit)
building **C**
3 ½-room apts., 99–110 m² (4 units)
4 ½-room apts., 131 m² (2 units)

Area per user
31.5–39.5 m²

Building depth
16.9 m

Access
double- and triple-loaded staircases
corridor in shorter slab

Open spaces
loggias, terraces

Parking
underground garage

Architect
Guignard & Saner Architekten AG
with Othmar Villiger
Zurich

Location
Illnau, Switzerland

Referring to floor plan 1:200

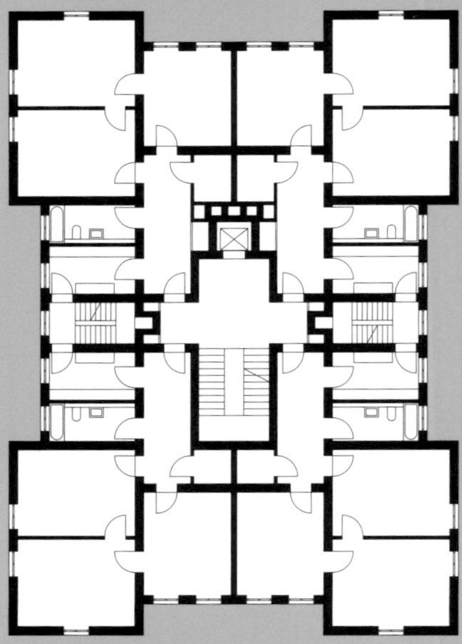

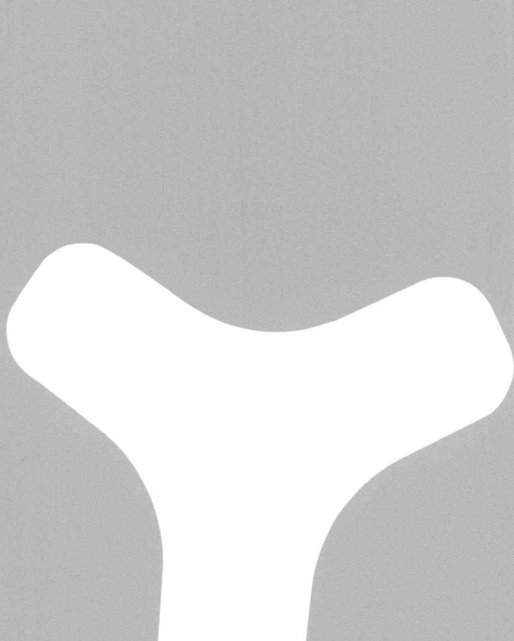

2.3
APARTMENT TOWER

The term apartment tower encompasses both urban high-rises and multi-story residential buildings. In building codes, a high-rise is defined as having occupied floors located above the height reachable by fire department vehicles; from this height onward, special fire protection measures and double fire escape routes must be considered in the design. The apartment tower follows the "law of series," stacking the highest possible number of identical or similar floor plans one above the other. In most cases, the apartments are grouped around a central core with several stairwells and elevators; due to the size of the core, they are often oriented toward one side only, sometimes around a corner. At times an undulating external skin is employed to provide the apartments with orientation in more than one direction. Apartment towers originated in 19th-century American cities, where real estate markets called for a building type that allowed for a significantly higher ratio of area utilization; this was only made possible through several technological advancements such as skeleton construction and increased safety features for passenger elevators. Apartment towers are thus a fundamentally urban building form, often furnished with luxurious lobbies reminiscent of those found in hotels because of the great number of occupants. The buildings combine a wide variety of highly different uses: retail, offices and apartments, parking and communal services are often incorporated in a single building for the comfort of the residents. The interiors of the apartment towers tend to be anonymous. With access areas that are often quite small and many residents per floor, there is little to inspire neighbourly encounters. Usually, the occupant's principal connection with his city is established by the stunning views afforded by apartments of this kind. From a certain height onward, high-rises become uneconomical and increasingly less sustainable due to the technical, structural, and energy requirements. Due to wind flow, the greater the height of the tower, the higher the cost and effort required to provide exterior spaces. Yet precisely these issues – the creation of communal and private outdoor spaces for the residents and an improvement of the energy balance – pose the challenges for the high-rises of the future.

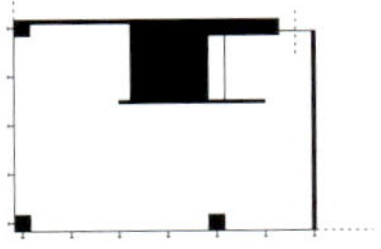

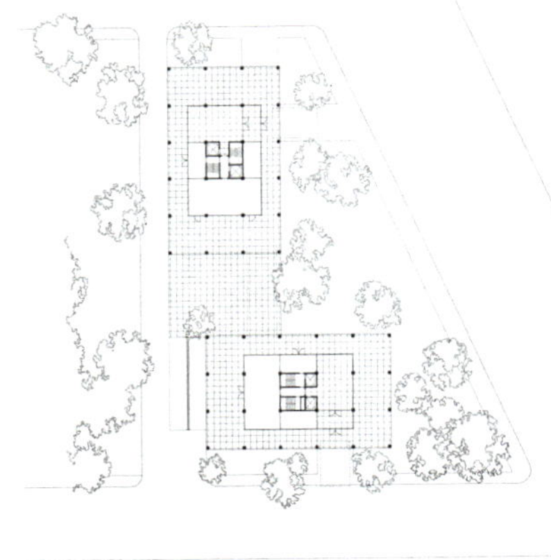

2.3
APARTMENT TOWER

Lake Shore Drive | Mies van der Rohe | 1951

The two apartment towers, the first to be realized almost exclusively in glass and steel, are placed on the trapezoid lot in such a manner that they appear to glide past one another and to provide all apartments with a panoramic vista of Lake Michigan. A roof between the recessed entrance stories links the two towers. Although based on an identical structural grid, the north tower accommodates eight 2-room apartments on each floor, and the south tower four 5-room apartments. The original floor plans were loosely based on the "open plan" of Farnsworth House. A service core with open kitchen and bathroom was to define living, dining and bedroom. After the financier objected that Americans preferred enclosed rooms, the floor plans had to be revised. The more conventional floor plan solution with separated bedrooms and enclosed kitchens was ultimately realized. Kitchen and bathroom are located at the center in both solutions. The other rooms benefit from the view.

Building type
apartment tower, 25 stories plus entrance level facing N/E/S/W

Date of construction
1951

Number of units
200 (north tower), 100 (south tower)

Size of units
north tower: 2-room apts., 66.5/68.5 m² (200 units)
south tower: 5-room apts., 133.5 m² (100 units)

Area per user
26.5–34.5 m²

Building depth
19.8/32.6 m

Access
8-strung stairs (north tower)/
4-strung stairs (south tower), two lifts in each tower and internal stairwells

Open spaces
none

Parking
116 parking spaces in underground garage

Architect
Mies van der Rohe, Chicago

Location
860/880 Lake Shore Drive, Chicago, USA

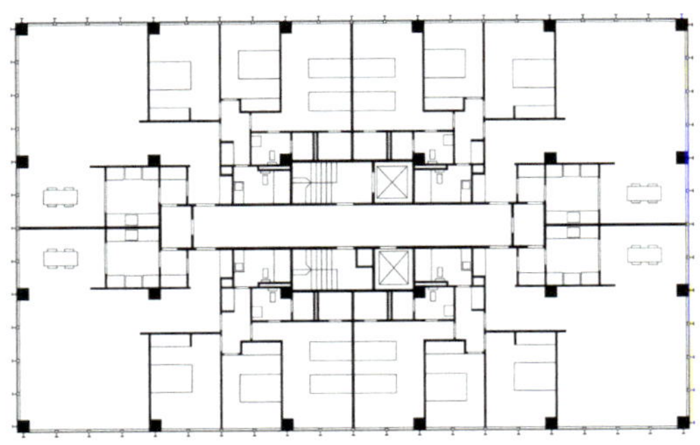

1

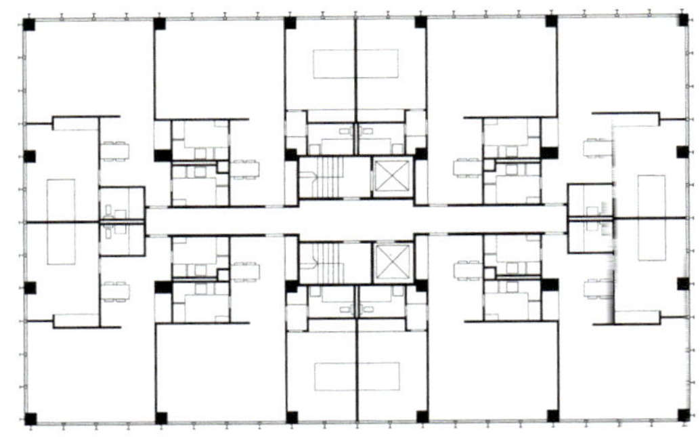

1

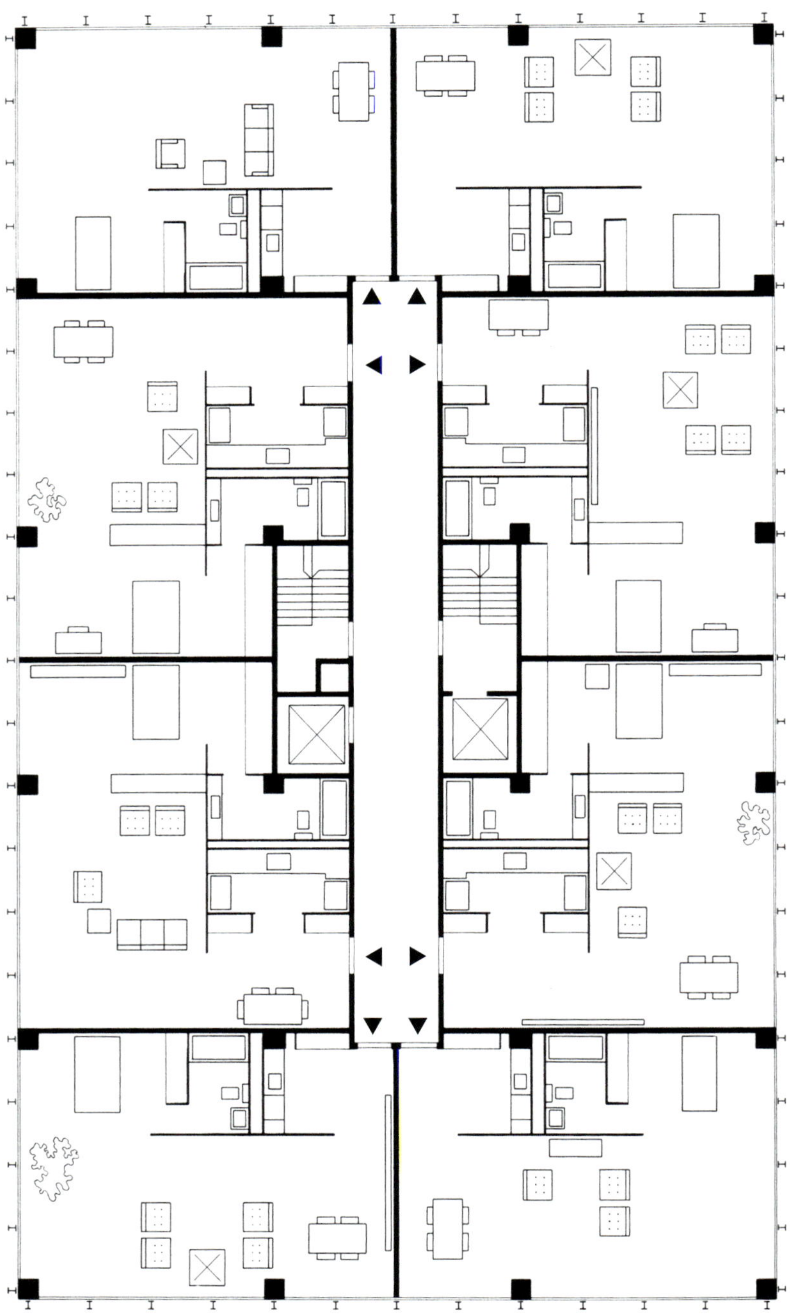

2

Standard floors, 1
as completed 1:500
Standard floor, 2
original design
1:200

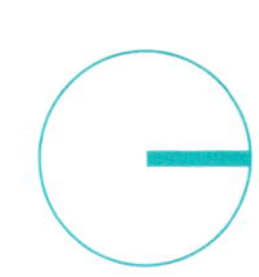

Referring to
floor plan 1:200

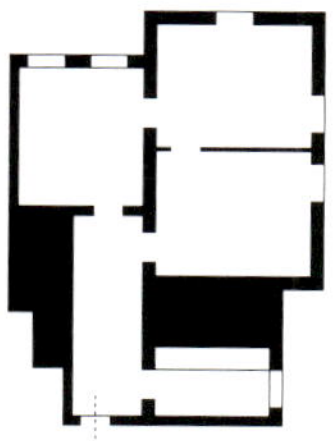

Weberwiese | Henselmann | 1952

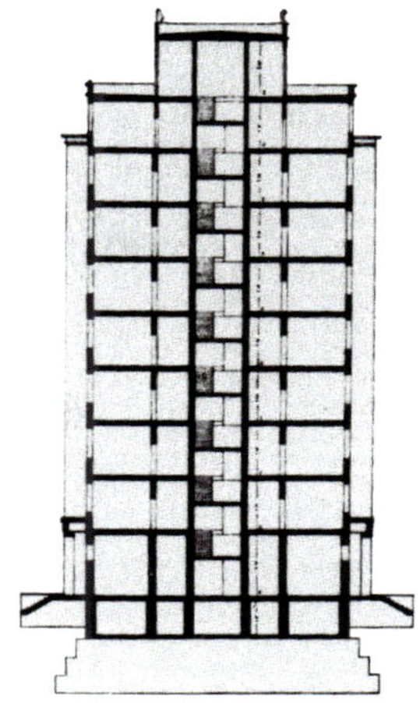

2.3 APARTMENT TOWER

"The first high-rise in the GDR." The structure, realized in brickwork, contains four identical units grouped around a central stairwell. A generous hallway leads to kitchen, bathroom, closet, and two of the three rooms. The latter, projecting from the building core, are arranged around the corner, with the corner room in the middle and lit from two sides. The attractive layout and dimensions of each of these rooms allows for a free choice of usage. Moreover, the non-load-bearing dividing wall suggests that two of these rooms could easily be combined into one. The kitchens open towards the auxiliary stairwells (stipulated by the building authorities) with garbage disposal chutes. The building is strongly defined by its "classic" image since it was intended to "satisfy the need for representation of the working people." On the other hand, the surprisingly neutral rooms are the reason why these apartments are once again in demand today.

Building type
apartment tower, 9 stories with retail stores on ground level, linked to a 5-story block on the south side of the square

Date of construction
1952

Number of units
32

Size of units
3-room apts., 96 m²

Area per user
32 m²

Building depth
21/29.3 m

Access
quadruple-loaded internal stairwell with two auxiliary stairwells

Open spaces
roof terrace, Weberwiese (public square with pond)

Parking
no parking spaces planned

Architect
Hermann Henselmann
Berlin

Location
Weberwiese
Berlin-Friedrichshain
Germany

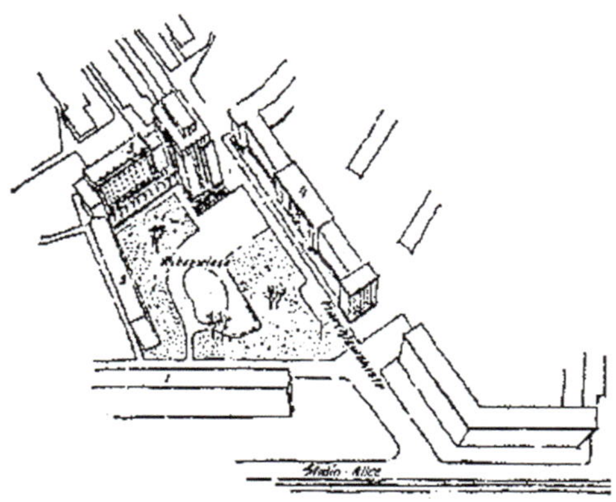

1

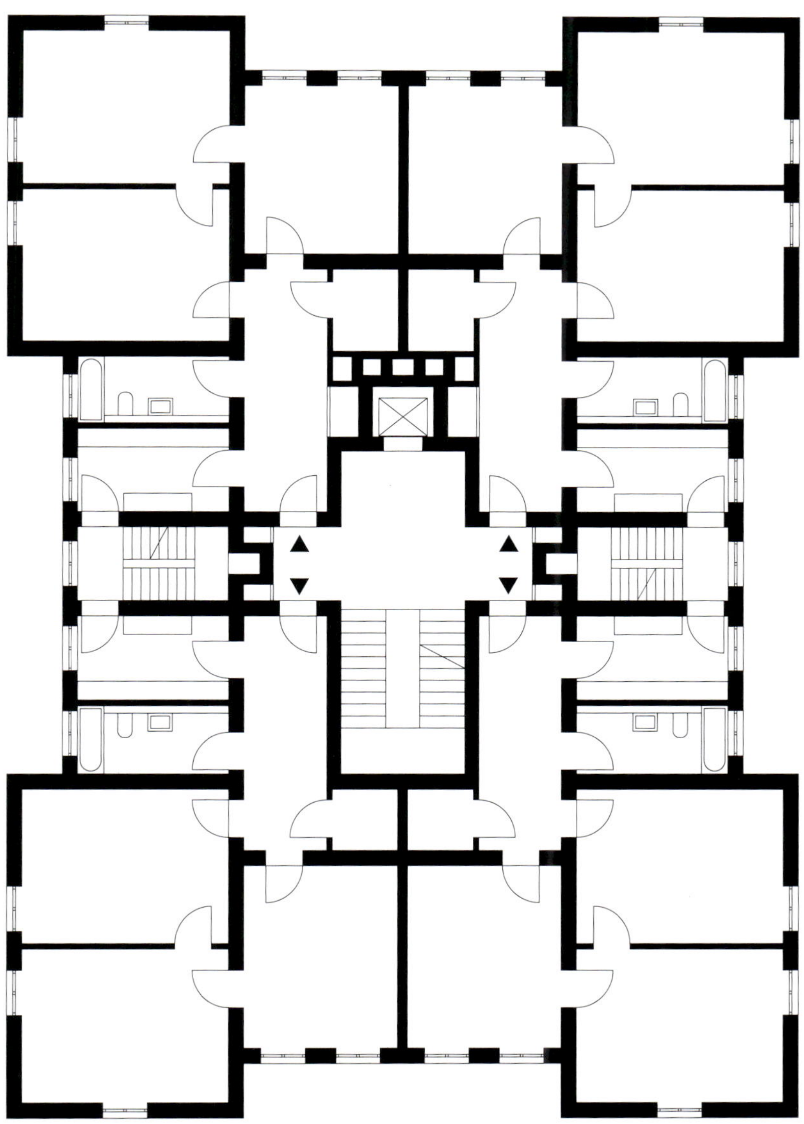

2

Site sketch of Weberwiese 1
with high-rise
and adjacent block
Standard floor 1:200 2

Referring to
floor plan 1:200

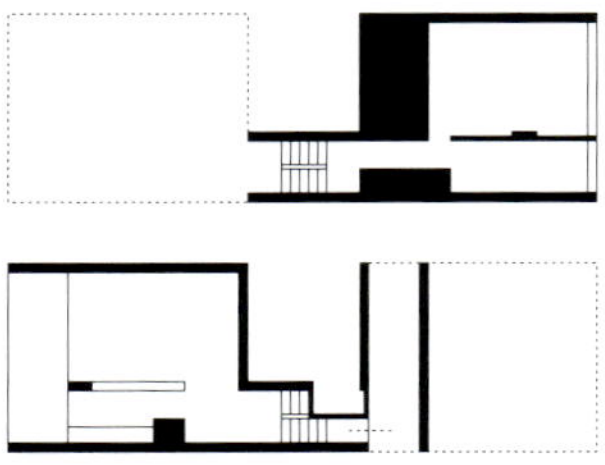

Hansaviertel | Van den Broek / Bakema | 1958

By virtue of a clever combination of section and plan, the building affords most units natural light on two sides and cross ventilation: three floors with the access corridor in the center and staggered split-level floors facing in the opposite direction are combined to form a greater unit, of which six are stacked one above the other in alternating fashion. The center corridor provides access to 1-room apartments on one side and – via split double span stairs – to two 3-room apartments on the other side: each set of stairs belongs to one apartment and leads up, or down, into the open living areas. The bedrooms, which are half a story up or down, are situated on the opposite side of the building. The apartments are thus continuous front-to-back over top of the stairs. All center corridors benefit from natural light on both sides. On the south side they open onto two-story loggias with sun and play decks for shared use. The six loggias define the south facade, which also reveals the staggered floor arrangement. The organization of the apartments is also reflected in the facade: the living rooms are fronted by room-high loggias, while the bedrooms lie behind horizontal bands of windows. In addition to a laundry facility, the attic story also features a roof garden suitable as a children's play area.

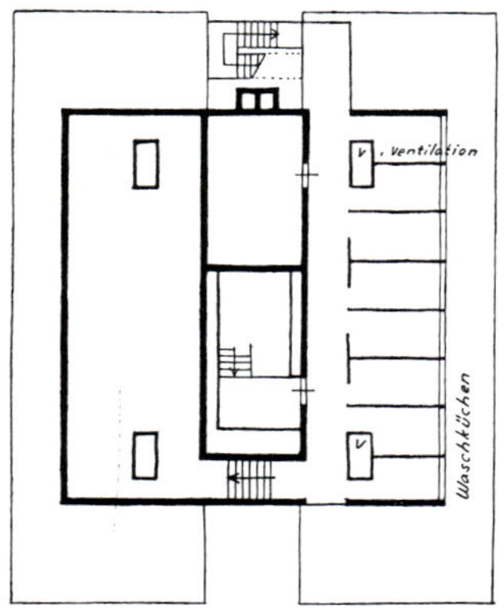

4

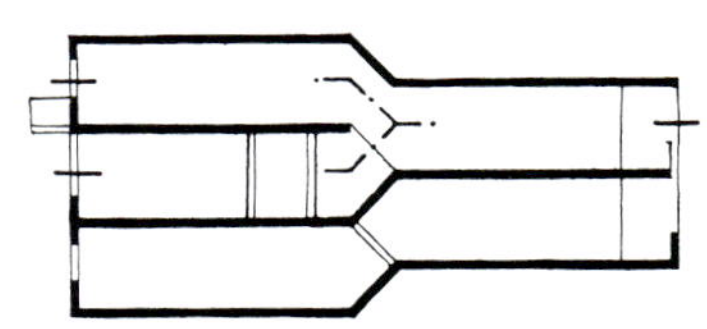

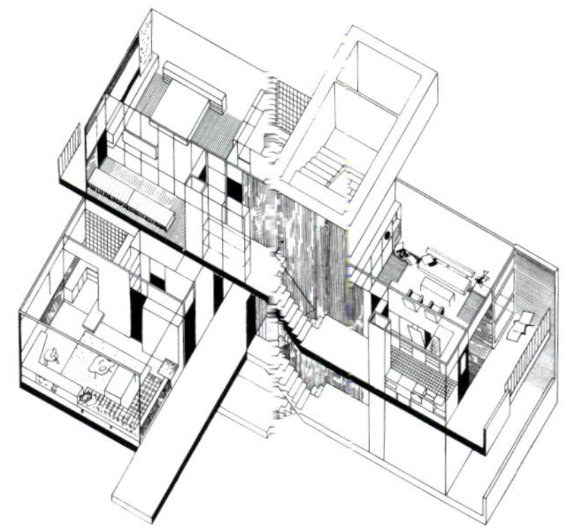

5

3

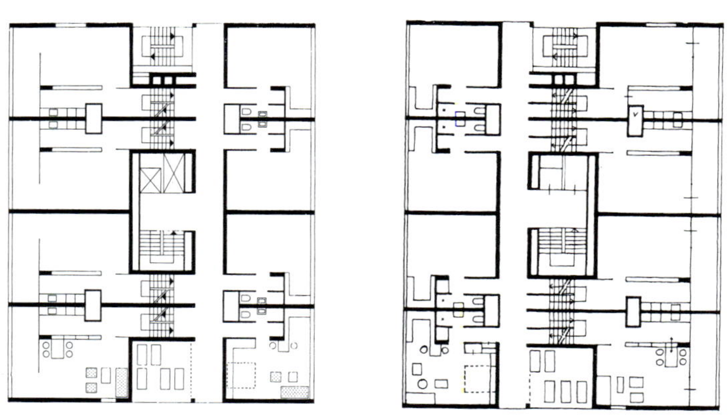

2

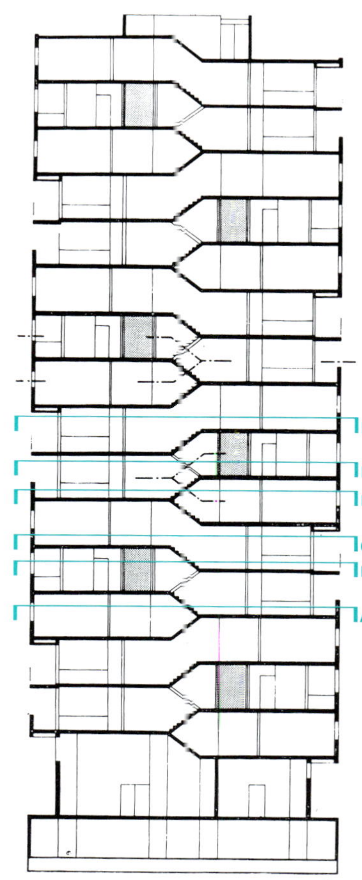

6

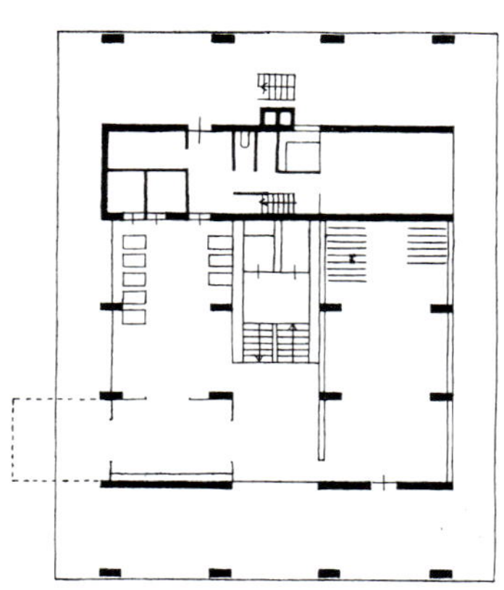

1

1 Ground floor: lobby, access core, and caretaker's apartment 1:500
2 Floor plans of access levels: central corridor on west i.e. east side, communal loggia at the southern end, 1-room apartments and living area of the split-level units 1:500
3 Floor plans with the two offset levels: living area and private rooms above i.e. below central corridors 1:500
4 Attic story: laundry rooms and roof terrace 1:500
5 Isometric drawing of 1-room apartment, central corridor and the 3-room split-level units
6 Section 1:500
7 Floor plans of the split-level units and 1-room apartments 1:200
8 Sketch of communal loggias

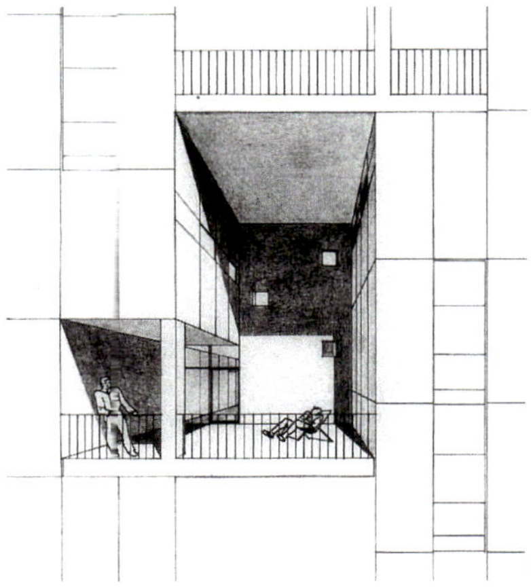

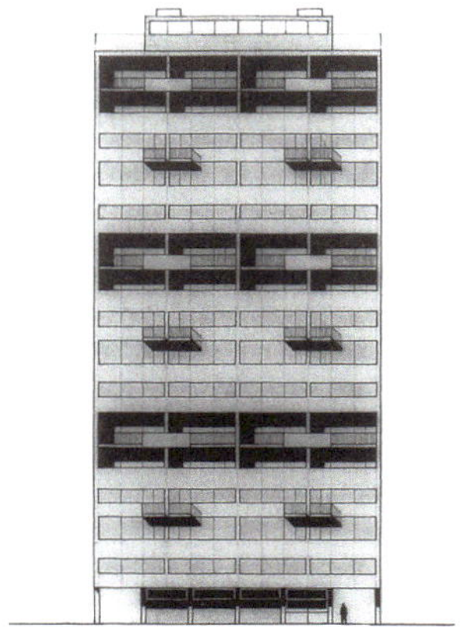

8

2.3
APARTMENT TOWER

FF
EE
DD
CC
BB
AA

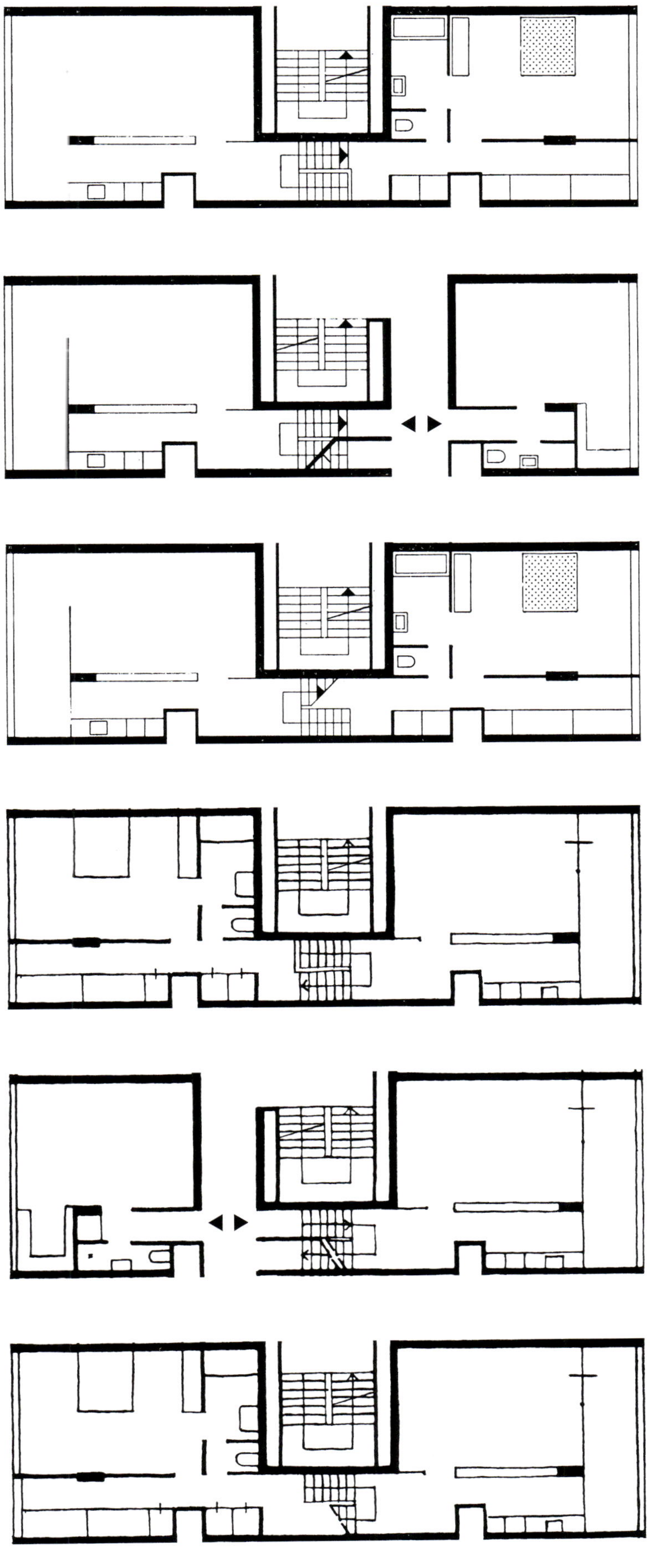

7

Building type
residential tower
with apartments and split-level units
16 stories
facing W/E
(common loggias facing S)

Date of construction
1958
for the Berlin
International Building Exhibition
(Interbau 1957)

Number of units
73

Size of units
1-room apts.: 36 m² (24 units)
3-room split-level apts.: 92 m² (48 units)
2-room apt.: 55 m²
(1 unit on ground floor for caretaker)

Area per user
27.5–36 m²

Building depth
20 m

Access core
access core
center corridors

Open spaces
private and common loggias
along center corridors
roof patio

Parking
no parking on lot

Architect
J.H. Van den Broek
J.B. Bakema
Rotterdam

Location
Hansaviertel
Berlin-Tiergarten
Germany

Referring to
floor plan 1:200

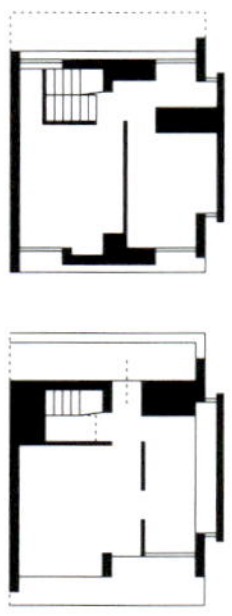

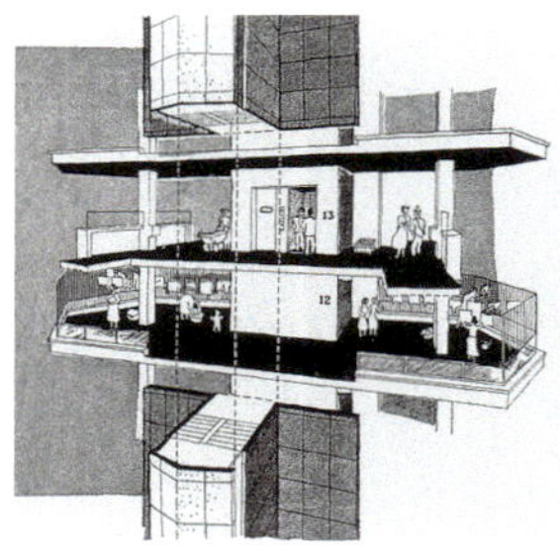

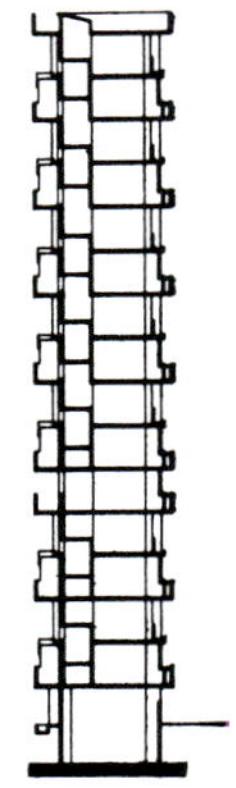

Cluster Block | Lasdun | 1958

Four individual towers are arranged as wings around a central access core; bridges lead to the apartment towers. Access from galleries on the north side (minimal shading). The core contains generous elevators, garbage disposals, landings for drying laundry, etc. The core is conceived as a kind of rear courtyard, originally envisioned with planted greenery. The galleries, each with two apartments, remain private. An installation zone, with kitchen and bath as buffer, is located on the gallery side. The apartments open out to south, south-west and south-east. With the exception of the sixth floor (bachelor units), all apartments are maisonettes, with the central living room (small balcony) and kitchen at bottom floor; two bedrooms on top floor. The outer corners have additional emergency exits (ancillary spaces).

Building type
quadruple tower (cluster)
16 stories
facing NNE/SSW, NNW/SSE

Date of construction
1955–1958

Number of units
64

Size of units
1-room apts., approx. 50 m² (8 units)
3-room mais., approx. 90 m² (56 units)

Area per user
30–50 m²

Building depth
7.5–9.25 m

Access
8-loaded stairwells
galleries for every 2 units

Open spaces
forecourt, balconies

Parking
street parking

Architect
Denys Lasdun + Partners
London
Architect in charge:
John Shaw
London

Location
Claredale Street
London-Bethnal Green
United Kingdom

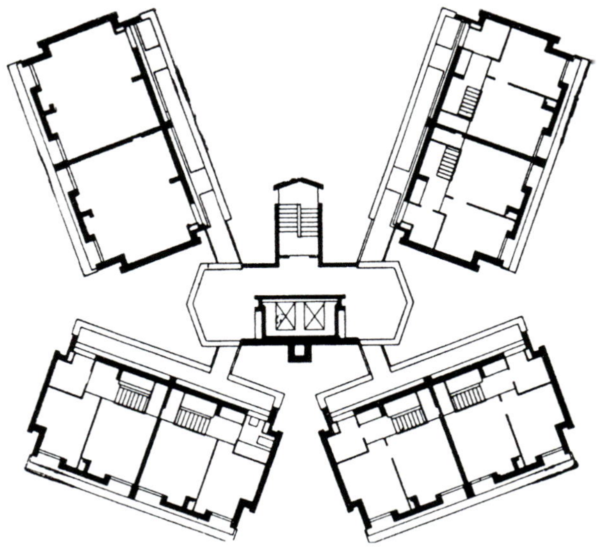

3

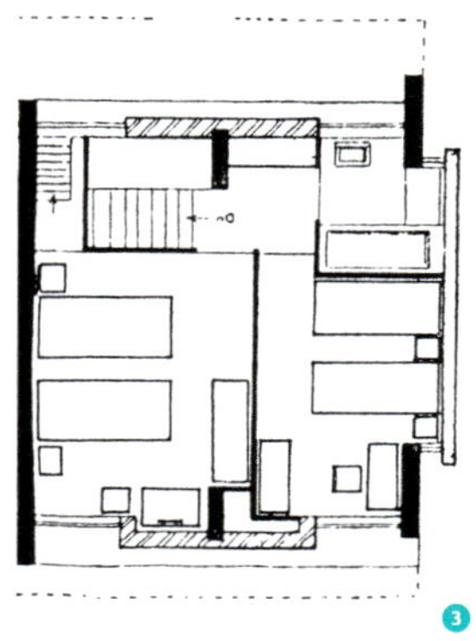

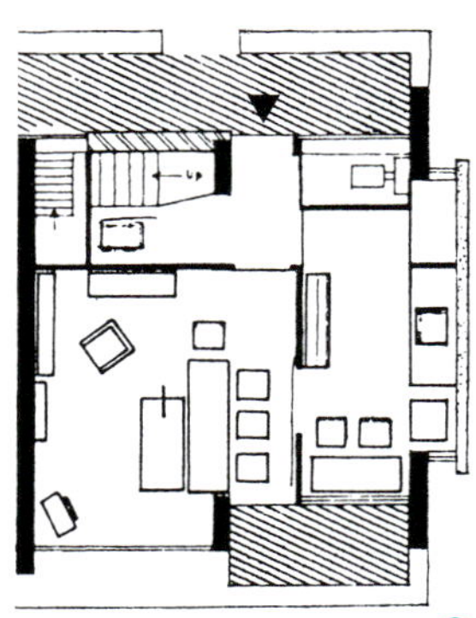

2

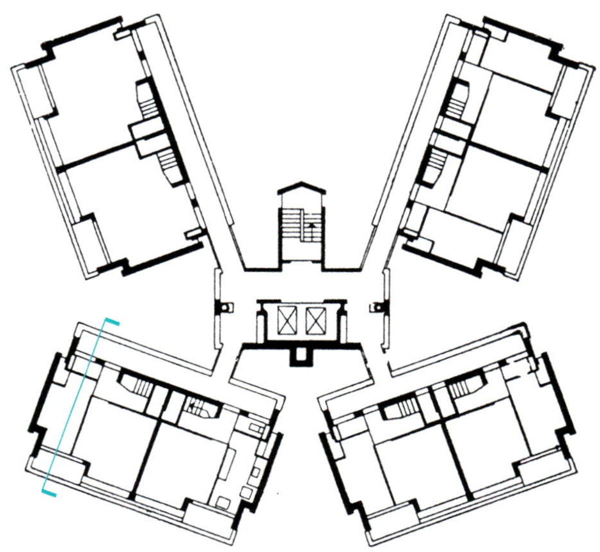

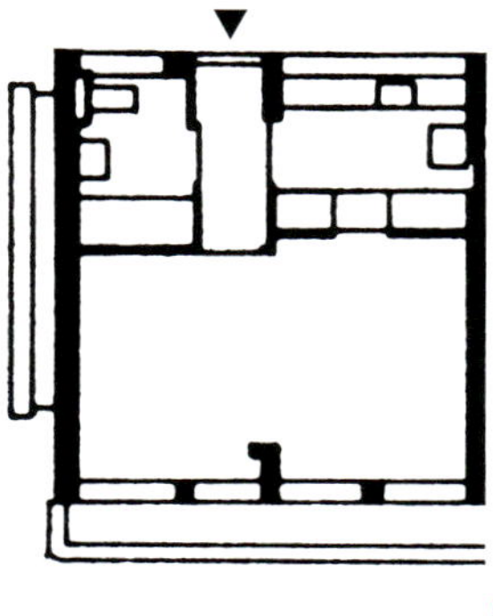

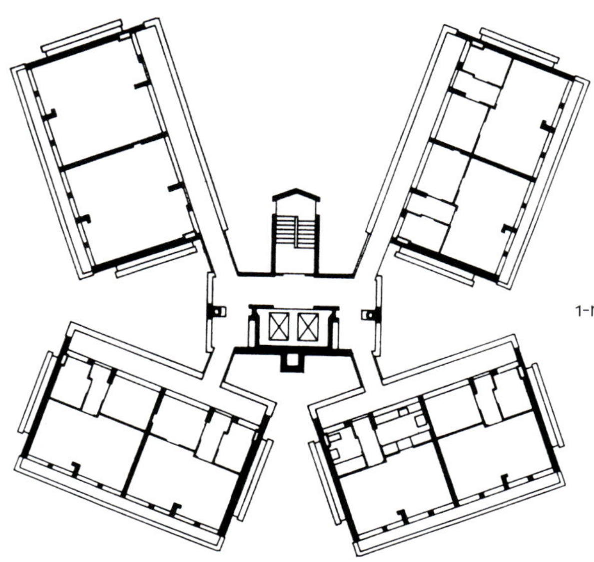

1

6th floor: 1
1-room apartments 1:200
Entrance floor of 2
maisonettes 1:200
Upper floor of 3
maisonettes 1:200

Referring to
floor plan 1:200

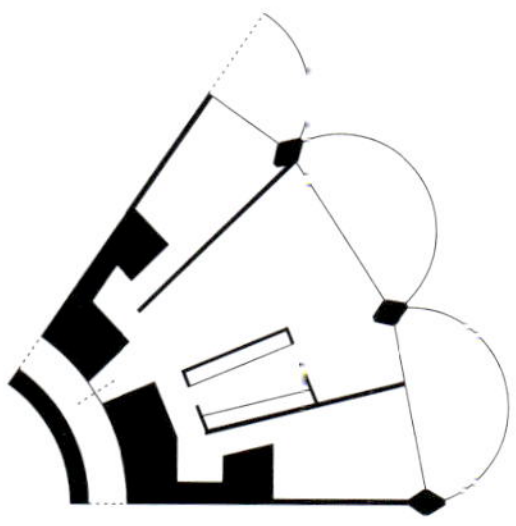

Marina City | Goldberg | 1963

A visionary complex, the first in the United States to combine living, working, and leisure. "Marina City" was to become a "city within the city" that would be alive 24 hours a day. In addition to two apartment towers, the development contains an office complex, theaters (cinemas), several restaurants, retail stores, etc., and a pier where yachts and boats can dock. The apartment towers are a "stacked city" themselves: the first fifteen floors are conceived as spiraling ramps and reserved for parking, followed by forty apartment floors above, wrapped around a central core, which houses the lift and building services. The dynamic segmental plan of the 1–3-room apartments leads from the narrow entrance hallway past dressing room, bathroom and kitchen, straight to the glass facades with wide balconies and the views they offer.

Building type
2 apartment towers
60 stories
40 apartment floors

Date of construction
1963

Number of units
2 × 448

Size of units
studios, approx. 40 m²
2-room apts., approx 70 m²
3-room apts., approx. 100 m²

Area per user
33.5–40 m²

Building depth
approx. 33 m

Access
elevator core
with surrounding hallway

Open spaces
balconies, public square
recreation facilities:
bowling lane, skating rink,
swimming pool, fitness club,
boat docking

Parking
15 parking levels

Architect
Bertrand Goldberg Associates
Chicago

Location
Marina City
Chicago
USA

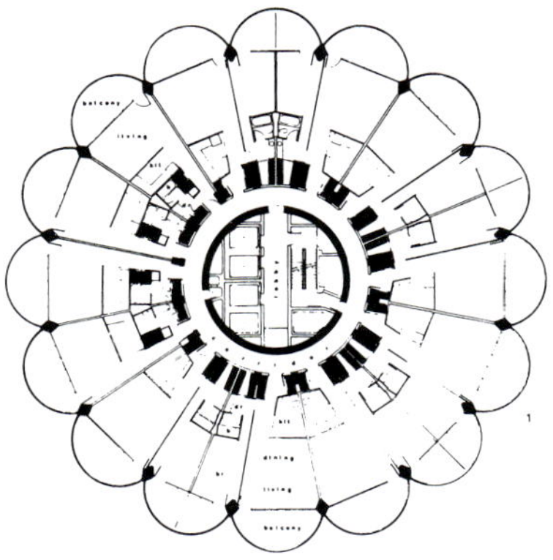

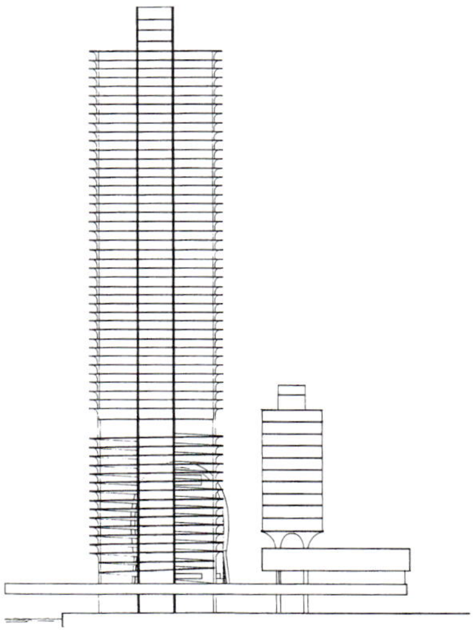

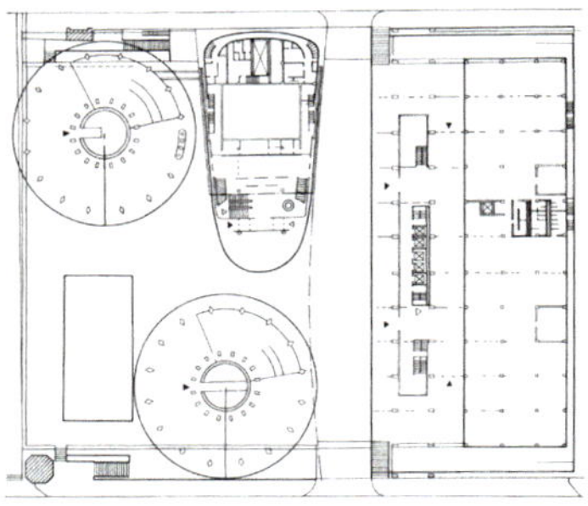

3-room apartment 1:200 ❶
1-room apartment 1:200 ❷
2-room apartment 1:200 ❸

Referring to
floor plan 1:200

Romeo and Juliet | Scharoun | 1959

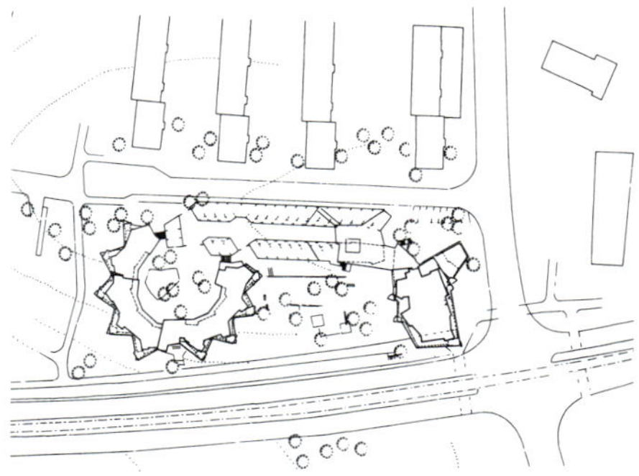

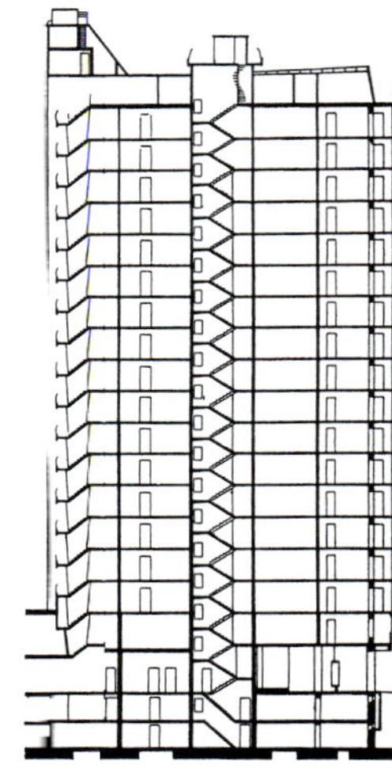

Two buildings on a lot, which enter into an inspired spatial dialog with one another. "Romeo" – a compact, tall apartment tower – occupies the eastern side of the shared lot. A central, naturally lit corridor provides access to six apartments on each floor. "Juliet," the lower building, on the other hand, appears more open and dissolved in volume, with its ring-shaped fabric divided into irregular steps from 11 to 4 stories. Covered walkways provide access to the apartments in this building, and are connected to a central stairwell. Penthouse units, so-called "studio" apartments with large terraces and cantilevered roofs generate a roof landscape. The crystalline, fanned-out form, which provides panoramic views in all directions, is common to both buildings. The projecting "teeth" in the building skin of "Juliet" increase the window surface in the individual rooms. Both buildings are strongly oriented toward the path of the sun: in "Romeo" all living rooms and balconies face south, while "Juliet's" circular plan is interrupted on the north side. Both buildings feature the same floor plan layout in principle, where a central hallway provides direct access to all rooms within a unit. There is always one room, which is additionally accessible from the living space, the separation of private and common area is rendered in a more flowing manner. The organic disposition of the internal dividing walls, which is also expressed in the facade, and the spatial effect of the two large fabrics on their own and in relation to each other, transforms the "Romeo and Juliet" complex into a differentiated housing experience. With this complex, Scharoun delivered proof in the postwar era that efficiency is not necessarily dependant on an orthogonal plan.

1

1 "Romeo":
typical floor plan with 1-room, 2-room, and 3-room apartments 1:200

2 "Juliet":
section of typical floor plan with 3-room apartments and one 4-room apartment 1:200

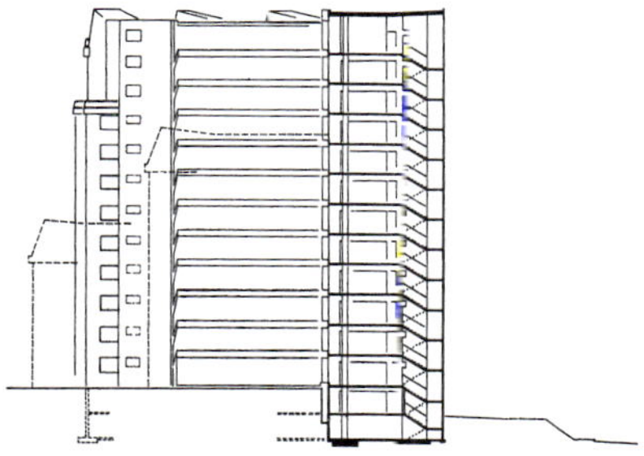

2.3
APARTMENT TOWER

Building type
residential high-rises
19 stories (Romeo)
11/7/4 stories (Juliet)
E/W and NW/W/S/SE

Date of construction
1954–1959

Number of units
104/82

Size of units
Romeo:
1-room apts., 38 m²
2-/3-/4-room apts., 88–96 m²
penthouse: 4 studios
Juliet:
3-room apts., 72 m²
4-room apts., 86 m²

Area per user
21.5–48 m²

Building depth
15–20 / 10–14 m

Access
6 apartments per story
on inner hallway (Romeo)
9 apartments
on gallery (Juliet)

Open spaces
balconies, rooftop terraces,
public green area

Parking
garage courtyard

Architect
Hans Scharoun
Berlin
with
Stefan Heise
Jo Zimmermann
Stuttgart
Project architect:
Wilhelm Frank

Location
Schozacher Straße /
Schwabbacher Straße
Stuttgart-Zuffenhausen
Germany

Referring to
floor plan 1:200

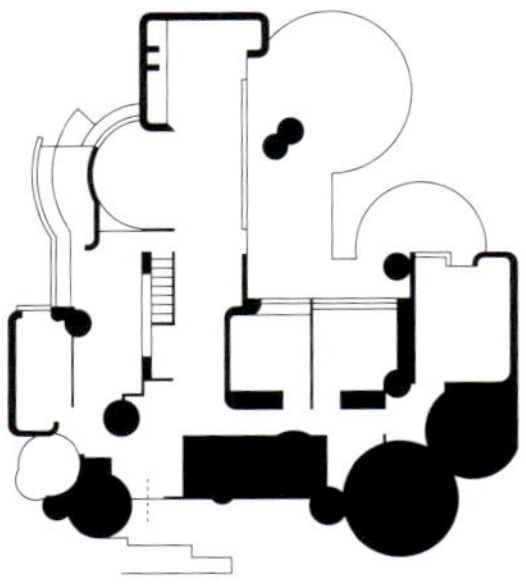

Torres Blancas | Sáenz de Oiza | 1969

The inspiration for this tower (the original plan envisioned two towers, hence the name "Torres Blancas," plural) was to create a vertical garden city that grows toward the sky like a tree. Cantilevered semicircular terraces envelope the building – like leaves on the branches of a tree. They provide shelter from the sun and noise and are the size of small gardens which can be extensively planted. It is through this green filter that the interiors communicate with the exterior space. The organic structure is continued on the inside of the units and determines the complex spatial construct: the geometry of the concrete structure, the layout of the access zones, as well as the internal organization of the apartments. A generous access area comprising an atrium, open and enclosed stairs, and cylindrical elevators provide access to four i. e. eight apartments per floor. The massive support columns with long-span ceilings make it possible to stack units with very different plans from floor to floor. Spacious open living spaces along the terraces are intentionally undefined to provide users with the utmost flexibility.

A two-story common area on the roof, which originally accommodated a pool, bar, and restaurant, now houses offices.

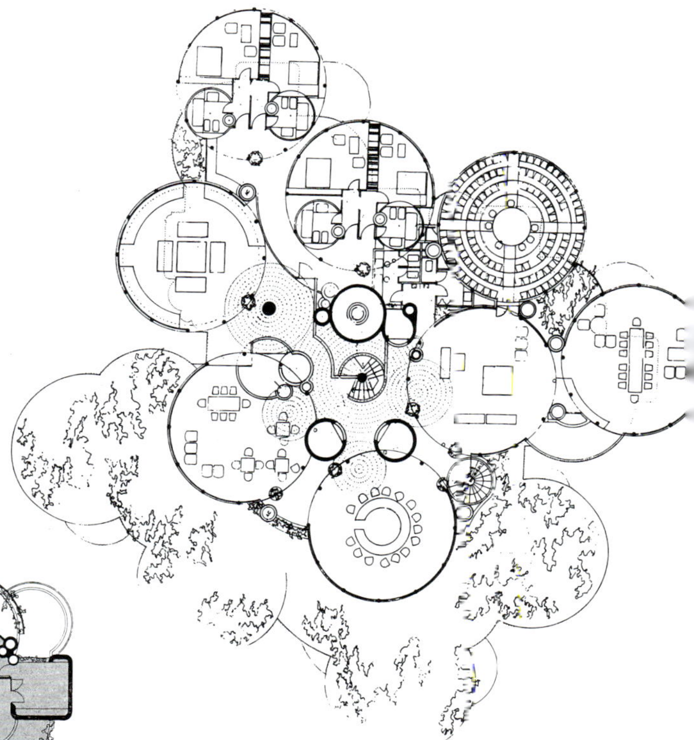

4

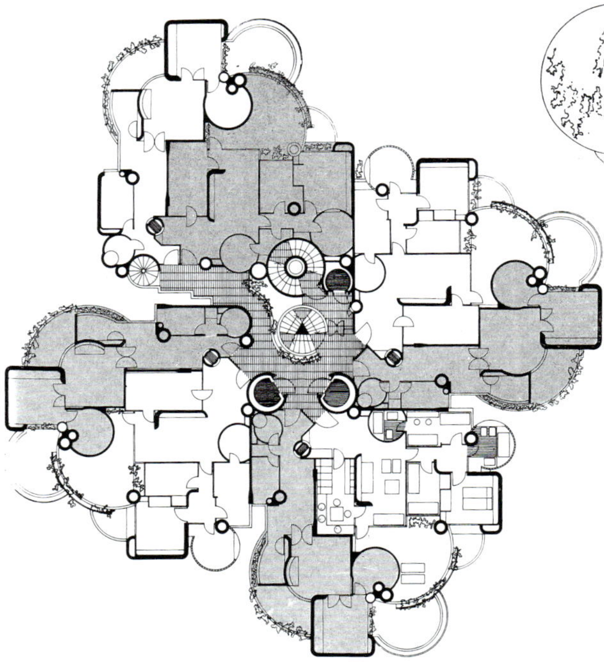

3

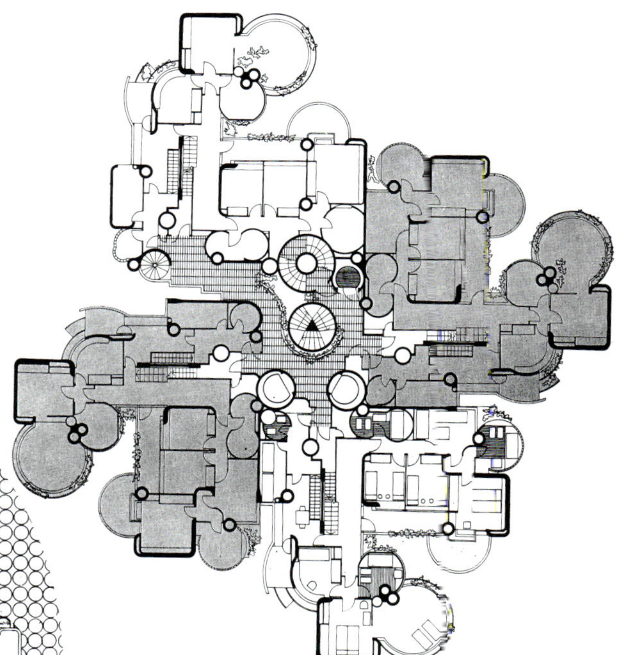

2

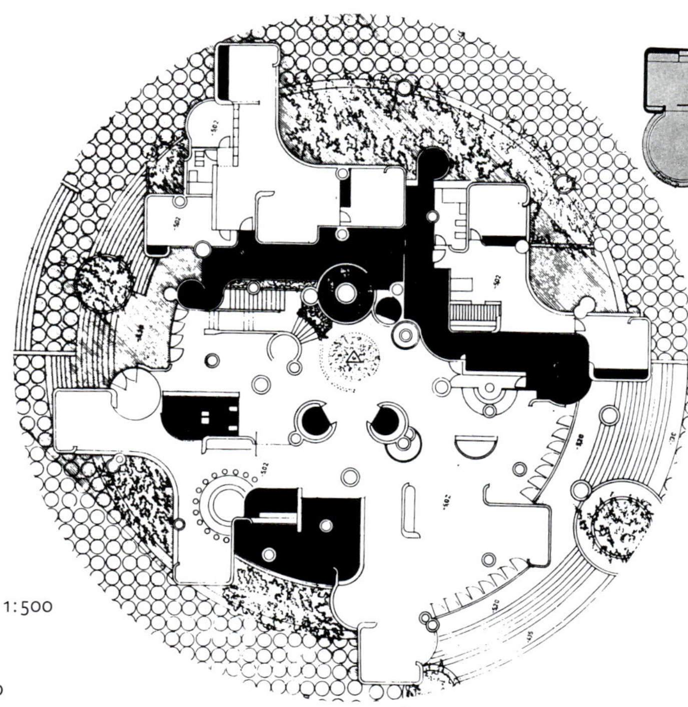

1

1 Ground floor: lobby 1:500
2 Total floor plan 6th and 12th floor 1:500
3 Total floor plan 8th–10th/20th–22nd floor 1:500
4 Public and communal areas in attic stories 1:500
5 Total floor plan of 7th and 13th floor 1:200
6 Section of attic stories 1:500

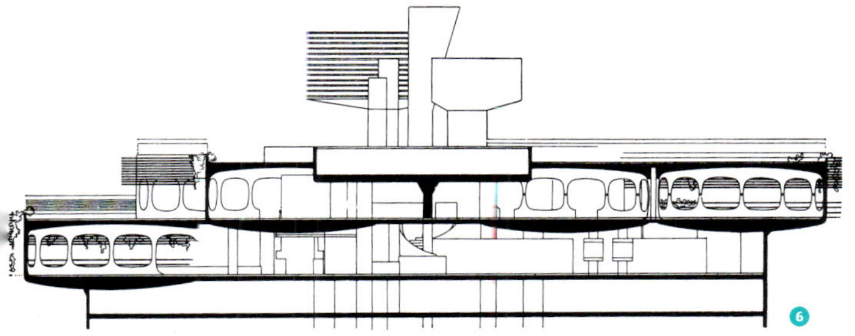

6

APARTMENT TOWER

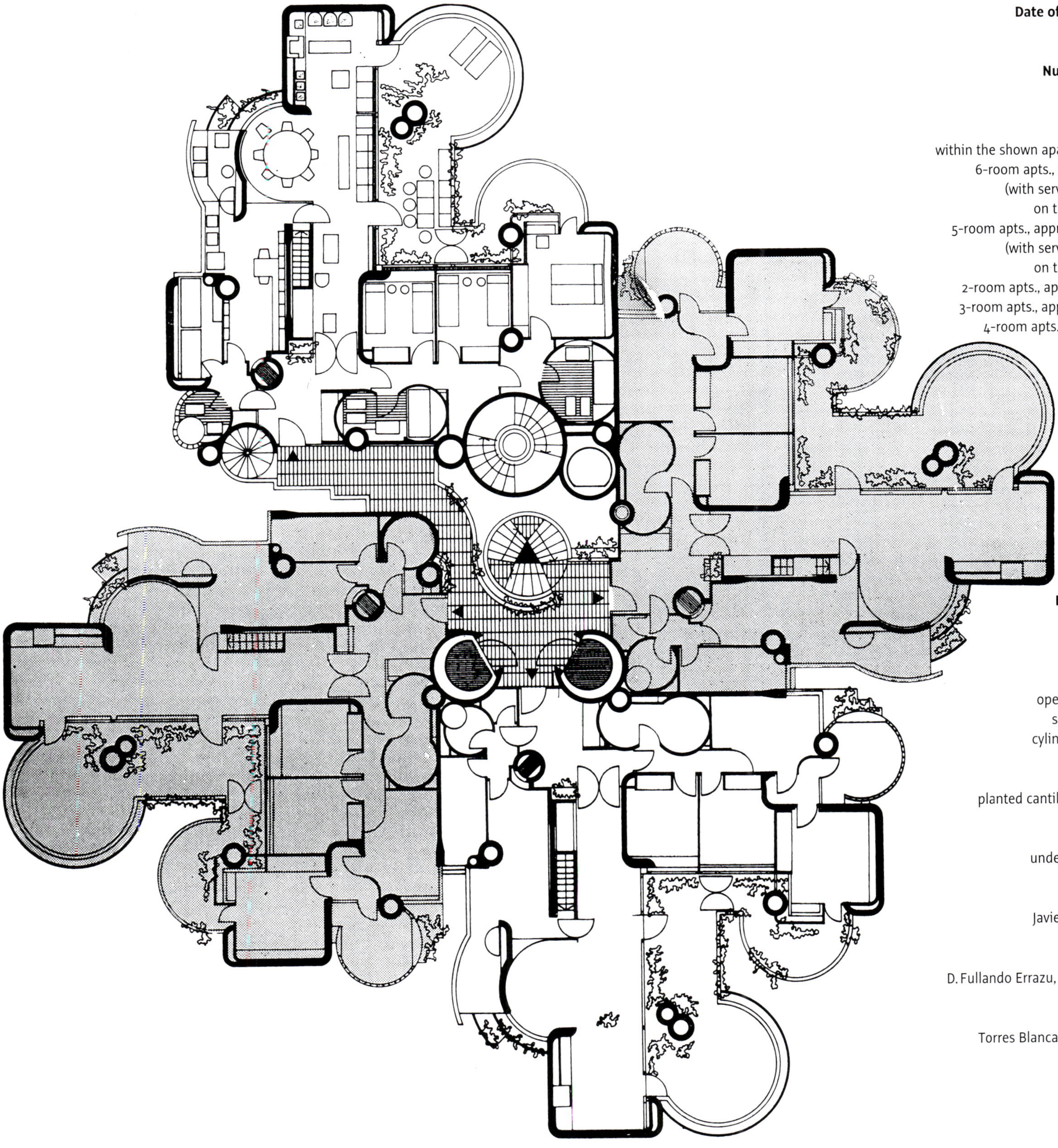

5

Building type
complex residential tower
composed of cylindrical volumes
21 stories, facing in all directions

Date of Construction
1964–1969

Number of units
160

Size of units
within the shown apartment sories:
6-room apts., approx. 150 m²
(with servants' entrance
on the level below)
5-room apts., approx. 139–147 m²
(with servants' entrance
on the level below)
2-room apts., approx. 64–67 m²
3-room apts., approx. 80–96 m²
4-room apts., approx. 78 m²

Area per user
20–33.5 m²

Building depth
approx. 40 m

Access
atrium with
open and enclosed
spiral staircases,
cylindrical elevators

Open spaces
planted cantilevered terraces

Parking
underground garage

Architect
Javier Sáenz de Oiza
Madrid
collaboration:
D. Fullando Errazu, R. Moneo Valles

Location
Torres Blancas, Madrid, Spain

Referring to
floor plan 1:200

Twin Parks Northwest | Prentice & Chan | 1970

A complex mixture of spare floors plans, combined in a variety of ways. Almost every apartment is oriented toward east and west, either as a single-floor apartment perpendicular to the access axis or as a maisonette. Only the two-room apartments and a few of the three-room apartments face only east or west. In the case of the maisonettes, the top floor is taken up by the living room, open kitchen, and dining area; the sleeping floor lies below, usually extending through the entire building. All kitchens are positioned as part of the living area along the access axis; all baths, WCs, and the double-return maisonette stairs also lie in this inner area. Access hallways are necessary on only two of three floors. The complicated arrangement of the apartments is also expressed in the facade: the large horizontal living-room windows appear on each third floor, with a row of narrow, almost floor-to-ceiling bedroom windows and a row of large and small windows mixed between them.

1

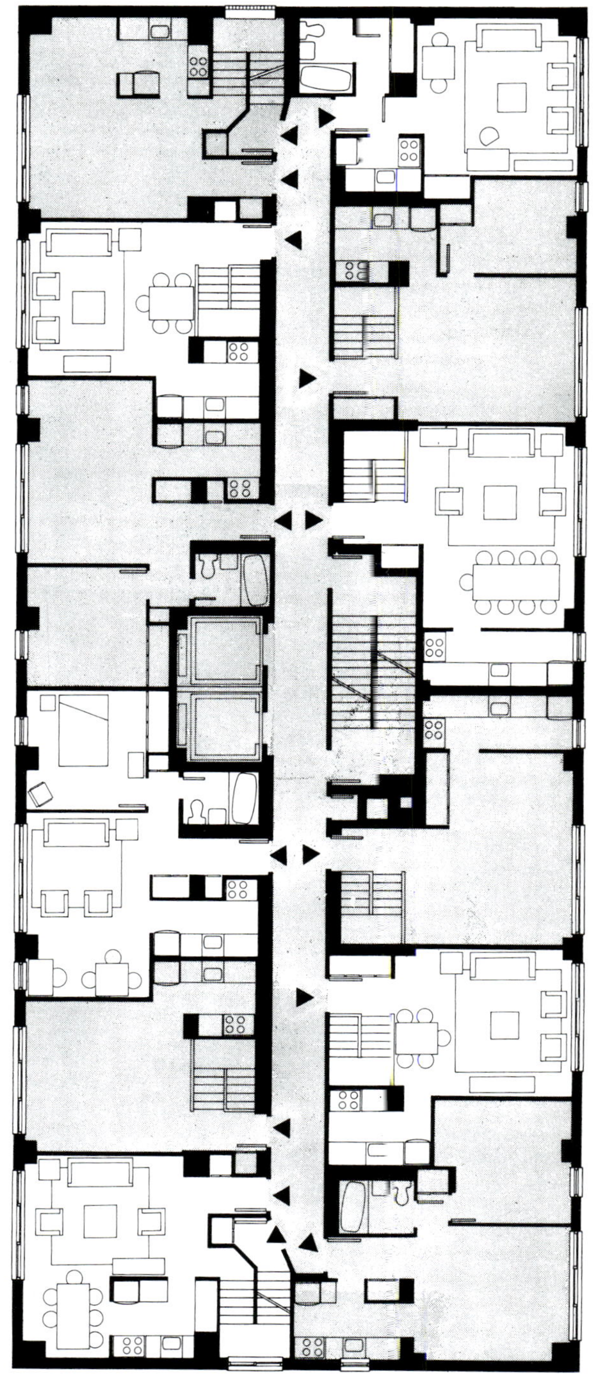

2

1 Typical floor plan,
2nd, 5th, 8th, 11th, 14th, 17th floor 1:200

2 Typical floor plan,
3rd, 6th, 9th, 12th, 15th, 18th floor 1:200

3 Typical floor plan,
4th, 7th, 10th, 13th, 16th, 19th floor 1:200

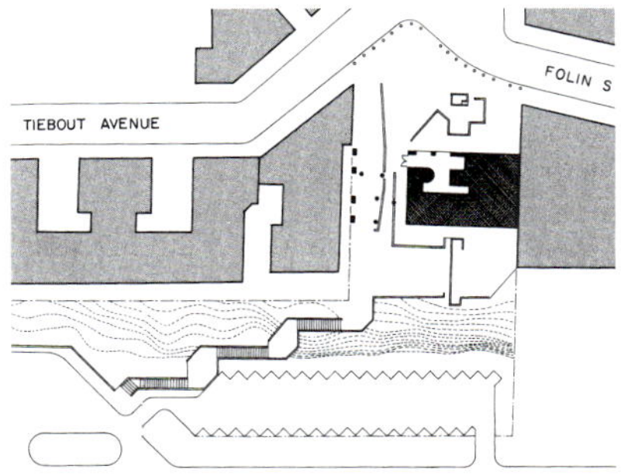

2.3
APARTMENT TOWER

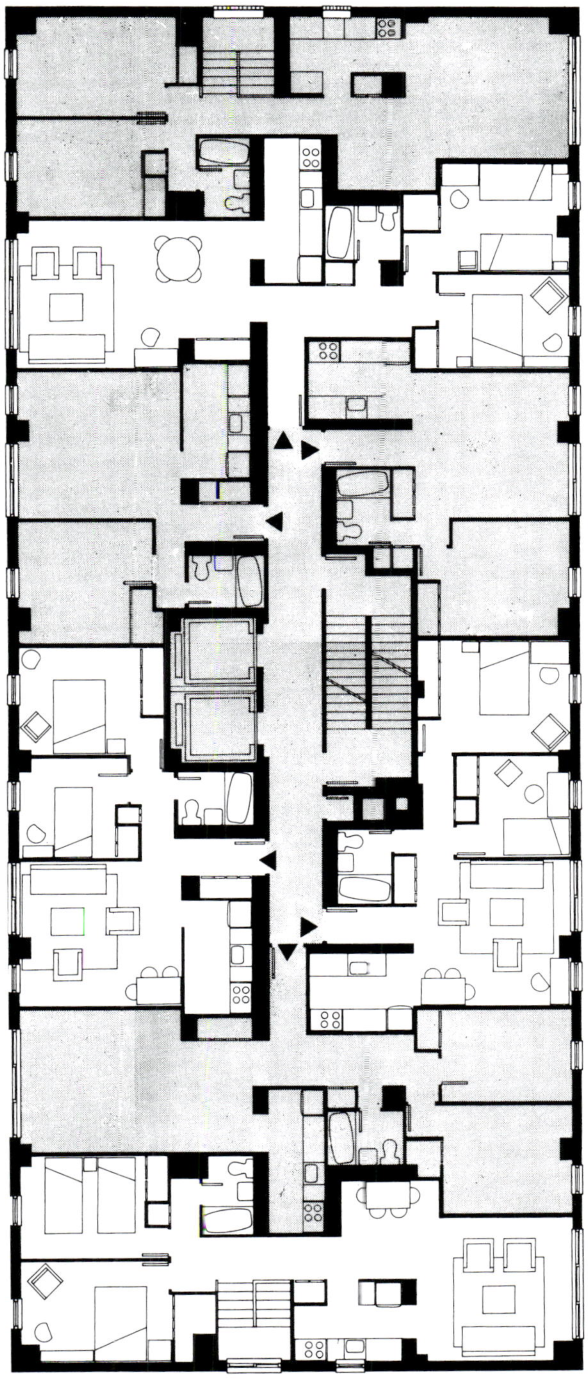

3

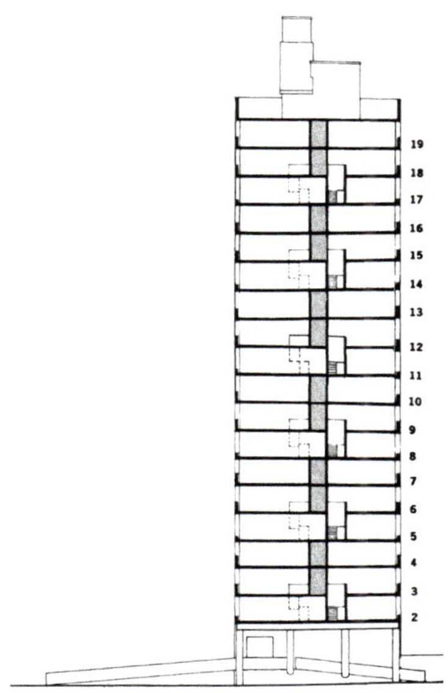

Building type
residential high-rise
19 stories
facing NW/SE

Date of construction
1970

Number of units
114

Size of units
2-room apts., (30 units)
3-room apts., (36 units)
4-room apts., (36 units)
5-room apts., (5 units)
6-room apts., (6 units)

Area per user
nondeterminable

Building depth
15 m

Access
internal hallway with
6 or 12 apts. on each floor

Open spaces
small park with play area

Parking
on street

Architect
Prentice & Chan
Ohlhausen
New York

Location
Twin Parks Northwest
Bronx
New York
USA

Referring to
floor plan 1:200

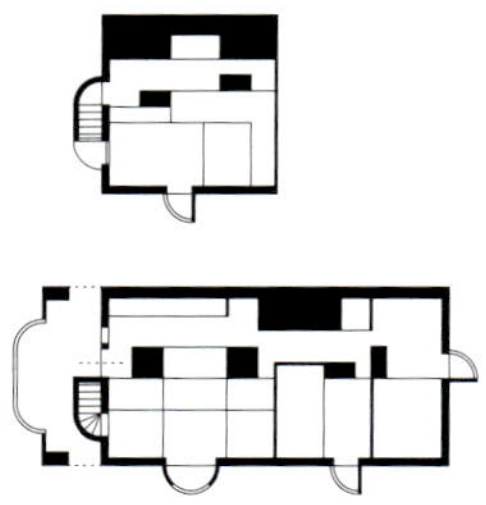

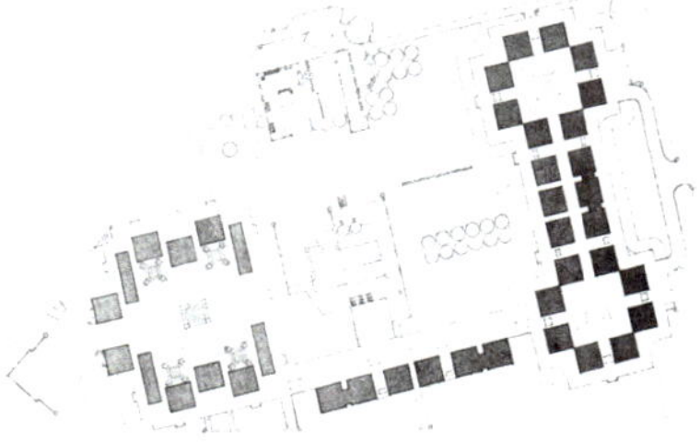

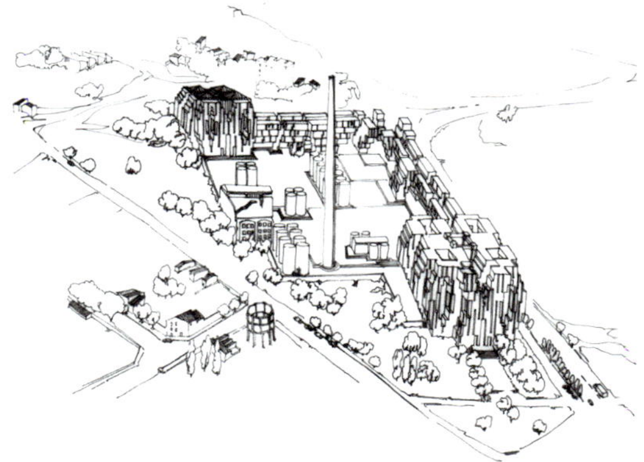

Walden 7 | Bofill | 1975

Located in a suburb of Barcelona, the project is a complex vertical conglomerate of 16 aggregated towers laid out in a grid around six courtyards. Tower pairs flanking the courtyards are not simply vertical, but widen alternately at their center or at their tops and bottoms. The pairs open toward the center around the four larger courtyards to form a "window to the city," the gap narrowing again at the top. In the central axis, they widen at the base to form a "city gate" connecting the open and linked courtyards with the surrounding neighborhood.

Bundled together in the central courtyards, the elevators form a junction for the residential community. Starting at each stop, a horizontal circulation network connects two levels and branches off around multiple courtyards, forming a space of chance encounters as it leads to the individual apartment entrances.

The floor plans are based on 30 m² modules, which can be combined horizontally or vertically to form apartments with different sizes and various configurations. The spatial configuration of each unit begins with a seemingly schematically zoned floor plan: bathrooms and kitchens are placed along the dividing wall between apartments, with the living spaces lined up on the opposite facade wall. A double-sided closet wall in between acts more like an interface than a boundary. The separation dissipates again and again. For instance, the bathtub opens up to the living room while one side of the dining table is in the kitchen and the other in the living room. The vision of living – as illustrated in the perspective sketches – focuses completely on a horizontal plane; to one side of the closet wall, a living landscape of seating and sleeping accommodations is sunk into the floor, devoid of any other furniture.

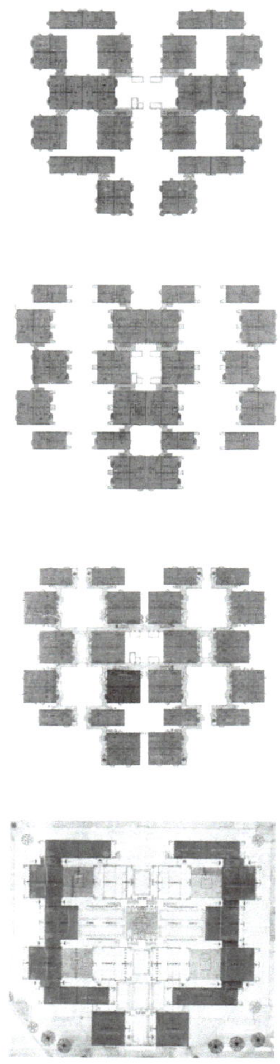

1

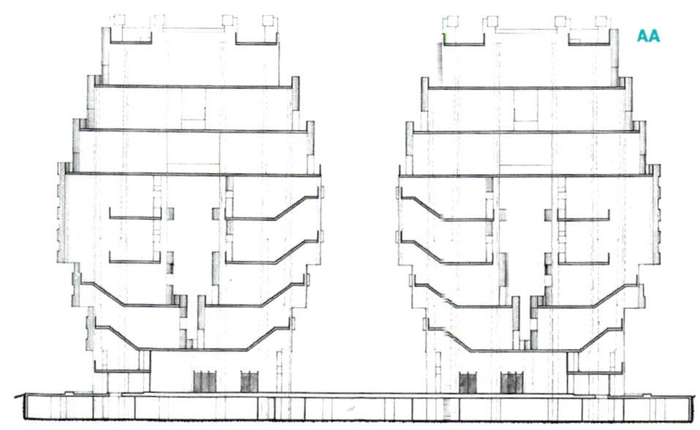

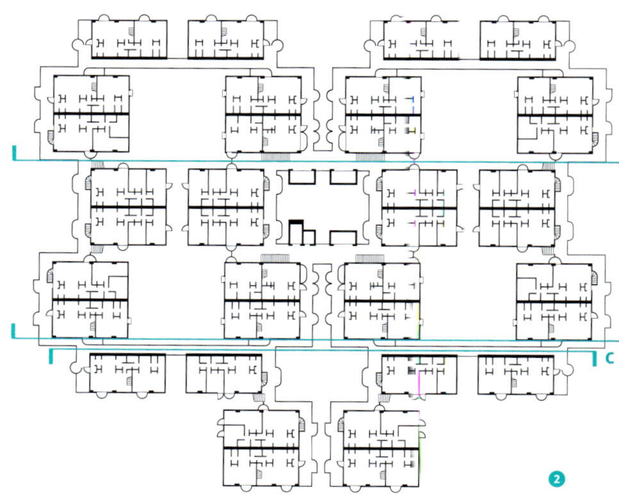

2

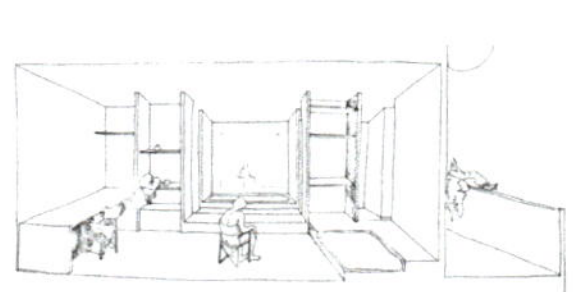

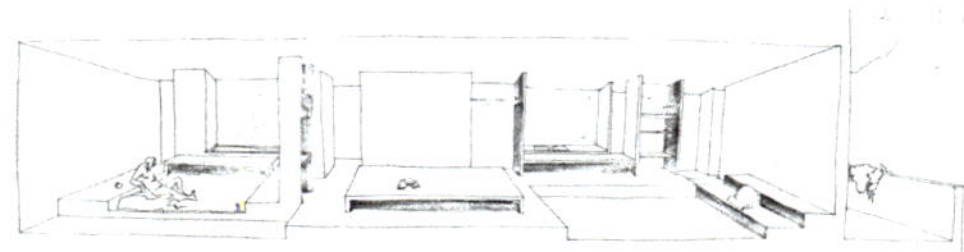

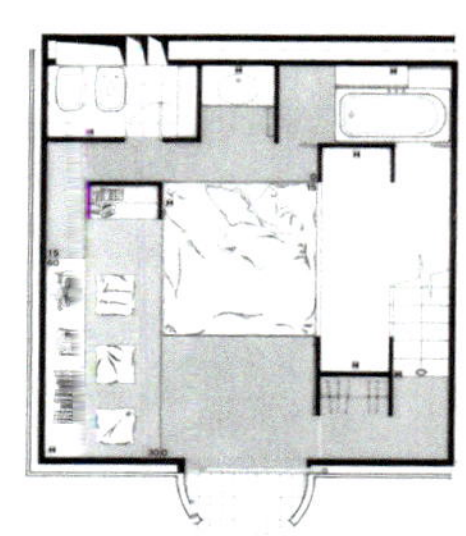

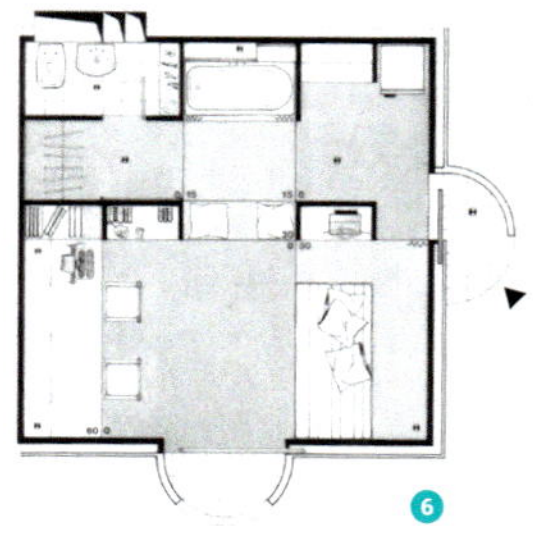

6

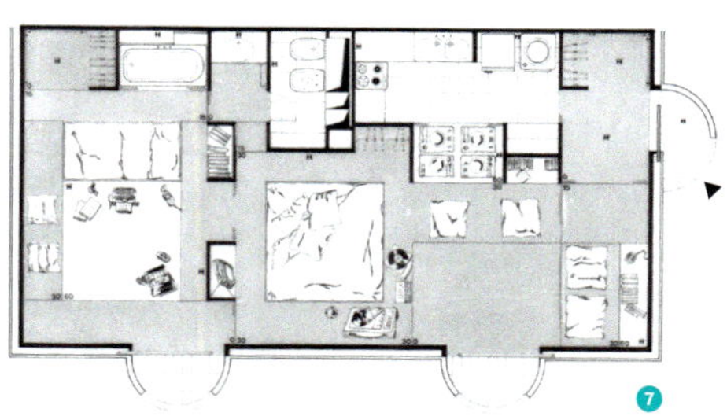

7

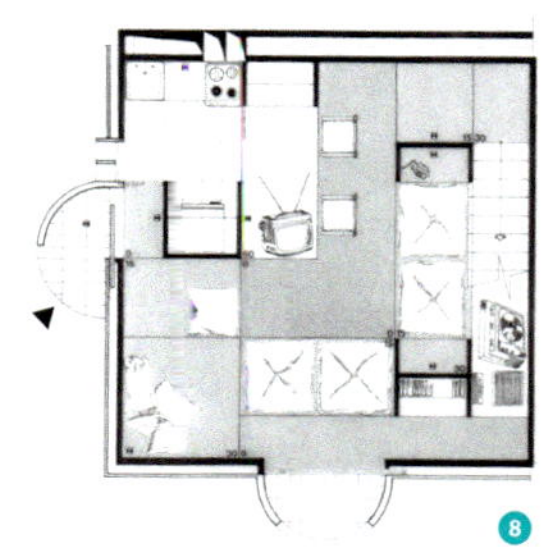

8

1 Total floor plans, from bottom to top ground floor, 5th floor, 10th floor, 16th floor
2 13th floor with central vertical access core and circulation network
3 – 5 Longitudinal sections through the complex high-rise cluster
6 1 module: studio apartment 1:200
7 2 modules: 2-room apartment 1:200
8 2 modules: studio-maisonette 1:200
9 3 modules: 4-room maisonette 1:200
10 4 modules: 5-room maisonette 1:200

2.3 APARTMENT TOWER

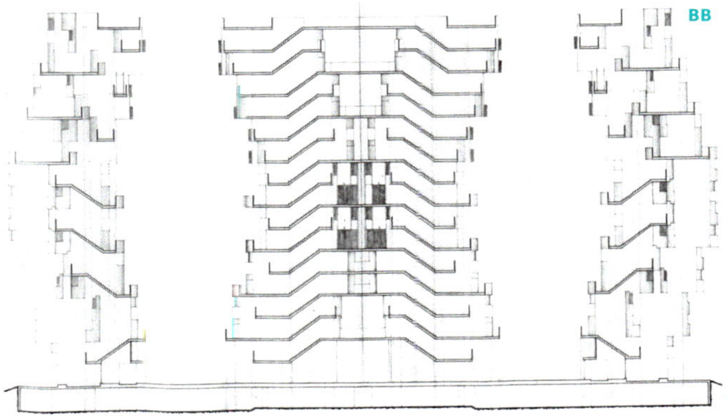

4

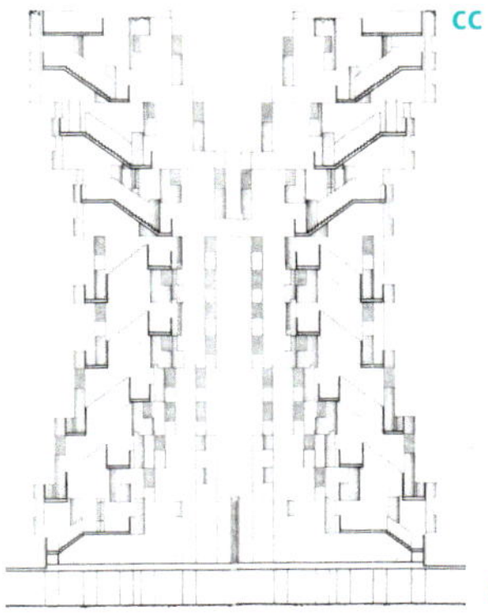

5

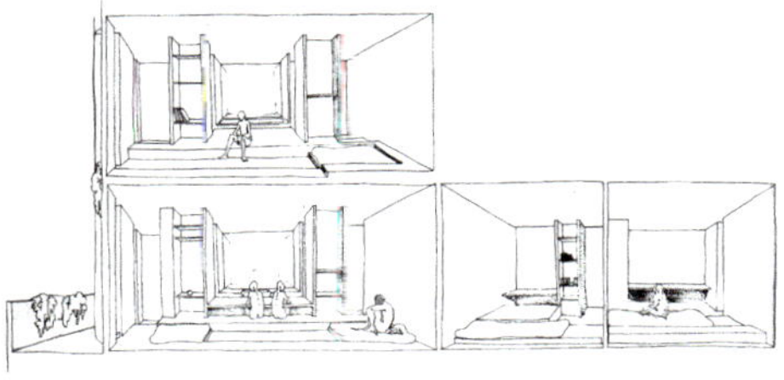

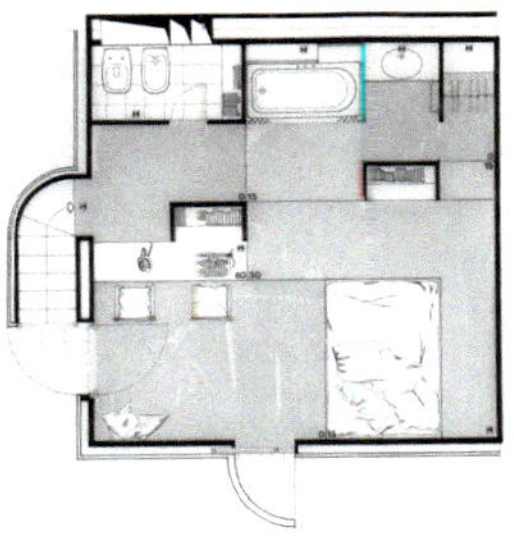

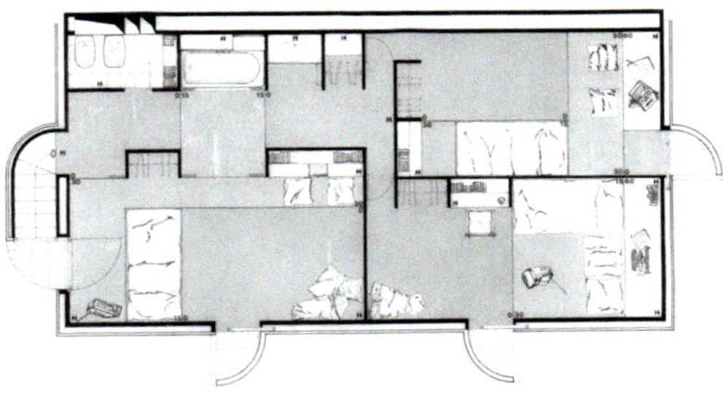

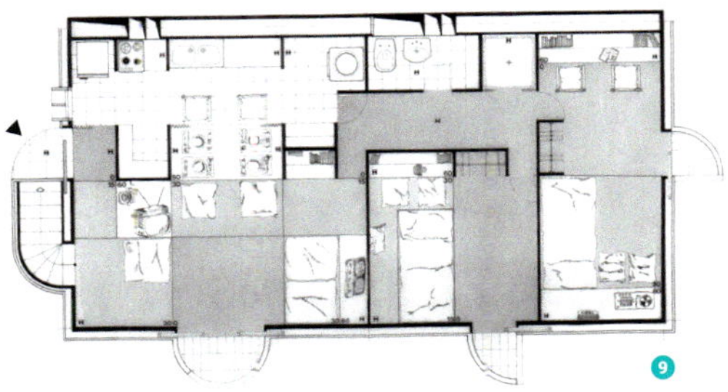

9

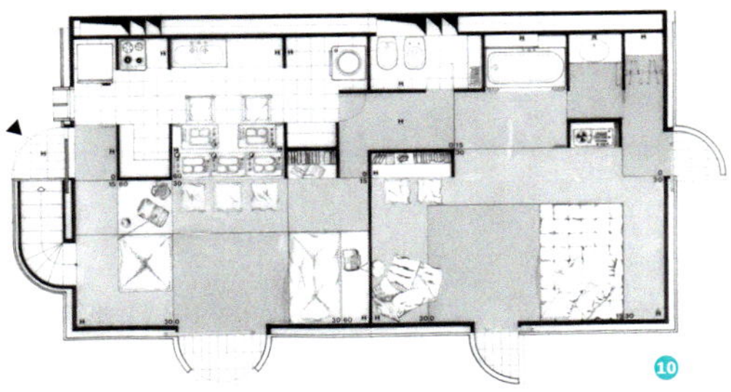

10

Building type
grid-based high-rise cluster with 6 courtyards and vertically bent tower segments
14 stories
facing NW/SE/NE/SW

Date of construction
1975

Number of units
368

Size of units
1-module studio apts., 30 m² (70 units)
2-module studio apts, 60 m² (23 units)
2-module apts., 60 m² (27 units)
3-module apts., 90 m² (174 units)
4-module apts., 120 m² (74 units)

Area per user
22.5–30 m²

Building depth
6/12 m per tower segment

Access
grid-like network of access galleries each comprising two levels, beginning at the vertical core in the central courtyard

Open spaces
small balconies, widened access galleries at apartment entrances, common courtyards and roof terraces with swimming pools

Parking
underground garage

Architect
Ricardo Bofill
Taller de Arquitectura
Barcelona

Location
Sant Just Desvern
Barcelona
Spain

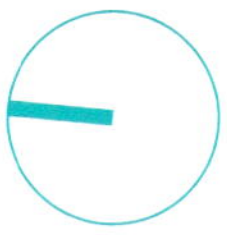

Referring to floor plan 1:200

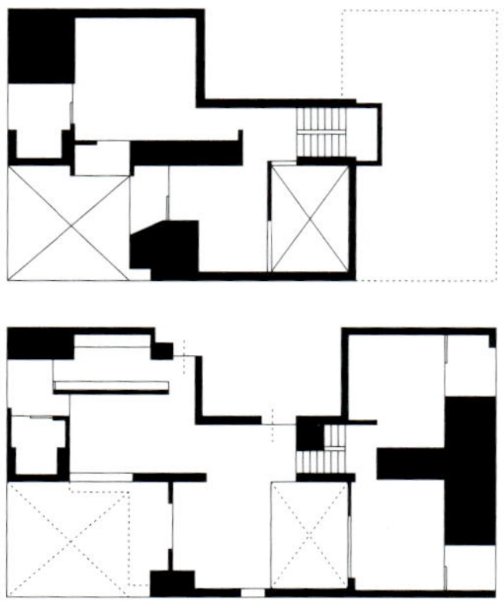

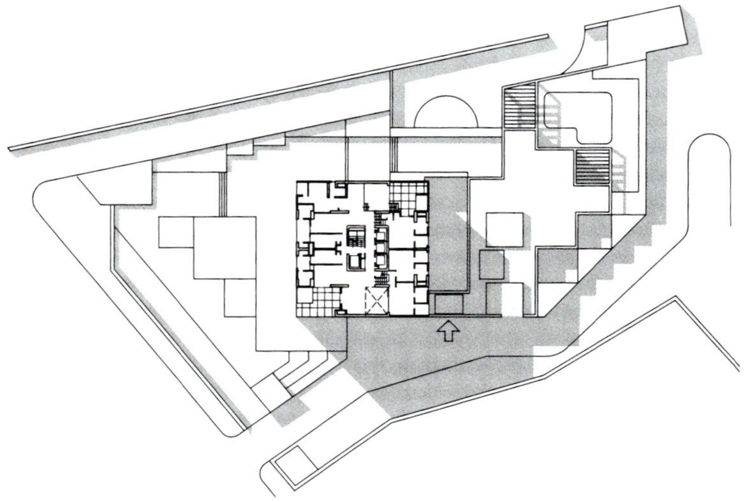

Kanchanjunga Apartments | Correa | 1983

The apartment tower opens in two directions to benefit from fresh breezes from the sea to the west and the harbor to the east. However, these orientations also expose the apartments to the hot afternoon sun and the monsoon rains. The tower is therefore wrapped in a protective layer of terraces; they are cut out from the volume at the corners and offer each apartment an exterior space on an urbane scale. Moreover, the position of the "cut-outs" alternates from floor to floor, thus defining the sculptural appearance of the tower. All 32 apartments occupy the entire depth of the volume and are wrapped around the central circulation core on three sides. All the large units occupy two stories, while the remaining apartments are single-story in the interior and only reach across two stories in the terrace area. All units are organized into split-levels, which facilitate uninterrupted sightlines throughout the interior. The differing apartment types are interlaced within the staggered levels that result from these split-levels.

In the largest units, the entrance area leads directly into a generous L-shaped living room wrapped around the double-height terrace. The stairs connect the living room to the individual rooms above and also to both levels behind the opposite facade. The bathrooms and bedroom loggias at the facade level form a buffer to the outside. There is also a two-story space in the interior. With its formal analogy to the terraces, this space seems to question its own placement within the interior – the entire construct appears as a light, open web of levels, connections, and spatial relationships, that run through the building and penetrate boundaries.

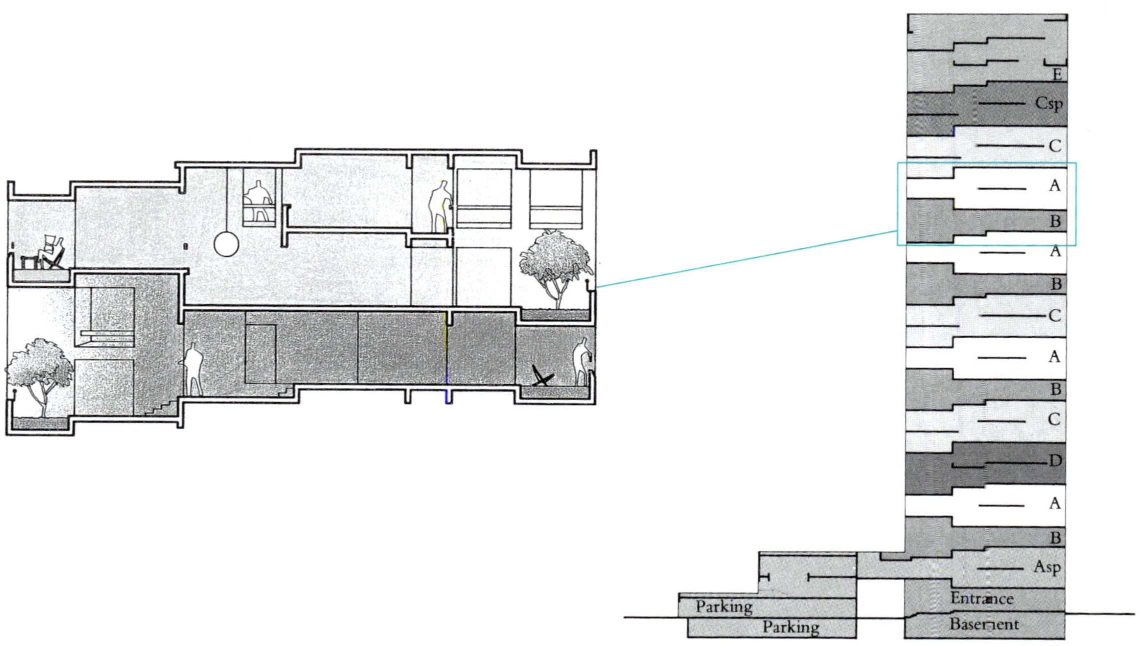

4

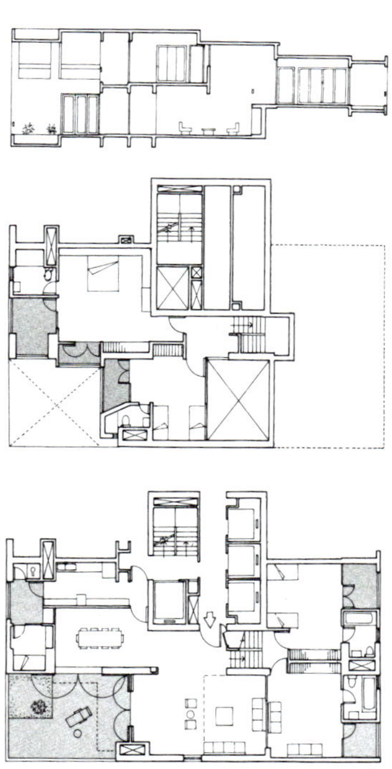

1

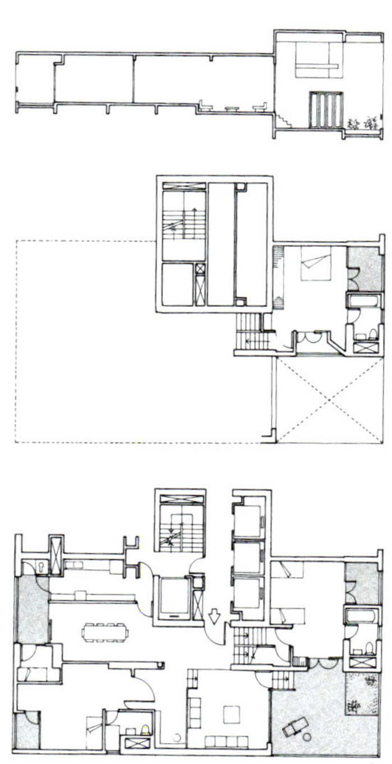

2

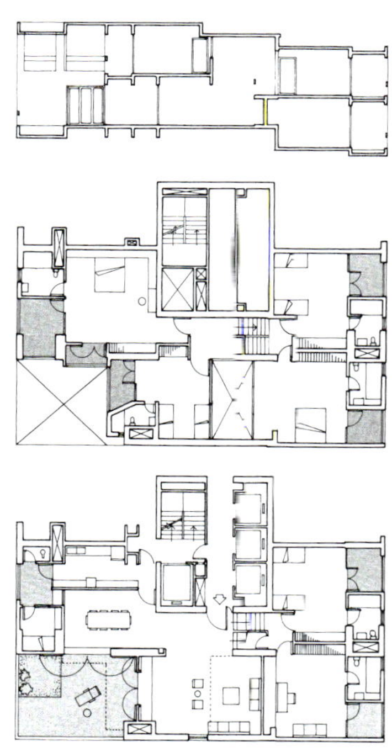

3

1 Type **A**: 5-room apartments: lower and upper level, section 1:500

2 Type **B**: 4-room apartments: lower and upper level, section 1:500

3 Type **C**: 7-room apartments: lower and upper level, section 1:500

4 Schematic section with apartment types **A, B, C, D** 1:1000

5 Type **D**: 5-room apartments: lower and upper level, section 1:200

2.3
APARTMENT TOWER

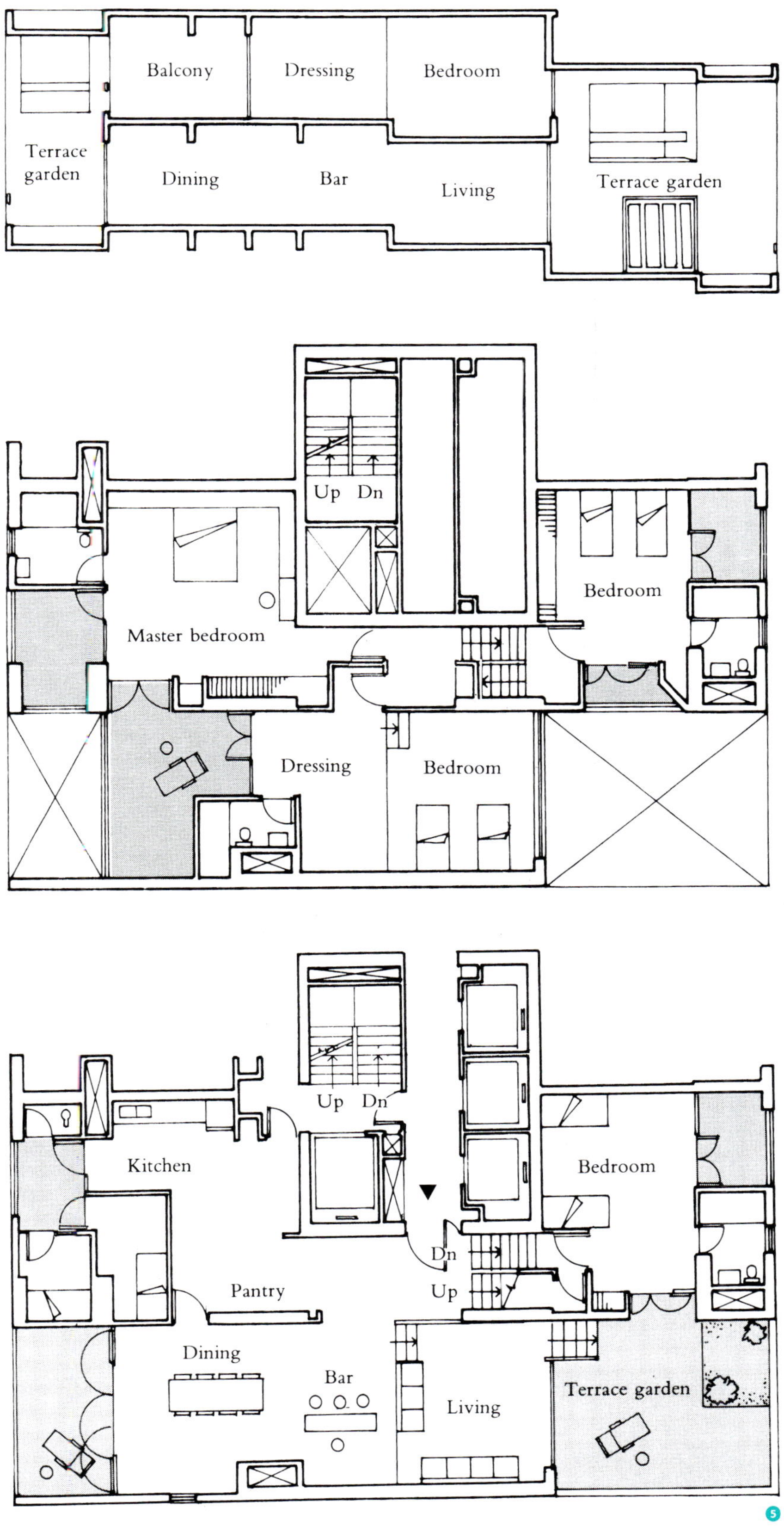

Building type
apartment tower with cut-out loggias
and split-level apartments
27 stories
facing SW/NE

Date of construction
1970–1983

Number of units
32

Size of units
Type **A**: 5-room apts., 198 m²
(plus 63 m² loggias)
Type **B**: 4-room apts., 164.5 m²
(plus 52 m² loggias)
Type **C**: 7-room apts., 258.5 m²
(plus 65.5 m² loggias)
Type **D**: 5-room apts., 237.5 m²
(plus 83.5 m² loggias)

Area per user
37–48 m²
(with loggias 46.5–64 m²)

Building depth
21 m

Access
central access core

Open spaces
cut-out double-height
terraces (loggias)

Parking
underground garage

Architect
Charles Correa
Mumbai

Location
Mumbai
India

Referring to
floor plan 1:200

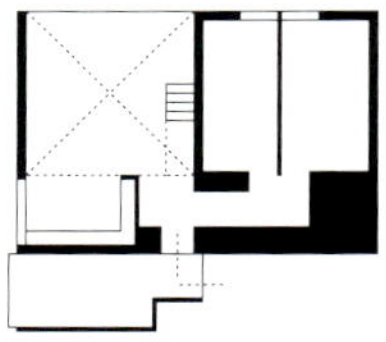

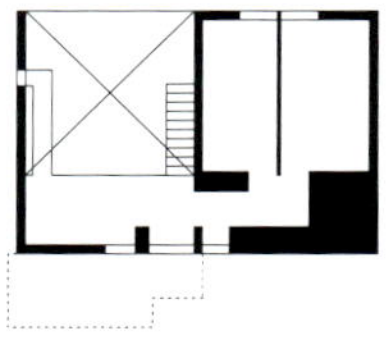

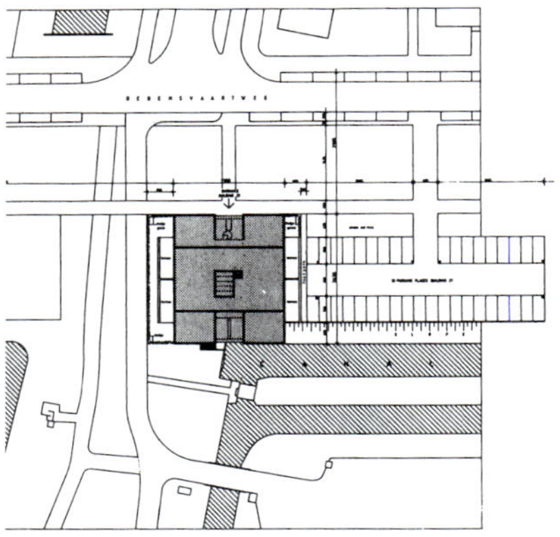

Morgenstond | Ciriani | 1994

The open entrance hall with access elements, galleries, and bridges forms the spine of the building. The terraces at both ends serve simultaneously as entrances to the apartments. This open area, at once terrace, entryway, and part of the hall, is the distinctive feature of this housing strategy. The ground floor and second floor, that is, the area of the two-story lobby, contain maisonettes on both sides that can be entered from the ground floor. On the upper floors, a central gallery along the longitudinal axis of the building plays a distributing role; this gallery gives way to two open terraces on both sides, forming this intermediate zone in front of the apartments. As a result, there are a wide variety of unobstructed views into all levels: clear space around the staircase, the elevator, the gallery. The apartments are organized on a classic pattern; installation zone on the inner side with bath, closet area, WC, kitchen. The kitchen is part of the living room and has a window to the intermediate zone in front of the apartment. All rooms are fronted by narrow terraces (wide terraces in front of the penthouses).

1

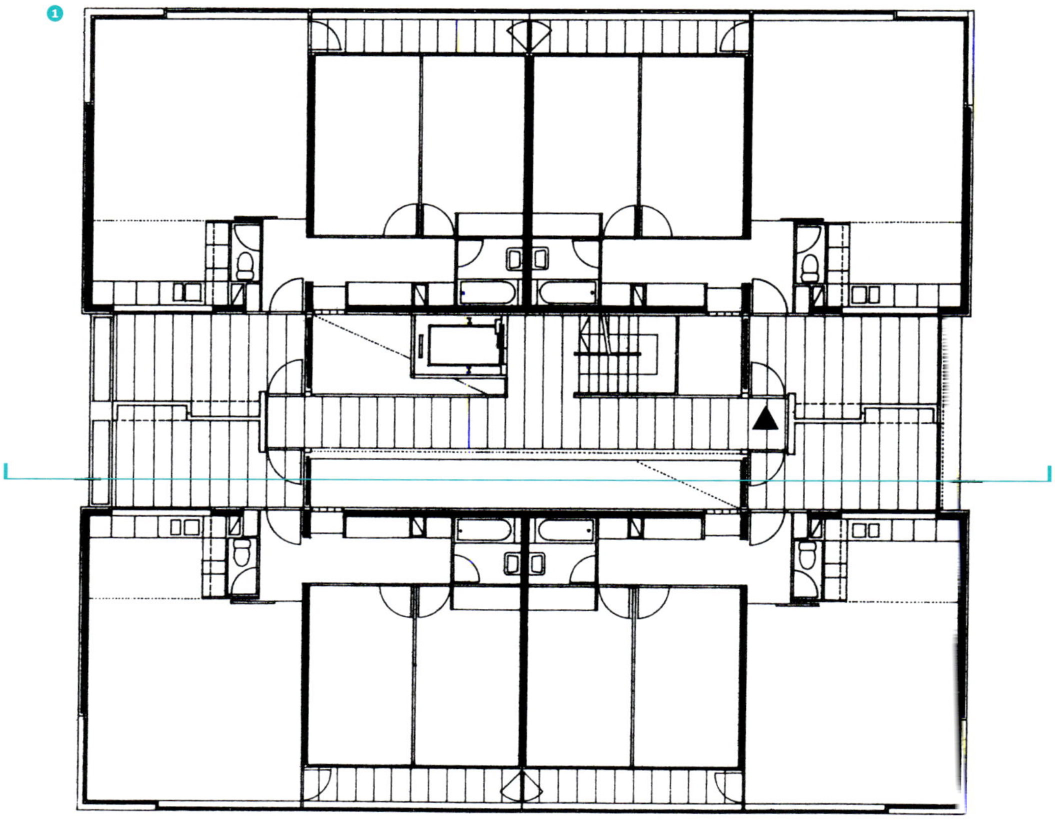

1 Typical floor plan with 3-room apartments 1:200
2 Ground floor plan with 5-room maisonettes 1:200
3 First floor with upper part of the maisonettes 1:200

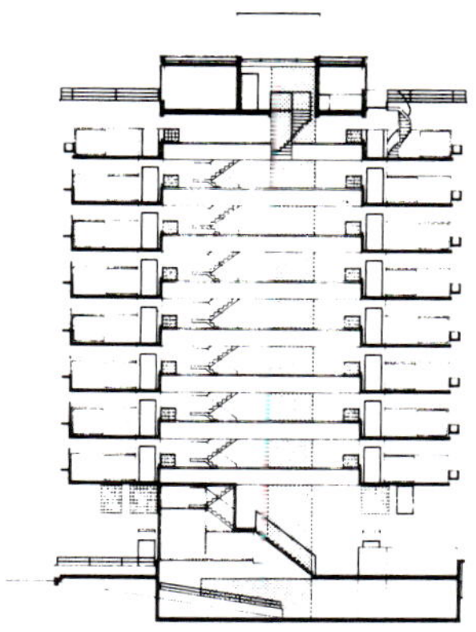

2.3
APARTMENT TOWER

3

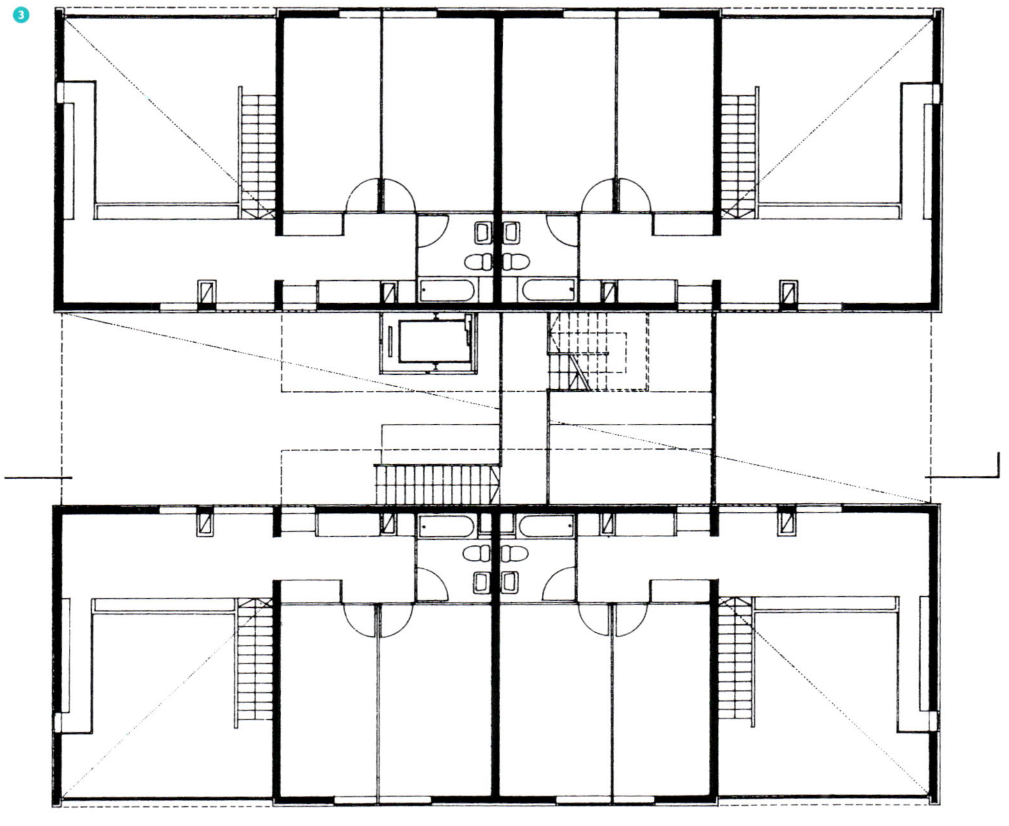

2

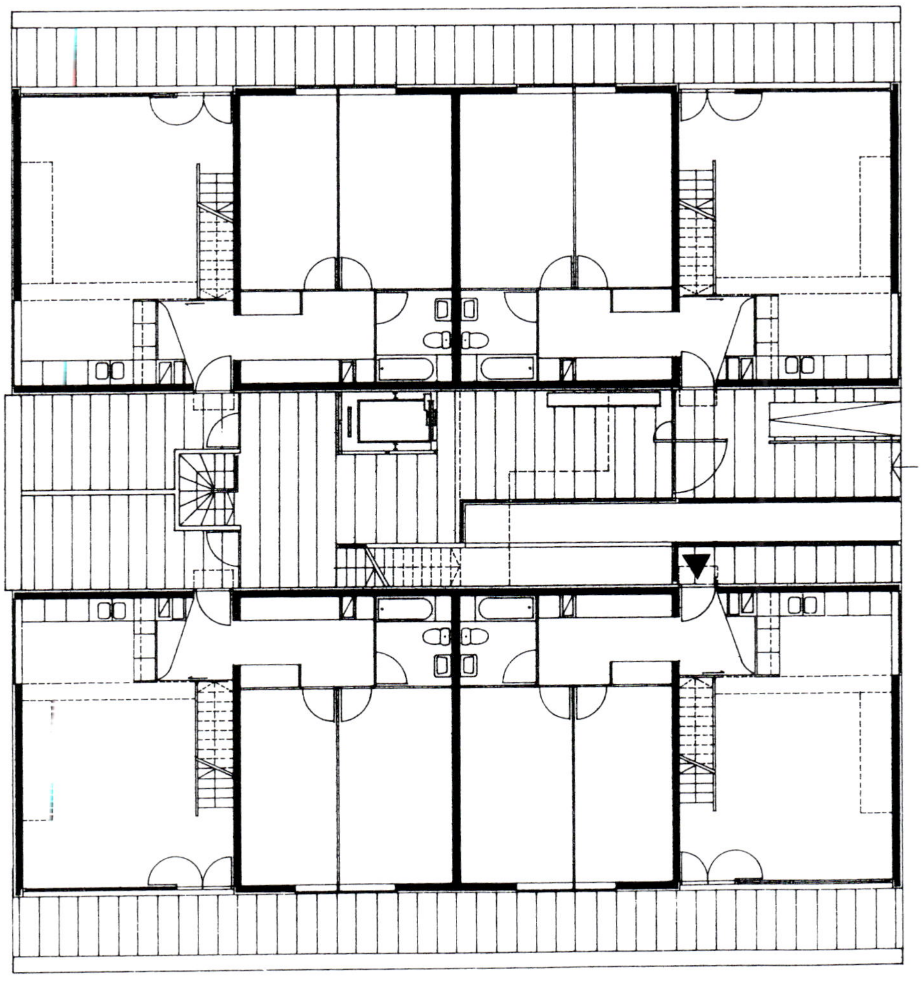

Building type
two high-rise slabs
around central hall
10 stories
facing NW/SE

Date of construction
1993–1994

Number of units
38

Size of units
3-room apts., 83 m² (32 units)
5-room mais., 138 m² (4 units)
2-/3½-room penthouse,
105 m² (2 units)

Area per user
27.5–52.5 m²

Building depth
21 m
(hall: 5 m)

Access
quadruple-loaded stairs,
entry to apartments
via private terraces

Open spaces
terraces to hall
garden

Parking
open parking spaces

Architect
Henri Ciriani
Paris
with Jean-Pierre Crousse
Town planning:
Rem Koolhaas

Location
Dedemsvaartweg
The Hague
The Netherlands

Referring to
floor plan 1:200

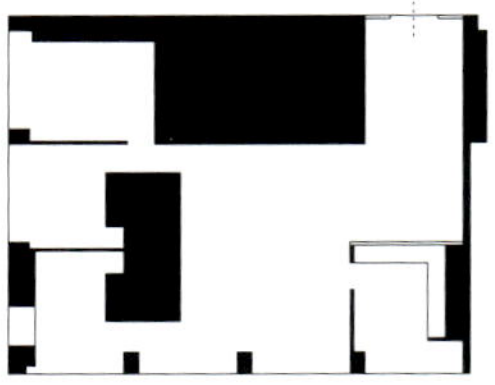

Jian Wai SOHO | Riken Yamamoto & Field Shop | 2005

3

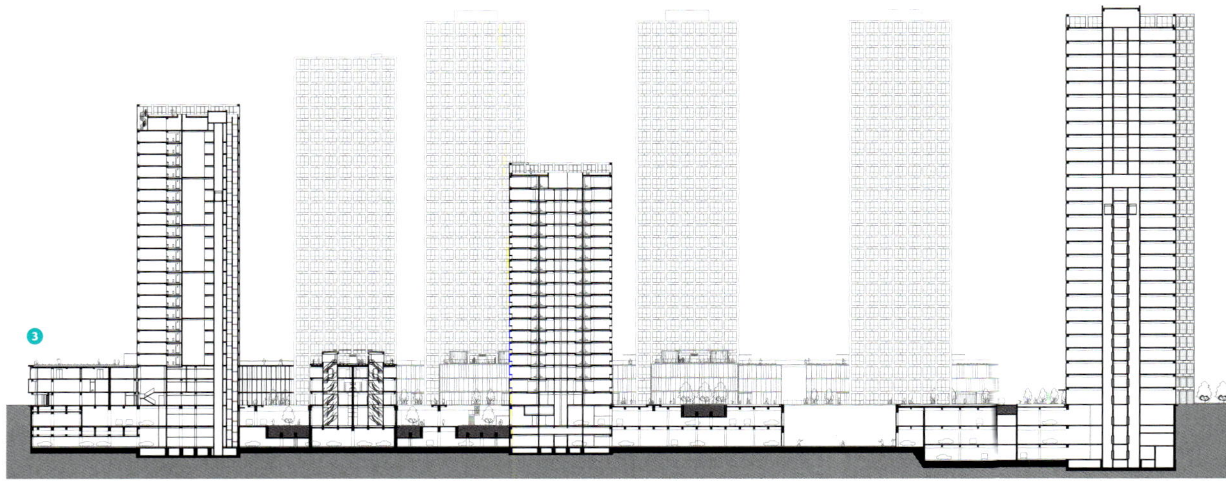

The development is a part of Beijing's new central business district east of Tiananmen Square and consists of an ensemble of nine residential high-rises and four mid-sized office buildings. The buildings and the open access grid are rotated 25 degrees to the surrounding city blocks, opening up the ensemble to increased sunlight and a view of the Tonghui River. A system of pedestrian walkways on multiple levels connects the buildings with sunken public cultural and green spaces. Together with the mixed-use areas – housing, restaurants and cafes, offices, and other commercial areas – the network of walkways makes a significant contribution to the density and urban livelihood of the development.

The square footprints of the high-rises contain various floor plan types, all of which have one similarity: they combine living and working areas within one unit. All apartments incorporate a so-called SOHO-zone ("Small Office/Home Office") and wide entrances along the spacious central corridor, and share in part multi-storied spaces which can be used privately or commercially. In the **L** type tower, three apartments naturally lit from two sides meet in the center. Inserted cores incorporating storage spaces, bathrooms, and kitchens as well as private corner rooms generate open living spaces with multiple zones, with the more public SOHO zone at the front and the more private rooms towards the rear along the facade. In the **MS** building type, a central corridor runs the length of the building, from which the SOHO areas of each apartment and the living spaces beyond branch off toward the facade.

2

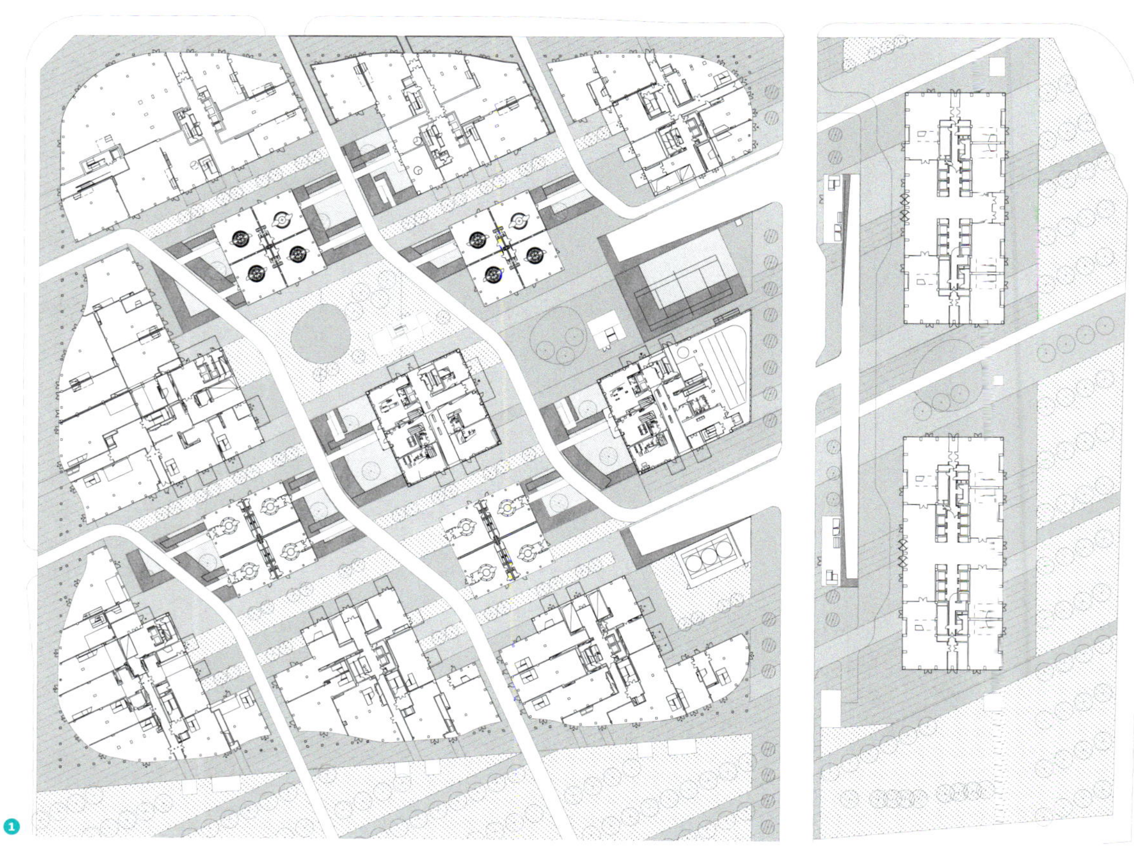

1

1 Ground floor, entire ensemble 1:2000
2 5th floor, entire ensemble 1:2000
3 Section through entire ensemble 1:2000
4 Partial section of base area
5 Typical floor high-rise type **MS** 1:500
6 SOHO apartment in type **MS** 1:200
7 Typical floor high-rise type **L** 1:500
8 SOHO apartment in type **L** 1:200

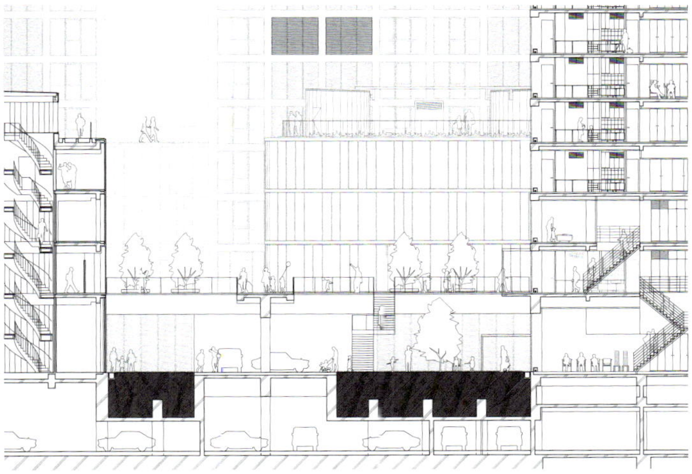

4

2.3
APARTMENT TOWER

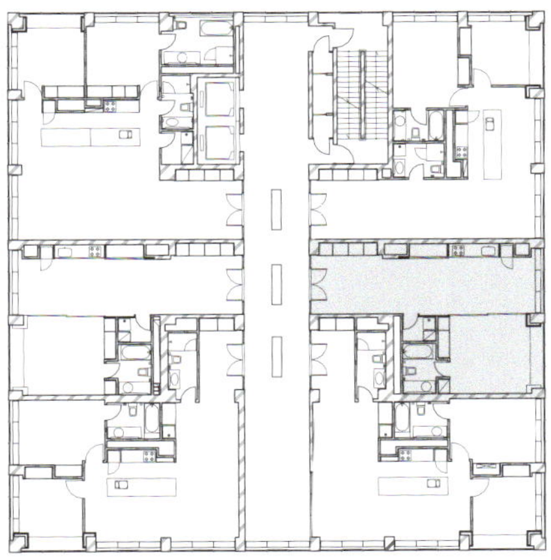

5

6

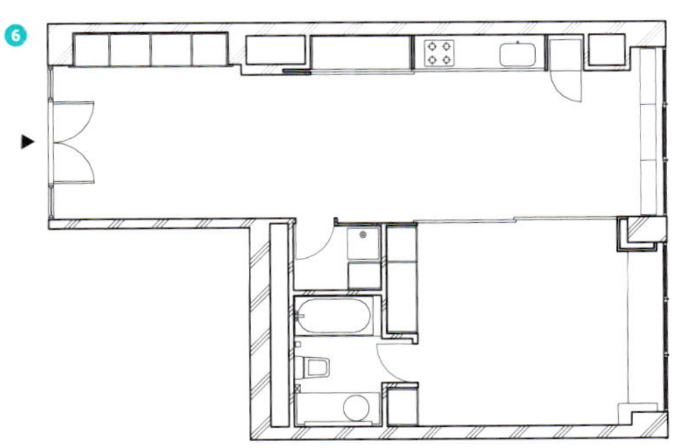

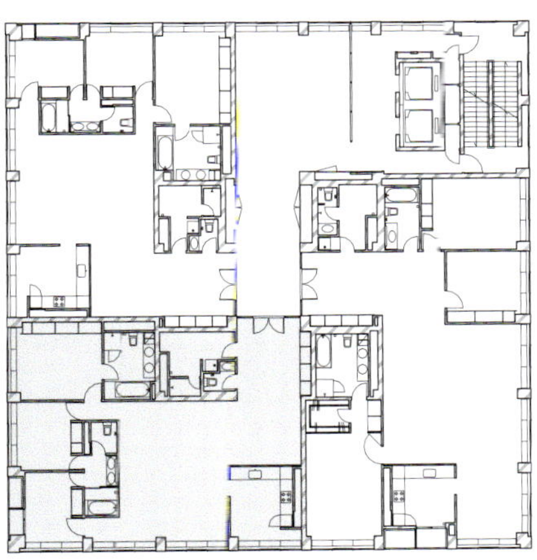

7

Building type
high-rise development with 9 towers
base of multi-level overlapping
public outdoor spaces and 4-story
commercial buildings above an
underground parking garage
19–28 stories
facing in all directions

Date of construction
2001–2005

Number of units
approx. 1954

Size of units
2- to 8-room apts. with home office,
approx. 68–311 m²

Area per user
approx. 25–68 m²

Building depth
27.3 m

Access
naturally lit corridor
with shared spaces

Open spaces
shared courtyards
and roof terraces

Parking
underground garage

Architect
Riken Yamamoto & Field Shop
Yokohama/Beijing

Location
Beijing
China

8

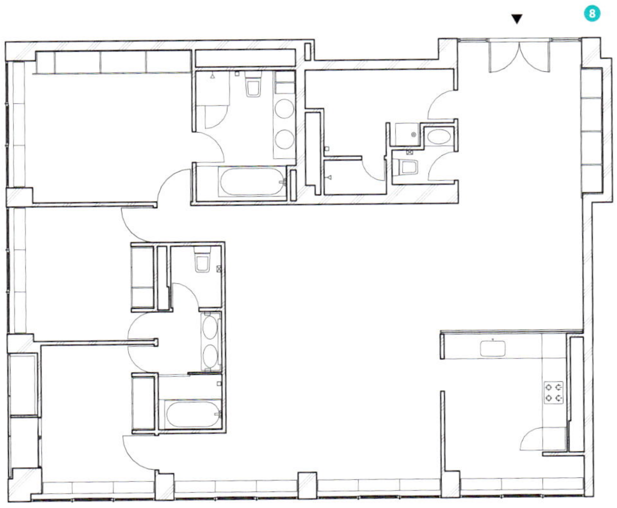

Referring to
floor plan 1:200

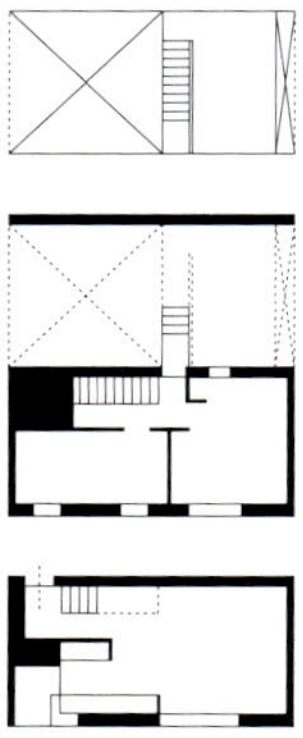

Mirador | MVRDV/ Blanca Lleó | 2005

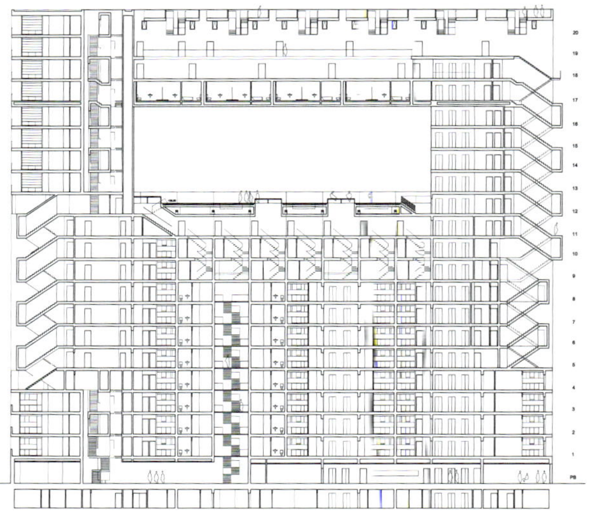

2

In an urban expansion area originally zoned for horizontal block developments, this typology is translated into the vertical: the result is a high-rise on an urban scale. The enclosed courtyard is transformed into an open communal garden at a height of 40 m, a framework for the ever-changing urban landscape. In deliberate contrast to the standard rows of identical apartment typologies, nine houses are stacked within each other in the Mirador complex, increasing density and each with their own apartment types. Connected by a complex, vertical and horizontal circulation and access sytem, they form a "super-block." Up to the 10th floor, the building features single-level front-to-back and corner units. The subsequent levels are reserved nearly exclusively for multi-story (2 to 3 stories) maisonette apartments. In and of themselves, the floor plans are not the unique characteristic of this building, rather it is the combination of these plans and the manner in which apartments and access are interwoven, and the fact that they are always oriented toward the communal areas – the loggias inserted into the corner, the large courtyard "in the sky" at the center and the projecting stairwells. The central corridors leading to the maisonettes are also integrated into this complex spatial layout. On the 19th and 20th floors, they are part of an open, multi-story atrium across the entire length of the building. From the upper maisonettes, external stairs bisect the atrium to the private roof patios.

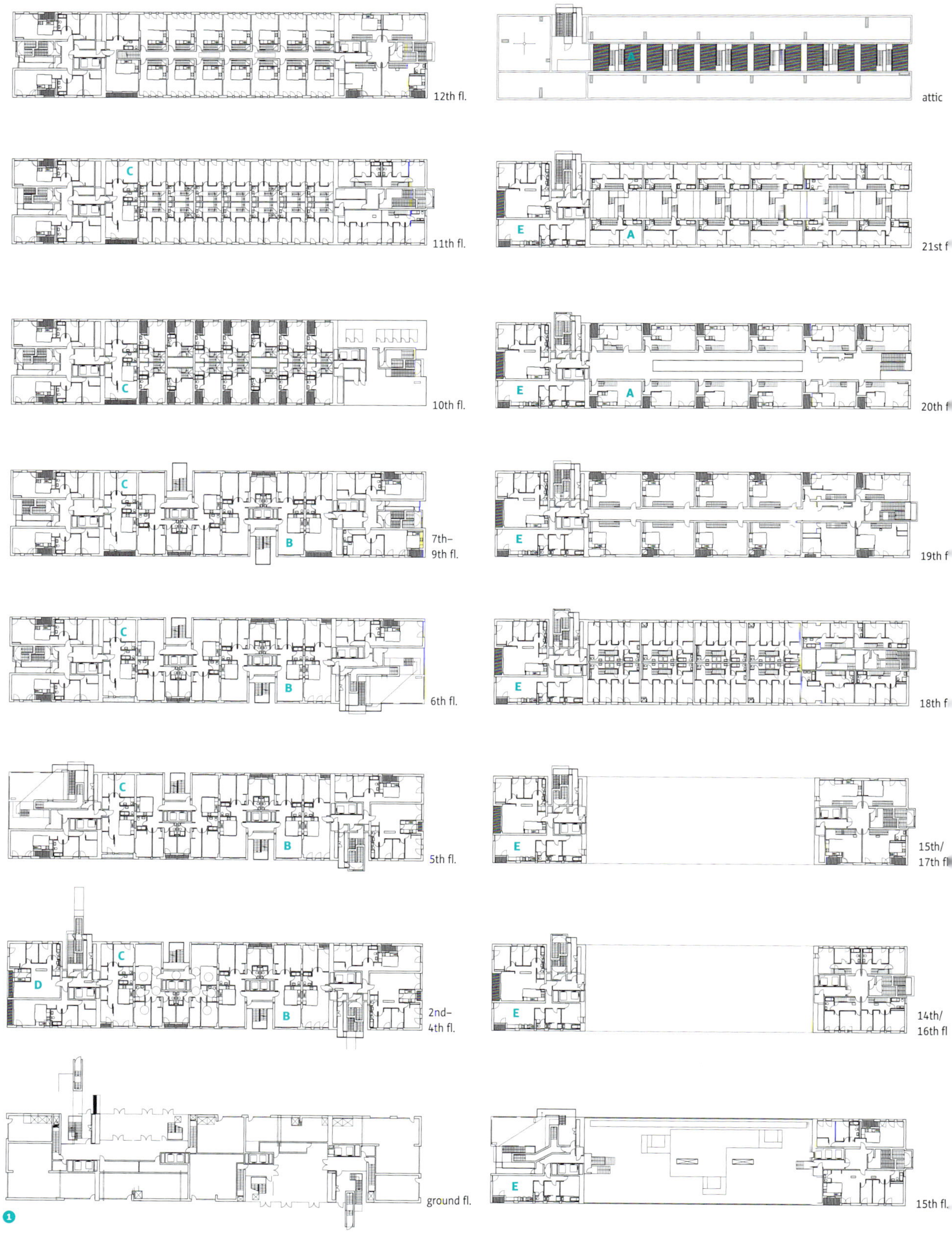

1

1 Total floor plan for ground floor–21st floor, attic story 1:1000
2 Longitudinal section with various access systems 1:1000
3 Concept of the urban block as propped-up urban block
4 Concept of nine houses stacked within one another
5 Floor plan examples 1:200
Type **A** 3-room maisonette on internal street, with roof patio
Type **B** Front-to-back 4-room apartment
Type **C** Front-to-back 4-room apartment
Type **D** 4-room corner apartment
Type **E** 3-room corner unit

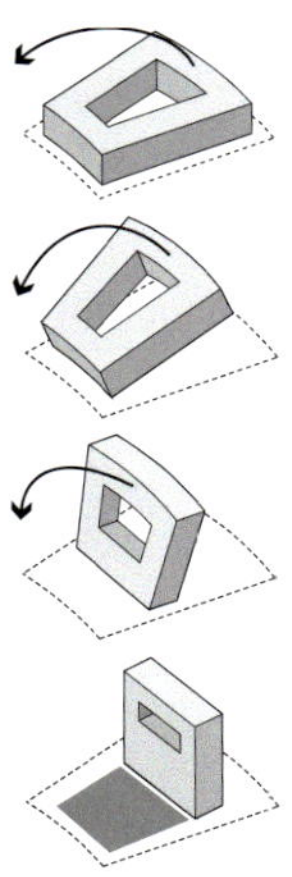

3

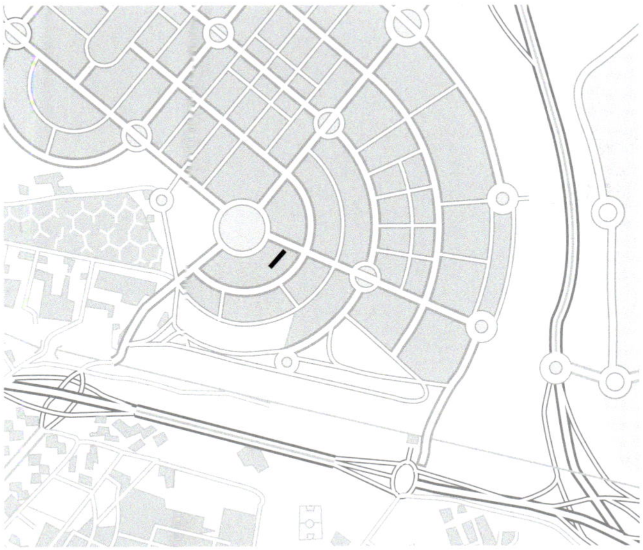

2.3
APARTMENT TOWER

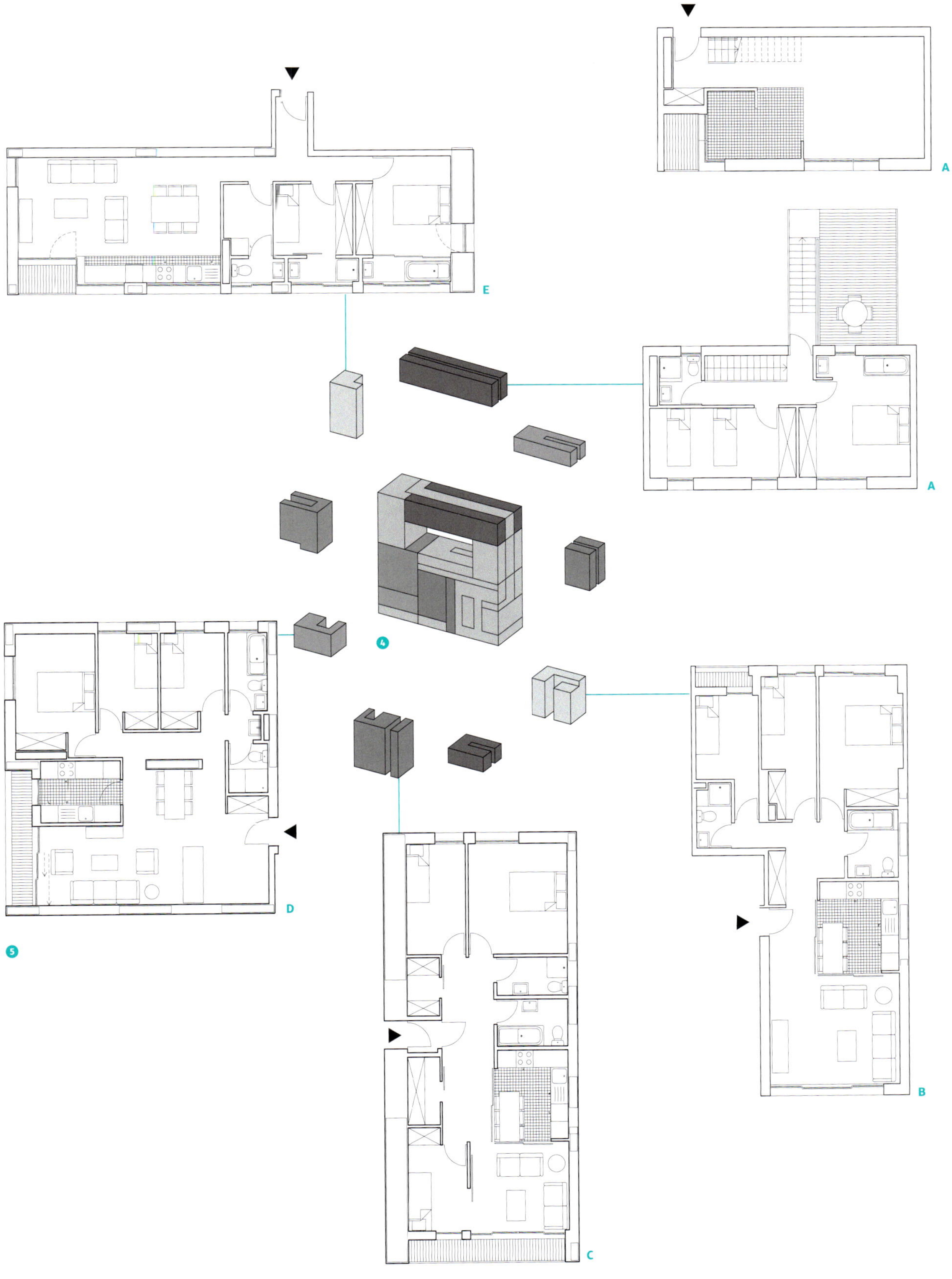

4

5

Building type
large residential block consisting of 9 building sections 21 stories with a 4-story open communal garden at a height of 40 m facing NW/SE

Date of construction
2001–2005

Number of units
157

Size of units
2-room apts., 55.5–74.5 m² (71 units)
3-room apts., 72.5–88 m² (77 units)
4-room apts., 101 m² (8 units)

Area per user
24–37.5 m²

Building depth
approx. 15 m

Access
access cores, cantilevered stairs stacked within each other, central corridors located in atria that rise to several stories in height in some areas

Open spaces
communal garden and loggias private loggias and roof patios

Parking
underground garage

Architect
MVRDV
Rotterdam
Blanca Lleó Associates
Madrid

Location
Madrid–Sanchinarro
Spain

Referring to floor plan 1:200

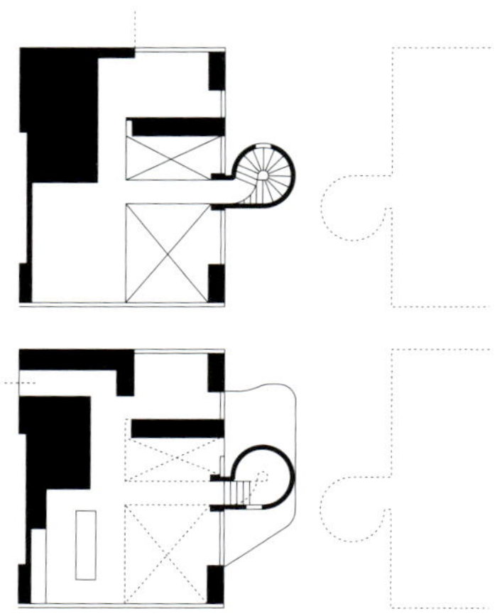

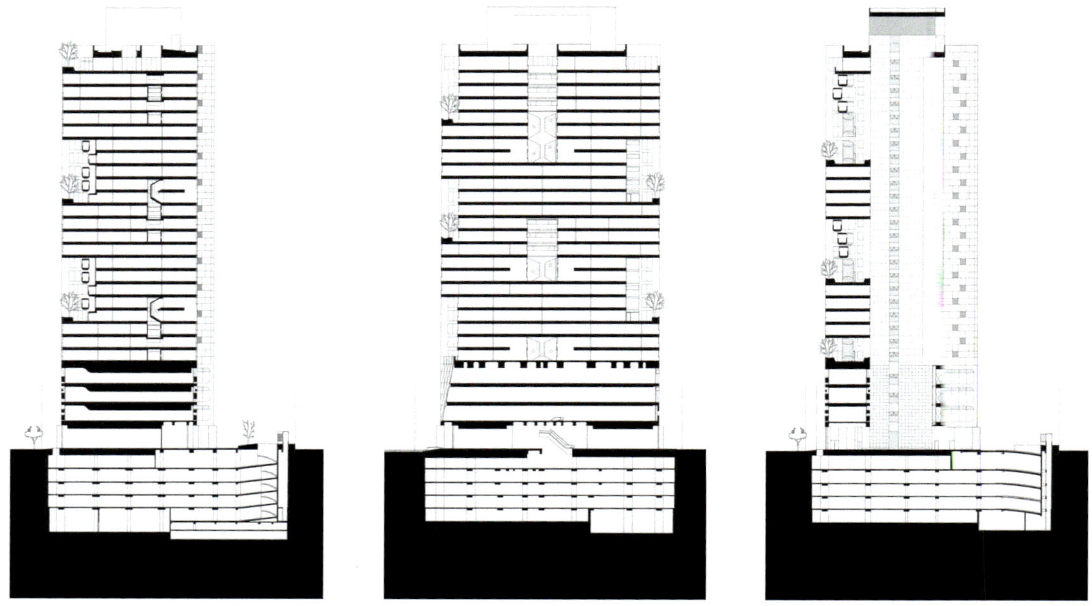

5

Boutique Monaco – Missing Matrix | Mass Studies | 2008

The four-story base of the C-shaped high-rise is reserved for cultural, commercial, and community uses; the 23 stories above are a complex structure housing 172 apartments of 49 different types, all accessed from the inner side of the two wings of the C-shaped plan. However, built to the maximum allowable height of 100 m, the skyscraper would have exceeded the permitted density by 10 percent, a problem that is solved with the so-called "missing matrices." These are voids cut out of the tower. They not only define the appearance of the tower, but also generate typologically surprising floor plans. The 15 voids augment the total facade surface and provide additional natural light and a better view for the apartments. This design also results in units that are open on three sides, with floor plans that are wrapped around and organized on either side of these voids. These units also benefit from having generous open spaces, in part planted with trees: in some two-story units the balcony reaches from one facade to the other and the only connection between the interior levels is a staircase volume projecting into the void; there are apartments in which the rooms are connected via a bridge, or corner units with generous terrace gardens, which divide the floor plan into private and common areas.

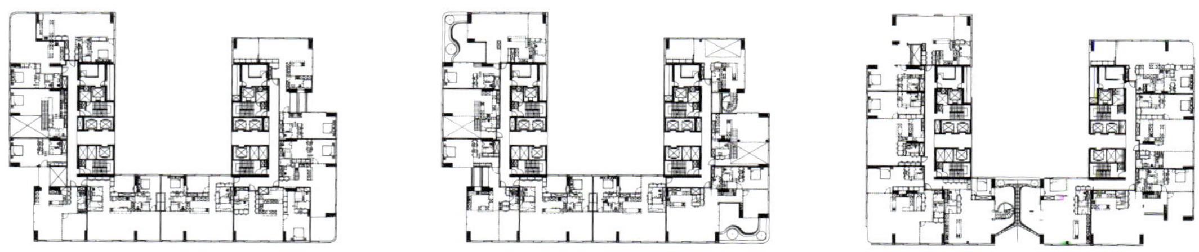

4

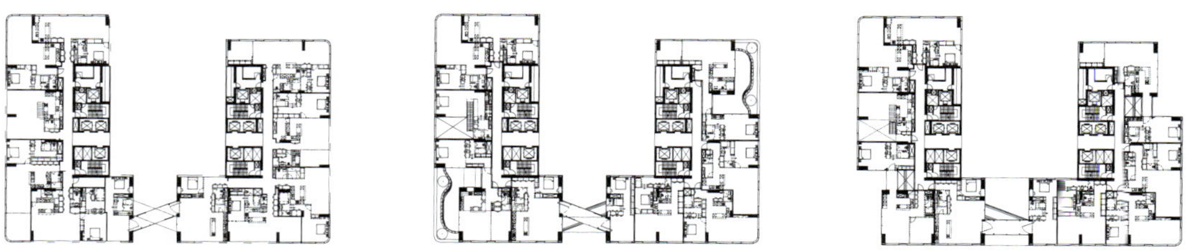

3

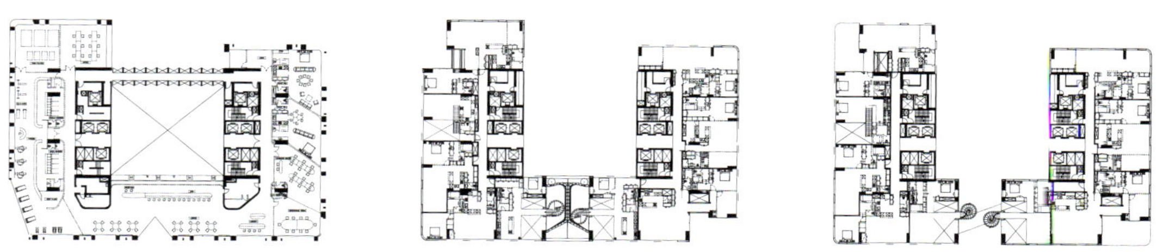

2

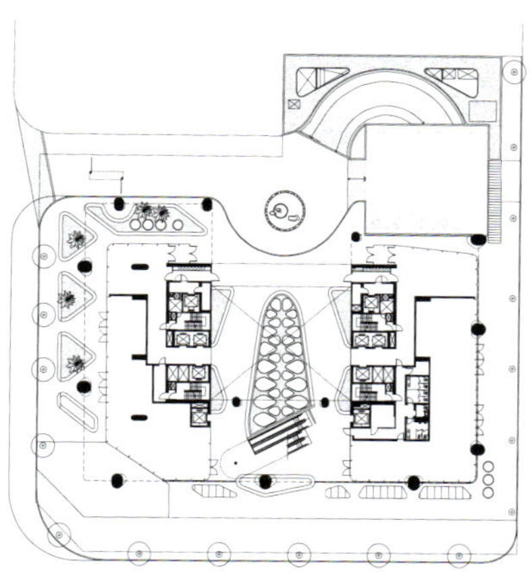

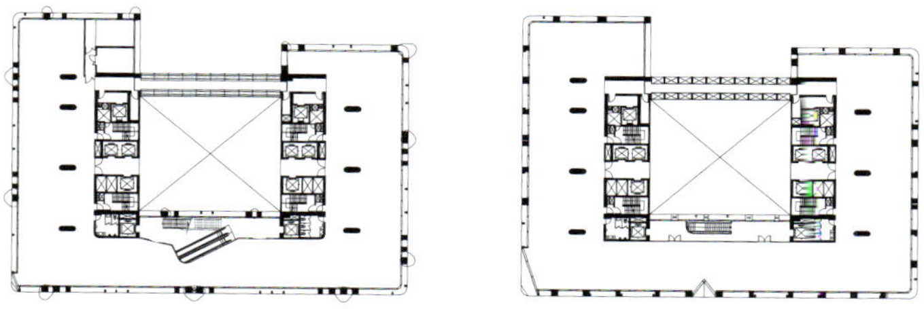

1

1 Ground floor, 2nd floor, 3rd floor (public areas) 1:1500
2 4th floor (public), 11th floor, 12th floor (residential) 1:1500
3 13th–15th floor (residential) 1:1500
4 16th floor, 17th floor, 20th floor (residential) 1:1500
5 Cross and longitudinal sections 1:2000
6 Apartment type "3 Bay" 1:500
7 Apartment type "Garden" 1:500
8 Apartment type "Minimal" 1:500
9 Apartment type "Duplex" 1:200
10 Apartment type "Bridge" 1:200

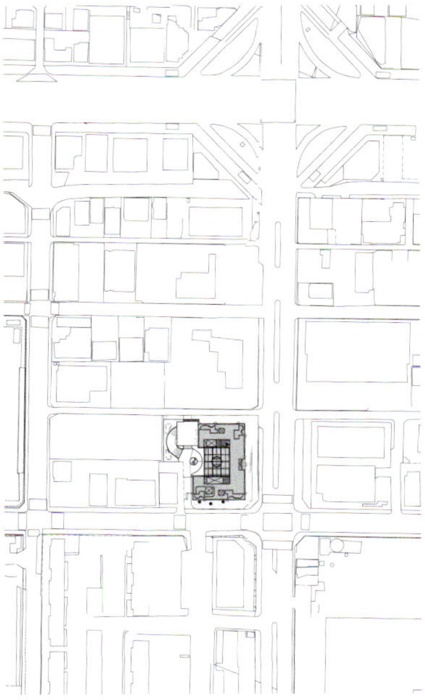

2.3 APARTMENT TOWER

6

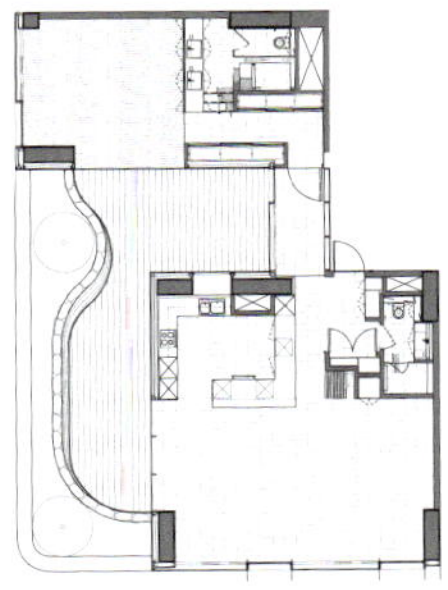

7

8

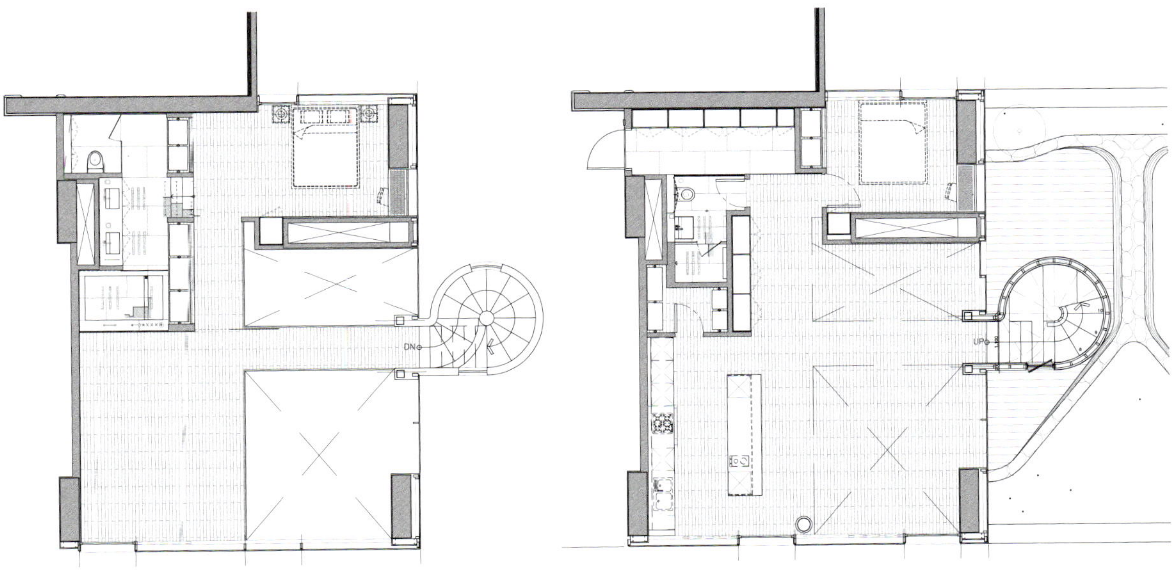

9

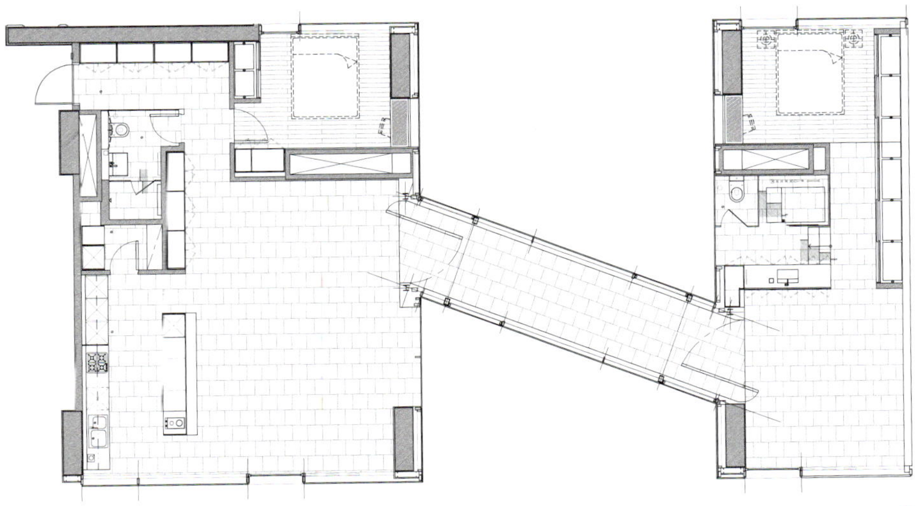

10

Building type
high-rise with voids (missing matrix)
1st–4th floor: commercial, cultural, and community uses
5th–27th floor: residential use
27 stories
facing in all directions

Date of construction
2004–2008

Number of units
172

Size of units
apt. type "3 Bay": 99.5–135.5 m² (66 units)
apt. type "Minimal": 141.5–172.5 m² (34 units)
apt. type "Garden": 141.5–161 m² (11 units)
apt. type "Bridge": 141.5–179 m² (41 units)
apt. type "Duplex": 147.5–186.5 m² (20 units)

Area per user
47–104.5 m²

Building depth
12/19 m

Access
mutual access cores on insides of C-shaped plan

Open spaces
terraces, balconies in cut-outs ("missing matrix")

Parking
underground garage

Architect
Mass Studies
Seoul

Location
Seoul
South Korea

Referring to floor plan 1:200

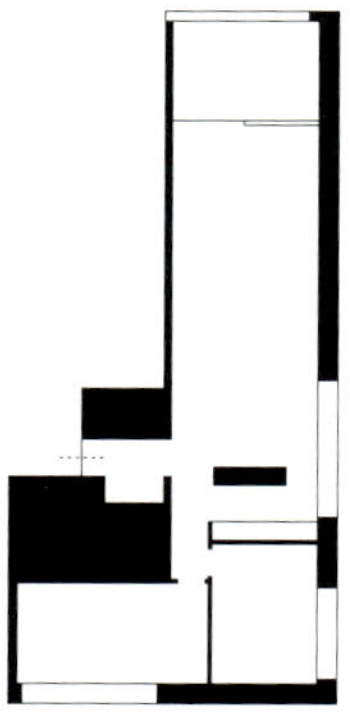

Escher Terraces | E2A | 2013

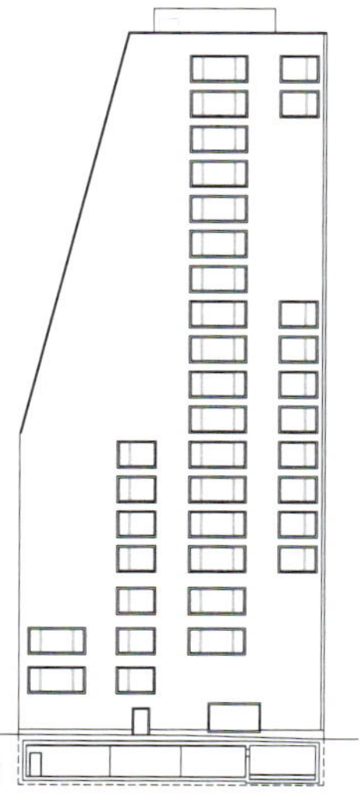

3

4

This high-rise building incorporates a former manufacturing facility into its base. With its scale and character, the base creates a visual connection to the surrounding production facilities. It is now used to house a rehearsal stage. While the tower rises vertically on the street side, on the opposite side it begins to taper above the roofline of the neighboring city blocks to create slightly terraced levels on its upper portion. Facing south, these floor-to-ceiling glazed loggias offer residents a panoramic view of the city center and Lake Zurich.

The floor plan organization and spatial quality of the residential levels is characterized by a structural system of parallel shear walls seemingly extending into the generous loggias on the terraced side of the building. Depending on the location, this spatial system generates a variety of apartment types and sizes. Along the side facades, the units (**A**) extend from one side of the building to the other and continue into the adjacent segment at the back of the building where the entrance, bedrooms, and bathrooms are located. The units in the center (**B**) of the building connect multiple segments with a deep living room bordered at the back by a kitchen and hallway and on the sides by more private areas. On the upper floors where the building depth begins to recede, three segments are combined, while on the lower floors, two duplex units are situated next to one another (**C**). Their lower level runs the length of the building, while on the floor above where the entrance is located, they extend only through the rear portion. Single-floor apartments running perpendicular through three structural segments (**D**) occupy the front portion of the floor above the duplexes' lower level.

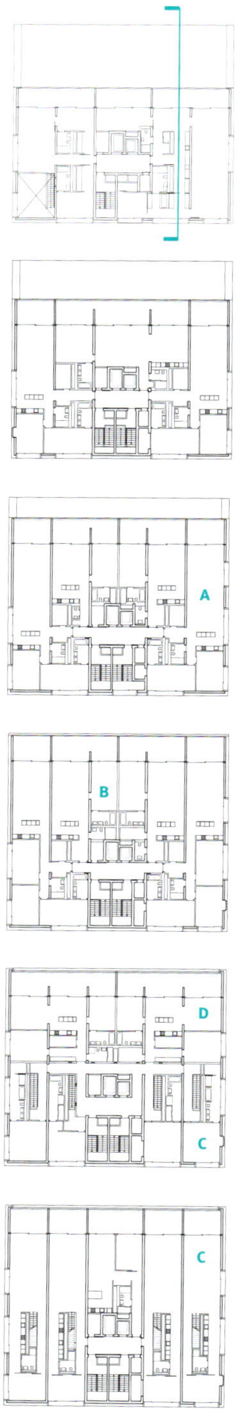

2

1 Ground floor with rehearsal stage 1:100
2 19th floor 1:1000 (top floor plan)
15th floor 1:1000
12th floor 1:1000
9th floor 1:1000
8th floor 1:1000
7th floor 1:1000 (bottom floor plan)
3 East elevation 1:1000
4 North elevation 1:1000
5 Cross section 1:1000
6 Type **C**: 3½-room maisonette, 7th and 8th floors 1:200
7 Type **A**: 3½-room apt., 12th floor 1:200
8 Type **B**: 2½-room apt., 9th floor 1:200
9 Type **D**: 3½-room apt., 8th floor 1:200

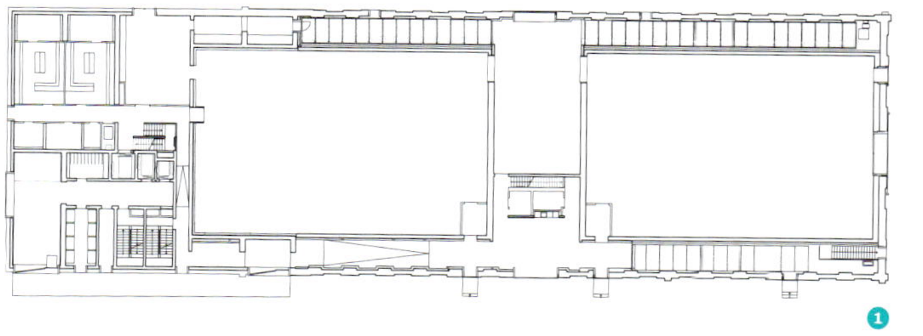

1

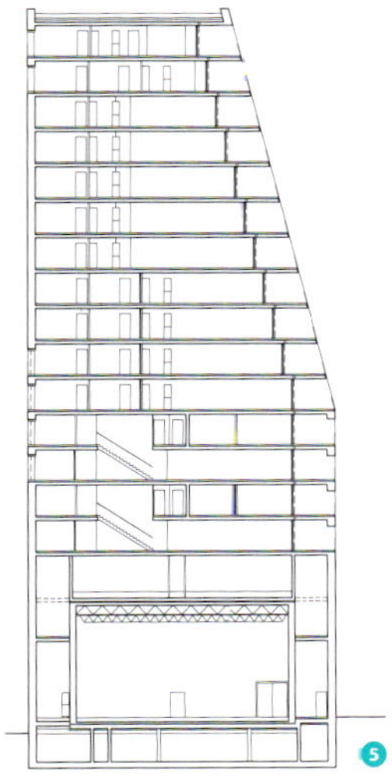

5

2.3 APARTMENT TOWER

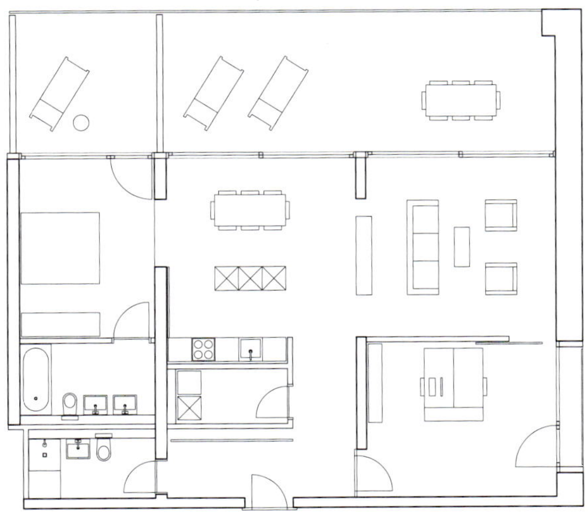

9

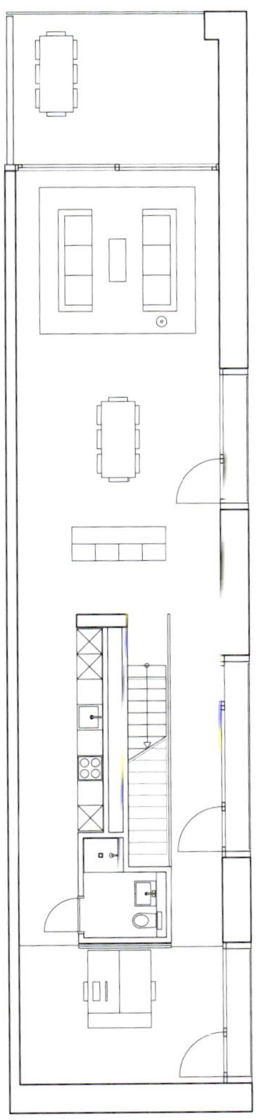

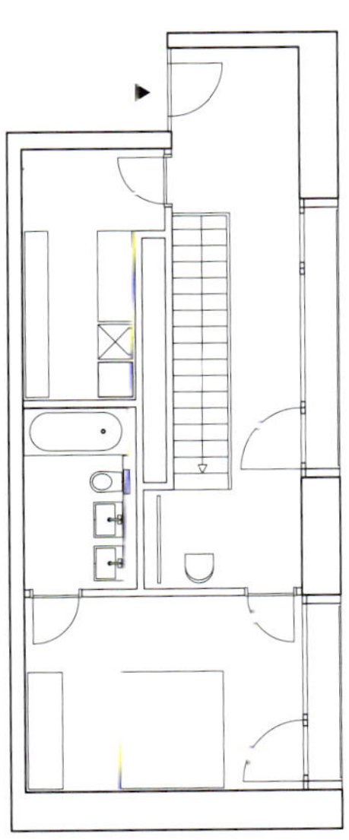

6

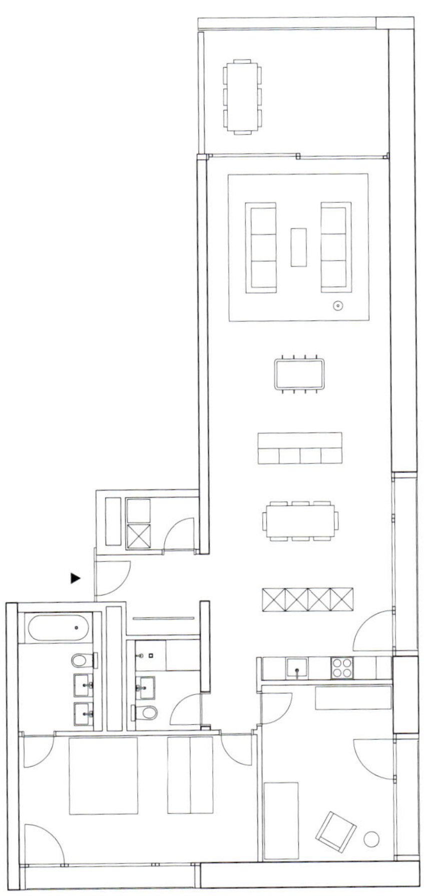

7

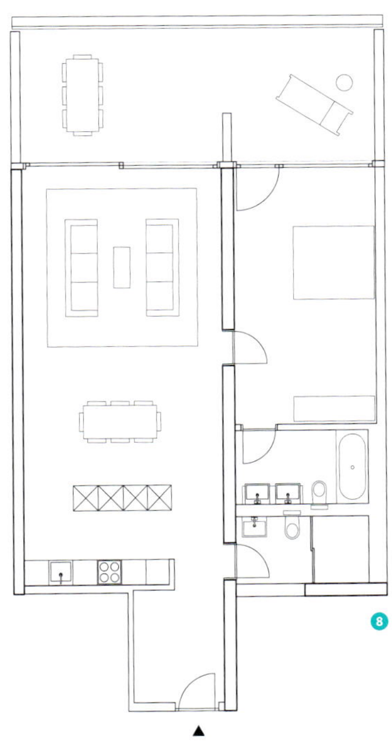

8

Building type
high-rise residential building above base with practice stage, tapered above the 10th floor to the south
19 stories
facing in all directions

Date of construction
2007–2013

Number of units
51

Size of units
2 ½-room apts., 56–106 m², loggias 23–57 m² (17 units)
3-room apt., 100 m², loggia 10 m² (1 unit)
3 ½-room apts., 111–180 m², loggias 16–44 m² (18 units)
3 ½-room mais., 138–157 m², loggia 16 m² (8 units)
4 ½-room apts., 136–149 m², loggias 16 m² (6 units)
5 ½-room mais., 251 m², loggia 49 m² (1 unit)

Area per user
38–67 m²

Building depth
17.8–26 m

Access
vertical access core on street facade
access to 3 to 6 apts. per floor

Open spaces
loggias

Parking
underground garage

Architect
E2A
Piet Eckert und Wim Eckert, Architekten ETH BSA SIA AG
Zurich

Location
Hardturmstrasse 5, 7
Zurich
Switzerland

Referring to floor plan 1:200

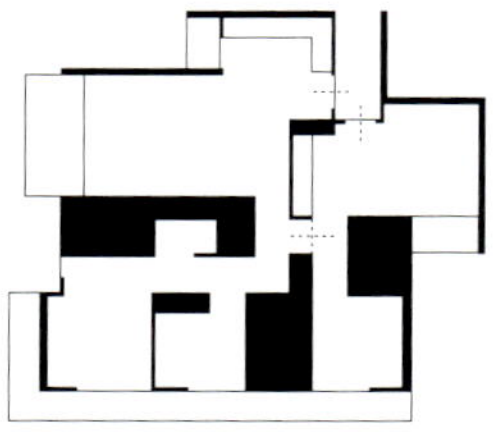

Kent Vale | MKPL | 2014

A high and spacious colonnade flanked by a parallel building with communal facilities forms the communal backbone for three residential towers positioned in a staggered row. The colonnade connects the towers with the entrance road and the surrounding university campus. From the colonnade, three two-level walkways branch off, framing patchwork-like arranged gardens and leading on the lower level to the multi-story portico-like entrance foyer. Their upper level connects the towers to shared amenities such as a swimming pool and playground on the roof of the parallel building. These communal spaces are complemented in the towers by roof gardens and large cantilevered balconies on every 8th floor.

Deep incisions into the building volumes, open access areas, and an exterior envelope with plant trellises and shading elements form an H-shaped cluster floor plan. The resulting open, multi-layered structure is green on all sides, naturally ventilated, and adequately shaded to facilitate a tropical lifestyle.

The apartments on the north side are laid out with an efficient floor plan subtly organized into three tiers. The sleeping quarters are directed to the north, while an open living room is located near the entrance. Three adjoining niches here form the hallway, kitchen, and dining area, and a deeply drawn-in loggia to the east or west can be connected with the apartment interior by opening folding doors. On the building's south side, an almost identical floor plan is treated slightly differently in the unit's first tier. Here, the living area is arranged in such a way to potentially allow the division of what would otherwise be a 5-room apartment into a 2- and a 3-room unit with separate entrances. The units can be easily connected to one another at will. In this manner, the often unpredictable needs of changing faculty members and their families can be taken into account.

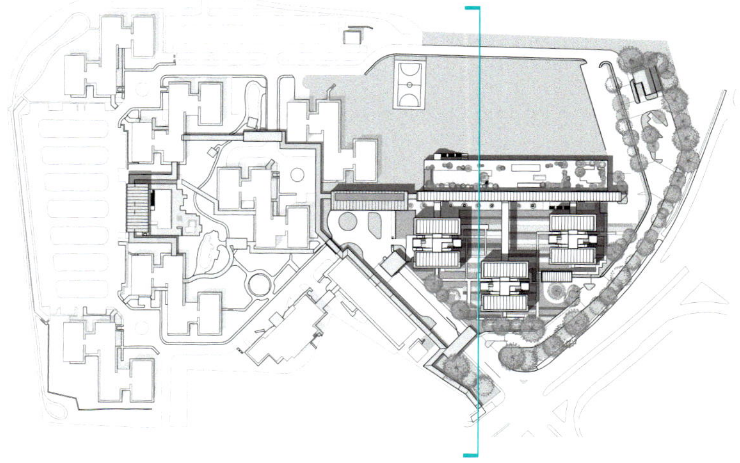

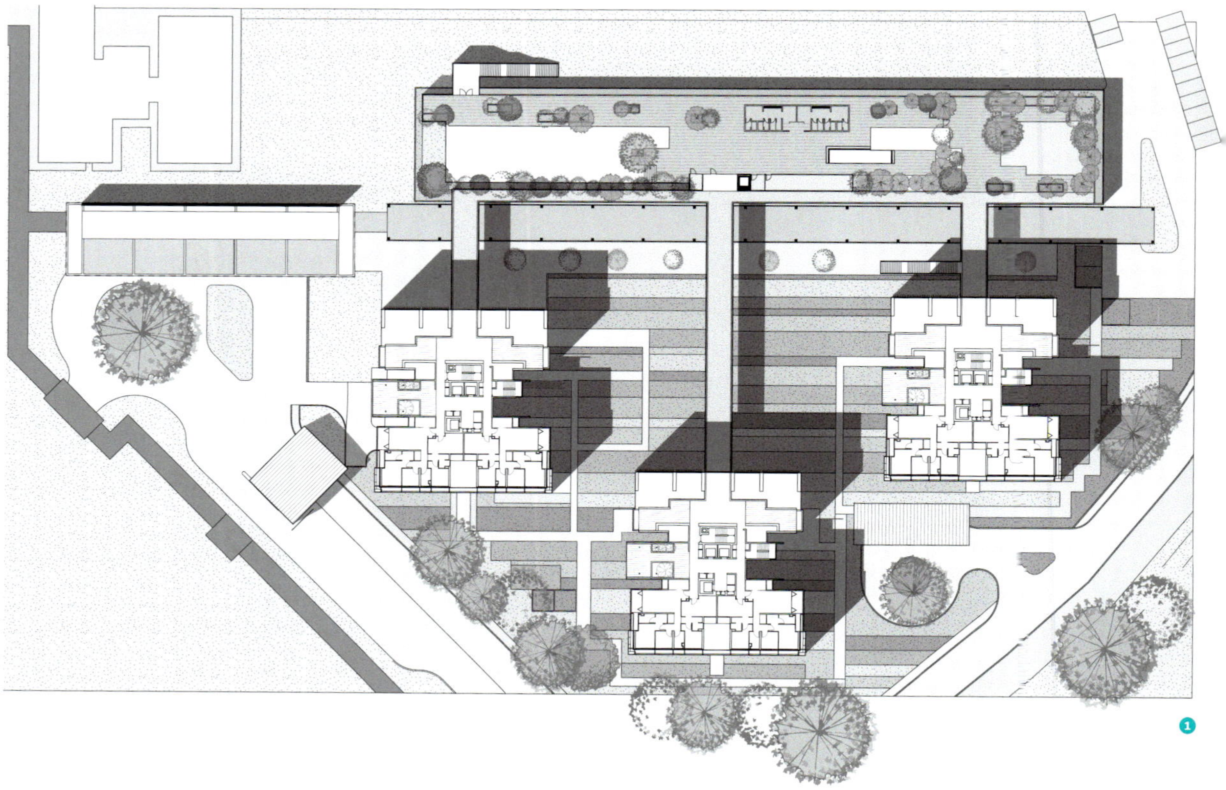

1

1 Site plan / 2nd floor with bridges connecting to communal facilities
2 Northeast elevation, section through colonnade and service building 1:2000
3 Northwest elevation 1:2000
4 Typical floor 1:200

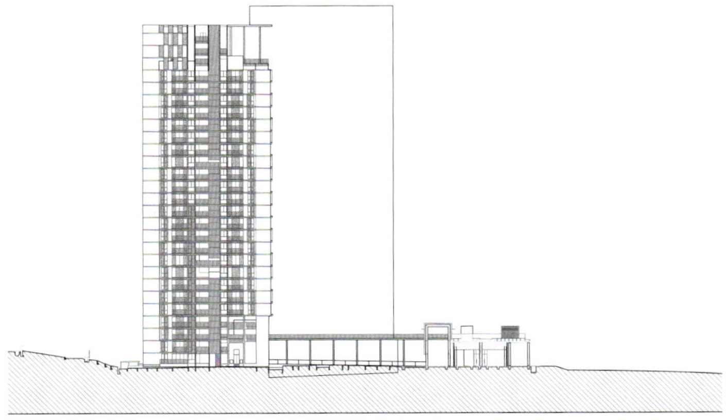

2

2.3 APARTMENT TOWER

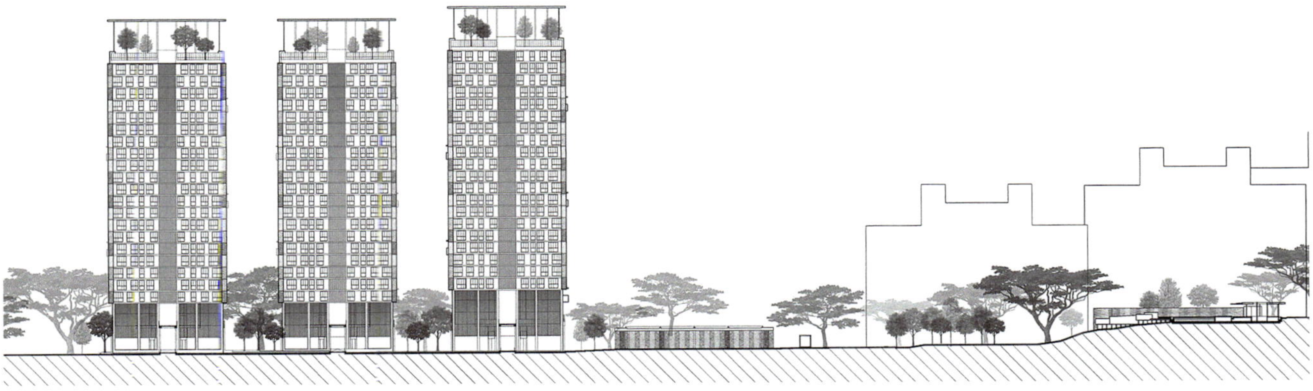

3

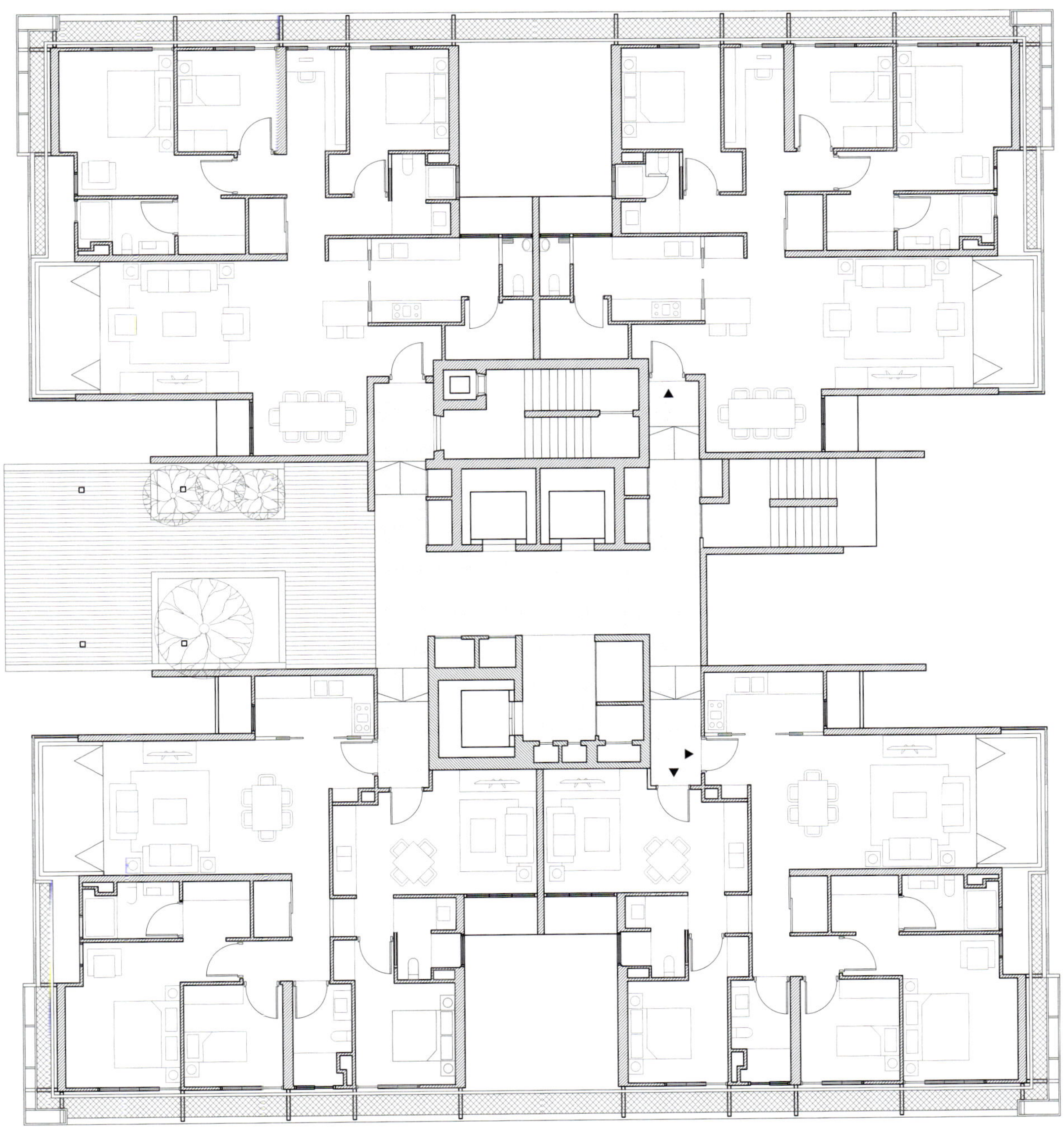

4

Building type
high-rise ensemble for faculty members on university campus, some serviced units
24–25 stories
facing in all directions

Date of construction
2010–2014

Number of units
352

Size of units
4-room apts., 125.5 m²
2-room apts., 38 m²
3-room apts., 87 m²
5-room apts., 125.5 m² (combination of 2- and 3-room apts.)

Area per user
19–31 m²

Building depth
13 m, 30.6/33.2 m

Access
open access area around central core, 4–6 apts. per floor, shared balconies every 8th floor, quadruple- to sextuple-loaded stairwell

Open spaces
private loggias, shared balconies and roof terraces in the towers, rooftop swimming pool and playground on building with communal facilities

Parking
parking lot

Architect
MKPL Architects
Singapore

Location
Clementi Road
Singapore

Referring to floor plan 1:200

2.4
TERRACED COMPLEX

The terraced complex is tantamount to a built utopia because it appears to interweave landscape and architecture, natural and built space, and aims to multiply the number of open, often verdant spaces. This typology is characterized by the design of the threshold between interior and exterior, by terraces as open spaces and topography. The occupants benefit from the direct relationship with the generous exterior spaces. The terraced complex is located either on a natural slope with orientation into one direction and a building and access configuration that follows the topography of the site. Alternatively, the complex can itself create an artificial topography with special functions located in the core areas, where natural light cannot penetrate. These functions may include access zones, storage areas, garage – and even a highway. Owing to their horizontality and complex staggering, terraced complexes are barely discernible as clearly circumscribable, sculptural building volumes; rather, they have the appearance of a structure that could occupy entire landscapes. However, this building type is economically and technically very demanding.

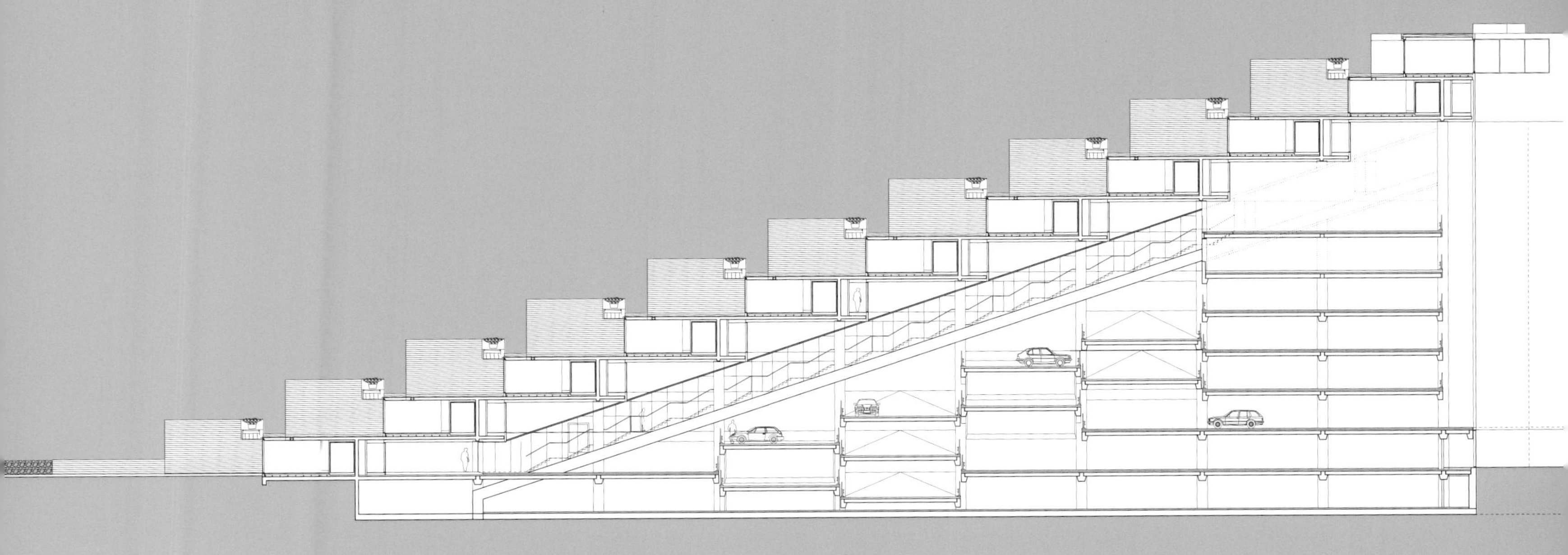

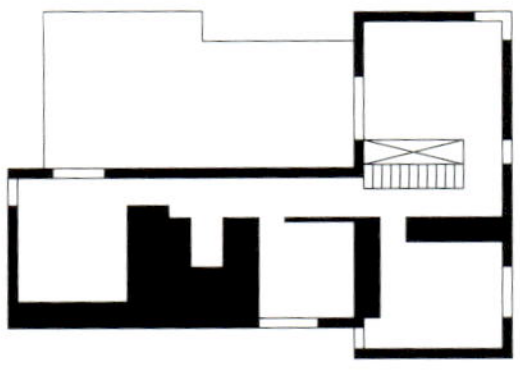

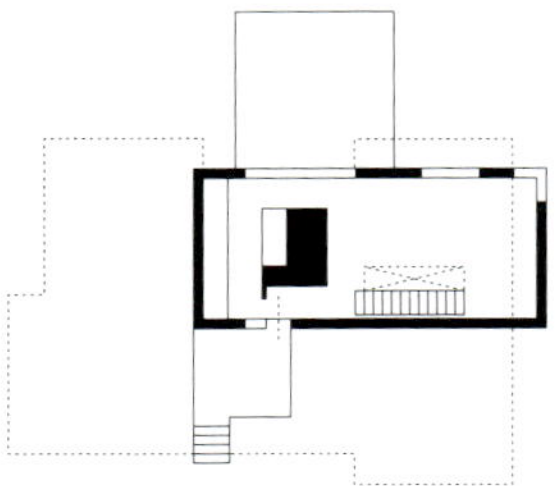

Habitat 67 | Safdie | 1967

Sculptural experimental housing project on the shores of the St. Lawrence River conceived on the occasion of the 1967 EXPO in Montreal. The conglomerate consist of 354 prefabricated, elongated concrete modules – each 55 m^2 in size – which were stacked in a staggered, orthogonal arrangement. The apartments can be composed of one or several modules: a total of 16 different configurations were realized. The rotation of the modules ensures that the rooms are oriented in different directions. One module generally accommodates a living room with kitchen and the internal stairs, and another one to two modules provide space for bedrooms and bathrooms. Each of the 1–5-room apartments is allocated to at least one large terrace on the roof of the unit below. The apartments are reached horizontally via covered pedestrian roads on every fourth level, and vertically via three lift cores. Short flights of stairs lead to the apartment above or below the corridor. All sections, play streets, prefabricated modules, and lift cores are linked in a manner that contributes to the overall structural support of the entire complex. Only 158 of a projected 1000 units were realized, making the project relatively expensive and ultimately preventing the development of the planned infrastructure with offices and retail stores to the full extent, which had been envisioned. Given the rigorous dedication to providing each unit with light, sunshine, and a generous terrace, Habitat 67 has no more density than a row-house development. It continues to function as a housing development to this day, having been bought by the tenants, who also assume the building management.

1

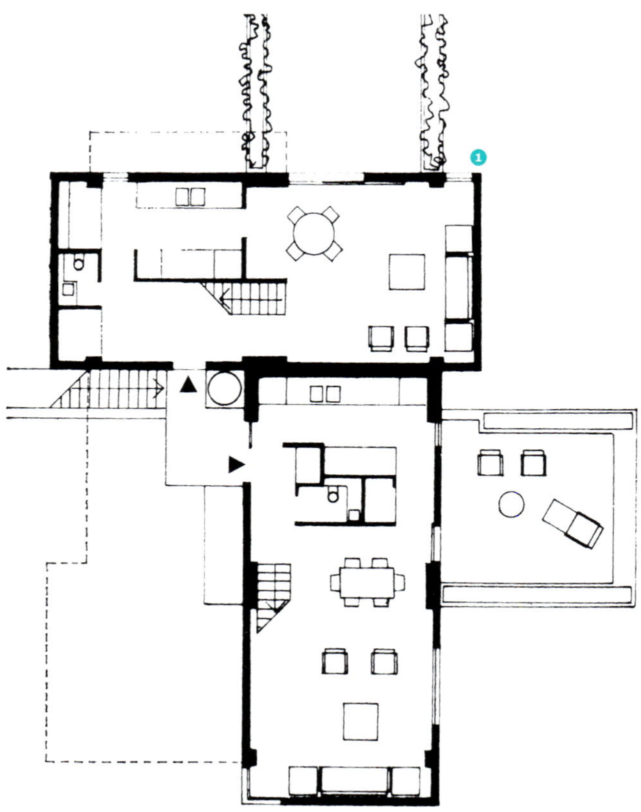

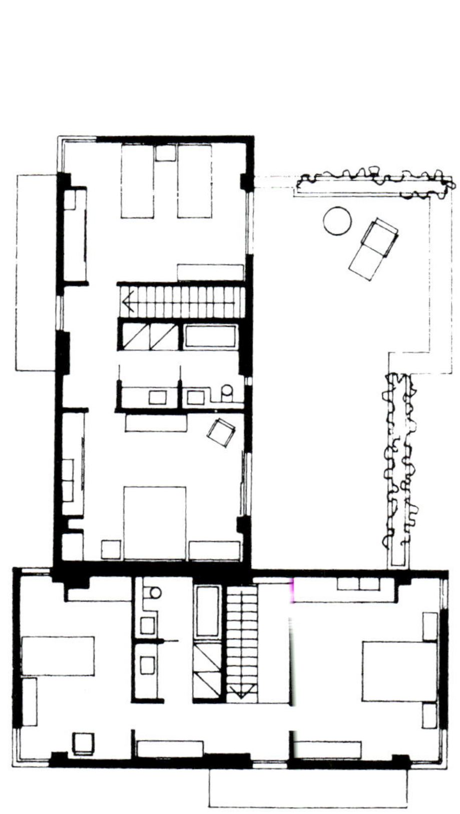

Examples of dwellings:
1. Two 3-room maisonettes 1:200
2. 5-room maisonette 1:200
3. 3-room apartment 1:200

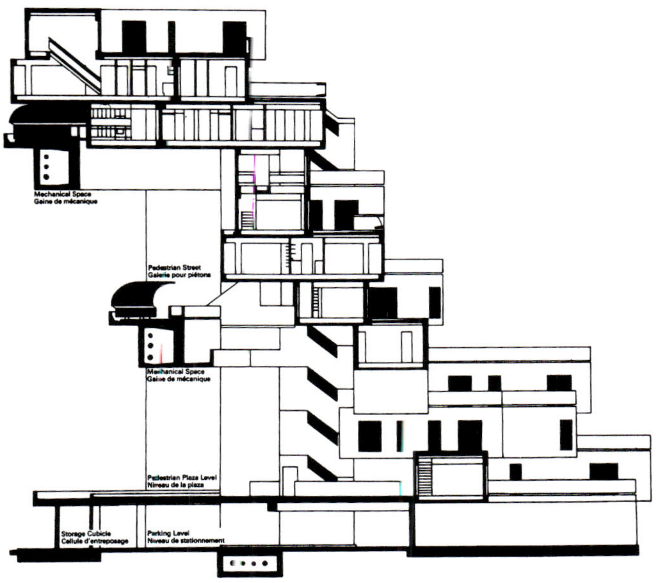

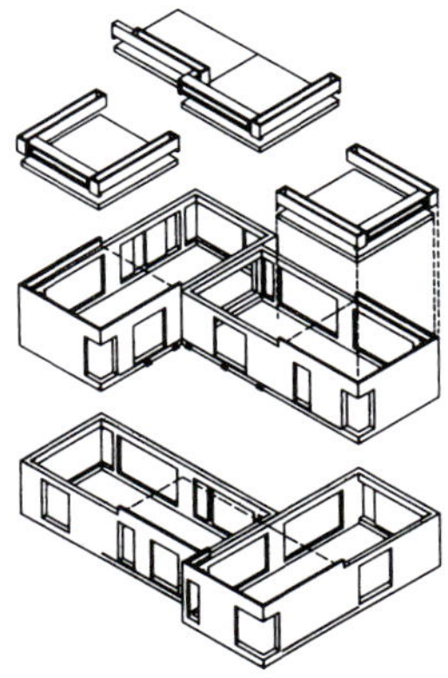

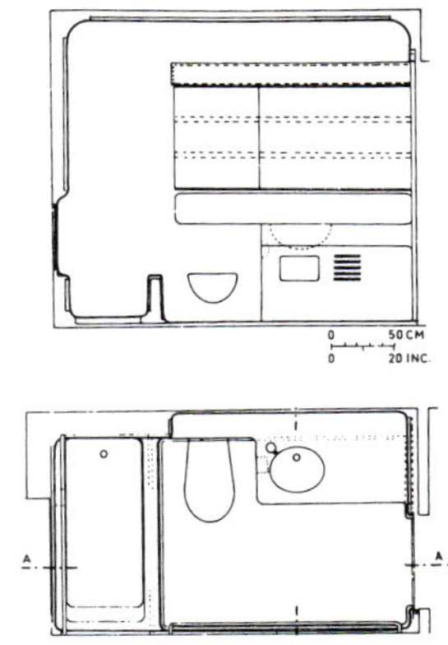

2.4
TERRACED COMPLEX

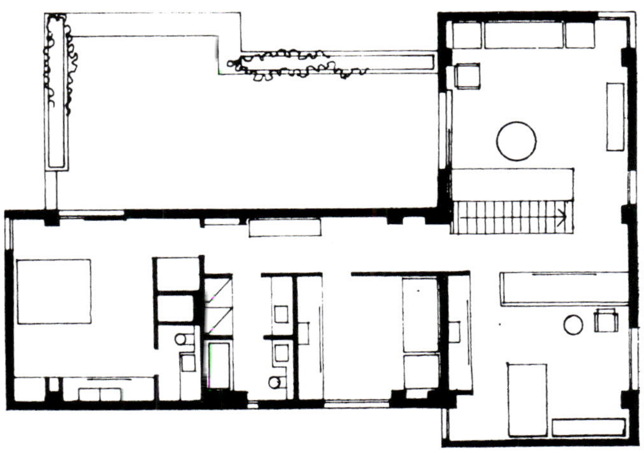

2

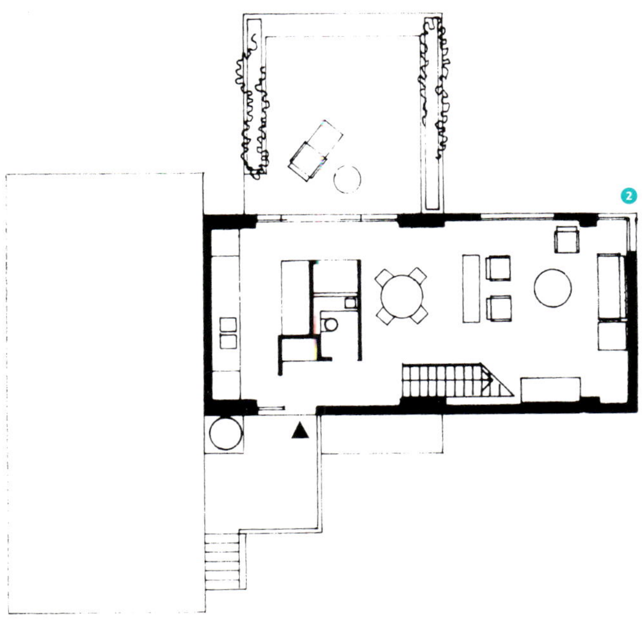

3

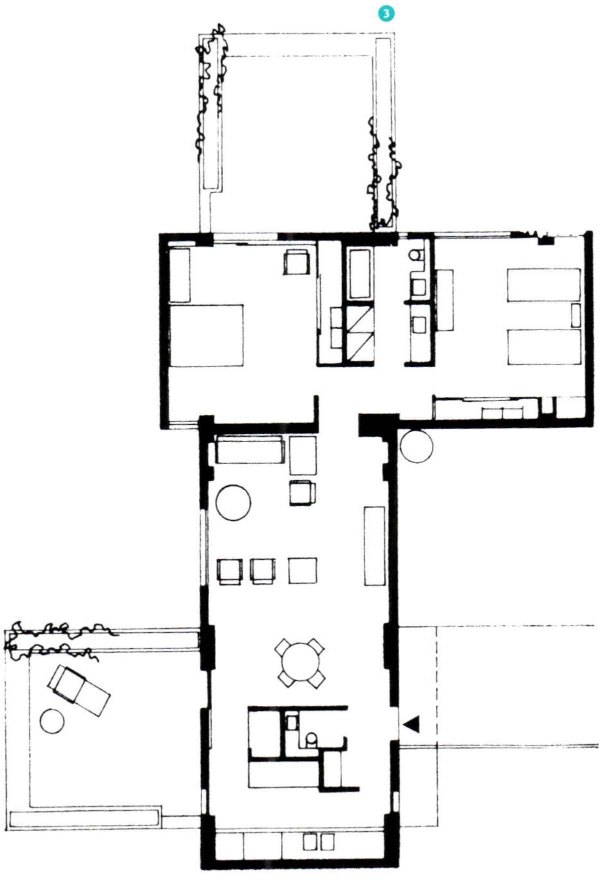

Building type
hill-shaped cluster building
22 stories

Date of construction
1967

Number of units
158

Size of units
15 different housing types,
1½–5-room apartments
with 55/110/165 m²,
basic module 5 × 11 m

Area per user
27.5–55 m²

Building depth
5/11 m

Access
walkways
galleries
elevator towers

Open spaces
terraces

Parking
covered parking

Architect
Moshe Safdie
Montreal
with
David, Barott, Boulva
Structural engineer:
A. E. Kommendant

Location
St. Lawrence River
Montreal
Canada

Referring to
floor plan 1:200

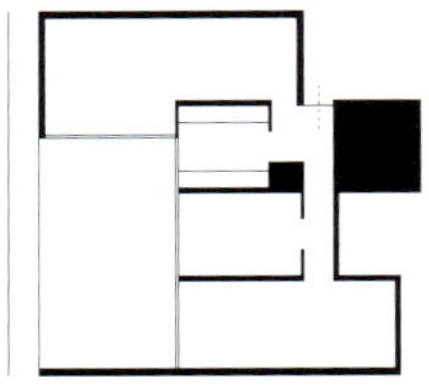

Brüderstraße |
Frey, Schröder, Schmidt | 1968

2.4
TERRACED COMPLEX

Artificial residential hill, symmetrically oriented toward east and west. The depth of the two lower floors allows apartments only in one point of the compass. Through living spaces on third and fourth floors. On the ground floor, the depth of the house is used by a central storage-room zone that runs all the way through. Within the fixed axial intervals and installation cores is a variety of apartments, with L-, U-, and T-shaped floor plans. This transforms the terraces (of different sizes) into inner courts enclosed on three sides. Clear zoning of living and sleeping areas. Large terrace in front of living room, often small ones in front of sleeping area. The terraces become patios on the ground floor.

Building type
artificial residential hill
4 stories
facing E/W

Date of construction
1968

Number of units
46

Size of units
2-room apts., 71 m² (8 units),
terraces 25 m²
3-room apts., 74/86/92 m² (18 units),
terraces 24–34 m²
4-room apts. 104/110 m² (16 units),
terraces 55/34 m²
5-room apts., 119 m² (4 units),
terraces 32 m²

Area per user
24–35.5 m²

Building depth
12–24 m

Access
single-, double- and quadruple-loaded stairwells

Open spaces
inner courts/terraces
patio gardens on ground floor

Parking
garages in basement story

Architect
Frey, Schröder, Schmidt
Stuttgart

Location
Brüderstraße 1
Marl
Germany

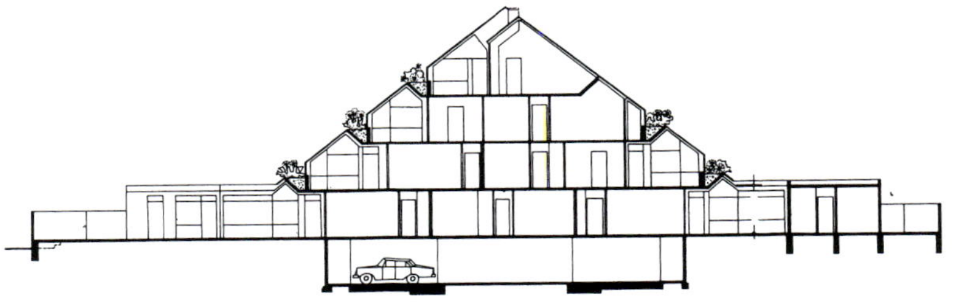

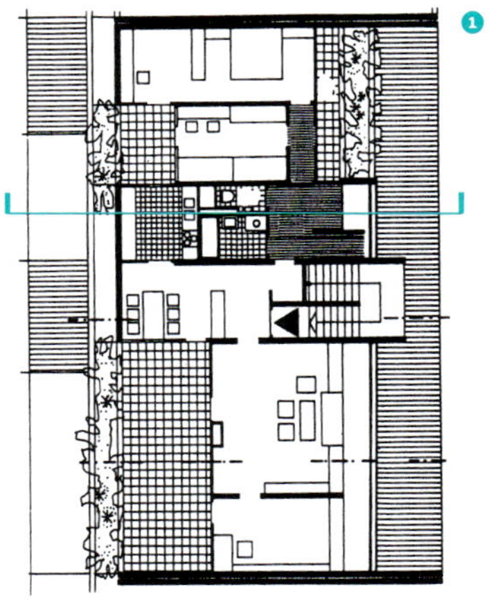

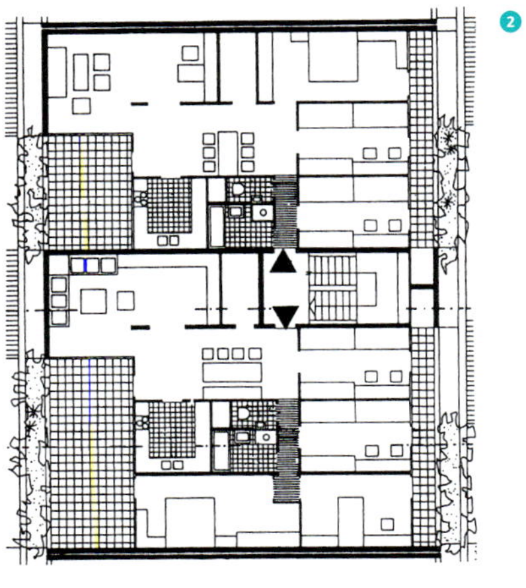

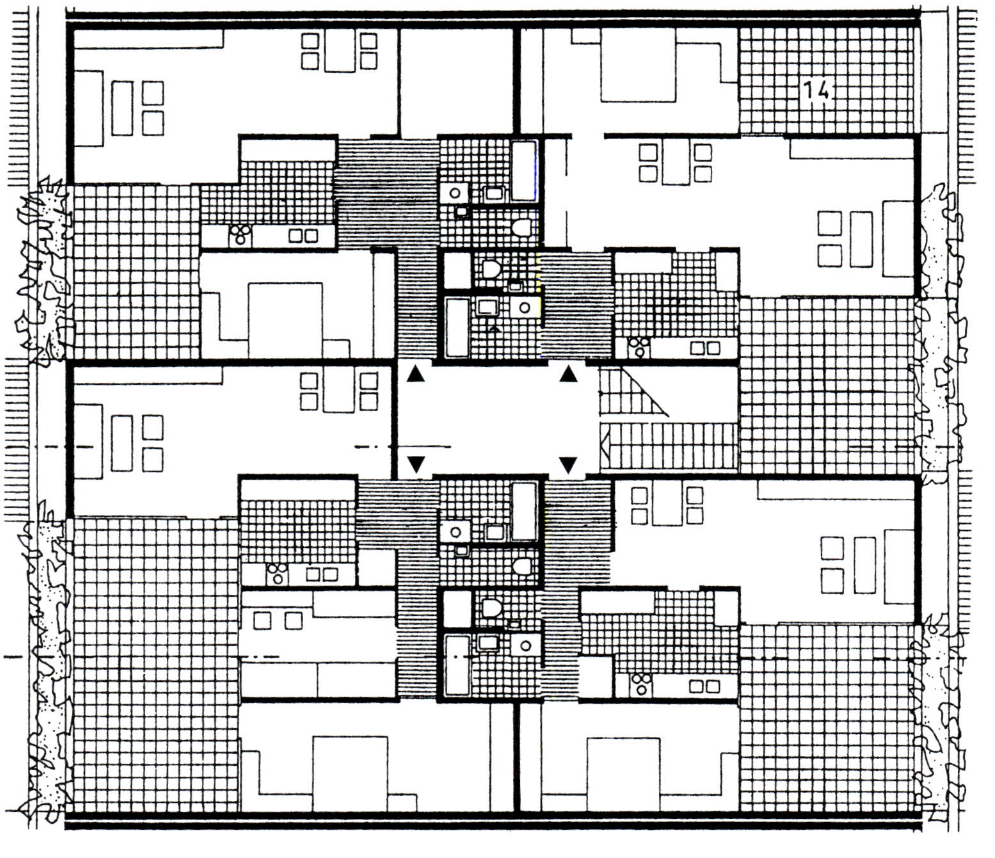

4th story: 1
2-room apt. 1:500
3rd story: 2
4-room and
5-room apts. 1:500
2nd story: 3
4 2-room apts. 1:200

Referring to
floor plan 1:200

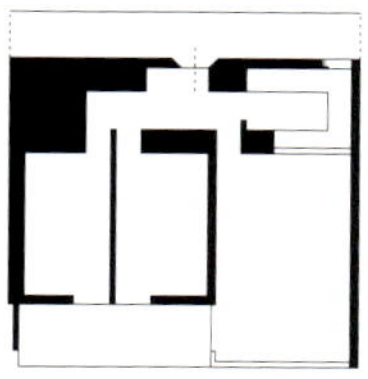

Brunswick Centre | Hodgkinson, Martin | 1972

2.4 TERRACED COMPLEX

A typical London street block is given a new interpretation. The basic approach: terraced apartments on both sides, gallery access in the hill interior and a grandiose five-story atrium. The apartments have simple layouts: the sanitary elements are placed along the gallery with the individual rooms lined up after them (the different apartment sizes result from adding up these rooms). The partition between kitchen and living room is designed as a counter. All living rooms end in a vertical and slanting winter garden section, entirely glassed-in; next to it a terrace with access from all rooms. Beneath the paved inner courtyard: neighbourhood uses (shops, restaurants, cinema).

Building type
terrace-house complex
in urban block
8 stories
facing SW/NE

Date of construction
1969–1972

Number of units
approx. 250

Size of units
1-room apts., approx. 40 m²
2-room apts., approx. 65 m²
3-room apts., approx. 90 m²
4-room apts., approx. 115 m²

Area per user
29–40 m²

Building depth
9 m

Access
inner galleries

Open spaces
semipublic terrace
in block interior,
winter gardens, terraces

Parking
garages in basement story

Architect
Patrick Hodgkinson
Sir Leslie Martin
London

Location
Foundling Court
London-Bloomsbury
United Kingdom

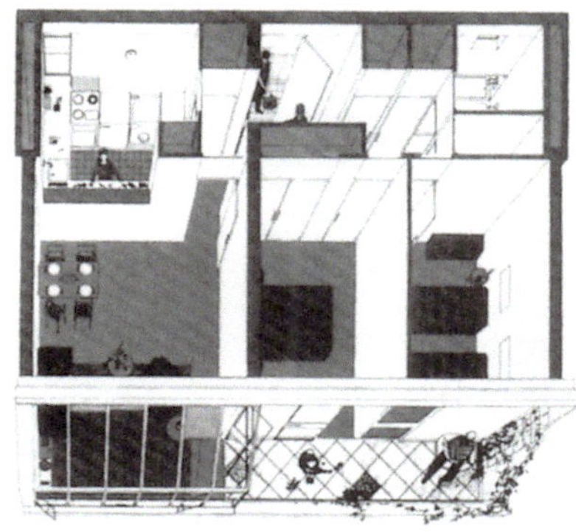

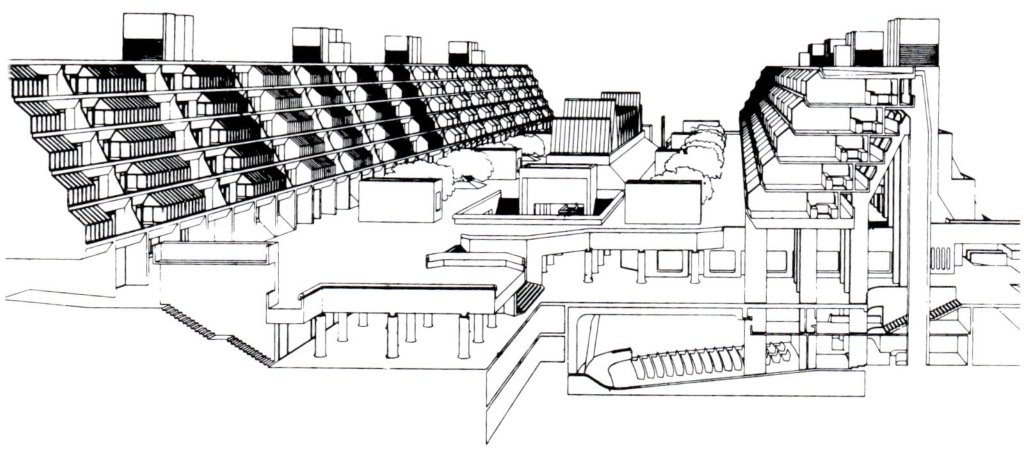

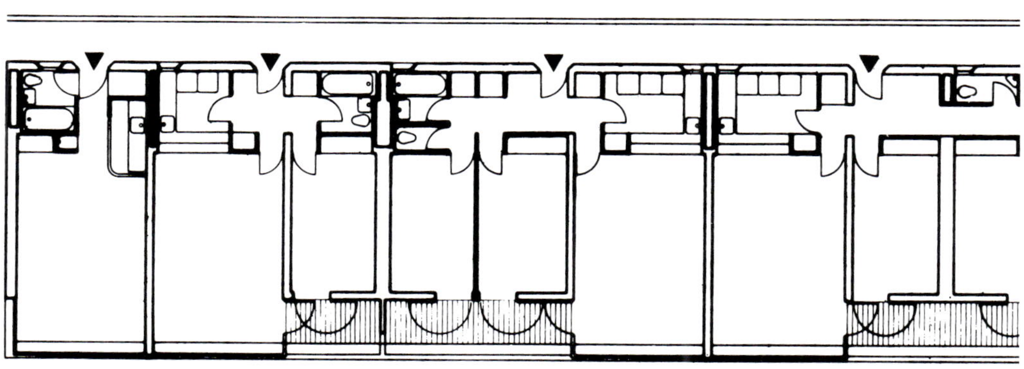

Isometric view of 1 typical apartment
Typical floor plan with 2 with 1-room, 2-room, 3-room and 4-room apt. (cut) 1:200

Referring to floor plan 1:200

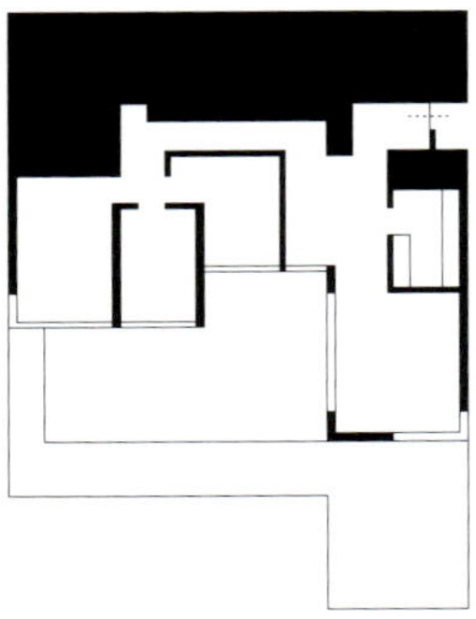

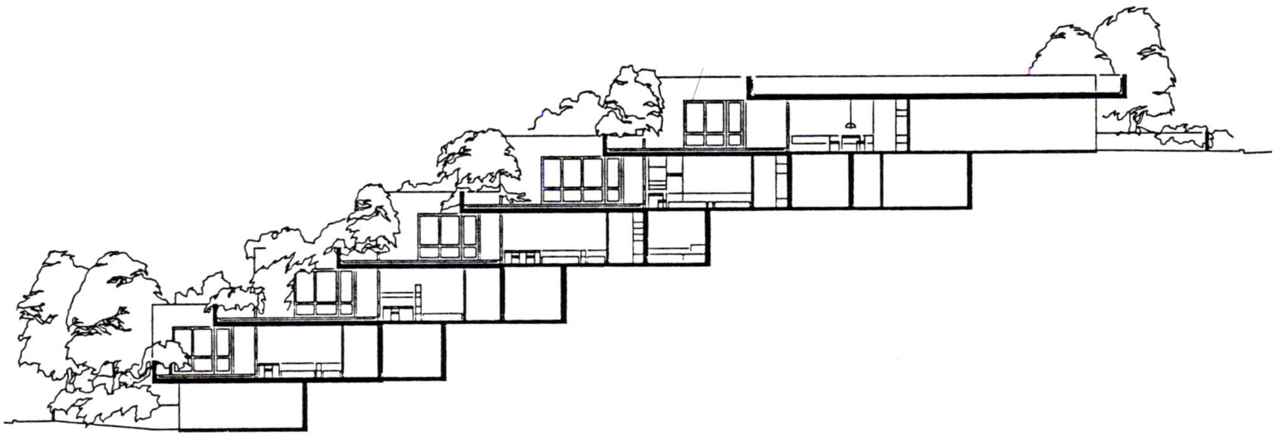

Trollingerweg | Kammerer, Belz | 1972

The housing complex contains two different building types, of which the apartments in the terrace houses are shown here (the three-story row houses appear only in the site plan). The typical terrace houses on angular floor plans are stacked on top of each other in strips of four to seven housing units. The intermediate zones between the development strips alternately contain either access paths – angular open stairways and landings filled with flowerbeds – or a second garden/service courtyard belonging to each apartment. The entrance to the apartment is set back (covered); the anteroom leads on one side to the small dining area, which, as an intermediate zone, provides access to living room and bedroom hallway, and on the other side to the large cellars, hobby rooms, and utility rooms excavated from the slope. The terrace on the southern side is enclosed by a deep plant trough that blocks the view into the apartment below. All rooms, including the dining area, have exits to the terrace. The small service courtyard lights either the spacious bath at the end of the hallway (with door to outside), the study, or one of the bedrooms.

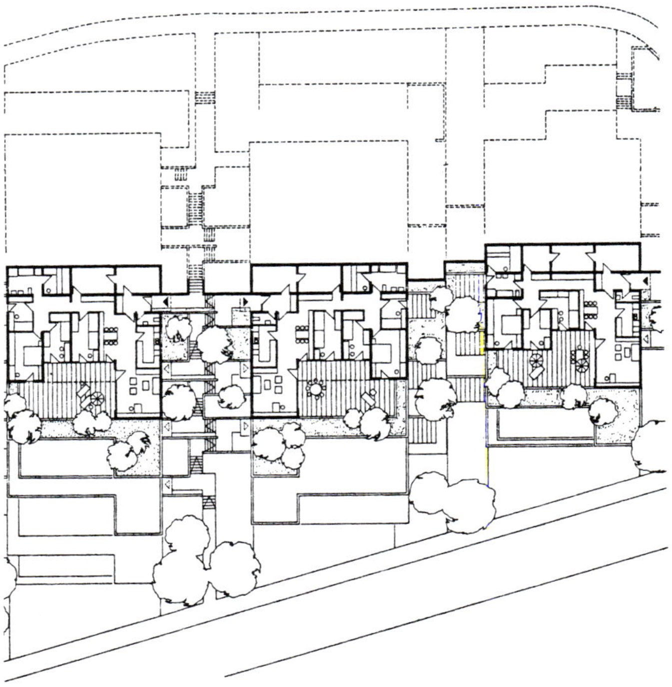

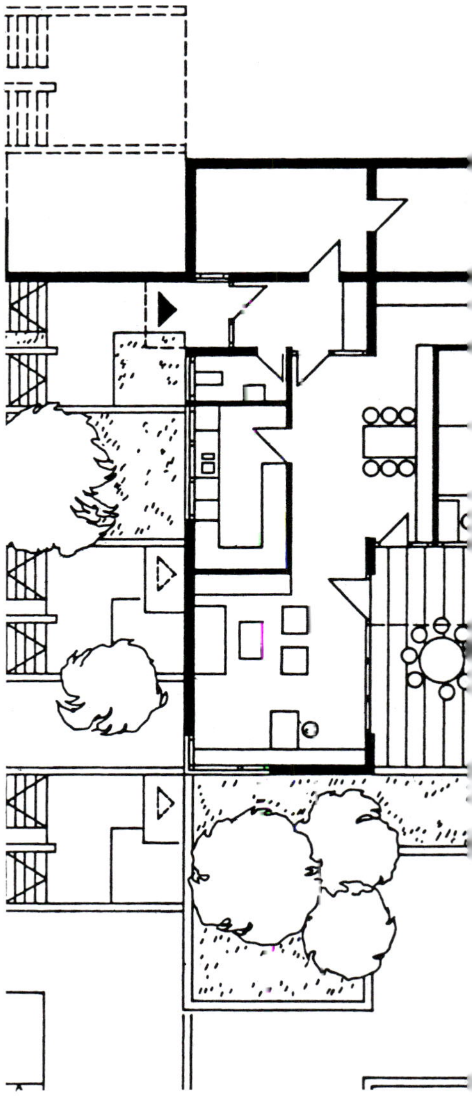

1. Layout of the building development
2. Section of floor plan 1:200 with 5-room apartments

1

2.4 TERRACED COMPLEX

Building type
terrace-house complex
5 stories
facing NNW/SSE

Date of construction
1972

Number of units
34
9 row houses

Size of units
5-room apts., aprox. 140 m²

Area per user
28 m²

Building depth
approx. 15–18 m

Access
directly from stair path

Open spaces
terrace courtyards

Parking
open parking spaces
along top access path

Architect
Hans Kammerer
Walter Belz
Stuttgart
Project architects:
Klaus Kucher
Josef Greitzke
Klaus Hallermann
Eberhard Munz
Roland Wittich

Location
Trollingerweg
Waiblingen-Neustadt
Germany

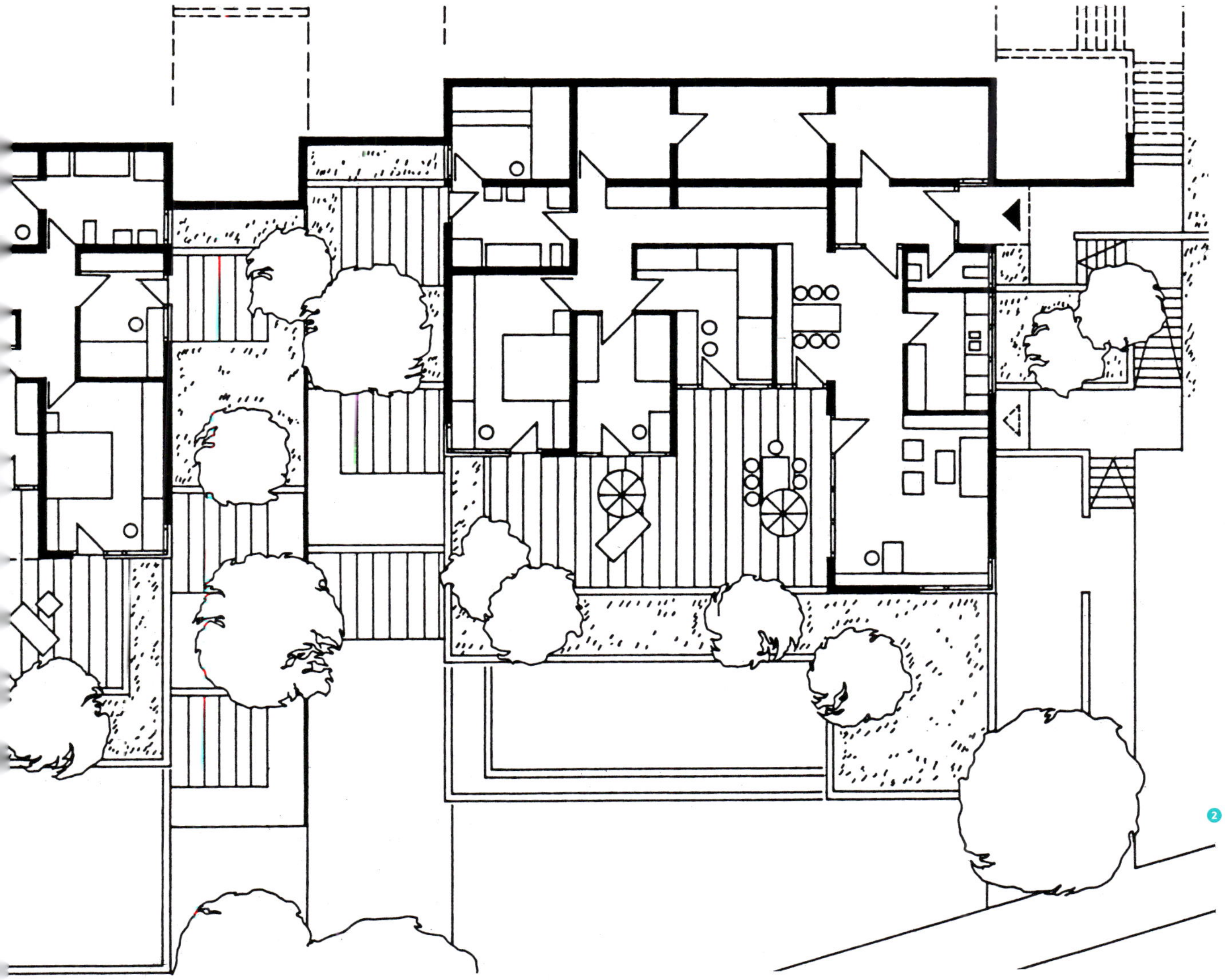

2

Referring to
floor plan 1:200

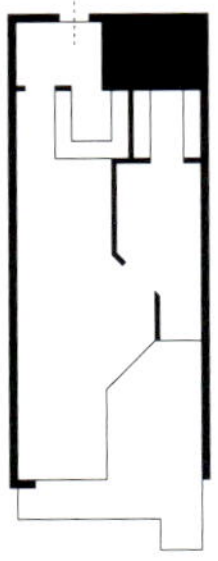

Schlangenbader Straße | Heinrichs | 1982

Built on top of an S-shaped double tunnel as part of one of the loop highways in Berlin: a 500-m-long, terraced residential development. A total of around 120 different apartment types were created: minimum-size flats, terraced apartments, standardized apartments, maisonettes. Generally the apartments are oriented to only one point of the compass, east or west; however, slanted glass doors toward oriels and terraces admit southeast or southwest light. In the minimum-size apartments, the kitchens are part of the living area (can be separated off by sliding doors) or are reduced to kitchen counters. Despite standard measurement of 6.10 m bay width, a wide variety of room layouts are offered. The great depth of the terraced apartments is exploited by inserted kitchens and closets; the slanted folding walls to the bedrooms create a greater sense of width. All larger apartments separate bath and WC; the bath, as a rule, is directly accessible from the bedroom; the second path leads back to the lobby. Large terraces and loggias supplement the housing strategy. On the fifth floor, an internal street with several community rooms: hobby rooms, playrooms, etc. Guest apartments on the third floor. On ground floor: laundries, bike rooms, etc.

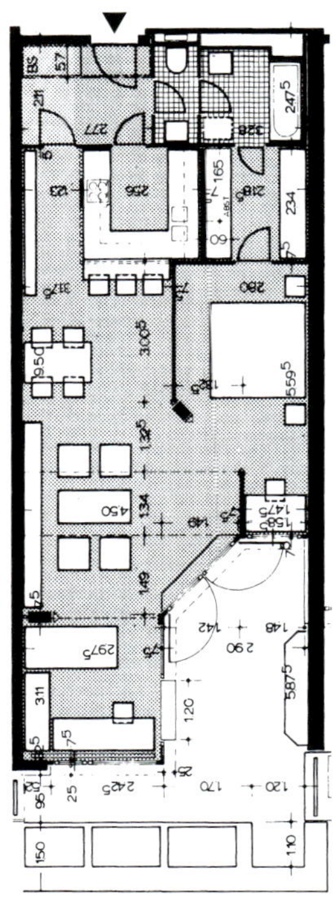

1

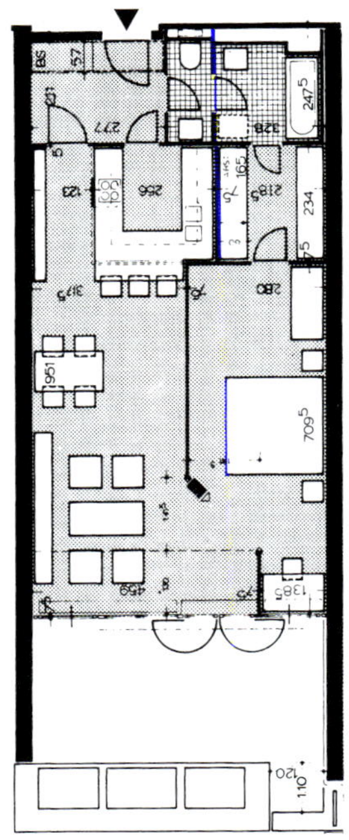

2

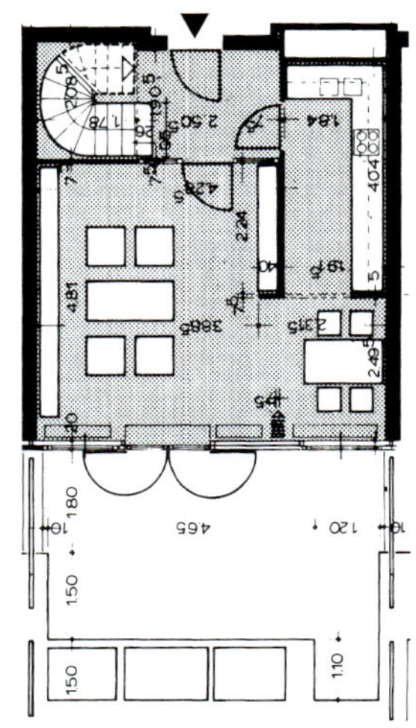

3

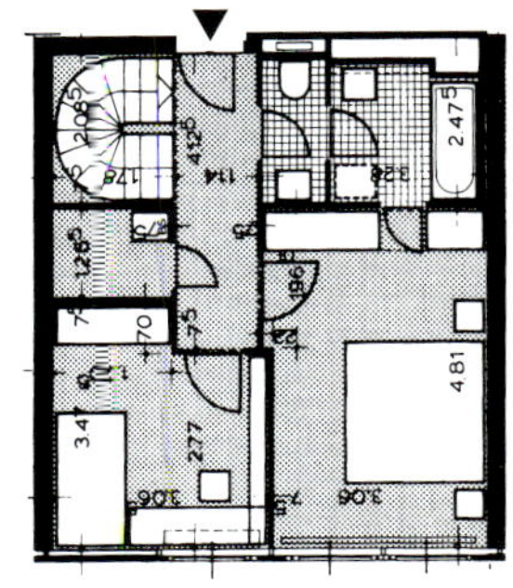

3

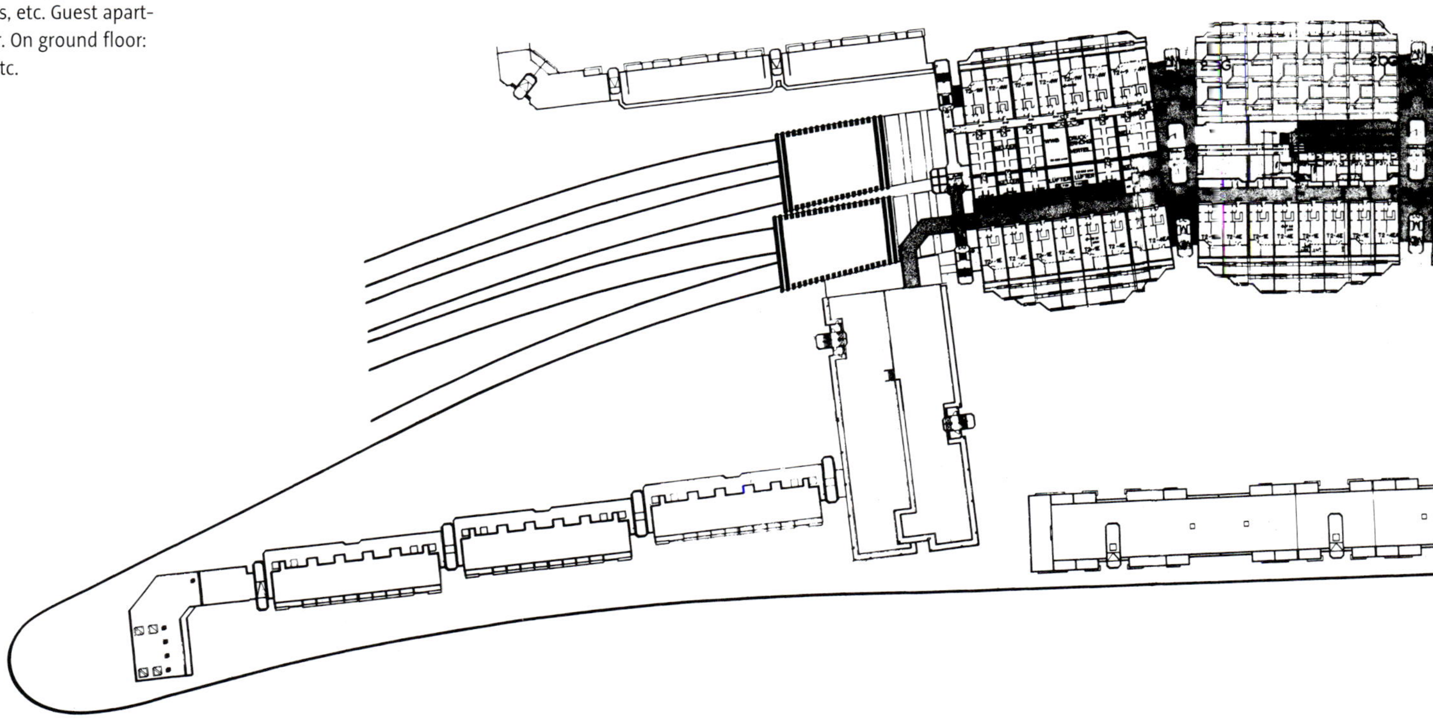

1 2-room apartment with terrace 1:200
2 2-room apartment with terrace 1:200
3 3-room maisonette with terrace 1:200
4 3 variations of 1-room flats 1:200

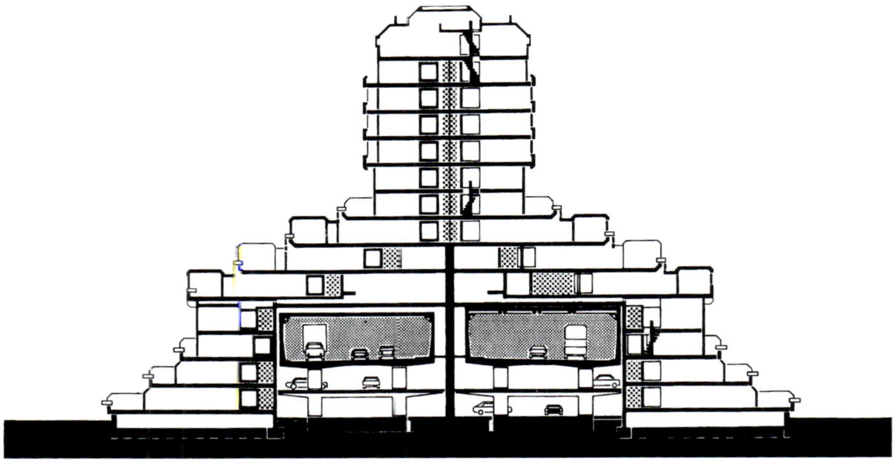

2.4
TERRACED COMPLEX

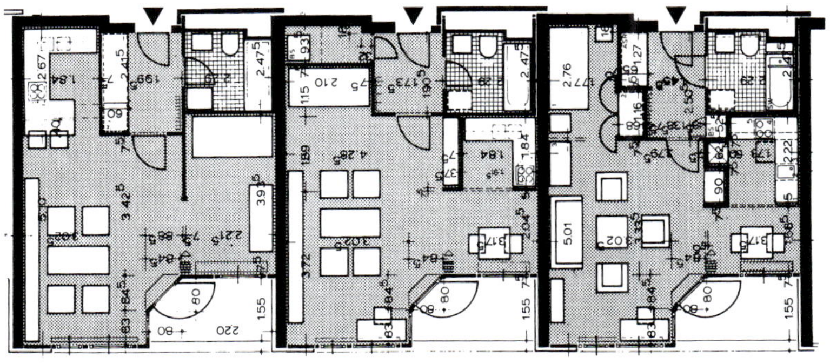

Building type
over-highway construction
14 stories
E/W

Date of construction
1976–1982

Number of units
1064

Size of units
1½-room apts., 42–52 m²
2-room apts., 67 m²
2½-room apts., 80–120 m²
3-room mais., 85 m²

Area per user
28.5–60 m²

Building depth
roughly 60 m in tunnel area, terraced back to 18 m

Access
inner hallways

Open spaces
terraces
loggias
gardens

Parking
parking spaces beneath the tunnel

Architekt
Georg Heinrichs
and G. and K. Krebs
Construction:
Georg Heinrichs and Partners

Location
Schlangenbader Straße/
Wiesbadener Straße
Berlin-Wilmersdorf
Germany

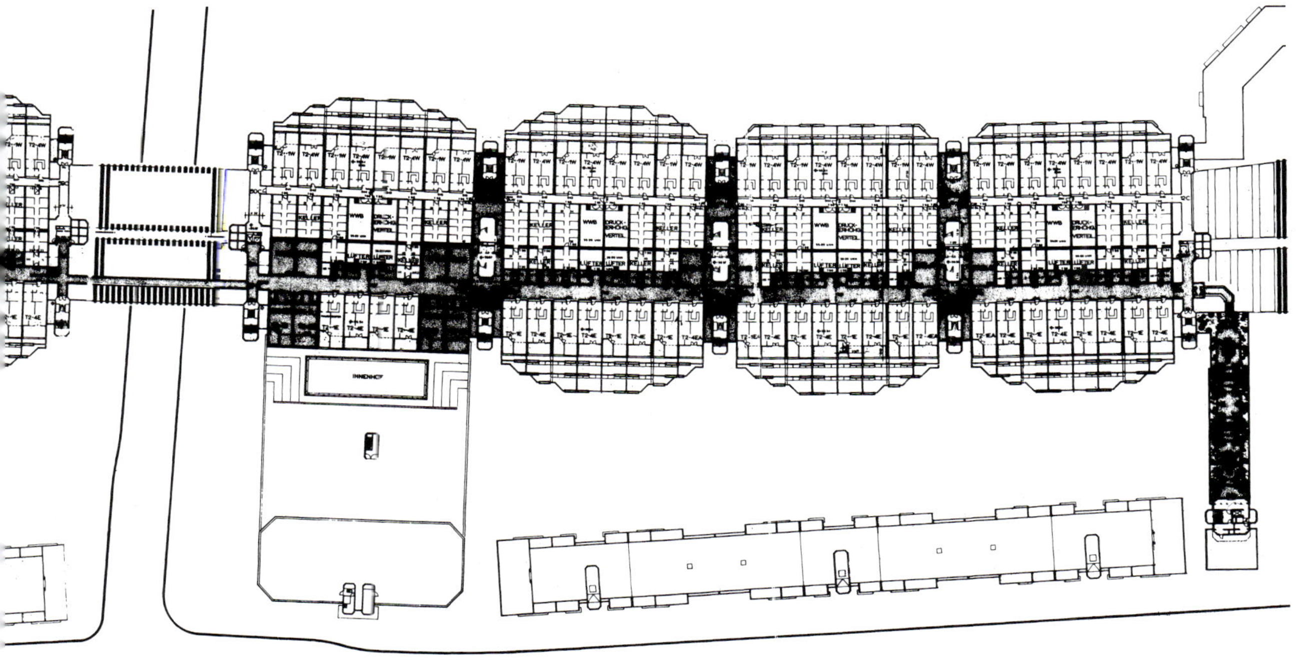

Referring to floor plan 1:200

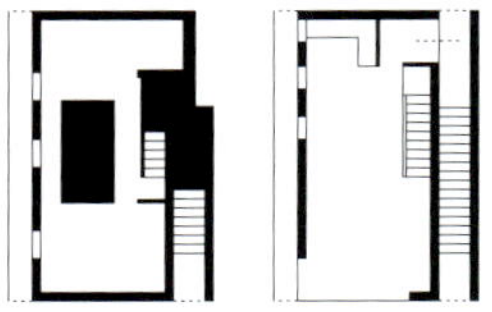

Living at the Lake | Baumschlager Eberle | 1988

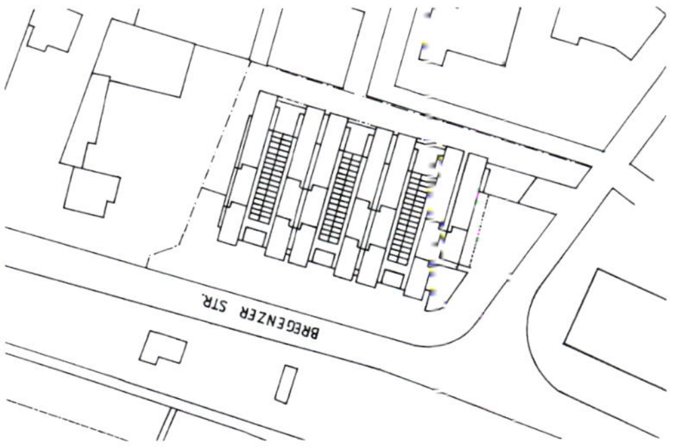

The terraced development on a slope is located between Lake Constance and the Bregenz Forest, although the site is separated from the lakeshore by a much-traveled road and a railway line. The configuration of the plinth story, which houses retail stores, commercial spaces, and the underground garage, ensures that the apartments on the upper levels lie above the acoustical shadow of the road. The atria at ground level, which provide access to the commercial units, continue on the upper floors as planted open spaces. The terraced building sections contain duplex and triplex apartments, staggered so that each apartment features a generous terrace with a splendid view of Lake Constance. With the exception of two small apartments in the rear section of the building, all living areas with kitchen open onto these terraces, while the bedrooms underneath overlook the elongated courtyards. Principal access is provided via glass-covered, cascading stairs with landings in front of sets of two apartments. The comb-like structure achieves a differentiated sequence of public (green spaces), semiprivate (pergolas) and private exterior spaces (terraces). The corner of the complex facing toward the city is marked by a 4-story office building, next to which the entrance ramp to the underground garage is located. The ingeniously stacked and staggered apartments create a typological variety that succeeds in bridging the gap between row-house development and dense low buildings.

1

2

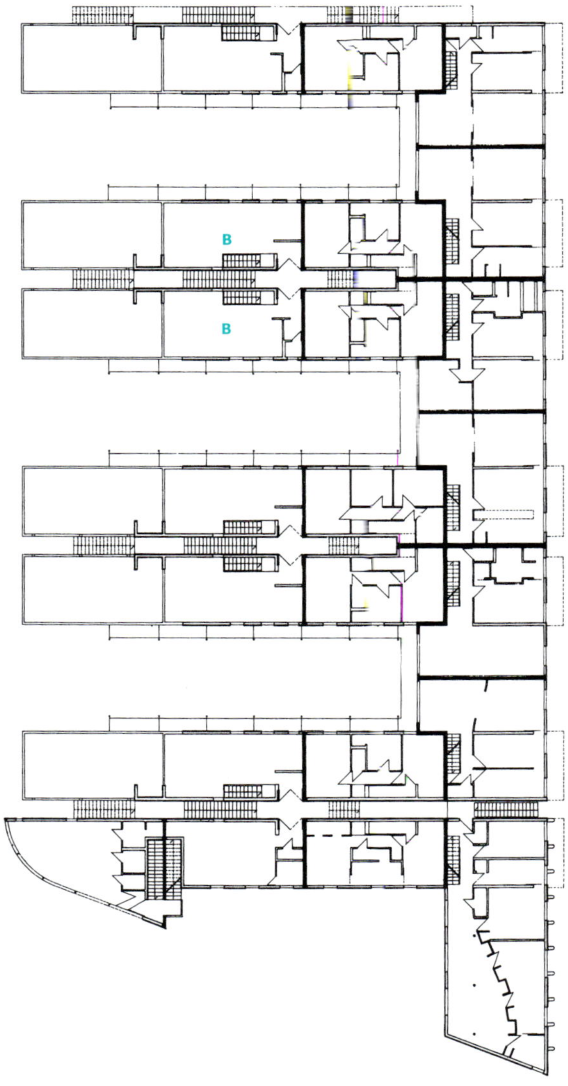

1. 2nd floor with green area for access 1:500
2. 3rd floor 1:500
3. 4th floor 1:500
4. 1-room apartment and 3-room maisonette 2nd/3rd floor 1:200

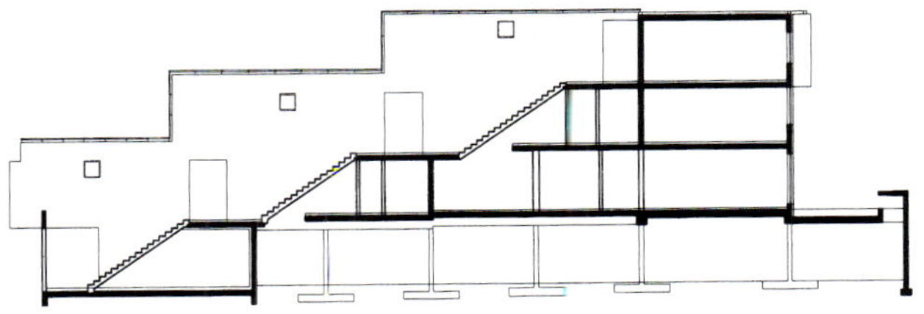

2.4
TERRACED COMPLEX

Building type
comb-like terrace house
4 stories
facing E/W, N/S
plinth story: offices, services

Date of construction
1986–1988

Number of units
37

Size of units
1-room apts., 33–39 m² (13 units)
2-room apts., 55.5 m² (3 units)
3-room apts./mais., 73.5–86 m² (15 units)
4-room mais., 95.5 m² (6 units)

Area per user
24–39 m²

Building depth
18/20/30 m

Access
via cascade of steps directly
or across large green area
on second floor

Open spaces
spacious roof terraces
rear gardens towards road
narrow verandas

Parking
underground car park

Architect
Baumschlager Eberle
Lochau

Location
Bregenzer Straße
Bregenz
Austria

3

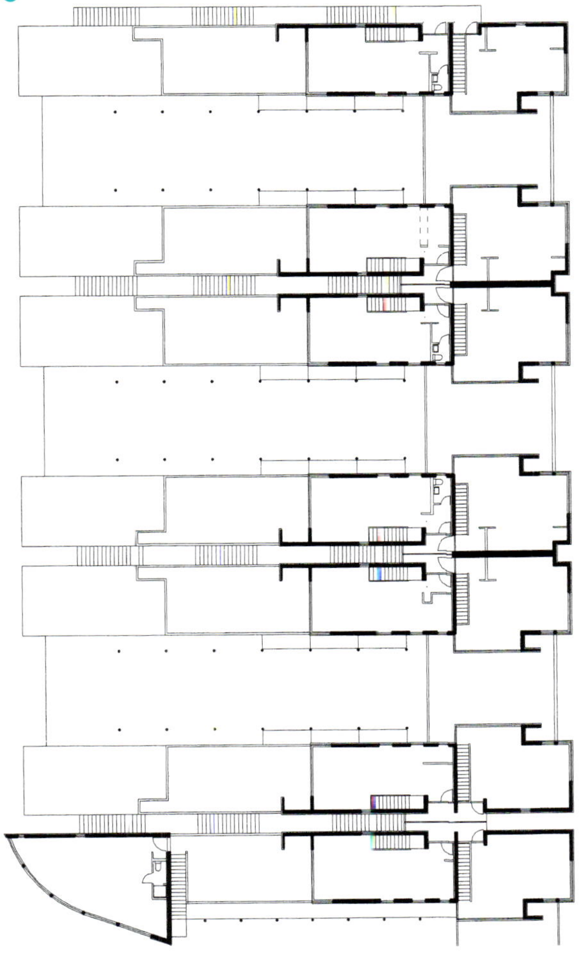

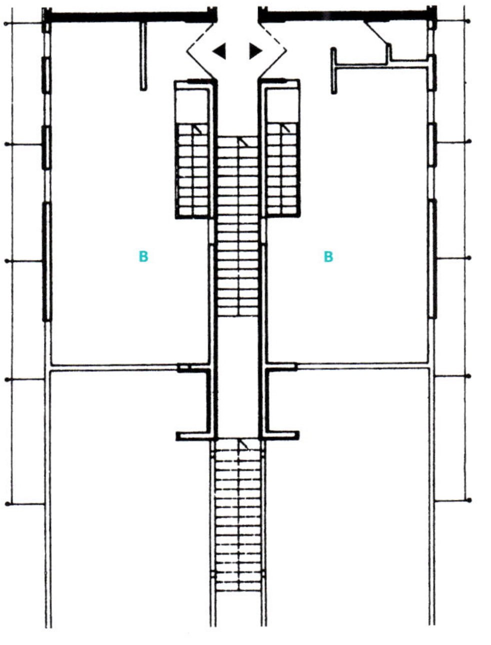

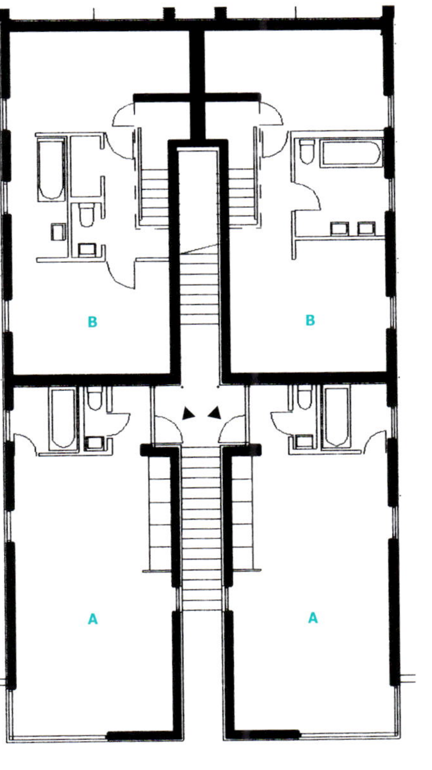

4

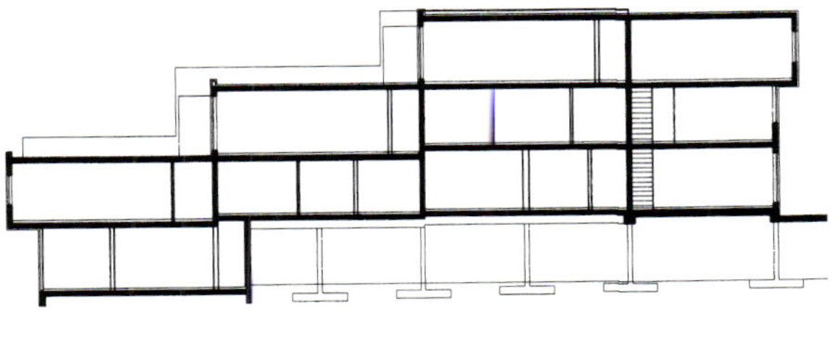

Referring to
floor plan 1:200

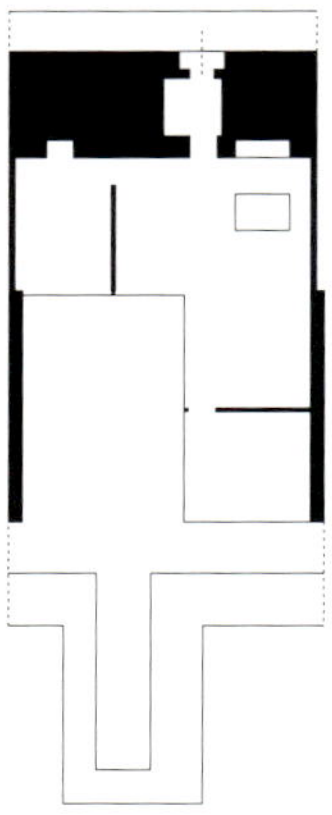

The Mountain | BIG Bjarke Ingels Group | 2008

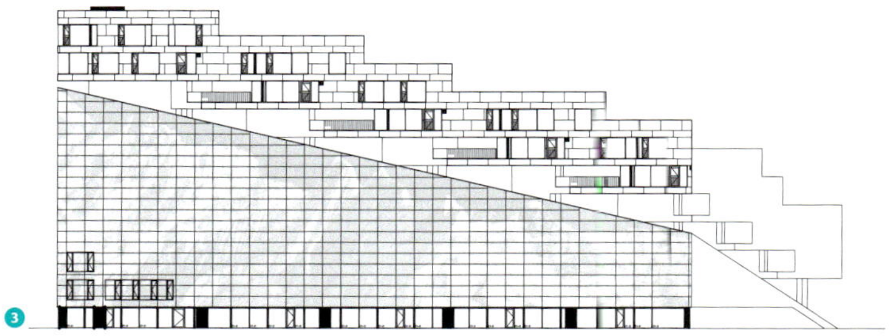

The program for "The Mountain" is summarized as two thirds parking and one third living: a 10-story parking garage serves as the base for a terraced housing development draped across the top like a patchwork quilt of gardens, patios, and apartments. The approach is one of combining the comforts of suburban living with the social intensity of urban density – residents drive almost to their front door and enter their apartment which affords them an expansive panoramic view across single-family homes nestled in a verdant suburban environment. A ramp with a sloping elevator and stairs placed between the parking and apartment decks, as well as bridges spanning the cascading atrium between the two layers, connect to the access corridors that lead to the apartments. In the interior, hallways, bathrooms, and walk-in closets create a transitional zone to the living room, which has an entirely different spatial sense with generous glazing affording a view onto the urban panorama to the southeast. The basic layout of the living room is L-shaped, wrapped around the deep roof patio, connected with lane-like gardens above the apartment on the terrace below. Wood-encased planters serve as balustrades and privacy screens. More complex units – some small and others larger, some two-stories high and others with two patios – are located along the edges of the complex, where the uniform terracing merges with the shape of the parcel. On the north and west facades, the walls of the parking decks are clad in perforated aluminum plates bearing an image of Mount Everest.

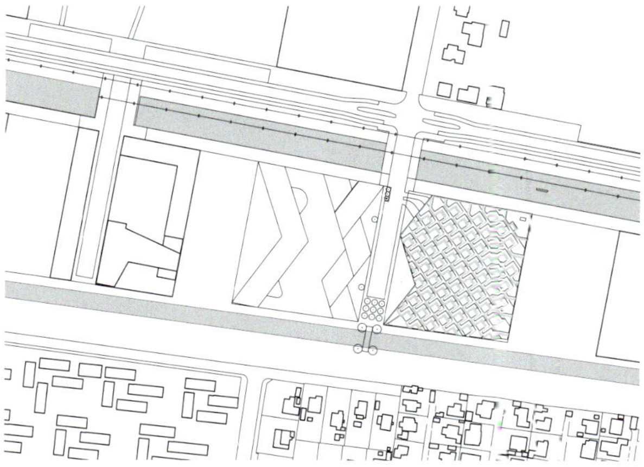

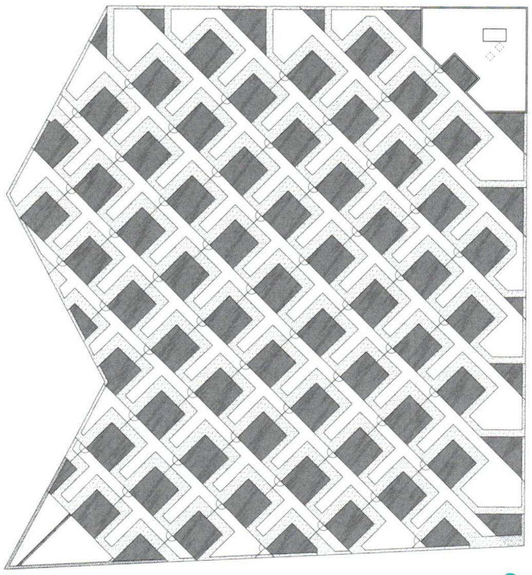

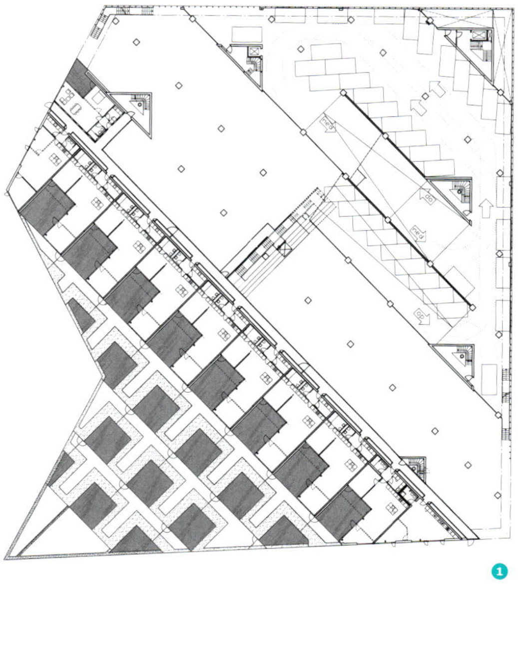

1 4th floor with apartments and parking deck 1:1250
2 Roof 1:1250
3 West elevation 1:1000
4 North elevation 1:1000
5 3-room apt. with terrace and roof garden 1:200
6 3-room apt. adjacent to facade, with terraces and roof garden 1:200
7 Cross section with access stairs and sloping elevator 1:500

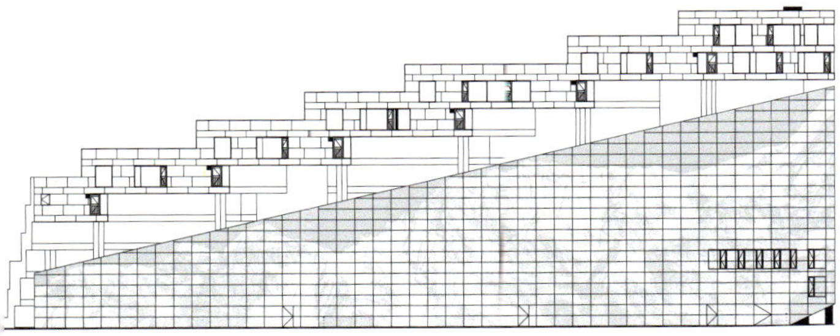

4

2.4
TERRACED COMPLEX

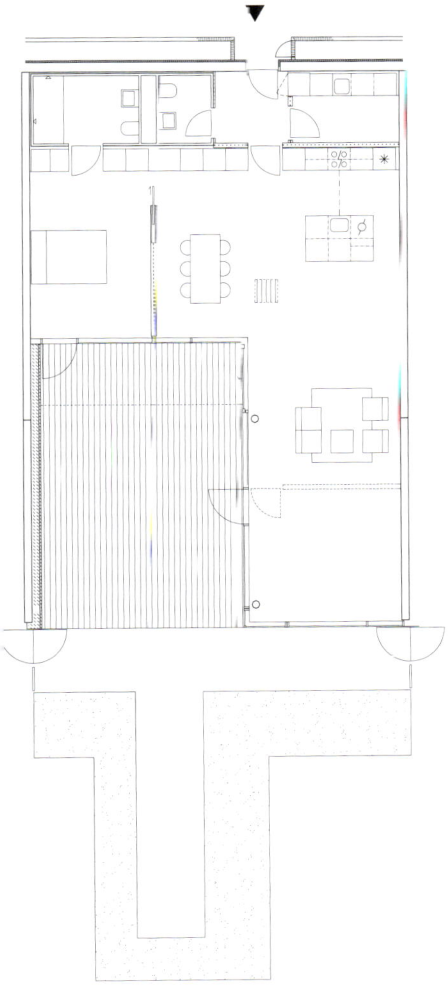

5

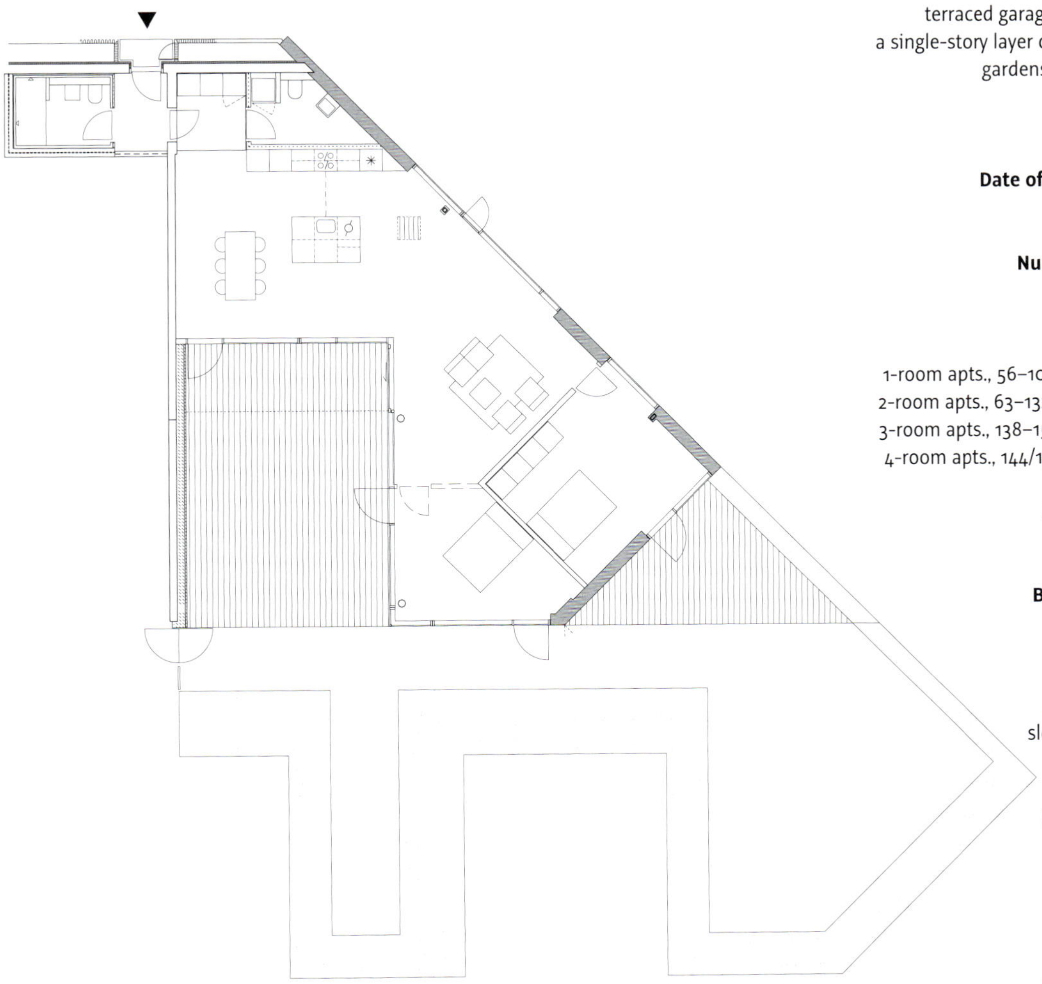

6

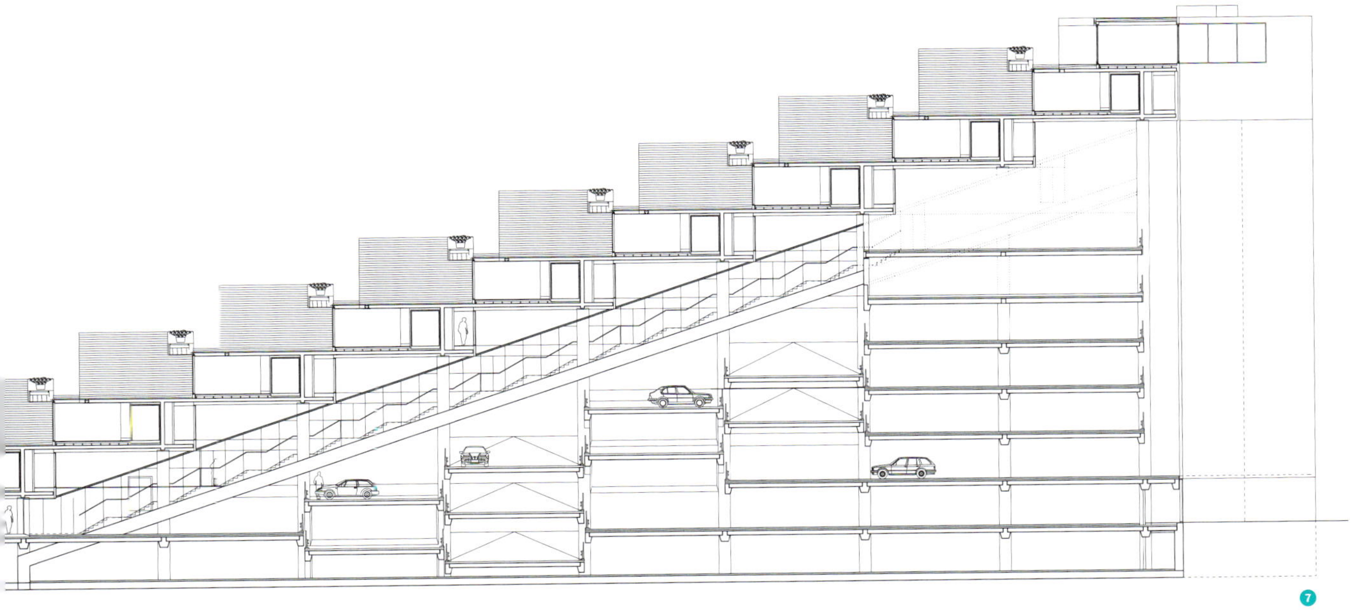

7

Building type
terraced garage as plinth for a single-story layer of apartments, gardens and terraces, 10 stories facing SE

Date of construction
2008

Number of units
80

Size of units
1-room apts., 56–103 m² (11 units)
2-room apts., 63–132 m² (64 units)
3-room apts., 138–150 m² (3 units)
4-room apts., 144/152 m² (2 units)

Area per user
31.5–103 m²

Building depth
10–94 m

Access
staircases and sloping elevator, internal corridors in parking garage

Open spaces
terraces

Parking
parking garage
parking spaces at dwelling level

Architect
BIG Bjarke Ingels Group
Copenhagen
In collaboration with:
JDS, Moe & Brødsgaard
Freddy Madsen, SLA

Location
Copenhagen-Ørestad
Denmark

Referring to floor plan 1:200

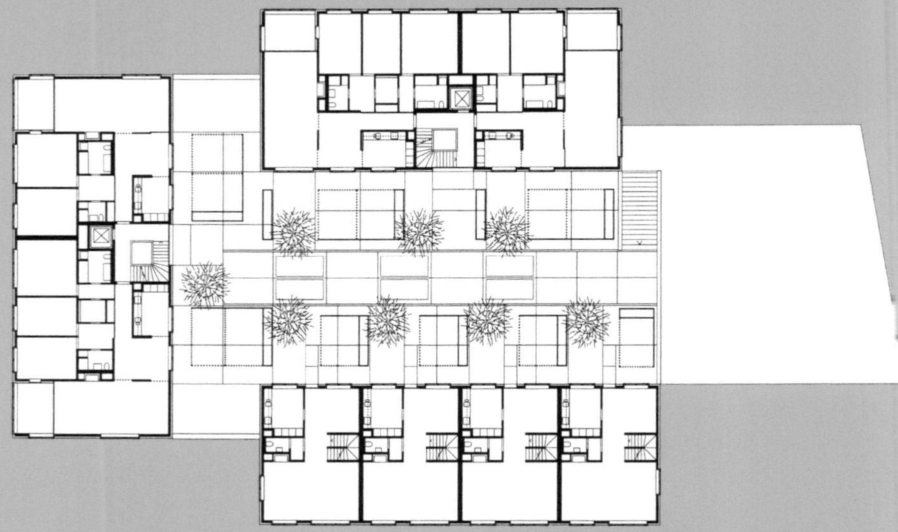

2.5
RESIDENTIAL COMPLEX / HOUSING ESTATE

This category is very heterogeneous: it covers single-family row houses in either linear or courtyard formations, multi-story conglomerates with interlaced and stacked apartment components, the meandering, space-enclosing "sculptures" of large-scale developments from the 1960s and 1970s, inner-city and suburban building ensembles, and even that autonomous urban building block – the housing estate.

Residential complexes and estates are structurally self-contained islands within the urban configuration and sufficient unto themselves. With the exception of large-scale developments, despite their density, they aim to fulfill the desire for a place of one's own: hence, efforts are undertaken to provide each occupant with their own access space, be it at ground level or on one of the upper floors. Similarly, the transition from public to private areas, and from public to private green is usually designed in a highly differentiated and varied manner. Clear boundaries to the surroundings, the often compact and homogeneous construction, common open spaces, and the deliberate restriction to a limited number of floor plan and house typologies – albeit often with subtle differences – contribute to the establishment of a shared identity among all occupants. In extreme cases, estates can become gated communities, which define themselves as communities for members of a uniform social class, usually affluent, and distance themselves from the outside world through security guards and barrier installations. By virtue of their scale, estates and residential complexes can contribute to stabilizing urban agglomerations and transition areas, giving them an identity and stimulus for further densification.

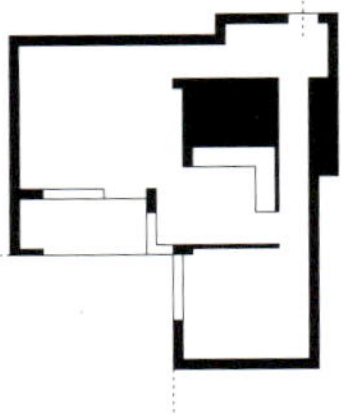

Märkisches Viertel | Fleig | 1966

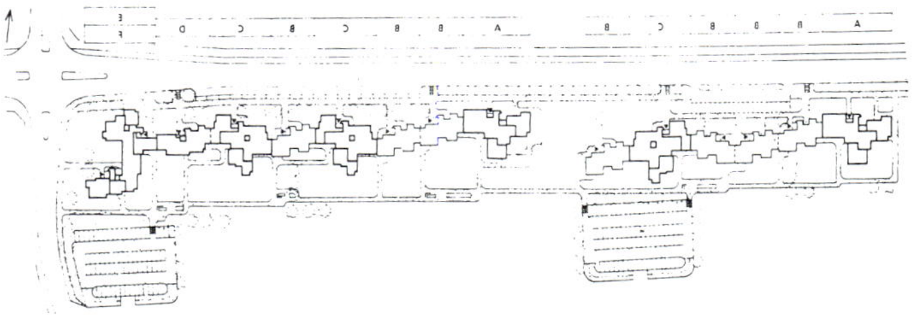

In response to the north-south orientation of the entire development, the architect developed a complex floor plan that allows for south-facing living rooms and balconies in all apartments. Kitchen and dining area are set diagonally to the eastern i.e. western exposure. The floor plans create a clear division between communal and individual areas: the sanitary block consisting of bathroom, WC, and kitchen was placed in the middle of the apartment space and thus creates two paths, one leading directly into the living room, the other to the sleeping area. In the two-room apartments the path becomes a circular path that leads via kitchen and dining area back into the living room; in the larger apartments the sleeping area can be separated off. All dining areas (but also other rooms) have corner windows.

Building type
meandering block,
4–8 stories, facing N/S, E/W

Date of construction
1965–1966

Number of units
283

Size of units
2-room apts., 79–85 m²
2½-room apts., 97 m²
2⅔-room apts., 110 m²
(type C)

Area per user
32.5–42.5 m²

Building depth
max. 32 m

Access
double- to
quintuple-loaded stairwells

Open spaces
loggias
greenspace to south

Parking
open parking spaces

Architect
Karl Fleig, Zurich

Location
Wilhelmsruher Damm 187–215
Märkisches Viertel
Berlin-Wittenau
Germany

1

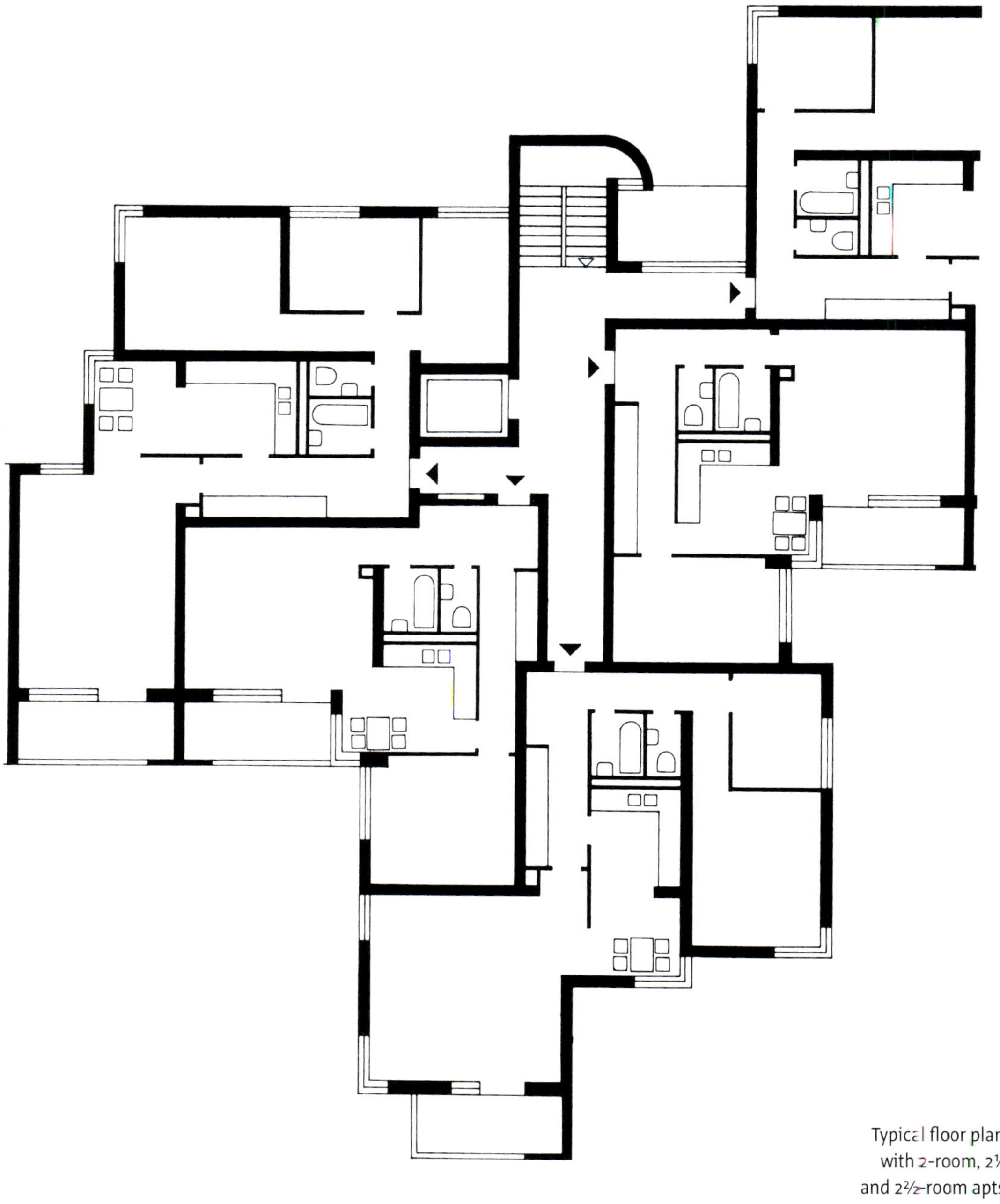

Typical floor plan, type C 1
with 2-room, 2½-room,
and 2⅔-room apts. 1:200

Referring to
floor plan 1:200

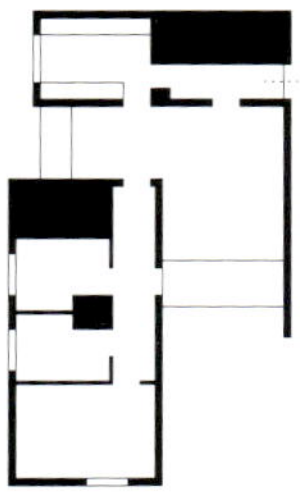

Märkisches Viertel | Ungers | 1969

The complex stacks identical spatial elements on top of each other. The stairwell is surrounded by autonomous units comprising hallways, guest WCs, kitchens, and storage rooms. The bedrooms and bathrooms form sleeping tracts of various lengths (depending on apartment size). From the outside they appear as bedroom towers. The positions of the kitchen towers and sleeping towers result in L-shaped living rooms lit from two sides and two loggias. The kitchen opens into the living room but can also be reached directly from the entry hallway. The floor plan is distinguishable from the outside. The last two floors are occupied by maisonettes composed of two superimposed sleeping tracts; the living-room space becomes a rooftop terrace.

Building type
meandering high-rise ensemble,
12–14 stories,
facing mainly E/W/S

Date of construction
1967–1969

Number of units
1305

Size of units
1-room apts., 49 m²
2-room apts., 73 m²
2½-room apts., 92–94 m²
on the top floors
maisonettes with rooftop terraces

Area per user
23.5–49 m²

Building depth
23–32 m

Access
quintuple-loaded stairwells
2 elevators

Open spaces
loggias
roof terraces
housing complex
forms green courtyards

Parking
garage

Architekt
Oswald Mathias Ungers
Berlin

Location
Wilhelmsruher Damm 165–185
Märkisches Viertel
Berlin-Wittenau
Germany

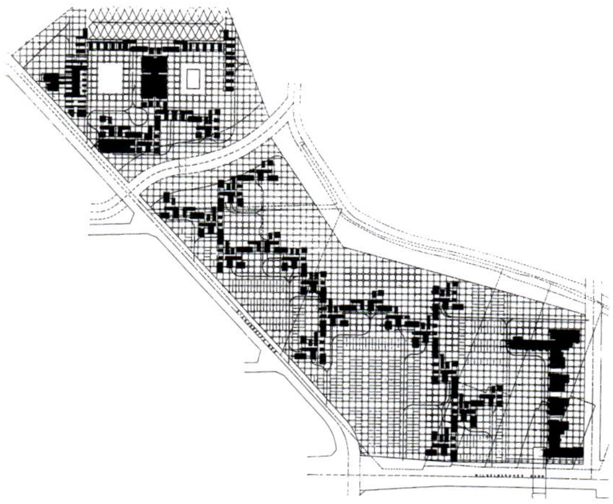

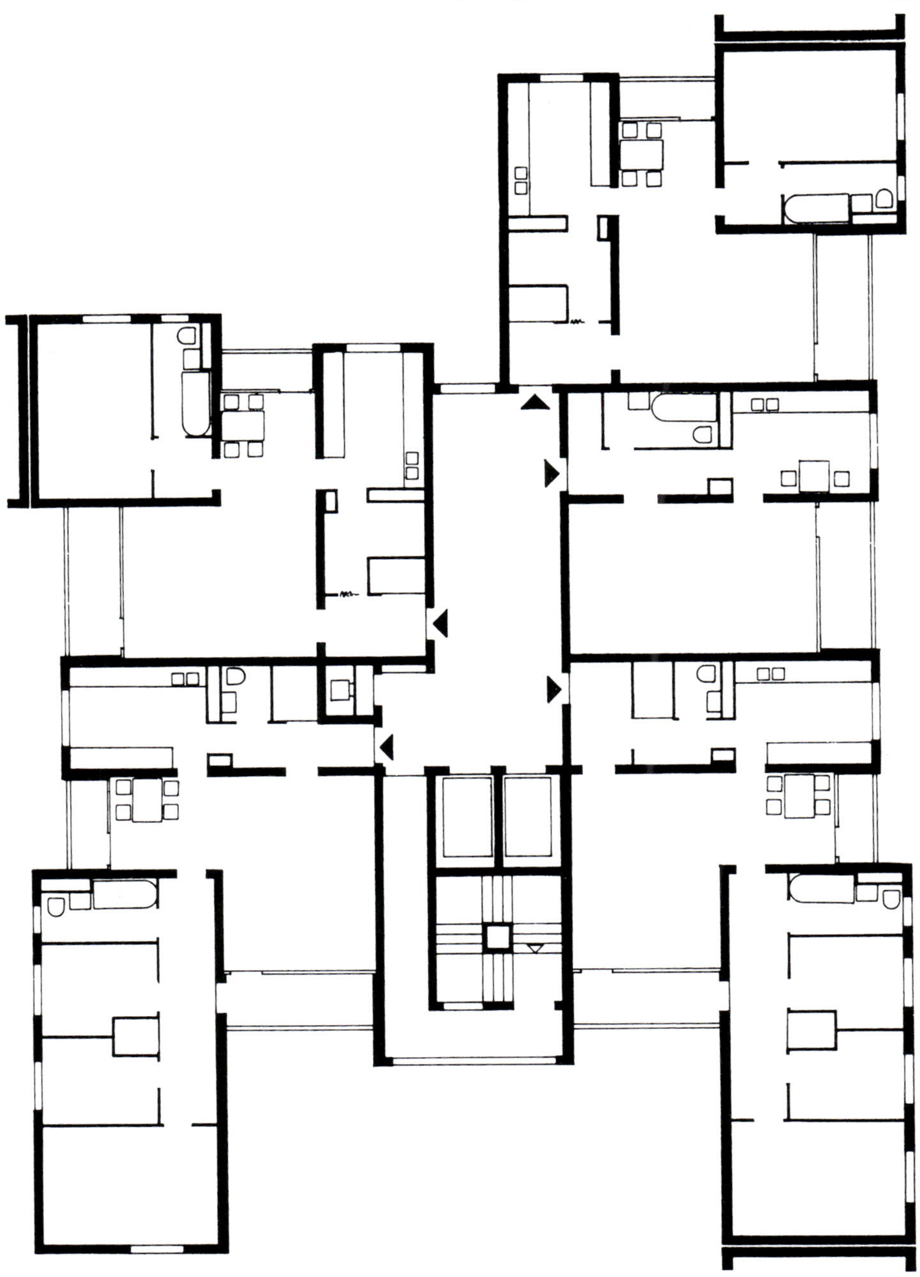

Typical floor plan, 1
with two 4-room,
two 2½-room,
one 1-room apt. 1:200

Referring to floor plan 1:200

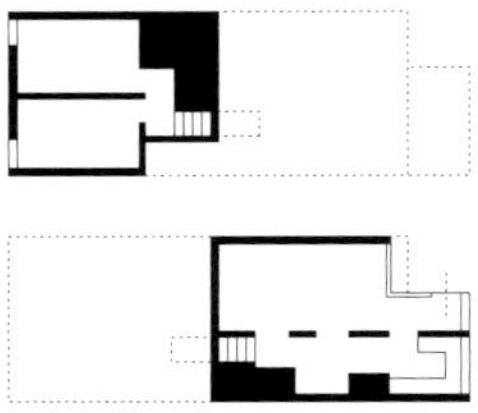

Marquess Road | Darbourne and Darke | 1977

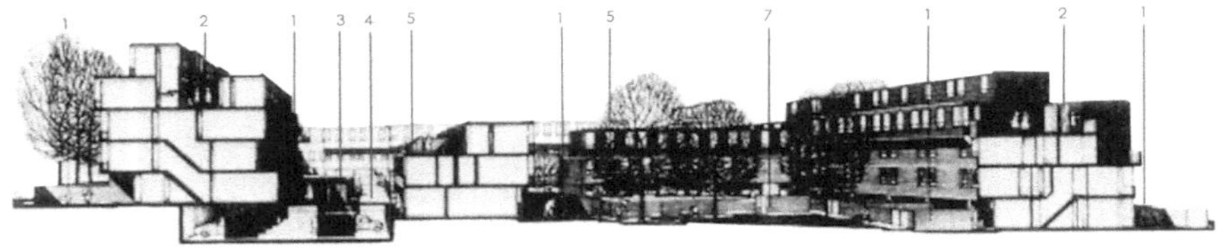

An introverted dense housing complex with row houses staggered in height. The larger apartments are accessed via courtyards, the smaller ones from the upper access path that runs through the center of the development. This path across the roofs serves as an open-air recreational area for residents. Transverse housing blocks divide these areas into easily identifiable sections. Between the house rows are residential paths and courtyards; the taller buildings provide noise protection along the street elevation. The apartments are mainly vertical in organization; the lower maisonettes have two to three floors following the "back-to-back crossover" scheme, i.e. the apartments are organized back to back, but the compass points are switched on the following floor so that each apartment and its living area profit from the more favourable direction. Access to the maisonettes is provided through the private green spaces in front; kitchen/dining, living, and sleeping are accommodated on separate levels. The third floor contains the pedestrian deck (footpath), to which the two-story maisonettes and/or flats are attached. Individual rooms can be assigned to various apartments; this provides additional flexibility in apartment sizes.

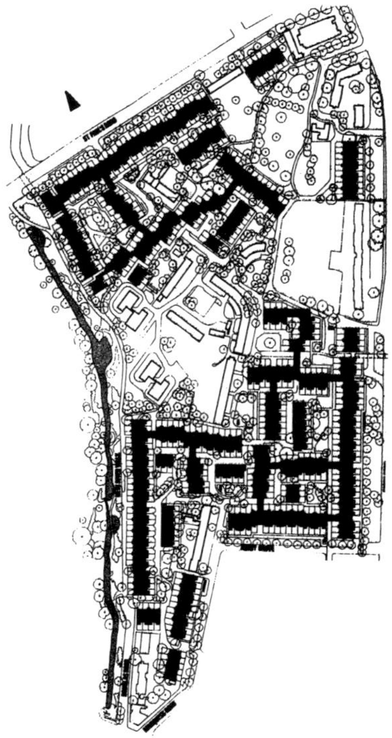

1 Basement 1:200
2 Ground floor 1:200
3 2nd floor 1:200
4 3rd floor 1:200
5 4th floor 1:200

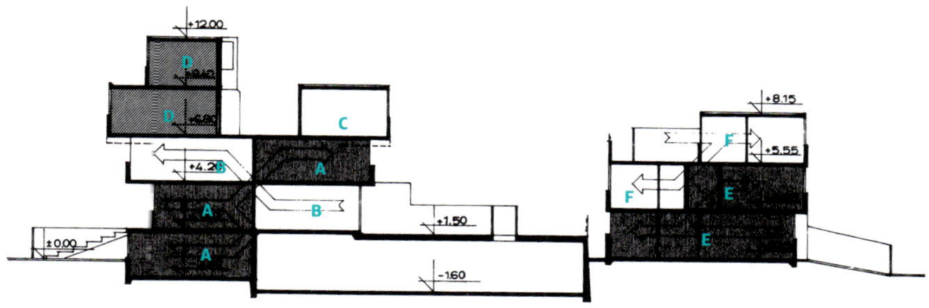

2.5
RESIDENTIAL COMPLEX / HOUSING ESTATE

Building type
row house development
2/3/4/5 stories
N/S, E/W, NW/SE, NE/SW

Date of construction
1968–1977

Number of units
993

Size of units
2-room apts., 49/58 m²
3-room apts., 85/90 m²
4-room apts., 95 m²
(60 % family and
40 % 1-/2-person apts.)

Area per user
24–29 m²

Building depth
11/13/15 m
(dim. betw. axis: 5.18 m)

Layout
row houses
(4/5 stories toward street,
2/3 stories inside),
bottom apartments
with ground-level access,
upper small apartments
accessible via pedestrian decks

Open spaces
private gardens for family
apartments (27 m²)
terraces
decks
public greenspace

Parking
garages in basement story

Architect
Darbourne and Darke
London

Location
Marquess Road
London-Islington
United Kingdom

5-story building:
- **A** 3-room maisonette basement: sleeping area
  ground floor: entrance, kitchen, dining area 1:200
  2nd floor: living area (crossed over) 1:200
- **B** 3-room maisonette ground floor: living area
  2nd floor: sleeping area (crossed over) 1:200
- **C** 2-room apartment along roof path 1:200
- **D** 2-room maisonette 1:200

Opposite building:
- **E** 4-room maisonette ground floor: living area
  2nd floor: sleeping area 1:200
- **F** 2-room maisonette 2nd floor: living area and entrance
  2nd floor, sleeping area 1:200

Referring to
floor plan 1:200

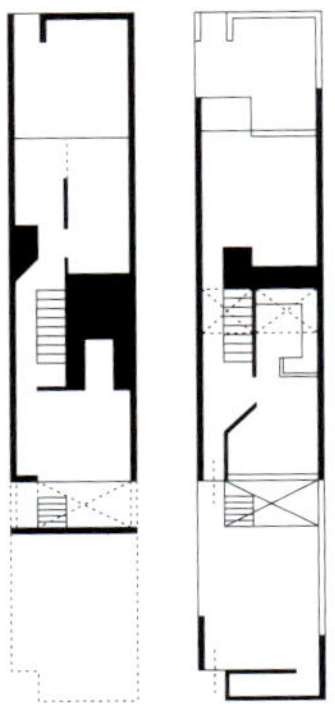

Maiden Lane | Benson, Forsyth | 1982

Despite the great urban density, most of the units in the Maiden Lane Estate feature individual entrances located at ground level or along the raised access roads, sheltered private open spaces and fairly generous, always deep floor plans lit from two sides, although the grid is only 4 m wide. The architects followed the model of the Halen development designed by the architects from Atelier 5 (see page 298). Similar to that model, the terraced site is utilized and manipulated to create different access levels for the pedestrian routes, to shift parking traffic onto another level and to endow the housing units with a high degree of privacy. To this end, similar row-house types, flats, and stacked maisonettes are linked in a specifically urban manner. The narrow row-house units (building **B**) are accessed via a courtyard-like and sheltered front garden. The bedrooms are located on this ground level and both are directly linked to a private open space of their own: the bedroom at the rear overlooks a patio, which is connected to the very large terrace on the upper floor via a steep set of stairs. Upstairs, the internal stairs and the kitchen/dining area face towards this open space, while the living room overlooks the entrance courtyard. The 2- to 3-story maisonettes adhere to the basic layout of these apartment units. The 2-room flats complete the buildings; they have balconies on both sides and an L-shaped kitchen, inviting front-to-back living despite the narrow hallway.

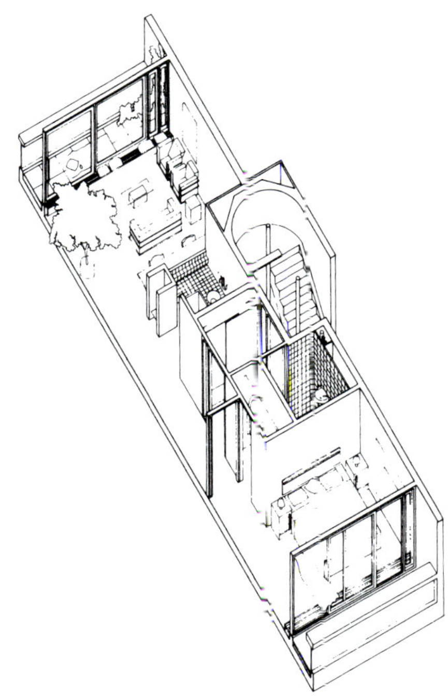

❶

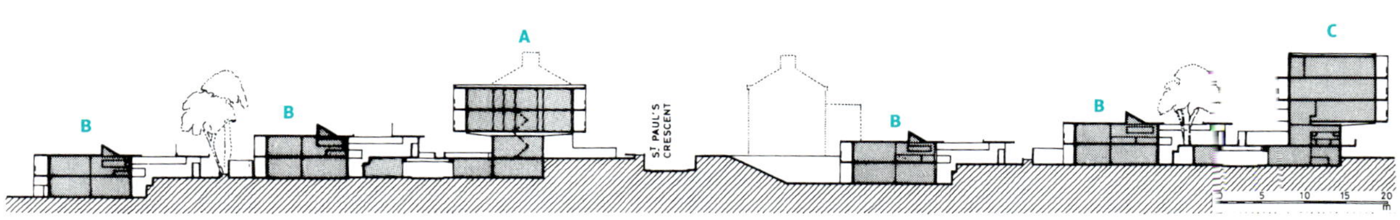

❶ Building **A**: 2-room apartment with stair access 1:200
❷ Building **B**: maisonette (5 persons) 1:200
❸ Building **C**: maisonette (4 persons) 1:200

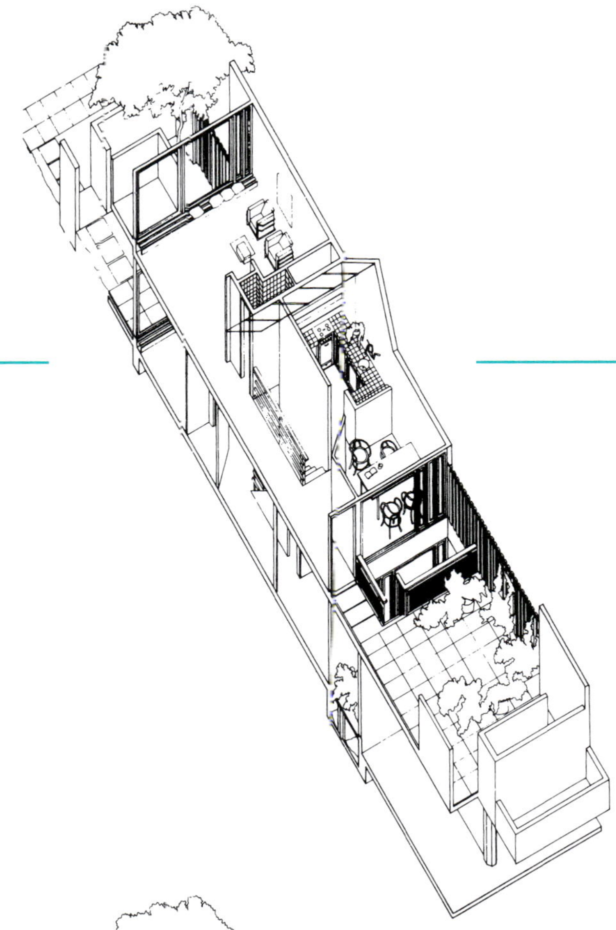

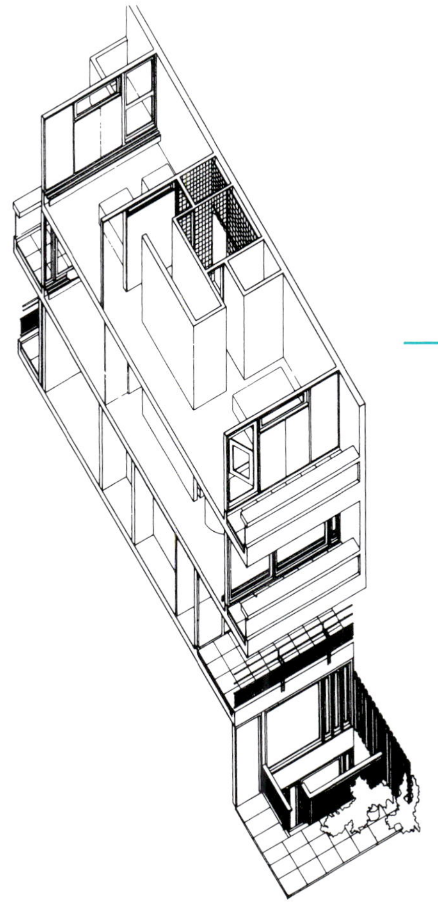

2.5
RESIDENTIAL COMPLEX / HOUSING ESTATE

Building type
row houses
residential slabs
2/4/5 stories
facing W/E

Date of construction
1976–1982

Number of units
500

Size of units
9 different types
of 2-room apts.
and 4-/5-room maisonettes
with private garden (50% of each)

Area per user
nondeterminable

Building depth
approx. 12/22 m

Layout
row houses on
artificially terraced ground,
pedestrian access on 3 levels,
communal facilities

Open spaces
private gardens, terraces,
garden courtyards,
park in the intermediate
zones and in the south

Parking
subterranean garage with
lightwells under pedestrian paths

Architect
Gordon Benson
Alan Forsyth
with F. Smith and
D. Usiskin, Q. Champion,
J. McCain
London
(London Borough of Camden)

Location
Maiden Lane
London
United Kingdom

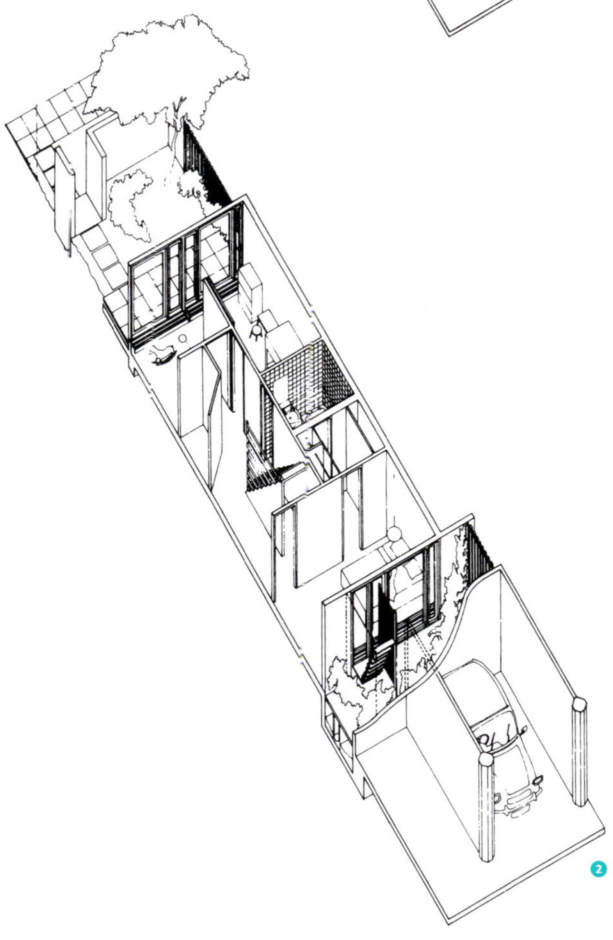

2

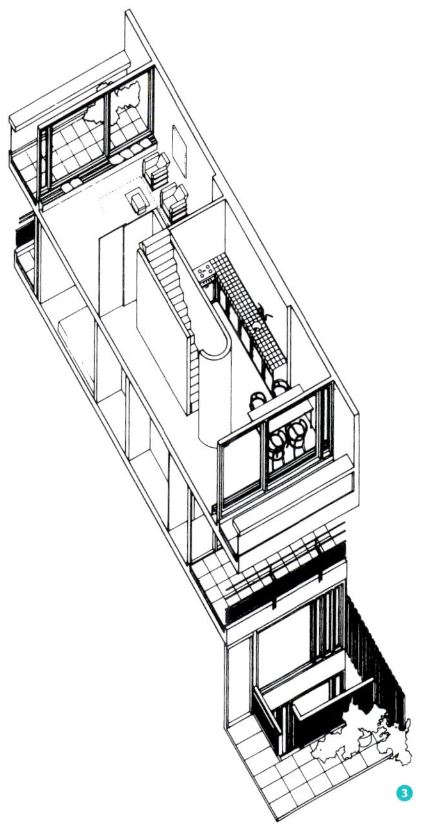

3

Referring to
floor plan 1:200

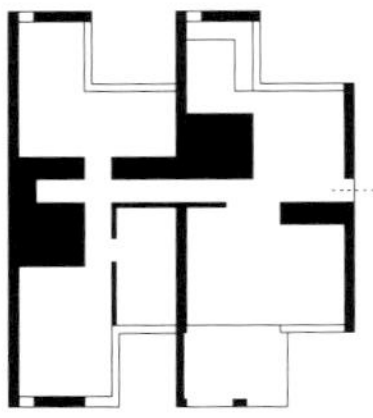

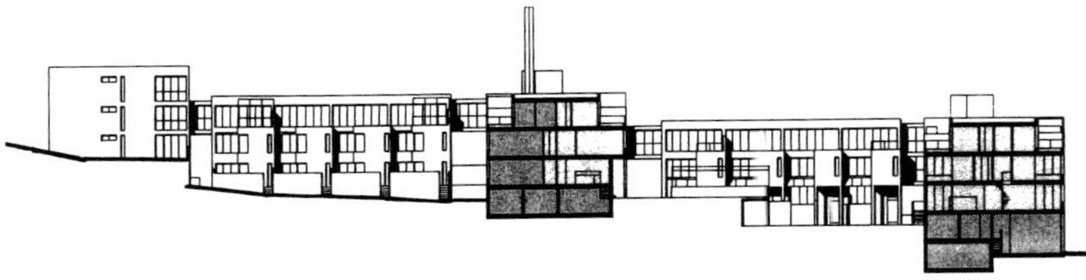

Ried 2 |
Atelier 5 | 1990

Apartments around two courtyards. Four row houses (three stories in N and W, four stories in E and S) make up one side of a courtyard; the square corner buildings, as joints, have five stories. Inside the courtyards: circulation, businesses, public outside space, communal institutions (laundry, club room). In outside ring, private greenery. The row houses consist of two floor plan types developing vertically or horizontally; a four-room maisonette over two or three levels (inside a bay) and a four-room apartment on the roof (over two bays). The corner buildings are double- or triple-loaded with one-, two-, or three-room apartments on one or two levels. All building types vary the same floor plan idea. Row houses: height differences between court level and entry level separate the public space (courtyard) from semipublic (entry platform, terrace) and private spaces. The stairway can be semipublic or private, inside or outside. The apartments are always accessed via the kitchen/dining area separated from the living area by the WC, cabinet, and inside stairway. From this level, access to garden (via stairway) or terrace. Above are three to four bedrooms, spatially separated by bath, walk-in closet, and stairway. Extra studios in the basement stories are assigned to some apartments. Three-room apartments on the roof: they are accessed from the corner buildings and distribute living and sleeping areas horizontally to one bay each. The corner houses mix all types.

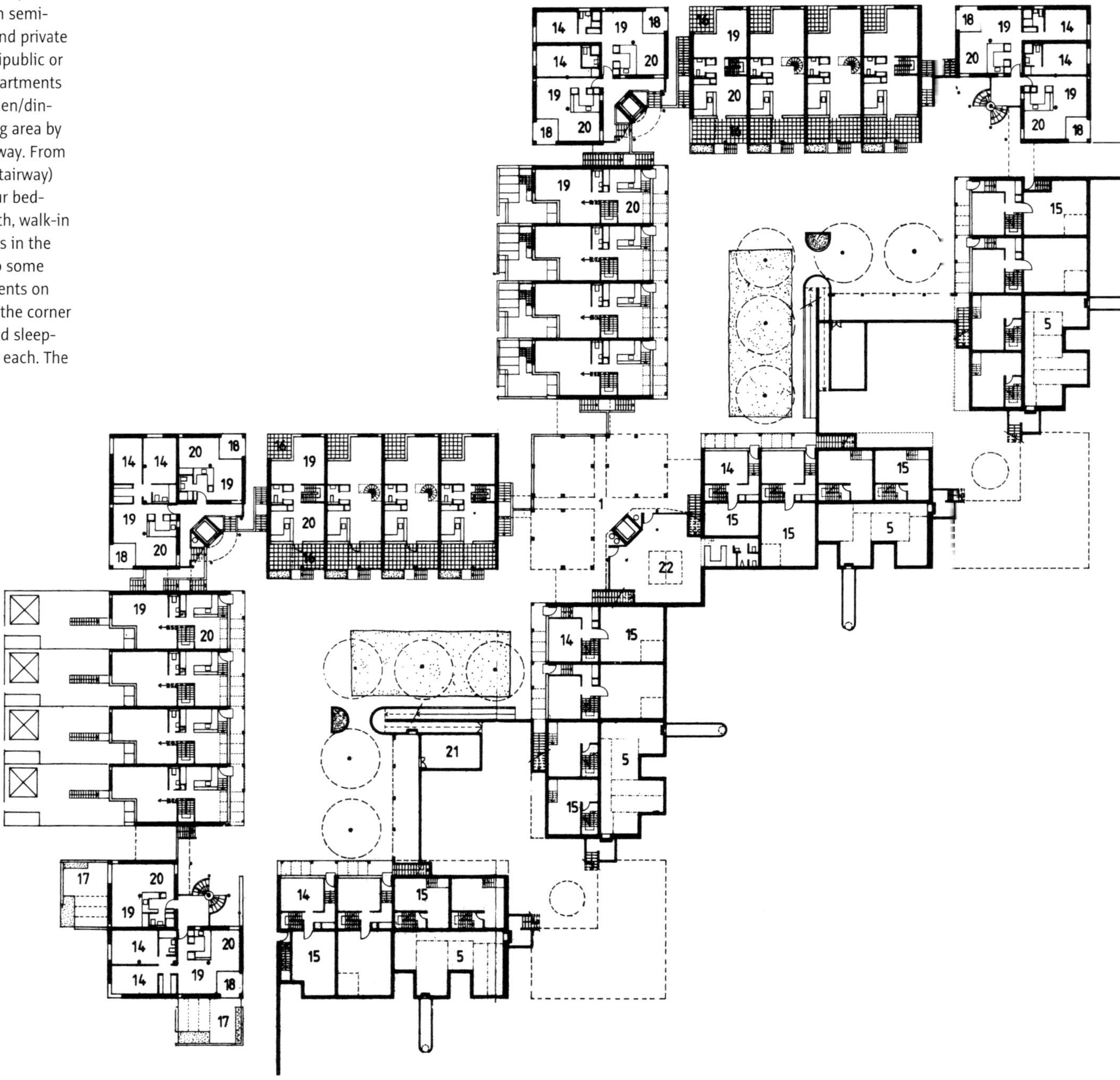

1 4½-room maisonette, levels 1, 2, 1:200
2 1½-room apt. 1:200
3 4½-room apt. 1:200
4 2½-room maisonette levels 1, 2, 1:200
5 3½-room apt. 1:200
6 3½-room apt. 1:200

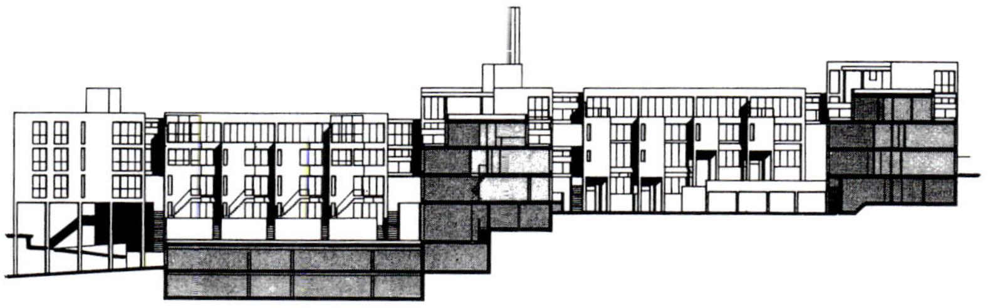

2.5
RESIDENTIAL COMPLEX / HOUSING ESTATE

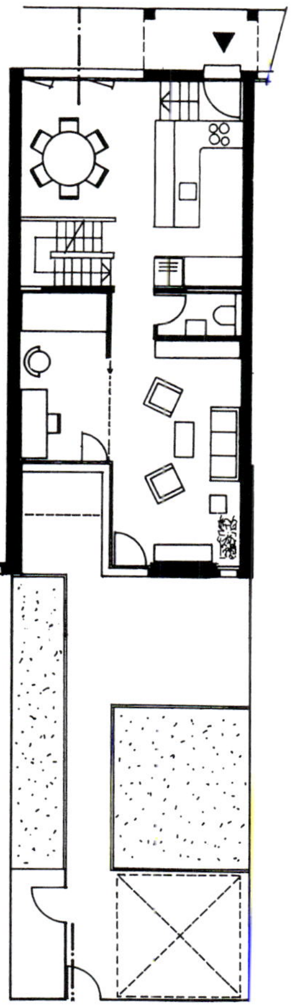

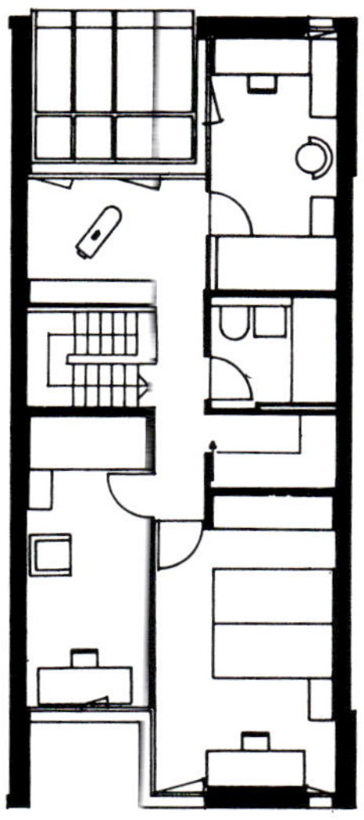

1

1

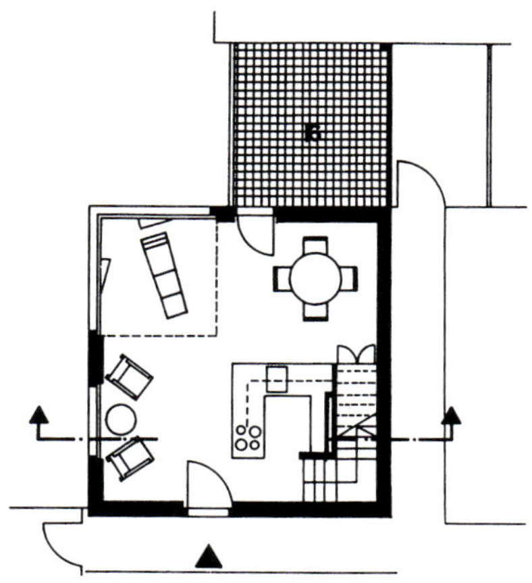

4

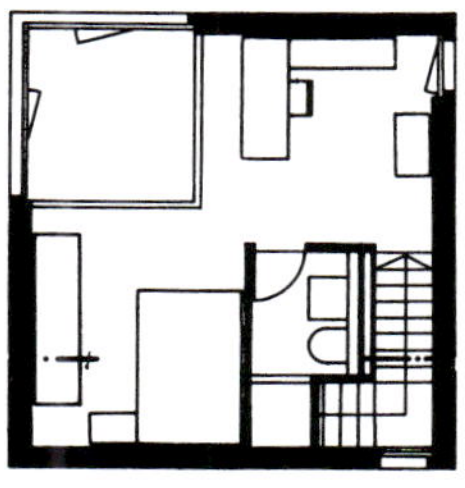

4

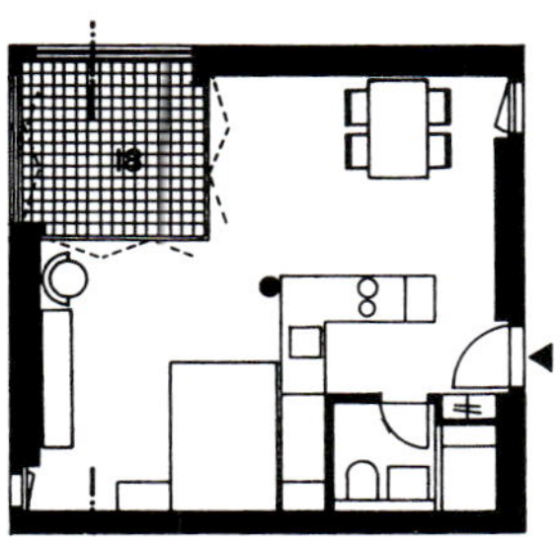

2

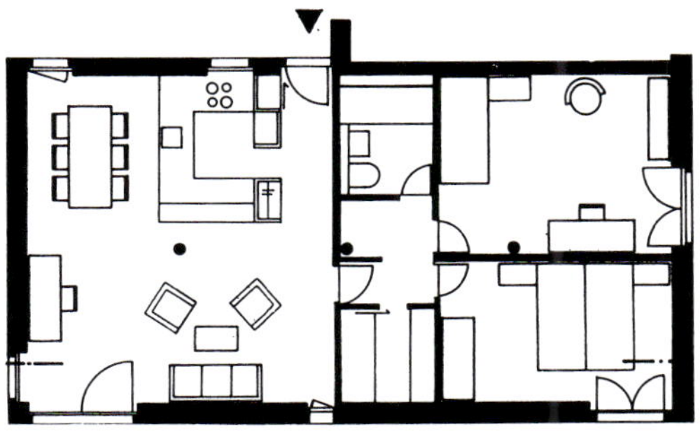

5

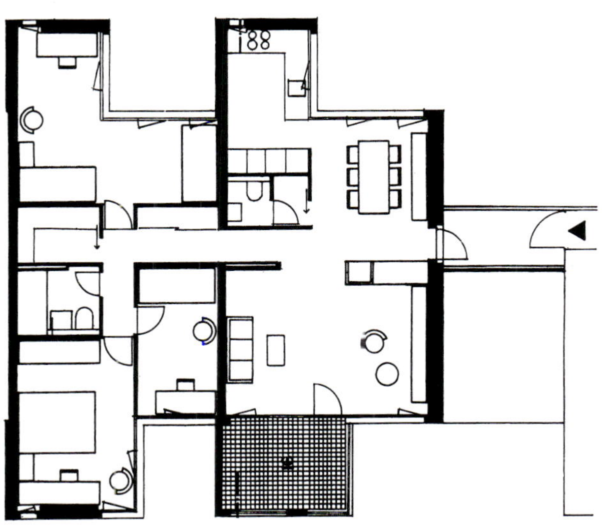

3

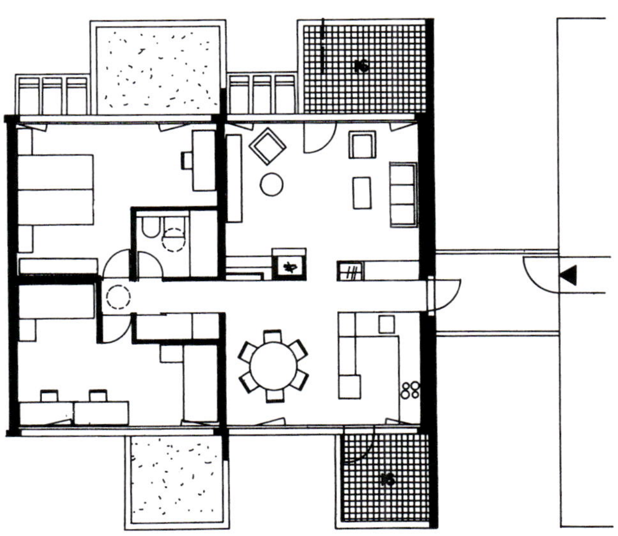

6

Building type
row houses, 3/4 stories
corner houses, 5 stories
facing N/S, E/W

Date of construction
1990

Number of units
93, 11 studios/
commercial spaces

Size of units
1½-room apts., 30/36 m²
2½-room apts., 50/54 m²
3½-room apts., 75/83 m²
4½-room apts., 100/104 m²
41½- and 5½-room mais.,
104/130 m²

Area per user
21–36 m²

Building depth
14 m

Layout
row houses and corner buildings
around two square courtyards

Open spaces
public outside space
with pergolas
private terraces and gardens

Parking
open parking spaces
subterranean garage

Architect
Atelier 5
Bern

Location
Brüggbühlstrasse
Niederwangen
Switzerland

Referring to
floor plan 1:200

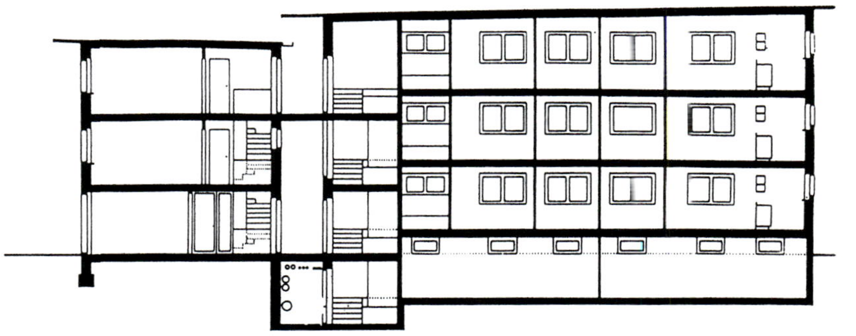

Vogelbach | Alder | 1992

The development forms four blocks, two on each side of a main path. The floor plans are extremely simple and clear. The southern wings contain mainly three-story maisonettes placed in a row like single-family houses. Kitchens, baths, stairwells, expanded corridor zones face north; all rooms open to the south. Small greenspaces lie behind a gravel path. The function of the living room on the ground floor is fixed by its proximity to the kitchen and its size; all other rooms are equal. On the third floor a gallery runs in front of the apartments on the north side. Above the two-story maisonettes, there is either a two-room apartment following the same floor plan scheme, or the building axis is cut up into two one-room studios that can be optionally connected. The apartments in the wings have east-west orientation; on the east or west sides (the floor plans are mirrored along the main axis) lie corridors with belt windows behind a zone with sanitary areas. The corridor, wide enough to work in, makes the rooms appear like display compartments. It leads into the living room with inserted open kitchen. A large terrace is located, orthogonal to the building axis, in front of the living room on each floor; this three-story terrace row closes the block to the north.

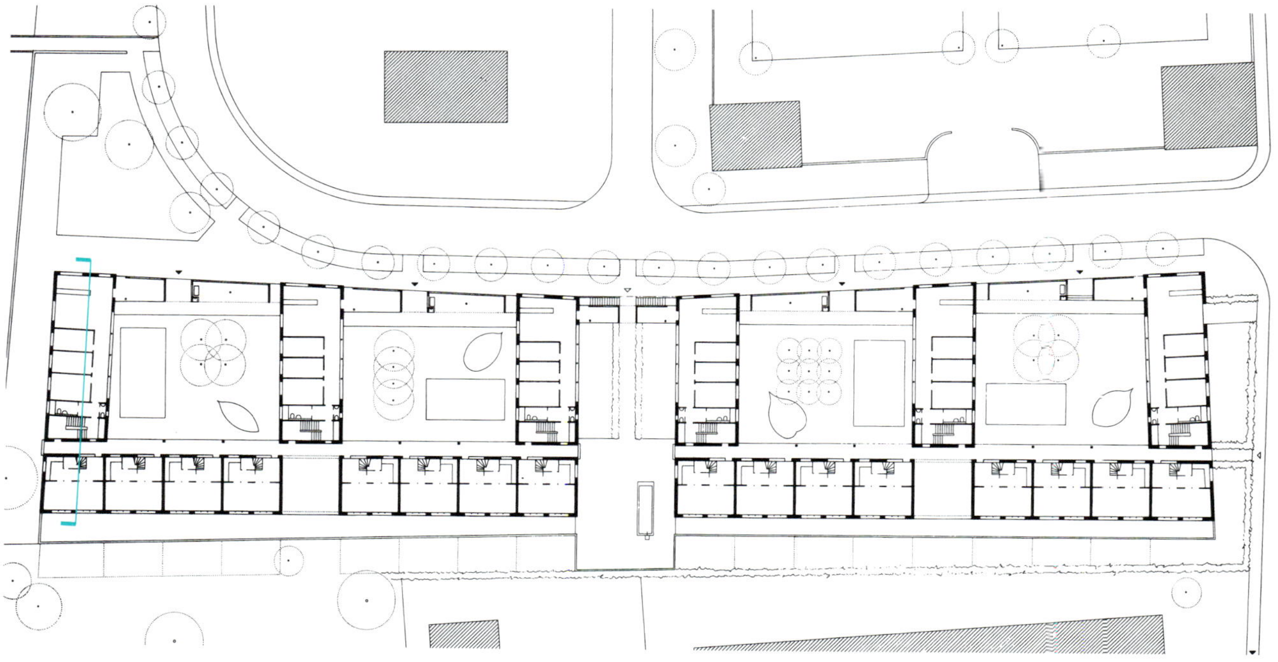

1

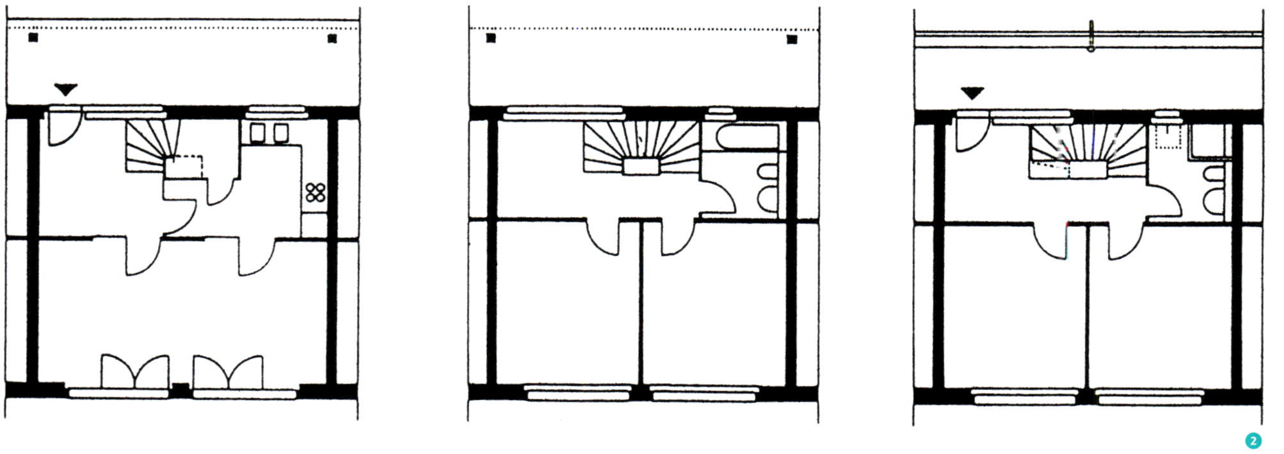

2

1 Layout of the building blocks
2 3-story maisonette 1:200
3 1-room studio (connectable room) 1:200
4 4-room apartment and 2-room apartment 1:200

2.5 RESIDENTIAL COMPLEX / HOUSING ESTATE

Building type
residential courtyards
3 stories
N/S, E/W

Date of construction
1991–1992
(competition 1988)

Number of units
38

Size of units
4½-room apts., 105–108 m² (18 units)
5½-room mais., 121 m² (14 units)
3½-room mais., 81 m² (2 units)
1-room studios, 22 m² connectable to apts. (4 units)

Area per user
22–27 m²

Building depth
8 m

Layout
block layout
mirrored along axis
2-/3-story mais. in south row,
apartments in cross wings

Open spaces
greenspaces inside courtyards,
private greenery
before maisonettes,
terraces for cross wings

Parking
underground parking in
adjacent commercial building

Architect
Michael Alder
Basel
partner:
Roland Naegelin

Location
Riehen
Basel
Switzerland

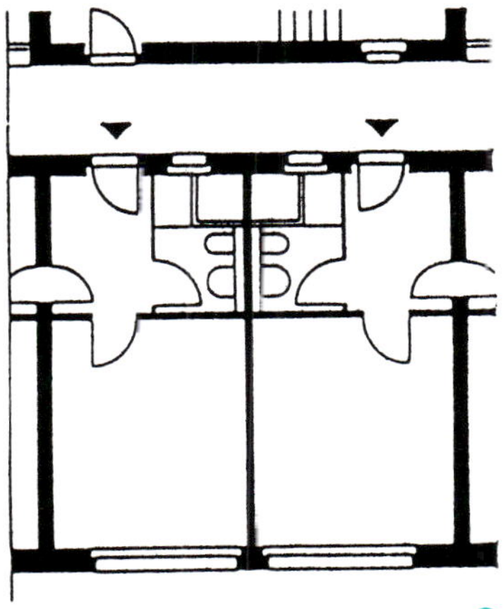
3

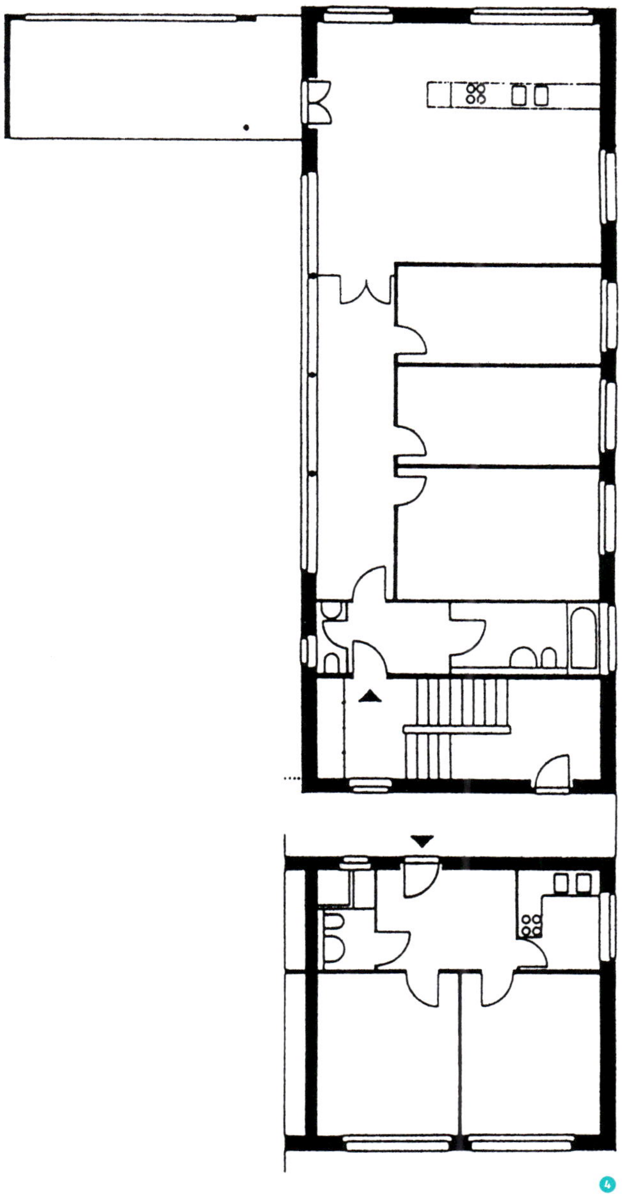
4

Referring to floor plan 1:200

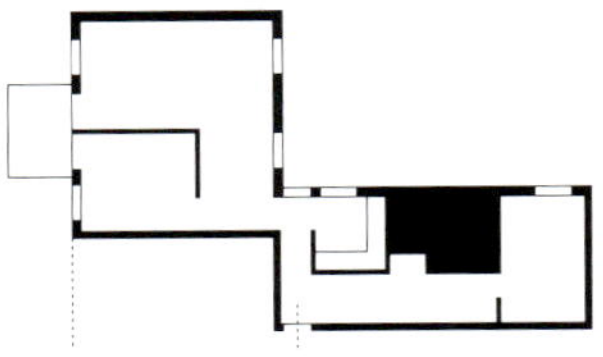

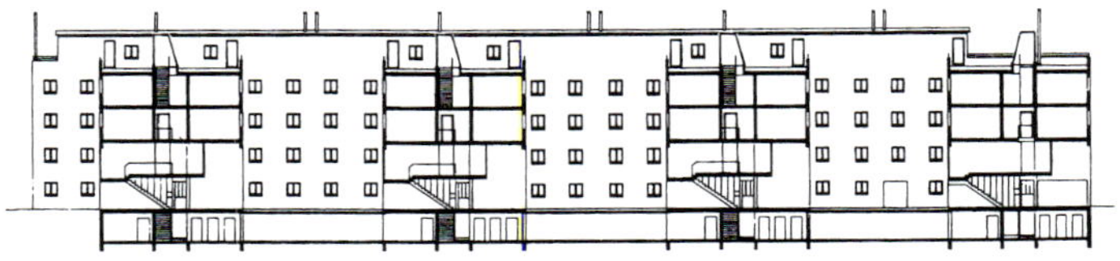

Wienerberggründe | Steidle + Partner | 1993

The two housing projects "Wien-Süd" and "Wien-Privat" were constructed following an urban plan by Otto Häuselmayer. Together they densify the double-row building type by weaving in perpendicular bars, creating courtyards within. "Wien-Süd": the living rooms and bedrooms are located in the main structures, while the cross rows hold stairwells and the apartments' ancillary functions. Passageways connect the courtyards through openings in the rows. At these openings are the long stairwells in which the stairs take a new direction on every floor. The apartments are oriented towards two sides: outwards to the garden and inwards to the courtyard. Most interesting are the corner units, which bend around and thus permit a visual relationship between one end of the apartment and the other. The entrance leads directly to the "elbow" (which has French windows) where living and bedrooms are to one side, while kitchen, bath, toilet, and sometimes a second bedroom lie along a broad corridor to the other side. On the 1st/2nd and 3rd/4th floor, maisonettes are tucked in between corner apartments. On the 5th/6th floor, the corner apartment becomes itself a maisonette. Where the living room was, one finds the stairs and three bedrooms. A bath, toilet, and large bedroom run along the corridor. Upstairs a long space is divided by the stairs, creating a kitchen/dining area and a living room.

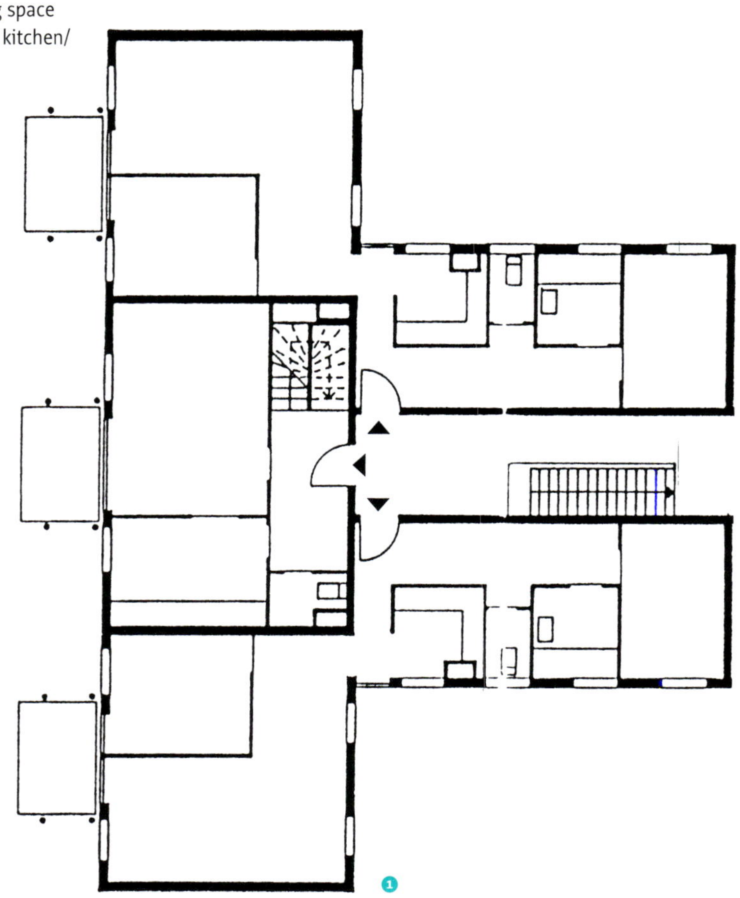

1

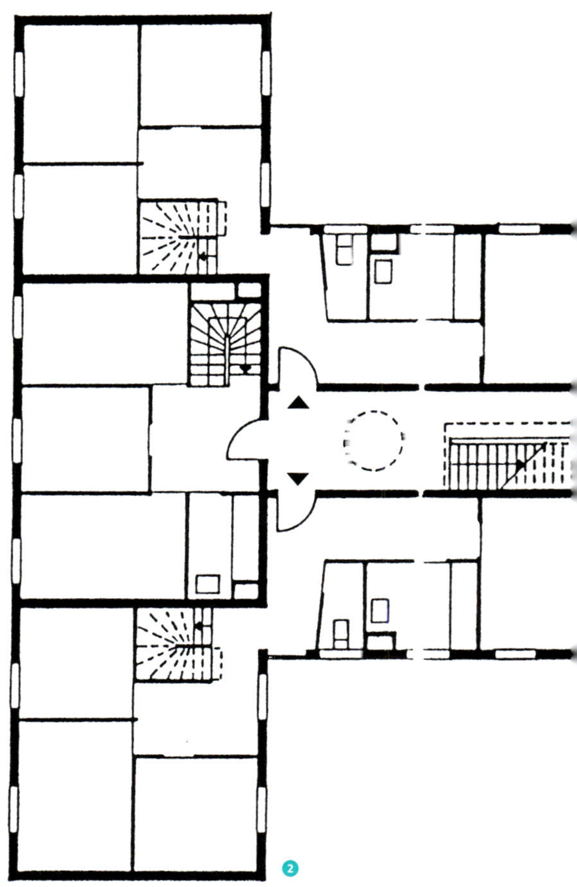

2

1 3rd floor: 3-room apt., 4-room mais. 1:200
2 4th floor: 5-room mais., 4-room mais. (upper part) 1:200
3 5th floor: 5-room mais. (upper part) 1:200
4 From left to right: Ground to 5th floor 1:500

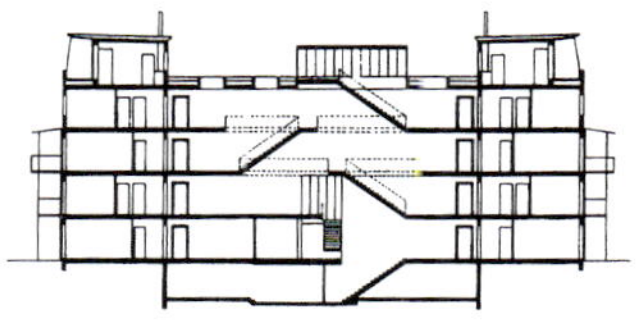

2.5 RESIDENTIAL COMPLEX / HOUSING ESTATE

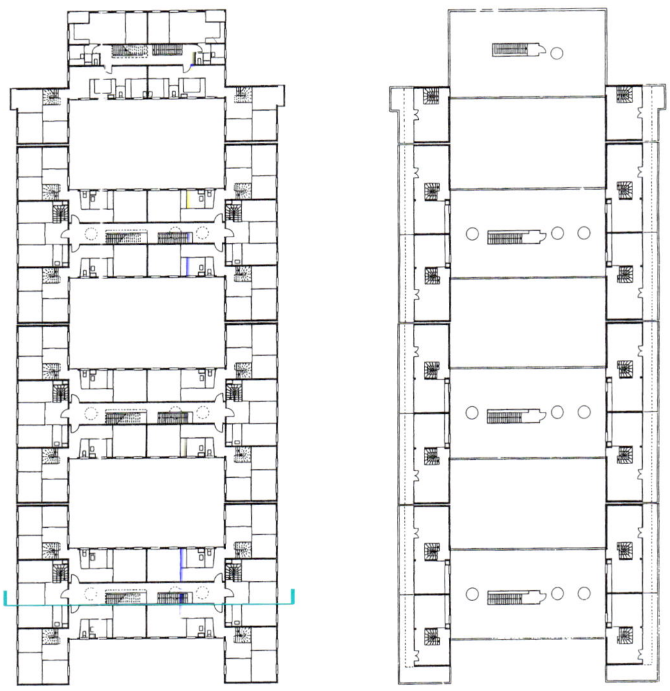

4

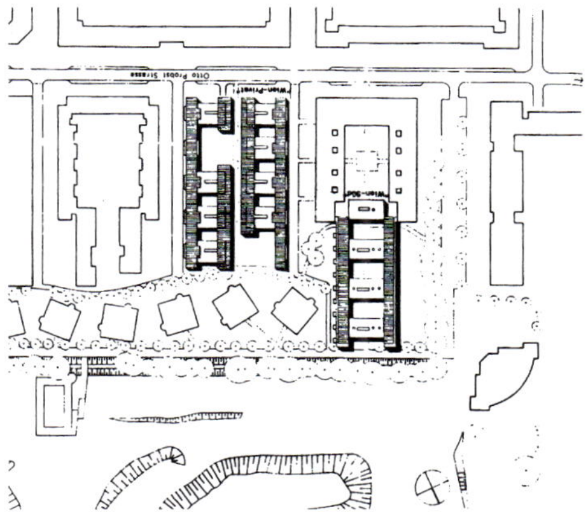

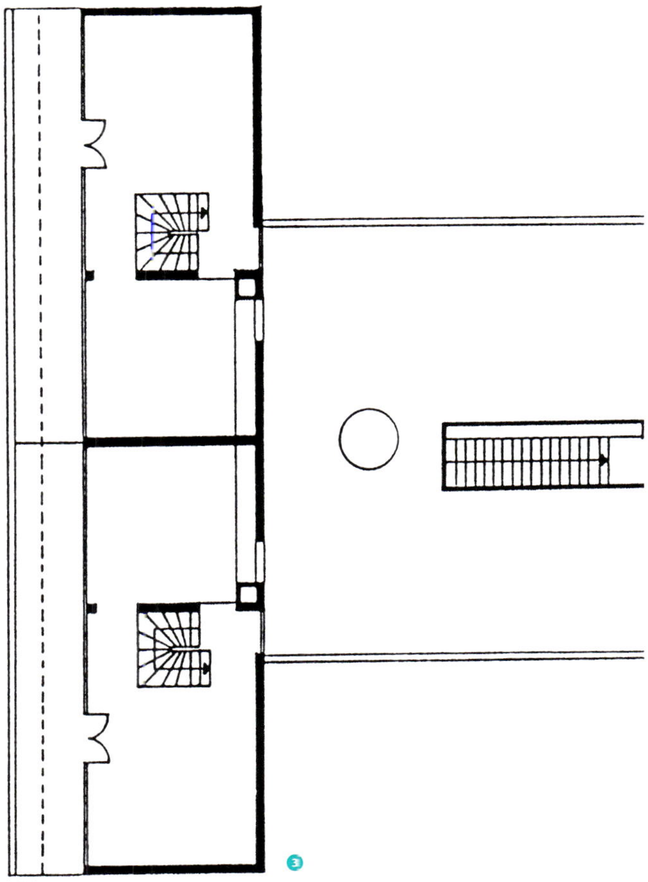

3

Building type
double block with courtyards
within cross bars
5 stories
facing SW/NE

Date of construction
1986–1993

Number of units
74

Size of units
2-room apts., 45/50/55 m² (8 units)
3-room apts., 82–87 m² (40 units)
4-room mais., 105 m² (12 units)
5-room mais., 135 m² (14 units)

Area per user
22.5–29 m²

Building depth
7/11.4 m

Layout
6-loaded stairwell

Open spaces
balconies
roof terraces

Parking
subterranean garage

Architect
Steidle + Partner
with Christian Kara
Celina Kress
Hans Kohl
Munich

Location
Otto-Probst-Straße
Vienna-Inzersdorf
Austria

Referring to
floor plan 1:200

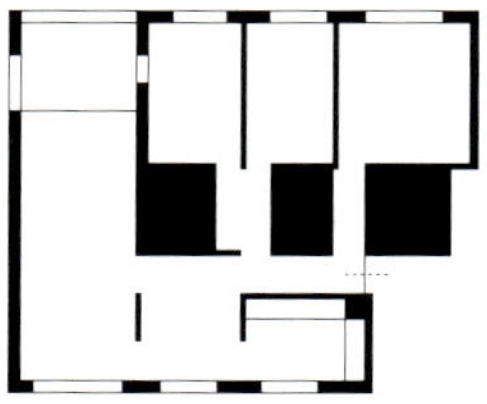

Kilchberg | Gigon/Guyer | 1996

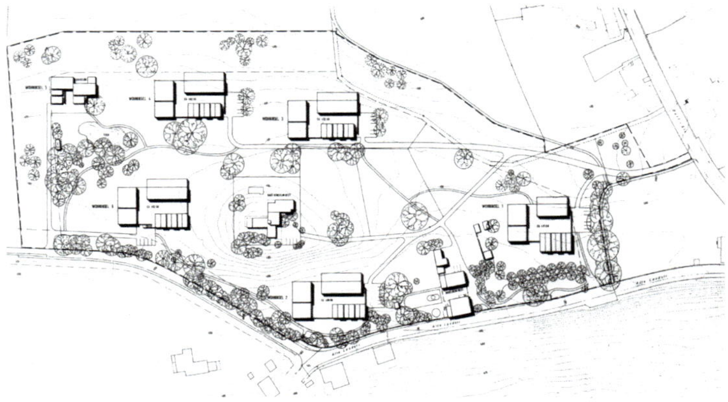

The first housing island, called "Kilchberg," is located in the park of a former estate. The architects were able to change the original plan for developing the entire open space with single-family homes into targeted densification with a total of six housing islands. The stunning vista of Lake Zurich and the Alps combined with proximity to the city suggested that this would be an ideal site for upscale apartments. The Housing Island I is introverted in nature. Three simple buildings are grouped around an interior courtyard. Two of the volumes contain flats, one is composed of row houses. The plinth story that connects the building houses contains parking places with direct access to the buildings. Above this level: the artificial courtyard with building entrances. In addition to paving and trees, the courtyard space is above all divided by gazebos constructed from boarding, which provide a semiprivate outside seating area for each house. The generous floor plans are oriented towards three directions. The rooms are arranged around a core with two bathrooms, which acts as a shield in front of the bedrooms. Kitchen and dining area overlook the courtyard, while private rooms and living room fronted by a winter garden benefit from a view of the park. The room proportions in combination with the sliding doors give the floor plan a strong character of flexibility. Across the facade, the floor-to-ceiling windows are distributed in such a free manner that even identical floor plans present an individual and different face on the outside. Thus, the window arrangement echoes the arrangement, on a larger scale, of the housing islands, which are scattered across the park like pavilions.

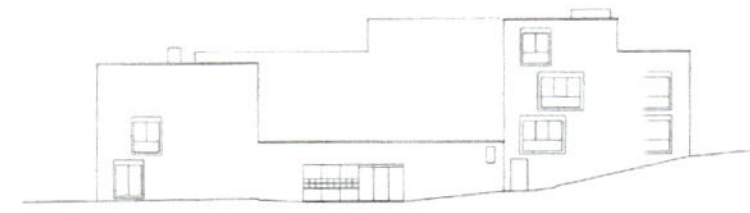

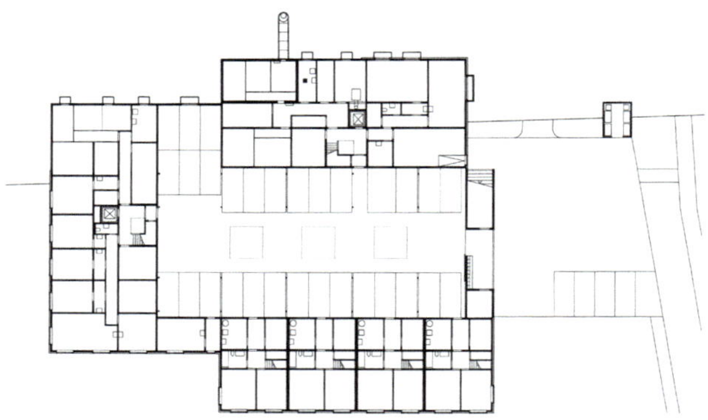

1

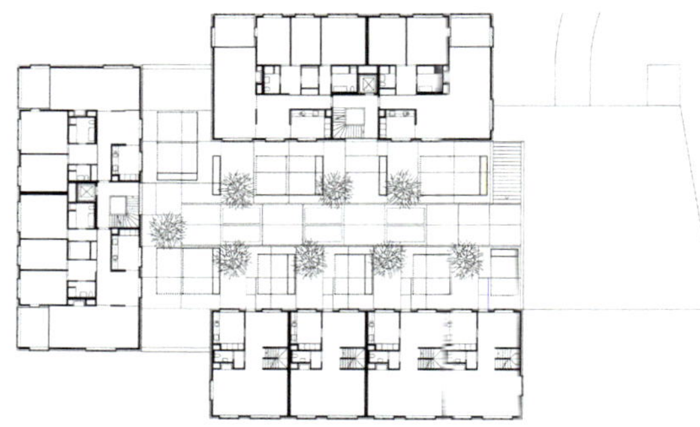

2

1 Base story 1:500
2 Entrance level 1:500
3 Standard floor 1:500
4 Top floor 1:500
5 Flats 1:200

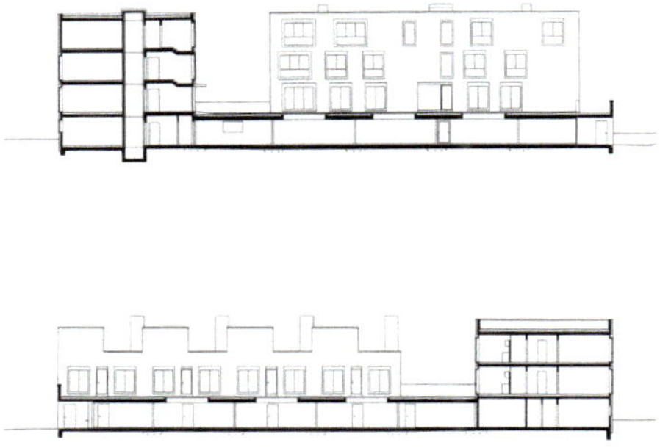

2.5 RESIDENTIAL COMPLEX / HOUSING ESTATE

Building type
housing complex in park setting
multifamily homes and
single-family row houses
3–4 stories
facing SE/NW

Date of construction
1996

Number of units
housing island I: 14

Size of units
4-room apts., 116,5 m² (4 units)
5-room apts., 139/146 m² (6 units)
5-room row house, 148 m² (4 units)

Area per user
28–29.5 m²

Building depth
12.2 m/10.8 m

Layout
housing development as part of targeted densification of parkland, plan for a total of six "housing islands," groups of three buildings create a courtyard

Open spaces
private winter gardens
and roof terraces

Parking
30 parking places in base story,
5 visitor parking spaces

Architect
Gigon/Guyer
Zurich

Location
Kilchberg
Zurich
Switzerland

3

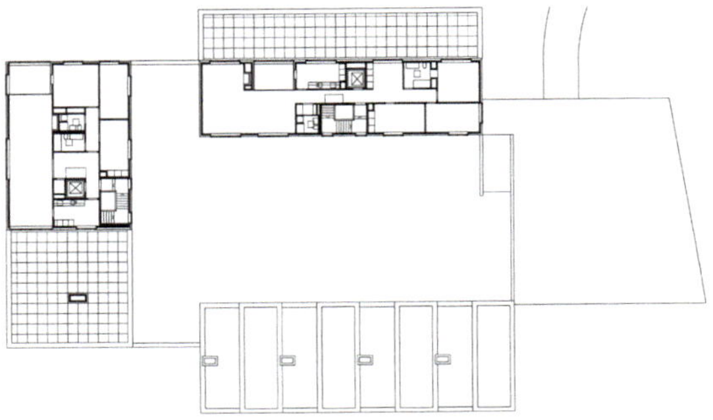
4

5

Referring to
floor plan 1:200

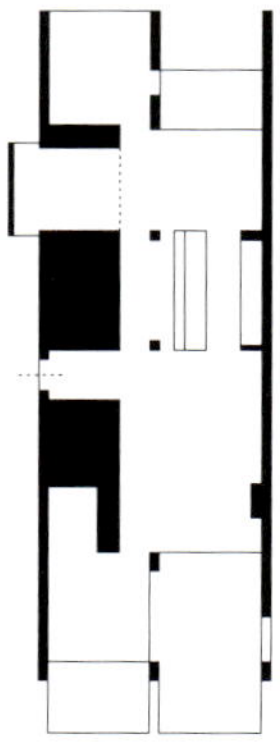

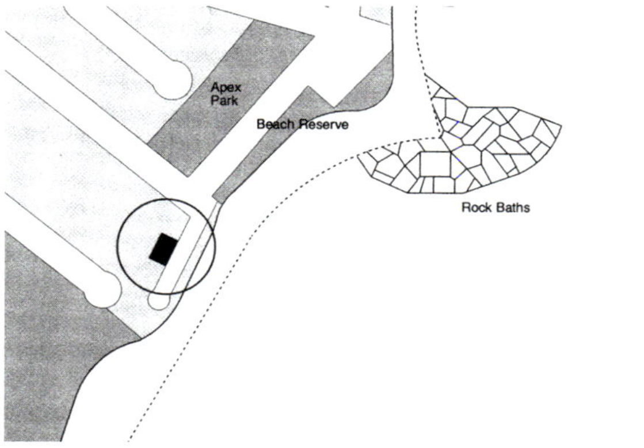

Rockpool | Popov | 1999

The development on the edge of Sydney in an area dominated by beach villas and characterized by a certain density, is distinguished by a balance between privacy and openness. The development consists of two rows: the front row features flats, the one to the rear maisonettes and row houses. A private path for the development runs between the rows debouching at the center of the complex into a communal square with pine trees. The individual units within the rows are set off from each other. This responds to the site and also creates a differentiated elevation in which each unit is specific in character. Each unit has balconies and generous terraces, those in the front row offering spectacular ocean views. The arrangement of the units within the overall structure of the complex varies while adhering to a principal idea: they are always divided into two stacked levels or two adjacent axial fields. The functions of living/sleeping are thus separated from one another; they are either accommodated one above the other, side by side or sequentially. The kitchens are set freely into the space or framed by short walls. Each unit has a study/guest room, equipped with folding doors and visually linked to the generous living room. This room is often shifted away from the main axis, tall and narrow glass louvers fill the resulting gap and light the space. The development is characterized by a playful treatment of filling the structure (diaphragm walls, framework with segmental arches), the resulting rhythm, and the deviation from this rhythm.

2

1

1 Entrance level: flats and maisonettes
2 Upper level: flats and maisonettes
3 Sections and street/beach elevation
4 Typical floor plan of flat 1:200

2.5
RESIDENTIAL COMPLEX / HOUSING ESTATE

3

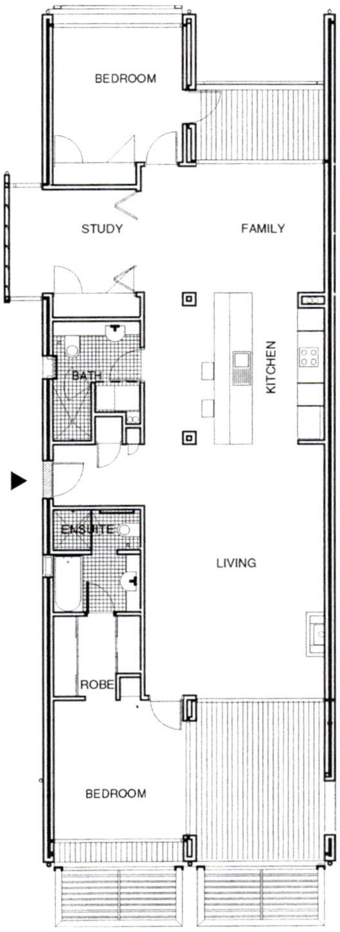

4

Building type
seaside development
2 stories
facing SE/NW

Date of construction
1995–1999

Number of units
17

Size of units
3–4-room apartments, 112–144 m²
2–3-room maisonettes, 122–156 m²

Area per user
37.5–78 m²

Building depth
18.3–25.3 m

Layout
1-/2-loaded stairwells
and private path

Open spaces
private gardens
balconies, terraces
landscaped communal path

Parking
40 parking spaces in
underground garage

Architect
Alex Popov Architects
Milsons Point
Australia

Location
Mona Vale Beach
Sydney
Australia

Referring to
floor plan 1:200

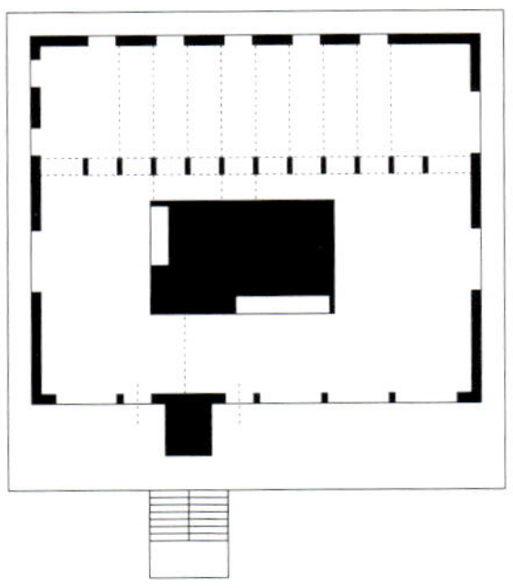

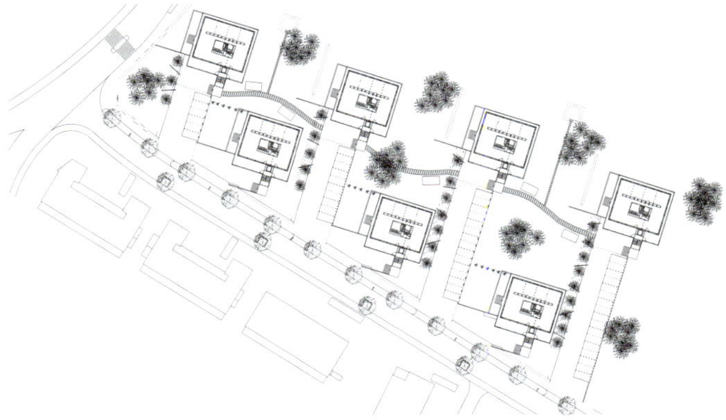

Balance Uster | Haerle Hubacher | 2001

The residential concept "Balance" has been realized on three different sites in the Zurich conurbation and was initially conceived as an answer to the phenomenon of increasingly changing and diverse living situations, as well as the desire of residents to take part in the design process for their own habitat. The basis of the design is a highly flexible floor plan covering an entire floor. Varied use and potential interior divisions were achieved by the precise placement of two spatial elements – a core containing the bathrooms and offering two potential niches for the kitchen unit or storage space and an open closet wall in the private areas. This closet wall defines a hallway on one side, while on the opposite side the individual segments enable the flexible configuration of room constellations and placement of access points. Taken together, they form a generous and divisible entrance area followed by two open living spaces with a zone at the back for individual rooms.

Fully glazed on all sides, each floor is surrounded by a wraparound balcony, meaning the residents can walk around their entire apartment as if it were a single-family home. This balcony deepens toward the southwest to form a wide loggia bay which is split by the vertical access core into two parts to enable the creation of an additional entrance should the floor be divided into two units.

By freely interpreting the spaces through the insertion of new walls to divide rooms or apartments, a variety of floor plan constellations are possible, which can be easily altered. In essence, all individual choices can be taken into account, with the exception of those affecting the load-bearing structure, the building envelope or the access and sanitary cores. In order to guarantee the success of this flexible concept, each floor was sold in its entirety.

1

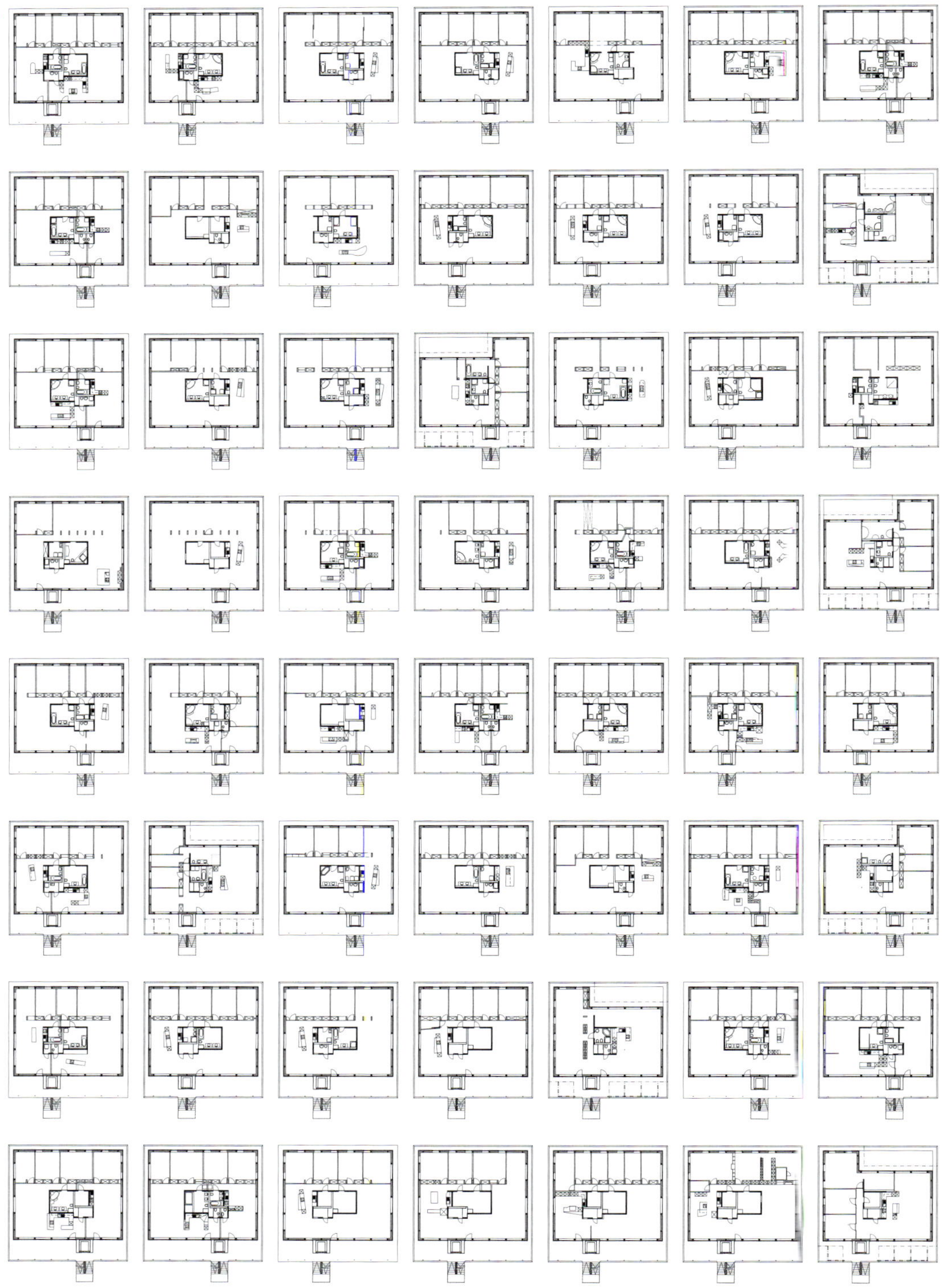

1 Floor plan solutions developed with the residents 1:1000
2 Southeast and southwest elevations 1:500
3 Cross section 1:500
4 Principal variations with optional room and apartment divisions
5 Basic floor plan: block with sanitary rooms and kitchen niches as well as closet wall system as the predetermined elements, optional room distributions and apartment dividing walls as the flexible elements 1:200

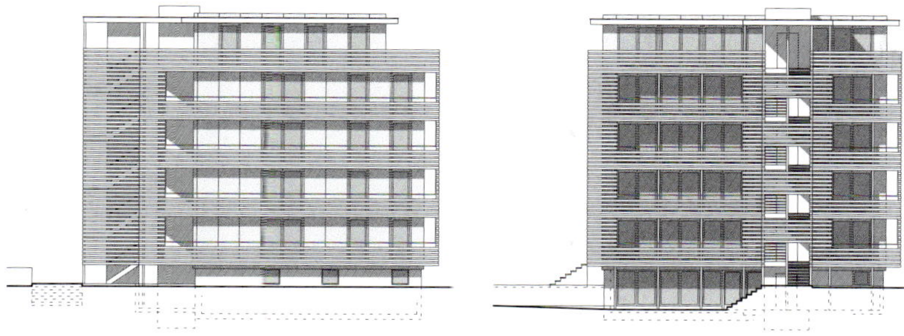

2

3

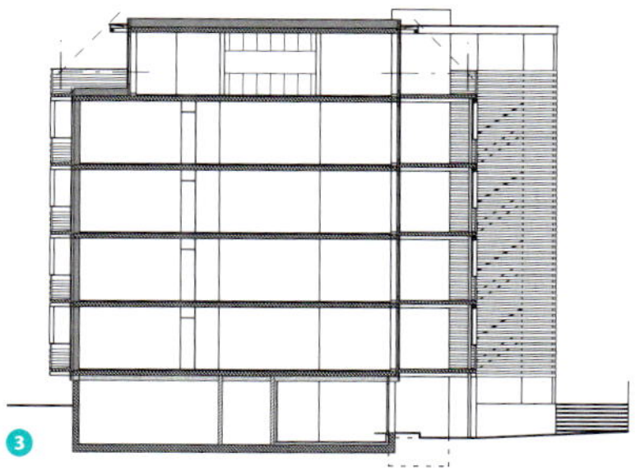

2.5 RESIDENTIAL COMPLEX / HOUSING ESTATE

4

5

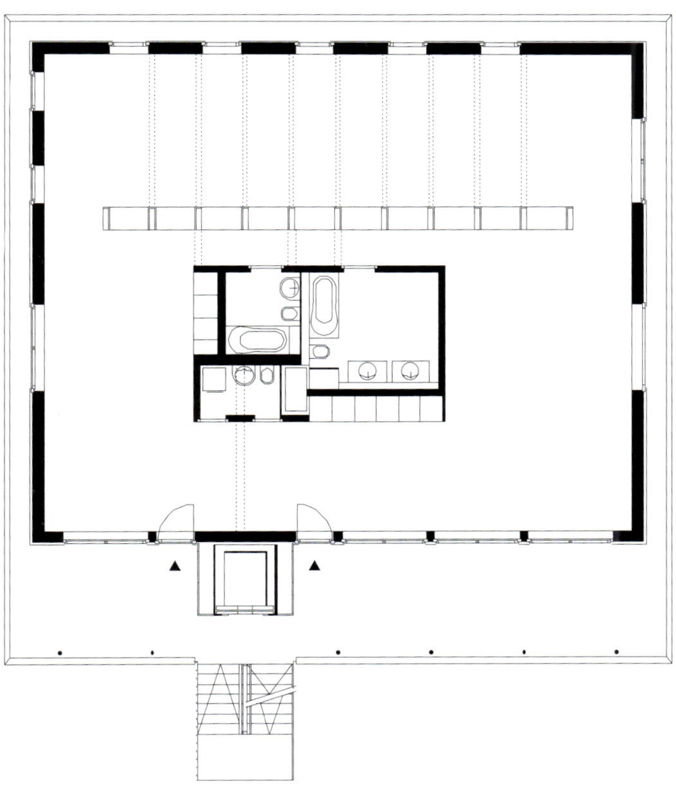

Building type
estate with seven identical cubic building volumes
5 stories
facing in all directions

Date of construction
1999–2001

Number of units
35

Size of units
182 m² per floor

Area per user
26–91 m²

Building depth
13/16 m

Layout
two rows of identical buildings offset to one another, with common areas in interstitial spaces
composite building structure with reinforced concrete floors and stiffening core and wooden columns along the perimeter
external staircase with access to the loggias, flexibly divisible floor plans

Open spaces
wraparound balconies, which widen to form loggia bays to the southwest

Parking
open parking spaces

Architect
Haerle Hubacher Architekten BSA GmbH
Zurich
construction planning:
Hans Andres Architekt ETH/SIA
Zurich

Location
Talweg 145–172
Uster
Switzerland

Referring to floor plan 1:200

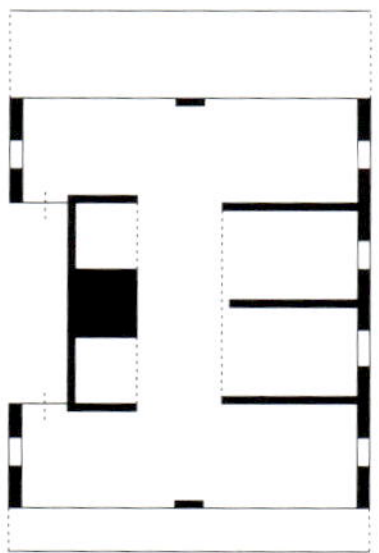

Steinfelsareal | Herczog Hubeli | 2002

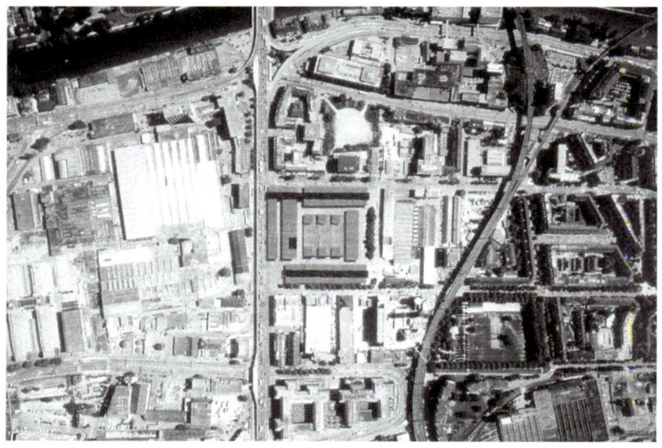

The "Steinfelsareal" is located in Zurich's former industrial district, where conversions have already created a strong mix of living and commercial uses. The existing mix was augmented to an even greater degree in the new complex by not only stacking the different uses but also creating many different, interchangeable floor plans. These floor plans are therefore capable of responding to shifts in demand for certain uses and make it possible to avoid vacant units. The ensemble is composed of very different building types: at the center, two rows of point blocks are grouped around an inner courtyard, which is sheltered by two rows that enclose it, the east row and the "basilica." The latter is a double row with a 5-story access hall in the middle, which can be reassigned to new uses according to the occupants' wishes. Depending on their orientation, these rows contain a variety of floor plans. In lieu of a roof, they are topped by the "cloud-hanger," a three-story, 165-m-long slab, which is cantilevered in a spectacular fashion over top of the existing Cinemax theater on the street side. Commercial and living spaces are arranged side by side – in the cloud-hanger, residential maisonettes are strung along commercial units that are twice as wide – or stacked one above the other: all plinth stories are used as 2-story commercial spaces with galleries, allowing for a wide variety of uses by virtue of their size and height. In some cases, flexibility of use is achieved through a free disposition of walls within the framework of the structural system, for example, on the north side of the basilica, where division forms a living or office unit as required. In other cases, the floor plans are reduced both in form and disposition to allow for residential or studio uses, for example, in the cloud-hanger.

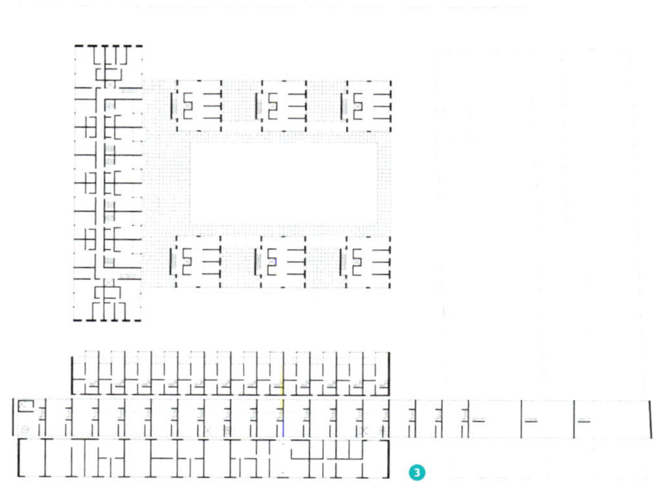

3

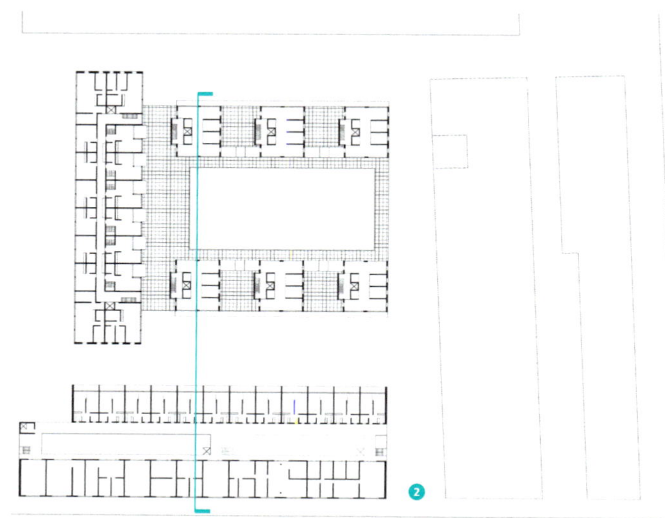

2

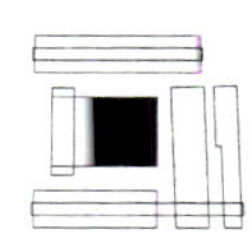

1

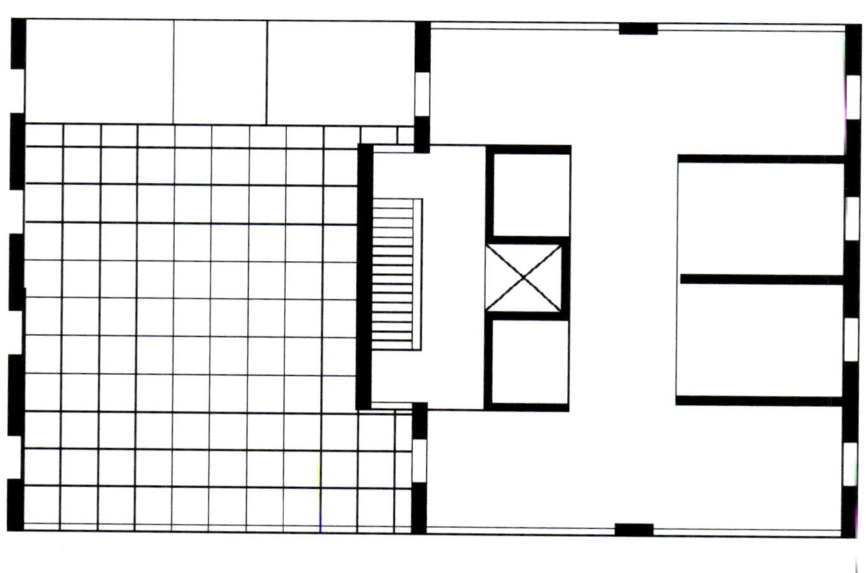

1. "Point blocks," floor plan for 2nd to 5th floor, apartment with highly flexible layout 1:200
2. Entire ensemble 4th floor, with access galleries for the "basilica"
3. Entire ensemble 9th floor, the "cloud-hanger" spans the access hall of the "basilica"
4. "Basilica," floor plans for 2nd/4th floor, 3rd, 5th, 6th floor from bottom to top: north-east row with commercial or living use, south-west row (2- or 4-room apartments)
5. "East row" floor plans for 2nd/4th and 3rd/5th floors, respectively: small 1-room apartments, Unité-like maisonettes along a naturally lit central corridor
6. "Cloud-hanger," floor plans for 9th to 12th floor from bottom to top: apartments in the smaller maisonette units, commercial spaces in the larger units

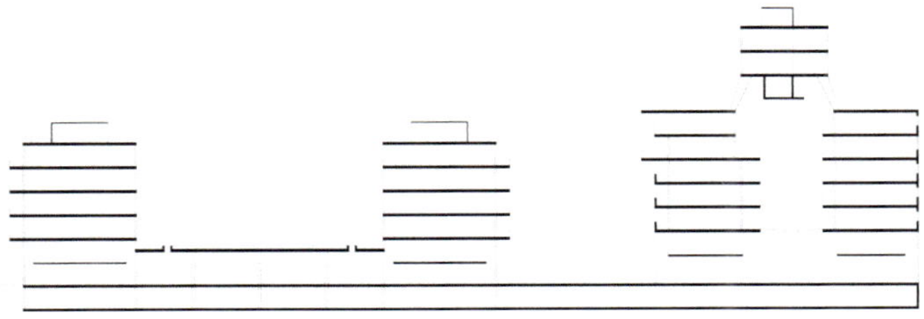

2.5
RESIDENTIAL COMPLEX / HOUSING ESTATE

4

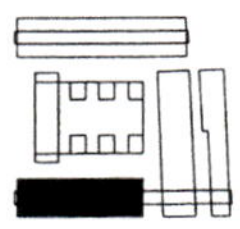

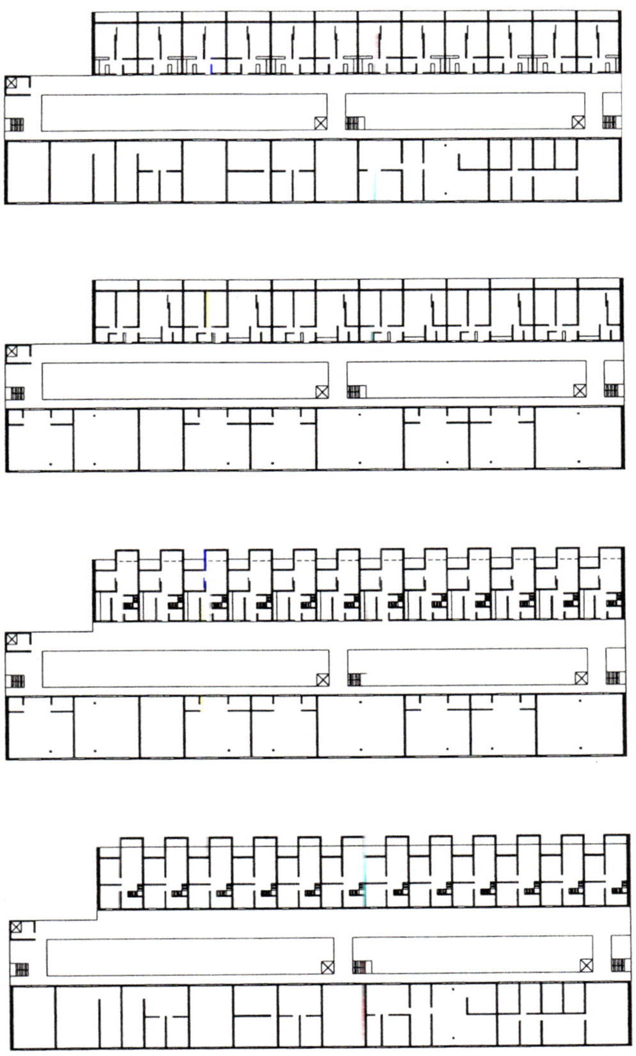

5

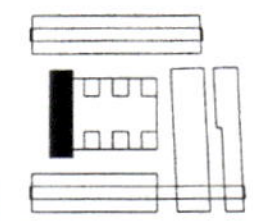

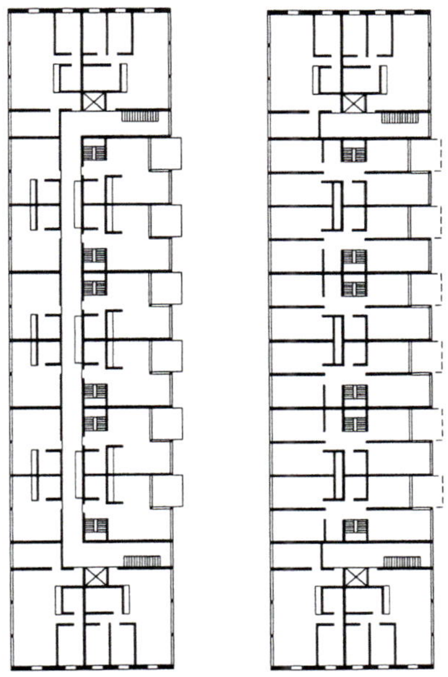

Building type
dense urban ensemble of typologically distinct long slabs and point blocks
5–9 stories
facing NE/SW, NW/SE

Date of construction
2002

Number of units
approx. 120 or more (depending on use of units)

Size of units
"basilica"
row 1: studios for living or commercial uses, small unit 70 m²
row 2: 5-room mais. 120 m² (12 u.)
2-/4-room apart. 60/120 m² (30 u.)
"cloud-hanger":
loft 150 m² (17 u.)
office 280 m² (3 units)
point blocks:
5-room apart. 130 m² (24 u.)

Area per user
26.5–70 m²

Building depth
"cloud-hanger"/"basilica" 10 m
point blocks 12/13.5 m
row to the east 17.5 m

Layout
ensemble composed of six point blocks, one double row with central access hall ("basilica"), with a 165-m-long floating slab ("cloud-hanger") above, row to the east

Open spaces
private and common roof patios, planted inner courtyard for detached houses and row to the east, loggias for detached houses within a separate structure between them

6

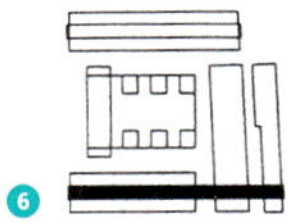

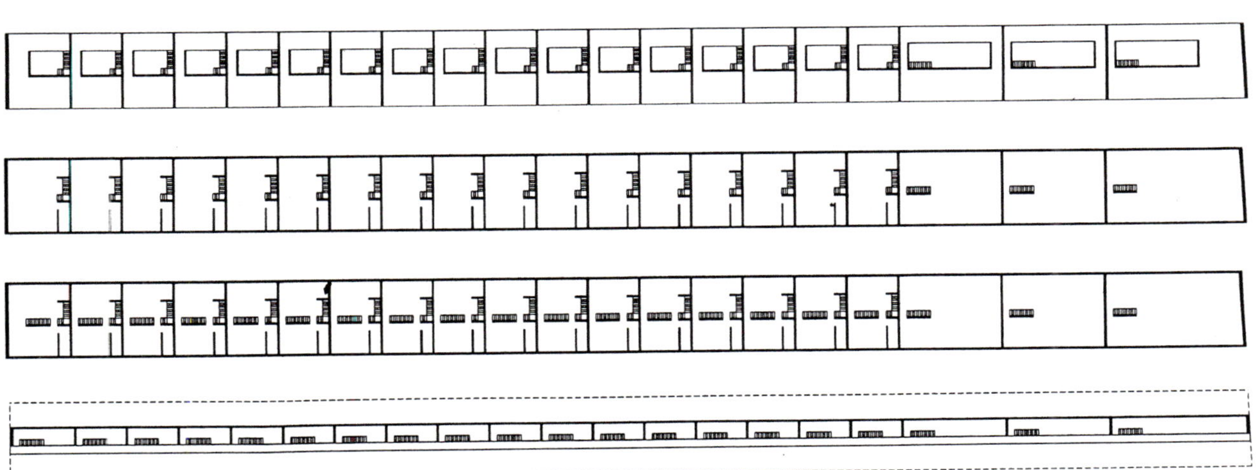

Parking
231 parking spaces in underground garage

Architect
Herczog Hubeli
Zurich

Location
Zurich
Switzerland

Referring to floor plan 1:200

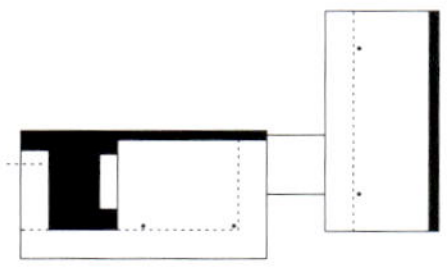

G-Flat | Koh Kitayama + Architecture Workshop | 2006

On a larger building lot, the building mass is dissolved into a series of compact volumes to provide better light and air flow and a scale more in line with the proportions of surrounding buildings. Each level in this connected chain of volumes is divided by a load-bearing wall running alternatingly length- or crosswise through its middle, thus dividing each floor into two halves. These floor halves are either connected to one another by bridges in the space between the volumes or separated by the vertical access cores placed here.

The different orientations of the volumes and connections between them are combined to generate different floor plan types. Studio apartments occupy one half of a floor. In type **A**, ground floor studios are located in the crosswise divided volumes and oriented towards the courtyard. Type **B** consists of a floor half divided longitudinally plus half of an outdoor bridge space. The somewhat larger units type **C** occupy an entire floor where one half contains the bedroom and auxiliary rooms such as the bathroom and kitchen, while the other remains a completely open living room. The largest unit type **D** is accessed via an open divisible living room (with the exception of the street-facing unit) in the longitudinally divided volume. It is connected by the bridge to a bedroom and bathroom in the adjacent, crosswise divided volume.

Near the transparent edges of the volumes, slender columns support the levels, enabling a visual connection between the apartment interior and the outdoor space beyond. A zone emerging between the columns and the glass facade can be used as an outlying access hallway between the different rooms of a single unit, or to slightly expand the size of the spaces.

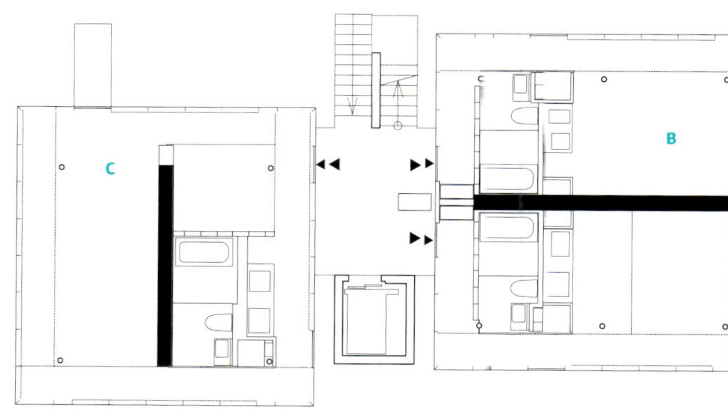

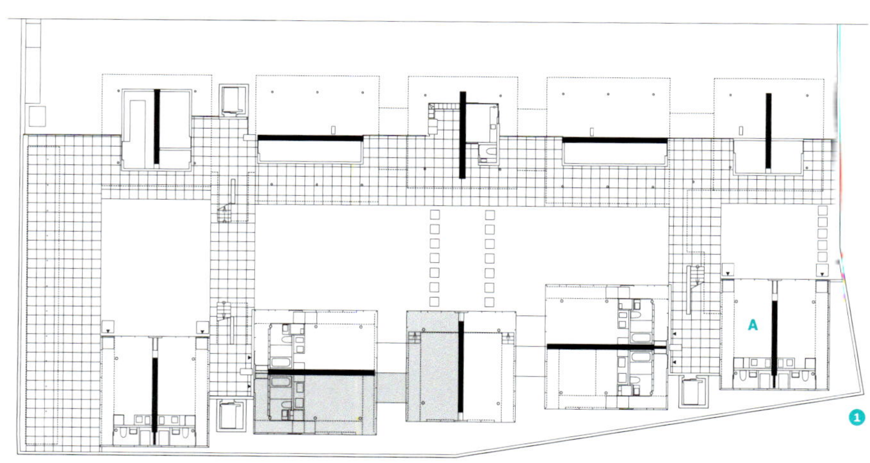

1 Ground floor 1:500
2 2nd floor 1:200
3 Street elevation 1:500
4 Northwest elevation 1:500

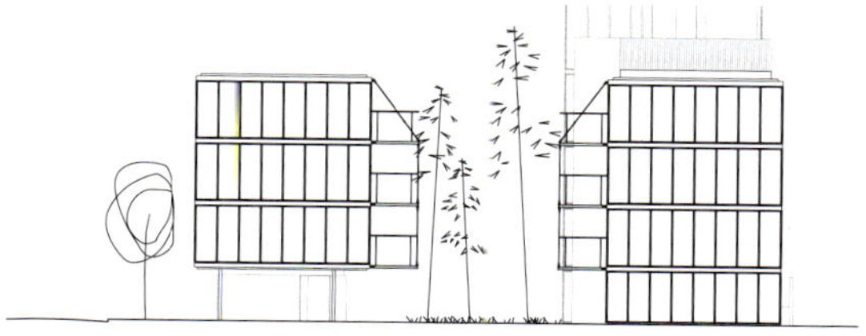

4

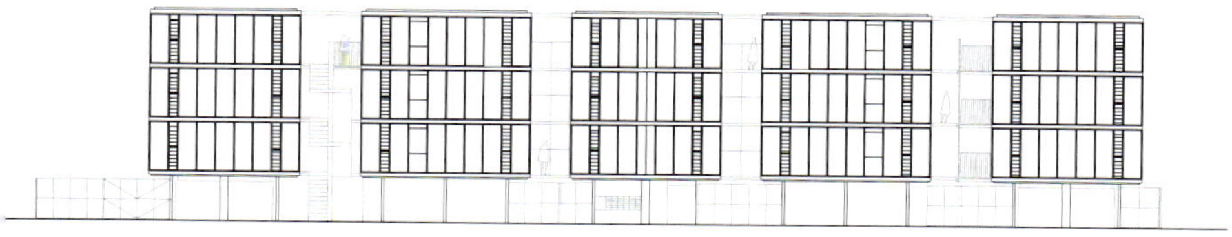

3

2.5
RESIDENTIAL COMPLEX / HOUSING ESTATE

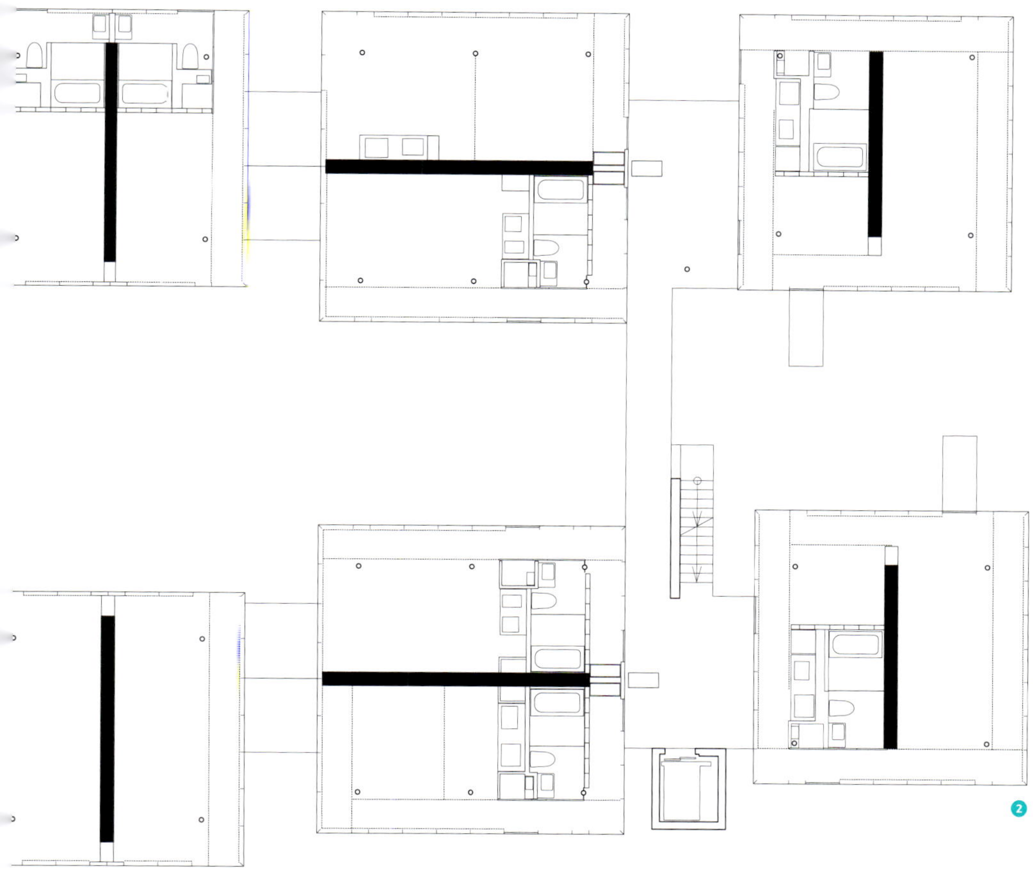

2

Building type
10 linked building volumes around an inner courtyard, punctuated by vertical access staircases and connecting bridge spaces
4 stories
facing in all directions but primarily NE/SW

Date of construction
2006

Number of units
47

Size of units
Type **A**: 26 m² (4 units)
Type **B**: 33 m² (14 units)
Type **C**: 52 m² (13 units)
Type **D**: 59 m² (14 units)

Area per user
26–33 m²

Building depth
7.3/8.2 m

Access
triple-loaded staircases, with bridge to three further apts. per floor

Open spaces
winter gardens, balconies, common courtyard with bamboo garden, roof terraces

Parking
street parking

Architect
Architecture WORKSHOP
Koh Kitayama
Hiroko Hasama
Mariko Hama
Yurie Kobayashi
Tokyo

Location
Kitasenzoku, Ota-ward
Tokyo
Japan

Referring to floor plan 1:200

Hunziker Areal Building A | Duplex Architekten | 2014

Instead of a typical residential development, a true urban district was the guiding principle for the design of this estate – dense, lively, and a place where public, communal, and commercial zones can intermingle with private spaces and residential units. Participatory dialogues between members of the building cooperative and the architectural offices involved resulted in an ensemble of 13 buildings with diverse living forms and floor plan constellations derived to help establish a heterogeneous community.

In Building A, each floor is divided into two large co-housing units containing five to six inserted small apartments as well as shared living spaces. The composition of the floor plan is strikingly similar to the urban layout of the entire ensemble: open spaces are framed by building volumes in such a way that they flow into one another. The living spaces flow between the chamber-like small apartments, offering plentiful space to cook, eat, and live together; they provide a visual connection to the circulation space in the central atrium, and open up to the facade on all sides in the form of quiet niches.

The chambers provide a quiet and private refuge for the residents. More than mere bedrooms, they can function as small, autonomous residential units. Each has its own bathroom, and some an additional living room, kitchen unit, or second bedroom, making it possible for smaller families to live together with others. An enclosed room group situated near the entrance houses a wardrobe, guest bathroom, and a storage and housekeeping room. Offices and studio spaces on the ground floor are almost identically configured, illustrating the neutrality and flexible potential of the floor plan composition.

1

2

1 Typical floor, entire ensemble 1:2000
2 Building A: ground floor 1:500
3 Building A: typical floor with two large co-housing units 1:200
4 Building A: section 1:600

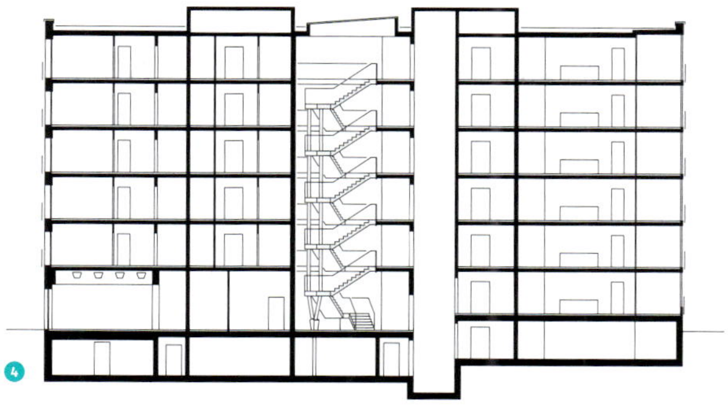

2.5
RESIDENTIAL COMPLEX / HOUSING ESTATE

Building type
solitary volume within an ensemble of 13 buildings of similar dimensions
ground floor with common and commercial spaces
6 stories
facing in all directions

Date of construction
2009–2014

Number of units
11 large shared units, each with 5 or 6 inserted small apartments

Size of units
10½-room apts., 323.5 m²
(6 units)
12½-room apts., 398 m²
(5 units)
individual inserted apts., 28.5–44 m²

Area per user
approx. 40 m²
(collective units)
20–30 m²
(individual apts.)

Building depth
23.6/31.4 m

Access
double-loaded staircase in central atrium

Open spaces
loggias
balconies

Parking
car-free living

Architect
Duplex Architekten
Zurich

Location
Leutschenbach
Zurich
Switzerland

Referring to floor plan 1:200

2.6
LOW RISE HIGH DENSITY

This building type evolved from criticism of the over-development of the landscape due to the construction of detached single-family homes. Partially inspired by vernacular courtyard typologies from North Africa – among other places – low rise high density developments achieve higher density not through increased height, but by reducing the unbuilt surface area to a minimum. Each resident is thus provided with spacious, private outdoor areas despite the overall horizontal density as well as direct, mostly ground-level entrances from the public sphere.

Following a specifically developed compositional logic, repetitive, sometimes varied building modules are grouped to create pattern-like configurations – for the most part with no dominant hierarchy or central focal point and with the potential to seemingly expand in width, height, and depth. Different formations are possible; for example, row houses in back-to-back arrangements, patchwork-like structures with courtyard houses – also called "carpet developments" – or multi-story conglomerates with intricately stacked or interlocked modules. Voids like patios or terraces are a necessary component to improve lighting conditions. In most cases, they become a distinctive design feature and influence the spatial organization of the residential units. The private nature of the individual, outdoor spaces resulting from incisions into the building mass help compensate for a lack of surrounding views.

The necessity to organize the circulation areas predominantly horizontally leads to specific solutions – systems of alleyways, courtyards that flow into one another, or partially or completely open ground levels. In multi-story building ensembles, maisonette units are the norm.

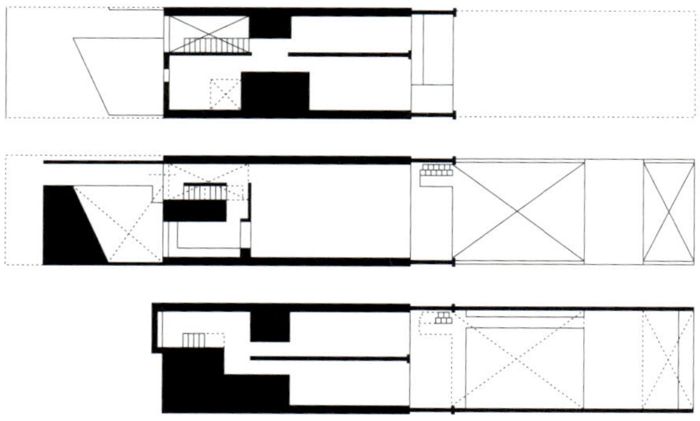

Halen | Atelier 5 | 1961

A dense, almost urban complex on deep, narrow properties on a slope. Highly efficient use of the construction site: Residents of the upper rows have their gardens on the roofs of the buildings that lie below; i.e., they have the legal right to use and even build on roofs which are not their property. Yet the private areas are absolutely protected, the view from each individual house cannot be blocked by buildings. Type 12: entrance at street level; covered path through front garden. The path axis continues in the house leading to the large living room; the kitchen lies behind the stairway. Living room with two-story loggia and steep stairway into the garden. The garden and upper levels contain two to three bedrooms with bathroom at the top and a living/lounge area below. Panoramic balcony on the upper floor. The garden floor opens onto the deep narrow garden with covered seating area. In the variant, the bedrooms on the garden floor are replaced by a multifunction room; the upper floor is shortened and creates room for a deep sun terrace. The narrower 380 type (again available in 2 options) is organized similarly except for the placement of the stairs. Additional studio buildings.

1

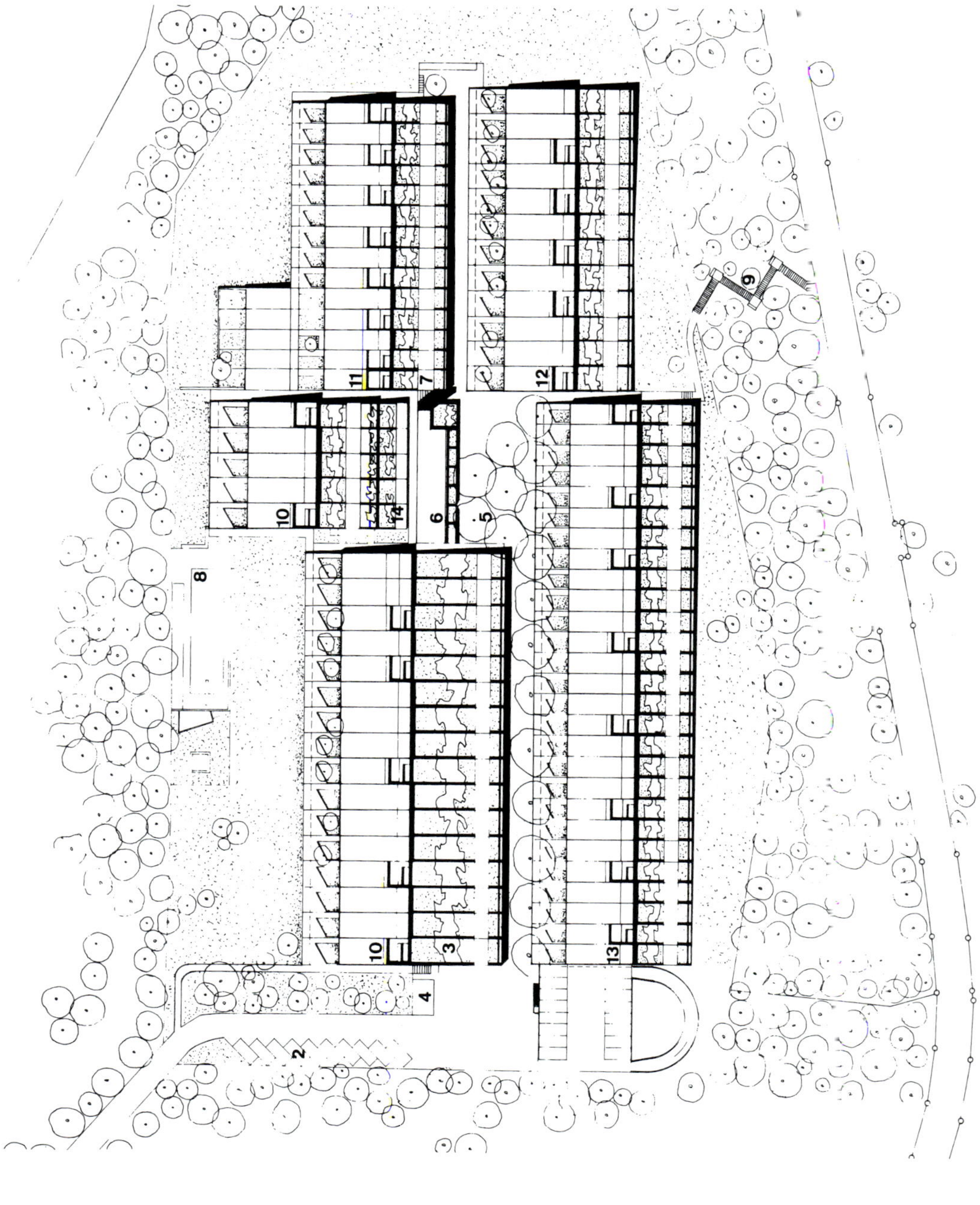

1 House type 380: 3 levels with studio building

2 Floor plans house type 12: garden level, street level with entrance, upper story 1:200

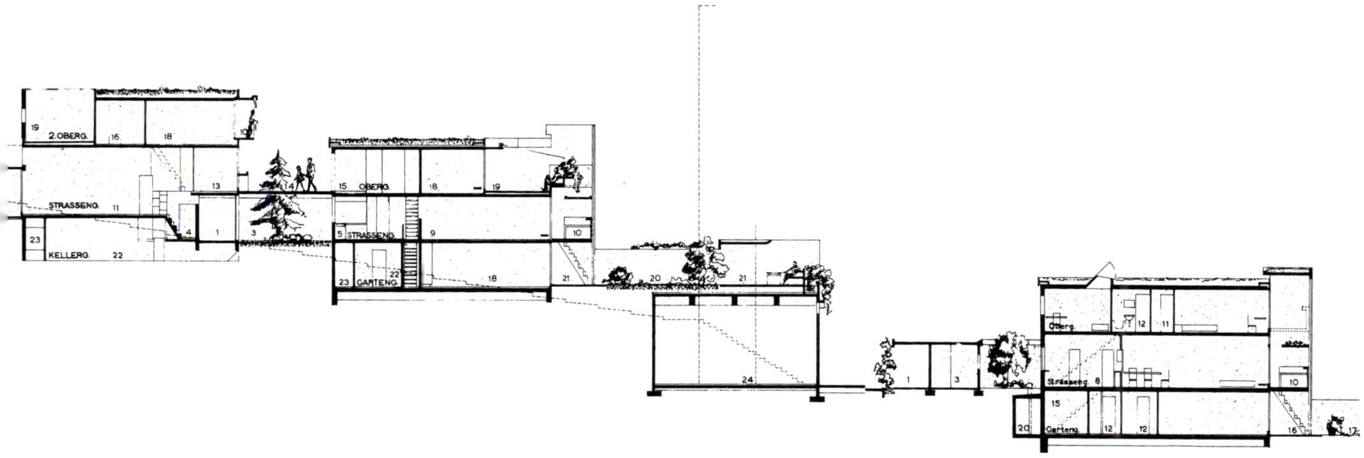

2.6
LOW RISE HIGH DENSITY

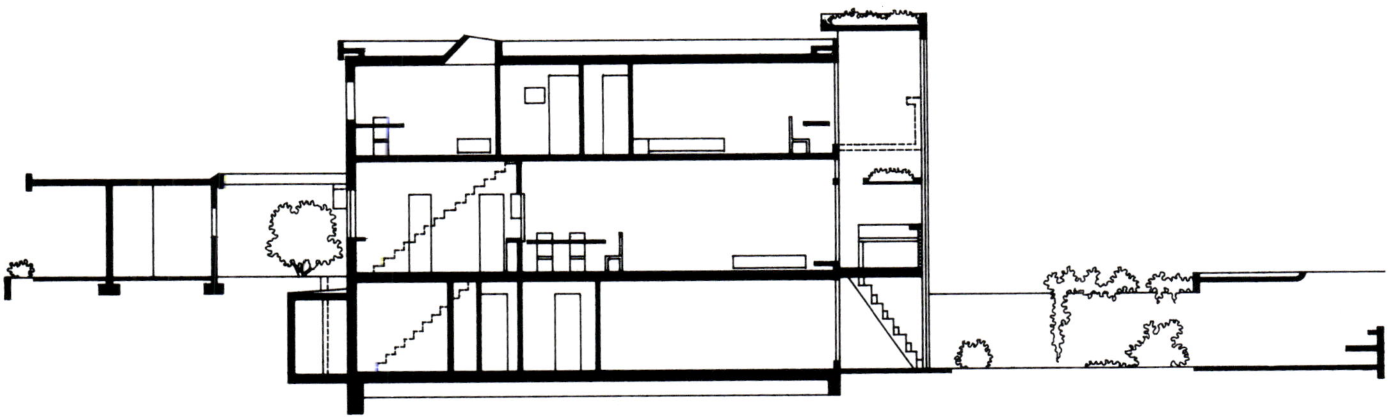

Building type
row houses on slope
3 stories
facing N/S

Date of construction
1955–1961

Number of units
79

Size of units
4 to 6 rooms,
approx. 130/190/220 m²

Area per user
32.5–36.5 m²

Building depth
12/13.5 m

Layout
79 units in terraced rows,
2 types, 4 and 5 m wide,
partly with studios,
paths closed to motor traffic,
many communal institutions

Open spaces
private gardens
shared forest lot
roof trerraces

Parking
parking garage

Architect
Atelier 5
Bern

Location
Herrenschwanden
Bern
Switzerland

2

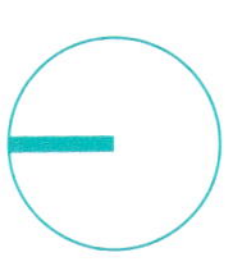

Referring to
floor plan 1:200

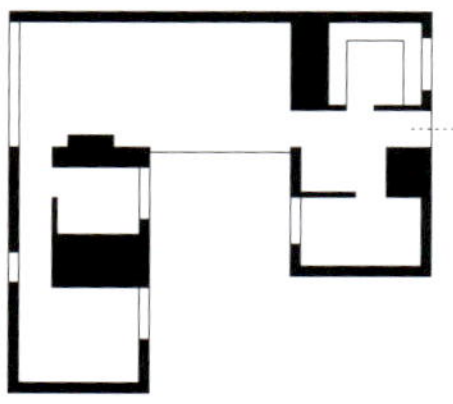

Ludwig-Windhorst-Straße | Gieselmann | 1961

2.6
LOW RISE HIGH DENSITY

Semiatrium houses: three house wings, originally conceived as prefabricated boxes, surround a paved interior courtyard which is supplemented by a garden (size depending on layout of property). The gardens are surrounded by perforated-masonry walls. Classical, spare floor plan: kitchen, dining room, and guest room (with WC), which can also be utilized as a study or a second children's bedroom, are located at the entrance. Narrow stairway to basement story (with heating system, supply room, hobby room). The elongated sunken living room (two steps down) features sliding doors that open onto the inner courtyard to the south, and, depending on position, a large window overlooking the garden. All windows of the sleeping tract face the courtyard. Only the entrance, kitchen window, and garden gate are located along the walkway or the square.

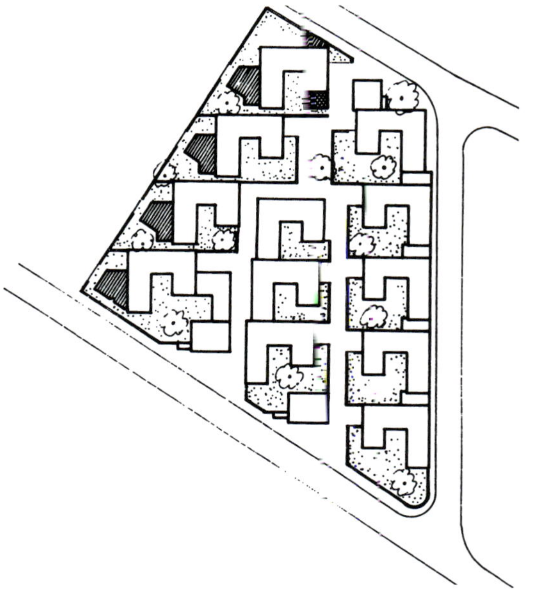

Building type
"carpet development" of semiatrium houses
1 story (with basement)
facing S, E/W

Date of construction
1960–1961

Number of units
12

Size of units
4 rooms, 95 m²
(expansion to 135 m²)

Area per user
24 m²

Building depth
12/14 m

Layout
development of semiatrium houses with broad paths, capable of expansion (small additions in 1969/1990)

Open spaces
terraces, gardens

Parking
individual and group garages

Architect
Reinhard Gieselmann
Karlsruhe / Vienna

Location
Ludwig-Windhorst-Straße
Karlsruhe
Germany

1
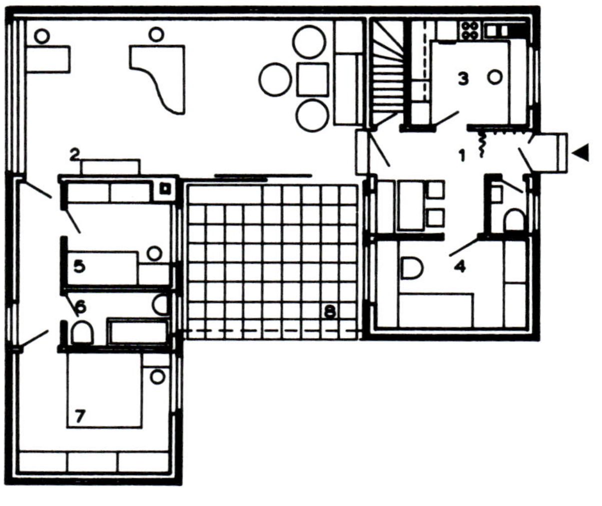

2
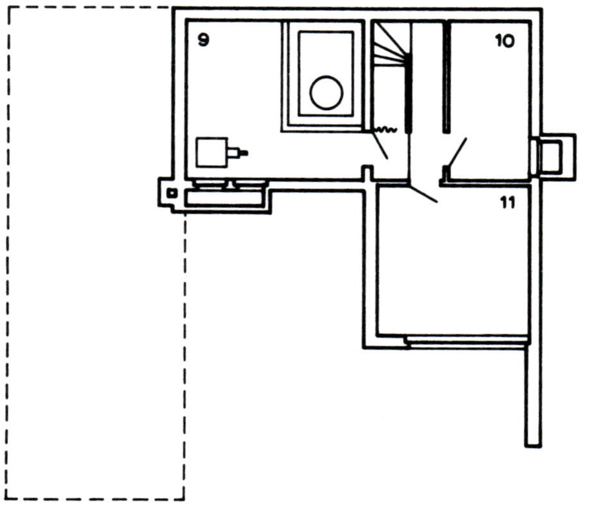

Ground floor 1:200 1
Basement 1:200 2

Referring to floor plan 1:200

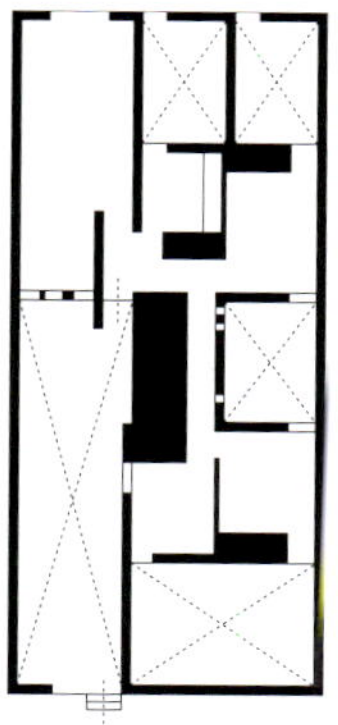

Villas en Bande | Zevaco | 1969

2.6 LOW RISE HIGH DENSITY

The single-story housing units, each arranged around four patios and a back garden, are divided into two blocks. The volumes of each unit are linked using homogenous materials, repetitive elements, and an all-enclosing perimeter wall, so that a patchwork emerges in which it is difficult to identify the single units. The common feature of the floor plans is the central section where the entrance, kitchen, corridors, and bathroom are situated, and from where three differently sized spatial volumes branch off to the sides. Due to the spatial organization, these appendices – a large living room as an extension of the entrance patio and the two perpendicular bedrooms in the private part of the house – always have outer walls and courtyards on two sides, thereby facilitating cross-ventilation for each room individually. The shady patios are a reminder of more traditional living forms where family life plays out in a space visually separate from the outside world and passersby.

Building type
patio house development
1 story
facing in all directions

Date of construction
1969

Number of units
17

Size of units
4- to 5-room apts., 105 m²

Area per user
21–26 m²

Building depth
4.5–13.8 m

Layout
patchwork of buildings with enclosed patio courtyards and gardens forming two separate blocks

Open spaces
patio courtyards and gardens

Parking
off-site parking spaces

Architect
Jean-François Zevaco
Casablanca

Location
Agadir
Morocco

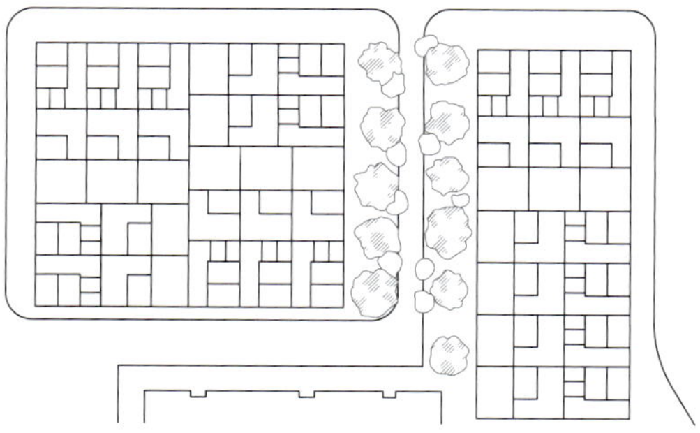

1 

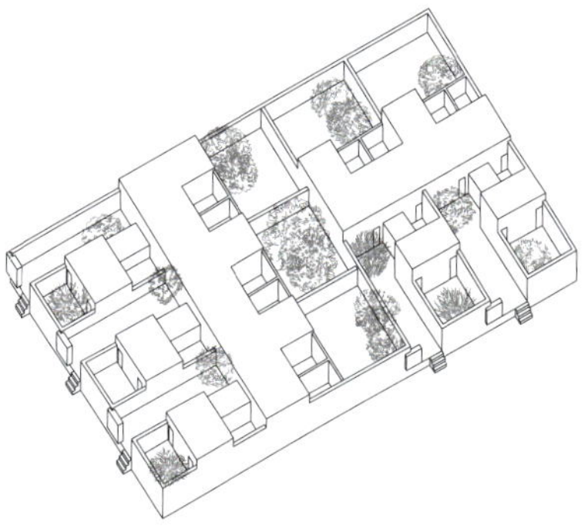

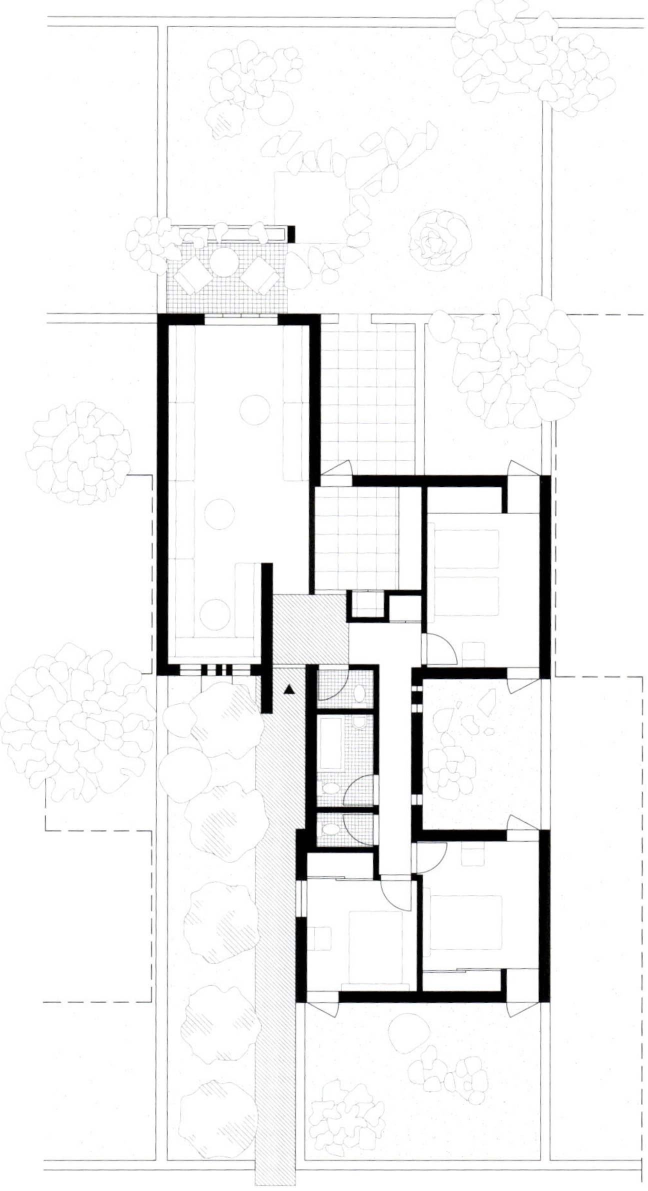

2

Floor plan and elevation of one of the two blocks 1:1000 (1)
Ground floor 1:200 (2)

Referring to floor plan 1:200

2.6

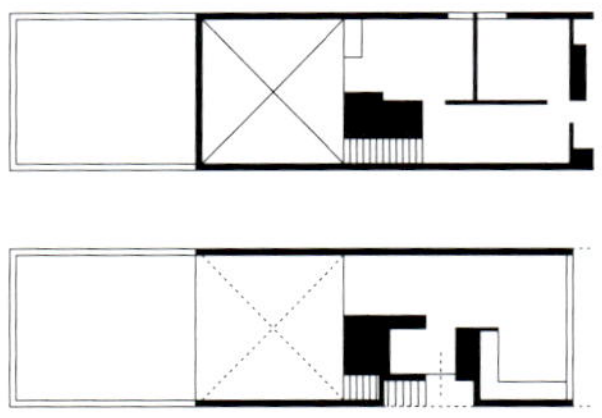

Kasbah | Blom | 1974

A shared roof over a patchwork-like cluster of houses is equally vision and program. Below the houses, all raised by an entire level, an open space is created to foster the development of a lively and social milieu. The lofty covered spaces connect laterally to the access paths and serve as parking spaces and access points to the open staircases. They also, however, can be adapted by individuals or groups of residents in the form of community organizations, shops, and small businesses – the pragmatic consideration being that under a single roof, anything can be built out or changed at will.

Starting on the second level on both sides of the open staircases are the two-story base units of all apartment types – in which the bathrooms, kitchens, internal staircases, and bedrooms above adjoin a double-height living room. Depending on the apartment type, additional modules connect to this base unit: in types **B** and **C** a roof garden that is enclosed on three sides, in type **C** an additional room above the street below, and an entire unit, when type **A** and **B** are combined, essentially forming type **D**.

To foster social diversity, roughly the same amount of each of these housing types is found throughout the complex. Type **A** is intended for use as a studio, or for students or young couples. Type **B** was developed as a low-income option, while type **C** is spacious enough for larger families. The upper floor under the gabled roof is described as a "sleeping loft" and can be subdivided if desired. In type **D**, in which the second upper floor room is transformed into a connecting hallway, the architect was not simply motivated by creating more rooms, but by increasing the spatial possibilities for additional activities.

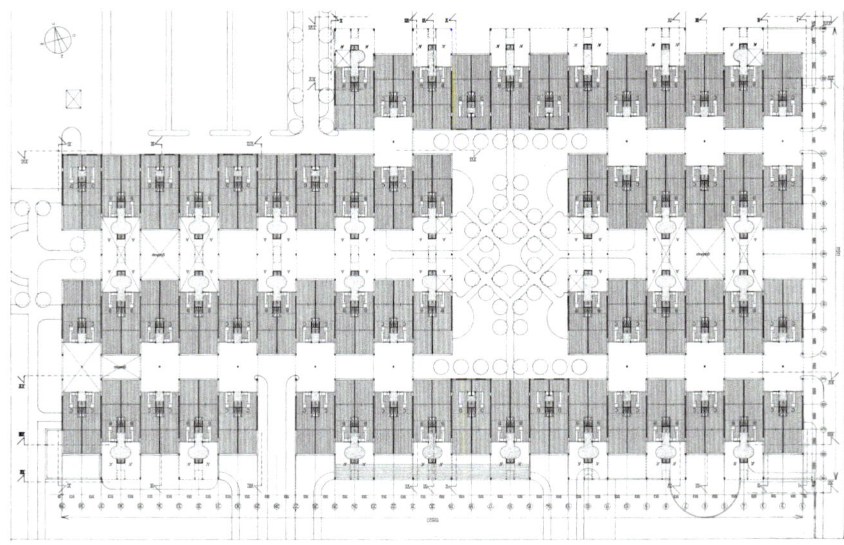

1

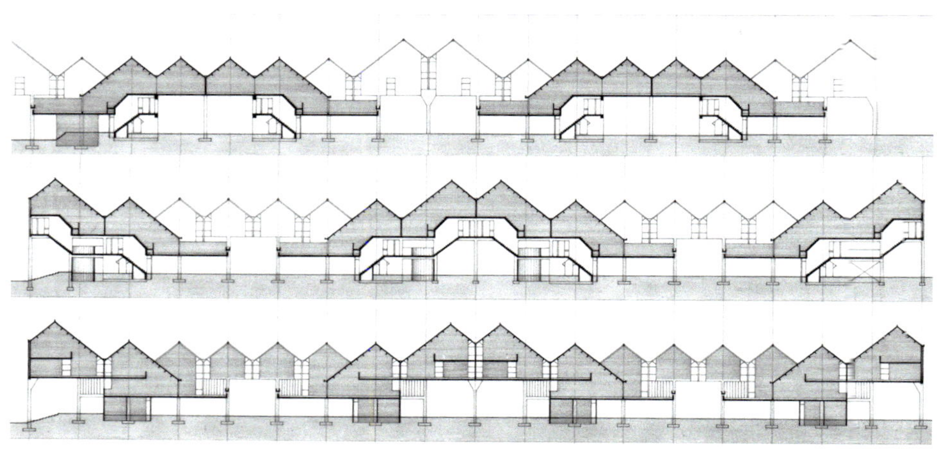

2

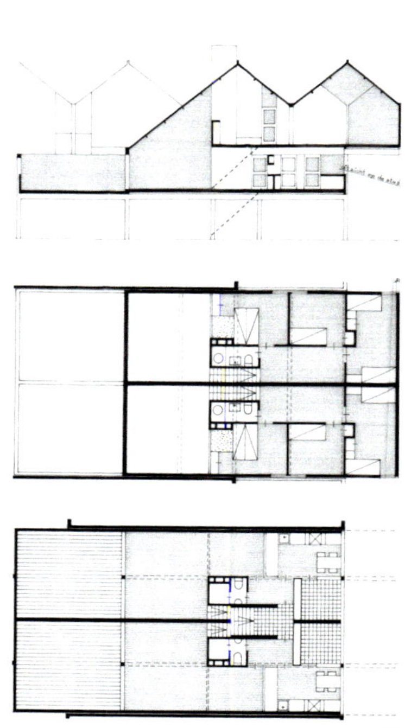

4

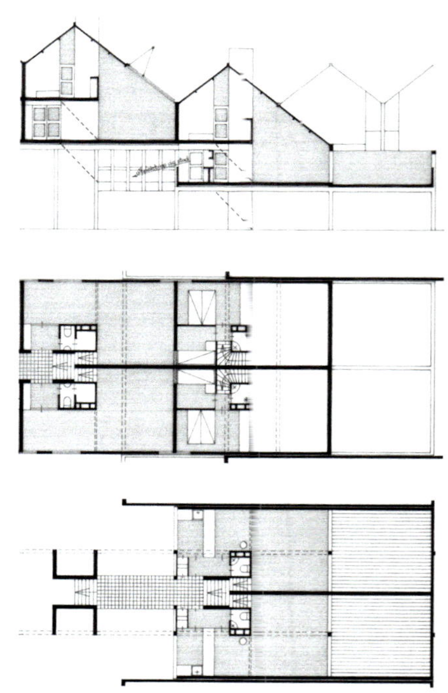

5

1 Entire ensemble, 2nd floor 1:2000
2 Sectional elevations with passageways 1:1000
3 Apartment types overview in section
4 Type **C**: 2nd floor with living area
3rd floor with "sleeping loft"
Section 1:500
5 Types **A** and **B**: 2nd floor with living area, type **A**
3rd floor with living area, type **B**
and "sleeping loft," type **A**
Section 1:500
6 Type **D** (combination of types **A** and **B**, living and studio):
2nd floor living area
3rd floor "sleeping loft" and studio (with a second gallery above)
Section 1:500
7 Type **C**: 2nd floor, 3rd floor with "sleeping loft"
Section 1:200
8 Visualization of the ground floor areas open to resident use and adaption

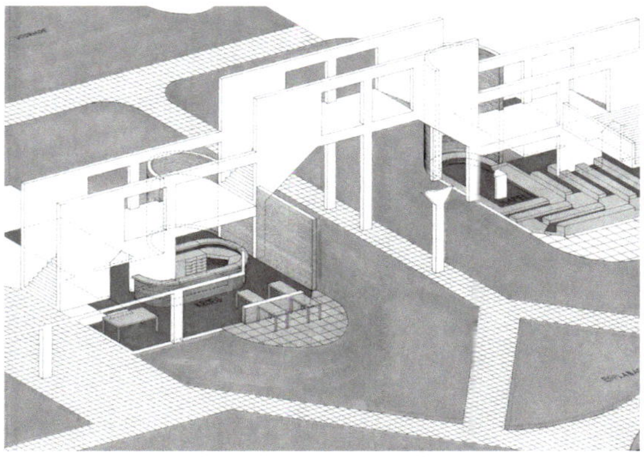

2.6
LOW RISE HIGH DENSITY

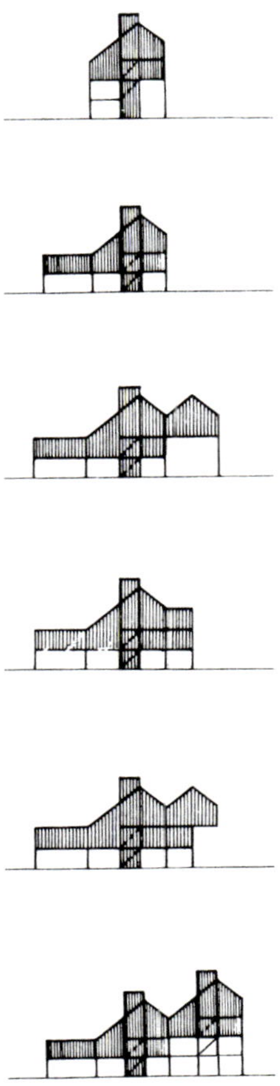

3

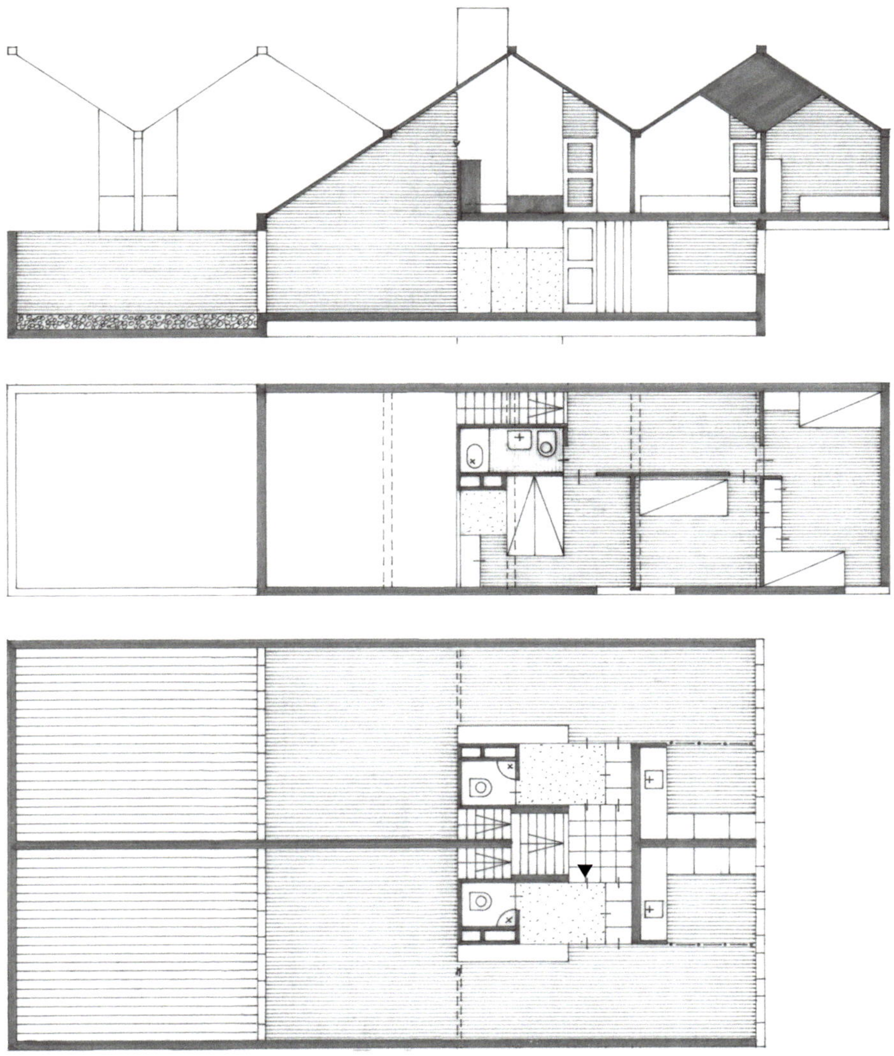

7

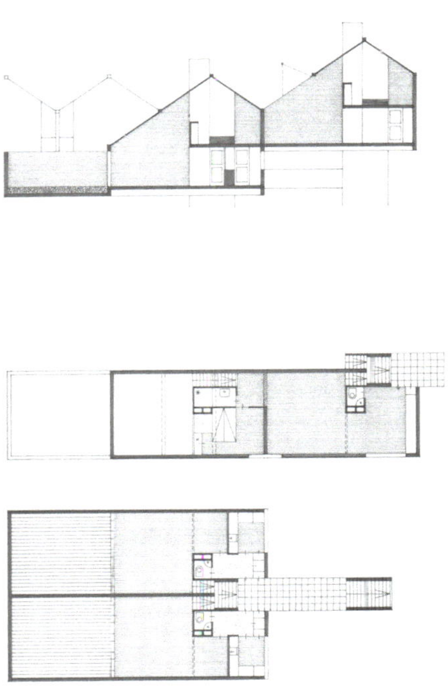

6

Building type
habitable urban roof on stilts
3 stories
facing NE/SW/NW/SE

Date of construction
1969–1974

Number of units
184

Size of units
Type **A**: 2 ½-room mais.,
approx. 69 m²
Type **B**: 2 ½-room mais.,
approx. 57 m²
Type **C**: 4-room mais.,
approx. 105 m²
Type **D**: 5-room mais.,
approx. 125 m²
(Types **A** and **B** combined)

Area per user
19–26 m²

Building depth
13.2–33 m

Access
North African kasbah as analogy for a building structure: patchwork of parallel rows of gabled roofs above a variety of housing types, carved out voids form common courtyards and private rooftop gardens
open ground floor throughout serves as an urban common space, for moving and stationary traffic and is adaptable to residents' commercial and social programs

Open spaces
roof terraces enclosed on three sides
patios

Parking
ground floor parking spaces

Architect
Piet Blom
Monnickendam

Location
Hengelo
The Netherlands

Referring to floor plan 1:200

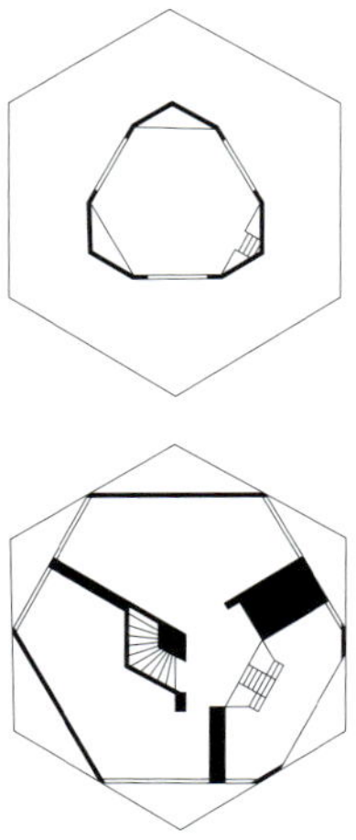

1

Cube house | Blom | 1984

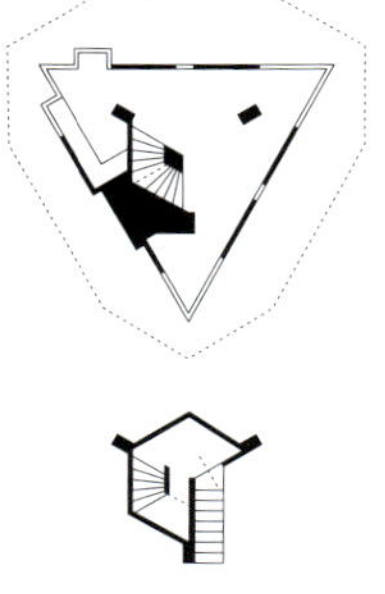

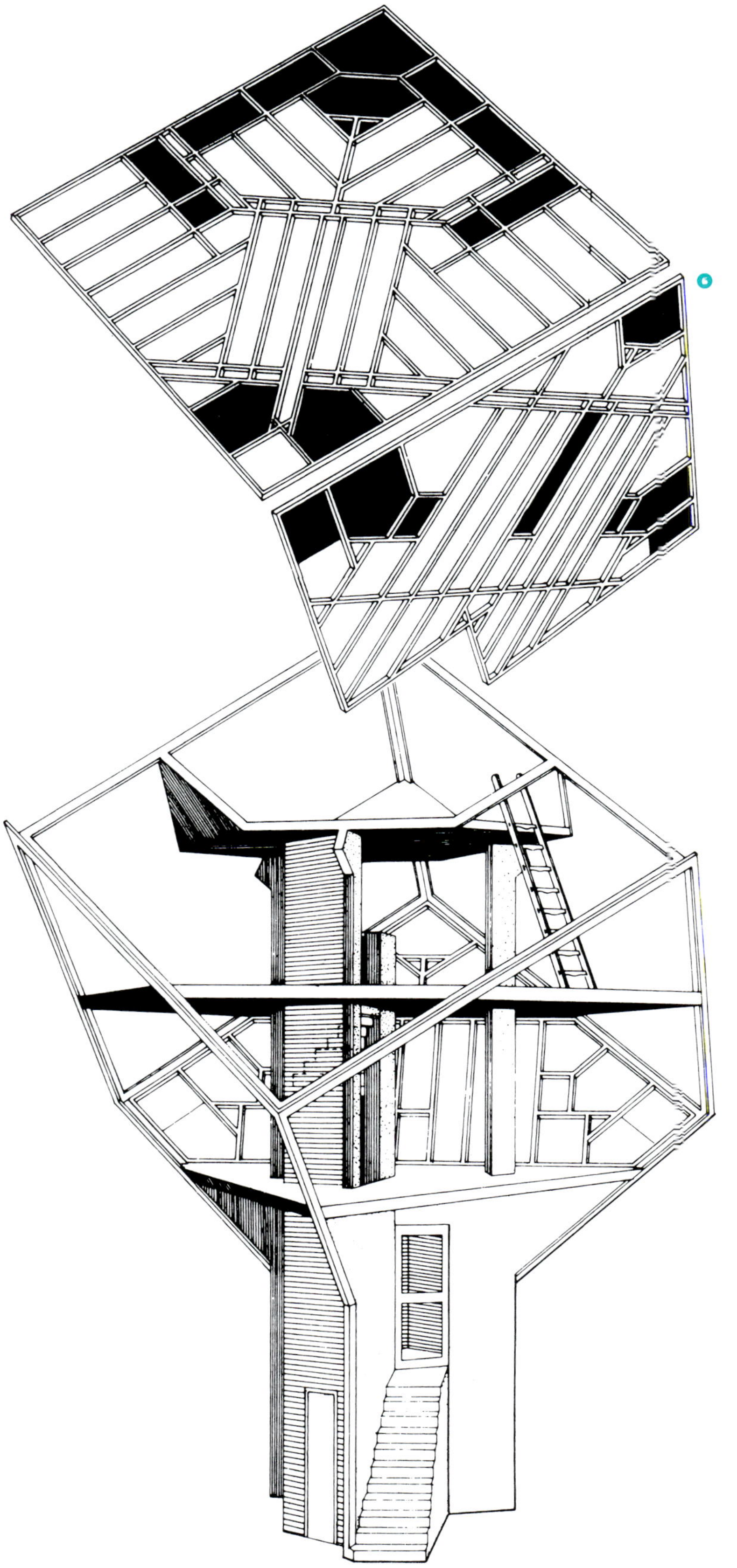

6

Experimental living: three levels in a tilted cube, supported by a hexagonal, 5-m-high central column. The result is a "cube house." The first level contains the "street house": a triangular living and dining room with kitchen alcove and WC behind the spiral staircase; the windows are slanted toward the street below. The second level contains the "sky house": living/sleeping zones with bathroom; here, the windows are slanted up, toward the sky. The attic story accommodates the "greenhouse," accessible via a ladder: this is the daylit studio with a vista through the glazed peak of the cube. The floor plan is highly flexible in theory, because each occupant can interpret the individual rooms at will – at the same time, however, the constraints imposed by the limited space within the slanted walls dictates specific uses and customized furnishings. A set of two cubes can be combined into a single unit. With the exception of the "trunk," which houses the private stairwell, the space below is available for public uses. The concept of this "forest of trees" arose from the desire to closely intertwine public and private spheres. Of the planned sixty units in the center of Helmond, three prototypes and the central "playhouse," which provides spaces for public events in a series of linked "trees," were initially realized in 1975. In 1984, thirty-eight additional tree units were built on a pedestrian bridge in downtown Rotterdam.

1 The sample houses in Helmond
2 The cube houses at port of Rotterdam
3 Cube houses as seen from the city in Rotterdam
4 5 Site plan cube houses in Rotterdam
6 Isometric view of the cube house's interior and surface
7 Floor plans of the various levels 1:200

3

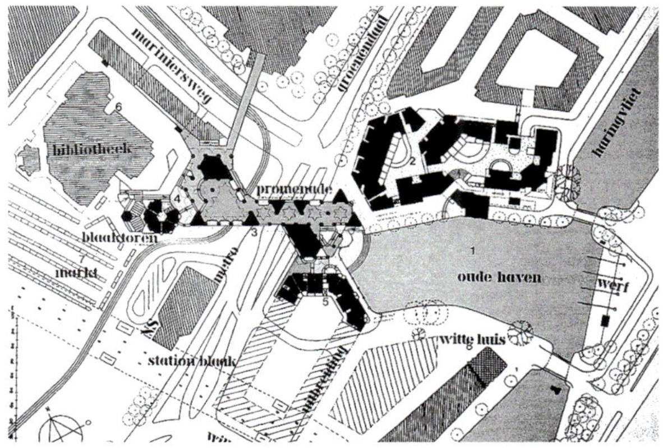

4

2.6
LOW RISE
HIGH DENSITY

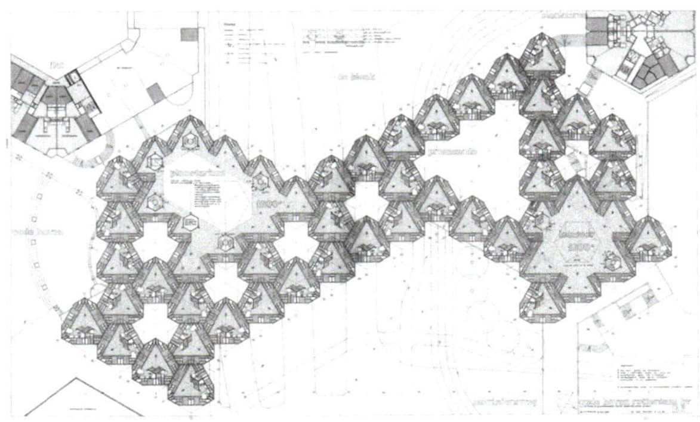

5

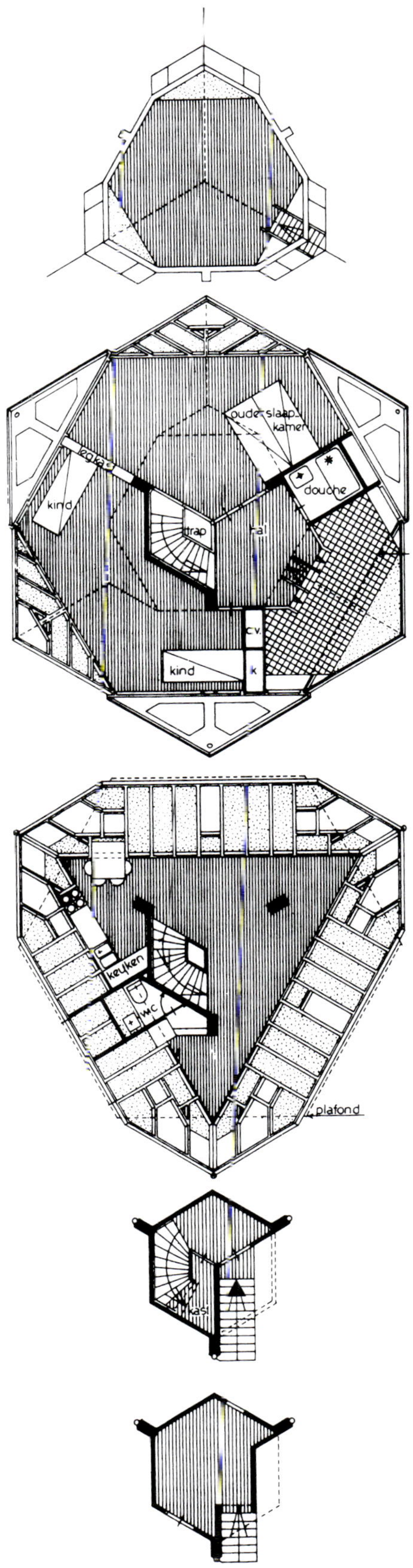

7

Building type
cube-shaped
individual houses
on trunks
put in row on tilted side
3 stories
orientation to all sides

Date of construction
1974–1976
1977–1984

Number of units
Helmond:
3 model houses,
"communal building"
Rotterdam:
38 units

Size of units
Helmond:
3–4 rooms, 90 m²
Rotterdam:
4-rooms, 102 m²
(2nd floor 24 m²/3rd floor 60 m²/
top floor 18 m²)

Area per user
22.5–34 m²

Building depth
approx. 11 m

Access
private stairs in "trunk"

Open spaces
beneath the
cube houses

Parking
on street

Architect
Piet Blom
Monnikendam

Location
Helmond/Oude Haven
Rotterdam-Blaak
The Netherlands

Referring to
floor plan 1:200

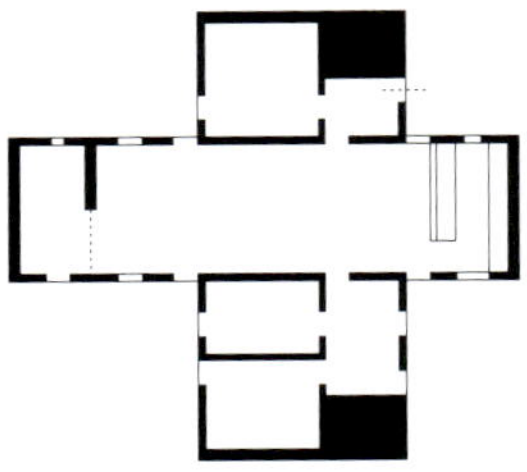

Galgebakken | Storgård, Orum-Nielsen, Marcussen | 1974

Car-free housing development near Copenhagen with row houses and garden courtyard houses. The design is primarily concerned with taking different housing needs into account through the provision of varied housing typologies, as well as enabling the residents to personally shape their immediate surroundings.

The single-story atrium houses (type **A**) are based on a cross-shaped plan. Groups of four houses form common garden courtyards to the rear and front courtyards accessed by pedestrian paths. The entrance, kitchen, bathroom, and play area face the path, while all bedrooms and the long living room overlook the garden. The central living room with linear kitchen and divisible (work/guest-)room serves as a distribution zone for the individual rooms complete with a separate hallway and bathroom. The layout makes it possible to sublet a wing with its own entrance, bathroom, kitchenette in the spacious hall, and garden access. The occupants are free to design the courtyards to their liking, separating them with walls or sharing them for common use.

In the same way, the green space adjacent to the pedestrian path and the garden to the rear offer residents of the two-story row houses (types **B** and **C**, depending on the house width) a spacious and open common area free to interpretation. The split-level row houses are characterized by a high kitchen and dining room lit by a slanted skylight, from which stairs lead up to a gallery level with bedrooms and down to the living room and an additional room. Scattered throughout the development is a further single-room house type (type **D**). Lined up in rows along courtyards, these houses are intended for either students, single elderly residents, or guests.

1

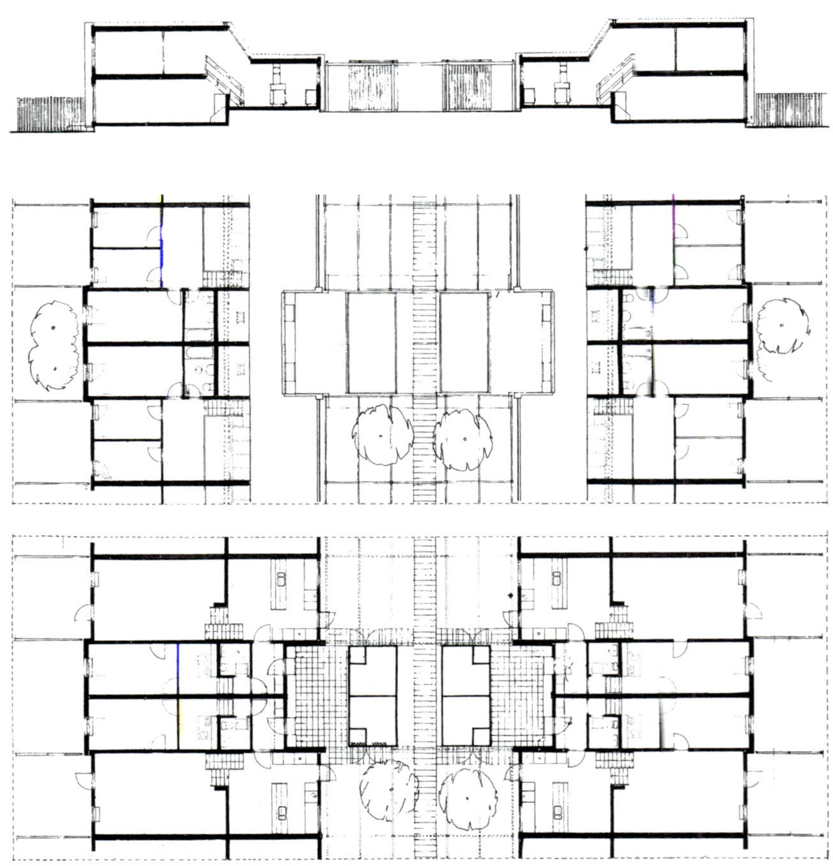

2

1. Site plan 1:3000
2. Split-level house, type **B**:
   Upper level open to kitchen below
   Lower level with living room and garden, kitchen and entrance along the raised footpath
   Floor plan, section 1:300
3. Atrium house, type **A**: four houses enclose a shared garden courtyard with pedestrian path
   Floor plan, section 1:300
4. 1-room apt., type **D**: "supplementary unit"
   Floor plan, section 1:300
5. Atrium building, type **A**, floor plan 1:200

2.6
LOW RISE
HIGH DENSITY

5

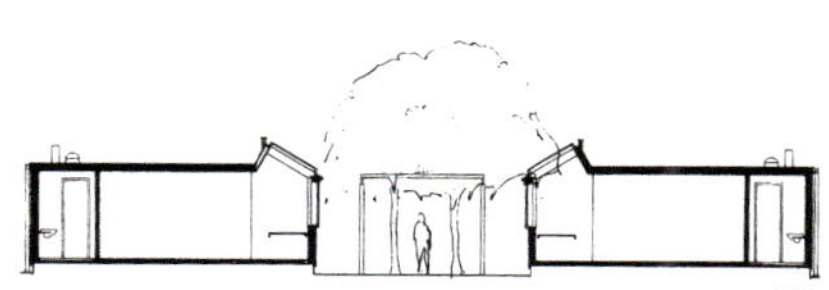

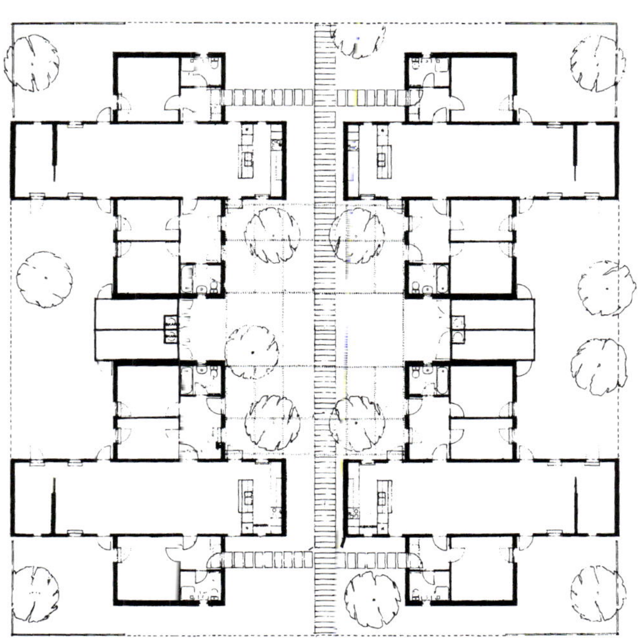

3

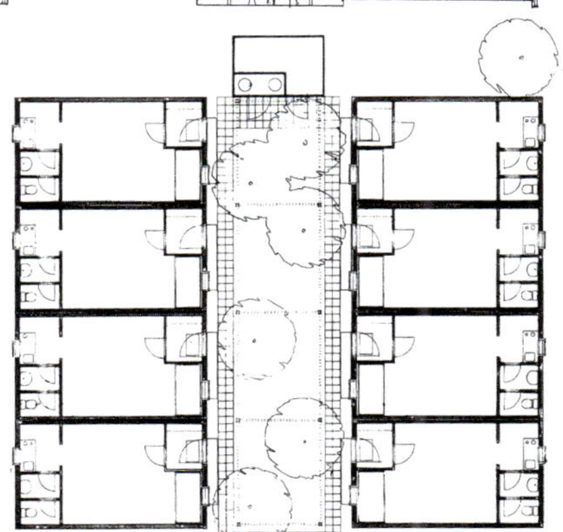

4

Building type
"carpet development" with split-level
row houses and patio houses
based on cross-shaped plan
1 or 2 stories
facing in all directions

Date of construction
1969–1974

Number of units
160 courtyard houses (Type **A**)
414 row houses (Types **B**, **C**)
70 1-room apts. (Type **D**)

Size of units
128 m² per unit (Type **A**)
125–170 m² per unit (Types **B**, **C**)
30 m² per unit (Type **D**)

Area per user
30 m²

Building depth
approx. 16/13.5 m (Type **A**)
approx. 12 m (Types **B**, **C**)
approx. 8 m (Type **D**)

Layout
courtyard houses (Type **A**):
groups of 8 houses,
4 buildings surrounding
a common garden area
split-level row houses (Types **B**, **C**):
row houses along pedestrian path,
rear gardens form common spaces
1-room houses (Type **D**):
2 rows, each 4 units form a courtyard

Open spaces
wide pedestrian paths (car-free),
private terraces,
shared garden courtyards
(can be subdivided into private
gardens upon request)

Parking
group parking spaces

Architect
J. P. Storgård
A. and J. Orum-Nielsen
H. Marcussen
Copenhagen

Location
Herstederne
Denmark

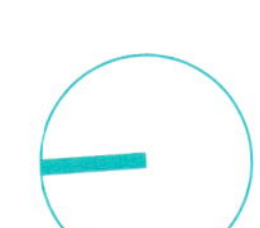

Referring to
floor plan 1:200

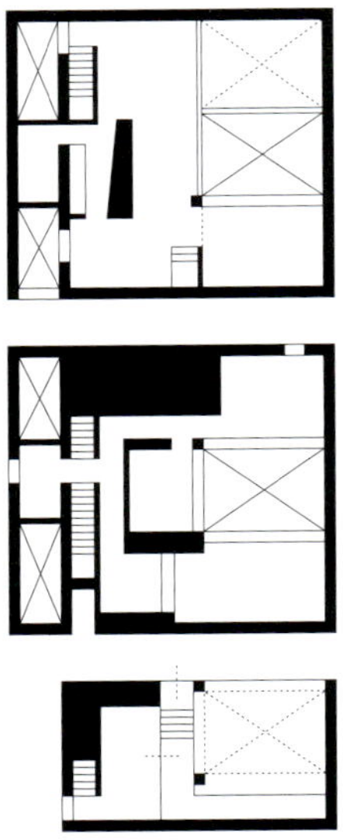

Nexus World | OMA/Koolhaas | 1991

Two building complexes, hermetically enclosed with black, structurally ornamented concrete walls, formulate the gate position leading into the building exhibition grounds "Nexus World" (see also, p. 198, Steven Holl). The large volumes are raised: the open ground floor, lit alternatively from above and from the side, houses several spaces for public use as well as the foundations for twelve of the integrated apartment units in each complex. The light joints, continuous from east to west, divide these units into individual rows. The units are accessed via an open, slightly inclined and network-like system of arcades. The entrances for most units are located in paved yards, which connect the floors above in the manner of patios. These patios create a separate open space for each apartment and a degree of privacy that would otherwise be impossible in a development of this density. In the interior, the apartments are linked by single-strung continuous stairs, behind which small chambers – in a zigzag arrangement – project into the building joints. On the second floor, the rooms and sanitary zones are lined up around the courtyard. The third floor accommodates the common area with kitchen; it overlooks the roof terrace, the patio reaching down to the paved garden and the sky. Akin to a wave-like shed roof, one half of this area is raised. Some of the units have an additional balcony on this uppermost level, accessible via a ladder.

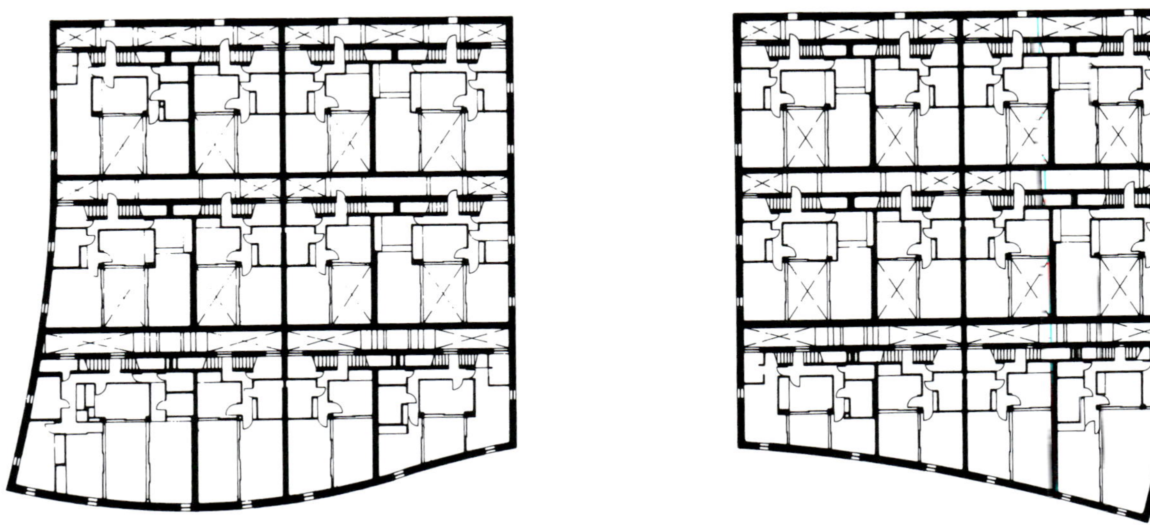

2

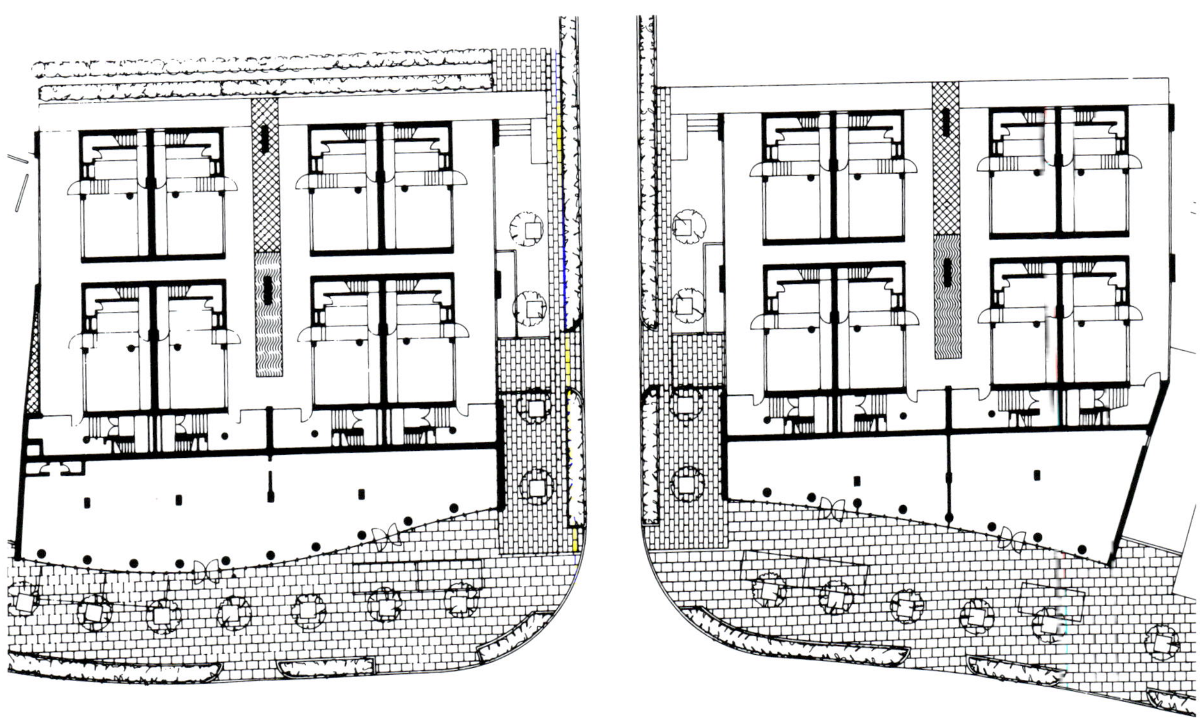

1

1 Ground floor of entire building
2 2nd floor of entire building
3 5-room unit on three levels 1:200
4 3-room unit on three levels 1:200
5 Section with light shafts at back of units
6 Section through patios

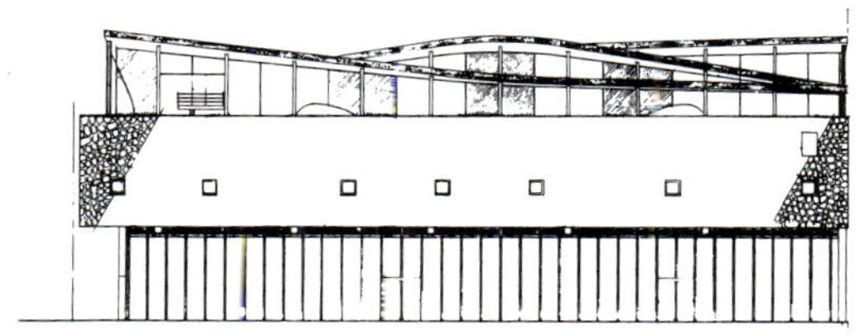

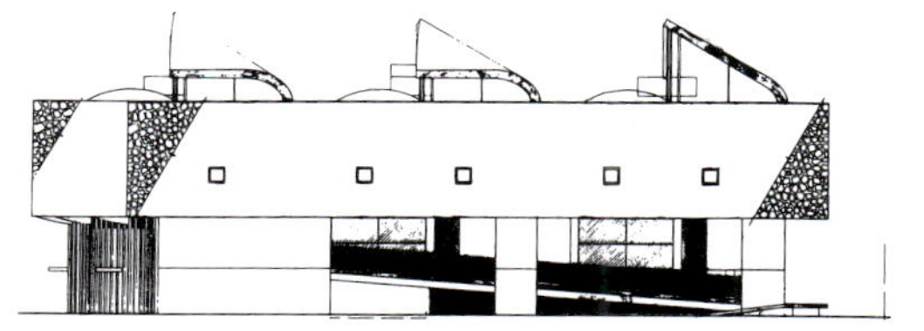

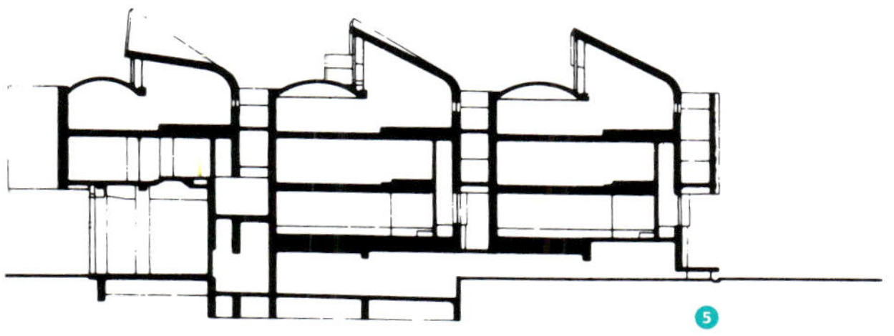

5

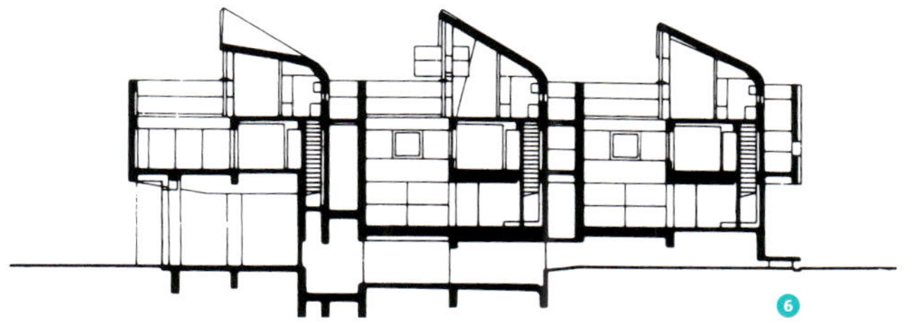

6

2.6
LOW RISE
HIGH DENSITY

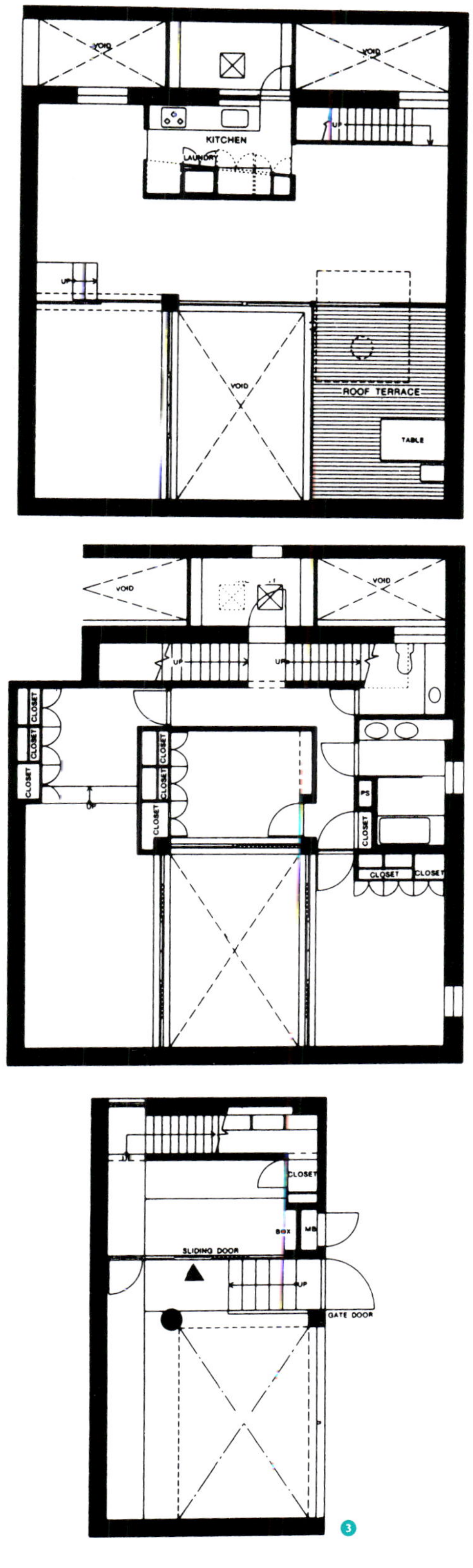

3

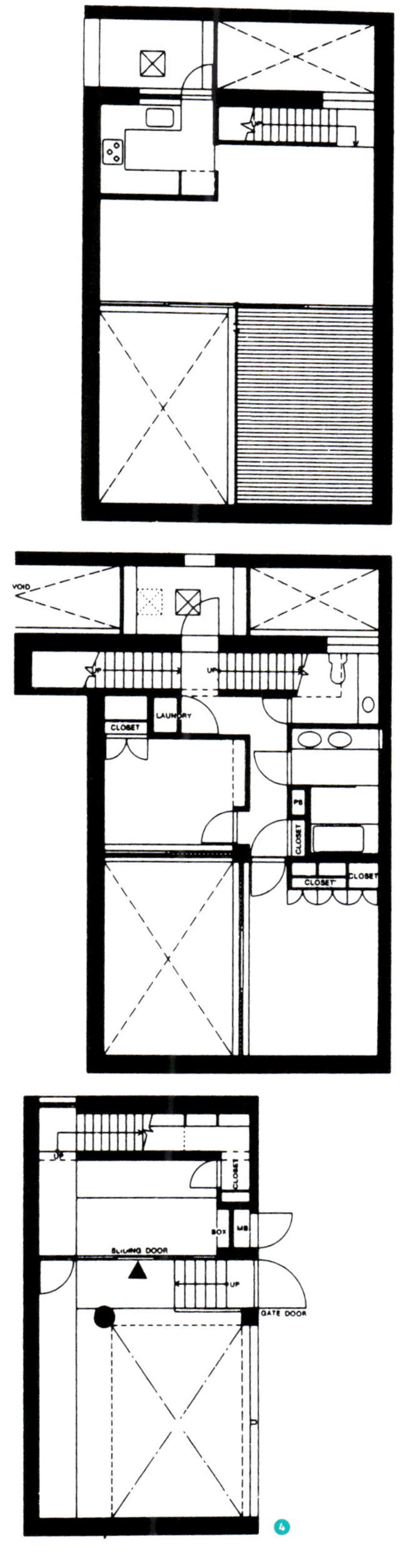

4

Building type
two blocks of packed-together patio houses
2/3 stories
facing NW/SE, NNE/SSW

Date of construction
1991

Number of units
2 × 12

Size of units
3-room apts., 102.5/108.5 m²
4-room apts., 156.5 m²
5-room apts., 150.5/179 m²
6-room apts., 223 m²

Area per user
30–37 m²

Building depth
approx. 12.5 m per unit
total depth approx. 35–40 m

Layout
2 full-surface housing developments, each with three separate house rows, individual apartment access

Open spaces
balconies
stone gardens in patios
roof terraces

Parking
parking spaces

Architect
OMA/Koolhaas
Amsterdam

Location
Fukuoka
Japan

Referring to floor plan 1:200

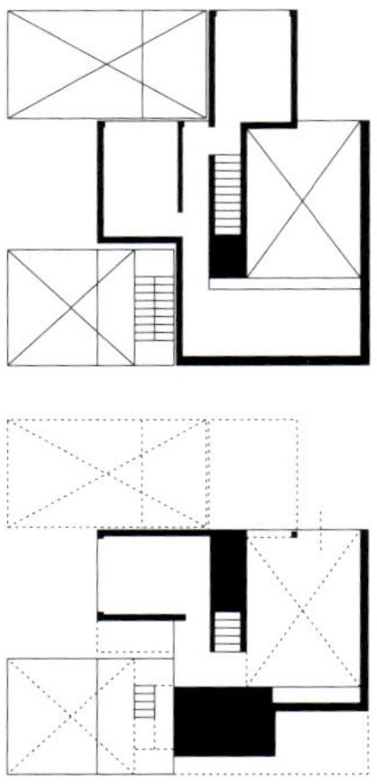

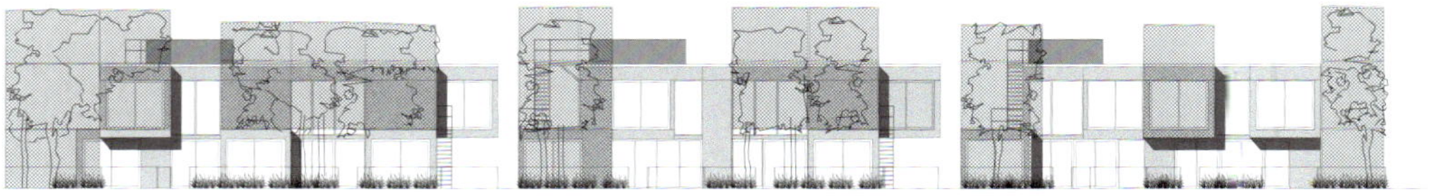

Cité Manifeste | Lewis / Block Architectes | 2005

On the 150th anniversary of the completion of France's first major workers' housing complex, the creation of a new and innovative housing development has been initiated by the same social housing association at the same location. The design is an adaptation of the historic building type and the process of ongoing modification to the latter by the residents. The so-called "Carré mulhousien" linked four houses on a cruciform plan into a single structure. Conversions and various expansions ensued over time. In the new project, four housing units – two of which are back-to-back where they have the kitchen and bathroom – form a single freestanding block. The groundfloor living room at the center of the unit is 5 m high; the bedrooms are inserted as cubes or "docked" at the outside on the upper or lower level and some on both levels. The sliding elements in front of the loft space above the kitchen/bathroom block facilitate variable uses: as an open space for socializing or a quiet, more private space. The result is a complex spatial construct, with additional densification of use on the outside in the form of wire-mesh enclosed gazebos for plants, open spaces, external stairs, and parking pads. The entire construct is rather like a complex parcours where spatial boundaries have been almost dissolved, than an enclosed apartment complex.

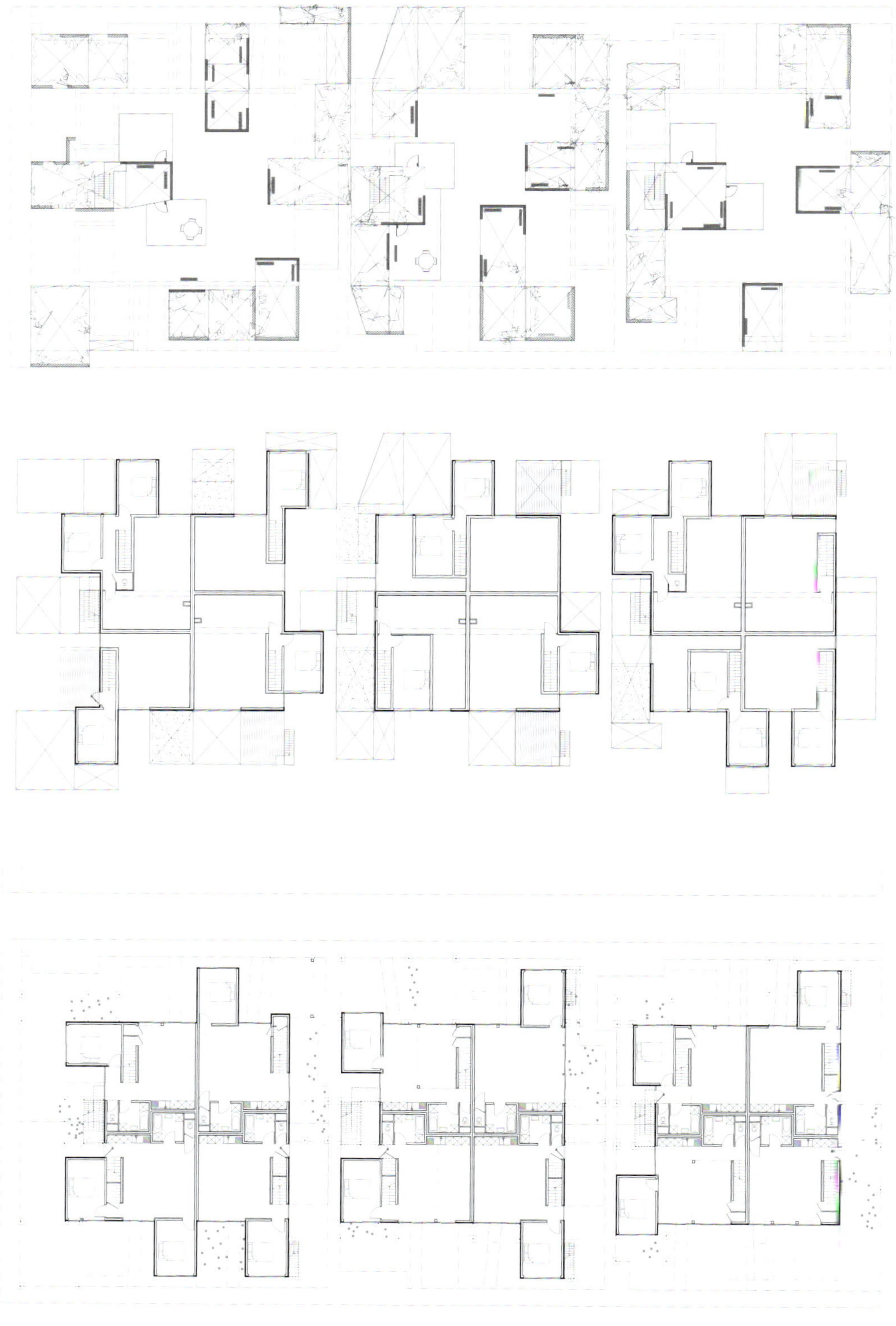

1

1 Floor plans of entire complex: ground floor, 2nd floor, roof with roof patios 1:500
2 Ground floor: two 4-room apartments with two-story living room, kitchen, and bathroom and attached "boxes" for private rooms 1:200
3 2nd floor: upper level with flexible gallery space and attached "boxes" for private rooms 1:200
4 Cross section of private, living, and gallery space / living room 1:500
5 Cross section of private rooms 1:500
6 West elevation 1:500

2.6
LOW RISE HIGH DENSITY

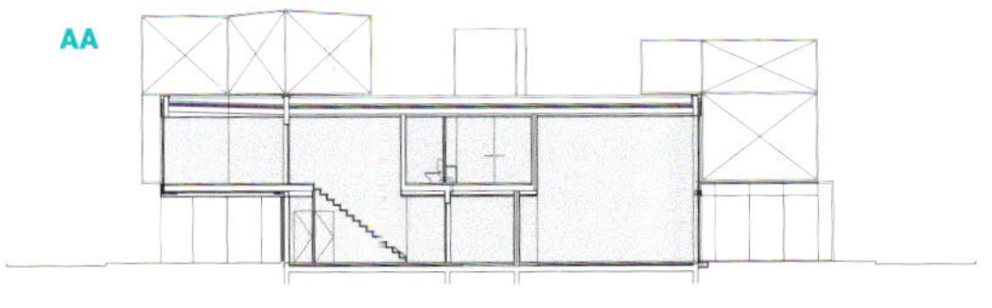

4

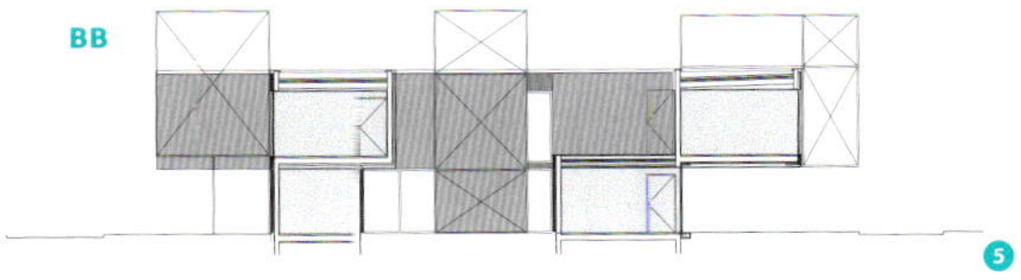

5

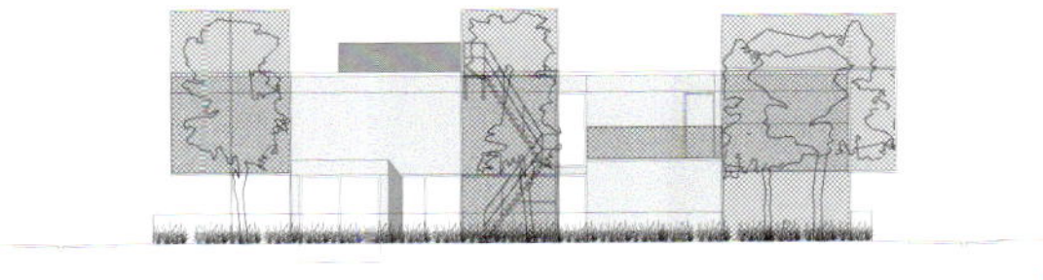

6

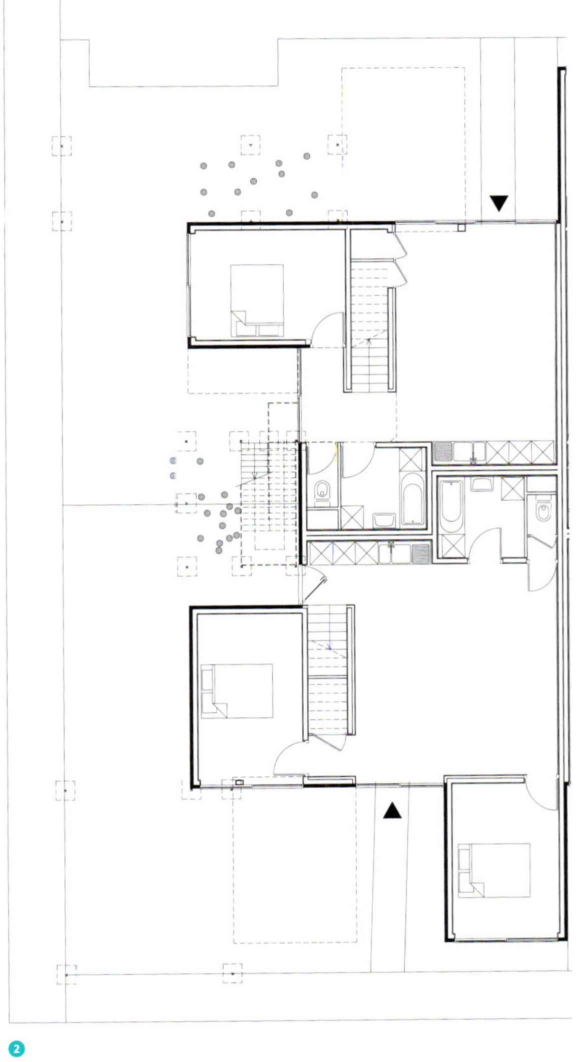

2

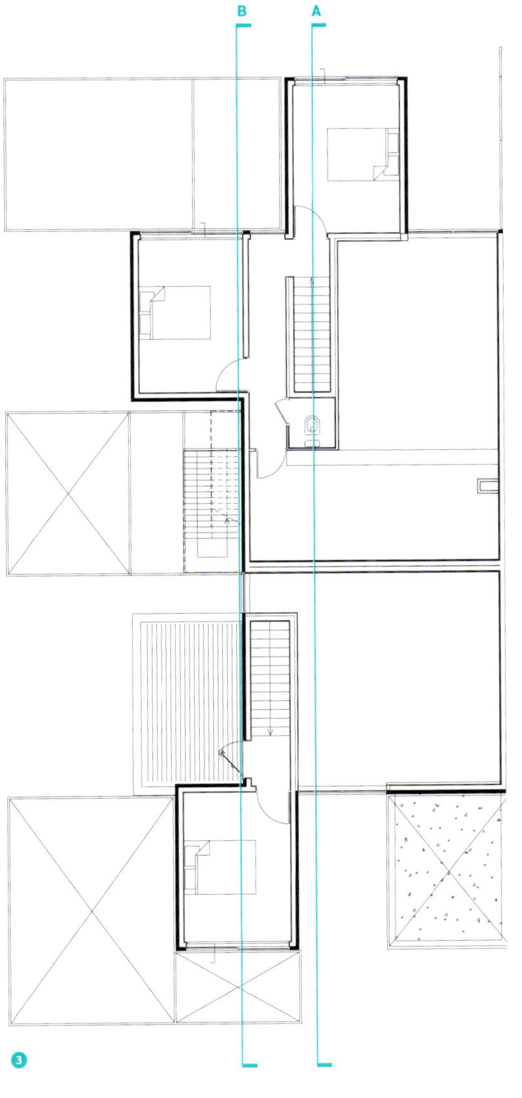

3

Building type
2-story housing units combined into blocks facing NE/ES/SW/WN

Date of construction
2005

Number of units
12

Size of units
2- to 4-room apts., 90 m²

Area per user
22.5–45 m²

Building depth
15–23 m

Layout
4 housing units arranged side-by-side i.e. back-to-back form a freestanding block with all-round orientation, attached cubes for individual spaces and with open spaces bordered in wire mesh

Open spaces
yards, gazebos, and roof patios

Parking
parking lot

Architect
Duncan Lewis
Scape Architecture
Bordeaux
Block Architectes
Nantes
Collaboration:
Tierry Maître
Tanguy Vermet

Location
Mulhouse
France

Referring to floor plan 1:200

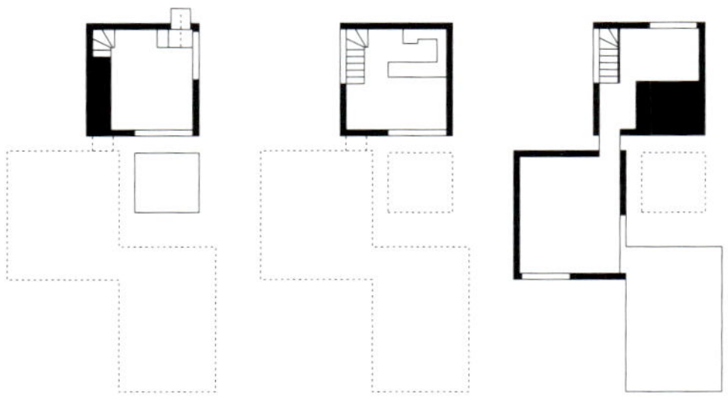

Seijo Townhouse | Sejima & Associates | 2007

Like on a board game, similar volumes are placed in relation to one another on a field, either linked at the corners or shifted away from each other. Complex and highly varied apartments, most occupying one room per volume, are created on four levels. Each apartment stretches horizontally and vertically across several volumes, forming a spatial continuum in the interior without its boundaries being recognizable from the outside. Thus there is an ambivalence as to whether the complex is a single large structure with apartments or a collection of individual houses. Up to four volumes are linked on one level, where the narrow passages at the interfaces without doors mark the transitions. As a result of the right-angled arrangement, the buildings frame intimate courtyards of similar size, which are treated as part of the spatial patchwork. Stairs usually connect to another room in the levels above and below, either providing an entrance into the apartment or leading to a quieter space. The interlocking creates different spatial scenarios, which inspire interpretations of one's own. Light and air flow in from alternating sides, the path through the apartment and the view leads to ever different yet similar exterior spaces. Every apartment has its own garden or a roof terrace. Although the gardens are separated, they remain readable as a continuous area. This creates a complex spatial configuration in which the living environments of the residents overlap.

2

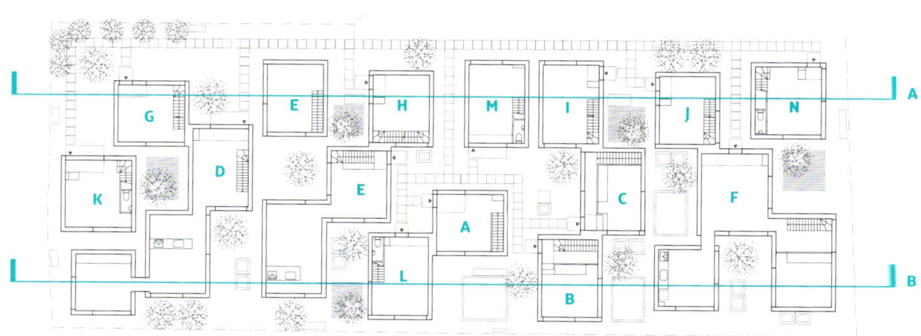

1

1. Total floor plans of basement, ground floor, 2nd and 3rd floor 1:500
2. Longitudinal sections 1:500
3. Floor plan detail basement 1:200
4. Floor plan detail ground floor 1:200
5. Floor plan detail 2nd floor 1:200
6. Floor plan detail 3rd floor 1:200
7. South elevation 1:500

2.6
LOW RISE HIGH DENSITY

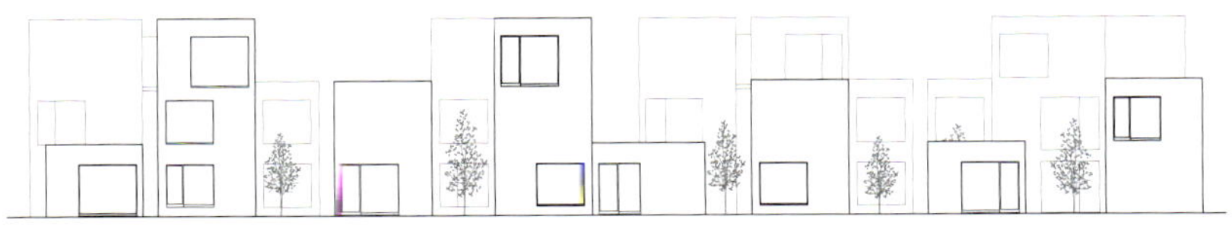

7

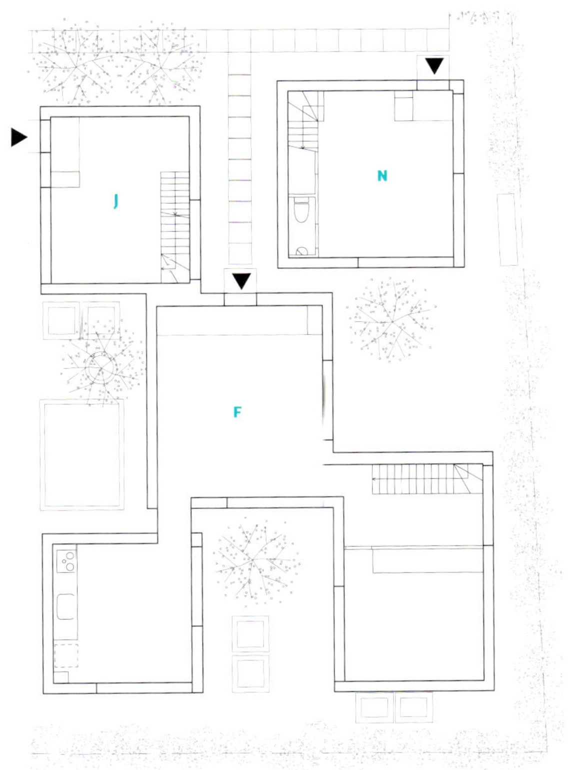

4

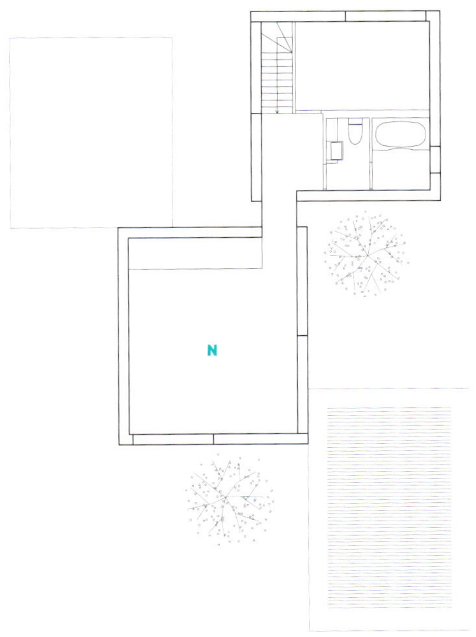

6

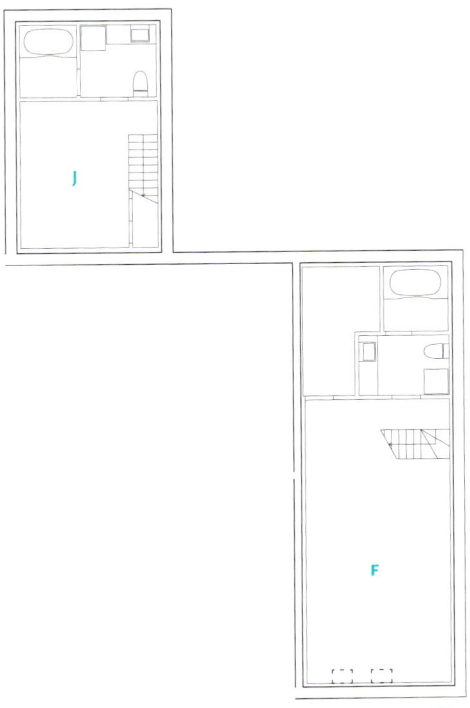

3

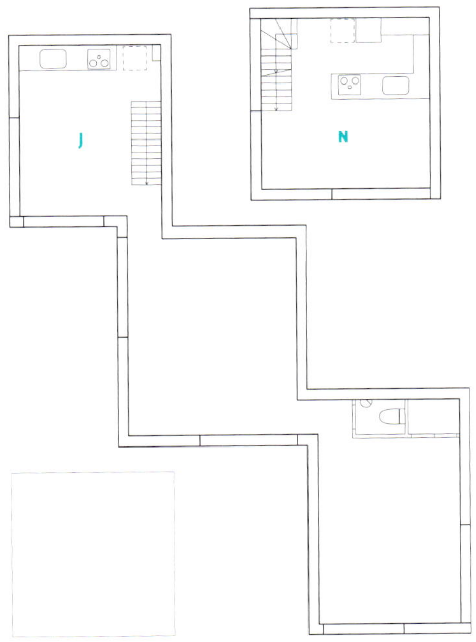

5

Building type
residential complex
4 stories
(incl. finished basement areas)
facing in all directions

Date of construction
2005–2007

Number of units
14

Size of unit
type **A**: 4-room apt., approx. 91 m²
type **B**: 4-room apt., approx. 105 m²
type **C**: 4-room apt., approx. 101 m²
type **D**: 3-room apt., approx. 95.5 m²
type **E**: 3-room apt., approx. 100 m²
type **F**: 3-room apt., approx. 103 m²
type **G**: 4-room apt., approx. 95.5 m²
type **H**: 4-room apt., approx. 91.5 m²
type **I**: 4-room apt., approx. 97.5 m²
type **J**: 4-room apt., approx. 98.5 m²
type **K**: 3-room apt., approx. 77.5 m²
type **L**: 3-room apt., approx. 74 m²
type **M**: 3-room apt., approx. 76.5 m²
type **N**: 3-room apt., approx. 79 m²

Area per user
23–33.5 m²

Building depth
6–16.5 m

Layout
patchwork-like conglomerate of 20 townhouse blocks and courtyards (partially lowered to basement level) with access paths, steel-frame construction, exposed brickwork and floor-to-ceiling glazing

Parking
no parking on lot

Architect
Kazuyo Sejima & Associates
Tokyo
design team:
Kazuyo Sejima, Mizuki Imamura
Takashige Yamashita, Sadaharu Ota
Tetsuo Kondo, Kansuke Kawashima

Location
Setagaya-ku
Tokyo
Japan

Referring to floor plan 1:200

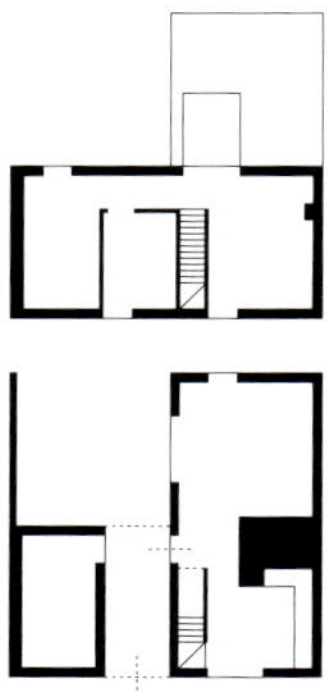

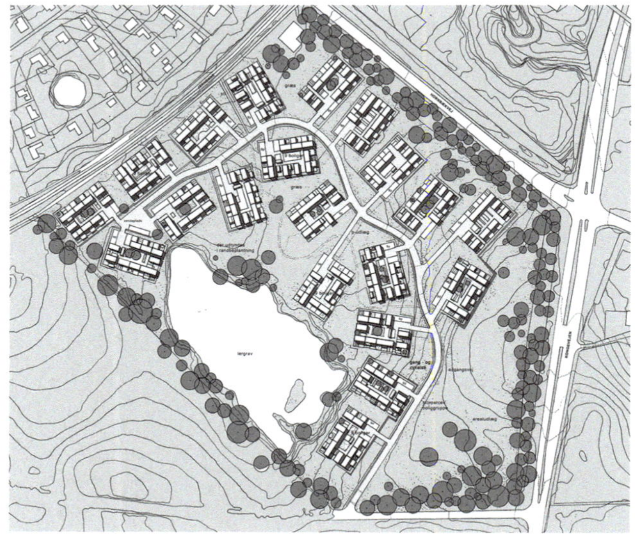

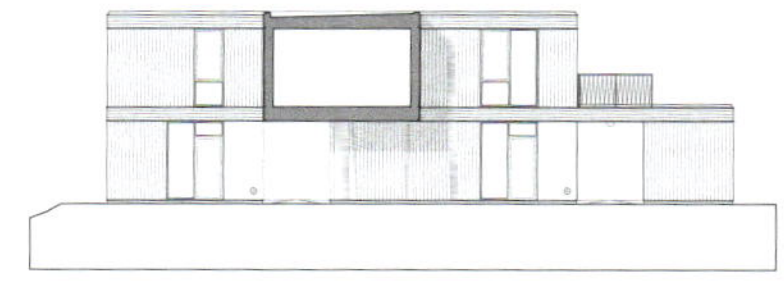

Kvistgård | Vandkunsten | 2008

The desire both to reduce building costs and engage in the design process determined the character and structure of this ensemble. Prefabricated wooden modules comprised of standardized building elements shortened the construction time and permitted residents to choose from different configurations.

In this organically laid-out ensemble nestled into the topography of the site, each group consists of nine residential units arranged around a shared courtyard protecting them from the wind and enabling a view for all into the surrounding landscaped park. From the outside spaces to the floor plan organization of the individual unit, the entire estate is characterized by a subtly formulated series of thresholds separating public from more private spaces. Branching walkways lead from the main path through the open corners of the more private courtyards. These are the interfaces of the small, manageable communities, each with its own connection to the landscape and passages to the individual residential units and their backyards. The kitchen of each unit is oriented toward the courtyard to foster an area of neighborly contact, while the bedrooms are all located on the top floors.

L-shaped constellations of three 5 × 5 m modules form the ground floors, which open up with an outdoor terrace to the surrounding landscape. An eat-in kitchen occupies the corner module where a staircase leads to the upper floor. Situated to one side of this is a covered passage to the courtyard and a large storage room behind, and on the other side, an open living room. For the upper floor, residents could choose between five options consisting of up to four modules. The smallest version uses only one module to provide a single bedroom with terrace above the kitchen. Two modules can be combined to create three bedrooms with an optional office or hallway as circulation space – with or without a bathroom. The four-module option incorporates an unused area of the neighboring unit.

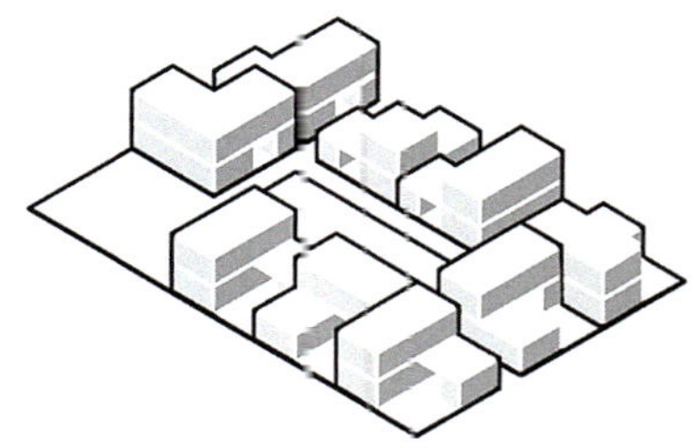

2

1

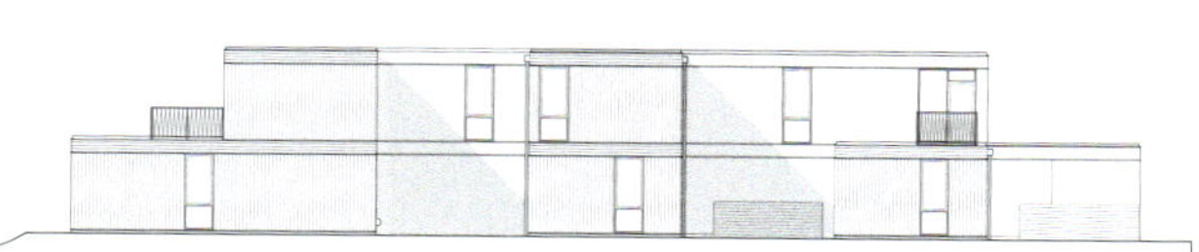

1 Courtyard with nine units, ground floor 1:500
2 2nd floor 1:500
3 Ground floor of base unit with three modules 1:200
4 2nd floor option with one module 1:200
5 2nd floor options with two modules 1:200
6 2nd floor options with 3 or 4 modules 1:200

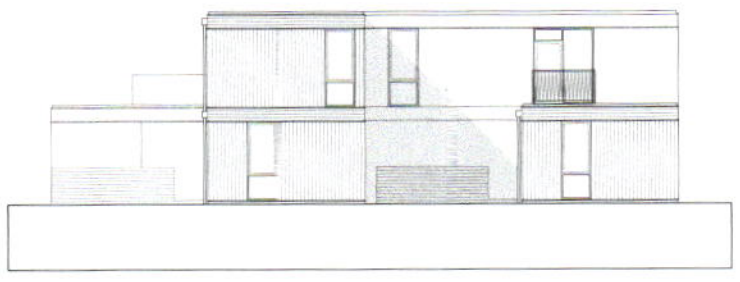

2.6
LOW RISE
HIGH DENSITY

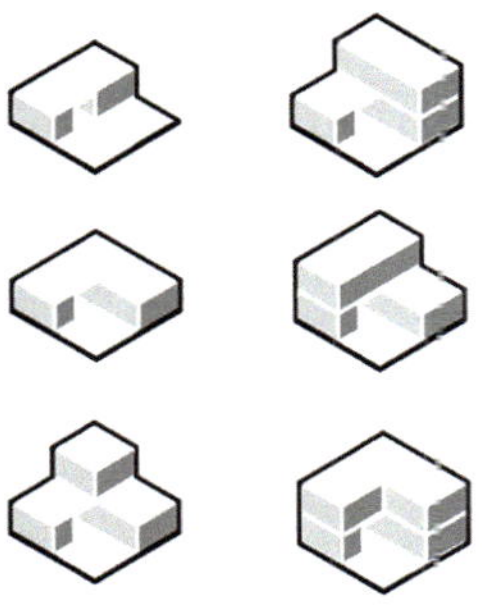

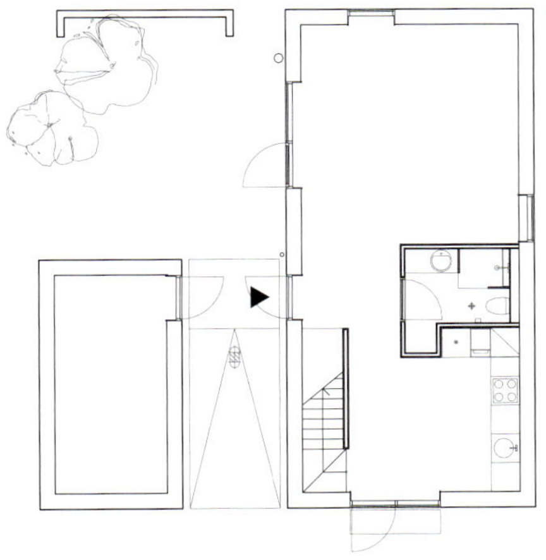

3

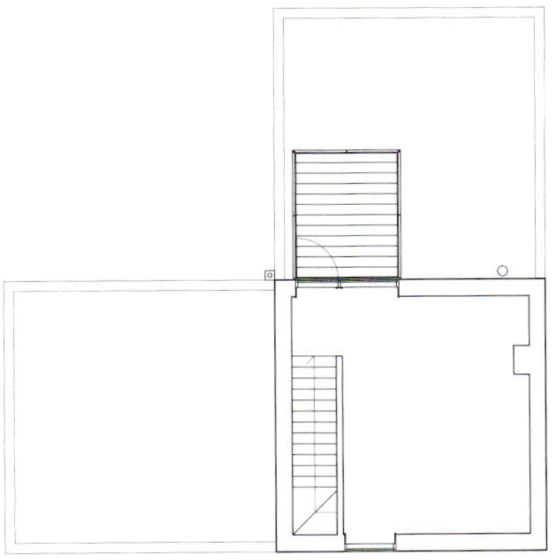

4

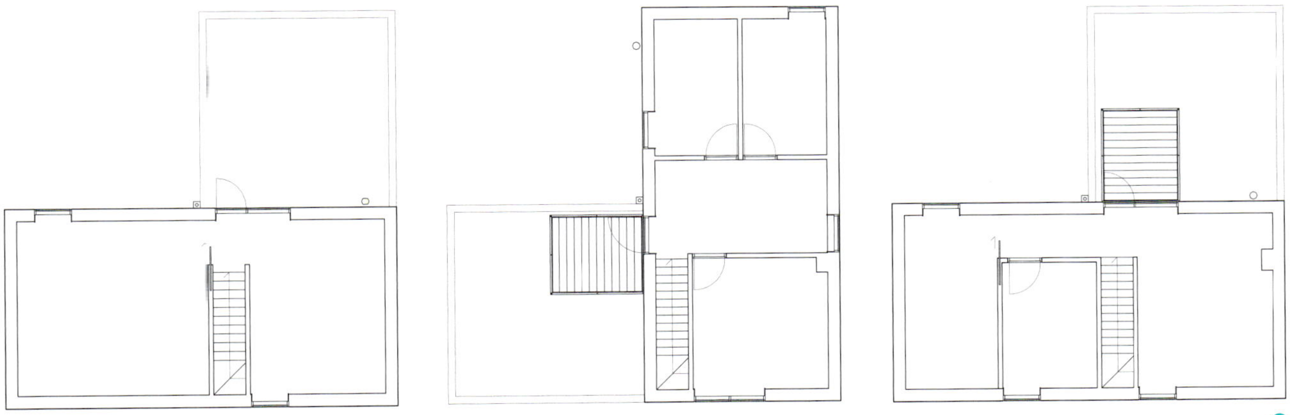

5

6

Building type
ensemble of 14 housing clusters, each with 9 units arranged around a courtyard
1–2 stories
oriented to all sides in various directions

Date of construction
2004–2008

Number of units
171

Size of units
2-room apts., 75 m²
4-room apts., 100 m²
4-room apts., 125 m²
6-room apts., 150 m²

Area per user
25–37.5 m²

Building depth
5.2 m/10.4 m

Layout
housing development in a landscaped park with a public pond and generous green spaces between housing clusters around courtyards
prefabricated wooden construction modules enabling a variety of configurations

Open spaces
private outdoor terraces, roof terraces, common courtyards

Parking
open parking spaces in courtyard corners

Architect
Vandkunsten Architects
Jan Albrechtsen
Jens Kristian Seier
Copenhagen

Location
Hornbækvej 115
Kvistgård
Denmark

Facing various directions

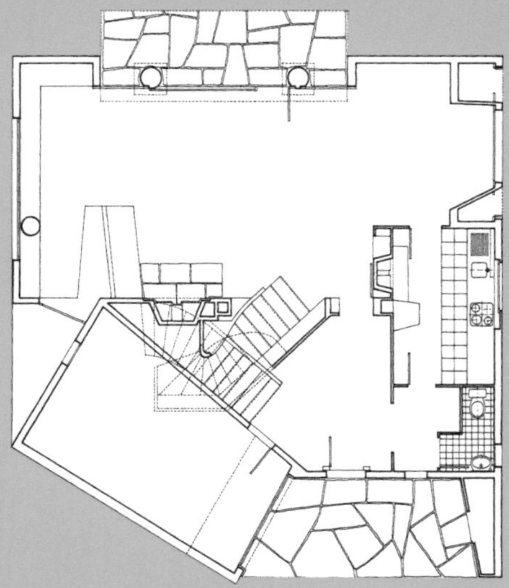

3.1
DETACHED HOUSE

As a morphological type, the detached house denotes a free-standing, small residential building predominantly in rural or suburban contexts with buildings of similar proportions. Generous amounts of space per occupant, orientation in all directions, and a private garden make detached houses a highly privileged form of building. A car is an indispensable requirement for this housing type, because it makes the dream of a house in a green setting possible. The almost unavoidable consequences of this building form therefore include urban sprawl, the paving of natural landscapes, and increased traffic.

In most cases – including most of the examples presented here – detached houses have a single leasing party, who enjoys a high degree of codetermination in the conception and design of the house. From the classic single-family home to the extreme form of a spacious villa, a multitude of radical and innovative approaches to habitation are possible, since each concept need function only for one, usually known, group of occupants. Thus the private home can serve as an opportunity for experimentation in residential architecture, often generating exciting, inspirational spatial concepts, even though they are rarely directly applicable to multifamily or multi-story dwellings.

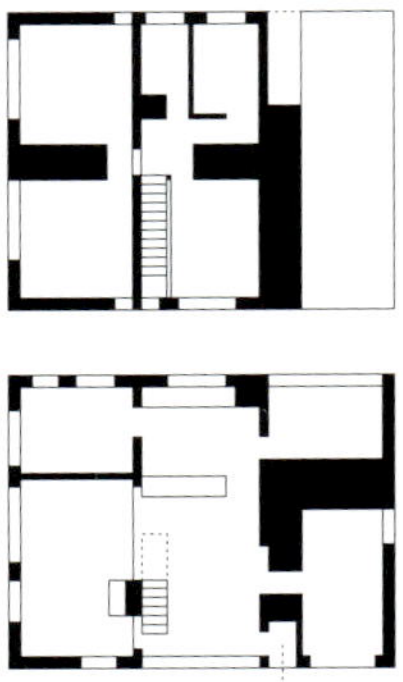

Sugden House | Smithson | 1956

The house for a couple, who were friends of the architects, was to be an "ordinary house," with the stipulation that it should feature masonry and a ridge roof. The house is still impressive today for the sensitive manner in which this simple program was met by characterizing the individual rooms instead of assigning them rigidly or leaving them completely without specification. A small entrance leads into the large middle zone, the heart of the house with (dining-)table and fenestrated facade. The stairs to the bedrooms on the upper floor and the chimney separate the living room on the ground floor from the central dining area, although they remain otherwise strongly linked. A customized piece of furniture, usable from both sides, defines the far end of the space as a kitchen, which provides access to the covered courtyard and the play-/bed-/study room and lies as a mediating space between these areas. The prefabricated windows are set freely into the masonry on all sides, at all points where a view of the exterior is desirable. The relationship to the garden is thus experienced in each of the rooms. The particular charm lies in the details: e.g., the manner in which the semitransparent trellis separates the small room above from the landing. This makes it possible to extend the common room upward. Or the small step, which sets the bedrooms apart from the common landing, thus separating them, and at the same time adding more space to the living room below.

Building type
single-family house
2 stories, facing SE/SW/NW

Date of construction
1956

Living area
approx. 94 m²

Area per user
31.5 m²

Layout
masonry construction with exposed timber beam ceiling, standard steel-frame windows, all doors customized

Architect
Alison + Peter Smithson
London

Location
Sugden House
Watford
United Kingdom

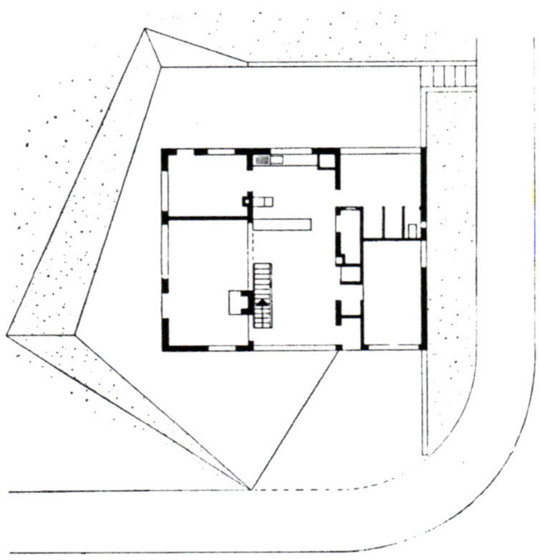

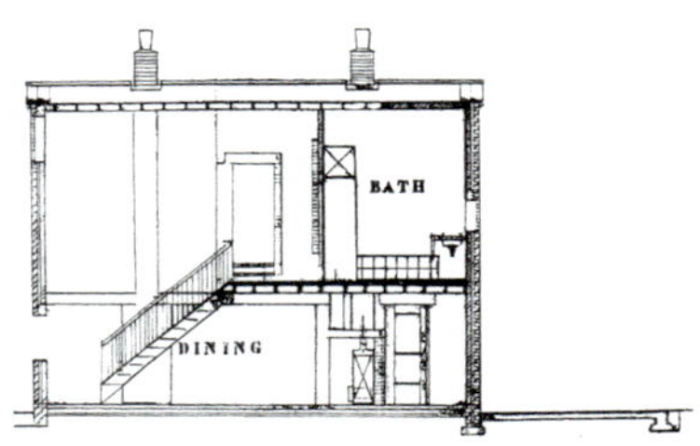

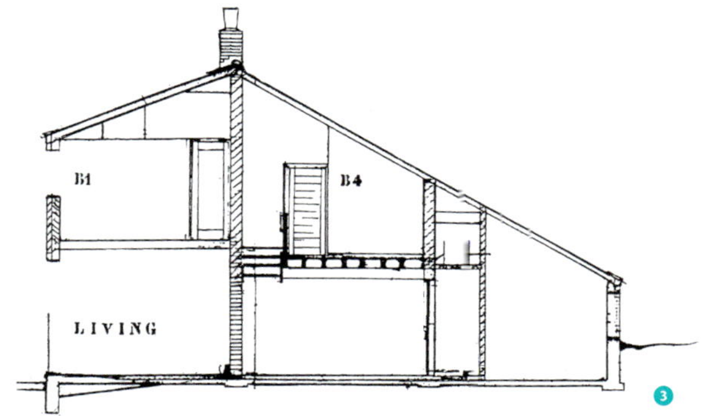

3

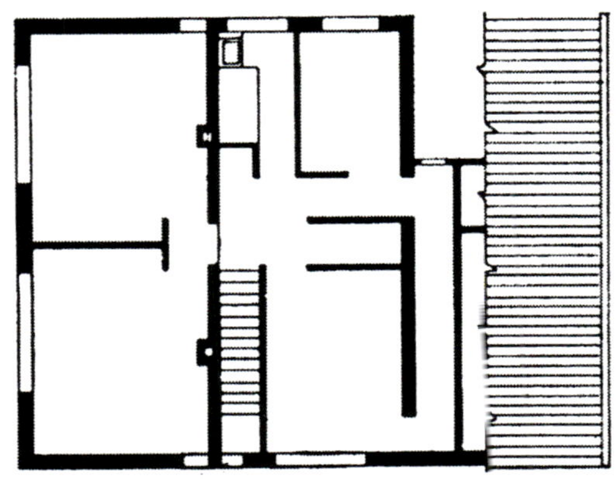

2

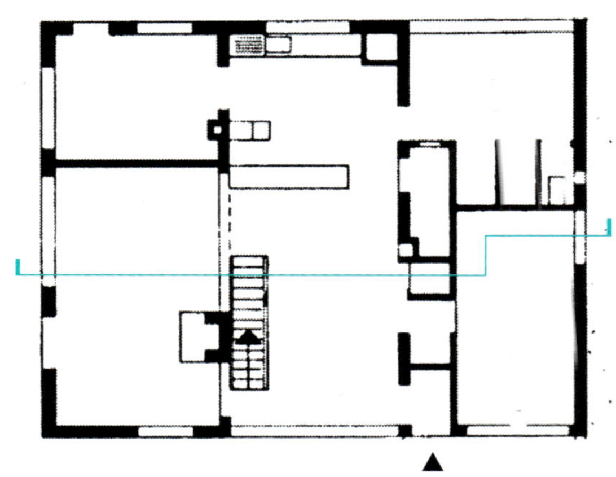

1

1 Entrance level with garage, covered courtyard, and living area 1:200
2 Upper level with private rooms 1:200
3 Longitudinal section of living and dining area and garage below, bedrooms above 1:200

Referring to floor plan 1:200

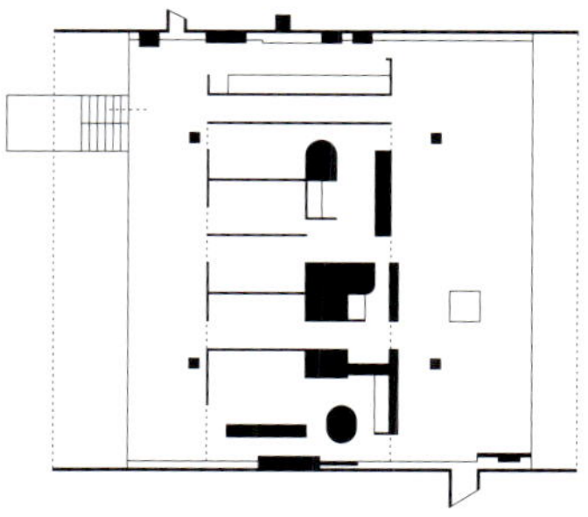

Casa Mendes da Rocha | Mendes da Rocha | 1960

All rooms are organized on a single raised level that is nearly square in plan and supported by only four pillars. Floor and ceiling with their ribbed slabs in fair-faced concrete are space-defining elements. To the north and south, the slabs transition into closed walls, perforated only by small, precisely placed openings for light and air. Bands of windows stretch along the entire east and west sides, interrupted only on the west side by the stairs and the house entrance. The garden is enclosed with walls and embankments and so densely planted that one has the sense of being perched above a horizon consisting solely of trees and shrubs, an impression that is enhanced by the screen wall on the cantilevered roof structure, which blocks out the sky like blinders.

The floor plan is organized into areas for living, dining, cooking, and working by subdividing a single large space with a series of inserted chamber-like rooms, which are treated as furnishings rather than parts of the house. The living room is wrapped around the bedroom area in the form of a "J" with the result that the interior space along the facade functions as a long and tranquil communal area, behind which the bedrooms are screened off. However, when the sliding doors are open, these rooms expand into the living area. Bathrooms, WC, and walk-in closets are arranged like a filter to the bedrooms in the direction of the east facade, where the living and dining areas extend through the entire breadth of the house. The bedrooms can also be opened toward this side. The floor plan can therefore be utilized and interpreted in multiple ways; it is brought to life entirely as a result of the sensual materiality, the diffuse light penetrating into the interior, and the view onto the verdant exterior.

Building type
single-family home, raised above open ground floor, 2 story facing WNW/ENE

Date of construction
1960

Floor area
Upper floor 180 m²

Area per user
30 m²

Layout
reinforced concrete structure on four pillars, side walls and cantilevered roof in fair-faced concrete, open living space with inserted room compartments property surrounded by enclosure wall i.e. embankment on all sides

Architect
Paulo Mendes da Rocha, São Paulo

Location
São Paulo, Brazil

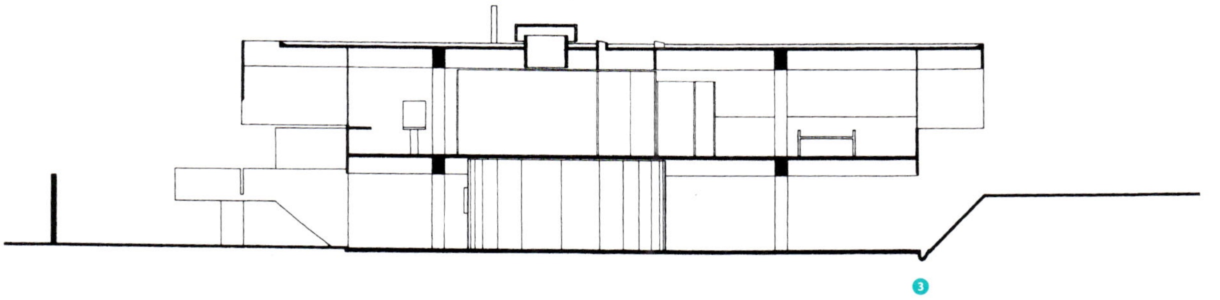

3

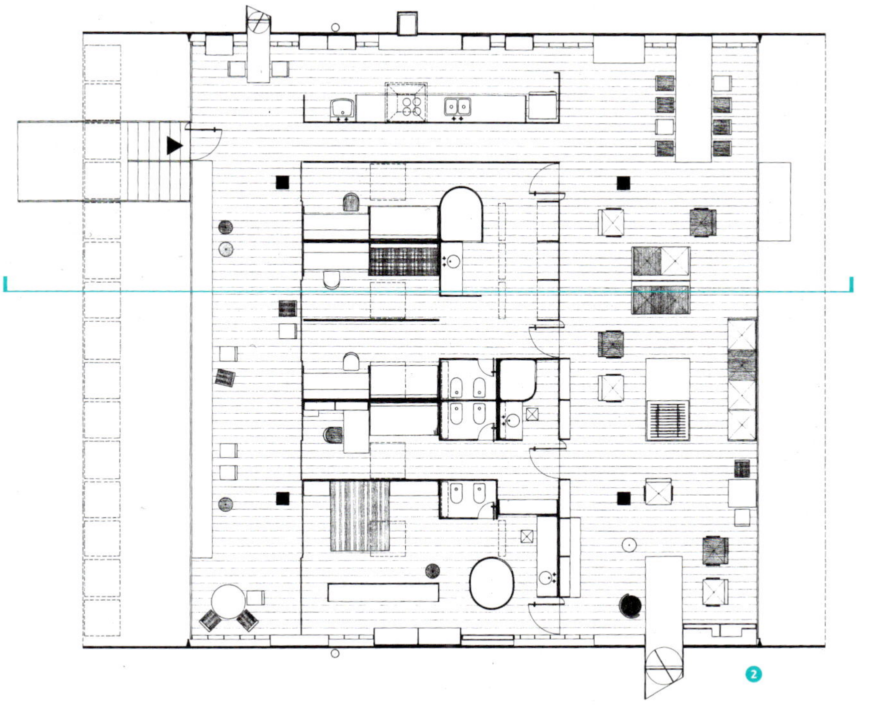

2

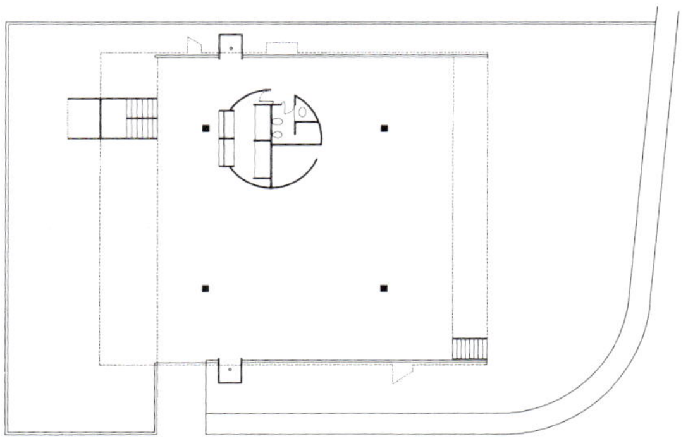

1

Open ground floor with staff apartment and garden 1:500 1
Upper floor: living area 1:200 2
Cross section 1:200 3

Referring to floor plan 1:200

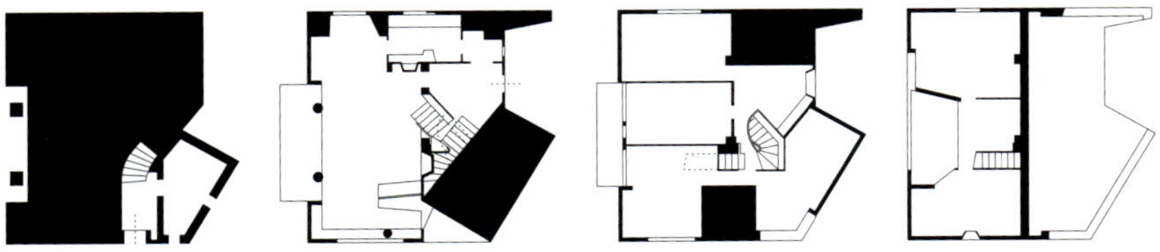

Witzig House | Olgiati | 1966

3.1 DETACHED HOUSE

The house is entirely laid out for a sensory experience. The sculptural and freely composed external wall is not only a facade, it turns toward the surroundings in an almost bodily manner: at times offering grandiose views of the mountain landscapes, at times closed off and inward. The double winding stairs tie all rooms together in a single dynamic movement. They are a spatial element that at times cuts into the building mass, at times appears as an object in space, winding around the chimney and thereby making it the center; being twined around the rooms, so to speak, thus defining the character of the floor plan. Between building skin and circulation, the open living space is organized into niches and various usage areas by the chimney, the inserted kitchen block and the bench facing in two directions. Every opening is devoted to a special situation on the interior or exterior: the recessed loggia, the funnel-like view to the outside from the eating nook, the inserted opening at the bench and the large window overlooking the valley. The room disposition continues throughout the house in similar fashion. The geometry of the two hallways, at the entrance and above in front of the bedrooms, mediates between the spaces and the stairs. The house itself is like a body and extraordinarily comfortable: here, functionality does not remain schematic, but is rather entirely conceived for the enjoyable use of the house.

Building type
detached single-family house
4 stories
orientation in all directions

Date of Construction
1966

Living area
approx. 226 m²

Area per user
approx. 37.5 m²

Layout
sculptural solid construction on a slope, organized in the interior around double winding stairs

Architect
Rudolf Olgiati
Flims-Dorf

Location
Flims-Waldhaus
Switzerland

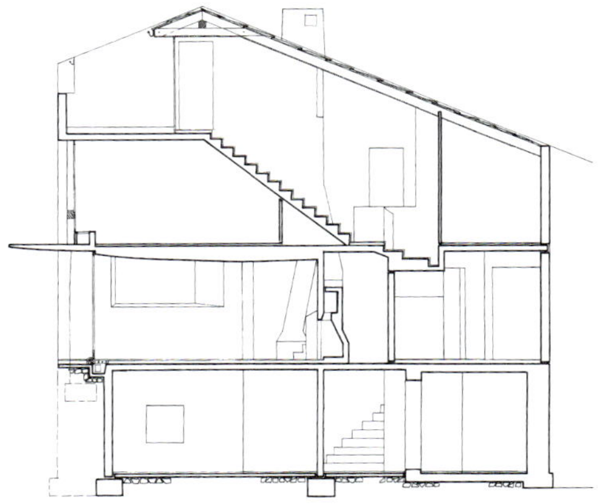

5

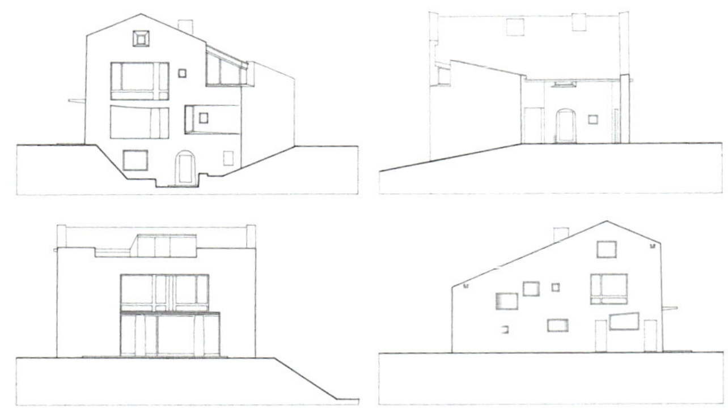

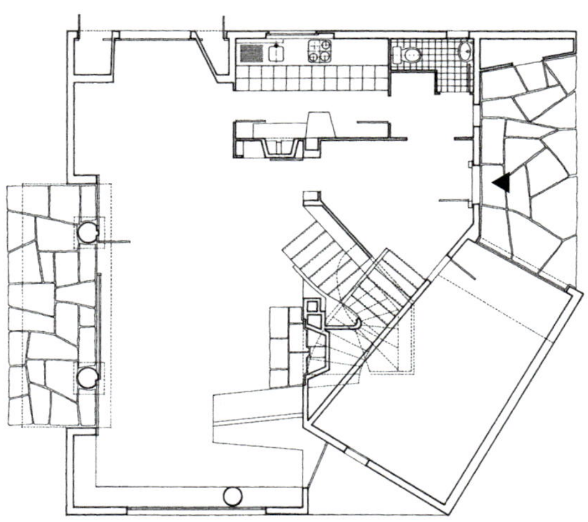

2

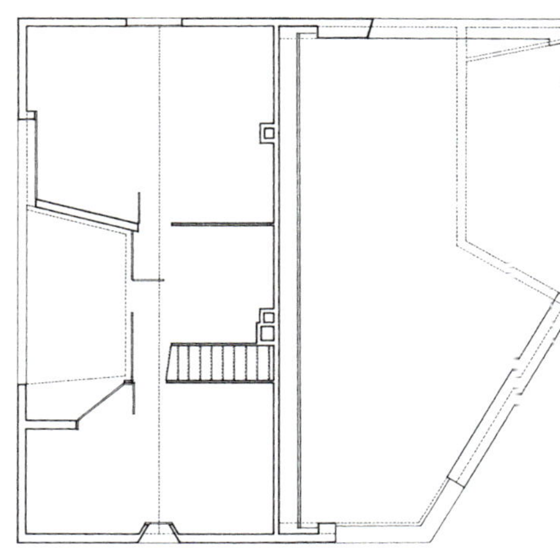

4

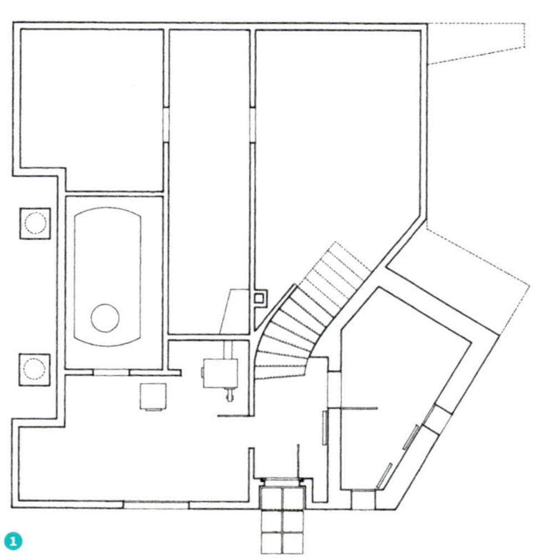

1

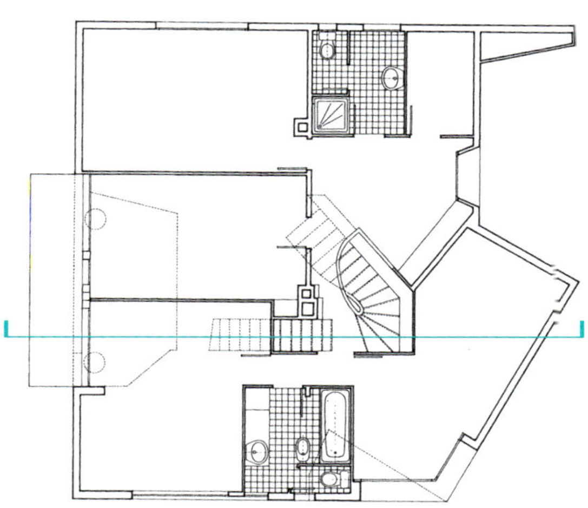

3

Basement with 2nd entrance 1:200 1
Ground floor with main entrance, living area and garage 1:200 2
Upper floor with private rooms 1:200 3
Attic story with private rooms 1:200 4
Section 1:200 5

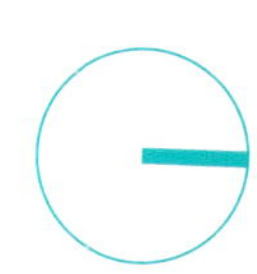

Referring to floor plan 1:200

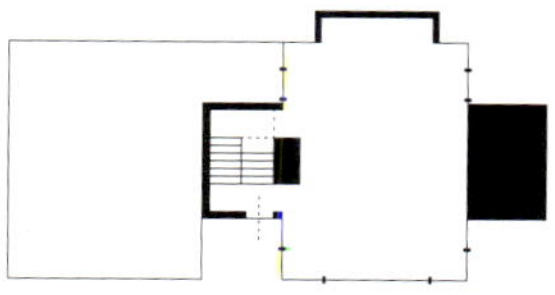

Kappe House | Kappe | 1967

This house was designed by the architect for his family, and, resting on six concrete cores due to weak near-surface soils, appears to hover over the inclined site. The U-shaped, niche-like rooms in these cores along the facade become a motif and are repeated on the other sides of the house, where they contain sanitary rooms or staircases, storage spaces or bay windows with integrated seating. The spatially dominant elements, however, are the floating wooden horizontal levels overlapping in an offset split-level formation over the terrain, which generate a near borderless space with numerous spatial and visual connections and open at multiple points onto outdoor terraces. The central portion of the Z-shaped floor plan incorporates a studio, living spaces, entrance and kitchen. Branching off to either side are the master and children's bedrooms.

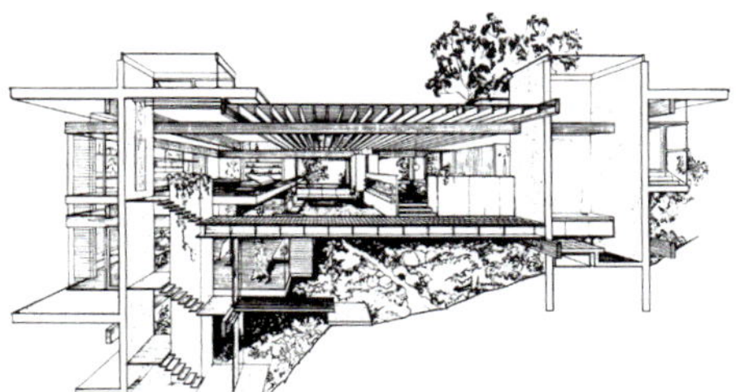

4

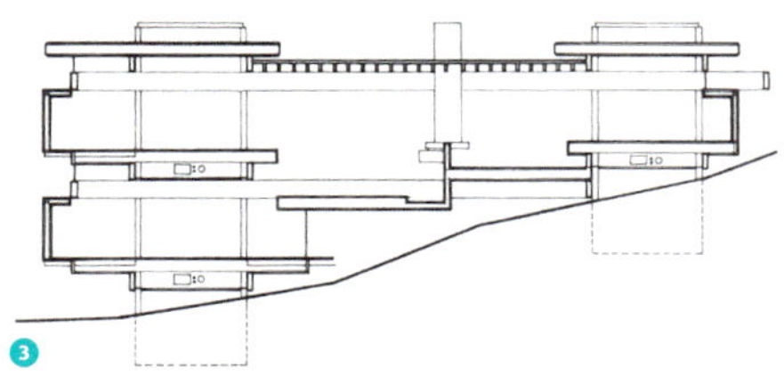

3

Building type
overlapping horizontal residential planes hovering over a steep slope surrounded by trees
2 stories with offset levels
facing in all directions

Date of construction
1965–1967

Number of units
1

Size of units
372 m²

Area per user
74.5 m²

Building depth
8–10 m

Layout
Z-shaped wooden frame construction spanned between reinforced concrete foundation cores
open, overlapping planes in the living areas, with flowing transitions to exterior spaces through large openings
access from below through one of the foundation cores.

Open spaces
terraces

Parking
open parking spaces below building

Architect
Ray Kappe
Los Angeles

Location
715 Brooktree Rd.
Los Angeles, USA

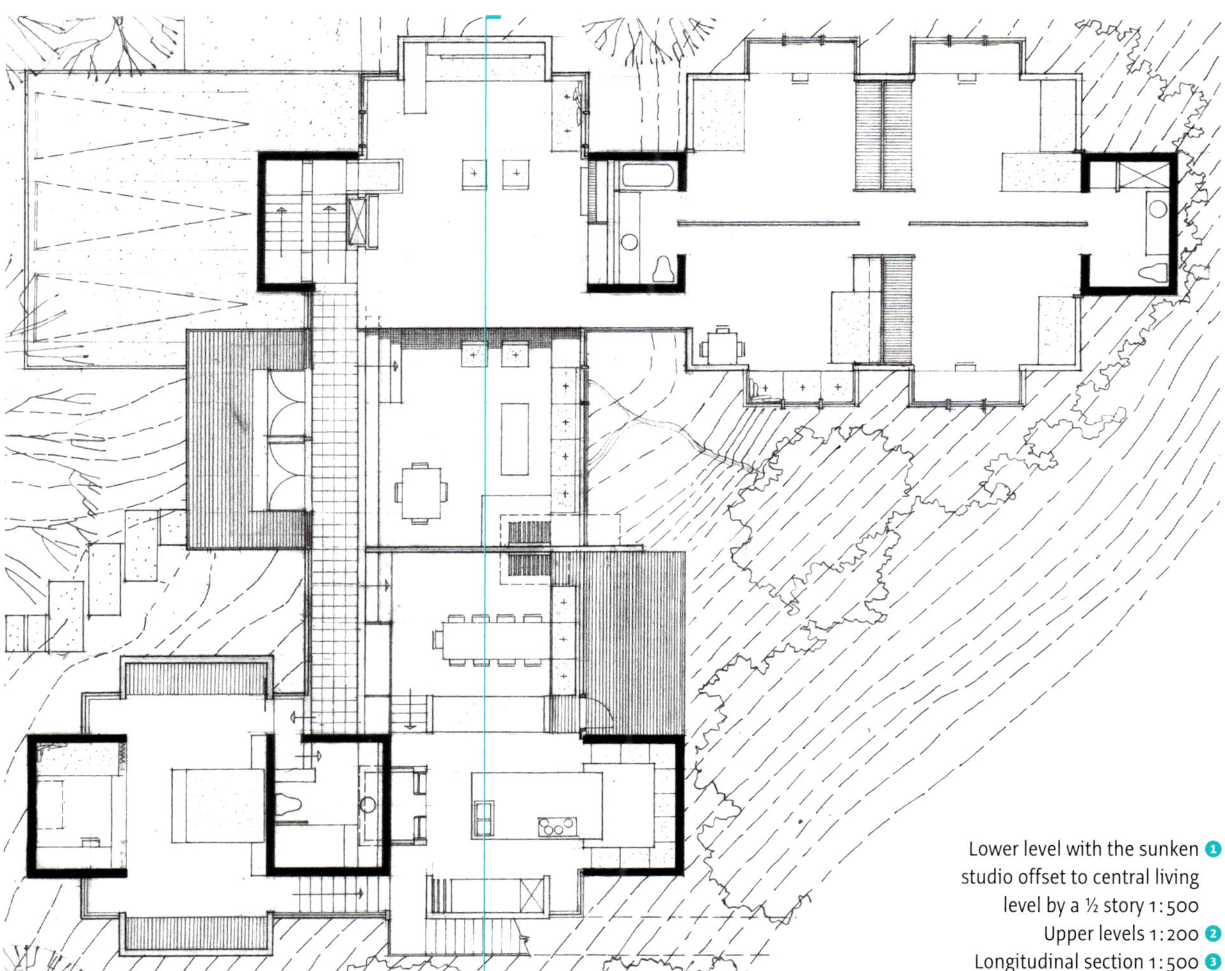

2

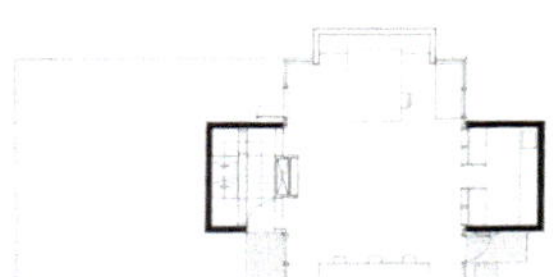

1

Lower level with the sunken studio offset to central living level by a ½ story 1:500 1
Upper levels 1:200 2
Longitudinal section 1:500 3
Sectional perspective 4

Referring to floor plan 1:200

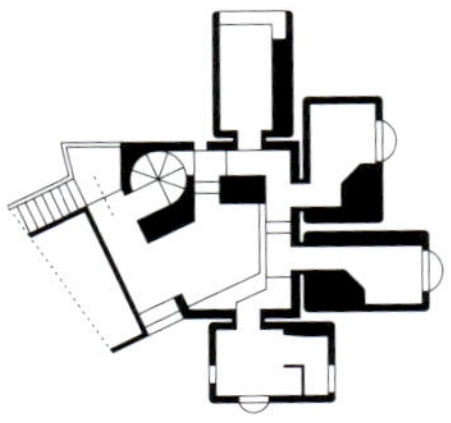

Karuizawa Capsule House | Kurokawa | 1973

3

One-room capsules are docked in all directions to a central concrete nucleus. The capsules are the same as those used for the famous larger residential tower in Nagakin. Their dimensions are an answer to the requirements for ease of transport and rapid assembly (on to the nucleus). This house on a hill develops from the top downwards: the garage and roof terrace are on the entrance level; the floor below is reached by an exterior flight of stairs and is composed of capsules containing a tea room, a kitchen, and two private rooms; the second floor down, consisting solely of the "trunk," is a living room with large circular window. A spiral staircase connects the stories. The capsules are ordered around the nucleus at increasing heights, suggesting a procession upwards to the tea room at the highest point. Kurokawa's idea of a "tea room in a space ship" was transformed into architectural reality here: state-of-the-art technology makes sense only when it is employed so as to enable people to partake of pleasures such as the tea ceremony.

Building type
summer house on a slope
2 stories, facing N/S/W

Date of construction
1973

Living area
103 m²

Area per user
34.5 m²

Layout
two-storied concrete core, with capsules made of steel (4 × 2.5 m) hung from it, exterior covering of Corten steel, PVC skylights as windows

Architect
Kisho Kurokawa & Partners
Tokyo

Location
Kita-saku
Nagano
Japan

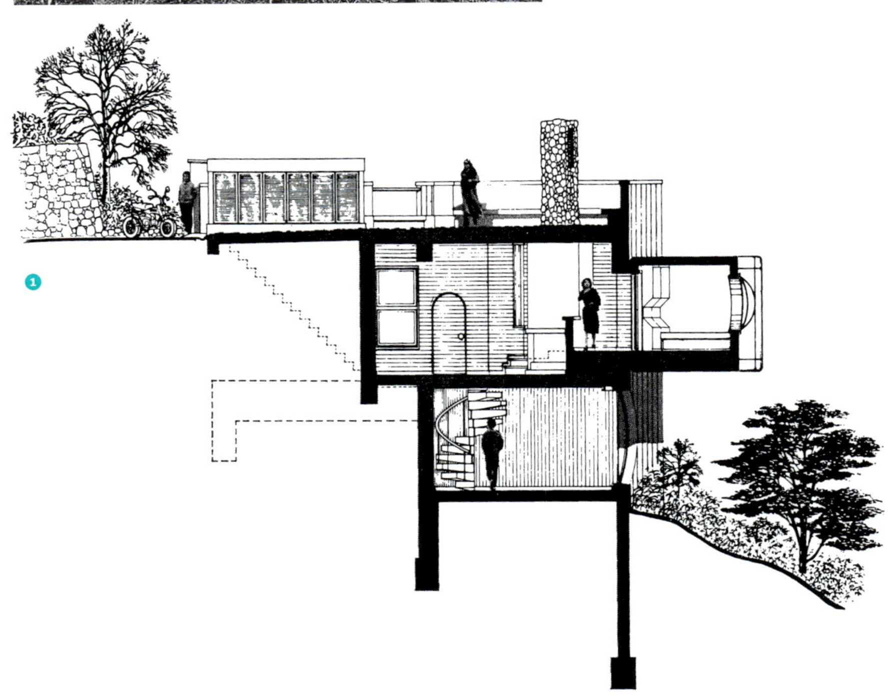

1

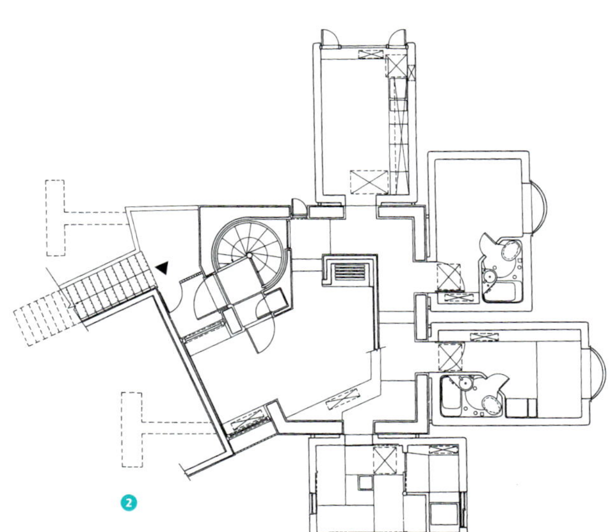

2

Section 1:200 1
Plan of main level 1:200 2
View into the tea room 3
in one of the capsules

Referring to
floor plan 1:200

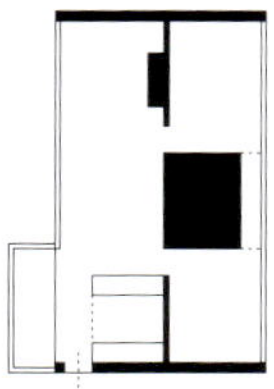

Kauf House | Märkli | 1989

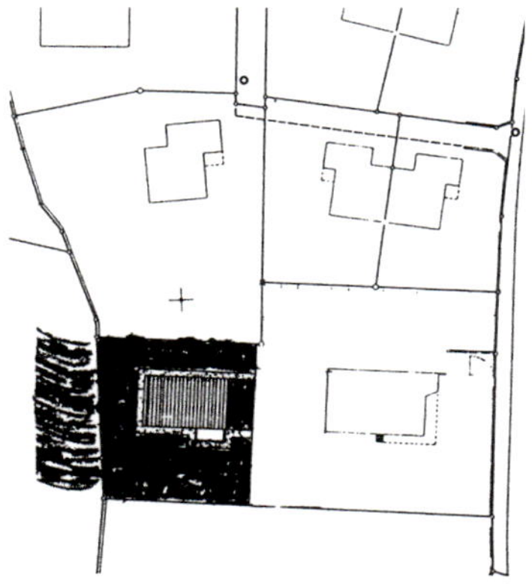

A simply unusual, unusually simple floor plan: per floor, the stairs provide access to only one of three identical units. The elongated apartment is divided into a private and a common area by a wall in the middle. The division of the floor area is effected simply by placing various elements along the middle wall. There is no need for a hallway. The kitchen and loggia are located at the entrance. The continuous windows give the kitchen a clear view of the living room, dominated to the rear by an open fireplace. A bathroom core serves as a dividing element for the private half of the unit. The bathroom's cabin-like sliding walls separate and connect the two bedrooms. The loggia is suspended in front of the facade like a separate room or body. The ribbons of windows along both longitudinal sides of the building bathe the interior in light.

Building type
detached apartment building
3 stories
facing N/S

Date of construction
1989

Number of units
3

Size of units
3-room apts., 71 m²

Area per user
23.5 m²

Building depth
6.9 m

Access
single-loaded stairwell

Open spaces
loggias

Parking
open parking places

Architect
Peter Märkli
in collaboration with
Gody Kühnis
Zurich

Location
Trübbach
Switzerland

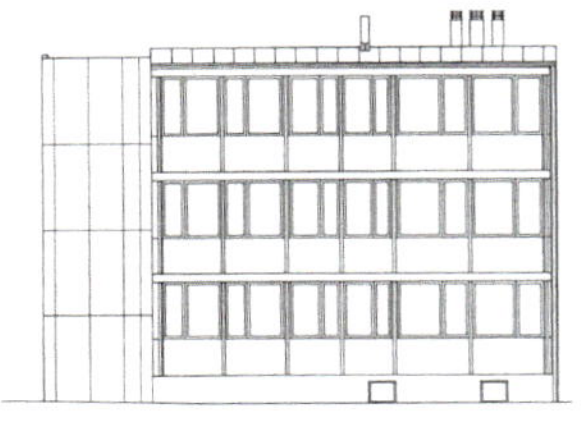

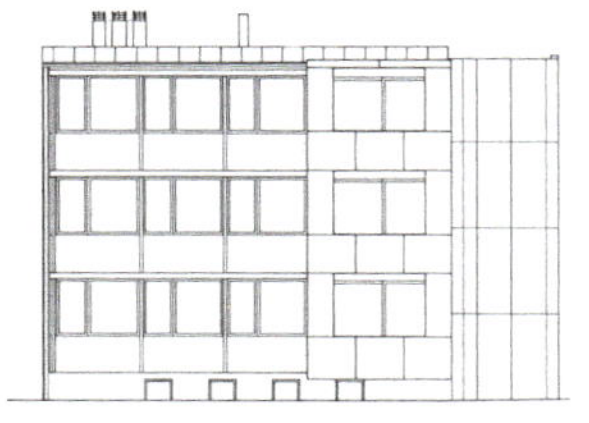

❸

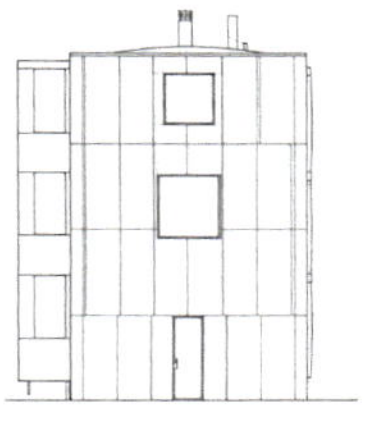

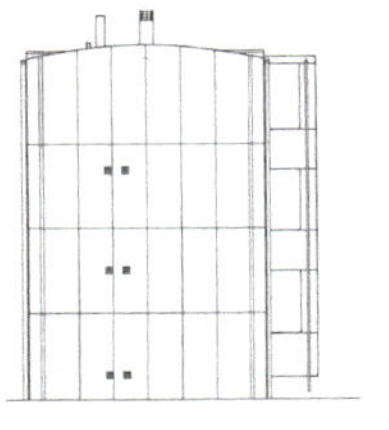

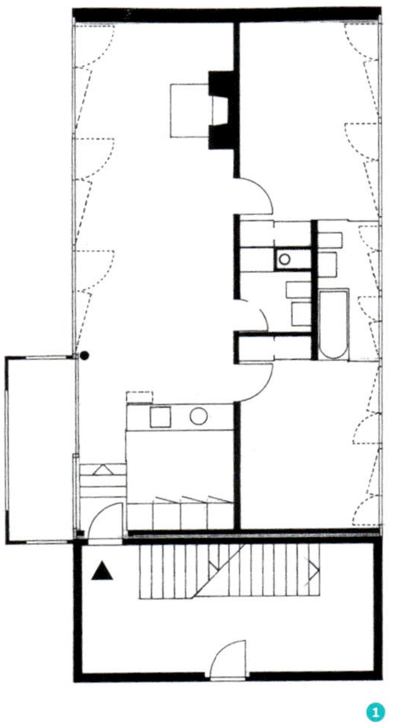

❶

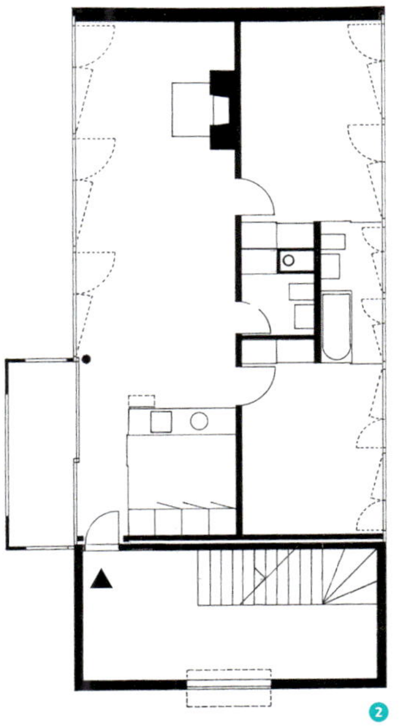

❷

Ground floor 1:200 ❶
Standard floor 1:200 ❷
Cross-section with load-bearing ❸
middle concrete wall,
wood construction of skin
and loggias

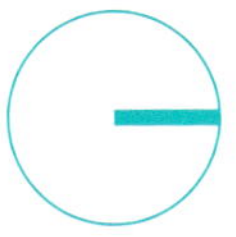

Referring to
floor plan 1:200

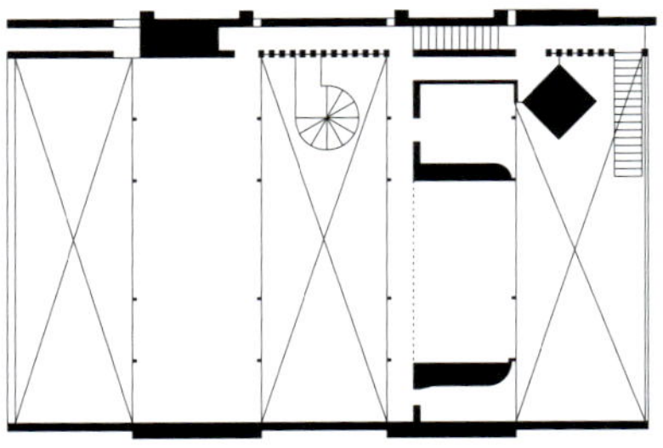

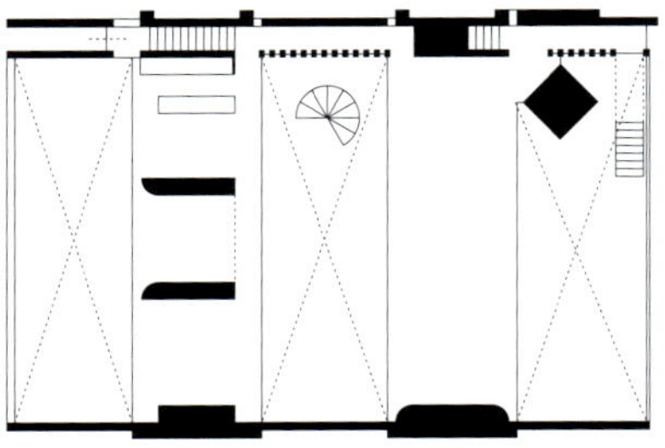

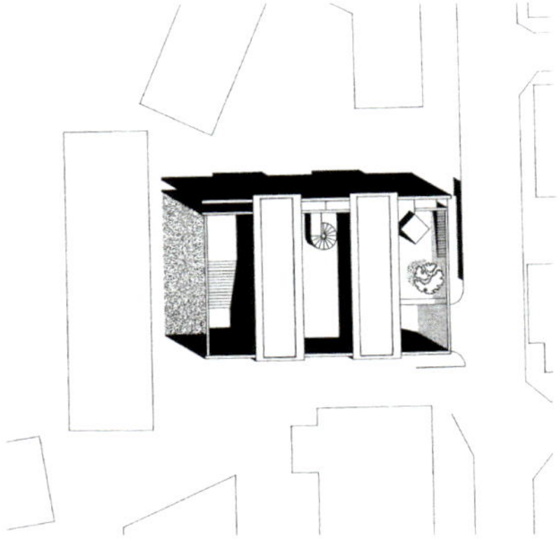

3.1
DETACHED
HOUSE

2/5 House | Ban | 1995

The 15 × 25 m lot is divided into five equal strips. These are allocated for use as: front garden, interior, central courtyard, interior, rear garden. The house consists of two glass containers raised on pilotis (reminiscent of Mies's Farnsworth House) that are linked. The area beneath the containers assumes the role of a "universal floor": sliding glass doors establish a visual link between interior and exterior and, when open, transform the two areas into a cohesive whole. The functional elements are arranged within this large space. In keeping with the tradition of the Japanese home, sliding doors are used to temporarily enclose bedrooms, bathrooms, and tatami rooms. In a similar vein, a tent roof can enclose the interior courtyard when desired. Openness and intimacy are options that can be chosen at will in this arrangement. One level of each half of the house is dedicated to a large common area, (one top left, and another bottom right), which is seen from the private rooms across the courtyard. The house is separated from the densely developed urban fabric by full-height concrete walls and planted trellises, and celebrates the expanse of the sky.

Building type
villa, two stories, N/S

Date of construction
1995

Living area
507 m²

Area per user
127 m²

Layout
2 two-story building struts 15 × 5 m in length, basement concrete and steel-frame construction

Architect
Shigeru Ban Architects
Tokyo

Location
Nishinomaiya, Hyogo
Osaka
Japan

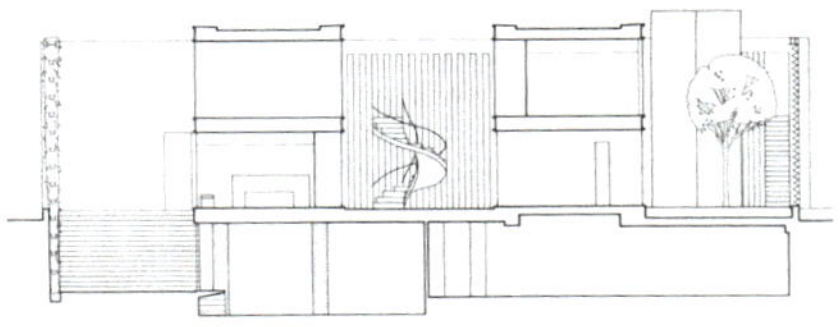

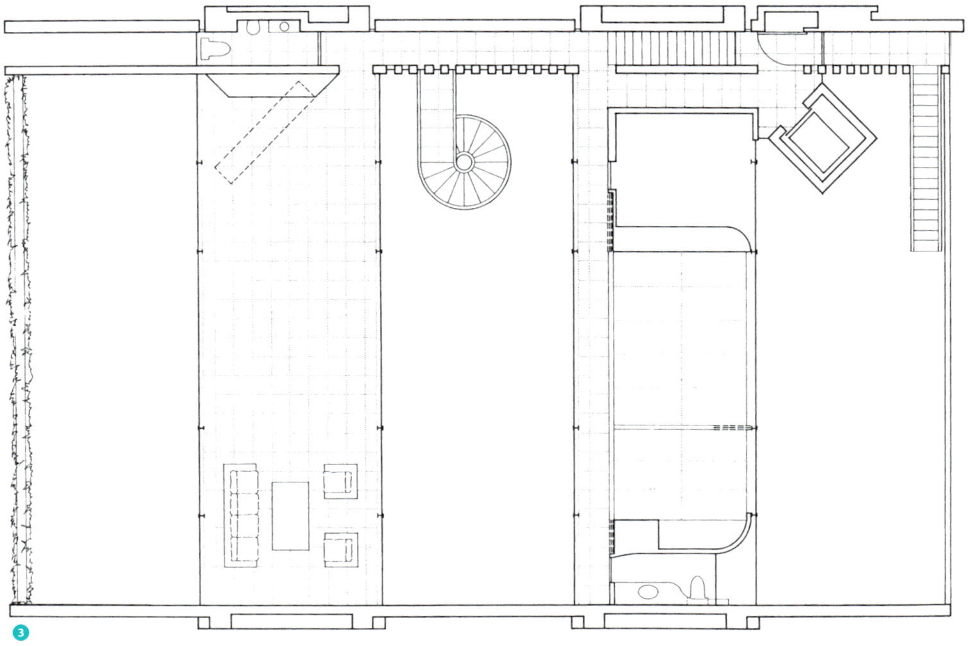

3

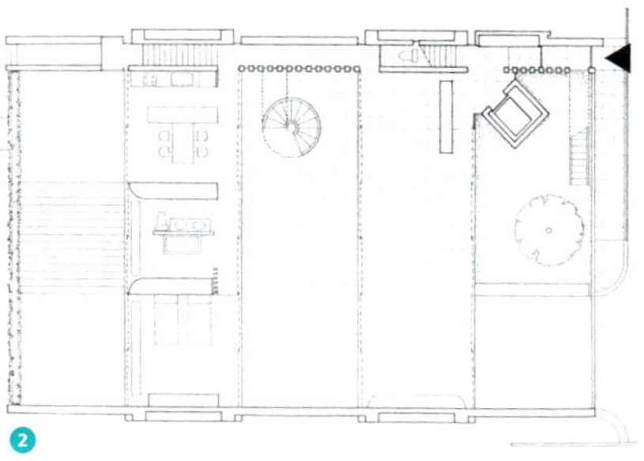

2

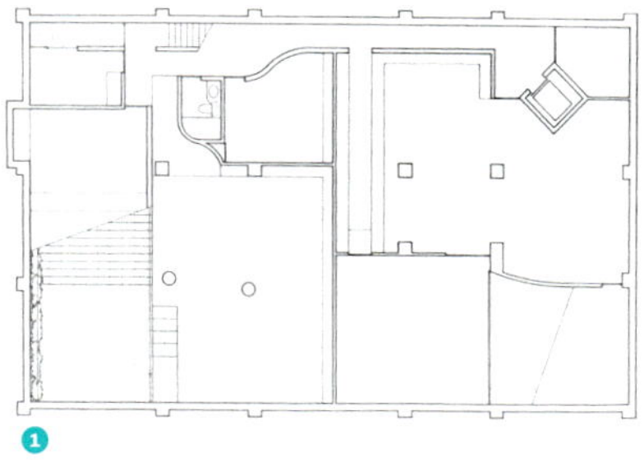

1

Basement with library, staff room, storage space, garage 1:500 ❶
Ground floor with living room, kitchen, bathroom, master bedroom 1:500 ❷
2nd floor with study, lounge, tatami rooms, children's rooms 1:200 ❸

Referring to floor plan 1:200

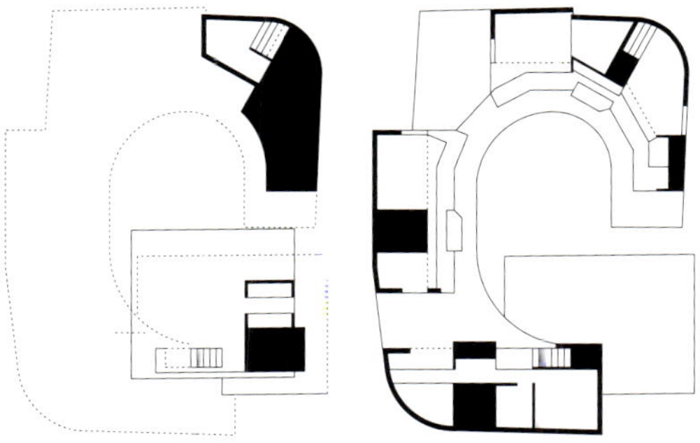

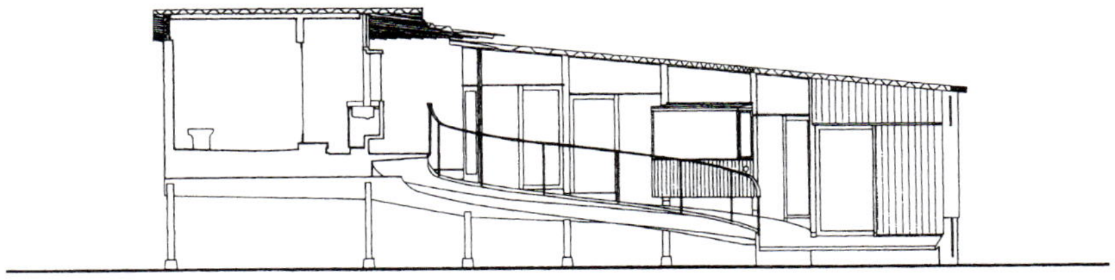

Aida-sou House | Miyamoto | 1995

A house for a mother, her two daughters, two lodgers, and six dogs. The design's point of departure was to analyze what kind of distances the occupants maintained and how they related to one another. The courtyard was to become the spatial and conceptual center of the project and was to remain open. The result was the arrangement of individual rooms clustered on a spiral ramp supported by steel columns. Each cluster has a bedroom, toilet, kitchen unit, and storage space, all grouped together behind a hall. The halls are separated from the courtyard by shoji screens. Each room is demarcated by its unique position along the ramp, from which it draws its individual character. Only the kitchen/dining area and service areas – on the ground level – are for shared use. By opening sliding doors, the dining room dissolves and becomes part of the courtyard with only the cantilevered roof left to trace its phantom outline. The dogs play in the kidney-shaped yards underneath the raised house.

Building type
ramped courtyard house
1–2 stories
oriented to all sides

Date of construction
1995

Living area
6 rooms, approx. 144 m²
including service spaces

Area per user
48 m²

Layout
house for 5 people, reinforced concrete ramp raised on steel columns around a courtyard, wood construction above with corrugated roofing

Architect
Katsuhiro Miyamoto +
Atelier Cinquième Architects
Kobe

Location
Takarazuka
Hyogo
Japan

1

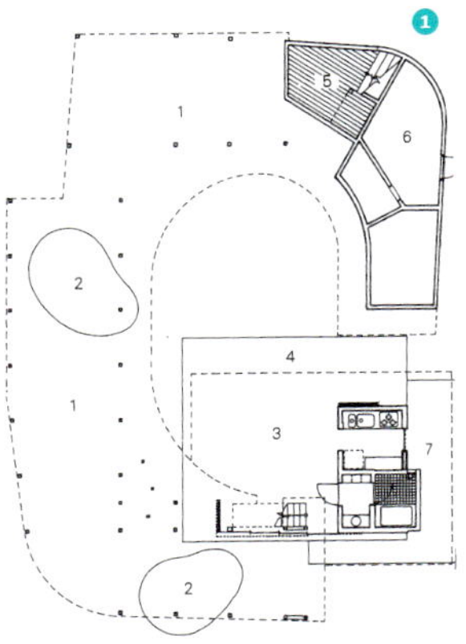

2

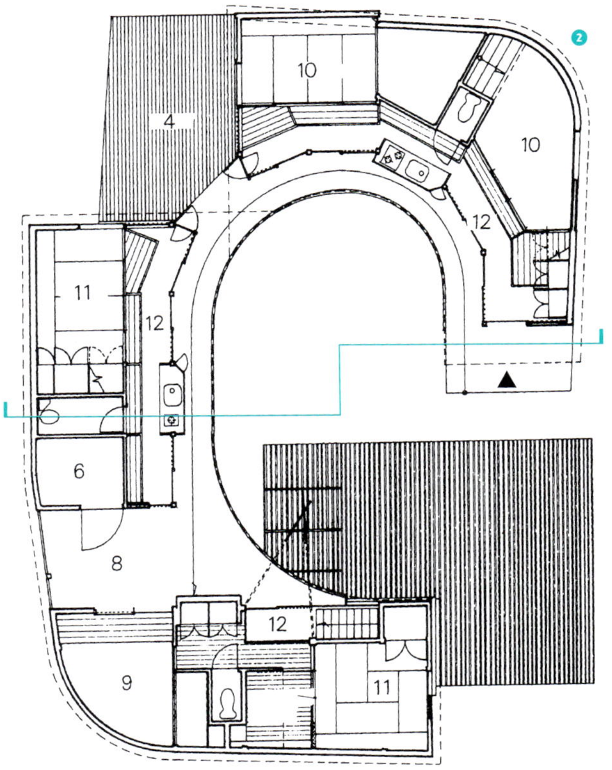

Pilotis 1
Dog yard 2
Dining room 3
Terrace 4
Closet 5
Storage 6
Laundry 7
Balcony 8
Mother's room 9
Daughter's room 10
Lodger's room 11
Earthen floor 12

Ground floor 1:400 1
Second floor 1:200 2

Referring to
floor plan 1:200

3.1

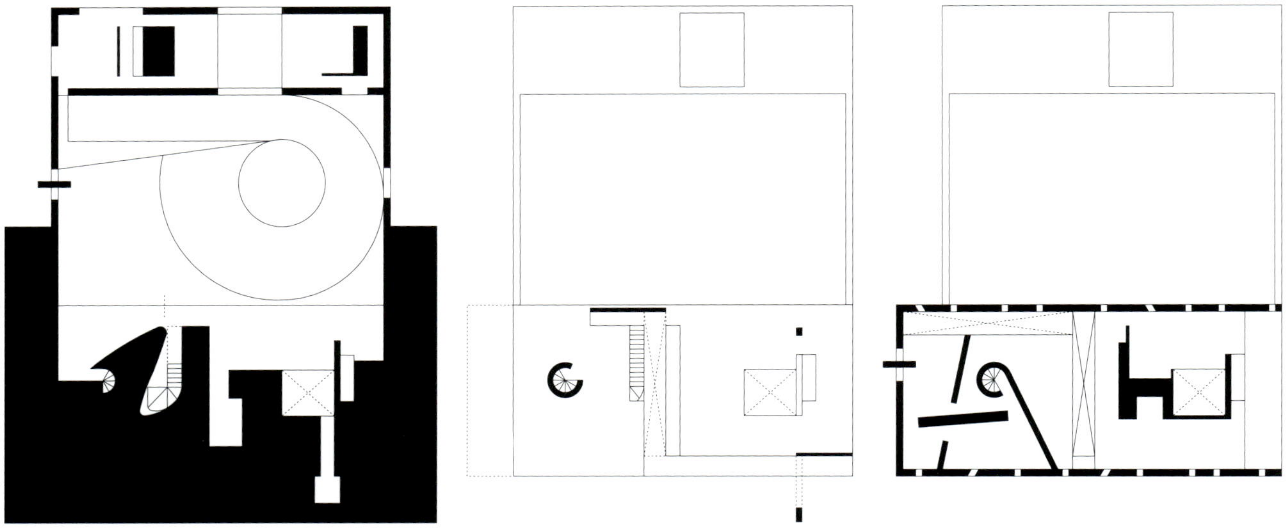

Floirac |
OMA / Koolhaas | 1998

The owner, who is confined to a wheelchair, expressed the wish for a complex house as a private cosmos. Rem Koolhaas designed not one, but three houses for the client and his family; the houses are stacked to promote a playful interaction of their individual characteristics. Thus the element that links the levels is transformed into the most prominent feature: the owner's moveable office. The central, hydraulic platform moves along bookshelves that reach from the basement to the attic and changes the configuration of spaces on each level depending on whether it is present or absent. Like all other floor levels, the basement is divided into two sections: the entrance with direct connections to the various vertical access links, the kitchen, and the family TV room are located in the glazed front half, overlooking the inner courtyard. The other half is carved into the hillside. This area houses the wine cellar (which can only be reached when the lift is stationed on this level) and the main entrance, staged much like a dark cave. The living space above merges with the landscape beyond – it is half interior, half terrace, with undefined boundaries since the walls are fully glazed. By contrast, the next level is pronounced in its solidity, with the concrete monolith perforated only by small portholes. This upper level contains the family's bedrooms, divided into parent house and children's house linked by a small bridge. The position of the portholes is carefully calculated. They offer views onto selected highlights in the landscape and are designed specifically for visual contact with the outside from a walking, standing, or reclining position. The house derives its strength from the effective collage of contrasts in form, material, and program.

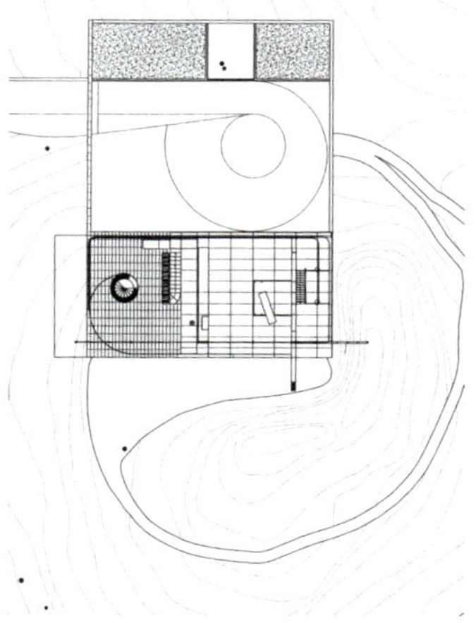

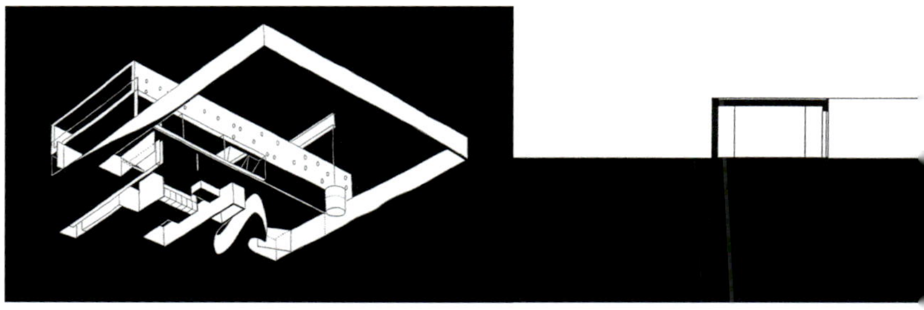

1

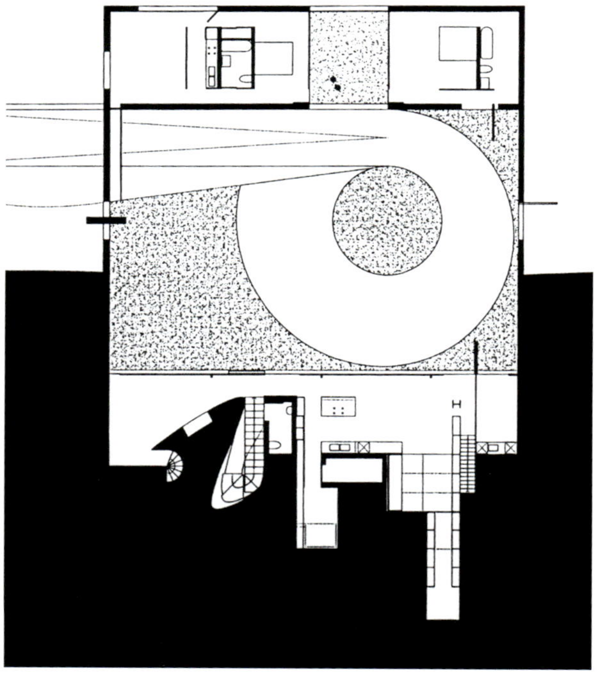

2

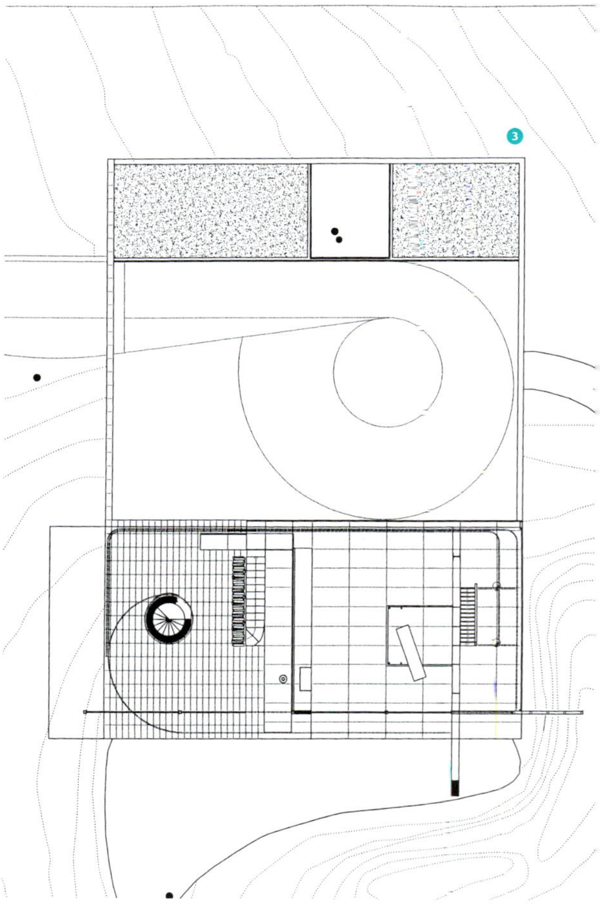

3

1 Cross-section and longitudinal section with moving platform
2 Plan of entrance level across from caretaker and guest house
3 Living area with dining and reception room, gallery, and the mother's home office

3.1
DETACHED HOUSE

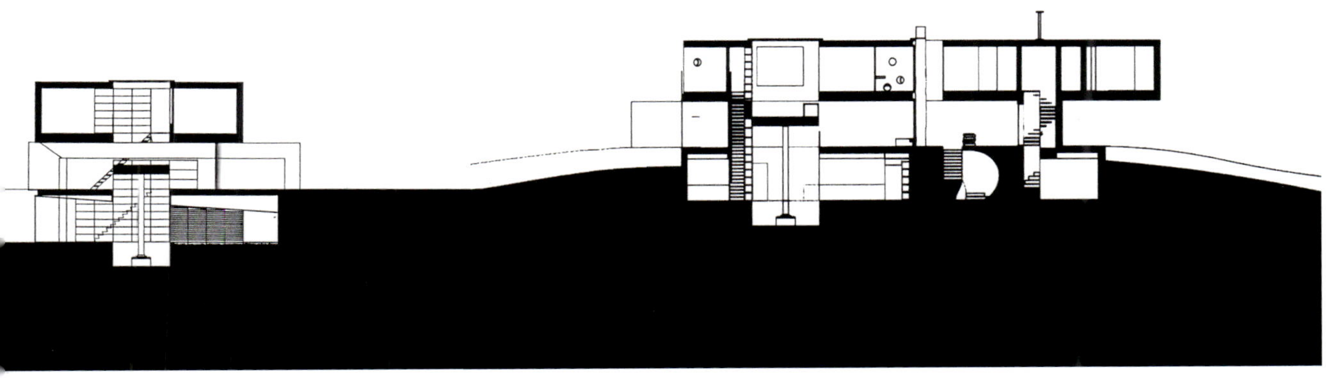

Building type
villa
3 stories
facing in all directions

Date of construction
1998

Living area
600 m² (125/215/260 m²)
caretaker and guest house 180 m²

Area per user
150 m²

Layout
concrete box (at top), suspended from steel girders on asymmetrically positioned, cylindrical external wall of spiral staircase, counterbalanced by means of tie rod, 3 different types of stairs (single-loaded, double-loaded, spiral), hydraulic platform, 3 × 3.5 m (with telephone and power connections) glass balustrades and aluminum doors, photocell-driven, as protective measure when lift is not in position, fully glazed sliding doors (ground floor, second floor), aluminum floor covering on interior and exterior (second floor)

Architect
OMA / Rem Koolhaas
Rotterdam

Location
Bordeaux
France

4

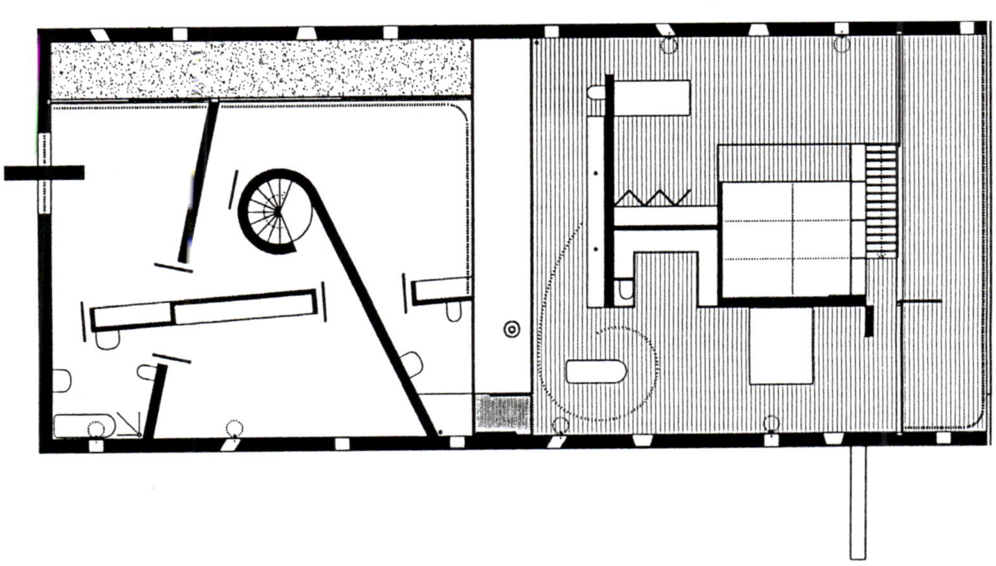

4 Upper floor with children's bedrooms (diagonal walls) and master bedroom (rectangular walls) 1:200

Referring to floor plan 1:200

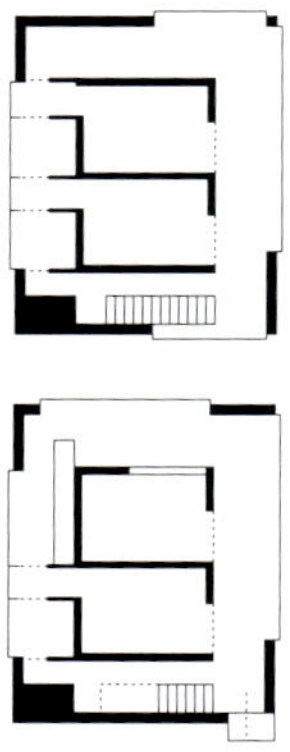

wunschhaus #1 | heide
von beckerath alberts | 1999

Dream house #1, conceived as a prefabricated house, is laid out around a core with two autonomous spaces and a perimeter circulation corridor, whose neutral character lends itself to a multiplicity of uses. The location of the core allows for an individual interpretation of the floor plan with regard to functions, the different door types (white double doors for rooms, dark wooden sliding doors for bathrooms) enable the occupants to establish varying relationships between rooms as needed. To the exterior, the floor-to-ceiling sliding glass doors also allow for extreme openness. The house has two levels, making it possible to divide the zones vertically if required. The stairs are located next to the entrance. The upper and lower level are identical in principle, with the ground floor also allowing for extending one room into the corridor (e.g., for cooking/dining). On the upper level, the rear section of the corridor can be divided off as an additional room for guests or as a library, etc. The corridor also accommodates the bathrooms, which can be used individually or linked, and which are directly accessible from the rooms. When the bathroom doors are closed, the room located at the core extends all the way to the exterior wall. With its surrounding corridor and variety of views even diagonally across the rooms, the house creates transparency and even a measure of chaos: it is up to the occupants to create quiet zones. The floor plan options demonstrate only some of the potential internal room arrangements corresponding to different patterns of living.

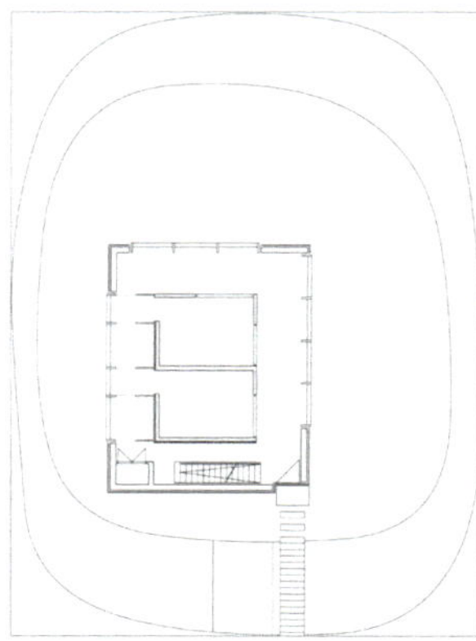

1

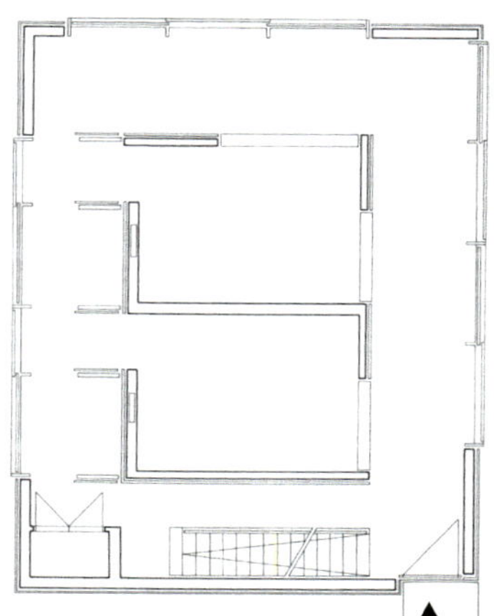

2

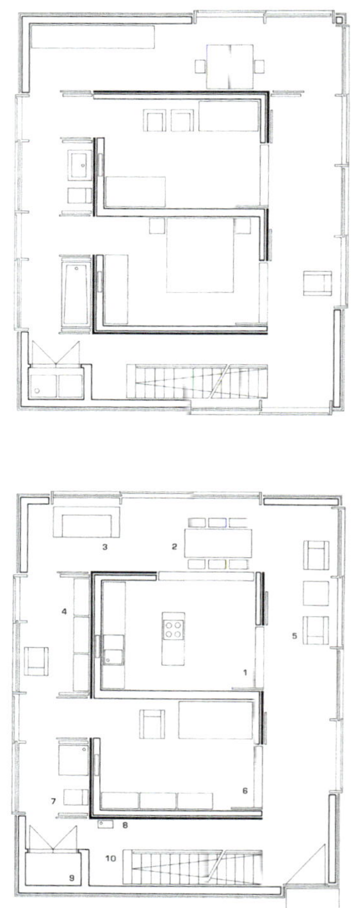

3

1 Site plan (ideal plan)
2 Basic plan 1:200
3 Option for a family
Below: living, dining
Above: children's and master bedrooms, additionally a children's playroom 1:200
4 Option for two families
(with or without child[ren]) 1:200
5 Option for couple working at home
Below: living and working
Above: bedroom and library/archive 1:200
6 Option for four equal partners
(co-op living) 1:200

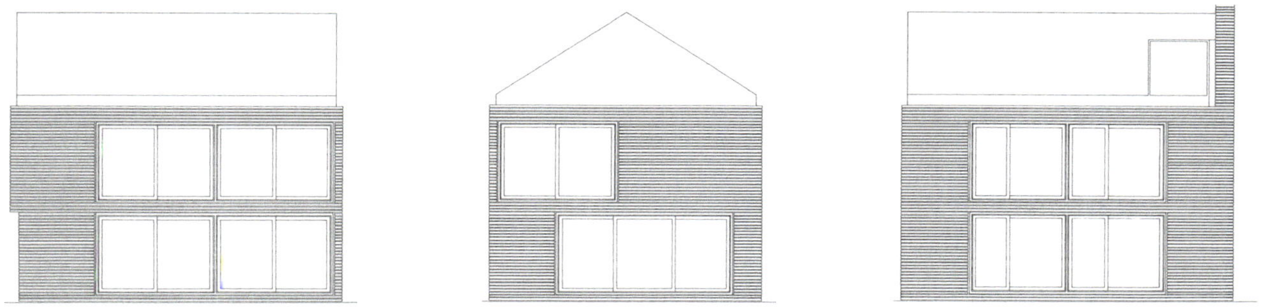

3.1
DETACHED HOUSE

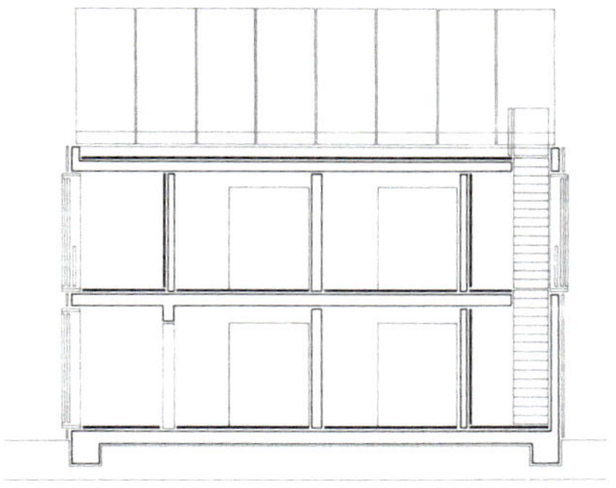

Building type
prefabricated house
2 stories

Date of construction
1999

Living area
144 m²

Area per user
29–48 m²

Layout
1996 competition "A dream house for Germans" by Schwäbisch Hall Savings Bank/"Stern" magazine, house sold by prefabricated home developer
8.8 × 10.6 m, timber frame construction with insulating plaster, large, sliding glass panels, timber ceilings, nonloadbearing interior walls

Architect
heide von beckerath alberts architects
Berlin

Location
Hamburg-Sülldorf
Germany

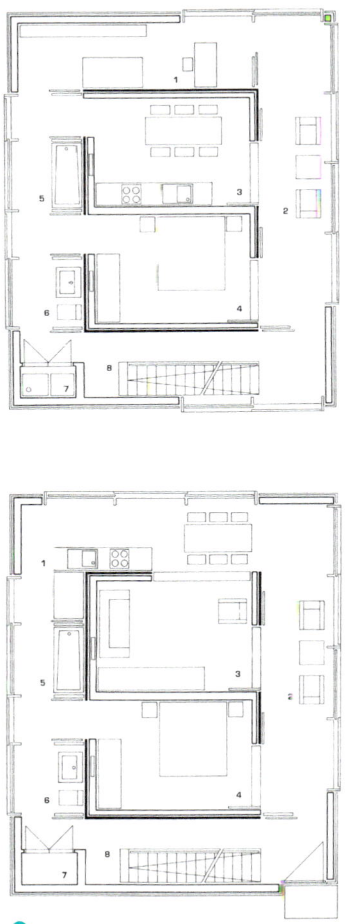

4

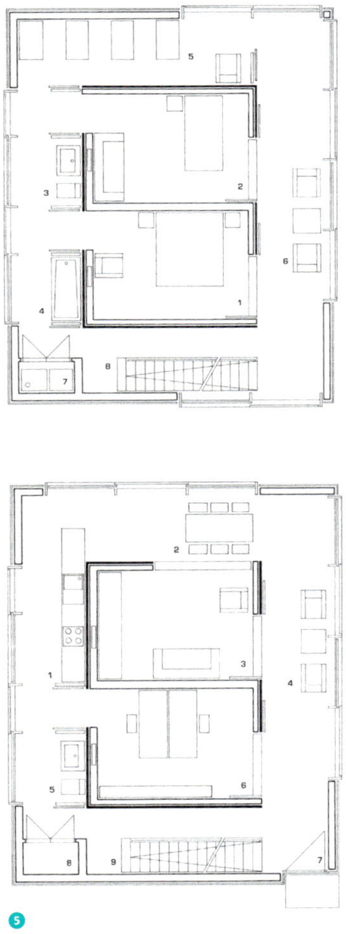

5

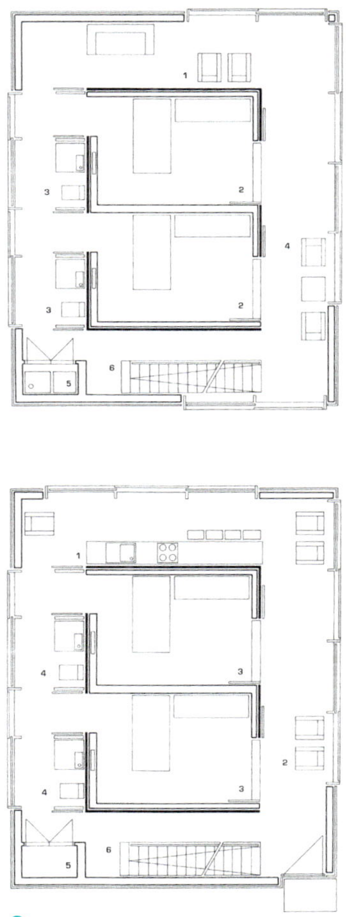

6

Referring to floor plan 1:200

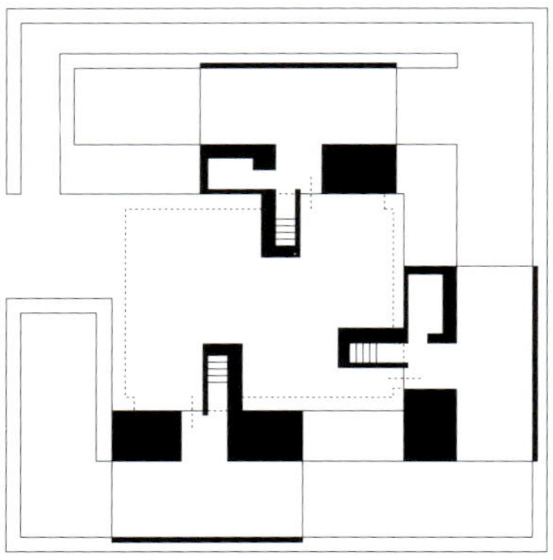

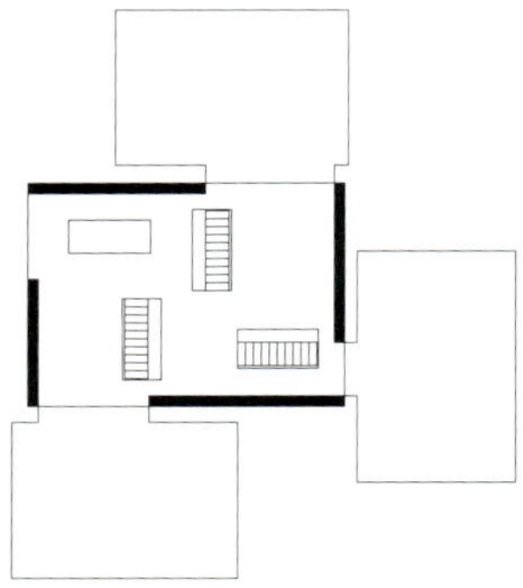

Haus der Gegenwart | Allmann Sattler Wappner | 2005

The experimental house, initiated by a magazine competition on housing for the future, reverses the spatial hierarchy of a conventional single-family home: the individual or private spaces on the ground floor are accessed at first. They are contained in three separate boxes surrounding a central forecourt and jointly support the shared living space above, which also connects the individual boxes with one another. Each of the three boxes has a separate entrance, additional ancillary rooms, and a private yard enclosed with box-hedges into which the personal living space extends nearly seamlessly. All boxes can be subdivided, if needed, by means of dividing walls at the center. Stairs lead from each of these boxes on the ground level to the large common space on the second floor, divided into kitchen, dining, and living area through strategic positioning of the stairs. The roofs of the three boxes provide space for generous roof patios with wooden decking. The core theme is individualization for all users: although the common area is generously proportioned as an interface for all inhabitants, it is essentially secondary in nature. Socializing is a conscious choice in this home, a reflection of the shift away from the single-family house designed solely for occupation by a typical nuclear family.

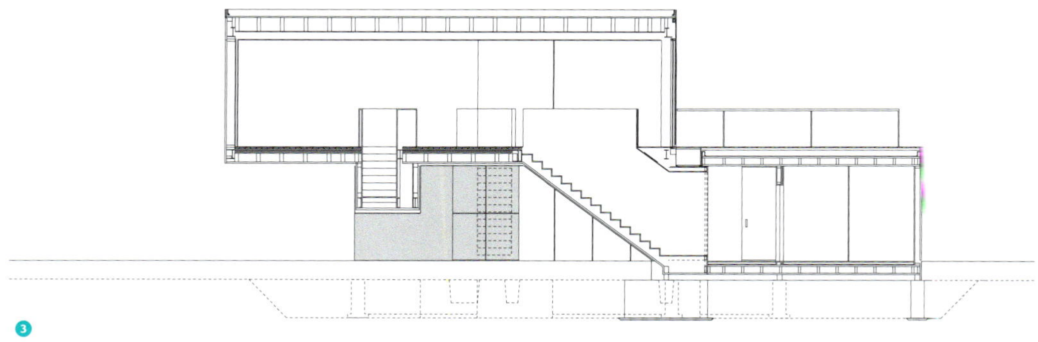

3

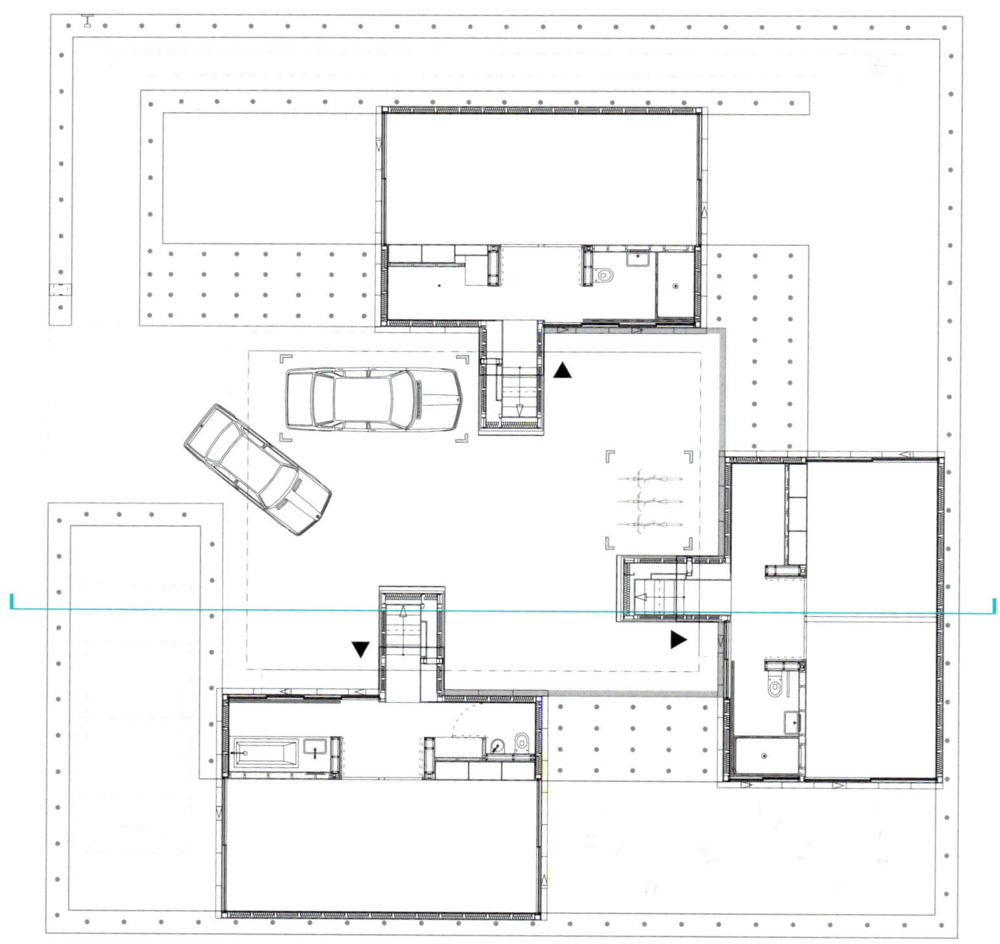

1

1 Ground floor with individual entrances, private spaces, and hedge-bordered gardens 1:200
2 Upper floor with common area and roof terraces 1:200
3 Longitudinal section 1:200
4 Cross-section 1:200

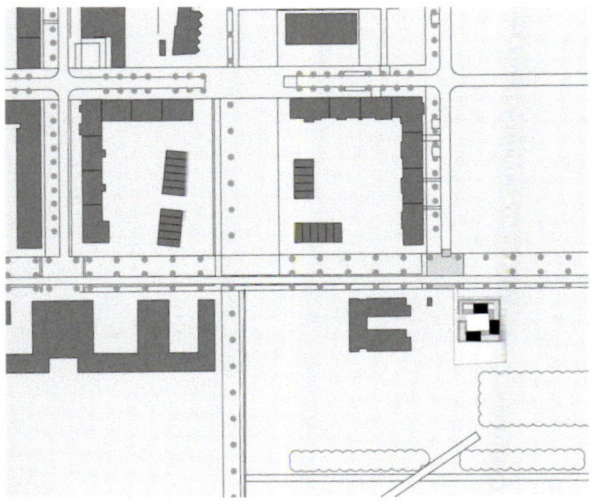

3.1
DETACHED HOUSE

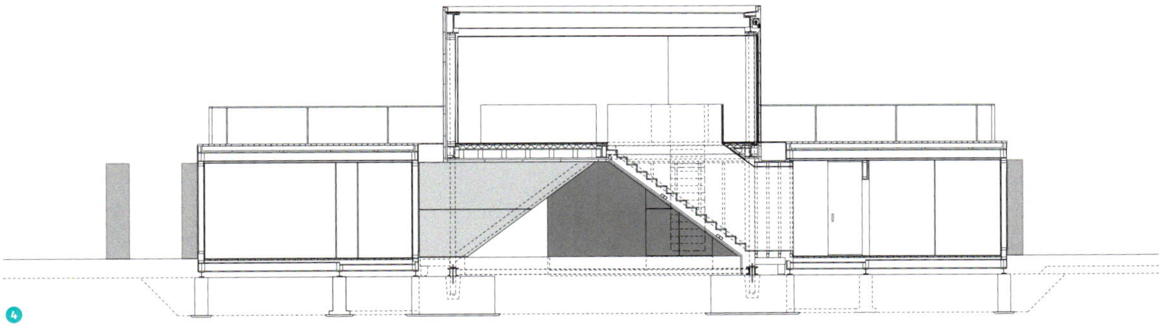

4

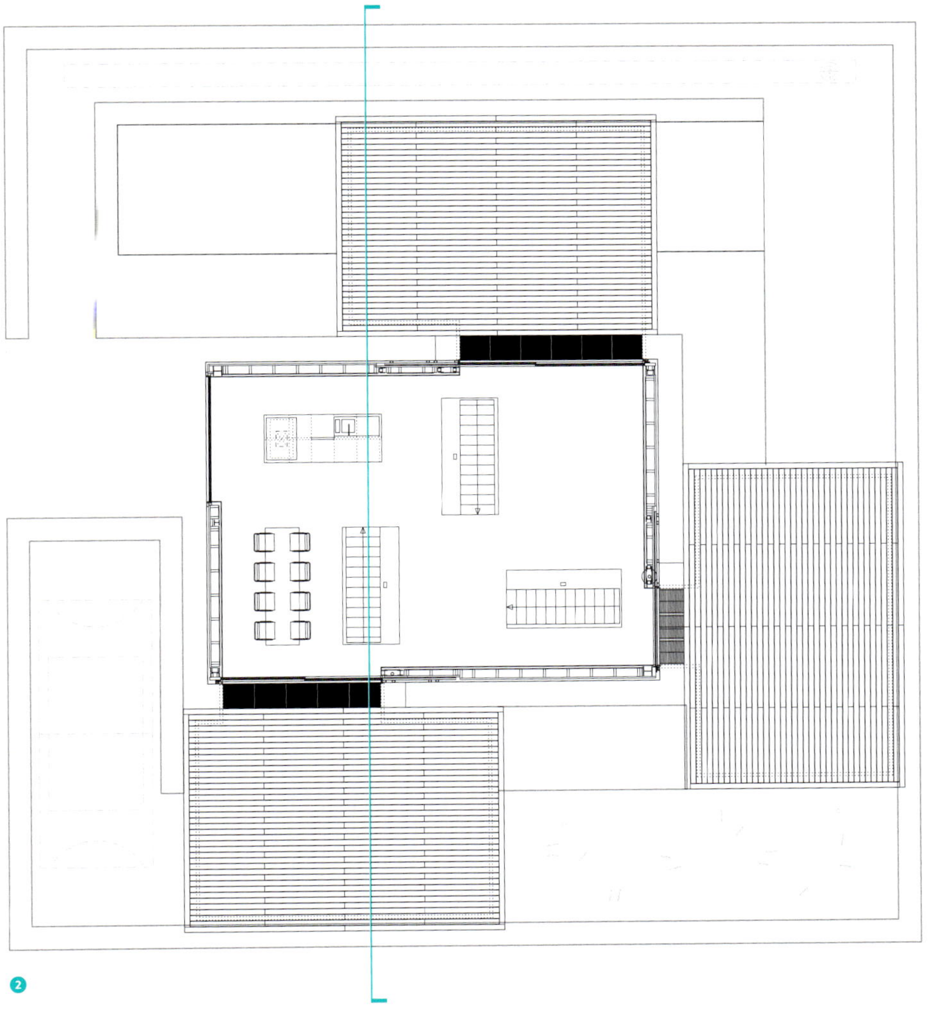

2

Building type
ensemble with three separate volumes on the ground floor and a common room above
2 stories
facing in all directions

Date of construction
2003–2005

Living area
238 m²

Area per user
59.5 m²

Layout
3 freestanding volumes on ground floor with individual living areas and a common area in a pillar-supported structure above, covered forecourt underneath as individual (outdoor) space, and parking, rooftops of single-story structures utilized as roof terraces, ensemble surrounded by hedges on all sides
individual cubes in post-and-beam construction with prefab wood panels, shared common area as steel-frame construction supported by pillars in staircases, with prefabricated wood panels and joist ceiling, floor-to-ceiling glazing

Architect
Allmann Sattler Wappner Architekten
Munich
Markus Allmann
Christian von Arenstorff
Kilian Jockisch
Frank Karlheim
Martin Plock
Olga Ritter
and Ulf Rössler

Location
Munich-Riem
Germany

Referring to floor plan 1:200

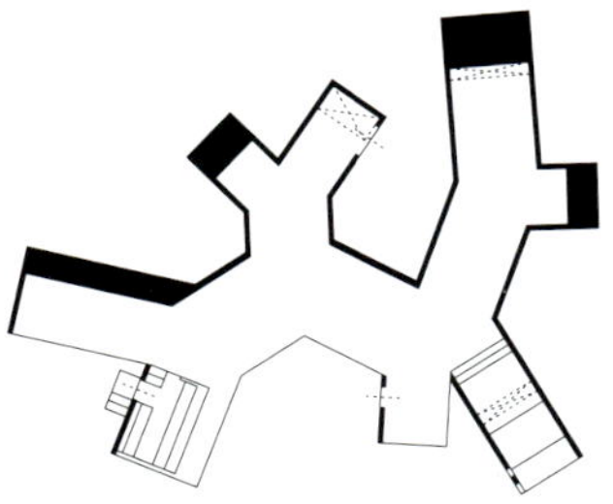

House O | Fujimoto | 2007

3.1 DETACHED HOUSE

A weekend house on the rocky Pacific coast for a couple that wish to be near the sea with a view to settling here permanently in the future – thus the brief. The sleeping, eating, working, and bathroom areas, the tatami room, and the porch are arranged in a ramified series like the branches of a tree. With the concrete wall facing the landside and the nearly fully glazed facade overlooking the sea, one moves through the house as if on a hiking path: from the sleeping and entrance area, one has a view of the coast as if from a deep cave; from the living area the panorama unfolds expansively, while from the bathtub it is clearly framed. The subtle deviations and branches in the floor plan enrich not only the perception; they prevent a direct sightline all the way through, thus defining the more intimate areas without them being separated by walls. The sleeping area only comes into view after several changes in direction; in this manner, the floor plan design imposes subtle rules of use without formulating them distinctly and hermetically. All functions, which would interrupt the spatial rigour and poetry of this spatial concept – such as toilets and cabinets – are discretely integrated into the wall.

Building type
branched-out 1-room house
1 story
facing in all directions

Date of construction
2007

Living area
129 m²

Area per user
64.5 m²

Layout
branched-out spatial structure without room division, closed off toward the landside and fully glazed toward the sea, external and internal walls in exposed concrete, with cedar structure on the interior, floor-to-ceiling frameless window glazing, flat roof in composite wood-steel construction, room depth approx. 3 m

Architect
Sou Fujimoto Architects, Tokyo

Location
Chiba
Japan

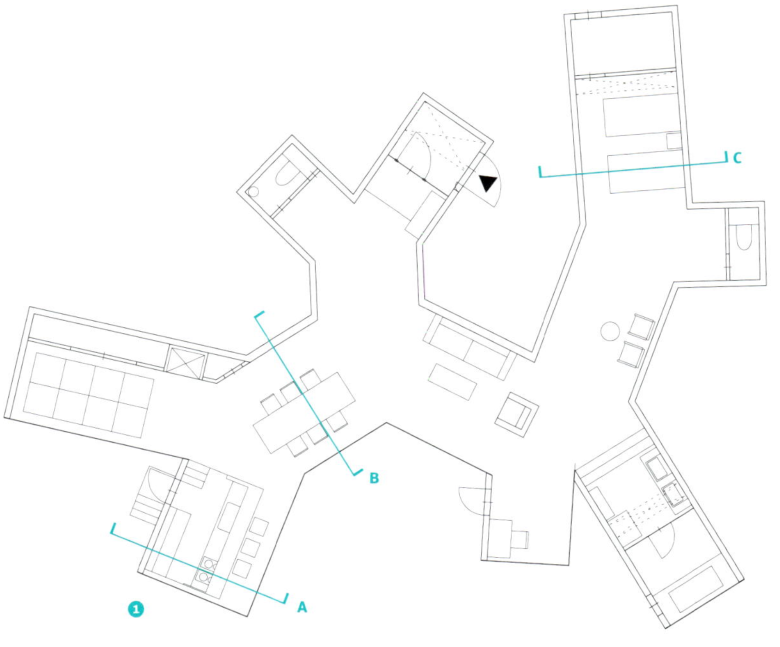

1

2

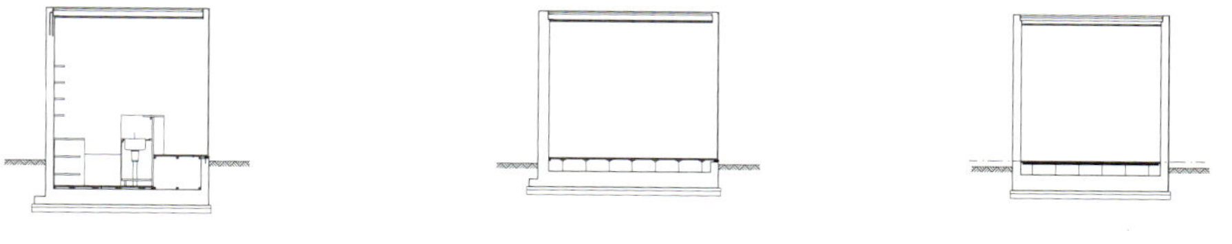

3

Floor plan 1:200 1
Sections 1:200 2
Development of the building form 3

Referring to floor plan 1:200

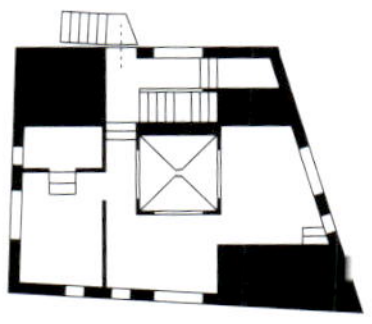

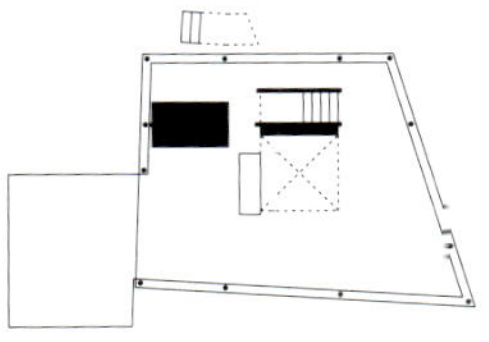

House W | Kraus Schönberg | 2007

The house is divided into two very distinct layers, which are also clearly differentiated typologically: the upper floor plan is compartmentalized and divided into smaller spaces, while the lower plan is entirely open and generously orientated toward the outdoors. The rooms on the upper floor are designed in various heights: master bedroom and children's bedrooms, bathrooms, anterooms, and dressing rooms project into the lower living area. The latter is structured by these differences in height without being subdivided through dividing walls. The horizontal layering of spaces, the cut into the ground, the eye-level perspective of the horizon on all four sides, the three-dimensional relief of the ceiling, which creates intimate zones despite the open plan, all these elements combine to give the house a very unique character. Another defining element is the atrium, – which opens to the lower level but not toward the top. With its bookshelves, soaring across the full height of the house, it serves as a connecting element between the two levels. The randomly placed openings of differing dimensions through all room layers emphasize the sculptural character of the house. The entrance lies between the two layers, with the internal stairs leading upward and downward. The semibasement has been cut into the ground in such a manner as to provide a sense of shelter when sitting while offering a direct view into the garden when standing. Seen from the outside, the light, sculptural volume appears to float above the terrain.

Building type
single-family house
2 stories facing in all directions

Date of construction
2006–2007

Living area
130 m²

Area per user
43.5 m²

Layout
partially prefabricated low-energy house, upper level as floating volume with individual functions, living area on lower level partially cut into ground, upper level: load-bearing timber structure resting on tubular steel pillars, in-ground structure constructed with concrete walls

Architect
Kraus Schönberg Architects
Constance/London

Location
Hamburg
Germany

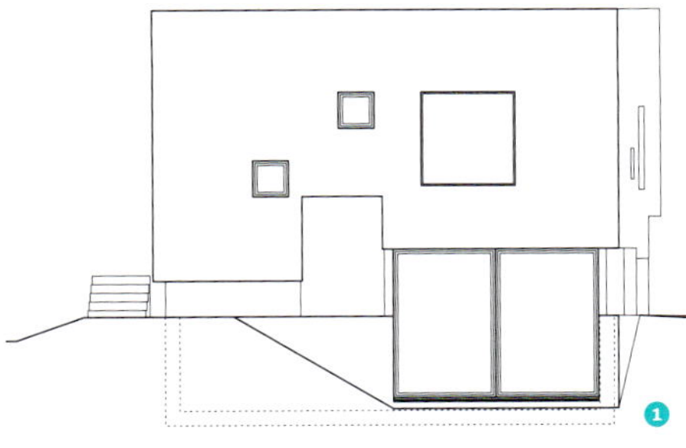

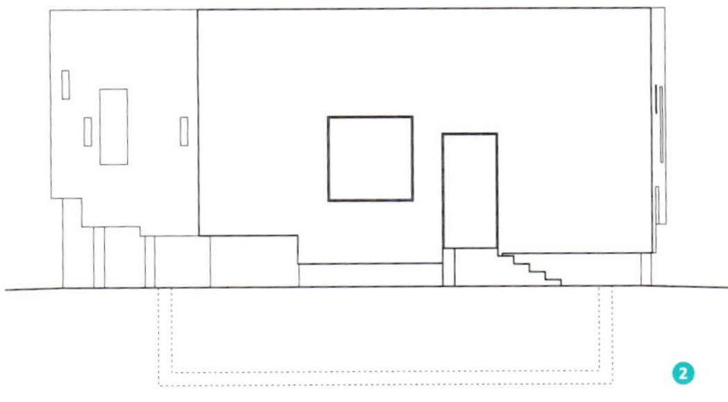

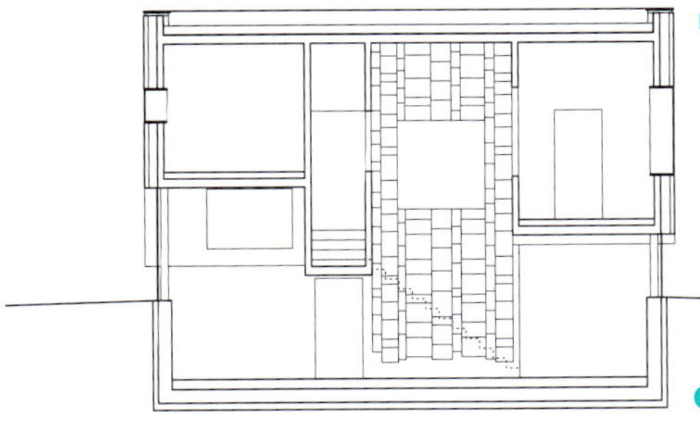

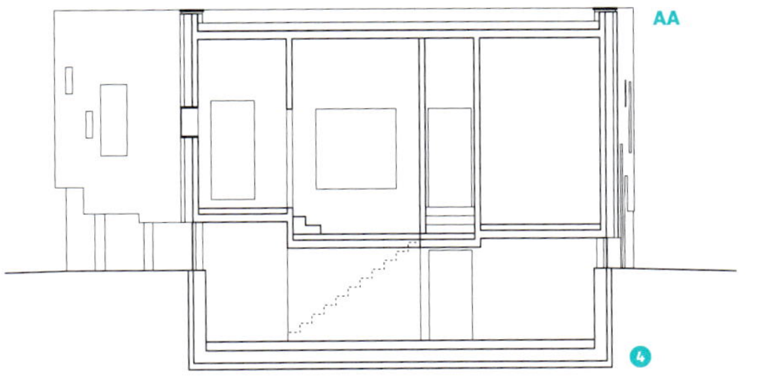

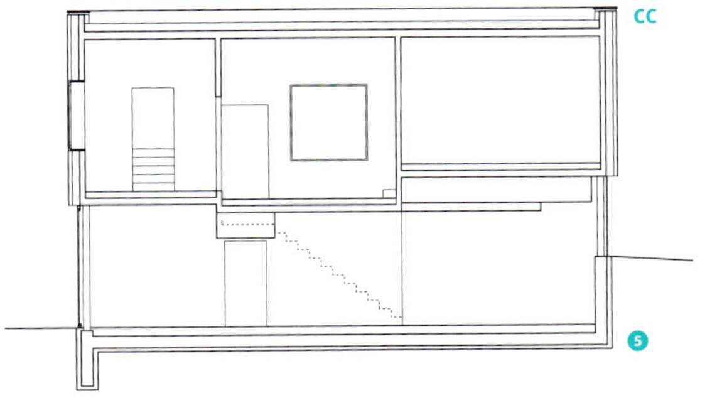

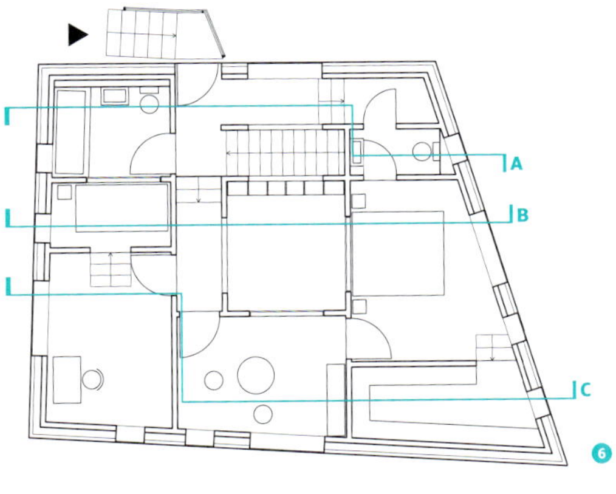

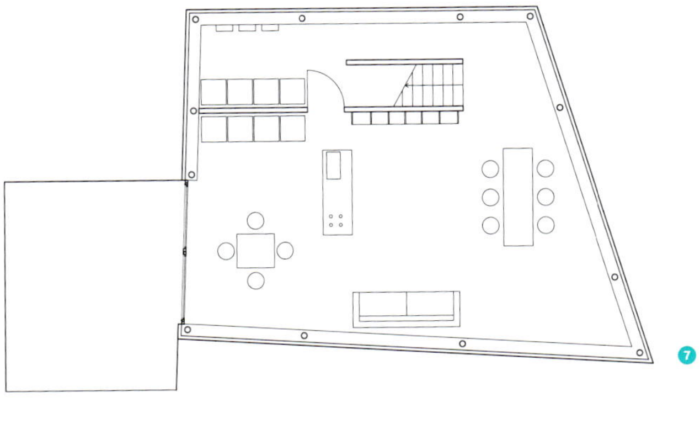

West elevation with sunken terrace 1:200 ①
North elevation with entrance 1:200 ②
Section with book shelf in atrium 1:200 ③
Section of entrance area 1:200 ④
Section of private and common area 1:200 ⑤
Upper floor with entrance and bedrooms arranged around atrium 1:200 ⑥
Basement with common area 1:200 ⑦

Referring to floor plan 1:200

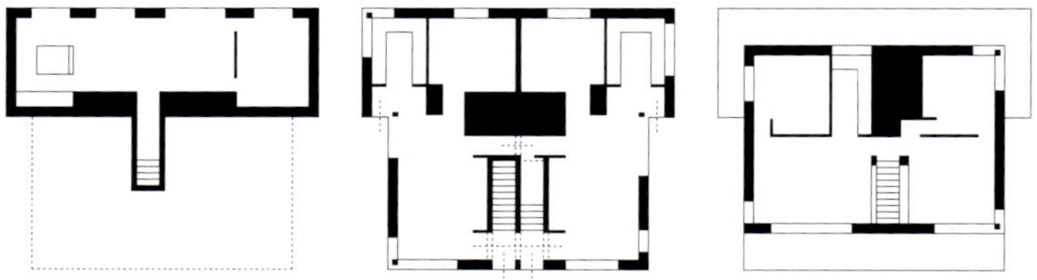

Am Ottersgraben | HAHOH | 2007

The clients requested a house with four compact rental units, designed in a manner that would also allow for entirely different floor plan configurations – one of the potential scenarios being the use as a multigenerational home. By virtue of their design and placement, the two single-loaded staircases function as an interface for various combinations: all areas adjoining the stairs – the apartments on the upper and lower levels, the units to the right and left as well as the landings – can be linked. The original layout comprises four small, compact yet open apartments, each with a private entrance. The apartment on the lower level of the site (garden apartment) and the attic story are accessed via separate sets of stairs. When the partition wall between bathrooms and stairwell is taken out, the resulting layout is an apartment that covers the entire ground floor with an open-plan eat-in kitchen and living area to one side, playroom and bedrooms to the other side. If, on the other hand, the partition wall between the bottom of the stairs and one adjoining apartment is opened up on the street elevation, the west-facing unit is combined with the attic story, and the east-facing unit with the garden unit. If the apartments on the ground floor are connected on both sides (front and rear), the result is a multigenerational home linked across all three levels with various private areas as well as a generous open living space on either side of the ground-floor stairs.

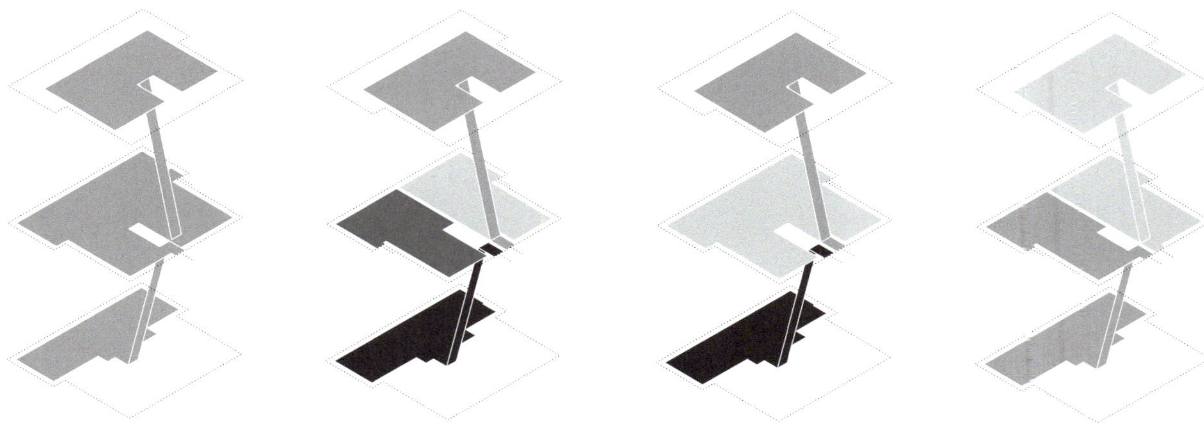

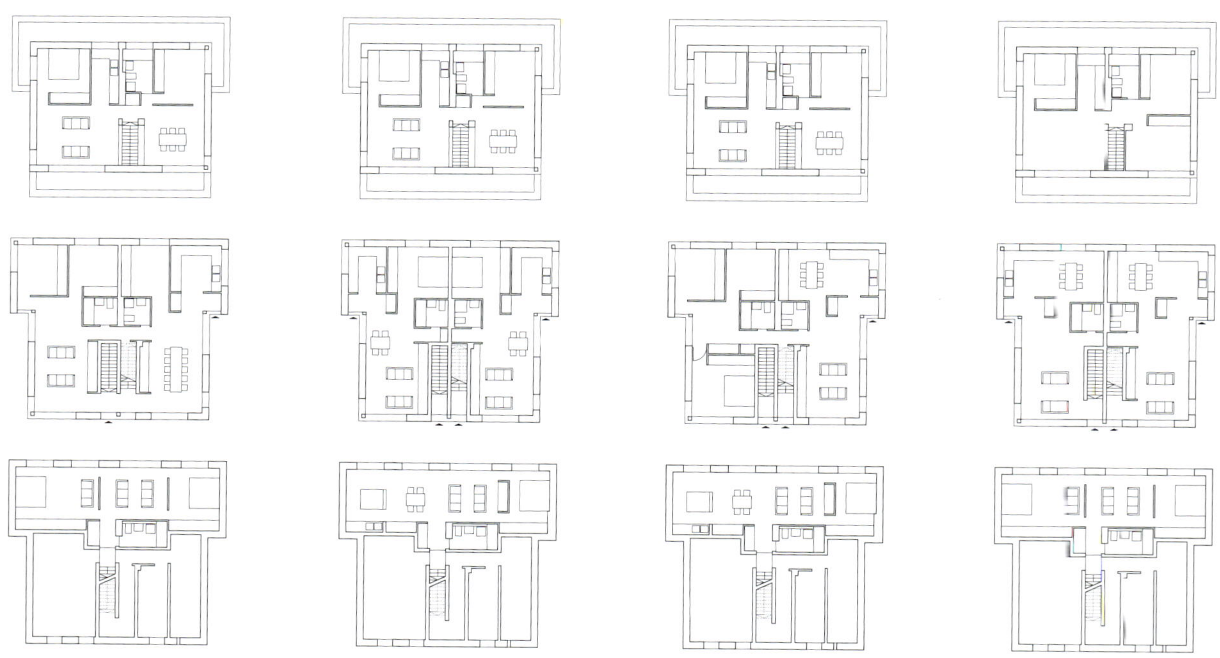

The interfaces on either side and behind the stairs allow for various combinations of floor plan units

1 Option 1: Multigenerational house 1:500

2 Option 2: Four individual 2- i. e. 3-room apartments with separate entrances 1:500

3 Option 3: 4-room apartment on ground floor, 2-room apartment in attic story, 3-room apartment on garden level 1:500

4 Option 4: 4-room maisonette apartment on ground and garden levels, 5-room maisonette on ground floor and in attic story 1:500

5 Option 5: Multigenerational house: garden level, ground floor and attic story with private and communal living areas and various access options 1:200

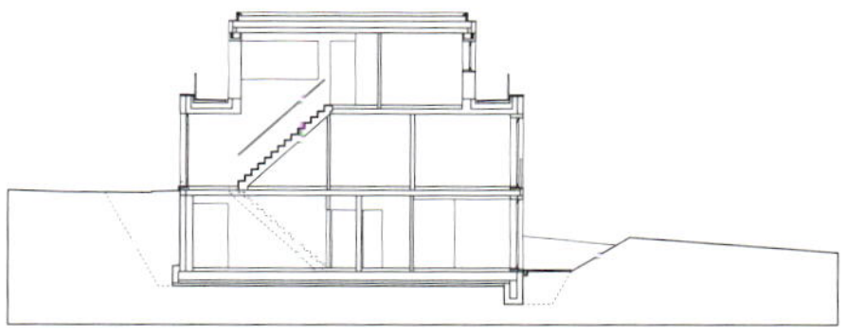

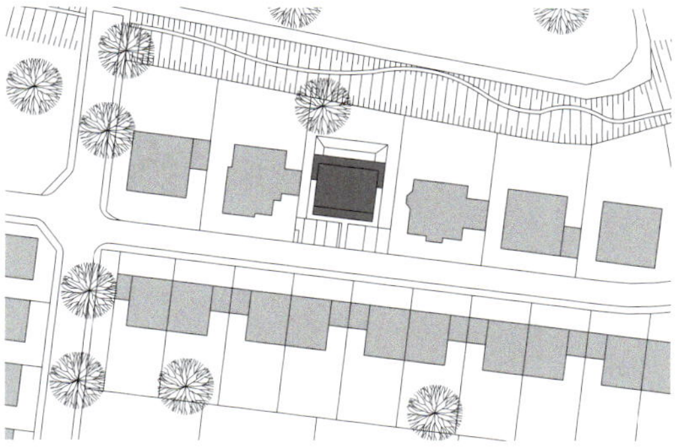

3.1
DETACHED HOUSE

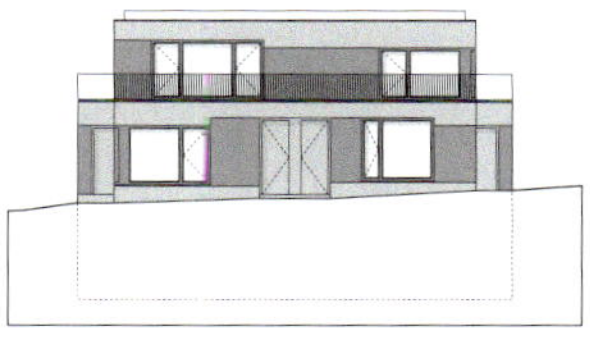

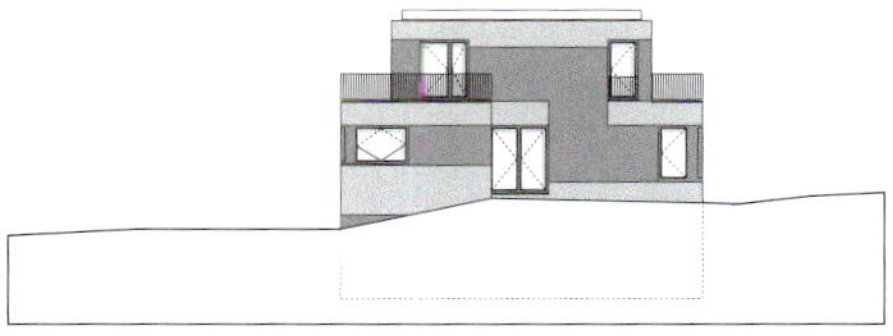

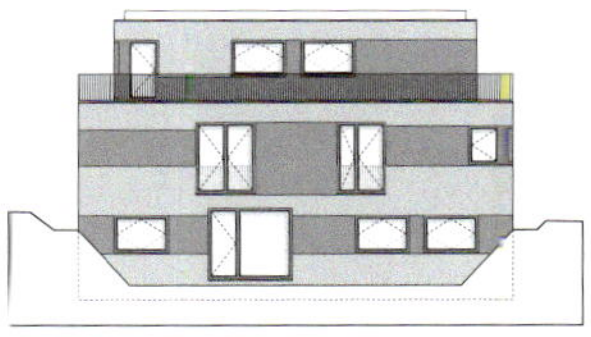

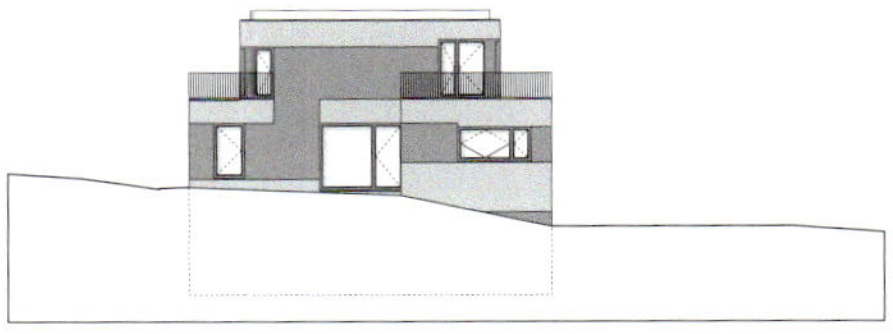

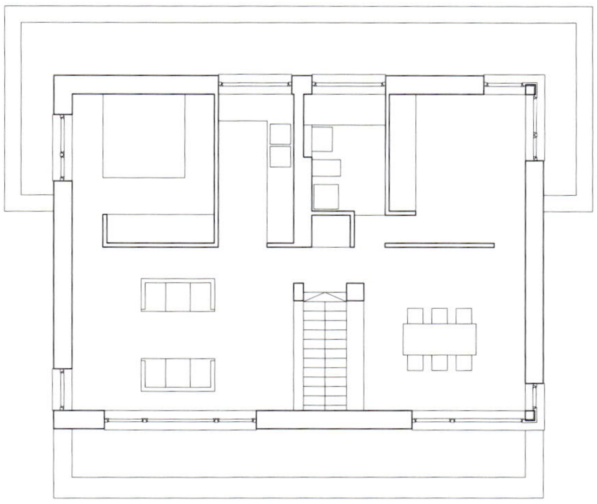

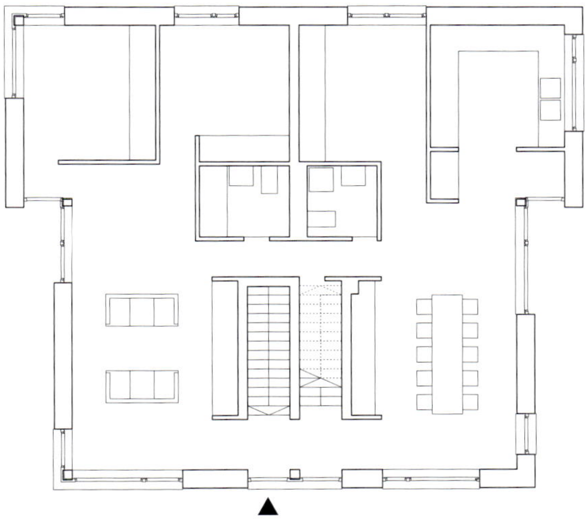

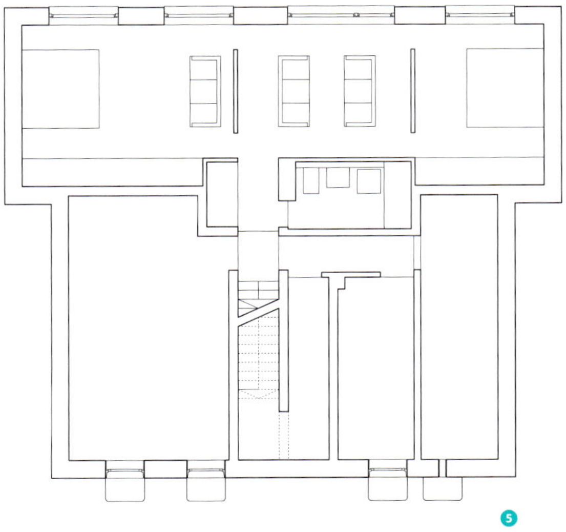

5

Building type
flexible multifamily house
on hillside
low-energy standard
3 stories
facing in all directions

Date of construction
2007

Number of units
1–4

Size of units
2-room apts., 58/59/61 m²
3-room apt., 83 m²
optional: 4-room apt., approx. 120 m²
optional: 5-room apt., approx. 141 m²
optional: multigenerational
home, 270 m²

Area per user
27.5–34 m²

Building depth
12 m

Access
two sets of stairs as interface for
various floor plan combinations

Open spaces
gardens
garden and roof terraces

Parking
parking in front of house

Architect
HAHOH Haas Heckmann Architekten
Berlin

Location
Zell am Harmersbach
Germany

Referring to
floor plan 1:200

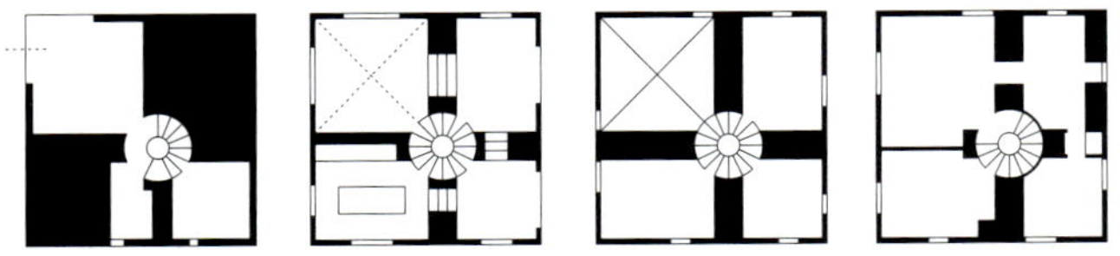

Casa Gago | Pezo von Ellrichshausen | 2012

3.1
DETACHED HOUSE

This house situated on a hillside differentiates itself from its neighbors through its clear cube-shaped volume and height in relation to its footprint. A spiral staircase placed asymmetrically within a square floor plan marks the point of intersection of a spatial cross figure, dividing the house into four different-sized quadrants for different uses. The staircase defines the very essence of the house by connecting the 12 levels as they transition from more public to more private rooms. The required dimensions of the steps and the columns in the cross axes determined the width of the double walls forming a cross around the round staircase, which incorporate different functions. On the lower, more public floors, stairs directly link the kitchen, dining areas, and the double-height living room to create an open enfilade. The remaining wall niches are used as built-in furniture so that the rooms themselves remain unoccupied. On the top floors, with the more private rooms and a one-and-a-half height workroom, these walls house WCs, bathrooms, and additional wall niches.

Building type
tower-like solitary house
4 stories
facing in all directions

Date of construction
2011–2013

Size of unit
241 m²

Area per user
60 m²

Layout
tower house with square floor plan, 12 levels in cross formation connected by a spiral staircase
reinforced concrete staircase and surrounding columns connecting every four steps to rooms
facade with exposed wooden structure

Open spaces
garden

Parking
ground floor garage

Architect
Mauricio Pezo, Sofia von Ellrichshausen
Concepción

Location
Cumbres de Andalue
San Pedro de la Paz
Chile

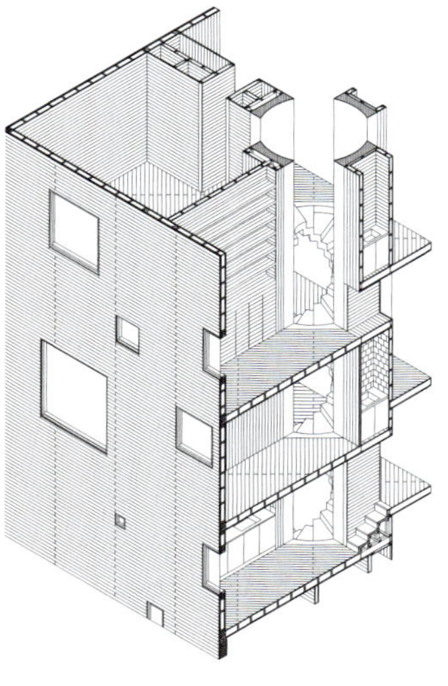

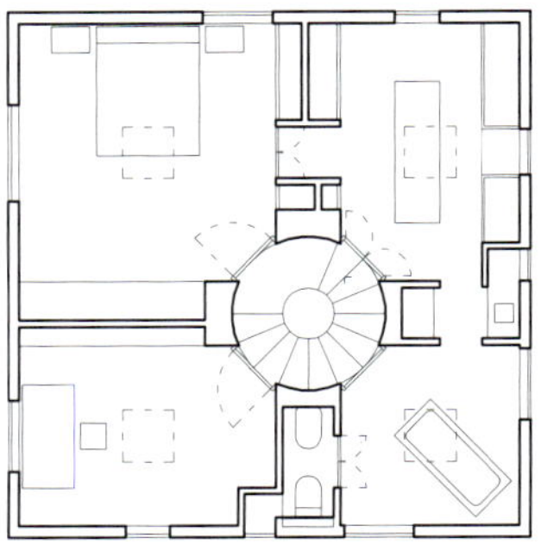

4

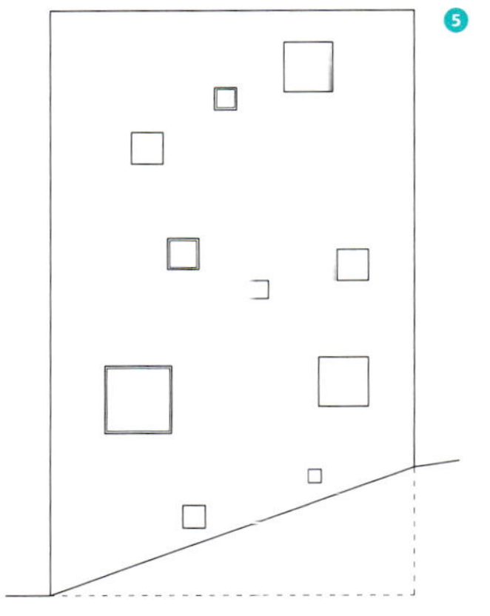

5

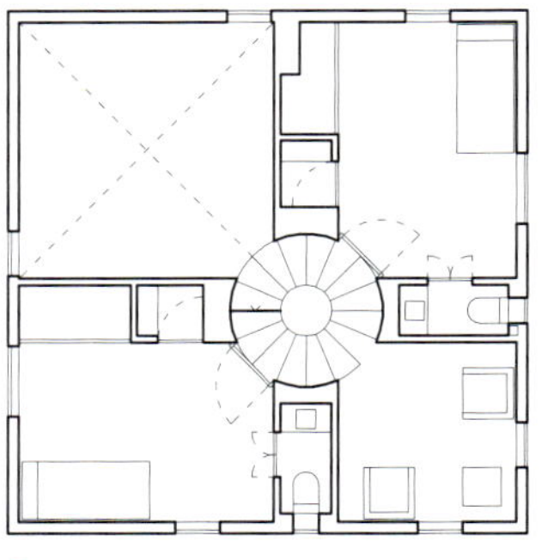

3

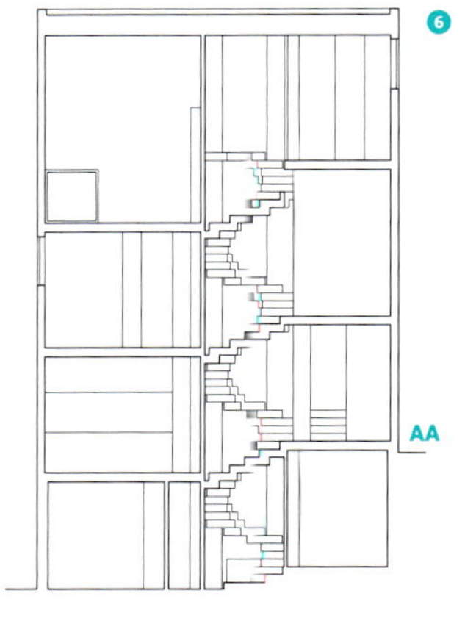

6

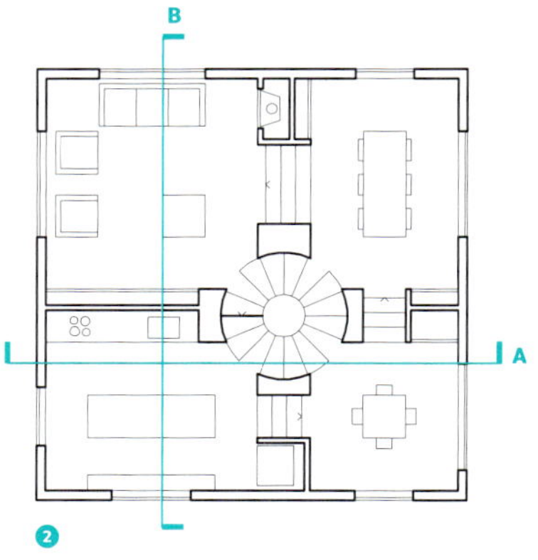

2

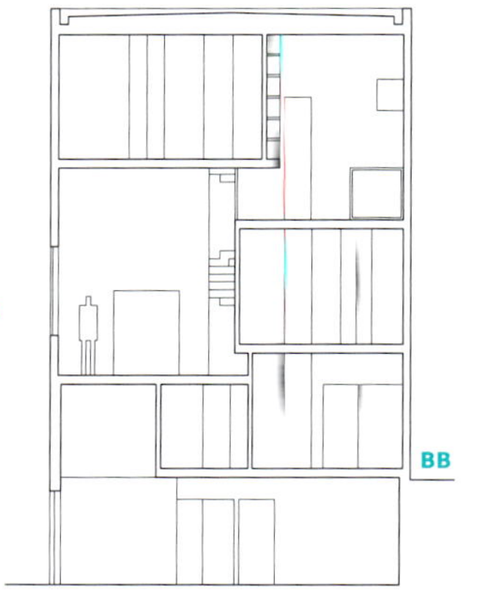

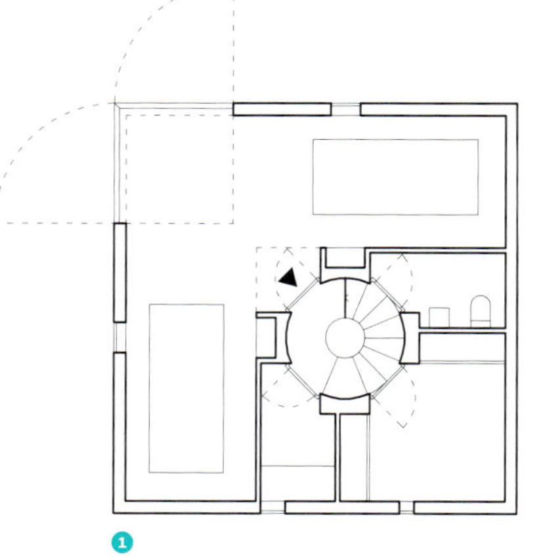

1

Entrance, parking spaces, ancillary rooms 1:200 1
Living area 1:200 2
Children's area, void above living room 1:200 3
Parents' area with work room 1:200 4
West elevation 1:250 5
Sections 1:250 6

Referring to floor plan 1:200

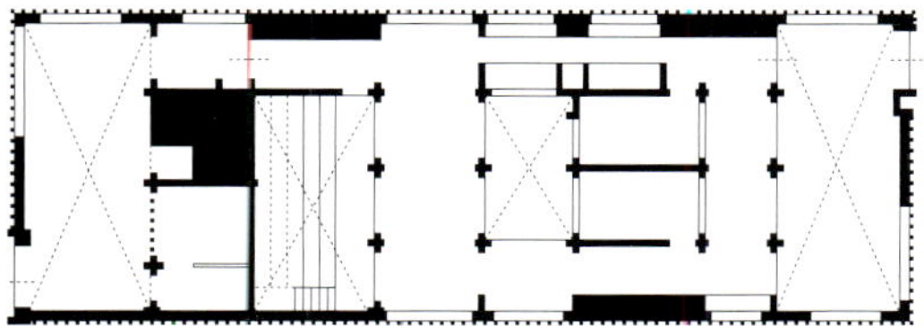

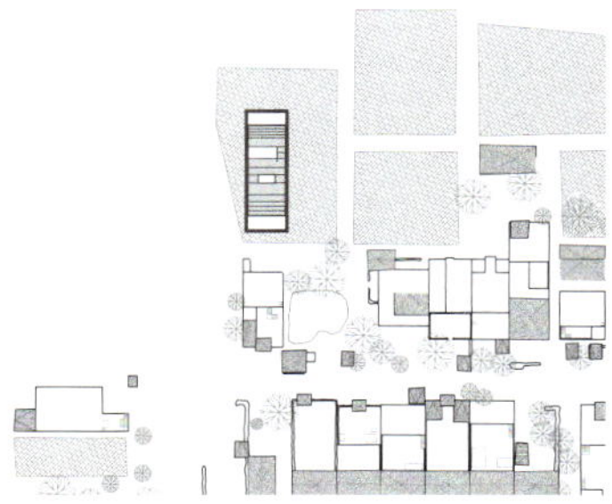

House in Shijia | John Lin / RUF | 2012

Wrapped in a partially open brick structure able to filter in light and wind, this house features a floor plan figure that can be described as a sequence of rooms meandering around four patios. The path is the space: in the center, it widens to form a living room extending from one side to the other and then branches off into spacious hallways on the sides. One hallway leads to the bedrooms, kitchen, and entrance, another one to the toilet, animal stall, and biogas unit. Each patio courtyard is given a specific use to complement the surrounding interior spaces: as a vestibule, washing, working, or planting courtyard. In the latter, a staircase ascends to the large roof terrace with stepped seating and rainwater reservoir. Against the backdrop of the massive population migrations to large Chinese cities, this model house attempts to lend new vibrancy and appeal to life in the countryside. By implementing sustainable technology such as a biogas unit powered by pig manure, a rainwater reservoir, a solar thermal water heater on the roof, planting courtyards, and a plant-based water filtration system, it is also an example of the trend toward supply autonomy. Simultaneously, its use for the production of local artisanal crafts reinforces both the village community and the local economy. The meandering floor plan of this house as a space-defining principle remains transferrable to other contexts.

Building type
patio building enclosed on all sides
single story with roof terrace
facing in all directions

Date of construction
2009–2012

Size of units
245 m²

Area per user
61.5 m²

Layout
self-sufficient house with shared workshop
sequence of rooms meandering around patio courtyards
ornamental, partially perforated, exposed brick structure, steeply terraced roof as common space and for rainwater collection

Open spaces
multi-functional patio courtyards and roof terrace

Architect
John Lin
Rural Urban Framework
Hong Kong

Location
Shijia, Shaanxi
China

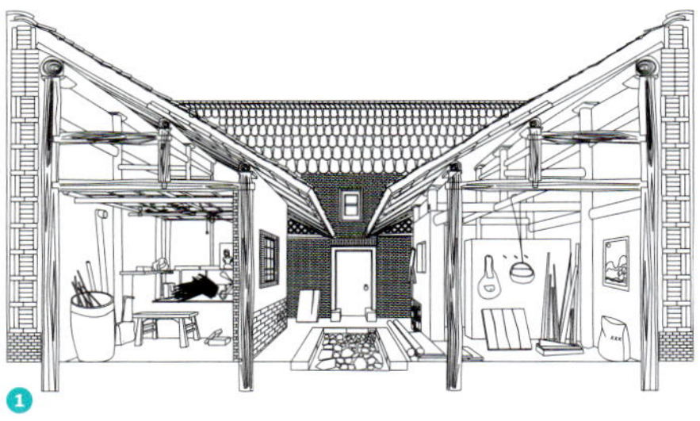

1

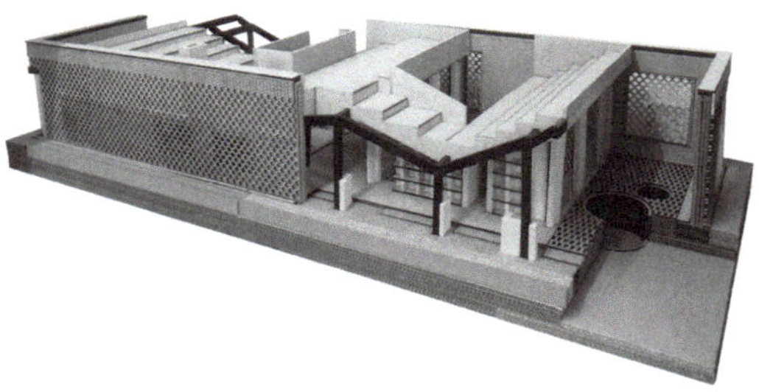

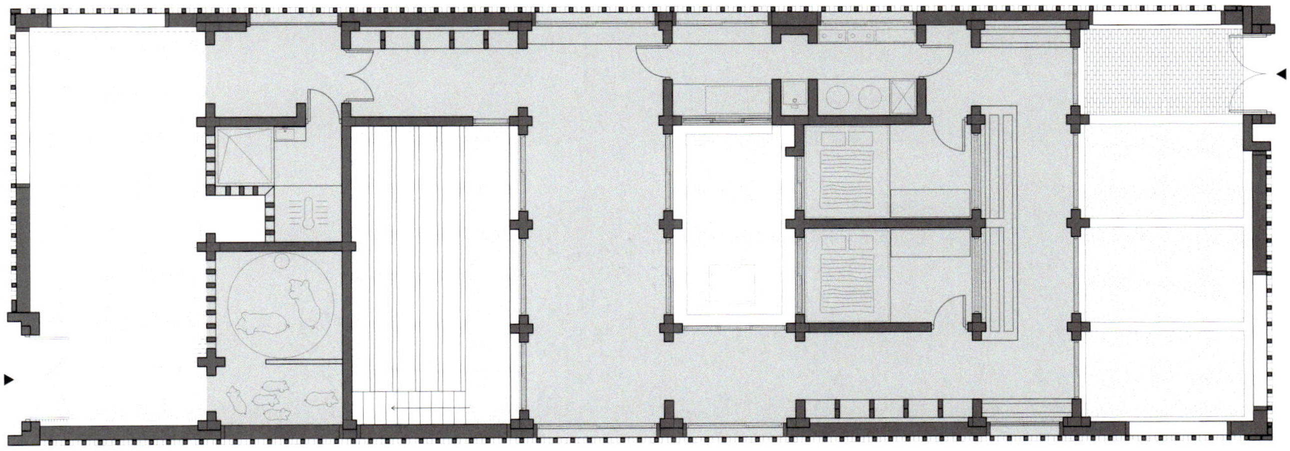

2

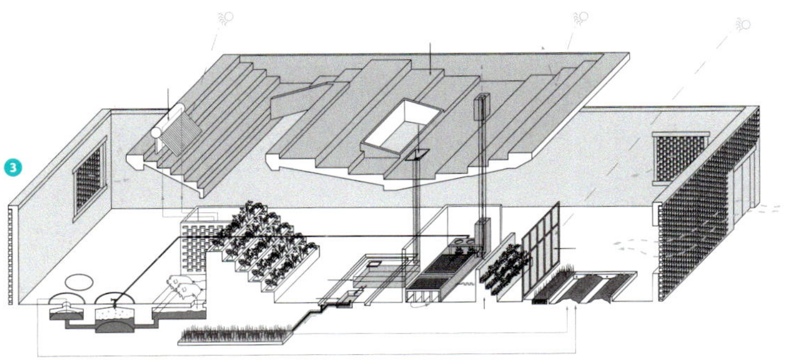

3

Traditional courtyard house type 1
Ground floor 1:200 2
Self-sufficiency systems: biogas unit, rainwater collection and reservoir, thermal solar system, plant-based filtration unit 3

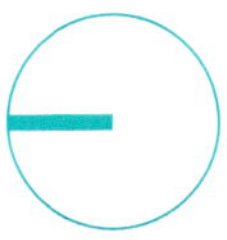

Referring to floor plan 1:200

3.2
DUPLEX

In many instances, duplex houses are created due to the desire of two parties to build and live jointly. The members of this type of coop seek to achieve a balance between separation and connectedness, which can become a model for a special form of cohabitation. Other duplex houses are not planned according to the wishes of the occupants, but are designed in series like a row house, frequently in the context of a planned housing scheme. The goal of the design is to protect the privacy of each neighbour, allowing everyone to fulfil their dream of a house of their own despite the close proximity. Unless the plan consists simply of two houses placed side-by-side and sharing a firewall, examples of duplex houses can often be spatially complex: in order to ensure that both units benefit from all orientations – instead of competing with one another –, they are partially intertwined. Occupants enjoy advantages similar to dwellers in single-family homes: they have their own lot, their own driveway, and a green space on several sides. However, the disadvantages are equally comparable: although this building type is somewhat less uneconomical, urban sprawl, soil sealing, and traffic are increased.

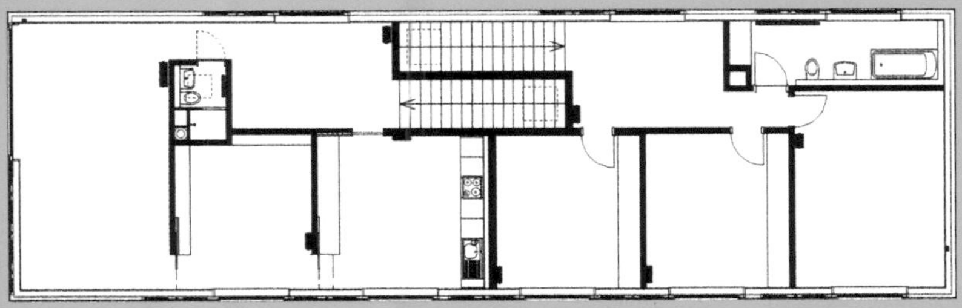

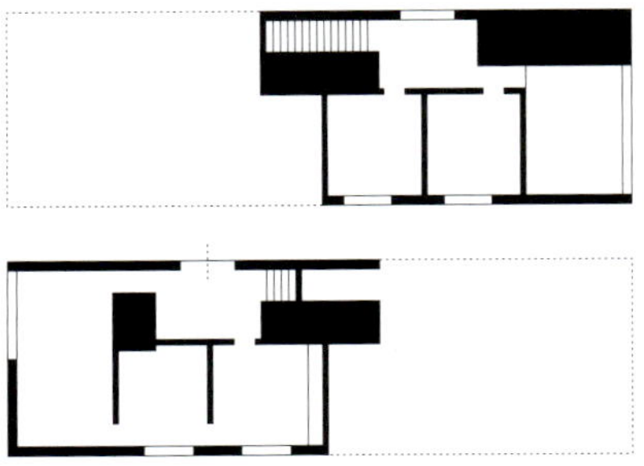

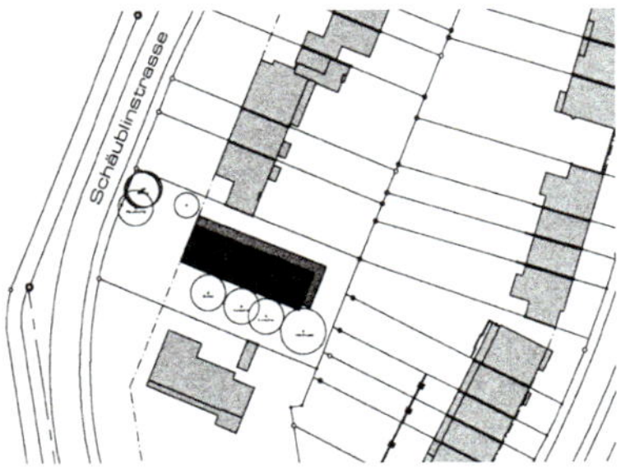

Bruderholz | Gugger | 1996

3.2
DUPLEX

The charm of the undeveloped lot, receding parallel to a row of old fruit trees, inspired the wish to preserve the character of the site and to place the building into the landscape as a solitary structure. Although not visible from the outside, the cube is composed of two identical apartments, which are stacked crosswise. This provides both units with the advantage of having different orientations and a direct relationship between the ground floor and the exterior. From the living area oriented toward the street, both occupants enjoy a spectacular panoramic vista, and in the sleeping area to the rear, the tranquility of the gardens. The full depth and height of the building volume is experienced in each apartment. This reinforces the impression of living in one's own house rather than in an apartment. The large sliding glass panels for the windows and the entrances are retracted into the wood boarding when opened. The result is a pure, unfettered opening, which transforms the rooms into loggias. As to the inhabitants: like the house itself, with its minimalist appearance, they enter into a direct dialog with the landscape.

Building type
detached duplex
apartments stacked crosswise
2 stories with studio in basement level
facing N/S/E/W

Date of construction
1996

Living area
each apartment 156 m²
basement studio 46 m²

Area per user
39 m²

Layout
cube with continuous vertical boarding of full-height, watered larch,
concrete plinth, projecting section covered with timber grating as footbridge

Size of lot
1097 m²

Architect
Harry Gugger
Basel

Location
Bruderholz
Basel
Switzerland

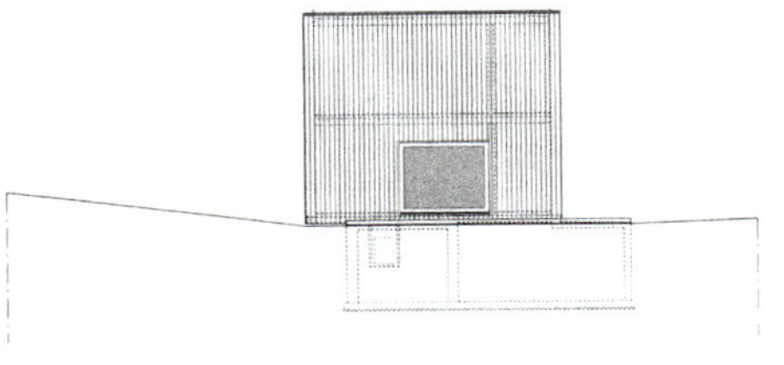
6

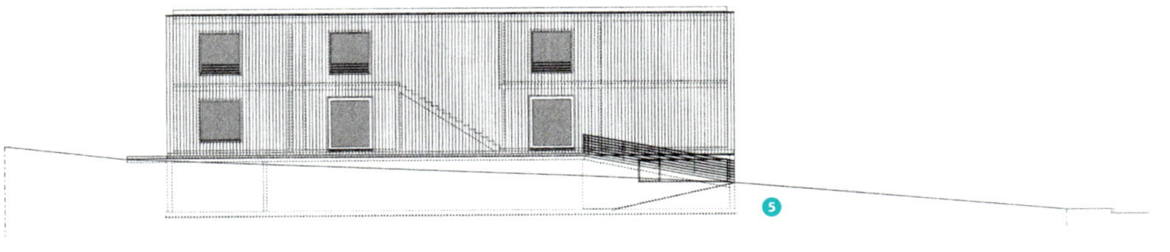
5

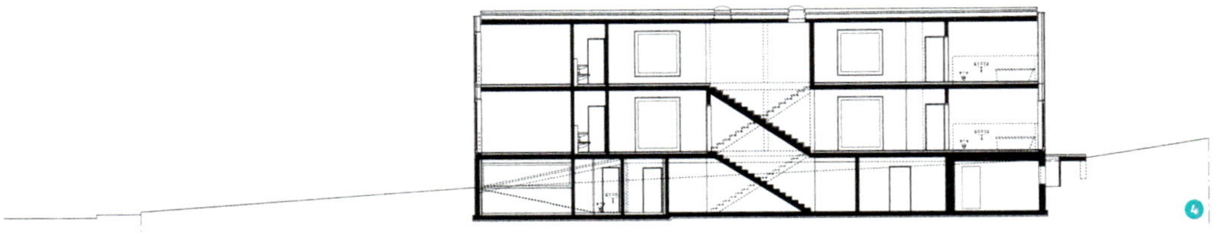
4

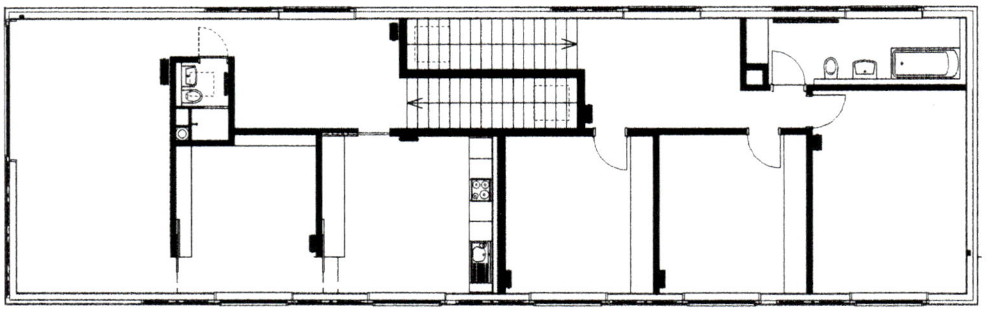
3

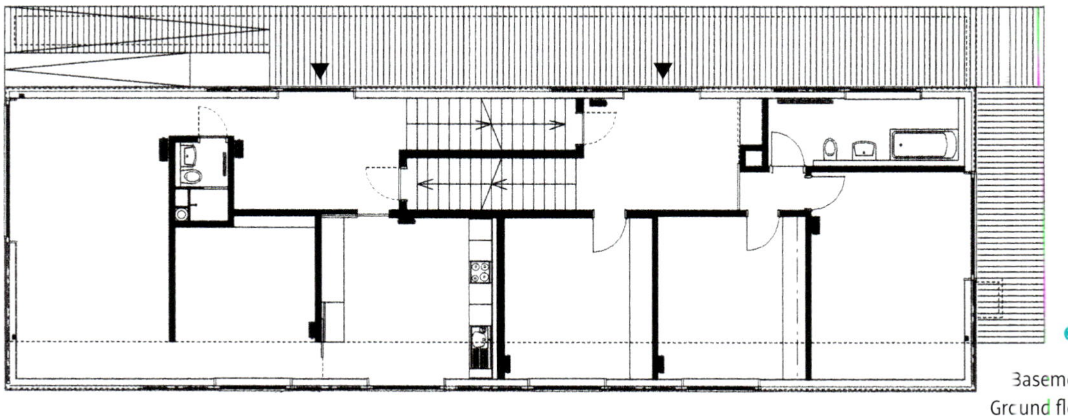
2

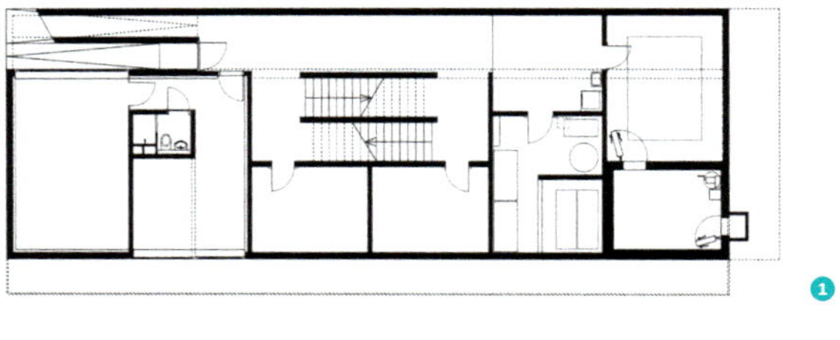
1

Basement 1:400 1
Ground floor 1:200 2
Second floor 1:200 3
Longitudinal section 1:400 4
North elevation with 5
access footbridge
East elevation 6

Referring to
floor plan 1:200

Vill | Noldin & Noldin | 2001

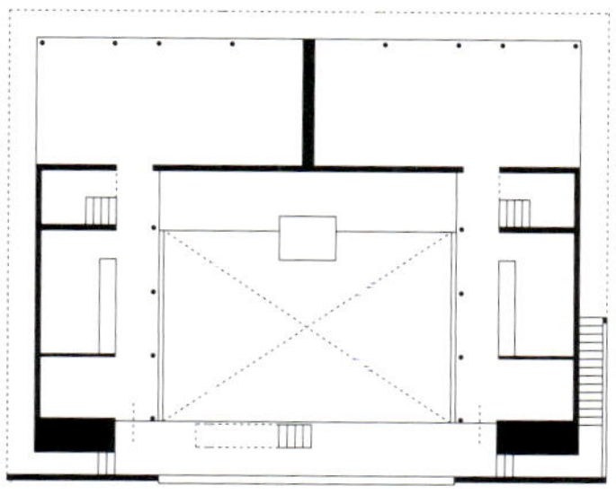

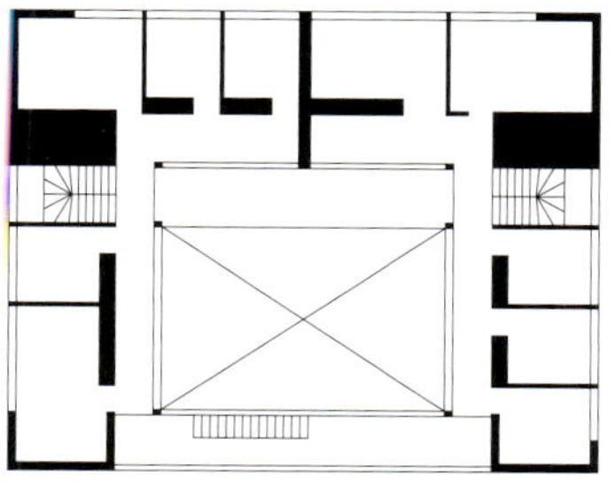

3.2
DUPLEX

Two families who are friends, wanted to create adjoining apartments with a shared semipublic space for socializing with neighbours and friends. To this end, the U-shaped duplex building surrounds a large shared courtyard, which is completed into a rectangle by a two-story verandah with external stairs. The internal plan in the two apartments is organized according to the same principle: a large open living area on the ground floor and individual rooms upstairs. The two kitchens face toward each other and overlook the courtyard. The living areas are oriented in the opposite direction, away from the courtyard. On the second floor, all rooms also face away from the courtyard, with the exception of the work areas, which are oriented in both directions. Both apartments can be divided on the upper floor, allowing for a flexible arrangement of rooms. The internal access areas for both apartments are turned toward each other whenever possible: access to both units is provided at the ground and upper level via the shared verandah, the internal stairs face one another and are linked on the ground floor by the wooden terrace. Upstairs, the corridors to the bedrooms on the courtyard side also face each other and are integrated via the verandah into a continuous ring connection. Opposite to this verandah, the balconies for both units lie side by side; like the verandah and the terrace, they are oriented toward the shared center in the manner of a theater box.

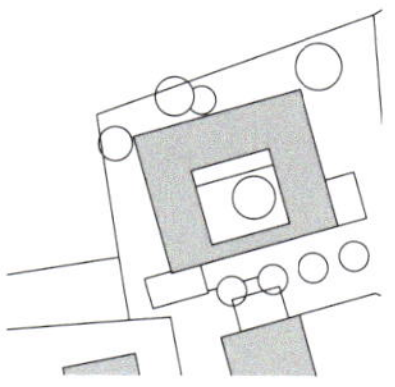

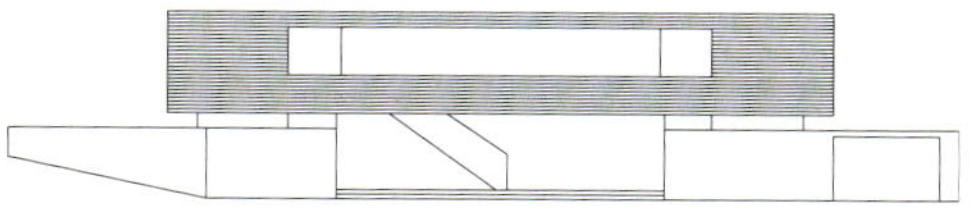

3

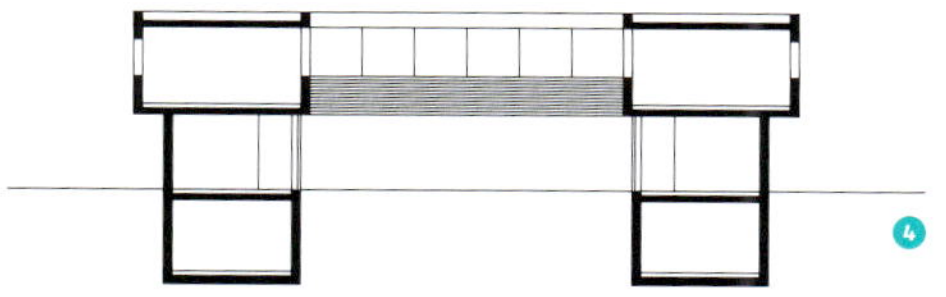

4

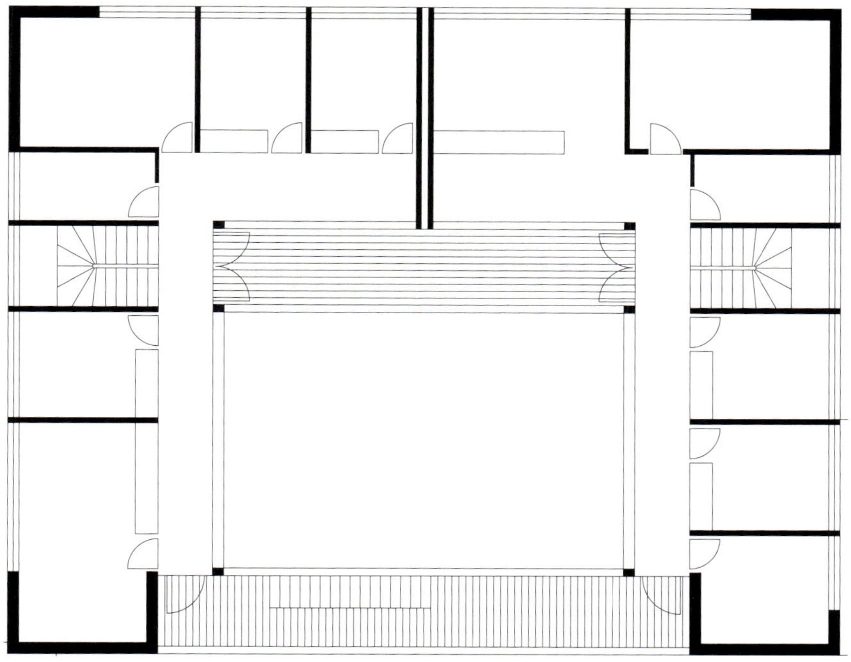

2

Building type
freestanding duplex around a courtyard
2 stories
facing in all directions

Date of construction
1991–2001

Living area
6-room apts., each 175 m²

Area per user
29–35 m²

Construction
low-energy house with shared courtyard, porch and terrace, plate structure constructed of glued laminated timber

Size of lot
approx. 690 m²

Architect
Noldin & Noldin Architekten
Innsbruck

Location
Vill
Austria

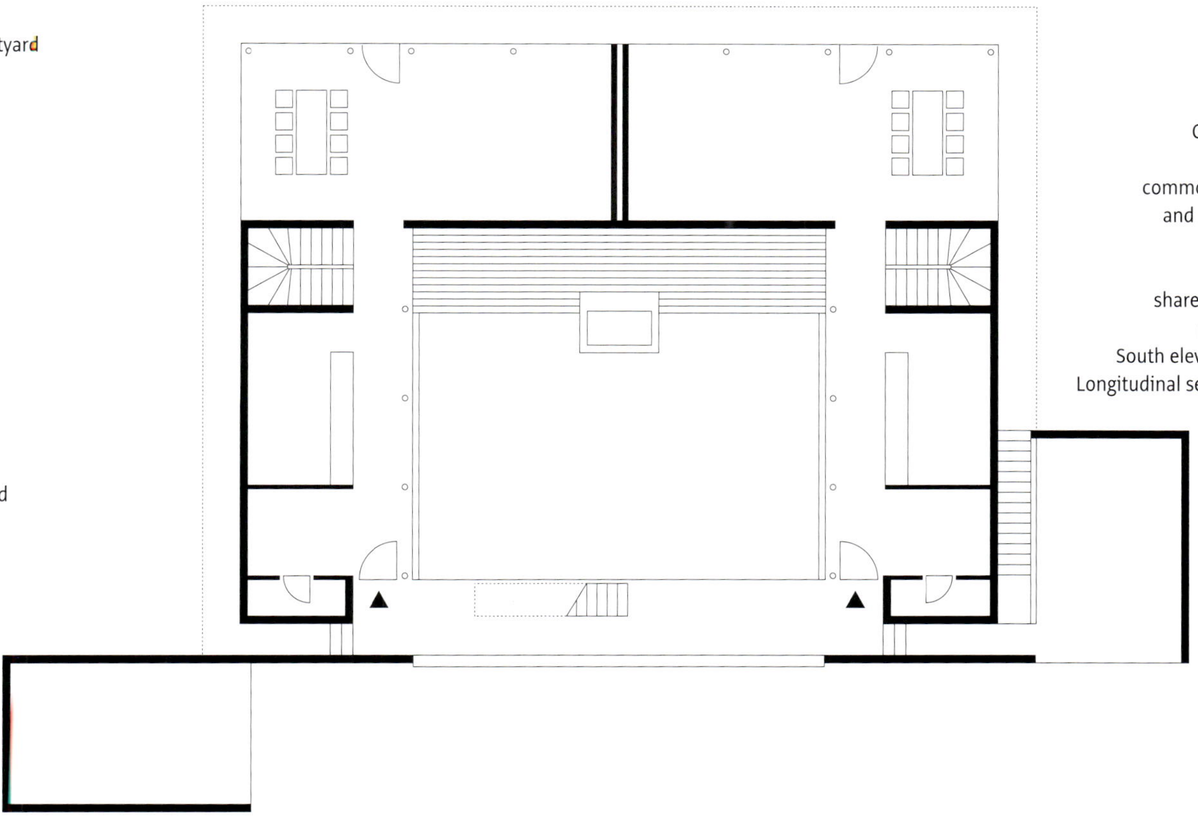

1

Ground floor: 1
living areas,
common courtyard
and porch 1:200
First floor: 2
bedroom,
shared loggia and
porch 1:200
South elevation 1:500 3
Longitudinal section 1:500 4

Referring to
floor plan 1:200

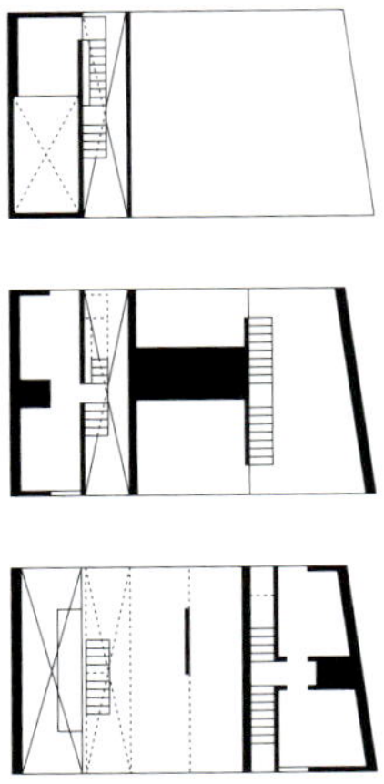

Villa KBWW |
De Architektengroep bv / MVRDV | 1997

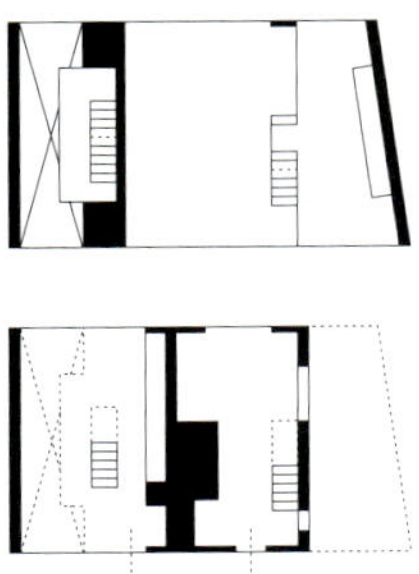

This building with two semidetached homes derives its structure from a highly individual development process, which it boldly displays. Two architectural firms were commissioned to accommodate the wishes of two very different clients under one roof. As different as their housing concepts were, both parties nevertheless asked for direct access to roof terrace and garden and, moreover, an unencumbered view of the park to the south. The living area was to be divided according to their unequal financial capacities, i.e., one-third to two-thirds. Following various developmental stages (see diagrams) the solution was found in a dividing wall, which does not constitute a continuous vertical separation between the halves but, instead, divides them in the guise of a wall, on the one hand, and then again as a staggered ceiling, all the while maintaining the same thickness and, what is more, with only minimal supporting columns, that is, in a very elaborate structural manner. The total building cube is therefore composed of different narrow and wide rooms, which are interlaced. The living areas extend far into the neighbouring semi and enjoy a view of the park across their full widths. The smaller semi on the left side, in particular, which is strongly articulated in the vertical direction, benefits from this elongated horizontal space. The rooms flow into one another and feature generous glazed surfaces; the bedrooms and bathrooms are the only fully enclosed volumes. The meandering course of the dividing wall is thus strongly visible on the facade.

Chronological evolution of design concept:

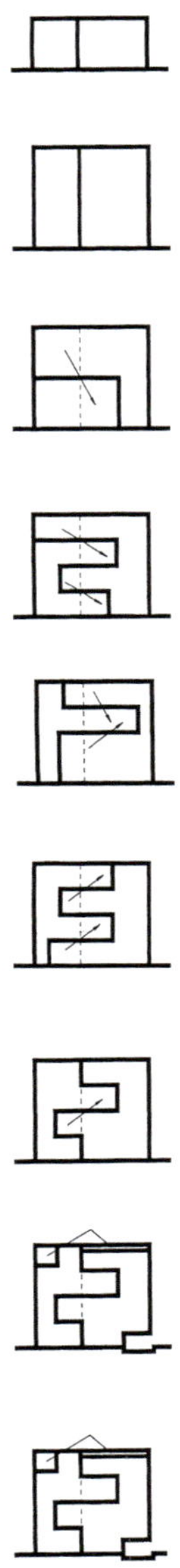

2 stories, 14 m building depth

4 stories, 7 m building depth

House-in-house,
no roof access for house 1

Interlacing 1,
no roof access for house 1

Interlacing 2, garden access
too narrow for house 1

Interlacing 3, garden access
too narrow for house 1

Interlacing 4, roof and
garden access for both houses

Roof terrace for house 1, roof terrace
and parking place for house 2

Final result, bedrooms for house 1
and 2, piano nobile of house 2
corresponds to fire-safety regulations

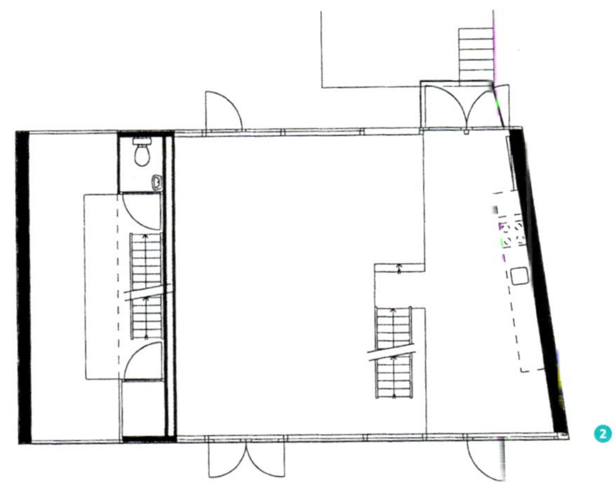

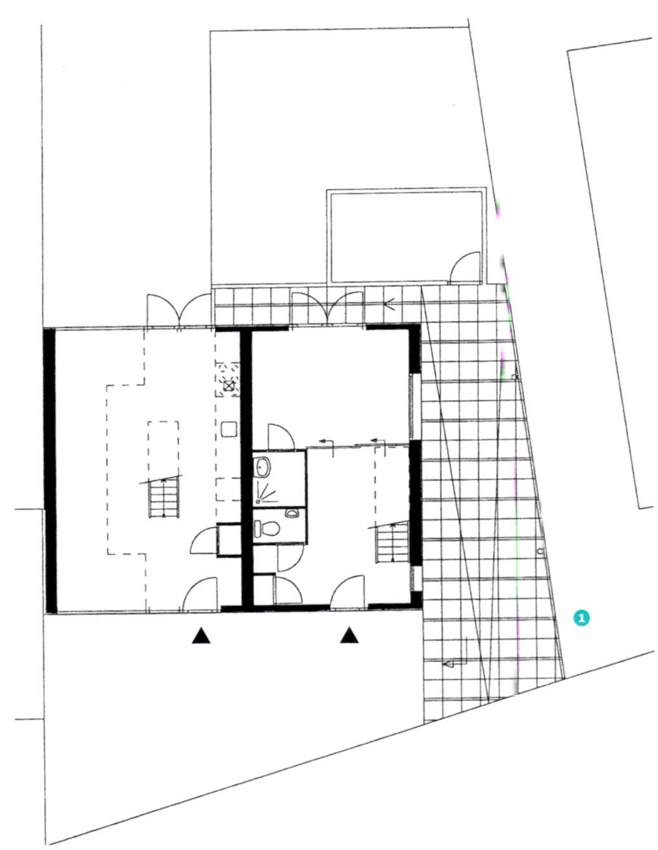

1 Ground floor
House 1: kitchen
House 2: parking place below house, guest room, and storage space in garden 1:200

2 2nd floor
House 1: library gallery
House 2: large living room 1:200

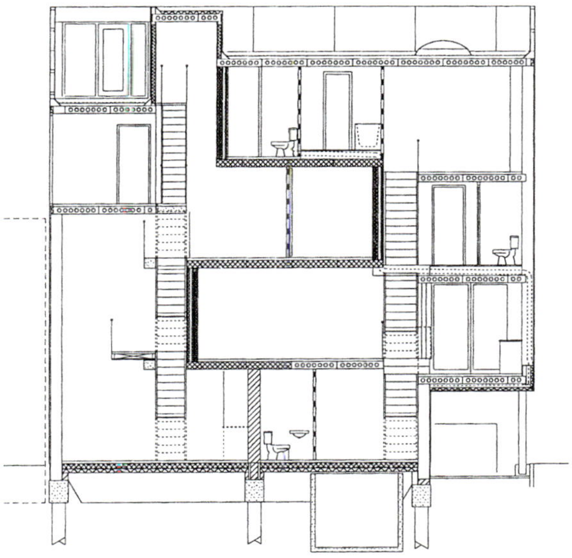

3.2
DUPLEX

Building type
semidetached houses
whose volumes are interlaced
5 stories
facing N/S

Date of construction
1997

Living area
110/160 m²

Area per user
40–55 m²

Layout
reinforced concrete construction,
clad on the inside
with different plywood panels
and on the outside with
boarding soaked in phenol

Size of lot
approx. 250 m²

Architect
Bjarne Mastenbroek /
De Architektengroep bv
Amsterdam
Winy Maas / MVRDV
Rotterdam

Location
Utrecht
The Netherlands

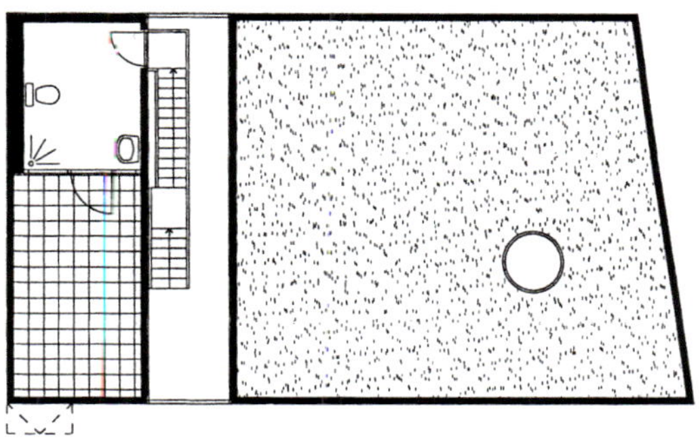
5

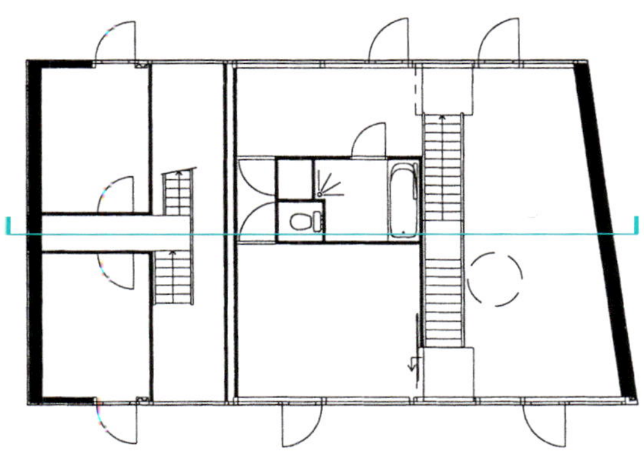
4

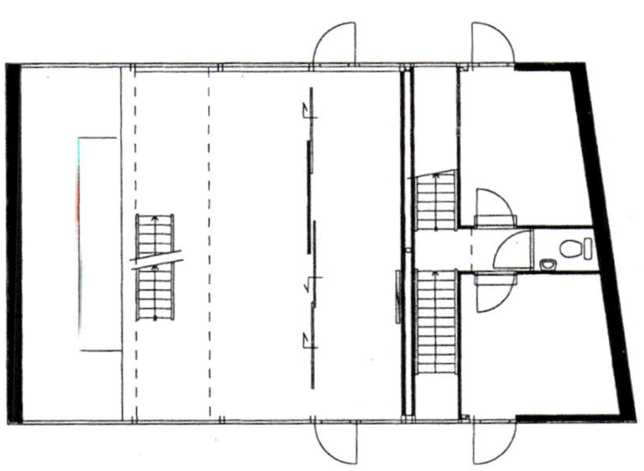
3

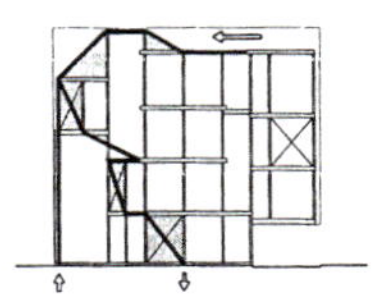
Distribution of forces

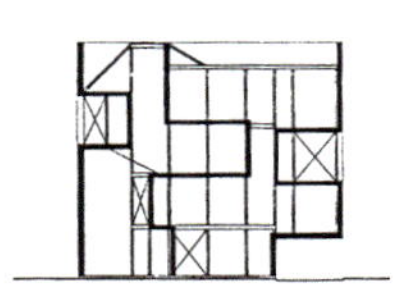
Division in apartment and face walls

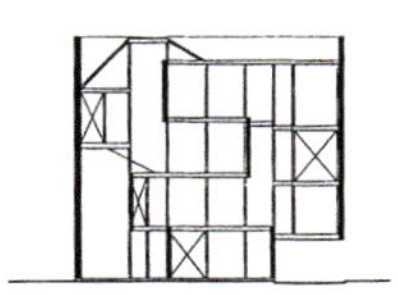
Bracing in north/south direction

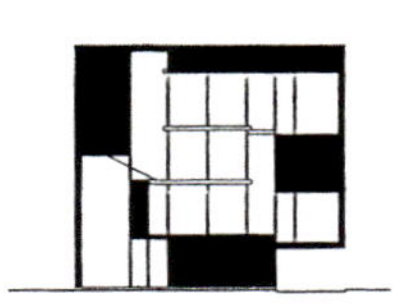
Closed facade areas

3 3rd floor
House 1: living room
House 2: bedroom 1:200
4 4th floor
House 1: bedroom
House 2: study/bedroom 1:200
5 5th floor
House 1: roof with bathroom and terrace
House 2: roof terrace 1:200

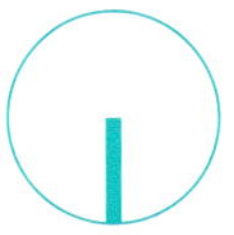
Referring to floor plan 1:200

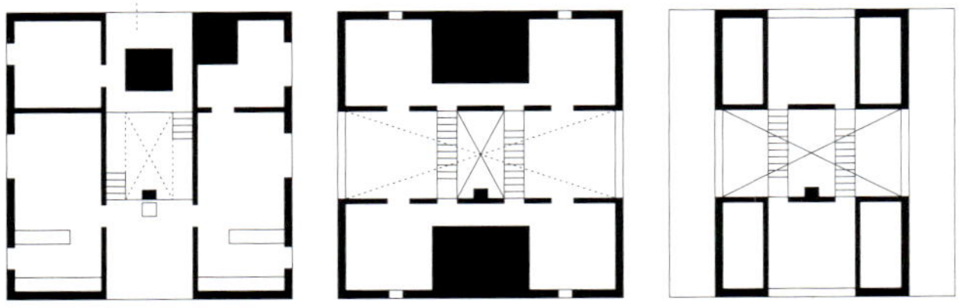

Patchwork House | Pfeifer, Roser, Kuhn | 2005

3.2 DUPLEX

Although conventional at first glance, the volume of this simple house reveals a surprising interior life: a house-within-a-house concept in which two different tenants occupy rooms that are in part stacked one above the other and in part facing each other, all connected via open stairs that are dedicated for the exclusive use of each tenant. The shared atrium, which rises across all three levels and in which the single-loaded stairs run past one another in a crisscross pattern, also serves as a space where the residents will naturally encounter one another, to stay awhile and socialize without restraint. Both units function in a fully autonomous manner. On the ground floor with its shared entrance, the two units unfold parallel to the stairs; this is where kitchen and living areas are located as well as separate spaces for other uses. On the upper levels, the stairs lead to the private rooms, now arranged crosswise to the stairs: the rooms for each unit alternate, lying first on one side and then on the other, with the result that both tenants benefit from all orientations. The atrium also serves as an "energy garden," which provides passive solar energy for the entire building.

Building type
Freestanding duplex house with optional in-law apartment, 3 stories facing in all directions

Date of construction
2003–2005

Living area
5 rooms each, shared living room on ground floor, total area 294 m²

Area per user
29.5 m²

Construction
Two units on opposite sides, alternating in orientation from floor to floor and connected by open, crisscrossing stairs in a joint atrium ("energy garden"). Gutter walls and roof surfaces constructed of wooden board stacking elements and transparent polycarbonate panels for solar energy gain. Solid, single-leaf end walls of porous concrete as storage mass, heating as building component activation in concrete floors

Size of lot
804 m²

Architect
Architektengemeinschaft Pfeifer, Roser, Kuhn Freiburg

Location
Am Lerchenbuch, Müllheim, Germany

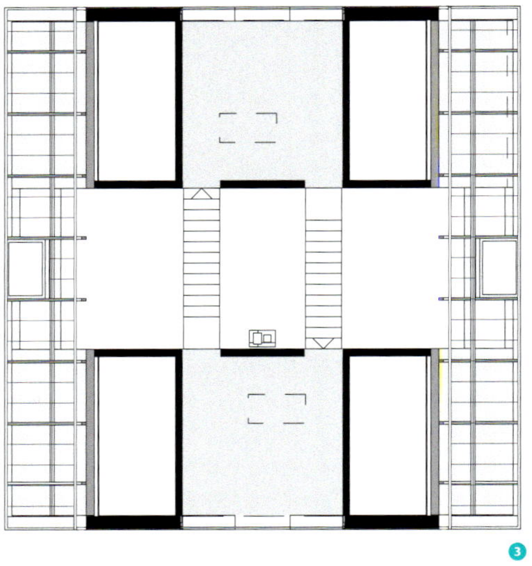

3

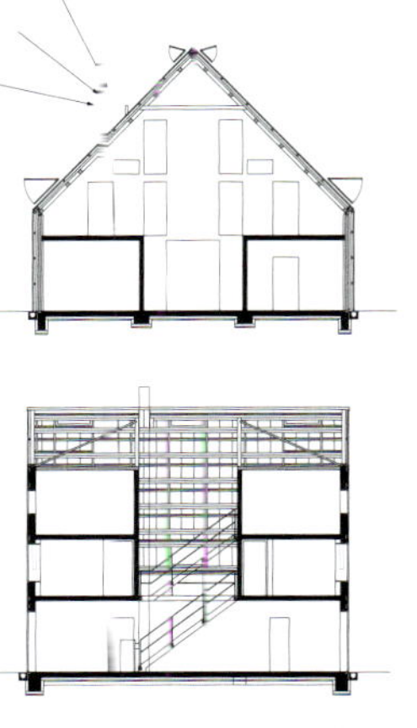

4

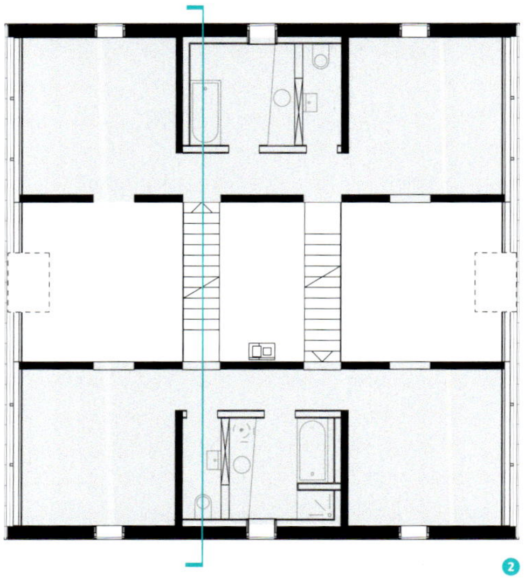

2

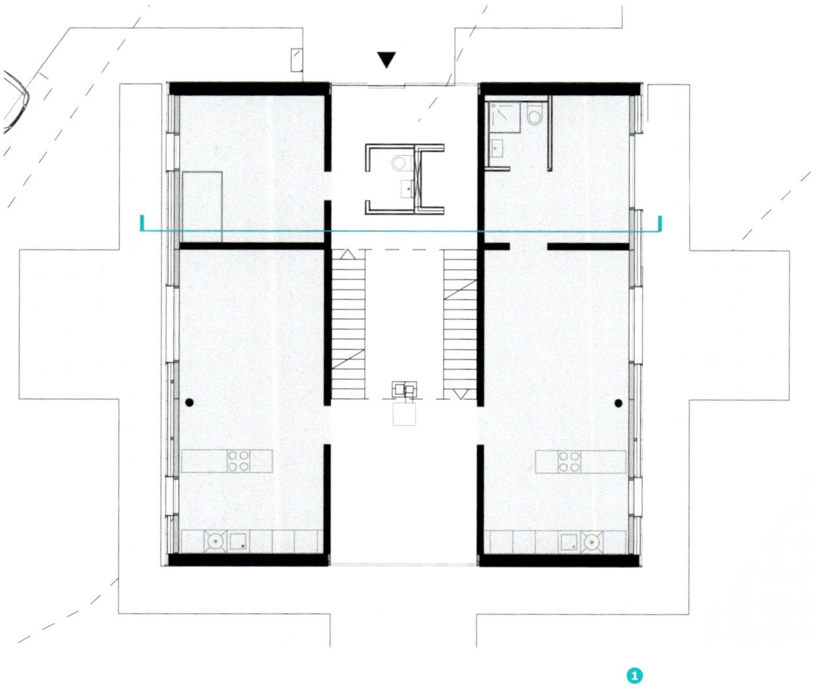

1

Ground floor with access in 1 lengthwise orientated hall, common and individual living areas 1:200
2nd floor with access in crosswise orientated 2 hall, bedrooms, and bathrooms 1:200
Attic story with one private room each 3
Cross and longitudinal section 1:500 4

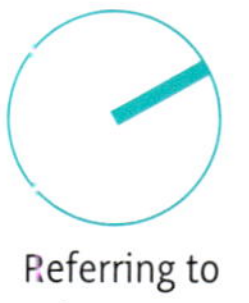

Referring to floor plan 1:200

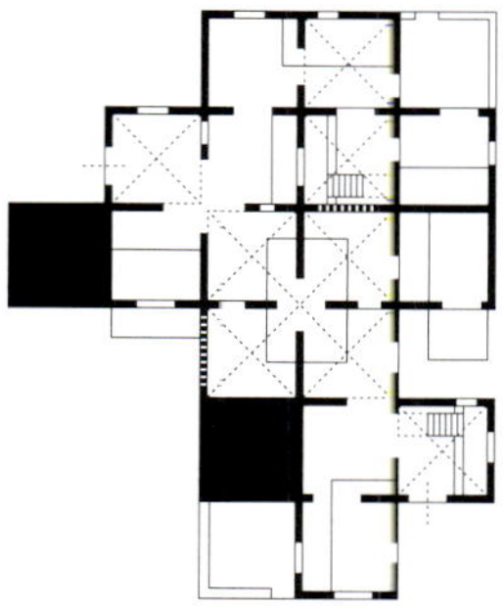

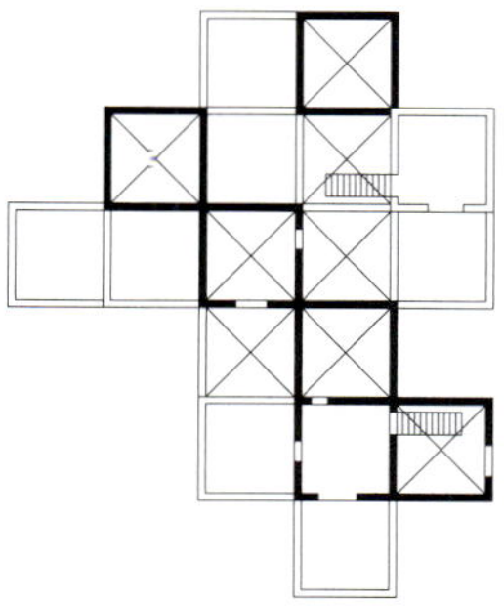

Minus K | KUU Architects | 2010

The floor plan is based on a square grid to form a three-dimensional conglomerate in which 3 x 3 m spaces are predominantly connecting rooms, linked axially or diagonally to form larger areas. Double-height volumes mark more important spaces or are used to access bedrooms or roof terraces above, and open patio courtyards seamlessly link the interior rooms to the outside.

This patchwork sets the stage for two intertwined residential units shared by neighbors – one is for the factory owner and his family, who occasionally reside here, the second for the warehouse supervisor and his child. Where the two units intersect, an open connection comprised of four quadrants – two in each apartment, one inside and one outside – houses a shared kitchen island stretching between the units, where the neighbors encounter one another from within their own space.

The concept is a reference to a living form common in China until not long ago where communal kitchens, bathrooms, and courtyards were shared among residents. What was once a necessity is here freely chosen.

Building type
bungalow-like cluster
1–2 stories
facing in all directions

Date of construction
2010

Size of units
5-room apt., 74 m² (1 unit)
6-room apt., 95 m² (1 unit)

Area per user
37 m²

Layout
patchwork-like brick building based on 3 × 3 m, partially multi-story spatial modules connected axially or diagonally
kitchens and patios act as interfaces between the two units

Size of lot
1,009 m²

Architect
KUU Architects
Kok-Meng TAN
Satoko SAEKI
Tokyo/Shanghai

Location
Nanhui
Shanghai
China

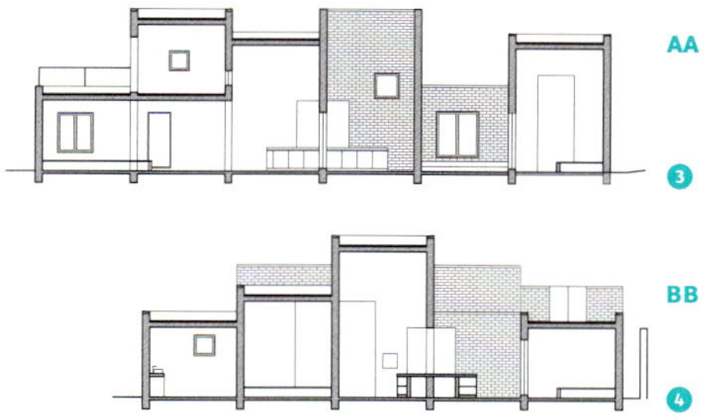

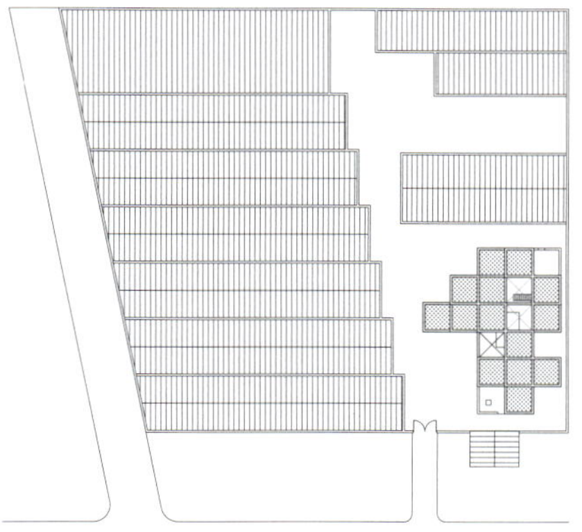

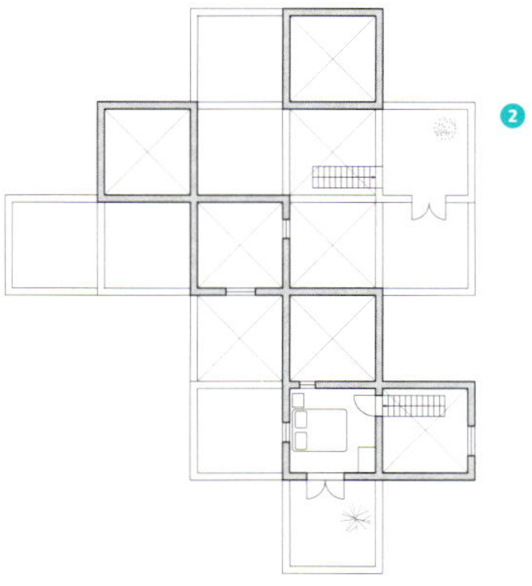

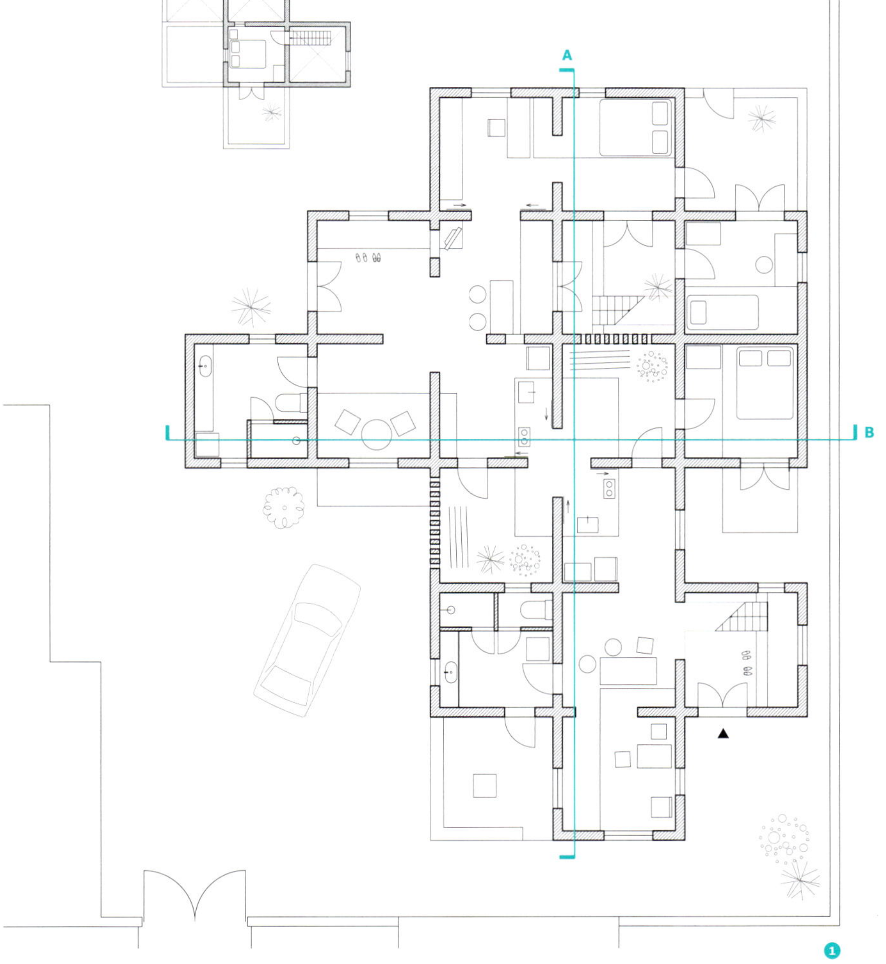

Ground floor 1:200 1
2nd floor/roof plan 1:400 2
Longitudinal section 1:400 3
Cross section 1:400 4

Referring to floor plan 1:200

3.3
ROW HOUSE

This house type is more economical in terms of construction and energy consumption than the detached house due to the large percentage of shared firewalls and the relatively small facade area. It is thus affordable to more people and corresponds to the desire for owning a house with a private garden. Row houses allow for a better degree of densification than residential developments with detached houses and are thus an appropriate building form for the urbanization of industrial wastelands close to the city center or for urban agglomerations. The classic row-house type is long and narrow, oriented toward the front and rear, and laid out between two closed walls. Most row houses are two stories high; the floor plan is divided into three sections with rooms and kitchen located at the facades and access and sanitary rooms in the middle. In principle, however, this type is highly flexible with regard to width, depth, and height. Innovative examples offer entirely new layouts: they achieve exciting floor plan configurations with patios, rotation, offset of levels, or change in orientation, thus relieving the potential monotony of long row-house developments.

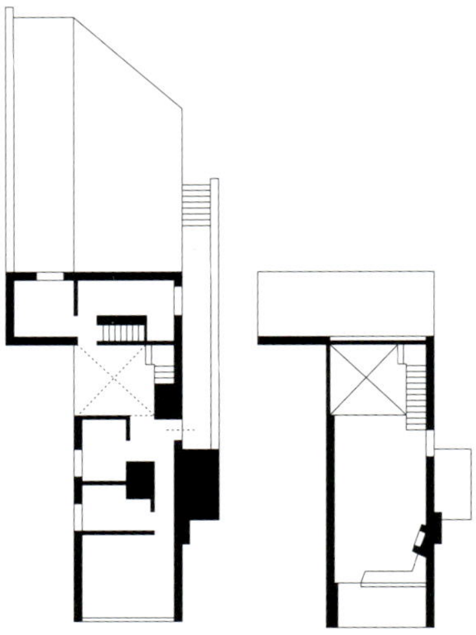

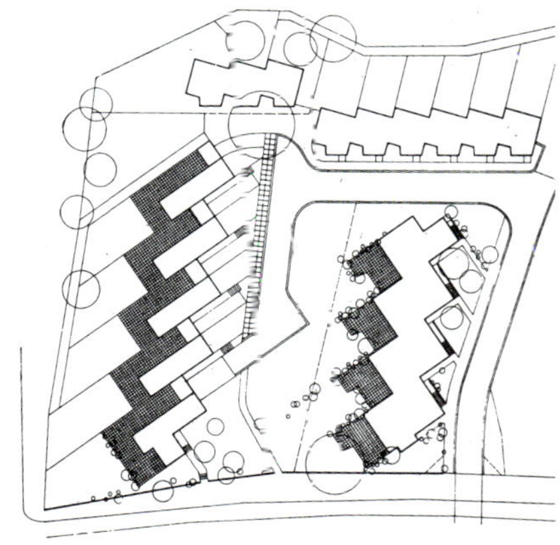

Søholm I – III | Jacobsen | 1954

The residential complex consists of Søholm I (5 linked houses, with the architect's private home at the end), Søholm II (9 row houses along the coastal path) and Søholm III (4 row houses, stepped back and only single-story in order to leave the view open for the other units). The shifted plan of the linked houses and the distinctive roofs and wall openings even on the longitudinal sides have the effect that each house of Søholm I is both part of the chain and an individual volume. The landscaping arises from the formation of the buildings comprising wraparound patios on the garden side and forecourts on the opposite side with entrance steps, front gardens, and a garage ramp. The external shape of the houses is closely harmonized with the internal layout. The dining area, for example, is embedded into the middle of the house both in cross-section and in plan. Located close to the entrance, connected via open stairs and its raised ceiling to the living room and oriented laterally toward the patio, this space is a key interface with the kitchen on one side and the bedrooms on the other. The kitchen leads to a set of stairs down to the garage, which is slightly lowered into the ground, and to the utility areas beyond as well as to an additional room, which defines the patio at right angles. Yet the living room occupies a tranquil position at the very top. The shape of the roof above it, with a band of windows facing northwest, guides the eye skyward, and also to the dining room below and out beyond the balcony to the sea. At this point, the built-in bench between fireplace and window beckons one to take everything in simultaneously – the surrounding landscape and the house.

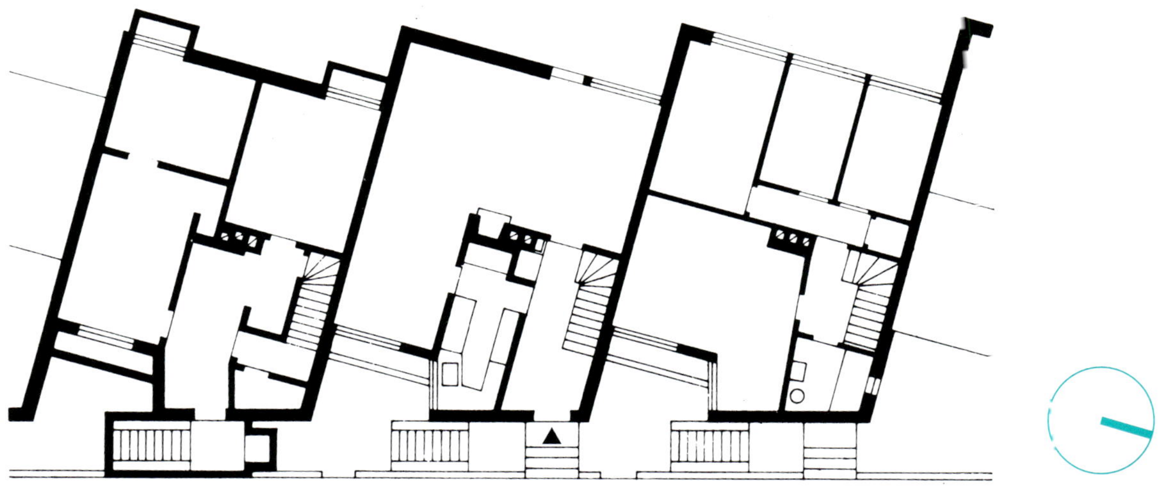

1

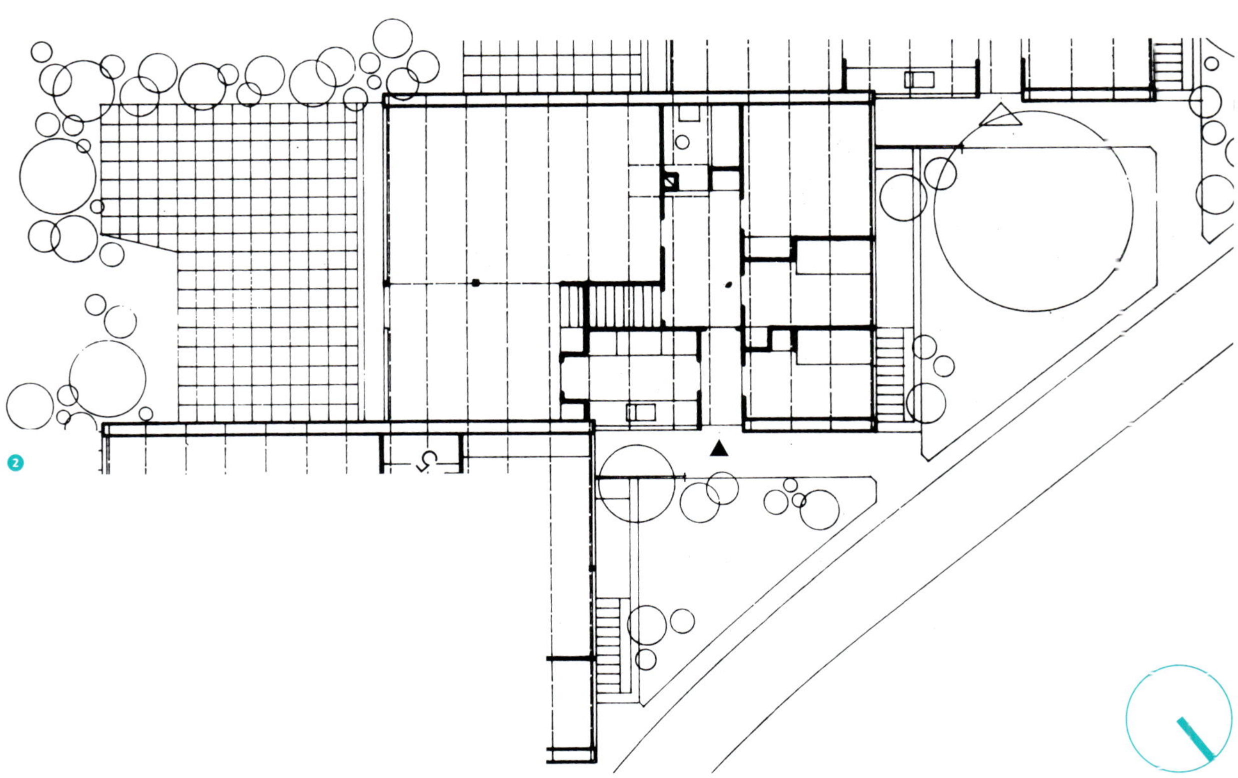

2

1 Søholm II: basement, ground floor, and upper floor 1:200
2 Søholm III: ground floor 1:200
3 Søholm I: basement, ground floor, and upper floor 1:200
4 Søholm I: longitudinal section 1:200

3.3 ROW HOUSE

Building type
Søholm I:
5 stepped-back, linked houses
2 stories
facing NW, NE/SW, SE
Søholm II:
9 row houses
2 stories facing W/E
Søholm III:
4 row bungalows
single-story
facing NW, NE/SW, SE

Date of construction
1951–1954

Number of units
18

Living area
Søholm I: 99.5 m²
Søholm II: 105.5 m²
Søholm III: 101 m²

Area per user
21–25.5 m²

Building depth
10–15 m

Layout
row-house complex
arranged in U-shaped formation
around common landscaped green
space with view of the Øresund

Open spaces
patios, balconies

Parking
garages
parking lot

Architect
Arne Jacobsen
Copenhagen

Location
Klampenborg
Gentofte
Denmark

Referring to
floor plan 1:200

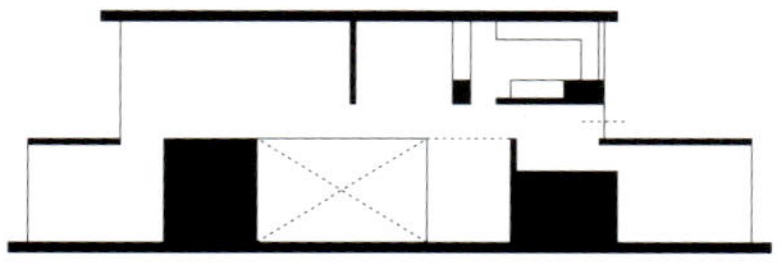

The Ryde | Phippen, Randall, Parkes | 1964

High residential quality despite reduced costs by means of narrow building widths and extremely simple structural features (concrete crosswalls, wooden roof frame). The development of twenty-eight staggered one-story homes is structured by the accentuated crosswalls that demarcate the houses and their open spaces (greens-paces to the front and rear of the house, as well as inner courtyards in the case of the larger houses). The open spaces are tightly interlocked with the interior spaces and, except for kitchen and bath, almost all interior rooms are defined by sliding walls and can be utilized with corresponding freedom. The entire length of the house, all the way to the garden, is immediately visible to anyone entering. The rooms to left and right bulge out allowing the house to appear much larger. The clever use of natural light – by means of floor-to-ceiling windows, glassed-in patio, and skylights – also enhances the atmosphere of spaciousness and brightness. The sizes of the houses are varied by changing the length of the buildings between the crosswalls.

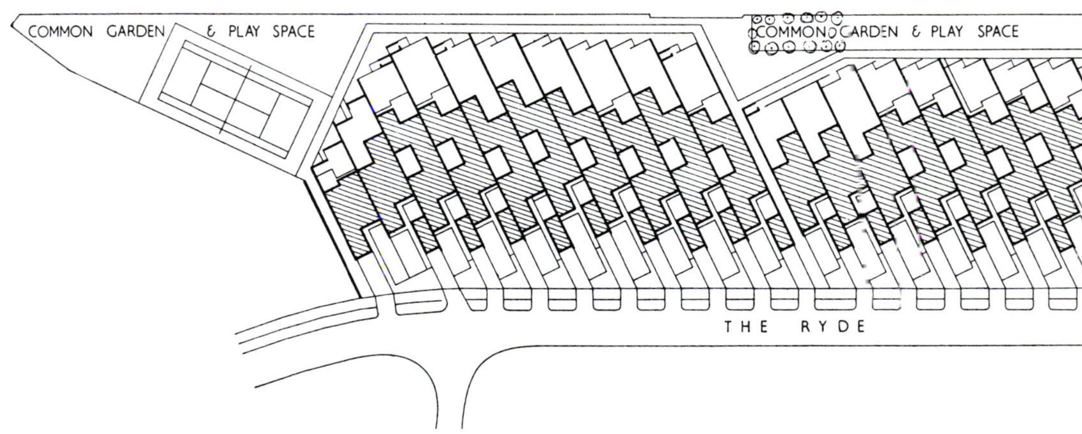

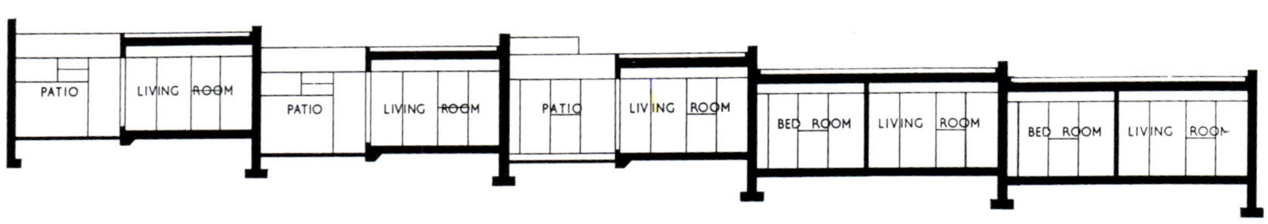

1

AA

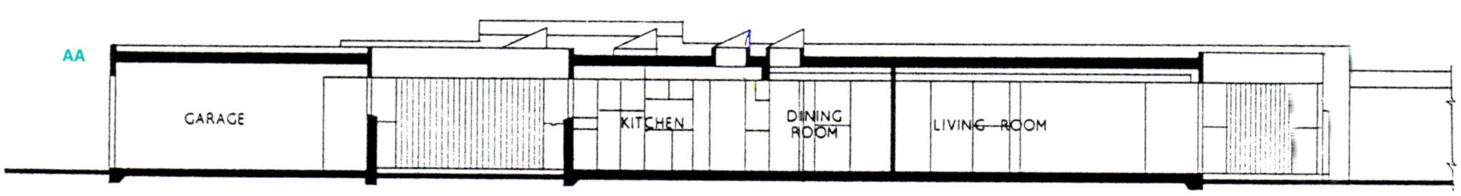

2

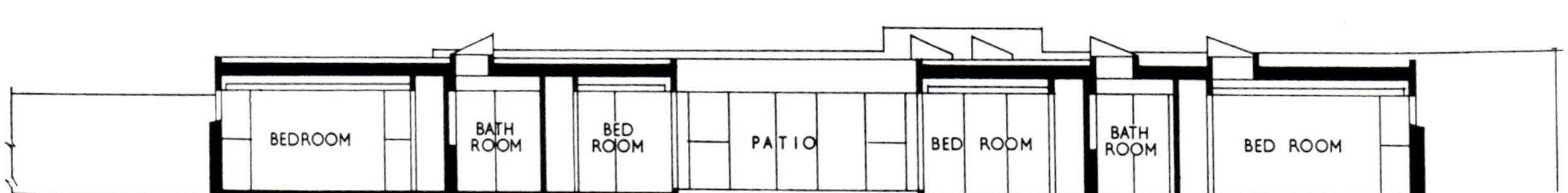

BB

2

1 Cross section through row of houses
2 Longitudinal sections through 5-room house 1:200
3 2-room house type with and without garage 1:200
4 3-room house type 1:200
5 4-room house type 1:200
6 5-room house type 1:200

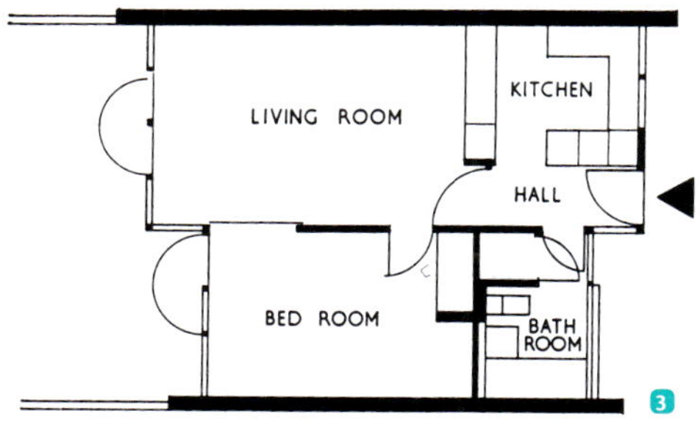

3

3.3 ROW HOUSE

Building type
row houses
1 story
facing NE/SW

Date of construction
1964

Living area
1-room apts., 58 m² (11 units)
2-room apts., 86 m² (10 units)
3-room apts., 112 m² (5 units)
4-room apts., 125 m² (2 units)

Area per user
31–58 m²

Building depth
9.9–25.3 m
(width 7.3 m)

Layout
compact development
consisting of deep,
28 parallel house units
with inside courts

Open spaces
patio, house gardens,
communal greenspace,
playgrounds

Parking
garages

Architect
Peter Phippen
Peter Randall
David Parkes
London

Location
London-Hatfield
United Kingdom

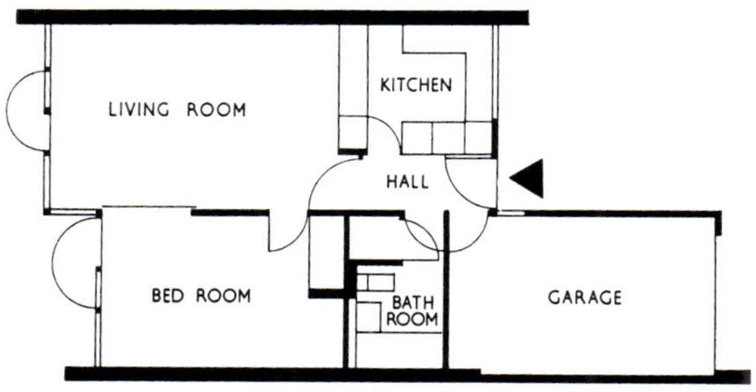

3

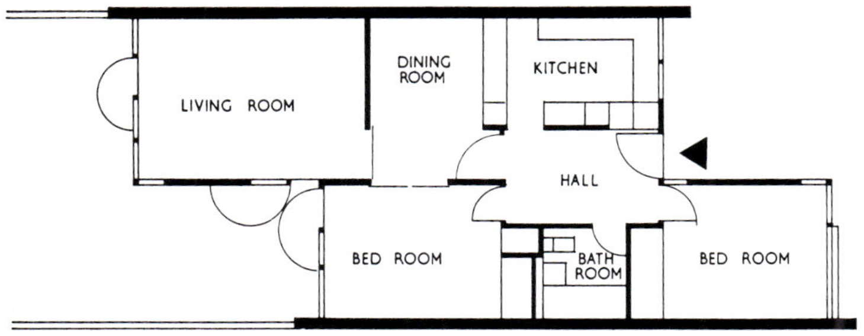

4

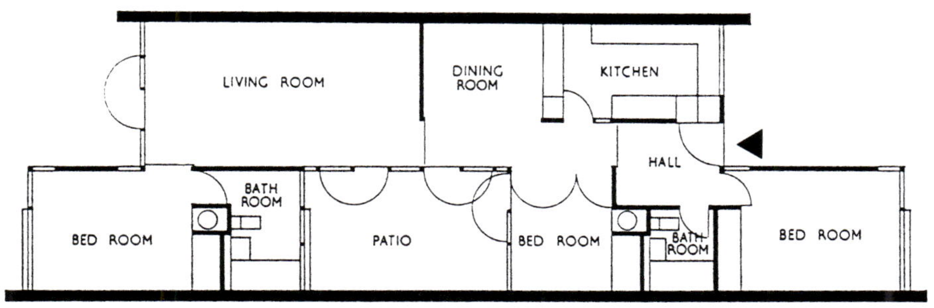

5

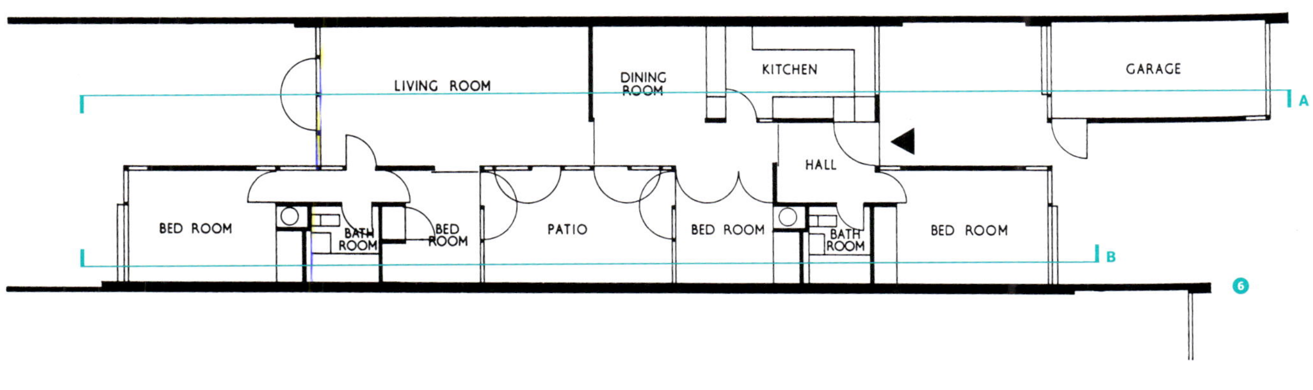

6

Referring to
floor plan 1:200

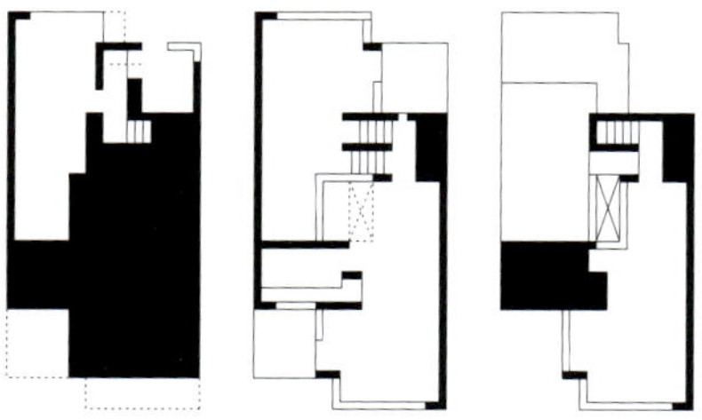

Diagoon Houses | Hertzberger | 1976

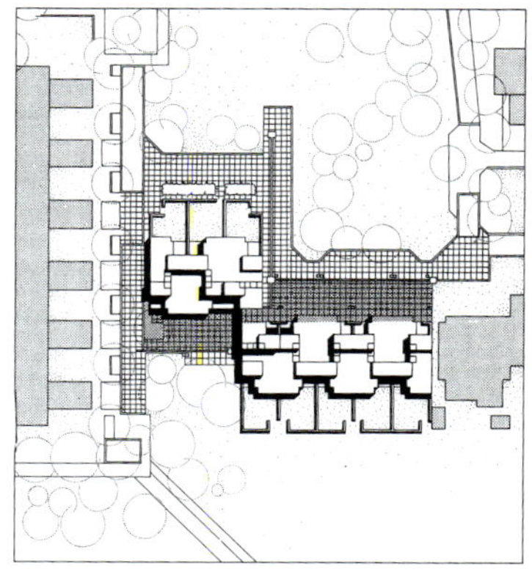

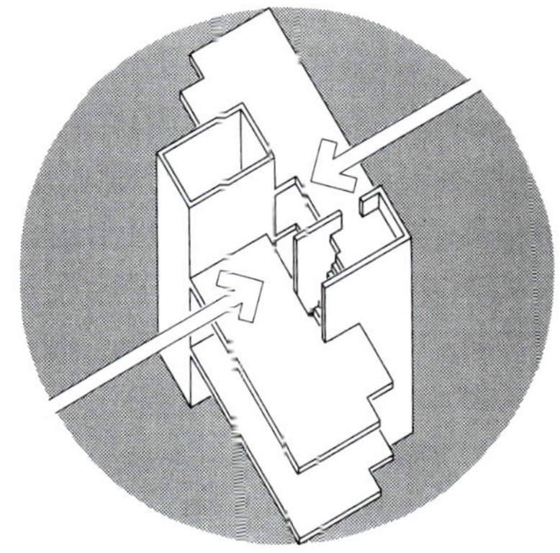

The idea behind the design is the intentional creation of "incomplete" houses that must be finished by their tenants according to their own needs and on their own responsibility. Installation and stairway cores are fixed, but the functions of the individual levels, which are staggered in split-levels, remain open. One room of each level can be separated off, with the rest remaining as a gallery on the central residential hall, which extends vertically through the entire house. These galleries are used for communal living. Many options for terraces; additions are also conceivable. An example: The room next to the entrance, which can be completed as desired. The reduced floor plan examples represent a series of possible divisions of the building floor plans. The basic layout is as follows: level a: entrance (with studio or garage); level b: kitchen with dining area (plus living room or bedroom); level c: living area or in-law apartment; level d: 1–2 bedroom(s) with bathroom; level e: roof terrace.

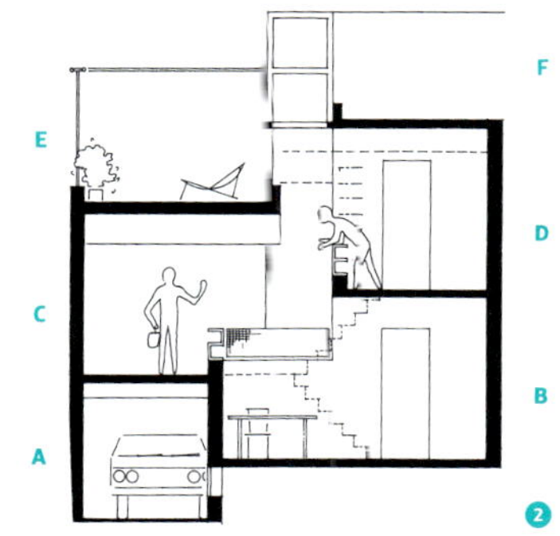

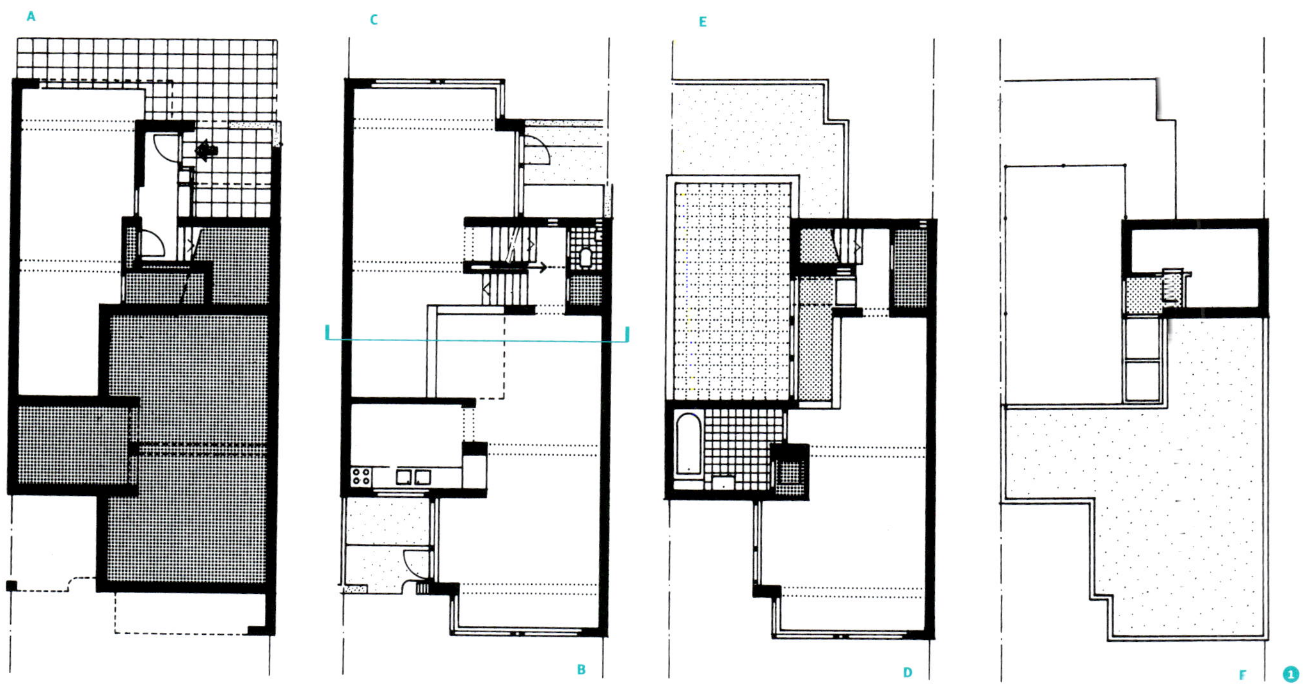

1 Floor plan schemes of levels **A–F** 1:200
2 Section 1:200
3 Variations of layout of levels **A–D**

D

3.3
ROW HOUSE

Building type
row houses
2/3 stories (4 levels)
facing NE/SW

Date of construction
1976

Living area
3–6 rooms,
approx. 145 m²

Area per user
24–48.5 m²

Building depth
15 m

Layout
8 units, 2 rows
of 5 or 3 house units,
each 2 units appear
as a single house

Open spaces
private gardens
rooftop terraces

Parking
lower entrance level

Architect
Herman Hertzberger
Amsterdam

Location
Delft
The Netherlands

C

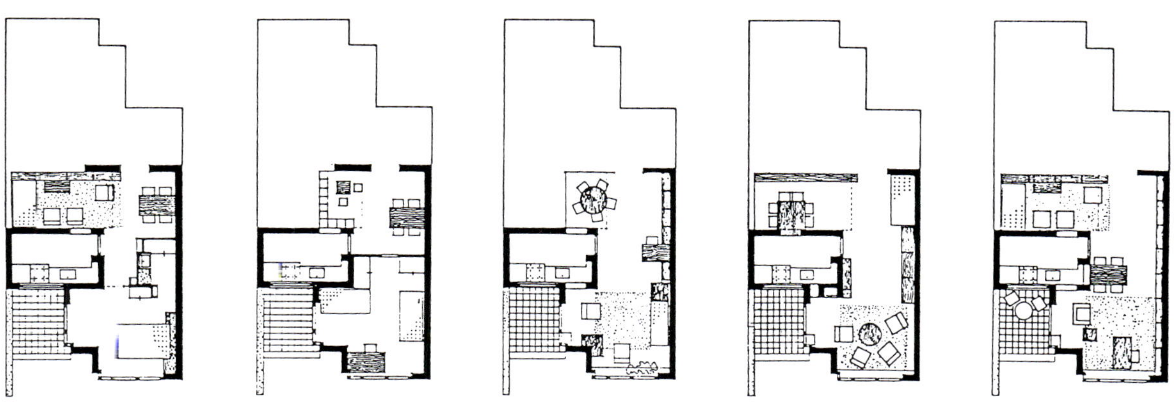

B

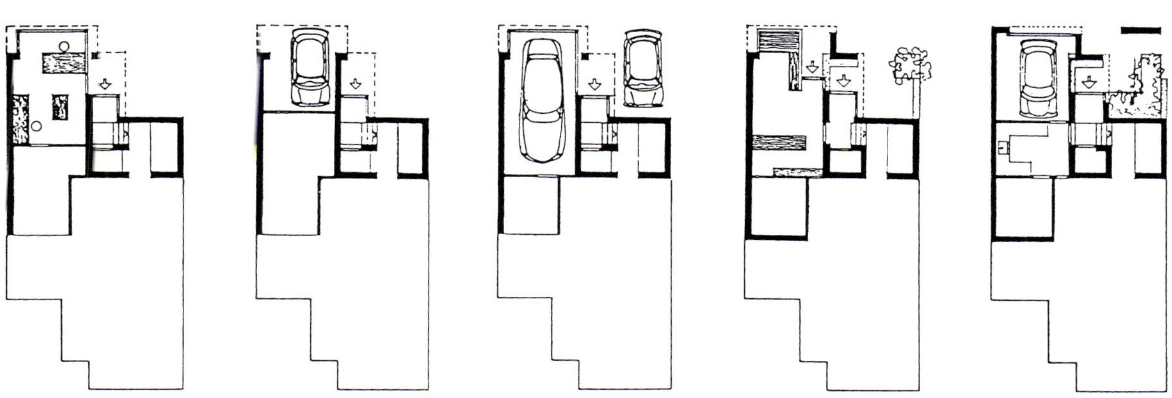

A

3

Referring to
floor plan 1:200

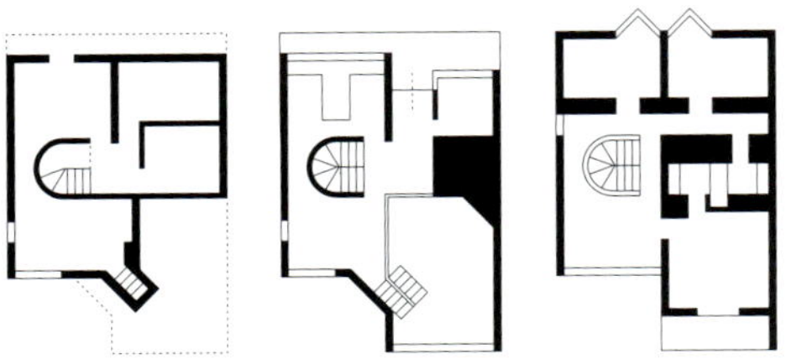

Altenbergstraße | Haas, Hermann | 1982

3.3
ROW HOUSE

Although part of a row, these houses are designed as individual homes. Their layouts organize living on various levels: the lobby leads directly to the parapet, and the viewer looks into the living room set a half-level lower. The living room is attached to the garden room (library), another half-level down. Located on the same level is a house workroom (with supply room). Kitchen and dining areas are located on the ground floor level; the inserted spiral staircase blocks them off without dividing them. A guest room is located at the entrance. The bedrooms are on the upper floor – with oriels to the street, terraces toward the garden. The spacious landing offers a large study area. An unusual aspect: the double bath, accessible from two sides, with two washing units and bathtub in the middle.

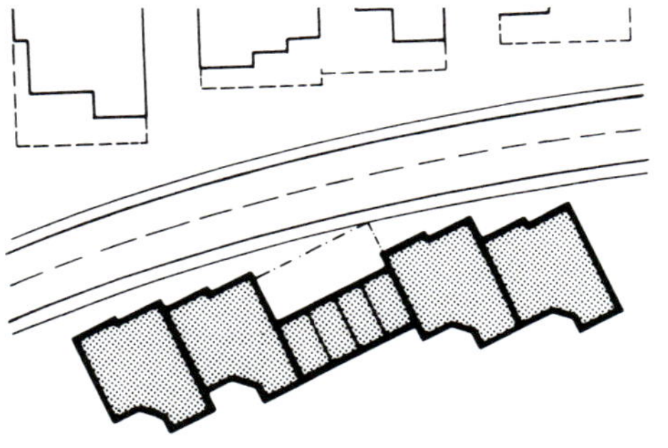

3

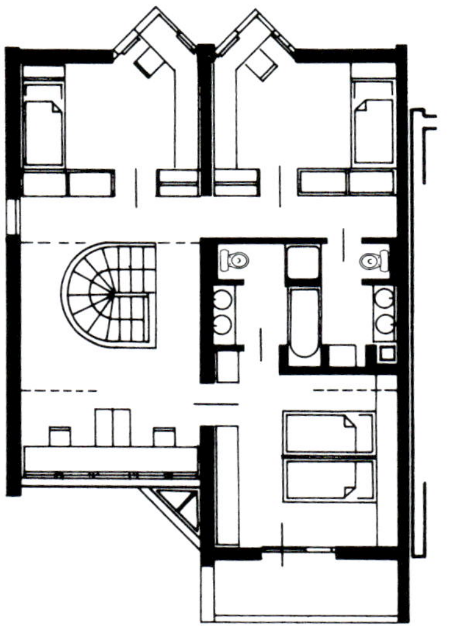

Building type
buildings on a slope
3 stories
facing NW/SE

Date of construction
1981–1982

Living area
6 rooms, 180 m²

Area per user
30 m²

Building depth
8.5/10.5 m

Layout
4 buildings, separated in middle into 2 double houses by garages, staggered towards street, brick masonry, stucco coated

Open spaces
gardens

Parking
garages

Architect
Peter Haas
Günter Hermann
Stuttgart
with G. Mayer

Location
Altenbergstraße
Stuttgart
Germany

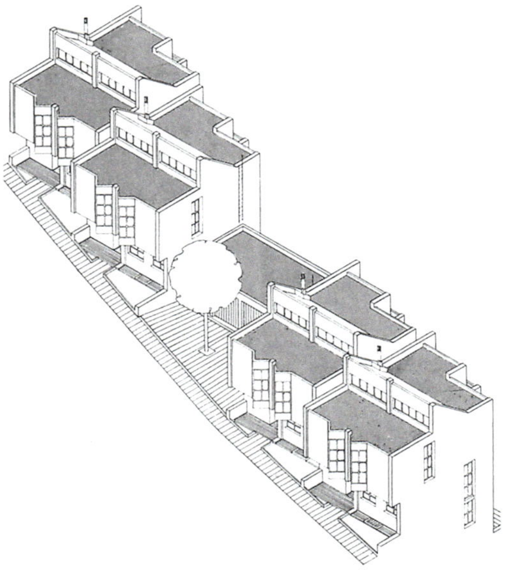

2

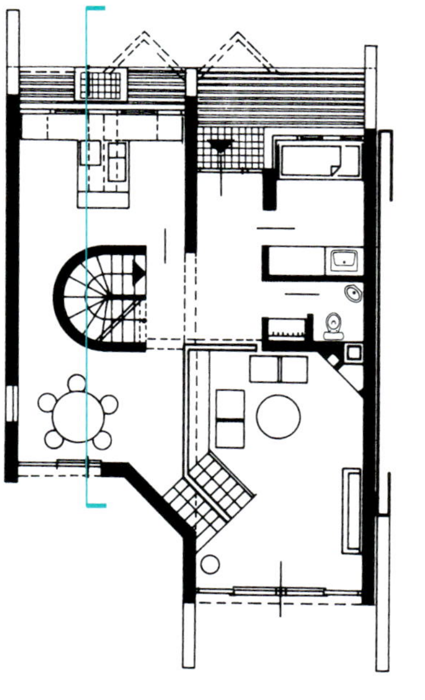

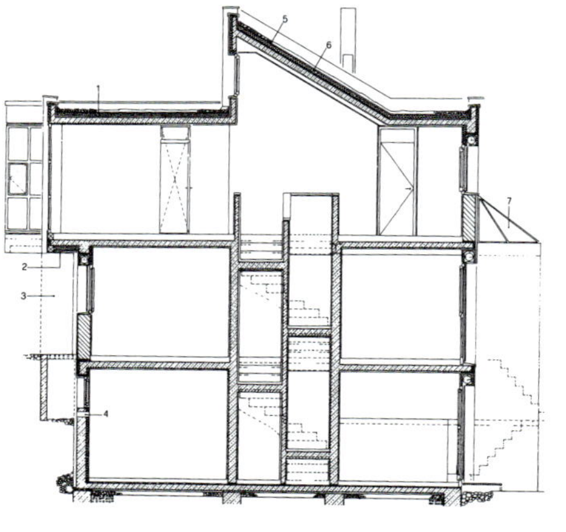

1

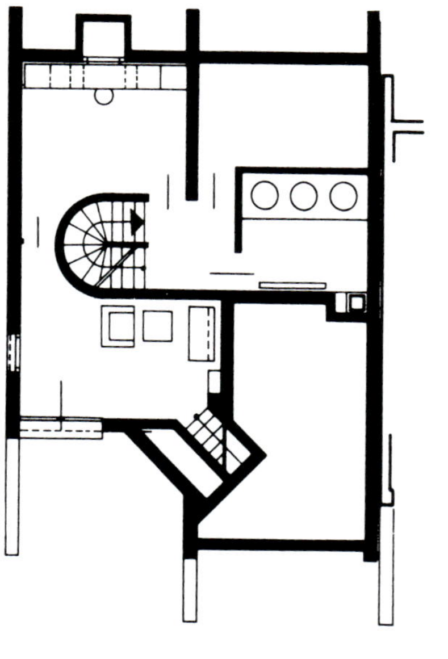

Basement 1:200 1
Ground floor 1:200 2
2nd floor with 3
bedrooms 1:200

Referring to
floor plan 1:200

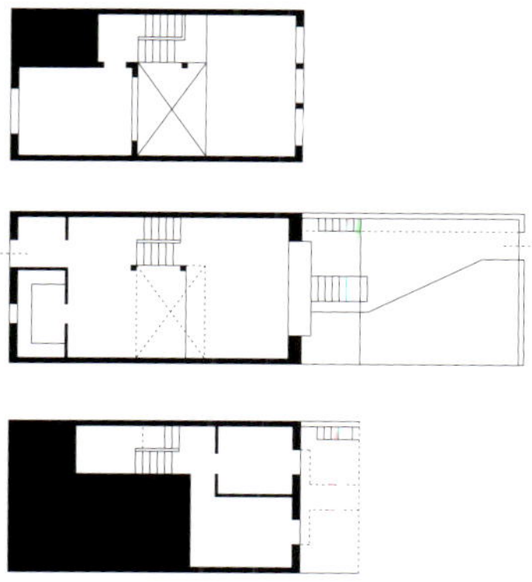

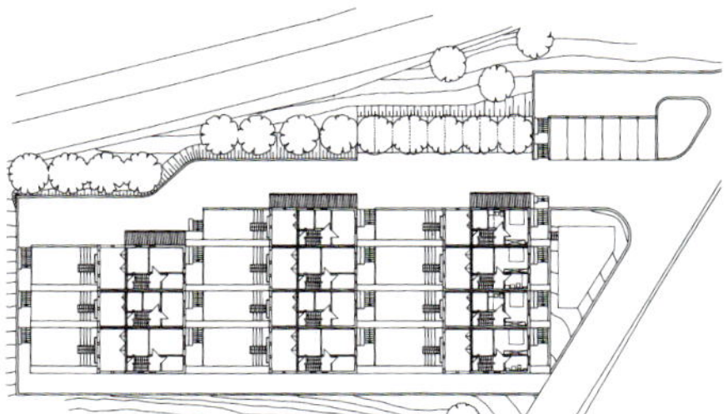

Kirchhölzle | GFP & Assoziierte | 1990

The property is a slope that rises from south to north and east to west. A highway runs along the north side of the valley, hence the east-west orientation with an emphasis on the west side (view of valley). The high density of the development allows only a garden courtyard of around 60 m² as the only open space for each house. However, the courtyards are supplemented by rooftop terraces of more than 20 m². Inside: a clearspace the height of two stories above the dining area. The entire layout of the buildings (split-level section, clearspace, balustrades, inside window openings) is designed to maximize lighting. The selected materials (light-filtering terrace roof, glass-brick walls on stairway and around courtyard) give different qualities to the light. Interior and exterior spaces are interwoven with great complexity; the terrace becomes a loggia, the courtyard is transformed into a garden room.

Building type
row houses, 2–3 stories
split-level, facing E/W

Date of construction
1986–1990

Living area
Type **A**: 4 rooms, 133 m²
Type **B**: 3 rooms, 107 m²

Area per user
34.5–36.5 m²

Building depth
12.5 m

Layout
11 units in small space, two-sided entrance, open space as garden courtyard, split-level

Open spaces
gardens, rooftop terrace

Parking
open parking places

Architect
GFP & Associates
Lörrach
Günter Pfeifer, Roland Mayer
with Harald Brutschin, Rolf Bühler
Ulrich Prutscher, Gerhard Zickenheimer

Location
Schopfheim-Wiechs
Germany

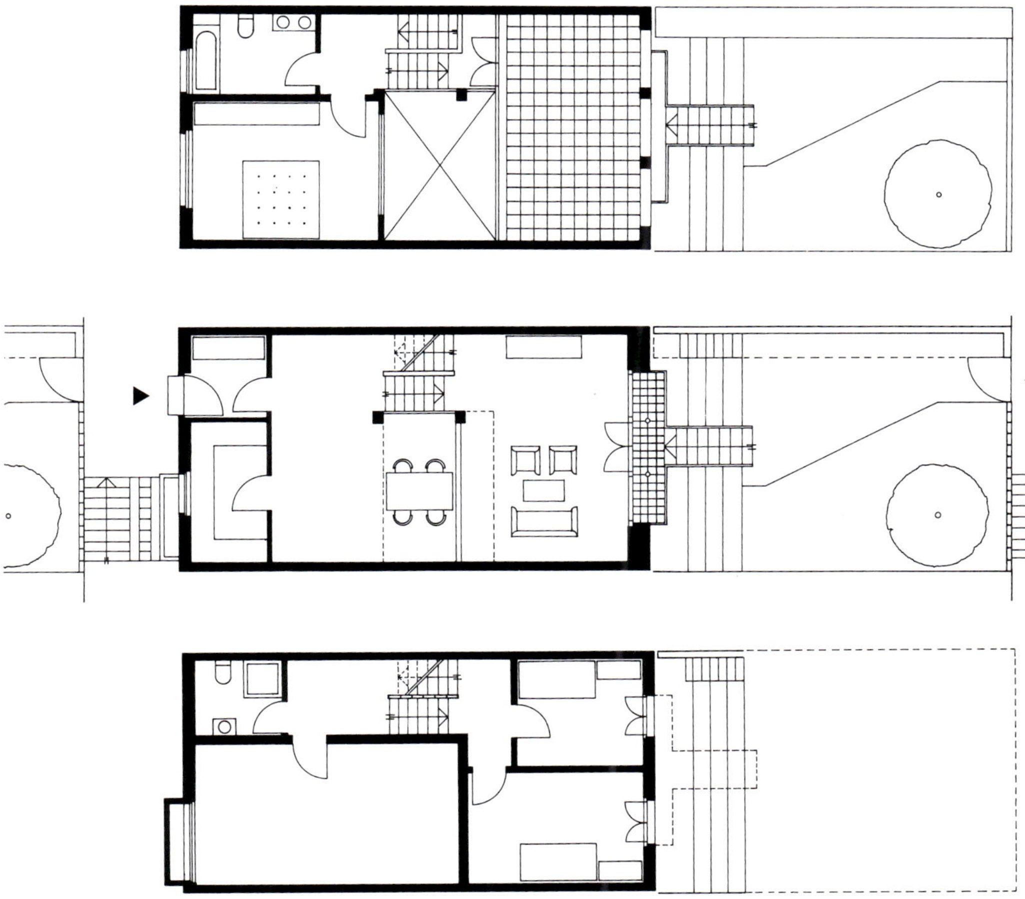

1

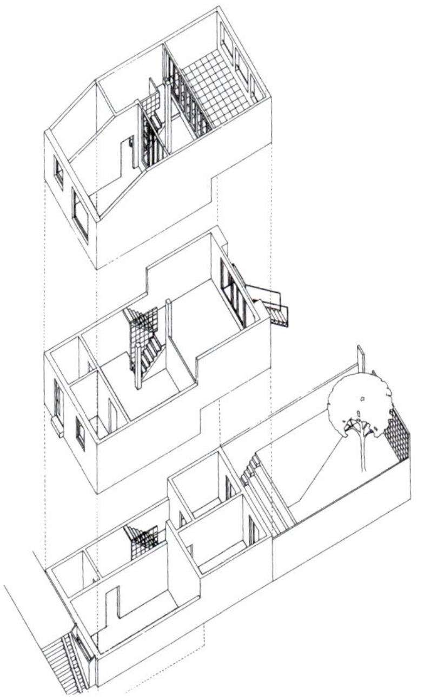

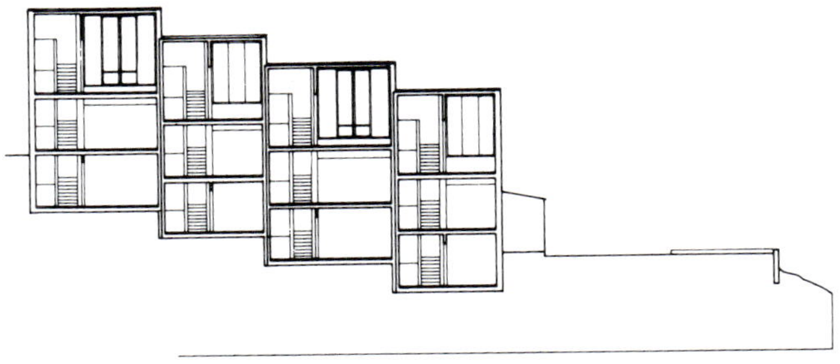

Type A: 4 rooms 1
on 6 levels 1:200

Referring to
floor plan 1:200

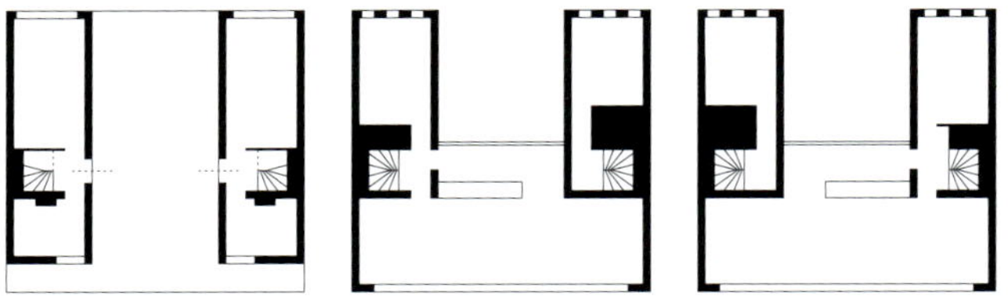

Huizen | Neutelings Riedijk | 1996

The first phase of the coastal development in Huizen consists of a series of duplex and single-family row houses, all facing north towards the water. To the south, the homes overlook private gardens and include a sunroom on the ground floor. The panorama units in the duplex homes are the most distinctive element in this design. Although the lot is only 6-m-wide, the clever stacked layout of the two duplex units makes it possible for each unit to have a living area stretching across 12 m, that is, across the full width of both lots. The continuous ribbon of windows offers a spectacular view of the Gooi Sea. While the living area of one apartment, on the second or third floor, occupies nearly the entire floor area on that level, space for a bedroom with bathroom and stairs is retained in one of the building wings. The recesses in the fabric allow sunlight to fall into the fully glazed kitchen on the south side and also create a kind of open courtyard at the entrance area. Cars are parked directly below the house. Generous roof terraces top the structure; the roof area also offers options for expansion. The single-family row houses with more conventional floor plans utilize the staggered design so that the oriel windows are oriented toward the water and the landscaped beach promenade.

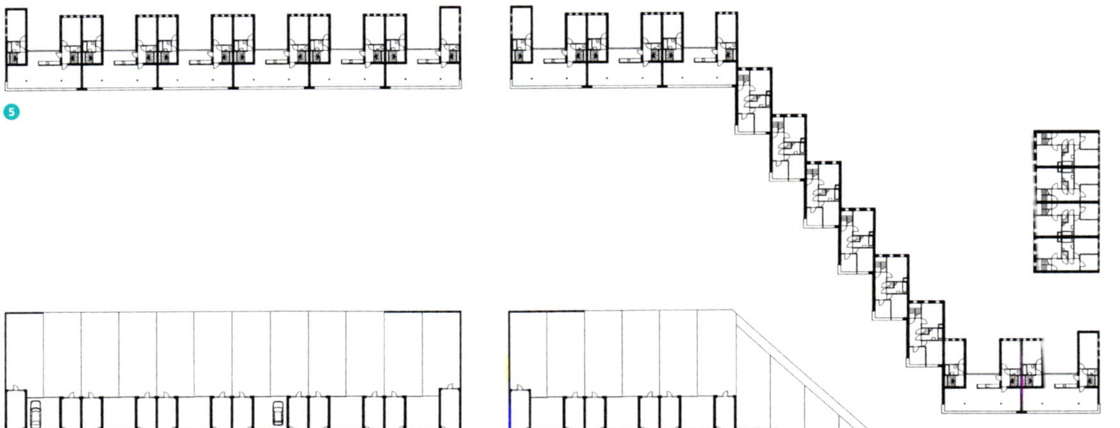

5

4

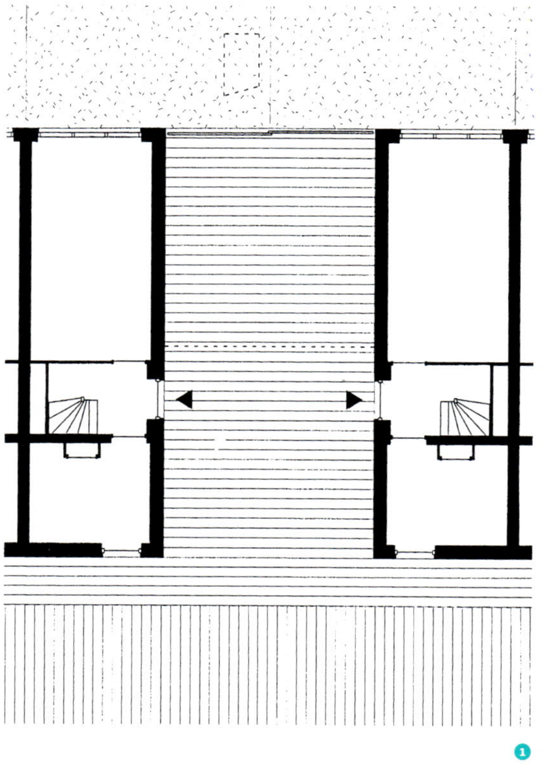

1

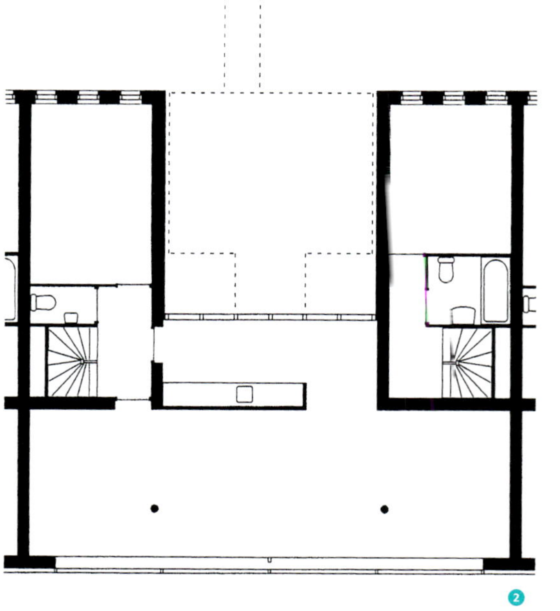

2

1 Ground floor 1:200
2 2nd floor with living area for apartment on left side 1:200
3 3rd floor with living area for apartment on right side 1:200
4 Entrance level for entire complex 1:1250
5 3rd floor of entire complex 1:1250
6 South elevation with garden facade
7 North elevation with weather facade overlooking the sea

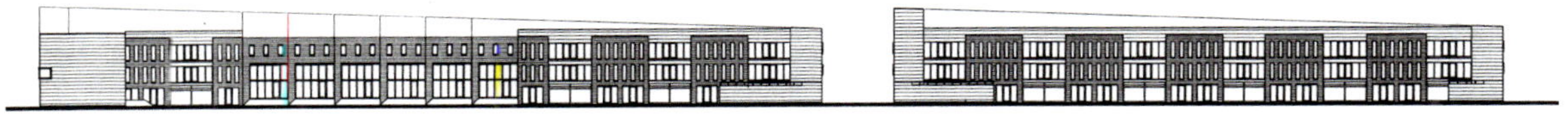

6

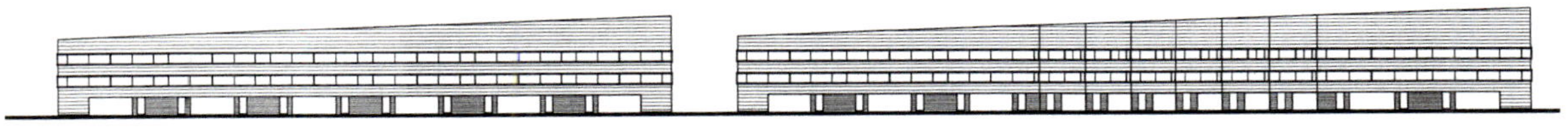

7

3.3
ROW HOUSE

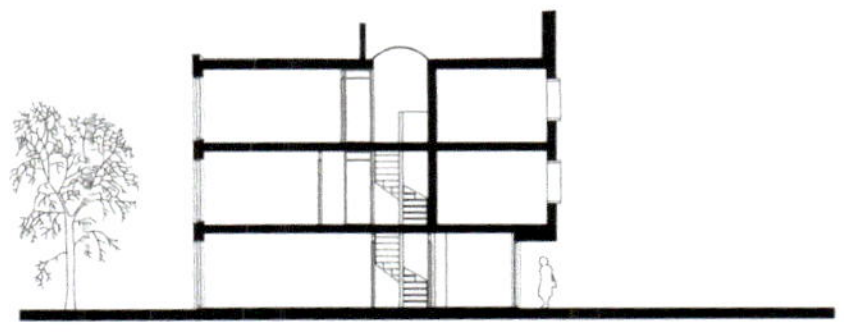

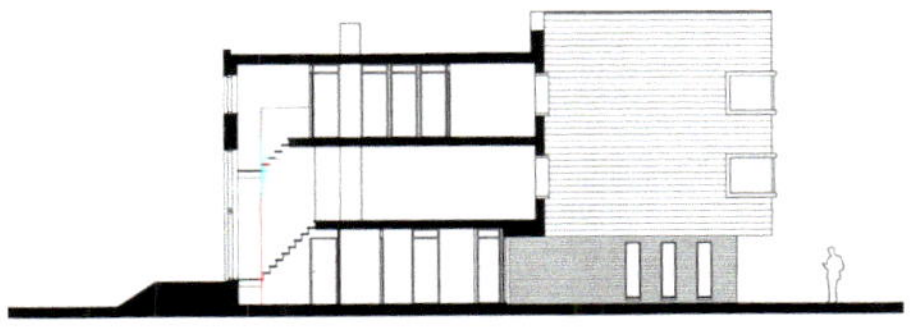

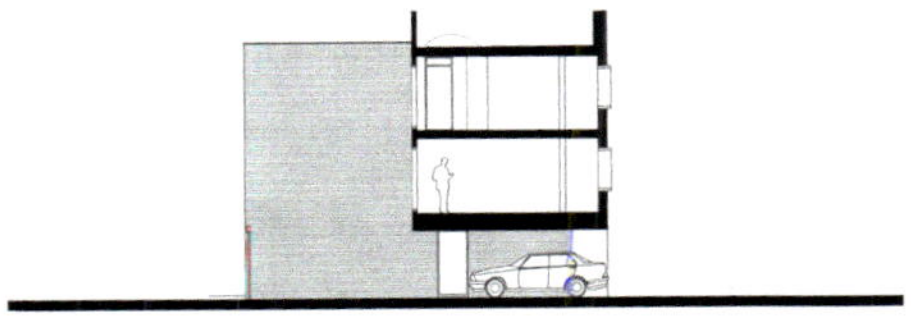

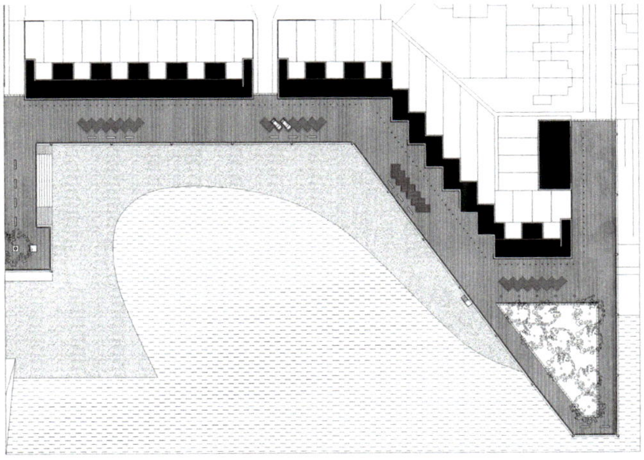

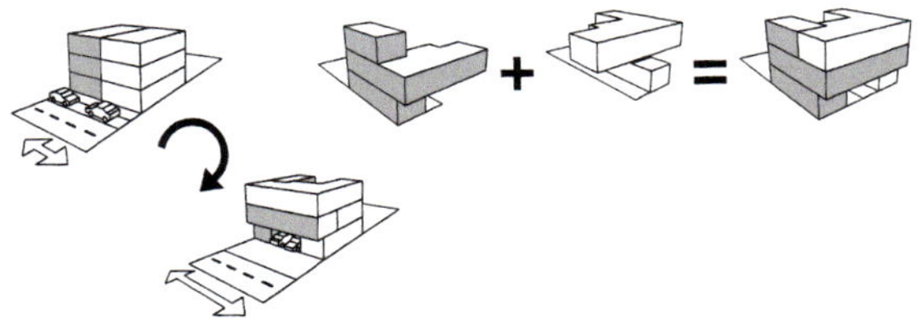

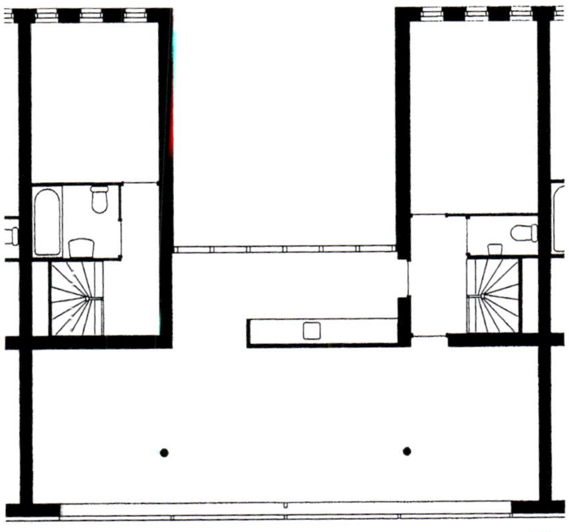

3

Building type
series of duplex
and single-family row houses
3 stories
facing N/S

Date of construction
phase 1 1994–1996

Living area
duplex and single-family houses,
each roughly 140 m²
max. 162 m²

Area per user
35 m²
(Duplex)

Building depth
6 and 12 m, respectively

Layout
11 duplex homes and
10 single-family row houses with
panorama views, carport and garden
staggered along the coastline of
a manmade bay on the sea

Open spaces
private roof terraces and gardens,
promenade along seashore with
landscaped green areas

Parking
private parking next to or
within the house,
24 guest parking spaces

Architect
Neutelings Riedijk Architects
Rotterdam

Location
Huizen
The Netherlands

Referring to
floor plan 1:200

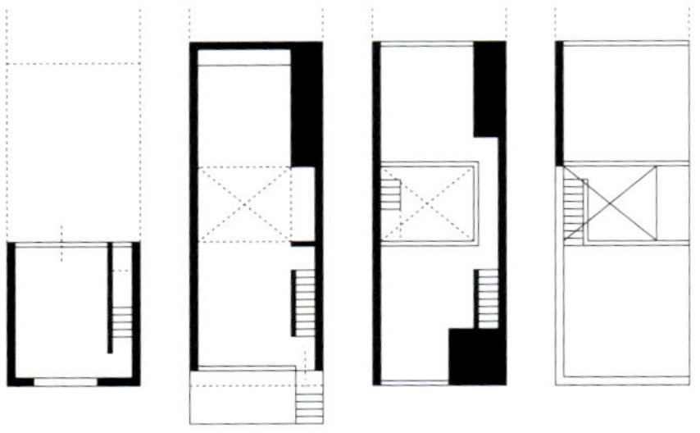

Borneo | MAP Architects / Mateo | 2000

The block at the end of the Borneo peninsula – monolithic in appearance at first glance – is divided into 26 "back-to-back" row houses with varying layouts. The basic crosswall construction is treated with great variety, resulting in eleven different 2–4-story unit types. The differing heights and recesses, introduced in the interest of lighting, are form-defining giving the complex an image of being simultaneously anonymous and animated. The complex is divided into two sections: the north elevation with an urban brick facade and the south elevation with a maritime timber facade. Each unit has direct access from the street and to the shared underground garage. The variety of open spaces, allocated to each unit, is especially notable. The homes on the south elevation have a verandah at the entrance, a large roof terrace, and a patio on the second floor. The patios are paved in glass-brick elements to allow light to penetrate into the 13-m-deep living area on the ground floor below. In the units along the north elevation, this glass ceiling is partially raised by one half story height, investing the living room with a lofty height of 5.7 m. The units on the north elevation are oriented toward the interior of the development, overlooking patios with small lawns. The 3–5-room end units are unique in layout; however, all feature a terrace on the 4th floor. One unit even rises to 5 levels, projecting above the remaining complex. All units benefit from a stunning view across the water that surrounds the development on three sides.

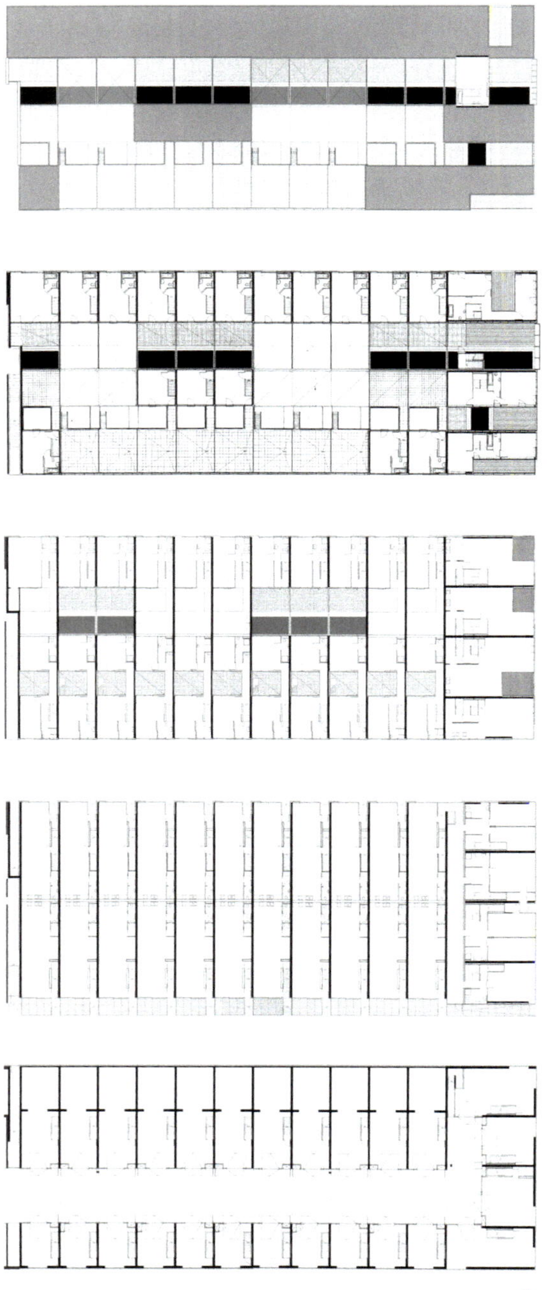

1

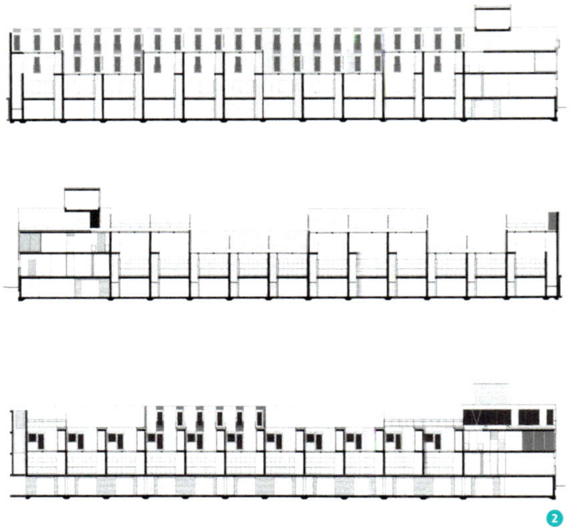

2

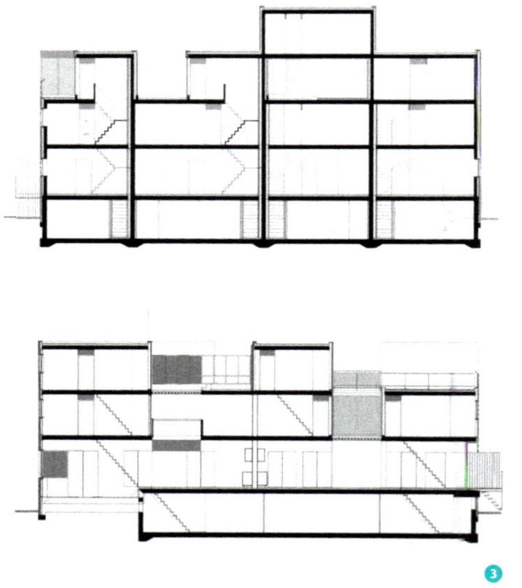

3

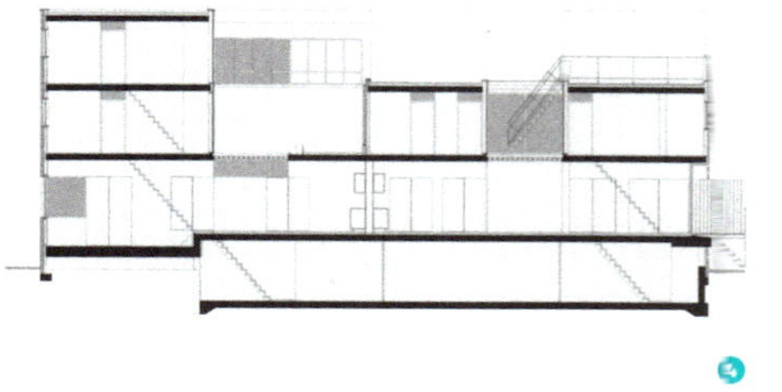

4

1 From bottom to top: entire scheme, basement to 4th floor 1:1000
2 Longitudinal sections 1:1000
3 Cross sections 1:500
4 Cross section matching floor plan details 1:500

3.3
ROW HOUSE

Building type
"back-to-back" row house development
2–4 stories
facing N/E/S

Date of construction
1995–2000

Living area
S: 3-/4-room hs. (4 types), 111–136 m² (11 u.)
N: 3-/4-room hs. (3 types), 118–175 m² (11 u.)
E: 3-/5-room hs. (4 types), 131–181 m² (4 u.)

Area per user
36–44 m²

Building depth
26 m
(house depth 13 m)

Layout
compact development of 26 patio row houses in "back-to-back" arrangement, surrounded by water on three sides, crosswall construction, 11 different house types, all units with one-sided orientation toward patio with the exception of corner units

Open spaces
verandah on south side, patios, small lawns, private roof terrace

Parking
26 parking spaces in underground garage

Architect
MAP Architects/Josep Lluís Mateo
Barcelona

Location
Borneo
Amsterdam
The Netherlands

5
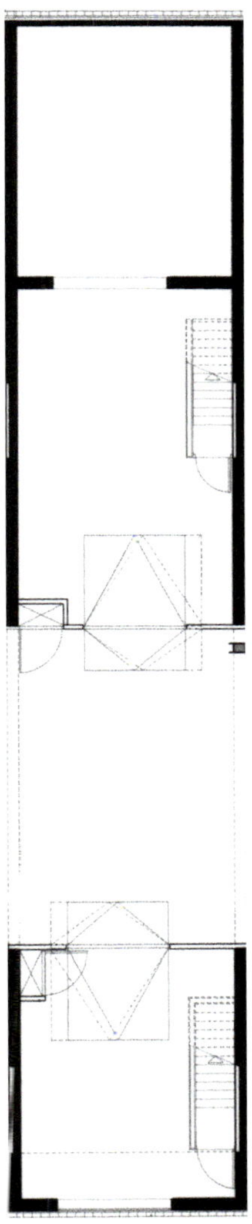

6
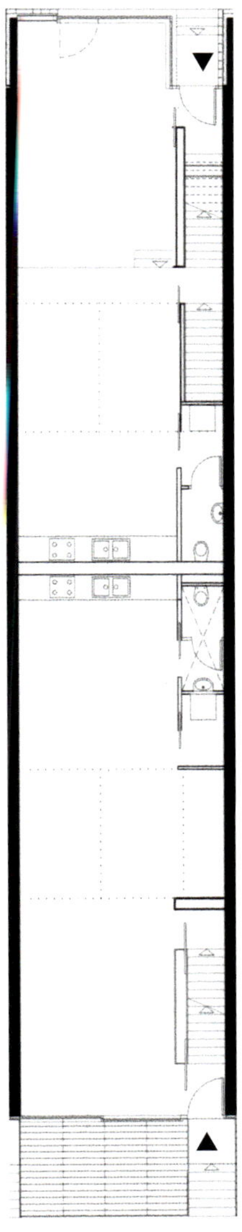

7
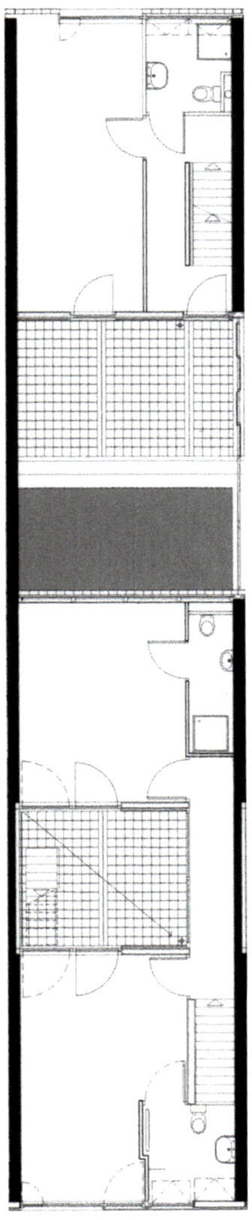

8
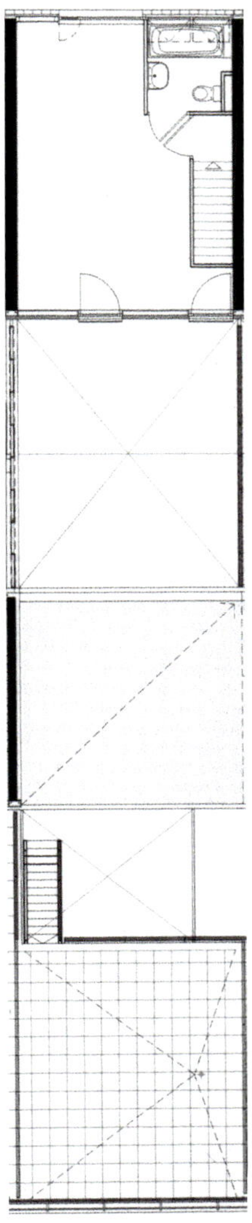

5 Floor plan detail basement with underground garage 1:200
6 Ground floor with living areas 1:200
7 2nd floor with bedrooms and patios 1:200
8 3rd floor with bedrooms/roof terrace 1:200

Referring to
floor plan 1:200

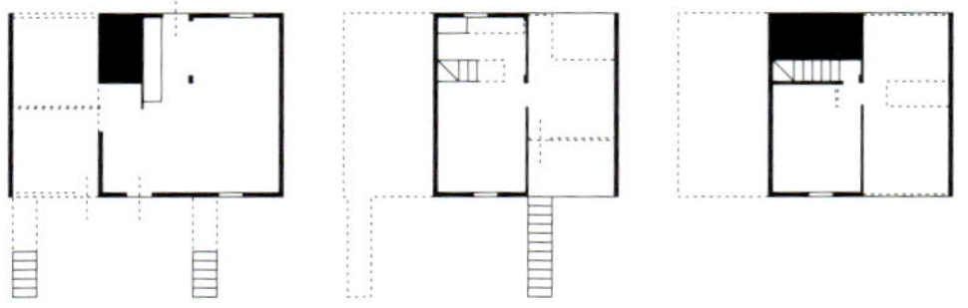

5

Quinta Monroy | Elemental | 2004

3.3 ROW HOUSE

The site on which 93 families had lived in an illegal squat, was to become their permanent legitimate home to ensure that they would not have to move away from their economic and social network. The solution was found in a dual densification strategy: instead of inefficient single-family homes, row houses with two stacked units were developed. Future additions by the inhabitants were part of the concept; the goal was to remain within the budget of the housing subsidy, and to offer growing families the option of enlarging their living space with their own efforts. The apartments are conceived to allow users to easily expand them themselves at any time in the future. The expansion spaces are an elemental component: they are the answer to the often uncontrollable growth of such housing developments. Elements such as load-bearing structure, dividing walls, stairs, kitchens, and baths form part of the basic finishes and are already scaled for future expansion. Unlike other projects, this housing complex is designed to ensure an increase in value; the building concept itself, the central location, the "room to grow" for changing family needs and the presence of a communal external space are all important prerequisites.

Building type
row houses designed for expansion by users
2 stories facing N/S i.e. S/E

Date of construction
2004

Number of units
93

Living area
ground floor apts: 36 m²/52.5 m²
maisonettes: 21 m²/59.5 m²
(at move in date/after expansion)

Area per user
11–18.5 m²

Building depth
6 m

Layout
row houses enclosing 4 communal courtyards
primary construction: reinforced concrete, masonry, temp. room enclosure, subsequent expansion by users: woods panels

Open spaces
communal courtyards and rear gardens

Parking
in front of house

Architect
Elemental
Santiago de Chile

Location
Iquique, Chile

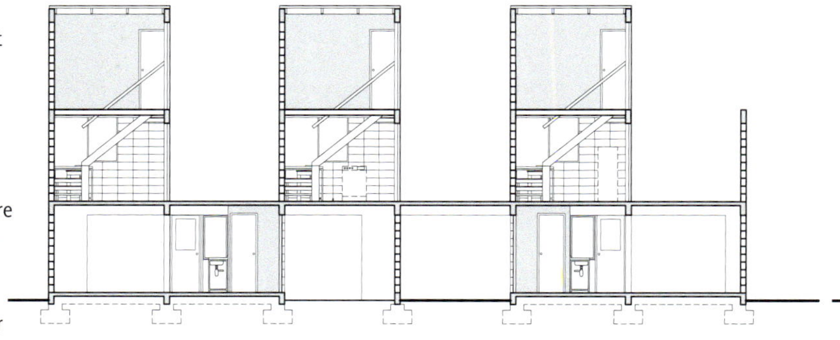

2

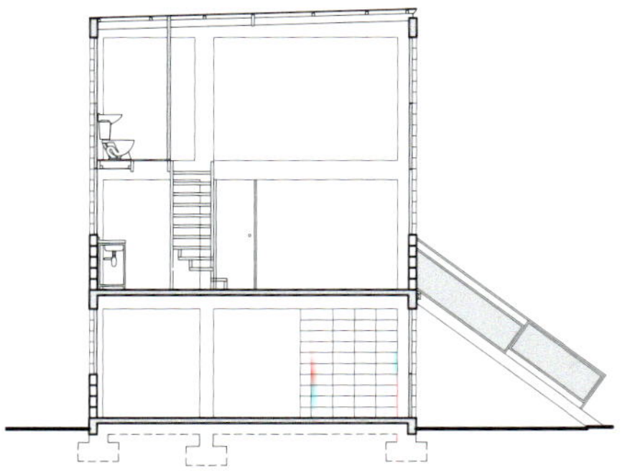

3

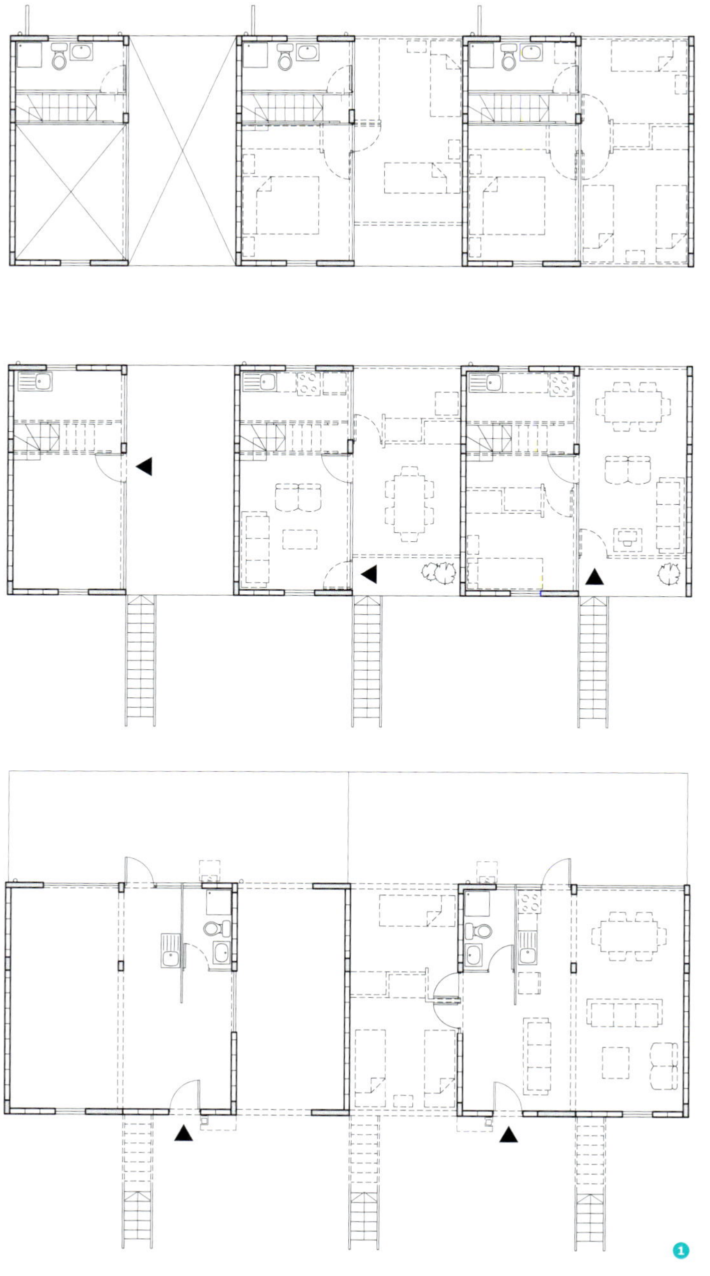

1

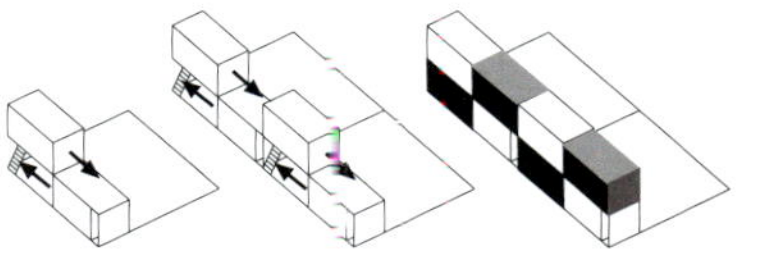

4

Ground floor, 2nd floor, 3rd floor of one unit – 1
to the left, the apartments as they were provided,
to the right the completion options 1:200
Longitudinal section prior to completion 1:200 2
Cross section prior to completion 1:200 3
Completion concept 4
Row houses before and after move-in and completion 5

Referring to
floor plan 1:200

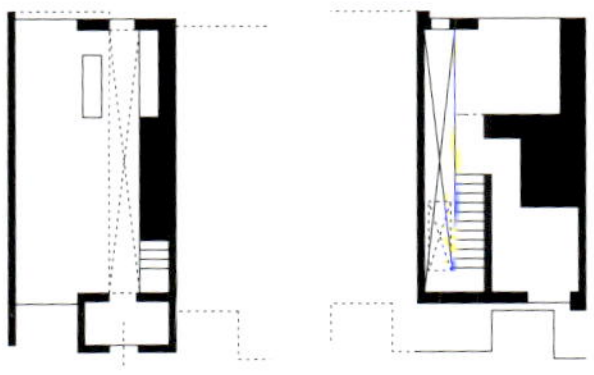

Skansen LIVING 2006+ | Dorte Mandrup Arkitekter | 2006

The row houses were created in the context of a housing exhibition. Arranged in two rows, each segment or terrace is slightly staggered in relation to its neighbour, in harmony with the gentle slope of the terrain. Although the atrium in each unit is only roughly 1 m in width, it runs front-to-back across the entire depth of the row house; the atrium terminates in a clerestory set into a triangular vault. However, the unique characteristic of this typology is that ground floor and the upper level are arranged on opposite sides of this atrium, with the result that all units are stacked within each other. The bedroom on the garden side is at right angles to this atrium, separated only by a glass wall; a glass sliding door leads from the upstairs landing into the bedroom. The atmosphere in the apartment is bright and open; the many sightlines resulting from this arrangement make it spacious and more complex in appearance than it is. The stairs are like a piece of furniture incorporated into the wall, into which a WC, storage space, and one of the two kitchen rows are also integrated on the ground floor. It is this densification of ancillary functions that creates the generous and open feel in the living room area. The staggered arrangement of the row houses creates additional space for another individual room, a bathroom, and a roof patio in the two end units. The row houses have front- and backyards, with entrance vestibules that also serve as a privacy screen.

Building type
2-story row houses facing WWS/EEN

Date of construction
2006

Living area
3-room apts., approx. 78 m² (3 units)
4-room apts., approx. 97 m² (4 units)

Area per user
24.5–26 m²

Building depth
9.5–11 m

Layout
7 units in two rows, staggered living spaces

Open spaces
patios

Parking
no parking on lot

Architect
Dorte Mandrup Arkitekter ApS, Copenhagen

Location
Ringsted
Denmark

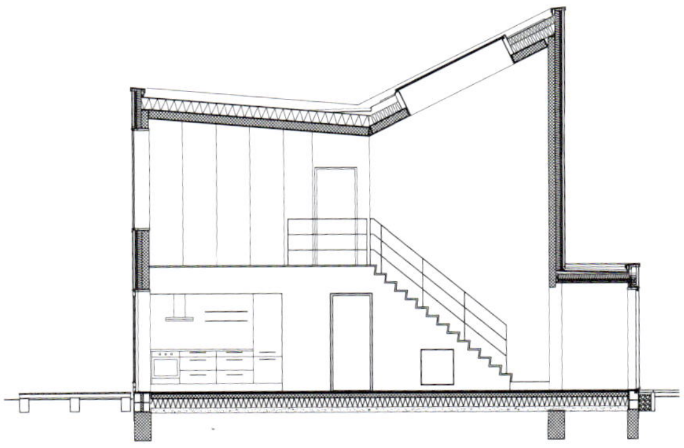

2

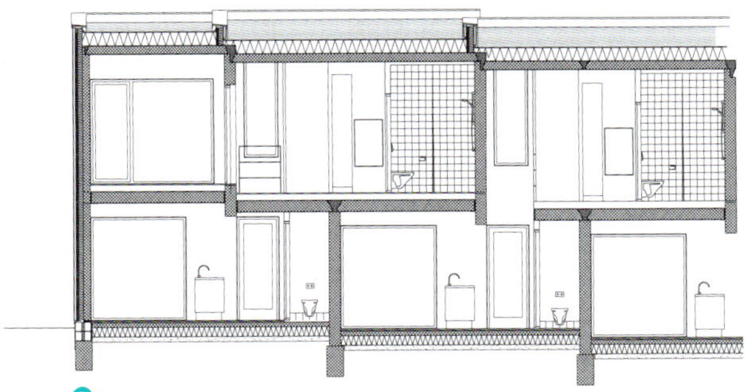

3

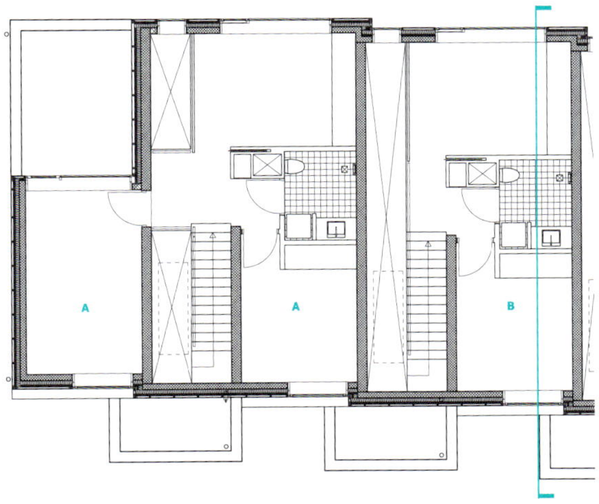

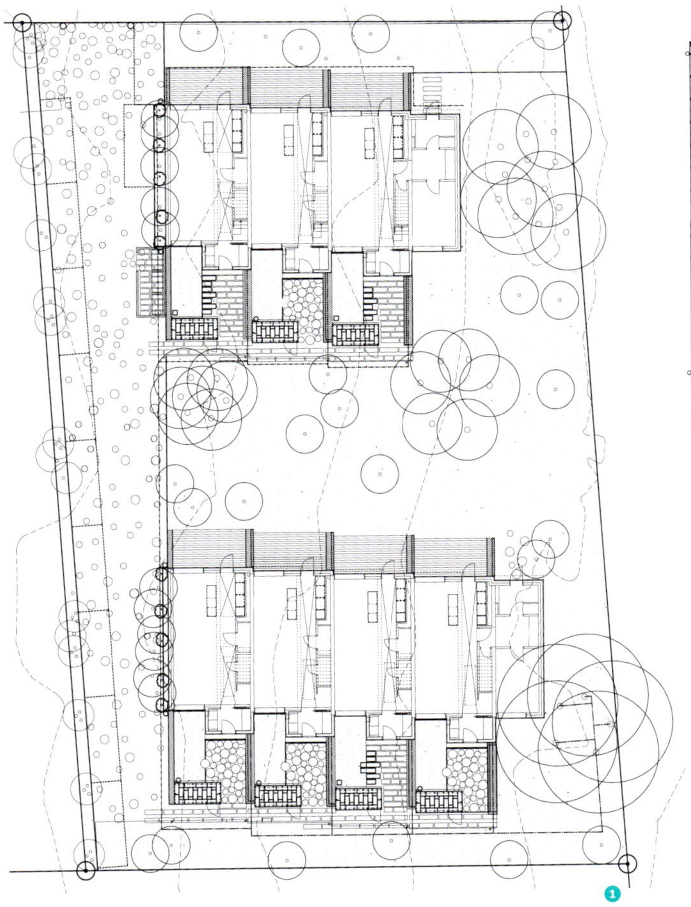

1

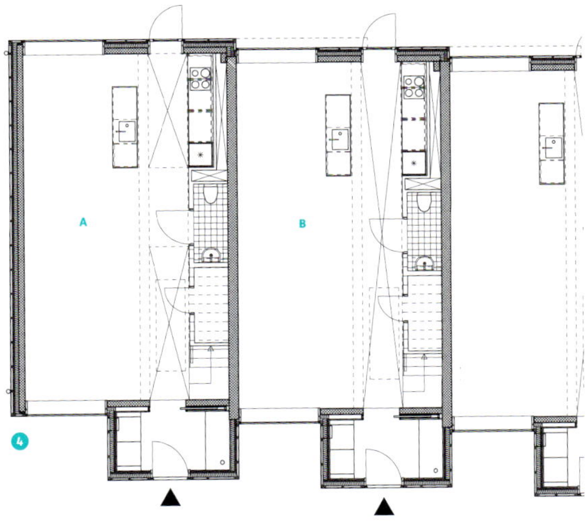

4

Ground floor of entire complex, 1
front and garden elevation 1:500
Cross section of void with stairs 2
and skylight 1:200
Longitudinal section with 3
offset housing units 1:200
Ground and upper floor of one 4- and 4
one 3-room row house 1:200

Referring to
floor plan 1:200

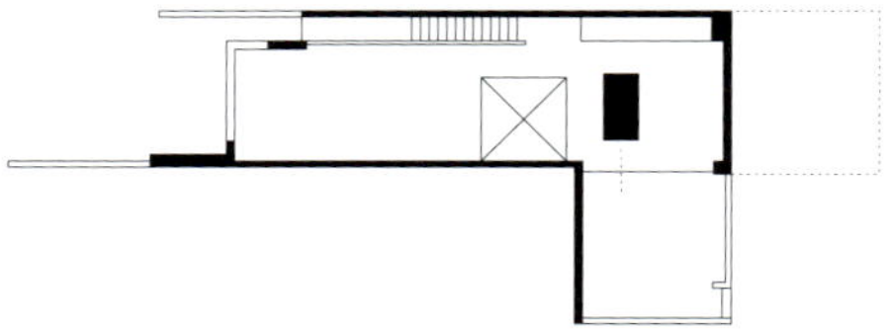

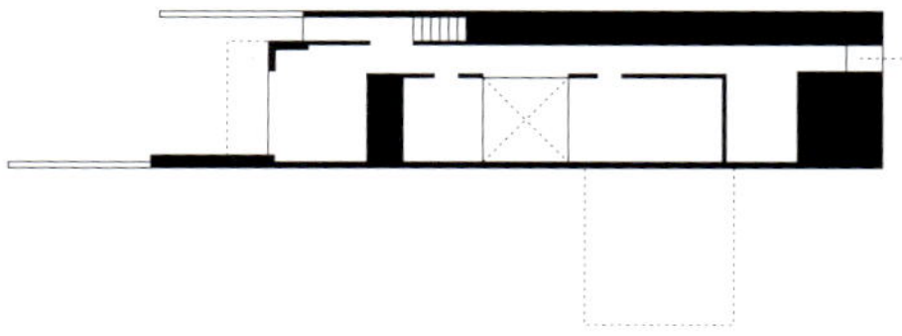

Bjørnveien |
Dahle, Dahle, Breitenstein | 2007

The compact housing development in a suburb of Oslo is both a combination of and a variation on row-house typologies. Here, the row houses are arranged in an unusual manner around a triangular communal courtyard at the center, which is often reserved for privacy in other row-house developments. The courtyard is bordered on one side by a series of rather conventional row houses and by a row of staggered patio houses on the other. A tall wall forms the third boundary of the courtyard garden, which is located above the underground garage and toward which nearly all patios are facing.

Due to the staggered layout of the patio houses, the individual units are oriented simultaneously to the west and to the south and benefit from three open spaces: the west-facing front garden at ground level, the two-story patio as the middle of the house and the large walled-in terrace to the south, which also marks the entrance. The directional shift from south to west is subtly reprised in the interior: the WC and wardrobe block and the patio indicate the way into the house and divide the interior into areas: the dining area with a view onto the patio, half-hidden behind it the kitchen and the internal corridor to the living area. Next, the floor plan turns in the other direction and the internal path leads either down the stairs or toward the living room overlooking the garden. On the lower level, the individual bedrooms are arranged in series, one behind the other, and overlook either the garden or the patio. The corridor, which reaches across the entire house and is bordered by built-in cabinets, connects with a second entrance, which leads directly to the garage.

1

1. Plan of entire complex ground floor, 2nd floor, 3rd floor 1:500
2. Floor plans of 2-story, 4-room patio row houses: lower level with private rooms, access to garden and garage 1:200
3. Upper level with living area and entrance from terrace and shared courtyard 1:200
4. Courtyard elevation of 2-story, 4-room patio row houses
5. Courtyard elevation of 3-story, 5-room row houses
6. Longitudinal section of 2- and 3-story row houses and parking garage beneath courtyard
7. Cross section of 2-story patio row houses

3.3
ROW HOUSE

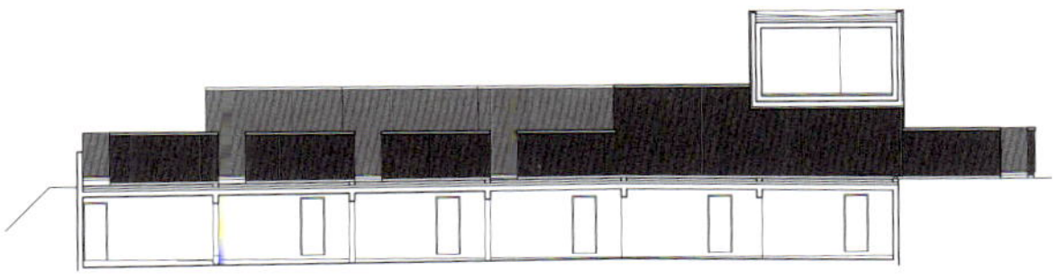

4

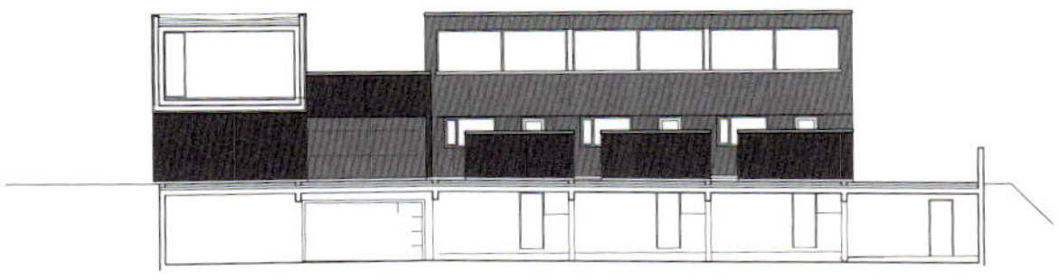

5

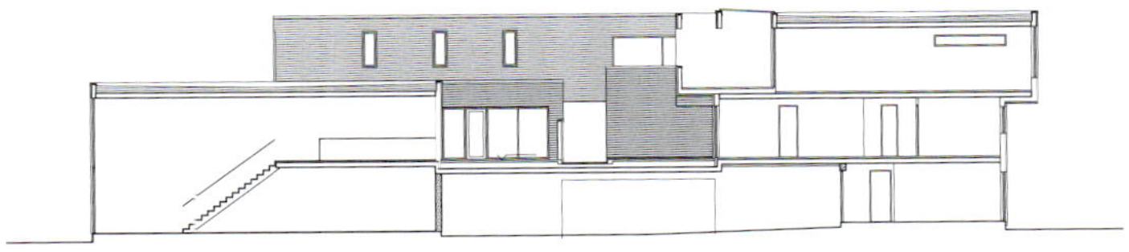

6

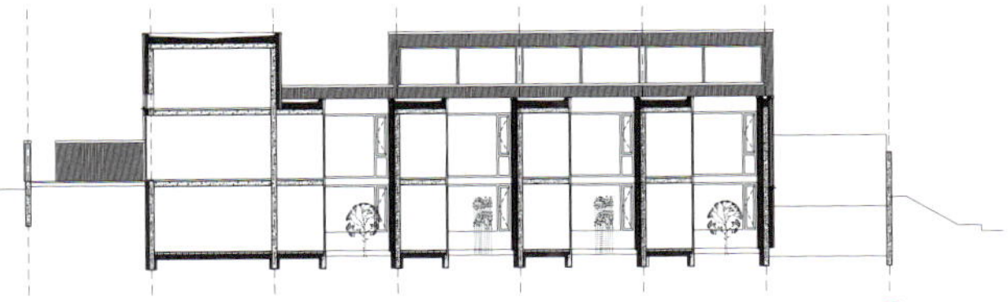

7

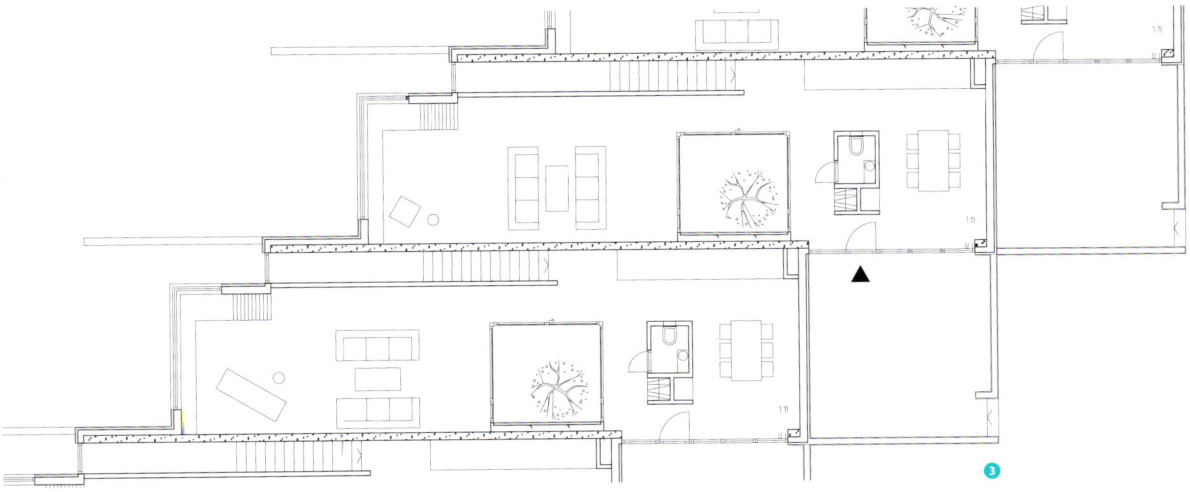

3

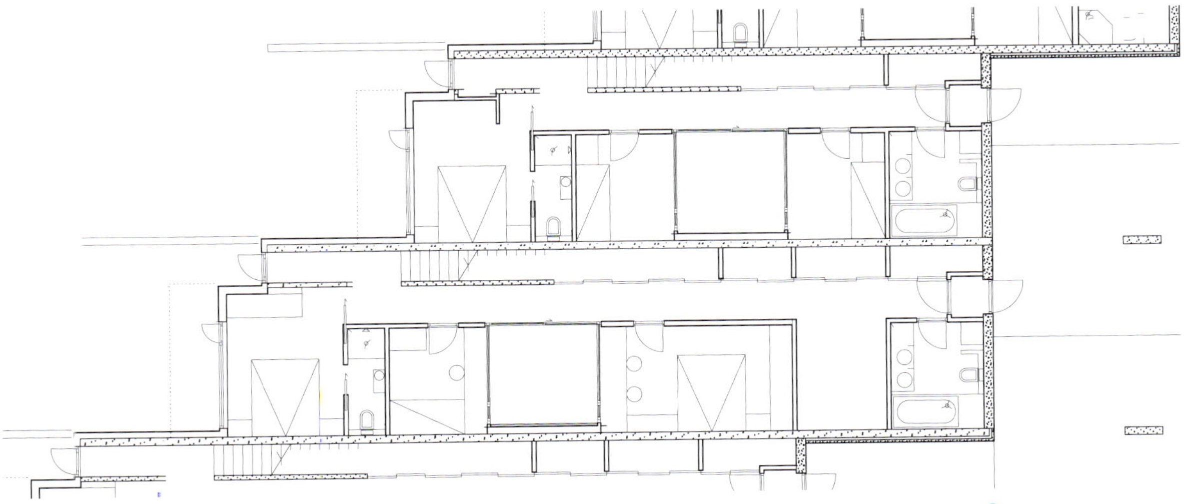

2

Building type
row housing
2- and 3-story houses
facing WNW/ESE/SSW

Date of construction
2007

Number of units
8

Living area
2-story patio houses:
154 m² (2 units)
130 m² (2 units)
2-story row house facing W:
161 m² (1 unit)
3-story row houses facing E:
136 m² (3 units)

Area per user
32.5–38.5 m²

Building depth
12 m, 17 m

Layout
courtyard layout,
access via courtyard terrace
and underground garage

Open spaces
front garden, terrace, patio

Parking
16 parking places in underground
garage beneath gardens with direct
access to units

Architect
Dahle, Dahle, Breitenstein
Oslo
Einar Dahle
Christian Dahle
Kurt Breitenstein

Location
Bjørnveien 119
Oslo
Norway

Referring to
floor plan 1:200

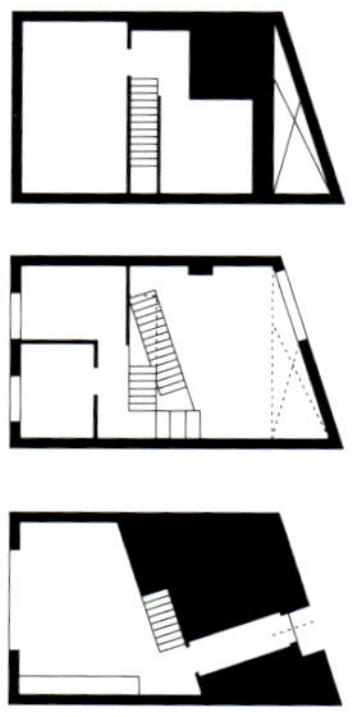

Regatta | Bovenbouw | 2016

Part of a larger new residential development, seven houses form a trapezoidal block, in which both the block as a whole and the individual houses are emphasized. All sides of the block – whether part of a building or the perimeter wall – are enclosed with the same brickwork. The gable facades are oriented to the narrow sides of the block while the eaves sides face its long sides. Two freestanding houses occupy the orthogonal corners while the remaining volumes translate the form of the block into five trapezoidal row houses. Orthogonally offset to one another on their back garden side, they adhere at the front to the oblique block edge, thus forming a striking, zigzagging roofline along the street. In this way, a correlation between the articulation of the block and the individual houses is created.

A feature of particular interest in the mid-block house type is a split-level living area that extends vertically through the entire building. It is arranged around the triangular void that occurs between the angled street facade and the orthogonal orientation of the rest of the house. The staircase ascends through this space, following the geometry of the surrounding walls and connecting the eat-in kitchen on the ground floor, with its large window to the private yard, to the living room above that overlooks the street. Offset by a few further steps, a corridor open to this living space leads to children's bedrooms, while the next flight of stairs goes up to the master bedroom and bathroom on the top floor. A maximum amount of spatial effect is thus generated in an extremely compact volume.

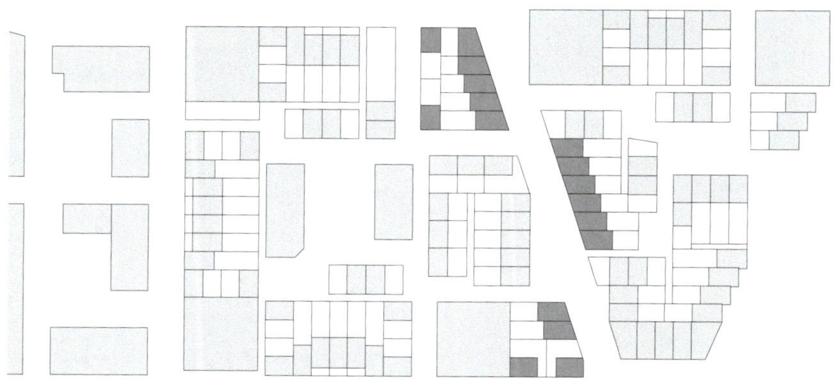

1

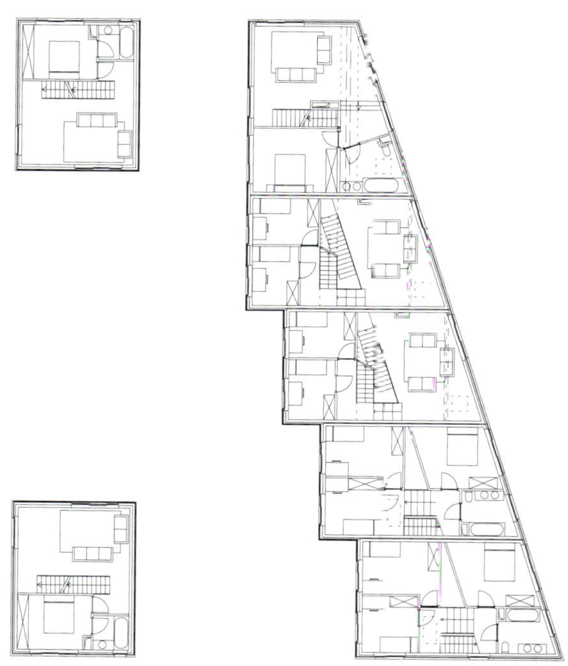

2

1 Entire block, ground floor 1:400
2 Entire block, 2nd floor 1:400
3 Entire block, attic floor 1:400
4 4-room row house, from bottom to top:
Ground floor, 2nd floor, attic floor, cross section 1:200

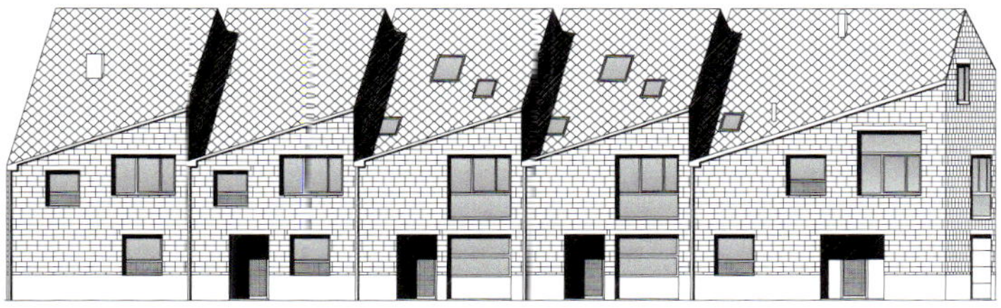

3.3
ROW HOUSE

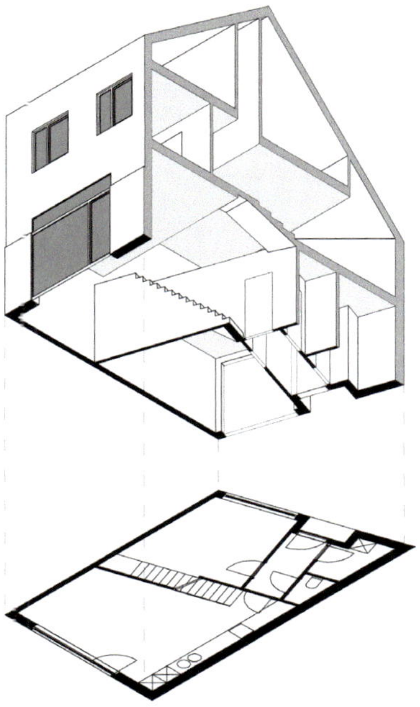

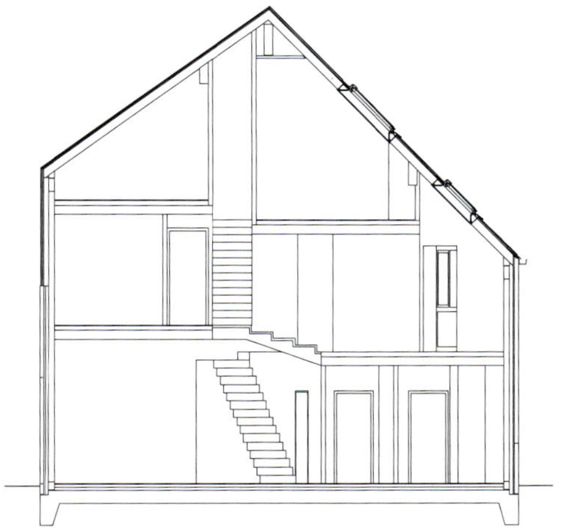

Building type
block ensemble in new residential area
2–3 stories
facing in all directions

Date of construction
2011–2016

Number of units
17

Size of units
row houses: 121–137 m²
(12 units)
corner houses: 104–108 m²
(5 units)

Area per user
26–34 m²

Building depth
5.9–11 m

Layout
single-family row houses and freestanding corner houses with private yards form a block with a surrounding perimeter wall, exposed brick structure with gabled roofs; roof shingles continue down the facades of some units

Open spaces
private yards

Parking
individual ground floor garages in some units

Architect
Bovenbouw Architectuur
Dirk Somers, Carole Boeckx, Wim Boesten, Annelien Grandry, Christopher Paesbrugghe
Antwerp

Location
Cornellielaan, Joris van Spiegelstraat
Linkeroever, Antwerp
Belgium

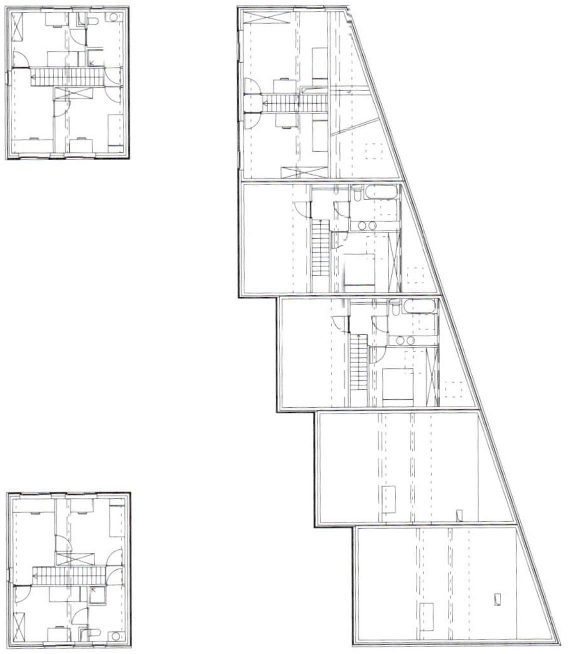

3

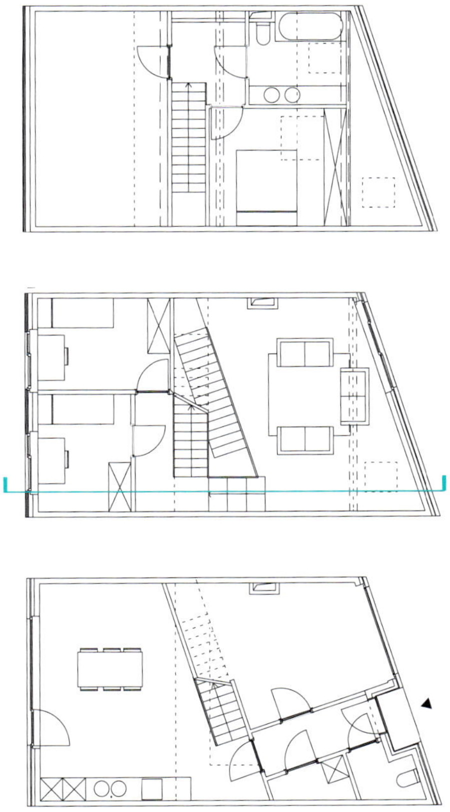

4

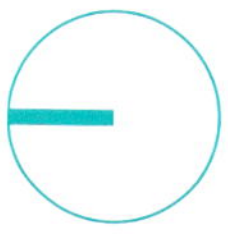

Referring to floor plan 1:200

Picture Credits

8/9 Evans, R.: "Figures, Doors and Passages", in: "Translations from Drawing to Building and Other Essays", AA Documents, London, 1997; Riegler Riewe Architekten; KUU Architects: Jeremy San

10/11 Märkli: Georg Gisel; b&k brandlhuber&kniess GbR: Michael Raske/Stefan Schneider; Alvar Aalto/Paul Baumgarten; Sou Fujimoto Architects; Field Design Architects

12/13 Smithson: Webster, H. (Ed.): "Modernism without Rhetoric", Academy Editions, London 1997/Rogier Hillier, photo/sections; "Alison and Peter Smithson, Changing the Art of Inhabitation", Artemis, London, Munich, 1994, Simon Smithson; "The Charged Void: Architecture", Monacelli Press, New York, 2001; Olgiati: Boga, T.: "Die Architektur von Rudolf Olgiati". Reprint of the original edition of 1977, Birkhäuser, Basel 2010/© original edition: 1977 ETH Zürich, Institut für Geschichte und Theorie der Architektur

14–35 1850-65: Gieselmann, R.: "Wohnbau", Vieweg & Sohn, Braunschweig/Wiesbaden 1979, 2nd edition, Institut für Wohnbau, TU Wien, 1991
1859: Stumberger, R.: "Die Geschichte einer 'Utopie réalisée'. Familistère Godin, Guise/Frankreich", Bauwelt Heft 27-28.04. 2004
1870,1: Firley, E., Stahl, C.: "The Urban Housing Handbook", John Wiley and Sons, Hoboken, 2009; 1870,2: Alpern, A.: "Apartments for the Affluent", McGraw-Hill, New York, 1975; 1870,3: Architekten- und Ingenieur-Verein Berlin (Ed.): Berlin und seine Bauten, 3rd edition, Ernst & Sohn, Berlin, 1964
1879: see 1850-65
1890: see 1870,1
1897: see 1870,3
1899: see 1850-65
1900,1: see 1850-65 and Gieselmann, R.: "Grundrißatlas Geschoßwohnungsbau", Prolegomena 52, Institut für Wohnbau, TU Wien, 1985; 1900,2: Gieselmann, R.: "Grundrißatlas Geschoßwohnungsbau", Prolegomena 52, Institut für Wohnbau, TU Wien, 1985
1902-08: Survey of London, Vol. 26, Lambeth, Southern Area, London County Council, London, 1956 and Zapel, E., redrawn from the original floor plan
1902: Howard, E.: "Garden Cities of To-morrow", Swan Sonnenschein & Co., London, 1902
1903: Rogers, E.N., "Auguste Perret", Il Balcone, Milan, 1955
1906,1: Gieselmann, R.: »Grundrißatlas Flachbau«, Prolegomena 56, Institut für Wohnbau, TU Wien 1987; 1906,2: Zapel, E., redrawn from the original floor plan; 1906,3: see 1850-65 and 1906,1
1907: Irma and Paul Milstein Division of United States History, Local History and Genealogy, The New York Public Library: "The Belnord Apartments. Floor plan - 2nd, 5th, 7th and 10th floors." The New York Public Library Digital Collections. 1910
http://digitalcollections.nypl.org/items/510d47da-d7e6-a3d9-e040-e00a18064a99 (accessed on 20.5.2017)
1917-25: Colquhoun, I., Fauset, P. G.: "Housing Design in Practice", Longman, Harlow, 1991
1918: Behrens, P., de Fries, H.: "Vom sparsamen Bauen: ein Beitrag zur Siedlungsfrage", Verlag der Bauwelt, Berlin, 1918
1919-22: Faller, P.: "Der Wohngrundriß: Entwicklungslinien 1920–1990", Deutsche Verlags-Anstalt, Stuttgart, 1969
1920: see 1906,1
1923: Faber, T.: "Kay Fisker", Arkitektens Forlag, Copenhagen, 1995
1924-28: Stein, C. S.: "Toward New Towns in America", MIT Press, Cambridge, 1978
1925,1 and 2: see 1906,1; 1925,3 and 4: Gieselmann, R.: "Der Wohngrundriß: Entwicklungslinien 1920–1990", Deutsche Verlags-Anstalt, Stuttgart, 1969
1926: see 1906,1
1927,1: Sherwood, R.: "Modern Housing Prototypes", Harvard University Press, Cambridge, 1978; 1927,2: see 1906,1; 1927,3: see 1900,2
1927-30: Reppé, S.: "Der Karl-Marx-Hof : Geschichte eines Gemeindebaus und seiner Bewohner", Picus Verlag, 1993
1928,1: Kellermüller, A., "Rationelle Gestaltung von Kleinwohnungsgrundrissen: Untersuchung von Architekt Alex. Klein, Berlin", Wohnen, Vol. 3, 1928, Zurich, 1928; 1928,2: see 1919-22; 1928,3: Zapel, E., redrawn from the original floor plan
1929,1: see 1924-28; 1929,2: Ruttenbaum, S., Roth, E.: "Mansions in the Clouds", Balsam Press, New York, 1986; 1929,3: Craib-Cox, J.: "Houses in the Sky", Architectural Review, 10/1977; 1929, 4: see 1906,1; 1929,5: Internationale Kongresse für Neues Bauen und Städtisches Hochbauamt Frankfurt/Main, "Die Wohnung für das Existenzminimum", Englert & Schlosser, Frankfurt, 1930
1930,1: Architectural Forum, Vol. 62, 1935; 1930,2: see 1900,2; 1930,3: Schneider, T., Till, J.: "Flexible Housing", Elsevier, Oxford, 2007; 1930,4: Johannes, H. : "Neues Bauen in Berlin", Deutscher Kunstverlag, Berlin, 1931; 1930,5: see 1927,1
1931-34: Yorke, F.R.S., Gibberd F.: "The Modern Flat - 3rd Edition", The Architectural Press, London, 1950
1932: see 1906,1
1933-35: see 1931-34
1934: Ledent, G.: "Potentiels relationnels: l'aptitude des dispositifs physiques de l'habitat à soutenir la sociabilité: Bruxelles, le cas des immeubles élevés et isolés de logement", Dissertation, UCL Brussels, Belgium 2014
1935,1: see 1931-34; 1935,2: Friberger, Erik; 1935,3: see 1927,1
1936-54: Fleig, K., Aalto, E.: "Alvar Aalto – The Complete Work", Birkhäuser, Basel, 1990
1943: www.omeka.design.ncsu.edu/omeka/items/show/567 (accessed on 3.5.2017); Zapel, E., redrawn from the original floor plan
1945: Fundacion Villanueva; Zapel, E., redrawn from the original floor plan
1945-48: Aga Khan Trust for Culture
1945-64: Britton, K., "Auguste Perret", Phaidon, London, 2001
1946: Architectural Journal, 10/1962
1947: Gibberd, F., "Harlow New Town : a plan prepared for the Harlow Development Corporation", Harlow Development Corporation, 1952; Opher, P., Bird, C.: "Architecture and Urban Design in Six British New Towns", Oxford Polytechnic, Oxford 1981
1947-51: House and Home 8/1954; Levitt and Sons, Inc.
1949-54: ArchiDiAP: http://www.archidiap.com/opera/quartiere-tiburtino/ (accessed on 20.5.2017)
1950-52: Franck, K.: "Affonso Eduardo Reidy", Hatje, Stuttgart, 1960
1952,1: Le Corbusier: Jenkins, D.: "Unité d'Habitation", London, Phaidon Press, 1993; 1952,2: Fry, M., Drew, J.: "Tropical Architecture in the Dry and Humid Zones", Batsford, London, 1964
1952-60: Schäche, W. (Ed.): "Hermann Henselmann. Ich habe Vorschläge gemacht", Ernst & Sohn, Berlin, 1995; Architektur der DDR 1/1989
1954,1: Joedicke, J.: "Candilis, Josic, Woods", Karl Kraemer, Stuttgart, 1978; 1954,2: Peters, P.: "Mehrgeschossige Wohnbauten", Callwey Verlag, Munich, 1978;1954,4: Meyer-Bohe, W.: "Apartments", Stuttgart, Verlagsanstalt Alexander Koch, 1970
1955: Atelier 5
1955-85: Projekt Russia, 73/2014
1955-83: Urban Housing Typologies, SUTD ASD Course Elective 2015, Singapore 2015
1955-57: see 1954,2
1956-60: Fils, A.: "Brasilia", Beton-Verlag, Düsseldorf, 1988
1957,1: see 1919-22; 1957,2: Dubor, B. F.: "Fernand Pouillon", Electa spa, Milan, 1987; 1957,3: The Builder, 4/1962
1957-61: Schmitt, K. W., "Multi-storey Housing", Hatje, Stuttgart, 1966; The Architectural Review, 130, 778, December 1961
1958,1: Hassenpflug, G.: "Scheibe Punkt und Hügel". Callwey Verlag, Munich, 1966; 1958,2: see 1900,2; 1958,3: Jianzhu Xuebao (Architectural Journal, Beijing), 10, 1958
1957-66: Casabella, 810, February 2012
1959: see 1956-60
1960: Zapel, E., redrawn from the original floor plan
1962: Stichting Architecten Research, S.A.R., 1970
1963-95: Bauwelt, 73, 8, 1982
1963,1: Phippen, Randall, Parkes; 1963,2: Zvi Hecker
1963-74: see 1850-65
1964: Aillaud, E.: "La Grande Borne à Grigny", Hachette, Paris, 1972
1967: Safdie, Moshe: AD, 3/1967
1967-70: Bauwelt, 7/1967
1968-75: Architectural Design (AD) 4, 1970; Correa, Ch.: "Charles Correa Housing and Urbanisation", Thames and Hudson, London, 2000
1968-77: Darbourne and Darke: AR 9/1974, AJ, 2 April 1975
1968: The Paul Rudolph Archive, Library of Congress, https://www.loc.gov/item/2007675744/ (Accessed on 13.5.2017)
1969: Ungers, Oswald Mathias: Bauwelt, 46/47, 1967
1969-82: Faller, P.: see 1919-22
1969-92: Salat, S.: "Fumihiko Maki", Rizzoli, New York, 1988
1970: Korean National Housing Corporation (1979, 306); Jung, I., "Architecture and Urbanism in Modern Korea", University of Hawai'i Press, Hawaii 2013
1971-73: Collection Het Nieuwe Instituut/ BLOM, 286
1971-81: Urban Housing Typologies, SUTD ASD Course Elective 2017, Singapore 2017
1972: www.pruitt-igoe.com/press-materials/ (accessed on 9.5.2017)
1972: Domus, 520, March 1973
1973: Nicolin, P.: "Carlo Aymonimo, Aldo Rossi: Housing Complex at the Gallaratese Quarter", A.D.A. Ed., Tokyo, 1977
1974,1: see 1955-83; 1974,2: Ricardo Bofill Taller de Arquitectura; 1974,3: Storgård, J. P.,Orum-Nielsen, A. and J.,Marcussen, H.: Bauen & Wohnen, 10/1974
1976-85: WBW, 7-8, 2016
1976,1: www.pearlbankapartments.com/ (accessed on 1.5.2017); 1976,2: Lüchinger, A.: "Herman Hertzberger, Bauten und Projekte 1959–1986", The Hague, ArchEdition-cop, 1987; 1976,3: Cruz, Antonio/Ortiz, Antonio
1977,2: Open House 2/1978
1979,1: Hilmer & Sattler: Werk, Bauen und Wohnen, 5/91; 1979,2: see 1954,2
1983: Charles Correa Associates
1984-87: Bauwelt 72, 1981
1985: Bauausstellung Berlin GmbH: "Internationale Bauausstellung Berlin 1987. Projektübersicht", updated and expanded edition, Berlin, 1991
1986: Rob Krier
1987: Nouvel, Ibos: GA Houses, 23
1988-93: Gast, K.P., "Modern Traditions. Contemporary Architecture in India", Birkhäuser, Basel, 2007
1988: Schwarzplan.eu
1991,1: Lotus International, 83, 1994; 1991,2: Steven Holl
1992: Schwarzplan.eu; Höhne & Rapp
1994,1: Riegler Riewe Architekten; 1994,2: Yositika Utida, SHU-KOH-SHA Architectural & Urban Design Studio; Osaka Gas Co. Ltd; 1994,3: Hans Kollhoff
1998: Kojima + Akamatsu
1998-2000: BKK-3 Architecture
1995-2003: MVRDV
2000: Kazuyo Sejima/Ryue Nishizawa

36-41 fig. 1: Dattner Architects; fig. 2: Michael Maltzan Architects; fig. 3: 5468796 Architecture + Cohlmeyer Architects; fig. 4: HAHOH Haas Heckmann Architekten; fig. 5: Archplan AG; fig. 6: Javier García-Solera; fig. 7: Naruse Inokuma Architects; fig. 8: SsD Architecture; fig. 9: nARCHITECTS; fig. 10: architecture WORKSHOP; fig. 11: Nakae, Takagi; fig. 12: Riken Yamamoto & Field Shop; fig. 13: BARarchitekten; fig. 14: MVRDV & Blanca Lleó Associates; fig. 15: Matti Ragaz Architekten AG; fig. 17: ANA architecten; fig. 18: Brandlhuber + NiehüserS; fig. 19: Duplex Architekten; fig. 20: Kaden Klingbeil Architekten; fig. 21: Case Architects; fig. 22: Urbanus: Bauwelt, 9, 2010; fig. 23: John Lin, Rural Urban Framework; fig. 24: Elemental; fig. 25: Kunihiro Ando + Satoyama Architecture Laboratory; fig. 26: Feldschnieders – Kister; fig. 27: Riken Yamamoto & Field Shop; fig. 28: Arc Studio Architecture + Urbanism, Housing Development Board (developer/owner)

42–47 "Mon Oncle de Jacques Tati (1958) © Les Films de Mon Oncle", www.tativille.com
Vázquez Consuegra: Lluìs Casals, Hisao Suzuki
Tessenow: Tessenow, H.: "Hausbau und dergleichen", Vieweg & Sohn, Braunschweig/Wiesbaden, 1953
Darbourne & Darke: Faller, P.: "Der Wohngrundriß: Entwicklungslinien 1920–1990", Deutsche Verlags-Anstalt, Stuttgart, 1996
Back-to-Back-Crossover Reihenhaustyp: Sting, H.: "Grundriß Wohnungsbau", Verlagsanstalt Alexander Koch GmbH, Stuttgart, 1975
Burkhard Meyer: Roger Frei, Zurich
zanderroth architekten
Ciriani
Kaden Klingbeil Architekten
Siza
Drawing by A. B. Walker in Life, March 1909: Koolhaas, R.: "Delirious New York", Arch+Verlag, Aachen, 1999
Elisha Otis, courtesy of Otis Elevator Company
Schipporeit & Heinrich
Smithson, A. and P.: "The Charged Void: Architecture", Monacelli Press, New York, 2001; Copyright Smithson Family Collection
querkraft architekten
van den Broek, Bakema
Maekawa: Altherr, A.: "Three Japanese Architects", Hatje, Stuttgart, 1968
Smithson, A. and P.: "The Charged Void: Architecture", Monacelli Press, New York, 2001; Copyright Smithson Family Collection
b&k+ brandlhuber&kniess GbR
Diener & Diener Architekten
MVRDV & Blanca Lleó Associates: Rob 't Hart
Ricardo Bofill Taller de Arquitectura
Vidal, Pau

54-57 All diagrams: Oliver Heckmann, Eric Zapel

72/73 Coderch/Valls: Carles Fochs, Arxiu Coderch E.T.S. d'Arquitectura del Vallès, Universitat Politècnica Catalunya, C/ Pere Serra, 1–15. 08173 Sant Cugat del Vallès [BCN] arxiu.coderch@etsav.upc.edu/José Antonio Coderch, photography [Girasol]. Madrid. 19[illegible]
74/75 Diener & Diener: Christian Richters
76/77 Diener & Diener: Christian Richters
78/79 Siza: Atelier 18, AR 10/1990 No 1124, Lotus, 96
80/81 Richter: Christian Richters

82/83 Alder: A. Helbling & T. Ineichen
84/85 Morger & Degelo: Ruedi Walti
86/87 Kollhoff: van der Vlugt, Schwendinger & Büttner
88/89 Spühler: Monika Bischof, Comet-photo AG
90/91 Neutelings Riedijk: Christian Richters
92/93 Aranguren & Gallegos: © Eduardo Sánchez López
94/95 Møller: Torben Eskerod
96/97 Office Winhov: Stefan Müller, Thomas Leden
98/99 Esch Sintzel Architekten: Walter Mair
100/101 ROBERTNEUN: Werner Huthmacher
104 Cruz, Antonio/Ortiz, Antonio
105 Duinker/van der Torre: van der Vlugt & Claus
106 Johannes Uhl
107 Rob Krier
108/109 Nylund, Puttfarken, Stürzebecher
110/111 OMA: Bauwelt Heft 15 4/1990, Architecture d'Aujourd'hui 9/1987
112/113 IBUS
114 Schnebli/Ammann: Lorenzo Bianda
115 Herzog & de Meuron
116 Gazeau: J. M. Monthiers
117 Hilde Léon/Konrad Wohlhage
118/119 Kojima + Akamatsu: Satoshi Asakawa
120 Nägeli/Zander: Stephanie Kiwitt
121 Höhne & Rapp: Kim Zwarts
122/123 Atelier Bow-Wow: Atelier Bow-Wow
124/125 Kaden Klingbeil Architekten: Kaden Klingbeil Architekten
126/127 BARarchitekten: www.janbitter.de
128 Naruse Inokuma: Masao Nishikawa
129 SsD Architecture; Jinhee Park; John Hong
130-131 Field Design Architects: Masao Nishikawa
134 Coderch: Fochs, C.: "J. A. Coderch 1913–1984", Barcelona, GG Gili, 1990
135 Baller: Reinhard Friedrich
136/137 Otto Steidle
138/139 Llinàs: J. Bernadó
140/141 Herzog & de Meuron: Margherita Spiluttini
142/143 Bedaux de Brouwer: Luuk Kramer
144/145 Caruso St John: Ioana Marinescu/Hélène Binet
146/147 zanderroth architekten: Michael Feser / Simon Menges
148-149 Stanley Saitowitz | Natoma Architects: Bruce Damonte
152/153 Caccia Dominioni: Christoph Rokitta, Berlin
154/155 Heinz Schudnagies
156/157 Yositika Utida; Shu-ko-sha Architectural and Urban Design Studio; Osaka Gas Co. Ltd.: Japan Architect 73/2009; Kendall, S. , Teicher, J.: "Residential Open Building", Taylor and Francis, London, 1999
158 Baumschlager Eberle: Eduard Hueber
159 Baumschlager Eberle: archphoto, inc.
160/161 Diener und Diener: Wim Ruigrok
162/163 De Architekten Cie./van Dongen: Oski Collado
164/165 Chiba, Manabu: Masao Nishikawa
166/167 Burkard Meyer: Roger Frei, Zurich
168/169 Graber Pulver Architekten: exterior photo: Graber Pulver Architekten/photo of stairwell: Walter Mair, Zurich
170/171 LOHA: Lawrence Anderson
172/173 NL Architects: Luuk Kramer/NL Architects
174/175 2b architectes: Roger Frei
176/177 5468796 Architecture: James Brittain Photography
178/179 Eric Lapierre Architecture: ELEX
180/181 Edelaar Mosayebi Inderbitzen: Roland Bernath
184/185 Le Corbusier: Jenkins, D.: "Unité d'Habitation, Marseille", London, Phaidon Press, 1993
186 Alvar Aalto/Paul Baumgarten
187 Kunio Maekawa: Altherr, A.: "Three Japanese Architects", Hatje, Stuttgart, 1968
188/189 Jäger, Müller, Wirth: Foto-Hatt
190/191 Smithson, A. + P.: "The Charged Void: Architecture", Monacelli Press, New York, 2001; Simon Smithson
192/193 Hotz: Peter Morf
194/195 Nouvel, Ibos: GA Houses 23
196 Tusquets Blanca: Lluís Casals
197 Cruz, Ortiz: Duccio Malagamba
198/199 Steven Holl
200/201 Zaaijer/Christiaanse: van der Vlugt & Claus
202/203 Riegler, Riewe: Margherita Spiluttini
204/205 Kovatsch: Angelo Kaunat
206/207 Dercon, T'Jonk, Van Broeck: Leo van Broeck
208/209 Sejima/Nishizawa: El Croquis 99, Shinkenchiku-sha
210/211 b&k+ brandlhuber&kniess GbR: Stefan Schneider
212/213 Rocha: Ferreira Alves
214/215 Morger & Degelo: Ulrike Ruh
216/217 Langenegger: Alexander Gempeler, Bern/SSWZ
218/219 Gmür & Steib Architekten AG: Roger Frei
220/221 Neff Neumann: Atelier Fontana, Michael Fontana
222/223 Matti Ragaz Hitz: Alexander Gempeler, Bern
224/225 Guignard & Saner: Roman Keller
228 Mies van der Rohe: Hedrich-Blessing Collection, The Chicago Historical Society, Florian Kessel, Werner Blaser
229 Schäche, W. (Ed.): "Henselmann, Hermann: Ich habe Vorschläge gemacht", Ernst & Sohn, 1995, Architektur der DDR 1/1989
230/231 van den Broek/Bakema: Oliver Heckmann
232 Lasdun, Denys
233 Goldberg, Bertrand: Bauen & Wohnen 1963
234/235 Scharoun: "Hans Scharoun", Akademie der Künste, Vol. 10; GA Houses 23
236/237 Sáenz de Oiza: Daniel Fernadez/Enrique Rojo Escoba
238/239 Prentice & Chan: AR, 9/1972
240/241 Ricardo Bofill Taller de Arquitectura
242/243 Correa: Charles Correa Associates
244/245 Ciriani: J. M. Monthiers
246/247 Riken Yamamoto & Field Shop: Shigeru Ohno
248/249 MVRDV/Blanca Lleó: Rob 'T Hart
250/251 Mass Studies: Photography © Yong-Kwan Kim
252/253 E2A Architekten: Georg Aerni/Jan Bitter
254/255 MKPL Architects: Robert Such/Tim Griffith
258/259 Safdie, Moshe: AD, 3/1967
260 Frey, Schröder, Schmidt: DB, 102, 4/1968
261 Hodgkinson, Patrick/Martin, Sir Leslie
262/263 Hans Kammerer/Walter Belz, Walter
264/265 Georg Heinrichs
266/267 Baumschlager Eberle: Gerhard Ullmann
268/269 BIG Bjarke Ingels Group: Carsten Kring/Dragor Luftfoto/Jakob Boserup
272 Fleig, Karl: Bauwelt, 46/47, 1967
273 Ungers, Oswald Mathias: Bauwelt, 46/47, 1967
274/275 Darbourne and Darke: AR, 9/1974, AJ, 2 April/1975
276/277 Benson, Gordon/Forsyth, Alan: AA, 234, 9/1984
278/279 Atelier 5: Terence du Fresne
280/281 Alder: A. Helbling & T. Ineichen
282/283 Steidle + Partner: Verena von Gagern
284/285 Gigon/Guyer: Heinrich Helfenstein
286/287 Popov: Kraig Carlstrom
288/289 Haerle Hubacher: Georg Aerni
290/291 Herczog Hubeli
292/293 Architecture WORKSHOP: Daici Ano
294/295 Duplex Architekten: Johannes Marburg
298/299 Atelier 5: Balthasar Burkhard
300 Reinhard Gieselmann
301 Zevaco: A. Goldenburg
302/303 Blom: Collection Het Nieuwe Instituut, BLOM, 286; Jeffrey Pardoen
304/305 Blom, Piet: a+u, 11/1985, GA Houses No. 3; Collection Het Nieuwe Instituut, BLOM, n286
306/307 Storgård, J. P., Orum-Nielsen, A. and J., Marcussen, H.: Bauen & Wohnen, 10/1974
308/309 OMA/Koolhaas, Rem
310/311 Duncan Lewis/Block Architectes: Architects Association Duncan Lewis + Block Architectes
312/313 Sejima & Associates: Iwan Baan
314/315 Vandkunsten Architects: Jens Kristian Seier
318 Smithson: Webster, H. (Ed.):"Modernism without Rhetoric", Academy Editions, London 1997, Simon Smithson
319 Mendes da Rocha: Spiro, A.: "Paulo Mendes da Rocha. Bauten und Projekte", Niggli Verlag, Sulgen 2002, Annette Spiro
320 Olgiati: Boga, T.: "Die Architektur von Rudolf Olgiati". Reprint of the original edition of 1977, Birkhäuser, Basel 2010/© original edition: 1977 ETH Zürich, Institut für Geschichte und Theorie der Architektur
321 J. Paul Getty Trust, Getty Research Institute, Los Angeles (2004.R.10)
322 Kurokawa, K.: "From metabolism to symbiosis/Kisho Kurokawa", Academy Editions, London, 1992
323 Peter Märkli;
324 Ban: Hiroyuki Hirai
325 Miyamoto, Katsuhiro + Atelier Cinquième: Shikenchiku-sha
326/327 OMA/Rem Koolhaas: Hans Werlemann, hectic pictures
328/329 heide von beckerath alberts architekten
330/331 Allmann Sattler Wappner: Ulrike Myrzik & Manfred Jarisch Fotografen
332 Fujimoto: Daici Ano
333 Kraus Schönberg Architekten: Ioana Marinescu/Kraus Schönberg Architekten
334/335 HAHOH: Hubert Haas
336 Pezo von Ellrichshausen: Christobal Palma
337 John Lin, Rural Urban Framework
340 Gugger: Margherita Spiluttini
341 Noldin & Noldin: Günter Richard Wett
342/343 De Architektengroep bv/MVRDV: Archis 10/1997; Werk, Bauen und Wohnen 3/1999; Christian Richters
344 Pfeifer, Roser, Kuhn: Ruedi Walti
345 KUU Architects: Jeremy San
348/349 Jacobsen: Per Munkgård Thorsen, Lars Degnbol/Realdania Byg
350/351 Phippen, Randall, Parkes
352/353 Lüchinger, A.: "Herman Hertzberger. Bauten und Projekte 1959–1986", Den Haag, ArchEdition-cop, 1987
354 Haas, P., Hermann, G.: Detail, 7/1982
355 GFP & Assoziierte
356/357 Neutelings Riedijk: Stijn Brakkee
358/359 MAP Architects/Mateo, Josep Lluís: Duccio Malagamba
360 Elemental: © Tadeuz Jalocha, 2004 © Elemental/Cristobal Palma
361 Dorte Mandrup Arkitekter: Adam Mørk
362/363 Dahle, Dahle, Breitenstein: ddb/Nils Petter Dale
364/365 Bovenbouw Architectuur: Filip Dujardin

Colophon

Floor plan diagrams: Oliver Heckmann and Eric Zapel

Layout, cover design and typesetting: Tom Unverzagt
(4th ed.: Alexander Müller)

Translation from German into English: Eric Zapel
(4th ed.: Elizabeth Schwaiger)

Production: Heike Strempel, Katja Jaeger

Editor for the Publisher: Andreas Müller
(4th ed.: Ulrike Ruh)

Paper: Condat matt Perigord, 135 g/m²

Printing: medialis Offsetdruck GmbH

Library of Congress Control Number: 2017051798

Bibliographic information published by the
German National Library
The German National Library lists this publication in the Deutsche Nationalbibliografie; detailed bibliographic data are available on the Internet at http://dnb.dnb.de.

This publication is also available as a hardcover (ISBN 978-3-0356-1143-4) and as an e-book (ISBN PDF 978-3-0356-1149-6) and in a German language edition (ISBN 978-3-0356-1142-7).

First edition 1994
Second, revised and expanded edition 1997
Third, revised and expanded edition 2004
Fourth, revised and expanded edition 2011
Fifth, revised and expanded edition 2018

P.O. Box 44, 4009 Basel, Switzerland
Part of Walter de Gruyter GmbH, Berlin/Boston

Printed on acid-free paper produced from chlorine-free pulp.
TCF ∞

Printed in Germany

ISBN 978-3-0356-1144-1

9 8 7 6 5 4 3
www.birkhauser.com